CLASSICAL AND COMPUTATIONAL

SOLID MECHANICS

Advanced Series in Engineering Science – Vol. 1

CLASSICAL AND COMPUTATIONAL
SOLID
MECHANICS

Y. C. FUNG

University of California, San Diego

PIN TONG

Hong Kong University of Science & Technology

 World Scientific
Singapore • New Jersey • London • Hong Kong

Published by

World Scientific Publishing Co. Pte. Ltd.

P O Box 128, Farrer Road, Singapore 912805

USA office: Suite 1B, 1060 Main Street, River Edge, NJ 07661

UK office: 57 Shelton Street, Covent Garden, London WC2H 9HE

British Library Cataloguing-in-Publication Data
A catalogue record for this book is available from the British Library.

CLASSICAL AND COMPUTATIONAL SOLID MECHANICS
Advanced Series in Engineering Science — Volume 1

ISBN 981-02-3912-2
ISBN 981-02-4124-0 (pbk)

Printed in Singapore by World Scientific Printers

DEDICATION

To my teacher Professor Fung and to Mrs. Fung who have been the constant source of caring, inspiration and guidance; to my parents who were the source of love and discipline; to my wife who has been the source of love, companionship and encouragement; to my children and grandchildren who have been the source of joy and love; to Denny who will be forever in my heart.

Pin Tong

To Luna, Conrad, and Brenda

Y. C.

PREFACE

The objective of this book is to offer students of science and engineering a concise, general, and easy-to-understand account of some of the most important concepts and methods of classical and computational solid mechanics. The classical part is mainly a re-issue of Fung's *Foundations of Solid Mechanics*, with a major addition to the modern theories of plasticity, and a major revision of the theory of large elastic deformation with finite strains. The computational part consists of five new chapters, which focus on numerical methods to solve many major linear and nonlinear boundary-value problems of solid mechanics.

We hold the principle of easy-to-understand for the readers as an objective of our presentation. We believe that to be easily understood, the presentation must be precise, the definitions and hypotheses must be clear, the arguments must be concise and with sufficient details, and the conclusion has to be drawn very carefully. We strive to pay strict attention to these requirements. We believe that the method must be general and the notations should be unified. Hence, we presented the tensor analysis in general coordinates, but kept the indicial notations for tensors in the first fifteen chapters. In Chapters 16–21, however, the dyadic notations of tensors and the notations for matrix operations are used to shorten the formulas.

This book was written for engineers who invent and design things for human kind and want to use solid mechanics to help implement their designs and applications. It was written also for engineering scientists who enjoy solid mechanics as a discipline and would like to help develop and advance the subject further. It was further designed to serve those physical and natural scientists and biologists and bioengineers whose activities might be helped by the classical and computational solid mechanics. For example, biologists are discovering that the functional behavior of cells depends on the stresses acting on the cell. It is widely recognized that the molecular mechanics of the cell must be developed as soon as possible.

Solid mechanics deals with deformation and motion of "solids." The displacement that connects the instantaneous position of a particle to its position in an "original" state is of general interest. The preoccupation about particle displacements distinguishes solid mechanics from that of fluids.

This book begins with an introductory chapter containing a brief sketch of history, an outline of some prototypes of theories, and a description of some more complex features of solid mechanics. In Chapter 2, an introduction to tensor analysis is given. The bulk of the text from Chapters 3 to 16 is concerned with the classical theory of elasticity, but the discussion also includes thermodynamics of solid, thermoelasticity, viscoelasticity, plasticity, and finite deformation theory. Fluid mechanics is excluded, but methods that are common to both fluid and solid mechanics are emphasized. Both dynamics and statics are treated; the concepts of wave propagation are introduced at an early stage. Variational calculus is emphasized, since it provides a unified point of view and is useful in formulating approximate theories and computational methods. The large deflection theory of plates presented in the concluding section of Chapter 16 illustrates the elegance of the general approach to the large deformation theory.

Chapters 17 to 21 are devoted to computational solid mechanics to deal with linear, nonlinear, and nonhomogeneous problems. It is recognized that the incremental approach is the most practical approach, hence the updated Lagrangian description. Chapter 17 develops the incremental theory in considerable detail. Chapter 18 is devoted to numerical methods, with the finite element methods singled out for a detailed discussion of the theory of elasticity. Chapter 19 presents methods of calculations based on the mixed and hybrid variational principles, illustrating the broadening of the computational power with a less restrictive (or weaker) hypothesis informulating the variational principle. Chapter 20 deals with finite element methods for plates and shells, making the computational methods accessible to the analysis of the structures of aircraft, marine architecture, land vehicles, and shell-like structures in human being, animals, plants, earth, globe, and space. Finally, the book concludes with Chapter 21 dealing with finite element modeling of nonlinear elasticity, viscoelasticity, plasticity, viscoplasticity, and creep. Thus, a broad sweep of modern, advanced topics are covered.

Overall, this book lays emphasis on general methodology. It prepares the students to tackle new problems. However, as it was said in the original preface of the *Foundations of Solid Mechanics*, no single path can embrace the broad field of mechanics. As in mountain climbing, some routes are safe to travel, others more perilous; some may lead to the summit, others to different vistas of interest; some have popular claims, others are less traveled. In choosing a particular path for a tour through the field, one is influenced by the curricular, the trends in literature, and the interest in engineering and science. Here, a particular way has been chosen to view some of the most beautiful vistas in classical and computational mechanics. In making this choice, we have aimed at straightforwardness and interest, and practical usefulness in the long run.

Holding the book to a reasonable length did not permit inclusion of many numerical examples, which have to be supplemented through problems and references. Fortunately, there are many excellent references to meet this demand. We have presented an extensive bibliography in this book, but we suggest that the reader consults the review journal *Applied Mechanics Reviews* (AMR) published by the American Society of Mechanical Engineers International since 1947, for the current information. The reader is referred to the periodic in-depth reviews of the literature in specific issues of AMR.

We are indebted to many authors and colleagues as acknowledged in the preface of the *Foundations of Solid Mechanics*. In the preparation of the present edition, we are especially indebted to Professors Satya Atluri and Theodore Pian. We would like to record our gratitude to many colleagues who wrote us to discuss various points and sent us errata in the *Foundations of Solid Mechanics*, especially to Drs. Pao-Show D. Cheng, Shun Cheng, Ellis H. Dill, Clive L. Dym, J. B. Haddow, Manohar P. Kamat, Hans Krumhaar, T. D. Leko, Howard A, Magrath, Sumio Murakami, Theodore Pian, R. S. Rivlin, William P. Rodden, Bertil Storakers, Howard J. White, Jr. and John C. Yao. We would also like to thank professors Y. Ohashi, S. Murakami and N. Kamiya for translating the *Foundations* book into Japanese, Professors Oyuang Zhang, Ma Wen-Hua and Wang Kai-Fu for translating the *Foundations* book into Chinese.

On the cover of this book, portraits of some pioneers of mechanics are presented. They are arranged, from left to right and top to bottom, according to their birthdays. They are: Galileo Galilei, 1564-1642, Isaac Newton, 1642-1727, Daniel Bernoulli, 1700-1782, Leonhard Euler, 1707-1783, Joseph Louis Lagrange, 1736-1813, Claude Louis Marie Henri Navier, 1785-1836, Augustin Louis Cauchy, 1789-1857, Lord Kelvin, William Thompson, 1824-1907, Gustave Robert Kirchhoff, 1824-1889, Lord Rayleigh, John William Strutt, 1842-1919, Ludwig Edward Boltsmann, 1844-1906, August Edward Hugh Love, 1863-1940, Stephen P. Timoshenko, 1878-1972. These portraits were supplied by Dr Stephen Juhasz, who was the editor of the *Applied Mechanics Reviews* for 30 years, from 1953-1983. In 1973, Dr Juhasz published an article entitled "Famous Mechanics

Scientists" in *App. Mech. Rev.* Vol. 26, No. 2. These portraits are from that article, in which acknowledgement to original sources is recorded. We thank you, Dr Juhasz.

We would like to take this opportunity to mention a few of editorial notes:

(1) The bibliography is given at the end of the book.

(2) Equations in each Section are numbered sequentially. When referring to equations in other Sections, we use the format (Sec. no: Eq. no), e.g. (5.6:4).

(3) Formulas are concise way of saying lots of things. The most important formulas are marked with a triangular star, ▲. They are worthy of being committed to memory.

Y. C. Fung

P. Tong

CONTENTS

17 INCREMENTAL APPROACH TO SOLVING SOME NONLINEAR PROBLEMS

18 FINITE ELEMENT METHODS

1

INTRODUCTION

Mechanics is the science of force and motion of matter. Solid mechanics is the science of force and motion of matter in the solid state. Physicists are of course interested in mechanics. The greatest advances in physics in the twentieth century are identified with mechanics: the theory of relativity, quantum mechanics, and statistical mechanics. Chemists are interested in the mechanics of chemical reaction, the formation of molecular aggregates, the formation of crystals, or the creation of new materials with desirable properties, or polymerization of larger molecules, etc. Biologists are interested in biomechanics that relates structure to function at all hierarchical levels: from biomolecules to cells, tissues, organs, and individuals. Although a living cell is not a homogeneous continuum, it is a protein machine, a protein factory, with internal machinery that moves and functions in an orderly way according to the laws of mechanics. Therefore, all scientists are interested in mechanics and mechanics is developed by scientists continuously.

Engineers, especially aeronautical, mechanical, civil, chemical, materials, biomedical, biotechnological, space, and structural engineers, are real developers and users of fluid and solid mechanics because of their professional needs. They design. They invent. They are concerned about the safety and economy of their products. They want to know the function of their products as precisely as possible. They want results fast. They experiment. They theorize. They test, compute, and validate. To them mechanics is a toy, a bread and butter, a feast or delicacy.

Engineering is quite different from science. Scientists try to understand nature. Engineers try to make things that do not exist in nature. Engineers stress invention. To embody an invention the engineer must put his idea in concrete terms, and design something that people can use. That something can be a device, a gadget, a material, a method, a computing program, an innovative experiment, a new solution to a problem, or an improvement on what is existing. Since a design has to be concrete, it must have its geometry, dimensions, and characteristic numbers. Almost all engineers working on new designs find that they do not have all the needed information. Most

1

often, they are limited by insufficient scientific knowledge. Thus they study mathematics, physics, chemistry, biology and mechanics. Often they have to add to the sciences relevant to their profession. Thus engineering sciences are born.

This book is written by engineering scientists, for engineering scientists, and this determines its style. The qualities we want are:

- Easy to read,
- Precise, concise, and practical,
- First priority on the formulation of problems,
- Presenting the classical results as gold standard, and
- Numerical approach as everyday tool to obtain solutions.

If the book is a banquet, we offer some hors d'oeuvres in this introductory chapter.

1.1. HOOKE'S LAW

Historically, the notion of elasticity was first announced in 1676 by Robert Hooke (1635–1703) in the form of an anagram, *ceiiinosssttuv*. He explained it in 1678 as

$$Ut\ tensio\ sic\ vis\,,$$

or "the power of any springy body is in the same proportion with the extension."[†]

As stated in the original form, Hooke's law is not very clear. Our first task is to give it a precise expression. Historically, this was done in two different ways. The first way is to make use of the common notion of "springs," and consider the load-deflection relationship. The second way is to state it as a tensor equation connecting the stress and strain. Although the second way is the proper way to start a general theory, the first, simpler and more restrictive, is not without interest. In this section, we develop the first alternative as a prototype of the theory of elasticity.

Let us consider the static equilibrium state of a solid body under the action of external forces (Fig. 1.1:1). Let the body be supported in some manner so that at least three points are fixed in a space which is described with respect to a rectangular Cartesian frame of reference. We shall make three basic hypothesis regarding the properties of the body under consideration.

(H1) *The body is continuous and remains continuous under the action of external forces.*

[†]Edme Mariotte enunciated the same law independently in 1680.

Under this hypothesis the atomistic structure of the body is ignored and the body is idealized into a geometrical copy in Euclidean space whose points are identified with the material particles of the body. Continuity is defined in mathematical sense as an isomorphism of the real number system. Neighboring points remain as neighbors under any loading condition. No cracks or holes may open up in the interior of the body under the action of external load.

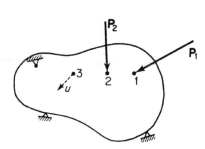

Fig. 1.1:1. Static equilibrium of a body under external forces.

A material satisfying this hypothesis is said to be a *continuum*. The study of the deformation or motion of a continuum under the action of forces is called the *continuum mechanics*.

To introduce the second hypothesis, let us consider the action of a set of forces on the body. Let every force be fixed in direction and in point of application, and let the magnitude of all the forces be increased or decreased together: always bearing the same ratio to each other. Let the forces be denoted by $\mathbf{P}_1, \mathbf{P}_2, \ldots, \mathbf{P}_n$ and their magnitude by $P_1, P_2, \ldots, P_n$. Then the ratios $P_1 : P_2 : \ldots : P_n$ remain fixed. When such a set of forces is applied on the body, the body deforms. Let the displacement at an arbitrary point in an arbitrary direction be measured with respect to a rectangular Cartesian frame of reference fixed with the supports. Let this displacement be denoted by u. Then our second hypothesis is

(H2) *Hooke's law;*

$$(1) \qquad\qquad u = a_1 P_1 + a_2 P_2 + \cdots + a_n P_n \,,$$

where $a_1, a_2, \ldots, a_n$ are constants independent of the magnitude of P_1, $P_2, \ldots, P_n$. The constants $a_1, a_2, \ldots, a_n$ depend, of course, on the location of the point at which the displacement component is measured and on the directions and points of application of the individual forces of the loading.

Hooke's law in the form (H2) is one that can be subjected readily to direct experimental examination.

To complete the formulation of the theory of elasticity, we need a third hypothesis:

(H3) *There exists a unique unstressed state of the body, to which the body returns whenever all the external forces are removed.*

A body satisfying these three hypothesis is called a *linear elastic solid.*

A number of deductions can be drawn from these assumptions. We shall list a few important ones.

(A) *Principle of superposition*

By a combination of (H2) and (H3), we can show that Eq. (1) is valid not only for systems of loads for which the ratios $P_1 : P_2 : \ldots : P_n$ remain fixed as originally assumed, but also for an arbitrary set of loads $\mathbf{P}_1, \mathbf{P}_2, \ldots, \mathbf{P}_n$. In other words, Eq. (1) holds regardless of the order in which the loads are applied.

Proof. If a proof of the statement above can be established for an arbitrary pair of loads, then the general theorem can be proved by mathematical induction.

Let $\mathbf{P}_1$ and $\mathbf{P}_2$ (with magnitudes P_1 and P_2) be a pair of arbitrary loads acting at points 1 and 2, respectively. Let the deflection in a specific direction be measured at a point 3 (see Fig. 1.1:1). According to (H2), if $\mathbf{P}_1$ is applied alone, then at the point 3 a deflection $u_3 = c_{31}P_1$ is produced. If $\mathbf{P}_2$ is applied alone, a deflection $u_3 = c_{32}P_2$ is produced. If $\mathbf{P}_1$ and $\mathbf{P}_2$ are applied together, with the ratio $P_1 : P_2$ fixed, then according to (H2) the deflection can be written as

(a)
$$u_3 = c'_{31}P_1 + c'_{32}P_2 \,.$$

The question arises whether $c'_{31} = c_{31}$, $c'_{32} = c_{32}$. The answer is affirmative, as can be shown as follows. After $\mathbf{P}_1$ and $\mathbf{P}_2$ are applied, we take away $\mathbf{P}_1$. This produces a change in deflection, $-c''_{31}P_1$, and the total deflection becomes

(b)
$$u_3 = c'_{31}P_1 + c'_{32}P_2 - c''_{31}P_1 \,.$$

Now only $\mathbf{P}_2$ acts on the body. Hence, upon unloading $\mathbf{P}_2$ we shall have

(c)
$$u_3 = c'_{31}P_1 + c'_{32}P_2 - c''_{31}P_1 - c_{32}P_2 \,.$$

Now all the loads are removed, and u_3 must vanish according to (H3). Rearranging terms, we have

(d)
$$(c'_{31} - c''_{31})P_1 = (c_{32} - c'_{32})P_2 \,.$$

Since the only possible difference of c'_{31} and c''_{31} must be caused by the action of $\mathbf{P}_2$, the difference $c'_{31} - c''_{31}$ can only be a function of P_2 (and not of P_1). Similarly, $c_{32} - c'_{32}$ can only be a function of P_1. If we write Eq. (d) as

(e)
$$\frac{c'_{31} - c''_{31}}{P_2} = \frac{c_{32} - c'_{32}}{P_1},$$

then the left-hand side is a function of P_2 alone, and the right-hand side is a function of P_1 alone. Since P_1 and P_2 are arbitrary numbers, the only possibility for Eq. (e) to be valid is for both sides to be a constant k which is independent of both P_1 and P_2. Hence,

(f)
$$c'_{32} = c_{32} - kP_1.$$

But a substitution of (f) into (a) yields

(g)
$$u_3 = c'_{31}P_1 + c_{32}P_2 - kP_1P_2.$$

The last term is nonlinear in P_1, P_2, and Eq. (g) will contradict (H2) unless k vanishes. Hence, $k = 0$ and $c'_{32} = c_{32}$. An analogous procedure shows $c'_{31} = c''_{31} = c_{31}$.

Thus the principle of superposition is established for one and two forces. An entirely similar procedure will show that if it is valid for m forces, it is also valid for $m + 1$ forces. Thus, the general theorem follows by mathematical induction. Q.E.D.

The constants c_{31}, c_{32}, etc., are seen to be of significance in defining the elastic property of the solid body. They are called *influence coefficients* or, more specifically, *flexibility influence coefficients*.

(B) *Corresponding forces and displacements and the unique meaning of the total work done by the forces.*

Let us now consider a set of external forces $\mathbf{P}_1, \ldots, \mathbf{P}_n$ acting on the body and *define the set of displacements at the points of application and in the direction of the loads as the displacements "corresponding" to the forces at these points.* The reactions at the points of support are considered as external forces exerted on the body and included in the set of forces.

Under the loads $\mathbf{P}_1, \ldots, \mathbf{P}_n$, the corresponding displacements may be written as

(2)
$$
\begin{aligned}
u_1 &= c_{11}P_1 + c_{12}P_2 + \cdots + c_{1n}P_n, \\
u_2 &= c_{21}P_1 + c_{22}P_2 + \cdots + c_{2n}P_n, \\
&\cdots\cdots\cdots\cdots\cdots\cdots\cdots\cdots\cdots\cdots \\
u_n &= c_{n1}P_1 + c_{n2}P_2 + \cdots + c_{nn}P_n.
\end{aligned}
$$

If we multiply the first equation by P_1, the second by P_2, etc., and add, we obtain

(3) $P_1 u_1 + P_2 u_2 + \cdots + P_n u_n = c_{11} P_1^2 + c_{12} P_1 P_2 + \cdots + c_{1n} P_1 P_n$

$$+ c_{21} P_1 P_2 + c_{22} P_2^2 + \cdots + c_{nn} P_n^2 \,.$$

The quantity above is independent of the order in which the loads are applied. It is the *total work done by the set of forces.*

(C) *Maxwell's reciprocal relation*

The influence coefficients for corresponding forces and displacements are symmetric.

(4) $c_{ij} = c_{ji} \,.$

In other words, *the displacement at a point i due to a unit load at another point j is equal to the displacement at j due to a unit load at i, provided that the displacements and forces "correspond," i.e., that they are measured in the same direction at each point.*

The proof is simple. Consider two forces $\mathbf{P}_1$ and $\mathbf{P}_2$ (Fig. 1.1:1). When the forces are applied in the order $\mathbf{P}_1$, $\mathbf{P}_2$, the work done by the forces is easily seen to be

$$W = \frac{1}{2}(c_{11} P_1^2 + c_{22} P_2^2) + c_{12} P_1 P_2 \,.$$

When the order of application of the forces is interchanged, the work done is

$$W' = \frac{1}{2}(c_{22} P_2^2 + c_{11} P_1^2) + c_{21} P_1 P_2 \,.$$

But according to (B) above, $W = W'$ for arbitrary P_1, P_2. Hence, $c_{12} = c_{21}$, and the theorem is proved.

(D) *Betti-Rayleigh reciprocal theorem*

Let a set of loads $\mathbf{P}_1, \mathbf{P}_2, \ldots, \mathbf{P}_n$ produce a set of corresponding displacements $u_1, u_2, \ldots, u_n$. Let a second set of loads $\mathbf{P}_1', \mathbf{P}_2', \ldots, \mathbf{P}_n'$, acting in the same directions and having the same points of application as those of the first, produce the corresponding displacements $u_1', u_2', \ldots, u_n'$. Then

(5) $P_1 u_1' + P_2 u_2' + \cdots + P_n u_n' = P_1' u_1 + \cdots + P_1' u_n \,.$

In other words, *in a linear elastic solid, the work done by a set of forces acting through the corresponding displacements produced by a second set of*

*forces is equal to the work done by the second set of forces acting through
the corresponding displacements produced by the first set of forces.*

A straightforward proof is furnished by writing out the u_i and u_i' in
terms of P_i and P_i', $(i = 1, 2, \ldots, n)$, with appropriate influence coefficients, comparing the results on both sides of the equation, and utilizing
the symmetry of the influence coefficients.

In the form of Eq. (5), the reciprocal theorem can be generalized to
include moments and rotations as the corresponding *generalized forces* and
generalized displacements. An illustration is given in Fig. 1.1:2. These
theorems are very useful in practical applications.

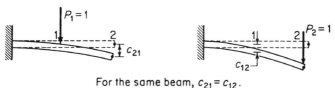

For the same beam, $c_{21} = c_{12}$.
(a) Forces and corresponding displacements.

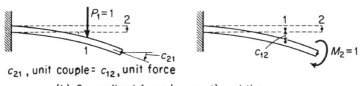

c_{21}, unit couple $= c_{12}$, unit force

(b) Generalized force (moment) and the
corresponding generalized displacement
(moment ~ rotation of angle).

Fig. 1.1:2. Illustration of the reciprocal theorem.

(E) *Strain energy*

Further insight can be gained from the first law of thermodynamics. When
a body is thermally isolated and thermal expansions are neglected the first
law states that the work done on the body by the external forces in a certain
time interval is equal to the increase in the kinetic energy and internal
energy in the same interval. If the process is so slow that the kinetic energy
can be ignored, the work done is seen to be equal to the change in internal
energy.

If the internal energy is reckoned as zero in the unstressed state, the
stored internal energy shall be called strain energy. Writing U for the

strain energy, we have, from (3) and (4),

$$(6) \qquad U = \frac{1}{2}\sum_{i=1}^{n}\sum_{j=1}^{n}c_{ij}P_iP_j = \frac{1}{2}\sum_{i=1}^{n}c_{ii}P_i^2 + \frac{1}{2}\sum_{i\neq j}\sum c_{ij}P_iP_j .$$

If we differentiate Eq. (6) with respect to P_i, we obtain

$$\frac{\partial U}{\partial P_i} = c_{ii}P_i + \sum_{j\neq i}c_{ij}P_j , \qquad\qquad i = 1, 2, \ldots, n .$$

But, the right-hand side is precisely u_i; hence, we obtain

(F) *Castigliano's theorem*

$$(7) \qquad\qquad\qquad \frac{\partial U}{\partial P_i} = u_i , \qquad\qquad i = 1, \ldots, n .$$

In other words, if a set of loads $P_1, \ldots, P_n$ is applied on a perfectly elastic body as described above and the strain energy is expressed as a function of the set $P_1, \ldots, P_n$, then the partial derivative of the strain energy, with respect to a particular load, gives the corresponding displacement at the point of application of that particular load in the direction of that load.

(E) *The principle of virtual work*

On the other hand, for a body in equilibrium under a set of external forces, the principle of virtual work can be applied to show that, *if the strain energy is expressed as a function of the corresponding displacements, then*

$$(8) \qquad\qquad\qquad \frac{\partial U}{\partial u_i} = P_i , \qquad\qquad i = 1, \ldots, n .$$

The proof consists in allowing a virtual displacement δu to take place in the body in such a manner that δu is continuous everywhere but vanishes at all points of loading except under P_i. Due to δu, the strain energy changes by an amount δU, while the virtual work done by the external forces is the product of P_i times the virtual displacement, i.e., $P_i\delta u_i$. According to the principle of virtual work, these two expressions are equal, $\delta U = P_i\delta u_i$. On rewriting it in the differential form, the theorem is established.

The important result (8) is established on the principle of virtual work as applied to a state of equilibrium under the additional assumption that a strain energy function that is a function of displacement exists. It is applicable also to elastic bodies that follow the nonlinear load-displacement relationship.

1.2. LINEAR SOLIDS WITH MEMORY: MODELS OF VISCOELASTICITY

Most structural metals are nearly linear elastic under small strain, as measurements of load-displacement relationship reveal. The existence of normal modes of free vibrations which are simple harmonic in time, is often quoted as an indication (although not as a proof) of the linear elastic character of the material. However, when one realizes that the vibration of metal instruments does not last forever, even in a vacuum, it becomes clear that metals deviate somewhat from Hooke's law. Thus, other constitutive laws must be considered. The need for such an extension becomes particularly evident when organic polymers are considered.

In this section we shall consider a simple class of materials which retains linearity between load and deflection, but the linear relationship depends on a third parameter, time. For this class of material, the present state of deformation cannot be determined completely unless the entire history of loading is known.

A linear elastic solid may be said to have a simple memory: it remembers only one configuration; namely, the unstrained natural state of the body. Many materials do not behave this way: they remember the past. Among such materials with memory there is one class that is relatively simple in behavior. This is the class of materials named above, for which the cause and effect are linearly related.

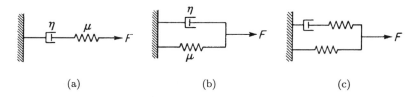

(a)   (b)   (c)

Fig. 1.2:1. Models of linear viscoelasticity: (a) Maxwell, (b) Voigt, (c) Kelvin.

Let us consider some simple examples. In Fig. 1.2:1 are shown three mechanical models of material behavior, namely, the Maxwell model, the Voigt model, and the Kelvin model, all of which are composed of combinations of linear springs with spring constant μ and dashpots (pistons moving in a viscous fluid) with coefficient of viscosity η. A *linear spring* is supposed to produce instantaneously a deformation proportional to the load. A *dashpot* is supposed to produce a velocity proportional to the load at any instant. The load-deflection relationship for these models are

(1)   Maxwell model: $\dot{u} = \dfrac{\dot{F}}{\mu} + \dfrac{F}{\eta}$, $u(0) = \dfrac{F(0)}{\mu}$,

(2)   Voigt model: $F = \mu u + \eta \dot{u}$, $u(0) = 0$,

(3)   Kelvin model: $F + \tau_\varepsilon \dot{F} = E_R(u + \tau_\sigma \dot{u})$, $\tau_\varepsilon F(0) = E_R \tau_\sigma u(0)$,

where τ_ε, τ_σ are two constants. When these equations are to be integrated, the initial conditions at $t = 0$ must be prescribed as indicated above.

The *creep functions* $c(t)$, which are the displacement $u(t)$ in response to a unit-step force $F(t) = \mathbf{1}(t)$ defined in Eq. (7) below, are the solution of Eqs. (1)–(3). They are:

(4) Maxwell solid:

$$c(t) = \left(\frac{1}{\mu} + \frac{1}{\eta} t \right) \mathbf{1}(t),$$

(5) Voigt solid:

$$c(t) = \frac{1}{\mu}(1 - e^{-(\mu/\eta)t}) \mathbf{1}(t),$$

(6) Kelvin solid:

$$c(t) = \frac{1}{E_R} \left[1 - \left(1 - \frac{\tau_e}{\tau_\sigma} \right) e^{-t/\tau_\sigma} \right] \mathbf{1}(t),$$

where the *unit-step function* $\mathbf{1}(t)$ is defined as

(7) $\mathbf{1}(t) = \begin{cases} 1 & \text{when} \quad t > 0, \\ \dfrac{1}{2} & \text{when} \quad t = 0, \\ 0 & \text{when} \quad t < 0. \end{cases}$

A body which obeys a load-deflection relation like that given by Maxwell's model is said to be a Maxwell solid. Similarly, Voigt and Kelvin solids are defined. These models are called models of *viscoelasticity*.

Interchanging the roles of F and u, we obtain the *relaxation function* as a response $F(t) = k(t)$ corresponding to an elongation $u(t) = \mathbf{1}(t)$, Eq. (7).

(8) Maxwell solid:

$$k(t) = \mu e^{-(\mu/\eta)t} \mathbf{1}(t),$$

(9) Voigt solid:

$$k(t) = \eta \delta(t) + \mu \mathbf{1}(t),$$

(10) Kelvin solid:

$$k(t) = E_R \left[1 - \left(1 - \frac{\tau_\sigma}{\tau_\varepsilon} \right) e^{-t/\tau_\varepsilon} \right] 1(t) \,.$$

Here we have used the symbol $\delta(t)$ to indicate the *unit-impulse function*, or *Dirac-delta function*, which is defined as a function with a singularity at the origin:

$$\delta(t) = 0 \qquad \text{for } t < 0, \text{ and } t > 0,$$

(11)
$$\int_{-\varepsilon}^{\varepsilon} f(t)\delta(t)dt = f(0) \,, \qquad\qquad\qquad \varepsilon > 0,$$

where $f(t)$ is an arbitrary function continuous at $t = 0$. These functions, $c(t)$ and $k(t)$, are illustrated in Figs. 1.2:2 and 1.2:3, respectively, for which we add the following comments.

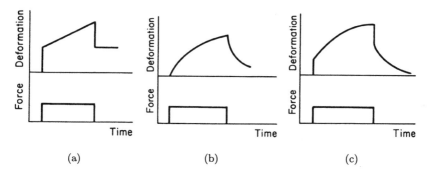

    (a)   (b)   (c)

Fig. 1.2:2. Creep function of (a) Maxwell, (b) Voigt, (c) Kelvin solid. A negative phase is superposed at the time of unloading.

For the Maxwell solid, a sudden application of a load induces an immediate deflection by the elastic spring, which is followed by "creep" of the dashpot. On the other hand, a sudden deformation produces an immediate reaction by the spring, which is followed by stress relaxation according to an exponential law Eq. (8). The factor η/μ, with dimensions of time, may be called a *relaxation time*: it characterizes the rate of decay of the force.

For the Voigt solid, a sudden application of force will produce no immediate deflection because the dashpot, arranged in parallel with the spring, will not move instantaneously. Instead, as shown by Eq. (5) and Fig. 1.2:2(b), a deformation will be gradually built up, while the spring takes a greater and greater share of the load. The dashpot displacement relaxes exponentially. Here the ratio η/μ is again a relaxation time: it characterizes the rate of decay of the deflection.

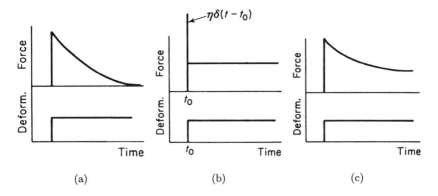

Fig. 1.2:3. Relaxation function, of (a) Maxwell, (b) Voigt, (c) Kelvin solid.

For the Kelvin solid, a similar interpretation is applicable. The constant τ_ε is the time of relaxation of load under the condition of constant deflection [see Eq. (10)], whereas the constant τ_σ is the time of relaxation of deflection under the condition of constant load [see Eq. (4)]. As $t \to \infty$, the dashpot is completely relaxed, and the load-deflection relation becomes that of the springs, as is characterized by the constant E_R in Eqs. (4) and (10). Therefore, E_R is called the *relaxed elastic modulus*.

Load-deflection relations such as (1)–(3) were proposed to extend the classical theory of elasticity to include *anelastic* phenomena. Lord Kelvin (Sir William Thomson, 1824–1907), on measuring the variation of the rate of dissipation of energy with frequency of oscillation in various materials, showed the inadequacy of the Maxwell and Voigt equations. A more successful generalization using mechanical models was first made by John H. Poynting (1852–1914) and Joseph John Thomson ("J.J.," 1856–1940) in their book *Properties of Matter* (London: C. Griffin and Co., 1902).

1.3. SINUSOIDAL OSCILLATIONS IN A VISCOELASTIC MATERIAL

It is interesting to examine the relationship between the load and the deflection in a body when it is forced to perform simple harmonic oscillations. A simple harmonic oscillation can be described by the real or imaginary part of a complex variable. Thus, if F_0 is a complex variable $F_0 = Ae^{i\phi} = A(\cos\phi + i\sin\phi)$, then a simple harmonic oscillatory force can be written as

(1) $$F(t) = F_0 e^{i\omega t} = A(\cos\phi + i\sin\phi)(\cos\omega t + i\sin\omega t)$$

$$= A\cos(\omega t + \phi) + iA\sin(\omega t + \phi).$$

Similarly, if $u_0 = Be^{i\psi}$, then a simple harmonic oscillatory displacement $u(t) = u_0 e^{i\omega t}$ is either the real part or the imaginary part of

$$(2) \qquad u(t) = u_0 e^{i\omega t} = B\cos(\omega t + \psi) + iB\sin(\omega t + \psi).$$

On substituting (1) and (2) into Eqs. (1)–(3) of Sec. 1.2, we can obtain the ratio u_0/F_0, which is a complex number. The inverse, F_0/u_0, is called the *complex modulus* of a viscoelastic material, and is often designated by $\mathcal{M}$:

$$(3) \qquad \mathcal{M} = F_0/u_0 = |\mathcal{M}|e^{i\delta},$$

where $|\mathcal{M}|$ is the magnitude and δ is the phase angle by which the strain lags behind the stress. The tangent of δ is often used as a measure of the *internal friction* of a linear viscoelastic material:

$$(4) \qquad \tan\delta = \frac{\text{imaginary part of } \mathcal{M}}{\text{real part of } \mathcal{M}}.$$

For the Kelvin model, we have

$$(5) \qquad \mathcal{M} = \frac{1 + i\omega\tau_\sigma}{1 + i\omega\tau_\varepsilon} E_R, \quad |\mathcal{M}| = \left(\frac{1 + \omega^2\tau_\sigma^2}{1 + \omega^2\tau_\varepsilon^2}\right)^{1/2} E_R,$$

$$(6) \qquad \tan\delta = \frac{\omega(\tau_\sigma - \tau_\varepsilon)}{1 + \omega^2(\tau_\sigma\tau_\varepsilon)} = \frac{(\tau_\sigma - \tau_\varepsilon)}{(\tau_\sigma\tau_\varepsilon)^{1/2}}\frac{\omega(\tau_\sigma\tau_\varepsilon)^{1/2}}{1 + \omega^2(\tau_\sigma\tau_\varepsilon)}.$$

When $|\mathcal{M}|$ and $\tan\delta$ in (5) and (6) are plotted against the logarithm of ω, curves as shown in Fig. 1.3:1 are obtained. Experiments with torsional

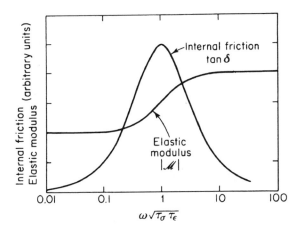

Fig. 1.3:1. Frequency dependence of internal friction and elastic modulus.

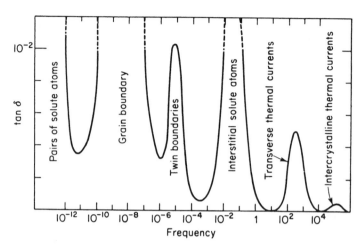

Fig. 1.3:2. A typical relaxation spectrum. (After C. M. Zener, *Elasticity and Anelasticity of Metals*, The University of Chicago Press, 1948.)

oscillations of metal wires at various temperatures, reduced to room temperature according to certain thermodynamic formula, yield a typical "relaxation spectrum" as shown in Fig. 1.3.2. Many peaks are seen in the internal-friction-versus-frequency curve. It has been suggested that each peak should be regarded as representing an elementary process as described above, with a particular set of relaxation times τ_σ, τ_ε. Each set of relaxation times τ_σ, τ_ε can be attributed to some process in the atomic or microscopic level. A detailed study of such a relaxation spectrum tells a great deal about the structures of metals; and the study of internal friction has provided a very effective key to metal physics.

1.4. PLASTICITY

Take a small steel rod. Bend it. When the deflection is small, the rod will spring back to its original shape when you release the load. This is elasticity. When the deflection is sufficiently large, a permanent deformation will remain when the load is released. That is plasticity. The load-deflection relationship of an *ideal plastic* material is shown in Fig. 1.4:1(a). For an ideal plastic material, the deflection is zero when the load is smaller than a critical value. Then, when the load reaches a critical value, the deflection continues to increase (the material flows) as long as the same load remains. If, at a given deflection, the load becomes smaller than the critical value, then the flow stops. There is no way the load can be increased beyond the critical value.

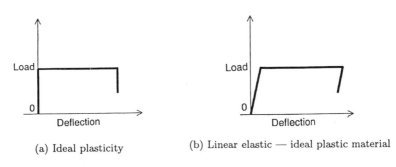

(a) Ideal plasticity   (b) Linear elastic — ideal plastic material

Fig. 1.4:1.

Structural steel behaves pretty much like an ideal plastic material, except that when the load (measured in terms of the maximum shear stress) is smaller than the critical load (called the *yield shear stress*), the load-deflection curve is an inclined straight line (Hooke's Law). Upon unloading, there is a small rebound. There are some details at the yield point that were ignored in the statement above. Other metals, such as copper, aluminum, lead, stainless steel, etc., behave in a somewhat similar, but more complex manner. Metals at a sufficiently high temperature may behave more like a fluid. Theories that deal with these features of materials are called the theories of *plasticity*, which are presented in Chapter 6.

1.5. VIBRATIONS

We know vibrations by experience while driving a car, flying an airplane, playing a musical instrument. The trees sway in the wind. A building shakes in an earthquake. Sometimes we want to know if a structure is safe in vibration. Sometimes we want to design a cushion that isolates an instrument from vibrations. A prototype of this kind of problem is shown in Fig. 1.5:1(a). A body with mass M is attached to an initially vertical massless spring, which has a spring constant k, and a damping constant c, and is "built-in" to a "ground" which moves horizontally with a displacement history $s(t)$. Let $x(t)$ denote the horizontal displacement of the mass and a dot over x or s denote a differentiation with respect to time. Then $\ddot{x}$ is the acceleration of the body, $k(x - s)$ is the spring force acting on the body, and $c(\dot{x} - \dot{s})$ is the viscous damping force acting on the body. Newton's second law requires that

(1) $$M\ddot{x} + c(\dot{x} - \dot{s}) + k(x - s) = 0 \,.$$

If we let

(2) $$y = x - s \,,$$

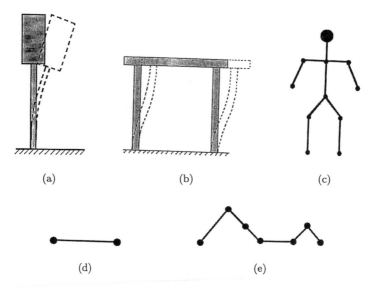

Fig. 1.5:1.

represent the displacement of the body relative to the ground, then Eq. (1) may be written as

$$(3) \qquad\qquad M\ddot{y} + c\dot{y} + ky = -M\ddot{s}.$$

The solution of Eq. (3) represents a forced vibration of a damped system. If the forcing function $M\ddot{s}(t)$ were zero, the equation

$$(4) \qquad\qquad M\ddot{y} + c\dot{y} + ky = 0,$$

describes the free vibration of the damped system. If the damping constant c vanishes, then the free vibration of the undamped system is described by the equation

$$(5) \qquad\qquad M\ddot{y} + ky = 0.$$

Equation (5) is satisfied by the solution

$$(6) \qquad\qquad y = A\cos\omega t + B\sin\omega t,$$

in which A and B are arbitrary constants, and

$$(7) \qquad\qquad \omega = \sqrt{k/M},$$

as can be verified by direct substitution of (6) into (5). When k and M are real values, ω is real. The motion $y(t)$ given by (6) is an oscillation of the body at a circular frequency of ω rad/sec. It can exist without external load. Hence ω is called the *natural frequency* of a *free vibration*. If the damping constant c were zero, and the forcing function $\ddot{s}$ is periodic with the same frequency as the natural one, then the amplitude of the oscillation is unbounded, and we have the phenomenon of *resonance*.

The solution of Eq. (4) must be an exponential function of time, because the derivative of an exponential function is another exponential function. Thus

(8) $\qquad y(t) = Ae^{\lambda t}$ implies $\dot{y}(t) = A\lambda e^{\lambda t}$, $\quad \ddot{y}(t) = A\lambda^2 e^{\lambda t}$.

On substituting (8) into Eq. (4), we have

(9) $$M\lambda^2 + c\lambda + k = 0.$$

If we write

(10) $$k/M = \omega^2, \quad \varepsilon = c/(2\sqrt{kM}),$$

then the two roots of Eq. (9) may be written as λ_1 and λ_2:

(11) $\qquad \lambda_1 = -\varepsilon\omega + i\omega\sqrt{1 - \varepsilon^2}, \qquad \lambda_2 = -\varepsilon\omega - i\omega\sqrt{1 - \varepsilon^2}.$

When k and M are real positive numbers, ω is real, and the solution of Eq. (9) under the initial conditions

(12) $$y(0) = y_o, \quad \dot{y}(0) = \dot{y}_o \quad \text{when } t = 0.$$

is

(13) $$y(t) = y_o e^{-\varepsilon\omega t} \cos\omega\sqrt{1 - \varepsilon^2}\, t$$
$$+ \frac{1}{\omega\sqrt{1 - \varepsilon^2}}(\dot{y}_o + y_o\varepsilon\omega)e^{-\varepsilon\omega t} \sin\omega\sqrt{1 - \varepsilon^2}\, t.$$

Now, we can return to the solution of Eq. (3). A particular solution of Eq. (3) can be obtained by Laplace or Fourier transformation or other methods. By direct substitution, it can be verified that a particular solution satisfying the initial conditions $y(t) = \dot{y}(t) = 0$ at $t = 0$ is:

(14) $\quad y(t, \omega, r) = -\dfrac{1}{\omega\sqrt{1 - \varepsilon^2}} \displaystyle\int_0^t \ddot{s}(\xi)e^{-\varepsilon\omega(t - \xi)} \sin\omega\sqrt{1 - \varepsilon^2}\,(t - \xi)d\xi.$

The general solution of Eq. (3) is the sum of the functions given in (13) and (14). From this solution we can examine the nature of the forced oscillation and its dependence on the parameters ω, ε, and the frequency spectrum of the forcing function $\ddot{s}(t)$.

The simple solution given by Eqs. (13) and (14) has important applications to the problems of isolation of delicate instruments in shipping, response of buildings to earthquake, impact of an airplane on landing, landing a robot on the moon. A frequently asked quention is: what is the maximum absolute value of the displacement y, or the acceleration $\ddot{y}$ as functions of the peak values of the ground displacement s, ground acceleration $\ddot{s}$, the time course of the ground motion, the system characteristis M, c, k, and the initial conditions y_o, $\dot{y}_o$ i.e.,

$$(15) \qquad \max |y(t)| / \max |s(t)| = \textit{function of } s(t), M, c, k, y_o, \dot{y}_o$$

$$\max |\ddot{y}(t)| / \max |\ddot{s}(t)| = \textit{functions of } s(t), M, c, k, y_o, \dot{y}_o \,.$$

These functions are called *shock spectra*. Studies of shock spectra are important not only for technological applications, but also for mathematics. Note that although Eq. (3) is linear in $y(t)$, the shock spectra are nonlinear functions of the parameters of the system. Qualitatively, we notice that the most important characteristic time of the system. Qualitatively, we notice that the most important characteristic time of the free oscillation is $1/\omega$. The most important characteristic time of the excitation $s(t)$ is the length of time it takes to rise from 0 to the peak value, t_r. We call t_r the *rise time of the signal*. Alluring to resonance, we see that a parameter of importance is the ratio of the two charateristic times $1/\omega$ and t_r. Hence the shock spectra are principally functions of ωt_r. See some references in the bibliography at the end of the book.

The picture shown in Fig. 1.5:1(b) may represent, in a very crude way, a concrete floor of a steel frame building during an earthquake. A system shown in Fig. 1.5:1(c) may represent an astronaut in flight. A dumbbell shown in Fig. 1.5:1(d) has been used to model a molecule. It is clear that fairly comprehensive models of dynamic systems can be devised by this approach, see, for example, Fig. 1.5:1(e). When there are too many particles of mass and springs of elasticity, however, simplicity may be lost, and one turns naturally to the continuum approach outlined in the next section.

1.6. PROTOTYPE OF WAVE DYNAMICS

A wire with an infinite number of particles of mass connected together elastically is an obvious candidate for the continuum approach. As an example, consider a wire as shown in Fig. 1.6:1. Let an axis x run along

Fig. 1.6:1.

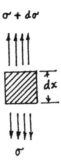

Fig. 1.6:2.

the length of the wire, with an origin O chosen at the lower end. When the wire is loaded, each particle in the wire will be displaced longitudinally from its original position by an amount u. We shall consider only axial loading and assume that the plane cross sections remain plane, so that u is parallel to the x-axis and is a function of x. We shall assume further that u is infinitesimal, so that the strain in the wire is

(1)
$$e = \frac{\partial u}{\partial x}.$$

The change in cross-sectional dimensions of the wire due to the axial load will be ignored. The wire is assumed to be elastic, and is so thin that all stress components other than the axial may be neglected. Then the axial stress is given by Hooke's law

(2)
$$\sigma = Ee = E\frac{\partial u}{\partial x},$$

in which E is a constant called the Young's modulus of elasticity. Now consider an element of the wire of a small length dx (see Fig. 1.6:2). The force acting at the lower end is σA, where A is the cross-sectional area of the wire. The force acting at the upper end is $A(\sigma + d\sigma)$. For a continuous function $\sigma(x)$ the differential $d\sigma$ is equal to $(\partial\sigma/\partial x)dx$. The acceleration of the element, $\partial^2 u/\partial t^2$, must be caused by the difference in axial forces. Now, the mass of the element is $\rho A dx$, with ρ denoting the density of the material. Hence, equating the mass times acceleration with the axial load $A d\sigma$, and canceling the nonvanishing factor $A dx$, we obtain the equation of motion

$$(3) \qquad\qquad \rho\frac{\partial^2 u}{\partial t^2} = \frac{\partial\sigma}{\partial x}.$$

A substitution of Eq. (2) yields the wave equation

$$(4) \qquad\qquad \frac{\partial^2 u}{\partial x^2} - \frac{1}{c^2}\frac{\partial^2 u}{\partial t^2} = 0,$$

in which the constant c is the wave speed:

$$(5) \qquad\qquad c = \sqrt{\frac{E}{\rho}}.$$

The wave speed c is a characteristic constant of the material.

The general solution of Eq. (4) is

$$(6) \qquad\qquad u = f(x - ct) + F(x + ct),$$

where f and F are two arbitrary functions. This can be verified by substituting Eq. (6) directly into Eq. (4). The function $u = f(x - ct)$ represents a wave propagating in the positive x direction. The function $u = F(x + ct)$ represents a wave propagating in the negative x direction. In either case we have

$$(7) \qquad\qquad \frac{\partial u}{\partial x} = \pm\frac{1}{c}\frac{\partial u}{\partial t} = \pm\frac{v}{c},$$

with $v = \partial u/\partial t$ denoting the particle velocity. A substitution of (7) and (5) into (2) yields the formula

$$(8) \qquad\qquad \sigma = \pm\frac{E}{c}v = \pm\rho c v,$$

where the $-$ sign applies to a wave propagating in the positive x direction, and the $+$ sign applies to a wave in the other direction. This result is remarkable. It says that the stress is equal to the product of the mass density of the material, the velocity of the sound wave, and the longitudinal velocity of the particles of the wire. This result has a very important

application to an experiment initiated by John Hopkinson (1872) who hang
a steel wire vertically from a ceiling, attached a stopper at the lower end as
shown in Fig. 1.6:1, then dropped a massive weight down the wire from a
measured height. When the weight struck the stopper, it stretches the wire
suddenly. Hopkinson's intention was to measure the strength of the wire
with this method. Using different weights dropped from different heights,
he found that the minimum height from which a weight had to be dropped
to break the wire was, within certain limits, almost independent of the
magnitude of the weight, and the diameter of the wire.

Now, when different solid bodies are dropped from a given height, the
velocity reached at any given time is independent of the weight. Thus
Hopkinson explains his result on the basis of elastic wave propagation.
Equation (8) shows that the stress is equal to $\rho c v$. The speed of sound of
longitudinal waves in the wire, given by Eq. (5) is 16,000 ft/sec for steel.
The velocity v in the wire, however, is not necessarily the largest at the
instant of impact at the lower end. To find v, we have to solve Eq. (4) with
the boundary conditions of the Hopkinson experiment (Fig. 1.6:1), which
has a clamped upper end, and a movable lower end:

$$(9) \qquad\qquad u = 0 \qquad \text{at } x = L \text{ for all } t, \text{ and} \qquad\qquad (16)$$

$$\text{when } t \le 0 \text{ for all } x,$$

$$(10) \qquad\qquad M\frac{dV}{dt} = A\sigma \qquad \text{at } x = 0 \text{ for all } t, \qquad\qquad (17)$$

$$(11) \qquad\qquad V = V_0 \qquad \text{at } x = 0 \text{ when } t = 0. \qquad\qquad (18)$$

Here V_0 is the velocity of the stopper, and M is the combined mass of the
dropped weight and the stopper. The mathematical problem is reduced to
finding the arbitrary function $f(x - ct)$ and $F(x + ct)$ so that $u(x, t)$ given
by Eq. (6) satisfies Eqs. (9)–(11).

Physically, the weight generated an elastic wave $u(x, t)$ in the wire initiated by a velocity V_0 at $x = 0$ when $t = 0$. The wave went up, reflected at
the top, went down, reflected at the lower end by the moving weight, and so
on. In John Hopkinson's test, a 27-foot long wire and weights ranging from
7 to 41 lbs. were used, and the absolute maximum of v and σ was reached
at the top after a number of reflections. The stress picture was very complicated. Bertram Hopkinson (1905) repeated his father's experiment with
a 1 lb. weight impacting on the lower end of a 30-foot long No. 10 gauge
wire weighing 1.3 lbs. Nevertheless, as shown by G. I Taylor (1946), the
maximum tensile stress in B. Hopkinson's experiment did not occur at the

first reflection, when the stress was $2\rho c V_0$, but at the third reflection, i.e., the second reflection at the top of the wire, when the tensile stress reached $2.15 \ \rho c V_0$.

If the mass of the wire were negligible, then the wave velocity is very high and the system of Fig. 1.6:1 would become a mass and spring system of Sec. 1.5.

Thus waves and vibrations are different features of the same dynamic system. Waves deliver extremely important information in a solid body. They are used in geophysics to probe the earth, to study earthquakes and to find oil and gas reserves. In medicine the pulse waves in the artery and the sound waves in the lung are used for diagnosis. In music the piano tuner and violin player listen to the vibration characteristics of their instruments. Some aspects of the general theory are discussed in Chapter 9.

1.7. BIOMECHANICS

Historically, the theory of solid mechanics was first developed along the lines of the linearized theory of elasticity, then it was expanded to the linearized theory of viscoelasticity. Then plasticity came on the scene. The nonlinear theory of large deformation and finite strain was developed. Further expansion was made possible by computational methods. Many significant problems in material science, aerospace structures, metal and plastics industries, geophysics, planetary physics, thermo-nuclear reactors, etc., were solved. Following this trend of expansion, it is natural to consider biological problems.

Many mechanics scientists had contributed to the understanding of physiology, e.g: Galileo Galilei (1564–1642), William Harvey (1578–1658), René Descartes (1596–1650), Giovanni Alfonso Borelli (1608–1679), Robert Boyle (1627–1691), and Robert Hooke (1635–1703) before Newton (1642–1727), and Leonhard Euler (1707–1783), Thomas Young (1773–1829), Jean Poiseuille (1797–1869), Herrmann von Helmholtz (1821–1894) after Newton. In fact, the Greek book *On the Parts of Animals* written by Aristotle (384–322 BC) and the Chinese book *Nei Jing* (or *Internal Classic*) attributed to Huangti (Yellow Emperor), but was believed to be written by anonymous authors in the Warring Period (472–221 BC), contain many concepts of biomechanics.

At the time when the manuscript of the first edition of this book was prepared, the molecule that is responsible for the genetics of the cells had been identified as DNA, and the double helix structure of DNA had been discovered. The beginning of a new age of understanding biology in terms of chemistry, physics and mechanics was recognized by many people.

Biomechanics has the following salient features:

(1) Material Constitution

Every living organism has a solid structure that gives it a unique shape and size, and an internal fluid flow that transports materials and keep the organism alive. Hence biosolid mechanics is inseparable from biofluid mechanics. The cells make new materials. Hence the composition, structure, and ultrastructure of biomaterials change dynamically.

(2) The Constitutive Equations

Blood is a non-Newtonian fluid. Synovial fluid in the knee joint, and body fluid in the abdominal cavity are also non-Newtonian. The bone obeys Hooke's law. The blood vessel, the skin and most other soft tissues in the body do not obey Hooke's law. We must know the constitutive equations of biological tissues, whose determination was a major task of the early biomechanics. When attention is focused on cells, the determination of the constitutive equations of DNA and other molecules becomes a primary task.

(3) Growth and Remodeling of Living Tissues Under Stress

Tissue is made of cells and extracellular matrices. The cell division, growth, hypertrophy, movement, or death are influenced by the cell geometry and the stress and strain in the cell, as well as by the chemical and physical environment. The tissue is, therefore, a changing, living entity. The determination of the growth law is a major task in biology.

(4) The Existence of Residual Stress

Organs such as the blood vessel, heart, esophagus and intestine have large residual strain and stress at the no-load condition. The residual strain changes in life because any birth or death of a cell in a continuum and the building and resorption of extracellular matrix by the cells create residual strains in the continuum. The changes of residual stress and strain cause changes of the stress state *in vivo*. The growth law and the residual stress are coupled.

(5) Concern about the Hierarchy of Sizes

In biology the hierarchy of sizes dominates the scene. Consider human. In studying gait, posture and sports, the length scale of interest is that of the whole body. In the study of the hemodynamics of the heart valves, a characteristic length is the size of the left ventricle. For coronary atherosclerosis studies, a characteristic dimension for hemodynamics in the vessel is the diameter of the coronary arteries, that for the shear stress on the vessel wall is the thickness of the endothelial cell, in the micrometers range, that for the molecular mechanism in the cells must be in the nanometers range. For microcirculation, features must be measured in the scale of the diameters of the red blood cells and the capillary blood vessels. In studying heart muscle contraction we must consider features in length scale of the sarcomere. In considering gene therapy we must think of the functions of the DNA molecule. In determining the active transport phenomenon we must be concerned with ion channels in the cell membrane. At each level of the hierarchy, there are important problems to be solved. Each hierarchy has its own appropriate mechanics. Integration of all the hierarchies is the objective of physiological studies.

(6) Perspectives

Life is motion. The science of motion is mechanics. Hence biology needs mechanics. Molecular biology needs molecular mechanics. Cell biology needs cell mechanics. Tissue biology needs tissue mechanics. Organ biology needs organ mechanics. Medicine, surgery, injury prevention, injury treatment, rehabilitation, and sports need mechanics. Solid and fluid mechanics have much to offer to biology. Yet, still, biomechanics seems to be a satellite, not in the main streams of biology and mechanics. Why? The reason probably lies in certain axiomatic differences between biology and mechanics. Classical continuum mechanics identifies a continuum as an isomorphism of the real number system. Biology studies molecules, cells, tissues, organs, and individuals in an orderly manner. The heirarchies are well defined. How can a biological entity be identified as a continuous body? The answer must lie in the clear recognition of the hierarchical structure. In each hierarchy, one must select a proper length scale as the starting point. No question with a characteristic length smaller than that selected dimension can be discussed. Then, in that hierarchy, one can identify a continuum to study it. This is the first axiom of biomechanics.

Secondly, solids of the classical solid mechanics have fixed structures and materials, an unchanging zero-stress state, and a set of fixed constitutive equations for the materials. Biomechanics, however, has to deal with DNA-controlled changes of materials, structures, zero-stress state, and constitutive equations. All these can be identified phenomenologically by suitably designed experiments. This is the second axiom of biomechanics.

Mechanics and biology can be unified only if these axiomatic differences are recognized. Knowing the theoretical structure of biomechanics, we can then develop classical and computational methods to cover it. Then biology becomes a new source of fresh water to irrigate our field of mechanics.

The present book will not deal with biomechanics any further. Some references are given in the Bibliography at the end of the book.

1.8. HISTORICAL REMARKS

The best-known constitutive equation for a solid is the Hooke's law of linear elasticity, which was discovered by Robert Hooke in 1660. This law furnishes the foundation for the mathematical theory of elasticity. By 1821, Louis M. H. Navier (1785–1836) had succeeded in formulating the general equations of the three-dimensional theory of elasticity. All questions of the small strain of elastic bodies were thus reduced to a matter of mathematical calculation. In the same year, 1821, Fresnel (1788–1827) announced his wave theory of light. The concept of transverse oscillations through an elastic medium attracted the attention of Cauchy and Poisson. Augustin L. Cauchy (1789–1857) developed the concept of stress and strain, and formulated the linear stress-strain relationship that is now called Hooke's law. Simeon D. Poisson (1781–1840) developed a molecular theory of elasticity and arrived at the same equation as Navier's. Both Navier and Poisson based their analysis on Newtonian conception of the constitution of bodies, and assumed certain laws of intermolecular forces. Cauchy's general reasoning, however, made no use of the hypothesis of material particles. In the ensuing years, with the contributions of George Green (1793–1841), George Stokes (1819–1903), Lord Kelvin (William Thomson, 1824–1907), and others, the mathematical theory was established. The fundamental questions of continuum mechanics have received renewed attention in recent years. Through the efforts of C. Truesdell, R. S. Rivlin, W. Noll, J. L. Erikson, A. E. Green, and others, theories of finite strain and nonlinear constitutive equations have been developed. Another focus of development is in the areas of micron and submicron-structures in which strain gradient effect can become dominant. In biomechical field, the study of the growth and remodeling of living tissues will lead continuum mechanics to another plateau. New developments in continuum mechanics since late 1940's have been truly remarkable.

The great impact of solid mechanics on civilization, however, is felt through its application to technology. In this respect a fine tradition was established by early masters. Galileo Galilei (1564–1642) considered the question of strength of beams and columns. James (or Jacques) Bernoulli (1654–1705) introduced the simple beam theory. Leonhard Euler (1707–1783) gave the column formula. Charles Augustin Coulomb (1736–1806) considered the failure criterion. Joseph Louis Lagrange (1736–1813) gave the equation that governs the bending and vibration of plates. Navier and Poisson gave numerous applications of their general theory to special problems. The best example was set by Barre de Saint-Venant (1797–1886), whose general solution of the problems of torsion and bending of prismatical bars is of great importance in engineering. With characteristic consideration for those who would use his results he solved these problems with a completeness that included many numerical coefficients and graphical presentations.

From this auspicious beginning, mechanics have developed into the mainstay of our civilization. Physics and chemistry are dominated by the theory of relativity, quantum mechanics, quantum electrodynamics. Geophysics, meteorology, and oceanography rest on the mechanics of seismic waves, atmospheric and ocean currents, and acoustic waves. Airplanes, ships, rockets, spacecraft, bathyscaphs, automobiles, trains, rails, highways, buildings, internal combustion engines, jet engines, are designed, constructed, and maintained on the basis of fluid and solid mechanics. New materials are invented for their mechanical properties. Artificial heart valves, hearts, kidneys, limbs, skin, pacemakers are mechanical marvels. Fundamental mechanics developed side by side with the developments in science and technology. In the first half of the 20th century, advancements were mainly concentrated on analytical solutions of the differential equations. In the second half of the 20th century, computational methods took center stage. These trends are reflected in this book.

PROBLEMS

1.1. Prove that it is possible to generalize Hooke's law to deal with moments and angles of rotations by considering a concentrated couple as the limiting case of two equal and opposite forces approaching each other but maintaining a constant moment.

1.2. A pin-jointed truss is shown in Fig. P1.2. Every member of the truss is made of the same steel and has the same cross-sectional area. Find the tension or

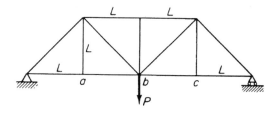

P1.2

compression in every member when a load P is applied at the point shown in the figure. *Note*: A joint is said to be a *pin joint* if no moment can be transmitted across it. This problem is statically determinate.

1.3. Find the vertical deflections at a, b, c of the truss in Fig. P1.2 due to a load P at point b. Assume that P is sufficiently small so that the truss remains linear elastic. Use the fact that for a single uniform bar in tension the total change in length of this bar is given by PL/AE, where P is the load in the bar, L is the bar length, A is the cross-sectional area, and E is the Young's modulus of the material. For structural steels E is about 3×10^7 lb/sq in, or 20.7×10^7 kPa. *Hint*: Use Castigliano's theorem.

1.4. A "rigid" frame of structural steel shown in Fig. P1.4 is acted on by a horizontal force P. Find the reactions at the points of support a, and d. The

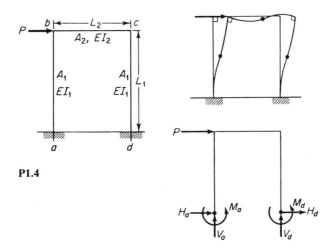

P1.4

cross-sectional area A and the bending rigidity constant EI of the various members are shown in the figure.

By a "rigid" frame is meant a frame whose joints are rigid; e.g., welded. Thus, when we say that corner b is a rigid joint, we mean that the two members meeting at b will retain the same angle at that corner under any loading. The frame under loading will deform. It is useful to sketch the deflection line of the frame. Try to locate the points of inflection on the members. The points of *inflection* are points where the bending moments vanish.

The reactions to be found are the vertical force V, the horizontal force H, and the bending moment M, at each support.

Use engineering beam theory for this problem. In this theory, the change of curvature of the beam is M/EI, where M is the local bending moment and EI is the local bending rigidity. The strain energy per unit length due to bending is, therefore, $M^2/2EI$. The strain energy per unit length due to a tensile force P is $P^2/2EA$. The strain energy due to transverse shear is negligible in comparison with that due to bending.

1.5. A circular ring of uniform linear elastic material and uniform cross section is loaded by a pair of equal and opposite forces at the ends of a diameter (Fig. P1.5). Find the change of diameters aa and bb. *Note:* This is a statically indeterminate problem. You have to determine the bending moment distribution in the ring.

1.6. Compare the changes in diameters aa, bb, and cc when a circular ring of uniform linear elastic material and uniform cross section is subjected to a pair of bending moments at the ends of a diameter (Fig. P1.6).

1.7. *Begg's Deformeter.* G. E. Beggs used an experimental model to determine the reactions of a statically indeterminate structure. For example, to determine the horizontal reaction H at the right-hand support b of an elastic arch under the load P, he imposes at b a small displacement δ in the horizontal

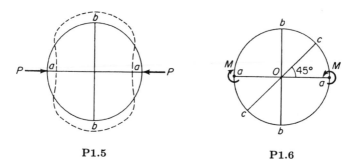

P1.5   **P1.6**

direction and measures at P the deflection δ' in the direction corresponding with P, while preventing the vertical displacement and rotation of the end b, Fig. P1.7. Show that $H = -P\delta'/\delta$. (Use the reciprocal theorem.) [G. E. Beggs, *J. Franklin Institute, 203* (1927), pp. 375–386.]

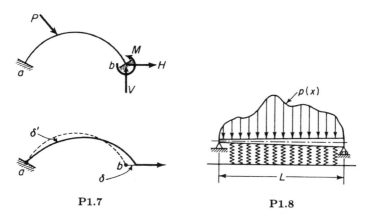

P1.7   **P1.8**

1.8. A simply supported, thin elastic beam of variable cross section with bending rigidity $EI(x)$ rests on an elastic foundation with spring constant k and is loaded by a distributed lateral load of intensity $p(x)$ per unit length (Fig. P1.8). Find an approximate expression for the deflection curve by assuming that it can be represented with sufficient accuracy by the expression

$$u(x) = \sum_{n=1}^{N} a_n \sin \frac{n\pi x}{L}.$$

Note: This expression satisfies the end conditions for arbitrary coefficients a_n. Use the minimum potential energy theorem.

1.9. Consider a uniform cantilever beam clamped at $x = 0$ (Fig. P1.9). According to Bernoulli-Euler theory of beams, the differential equation governing the deflection of the beam is

$$\frac{d^2 w}{dx^2} = \frac{M}{EI},$$

P1.9

where w is the deflection parallel to the z-axis and M is the bending moment at station x. This equation is valid if the beam is straight and slender and if the loading acts in a plane containing a principal axis of all cross sections of the beam, in which case the deflection w occurs also in that plane.

Use this differential equation and the boundary conditions:

$$At\ clamped\ end: \quad w = \frac{dw}{dx} = 0.$$

$$At\ free\ end: \quad EI\frac{d^2 w}{dx^2} = EI\frac{d^3 w}{dx^3} = 0.$$

Derive the deflection curve of the beam when it is loaded by a unit force located at $x = \xi$.

$$Ans. \quad For\ 0 \le x \le \xi \le l: \quad w = \frac{x^2}{6EI}\left(3\xi - x\right).$$

$$For\ 0 \le \xi \le x \le l: \quad w = \frac{\xi^2}{6EI}\left(3x - \xi\right).$$

The solution fits the definition of influence coefficients and is known as an influence function.

1.10. Let the beam of Problem 1.9 be divided into six equidistant sections as marked in Fig. P1.9. Compute the influence coefficients c_{ij}, where $i, j = 1, 2, \ldots, 6$.

2

TENSOR ANALYSIS

In attempting to develop the theories outlined in the previous chapter rigorously and succinctly, we shall first learn to use the powerful tool of tensor calculus. At first sight, the mathematical analysis may seem involved, but a little study will soon reveal its simplicity.

2.1. NOTATION AND SUMMATION CONVENTION

Let us begin with the matter of notation. In tensor analysis one makes extensive use of indices. A set of n variables $x_1, x_2, \ldots, x_n$ is usually denoted as $x_i, i = 1, \ldots, n$. A set of n variables $y^1, y^2, \ldots, y^n$ is denoted by $y^i, i = 1, \ldots, n$. We emphasize that $y^1, y^2, \ldots, y^n$ are n independent variables and *not* the first n powers of the variable y.

Consider an equation describing a plane in a three-dimensional space x^1, x^2, x^3,

$$(1) \qquad a_1 x^1 + a_2 x^2 + a_3 x^3 = p,$$

where a_i and p are constants. This equation can be written as

$$(2) \qquad \sum_{i=1}^{3} a_i x^i = p.$$

However, we shall introduce the *summation convention* and write the equation above in the simple form

$$(3) \qquad a_i x^i = p.$$

The convention is as follows: *The repetition of an index (whether superscript or subscript) in a term will denote a summation with respect to that index over its range.* The *range* of an index i is the set of n integer values 1 to n. A lower index i, as in a_i, is called a *subscript*, and upper index i, as in x^i, is called a *superscript*. An index that is summed over is called a *dummy index*; one that is not summed out is called a *free index*.

Since a dummy index just indicates summation, it is immaterial which symbol is used. Thus, $a_i x^i$ may be replaced by $a_j x^j$, etc. This is analogous to the dummy variable in an integral

$$\int_a^b f(x)dx = \int_a^b f(y)dy\,.$$

The use of index and summation convention may be illustrated by other examples. Consider a unit vector $\boldsymbol{\nu}$ in a three-dimensional Euclidean space with rectangular Cartesian coordinates x, y, and z. Let the direction cosines α_i be defined as

$$\alpha_1 = \cos(\boldsymbol{\nu},\mathbf{x})\,,\quad \alpha_2 = \cos(\boldsymbol{\nu},\mathbf{y})\,,\quad \alpha_3 = \cos(\boldsymbol{\nu},\mathbf{z})\,,$$

where $(\boldsymbol{\nu},\mathbf{x})$ denotes the angle between $\boldsymbol{\nu}$ and the x-axis, etc. The set of numbers $\alpha_i(i = 1,2,3)$ represents the projections of the unit vector on the coordinate axes. The fact that the length of the vector is unity is expressed by the equation

$$(\alpha_1)^2 + (\alpha_2)^2 + (\alpha_3)^2 = 1\,,$$

or simply

(4) $$\alpha_i\alpha_i = 1\,.$$

As a further illustration, consider a line element (dx, dy, dz) in a three-dimensional Euclidean space with rectangular Cartesian coordinates x, y, z. The square of the length of the line element is

(5) $$ds^2 = dx^2 + dy^2 + dz^2\,.$$

If we define

(6) $$dx^1 = dx\,,\quad dx^2 = dy\,,\quad dx^3 = dz\,,$$

and

(7) ▲
$$\delta_{11} = \delta_{22} = \delta_{33} = 1\,,$$
$$\delta_{12} = \delta_{21} = \delta_{13} = \delta_{31} = \delta_{23} = \delta_{32} = 0\,.$$

Then (5) may be written as

(8) $$ds^2 = \delta_{ij}dx^i dx^j\,,$$

with the understanding that the range of the indices i and j is 1 to 3. Note that there are two summations in the expression above, one over i and one over j. The symbol δ_{ij} as defined in Eq. (7) is called the *Kronecker delta*.

The following determinant illustrates another application

$$
\begin{vmatrix} a_{11} & a_{12} & a_{13} \\ a_{21} & a_{22} & a_{23} \\ a_{31} & a_{32} & a_{33} \end{vmatrix} = \begin{array}{l} a_{11}a_{22}a_{33} + a_{21}a_{32}a_{13} + a_{31}a_{12}a_{23} \\[4pt] - a_{11}a_{32}a_{23} - a_{12}a_{21}a_{33} - a_{13}a_{22}a_{31} \, . \end{array}
$$

If we denote the general term in the determinant by a_{ij} and write the determinant as $|a_{ij}|$, then the equation above can be written as

$$
(9) \qquad |a_{ij}| = e_{rst}a_{r1}a_{s2}a_{t3} \, ,
$$

where e_{rst}, the *permutation symbol*, is defined by the equations

$$
(10) \quad \blacktriangle \qquad
\begin{aligned}
e_{111} &= e_{222} = e_{333} = e_{112} = e_{113} = e_{221} = e_{223} \, , \\
&= e_{331} = e_{332} = 0 \, , \\
e_{123} &= e_{231} = e_{312} = 1 \, , \\
e_{213} &= e_{321} = e_{132} = -1 \, .
\end{aligned}
$$

In other words, e_{ijk} vanishes whenever the values of any two indices coincide; $e_{ijk} = 1$ when the subscripts permute like $1, 2, 3$; and $e_{ijk} = -1$ otherwise.

The Kronecker delta and the permutation symbol are very important quantities which will appear again and again in this book. They are connected by the identity

$$
(11) \quad \blacktriangle \qquad e_{ijk}e_{ist} = \delta_{js}\delta_{kt} - \delta_{jt}\delta_{ks} \, .
$$

This e-δ identity is used sufficiently frequently to warrant special attention here. It can be verified by actual trial.

Finally, we shall extend the summation convention to differentiation formulas. Let $f(x^1, x^2, \ldots, x^n)$ be a function of n variables $x^1, x^2, \ldots, x^n$. Then its differential shall be written as

$$
(12) \qquad df = \frac{\partial f}{\partial x^i} \, dx^i \, .
$$

The Kronecker delta and the permutation symbol play important roles in vector or tensor operations in rectangular Cartesian coordinate system. The dot product of two vectors $\boldsymbol{x} = (x_1, x_2, x_3)$ and $\boldsymbol{y} = (y_1, y_2, y_3)$ is a scalar

$$
c = \boldsymbol{x} \cdot \boldsymbol{y} = x_1 y_1 + x_2 y_2 + x_3 y_3 \, ,
$$

DRAWING: C 1022828

DRAWING #   ISS# TITLE   PPS#   ASN-DATE
-------   ---- -----   ----   --------
C -1022828   2 FINDER SHEET IMT/CMT WINDER & CORE DELIVER   06/17/98
DRAWING ORIGINAL CHARGED TO: ALEXANDER & ASSOC   06/17/98 WICKEN   .DED
                 CONTACT: ROSE WICKENHOFER (513)731-7800

MICROFILM CUSTODIAN: REPRO PE
MICROFILM STATUS: YEAR AUDITED   ISS#
     USAGE DECK:
     SECURITY DECK:

                      DRAWING ROOTS

TECH   PLANT PRES   AREA   DESC   BLDG   LINE   LEVEL   EQUIP#
----   ----- ----   ----   ----   ----   ----   -----   ------
CONV   GLBT   FS   KRT   WIND
CONV   GLBT   FS   ALL

          DRAWINGS - NEXT LEVEL UP

DRAWING #   ISS# TITLE   PRES   BOM IT#
-------   ---- -----   ----   -------
DL-27260   0 IMT/CMT WINDER & CORE DELIVERY   DL
          DRAWINGS - NEXT LEVEL DOWN

NONE

which can be written as

(13) $$c = \delta_{ij} x_i y_j = x_i y_i\,.$$

The vector product of two vectors x and y is the vector $z = x \times y$ whose three components are

(14) $z_1 = x_2 y_3 - x_3 y_2\,,\quad z_2 = x_3 y_1 - x_1 y_3\,,\quad z_3 = x_1 y_2 - x_2 y_1\,.$

This can be shortened by writing

(15) $$z_i = e_{ijk} x_j y_k\,.$$

PROBLEMS

2.1.   Show that, when i, j, k range over $1, 2, 3$,

(a) $\delta_{ij}\delta_{ij} = 3$ (c) $e_{ijk} A_j A_k = 0$

(b) $e_{ijk} e_{jki} = 6$ (d) $\delta_{ij}\delta_{jk} = \delta_{ik}$

2.2.   Verify the following identity connecting three arbitrary vectors by means of the e-δ identity.

$$\mathbf{A} \times (\mathbf{B} \times \mathbf{C}) = (\mathbf{A} \cdot \mathbf{C})\mathbf{B} - (\mathbf{A} \cdot \mathbf{B})\mathbf{C}\,.$$

Note: The last equation is well known in vector analysis. After identifying the quantities involved as Cartesian tensors, this verification may be construed as a proof of the e-δ identity.

2.2. COORDINATE TRANSFORMATION

The central point of view of tensor analysis is to study the change of the components of a quantity such as a vector with respect to coordinate transformations.

A set of independent variables x_1, x_2, x_3 may be thought of as specifying the coordinates of a point in a frame of reference. A transformation from x_1, x_2, x_3 to a set of new variables $\bar{x}_1, \bar{x}_2, \bar{x}_3$ through the equations

(1) $$\bar{x}_i = f_i(x_1, x_2, x_3)\,,\quad i = 1, 2, 3,$$

specifies a transformation of coordinates. The inverse transformation

(2) $$x_i = g_i(\bar{x}_1, \bar{x}_2, \bar{x}_3)\quad i = 1, 2, 3,$$

proceeds in the reverse direction. In order to insure that such a transformation is reversible and in one-to-one correspondence in a certain region R of the variables (x_1, x_2, x_3), i.e., in order that each set of numbers (x_1, x_2, x_3) defines a unique set of numbers $(\bar{x}_1, \bar{x}_2, \bar{x}_3)$, for (x_1, x_2, x_3) in the region R, and vice versa, it is sufficient to meet the following conditions:

(a) The functions f_i are single-valued, continuous, and possess continuous first partial derivatives in the region R, and

(b) The *Jacobian determinant* $J = |\partial \bar{x}_i / \partial x_j|$ does not vanish at any point of the region R, i.e.,

(3)
$$
J = \left| \frac{\partial \bar{x}_i}{\partial x_j} \right| \equiv
\begin{vmatrix}
\dfrac{\partial \bar{x}_1}{\partial x_1} & \dfrac{\partial \bar{x}_1}{\partial x_2} & \dfrac{\partial \bar{x}_1}{\partial x_3} \\[2ex]
\dfrac{\partial \bar{x}_2}{\partial x_1} & \dfrac{\partial \bar{x}_2}{\partial x_2} & \dfrac{\partial \bar{x}_2}{\partial x_3} \\[2ex]
\dfrac{\partial \bar{x}_3}{\partial x_1} & \dfrac{\partial \bar{x}_3}{\partial x_2} & \dfrac{\partial \bar{x}_3}{\partial x_3}
\end{vmatrix}
\neq 0 \quad \text{in } R.
$$

Coordinate transformations with the properties (a), (b) named above are called *admissible transformations*. If the Jacobian is positive everywhere, then a right-handed set of coordinates is transformed into another right-handed set, and the transformation is said to be *proper*. If the Jacobian is negative everywhere, a right-handed set of coordinates is transformed into a left-handed one, and the transformation is said to be *improper*. *In this book, we shall tacitly assume that our transformations are admissible and proper.*

2.3. EUCLIDEAN METRIC TENSOR

The first thing we must know about any coordinate system is how to measure length in that reference system. This information is given by the metric tensor.

Consider a three-dimensional Euclidean space, with the range of all indices 1, 2, 3. Let

(1)
$$
\theta_i = \theta_i(x_1, x_2, x_3),
$$

be an admissible transformation of coordinates from the rectangular Cartesian (in honor of Cartesius, i.e., Descartes) coordinates x_1, x_2, x_3 to some general coordinates $\theta_1, \theta_2, \theta_3$. The inverse transformation

(2)
$$
x_i = x_i(\theta_1, \theta_2, \theta_3),
$$

is assumed to exist, and the point (x_1, x_2, x_3) and $(\theta_1, \theta_2, \theta_3)$ are in one-to-one correspondence.

Consider a line element with three components given by the differentials dx^1, dx^2, dx^3. Since the coordinates x_1, x_2, x_3 are assumed to be rectangular Cartesian, the length of the element ds is determined by Pythagoras' rule:

(3) ▲ $$ds^2 = dx^i dx^i = \delta_{ij} dx^i dx^i \,.$$

Here ds^2 is the square of ds, not ds with a superscript 2. When a coordinate transformation (1) is effected, we obtain from Eq. (2), according to the ordinary rules of differentiation,

(4) $$dx^i = \frac{\partial x_i}{\partial \theta_k} d\theta^k \,.$$

Substituting (4) into (3), we have

$$ds^2 = \frac{\partial x_i}{\partial \theta_k} \frac{\partial x_i}{\partial \theta_m} d\theta^k d\theta^m \,.$$

If we define the functions $g_{km}(\theta_1, \theta_2, \theta_3)$ by

(5) ▲ $$g_{km}(\theta_1, \theta_2, \theta_3) = \frac{\partial x_i}{\partial \theta_k} \frac{\partial x_i}{\partial \theta_m} \,,$$

then the square of the line element in the general $\theta_1, \theta_2, \theta_3$ coordinates takes the form

(6) ▲ $$ds^2 = g_{km} d\theta^k d\theta^m \,.$$

This, of course, stands for

(7) $$\begin{aligned}
ds^2 = {} & g_{11}(d\theta^1)^2 + g_{12} d\theta^1 d\theta^2 + g_{13} d\theta^1 d\theta^3 \\
& + g_{21} d\theta^1 d\theta^2 + g_{22}(d\theta^2)^2 + g_{23} d\theta^2 d\theta^3 \\
& + g_{31} d\theta^1 d\theta^3 + g_{32} d\theta^2 d\theta^3 + g_{33}(d\theta^3)^2 \,.
\end{aligned}$$

It is apparent from (5) that

(8) ▲ $$g_{km} = g_{mk} \qquad \text{for each } k \text{ and } m \,,$$

so the functions g_{km} are symmetric in k and m. The functions g_{km} are called the components of the *Euclidean metric tensor* in the coordinate system $\theta_1, \theta_2, \theta_3$.

Let $\bar{\theta}_1, \bar{\theta}_2, \bar{\theta}_3$ be another general coordinate system. Let

$$(9) \qquad\qquad \theta_i = \theta_i(\bar{\theta}_1, \bar{\theta}_2, \bar{\theta}_3),$$

be the transformation of coordinates from $\bar{\theta}_1, \bar{\theta}_2, \bar{\theta}_3$ to $\theta_1, \theta_2, \theta_3$. Now

$$(10) \qquad\qquad d\theta^k = \frac{\partial \theta_k}{\partial \bar{\theta}_l} d\bar{\theta}^l.$$

Hence, from Eq. (6),

$$(11) \qquad\qquad ds^2 = g_{km} \frac{\partial \theta_k}{\partial \bar{\theta}_l} \frac{\partial \theta_m}{\partial \bar{\theta}_n} d\bar{\theta}^l d\bar{\theta}^n.$$

If we define
$$(12) \qquad \bar{g}_{ln}(\bar{\theta}_1, \bar{\theta}_2, \bar{\theta}_3) = g_{km}(\theta_1, \theta_2, \theta_3) \frac{\partial \theta_k}{\partial \bar{\theta}_l} \frac{\partial \theta_m}{\partial \bar{\theta}_n}.$$

Then Eq. (11) assumes a form which is the same as Eq. (3) or (6):

$$(13) \qquad\qquad ds^2 = \bar{g}_{ln} d\bar{\theta}^l d\bar{\theta}^n.$$

Accordingly, we call $\bar{g}_{ln}$ the components of the Euclidean metric tensor in the coordinate system $\bar{\theta}_1, \bar{\theta}_2, \bar{\theta}_3$.

The quadratic differential forms (3), (6), and (13) are of fundamental importance since they define the length of any line element in general coordinate systems. We conclude that if $\theta_1, \theta_2, \theta_3$ and $\bar{\theta}_1, \bar{\theta}_2, \bar{\theta}_3$ are two sets of general coordinates, then Euclidean metric tensors $g_{km}(\theta_1, \theta_2, \theta_3)$ and $\bar{g}_{km}(\bar{\theta}_1, \bar{\theta}_2, \bar{\theta}_3)$ are related by means of the law of transformation (12).

The law of transformation of the components of a quantity with respect to coordinate transformation is an important property of that quantity. In the following section, we shall see that a quantity shall be called a tensor if and only if it follows certain specific laws of transformation.

All the results above apply as well to the plane (a two-dimensional Euclidean space), as can be easily verified by changing the range of all indices to 1, 2.

P R O B L E M S

2.3. Find the components of the Euclidean metric tensor in plane polar coordinates ($\theta_1 = r, \theta_2 = \theta$; see Fig. P2.3) and the corresponding expression for the length of a line element
Ans. Let x_1, x_2 be a set of rectangular Cartesian coordinates.

Then

$$x_1 = \theta_1 \cos \theta_2, \quad \theta_1 = \sqrt{(x_1)^2 + (x_2)^2},$$

$$x_2 = \theta_1 \sin \theta_2, \quad \theta_2 = \sin^{-1}\left(\frac{x_2}{\sqrt{(x_1)^2 + (x_2)^2}}\right),$$

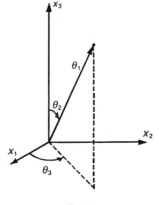

$$g_{11} = \cos^2 \theta_2 + \sin^2 \theta_2 = 1,$$

$$g_{12} = (\cos \theta_2)(-\theta_1 \sin \theta_1) + (\sin \theta_2)(\theta_1 \cos \theta_2) = 0 = g_{21},$$

$$g_{22} = (-\theta_1 \sin \theta_2)^2 + (\theta_1 \cos \theta_2)^2 = (\theta_1)^2.$$

P2.3

The line element is

$$ds^2 = (d\theta_1)^2 + (\theta_1)^2 (d\theta_2)^2.$$

2.4. Let x_1, x_2, x_3 be rectangular Cartesian coordinates and $\theta_1, \theta_2, \theta_3$ be spherical polar coordinates. (See Fig. P2.4.) Then

$$\theta_1 = \sqrt{(x_1)^2 + (x_2)^2 + (x_3)^2},$$

$$\theta_2 = \cos^{-1} \frac{x_3}{\sqrt{(x_1)^2 + (x_2)^2 + (x_3)^2}},$$

$$\theta_3 = \tan^{-1} \frac{x_2}{x_1},$$

and the inverse transformation is

$$x_1 = \theta_1 \sin \theta_2 \cos \theta_3,$$

$$x_2 = \theta_1 \sin \theta_2 \sin \theta_3,$$

$$x_3 = \theta_1 \cos \theta_2.$$

P2.4

Show that the components of the Euclidean metric tensor in the spherical polar coordinates are

$$g_{11} = 1, \quad g_{22} = (\theta_1)^2, \quad g_{33} = (\theta_1)^2 (\sin \theta_2)^2,$$

and all other $g_{ij} = 0$. The square of the line element is, therefore,

$$ds^2 = (d\theta_1)^2 + (\theta_1)^2 (d\theta_2)^2 + (\theta_1)^2 (\sin \theta_2)^2 (d\theta_3)^2.$$

2.5.   Show that the length of the line element $(d\theta^1, 0, 0)$ is $\sqrt{g_{11}}\,|d\theta^1|$; that of the line element $(0, d\theta^2, 0)$ is $\sqrt{g_{22}}\,|d\theta^2|$.

2.6.   Let the angle between the line elements $(d\theta^1, 0, 0)$ and $(0, d\theta^2, 0)$ be denoted by α_{12}. Show that

$$\cos\alpha_{12} = \frac{g_{12}}{\sqrt{g_{11}}\sqrt{g_{22}}}\,.$$

Hint: Find the components of the line elements with respect to a system of rectangular Cartesian coordinates in which we know how to compute the angle between two vectors.

2.4. SCALARS, CONTRAVARIANT VECTORS, COVARIANT VECTORS[†]

In nonrelativistic physics there are quantities like mass and length which are independent of reference coordinates, and there are quantities like displacement and velocity whose components do depend on reference coordinates. The former are the scalars; the latter, vectors. Mathematically, we define them according to the way their components change under admissible transformations. In the following discussions we consider a system whose components are defined in the general set of variables θ^i and are functions of θ^1, θ^2, θ^3. If the variables θ^i can be changed to $\bar{\theta}^i$ by an admissible and proper transformation, then we can define new components of the system in the new variables $\bar{\theta}^i$. The system will be given various names according to the way in which the new and the old components are related.

A system is called a *scalar* if it has only a single component ϕ in the variables θ^i and a single component $\bar{\phi}$ in the variables $\bar{\theta}^i$, and if ϕ and $\bar{\phi}$ are numerically equal at the corresponding points,

$$(1) \qquad\qquad \phi(\theta^1, \theta^2, \theta^3) = \bar{\phi}(\bar{\theta}^1, \bar{\theta}^2, \bar{\theta}^3)\,.$$

A system is called a *contravariant vector field* or a *contravariant tensor field of rank one*, if it has three components ξ^i in the variables θ^i and three components $\bar{\xi}^i$ in the variables $\bar{\theta}^i$, and if the components are related by the law

$$(2) \quad\blacktriangle\qquad\qquad \bar{\xi}^i(\bar{\theta}^1, \bar{\theta}^2, \bar{\theta}^3) = \xi^k(\theta^1, \theta^2, \theta^3)\frac{\partial\bar{\theta}^i}{\partial\theta^k}\,.$$

[†]One may pass over the rest of this chapter and proceed directly to Chapter 3 on first reading.

A contravariant vector is indicated by a superscript, called a *contravariant index*.

A differential $d\theta^i$ is a prototype of a contravariant vector:

$$(3) \qquad\qquad d\bar{\theta}^i = \frac{\partial \bar{\theta}^i}{\partial \theta^k}\, d\theta^k \,.$$

Hence, a vector is contravariant if it transforms like a differential.[†]

A system is called a *covariant vector field*, or a *covariant tensor field of rank one*, if it has three components η_i in the variables θ^1, θ^2, θ^3 and three components $\bar{\eta}_i$ in the variables $\bar{\theta}^1$, $\bar{\theta}^2$, $\bar{\theta}^3$, and if the components in these two coordinate systems are related by the law

$$(4) \quad \blacktriangle \qquad\qquad \bar{\eta}_i(\bar{\theta}^1, \bar{\theta}^2, \bar{\theta}^3) = \eta_k(\theta^1, \theta^2, \theta^3) \frac{\partial \theta^k}{\partial \bar{\theta}^i} \,.$$

A covariant vector is indicated by a subscript, called a *covariant index*.

A gradient of a scalar potential ϕ transforms like (4),

$$(5) \qquad\qquad \frac{\partial \phi}{\partial \bar{\theta}^i} = \frac{\partial \phi}{\partial \theta^k}\frac{\partial \theta^k}{\partial \bar{\theta}^i} \,.$$

Hence, a vector is covariant if it transforms like a gradient of a scalar potential.

2.5. TENSOR FIELDS OF HIGHER RANK

A system such as the three-dimensional Euclidean metric tensor g_{ij} has nine components when i and j range over 1, 2, 3. Such quantities are given special names when their components in any two coordinate systems are related by specific transformation laws.

Covariant tensor field of rank two, t_{ij}:

$$(1) \qquad\qquad \bar{t}_{ij}(\bar{\theta}^1, \bar{\theta}^2, \bar{\theta}^3) = t_{mn}(\theta^1, \theta^2, \theta^3)\frac{\partial \theta^m}{\partial \bar{\theta}^i}\frac{\partial \theta^n}{\partial \bar{\theta}^j} \,.$$

Contravariant tensor field of rank two, t^{ij}:

$$(2) \qquad\qquad \bar{t}^{ij}(\bar{\theta}^1, \bar{\theta}^2, \bar{\theta}^3) = t^{mn}(\theta^1, \theta^2, \theta^3)\frac{\partial \bar{\theta}^i}{\partial \theta^m}\frac{\partial \bar{\theta}^j}{\partial \theta^n} \,.$$

[†]The variables θ^i and $\bar{\theta}^i$, in general, are not so related [see Eq. (2.3:9)]. Thus, although the differential $d\theta^i$ is a contravariant vector, the set of variables θ^i itself does not transform like a vector. Hence, in this instance, the position of the index of θ^i must be regarded as without significance.

Mixed tensor field of rank two, t^i_j:

(3)
$$\bar{t}^i_j(\bar{\theta}^1, \bar{\theta}^2, \bar{\theta}^3) = t^m_n(\theta^1, \theta^2, \theta^3)\frac{\partial \bar{\theta}^i}{\partial \theta^m}\frac{\partial \theta^n}{\partial \bar{\theta}^j}\,.$$

Generalization to tensor fields of higher ranks is immediate. Thus, we call a quantity $t^{\alpha_1\cdots\alpha_p}_{\beta_1\cdots\beta_q}$ a *tensor field of rank* $r = p + q$, *contravariant of rank p and covariant of rank q*, if the components in any two coordinate systems are related by

(4)
$$\bar{t}^{\;\alpha_1\cdots\alpha_p}_{\;\beta_1\cdots\beta_q} = \frac{\partial \bar{\theta}^{\alpha_1}}{\partial \theta^{k_1}}\cdots\frac{\partial \bar{\theta}^{\alpha_p}}{\partial \theta^{k_p}}\cdot\frac{\partial \theta^{m_1}}{\partial \bar{\theta}^{\beta_1}}\cdots\frac{\partial \theta^{m_q}}{\partial \bar{\theta}^{\beta_q}}\,t^{k_1\cdots k_p}_{m_1\cdots m_q}\,.$$

Thus, the location of an index is important in telling whether it is contravariant or covariant. Again, if only *rectangular Cartesian coordinates* are considered, the distinction disappears.

These definitions can be generalized in an obvious manner if the range of the indices are $1, 2, \ldots, n$.

PROBLEMS

2.7. Show that, *if all components of a tensor vanish in one coordinate system, then they vanish in all other coordinate systems which are in one-to-one correspondence with the given system.*

This is perhaps the most important property of tensor fields.

2.8. Prove the theorem: *The sum or difference of two tensors of the same type and rank (with the same number of covariant and the same number of contravariant indices) is again a tensor of the same type and rank.*

Thus, any linear combination of tensors of the same type and rank is again a tensor of the same type and rank.

2.9. *Theorem. Let* $A^{\beta_1\cdots\beta_s}_{\alpha_1\cdots\alpha_r}$, $B^{\beta_1\cdots\beta_s}_{\alpha_1\cdots\alpha_r}$ *be tensors. The equation*

$$A^{\beta_1\cdots\beta_s}_{\alpha_1\cdots\alpha_r}(\theta^1, \theta^2, \ldots, \theta^n) = B^{\beta_1\cdots\beta_s}_{\alpha_1\cdots\alpha_r}(\theta^1, \theta^2, \ldots, \theta^n),$$

is a tensor equation; i.e., if this equation is true in some coordinate system, then it is true in all coordinate systems which are in one-to-one correspondence with each other.

Hint: Use the results of the previous problems.

2.6. SOME IMPORTANT SPECIAL TENSORS

If we define the Kronecker delta and the permutation symbol introduced in Sec. 2.1 as components of covariant, contravariant, and mixed tensors of ranks 2 and 3 in rectangular Cartesian coordinates x_1, x_2, x_3,

(1) $\delta_{ij} = \delta^{ij} = \delta^i_j = \delta^j_i \begin{cases} = 0 & \text{when } i \neq j\,, \\ = 1 & \text{when } i = j, j \text{ not summed}\,, \end{cases}$

(2) $e_{ijk} = e^{ijk} \begin{cases} = 0 & \text{when any two indices are equal}\,, \\ = 1 & \text{when } i, j, k \text{ permute like } 1, 2, 3\,, \\ = -1 & \text{when } i, j, k \text{ permute like } 1, 3, 2\,, \end{cases}$

what will be their components in general coordinates θ^i? The answer is provided immediately by the tensor transformation laws. Thus,

(3) $$g_{ij} = \frac{\partial x^m}{\partial \theta^i} \frac{\partial x^n}{\partial \theta^j} \delta_{mn} = \frac{\partial x^m}{\partial \theta^i} \frac{\partial x^m}{\partial \theta^j}\,,$$

(4) $$g^{ij} = \frac{\partial \theta^i}{\partial x^m} \frac{\partial \theta^j}{\partial x^n} \delta^{mn} = \frac{\partial \theta^i}{\partial x^m} \frac{\partial \theta^j}{\partial x^m}\,,$$

(5) $$g^i_j = \frac{\partial \theta^i}{\partial x^m} \frac{\partial x^n}{\partial \theta^j} \delta^m_n = \frac{\partial \theta^i}{\partial x^m} \frac{\partial x^m}{\partial \theta^j} = \delta^i_j\,,$$

(6) $$\epsilon_{ijk} = \frac{\partial x^r}{\partial \theta^i} \frac{\partial x^s}{\partial \theta^j} \frac{\partial x^t}{\partial \theta^k} e_{rst} = e_{ijk} \left| \frac{\partial x^m}{\partial \theta^n} \right| = e_{ijk} \sqrt{g}\,,$$

(7) $$\epsilon^{ijk} = \frac{\partial \theta^i}{\partial x^r} \frac{\partial \theta^j}{\partial x^s} \frac{\partial \theta^k}{\partial x^t} e^{rst} = e^{ijk} \left| \frac{\partial \theta^m}{\partial x^n} \right| = \frac{e^{ijk}}{\sqrt{g}}\,,$$

where g is the value of the determinant $|g_{ij}|$ and is positive for any proper coordinate system, i.e.,

(8) $$g = |g_{ij}| > 0\,.$$

We see that *the proper generalizations of the Kronecker delta are the Euclidean metric tensors, and those of the permutation symbol are ϵ_{ijk} and ϵ^{ijk}, which are called permutation tensors or alternators.* As defined by Eqs. (6) and (7), the components of ϵ_{ijk} and ϵ^{ijk} do not have values $1, -1$, or 0 in general coordinates, they are $(\sqrt{g}, -\sqrt{g}, 0)$ and $(1/\sqrt{g}, -1/\sqrt{g}, 0)$ respectively.

Note that the mixed tensor g^i_j is identical with δ^i_j and is constant in all coordinate systems. From Eqs. (3) and (4) we see that

(9) $$g_{im} g^{mj} = \delta^j_i\,.$$

Hence, the determinant

$$|g_{im} g^{mj}| = |g_{ij}| \cdot |g^{ij}| = |\delta^i_j| = 1\,.$$

Using (3) and (4), we can write

(10)
$$g = |g_{ij}| = \left|\frac{\partial x^i}{\partial \theta^j}\right|^2 ,$$

$$\frac{1}{g} = |g^{ij}| = \left|\frac{\partial \theta^i}{\partial x^j}\right|^2 .$$

Since g is positive, Eq. (9) may be solved to give

(11)
$$g^{ij} = \frac{D^{ij}}{g} ,$$

where D^{ij} is the cofactor of the term g_{ij} in the determinant g. *The tensor g^{ij} is called the associated metric tensor. It is as important as the metric tensor itself in the further development of tensor analysis.*

PROBLEM

2.10. Prove Eqs. (6), (7) and (10). Write down explicitly D^{ij} in Eq. (11) in terms of $g_{11}, g_{12}, \ldots, g_{33}$.

2.7. THE SIGNIFICANCE OF TENSOR CHARACTERISTICS

The importance of tensor analysis may be summarized in the following statement. The form of an equation can have general validity with respect to any frame of reference only if every term in the equation has the same tensor characteristics. If this condition is not satisfied, a simple change of the system of reference will destroy the form of the relationship. This form is, therefore, only fortuitous and accidental.

Thus, tensor analysis is as important as dimensional analysis in any formulation of physical relations. Dimensional analysis considers how a physical quantity changes with the particular choice of fundamental units. Two physical quantities cannot be equal unless they have the same dimensions; any physical equation cannot be correct unless it is invariant with respect to change of fundamental units.

Whether a physical quantity should be a tensor or not is a decision for the physicist to make. Why is a force a vector, a stress tensor a tensor? Because we say so! It is our judgement that endowing tensorial character to these quantities is in harmony with the world.

Once we decided upon the tensorial character of a physical quantity, we may take as the components of a tensor field in a given frame of reference any set of functions of the requisite number. A tensor field thus assigned in a given frame of reference then transforms according to the tensor transformation law when admissible transformations are considered. In other words, once the values of the components of a tensor are assigned in one particular coordinate system, the values of the components in any general coordinate system are fixed.

Why are the tensor transformation laws in harmony with physics? Because tensor analysis is designed so. For example, a tensor of rank one is defined in accordance with the physical idea of a vector. The only point new to the student is perhaps the distinction between contravariance and covariance. In elementary physics, natural laws are studied usually only in rectangular Cartesian coordinates of reference, in which the distinction between the contravariance and covariance disappears. When curvilinear coordinates are used in elementary physics, the vectorial components must be defined specifically in each particular case, and mathematical expressions of physical laws must be derived anew for each particular coordinate system. These derivations are usually quite tedious. Now, what is achieved by the definition of a tensor is a unified treatment, good for any curvilinear coordinates, orthogonal or nonorthogonal. This simplicity is obtained, however, at the expense of recognizing the distinction between contravariance and covariance.

In Sec. 2.14 we discuss the geometric interpretation of the tensor components of a vector in curvilinear coordinates. It will become clear that each physical vector has two tensor images: one contravariant and one covariant, depending on how the components are resolved.

2.8. RECTANGULAR CARTESIAN TENSORS

We have seen that it is necessary to distinguish the contravariant and covariant tensor transformation laws. However, *if only transformations between rectangular Cartesian coordinate systems are considered, the distinction between contravariance and covariance disappears.* To show this, let x_1, x_2, x_3 and $\bar{x}_1, \bar{x}_2, \bar{x}_3$ be two sets of rectangular Cartesian coordinates of reference. The transformation law must be simply

$$(1) \qquad \bar{x}_i = \beta_{ij} x_j + a_i ,$$

where a_1, a_2, a_3 are constants and β_{ij} are the direction cosines of the angles between unit vectors along the coordinate axes $\bar{x}_i$ and x_j. Thus,

$$(2) \qquad \beta_{21} = \cos(\bar{x}_2, x_1) ,$$

etc. The inverse transform is

(3)
$$x_i = \cos(x_i, \bar{x}_j)\bar{x}_j + b_i = \cos(\bar{x}_j, x_i)\bar{x}_j + b_i$$
$$= \beta_{ji}\bar{x}_j + b_i ,$$

where

$$b_i = \beta_{ki}a_k .$$

Hence,

(4)
$$\frac{\partial x_k}{\partial \bar{x}_i} = \beta_{ik} = \frac{\partial \bar{x}_i}{\partial x_k} ,$$

and the distinction between the transformation laws (2.4:2) and (2.4:4) disappears.

The components of a vector in the two coordinate systems are related by

$$\bar{\eta}_i(\bar{x}_1, \bar{x}_2, \bar{x}_3) = \beta_{ij}\eta_j(x_1, x_2, x_3) .$$

The components of a tensor of rank two are related by

$$\bar{\xi}_{ij}(\bar{x}, \bar{x}_2, \bar{x}_3) = \beta_{ik}\beta_{js}\xi_{ks}(x_1, x_2, x_3) .$$

When only rectangular Cartesian coordinates are considered, we shall write all indices as subscripts. This convenient practice will be followed throughout this book.

2.9. CONTRACTION

We shall now consider some operations on tensors that generate new tensors.

Let A^i_{jkl} be a mixed tensor so that, in a transformation from the coordinates x^α to $\bar{x}^\alpha (\alpha = 1, 2, \ldots, n)$, we obtain

$$\bar{A}^i_{jkl}(\bar{x}) = \frac{\partial \bar{x}^i}{\partial x^\alpha} \frac{\partial x^\beta}{\partial \bar{x}^j} \frac{\partial x^\gamma}{\partial \bar{x}^k} \frac{\partial x^\delta}{\partial \bar{x}^l} A^\alpha_{\beta\gamma\delta}(x) .$$

If we equate the indices i and k and sum, we obtain the set of quantities

$$\bar{A}^i_{jil}(\bar{x}) = \frac{\partial \bar{x}^i}{\partial x^\alpha} \frac{\partial x^\beta}{\partial \bar{x}^j} \frac{\partial x^\gamma}{\partial \bar{x}^i} \frac{\partial x^\delta}{\partial \bar{x}^l} A^\alpha_{\beta\gamma\delta}(x)$$

$$= \frac{\partial x^\beta}{\partial \bar{x}^j} \frac{\partial x^\delta}{\partial \bar{x}^l} \delta^\gamma_\alpha A^\alpha_{\beta\gamma\delta}(x)$$

$$= \frac{\partial x^\beta}{\partial \bar{x}^j} \frac{\partial x^\delta}{\partial \bar{x}^l} A^\alpha_{\beta\alpha\delta}(x) .$$

Let us write $A^{\alpha}_{\beta\alpha\delta}$ as $B_{\beta\delta}$. Then the equation above shows that

$$\bar{B}_{jl} = \frac{\partial x^{\beta}}{\partial \bar{x}^{j}} \frac{\partial x^{\delta}}{\partial \bar{x}^{l}} B_{\beta\delta}.$$

Hence, $B_{\beta\delta}$ satisfies the tensor transformation law and is therefore a tensor.

The process of equating and summing a covariant and a contravariant index of a mixed tensor is called a *contraction*. It is easy to see that the example above can be generalized to mixed tensors of other ranks. *The result of a contraction is another tensor.* If, as a result of contraction, there is no free index left, the resulting quantity is a scalar.

The following problem shows that, in general, equating and summing two covariant indices or two contravariant indices does not yield a tensor of lower order and is not a proper contraction. However, when only Cartesian coordinates are considered, we write all indices as subscripts and contract them by equating two subscripts and summing.

PROBLEM

2.11. If A^i_j is a mixed tensor of rank two, show that A^i_i is a scalar. If A^{ij} is a contravariant tensor of rank two show that in general A^{ii} is not an invariant. Similarly, if A_{ij} is a covariant tensor of rank two, A_{ii} is not, in general, an invariant.

2.10. QUOTIENT RULE

Consider a set of n^3 functions $A(111), A(112), A(123)$, etc., or $A(i, j, k)$ for short, with the indices i, j, k each ranging over $1, 2, \ldots, n$. Although the set of functions $A(i, j, k)$ has the right number of components, we do not know whether it is a tensor or not. Now, suppose that we know something about the nature of the product of $A(i, j, k)$ with an arbitrary tensor. Then there is a theorem which enables us to establish whether $A(i, j, k)$ is a tensor without going to the trouble of determining the law of transformation directly.

For example, let $\xi^{\alpha}(x)$ be an arbitrary tensor of rank 1 (a vector). Let us suppose that the product $A(\alpha, j, k)\xi^{\alpha}$ (summation convention used over α) is known to yield a tensor of the type $A^j_k(x)$,

$$A(\alpha, j, k,)\xi^{\alpha} = A^j_k.$$

Then we can prove that $A(i, j, k)$ is a tensor of the type $A^j_{ik}(x)$.

The proof is very simple. Since $A(\alpha, j, k)\xi^\alpha$ is of type A_k^j, it is transformed into $\bar{x}$-coordinates as

$$\bar{A}(\alpha, j, k)\bar{\xi}^\alpha = \bar{A}_k^j = \frac{\partial \bar{x}^j}{\partial x^r} \frac{\partial x^s}{\partial \bar{x}^k} A_s^r$$

$$= \frac{\partial \bar{x}^j}{\partial x^r} \frac{\partial x^s}{\partial \bar{x}^k} [A(\beta, r, s)\xi^\beta].$$

But $\xi^\beta = (\partial x^\beta / \partial \bar{x}^\alpha)\bar{\xi}^\alpha$. Inserting this in the right-hand side of the equation above and transposing all terms to one side of the equation, we obtain

$$\left[\bar{A}(\alpha, j, k) - \frac{\partial \bar{x}^j}{\partial x^r} \frac{\partial x^s}{\partial \bar{x}^k} \frac{\partial x^\beta}{\partial \bar{x}^\alpha} A(\beta, r, s) \right] \bar{\xi}^\alpha = 0.$$

Now $\bar{\xi}^\alpha$ is an arbitrary vector. Hence, the bracket must vanish and we have

$$\bar{A}(\alpha, j, k) = \frac{\partial \bar{x}^j}{\partial x^r} \frac{\partial x^s}{\partial \bar{x}^k} \frac{\partial x^\beta}{\partial \bar{x}^\alpha} A(\beta, r, s),$$

which is precisely the law of transformation of the tensor of the type A_{ik}^j.

The pattern of the example above can be generalized to prove the theorem that, if $[A(i_1, i_2, \ldots, i_r)]$ *is a set of functions of the variables* x^i, *and if the product* $A(\alpha, i_2, \ldots, i_r)\xi^\alpha$ *with an arbitrary vector* ξ^α *be a tensor of the type* $A_{k_1 \cdots k_p}^{j_1 \cdots j_q}(x)$, *where* $p + q = r$, *then the set* $A(i_1, i_2, \ldots, i_r)$ *represents a tensor of the type* $A_{\alpha k_1 \cdots k_p}^{j_1 \cdots j_q}(x)$.

Similarly, if the product of a set of n^2 *functions* $A(\alpha, j)$ *with an arbitrary tensor* $B_{\alpha k}$ *(and is summed over* α*) is a covariant tensor of rank 2, then* $A(i, j)$ *represents a tensor of the type* A_j^i.

These and similar theorems that can be derived are called *quotient rules*. Numerous applications of these rules follow. See, for example, Sec. 3.3, following Eq.(3.3:3); Sec. 3.12, following Eq. (3.12:11); Sec. 4.1, following Eqs. (4.1:11) and (4.1.:12).

2.11. PARTIAL DERIVATIVES IN CARTESIAN COORDINATES

The generation of a tensor of rank one from a tensor of rank zero by differentiation, such as the gradient of a scalar potential, indicates a way of generating tensors of higher rank. But, in general, the set of partial derivatives of a tensor does not behave like a tensor field. However, *if only Cartesian coordinates are considered, then the partial derivatives of any tensor field behave like the components of a tensor field under a transformation from Cartesian coordinates to Cartesian coordinates.* To show this, let us consider two Cartesian coordinates (x_1, x_2, x_3) and $(\bar{x}_1, \bar{x}_2, \bar{x}_3)$ related by

(1) $$\bar{x}_i = a_{ij} x_j + b_i ,$$

where a_{ij} and b_i are constants. From Eq. (1), we have

(2) $$\frac{\partial \bar{x}_i}{\partial x_j} = a_{ij} ,$$

(3) $$\frac{\partial^2 \bar{x}_i}{\partial x_j \partial x_k} = 0 .$$

Now, if $\xi^i(x_1, x_2, x_3)$ is a contravariant tensor, so that

(4) $$\bar{\xi}^i(\bar{x}_1, \bar{x}_2, \bar{x}_3) = \xi^\alpha(x_1, x_2, x_3) \frac{\partial \bar{x}_i}{\partial x_\alpha} .$$

Then, on differentiating both sides of the equation, one obtains

(5) $$\frac{\partial \bar{\xi}^i}{\partial \bar{x}^j} = \left(\frac{\partial \xi^\alpha}{\partial x_\beta} \frac{\partial x_\beta}{\partial \bar{x}_j} \right) \frac{\partial \bar{x}_i}{\partial x_\alpha} + \xi^\alpha \left(\frac{\partial^2 \bar{x}_i}{\partial x_\alpha \partial x_\beta} \right) \frac{\partial x_\beta}{\partial \bar{x}_j} .$$

When x_i and $\bar{x}_i$ are Cartesian coordinates, the last term vanishes according to Eq. (3). Hence,

(6) $$\frac{\partial \bar{\xi}^i}{\partial \bar{x}_j} = \frac{\partial \xi^\alpha}{\partial x_\beta} \frac{\partial x_\beta}{\partial \bar{x}_j} \frac{\partial \bar{x}_i}{\partial x_\alpha} .$$

Thus, the set of partial derivatives $\partial \xi^\alpha / \partial x_\beta$ follows the transformation law for a mixed tensor of rank two under a transformation *from Cartesian coordinates to Cartesian coordinates*. However, the presence of the second derivative terms in Eq. (5) which does not vanish in curvilinear coordinates shows that $\partial \xi^\alpha / \partial x_\beta$ are not really the components of a tensor field in general coordinates. A similar situation holds obviously also for tensor fields of higher ranks. See Sec. 2.12 below.

When Cartesian coordinates are used, *we shall use a comma to denote partial differentiation*. Thus,

$$\xi_{i,j} \equiv \frac{\partial \xi_i}{\partial x_j} , \quad \phi_{,i} \equiv \frac{\partial \phi}{\partial x_i} , \quad \sigma_{ij,k} \equiv \frac{\partial \sigma_{ij}}{\partial x_k} .$$

When we restrict ourselves to Cartesian coordinates, $\phi_{,i}, \xi_{i,j}, \sigma_{ij,k}$ are tensors of rank one, two, three, respectively, provided that ϕ, ξ_i, σ_{ij} are tensors.

Warning: Cartesian tensor equations derived through the use of differentiation are in general not valid in curvilinear coordinates. This important point is discussed in the following five sections.

2.12. COVARIANT DIFFERENTIATION OF VECTOR FIELDS

The generalization of the concept of partial derivatives to the concept of *covariant derivative*, so that the covariant derivative of a tensor field is another tensor field, is the most important milestone in the development of tensor calculus. It is natural to search for such an extension in the form of a correction term that depends on the vector itself. Thus, if ξ^i is a vector, we might expect the combination

$$\frac{\partial \xi^i}{\partial x^j} + \Gamma(i,j,\alpha)\xi^\alpha \,,$$

to be a tensor. Here the suggested correction is linear function of ξ^i, and $\Gamma(i,j,\alpha)$ is some function with three indices. The success of this scheme hinges on the Euclidean Christoffel symbols, which are certain linear combinations of the derivatives of the metric tensor g_{ij}. This subject is beautiful, and the results are powerful in handling curvilinear coordinates. However, since the topic is not absolutely necessary for the development of solid mechanics, we shall not discuss it in detail, but merely outline below some of the salient results.

We discussed in Sec. 2.3 the metric tensor g_{ij} in a set of general coordinates (x^1, x^2, x^3), and in Sec. 2.6 the associated metric tensor g^{ij}. By means of these metric tensors, the *Euclidean Christoffel symbols* $\Gamma^i_{\alpha\beta}(x^1, x^2, x^3)$ *are defined as follows:*

(1) $$\Gamma^i_{\alpha\beta}(x^1, x^2, x^3) = \frac{1}{2}g^{i\sigma}\left(\frac{\partial g_{\sigma\beta}}{\partial x^\alpha} + \frac{\partial g_{\alpha\sigma}}{\partial x^\beta} - \frac{\partial g_{\alpha\beta}}{\partial x}\right).$$

The $\Gamma^i_{\alpha\beta}$ is not a tensor. It transforms under a coordinate transformation $\bar{x}^i = f^i(x^1, x^2, x^3)$ as follows (see Prob. 2.24, p. 55)

(2) $$\bar{\Gamma}^i_{\alpha\beta}(\bar{x}^1, \bar{x}^2, \bar{x}^3) = \Gamma^\lambda_{\mu\nu}(x^1, x^2, x^3)\frac{\partial x^\mu}{\partial \bar{x}^\alpha}\frac{\partial x^\nu}{\partial \bar{x}^\beta}\frac{\partial \bar{x}^i}{\partial x^\lambda} + \frac{\partial^2 x^\lambda}{\partial \bar{x}^\alpha \partial \bar{x}^\beta}\frac{\partial \bar{x}^i}{\partial x^\lambda}.$$

This equation can be solved for $\partial^2 x^\lambda / \partial \bar{x}^\alpha \partial \bar{x}^\beta$ by multiplying (2) with $\partial x^m / \partial \bar{x}^i$ and sum over i to obtain

(3) $$\frac{\partial^2 x^\lambda}{\partial \bar{x}^\alpha \partial \bar{x}^\beta} = \bar{\Gamma}^i_{\alpha\beta}(\bar{x})\frac{\partial x^\lambda}{\partial \bar{x}^i} - \Gamma^\lambda_{\mu\nu}(x)\frac{\partial x^\mu}{\partial \bar{x}^\alpha}\frac{\partial x^\nu}{\partial \bar{x}^\beta}.$$

Interchanging the roles of x_i and $\bar{x}_i$ and with suitable changes in indices, we can substitute (3) into Eq. (2.11:5) to obtain

$$\frac{\partial \bar{\xi}^i}{\partial \bar{x}^\alpha} = \frac{\partial \xi^\lambda}{\partial x^\mu}\frac{\partial x^\mu}{\partial \bar{x}^\alpha}\frac{\partial \bar{x}^i}{\partial x^\lambda} + \xi^\lambda\frac{\partial x^\mu}{\partial \bar{x}^\alpha}\left[\Gamma^s_{\lambda\mu}(x)\frac{\partial \bar{x}^i}{\partial x^s} - \bar{\Gamma}^i_{mn}(\bar{x})\frac{\partial \bar{x}^m}{\partial x^\lambda}\frac{\partial \bar{x}^n}{\partial x^\mu}\right],$$

which can be reduced to

$$(4) \qquad \frac{\partial \bar{\xi}^i}{\partial \bar{x}^\alpha} + \bar{\Gamma}^i_{m\alpha} \bar{\xi}^m = \left(\frac{\partial \xi^\lambda}{\partial x^\mu} + \Gamma^\lambda_{s\mu} \xi^s \right) \frac{\partial x^\mu}{\partial \bar{x}^\alpha} \frac{\partial \bar{x}^i}{\partial x^\lambda}.$$

But this states that the functions $\partial \xi^\lambda / \partial x^\mu + \Gamma^\lambda_{s\mu} \xi^s$ are the components of a mixed tensor of rank two. Hence, the *functions*

$$(5) \quad \blacktriangle \qquad \xi^i|_\alpha \equiv \frac{\partial \xi^i(x^1, x^2, x^3)}{\partial x^\alpha} + \Gamma^i_{\sigma\alpha}(x^1, x^2, x^3) \xi^\sigma(x^1, x^2, x^3),$$

are the components of a mixed tensor field of rank two, called the covariant derivative of the contravariant vector ξ^i. We shall use *the notation* $\xi^i|_\alpha$ *for the covariant derivative of* ξ^i.

By a slight variation in the derivation, it can be shown that the *functions*

$$(6) \quad \blacktriangle \qquad \xi_i|_\alpha \equiv \frac{\partial \xi_i}{\partial x^\alpha} - \Gamma^\sigma_{i\alpha} \xi_\sigma,$$

are the components of a covariant tensor field of rank two whenever ξ_i *are the components of a covariant vector field.* This is called the *covariant derivative of* ξ_i, and is denoted by $\xi_i|_\alpha$.

More generally, a long but quite straightforward calculation analogous to the above can be made to establish the *covariant derivative of a tensor* $T^{\alpha_1 \cdots \alpha_p}_{\beta_1 \cdots \beta_q}$ of rank $p + q$, contravariant of rank p, covariant of rank q:

$$T^{\alpha_1 \cdots \alpha_p}_{\beta_1 \cdots \beta_q}|_\gamma = \frac{\partial T^{\alpha_1 \cdots \alpha_p}_{\beta_1 \cdots \beta_q}}{\partial x^\gamma} + \Gamma^{\alpha_1}_{\sigma\gamma} T^{\sigma \alpha_2 \cdots \alpha_p}_{\beta_1 \beta_2 \cdots \beta_q} + \cdots$$

$$+ \Gamma^{\alpha_p}_{\sigma\gamma} T^{\alpha_1 \cdots \alpha_{p-1}\sigma}_{\beta_1 \cdots \beta_{q-1}\beta_q} - \Gamma^\sigma_{\beta_1\gamma} T^{\alpha_1 \alpha_2 \cdots \alpha_p}_{\sigma \beta_2 \cdots \beta_q} - \cdots - \Gamma^\sigma_{\beta_q\gamma} T^{\alpha_1 \cdots \alpha_{p-1}\alpha_p}_{\beta_1 \cdots \beta_{q-1}\sigma}.$$

This derivative is contravariant of rank p, and covariant of rank $q + 1$.

Since the components of the metric tensor g_{ij} are constant in Cartesian coordinates, we see from Eq. (1) that the Euclidean Christoffel symbols are zero in Cartesian coordinates. The covariant derivative of a tensor field reduces to partial derivatives of the tensor field when the tensor field and the operations are evaluated in Cartesian coordinates.

2.13. TENSOR EQUATIONS

The theorems included in the problems at the end of Sec. 2.5 contain perhaps the most important property of tensor fields: *if all the components of a tensor field vanish in one coordinate system, they vanish likewise in*

all coordinate systems which can be obtained by admissible transformations.
Since the sum and difference of tensor fields of a given type are tensors of
the same type, we deduce that *if a tensor equation can be established in one
coordinate system, then it must hold for all coordinate systems obtained by
admissible transformations.*

The last statement affords a powerful method for establishing equations
in mathematical physics. For example, if a certain tensor relationship can
be shown to be true in rectangular Cartesian coordinates, then it is also true
in general curvilinear coordinates in Euclidean space. Thus, once an equation is established in rectangular Cartesian coordinates, the corresponding
equation (stating a physical fact, such as a condition of equilibrium, of conservation of energy, etc.) in any specific curvilinear coordinates in Euclidean
space can be obtained by a straightforward "translation" in the language of
tensors. The key word here is "tensor." Every term in the equation must
be a tensor. Partial derivatives are tensors only in Cartesian cordinates.
They are not tensors in curvilinear coordinates. Covariant differentiation
must be used to generate new tensors in curvilinear coordinates.

As an example of the application of these remarks, let us consider the
successive covariant derivatives of a tensor field. Since the covariant derivative of a tensor field is a tensor field, we can form the covariant derivative
of the latter, which is called the *second covariant derivative of the original
tensor field.* If T^{ij}_{kl} denotes the original tensor field, we can consider the sec-
ond covariant derivative $T^{ij}_{kl}|_{\gamma\delta}$. Now, if the space is Euclidean, then it can
be described by a rectangular Cartesian coordinate system. In Cartesian
coordinates, the Euclidean Christoffel symbols are all zero and the covariant
derivatives of a tensor field reduce to partial derivatives of the tensor field.
But partial derivatives are *commutative* if they are continuous. Hence, we
see that the following equation is true in Cartesian coordinates if every
component is continuous,

$$(1) \qquad\qquad T^{ij}_{kl}|_{\gamma\delta} = T^{ij}_{kl}|_{\delta\gamma}\,,$$

and, therefore, it is true in all coordinates that can be obtained by admissible transformations from a Cartesian system.

We must remark that the commutativeness of the covariant differentiation operation is established above only in Euclidean space. A coordinate
system in a more general Riemannian space may not be transformed into
Cartesian coordinates and the method of proof used above cannot be applied. In fact, the theorem expressed in Eq. (1) is, in general, untrue
in Riemannian space: successive covariant derivatives are, in general, not
commutative in the Riemannian space. (See Prob. 2.31 below.)

As a second application we can prove the following theorem: *the covariant derivatives of the Euclidean metric tensor g_{ij} and the associated contravariant tensor g^{ij} are zero:*

(2) $$g_{ij}|_k = 0, \quad g^{ij}|_k = 0.$$

Since $g_{ij}|_k$ and $g^{ij}|_k$ are tensors, the truth of the theorem can be established if we can demonstrate Eq. (2) in one particular coordinate system. But this is exactly the case in Cartesian coordinates, in which g_{ij} and g^{ij} are constants and, hence, their derivative vanish. Thus, the proof is completed. In contrast to Eq. (1), however, it can be proved that Eqs. (2) remain true in Riemannian space.

Further examples are furnished in Probs. 2.17 to 2.23.

To apply this powerful procedure, one must make sure that all quantities involved are tensors. In particular, we must ascertain that all scalars are "absolute" constants. This remark is very important because in physics we also use quantities that transform like *relative tensors*. A *relative tensor of weight w* is an object with components whose transformation law differs from the tensor transformation law by the appearance of the Jacobian to the wth power as a factor. Thus,

(3) $$\bar{\phi}(\theta) = \left| \frac{\partial x_j}{\partial \theta_\beta} \right|^w \phi(x),$$

(4) $$\bar{\xi}^i(\theta) = \left| \frac{\partial x_j}{\partial \theta_\beta} \right|^w \xi^\alpha(x) \frac{\partial \theta_i}{\partial x_\alpha},$$

are the transformation laws for a relative scalar field of weight w and a relative contravariant vector field of weight w, respectively. If $w = 0$, we have the previous notion of a tensor field. Whether an object is a tensor or a relative tensor is often a matter of definition.

As an example, consider the total mass enclosed in a volume expressed in terms of density. Let x_i be rectangular coordinates which are transformed into curvilinear coordinates θ_j. We have

(5) $$M = \iiint_V \rho_0(x_1, x_2, x_3) dx_1, dx_2, dx_3$$

$$= \iiint_V \rho_0[x(\theta)] \left| \frac{\partial x_i}{\partial \theta_j} \right| d\theta_1 d\theta_2 d\theta_3$$

$$= \iiint_V \bar{\rho}(\theta_1, \theta_2, \theta_3) d\theta_1 d\theta_2 d\theta_3.$$

If $\bar{\rho}(\theta)$ in the last term is defined as the density distribution in the θ-coordinates, then it is a relative scalar of weight one. On the other hand, $\rho_0[x(\theta)] = \rho_0(x)$ is an absolute scalar which defines the (physical) density of the medium.

As another example, consider the determinant g of the Euclidean metric tensor g_{ij} whose transformation law is

$$(6) \qquad \bar{g}_{\alpha\beta}(\theta) = g_{ij}(x)\frac{\partial x_i}{\partial \theta_\alpha}\frac{\partial x_j}{\partial \theta_\beta}.$$

Let $\bar{g} = |\bar{g}_{\alpha\beta}|$, the determinant of $\bar{g}_{\alpha\beta}$. By a double use of the formula for the product of two determinants when applied to Eq. (6), it is easy to prove that

$$(7) \qquad \bar{g}(\theta) = \left|\frac{\partial x_i}{\partial \theta_\alpha}\right|^2 g(x),$$

which shows that g is a relative scalar of weight two. It follows that $\sqrt{g}$ is a relative scalar of weight one. We note that if x_i are rectangular coordinates, $g = 1$. Hence Eq. (5) shows that $\sqrt{g[x(\theta)]} = |\partial x_i/\partial \theta_j|$, the Jacobian of the transformation. Thus the volume enclosed by a closed surface can be written as

$$(8) \qquad V = \iiint dx_1 dx_2 dx_3$$

$$= \iiint \left|\frac{\partial x_i}{\partial \theta_\alpha}\right| d\theta_1 d\theta_2 d\theta_3$$

$$= \iiint \sqrt{g[x(\theta)]} d\theta_1 d\theta_2 d\theta_3.$$

The last integral shows the importance of $\sqrt{g}$ in mechanics.

The method of tensor equations does not apply to relative tensors. Therefore it is important to properly define all quantities involved in an equation to be tensors.

2.14. GEOMETRIC INTERPRETATION OF TENSOR COMPONENTS

Before we conclude this chapter we shall consider briefly the geometric interpretation of tensor components. For this purpose we must use the concept of *base vectors*. We know that in a three-dimensional Euclidean space any three linearly independent vectors form a *basis* with which any other vectors can be expanded as a linear combination of these three vectors. When a rectangular Cartesian frame of reference is chosen, we can choose as base vectors the unit vectors $\mathbf{i}_1, \mathbf{i}_2, \mathbf{i}_3$ parallel to the coordinate axes: thus, if a vector $\mathbf{A}$ has three components (A_1, A_2, A_3), we can write

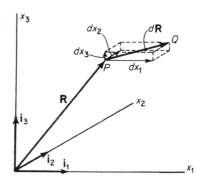

Fig. 2.14:1. Base vectors.

$$\mathbf{A} = A_i \mathbf{i}_1 + A_2 \mathbf{i}_2 + A_3 \mathbf{i}_3 \,.$$

In a curvilinear coordinate system in a Euclidean space, we shall introduce the base vectors from the following consideration.

Let $d\mathbf{R}$ denote an infinitesimal vector PQ joining a point $P = (x_1, x_2, x_3)$ to a point $Q = (x_1 + dx_1, x_2 + dx_2, x_3 + dx_3)$ where x_1, x_2, x_3 are referred to a rectangular Cartesian frame of reference. (See Figure 2.14:1.) Then, obviously,

$$(1) \qquad d\mathbf{R} = dx^r \mathbf{i}_r = dx_r \mathbf{i}^r \,,$$

where $\mathbf{i}_1, \mathbf{i}_2, \mathbf{i}_3$, or $\mathbf{i}^1, \mathbf{i}^2, \mathbf{i}^3$ denote the base vectors along coordinate axes. Here, since a rectangular Cartesian coordinate system is used, we can assign arbitrarily an index as contravariant or covariant. In the assignment chosen above, dx^r and dx_r are, respectively, contravariant and covariant differentials.

Now let us consider a transformation from the rectangular Cartesian coordinates x^i to general coordinates θ^i.

$$(2) \qquad \theta^i = \theta^i(x_1, x_2, x_3) \,.$$

According to the tensor transformation law, when dx^r and dx_r are regarded as tensors of order one, their components in the general coordinates become

$$(3) \qquad d\theta^j = \frac{\partial \theta^j}{\partial x^i} dx^i \,, \quad dx^i = \frac{\partial x^i}{\partial \theta^j} d\theta^j \,,$$

$$(4) \qquad d\theta_j = \frac{\partial x^i}{\partial \theta^j} dx_i \,, \quad dx_i = \frac{\partial \theta^j}{\partial x^i} d\theta_j \,.$$

Now in (3), $d\theta^j$ can be identified as the usual differential of the variable θ^j, as specified in Eq. (2); hence, the superscript is justified. But in Eq. (4), $d\theta_j$, is not to be identified with the usual differential; $d\theta_j$ is a *covariant* differential that will have a different geometric meaning as will be seen later.

By Eqs. (3) and (4), we may write Eq. (1) as

(5) ▲ $$d\mathbf{R} = \mathbf{g}_r \, d\theta^r = \mathbf{g}^r d\theta_r,$$

where

(6) $$\mathbf{g}_r = \frac{\partial x^s}{\partial \theta^r} \, \mathbf{i}_s \, , \quad \mathbf{g}^r = \frac{\partial \theta^r}{\partial x^s} \, \mathbf{i}^s \, .$$

Since $\mathbf{g}_r$ and $\mathbf{g}^r$ are linear combinations of unit vectors, they are themselves vectors; they are known as the *covariant* and *contravariant base vectors*, respectively, or as the *base vectors* and *reciprocal base vectors*. Equation (5) shows that

(7) ▲ $$\mathbf{g}_i = \frac{\partial \mathbf{R}}{\partial \theta^i} \, .$$

Hence $\mathbf{g}_i$ characterizes the change of the position vector $\mathbf{R}$ as θ^i varies. In other words, $\mathbf{g}_i$ *is directed tangentially along the coordinate curve* θ^i. These vectors are illustrated in Fig. 2.14:2.

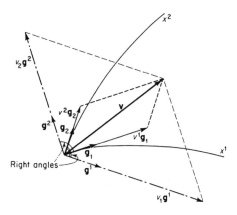

Fig. 2.14:2. Contravariant and covariant components of a vector $\mathbf{v}$ in two dimensions.

It is easily verified that

(8) ▲ $$\mathbf{g}_r \cdot \mathbf{g}_s = g_{rs} \, , \quad \mathbf{g}^r \cdot \mathbf{g}^s = g^{rs} \, ,$$

(9) ▲ $$\mathbf{g}^r \cdot \mathbf{g}_s = g^r_s = \delta^r_s \, ,$$

(10) ▲ $$\mathbf{g}^r = g^{rs} \mathbf{g}_s \, , \quad \mathbf{g}_r = g_{rs} \mathbf{g}^s \, ,$$

where g_{rs} is the Euclidean metric tensor of the coordinate system, and g^{rs} is the *associated*, or *conjugate metric tensor*. From Eq. (9) it is clear that *the contravariant base vectors* $\mathbf{g}^1, \mathbf{g}^2, \mathbf{g}^3$ *are, respectively, perpendicular to the planes of* $\mathbf{g}_2\mathbf{g}_3, \mathbf{g}_3\mathbf{g}_1, \mathbf{g}_1\mathbf{g}_2$. (See also Prob. 2.28.) For orthogonal coordinates, that means $\mathbf{g}_i \cdot \mathbf{g}_j = 0$ for $i \neq j$, it can be shown that $\mathbf{g}^i$ is in the same direction as $\mathbf{g}_i$ and that $\mathbf{g}^i \cdot \mathbf{g}^j = 0$ for $i \neq j$.

We can now deal with the contravariant and covariant components of a vector. Consider the expression $v^r \mathbf{g}_r$, where v^r is a contravariant tensor of rank one, and $\mathbf{g}_r$ are the covariant base vectors at a generic point. This expression remains *invariant* under coordinate transformations, and, since it is a linear combination of the base vectors $\mathbf{g}_r$, it is a vector, which may be designated $\mathbf{v}$:

$$(11) \qquad\qquad \mathbf{v} = v^r \mathbf{g}_r .$$

By Eq. (10), replacing $\mathbf{g}_r$ by $g_{rs}\mathbf{g}^s$, we also have

$$(12) \quad \blacktriangle \qquad\qquad \mathbf{v} = v^r \mathbf{g}_r = v_s \mathbf{g}^s ,$$

where

$$(13) \quad \blacktriangle \qquad\qquad v_s = g_{rs}v^r , \qquad v^r = g^{rs}v_s .$$

According to Eq. (12), if we represent $\mathbf{v}$ by a directed line, then the contravariant components v^r are the components of $\mathbf{v}$ in the direction of the covariant base vectors, while the covariant components v_r are the components of $\mathbf{v}$ in the direction of the contravariant base vectors. A two-dimensional illustration is shown in Fig. 2.14:2.

Equation (12) justifies naming v^r the *contravariant components* of the vector $\mathbf{v}$, and v_r the *covariant components* of $\mathbf{v}$. Thus, in our Euclidean space, the tensors v^r and v_r are two different representations of the same vector $\mathbf{v}$. Equations (13) establish the process of raising and lowering of indices.

Having recognized the covariant and contravariant base vectors of each coordinate system, we can use them to derive the transformation laws of vectors and tensors from one coordinate system to another. Let $\mathbf{g}^i$, $\bar{\mathbf{g}}^i$ and $\mathbf{g}_i$, $\bar{\mathbf{g}}_i$ be, respectively, the contravariant and covariant base vectors of two coordinate systems of reference. We define

$$(14) \qquad \begin{aligned} \beta^{\cdot r}_m &= \bar{\mathbf{g}}_m \cdot \mathbf{g}^r = |\bar{g}_{mm}g^{rr}| \cos(\bar{\mathbf{g}}_m, \mathbf{g}^r) , \\ \beta^m_{\cdot r} &= \bar{\mathbf{g}}^m \cdot \mathbf{g}_r = |\bar{g}^{mm}g_{rr}| \cos(\bar{\mathbf{g}}^m, \mathbf{g}_r) , \end{aligned}$$

in which m and r are not summed. We use a dot in the superscript or subscript to emphasis the difference between $\beta^{\cdot r}_m$ and $\beta^r_{\cdot m} [= |g_{mm}\bar{g}^{rr}| \cos(\mathbf{g}_m, \bar{\mathbf{g}}^r)]$

(m, r not summed). In general, $\beta^r_{\cdot m} \neq \beta^{\cdot m}_r \neq \beta^{\cdot r}_m \neq \beta^m_{\cdot r}$. Here the first indices of the superscript and the subscript are associated with the coordinates with base vectors $\bar{\mathbf{g}}$'s and the second indices with $\mathbf{g}$'s. The distinction between $\beta^{\cdot r}_m$ and $\beta^r_{\cdot m}$ disappear in Cartesian coordinates.

One can easily show that

(15) $$\bar{\mathbf{g}}_r = (\bar{\mathbf{g}}_r \cdot \mathbf{g}^m)\mathbf{g}_m = \beta^{\cdot m}_r \mathbf{g}_m \,,$$

(16) $$\mathbf{g}_m = (\mathbf{g}_m \cdot \bar{\mathbf{g}}^k)\bar{\mathbf{g}}_k = (\bar{\mathbf{g}}^k \cdot \mathbf{g}_m)\bar{\mathbf{g}}_k = \beta^k_{\cdot m}\bar{\mathbf{g}}_k$$

and that the β's are related by the metric tensors of the two coordinate systems:

$$\beta^{\cdot r}_m = \bar{g}_{mp}\beta^p_{\cdot q}\, g^{rq} \,, \qquad \beta^r_{\cdot m} = \bar{g}^{rp}\beta^{\cdot q}_p\, g_{qm} \,.$$

From Eqs. (6) and (14), it can be seen that

(17) $$\beta^{\cdot m}_r = \bar{\mathbf{g}}_r \cdot \mathbf{g}^m = \frac{\partial x^m}{\partial \bar{x}^r} \,, \qquad \beta^k_{\cdot m} = \bar{\mathbf{g}}^k \cdot \mathbf{g}_m = \frac{\partial \bar{x}^k}{\partial x^m} \,.$$

Thus, Eqs. (15) and (16) establish that the transformation between $\bar{\mathbf{g}}$'s and $\mathbf{g}$'s follows the tensor transformation law. A substitution of Eq. (16) into Eq. (15) and *vice versa* yield

$$\bar{\mathbf{g}}_r = \beta^{\cdot k}_r \mathbf{g}_k = \beta^{\cdot k}_r \beta^m_{\cdot k}\bar{\mathbf{g}}_m \,, \qquad \mathbf{g}_m = \beta^k_{\cdot m}\bar{\mathbf{g}}_k = \beta^k_{\cdot m}\beta^{\cdot r}_k \mathbf{g}_r \,,$$

which implies that

(18) $$\beta^{\cdot k}_r \beta^m_{\cdot k} = \delta_{rm} \,. \qquad \beta^k_{\cdot m}\beta^{\cdot r}_k = \delta_{rm} \,.$$

We can now determine the transformation law for the components of $\mathbf{v}$ between different coordinate systems. Dotting both sides of the equation

$$\bar{v}^r \bar{\mathbf{g}}_r = v^m \mathbf{g}_m$$

with $\bar{\mathbf{g}}^i$ gives the transformation law

(19) ▲ $$\bar{v}^i = (\bar{\mathbf{g}}^i \cdot \mathbf{g}_m)v^m = \beta^i_{\cdot m}v^m \,.$$

Similarly, dotting the same equation with $\mathbf{g}^j$ gives the inverse transformation

(20) ▲ $$v^j = (\bar{\mathbf{g}}_r \cdot \mathbf{g}^j)\bar{v}^r = \beta^{\cdot j}_r \bar{v}^r \,.$$

Using the relations of the contravariant base vectors

(21) ▲ $$\bar{\mathbf{g}}^r = (\bar{\mathbf{g}}^r \cdot \mathbf{g}_m)\mathbf{g}^m = \beta^r_{\cdot m}\mathbf{g}^m \,, \qquad \mathbf{g}^m = (\bar{\mathbf{g}}_r \cdot \mathbf{g}^m)\bar{\mathbf{g}}^r = \beta^{\cdot m}_r \bar{\mathbf{g}}^r \,,$$

together with $\mathbf{v} = \bar{v}_r \bar{\mathbf{g}}^r = v_m \mathbf{g}^m$, we obtain

(22)   ▲ $\bar{v}_r = (\bar{\mathbf{g}}_r \cdot \mathbf{g}^m) v_m = \beta_{.r}^{.m} v_m$, $v_m = (\bar{\mathbf{g}}^r \cdot \mathbf{g}_m) \bar{v}_r = \beta_{.m}^r \bar{v}_r$.

If both coordinate systems are rectangular Cartesian coordinates, $\mathbf{g}_i = \mathbf{g}^i$ and $\bar{\mathbf{g}}_i = \bar{\mathbf{g}}^i$. There is no need to distinguish the superscripts and the subscripts of all quantities and $\bar{\mathbf{g}}$'s, $\mathbf{g}$'s are unit base vectors. Then Eq. (14) reduces to

(23) $\beta_{rm} = \cos(\bar{\mathbf{g}}_r, \mathbf{g}_m) = \bar{\mathbf{g}}_r \cdot \mathbf{g}_m = \mathbf{g}_m \cdot \bar{\mathbf{g}}_r$,

that β_{rm} is the direction cosine of the angle between $\bar{\mathbf{g}}_r$ and $\mathbf{g}_m$. Equation (18) becomes

(24) $\beta_{rk} \beta_{mk} = \beta_{kr} \beta_{km} = \delta_{rm}$,

i.e., the transpose of the Cartesian tensor β_{rm} is the inverse of β_{rm}. Equations (19), (20) and (22) become

(25) $\bar{v}_r = (\bar{\mathbf{g}}_r \cdot \mathbf{g}_m) v_m = \beta_{rm} v_m$, $v_m = (\bar{\mathbf{g}}_r \cdot \mathbf{g}_m) \bar{v}_r = \beta_{rm} \bar{v}_r$.

Note that in general we still have $\beta_{rm} \neq \beta_{mr}$ for $r \neq m$.

If $\bar{v}$'s and v's are the coordinates of the two rectangular Cartesian coordinate systems, Eq. (25) gives the transformation law between the two coordinate systems as stated in Eqs. (2.8:2,3) for zero translation.

Similar concept can be generalized to deal with the contravariant and covariant components of tensors of rank two or higher. Consider the expression

$$\mathbf{A} = A^{rs} \mathbf{g}_r \mathbf{g}_s ,$$

where A^{rs} is a contravariant tensor of rank 2. By Eq. (10), replacing $\mathbf{g}_r$ by $g_{rm} \mathbf{g}^m$, we also have

$$\mathbf{A} = A^{rs} \mathbf{g}_r \mathbf{g}_s = A_{mn} \mathbf{g}^m \mathbf{g}^n .$$

The expressions above are invariant under coordinate transformation, i.e.,

(26) $\mathbf{A} = \bar{A}^{rs} \bar{\mathbf{g}}_r \bar{\mathbf{g}}_s = A^{mn} \mathbf{g}_m \mathbf{g}_n = \bar{A}_{rs} \bar{\mathbf{g}}^r \bar{\mathbf{g}}^s = A_{mn} \mathbf{g}^m \mathbf{g}^n$,

in two different coordinate systems of reference. To examine the relations between $\bar{A}^{rs}$ and A^{rs}, we substitute Eq. (16) into Eq. (26) and get

(27) $\mathbf{A} = \bar{A}^{rs} \bar{\mathbf{g}}_r \bar{\mathbf{g}}_s = A^{mn} \beta_{.m}^r \beta_{.n}^s \bar{\mathbf{g}}_r \bar{\mathbf{g}}_s$,   or $\bar{A}^{rs} = \beta_{.m}^r \beta_{.n}^s A^{mn}$.

The substitution of Eq. (15) into Eq. (26) yields

(28) $\mathbf{A} = A^{mn} \mathbf{g}_m \mathbf{g}_n = \bar{A}^{rs} \beta_r^{.m} \beta_s^{.n} \mathbf{g}_m \mathbf{g}_n$ or $A^{mn} = \beta_r^{.m} \beta_s^{.n} \bar{A}^{rs}$.

Similarly, using the relations of the contravariant base vectors given in Eq. (21), we obtain

(29) $\bar{A}_{rs} = \beta_{\cdot r}^{\cdot m}\beta_{s}^{\cdot n}A_{mn}$ and $A_{mn} = \beta_{\cdot m}^{r}\beta_{\cdot n}^{s}\bar{A}_{rs}$

If both coordinate systems are rectangular Cartesian, Eqs. (27), (28) and (29) becomes

(30) $\bar{A}_{rs} = \beta_{rm}\beta_{sn}A_{mn}$ and $A_{mn} = \beta_{rm}\beta_{sn}\bar{A}_{rs}$.

These formulas tell us how base vectors and tensor components are changed in coordinate transformation. Coordinate transformation occupies a uniquely important position in mechanics for many reasons. One reason is exemplified by Einstein's using it to develop the theory of relativity. Another reason is that the shape of a natural object of interest to science and engineering often has a natural, preferred coordinate system for its description, e.g., the earth is round, the rail is straight, and the egg is egg-shaped. Whichever be the reason, we often find it desirable to transform an equation written in one coordinate system to one valid in another cordinate system. For these tasks the method illustrated above can be helpful.

2.15. GEOMETRIC INTERPRETATION OF COVARIANT DERIVATIVES

Consider now a vector field $\mathbf{v}$ defined at every point of space in a region R. Let the vector at the point $P(\theta^1, \theta^2, \theta^3)$ be

(1) $\mathbf{v}(P) = v^i(P)\mathbf{g}_i(P)$.

At a neighboring point $P'(\theta^1 + d\theta^1, \theta^2 + d\theta^2, \theta^3 + d\theta^3)$, the vector becomes

$$\mathbf{v}(P') = \mathbf{v}(P) + d\mathbf{v}(P)$$
$$= [v^i(P) + dv^i(P)][\mathbf{g}_i(P) + d\mathbf{g}_i(P)] .$$

On passing to the limit $d\theta^i \to 0$, we obtain the principal part of the difference

(2) $d\mathbf{v} = (v^i + dv^i)(\mathbf{g}_i + d\mathbf{g}_i) - v^i\mathbf{g}_i = \mathbf{g}_i dv^i + v^i d\mathbf{g}_i$,

and the derivative

(3) ▲ $\dfrac{\partial \mathbf{v}}{\partial \theta^j} = \mathbf{g}_i \dfrac{\partial v^i}{\partial \theta^j} + \dfrac{\partial \mathbf{g}_i}{\partial \theta^j} v^i$.

Thus the derivative of the vector $\mathbf{v}$ is resolved into two parts: one arising from the variation of the components v^i as the coordinates $\theta^1, \theta^2, \theta^3$ are changed, the other arising from the change of the base vector $\mathbf{g}_i$ as the position of the point θ^i is changed. It is shown below that, in a Euclidean space,

(4)   ▲
$$\frac{\partial \mathbf{g}_i}{\partial \theta^j} = \Gamma_{ij}^\alpha \mathbf{g}_\alpha \,.$$

Hence

$$\frac{\partial \mathbf{v}}{\partial \theta^j} = \mathbf{g}_\alpha \frac{\partial v^\alpha}{\partial \theta^j} + v^i \Gamma_{ij}^\alpha \mathbf{g}_\alpha$$

(5)   ▲
$$= \left(\frac{\partial v^\alpha}{\partial \theta^j} + v^i \Gamma_{ij}^\alpha \right) \mathbf{g}_\alpha = v^\alpha|_j \mathbf{g}_\alpha \,.$$

Thus, the covariant derivative $v^\alpha|_j$ represents the components of $\partial \mathbf{v}/\partial \theta^j$ referred to the base vectors $\mathbf{g}_\alpha$.

To establish Eq. (4), we differentiate the equation $g_{ij} = \mathbf{g}_i \cdot \mathbf{g}_j$ to obtain

$$\frac{\partial g_{ij}}{\partial \theta^k} = \frac{\partial \mathbf{g}_i}{\partial \theta^k} \cdot \mathbf{g}_j + \frac{\partial \mathbf{g}_j}{\partial \theta^k} \cdot \mathbf{g}_i \,.$$

On permuting the indices i, j, k, we can obtain the derivatives $\partial g_{ij}/\partial \theta^j$, $\partial g_{ik}/\partial \theta^i$. Furthermore, since $\mathbf{g}_i = \partial \mathbf{R}/\partial \theta^i$, we have

$$\frac{\partial \mathbf{g}_i}{\partial \theta^j} = \frac{\partial}{\partial \theta^j} \left(\frac{\partial \mathbf{R}}{\partial \theta^i} \right) = \frac{\partial}{\partial \theta^i} \left(\frac{\partial \mathbf{R}}{\partial \theta^j} \right) = \frac{\partial \mathbf{g}_j}{\partial \theta^i} \,.$$

Hence, by substitution, it is easy to show that

$$\frac{1}{2} \left(\frac{\partial g_{ik}}{\partial \theta^j} + \frac{\partial g_{jk}}{\partial \theta^i} - \frac{\partial g_{ij}}{\partial \theta^k} \right) = \frac{\partial \mathbf{g}_i}{\partial \theta^j} \cdot \mathbf{g}_k \,.$$

On multiplying the two sides of the equation by $g^{\alpha k}$ and summing over k, the left-hand side becomes Γ_{ij}^α, according to the definition on Christoffel symbols. We then multiply the scalar quantity on both sides of the equation by the vectors $\mathbf{g}_\alpha$ and sum over α to obtain

(6)
$$\Gamma_{ij}^\alpha \mathbf{g}_\alpha = \left(\frac{\partial \mathbf{g}_i}{\partial \theta^j} \cdot \mathbf{g}_k \right) g^{\alpha k} \mathbf{g}_\alpha$$

$$= \left(\frac{\partial \mathbf{g}_i}{\partial \theta^j} \cdot \mathbf{g}_k \right) \mathbf{g}^k \,.$$

The right-hand side, the sum of the vectors $\mathbf{g}^k$ multiplied by the scalar quantities $(\frac{\partial \mathbf{g}_i}{\partial \theta^j} \cdot \mathbf{g}_k)$ is exactly equal to the vector $\partial \mathbf{g}_i / \partial \theta^j$ itself. To see this, we note that since the set of vectors $\mathbf{g}^1, \mathbf{g}^2, \mathbf{g}^3$ are linearly independent and form a basis of the vector space, $\partial \mathbf{g}_i / \partial \theta^j$ can be expressed as a linear combination

(7)
$$\frac{\partial \mathbf{g}_i}{\partial \theta^j} = \lambda_1 \mathbf{g}^1 + \lambda_2 \mathbf{g}^2 + \lambda_3 \mathbf{g}^3 ,$$

where $\lambda_1, \lambda_2, \lambda_3$ are scalars. If we multiply this equation by $\mathbf{g}_1$, we obtain

(8)
$$\lambda_1 = \frac{\partial \mathbf{g}_i}{\partial \theta^j} \cdot \mathbf{g}_1 .$$

Similarly, λ_2, λ_3 can be evaluated. A comparison of Eqs. (6), (8), and (7) thus shows the truth of Eq. (2.15:4):

$$\Gamma_{ij}^\alpha \mathbf{g}_\alpha = \frac{\partial \mathbf{g}_i}{\partial \theta^j} . \qquad\qquad \text{Q.E.D.}$$

2.16. PHYSICAL COMPONENTS OF A VECTOR

The base vectors $\mathbf{g}_r$ and $\mathbf{g}^r$ are in general not unit vectors. In fact, their lengths are

$$|\mathbf{g}_r| = \sqrt{g_{rr}} , \quad |\mathbf{g}^r| = \sqrt{g^{rr}} , \qquad\qquad r \text{ not summed.}$$

Let us write Eq. (2.14:12) as

(1)
$$\mathbf{v} = \sum_{r=1}^{3} v^r \sqrt{g_{rr}} \, \frac{\mathbf{g}_r}{\sqrt{g_{rr}}} = \sum_{r=1}^{3} v_r \sqrt{g^{rr}} \, \frac{\mathbf{g}^r}{\sqrt{g^{rr}}} .$$

Then, since $\mathbf{g}_r / \sqrt{g_{rr}}$ and $\mathbf{g}^r / \sqrt{g^{rr}}$ are unit vectors, all components $v^r \sqrt{g_{rr}}$ and $v_r \sqrt{g^{rr}}$ (r not summed) will have the same physical dimensions. It is seen that $v^r \sqrt{g_{rr}}$ are the components of $\mathbf{v}$ resolved in the direction of unit vectors $\mathbf{g}_r / \sqrt{g_{rr}}$ which are tangent to the coordinate lines; and that $v_r \sqrt{g^{rr}}$ are the components of $\mathbf{v}$ resolved in the direction of unit vectors $\mathbf{g}^r / \sqrt{g^{rr}}$ which are perpendicular to the coordinate planes. The components

$$v^r \sqrt{g_{rr}} , \quad v_r \sqrt{g^{rr}} , \qquad\qquad r \text{ not summed,}$$

are called the *physical components of the vector* $\mathbf{v}$. They do not transform according to the tensor transformation law and are not components of tensors. Note that, in general $v^r \sqrt{g_{rr}} \neq v_r \sqrt{g^{rr}}$ (r not summed) except for orthogonal coordinates.

We should remember that the tensor components of a physical quantity which is referred to a particular curvilinear coordinate system may or may not have the same physical dimensions. This difficulty (and it is also a great convenience!) arises because we would like to keep our freedom in choosing arbitrary curvilinear coordinates. Thus, in spherical polar coordinates for a Euclidean space, the position of a point is expressed by a length and two angles. In a four-dimensional space a point may be expressed in three lengths and a time. For this reason, we must distinguish the tensor components from the "physical components," which must have uniform physical dimensions.

PROBLEMS

2.12. Show that in an orthogonal n-dimensional coordinates system, we have for each component $i = j$, $g^{ij} = 1/g_{ij}$.

2.13. If a_{ij} is a tensor, and the components $a_{ij} = a_{ji}$, then the tensor a_{ij} is called a *symmetric tensor*. If the components $a_{ij} = -a_{ji}$, then the tensor a_{ij} is said to be *skew-symmetric*, or *antisymmetric*. *Show that the symmetry of a tensor with respect to two indices at the same level is conserved under coordinate transformations.* Since $a_{ij} = \frac{1}{2}(a_{ij} + a_{ji}) + \frac{1}{2}(a_{ij} - a_{ji})$, *any covariant (or contravariant) second-order tensor can be written as the sum of a symmetric and a skew-symmetric tensor.*

2.14. Show that the scalar product of a symmetric tensor S^{ij} and a skew-symmetric tensor W_{ij} vanishes indentically.

2.15. Show that the Cartesian tensor $\omega_{ik} = e_{ijk}u_j$ is skew-symmetric, where u_j is a vector.

2.16. Show that if A_{jk} is a skew-symmetric Cartesian tensor, then the unique solution of the equation $\omega_i = \frac{1}{2}e_{ijk}A_{jk}$ is $A_{mn} = e_{mni}\omega_i$.

2.17. Let ∇ be the operator

$$\nabla = \mathbf{g}^r \frac{\partial}{\partial\theta^r}.$$

Show that

$$\operatorname{grad} \phi = \nabla\phi = \mathbf{g}^r \frac{\partial\phi}{\partial\theta^r},$$

$$\operatorname{div} \mathbf{F} = \nabla \cdot \mathbf{F} = F^r|_r,$$

$$\operatorname{curl} \mathbf{A} = \nabla \times \mathbf{A} = \epsilon^{rst} A_s|_r \mathbf{g}_t.$$

Show that these functions are invariant under coordinate transformations.

2.18. Let $g^{\alpha\beta}$ be the associated contravariant Euclidean metric tensor and $\psi(x^1, x^2, x^3)$ be a scalar field. Show that

(a) $g^{\alpha\beta}\psi|_{\alpha\beta}$ is a scalar field.

(b) In rectangular Cartesian coordinates $g_{\alpha\beta} = \delta_{\alpha\beta}$, the scalar $g^{\alpha\beta}\psi|_{\alpha\beta}$ reduces to the form (writing $x^1 = x, x^2 = y, x^3 = z$)

$$\frac{\partial^2\psi}{\partial x^2} + \frac{\partial^2\psi}{\partial y^2} + \frac{\partial^2\psi}{\partial z^2}.$$

(c) Hence, the Laplace equation in curvilinear coordinates with the scalar field $\psi(x^1, x^2, x^3)$ is given by

$$g^{\alpha\beta}(x^1, x^2, x^3)\psi(x^1, x^2, x^3)|_{\alpha\beta} = 0.$$

2.19. Let y^1, y^2, y^3 (or x, y, z) be rectangular Cartesian coordinates and x^1, x^2, x^3 (or r, ϕ, θ) be the spherical polar coordinates. Show that the Laplace equation in spherical polar coordinates, when the unknown function is a scalar field $\psi(r, \phi, \theta)$, is

$$\frac{\partial^2\psi}{\partial r^2} + \frac{1}{r^2}\frac{\partial^2\psi}{\partial\phi^2} + \frac{1}{r^2\sin^2\phi}\frac{\partial^2\psi}{\partial\theta^2} + \frac{2}{r}\frac{\partial\psi}{\partial r} + \frac{\cot\phi}{r^2}\frac{\partial\psi}{\partial\phi} = 0.$$

2.20. Let $\xi^i(y^1, y^2, y^3)$ be the components of an unknown vector in rectangular Cartesian coordinates. Let each component satisfy Laplace's equation in rectangular coordinates,

$$\frac{\partial^2\xi^i}{(\partial y^1)^2} + \frac{\partial^2\xi^i}{(\partial y^2)^2} + \frac{\partial^2\xi^i}{(\partial y^3)^2} = 0.$$

Show that a generalization of this equation is that $\xi^i(x^1, x^2, x^3)$ in curvilinear coordinates x^1, x^2, x^3 will satisfy the system of three differential equations

$$g^{\alpha\beta}\xi^i|_{\alpha\beta} = 0,$$

where $\xi^i|_{\alpha\beta}$ is the second covariant derivative of $\xi^i(x^1, x^2, x^3)$.

2.21. Prove that

$$F^r|_r = \frac{1}{\sqrt{g}}\frac{\partial(\sqrt{g}\,F^r)}{\partial x^r}.$$

Hint: Since $g = g_{i\alpha}G^{i\alpha}$ (i not summed), G^{ij} is the cofactor of the element g_{ij} in $g = |g_{ij}|$, show that

$$\frac{\partial g}{\partial x^i} = gg^{\alpha\beta}\frac{\partial g_{\alpha\beta}}{\partial x^i},$$

and that

$$\frac{\partial g}{\partial x^i} = 2g\Gamma^{\alpha}_{\alpha i}.$$

Hence,

$$\Gamma^{\alpha}_{\alpha i} = \frac{\partial}{\partial x^i}\log\sqrt{g}.$$

2.22. Prove that the Laplacian of Prob. 2.18 can be written

$$g^{ij}\psi|_{ij} = \frac{1}{\sqrt{g}}\frac{\partial[\sqrt{g}\,g^{ij}(\partial\psi/\partial x^j)]}{\partial x^i}.$$

Hint: Use the results of Prob. 2.21.

2.23. Show that the covariant differentiation of sums and products follows the usual rules for partial differentiation. Thus,

$$(\phi v^r)|_i = \phi,_i v^r + \phi v^r|_i,$$

$$(v^r v_r)|_i = (v^r v_r),_i = v^r|_i v_r + v^r v_r|_i,$$

$$(A_{ij}B^{mn})|_r = A_{ij}|_r B^{mn} + A_{ij}B^{mn}|_r,$$

where a comma indicates partial differentiation. Remember that the covariant derivatives of a scalar are the same as the partial derivatives.

2.24. Derive the transformation law for the Euclidean Christoffel symbols $\Gamma_{\alpha\beta}^i(x^1,x^2,x^3)$. *Ans.* Under a transformation of coordinates from x^i to $\bar{x}^i$, the Euclidean metric tensor transforms as

$$\bar{g}_{\sigma\beta}(\bar{x}^1,\bar{x}^2,\bar{x}^3) = g_{\mu\nu}(x^1,x^2,x^3)\frac{\partial x^\mu}{\partial\bar{x}^\sigma}\frac{\partial x^\nu}{\partial\bar{x}^\beta}.$$

Differentiating both sides of the equation with respect to $\bar{x}^\alpha$, considering $\bar{x}^\alpha$ as an independent variable, we obtain

$$\frac{\partial\bar{g}_{\sigma\beta}}{\partial\bar{x}^\alpha} = \left(\frac{\partial g_{\mu\nu}}{\partial x^\rho}\frac{\partial x^\rho}{\partial\bar{x}^\alpha}\right)\frac{\partial x^\mu}{\partial\bar{x}^\sigma}\frac{\partial x^\nu}{\partial\bar{x}^\beta} + g_{\mu\nu}\frac{\partial^2 x^\mu}{\partial\bar{x}^\sigma\partial\bar{x}^\alpha}\frac{\partial x^\nu}{\partial\bar{x}^\beta} + g_{\mu\nu}\frac{\partial x^\mu}{\partial\bar{x}^\sigma}\frac{\partial^2 x^\nu}{\partial\bar{x}^\beta\partial\bar{x}^\alpha}.$$

On first interchanging α and σ in this formula and then interchanging β and α, two similar expressions for $\partial\bar{g}_{\alpha\beta}/\partial\bar{x}^\sigma$ and $\partial\bar{g}_{\alpha\sigma}/\partial\bar{x}^\beta$ are obtained. Adding and subtracting, with several further interchange of dummy indices, and recalling $g_{\nu\mu} = g_{\mu\nu}$, we obtain

$$\frac{1}{2}\left(\frac{\partial\bar{g}_{\sigma\beta}}{\partial\bar{x}^\alpha} + \frac{\partial\bar{g}_{\alpha\sigma}}{\partial\bar{x}^\beta} - \frac{\partial\bar{g}_{\alpha\beta}}{\partial\bar{x}^\sigma}\right) = \frac{1}{2}\left(\frac{\partial g_{\rho\nu}}{\partial x^\mu} + \frac{\partial g_{\mu\rho}}{\partial x^\nu} - \frac{\partial g_{\mu\nu}}{\partial x^\rho}\right)\frac{\partial x^\mu}{\partial\bar{x}^\alpha}\frac{\partial x^\nu}{\partial\bar{x}^\beta}\frac{\partial x^\rho}{\partial\bar{x}^\sigma}$$

$$+ g_{\mu\nu}\frac{\partial x^\mu}{\partial\bar{x}^\sigma}\frac{\partial^2 x^\nu}{\partial\bar{x}^\beta\partial\bar{x}^\alpha}.$$

Now,

$$\bar{g}^{i\sigma} = g^{\lambda\omega}\frac{\partial\bar{x}^i}{\partial x^\lambda}\frac{\partial\bar{x}^\sigma}{\partial x^\omega}.$$

A multiplication of the corresponding sides of the last two equations leads to the desired result.

2.25. All the results obtained above can be applied to two-dimensional spaces by assigning the range of indices to be 1 to 2 instead of 1 to 3. Compute the Euclidean Christoffel symbols for the plane polar coordinates $(x^1 = r, x^2 = \theta)$

Ans. $\Gamma_{22}^1 = -x^1$, $\Gamma_{12}^2 = 1/x^1$, all other components $= 0$.

2.26.   Show that Γ_{mm}^r is symmetric in m and n; i.e., $\Gamma_{mn}^r = \Gamma_{nm}^r$.

2.27.   Show that the necessary and sufficient condition that a given curvilinear coordinate system be orthogonal is that $g_{ij} = 0$, if $i \neq j$, throughout the domain.

2.28.   Prove that $\mathbf{g}_r \times \mathbf{g}_s = \epsilon_{rst}\mathbf{g}^t$, $\mathbf{g}^r \times \mathbf{g}^s = \epsilon^{rst}\mathbf{g}_t$, where ϵ_{rst}, ϵ^{rst} are the permutation tensor of Sec. 2.6. Hence, if we denote the scalar product of the vectors $\mathbf{g}_1$ and $\mathbf{g}_2 \times \mathbf{g}_3$ by $[\mathbf{g}_1\mathbf{g}_2\mathbf{g}_3]$ or $(\mathbf{g}_1, \mathbf{g}_2 \times \mathbf{g}_3)$, we have

$$[\mathbf{g}_1\mathbf{g}_2\mathbf{g}_3] = (\mathbf{g}_1, \mathbf{g}_2 \times \mathbf{g}_3) = \sqrt{g}, \quad [\mathbf{g}^1\mathbf{g}^2\mathbf{g}^3] = (\mathbf{g}^1, \mathbf{g}^2 \times \mathbf{g}^3) = 1/\sqrt{g}.$$

2.29.   The element of area of a parallelogram with two adjacent edges $d\mathbf{s}_2 = \mathbf{g}_2 d\theta^2$ and $d\mathbf{s}_3 = \mathbf{g}_3 d\theta^3$ is

$$dS_1 = |d\mathbf{s}_2 \times d\mathbf{s}_3| = |\mathbf{g}_2 \times \mathbf{g}_3| d\theta^2 d\theta^3.$$

Show that $dS_1 = \sqrt{(gg^{11})}\, d\theta^2 d\theta^3$. In general, the element of area dS_i of a parallelogram formed by the elements $\mathbf{g}_j d\theta^j$ and $\mathbf{g}_k d\theta^k$, $(j, k$ not summed), on the θ_i-surface is $dS_i = \sqrt{(gg^{ii})}\, d\theta^j d\theta^k$ (i not summed, $i \neq j \neq k$).

2.30.   With reference to Prob. 2.28, show that the volume element

$$dV = d\mathbf{s}_1 \cdot (d\mathbf{s}_2 \times d\mathbf{s}_3) = [\mathbf{g}_1\mathbf{g}_2\mathbf{g}_3]\, d\theta^1 d\theta^2 d\theta^3 = \sqrt{g}\, d\theta^1 d\theta^2 d\theta^3.$$

2.31.   If the space is non-Euclidean, we cannot find a coordinate system in which the metric tensor g_{ij} has constant components everywhere throughout the space. (In a Euclidean space there do exist just such coordinate systems, namely, Cartesian coordinate systems.) In this case, we must compute the successive derivatives $\xi_r|_{st}$ and $\xi_r|_{ts}$ according to the covariant differentiation rules. Show that

$$\xi^i|_{st} - \xi^i|_{ts} = R_{pst}^i \xi^p,$$

where

$$R_{pst}^i = \frac{\partial \Gamma_{ps}^i}{\partial x^t} - \frac{\partial \Gamma_{pt}^i}{\partial x^s} + \Gamma_{ps}^r \Gamma_{rt}^i - \Gamma_{pt}^r \Gamma_{rs}^i.$$

Show that R_{pst}^i is a tensor of rank 4, which is the famous *Riemann-Christoffel curvature tensor*. It is not a zero tensor in a general Riemannian space. Hence, in general,

$$\xi^i|_{st} \neq \xi^i|_{ts},$$

in a Riemannian space.

Note: The results obtained above hold true also for two-dimensional spaces, provided all indices range over 1 and 2 only. A curved shell in a three-dimensional Euclidean space appears, in general, to be a two-dimensional non-Euclidean space to a "two-dimensional" animal who has to measure distances right on the shell surface and is never allowed to leave the shell surface to view the third dimension. For a spherical surface, the nonvanishing components of the two-dimensional

Riemann-Christoffel curvature tensor are all equal to a constant, which may be written as 1. For a flat plate, they are zero. For certain hyperboloidal surface all the nonvanishing components of curvature may take on the value -1. Since the spirit of the theory of thin elastic shells is to reduce all the properties of the shells into differential equations describing the middle surface of the shell, an engineer deals with non-Euclidean geometry rather frequently.

2.32. In cylindrical coordinates with unit base vectors $(\mathbf{e}_r, \mathbf{e}_\theta, \mathbf{e}_z)$, a vector $\mathbf{u}$ and its gradient, gradient transpose and divergent are, respectively,

$$\mathbf{u} = u_r \mathbf{e}_r + u_\theta \mathbf{e}_\theta + u_z \mathbf{e}_z ,$$

$$\nabla \mathbf{u} = \frac{\partial \mathbf{u}}{\partial r} \mathbf{e}_r + \frac{1}{r} \frac{\partial \mathbf{u}}{\partial \theta} \mathbf{e}_\theta + \frac{\partial \mathbf{u}}{\partial z} \mathbf{e}_z , \qquad (\nabla \mathbf{u})^T = \mathbf{e}_r \frac{\partial \mathbf{u}}{\partial r} + \frac{\mathbf{e}_\theta}{r} \frac{\partial \mathbf{u}}{\partial \theta} + \mathbf{e}_z \frac{\partial \mathbf{u}}{\partial z} ,$$

$$\nabla \cdot \mathbf{u} = u^i|_i = \mathbf{e}_r \cdot \frac{\partial \mathbf{u}}{\partial r} + \frac{\mathbf{e}_\theta}{r} \cdot \frac{\partial \mathbf{u}}{\partial \theta} + \mathbf{e}_z \cdot \frac{\partial \mathbf{u}}{\partial z} = \frac{1}{r} \frac{\partial r u_r}{\partial r} + \frac{1}{r} \frac{\partial u_\theta}{\partial \theta} + \frac{\partial u_z}{\partial z} .$$

Show that, $\nabla \cdot (\nabla \mathbf{u}) = \nabla (\nabla \cdot \mathbf{u}) = u^i|_{ij}$ in general, and for cylindrical coordinates,

$$\nabla \cdot (\nabla \mathbf{u}) = u^i|_{ij} = \frac{\partial (\nabla \cdot \mathbf{u})}{\partial r} \mathbf{e}_r + \frac{1}{r} \frac{\partial (\nabla \cdot \mathbf{u})}{\partial \theta} \mathbf{e}_\theta + \frac{\partial (\nabla \cdot \mathbf{u})}{\partial z} \mathbf{e}_z ,$$

$$\nabla \cdot (\nabla \mathbf{u})^T = u^i|_{kj} g^{jk} = \frac{1}{r} \frac{\partial}{\partial r} \left(r \frac{\partial \mathbf{u}}{\partial r} \right) + \frac{1}{r^2} \frac{\partial^2 \mathbf{u}}{\partial \theta^2} + \frac{\partial^2 \mathbf{u}}{\partial z^2} .$$

2.33. In spherical coordinates with unit base vectors $(\mathbf{e}_r, \mathbf{e}_\phi, \mathbf{e}_\theta)$, we have

$$\mathbf{u} = u_r \mathbf{e}_r + u_\phi \mathbf{e}_\phi + u_\theta \mathbf{e}_\theta , \qquad \nabla \mathbf{u} = \frac{\partial \mathbf{u}}{\partial r} \mathbf{e}_r + \frac{1}{r} \frac{\partial \mathbf{u}}{\partial \phi} \mathbf{e}_\phi + \frac{1}{r \sin \phi} \frac{\partial \mathbf{u}}{\partial \theta} \mathbf{e}_\theta ,$$

$$(\nabla \mathbf{u})^T = \mathbf{e}_r \frac{\partial \mathbf{u}}{\partial r} + \frac{\mathbf{e}_\theta}{r} \frac{\partial \mathbf{u}}{\partial \phi} + \frac{\mathbf{e}_\theta}{r \sin \phi} \frac{\partial \mathbf{u}}{\partial \theta} ,$$

$$\nabla \cdot \mathbf{u} = \mathbf{e}_r \cdot \frac{\partial \mathbf{u}}{\partial r} + \frac{\mathbf{e}_\phi}{r} \cdot \frac{\partial \mathbf{u}}{\partial \phi} + \frac{\mathbf{e}_\theta}{r \sin \phi} \cdot \frac{\partial \mathbf{u}}{\partial \theta} = \frac{1}{r^2} \frac{\partial r^2 u_r}{\partial r} + \frac{1}{r \sin \phi} \frac{\partial \sin \phi u_\phi}{\partial \phi}$$

$$+ \frac{1}{r \sin \phi} \frac{\partial u_\theta}{\partial \theta} .$$

Show that

$$\nabla \cdot (\nabla \mathbf{u}) = u^i|_{ij} = \frac{\partial (\nabla \cdot \mathbf{u})}{\partial r} \mathbf{e}_r + \frac{1}{r} \frac{\partial (\nabla \cdot \mathbf{u})}{\partial \phi} \mathbf{e}_\phi + \frac{1}{r \sin \phi} \frac{\partial (\nabla \cdot \mathbf{u})}{\partial \theta} \mathbf{e}_\theta .$$

$$\nabla \cdot (\nabla \mathbf{u})^T = u^i|_{kj} g^{jk} = \frac{1}{r^2} \left[\frac{\partial}{\partial r} \left(r^2 \frac{\partial \mathbf{u}}{\partial r} \right) + \frac{1}{\sin \phi} \frac{\partial}{\partial \phi} \left(\sin \phi \frac{\partial \mathbf{u}}{\partial \phi} \right) \right.$$

$$\left. + \frac{1}{\sin^2 \phi} \frac{\partial^2 \mathbf{u}}{\partial \theta^2} \right] .$$

These equations are needed in the Navier equations in later chapters.

3

STRESS TENSOR

The definitions of stress vector and stress components will be given and the equations of equilibrium will be derived. We shall then show how the stress components change when the frames of reference are changed from one rectangular Cartesian frame of reference to another, and in this way we will prove from physical standpoint that the stress components transform according to the tensor transformation rules. The symmetry of the stress tensor will then be discussed, and the consequences of the symmetry property will be derived. The principal stresses, the stress deviations, the octahedral stresses, and finally, the stress tensor in general curvilinear coordinates, form the material for the remainder of the chapter.

Except for Sec. 3.14 *et seq.*, we shall use only rectangular Cartesian frames of reference, whose coordinate axes will be denoted by x_1, x_2, x_3 and are rectilinear and orthogonal to each other. We shall use subscripts for all components unless stated otherwise.

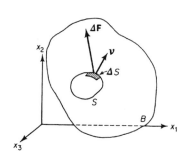

Fig. 3.1:1. Stress principle.

3.1. STRESSES

Consider a configuration occupied by a body B at some time (Fig. 3.1:1). Imagine a closed surface S within B. We would like to know the interaction between the material exterior to this surface and that in the interior. In this consideration, there arises the basic defining concept of continuum mechanics — the *stress principle* of Euler and Cauchy.

Consider a small surface element of area ΔS on our imagined surface S. Let us draw a unit vector ν normal to ΔS, with its direction outward from the interior of S. Then we can distinguish the two sides of ΔS according to the direction of ν. Consider the part of material lying on the positive side of the normal. This part exerts a force $\Delta \mathbf{F}$ on the other part, which is situated on the negative side of the normal. The force $\Delta \mathbf{F}$ is a function of the area and the orientation of the surface. We introduce *the assumption that as ΔS tends to zero, the ratio $\Delta \mathbf{F}/\Delta S$ tends to a definite limit $d\mathbf{F}/dS$ and that the moment of the forces acting on the surface ΔS about any point within the area vanishes in the limit.* The limiting vector will be written as

$$(1) \qquad \overset{\nu}{\mathbf{T}} = \frac{d\mathbf{F}}{dS},$$

where an overhead ν is introduced to denote the direction of the normal ν of the surface ΔS. The limiting vector $\overset{\nu}{\mathbf{T}}$ is called the *stress vector*, or *traction*, and represents the force per unit area acting on the surface.

The assertion that there is defined upon any imagined closed surface S in the interior of a continuum a stress vector field whose action on the material occupying the space interior to S is equipollent to the action of the exterior material upon it, is the stress principle of Euler and Cauchy.

Consider now a special case in which the surface ΔS_k is parallel to one of the coordinate planes. Let the normal of ΔS_k be in the positive direction of the x_k-axis. Let the stress vector acting on ΔS_k be denoted by $\overset{k}{\mathbf{T}}$, with three components $\overset{k}{\mathbf{T}}_1, \overset{k}{\mathbf{T}}_2, \overset{k}{\mathbf{T}}_3$ along the direction of the coordinate axes x_1, x_2, x_3, respectively. The index i of $\overset{k}{\mathbf{T}}_i$ denotes the components of the force, and the symbol k indicates the surface on which the force acts. In this special case, we introduce a new set of symbols for the stress components,

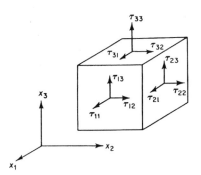

Fig. 3.1:2. Notations of stress components.

$$(2) \qquad \overset{k}{\mathbf{T}}_1 = \tau_{k1}, \quad \overset{k}{\mathbf{T}}_2 = \tau_{k2}, \quad \overset{k}{\mathbf{T}}_3 = \tau_{k3}.$$

If we arrange the components of the tractions acting on the surfaces $k = 1$, $k = 2$, $k = 3$ in a square matrix, we obtain

| | Components of Stresses | | |
	1	2	3
Surface normal to x_1	τ_{11}	τ_{12}	τ_{13}
Surface normal to x_2	τ_{21}	τ_{22}	τ_{23}
Surface normal to x_3	τ_{31}	τ_{32}	τ_{33}

This is illustrated in Fig. 3.1:2. The components τ_{11}, τ_{22}, τ_{33} are called *normal stresses*, and the remaining components τ_{12}, τ_{13}, etc., are called *shearing stresses*. Each of these components has the dimension of force per unit area, or that of [Mass] [Length]$^{-1}$ [Time]$^{-2}$.

A great diversity in notations for stress components exists in the literature. The most widely used notations in American literature are, in reference to a system of rectangular Cartesian coordinates x, y, z.

(3)
$$\sigma_x \ \tau_{xy} \ \tau_{xz} ,$$
$$\tau_{yx} \ \sigma_y \ \tau_{yz} ,$$
$$\tau_{zx} \ \tau_{zy} \ \sigma_z .$$

Love writes X_x, Y_x for σ_x and τ_{xy}, and Todhunter and Pearson use $\widehat{xx}$, $\widehat{xy}$. *In this book we shall use both τ_{ij} and σ_{ij}. We use τ_{ij} to denote stress tensors in general, and we use σ_{ij} to denote the physical components of stress tensors in curvilinear coordinate (see p. 86). In rectangular Cartesian coordinates, the tensor components and the physical components coincide. Hence, we use σ_{ij} for Cartesian stress tensors.* Although the lack of uniformity may seem awkward, little confusion will arise, and in many instances different notations actually result in clarity.

It is important to emphasize again that a stress will always be understood to be the force (per unit area) which the part lying on the positive side of a surface element (the side on the positive side of the outer normal) exerts on the part lying on the negative side. Thus, if the outer normal of a surface element points in the positive direction of the x_1-axis and τ_{11} is positive, the vector representing the component of normal stress acting on the surface element will point in the positive x_1-direction. But if τ_{11} is positive while the outer normal points in the negative x_1-axis direction, then the stress vector acting on the element also points to the negative x_1-axis direction (see Fig. 3.1:3).

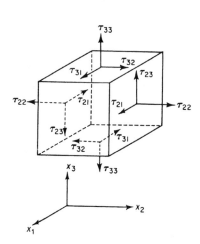

Fig. 3.1:3. Senses of positive stresses.

Similarly, positive values of τ_{12}, τ_{13} will imply shearing stress vectors pointing to positive x_2, x_3-axes if the outer normal agrees in sense with

x_1-axis, whereas they point to the negative x_2, x_3-direction if the outer normal disagrees in sense with the x_1-axis, as illustrated in Fig. 3.1:3. A careful study of the figure is essential. Naturally, these rules agree with the usual notions of tension, compression, and shear.

3.2. LAWS OF MOTION

The fundamental laws of mechanics for bodies of all kinds are Euler's equations, which extend Newton's laws of motion for particles. Let the coordinate system x_1, x_2, x_3 be an inertial frame of reference. Let the space occupied by a material body at any time t be denoted by $B(t)$. Let **r** be the position vector of a particle with respect to the origin of the coordinate system. Let **V** be the velocity vector of a particle at the point x_1, x_2, x_3. Then

$$(1) \qquad\qquad \boldsymbol{\mathcal{P}} = \int_{B(t)} \mathbf{V}\rho dv\,,$$

is called the *linear momentum* of the body in the configuration $B(t)$, and let

$$(2) \qquad\qquad \boldsymbol{\mathcal{H}} = \int_{B(t)} \mathbf{r} \times \mathbf{V}\rho dv\,,$$

is called the *moment of momentum*. In these formulas ρ is the density of the material and the integration is over the volume $B(t)$. *Newton's laws, as stated by Euler for a continuum, assert that the rate of change of linear momentum is equal to the total applied force $\boldsymbol{\mathcal{F}}$ acting on the body,*

$$(3) \qquad\qquad \dot{\boldsymbol{\mathcal{P}}} = \boldsymbol{\mathcal{F}}\,,$$

and that the rate of change of moment of momentum is equal to the total applied torque $\boldsymbol{\mathcal{L}}$,

$$(4) \qquad\qquad \dot{\boldsymbol{\mathcal{H}}} = \boldsymbol{\mathcal{L}}\,.$$

The torque $\boldsymbol{\mathcal{L}}$ is taken with respect to the same point as the origin of the position vector **r**. It is easy to verify that if (3) holds, then if (4) holds for one choice of origin, it holds for all choices of origin.[†]

It is assumed that force and torque are quantities about which we have *a priori* information in certain frames of reference.

On material bodies considered in mechanics of continuous media, two types of external forces act:

[†] The derivatives $\dot{\boldsymbol{\mathcal{P}}}$ and $\dot{\boldsymbol{\mathcal{H}}}$ are *material derivatives*; i.e., the time rate of change of $\boldsymbol{\mathcal{P}}$ and $\boldsymbol{\mathcal{H}}$ of a fixed set of particles (cf. Secs. 5.2 and 5.3).

1. Body forces, acting on elements of volume of the body.
2. Surface forces or stresses, acting on surface elements.

Examples of body forces are gravitational forces and electromagnetic forces. Examples of surface forces are aerodynamic pressure acting on a body and pressure due to mechanical contact of two bodies.

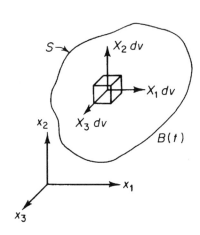

To specify a body force, we consider a volume bounded by an arbitrary surface S (Fig. 3.2:1). The resultant force vector contributed by the body force is assumed to be representable in the form of a volume integral

$$\int_B \mathbf{X} dv \, .$$

Fig. 3.2:1. Body forces.

The three components of $\mathbf{X}$, namely, X_1, X_2, X_3, all of dimensions force per unit volume $M(LT)^{-2}$, are called the body force per unit volume. For example, in a gravitational field,

$$X_i = \rho g_i \, ,$$

where g_i are components of a gravitational acceleration field and ρ is the density (mass per unit volume) at a given point of the body.

The surface force acting on an imagined surface in the interior of a body is the stress vector conceived in Euler and Cauchy's stress principle. According to this concept, the total force acting upon the material occupying the region B interior to a closed surface S is

$$(5) \qquad \mathcal{F} = \oint_S \overset{\nu}{\mathbf{T}} dS + \int_B \mathbf{X} dv \, ,$$

where $\overset{\nu}{\mathbf{T}}$ is the stress vector acting on dS whose outer normal vector is $\boldsymbol{\nu}$. Similarly, the torque about the origin is given by the expression

$$(6) \qquad \mathcal{L} = \oint_S \mathbf{r} \times \overset{\nu}{\mathbf{T}} dS + \int_B \mathbf{r} \times \mathbf{X} dv \, .$$

In the following section we shall make some elementary applications of these equations to obtain the fundamental properties of the stress tensor.

3.3. CAUCHY'S FORMULA

With the equations of motion, we shall first derive a simple result which states that *the stress vector* $\mathbf{T}^{(+)}$ *representing the action of material exterior to a surface element on the interior is equal in magnitude and opposite in direction to the stress vector* $\mathbf{T}^{(-)}$ *which represents the action of the interior material on the exterior across the same surface element*:

(1) ▲ $$\mathbf{T}^{(-)} = -\mathbf{T}^{(+)}.$$

To prove this, we consider a small "pill box" with two parallel surfaces of area ΔS and thickness δ, as shown in Fig. 3.3:1. When δ shrinks to zero, while ΔS remains small but finite, the volume forces and the linear momentum and its rate of change with time vanish, as well as the contribution of surface forces on the sides of the pill box. The equation of motion (3.2:3) implies, therefore, for small ΔS,

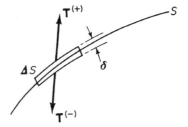

$$\mathbf{T}^{(+)}\Delta S + \mathbf{T}^{(-)}\Delta S = 0.$$

Equation (1) then follows.

Another way of stating this result is that the stress vector is a function of

Fig. 3.3:1. Equilibrium of a "pill box" across a surface S.

the normal vector to a surface. When the sense of direction of the normal vector reverses, the stress vector reverses also.

Now we shall show that *knowing the components* τ_{ij}, *we can write down at once the stress vector acting on any surface with unit outer normal vector* $\boldsymbol{\nu}$ *whose components are* ν_1, ν_2, ν_3. *This stress vector is denoted by* $\overset{\nu}{\mathbf{T}}$, *with components* $\overset{\nu}{T}_1, \overset{\nu}{T}_2, \overset{\nu}{T}_3$ *given by Cauchy's formula*

(2) ▲ $$\overset{\nu}{T}_i = \nu_j \tau_{ji},$$

which can be derived in several ways. We shall give first an elementary derivation.

Let us consider an infinitesimal tetrahedron formed by three surfaces parallel to the coordinate planes and one normal to the unit vector $\boldsymbol{\nu}$ (see Fig. 3.3:2). Let the area of the surface normal to $\boldsymbol{\nu}$ be dS. Then the area

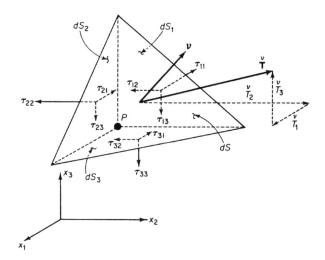

Fig. 3.3:2. Surface tractions on a tetrahedron.

of the other three surfaces are

$$dS_1 = dS\cos(\boldsymbol{\nu}, \mathbf{x}_1)$$

$$= \nu_1 dS = \text{area of surface} \parallel \text{to the } x_2 x_3\text{-plane},$$

$$dS_2 = \nu_2 dS = \text{area of surface} \parallel \text{to the } x_3 x_1\text{-plane},$$

$$dS_3 = \nu_3 dS = \text{area of surface} \parallel \text{to the } x_1 x_2\text{-plane},$$

and the volume of the tetrahedron is

$$dv = \frac{1}{3}h\,dS\,,$$

where h is the height of the vertex P from the base dS. The forces in the positive direction of $\mathbf{x}_1$ acting on the three coordinate surfaces can be written as

$$(-\tau_{11} + \varepsilon_1)dS_1\,, \quad (-\tau_{21} + \varepsilon_2)dS_2\,, \quad (-\tau_{31} + \varepsilon_3)dS_3\,,$$

where τ_{11}, τ_{21}, τ_{31} are the stresses at the point P. The negative sign is obtained because the outer normals to the three surfaces are opposite in sense with respect to the coordinate axes, and the ε's are inserted because the tractions act at points slightly different from P. If we assume that the stress field is continuous, then ε_1, ε_2, ε_3 are infinitesimal quantities. On the other hand, the force acting on the triangle normal to $\boldsymbol{\nu}$ has a component $(\overset{\nu}{T}_1 + \varepsilon)dS$ in the x_1-axis direction, the body force has an x_1-component equal to $(X_1 + \varepsilon')dv$, and the rate of change of linear momentum has a component $\rho\dot{V}_L dv$. Here $\overset{\nu}{T}_1$ and X_1 refer to the point P and ε, ε' are again infinitesimal. The first equation of motion is thus

$$(-\tau_{11} + \varepsilon_1)\nu_1 dS + (-\tau_{21} + \varepsilon_2)\nu_2 dS + (-\tau_{31} + \varepsilon_3)\nu_3 dS$$
$$+ (\overset{\nu}{T}_1 + \varepsilon)dS + (X_1 + \varepsilon')\frac{1}{3}hdS = \rho\dot{V}_1\frac{1}{3}hdS.$$

Dividing through by dS and taking the limit $h \to 0$, one obtains

(3) $$\overset{\nu}{T}_1 = \tau_{11}\nu_1 + \tau_{21}\nu_2 + \tau_{31}\nu_3,$$

which is the first component of Eq. (2). Other components follow similarly.

Cauchy's formula assures us that the nine components of stresses τ_{ij} are necessary and sufficient to define the traction across any surface element in a body. Hence the stress state in a body is characterized completely by the set of quantities τ_{ij}. Since $\overset{\nu}{T}_i$ is a vector and Eq. (2) is valid for an arbitrary vector ν_j, it follows from the quotient rule (Sec. 2.10) that τ_{ij} is a tensor. Henceforth τ_{ij} will be called a stress tensor.

We note again that in the theoretical development up to this point we have assumed, first, that stress can be defined everywhere in a body, and, second, that the stress field is continuous. The same assumption will be made later with respect to strain. These are characteristic assumptions of continuum mechanics. Without these assumptions we can do very little indeed. However, in the further development of the theory, certain mathematical discontinuities will be permitted — often they are very useful tools — but one should always view these discontinuities with great care against the general basic assumptions of continuity of the stress and strain fields.

3.4. EQUATIONS OF EQUILIBRIUM

We shall now transform the equations of motion (3.2:3), (3.2:4) into differential equations. This can be done elegantly by means of Gauss' theorem and Cauchy's formula. But we shall pursue here an elementary course to assure physical clarity.

Consider the static equilibrium state of an infinitesimal parallelepiped with surfaces parallel to the coordinate planes. The stresses acting on the various surfaces are shown in Fig. 3.4:1. The force $\tau_{11}dx_2dx_3$ acts on the left-hand side, the force $(\tau_{11} + \frac{\partial\tau_{11}}{\partial x_1} dx_1)dx_2dx_3$ acts on the right-hand side, etc. These expressions are based on the assumption of continuity of the stresses. The body force is $X_i dx_1 dx_2 dx_3$.

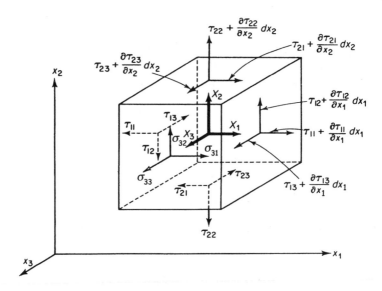

Fig. 3.4:1. Equilibrating stress components on an infinitesimal parallelepiped.

Now, the equilibrium of the body demands that the resultant forces vanish. Consider the forces in the x_1-direction. As shown in Fig. 3.4:2, we have six components of surface forces and one component of body force. The sum is

$$\left(\tau_{11} + \frac{\partial \tau_{11}}{\partial x_1} dx_1\right) dx_2 dx_3 - \tau_{11} dx_2 dx_3$$

$$+ \left(\tau_{21} + \frac{\partial \tau_{21}}{\partial x_2} dx_2\right) dx_3 dx_1 - \tau_{21} dx_3 dx_1$$

$$+ \left(\tau_{31} + \frac{\partial \tau_{31}}{\partial x_3} dx_3\right) dx_1 dx_2 - \tau_{31} dx_1 dx_2 + X_1 dx_1 dx_2 dx_3 = 0.$$

Dividing by $dx_1 dx_2 dx_3$, we obtain

(1) $$\frac{\partial \tau_{11}}{\partial x_1} + \frac{\partial \tau_{21}}{\partial x_2} + \frac{\partial \tau_{31}}{\partial x_3} + X_1 = 0.$$

A cyclic permulation of subscripts 1, 2, 3 leads to similar equations of equilibrium of forces in x_2, x_3-directions. The whole set, written concisely, is

(2) ▲ $$\frac{\partial \tau_{ji}}{\partial x_j} + X_i = 0.$$

This is an important result. A shorter derivation will be given later in Sec. 5.5.

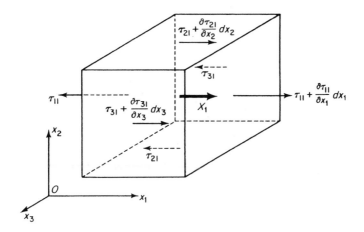

Fig. 3.4:2. Components of tractions in x_1 direction.

The equilibrium of an element requires also that the resultant moment vanishes. If there do not exist external moments proportional to a volume, the consideration of moments will lead to the important conclusion that *the stress tensor is symmetric*,

$$(3) \quad \blacktriangle \qquad\qquad\qquad \tau_{ij} = \tau_{ji} \,.$$

This is demonstrated as follows. Referring to Fig. 3.4:3 and considering the moment of all the forces about the axis Ox_3, we see that those components

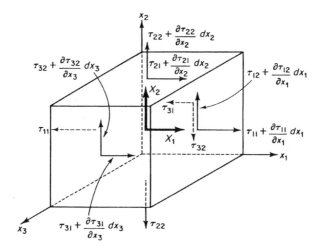

Fig. 3.4:3. Components of tractions that contribute moment about Ox_3-axis.

of forces parallel to Ox_3 or lying in planes containing Ox_3 do not contribute any moment. The components that do contribute a moment about the Ox_3-axis are shown in Fig. 3.4:3. Therefore, properly taking care of the moment arm, we have

$$- \left(\tau_{11} + \frac{\partial \tau_{11}}{\partial x_1} dx_1 \right) dx_2 dx_3 \cdot \frac{dx_2}{2} + \tau_{11} dx_2 dx_3 \cdot \frac{dx_2}{2}$$

$$+ \left(\tau_{12} + \frac{\partial \tau_{12}}{\partial x_1} dx_1 \right) dx_2 dx_3 \cdot dx_1 - \left(\tau_{21} + \frac{\partial \tau_{21}}{\partial x_2} dx_2 \right) dx_1 dx_3 \cdot dx_2$$

$$+ \left(\tau_{22} + \frac{\partial \tau_{22}}{\partial x_2} dx_2 \right) dx_1 dx_3 \cdot \frac{dx_1}{2} - \tau_{22} dx_1 dx_3 \cdot \frac{dx_1}{2}$$

$$+ \left(\tau_{32} + \frac{\partial \tau_{32}}{\partial x_3} dx_3 \right) dx_1 dx_2 \cdot \frac{dx_1}{2} - \tau_{32} dx_1 dx_2 \cdot \frac{dx_1}{2}$$

$$- \left(\tau_{31} + \frac{\partial \tau_{31}}{\partial x_3} dx_3 \right) dx_1 dx_2 \cdot \frac{dx_2}{2} + \tau_{31} dx_1 dx_2 \cdot \frac{dx_2}{2}$$

$$- X_1 dx_1 dx_2 dx_3 \cdot \frac{dx_2}{2} + X_2 dx_1 dx_2 dx_3 \cdot \frac{dx_1}{2} = 0 \, .$$

On dividing through by $dx_1 dx_2 dx_3$ and passing to the limit $dx_1 \to 0$, $dx_2 \to 0$, $dx_3 \to 0$, we obtain

(4) $$\tau_{12} = \tau_{21} \, .$$

Similar considerations of resultant moments about Ox_1, Ox_2 lead to the general result given by Eq. (3). Again a shorter derivation will be given later in Sec. 5.5.

It should be noted that if an external moment proportional to the volume does exist, then the symmetry condition does not hold. For example, if there is a moment $c_3 dx_1 dx_2 dx_3$ about the axis Ox_3, then we obtain in place of Eq. (4) the result

(5) $$\tau_{12} - \tau_{21} + c_3 = 0 \, .$$

Maxwell pointed out that nonvanishing body moments exist in a magnet in a magnetic field and in a dielectric material in an electric field with different planes of polarization. If the electromagnetic field is so intense and the stress level is so low that τ_{12} and c_3 are of the same order of magnitude, then, according to (5), τ_{12} cannot be equated to τ_{21}. In this case, we have to admit the stress tensor τ_{ij} as asymmetric. If c_3 is very much smaller in comparison with τ_{12}, then we can omit c_3 in (5) and consider (4) as valid approximately.

In developing a physical theory, particularly for the purpose of engineering, one of the most important objectives is to obtain the simplest formulation consistent with the desired degree of accuracy. A decision of whether or not we shall treat the stress tensor as symmetric must be based on the purpose of the theory. Since electromagnetic fields pervade the universe, the stress tensor is in general unsymmetric. But, if the theory is formulated for the purpose of a structural or mechanical engineer who studies the stress distribution in a structure or a machine with a view towards assessing its strength, stability, or rigidity, then a stress is important when it is of the order of the yielding stress of the material. Even for a structure which is designed primarily on the basis of stability, such as a column, an arch, or a thin-walled shell, a good design should produce a critical stress of the order of the yielding stress under the critical conditions, for otherwise the material is not economically used. When τ_{ij} is comparable to the yielding stress in magnitude (of order 10,000 to 100,000 lb/sq in. or 70 to 700 mPa for a steel, or 50 to 5,000 lb/sq in. or 0.35 to 35 mPa for a concrete), there are few circumstances in which the assumption of symmetry in stress tensor should cause concern.

However, if one wants to study the influence of a strong electromagnetic field on the propagation of elastic waves, or such influence on some high-frequency phenomenon in the material, then the stress level may be very low and the body moment may be significant. In such problems the stress tensor may not be assumed symmetric.

In the rest of this book the stress tensor will be assumed to be symmetric unless stated otherwise.

Notes on Couple-stresses

If, following Voigt, we assume that across any infinitesimal surface element in a solid the action of the exterior material upon the interior is equipollent to a force *and a couple* (in contrast to the assumption made in Sec. 3.1) then in addition to the traction $\overset{\nu}{\mathbf{T}}$ that acts on the surface we must have also a *couple-stress* vector $\overset{\nu}{\mathbf{M}}$. These two vectors $\overset{\nu}{\mathbf{T}}$ and $\overset{\nu}{\mathbf{M}}$, together, are now equipollent to the action of the exterior upon the interior. Similarly, one might have body couples as pointed out by Maxwell, i.e., couple per unit mass, $\mathbf{c}$, with components c_i, $(i = 1, 2, 3)$. If we accept these possibilities, then we must define a couple-stress tensor, $\mathcal{M}_{ij}$, in addition to the stress tensor τ_{ij}. The tensor $\mathcal{M}_{ij}$ is related to the couple-stress vector by a linear transformation like Eq. (3.3:2):

$$\overset{\nu}{M}_i = \mathcal{M}_{ji}\nu_j\,.$$

An analysis of the angular momentum then leads to the equation

$$\frac{\partial \mathcal{M}_{ji}}{\partial x_j} + \rho c_i = e_{ijk}\tau_{jk}\,,$$

i.e.,

$$\frac{\partial \mathcal{M}_{xx}}{\partial x} + \frac{\partial \mathcal{M}_{yx}}{\partial y} + \frac{\partial \mathcal{M}_{zx}}{\partial z} + \rho c_x = \tau_{yz} - \tau_{zy}\,, \quad \text{etc}\,.$$

Thus, the antisymmetric part of the stress tensor is determined by the body couples and the divergence of the couple-stress tensor. When couples of both kinds are absent, the stress tensor must be symmetric.

Couple-stresses and body couples are useful concepts in dealing with submicron structures and molecular mechanics of materials, and in the dislocation theory of metals.

3.5. TRANSFORMATION OF COORDINATES

In the previous section, the components of stress τ_{ij} are defined with respect to a rectangular Cartesian system x_1, x_2, x_3. Let us now take a second set of rectangular Cartesian coordinates x_1', x_2', x_3' with the same origin but oriented differently, and consider the stress components in the new reference system (Fig. 3.5:1). Let these coordinates be connected by the linear relations

$$(1) \qquad x_k' = \beta_{ki}x_i = \frac{\partial x_k'}{\partial x_i}x_i\,, \quad k = 1,2,3\,,$$

where $\beta_{ki} = \cos(x_k', x_i)$ are the direction cosines of the x_k'-axis with respect to the x_i-axis. Since τ_{ij} is a tensor (Sec. 3.3) we can write down the transformation law at once. However, in order to emphasize the importance

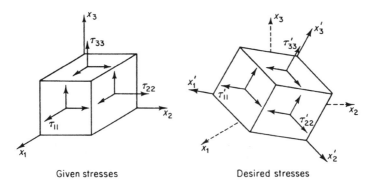

Fig. **3.5:1.** Transformation of stress components under rotation of coordinates system.

of the result we shall insert an elementary derivation based on Cauchy's formula derived in Sec. 3.3, which states that if dS is a surface element whose unit outer normal vector $\boldsymbol{\nu}$ has components ν_i, then the force per unit area acting on dS is a vector $\overset{\nu}{\mathbf{T}}$ with components

$$\overset{\nu}{T}_i = \tau_{ji}\nu_j \,.$$

If the normal $\boldsymbol{\nu}$ is chosen to be parallel to the axis $\mathbf{x}'_k$, so that

$$\nu_1 = \beta_{k1}\,, \quad \nu_2 = \beta_{k2}\,, \quad \nu_3 = \beta_{k3}\,,$$

then denoting $\overset{\nu}{T}_i$ as $\overset{k}{\mathbf{T}}_i$, we have

$$\overset{\nu}{T}_i = \overset{k}{\mathbf{T}}_i = \tau_{ji}\beta_{kj}\,.$$

Since the component of the vector $\overset{k}{\mathbf{T}}(=\overset{\nu}{\mathbf{T}})$ in the direction x'_m is τ'_{km}, i.e.,

$$\overset{k}{\mathbf{T}} = \tau'_{km}\mathbf{g}'_m\,,$$

then

$$\overset{k}{\mathbf{T}} = \tau'_{km}\mathbf{g}'_m = \overset{k}{\mathbf{T}}_i\mathbf{g}_i = \tau_{ji}\beta_{kj}\mathbf{g}_i\,,$$

where $\mathbf{g}'_m$ and $\mathbf{g}_i$ are the unit base vectors of the two coordinate systems. Using the relation $\mathbf{g}_i = \beta_{mi}\mathbf{g}'_m$ [Eq. (2.14:16)] and Eq. (1) for rectangular Cartesian coordinates, we obtain

$$\overset{k}{\mathbf{T}} = \tau'_{km}\mathbf{g}'_m = \tau_{ji}\beta_{kj}\beta_{mi}\mathbf{g}'_m\,,$$

i.e.,

(3) ▲ $$\tau'_{km} = \tau_{ji}\beta_{kj}\beta_{mi} = \tau_{ji}\frac{\partial x'_k}{\partial x_j}\frac{\partial x'_m}{\partial x_i}\,.$$

If we compare Eq. (3) and Eq. (2.5:2) we see that the stress components transform like a Cartesian tensor of rank two. Thus, the physical concept of stress which is described by τ_{ij} agrees with the mathematical definition of a tensor of rank two in a Euclidean space.

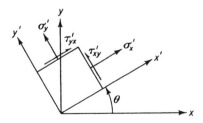

Fig. 3.6:1. Change of coordinates in plane state of stress.

3.6. PLANE STATE OF STRESS

A state of stress in which

(1) $$\tau_{33} = \tau_{31} = \tau_{32} = 0\,,$$

is called a *plane state of stress* in the x_1x_2-plane. In this case, the direction

cosines between two systems of rectangular Cartesian coordinates can be
expressed in terms of a single angle θ, as shown in Fig. 3.6:1. We have,

$$(\beta_{ij}) = \begin{pmatrix} \cos\theta & \sin\theta & 0 \\ -\sin\theta & \cos\theta & 0 \\ 0 & 0 & 1 \end{pmatrix}.$$

Writing x, y and x', y' in place of x_1, x_2 and x'_1, x'_2; σ_x for τ_{11}; τ_{xy} for τ_{12},
etc., we have

$$\sigma'_x = \sigma_x \cos^2\theta + \sigma_y \sin^2\theta + 2\tau_{xy}\sin\theta\cos\theta,$$

$$\sigma'_y = \sigma_x \sin^2\theta + \sigma_y \cos^2\theta - 2\tau_{xy}\sin\theta\cos\theta,$$

$$\tau'_{xy} = (-\sigma_x + \sigma_y)\sin\theta\cos\theta + \tau_{xy}(\cos^2\theta - \sin^2\theta).$$

Since

$$\sin^2\theta = \frac{1}{2}(1 - \cos 2\theta), \quad \cos^2\theta = \frac{1}{2}(1 + \cos 2\theta),$$

we may also write

$$\sigma'_x = \frac{\sigma_x + \sigma_y}{2} + \frac{\sigma_x - \sigma_y}{2}\cos 2\theta + \tau_{xy}\sin 2\theta,$$

(3) $$\sigma'_y = \frac{\sigma_x + \sigma_y}{2} - \frac{\sigma_x - \sigma_y}{2}\cos 2\theta - \tau_{xy}\sin 2\theta,$$

$$\tau'_{xy} = -\frac{\sigma_x - \sigma_y}{2}\sin 2\theta + \tau_{xy}\cos 2\theta.$$

Note that

(4) $$\sigma'_x + \sigma'_y = \sigma_x + \sigma_y,$$

(5) $$\frac{\partial \sigma'_x}{\partial \theta} = 2\tau'_{xy},$$

(6) $$\tau'_{xy} = 0 \quad \text{when} \quad \tan 2\theta = \frac{2\tau_{xy}}{\sigma_x - \sigma_y}.$$

The directions given by the particular values of θ given by (6) are called
the *principal directions*; the corresponding normal stresses are called the
principal stresses (see Sec. 3.7). Following (5) and (6), the principal stresses
are extreme values of the normal stresses,

(7) $$\left.\begin{array}{c}\sigma_{max}\\\sigma_{min}\end{array}\right\} = \frac{\sigma_x + \sigma_y}{2} \pm \sqrt{\left(\frac{\sigma_x - \sigma_y}{2}\right)^2 + \tau_{xy}^2}.$$

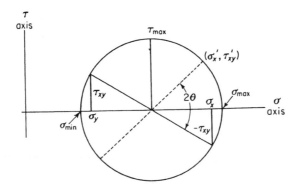

Fig. 3.6:2. Mohr's circle in plane state of stress.

Differentiating τ'_{xy} with respect to θ and setting the derivative to zero, we can find the angle θ at which τ'_{xy} attends its extreme value. This angle is easily seen to be $\pm 45°$ from the principal directions given by (6), and the maximum value of τ'_{xy} is

$$(8) \qquad \tau_{max} = \frac{\sigma_{max} - \sigma_{min}}{2} = \sqrt{\left(\frac{\sigma_x - \sigma_y}{2}\right)^2 + \tau_{xy}^2}\,.$$

Figure 3.6:2 is a geometric representation of the relations above. It is the well-known *Mohr's circle*.

Problem 3.1. Find the transformation law for the moments of inertia and products of inertia of an area about a set of rectangular Cartesian coordinates in a plane,

$$I_{xx} = \int y^2 dA\,, \quad I_{xy} = \int xy dA\,, \quad I_{yy} = \int x^2 dA\,,$$

with respect to rotation of the coordinate axes about the origin. Mohr's circle was invented in 1887 for the transformation of the inertia tensor.

Problem 3.2. Let v_i, $i = 1, 2, 3$, be the velocity vector field of a continuum, and let $\overline{v_i v_j}$ be the average value of the product $v_i v_j$ over a period of time. Show that the correlation function $\overline{v_i v_j}$, with components $\overline{u^2}$, $\overline{uv}$, $\overline{uw}$, $\overline{v^2}$, etc., in unabridged notations, is a symmetric tensor of the second order.

Problem 3.3. Show that the mass moment of inertia of a set of particles,

$$I_{ij} = e_{ipq} e_{jkq} \int x_p x_k dm\,, \quad i, j = 1, 2, 3\,,$$

is a tensor, where dm is an element of mass and the integration is extended over the entire set of particles. Write out the matrix of the inertia tensor I_{ij}. Show that I_{ii} (i not summed) is the moment of inertia about the x_i axis, whereas I_{ij} ($i \neq j$) is equal to the *negative* of the product of inertia about the axes x_i and x_j. Show that for a rigid body rotating at an angular velocity ω_j, the angular momentum vector of the body is $I_{ij}\omega_j$.

3.7. PRINCIPAL STRESSES

The results of the previous section are restricted to the plane state of stress. Let us now generalize them to general three dimensional problems.

In a general state of stress, the stress vector acting on a surface with outer normal ν depends on the direction of ν. Let us ask in what direction ν the stress vector becomes normal to the surface, on which the shearing stress vanishes. Such a surface is called a *principal plane*, its normal a *principal axis*, and the value of the normal stress acting on the principal plane is called a *principal stress*.

Let ν define a principal axis and let σ be the corresponding principal stress. Then the stress vector acting on the surface normal to ν has components $\sigma\nu_i$. On the other hand, this same vector is given by the expression $\tau_{ji}\nu_j$. Hence, writing $\nu_i = \delta_{ji}\nu_j$, we have, on equating these two expressions and transposing them to the same side,

$$(1) \qquad (\tau_{ji} - \sigma\delta_{ji})\nu_j = 0 \,.$$

The three equations, $i = 1, 2, 3$, are to be solved for ν_1, ν_2, ν_3. Since ν is a unit vector, we must find a set of nontrivial solutions for which $\nu_1^2 + \nu_2^2 + \nu_3^2 = 1$. Thus, Eq. (1) poses an eigenvalue problem. Since τ_{ij} as a matrix is real and symmetric, we need only to recall a result in the theory of matrices to assert that *there exist three real-valued principal stresses and a set of orthonormal principal axes*. Because of the importance of these results, we shall give the details of the reasoning below.

Equation (1) has a set of nonvanishing solutions ν_1, ν_2, ν_3 if and only if the determinant of the coefficients vanishes, i.e.,

$$(2) \qquad |\tau_{ij} - \sigma\delta_{ij}| = 0 \,.$$

Equation (2) is a cubic equation in σ; its roots are the principal stresses. For each value of the principal stress, a unit normal vector ν can be determined. On expanding Eq. (2), we have

$$(3) \qquad |\tau_{ij} - \sigma\delta_{ij}| = \begin{vmatrix} \tau_{11} - \sigma & \tau_{12} & \tau_{13} \\ \tau_{21} & \tau_{22} - \sigma & \tau_{23} \\ \tau_{31} & \tau_{32} & \tau_{33} - \sigma \end{vmatrix}$$

$$= -\sigma^3 + I_1\sigma^2 - I_2\sigma + I_3 = 0 \,,$$

where

$$I_1 = \tau_{11} + \tau_{22} + \tau_{33} \,,$$

$$I_2 = \begin{vmatrix} \tau_{22} & \tau_{23} \\ \tau_{32} & \tau_{33} \end{vmatrix} + \begin{vmatrix} \tau_{11} & \tau_{13} \\ \tau_{31} & \tau_{33} \end{vmatrix} + \begin{vmatrix} \tau_{11} & \tau_{12} \\ \tau_{21} & \tau_{22} \end{vmatrix} \,,$$

(4)

$$I_3 = \begin{vmatrix} \tau_{11} & \tau_{12} & \tau_{13} \\ \tau_{21} & \tau_{22} & \tau_{23} \\ \tau_{31} & \tau_{32} & \tau_{33} \end{vmatrix} \,.$$

On the other hand, if σ_1, σ_2, σ_3 are the roots of Eq. (3), which can be written as

$$(\sigma - \sigma_1)(\sigma - \sigma_2)(\sigma - \sigma_3) = 0 \,,$$

it can be seen that the following relations between the roots and the coefficients must hold:

(5)
$$I_1 = \sigma_1 + \sigma_2 + \sigma_3 \,,$$
$$I_2 = \sigma_1\sigma_2 + \sigma_2\sigma_3 + \sigma_3\sigma_1 \,,$$
$$I_3 = \sigma_1\sigma_2\sigma_3 \,.$$

Since the principal stresses characterize the physical state of stress at a point, they are independent of any coordinates of reference. Hence, Eq. (3) and the coefficients I_1, I_2, I_3 are invariant with respect to the coordinate transformation; I_1, I_2, I_3 are the *invariants* of the stress tensor. The importance of invariants will become evident when physical laws are formulated (see, for example, Chapter 6).

We shall show now that for a symmetric stress tensor the three principal stresses are all real and that the three principal planes are mutually orthogonal. These important properties can be established when the stress tensor is symmetric,

(6) $$\tau_{ij} = \tau_{ji} \,.$$

The proof is as follows. Let $\overset{1}{\nu}$, $\overset{2}{\nu}$, $\overset{3}{\nu}$, be unit vectors in the direction of the principal axes, with components $\overset{1}{\nu_i}$, $\overset{2}{\nu_i}$, $\overset{3}{\nu_i}$ ($i = 1, 2, 3$) which are the solutions of Eq. (1) corresponding to the roots σ_1, σ_2, σ_3, respectively;

(7)
$$(\tau_{ij} - \sigma_1\delta_{ij})\overset{1}{\nu_j} = 0 \,,$$
$$(\tau_{ij} - \sigma_2\delta_{ij})\overset{2}{\nu_j} = 0 \,,$$
$$(\tau_{ij} - \sigma_3\delta_{ij})\overset{3}{\nu_j} = 0 \,.$$

Multiplying the first equation by $\overset{2}{\nu}_i$, the second by $\overset{1}{\nu}_i$, summing over i and subtracting the resulting equations, we obtain

$$(8) \qquad\qquad (\sigma_2 - \sigma_1)\overset{1}{\nu}_i\overset{2}{\nu}_i = 0\,,$$

on account of the symmetry condition (6), which implies that

$$(9) \qquad\qquad \tau_{ij}\overset{1}{\nu}_j\overset{2}{\nu}_i = \tau_{ji}\overset{1}{\nu}_j\overset{2}{\nu}_i = \tau_{ij}\overset{2}{\nu}_j\overset{1}{\nu}_i\,.$$

The last equality is obtained by interchanging the dummy indices i and j.

Now, if we assume tentatively that Eq. (3) has a complex root, then, since the coefficients in Eq. (3) are real-valued, a complex conjugate root must also exist and the set of roots may be written as

$$\sigma_1 = \alpha + i\beta\,, \qquad \sigma_2 = \alpha - i\beta\,, \qquad \sigma_3\,,$$

where α, β, σ_3 are real numbers and i stands for the imaginary number $\sqrt{-1}$. In this case, Eqs. (7) show that $\overset{1}{\nu}_j$ and $\overset{2}{\nu}_j$ are complex conjugate to each other and can be written as

$$\overset{1}{\nu}_j \equiv a_j + ib_j\,, \qquad \overset{2}{\nu}_j \equiv a_j - ib_j\,,$$

in which a_j and b_j are real numbers and at least one of them is not zero. Therefore,

$$\overset{1}{\nu}_j\overset{2}{\nu}_j = (a_j + ib_j)(a_j - ib_j)$$
$$= a_1^2 + a_2^2 + a_3^2 + b_1^2 + b_2^2 + b_3^2 \neq 0\,.$$

It follows from (8) that $\sigma_1 - \sigma_2 = 2i\beta = 0$ or $\beta = 0$. This contradicts the original assumption that the roots are complex. Thus, the assumption of the existence of complex roots is untenable, and the roots σ_1, σ_2, σ_3 are all real.

When $\sigma_1 \neq \sigma_2 \neq \sigma_3$, Eq. (8) implies

$$(10) \qquad\qquad \overset{1}{\nu}_i\overset{2}{\nu}_i = 0\,, \quad \overset{2}{\nu}_i\overset{3}{\nu}_i = 0\,, \quad \overset{3}{\nu}_i\overset{1}{\nu}_i = 0\,;$$

i.e., the principal vectors are mutually orthogonal to each other. If $\sigma_1 = \sigma_2 \neq \sigma_3$, we can determine an infinite number of pairs of orthogonal vectors $\overset{1}{\nu}_i$ and $\overset{2}{\nu}_i$ and define $\overset{3}{\nu}_i$ as a vector orthogonal to $\overset{1}{\nu}_i$ and $\overset{2}{\nu}_i$. If $\sigma_1 = \sigma_2 = \sigma_3$, then any set of orthogonal axes may be taken as the principal axes.

If the reference axes x_1, x_2, x_3 are chosen to coincide with the principal axes, then the matrix of stress components becomes

(11)
$$(\tau_{ij}) = \begin{pmatrix} \sigma_1 & 0 & 0 \\ 0 & \sigma_2 & 0 \\ 0 & 0 & \sigma_3 \end{pmatrix}.$$

3.8. SHEARING STRESSES

We have seen that on an element of surface with a unit outer normal $\boldsymbol{\nu}$, (ν_i), there acts a traction $\overset{\nu}{\mathbf{T}}$, $(\overset{\nu}{T}_i = \tau_{ji}\nu_j)$. The component of $\overset{\nu}{\mathbf{T}}$ in the direction of $\boldsymbol{\nu}$ is the normal stress acting on the surface element. Let this normal stress be denoted by $\sigma_{(n)}$. Since the component of a vector in the direction of another vector is given by the scalar product of the two vectors, we obtain

(1)
$$\sigma_{(n)} = \tau_{ij}\nu_i\nu_j.$$

The magnitude of the resultant shearing stress on a surface element having the normal ν_i is given by the equation

(2)
$$\tau^2 = |\overset{\nu}{T}_i|^2 - \sigma_{(n)}^2,$$

(see Fig. 3.8:1). Let the principal axes be chosen as the coordinate axes,

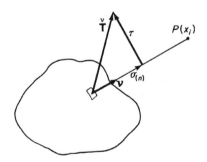

Fig. 3.8:1. Notations.

and let σ_1, σ_2, σ_3 be the principal stresses. Then

$$|\overset{\nu}{T}_i|^2 = (\sigma_1\nu_1)^2 + (\sigma_2\nu_2)^2 + (\sigma_3\nu_3)^2,$$

and, from Eq. (1)

$$\sigma_{(n)}^2 = [\sigma_1(\nu_1)^2 + \sigma_2(\nu_2)^2 + \sigma_3(\nu_3)^2]^2.$$

On substituting into Eq. (2) and noting that

$$(\nu_1)^2 - (\nu_1)^4 = (\nu_1)^2[1 - (\nu_1)^2] = (\nu_1)^2[(\nu_2)^2 + (\nu_3)^2],$$

we see that

(3) $$\tau^2 = (\nu_1)^2(\nu_2)^2(\sigma_1 - \sigma_2)^2 + (\nu_2)^2(\nu_3)^2(\sigma_2 - \sigma_3)^2$$
$$+ (\nu_3)^2(\nu_1)^2(\sigma_3 - \sigma_1)^2.$$

If $\nu_1 = \nu_2 = 1/\sqrt{2}$ and $\nu_3 = 0$, then $\tau = \pm\frac{1}{2}(\sigma_1 - \sigma_2)$ and $\sigma_n = \frac{1}{2}(\sigma_1 + \sigma_2)$.

Problem 3.4.   Show that $\tau_{max} = \frac{1}{2}(\sigma_{max} - \sigma_{min})$ and that the plane on which τ_{max} acts makes an angle of 45° with the directions of the largest and the smallest principal stresses.

3.9. MOHR'S CIRCLES

Let σ_1, σ_2, σ_3 be the principal stresses at a point. The stress components acting on any other surface elements can be obtained by the tensor transformation laws, Eq. (3.5:3). Otto Mohr, in papers published in 1882 and 1900, has shown the interesting result that if the normal stress $\sigma_{(n)}$ and the shearing stress τ acting on any surface element be plotted on a plane,

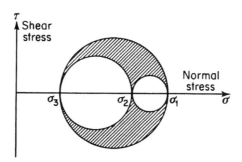

Fig. 3.9:1. Mohr's circles.

with σ and τ as coordinates as shown in Fig. 3.9:1, the locus necessarily falls in a closed domain represented by the shaded area bounded by the three semicircles with centers on the σ-axis. A detailed proof can be found in Westergaard,[1.2] *Elasticity and Plasticity*, pp. 61–64; or Sokolnikoff,[1.2] *Elasticity*, p. 52. The practical problem of graphical construction of Mohr's circle from strain-gage data is discussed in Biezeno and Grammel,[1.2] *Engineering Dynamics*, Vol. 1, p. 245; Pearson,[1.4] *Theoretical Elasticity*, p. 64. See Bibliography on pp. 873–876.

3.10. STRESS DEVIATIONS

The tensor

(1) ▲ $$\tau'_{ij} = \tau_{ij} - \sigma_0 \delta_{ij},$$

is called the *stress deviation tensor*, where δ_{ij} is the Kronecker delta and σ_0 is the mean stress

(2) $$\sigma_0 = \frac{1}{3}(\sigma_1 + \sigma_2 + \sigma_3) = \frac{1}{3}(\tau_{11} + \tau_{22} + \tau_{33}) = \frac{1}{3}I_1,$$

where I_1 is the first invariant of Sec. 3.7 and τ'_{ij} specifies the deviation of the state of stress from the means stress.

The first invariant of the stress deviation tensor always vanishes:

(3) $$I'_1 = \tau'_{11} + \tau'_{22} + \tau'_{33} = 0.$$

To determine the principal stress deviations, the procedure of Sec. 3.7 may be followed. The determinental equation

(4) $$|\tau'_{ij} - \sigma' \delta_{ij}| = 0,$$

may be expanded in the form

(5) $$\sigma'^3 - J_2 \sigma' - J_3 = 0.$$

It is easy to verify the following equations relating J_2, J_3 to the invariants I_2, I_3 defined in Sec. 3.7,

(6) $$J_2 = 3\sigma_0^2 - I_2,$$

(7) $$J_3 = I_3 - I_2\sigma_0 + 2\sigma_0^3 = I_3 + J_2\sigma_0 - \sigma_0^3,$$

and the alternative expressions below on account of Eq. (3),

(8) $$J_2 = -\tau'_{11}\tau'_{22} - \tau'_{22}\tau'_{33} - \tau'_{33}\tau'_{11} + (\tau_{12})^2 + (\tau_{23})^2 + (\tau_{31})^2$$

$$= \frac{1}{2}[(\tau'_{11})^2 + (\tau'_{22})^2 + (\tau'_{33})^2] + (\tau_{12})^2 + (\tau_{23})^2 + (\tau_{31})^2$$

$$= \frac{1}{6}[(\tau_{11} - \tau_{22})^2 + (\tau_{22} - \tau_{33})^2 + (\tau_{33} - \tau_{11})^2]$$

$$+ (\tau_{12})^2 + (\tau_{23})^2 + (\tau_{31})^2$$

$$= \frac{3}{2}\tau_0^2.$$

The τ_0 in the last equation is the *octahedral stress,* which will be defined in the next section.

We note also the simple expressions

(9) ▲
$$J_2 = \frac{1}{2}\tau'_{ij}\tau'_{ij} ,$$

(10) ▲
$$J_3 = \frac{1}{3}\tau'_{ij}\tau'_{jk}\tau'_{ki} .$$

It can be easily shown that the principal stress deviations are

(11)
$$\sigma'_i = \sigma_i - \sigma_0 .$$

Problem 3.5.   Show that the principal stresses as given by the three roots of Eq. (5) can be written as

$$\sigma'_1 = \tau_0\sqrt{2}\,\cos\alpha, \quad \sigma'_2 = \tau_0\sqrt{2}\,\cos\left(\alpha + \frac{2\pi}{3}\right), \quad \sigma'_3 = \tau_0\sqrt{2}\,\cos\left(\alpha - \frac{2\pi}{3}\right),$$

where $\cos 3\alpha = J_3\sqrt{2}/\tau_0^3$, and $J_2 = 3\tau_0^2/2$.

3.11. OCTAHEDRAL SHEARING STRESS

The octahedral shearing stress τ_0 is the resultant shearing stress on a plane that makes the same angle with the three principal directions. Such a plane is called an *octahedral plane;* eight such planes can form an octahedron. See Fig. 3.11:1. The direction cosines ν_i of a normal to the octahedral

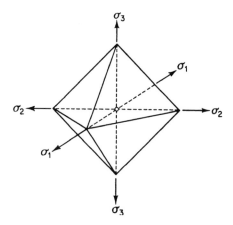

Fig. 3.11:1. Octahedral planes.

plane relative to the principal axes are such that

$$(\nu_1)^2 = (\nu_2)^2 = (\nu_3)^2 = \frac{1}{3}.$$

Hence, Eq. (3.8:3) gives

$$9\tau_0^2 = (\sigma_1 - \sigma_2)^2 + (\sigma_2 - \sigma_3)^2 + (\sigma_3 - \sigma_1)^2,$$

which is proportional to the sum of the areas of Mohr's three semicircles. From Eqs. (3.10:6) and (3.10:8), it can be easily verified that the octahedral stress can be expressed in terms of the two invariants I_1 and I_2 of Sec. 3.7,

$$9\tau_0^2 = 2I_1^2 - 6I_2.$$

The square of the octahedral stress happens to be proportional to the second invariant J_2 of the stress deviation, Eqs. (3.10:6) and (3.10:8). In 1913, Richard von Mises proposed the hypothesis that yielding of some of the most important materials occurs at a constant value of the quantity J_2. Nadai then introduced the interpretation of J_2 as proportional to the octahedral shearing stress. In this way, J_2 or τ_0 enters into the basic equations of plasticity. Note that the normal stress on the octohedral plane equals the mean stress.

Problem 3.6. If $\sigma_1 > \sigma_2 > \sigma_3$ and σ_1, σ_3 are given, at what values of σ_2 does τ_0 attain its extreme values?

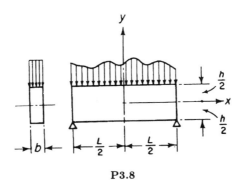

P3.8

Problem 3.7. Let $\sigma_x = -5c$, $\sigma_y = c$, $\sigma_z = c$, $\tau_{xy} = -c$, $\tau_{yz} = \tau_{zx} = 0$, where $c = 1,000$ lb/sq in. Determine the principal stresses, the principal stress deviations, the direction cosines of the principal directions, the greatest shearing stress, and the octahedral stress.

Problem 3.8. Consider a horizontal beam as shown in Fig. P3.8. According to the usual elementary theory of bending, the "fiber stress" is $\sigma_{xx} = -12My/bh^3$, where M is the bending moment which is a function of x. Assume this value of σ_{xx}, and assume further that $\sigma_{zz} = \sigma_{zx} = \sigma_{zy} = 0$, that the body force is absent, that $\sigma_{xy} = 0$ at the top and bottom of the beam $(y = \pm h/2)$, and that $\sigma_{yy} = 0$ at the bottom. Derive σ_{xy} and σ_{yy} from the equations of equilibrium. Compare the results with those derived in elementary mechanics of materials.

3.12. STRESS TENSOR IN GENERAL COORDINATES

So far we have discussed the stress tensor in rectangular Cartesian coordinates, in which there is no necessity to distinguish the contravariant and covariant transformations. The necessary distinction arises in curvilinear coordinates.

Just as a vector in three-dimensional Euclidean space may assume either a contravariant or a covariant form, a stress tensor can be either contravariant, τ^{ij}, or mixed, τ^i_j, or covariant, τ_{ij}. The tensors τ^{ij}, τ^i_j, and τ_{ij} are related to each other by raising or lowering of the indices by forming inner products with the metric tensors g_{ij} and g^{ij}:

$$\begin{aligned} \tau^i_j &= g_{\alpha j}\tau^{i\alpha} = g_{\alpha j}\tau^{\alpha i} \,, \\ \tau_{ij} &= g_{i\alpha}\tau^\alpha_j \,, \\ \tau^{ij} &= g^{i\alpha}\tau^j_\alpha \,. \end{aligned}$$

(1)

The correctness of these tensor relations are again seen by specializing them into rectangular Cartesian coordinates. We have seen that τ_{ij} is symmetric in rectangular Cartesian coordinates. It is easy to see from the tensor transformation law that *the symmetry property remains invariant in coordinate transformation.* Hence, $\tau_{ij} = \tau_{ji}$ in all admissible coordinates,[†] and we are allowed to write the mixed tensor as τ^i_j and not $\tau^i{}_j$ or $\tau_j{}^i$. But what is the physical meaning of each of these components?

To clarify the meaning of the stress components in an arbitrary curvilinear coordinates system, let us first consider a geometric relationship. Let us form an infinitesimal tetrahedron whose edges are formed by the coordinate curves PP_1, PP_2, PP_3 and the curves P_1P_2, P_2P_3, P_3P_1, as shown in Fig. 3.12:1. Let us write, vectorially,

$$\overline{PP_1} = \mathbf{r}_1 \,, \quad \overline{PP_2} = \mathbf{r}_2 \,, \quad \overline{PP_3} = \mathbf{r}_3 \,.$$

Then

$$\overline{P_1P_2} = \mathbf{r}_2 - \mathbf{r}_1 \,, \quad \overline{P_1P_3} = \mathbf{r}_3 - \mathbf{r}_1 \,, \quad \overline{P_2P_3} = \mathbf{r}_3 - \mathbf{r}_2 \,,$$

[†]Similarly, *the contravariant stress tensor τ^{ij} is symmetric, $\tau^{ij} = \tau^{ji}$.* It does not make sense, however, to say that the mixed tensor τ^i_j is symmetric, since an equation like $\tau^i_j = \tau^j_i$, with the indices switching roles on the two sides of the equation, is not a tensor equation.

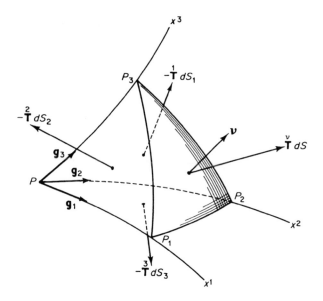

Fig. 3.12:1. Geometric relationship.

and we have

(2)
$$\overline{P_1P_2} \times \overline{P_1P_3} = (\mathbf{r}_2 - \mathbf{r}_1) \times (\mathbf{r}_3 - \mathbf{r}_1)$$

$$= -\mathbf{r}_1 \times \mathbf{r}_3 - \mathbf{r}_2 \times \mathbf{r}_1 + \mathbf{r}_2 \times \mathbf{r}_3$$

$$= \mathbf{r}_2 \times \mathbf{r}_3 + \mathbf{r}_3 \times \mathbf{r}_1 + \mathbf{r}_1 \times \mathbf{r}_2 \,.$$

Now the vector product $\mathbf{A} \times \mathbf{B}$ of any two vectors $\mathbf{A}$ and $\mathbf{B}$ is a vector perpendicular to $\mathbf{A}$ and $\mathbf{B}$, whose positive sense is determined by the *right-hand screw rule* from $\mathbf{A}$ to $\mathbf{B}$, and whose length is equal to the area of a parallelogram formed by $\mathbf{A}, \mathbf{B}$ as two sides. Hence, if we denote by ν, ν_1, ν_2, ν_3 the unit vectors normal to the surfaces $P_1P_2P_3$, PP_2P_3, PP_3P_1, PP_1P_2, respectively, and by dS, dS_1, dS_2, dS_3 their respective areas, Eq. (2) may be written as

(3)
$$\nu dS = \nu_1 dS_1 + \nu_2 dS_2 + \nu_3 dS_3 \,.$$

Now let us recall that in Sec. 2.14, we defined the reciprocal base vectors $\mathbf{g}^1$, $\mathbf{g}^2$, $\mathbf{g}^3$ which are perpendicular to the coordinate planes and are of length $\sqrt{g^{11}}$, $\sqrt{g^{22}}$, $\sqrt{g^{33}}$, respectively. We see that the unit vectors ν_1, ν_2, ν_3 are exactly $\mathbf{g}^1/\sqrt{g^{11}}$, $\mathbf{g}^2/\sqrt{g^{22}}$, $\mathbf{g}^3/\sqrt{g^{33}}$, respectively. Hence,

(4)
$$\nu dS = \sum_{i=1}^{3} \frac{dS_i}{\sqrt{g^{ii}}} \mathbf{g}^i \,.$$

If the unit normal vector $\boldsymbol{\nu}$ is resolved into its covariant components with respect to the reciprocal base vectors, then

(5) $$\boldsymbol{\nu} = \nu_i \mathbf{g}^i \,.$$

We see from the last two equations that

(6) ▲ $$\nu_i \sqrt{g^{ii}} \, dS = dS_i \,, \quad i \text{ not summed} \,.$$

This is the desired result.

Let us now consider the forces acting on the external surface of the infinitesimal tetrahedron. In Sec. 3.1, the inside and outside of a volume are distinguished by drawing an outward pointing normal vector. The stress vector $\overset{\nu}{\mathbf{T}}$ is defined as the limit of force acting on the outside surface with a normal $\boldsymbol{\nu}$ divided by the area of the surface. On the other side, the stress vector is $-\overset{\nu}{\mathbf{T}}$, see Sec. 3.3, Fig. 3.3:1. Now, for the tetrahedron shown in Fig. 3.12:1, the normal vector $\boldsymbol{\nu}$ of the triangle $P_1 P_2 P_3$ is outward, the stress vector is $\overset{\nu}{\mathbf{T}}$, the area is dS, the force is $\overset{\nu}{\mathbf{T}} dS$. On the triangle $P\,P_2\,P_3$, the normal vector $\boldsymbol{\nu}_i$ points inward, the stress vector is $\overset{i}{\mathbf{T}}$, the area is dS_i, the force is, therefore, $-\overset{i}{\mathbf{T}} dS_i$ as shown in Fig. 3.12:1. The forces on the other surfaces are determined similarly. The equation of motion of this infinitesimal tetrahedron is, in the limit,

(7) $$\overset{\nu}{\mathbf{T}} dS = \overset{i}{\mathbf{T}} dS_i \,.$$

Volume forces and inertia (mass × acceleration) forces acting on the tetrahedron do not enter into this equation, because they are of higher order of smallness than the surface forces. On substituting (6) into (7) and canceling the nonvanishing factor dS, we have

(8) $$\overset{\nu}{\mathbf{T}} = \sum_{i=1}^{3} \overset{i}{\mathbf{T}} \nu_i \sqrt{g^{ii}} \,.$$

If the coordinates x^i are changed to a new set $\bar{x}^i$ while the surface $P_1 P_2 P_3$ and the unit outer normal $\boldsymbol{\nu}$ remain unchanged, the stress vector $\overset{\nu}{\mathbf{T}}$ is invariant, but the vectors $\overset{i}{\mathbf{T}}$ will change because they will be associated with the new coordinate surfaces. (See Fig. 3.12:2, in which only one pair of vectors $\overset{2}{\mathbf{T}}$, $\overset{2}{\mathbf{T}}'$ are shown.) The covariant components ν_i will change also. Inasmuch as $\overset{\nu}{\mathbf{T}}$ is invariant and ν_i is a covariant tensor, Eq. (8) shows that $\overset{i}{\mathbf{T}} \sqrt{g^{ii}}$ transforms according to a contravariant type of transformation.

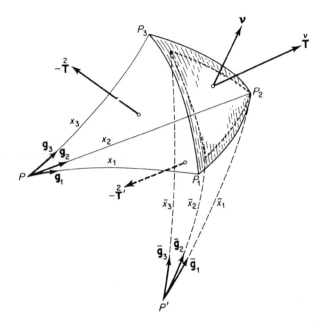

Fig. 3.12:2. Tractions referred to two different elementary tetrahedron with a common surface.

Resolving the vectors $\overset{i}{\mathbf{T}}\sqrt{g^{ii}}$ into their components with respect to the base vectors $\mathbf{g}_i$ and the reciprocal base vectors $\mathbf{g}^i$, we have

(9) ▲ $$\sqrt{g^{ii}}\,\overset{i}{\mathbf{T}} = \tau^{ij}\mathbf{g}_j = \tau^i_j\mathbf{g}^j\,, \quad i \text{ not summed}.$$

On the other hand, the components of $\overset{\nu}{\mathbf{T}}$ may be written as

(10) $$\overset{\nu}{\mathbf{T}} = \overset{\nu}{T}{}^j\mathbf{g}_j = \overset{\nu}{T}_j\mathbf{g}^j\,.$$

Substitution of (9) and (10) into (8) shows that

(11) ▲ $$\overset{\nu}{T}{}^j = \tau^{ij}\nu_i\,, \quad \overset{\nu}{T}_j = \tau^i_j\nu_i\,.$$

The tensorial character of τ^{ij} and τ^i_j are demonstrated both by (9) and by (11) according to the quotient rule. Following the tensor transformation rules for τ^{ij}, τ^i_j and ν_i, we can rewrite Eq. (11) in the form (12) $\overset{\nu}{T}{}^j = \tau^j_i\nu^i$, $\overset{\nu}{T}_j = \tau_{ij}\nu^i$ where ν^i are the contravariant components of $\boldsymbol{\nu}$.

Let us recapitulate the important points. With the scaling factor $\sqrt{g^{ii}}$, the vectors $\overset{i}{\mathbf{T}}\sqrt{g^{ii}}$ ($i = 1, 2, 3$, not summed) transform according to the contravariant rule. When the vectors $\overset{i}{\mathbf{T}}\sqrt{g^{ii}}$ are resolved into components along the base vectors $\mathbf{g}_j$, the tensor components τ^{ij} are obtained. When the resolution is made with respect to the reciprocal base vectors $\mathbf{g}^j$, then the mixed tensor τ^i_j is obtained. On the other hand, if the stress vector $\overset{\nu}{\mathbf{T}}$ (acting on the triangle $P_1P_2P_3$ with outer normal ν) is resolved into the contravariant components $\overset{\nu}{T^j}$ along the base vectors $\mathbf{g}_j$ and the covariant components $\overset{\nu}{\mathbf{T}}_j$ along the reciprocal base vectors $\mathbf{g}^j$, then we obtain Eq. (11).

The contravariant stress tensor τ^{ij} and the mixed stress tensor τ^i_j are related to the stress vectors $\overset{i}{\mathbf{T}}$ by the Eq. (9). The covariant stress tensor τ_{ij} defined in Eq. (1) cannot be so simply related to the vectors $\overset{i}{\mathbf{T}}$ and are therefore of less importance.

3.13. PHYSICAL COMPONENTS OF A STRESS TENSOR IN GENERAL COORDINATES

If we write Eq. (3.12:9) as

(1) $$\overset{i}{\mathbf{T}} = \sum_{j=1}^{3}\sqrt{\frac{g_{jj}}{g^{ii}}}\,\tau^{ij}\frac{\mathbf{g}_j}{\sqrt{g_{jj}}} = \sum_{j=1}^{3}\sigma^{ij}\frac{\mathbf{g}_j}{\sqrt{g_{jj}}}\,, \qquad i \text{ not summed}.$$

Then, since $\mathbf{g}_j/\sqrt{g_{jj}}$ (j not summed) are unit vectors along the coordinate curves, the components σ^{ij} are uniform in physical dimensions and represent the *physical components* of the stress vector $\overset{i}{\mathbf{T}}$ in the direction of the unit vectors $\mathbf{g}_j/\sqrt{g_{jj}}$,

(2) $$\sigma^{ij} = \sqrt{\frac{g_{jj}}{g^{ii}}}\tau^{ij}\,, \qquad i,j \text{ not summed}.$$

But σ^{ij} is not a tensor.

On the other hand, if we use the mixed tensor τ^i_j in (3.12:1), we have

(3) $$\overset{i}{\mathbf{T}} = \sum_{j=1}^{3}\tau^i_j\frac{\mathbf{g}^j}{\sqrt{g^{ii}}} = \sum_{j=1}^{3}\tau^i_j\sqrt{\frac{g^{jj}}{g^{ii}}}\cdot\frac{\mathbf{g}^j}{\sqrt{g^{jj}}}\,, \qquad i \text{ not summed}.$$

Thus

(4) $$\sigma^i_j = \sqrt{\frac{g^{jj}}{g^{ii}}}\tau^i_j\,, \qquad i,j \text{ not summed},$$

are the physical components of the tensor τ^i_j, of uniform physical dimensions, representing the components of the stress vector $\overset{i}{\mathbf{T}}$ resolved in the directions of the reciprocal base vectors. Note that, in general, $\sigma^{ij} \neq \sigma^i_j \neq \sigma^j_i$, except for orthogonal coordinates. If the coordinates are orthogonal, $\mathbf{g}^i$ and $\mathbf{g}_i$ are in the same direction.

When curvilinear coordinates are used, we like to retain the liberty of choosing coordinates without regard to dimensions. Thus, in cylindrical polar coordinates (r, θ, z), r and z have the dimensions of length and θ is an angle. The corresponding tensor components of a vector referred to polar coordinates will have different dimensions. For physical understanding it is desirable to employ the physical components, but for the convenience of analysis it is far more expedient to use the tensor components.

3.14. EQUATIONS OF EQUILIBRIUM IN CURVILINEAR COORDINATES

In Sec. 3.4 we discussed the equations of equilibrium in terms of Cartesian tensors in rectangular Cartesian coordinates. To obtain these equations in any curvilinear coordinates, it is only necessary to observe that the equilibrium conditions must be expressed in a tensor equation. Thus the equations of equilibrium must be

(1) $$\tau^{ij}|_j + X^i = 0, \quad \text{in volume},$$

(2) $$\tau^{ji}\nu_i = \overset{\nu}{T}^i, \quad \text{on surface}.$$

The truth is at once proved by observing that these are truly tensor equations and that they hold in the special case of rectangular Cartesian coordinates. Hence, they hold in any coordinates that can be derived from the Cartesian coordinates through admissible transformations.

The practical application of tensor analysis in the derivation of the equations of equilibrium in particular curvilinear coordinates will be illustrated in Secs. 4.11 and 4.12. It will be seen that these lengthly equations can be obtained in a routine manner without too must effort. Because the manipulation is routine, chances of error are minimized. This practical application may be regarded as the first dividend to be paid for the long process of learning the tensor analysis.

Problem 3.9. Let us recast the principal results obtained above into tensor equations in general coordinates of reference. In rectangular Cartesian coordinates, there is no difference in contravariant and covariant transformations. Hence, the Cartesian stress tensor may be written as τ_{ij}, or τ^{ij}, or τ^i_j. In general frames of reference, τ_{ij}, τ^{ij}, τ^i_j are different. Their components may have different values. They are different versions of the same physical entity. Now prove the following results in general coordinates.

(a) The tensors τ^{ij}, τ_{ij} are symmetric if there is no body moment acting on the medium; i.e.,

$$\tau^{ij} = \tau^{ji}, \quad \tau_{ij} = \tau_{ji}.$$

(b) Principal planes are planes on which the stress vector $\overset{\nu}{T}$ is parallel to the normal vector ν. If we use contravariant components, we have, on a principal plane, $\overset{\nu}{T}{}^j = \sigma\nu^j = \sigma g^{ij}\nu_i = \tau^{ij}\nu_i$, where σ is a scalar. If we use covariant components, we have correspondingly $\overset{\nu}{T}_j = \sigma\nu_j = \sigma g^i_j\nu_i = \tau^i_j\nu_i$. Show that σ must satisfy the characteristic determinantal equation

$$|\tau^i_j - \sigma\delta^i_j| = 0 \quad \text{or its equivalent} \quad |\tau^{ij} - \sigma g^{ij}| = 0.$$

(c) The first invariant of the stress tensor is τ^i_i, or $\tau^{ij}g_{ij}$. However, τ_{ii} and τ^{ii} are in general not invariants.

(d) The stress deviation tensor s^i_j is defined as

$$s^i_j = \tau^i_j - \frac{1}{3}\tau^\alpha_\alpha g^i_j.$$

The first invariant of s^i_j zero. The second invariant has the convenient form $J_2 = \frac{1}{2}s^i_k s^k_i$.

(e) The octahedral shearing stress has the same value in any coordinates system.

4

ANALYSIS OF STRAIN

In this chapter we shall consider the deformation of a body as a "mapping" of the body from the original state to the deformed state. Strain tensors which are useful for finite strains as well as infinitesimal strains are then defined.

4.1. DEFORMATION

In the formulation of continuum mechanics the configuration of a solid body is described by a continuous mathematical model whose geometrical points are identified with the place of the material particles of the body. When such a continuous body changes its configuration under some physical action, we impose the assumption that the change is continuous; i.e., neighborhoods are changed into neighborhoods. Any introduction of new boundary surfaces, such as caused by tearing of a membrane or fracture of a test specimen, must be regarded as an extraordinary circumstance requiring special attention and explanation.

Let a system of coordinates a_1, a_2, a_3 be chosen so that a point P of a body at a certain instant of time is described by the coordinates $a_i(i = 1, 2, 3)$. At a later instant of time, the body is moved (deformed) to a new configuration; the point P is moved to Q with coordinates $x_i(i = 1, 2, 3)$ with respect to a new coordinate system x_1, x_2, x_3. The coordinate systems a_1, a_2, a_3 and x_1, x_2, x_3 may be curvilinear and need not be the same (Fig. 4.1:1), but they both describe a Euclidean space.

The change of configuration of the body will be assumed to be continuous, and the *point transformation* (*mapping*) from P to Q is assumed to be one-to-one. The equation of transformation can be written as

$$(1) \qquad x_i = \hat{x}_i(a_1, a_2, a_3),$$

which has a unique inverse

$$(2) \qquad a_i = \hat{a}_i(x_1, x_2, x_3),$$

for every point in the body. The functions $\hat{x}_i(a_1, a_2, a_3)$ and $\hat{a}_i(x_1, x_2, x_3)$ are assumed to be continuous and differentiable.

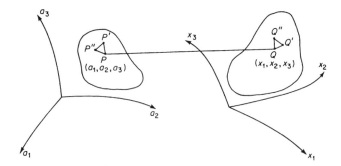

Fig. 4.1:1. Deformation of a body.

We shall be concerned with the description of the strain of the body, i.e., with the stretching and distortion of the body. If P, P', P'' are three neighboring points forming a triangle in the original configuration, and if they are transformed to points Q, Q', Q'' in the deformed configuration, the change in area and angles of the triangle is completely determined if we know the change in length of the sides. But the "location" of the triangle is undetermined by the change of the sides. Similarly, if the change of length between any two arbitrary points of the body is known, the new configuration of the body will be completely defined except for the location of the body in space. In the following discussions our attention will be focused on the strain of the body, because it is the strain that is related to the stress. The description of the change in distance between any two points of the body is the key to the analysis of deformation.

Consider an infinitesimal line element connecting the point $P(a_1, a_2, a_3)$ to a neighboring point $P'(a_1 + da^1, a_2 + da^2, a_3 + da^3)$.[†] The square of the length ds_0 of PP' in the original configuration is given by

$$(3) \qquad ds_0^2 = a_{ij} da^i da^j \,,$$

where a_{ij}, evaluated at the point P, is the Euclidean metric tensor for the coordinate system a_i. When P, P' is deformed to the points $Q(x_1, x_2, x_3)$ and $Q'(x_1 + dx^1, x_2 + dx^2, x_3 + dx^3)$, respectively, the square of the length ds of the new element QQ' is

$$(4) \qquad ds^2 = g_{ij} dx^i dx^j \,,$$

where g_{ij} is the Euclidean metric tensor for the coordinate system x_i.

[†] As remarked before in a footnote in Sec. 2.4, the point transformation between a_i and x_i does not follow tensor transformation law and the position of the indices has no significance. But the *differentials* da^i and dx^i do transform according to the contravariant tensor law and are contravariant vectors.

By Eqs. (1) and (2), we may also write

(5) $$ds_0^2 = a_{ij}\frac{\partial a_i}{\partial x_l}\frac{\partial a_j}{\partial x_m}dx^l dx^m ,$$

(6) $$ds^2 = g_{ij}\frac{\partial x_i}{\partial a_l}\frac{\partial x_j}{\partial a_m}da^l da^m .$$

The difference between the squares of the length elements may be written, after several changes in the symbols for dummy indices, either as

(7) $$ds^2 - ds_0^2 = \left(g_{\alpha\beta}\frac{\partial x_\alpha}{\partial a_i}\frac{\partial x_\beta}{\partial a_j} - a_{ij} \right) da^i da^j ,$$

or as

(8) $$ds^2 - ds_0^2 = \left(g_{ij} - a_{\alpha\beta}\frac{\partial a_\alpha}{\partial x_i}\frac{\partial a_\beta}{\partial x_j} \right) dx^i dx^j .$$

We define the *strain tensors*

(9) ▲ $$E_{ij} = \frac{1}{2}\left(g_{\alpha\beta}\frac{\partial x_\alpha}{\partial a_i}\frac{\partial x_\beta}{\partial a_j} - a_{ij} \right) ,$$

(10) ▲ $$e_{ij} = \frac{1}{2}\left(g_{ij} - a_{\alpha\beta}\frac{\partial a_\alpha}{\partial x_i}\frac{\partial a_\beta}{\partial x_j} \right) ,$$

so that

(11) ▲ $$ds^2 - ds_0^2 = 2E_{ij}da^i da^j ,$$

(12) ▲ $$ds^2 - ds_0^2 = 2e_{ij}dx^i dx^j .$$

The strain tensor E_{ij} *was introduced by Green and St. Venant, and is called Green's strain tensor. The strain tensor* e_{ij} *was introduced by Cauchy for infinitesimal strains and by Almansi and Hamel for finite strains, and is known as Almansi's strain tensor.* In analogy with a terminology in hydrodynamics, E_{ij} is often referred to as a strain tensor in Lagrangian coordinates and e_{ij} as a strain tensor in Eulerian coordinates.

That E_{ij} and e_{ij} thus defined are tensors in the coordinate systems a_i and x_i, respectively, follows from the quotient rule when it is applied to Eqs. (11) and (12). The tensorial character of E_{ij} and e_{ij} can also be verified directly from their definitions (9), (10) by considering further coordinate transformations in either the original configuration (from a_1, a_2, a_3 to $\bar{a}_1, \bar{a}_2, \bar{a}_3$), or the deformed configuration (from x_1, x_2, x_3 to $\bar{x}_1, \bar{x}_2, \bar{x}_3$).

The details are left to the reader. The tensors E_{ij} and e_{ij} are obviously *symmetric*, i.e.,

(13) ▲ $$E_{ij} = E_{ji}, \quad e_{ij} = e_{ji}.$$

An immediate consequence of Eqs. (11) and (12) is the fundamental result that a *necessary and sufficient condition that a deformation of a body be a rigid-body motion (consists merely of translation and rotation without changing distances between particles) is that all the components of the strain tensor* E_{ij} *or* e_{ij} *be zero throughout the body.*

In the discussion above we have used two sets of curvilinear coordinates to describe the position of each particle. One, a_1, a_2, a_3, is used in the original configuration, the other, x_1, x_2, x_3, is used in the deformed configuration.

Now, *there are two particularly favored choices of coordinates*:

I. *We use the same rectangular Cartesian coordinates for both the original and the deformed configurations of the body.* In this case, the metric tensors are extremely simple:

(14) $$g_{ij} = a_{ij} = \delta_{ij}.$$

II. *We distort the frame of reference in the deformed configuration in such a way that the coordinates* x_1, x_2, x_3, *of a particle have the same numerical values* a_1, a_2, a_3 *as in the original configuration.* In this case, $x_i = a_i$, $\partial x_\alpha / \partial a_i = \delta_{\alpha i}$, $\partial a_\alpha / \partial x_i = \delta_{\alpha i}$, and Eqs. (9) and (10) are reduced to

(15) $$E_{ij} = e_{ij} = \frac{1}{2}(g_{ij} - a_{ij}).$$

Thus all the information about strain is contained in the change of the metric tensor as the frame of reference is distorted from the original configuration to the deformed configuration. In many ways this is the most convenient choice in the study of large deformations. The coordinates so chosen are called *convected* or *intrinsic* coordinates.

In the following sections we wish to discuss the meaning of the individual components of the strain tensor. For this purpose the choice (I) above is the most appropriate.

4.2. STRAIN TENSORS IN RECTANGULAR CARTESIAN COORDINATES

If we use the *same rectangular Cartesian* (rectilinear and orthogonal) coordinate system to describe both the original and the deformed

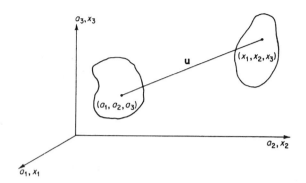

Fig. 4.2:1. Displacement vector.

configurations of the body, then

(1)
$$g_{ij} = a_{ij} = \delta_{ij} = \begin{cases} 1 & \text{for } i = j, \\ 0 & \text{for } i \neq j. \end{cases}$$

Furthermore, if we introduce the *displacement vector* **u** with components

(2)
$$u_i = x_i - a_i, \quad i = 1, 2, 3,$$

(see Fig. 4.2:1) then

$$\frac{\partial x_\alpha}{\partial a_i} = \frac{\partial u_\alpha}{\partial a_i} + \delta_{\alpha i}, \quad \frac{\partial a_\alpha}{\partial x_i} = \delta_{\alpha i} - \frac{\partial u_\alpha}{\partial x_i}.$$

The strain tensors reduce to the simple form

(3) ▲
$$E_{ij} = \frac{1}{2} \left[\delta_{\alpha\beta} \frac{\partial x_\alpha}{\partial a_i} \frac{\partial x_\beta}{\partial a_j} - \delta_{ij} \right]$$

$$= \frac{1}{2} \left[\delta_{\alpha\beta} \left(\frac{\partial u_\alpha}{\partial a_i} + \delta_{\alpha i} \right) \left(\frac{\partial u_\beta}{\partial a_j} + \delta_{\beta j} \right) - \delta_{ij} \right]$$

$$= \frac{1}{2} \left[\frac{\partial u_j}{\partial a_i} + \frac{\partial u_i}{\partial a_j} + \frac{\partial u_\alpha}{\partial a_i} \frac{\partial u_\alpha}{\partial a_j} \right],$$

and

(4) ▲
$$e_{ij} = \frac{1}{2} \left[\delta_{ij} - \delta_{\alpha\beta} \frac{\partial a_\alpha}{\partial x_i} \frac{\partial a_\beta}{\partial x_j} \right]$$

$$= \frac{1}{2} \left[\delta_{ij} - \delta_{\alpha\beta} \left(-\frac{\partial u_\alpha}{\partial x_i} + \delta_{\alpha i} \right) \left(-\frac{\partial u_\beta}{\partial x_j} + \delta_{\beta j} \right) \right]$$

$$= \frac{1}{2} \left[\frac{\partial u_j}{\partial x_j} + \frac{\partial u_i}{\partial x_j} - \frac{\partial u_\alpha}{\partial x_i} \frac{\partial u_\alpha}{\partial x_j} \right].$$

In unabridged notations (x, y, z for x_1, x_2, x_3; a, b, c for a_1, a_2, a_3; and u, v, w for u_1, u_2, u_3), we have the typical terms

$$E_{aa} = \frac{\partial u}{\partial a} + \frac{1}{2}\left[\left(\frac{\partial u}{\partial a}\right)^2 + \left(\frac{\partial v}{\partial a}\right)^2 + \left(\frac{\partial w}{\partial a}\right)^2\right],$$

$$e_{xx} = \frac{\partial u}{\partial x} - \frac{1}{2}\left[\left(\frac{\partial u}{\partial x}\right)^2 + \left(\frac{\partial v}{\partial x}\right)^2 + \left(\frac{\partial w}{\partial x}\right)^2\right],$$

(5)

$$E_{ab} = \frac{1}{2}\left[\frac{\partial u}{\partial b} + \frac{\partial v}{\partial a} + \left(\frac{\partial u}{\partial a}\frac{\partial u}{\partial b} + \frac{\partial v}{\partial a}\frac{\partial v}{\partial b} + \frac{\partial w}{\partial a}\frac{\partial w}{\partial b}\right)\right],$$

$$e_{xy} = \frac{1}{2}\left[\frac{\partial u}{\partial y} + \frac{\partial v}{\partial x} - \left(\frac{\partial u}{\partial x}\frac{\partial u}{\partial y} + \frac{\partial v}{\partial x}\frac{\partial v}{\partial y} + \frac{\partial w}{\partial x}\frac{\partial w}{\partial y}\right)\right].$$

Note that u, v, w are considered as functions of a, b, c, the position of points in the body in unstrained configuration, when the Lagrangian strain tensor is evaluated; whereas they are considered as functions of x, y, z, the position of points in the strained configuration, when the Eulerian strain tensor is evaluted.

If the components of displacement u_i are such that their first derivatives are so small that the squares and products of the partial derivatives of u_i are negligible, then e_{ij} is reduced to Cauchy's *infinitesimal strain tensor*,

(6) ▲
$$e_{ij} = \frac{1}{2}\left[\frac{\partial u_j}{\partial x_i} + \frac{\partial u_i}{\partial x_j}\right].$$

In unabridged notation,

$$e_{xx} = \frac{\partial u}{\partial x}, \quad e_{xy} = \frac{1}{2}\left(\frac{\partial u}{\partial y} + \frac{\partial v}{\partial x}\right) = e_{yx},$$

(7)
$$e_{yy} = \frac{\partial v}{\partial y}, \quad e_{xz} = \frac{1}{2}\left(\frac{\partial u}{\partial z} + \frac{\partial w}{\partial x}\right) = e_{zx},$$

$$e_{zz} = \frac{\partial w}{\partial z}, \quad e_{yz} = \frac{1}{2}\left(\frac{\partial v}{\partial z} + \frac{\partial w}{\partial y}\right) = e_{zy}.$$

In the infinitesimal displacement case, the distinction between the Lagrangian and Eulerian strain tensors disappears, since then it is immaterial whether the derivatives of the displacements are calculated at the position of a point before or after deformation.

4.3. GEOMETRIC INTERPRETATION OF INFINITESIMAL STRAIN COMPONENTS

Let x, y, z be a set of rectangular Cartesian coordinates. Consider a line element of length dx parallel to the x-axis ($dy = dz = 0$). The change of the square of the length of this element due to deformation is

$$ds^2 - ds_0^2 = 2e_{xx}(dx)^2 \, .$$

Hence,

$$ds - ds_0 = \frac{2e_{xx}(dx)^2}{ds + ds_0} \, .$$

But $ds = dx$ in this case, and ds_0 differs from ds only by a small quantity of the second order, if we assume the displacements u, v, w and the strain

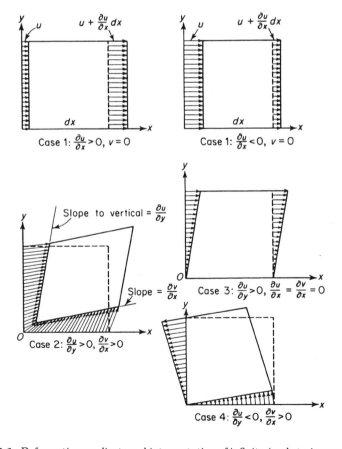

Fig. 4.3:1. Deformation gradients and interpretation of infinitesimal strain components.

components e_{ij} to be infinitesimal. Hence,

$$\frac{ds - ds_0}{ds} = e_{xx} ,$$

and it is seen that e_{xx} represents the *extension*, or change of length per unit length of a vector parallel to the x-axis. An application of the above discussion to a volume element is illustrated in Fig. 4.3:1, Case 1.

To see the meaning of the component e_{xy}, let us consider a small rectangle in the body with edges dx, dy. It is evident from Fig. 4.3:1, Cases 2, 3, and 4 that the sum $\partial u/\partial y + \partial v/\partial x$ represents the change of angle xOy which was originally a right angle. Thus,

$$e_{xy} = \frac{1}{2}\left(\frac{\partial u}{\partial y} + \frac{\partial v}{\partial x}\right) = \frac{1}{2}\ (\text{change of angle } xOy) .$$

In engineering usage, the strain components e_{ij} $(i \neq j)$ doubled, i.e., $2e_{ij}$, are called the *shearing strains*, or *detrusions*. The name is perhaps particularly suggestive in Case 3 of Fig. 4.3:1, which is called the case of *simple shear*.

The quantity

$$\omega_z = \frac{1}{2}\left(\frac{\partial v}{\partial x} - \frac{\partial u}{\partial y}\right) ,$$

is called the *infinitesimal rotation* of the element $dxdy$. This terminology is suggested by Case 4 in Fig. 4.3:1. If

$$\frac{\partial v}{\partial x} = -\frac{\partial u}{\partial y} ,$$

then $e_{xy} = 0$ and ω_z is indeed the angle through which the rectangular element is rotated as a rigid body.

4.4. ROTATION

Consider an infinitesimal displacement field $u_i(x_1, x_2, x_3)$. From u_i, form the Cartesian tensor

$$(1) \qquad\qquad \omega_{ij} = \frac{1}{2}\left(\frac{\partial u_j}{\partial x_i} - \frac{\partial u_i}{\partial x_j}\right) ,$$

which is antisymmetric, i.e.,

$$(2) \qquad\qquad \omega_{ij} = -\omega_{ji} .$$

In three dimensions, from an antisymmetric tensor we can always build a *dual vector*,

$$(3) \qquad\qquad \omega_k = \frac{1}{2}e_{kij}\omega_{ij} , \quad \text{i.e.,} \quad \boldsymbol{\omega} = \frac{1}{2}\ \text{curl } \mathbf{u} ,$$

where e_{kij} is the permutation tensor (Secs. 2.1 and 2.6). On the other hand, from Eq. (3) and the $e - \delta$ identity [Eq. (2.1:11)] it follows that $e_{ijk}\omega_k = \frac{1}{2}(\omega_{ij} - \omega_{ji})$. Hence, if ω_{ij} is antisymmetric, the relation Eq. (3) has a unique inverse,

$$(4) \qquad\qquad\qquad\qquad \omega_{ij} = e_{ijk}\omega_k \,.$$

Thus, ω_{ij} may be called the *dual* (antisymmetric) *tensor* of a vector ω_k. We shall call ω_k and ω_{ij}, respectively, the *rotation vector* and *rotation tensor* of the displacement field u_i.

We shall consider the physical meaning of these quantities below.

At the end of Sec. 4.3 we saw that ω_z represents the (infinitesimal) rotation of the element as a rigid body if e_{xy} vanishes. Now we have the general theorem that *the vanishing of the symmetric strain tensor E_{ij} or e_{ij} is the necessary and sufficient condition for a neighborhood of a particle to be moved like a rigid body.* This follows at once from the definitions of strain tensors, Eqs. (4.1:11) and (4.1:12). For, if a neighborhood of a particle P moves like a rigid body, the length of any element in the neighborhood will not change, so that $ds = ds_0$. It follows that $E_{ij} = e_{ij} = 0$, because, for a symmetric strain tensor E_{ij}, there exists a coordinates system that

$$ds^2 - ds_0^2 = \Lambda_1 da_1^2 + \Lambda_2 da_2^2 + \Lambda_3 da_3^2 \,,$$

where Λ_i are the eigenvalues of E_{ij}. If $ds = ds_0$ for any line element i.e., ds_0 can just equal da_1, da_2 or da_3, it implies Λ_i $(i = 1, 2, 3)$ are zero which in turn implies that $E_{ij} = 0$. The proof for $e_{ij} = 0$ is the same.

Conversely, if E_{ij} or e_{ij} vanishes at P, the length of line elements joining any two points in a neighbourhood of P will not change and the neighborhood moves like a rigid body.

We can show that an infinitesimal displacement field in the neighborhood of a point P can be decomposed into a stretching deformation and a rigid-body rotation. To show this, consider a point P' in the neighborhood of P. Let the coordinates of P and P' be x_i and $x_i + dx_i$, respectively. The relative displacement of P' with respect to P is

$$(5) \qquad\qquad\qquad\qquad du_i = \frac{\partial u_i}{\partial x_j} dx_j \,.$$

This can be written as

$$du_i = \frac{1}{2}\left(\frac{\partial u_i}{\partial x_j} + \frac{\partial u_j}{\partial x_i}\right) dx_j + \frac{1}{2}\left(\frac{\partial u_i}{\partial x_j} - \frac{\partial u_j}{\partial x_i}\right) dx_j$$

$$= e_{ij} dx_j - \omega_{ij} dx_j \,.$$

The first term in the right-hand side represents a stretching deformation which involves both stretching and rotation. The second term can be written as

(6)
$$-\omega_{ij}dx_j = \omega_{ji}dx_j$$
$$= -e_{ijk}\omega_k dx_j$$
$$= (\boldsymbol{\omega} \times d\mathbf{x})_i\,.$$

Thus the second term is the vector product of $\boldsymbol{\omega}$ and $d\mathbf{x}$. This is exactly what would have been produced by an infinitesimal rotation $|\boldsymbol{\omega}|$ about an axis through P in the direction of $\boldsymbol{\omega}$.

It should be noted that we have restricted ourselves to infinitesimal displacements. Angular and strain measures for finite displacements are related to the deformation gradient in a more complicated way.

4.5. FINITE STRAIN COMPONENTS

When the strain components are not small, it is no longer possible to give simple geometric interpretations of the components of the strain tensors.

Consider a set of rectangular Cartesian coordinates with respect to which the strain components are defined as in Sec. 4.1. Let a line element before deformation be $da^1 = ds_0$, $da^2 = 0$, $da^3 = 0$. Let the extension E_1 of this element be defined by

$$E_1 = \frac{ds - ds_0}{ds_0}\,,$$

or

(1)
$$ds = (1 + E_1)ds_0\,.$$

From Eq. (4.1:11), we have

(2)
$$ds^2 - ds_0^2 = 2E_{ij}da^i da^j = 2E_{11}(da^1)^2\,.$$

Combining (1) and (2), we obtain

$$(1 + E_1)^2 - 1 = 2E_{11}\,,$$

or

(3)
$$E_1 = \sqrt{1 + 2E_{11}} - 1\,.$$

This reduces to

(4)
$$E_1 \doteq E_{11}\,,$$

if E_{11} is small.

To get the physical significance of the component E_{12}, let us consider two line elements ds_0 and $d\bar{s}_0$ which are at a right angle in the original state:

(5)
$$ds_0 : \quad da^1 = ds_0, \quad da^2 = 0, \quad da^3 = 0;$$
$$d\bar{s}_0 : \quad d\bar{a}^1 = 0, \quad d\bar{a}^2 = d\bar{s}_0, \quad da^3 = 0.$$

After deformation these line elements become ds, (dx^i) and $d\bar{s}$, $(d\bar{x}^i)$. Forming the scalar product of the deformed elements, we obtain

$$ds\, d\bar{s} \cos\theta = dx^k d\bar{x}^k = \frac{\partial x_k}{\partial a_i} da^i \frac{\partial x_k}{\partial a_j} d\bar{a}^j$$

$$= \frac{\partial x_k}{\partial a_1} \frac{\partial x_k}{\partial a_2} ds_0 d\bar{s}_0.$$

The right-hand side is related to the strain component E_{12} according to Eq. (4.1:9), on specializing to rectangular Cartesian coordinates. Hence

(6)
$$ds\, d\bar{s} \cos\theta = 2E_{12} ds_0 d\bar{s}_0.$$

But, from Eqs. (1) and (3), we have

$$ds = \sqrt{1 + 2E_{11}}\, ds_0, \quad d\bar{s} = \sqrt{1 + 2E_{22}}\, d\bar{s}_0.$$

Hence, Eq. (6) yields

(7)
$$\cos\theta = \frac{2E_{12}}{\sqrt{1 + 2E_{11}}\, \sqrt{1 + 2E_{22}}}.$$

The angle θ is the angle between the line elements ds and $d\bar{s}$ after deformation. The change of angle between the two line elements, which in the original state ds_0 and $d\bar{s}_0$ are orthogonal, is $\alpha_{12} = \pi/2 - \theta$. From Eq. (7) we therefore obtain

(8)
$$\sin\alpha_{12} = \frac{2E_{12}}{\sqrt{1 + 2E_{11}}\, \sqrt{1 + 2E_{22}}}.$$

This reduces, in the case of infinitesimal strain, to

(9)
$$\alpha_{12} \doteq 2E_{12}.$$

A completely analogous interpretation can be made for the Eulerian strain components. Defining the extension e_1 per unit *deformed* length as

(10)
$$e_1 = \frac{ds - ds_0}{ds},$$

we find

(11) $$e_1 = 1 - \sqrt{1 - 2e_{11}}.$$

Furthermore, if the deviation from a right angle between two elements in the original state which after deformation become orthogonal be denoted by β_{12}, we have

(12) $$\sin \beta_{12} = \frac{2e_{12}}{\sqrt{1 - 2e_{11}} \sqrt{1 - 2e_{22}}}.$$

These again reduce to the familiar results

$$e_1 \doteq e_{11}, \quad \beta_{12} \doteq 2e_{12},$$

in the case of infinitesimal strain.

4.6. COMPATIBILITY OF STRAIN COMPONENTS

The question of how to determine the displacements u_i when the components of strain tensor are given naturally arises. In other words, how do we integrate the differential equations (in rectangular Cartesian coordinates)

(1) $$e_{ji} = e_{ij} = \frac{1}{2} \left[\frac{\partial u_j}{\partial x_i} + \frac{\partial u_i}{\partial x_j} - \frac{\partial u_\alpha}{\partial x_i} \frac{\partial u_\alpha}{\partial x_j} \right], \qquad (i, j = 1, 2, 3)$$

to determine the three unknowns u_1, u_2, u_3?

Inasmuch as there are six equations for three unknown functions u_i, the system of Eqs. (1) will not have a single-valued solution in general, if the functions e_{ij} were arbitrarily assigned. One must expect that a solution may exist only if the functions e_{ij} satisfy certain conditions.

Since strain components only determine the relative positions of points in the body, and since any rigid-body motion corresponds to zero strain, we expect the solution u_i can be determined only up to an arbitrary rigid-body motion. But, if e_{ij} were specified arbitrarily, we may expect that something like the cases shown in Fig. 4.6:1 may happen. Here a continuous triangle (portion of material in a body) is given. If we deform it by following an arbitrarily specified strain field starting from the point A, we might end at the points C and D either with a gap between them or with overlapping of material. For a single-valued continuous solution to exist (up to a rigid-body motion), the ends C and D must meet perfectly in the strained configuration. This cannot be guaranteed unless the specified strain field along the edges of the triangle obeys certain conditions.

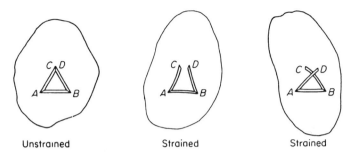

Unstrained   Strained   Strained

Fig. 4.6:1. Illustrations for the requirement of compatibility.

The conditions of integrability of Eqs. (1) are called the *compatibility conditions*. They are conditions to be satisfied by the strain components e_{ij} and can be obtained by eliminating u_i from Eqs. (1)

The nonlinear differential Eqs. (1) are difficult to handle, so first *let us consider the linear infinitesimal strain case*

$$(2) \qquad e_{ij} = \frac{1}{2} \left(\frac{\partial u_i}{\partial x_j} + \frac{\partial u_j}{\partial x_i} \right) , \quad \text{i.e.,} \quad e_{ij} = \frac{1}{2}(u_{i,j} + u_{j,i}) ,$$

where an index i following a comma indicates partial differentiation with respect to x_i. By differentiation of Eq. (2), we have

$$e_{ij,kl} = \frac{1}{2}(u_{i,jkl} + u_{j,ikl}) .$$

Interchanging subscripts, we have

$$e_{kl,ij} = \frac{1}{2}(u_{k,lij} + u_{l,kij}) ,$$

$$e_{jl,ik} = \frac{1}{2}(u_{j,lik} + u_{l,jik}) ,$$

$$e_{ik,jl} = \frac{1}{2}(u_{i,kjl} + u_{k,ijl}) .$$

From these we verify at once that

$$(3) \quad \blacktriangle \qquad\qquad e_{ij,kl} + e_{kl,ij} - e_{ik,jl} - e_{jl,ik} = 0 .$$

This is the *equation of compatibility* of St. Venant, first obtained by him in 1860.

Of the 81 equations represented by Eq. (3), only six are essential. The rest are either identities or repetitions on account of the symmetry of e_{ij} and of e_{kl}. The six equations written in unabridged notations are

$$\frac{\partial^2 e_{xx}}{\partial y \partial z} = \frac{\partial}{\partial x} \left(-\frac{\partial e_{yz}}{\partial x} + \frac{\partial e_{zx}}{\partial y} + \frac{\partial e_{xy}}{\partial z} \right),$$

$$\frac{\partial^2 e_{yy}}{\partial z \partial x} = \frac{\partial}{\partial y} \left(-\frac{\partial e_{zx}}{\partial y} + \frac{\partial e_{xy}}{\partial z} + \frac{\partial e_{yz}}{\partial x} \right),$$

$$\frac{\partial^2 e_{zz}}{\partial x \partial y} = \frac{\partial}{\partial z} \left(-\frac{\partial e_{xy}}{\partial z} + \frac{\partial e_{yz}}{\partial x} + \frac{\partial e_{zx}}{\partial y} \right),$$

(4) ▲
$$2\frac{\partial^2 e_{xy}}{\partial x \partial y} = \frac{\partial^2 e_{xx}}{\partial y^2} + \frac{\partial^2 e_{yy}}{\partial x^2},$$

$$2\frac{\partial^2 e_{yz}}{\partial y \partial z} = \frac{\partial^2 e_{yy}}{\partial z^2} + \frac{\partial^2 e_{zz}}{\partial y^2},$$

$$2\frac{\partial^2 e_{zx}}{\partial z \partial x} = \frac{\partial^2 e_{zz}}{\partial x^2} + \frac{\partial^2 e_{xx}}{\partial z^2}.$$

These conditions are derived for infinitesimal strains referred to rectangular Cartesian coordinates. If the general curvilinear coordinates are used, we define the infinitesimal strain by

(5)
$$e_{ij} = \frac{1}{2}(u_i|_j + u_j|_i),$$

where $u_i|_j$ is the covariant derivative of u_i, etc. The corresponding compatibility condition is then

(6)
$$e_{ij}|_{kl} + e_{kl}|_{ij} - e_{ik}|_{jl} - e_{jl}|_{ik} = 0.$$

If, however, the strain is finite so that e_{ij} is given by Eq. (1) in rectangular Cartesian coordinates, or by

(7)
$$e_{ij} = \frac{1}{2}[u_i|_j + u_j|_i - u_\alpha|_i u^\alpha|_j],$$

in general coordinates, then it is necessary to use a new method of derivation. A successful method uses the basic concept that the compatibility conditions say that our body initially situated in a three-dimensional Euclidean space, must remain in the Euclidean space after deformation. A mathematical statement to this effect, expressed in terms of the strain components, gives the compatibility conditions. (See Probs. 2.31 and 4.9.

Full expressions can be found in Green and Zerna,[1,2] *Theoretical Elasticity*, p. 62.)

Let us now return to the question posed at the beginning of this section and inquire whether conditions (3) and (4) are sufficient to assure the existence of a single-valued continuous solution of the differential Eqs. (2) up to a rigid-body motion. The answer is affirmative. Various proofs are available, the simplest having been given by E. Cesaro[4.1] in 1906. (See notes in Bibliography 4.1.) The proof may proceed as follows.

Let $P_0(x_1^0, x_2^0, x_3^0)$ be a point at which the displacements u_i^0 and the components of rotation w_{ij}^0 are specified. The displacement u_i at an arbitrary point $\bar{P}$ $(\bar{x}_1, \bar{x}_2, \bar{x}_3)$ is obtained by a line integral along a continuous rectifiable curve C joining P_0 and $\bar{P}$,

$$(8) \qquad u_i(\bar{x}_1, \bar{x}_2, \bar{x}_3) = u_i^0 + \int_{P_0}^{\bar{P}} du_i = u_i^0 + \int_{P_0}^{\bar{P}} \frac{\partial u_i}{\partial x_k} dx^k .$$

But

$$(9) \qquad \frac{\partial u_i}{\partial x_k} = \frac{1}{2} \left[\left(\frac{\partial u_i}{\partial x_k} + \frac{\partial u_k}{\partial x_i} \right) + \left(\frac{\partial u_i}{\partial x_k} - \frac{\partial u_k}{\partial x_i} \right) \right] = e_{ik} - w_{ik} ,$$

according to the definitions of the infinitesimal strain and rotation tensors, e_{ij} and w_{ij}, respectively, as given by Eqs. (4.2:6) and (4.4:1). Hence, Eq. (8) becomes

$$(10) \qquad u_i(\bar{x}_1, \bar{x}_2, \bar{x}_3) = u_i^0 + \int_{P_0}^{\bar{P}} e_{ik} dx^k - \int_{P_0}^{\bar{P}} w_{ik} dx^k .$$

We must eliminate w_{ik} in terms of e_{ik}. To achieve this, the last integral is integrated by parts to yield

$$(11) \qquad \int_{P_0}^{\bar{P}} w_{ik} dx^k = \int_{P_0}^{\bar{P}} w_{ik}(x)(dx^k - d\bar{x}^k)$$

$$= (\bar{x}_k - x_k^0) w_{ik}^0 + \int_{P_0}^{\bar{P}} (\bar{x}_k - x_k) \frac{\partial w_{ik}}{\partial x_l} dx^l ,$$

where w_{ik}^0 is the value of w_{ik} at P_0. In Eq. (11) we have replaced dx^k by $dx^k - d\bar{x}^k$, to have w_{ik}^0 appear outside the integral instead of w_{ik} at $\bar{P}$. This is permissible, since $\bar{P}$ is fixed with respect to the integration so that

$d\bar{x}^k = 0$. Now,

$$(12) \quad -\frac{\partial \omega_{ik}}{\partial x_l} = \frac{1}{2}\left(\frac{\partial^2 u_i}{\partial x_l \partial x_k} - \frac{\partial^2 u_k}{\partial x_l \partial x_i}\right)$$

$$= \frac{1}{2}\left(\frac{\partial^2 u_i}{\partial x_k \partial x_l} + \frac{\partial^2 u_l}{\partial x_k \partial x_i}\right) - \frac{1}{2}\left(\frac{\partial^2 u_k}{\partial x_i \partial x_l} + \frac{\partial^2 u_l}{\partial x_i \partial x_k}\right)$$

$$= \frac{\partial}{\partial x_k}\frac{1}{2}\left(\frac{\partial u_i}{\partial x_l} + \frac{\partial u_l}{\partial x_i}\right) - \frac{\partial}{\partial x_i}\frac{1}{2}\left(\frac{\partial u_k}{\partial x_l} + \frac{\partial u_l}{\partial x_k}\right)$$

$$= \frac{\partial e_{il}}{\partial x_k} - \frac{\partial e_{kl}}{\partial x_i}.$$

Substitution of Eqs. (11) and (12) into Eq. (10) yields

$$(13) \quad u_i(\bar{x}_1, \bar{x}_2, \bar{x}_3) = u_i^0 - (\bar{x}_k - x_k^0)\omega_{ik}^0 + \int_{P_0}^{\bar{P}} U_{il}dx^l ,$$

where

$$(14) \quad U_{il} = e_{il} + (\bar{x}_k - x_k)\left(\frac{\partial e_{il}}{\partial x_k} - \frac{\partial e_{kl}}{\partial x_i}\right).$$

Now, if $u_i(\bar{x}_1, \bar{x}_2, \bar{x}_3)$ is to be single-valued and continuous, the last integral in Eq. (13) must depend only on the end points P_0, $\bar{P}$ and must not

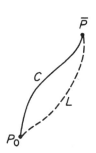

Fig. 4.6:2. Paths of integration.

be dependent on the path of integration C. Therefore, the integrand must be an exact differential. (See Sec. 4.7 below.) The necessary condition that $U_{il}dx^l$ be an exact differential is (see Sec. 4.7)

$$(15) \quad \frac{\partial U_{il}}{\partial x_m} - \frac{\partial U_{im}}{\partial x_l} = 0 .$$

This condition also suffices in assuring the single-valuedness of the integral if the region is simply connected. However, for a multiply connected region, some additional conditions must be imposed; they will be discussed in the next section.

Now we obtain, from Eqs. (14) and (15),

$$\frac{\partial e_{il}}{\partial x_m} - \delta_{km}\left(\frac{\partial e_{il}}{\partial x_k} - \frac{\partial e_{kl}}{\partial x_i}\right) + (\bar{x}_k - x_k)\frac{\partial}{\partial x_m}\left(\frac{\partial e_{il}}{\partial x_k} - \frac{\partial e_{kl}}{\partial x_i}\right) - \frac{\partial e_{im}}{\partial x_l}$$

$$+ \delta_{kl}\left(\frac{\partial e_{im}}{\partial x_k} - \frac{\partial e_{km}}{\partial x_i}\right) - (\bar{x}_k - x_k)\frac{\partial}{\partial x_l}\left(\frac{\partial e_{im}}{\partial x_k} - \frac{\partial e_{km}}{\partial x_i}\right) = 0 .$$

The factors not multiplied by $\bar{x}_k - x_k$ cancel each other, and the factors multiplied by $\bar{x}_k - x_k$ form exactly the compatibility condition given by Eq. (3). Hence, Eq. (15) is satisfied if the compatibility conditions (3) are satisfied. Thus, we have proved that the satisfaction of the compatibility conditions (3) is necessary and sufficient for the displacement to be single-valued in a simply connected region. For a multiply connected region, Eq. (3) is necessary, but no longer sufficient. To guarantee single-valuedness of displacement for an assumed strain field, some additional conditions as described in Sec. 4.7 must be imposed.

Problem 4.1. Consider the two-dimensional case in unabridged notations,

$$\frac{\partial u}{\partial x} = e_{xx}, \quad \frac{\partial v}{\partial y} = e_{yy}, \quad \frac{\partial u}{\partial y} + \frac{\partial v}{\partial x} = 2e_{xy}.$$

Prove that in order to guarantee the solutions $u(x,y)$, $v(x,y)$ to be single-valued in a simply connected domain $D(x,y)$, the functions $e_{xx}(x,y)$, $e_{yy}(x,y)$, $e_{xy}(x,y)$ must satisfy the compatibility condition

$$\frac{\partial^2 e_{xx}}{\partial y^2} + \frac{\partial^2 e_{yy}}{\partial x^2} = 2\frac{\partial^2 e_{xy}}{\partial x \partial y}.$$

Prove the sufficiency of this condition in unabridged notations.

4.7. MULTIPLY CONNECTED REGIONS

As necessary conditions, the compatibility equations derived in Sec. 4.6 apply to any continuum. But the sufficiency proof at the end of the preceding section requires that the region be simply connected. For a multiply connected continuum, additional conditions must be imposed.

A region is *simply connected* if any simple closed contour drawn in the region can be shrunk continuously to a point without leaving the region; otherwise the region is said to be *multiply connected.*

Figure 4.7:1(a) illustrates a simply connected region $\mathcal{R}$ in which an arbitrary closed curve C can be shrunk continuously to a point without leaving $\mathcal{R}$. Figures 4.7:1(b) and 4.7:2 illustrate doubly connected regions in two and three dimensions, respectively. Figures 4.7:1(c) and (d) show how multiply connected regions can be made simply connected by introducing cuts — imaginary boundaries.

The condition (4.6:15) is based on Stokes' theorem, and we shall review the reasoning in preparation for discussing multiply connected regions. In

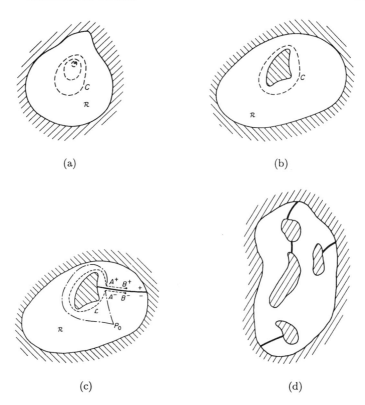

(a)   (b)

(c)   (d)

Fig. 4.7:1. Two-dimensional examples of simply and multiply connected regions. In (a), region $\mathcal{R}$ is simply connected; in (b), $\mathcal{R}$ is doubly connected.

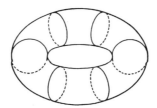

Fig. 4.7:2. A torus doubly connected in three dimensions.

two dimensions, Stokes' theorem for continuous and differentiable functions $P(x, y), Q(x, y)$ may be written as

$$(1) \qquad \int_C (P\,dx + Q\,dy) = \int\int_{\mathcal{R}} \left(\frac{\partial Q}{\partial x} - \frac{\partial P}{\partial y} \right) dx\,dy \,,$$

for a simply connected region $\mathcal{R}$ on a plane x, y bounded by a closed contour C. In three dimensions, Stokes' theorem states, for three functions P, Q, R,

$$(2) \quad \int_C (Pdx + Qdy + Rdz) = \int \int_{\mathcal{R}} \left\{ \left(\frac{\partial R}{\partial y} - \frac{\partial Q}{\partial z} \right) \cos(\nu, \mathbf{x}) \right.$$

$$\left. + \left(\frac{\partial P}{\partial z} - \frac{\partial R}{\partial x} \right) \cos(\nu, \mathbf{y}) + \left(\frac{\partial Q}{\partial x} - \frac{\partial P}{\partial y} \right) \cos(\nu, \mathbf{z}) \right\} dS ,$$

for a simply connected region $\mathcal{R}$ on a surface S bounded by a closed contour C. The factor $\cos(\nu, \mathbf{x})$ is the cosine of the angle between the x-axis and the normal vector ν normal to the surface S. When the conditions

$$(3) \qquad \frac{\partial R}{\partial y} - \frac{\partial Q}{\partial z} = 0 , \quad \frac{\partial P}{\partial z} - \frac{\partial R}{\partial x} = 0 , \quad \frac{\partial Q}{\partial x} - \frac{\partial P}{\partial y} = 0 ,$$

are imposed, the line integral on the left-hand side of Eq. (2) vanishes for every possible closed contour C,

$$\int_C (Pdx + Qdy + Rdz) = 0 .$$

When this result is applied to any two possible paths of integration between a fixed point P_0 and a variable point $\bar{P}$, as in Fig. 4.6:2, we see that the integral

$$(4) \qquad F(\bar{P}) = \int_{P_0}^{\bar{P}} (Pdx + Qdy + Rdz) ,$$

must be a single-valued function of the point $\bar{P}$, independent of the path of integration; because in Fig. 4.6:2 the curve $P_0 C \bar{P} L P_0$ forms a simple contour. Hence,

$$\int_{P_0 C \bar{P}} + \int_{\bar{P} L P_0} = 0 = \int_{P_0 C \bar{P}} - \int_{P_0 L \bar{P}} .$$

Thus, the last two integrals are equal. This is the result used in Sec. 4.6 to conclude that the compatibility conditions as posed are sufficient for uniqueness of displacements in a simply connected region.

Let us now consider a doubly connected region on a plane as shown in Fig. 4.7:1(b). After a cut is made as in Fig. 4.7:1(c), the region $\mathcal{R}$ is simply connected. Let the two sides of the cut be denoted by $(+)$ and $(-)$ and let the points A^+ and A^- be directly opposite to each other on the two sides of the cut. The line integrals

$$F(A^-) = \int_{P_0}^{A^-} (Pdx + Qdy) , \quad F(A^+) = \int_{P_0}^{A^+} (Pdx + Qdy) ,$$

integrated from a fixed point P_0 along paths in the cut region, are both single-valued. But, since A^+ and A^- are on the opposite sides of the cut, the values of the integrals need not be the same. Let the chain-dot path P_0A^+ in Fig. 4.7:1(c) be deformed into P_0A^- plus a line $\mathcal{L}$ connecting A^- to A^+, encircling the inner boundary. Then

$$\int_{P_0}^{A^+} = \int_{P_0}^{A^-} + \int_{\mathcal{L}} .$$

Hence,

(5) $$F(A^+) - F(A^-) = \int_{\mathcal{L}} (P\,dx + Q\,dy) .$$

Similar consideration for another pair of arbitrary points B^+ and B^- across the boundary leads to the same result,

(6) $$F(B^+) - F(B^-) = \int_{\mathcal{L}'} (P\,dx + Q\,dy) .$$

The right-hand sides of Eqs. (5) and (6) are the same if P and Q are single-valued in the region $\mathcal{R}$, because a line $\mathcal{L}'$ connecting B^- to B^+, encircling the inner boundary, may be deformed into one that goes from B^- to A^- along the cut, then follows $\mathcal{L}$ from A^- to A^+, and finally moves from A^+ to B^+ along the $+$ side of the cut. If P and Q were single-valued, the integral from B^- to A^- cancels exactly the one from A^+ to B^+.

From these considerations, it is clear that function $F(\bar{P})$ defined by the line integral (4) will be single-valued in a doubly connected region if the supplementary condition

(7) $$\int_{\mathcal{L}} (P\,dx + Q\,dy) = 0 ,$$

is imposed, where $\mathcal{L}$ is a closed contour that goes from one side of a cut to the other, without leaving the region.

The same consideration can be applied generally to multiply connected regions in two or three dimensions. An $(m+1)$-ply connected region can be made simply connected by m cuts. In the cut, simply-connected region, m independent simple contours $\mathcal{L}_1, \mathcal{L}_2, \ldots, \mathcal{L}_m$ can be drawn. Each $\mathcal{L}_i$ starts from one side of a cut, and ends on the other side of the same cut. All cuts are thus embraced by the $\mathcal{L}$'s. Then the single-valuedness of the function $F(\bar{P})$ defined by the line integral (4) can be assured by imposing, in addition to the conditions (3), m supplementary conditions

(8) $$\int_{\mathcal{L}_1} (P\,dx + Q\,dy + R\,dz) = \int_{\mathcal{L}_2} (\quad) = \cdots = \int_{\mathcal{L}_m} (\quad) = 0 .$$

When these results are applied to the compatibility problem of Sec. 4.6 we see that if the body is $(m+1)$-ply connected, the single-valuedness of the displacement function $u_i(x_1, x_2, x_3)$ given by Eq. (4.6:13) requires $m \times 3$ supplementary conditions:

(9) ▲
$$\int_{\mathcal{L}_1} U_{il} dx^l = \cdots = \int_{\mathcal{L}_m} U_{il} dx^l = 0, \quad i = 1, 2, 3,$$

where

(10) ▲
$$U_{il} = e_{il} + (\bar{x}_k - x_k) \left(\frac{\partial e_{il}}{\partial x_k} - \frac{\partial e_{kl}}{\partial x_i} \right),$$

and $\mathcal{L}_1, \ldots, \mathcal{L}_m$ are m contours in $\mathcal{R}$ as described above.

4.8. MULTIVALUED DISPLACEMENTS

Some problems of thermal stress, initial strain, rigid inclusions, etc., can be formulated in terms of multivalued displacements. For example, if the horseshoe in Fig. 4.8:1(a) is strained so that the faces S_1 and S_2 come into contact and are then welded together, the result is the doubly connected body shown in Fig. 4.8:1(b). If the strain of the body (b) is known, we

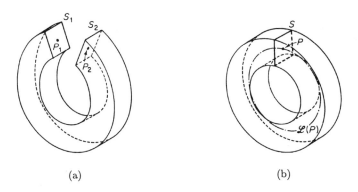

(a)   (b)

Fig. 4.8:1. Possibility and use of multivalued displacements.

can use the procedure described in Sec. 4.6 to compute displacements and generally we will obtain the configuration of Fig. 4.8:1(a). In this case a single point P on the welded surface S will open up into two points P_1 and P_2 on the two open ends. Thus the displacement at P is double-valued.

In the example posed here, the strain components and their derivatives may be discontinuous on the two sides of the interface S of Fig. 4.8:1(b).

Hence, the value of a line integral around a contour in the body that crosses the interface S will depend on where the crossing point is. In other words,

$$(1) \qquad u_i(P_2) - u_i(P_1) = \int_{\mathcal{L}(P)} U_{il} dx^l \,,$$

depends on the location of P, where $\mathcal{L}(P)$ is a contour passing through the point P as indicated in the figure and U_{il} is given in Eq. (4.6:14) or (4.7:10).

On the other hand, if we have a ring as in Fig. 4.8:1(b) and apply some load on it and ask what is the corresponding deformation, we must first decide whether some crack is allowed to be opened or not. If the ring is to remain integral, the deformation must be such that the displacement field is single-valued. In this case the strain due to the load must be such that

$$(2) \qquad \int_{\mathcal{L}(P)} U_{il} dx^l = 0 \,,$$

for a contour $\mathcal{L}(P)$ as shown in Fig. 4.8:1(b). Of course, Eq. (2) in conjunction with the compatibility conditions (4.6:4) assures that the line integral $\int U_{il} dx^l$ will vanish on any contour in the ring. Thus the point P in Eq. (2) is arbitrary.

These examples show that both single-valued displacement fields and multivalued displacement fields have proper use in answering appropriate question.

4.9. PROPERTIES OF THE STRAIN TENSOR

The symmetric strain tensor e_{ij} has many properties in common with the stress tensor. Thus, the existence of real-valued principal strains and principal planes, the Mohr circle for strain, the octahedral strain, strain deviations, and the strain invariants need no further discussion.

The first strain invariant,

$$(1) \qquad e = e_{11} + e_{22} + e_{33} = \text{sum of principal strains} \,,$$

has a simple geometrical meaning in the case of infinitesimal strain. Let a volume element consist of a rectangular parallelepiped with edges parallel to the principal directions of strain. Let the length of the edges be l_1, l_2, l_3 in the unstrained state. Let e_{11}, e_{22}, e_{33} be the principal strains. In the strained configuration, the edges become of lengths $l_1(1 + e_{11})$, $l_2(1 + e_{22})$, $l_3(1 + e_{33})$ and remain orthogonal to each other. Hence, for small strain the change of volume is

$$\Delta V = l_1 l_2 l_3 (1 + e_{11})(1 + e_{22})(1 + e_{33}) - l_1 l_2 l_3$$

$$\doteq l_1 l_2 l_3 (e_{11} + e_{22} + e_{33}) \,.$$

Therefore,

(2) $$e = e_{ii} = \Delta V / V .$$

Thus, in the infinitesimal strain theory the first invariant represents the expansion in volume per unit volume. For this reason, e is called the *dilatation.*

If two-dimensional strain state (plane strain) is considered u_3 or $w \equiv 0$, the first invariant $e_{11} + e_{22}$ represents the change of area per unit area of the surface under strain.

For finite strain, the sum of principal strains does not have such a simple interpretation.

P R O B L E M S

4.2. A state of deformation in which all the strain components are constant throughout the body is called a *homogeneous deformation.* What is the equation, of the type $f(x, y, z) = 0$, of a surface which becomes a sphere $x^2 + y^2 + z^2 = r^2$ *after* a homogeneous deformation? What kind of surface is it? (x, y, z are rectangular Cartesian coordinates.)

4.3. Show that a strain state described in rectangular Cartesian coordinates

$$e_{xx} = k(x^2 + y^2) , \quad e_{yy} = k(y^2 + z^2) , \quad e_{xy} = k'xyz ,$$

$$e_{xz} = e_{yz} = e_{zz} = 0 ,$$

where k, k' are small constants, is not a possible state of strain for a continuum.

4.4. A solid is heated nonuniformly to a temperature $T(x, y, z)$. If each element has unrestrained thermal expansion, the strain components will be

$$e_{xx} = e_{yy} = e_{zz} = \alpha T ,$$

$$e_{xy} = e_{yz} = e_{zx} = 0 ,$$

where x, y, z, are rectangular Cartesian coordinates and α is the thermal expansion coefficient (a constant). Prove that this can only occur when T is a linear function of x, y, z; i.e., $T = c_1 x + c_2 y + c_3 z + c_4$, where $c_1, \ldots, c_4$ are constants.

4.5. A soap-film like membrane stretched over a ring is deformed by uniform pressure into a hemisphere of the same radius (Fig. P4.5). In so doing, a point P on the flat surface over the ring is deformed into a point Q on the sphere. Determine a mathematical transformation from P to Q. *Note:* A "soap-film like" membrane shall be defined as a membrane in which the tension is a constant, isotropic, and independent of the stretching of the membrane.

Let P be referred to plane polar coordinates (r, θ) and Q be referred to spherical polar coordinates (R, θ, ϕ) with the same origin. We assume that $R = a$, $\theta = \theta$. The problem is to determine the angle ϕ as a function of r.

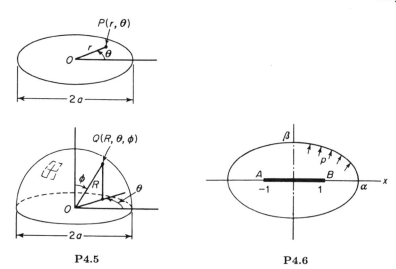

P4.5   P4.6

If the thickness of the film is uniform originally, it will not remain so after being blown into spherical shape. Determine the thickness distribution in the final configuration under the assumption of material incompressibility.

4.6. Consider a two-dimensional inflatable structure which consists of an infinitely long tube. When folded flat, the cross section of the tube appears like a line segment AB of length 2 in Fig. P4.6. When blown up with an internal gas pressure p, the tube assumes the form of an ellipse with major axis α and minor axia β.

Questions:

(a) Devise (arbitrarily, but specifically) a law of transformation which transforms the folded tube into the elliptic cylinder.

(b) Compute the components of membrane strain in the midsurface of the tube referred to the original folded tube (components of Green's strain tensor).

(c) Compute the corresponding components of strain referred to the blown up tube (components of Almansi's strain tensor).

(d) For equilibrium, what must be the distribution of membrane tension force T in the blown-up tube? (The internal pressure is uniform.)

(e) Assume that $T = khe_1$, where T is the membrane tension, e_1 is the strain component of Almansi on the midsurface of the tube wall, h is the wall thickness of the blown up tube, and k is a constant. Determine h.

(f) If the tube material is incompressible (like natural rubber), what would have to be the initial thickness distribution?

4.10. PHYSICAL COMPONENTS

In Sec. 2.16 we defined the physical components of a vector $\mathbf{v}$ as the components of $\mathbf{v}$ resolved in the directions of a set of unit vectors which are parallel either to the set of base vectors or to the set of reciprocal base vectors. Thus,

$$
\begin{array}{lll}
\text{Tensor components:} & v^r & v_r \\
\text{Physical components:} & v^r \sqrt{g_{rr}} & v_r \sqrt{g^{rr}} \\
\text{Reference base vectors:} & \mathbf{g}_r & \mathbf{g}^r
\end{array}
$$

(1)

All the physical components have uniform physical dimensions; but they do not transform conveniently under coordinate transformations.

In Sec. 3.13 we defined the physical components of the stress tensors τ^{ij} as the components of the stress vector $\overset{i}{\mathbf{T}}$ resolved in the directions of unit vectors parallel to the base vectors. If the physical components of τ^{ij} are denoted by σ^{ij}, we have

$$
(2) \qquad \sigma^{ij} = \sqrt{\frac{g_{jj}}{g^{ii}}}\, \tau^{ij}, \quad i,j \text{ not summed}.
$$

Correspondingly, when $\overset{i}{\mathbf{T}}$ is resolved with respect to the reciprocal base vectors, the physical components are

$$
(3) \qquad \sigma^i_j = \sqrt{\frac{g^{jj}}{g^{ii}}}\, \tau^i_j, \quad i,j \text{ not summed}.
$$

All components σ^{ij}, σ^i_j have uniform physical dimensions. The physical components of the contravariant, covariant, and mixed tensors are the same if the coordinates are orthogonal.

Now consider the strain tensor e_{ij}, which is covariant in the most natural form of definition (Sec. 4.1). If ds is the length of an element $(d\theta^1, d\theta^2, d\theta^3)$ at a point $(\theta^1, \theta^2, \theta^3)$, and ds_0 is the length of the same element before the deformation takes place, then

$$
(4) \qquad ds^2 - ds_0^2 = 2e_{ij}\, d\theta^i d\theta^j .
$$

The differential element $(d\theta^i)$ is a vector, whose physical components are $\sqrt{g_{ii}}\, d\theta^i$, ($i$ not summed). We may rewrite (4) as

$$
(5) \qquad ds^2 - ds_0^2 = \sum_{i=1}^{3}\sum_{j=1}^{3} e_{ij} \frac{1}{\sqrt{g_{ii}}\sqrt{g_{jj}}} \left(\sqrt{g_{ii}}\, d\theta^i\right)\left(\sqrt{g_{jj}}\, d\theta^j\right).
$$

Since all the components $\sqrt{g_{ii}}\,d\theta^i$, $\sqrt{g_{jj}}\,d\theta^j$ have the dimension of length, we may define the *physical strain components*

(6)
$$\epsilon_{ij} = \frac{1}{\sqrt{g_{ii}}\,\sqrt{g_{jj}}}\,e_{ij}\,, \quad i, j \text{ not summed}\,,$$

which are all dimensionless.

The rules of forming physical components now appear rather complicated. The appearance can be made more systematic if we utilize the relation given in Eq. (2.14:9), i.e.,

(7)
$$\mathbf{g}^i \cdot \mathbf{g}_j = \delta^i_j\,,$$

which implies

(8)
$$g^{ii} = (g_{ii})^{-1}\,, \quad i \text{ not summed}\,.$$

Then Eqs. (2), (3), and (6) may be written as

(9)
$$\sigma^{ij} = \sqrt{g_{ii}}\,\sqrt{g_{jj}}\,\tau^{ij}\,,$$

(10)
$$\sigma^i_j = \sqrt{g^{jj}}\,\sqrt{g_{ii}}\,\tau^i_j\,,$$

(11)
$$\epsilon_{ij} = \sqrt{g^{ii}}\,\sqrt{g^{jj}}\,e_{ij}\,,$$

where i, j are not summed. These formulas, similar to Eq. (1), may serve as a pattern for defining the physical components of any tensor in any curvilinear coordinates. In orthogonal curvilinear coordinates, $\sigma^{ij} = \sigma^i_j$.

Let us repeat. The physical components and the tensor components of a physical quantity have the same geometric and physical interpretations, except for physical dimensions. The unification in dimensions is achieved by multiplying the tensor components with appropriate scale factors, which are related to the components of the metric tensor. The physical components do not transform conveniently under general transformation of coordinates. In practical applications, therefore, we generally write basic equations in the tensor form and then substitute the individual tensor components by their physical counterpart, if it is so desired.

We shall illustrate these applications in the following sections.

4.11. EXAMPLE — SPHERICAL COORDINATES

Let $x^1 = r$, $x^2 = \phi$, $x^3 = \theta$ (Fig. 4.11:1). Then

$$ds^2 = dr^2 + r^2 d\phi^2 + r^2 \sin^2 \phi d\theta^2 \,,$$

$$g_{11} = 1 \,, \quad g_{22} = r^2 \,,$$

$$g_{33} = r^2 \sin^2 \phi \,, \quad g_{ij} = 0 \quad \text{if} \quad i \neq j \,.$$

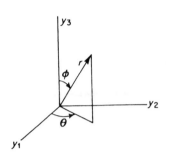

Fig. 4.11:1. Spherical polar coordinates.

Thus the coordinate system is orthogonal. From the definition of $g^{\alpha\beta}$, we have

$$g^{11} = 1 \,, \quad g^{22} = \frac{1}{r^2} \,,$$

$$g^{33} = \frac{1}{r^2 \sin^2 \phi} \,, \quad g^{ij} = 0 \quad \text{if} \quad i \neq j \,.$$

The Euclidean Christoffel symbols are

$$\Gamma^1_{22} = -r \,, \qquad \Gamma^1_{33} = -r \sin^2 \phi \,,$$

$$\Gamma^2_{12} = \Gamma^2_{21} = \frac{1}{r} \,, \qquad \Gamma^2_{33} = -\sin \phi \cos \phi \,,$$

$$\Gamma^3_{13} = \Gamma^3_{31} = \frac{1}{r} \,, \qquad \Gamma^3_{23} = \Gamma^3_{32} = \cot \phi \,,$$

all other $\Gamma^i_{jk} = 0$.

Let u_i be the covariant components of the displacement vector. We have the infinitesimal strain tensor components

$$e_{ij} = \frac{1}{2}(u_i|_j + u_j|_i) = \frac{1}{2}\left(\frac{\partial u_i}{\partial x^j} + \frac{\partial u_j}{\partial x^i}\right) - \Gamma^\sigma_{ij} u_\sigma \,,$$

since $\Gamma^\sigma_{ij} = \Gamma^\sigma_{ji}$. Hence,

$$e_{11} = \frac{\partial u_1}{\partial r} \,,$$

$$e_{22} = \frac{\partial u_2}{\partial \phi} + r u_1 \,,$$

$$e_{33} = \frac{\partial u_3}{\partial \theta} + r \sin^2 \phi \, u_1 + \sin \phi \cos \phi \, u_2 \,,$$

$$e_{12} = \frac{1}{2}\left(\frac{\partial u_1}{\partial \phi} + \frac{\partial u_2}{\partial r}\right) - \frac{1}{r} u_2 \,,$$

$$e_{23} = \frac{1}{2}\left(\frac{\partial u_2}{\partial \theta} + \frac{\partial u_3}{\partial \phi}\right) - \cot \phi u_3 \,,$$

$$e_{31} = \frac{1}{2}\left(\frac{\partial u_3}{\partial r} + \frac{\partial u_1}{\partial \theta}\right) - \frac{1}{r}u_3 \,.$$

The u_i and e_{ij} are tensor components. Let the corresponding physical components of the displacement vector be written as ξ_r, ξ_ϕ, ξ_θ and that of the strain tensor be written as ϵ_{ij}. Then, since the spherical coordinates are orthogonal,

$$\xi_r = \sqrt{g^{11}}\, u_1 = u_1\,, \qquad \xi_\phi = \sqrt{g^{22}}\, u_2 = \frac{u_2}{r}\,,$$

$$\xi_\theta = \sqrt{g^{33}}\, u_3 = \frac{u_3}{r\sin\phi}\,, \qquad \epsilon_{ij} = \sqrt{g^{ii}g^{jj}}\, e_{ij}\,.$$

Therefore,

$$\epsilon_{11} = \frac{\partial \xi_r}{\partial r}\,,$$

$$\epsilon_{22} = \frac{\partial \xi_\phi}{r\partial \phi} + \frac{\xi_r}{r}\,,$$

$$\epsilon_{33} = \frac{1}{r\sin\phi}\frac{\partial \xi_\theta}{\partial \theta} + \frac{1}{r}\xi_r + \frac{\cot\phi}{r}\xi_\phi\,,$$

$$\epsilon_{12} = \frac{1}{2}\left(\frac{1}{r}\frac{\partial \xi_r}{\partial \phi} + \frac{1}{r}\frac{\partial(r\xi_\phi)}{\partial r}\right) - \frac{\xi_\phi}{r} = \frac{1}{2}\left(\frac{1}{r}\frac{\partial \xi_r}{\partial \phi} + \frac{\partial \xi_\phi}{\partial r} - \frac{\xi_\phi}{r}\right)\,,$$

$$\epsilon_{23} = \frac{1}{2}\left(\frac{1}{r\sin\phi}\frac{\partial \xi_\phi}{\partial \theta} + \frac{1}{r}\frac{\partial \xi_\theta}{\partial \phi} - \frac{\cot\phi}{r}\xi_\theta\right)\,,$$

$$\epsilon_{31} = \frac{1}{2}\left(\frac{1}{r\sin\phi}\frac{\partial \xi_r}{\partial \theta} + \frac{\partial \xi_\theta}{\partial r} - \frac{\xi_\theta}{r}\right)\,.$$

The equations of equilibrium now become, with F_r, F_ϕ, F_θ denoting the physical components of the body force vector,

$$-F_r = \frac{1}{r^2}\frac{\partial}{\partial r}(r^2\sigma_{rr}) + \frac{1}{r\sin\phi}\frac{\partial}{\partial \phi}(\sin\phi\,\sigma_{r\phi})$$

$$+ \frac{1}{r\sin\phi}\frac{\partial}{\partial \theta}\sigma_{r\theta} - \frac{1}{r}(\sigma_{\theta\theta} + \sigma_{\phi\phi})\,,$$

$$-F_\phi = \frac{1}{r^3}\frac{\partial}{\partial r}(r^3\sigma_{r\phi}) + \frac{1}{r\sin\phi}\frac{\partial}{\partial \phi}(\sin\phi\,\sigma_{\phi\phi}) + \frac{1}{r\sin\phi}\frac{\partial}{\partial \theta}\sigma_{\theta\phi} - \frac{\cot\phi}{r}\sigma_{\theta\theta}\,.$$

$$-F_\theta = \frac{1}{r^3}\frac{\partial}{\partial r}(r^3\sigma_{\theta r}) + \frac{1}{r\sin^2\phi}\frac{\partial}{\partial \phi}(\sin^2\phi\,\sigma_{\theta\phi}) + \frac{1}{r\sin\phi}\frac{\partial}{\partial \theta}\sigma_{\theta\theta}\,,$$

4.12. EXAMPLE — CYLINDRICAL POLAR COORDINATES

Letting ξ_r, ξ_θ, ξ_z be the physical components of the displacement and ϵ_{rr}, $\epsilon_{r\theta}$, ..., σ_{rr}, $\sigma_{r\theta}$, ..., etc., be the physical components of the strain and stress, respectively, we obtain the following results:

$$x^1 = r, \quad x^2 = \theta, \quad x^3 = z,$$

$$ds^2 = dr^2 + r^2 d\theta^2 + dz^2,$$

$$g_{11} = 1, \quad g_{22} = r^2, \quad g_{33} = 1, \quad \text{all other} \quad g_{ij} = 0,$$

$$g^{11} = 1, \quad g^{22} = \frac{1}{r^2}, \quad g^{33} = 1, \quad \text{all other} \quad g^{ij} = 0,$$

$$\Gamma^1_{22} = -r, \quad \Gamma^2_{21} = \Gamma^2_{21} = \frac{1}{r}, \quad \text{all other} \quad \Gamma^i_{jk} = 0.$$

Hence,

$$\epsilon_{rr} = \frac{\partial \xi_r}{\partial r}, \quad \epsilon_{\theta\theta} = \frac{1}{r}\frac{\partial \xi_\theta}{\partial \theta} + \frac{\xi_r}{r}, \quad \epsilon_{zz} = \frac{\partial \xi_z}{\partial z},$$

$$\epsilon_{\theta z} = \frac{1}{2}\left(\frac{\partial \xi_\theta}{\partial z} + \frac{1}{r}\frac{\partial \xi_z}{\partial \theta}\right), \quad \epsilon_{zr} = \frac{1}{2}\left(\frac{\partial \xi_z}{\partial r} + \frac{\partial \xi_r}{\partial z}\right),$$

$$\epsilon_{r\theta} = \frac{1}{2}\left(\frac{1}{r}\frac{\partial \xi_r}{\partial \theta} + \frac{\partial \xi_\theta}{\partial r} - \frac{\xi_\theta}{r}\right).$$

The equations of equilibrium are

$$-F_r = \frac{1}{r}\frac{\partial}{\partial r}(r\sigma_{rr}) + \frac{1}{r}\frac{\partial}{\partial \theta}\sigma_{r\theta} + \frac{\partial}{\partial z}\sigma_{rz} - \frac{\sigma_{\theta\theta}}{r},$$

$$-F_\theta = \frac{1}{r^2}\frac{\partial}{\partial r}(r^2\sigma_{\theta r}) + \frac{1}{r}\frac{\partial}{\partial \theta}\sigma_{\theta\theta} + \frac{\partial}{\partial z}\sigma_{\theta z},$$

$$-F_z = \frac{1}{r}\frac{\partial}{\partial r}(r\sigma_{zr}) + \frac{1}{r}\frac{\partial}{\partial \theta}\sigma_{z\theta} + \frac{\partial}{\partial z}\sigma_{zz}.$$

PROBLEMS

4.7. Give proper definitions of strain tensors e^{ij}, e^i_j in general frames of reference in terms of e_{ij}. Show that e_{ij}, e^{ij} are symmetric tensors. Show that the principal strains e_1, e_2, e_3 are the roots of the characteristic equation

$$|e^i_j - e\delta^i_j| = 0$$

and that the first invariant is e^i_i.

4.8. The generalization of the expressions of strain tensors in terms of the displacement vector field **u**, which represents the displacement of any particle as the body deforms as explained in Sec. 4.2, can be done as follows. The displacement vector **u** can be resolved into covariant and contravariant components u_i and u^i, respectively, along base vectors and reciprocal base vectors defined in the original coordinates a_1, a_2, a_3. The Green's strain tensor is

$$E_{ij} = \frac{1}{2}[u_j|_i + u_i|_j + u^k|_i u_k|_j],$$

where $u_i|_j$ is the covariant derivative of u_i with respect to a_j.

The displacement vector **u** can be resolved also into covariant and contravariant components u_i and u^i along base vectors and reciprocal base vectors defined in the Eulerian coordinates x_1, x_2, x_3. Then the Almansi strain tensor is

$$e_{ij} = \frac{1}{2}[u_j\|_i + u_i\|_j - u^k\|_i u_k\|_j],$$

where a double bar is used to indicate that u_i, u^j and the covariant derivatives are referred to the coordinates x_i.

Prove these statements.

4.9. The condition of compatibility for finite strain can be derived easily if convected coordinates are used [see Eq. (4.1:15)]. By the fact that the space of the deformed body is Euclidean, derive the compatibility conditions. *Hint*: See comments on p. 101 and Prob. 2.31. *Note*: however, the properties $g^{ij}|_k = 0$, $g_{ij}|_k = 0$ are valid in Riemannian space as well as Euclidean space. The feature that distinguishes an Euclidean space is the vanishing of the Riemann-Christoffel curvature tensor R^i_{pst} of Prob. 2.31. Expressing $R^i_{pst} = 0$ in terms of the metric tensors before and after deformation leads to the compatibility conditions.

5

CONSERVATION LAWS

We shall discuss in this chapter the basic laws of conservation of mass and momentum. A rectangular Cartesian frame of reference will be used throughout. All tensors are Cartesian tensors.

5.1. GAUSS' THEOREM

Consider a convex region V bounded by a surface S that consists of a finite number of parts whose outer normals form a continuous vector field (Fig. 5.1:1). Let a function $A(x_1, x_2, x_3)$ be defined and differentiable in the region $V + S$. Then, by the usual process of integration, we obtain

$$(1) \qquad \iiint_V \frac{\partial A}{\partial x_1} dx_1 dx_2 dx_3 = \iint_S (A^* - A^{**}) dx_2 dx_3 ,$$

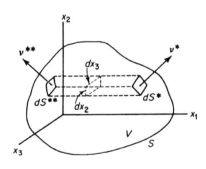

Fig. 5.1:1.

where A^* and A^{**} are the values of A on the surface S at the right and left ends of a line parallel to the x_1-axis, respectively. The factors $\pm dx_2 dx_3$ in the surface integral in Eq. (1) are the projections of the x_2, x_3-plane of the areas dS^* and dS^{**} at the ends of a line parallel to the x_1-axis. Let $\boldsymbol{\nu} = (\nu_1, \nu_2, \nu_3)$ be the unit vector along the outer normal of S. Then $dx_2 dx_3 = \nu_1^* dS^*$ at the right end and $dx_2 dx_3 = -\nu_1^{**} dS^{**}$ at the left end. Therefore, the surface integral in Eq. (1) can be written as

$$\int_S (A^* \nu_1^* dS^* + A^{**} \nu_1^{**} dS^{**}) = \int_S A\nu_1 dS .$$

Thus, Eq. (1) may be written as

$$\int_V \frac{\partial A}{\partial x_1} dV = \int_S A\nu_1 dS ,$$

127

where dV and dS denote the elements of V and S, respectively. A similar argument applies to the volume integral of $\partial A/\partial x_2$, or $\partial A/\partial x_3$. In summary, we obtain Gauss' theorem

$$(2) \qquad \int_V \frac{\partial A}{\partial x_i}\, dV = \int_S A\nu_i\, dS \,.$$

This formula holds for any convex regular region or for any region that can be decomposed into a finite number of convex regular regions, as can be seen by summing Eq. (2) over these component regions.

Now let us consider a tensor field $A_{jkl\ldots}$. Let the region V with boundary surface S be within the region of definition of $A_{jkl\ldots}$. Let every component of $A_{jkl\ldots}$ be continuously differentiable. Then Eq. (2) is applicable to every component of the tensor, and we may write

$$(3) \quad \blacktriangle \qquad \int_V \frac{\partial}{\partial x_i} A_{jkl\ldots}\, dV = \int_S \nu_i A_{jkl\ldots}\, dS \,.$$

This is Gauss' theorem in a general form.

Problem 5.1. Show that

$$\int \phi_{,i}\, dV = \int \phi \nu_i\, dS \quad \text{or} \quad \int \operatorname{grad} \phi\, dV = \int \nu \phi\, dS \,,$$

$$\int u_{i,i}\, dV = \int u_i \nu_i\, dS \quad \text{or} \quad \int \operatorname{div} \mathbf{u}\, dV = \int \nu \cdot \mathbf{u}\, dS \,,$$

$$\int e_{ijk} u_{k,j}\, dV = e_{ijk} \int u_{k,j}\, dV = e_{ijk} \int u_k \nu_j\, dS = \int e_{ijk} u_k \nu_j\, dS \,,$$

or

$$\int \operatorname{curl} \mathbf{u}\, dV = \int \nu \times \mathbf{u}\, dS \,,$$

where e_{ijk} is the permutation tensor. Verify that these formulas are also valid in two dimensions, in which case the range of indices is 1, 2 and the volume and surface integrals are replaced by surface and line integrals, respectively.

5.2. MATERIAL AND SPATIAL DESCRIPTIONS OF CHANGING CONFIGURATIONS

We shall speak of a *particle* in the sense of a material particle as we know it in Newtonian particle mechanics. The instantaneous geometric location of a particle will be spoken of as a *point*. A body is composed of

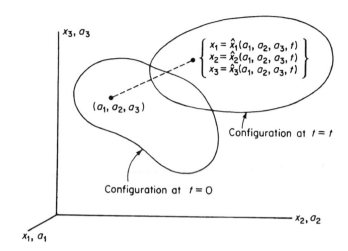

Fig. 5.2:1. Labeling of particles.

particles. To label the particles of a body we choose a Cartesian frame of reference and identify the coordinates of the particles at a time $t = 0$ as (a_1, a_2, a_3) (Fig. 5.2:1). At a later time the particle moves to another point whose coordinates are (x_1, x_2, x_3), *referred to the same coordinate system.* The relation

$$(1) \qquad x_i = \hat{x}_i(a_1, a_2, a_3, t), \quad i = 1, 2, 3,$$

links the configurations of the body at different instants of time. The functions $\hat{x}_i$ are single-valued continuous functions whose Jacobian does not vanish.

A basic property of bodies is that they have mass. In classical mechanics, mass is assumed to be conserved; i.e., the mass of a material body is the same at all times. In continuum mechanics it is further assumed that the mass is a continuous function of volume. In other words, it is assumed that a positive quantity ρ, called *density*, can be defined at every point in the body as the limit

$$(2) \qquad \rho(\mathbf{x}) = \lim_{k \to \infty} \frac{\text{mass of } B_k}{\text{volume of } B_k},$$

where B_k is a suitably chosen infinite sequence of particle sets shrinking down to the point $\mathbf{x}$; the symbol $\mathbf{x}$ stands for (x_1, x_2, x_3). At time $t = 0$, the density at the point (a_1, a_2, a_3) is denoted by $\rho_0(\mathbf{a})$.

The conservation of mass is expressed in the formula

(3)
$$\int \rho(\mathbf{x})dx_1 dx_2 dx_3 = \int \rho_0(\mathbf{a})da_1 da_2 da_3 ,$$

where the integrals extend over the same particles. Since

$$\int \rho(\mathbf{x})dx_1 dx_2 dx_3 = \int \rho(\mathbf{x}) \left| \frac{\partial x_i}{\partial a_j} \right| da_1 da_2 da_3 ,$$

and since this relation must hold for all bodies, we have

$$\rho_0(\mathbf{a}) = \rho(\mathbf{x}) \left| \frac{\partial x_i}{\partial a_j} \right| ,$$

(4) ▲
$$\rho(\mathbf{x}) = \rho_0(\mathbf{a}) \left| \frac{\partial a_i}{\partial x_j} \right| ,$$

where $|\partial a_i/\partial x_j|$ denotes the determinant of the matrix $\{\partial a_i/\partial x_j\}$. These equations relate the density in different configurations of the body to the transformation that leads from one configuration to the other.

For the particle (a_1, a_2, a_3) whose trajectory is given by Eq. (1), the velocity is

(5) ▲
$$v_i(\mathbf{a}, t) = \frac{\partial}{\partial t} x_i(\mathbf{a}, t) ,$$

and the acceleration is

(6) ▲
$$\dot{v}_i(\mathbf{a}, t) = \frac{\partial^2}{\partial t^2} x_i(\mathbf{a}, t) = \frac{\partial}{\partial t} v_i(\mathbf{a}, t) ,$$

where $\mathbf{a}$ stands for (a_1, a_2, a_3) and is held constant.

A description of mechanical evolution which uses (a_1, a_2, a_3) and t as independent variables is called a *material description*. In hydrodynamics, traditionally, a different description, called the *spatial description*, is used. In the spatial description, the location (x_1, x_2, x_3) and time t are taken as independent variables. This is convenient because measurements in many kinds of materials are more directly interpreted in terms of what happens at a certain place, rather than following the particles. These two methods of description are commonly designated as the *Lagrangian* and the *Eulerian descriptions*, respectively, although both are due to Euler. The variables a_1, a_2, a_3, t are usually called the *Lagrangian variables*, whereas x_1, x_2, x_3, t are called the *Eulerian variables*; they are related by Eq. (1). *For a given particle, it is convenient to speak of* (a_1, a_2, a_3) *as the Lagrangian coordinates of the particle at* (x_1, x_2, x_3).

In a spatial description, the instantaneous motion of the body is described by the velocity vector field $v_i(x_1, x_2, x_3, t)$ associated with the instantaneous location of each particle. The acceleration of the particle is given by the formula

(7) ▲ $$\dot{v}_i(\mathbf{x}, t) = \frac{\partial v_i}{\partial t}(\mathbf{x}, t) + v_j \frac{\partial v_i}{\partial x_j}(\mathbf{x}, t),$$

where $\mathbf{x}$ again stands for the variables x_1, x_2, x_3, and every quantity in this formula is evaluated at $(\mathbf{x}, t)$. The proof follows the fact that a particle located at (x_1, x_2, x_3) at time t is moved to a point with coordinates $x_i + v_i dt$ at the time $t + dt$; and that, according to Taylor's theorem,

$$\dot{v}_i(\mathbf{x}, t)dt = v_i(x_j + v_j dt, t + dt) - v_i(\mathbf{x}, t)$$

$$= v_i + \frac{\partial v_i}{\partial t} dt + \frac{\partial v_i(\mathbf{x}, t)}{\partial x_j} v_j dt - v_i,$$

which reduces to Eq. (7). The first term in Eq. (7) may be interpreted as arising from the time dependence of the velocity field; the second term as the contribution of the motion of the particle in the instantaneous velocity field. Accordingly, these terms are called the *local* and the *convective* parts of the acceleration, respectively.

The same reasoning that led to Eq. (7) is applicable to any function $F(x_1, x_2, x_3, t)$ that is attributable to the moving particles, such as the temperature. A convenient terminology is the *material derivative*, and it is denoted by a dot or the symbol D/Dt. Thus

(8) ▲ $$\dot{F} \equiv \frac{DF}{Dt} \equiv \left(\frac{\partial F}{\partial t}\right)_{\mathbf{x}=\text{const.}} + v_1 \frac{\partial F}{\partial x_1} + v_2 \frac{\partial F}{\partial x_2} + v_3 \frac{\partial F}{\partial x_3}$$

$$\equiv \left(\frac{\partial F}{\partial t}\right)_{\mathbf{a}=\text{const.}},$$

where $\mathbf{a} = (a_1, a_2, a_3)$ is the Lagrangian coordinate of the particle which is located at $\mathbf{x}$ at the time t, connected by Eq. (1)

5.3. MATERIAL DERIVATIVE OF VOLUME INTEGRAL

Consider a volume integral taken over the body

(1) $$I = \int_V A(\mathbf{x}, t)\, dV,$$

where $A(\mathbf{x}, t)$ denotes a property of the continuum and the integral is evaluated at an instant of time t. We may wish to know how fast the body itself sees its own value of I is changing, so it is of interest to know *the derivative of I with respect to time for a given set of particles.* Now the particle at x_i at the instant t will have the coordinates $x_i' = x_i + v_i dt$ at the time $t+dt$. The boundary S of the body at the instant t will have moved at time $t+dt$ to a neighboring surface S', which bounds the volume V' (Fig. 5.3:1). The material derivative of I is defined as

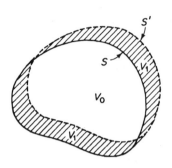

Fig. 5.3:1. Continuous change of the boundary of a region.

$$(2) \qquad \frac{DI}{Dt} = \lim_{dt \to 0} \frac{1}{dt} \left[\int_{V'} A(\mathbf{x}', t + dt) \, dV' - \int_{V} A(\mathbf{x}, t) \, dV \right].$$

Now there are two contributions to the difference on the right-hand side of Eq. (2); one over the region V_0 where V and V' share in common, and another over the region V_1 where V and V' differ. The former contribution to DI/Dt is evidently

$$\int_{V_0} \frac{\partial A}{\partial t} \, dV .$$

The latter contribution comes from the value of A on the boundary multiplied by the volumes swept by the particles on the boundary in the time interval dt. If ν_i is the unit vector along the exterior normal of S, then, since the displacement of a particle on the boundary is $v_i dt$, the volume swept by particles occupying an element of area dS on the boundary S is $dV = v_i \nu_i dS dt$. The contribution of this element to DI/Dt is $A v_i \nu_i dS$. The total contribution is obtained by an integration over S. Therefore,

$$(3) \quad \blacktriangle \qquad \frac{D}{Dt} \int_V A dV = \int_V \frac{\partial A}{\partial t} \, dV + \int_S A v_j \nu_j \, dS .$$

Transforming the last integral by Gauss' theorem and using Eq. (5.2:8), we have

(4)   ▲
$$\frac{D}{Dt}\int_V A\,dV = \int_V \frac{\partial A}{\partial t}\,dV + \int_V \frac{\partial}{\partial x_j}(Av_j)\,dV$$

$$= \int_V \left(\frac{\partial A}{\partial t} + v_j\frac{\partial A}{\partial x_j} + A\frac{\partial v_j}{\partial x_j}\right)dV$$

$$= \int_V \left(\frac{DA}{Dt} + A\frac{\partial v_j}{\partial x_j}\right)dV\,.$$

This important formula will be used over and over again below. It should be noted that *the operation of forming the material derivative and that of spatial integration is noncommutative* in general.

5.4. THE EQUATION OF CONTINUITY

The law of conservation of mass has been discussed in Sec. 5.2. With the results of Sec. 5.3, we can now give some alternative forms.

The mass contained in a region V at a time t is

(1)
$$m = \int_V \rho\,dV\,,$$

where $\rho = \rho(\mathbf{x}, t)$ is the density field of the continuum. Conservation of mass requires that $Dm/Dt = 0$. The derivative Dm/Dt is given by Eqs. (5.3:3) and (5.3:4), when A is replaced by ρ. Since the result must hold for arbitrary V, the integrand must vanish. Hence, we obtain the following alternative forms of the law of conservation of mass.

(2)   ▲
$$\int_V \frac{\partial \rho}{\partial t}\,dV + \int_S \rho v_j \nu_i\,dS = 0\,.$$

(3)   ▲
$$\frac{\partial \rho}{\partial t} + \frac{\partial \rho v_j}{\partial x_j} = 0\,.$$

(4)   ▲
$$\frac{D\rho}{Dt} + \rho\frac{\partial v_j}{\partial x_j} = 0\,.$$

These are called the *equations of continuity*. The integral form (2) is useful when the differentiability of ρv_j cannot be assumed.

In problems of statics, these equations are identically satisfied. Then the conservation of mass must be expressed by Eq. (5.2:3), or (5.2:4).

5.5. THE EQUATIONS OF MOTION

Newton's laws of motion state that, in an inertial frame of reference, the material rate of change of the linear momentum of a body is equal to the resultant of applied forces and that the material rate of change of the moment of momentum with respect to the coordinate origin is equal to the resultant moment of applied forces about the same origin.

At an instant of time t, a regular region V of space contains the linear momentum

$$(1) \qquad \mathcal{P}_i = \int_V \rho v_i \, dV .$$

If the body is subjected to surface traction $\overset{\nu}{T}_i$ and body force per unit volume X_i, the resultant force is

$$(2) \qquad \mathcal{F}_i = \int_S \overset{\nu}{T}_i \, dS + \int_V X_i \, dV .$$

According to the stress principle of Euler and Cauchy (Secs. 3.2 and 3.3), $\overset{\nu}{T}_i = \sigma_{ji} \nu_j$, where σ_{ij} is the stress field and ν_j is the unit vector along the exterior normal to the boundary surface S of the region V. Substituting into Eq. (2) and transforming into a volume integral by Gauss' theorem, we have

$$(3) \qquad \mathcal{F}_i = \int_V \left(\frac{\partial \sigma_{ij}}{\partial x_j} + X_i \right) dV .$$

Newton's law states that

$$(4) \qquad \frac{D}{Dt} \mathcal{P}_i = \mathcal{F}_i .$$

Hence, according to Eq. (5.3:4), with A identified with ρv_i, we have

$$(5) \qquad \int_V \left[\frac{\partial \rho v_i}{\partial t} + \frac{\partial}{\partial x_j} (\rho v_i v_j) \right] dV = \int_V \left(\frac{\partial \sigma_{ij}}{\partial x_j} + X_i \right) dV .$$

Since this equation must hold for an arbitrary region V, the integrand on the two sides must be equal. Thus

$$(6) \qquad \frac{\partial \rho v_i}{\partial t} + \frac{\partial}{\partial x_j} (\rho v_i v_j) = \frac{\partial \sigma_{ij}}{\partial x_j} + X_i .$$

The left-hand side of Eq. (6) is equal to

$$v_i \left(\frac{\partial \rho}{\partial t} + \frac{\partial \rho v_j}{\partial x_j} \right) + \rho \left(\frac{\partial v_i}{\partial t} + v_j \frac{\partial v_i}{\partial x_j} \right) .$$

The first parenthesis vanishes by the equation of continuity (5.4:3), while the second is the acceleration Dv_i/Dt. Hence

(7) ▲
$$\rho \frac{Dv_i}{Dt} = \frac{\partial \sigma_{ij}}{\partial x_j} + X_i .$$

This is the *Eulerian equation of motion* of a continuum. The equation of equilibrium discussed in Sec. 3.4 is obtained by setting all velocity components v_i equal to zero.

If differentiability of the stress field or the momentum field cannot be assumed, we may use Eq. (5.3:3) to compute $D\mathcal{P}/Dt$. Then Eqs. (2) and (4) give *Euler's equation in the integral form*,

(8)
$$\int_V \frac{\partial \rho v_i}{\partial t} \, dV = \int_S (\sigma_{ij} - \rho v_i v_j) \nu_j \, dS + \int_V X_i \, dV .$$

The corresponding equations of static equilibrium are obtained, of course, by setting all velocity components to zero.

5.6. MOMENT OF MOMENTUM

An application of the law of balance of angular momentum to the particular case of *static equilibrium* leads to the conclusion that stress tensors are symmetric tensors (see Sec. 3.4). We shall now show that no additional restriction to the motion of a continuum is introduced in dynamics by the angular momentum postulate.

At an instant of time t, a body occupying a regular region V of space with boundary S has the moment of momentum [Eq. (3.2:2)]

(1)
$$\mathcal{H}_i = \int_V e_{ijk} x_j \rho v_k \, dV ,$$

with respect to the origin. If the body is subjected to surface traction $\overset{\nu}{T}_i$ and body force per unit volume X_i, the resultant moment about the origin is

(2)
$$\mathcal{L}_i = \int_V e_{ijk} x_j X_k \, dV + \int_S e_{ijk} x_j \overset{\nu}{T}_k \, dS .$$

Introducing Cauchy's formula $\overset{\nu}{T}_k = \sigma_{lk} \nu_l$ into the last integral and transforming the result into a volume integral by Gauss' theorem, we obtain

(3)
$$\mathcal{L}_i = \int_V e_{ijk} x_j X_k \, dV + \int_V (e_{ijk} x_j \sigma_{lk})_{,l} \, dV .$$

Euler's law states that for any region V

(4) $$\frac{D}{Dt}\mathcal{H}_i = \mathcal{L}_i .$$

Evaluating the material derivative of $\mathcal{H}_i$ according to (5.3:4), and using Eq. (3), we obtain

(5) $$e_{ijk}x_j\frac{\partial}{\partial t}(\rho v_k) + \frac{\partial}{\partial x_l}(e_{ijk}x_j\rho v_k v_l) = e_{ijk}x_j X_k + e_{ijk}(x_j\sigma_{lk}),_l .$$

The second term in Eq. (5) can be written as

$$e_{ijk}\rho v_j v_k + e_{ijk}x_j\frac{\partial}{\partial x_l}(\rho v_k v_l) = e_{ijk}x_j\frac{\partial}{\partial x_l}(\rho v_k v_l) .$$

The last term in Eq. (5) can be written as $e_{ijk}\sigma_{jk} + e_{ijk}x_j\sigma_{lk,l}$. Hence, Eq. (5) becomes

(6) $$e_{ijk}x_j\left[\frac{\partial}{\partial t}\rho v_k + \frac{\partial}{\partial x_l}(\rho v_k v_l) - X_k - \sigma_{lk,l}\right] - e_{ijk}\sigma_{jk} = 0 .$$

The sum in the square brackets vanishes by the equation of motion (5.5:6). Hence, Eq. (6) is reduced to

(7) $$e_{ijk}\sigma_{jk} = 0 ;$$

i.e., $\sigma_{jk} = \sigma_{kj}$. Thus, if the stress tensor is symmetric, the law of balance of moment of momentum is identically satisfied.

5.7. OTHER FIELD EQUATIONS

The motion of a continuum must be governed further by the law of conservation of energy. If mechanical energy is the only form of energy of interest in a problem, then the energy equation can be obtained by suitable integration of the equation of motion, Eq. (5.5:7). If the interaction of thermal process and mechanical process is significant, then the equation of energy contains a thermal energy term and is an independent equation to be satisfied. We shall discuss the energy equation in Sec. 12.2.

The equations of continuity and motion, Eqs. (5.4:3) and (5.5:7), constitute four equations for ten unknown functions of time and position; namely, the density ρ, the three velocity components v_i (or displacements u_i) and the six independent stress components σ_{ij}. Further restrictions would have to be introduced before the motion of a continuum can be determined. One group of such additional restrictions comes from a statement about the mechanical property of the medium in the form of a specification of stress-strain relationship. These specifications are called *constitutive equations*.

The constitutive equations of elastic and plastic materials are discussed in Chapter 6. The constitutive equations of thermomechanical materials are discussed in Chapters 12 and 14. The constitutive equations of viscoelastic materials are discussed in Chapters 13 and 15. The constitutive equations of solids with large deformation are discussed in Chapter 16. The constitutive equations of plates and shells are discussed in Chapter 20. A field that is very rich in a variety of constitutive equations is biomechanics. See Fung *Biomechanics: Mechanical Properties of Living Tissues, 1993, Biomechanics: Circulation, 1996, Biomechahics: Motion, Flow, Stress, and Growth,* 1990, Springer Verlag, and other references listed in the Bibliography.

PROBLEMS

5.2. Express the following statements in tensor equations. Define your symbols.

(a) The force of gravitational attraction between two particles of inertial masses m_1 and m_2, respectively, and separated by a distance r, is equal to Gm_1m_2/r^2 and is directed toward each other.

(b) The components of stress is a linear function of the components of strain.

(c) A normal vector of a surface is perpendicular to any two line elements tangent to the surface (in particular, to tangents of the parametric curves).

5.3. If the stress-strain law in rectangular Cartesian coordinates is

$$\tau_{ij} = \lambda\theta\delta_{ij} + 2Ge_{ij}, \quad \theta = e_{\alpha\alpha},$$

what is the proper form of tensor relation between τ^{ij} and e_{ij} in general coordinates of reference?

5.4. Discuss the appropriateness, from the tensorial point of view, of the following proposals for the constitutive equations in Cartesian tensors for certain materials.

(a) $\tau_{ij} = P(e_{mn})e_{ij}$, where $P(e_{mn}) = ae_{11} + be_{11}^2$ (a, b, constants).

(b) $\tau_{ij} = P(e_{mn})e_{ij}$, where $P(e_{mn}) = a + b|e_1|^2$ (a, b, constants), and e_1, e_2, e_3 are the three principal strains satisfying the relation $e_1 \geq e_2 \geq e_3$.

(c) $\tau_{ij} = Q(I_1, I_2, I_3)e_{ij}$, where I_1, I_2, I_3 are the first, second, and third invariants of e_{ij}, respectively.

(d) $\tau_{ij} = \alpha\delta_{ij} + \beta e_{ij} + \gamma e_{ik}e_{kj} + \lambda e_{ik}e_{km}e_{mj}$, where α, β, γ, λ are constants.

(e) If in (d), α, β, γ, λ are permitted to depend on the stress components, what kind of combinations of the stress components would be allowed for α, β, γ, λ to be functionally dependent upon?

6

ELASTIC AND PLASTIC BEHAVIOR
OF MATERIALS

In this chapter, some commonly used constitutive laws are considered. These laws describe the material behavior at isothermal condition in the range of relatively small strains, and with slow rates of flow. The plasticity theories discussed in this chapter are mathematical formulations of experimental observations. The approach is phenomenological. We shall discuss first the strain-rate-independent plasticity theory considering the yield surface, the flow rule, the hardening behavior, and the loading and unloading criterion. We then consider cyclic loading reversals and strain softening. It is, however, beyond the scope of this book to consider the atomic, crystalline or amorphous structural changes in plastic deformation, atomic reasons of how and why plasticity occurs, and plasticity at large strain. The plasticity considered here is appropriate to most problems in structural engineering, in which excessive plastic flow is undesirable; but it does not meet the needs to solve problems in metal forming, wire drawing, rolling, etc.

We shall use Cartesian tensors in this chapter. In rectangular Cartesian coordinates the physical components of a tensor are the same as the tensor components, so stresses can be denoted by σ_{ij} or σ^{ij}. The reader should try to put all the equations in this chapter into general tensor equations in curvilinear coordinates.

6.1. GENERALIZED HOOKE'S LAW

With the introduction of the concepts of stress and strain, Cauchy generalized Hooke's law into the statement that the components of stress are linearly related to the components of strain. As a tensor equation, the generalized Hooke's law may be written in the form

(1) ▲
$$\sigma^{ij} = D^{ijkl} e_{kl}$$

where σ_{ij} is the stress tensor, e_{ij} is the strain tensor,[1] and D^{ijkl} is the tensor of the *elastic constants*, or *moduli*, of the material and is called the *elastic modulus tensor*. Inasmuch as $\sigma^{ij} = \sigma^{ji}$, we must have

(2) $$D^{ijkl} = D^{jikl} .$$

Furthermore, since $e_{kl} = e_{lk}$, and in Eq. (1) the indices k and l are dummy variables, we can always symmetrize D^{ijkl} with respect to k and l without altering the sum. Hence, without loss of generality, we may assume that

(3) $$D^{ijkl} = D^{ijlk} .$$

According to these symmetry properties, the maximum number of independent elastic constants is 36.

If there exists a strain energy function W,

(4) $$W = \frac{1}{2} D^{ijkl} e_{ij} e_{kl} ,$$

with the property

(5) $$\frac{\partial W}{\partial e_{ij}} = \sigma^{ij} ,$$

then we can always assume that the quadratic form Eq. (4) is symmetric, and it follows that

(6) $$D^{ijkl} = D^{klij} .$$

Under the symmetry condition, Eq. (6), the number of independent elastic constants is reduced to 21. The question of the existence of the strain energy function is discussed in Chapter 12.

If a material possesses further symmetry in its elastic property, the number of independent elastic constants will be reduced. For example, if the material exhibits symmetry with respect to a plane, the number of independent elastic constants becomes 13. If there is symmetry with respect to three mutually perpendicular planes, the number becomes 9. However, if the material is elastically *isotropic*, i.e., if its elastic properties are identical in all directions, then the number of independent elastic constants reduces to 2.

The study of crystal symmetry is a very interesting subject. Many excellent references exist. See Love,[1.1] Green and Adkins,[1.2] Sokolnikoff.[1.2]

In the following discussion, we shall limit our attention first to *isotropic* materials. Since we are going to use Cartesian tensors in the remaining

---

[1]If the displacement u_i is infinitesimal, $e_{ij} = (u_{i,j} + u_{j,i})/2$. If u_i is finite, e_{ij} is the Almansi strain tensor of Sec. 4.1. See also Sec. 16.6, 16.7.

chapter, we shall just use subscripted indices to indicate tensorial components.

6.2. STRESS-STRAIN RELATIONSHIP FOR ISOTROPIC ELASTIC MATERIALS

For an *isotropic elastic material* in which there is no change of temperature, Hooke's law referred to a set of rectangular Cartesian coordinates may be stated in the form

$$(1) \qquad \sigma_{\alpha\alpha} = 3Ke_{\alpha\alpha} \,,$$

$$(2) \qquad \sigma'_{ij} = 2Ge'_{ij} \,,$$

where K and G are constants and σ'_{ij} and e'_{ij} are the *stress deviation* and *strain deviation*, respectively; i.e.,

$$(3) \qquad \sigma'_{ij} = \sigma_{ij} - \frac{1}{3}\sigma_{\alpha\alpha}\delta_{ij} \,,$$

$$(4) \qquad e'_{ij} = e_{ij} - \frac{1}{3}e_{\alpha\alpha}\delta_{ij} \,.$$

We have seen before that $\sigma_{\alpha\alpha}/3$ is the mean stress at a point and that, if the strain were infinitesimal, $e_{\alpha\alpha}$ is the change of volume per unit volume. Both $\sigma_{\alpha\alpha}$ and $e_{\alpha\alpha}$ are *invariants*. Thus, Eq. (1) states that the change of volume of the material is proportional to the mean stress. In the special case of hydrostatic compression,

$$\sigma_{xx} = \sigma_{yy} = \sigma_{zz} = -p, \qquad \sigma_{xy} = \sigma_{yz} = \sigma_{zx} = 0 \,,$$

we have $\sigma_{\alpha\alpha} = -3p$, and Eq. (1) may be written, in the case of infinitesimal strain, with v and Δv denoting volume and change of volume, respectively,

$$(5) \qquad \frac{\Delta v}{v} = -\frac{p}{K} \,.$$

Thus, the coefficient K is appropriately called the *bulk modulus* of the material.

The strain deviation e'_{ij} describes a deformation without volume change. Equation (2) states that the stress deviation is simply proportional to the strain deviation. The constant G is called the *modulus of elasticity in shear*, the *shear modulus*, or the *modulus of rigidity*. In the special case in which $e_{xy} \neq 0$, but all other strain components vanish, we have

$$(6) \qquad \sigma_{xy} = 2Ge_{xy} \,,$$

whereas all other stress components vanish. The coefficient 2 is included because, before the tensor concept was introduced, it was customary to define the engineering shear strain as $\gamma_{xy} = 2e_{xy}$.

If we substitute Eqs. (3) and (4) into Eq. (2) and make use of Eq. (1), the result may be written in the form

(7) $$\sigma_{ij} = \lambda e_{\alpha\alpha}\delta_{ij} + 2Ge_{ij} ,$$

or

(8) $$e_{ij} = \frac{1+\nu}{E}\sigma_{ij} - \frac{\nu}{E}\sigma_{\alpha\alpha}\delta_{ij} .$$

The constants λ and G are called *Lamé's constants* (G. Lamé, 1852). In many books the symbol μ is used in place of G. The constant E is called the modulus of elasticity, or Young's modulus (Thomas Young, 1807). The constant ν is called Poisson's ratio. The relationships between these constants are

(9)
$$\lambda = \frac{2G\nu}{1-2\nu} = \frac{G(E-2G)}{3G-E} = K - \frac{2}{3}G = \frac{E\nu}{(1+\nu)(1-2\nu)}$$
$$= \frac{3K\nu}{1+\nu} = \frac{3K(3K-E)}{9K-E} ,$$
$$G = \frac{\lambda(1-2\nu)}{2\nu} = \frac{3}{2}(K-\lambda) = \frac{E}{2(1+\nu)} = \frac{3K(1-2\nu)}{2(1+\nu)} = \frac{3KE}{9K-E} ,$$
$$\nu = \frac{\lambda}{2(\lambda+G)} = \frac{\lambda}{(3K-\lambda)} = \frac{E}{2G} - 1 = \frac{3K-2G}{2(3K+G)} = \frac{3K-E}{6K} ,$$
$$E = \frac{G(3\lambda+2G)}{\lambda+G} = \frac{\lambda(1+\nu)(1-2\nu)}{\nu} = \frac{9K(K-\lambda)}{3K-\lambda} = 2G(1+\nu)$$
$$= \frac{9KG}{3K+G} = 3K(1-2\nu) ,$$
$$K = \lambda + \frac{2}{3}G = \frac{\lambda(1+\nu)}{3\nu} = \frac{2G(1+\nu)}{3(1-2\nu)} = \frac{GE}{3(3G-E)} = \frac{E}{3(1-2\nu)} .$$

To these we may add the following combinations that appear frequently.

(10) $$\frac{G}{\lambda+G} = 1 - 2\nu , \qquad \frac{\lambda}{\lambda+2G} = \frac{\nu}{1-\nu} .$$

In unabridged notation, Eq. (8) reads

$$e_{xx} = \frac{1}{E}[\sigma_{xx} - \nu(\sigma_{yy} + \sigma_{zz})],$$

$$e_{yy} = \frac{1}{E}[\sigma_{yy} - \nu(\sigma_{xx} + \sigma_{zz})],$$

$$e_{zz} = \frac{1}{E}[\sigma_{zz} - \nu(\sigma_{xx} + \sigma_{yy})],$$

(11)

$$e_{xy} = \frac{1+\nu}{E}\sigma_{xy} = \frac{1}{2G}\sigma_{xy},$$

$$e_{yz} = \frac{1+\nu}{E}\sigma_{yz} = \frac{1}{2G}\sigma_{yz},$$

$$e_{zx} = \frac{1+\nu}{E}\sigma_{zx} = \frac{1}{2G}\sigma_{zx}.$$

One can express σ's in terms of e's

(12)
$$\sigma_{xx} = \frac{E}{(1+\nu)(1-2\nu)}[(1-\nu)e_{xx} + \nu(e_{yy} + e_{zz})],$$

$$\sigma_{xy} = 2Ge_{xy},$$

with corresponding expressions for σ_{yy}, σ_{zz}, σ_{yz}, σ_{zx}. Table 6.2:1 gives the average values of E, G, and ν at room temperature for several engineering materials which are approximately isotropic.

Table 6.2:1

	E, 10^6 lb/sq in.	G, 10^6 lb/sq in.	ν	Speed of sound (Dilatational wave) 10^3 ft/sec
Metals:				
Steels	30	11.5	0.29	16.3
Aluminum alloys	10	2.4	0.31	16.5
Magnesium alloys	6.5	2.4	0.35	16.6
Copper (hot rolled)	15.0	5.6	0.33	—
Plastics:				
Cellulose acetate	0.22			0.36
Vinylchloride acetate	0.46			5.1
Phenolic laminates	1.23		0.25	8.2
Glass	8	3.2	0.25	
Concrete	4		0.2	

In 1829, Poisson advanced arguments, which were found untenable later, that the value of ν should be $1/4$. The special value of Poisson's ratio $\nu = 1/4$ makes

(13) $$\lambda = G$$

and simplifies the equations of elasticity considerably. Consequently, this assumption is often used, particularly in geophysics, in the study of complicated wave-propagation problems. The special value $\nu = 1/2$ implies that

(14) $$G = \frac{1}{3}E, \qquad \frac{1}{K} = 0, \qquad \frac{\Delta v}{v} = e_{\alpha\alpha} = 0.$$

So far in this chapter we have discussed the simplest of all stress-strain relationships for elastic materials — the *Hooke's law*, which are the mechanical properties of a class of materials like steel. There are of course also many things in the world whose mechanical properties cannot be described by the simple elasticity discussed above. For example, almost all biological materials lie beyond the reach of these simple laws. Even rubber, plastics, and metals like aluminum, magnesium, and lead cannot be so described. Many materials obey Hooke's law when the stress and strain are sufficiently small, but yield and flow plastically when a critical condition of yield is reached and maintained. The behavior of materials beyond their elastic limit is complicated. In the rest of this book, we first demonstrate the classical methods that can solve boundary-value problems of bodies whose materials obey Hooke's law; then consider the world beyond. We will study plasticity, thermodynamics, thermoelasticity, irreversible thermodynamics, viscoelasticity, finite strains, large deformation, elasticity with nonlinear stress-strain relationship, plates and shells, and numerical methods to handle boundary-value problems involving materials with these mechanical properties. The final mathematical structure of the theories renders boundary-value problems solvable by the numerical methods that will be proposed.

6.3. IDEAL PLASTIC SOLIDS

Metals obey Hooke's law only in a certain range of small strain. When a metal is strained beyond an *elastic limit*, Hooke's law no longer applies. The behavior of metals beyond their elastic limit is rather complicated. We shall give a very brief outline of some experimental facts in the next section, and we will discuss the formulation of the constitutive relations in the plastic regime later. However, before engaging in a long and involved discussion, let us present a set of laws which defines the simplest plastic materials — an *ideal plastic solid obeying von Mises' yield criterion and flow rule.* That such a set of laws is a reasonable abstraction of the behavior of certain real materials will be discussed later. At the moment, we shall just exhibit the minimum ingredients that constitute a theory of plasticity.

In Secs. 3.12 and 3.13, we defined the *stress deviation* σ'_{ij} which has a second invariant J_2 defined as

$$(1) \qquad J_2 = \frac{1}{2}\sigma'_{ij}\sigma'_{ij} .$$

In a similar way, the *strain deviation* e'_{ij} is defined by subtracting the mean strain from the strain tensor e_{ij},

$$(2) \qquad e'_{ij} = e_{ij} - \frac{1}{3}e_{\alpha\alpha}\delta_{ij} .$$

If the material is isotropic and obeys Hooke's law, then

$$(3) \qquad e'_{ij} = \frac{1}{2G}\sigma'_{ij} .$$

When an increasing deformation reaches a certain limit, called the *elastic limit*, the material starts to deform plastically, and e'_{ij} is no longer given by Hooke's law. In this case we define the *plastic strain increment* $de^{(p)}_{ij}$ as the *increment* of the actual strain deviation de'_{ij} minus the *increment* of the elastic strain $de'^{(e)}_{ij}$ computed from Hooke's law as if it still applied. $de'^{(e)}_{ij}$ is called the *elastic strain increment.* For metallic materials experimental evidence indicates that the hydrostatic pressure has little effect on plastic yielding and that the volumetric change of the material is negligible during plastic yielding. Hence one can assume $de^{(p)}_{ii} = 0$. Thus $de^{(p)}_{ij}$ can be considered as the increment of the plastic strain deviation and be written in the form

$$(4) \qquad de^{(p)}_{ij} = de'_{ij} - de'^{(e)}_{ij} = de'_{ij} - \frac{d\sigma'_{ij}}{2G} , \qquad de^{(p)}_{ii} = 0 .$$

where $(\cdot)'$ denotes the deviation of the associated quantity in the parenthesis. The total strain deviation increment may now be written as

$$(5) \qquad de'_{ij} = de^{(p)}_{ij} + \frac{d\sigma'_{ij}}{2G} .$$

Equation (5) becomes a rate equation of deformation when all incremental quantities $d(\cdot)$ are replaced by rate quantities $(\cdot)$,

$$(5a) \qquad \dot{e}_{ij} = \dot{e}^{(p)}_{ij} + \frac{\dot{\sigma}'_{ij}}{2G} .$$

The rate of plastic deformation is $\dot{e}_{ij}^{(p)}$ and the increment of the plastic strain in the time interval dt is $\dot{e}_{ij}^{(p)} dt$. Then the total plastic strain deviation after successive stages of yielding is the algebraic sum of the deformations that occur at all stages:

$$(6) \qquad\qquad e_{ij}^{(p)} = e_{ij}^{(p)}(0) + \int_0^t \dot{e}_{ij}^{(p)}(t)\, dt\,,$$

where $e_{ij}^{(p)}(0)$ is the initial value of $e_{ij}^{(p)}$ at time $t = 0$. A theory of plasticity is formulated by specifying how $e_{ij}^{(p)}$ can be computed.

For an *ideal plastic solid obeying von Mises' yield criterion and flow rule* the specifications are as follows:

(a) Hooke's law holds for the mean stress and the mean strain at all times, i.e.,

$$\sigma_{\alpha\alpha} = 3K e_{\alpha\alpha}\,.$$

Hence, the plastic strain is incompressible $(e_{ii}^{(p)} = 0)$ and the plastic strain deviation tensor is the same as the plastic strain tensor. This justifies the notation $e_{ij}^{(p)}$ in places where normally we would write $e_{ij}'^{(p)}$ to indicate a plastic strain deviation.

(b) The material is elastic and obeys Hooke's law as long as the second invariant J_2 is less than a constant k^2. In other words, no change in plastic strain can occur as long as $J_2 < k^2$.

$$\dot{e}_{ij}^{(p)} = 0 \qquad \text{when} \qquad J_2 < k^2\,.$$

(c) Yielding occurs (the elastic limit is reached) when and only when $J_2 = k^2$. When the yield condition $J_2 - k^2 = 0$ prevails, the rate of change of the plastic strain is proportional to the stress deviation.

$$\dot{e}_{ij}^{(p)} = \frac{1}{\mu}\sigma_{ij}'\,, \qquad \mu > 0\,,$$

where μ is a *positive* factor of proportionality.

(d) Stress state corresponding to $J_2 > k^2$ cannot be realized in the material.

The set of laws above contains two essential parts: the criterion for yield, and the stress-strain relation in the elastic and plastic regimes. In the specifications above, the yield condition is based on the second invariant of the stress deviation tensor. This yield criterion was first proposed by von Mises. The constant k can be identified with the yield stress in simple shear or simple tension.

We notice that only *the rate* of plastic strain is specified by these laws. Under a varying loading program, successive plastic strain increments must be added together algebraically according to Eq. (6). Such a theory is called an *incremental theory*.

In many applications of the theory of plasticity the rate at which plastic flow occurs is relatively small. The concern is mostly with the total amount of plastic flow. In such applications we often assume that the plastic deformation is rate insensitive and that we can replace the equation in (c) by the incremental law

$$de_{ij}^{(p)} = \lambda \sigma'_{ij}, \qquad \lambda > 0,$$

where λ is a factor of proportionality and not a characteristic constant of the material. The factor λ can be a function of $e_{ij}^{(p)}$ and σ_{ij}, but not $\dot{\sigma}_{ij}$. The sign of λ is determined by the fact that plastic flow involves dissipation of energy. Under a given set of loading conditions the value of λ, and hence the total plastic flow, is determined by the total work done by the external load.

Actual materials may exhibit much more complicated plastic behavior as specified above. Many yield conditions and plastic flow rules have been proposed. To prepare for the study of these formulations, we shall review briefly below some basic experimental facts.

6.4. SOME EXPERIMENTAL INFORMATION

Simple Tension Tests. If a rod of a ductile metal is pulled in a testing machine at room temperature, the load applied on the test specimen may be plotted against the elongation,

$$\epsilon = \frac{l - l_0}{l_0},$$

where l_0 is the original length of the rod and l is the length under load. Numerous experiments show typical load-elongation relationships, as indicated in the diagrams of Fig. 6.4:1. The initial region appears as a straight line. This is the region in which the law of linear elasticity is expected to hold. Mild steel shows an *upper yield point* A*, a *lower yield point* A, and a flat *yield region* AB [Fig. 6.4:1(a)], which is caused by many discontinuous steps of microscopic slip along slip planes of the crystals. Most other metals do not have such a flat yield region [Fig. 6.4:1(b)]. Metals such as aluminum, copper and stainless steel exhibit a gradual transition from

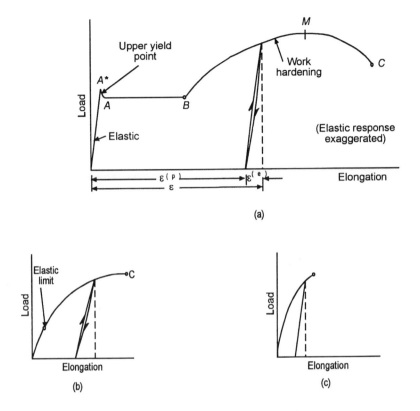

Fig. 6.4:1. Typical load-elongation curves in simple tension tests.

a linear elastic to a nonlinear plastic behavior instead having pronounced yield points as shown in Fig. 6.4:1(a).

Upon unloading at any stage in the deformation, the strain is reduced along an elastic unloading line. Reloading retraces the unloading curve with relatively minor deviation and then produces further plastic deformation when approximately the previous maximum stress is exceeded. During the loading process, the test specimen may "neck" at certain strain, so that the cross section is reduced in a small region. When necking occurs under continued elongation, the load reaches a maximum and then drops down, although the actual average stress in the neck region (load divided by the true area of the neck) continues to increase. The maximum M in Fig. 6.4:1(a) is the *ultimate load*. Beyond the ultimate load the metal continues to flow. At point C in the curves of Fig. 6.4:1 the specimen breaks.

Materials like cast iron, titanium carbide, beryllium, or any rock material, which allow very little plastic deformation before reaching the breaking point, are called *brittle* materials. The load-strain curve for a brittle material is given in Fig. 6.4:1(c). Point C is the *breaking point*.

A fact of great importance for geology is that brittle materials such as rocks tend to become ductile when subjected to large hydrostatic pressure (large negative mean stress). Theodore von Karman (1911) demonstrated this in his classical experiments on marbles. Simple compression and simple shear tests of cylindrical specimens give load-strain diagrams similar to those of Fig. 6.4:1. Bending of a beam with a shear load is often used to test the behavior of a material in tension, compression, and shear combined in a specific way.

Bauschinger Effect. When a metal specimen is subjected to repeated tension-compression loads with deformation exceeding the elastic limit, the load-deflection curve sometimes appears as in Fig. 6.4:2. The tension stroke and the compression stroke are dissimilar. This is referred to as the *Bauschinger effect*, after J. Bauschinger's basic paper on strain hardening published in 1886.

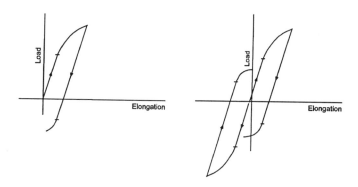

Fig. 6.4:2. Bauschinger effect in a simple tension-elongation test.

Anisotropy. Plastic deformation in metal is a result of slip along certain crystallographic planes. The process is clearly directional. As a consequence, any initial isotropy, which may have been present is usually destroyed by plastic deformation. From the point of view of the dislocation theory, slip is an irreversible process; every slip produces a new material. These changes are revealed in the Bauschinger effect and in the anisotropy of materials after plastic deformation.

Time and Temperature Effects. The results described above are typically obtained by slow application of the load in a testing machine. Higher strain rate or loading rate can have a pronounced effect on the material behavior in the plastic regime. Increasing loading rate generally decreases the ductility of the material, and increases the initial and subsequent yield stresses. Under rapid loading, the hereditary nature of the material usually reveals itself. If the hereditary stress-strain law is linear, it is often described as *anelasticity* or *viscoelasticity* (see Chapters 1 and 15). If the strain is large, the hereditary stress-strain laws for metals are generally nonlinear.

The time-dependent plastic flow is often described as *creep*. A simple tension specimen of lead wire under a constant tension load shows a creep curve as illustrated in Fig. 6.4:3. Following an initial extension, the rate of strain first decreases gradually, then remains nearly constant for a while, and finally accelerates until the specimen breaks. The *three stages of creep* are called, respectively, the *primary*, the *secondary*, and the *tertiary creep*.

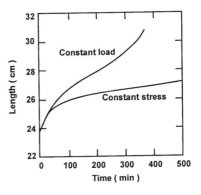

Fig. 6.4:3. Creep test of a lead wire in tension. The initial lengths and initial loads were the same in both tests. (After Andrade, 1910; courtesy of the Royal Society, London.).

Andrade[6.4] has pointed out, however, that if a creep test is performed at constant stress, the final stage of accelerated elongation may or may not appear, depending on whether the material deteriorates or not. As shown in Fig. 6.4:3, the creep curve at constant load differs considerably from that at constant stress.

The creep rate and length of the primary and secondary stages are dependent on temperature and stress level. At high temperature and high stresses the secondary creep stage may not be appreciable.

The phenomenon of creep is important in geophysics and in those engineering problems in which accurate dimensions must be maintained over a long period of time. It is a controlling factor for the design of machinery that operates at high temperature. On the other hand, plastic deformation generates heat, and a very rapid application of load is usually associated with a large temperature gradient, localized plastic flow and anisotropic effects. Continuous metal cutting is a good example of large localized plastic deformation. Temperature, as strain rate, generally has a very significant influence on material properties. As temperature increases, the ductility of metals usually increases while its stiffness decreases.

Size Effect and Stress Inhomogeneity. Since practically all materials have a nonhomogeneous microscopic structure, plastic properties of a small

region may be different from those of a larger region. This consideration becomes important in the question of stress concentration in notched
specimens. Size effect is also important for micron- and submicron structures in micro electronic applications as the strain gradient can strongly
influence the structural behaviors.

Combined Stress. One of the most important tasks of mathematical
theory is to extend and generalize the experiences of simple experiments
and to suggest crucial tests to verify the basic assumptions of the theory. In
plasticity, this function of mathematics is particularly evident, because here
one must consider the stress and strain tensors; yet no direct observation
of all components of these tensors is possible.

A relatively simple combined-stress experiment employs a thin-walled
circular cylinder. With the combination of an axial tension, a twisting
moment about the cylinder axis, and an internal or external pressure, a
good approximation to a biaxial state of stress (involving σ_{11}, σ_{12}, σ_{22}
with $\sigma_{31} = \sigma_{32} = \sigma_{33} = 0$) can be produced. Much information has been
gathered from such experiments. Among other things, it is found that the
shear stress is by far the major cause of yielding.

P. W. Bridgman's renowned experiments on the *influence of hydrostatic
pressure on yielding* and other mechanical and metallographical characteristics of metals have shown that hydrostatic pressures of one order higher
than the yield stress have practically no influence on yielding of metals. A
tensile test or shear test was run at atmospheric pressure in the standard
manner, with load and deformation recorded; and then was repeated by im-
mersion of the whole setup in a chamber under high hydrostatic pressure.
Bridgman found that the stress-strain curves in the small strain range did
not change, but the ductility of the material increased greatly, thus permit-
ting a much larger deformation prior to fracture.

Conversely, very small changes in material density are found when a
metal is subjected to repeated plastic deformation, indicating that the plastic volume change is small.

In the subsequent sections, we shall present some of the best known
theories of plasticity and derive the mathematical relationships between
stress and strain for plastically deformed solids.

6.5. A BASIC ASSUMPTION OF THE MATHEMATICAL THEORY OF PLASTICITY: THE EXISTENCE OF A YIELD FUNCTION

The theory of plasticity descirbes the mechanical behavior of materials
in the plastic range including *energy dissipation, irreversible and history*

or path dependent process, the initial yield surface, its subsequent growth, the constitutive equations for plastic deformation, and the criteria for loading and unloading. Throughout this chapter, we assume that the *plastic deformation is not rate sensitive*, so that the constitutive equations are *invariant with respect to time scale* and that *the rate form and the incremental form are equivalent to each other*.

By comparing the laws of linear elasticity and ideal plasticity as described in Secs. 6.1 and 6.3 with the typical experimental features described in Sec. 6.4, we can determine the applicability and limitations of these mathematical laws. We see that near the origin of the stress-strain diagrams of Fig. 6.4:1, the linear relationship is good in nearly all cases. This is the range over which the material behavior follows the laws of linear elasticity as described in Secs. 6.1 and 6.2. For the case of mild steel represented by Fig. 6.4:1(a), there is a flat region A-B. This is the feature of ideal plasticity that the load remains constant as the elongation increases under uniaxial loading. The slightly unstable region on the yield curve at the peak A* before the flat range is normally ignored by plasticity laws. The rise beyond point B in Fig. 6.4:1(a), as well as the curved portion of the curves in Figs. 6.4:1(b) and 1(c), obviously cannot be represented by the linear elasticity or ideal plasticity. The features of these curved portions of stress-strain curves are said to be features of *strain hardening*.[2] It is evident that *the stress-strain laws of strain-hardening will be basically different from those of ideal plasticity*. If the material is unloaded in this region beyond point A, it will follow a path parallel to the initial loading path OA*. As a result only part of the strain, called the *elastic strain* $e^{(e)}$, is recovered, while the other part of the strain will remain as the permanent or *plastic strain* $e^{(p)}$.

For the purpose of describing the state of stress at any point in the material, it is convenient to represent each state of stress by a point in a nine-dimensional *stress space*, with axes $\sigma_{ij}(i, j = 1, 2, 3)$. Similarly, a state of strain is a point in a nine-dimensional *strain space*, with components e_{ij}. In the theory of plasticity, a state of plastic strain $e_{ij}^{(p)}$ may be so represented

---

[2]The material that is most qualified to be called *elastic* is probably natural rubber. But if the stress-strain relationship of natural rubber is plotted in the manner of Fig. 6.4:1, the resulting curve will not be a straight line. In some rubbers this can be corrected by plotting the stress components versus the (finite) Eulerian (Almansi) strains or, equivalently, by plotting the Kirchhoff stress components versus Green's strain components. In such cases, the material is said to be *linearly elastic* with respect to finite strain. In most cases the nonlinearity in the stress-strain relationship remains, although the material returns to its original state along the same stress-strain curve when all the loads are removed. In this case, the material is said to be *nonlinearly elastic*. In this chapter, we discuss the strain hardening only under the assumption of small strain. Finite deformation is discussed in Chapter 16.

also. A program of loading may be regarded as a path in the stress space. The corresponding deformation history is a path in the strain space.

To develop a mathematical theory of plasticity, a basic assumption is made that there exists a continuous scalar *yield function* $f(\sigma_{ij}, T, \xi_i)$, which has the following properties:

- The equation $f(\sigma_{ij}, T, \xi_i) = 0$ represents a closed surface, called the *yield* or *loading surface*, in the stress space σ_{ij} for a given temperature T and an array of internal variables $\xi_1, \ldots, \xi_n$.
- The plastic strain-rate $\dot{e}_{ij}^{(p)}$ and all internal-variable rates $\dot{\xi}_i$ vanish in the region in which $f(\sigma_{ij}, T, \xi_i) < 0$. This region is called the *elastic region*, which occupies the interior of the yield surface.
- The plastic strain rate $\dot{e}_{ij}^{(p)}$ can be nonzero in the region where $f(\sigma_{ij}, T, \xi_i) = 0$.
- No meaning is associated with $f(\sigma_{ij}, T, \xi_i) > 0$.

There are two types of internal variables $\xi_1, \ldots, \xi_n$. One type consists of "physical" variables, which describe chemical reaction, phase changes, or structural defects. Another type consists of phenomenological variables including $e_{ij}^{(p)}$ itself. Together they characterize the hardening properties of the material and, therefore, are also called *work-hardening parameters*. These parameters may depend on the plastic deformation history and their rates may be functions of σ_{ij}, T and ξ_i.

Work hardening will be considered later. Here let us consider several simple examples to see the plausibility of the yield assumption.

Von Mises Criterion. The *von Mises yield condition* is defined by the yield function

(1) ▲ $$f(\sigma_{ij}) = J_2 - k^2,$$

where k is a parameter and J_2 is the second invariant of the stress deviation, which is in the form

(2) $$J_2 = \frac{1}{2}\sigma_{ij}'\sigma_{ij}'.$$

Alternative expressions of J_2 are given in Eq. (3.10:8). For an ideal plastic material obeying von Mises's condition, k is a constant independent of strain history. The stress state is characterized by elastic deformation when the condition $J_2 \leq k^2$, with plastic flow possible only when $J_2 = k^2$, whereas the condition $J_2 > k^2$ can never be realized. These are the conditions specified in Sec. 6.3. For a work-hardening material, k will be allowed to change with strain history.

If the material is subjected to a simple shear σ_{12} while all other stress components vanish (a state of stress realizable in torsion of a thin-walled tube), then $J_2 = \sigma_{12}^2$, and yielding should occur when

$$\sigma_{12} = k.$$

Hence *the constant k means the yield stress in simple shear.*

Tresca Criterion. Tresca[6.3] (1868) first advanced the idea of yield criterion. Through his work on metal forming in an armory, he concluded that the decisive factor for yielding is the maximum shear stress in the material. Tresca proposed the criterion stipulating that the maximum shear stress has a constant value during plastic flow.

To express Tresca's idea analytically, it is the simplest to use the principal stresses σ_1, σ_2, σ_3. If the principal axes of stress are so labeled that

$$(3) \qquad\qquad \sigma_1 \geq \sigma_2 \geq \sigma_3,$$

then *Tresca's yield condition* is

$$(4) \; \blacktriangle \qquad\qquad f \equiv \sigma_1 - \sigma_3 - 2k = 0.$$

However, f in this form is not analytic; it violates the rule that the manner in which the principal axes are labeled 1, 2, 3 should not affect the form of the yield function. To obey this rule, we observe that Tresca's condition states that during plastic flow one of the differences $|\sigma_1 - \sigma_2|$, $|\sigma_2 - \sigma_3|$, $|\sigma_3 - \sigma_1|$ has the value $2k$. Hence we may write

$$(5) \quad f \equiv [(\sigma_1 - \sigma_2)^2 - 4k^2][(\sigma_2 - \sigma_3)^2 - 4k^2][(\sigma_3 - \sigma_1)^2 - 4k^2] = 0.$$

This equation is now symmetrical with respect to the principal stresses, and can be put into an invariant form, which is due to Reuss:

$$(6) \; \blacktriangle \qquad f = 4J_2^3 - 27J_3^2 - 36k^2 J_2^2 + 96k^4 J_2 - 64k^6 = 0,$$

where J_2 and J_3 are the second and third invariants of the stress deviation tensor with J_2 as given in Eq. (2) and

$$J_3 = \det(\sigma'_{ij}) = \frac{1}{3}\sigma'_{ij}\sigma'_{jk}\sigma'_{ki}$$

$$= \sigma'_{11}\sigma'_{22}\sigma'_{33} + 2\sigma_{12}\sigma_{23}\sigma_{31} - \sigma'_{11}\sigma^2_{23} - \sigma'_{22}\sigma^2_{31} - \sigma'_{33}\sigma^2_{12}.$$

Using the results of Problem 3.5 in Sec. 3.10, we can write the Tresca yield surface in the form,

$$f = \sqrt{J_2}\,\sin(\alpha + 60°) - k = 0,$$

where

$$\alpha = \frac{1}{3}\cos^{-1}\left(\frac{3\sqrt{3}}{2}\frac{J_3}{J_2^{3/2}}\right) \qquad \text{with} \qquad -\frac{\pi}{6} < \alpha < \frac{\pi}{6}\,.$$

It can be shown that the absolute value of $\frac{3\sqrt{3}}{2}\frac{J_3}{J_2^{3/2}}$ is always less or equal to one.

In certain problems, the direction of the principal axes and the relative magnitudes of the principal stresses are known by symmetry considerations or by intuition. Then it is possible to use the simple form Eq. (4) for Tresca's criterion. In general, however, we cannot tell *a priori* the relative magnitude of the principal stresses, and would be obliged to use the general form Eq. (6).

Example 1. *Comparison of the Tresca and von Mises Criteria.*

Consider a thin-walled tube subjected to the combined uniaxial tensile stress $\sigma (\geq 0)$ and shear stress τ. From Eq. (3.6:7), the principal stresses are

$$(7) \qquad \left\{\begin{matrix} \sigma_1 \\ \sigma_3 \end{matrix}\right\} = \frac{\sigma}{2} \pm \sqrt{\left(\frac{\sigma}{2}\right)^2 + \tau^2}\,, \qquad \sigma_2 = 0\,.$$

The Tresca and von Mises yield functions can be simplified as

$$f = \bar{\sigma} - \sigma_Y\,,$$

where σ_Y is the yield stress in uniaxial tension, and

$$(8) \qquad \bar{\sigma} = \sigma_1 - \sigma_3 = \sqrt{\sigma^2 + 4\tau^2} = 2k \qquad \text{(Tresca)}\,,$$

$$(9) \qquad \bar{\sigma} = \sqrt{3J_2} = \sqrt{\frac{1}{2}[(\sigma_1 - \sigma_3)^2 + \sigma_1^2 + \sigma_3^2]}$$

$$= \sqrt{\sigma^2 + 3\tau^2} = \sqrt{3}k \qquad \text{(von Mises)}\,.$$

If k is the yield stress in simple shear ($\sigma = 0, \tau = k$), then $k = \sigma_Y/\sqrt{3}$ according to the von Mises criterion in Eq. (9) and $k = \sigma_Y/2$ according to the Tresca criterion in Eq. (8).

Example 2. *Experimental Investigation of the Tresca and von Mises Criteria.*

Lode[6.3] (1925) investigated the influence of the intermediate principal stress on yielding. He used the parameter, called the *Lode parameter*,

$$(10) \qquad \eta = \frac{2\sigma_2 - \sigma_1 - \sigma_3}{\sigma_1 - \sigma_3}$$

to characterize the stress state where the principal stresses σ_i satisfy the condition

$$(11) \qquad \sigma_1 > \sigma_2 > \sigma_3 .$$

According to the Tresca criterion

$$(12) \qquad \frac{\sigma_1 - \sigma_3}{\sigma_Y} = 1 ,$$

thus yielding does not depend on η. On the other hand, from Eq. (10), we have

$$\sigma_2 - \sigma_1 = \frac{\sigma_1 - \sigma_3}{2}(\eta - 1) ,$$

$$\sigma_2 - \sigma_3 = \frac{\sigma_1 - \sigma_3}{2}(\eta + 1) .$$

The von Mises criterion

$$\frac{1}{2}[(\sigma_1 - \sigma_2)^2 + (\sigma_2 - \sigma_3)^2 + (\sigma_3 - \sigma_1)^2] = \sigma_Y^2$$

can be reduced to

$$(13) \qquad \frac{\sigma_1 - \sigma_3}{\sigma_Y} = \frac{2}{\sqrt{3 + \eta^2}} ,$$

which is a function of the Lode parameter η. Lode (1925) tested thin-walled tubes of steel, copper, and nickel subjected to a combined loading of uniaxial tensile force F and internal pressure p. The induced stresses are

$$\sigma_1 = \sigma_\theta = \frac{pR}{t} ,$$

$$\sigma_2 = \sigma_z = \frac{F}{2\pi Rt} ,$$

$$\sigma_3 = \sigma_r \approx 0 ,$$

where t is the wall thickness and R the mean tube radius. From Eq. (10), the Lode parameter is

$$\eta = \frac{F - \pi R^2 p}{\pi R^2 p} ,$$

with $2\pi R^2 p > F > 0$ to assure that Eq. (11) is satisfied. Experimental results in Fig. 6.5:1 seem to favor the von Mises yield criterion as the results show that the value of $(\sigma_1 - \sigma_3)/\sigma_Y$ depends on η.

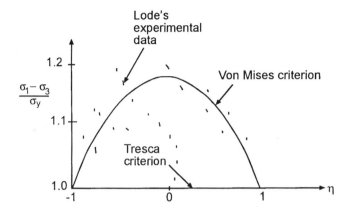

Fig. 6.5:1. Lode's comparison of von Mises and Tresca yield criteria (Data from Khan and Huang 1995). Dots are Lode's experimental data.

Problem 6.1.   Apply the yield criteria of von Mises and Tresca to simple tension and to simple shear. Show that, if they give the same yield stress in simple shear, the tensile yield stress predicted by von Mises' criterion is smaller than that predicted by Tresca's criterion by a factor $\sqrt{3}/2$.

Problem 6.2.   Show that in no other type of stress is the discrepancy between the predictions of the yield stress by von Mises, and by Tresca's criteria as large as it is in simple tension (von Mises[6.3] 1913), if they give the same yield stress in simple shear.

6.6. LOADING AND UNLOADING CRITERIA

Let us first clarify what loading and unloading in a plastic state mean. Consider a plastic state at which the yield function introduced in Sec. 6.5 vanishes, i.e.,

$$(1) \qquad\qquad f(\sigma_{ij}, T, \xi_i) = 0\,.$$

The time rate of f is

$$(2) \qquad\qquad \dot{f} = \frac{\partial f}{\partial \sigma_{ij}}\dot{\sigma}_{ij} + \frac{\partial f}{\partial \xi_i}\dot{\xi}_i + \frac{\partial f}{\partial \xi_i}\dot{T}\,.$$

Obviously, $f = 0$ and $\dot{f} < 0$ at a time t would imply $f < 0$, the next instant of time. Such a change leads to an elastic state and is a natural attribute to the term *unloading*. However, we also require that in an unloading process there is no change in the internal-variables and temperature, i.e., $\dot{\xi}_i = \dot{T} = 0$. Hence, by Eq. (2) we stipulate that the criterion for unloading from a plastic state at constant temperature is

$$(3) \qquad\qquad \frac{\partial f}{\partial \sigma_{ij}}\dot{\sigma}_{ij} < 0\,, \qquad f = 0\,, \quad \text{during unloading}\,.$$

Otherwise it is said to be loading or neutral loading. Thus,

(4) $$\frac{\partial f}{\partial \sigma_{ij}} \dot{\sigma}_{ij} = 0 , \qquad f = 0 , \qquad \text{during neutral loading} ;$$

(5) $$\frac{\partial f}{\partial \sigma_{ij}} \dot{\sigma}_{ij} > 0 , \qquad f = 0 , \qquad \text{during loading} .$$

The function f is also called a *loading function* because of its prominence in these loading criteria.

A simple geometric interpretation of these criteria exists. Since the yield surface, i.e., the loading surface, is assumed to be a closed surface, we can speak of its inside and outside. Then, for a state of stress on the loading surface, loading, unloading, or neutral loading takes place, according to whether the stress increment vector is directed outward, inward, or along the tangent to the loading surface, respectively. Because of this geometric interpretation (also for reasons to be discussed in Sec. 6.9), it is important to obey the sign convention in writing Eq. (1) so that $\partial f / \partial \sigma_{ij}$ be directed outwardly normal to the surface $f = 0$.

6.7. ISOTROPIC STRESS THEORIES OF YIELD FUNCTION

A material is said to be *isotropic* if there is no orientation effect in the material. If the yield function f depends only on the *invariants* (with respect to rotation of coordinates) of stress, strain, and strain history, then the plastic characteristics of the material is isotropic. If the yield function f is an isotropic function of the stress alone, then the theory of plasticity is called an *isotropic stress theory*. In such theories

(1) $$f = f(I_1, I_2, I_3) ,$$

where I_1, I_2, I_3 are the three invariants of the stress tensor σ_{ij}. Equivalently, we may write Eq. (1) in terms of principal stresses,

(2) $$f = f(\sigma_1, \sigma_2, \sigma_3) .$$

If the principal stresses σ_1, σ_2, σ_3 are taken as the coordinate axes, the surface $f(\sigma_1, \sigma_2, \sigma_3) = 0$ can be plotted in a three-dimensional stress space. For example, if we take $f = J_2 - k^2$, with k being a constant, the surface $f = 0$ appears as a circular cylinder whose axis is equally inclined to the coordinate axes, as shown in Fig. 6.7:1.

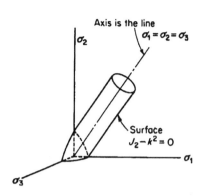

Fig. 6.7:1. A yield surface in the principal stress plane.

According to P. W. Bridgman[6.3] (1923), the plastic deformation of metals essentially is independent of hydrostatic pressure; Crossland[6.3] (1954) and many others have verified this result. If this conclusion is accepted, the yield function will be independent of $I_1 = \sigma_1 + \sigma_2 + \sigma_3$. Then it is advantageous to introduce the stress deviations $\sigma'_{ij} = \sigma_{ij} - \delta_{ij}\sigma_{kk}/3$. Let J_2, J_3 be the second and third invariants of the stress deviation σ'_{ij}. Then we have $f = f(J_2, J_3)$. Von Mises took the simplest of such functions, assuming $f = J_2 - \text{const}$. The surface $f(J_2) = 0$, when plotted in the space of principal stresses, will be a cylinder perpendicular to the plane

$$(3) \qquad\qquad I_1 = \sigma_1 + \sigma_2 + \sigma_3 = 0 .$$

Its axis is equally inclined to the coordinate axes σ_1, σ_2, σ_3, but its cross section is no longer circular if J_3 participates in f.

When f depends on J_2, J_3 alone, it can be written in the form $f(\sigma_1 - \sigma_3, \sigma_2 - \sigma_3)$. Then a two-dimensional plot of the surface $f = 0$ is possible, with $\sigma_1 - \sigma_3$, $\sigma_2 - \sigma_3$ as coordinate axes (see Fig. 6.7:2).

Another way of representing a yield surface when it is unaffected by hydrostatic pressure is to *project the yield surface on the so-called π-plane,*

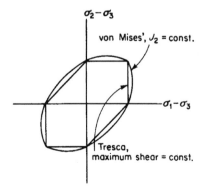

Fig. 6.7:2. Yield surface plotted on the plane of $(\sigma_1 - \sigma_3)$, $(\sigma_2 - \sigma_3)$.

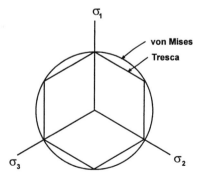

Fig. 6.7:3. Projection of yield surfaces on the π-plane.

$\sigma_1 + \sigma_2 + \sigma_3 = 0$. For example, von Mises's criterion would appear as a circle on this plane, and Tresca's criterion would appear as a regular hexagon, as shown in Fig. 6.7:3. The π-plane projection will be used extensively below in discussing the flow and the hardening rules in plasticity.

6.8. FURTHER EXAMPLES OF YIELD FUNCTIONS

Let σ'_{ij} stand for the stress deviation and J_2, J_3 for the second and third invariants of σ'_{ij}, respectively. Von Mises' and Tresca's yield functions discussed in Sec. 6.5 imply initial isotropy, and equality of tensile and compressive yield stresses at all stages of the deformation. The more general expression

$$(1) \qquad\qquad f = F(J_2, J_3) - k^2$$

still contains no *Bauschinger effect*. A simple example is

$$(2) \qquad\qquad F = J_2^3 - cJ_3^2 \,.$$

A good correlation with Osgood's experimental data[6.3] (1947) is obtained by taking $c = 2.25$.

To include Bauschinger effect, but preserve isotropy, the following yield functions have been suggested.

$$(3) \qquad f = F(J_2) - m\sigma'_{ij}e_{ij}^{(p)} - k^2 \,, \qquad m, \text{ a constant} \,,$$

$$(4) \qquad f = F(J_2, J_3) - m\sigma'_{ij}e_{ij}^{(p)} - k^2 \,,$$

$$(5) \qquad f = F(J_2, J_3) - H(J_2, J_3)\sigma'_{ij}e_{ij}^{(p)} - k^2 \,,$$

$$(6) \qquad f = F(J_2, J_3) - [P(J_2, J_3)\sigma'_{ij} + Q(J_2, J_3)t_{ij}]e_{ij}^{(p)} - k^2 \,,$$

where
$$(7) \qquad\qquad t_{ij} = \sigma'_{ik}\sigma'_{kj} - \frac{2}{3}J_2\delta_{ij} \,.$$

Anisotropic Materials. As mentioned before, for general *anisotropic* materials, the yield criterion has to be expressed in terms of the nine components of the stress tensor σ. Following Betten[6.6] (1982, 1988), we assume the *initial yield function* in a quadratic form

$$(8) \qquad\qquad f(H_{ijkl}\sigma_{ij}\sigma_{kl}) = 0 \,,$$

where H_{ijkl} are the components of a fourth-order material tensor. Bauschinger effect is still excluded. Due to the symmetry conditions

$$H_{ijkl} = H_{klij} = H_{jikl} = H_{ijlk} \,,$$

then H_{ijkl} are composed of 21 independent constants. If we assume incompressibility for the plastic deformation, the number of independent constants reduces to 15 and the yield criterion can be written as

$$(9) \qquad\qquad f(H_{ijkl}\sigma'_{ij}\sigma'_{kl}) = 0\,.$$

Several examples follow.

Hill's Criterion for Anisotropic Materials. Equation (9) reduces to Hill's yield criterion, if

$$(10) \qquad\qquad 2f(H_{ijkl}\sigma'_{ij}\sigma'_{kl}) = H_{ijkl}\sigma'_{ij}\sigma'_{kl} - 1 = 0\,,$$

where

$$
\begin{aligned}
& H_{1111} = G + H\,, \qquad H_{2222} = H + F\,, \qquad H_{3333} = F + G\,,\\
(11)\quad & H_{1122} = -H\,, \qquad\;\; H_{2233} = -F\,, \qquad\;\; H_{3311} = -G\,,\\
& H_{2323} = \frac{L}{2}\,, \qquad\;\; H_{3131} = \frac{M}{2}\,, \qquad\;\; H_{1212} = \frac{N}{2}\,,
\end{aligned}
$$

in which F, G, H, L, M, and N are material constants. Written out *in extenso*, Hill's equation is in the form

$$
\begin{aligned}
(12)\quad 2f(\sigma_{ij}) = & F(\sigma_{yy} - \sigma_{zz})^2 + G(\sigma_{zz} - \sigma_{xx})^2 + H(\sigma_{xx} - \sigma_{yy})^2\\
& + 2L\sigma_{yz}^2 + 2M\sigma_{zx}^2 + 2N\sigma_{xy}^2 - 1\,.
\end{aligned}
$$

Mohr–Coulomb Criterion for Pressure-Sensitive Materials. A yield criterion depending on the mean stress is necessary when it applies to soils, rocks, concrete, or porous materials. One such criterion occurs in the Mohr-Coulomb theory of rupture. This criterion postulates that yield occurs in a body on a plane on which the normal and shear stresses reach a critical combination, as in dry friction between surfaces. The critical condition can be expressed by two bounding curves represented by the equations

$$(13) \qquad\qquad f = \tau \mp g(\sigma) = 0\,,$$

where σ and τ are the normal and shear stresses on the failure plane. If the state of stress (σ, τ) is described by points in a shaded area bounded by the three Mohr's circles shown in Fig. 3.9:1 in Chapter 3, then the bounding curves $\tau = \pm g(\sigma)$ can be drawn on the (σ, τ) plane. A material is safe if its state of stress lies within these bounding curves. The state of stress is critical if the bounding curves become tangent to the largest of the three Mohr's circles.

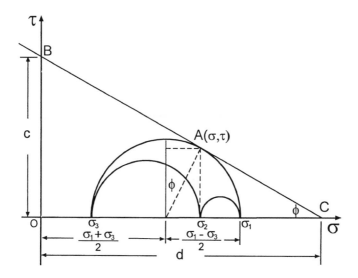

Fig. 6.8:1. Mohr–Coulomb yield criterion.

A special case of the Mohr–Coulomb criterion is that the bounding curves are straight lines,

$$(14) \qquad\qquad g(\sigma) = c - \sigma \tan \phi$$

with c being a material constant and ϕ the angle of internal friction. On the (σ, τ) plane, we can represent the criterion graphically by the straight line BC tangent to the largest Mohr's circle. See Fig. 6.8:1. Let σ_1, σ_2, σ_3 be the principal stresses with $\sigma_1 \geq \sigma_2 \geq \sigma_3$, then

$$\frac{\sigma_1 - \sigma_3}{2} = \left(d - \frac{\sigma_1 + \sigma_3}{2}\right) \sin \phi = \left(c \cot \phi - \frac{\sigma_1 + \sigma_3}{2}\right) \sin \phi .$$

Hence the yield surface Eq. (13) can be expressed in terms of the maximum and minimum principal stresses σ_1, σ_3 as follows:

$$(15) \qquad\qquad f = \frac{\sigma_1 - \sigma_3}{2} + \frac{\sigma_1 + \sigma_3}{2} \sin \phi - c \cos \phi = 0 .$$

From the Mohr's circle, we have

$$(16) \qquad\qquad \sigma = \frac{\sigma_1 + \sigma_3}{2} + \frac{\sigma_1 - \sigma_3}{2} \sin \phi ,$$

$$(17) \qquad\qquad \tau = \frac{\sigma_1 - \sigma_3}{2} \cos \phi .$$

Eliminating σ_1, σ_3, we can rewrite the yield surface Eq. (15) in the form

$$\tau = c - \sigma \tan\phi,$$

which reduces to the Tresca criterion if $\phi = 0$ and $c = \tau_Y = \sigma_Y/\sqrt{3}$. Let

(18)
$$\sigma_t = \frac{2c\cos\phi}{1 + \sin\phi},$$

(19)
$$\sigma_c = \frac{2c\cos\phi}{1 - \sin\phi}.$$

Equation (15) becomes

(20)
$$\frac{\sigma_1}{\sigma_t} - \frac{\sigma_3}{\sigma_c} = 1,$$

in which σ_t, σ_c are identified as the yield strengths in tension and in compression, respectively. Note that $\sigma_t < \sigma_c$. From Eqs. (18) and (19), we can express the parameters c, ϕ in terms of σ_t, σ_c:

$$\phi = \sin^{-1}\left(\frac{\sigma_c - \sigma_t}{\sigma_c + \sigma_t}\right) \quad \text{with} \quad 0 \le \phi < \frac{\pi}{2}, \quad c = \frac{1}{2}\sqrt{\sigma_c\sigma_t}.$$

Thus c, ϕ can be obtained from measured values of σ_t, σ_c in uniaxial tension and compression tests.

Equation (20) is a straight line between the positive σ_1-axis and the negative σ_3-axis on the octahedral plane for a fixed first stress invariant I_1. The line intersects the positive σ_1-axis at

(21)
$$\sigma_1 = (\sigma_1)_0 = \frac{4c\cos\phi + I_1(1 - \sin\phi)}{3 + \sin\phi}(> 0),$$

which is derived by a substitution of $2\sigma_3 = I_1 - \sigma_1$ into Eq. (15). In other words, at the intersection, $\sigma_2 = \sigma_3$ and $\sigma_1 + \sigma_2 + \sigma_3 = I_1$. Similarly we find the intersection of the line with the negative σ_3-axis to be

(22)
$$\sigma_3 = -(\sigma_3)_0 = -\frac{4c\cos\phi - I_1(1 + \sin\phi)}{3 - \sin\phi}(< 0),$$

which is obtained by substituting $2\sigma_1 = I_1 - \sigma_3$ in Eq. (15), i.e. $\sigma_2 = \sigma_1$ and $\sigma_1 + \sigma_2 + \sigma_3 = I_1$. It can be shown that $(\sigma_3)_0 \ge (\sigma_1)_0$.

The yield surface is an irregular hexagon on the *octahedral plane* or *π-plane* as shown in Fig. 6.8:2. The yield locus is symmetric about the $\sigma_1, \sigma_2, \sigma_3$ axes and can be obtained by the symmetry condition.

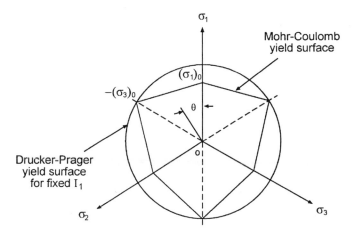

Fig. 6.8:2. Mohr–Coulomb and Drucker–Prager yield surfaces [Eq. (6.8:27)] on the π-plane for $\kappa - c_1 I_1 = (\sigma_3)_0/\sqrt{3}$.

For two-dimensional stress with $\sigma_3 = 0$, the yield criterion can be written

$$\sigma_1 = \sigma_t \qquad \text{if } \sigma_1 > \sigma_2 > 0,$$

$$\sigma_2 = -\sigma_c \qquad \text{if } \sigma_2 < \sigma_1 < 0,$$

$$\frac{\sigma_1}{\sigma_t} - \frac{\sigma_2}{\sigma_c} = 1 \qquad \text{if } \sigma_1 > 0 > \sigma_2.$$

These equations represent an irregular hexagon as shown in Fig. 6.8:3.

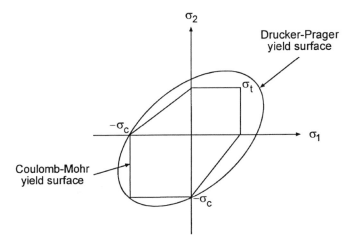

Fig. 6.8:3. Two-dimensional Mohr–Coulomb and Drucker–Prager yield surfaces [Eq. (6.8:27)] for $\kappa - c_1 I_1 = \sigma_c/\sqrt{3}$.

Using the results of Problem 6.3 at the end of this section, we can write the yield surface Eq. (15) in the form

$$(23) \quad f = \frac{I_1}{3}\sin\phi + \sqrt{J_2}\left[\sin(\theta + 60°) + \frac{1}{\sqrt{3}}\cos(\theta + 60°)\sin\phi\right] - c\cos\phi = 0$$

where I_1 is the first stress invariant, J_2 is second deviatoric stress invariant given by Eq. (3.10:8), and θ is the angle between the principal deviatoric stress axis σ_1' and the projection vector of the stress state on the octahedral plane. Equation (23) is valid only for $0 \le \theta \le 60°$ (in the region between the positive σ_1-axis and the negative σ_3-axis where $\sigma_1 \ge \sigma_2 \ge \sigma_3$). For other values of θ, the yield locus is obtained by the symmetric condition about the principal axes $\sigma_1, \sigma_2, \sigma_3$.

Betten's Form for Pressure Sensitive Materials. Betten (1982) introduced the following form of yield function for pressure sensitive materials,

$$(24) \qquad f(H_{ijkl}\sigma_{ij}\sigma_{kl}) = H_{ijkl}(\sigma_{ij} - \delta_{ij}b)(\sigma_{kl} - \delta_{kl}b) - \kappa^2\,.$$

If the material is isotropic, H_{ijkl} can be written as

$$(25) \qquad H_{ijkl} = a_1\delta_{ij}\delta_{kl} + a_2(\delta_{ik}\delta_{jl} + \delta_{il}\delta_{jk})\,.$$

A substitution of H_{ijkl} above into Eq. (24) yields

$$(26) \qquad f = a_2 J_2 + (3a_1 + 2a_2)\left[\frac{1}{3}I_1^2 - 2bI_1 + 3b^2\right] - \kappa^2$$

$$= a_2(J_2 + b_1 I_1^2 + b_2 I_1 + b_3) = 0\,,$$

where b_1, b_2, and b_3 are material constants and I_1 is the first stress invariant.

Drucker–Prager Criterion. Equation (26) reduces to the *Drucker–Prager yield function*

$$(27) \qquad\qquad f = \sqrt{J_2} + c_1 I_1 - \kappa = 0\,,$$

if $b_1 = -c_1^2$, $b_2 = 2c_1\kappa$, and $b_3 = -\kappa^2$. The yield locus on any octahedral plane (fixed I_1) is a circle. If the radius of the circle is $(\sigma_3)_0$ as given in Eq. (22) on all octahedral planes, we have

$$\kappa = \frac{4c\cos\phi}{\sqrt{3}(3 - \sin\phi)}\,, \qquad c_1 = \frac{1}{\sqrt{3}}\frac{\sin\phi + 1}{3 - \sin\phi}\,.$$

If the radius equals to $(\sigma_1)_0$ as defined in Eq. (21) on all octahedral planes, we find

$$\kappa = \frac{4c\cos\phi}{\sqrt{3}(3+\sin\phi)}, \qquad c_1 = \frac{1}{\sqrt{3}}\frac{\sin\phi - 1}{3+\sin\phi}.$$

The yield functions discussed up to this point contain no Bauschinger effect. However, the yield function

(28) $$f(H_{ijkl}\sigma_{ij}\sigma_{kl}) = H_{ijkl}(\sigma_{ij} - me_{ij}^{(p)})(\sigma_{kl} - me_{kl}^{(p)}) - \kappa^2$$

exhibits the Bauschinger effect, which is controlled by the constant m. The initial anisotropy is not preserved during deformation. Yield functions such as this will be discussed more fully in Sec. 6.12.

All these examples use a single analytic function to represent the entire yield surface. If a yield surface is composed of piecewise smooth surfaces, which meet to form corners, it would be convenient to use a separate expression for each of these piecewise smooth surfaces. This concept leads to Koiter's generalization (see Sec. 6.10).

Problem 6.3.   Let σ_1', σ_2', σ_3' be the projections of the principal-stress axes $\sigma_1, \sigma_2, \sigma_3$ on an octahedral plane and n_1' be the unit vector in σ_1' direction. The components of n_1' with respect to the principal axes can be expressed in the form $[\cos(90° - \alpha), -a, -a]$, where $90° - \alpha$ is the angle between the σ_1- and σ_1'-axes with $\cos\alpha = 1/\sqrt{3}$ and "a" is a constant. (An octahedral plane is the plane whose normal makes equal angles α with each of the principal axes. Also n_1' makes equal angles with the σ_2', σ_3'-axes.)

(a) Show that $n_1' = \frac{1}{\sqrt{6}}(2, -1, -1)$.

(b) Let S_i be the principal deviatoric stress corresponding to σ_i. Show that

$$\begin{bmatrix} \sigma_1 \\ \sigma_2 \\ \sigma_3 \end{bmatrix} = \frac{I_1}{3}\begin{bmatrix} 1 \\ 1 \\ 1 \end{bmatrix} + \begin{bmatrix} S_1 \\ S_2 \\ S_3 \end{bmatrix} = \frac{I_1}{3}\begin{bmatrix} 1 \\ 1 \\ 1 \end{bmatrix} + 2\sqrt{\frac{J_2}{3}}\begin{bmatrix} \cos\theta \\ \cos(120° - \theta) \\ \cos(120° + \theta) \end{bmatrix}$$

where I_1 is the first stress invariant, $J_2 = (S_1^2 + S_2^2 + S_3^2)/2$ is the second deviatoric-stress invariant, and θ is the angle between the σ_1'-axes and the vector representing the projection of the stress state on the octahedral plane. This result shows that one can choose I_1, J_2, θ as the coordinate axes to represent a stress state. [*Hint*: The components of the projection of a stress state on the octahedral plane with respect to the $\sigma_1, \sigma_2, \sigma_3$-axes are (S_1, S_2, S_3) and the magnitude of the projection is $\sqrt{S_1^2 + S_2^2 + S_3^2}$ $(= \sqrt{2J_2})$. Also $\sqrt{2J_2}\cos\theta = (S_1, S_2, S_3) \cdot n_1'$.]

Problem 6.4. Show that the line given by Eq. (20) intersects the σ_1, σ_3-axes at $(\sigma_1)_0, (\sigma_3)_0$ as given in Eqs. (21) and (22), respectively. [*Hint*: At $(\sigma_1)_0$, the deviatoric stress $S_3 = -S_1/2$. Thus, $\sigma_1 = \frac{I_1}{3} + S_1$ and $\sigma_3 = \frac{I_1}{3} - \frac{S_1}{2}$ at $\sigma_1 = (\sigma_1)_0$. Show that $S_1 = \frac{4c \cos \phi}{3 + \sin \phi} - \frac{4I_1 \sin \phi}{3(3 + \sin \phi)}$ from Eq. (6.8:15). Similarly at $(\sigma_3)_0$, $S_1 = -\frac{S_3}{2}(< 0)$, $\sigma_1 = \frac{I_1}{3} - \frac{S_3}{2}$ and $\sigma_3 = \frac{I_1}{3} + S_3$. Show that $S_3 = -\frac{4c \cos \phi}{3 - \sin \phi} + \frac{4I_1 \sin \phi}{3(3 - \sin \phi)}$.]

Problem 6.5. Show that the yield stresses for the Drucker–Prager criterion in simple shear, tension and compression are, c_2, $\sqrt{3}c_2/(1 + \sqrt{3}c_1)$ and $\sqrt{3}c_2/(1 - \sqrt{3}c_1)$, respectively. For this criterion to be physically meaningful, we must have $c_2 > 0$ and $\sqrt{3}c_1 < 1$.

Problem 6.6. The Mises–Schleicher criterion is

$$f(\sigma_{ij}) = 3J_2 + (\sigma_c - \sigma_t)I_1 - \sigma_c \sigma_t = 0 \,.$$

Show that σ_c and σ_t are, respectively, the compressive and tensile yield stresses in uniaxial tests.

6.9. WORK HARDENING — DRUCKER'S HYPOTHESIS AND DEFINITION

Work hardening in a simple tension experiment means that the stress is a monotonically increasing function of increasing strain. To generalize this concept, D. C. Drucker[6.3] (1951) considers the work done on a material element in equilibrium by an external agency, which slowly applies a set of self-equilibrating forces and then slowly removes them. This external agency is to be understood as entirely separate and distinct from the agency which causes the existing state of stress. This process of application and removal of the additional stress is called a *stress cycle*. Removing the additional stress enables the stress of the body to return to the original stress state, but the strain state can be different if plastic deformation occurred during the stress cycle. *Work hardening is then defined to mean that, for all such added sets of stresses, positive work is done by the external agency during the application of the stresses, and the net work performed by it over the cycle of application and removal is either zero or positive.* Rephrased, Drucker's definition of work hardening means that *useful net energy over and above the elastic energy cannot be extracted from the material and the system of forces acting upon it.*

Consider a volume of material in which there is a homogeneous state of stress σ_{ij} and strain e_{ij}. Suppose that an external agency applies a small surface traction, which alters the stress at each point by $d\sigma_{ij}$, and the strain by de_{ij}. On removal of the small traction, $d\sigma_{ij}$ returns to zero, and a strain de_{ij}^e is recovered. Then, according to the definition above, *the material is said to be work-hardening if the following two conditions hold true.*

(1) $d\sigma_{ij} de_{ij} > 0$,   upon loading;

(2) $d\sigma_{ij}(de_{ij} - de_{ij}^{(e)}) \geq 0$,   on completing a cycle.

Let $de_{ij}^{(p)}$ denote the *plastic strain increment*, which is not recovered by the process named above. Then, since $de_{ij}^{(p)} = de_{ij} - de_{ij}^{(e)}$, the second condition may be written as

(3) $d\sigma_{ij} de_{ij}^{(p)} \geq 0$.

Drucker extended the definition of work-hardening to allow for a finite $d\sigma_{ij}$ produced by the external agency. In fact, the initial stress, say $\bar{\sigma}_{ij}$, may at any point be inside or on the yield surface far away from σ_{ij}. The work per unit volume done by the external agency is $(\sigma_{ij} - \bar{\sigma}_{ij})de_{ij}^{(p)}$. Equation (3) is replaced by

(4) $(\sigma_{ij} - \bar{\sigma}_{ij})\dot{e}_{ij}^{(p)} \geq 0$.

Drucker's hypothesis Eq. (3) or Eq. (4) can be satisfied only by materials whose subsequent yield strength increases with deformation. Any material on which an external agency does positive work during an elastic-plastic stress cycle is called a *hardening material*. Otherwise, it is considered as *nonhardening* or *work-softening material*. It can be shown that Eq. (3) or Eq. (4) holds for work-softening and perfectly plastic materials [von Mises (1928), Bishop and Hill (1951)] under Il'iushin's postulate of plasticity in strain space to be considered in Sec. 6.14.

Equation (4) is also called the *principle of maximum plastic dissipation*. It can be written in the form

(5) $\sigma_{ij}\dot{e}_{ij}^{(p)} = D(\dot{e}_{ij}^{(p)}, \xi_i) \geq \bar{\sigma}_{ij}\dot{e}_{ij}^{(p)}$,

where $D(\dot{e}_{ij}^{(p)}, \xi_i)$ depends on the plastic strain-rates $\dot{e}_{ij}^{(p)}$ and the internal variables ξ_i only. Equation (5) will be used for limit analysis to be discussed in Chapter 10.

6.10. IDEAL PLASTICITY

According to Drucker's definition of strain hardening, we can define *ideal plasticity* as a plastic deformation without strain hardening. It is mathematically specified by the condition that, when plastic deformation occurs, the equality sign prevails in Eq. (6.9:3):

(1) $d\sigma_{ij} de_{ij}^{(p)} = 0$,

and that the yield function is unaffected by $e_{ij}^{(p)}$. The differentials $d\sigma_{ij}$ and $de_{ij}^{(p)}$ must be interpreted as in Sec. 6.9.

As an application of this definition, we shall derive the *flow rule* during an ideal plastic deformation. We notice that the yield function furnishes a criterion to tell whether yielding occurs or not. If yielding does occur, we need further information concerning the increment or rate of deformation in order to complete the description of the material behavior. In other words, we need a flow rule.

Now, for an ideal plastic material, we assume that a yield function $f(\sigma_{ij})$ exists, which is a function of the stresses σ_{ij} and not of the strains $e_{ij}^{(p)}$, such that $f(\sigma_{ij}) \leq 0$ prevails; and

$$(2) \qquad \dot{e}_{ij}^{(p)} \neq 0 \qquad \text{only if} \qquad f(\sigma_{ij}) = 0 \,.$$

Since f is assumed to be a function of σ_{ij} only, any change in stresses during plastic flow must satisfy the relation

$$(3) \qquad df = \frac{\partial f}{\partial \sigma_{ij}} d\sigma_{ij} = 0 \,.$$

Equation (3) is often called *the consistency condition for ideal plasticity.* When plastic flow occurs, Eq. (1) holds. A comparison of Eqs. (1) and (3) shows that

$$(4) \; \blacktriangle \qquad de_{ij}^{(p)} = d\Lambda \frac{\partial f}{\partial \sigma_{ij}} \,,$$

where $d\Lambda$ is an arbitrary constant of proportionality. We can also write Eq. (4) in the rate form:

$$(5) \qquad \dot{e}_{ij}^{(p)} = \frac{1}{\mu} \frac{\partial f}{\partial \sigma_{ij}} \,.$$

If $\partial f / \partial \sigma_{ij}$ has the dimension of stress, then μ has the physical dimensions of the coefficient of viscosity, but it is certainly not a material constant, and may vary during the deformation. The sign of $d\Lambda$ or μ is restricted by the condition that plastic flow always involves dissipation of mechanical energy, a condition which may be written as

$$(6) \qquad \dot{W} = \sigma_{ij}\dot{e}_{ij}^{(p)} > 0 \,.$$

Equation (4) gives *the rule of plastic flow* (plastic strain increments) in ideal plasticity. It is a prototype of *the theory of plastic potential* developed by Richard von Mises[6.3] (1928). The general theory is discussed in Sec. 6.11 *et seq.*

Prager's Geometric Interpretation.   An interesting geometric interpretation of Eqs. (4) and (5) was given by Prager. The formula $f(\sigma_{ij}) = 0$

defines a surface in the nine-dimensional stress space with $\sigma_{ij}(i, j = 1, 2, 3)$ as coordinates. The outward normal vector to this surface has the components $\partial f/\partial \sigma_{ij}$. Equation (5) states thus that the vector of plastic deformation rate $\dot{e}_{ij}^{(p)}$ is normal to the surface $f = 0$ in the stress space. So, during unloading,

$$f = 0, \qquad \frac{\partial f}{\partial \sigma_{ij}} d\sigma_{ij} < 0,$$

the stress increment $d\sigma_{ij}$ is pointing inward from the yield surface, while during loading or neutral loading,

$$f = 0, \qquad \frac{\partial f}{\partial \sigma_{ij}} d\sigma_{ij} = 0,$$

i.e., the stress increment $d\sigma_{ij}$ is on the tangential plane at a stress point on the yield surface. Because the yield surface is fixed for ideal plasticity, $d\sigma_{ij}$ cannot point outward.

Incremental Stress-Strain Relationship in Plastic Flow. Using the basic assumption that the total strain increment can be decomposed into an elastic and a plastic component,

$$(7) \qquad de_{ij} = de_{ij}^{(e)} + de_{ij}^{(p)},$$

together with the flow rule and the consistency condition, we can determine $d\Lambda$ in Eqs. (4) and (5) in terms of the total strain increment and the current state of stress. We have

$$(8) \qquad d\sigma_{ij} = D_{ijkl}[de_{kl} - de_{kl}^{(p)}] = D_{ijkl}de_{kl} - D_{ijkl}\frac{\partial f}{\partial \sigma_{kl}}d\Lambda.$$

Multiplying both sides of Eq. (8) by $\frac{\partial f}{\partial \sigma_{ij}}$, we obtain

$$(9) \qquad \frac{\partial f}{\partial \sigma_{ij}}d\sigma_{ij} = \frac{\partial f}{\partial \sigma_{ij}}D_{ijkl}de_{kl} - D_{ijkl}\frac{\partial f}{\partial \sigma_{ij}}\frac{\partial f}{\partial \sigma_{kl}}d\Lambda.$$

Since $\frac{\partial f}{\partial \sigma_{ij}}d\sigma_{ij} = 0$ from the consistency condition, we can solve for $d\Lambda$

$$(10) \; \blacktriangle \qquad d\Lambda = D_{ijkl}de_{kl}\frac{\partial f}{\partial \sigma_{ij}}\left(D_{mnrs}\frac{\partial f}{\partial \sigma_{mn}}\frac{\partial f}{\partial \sigma_{rs}}\right)^{-1}.$$

A substitution of Eq. (10) into Eq. (8) establishes the *general stress-strain increment relationship*

$$(11) \qquad d\sigma_{ij} = D_{ijkl}[de_{kl} - de_{kl}^{(p)}] = D_{ijkl}^{ep}de_{kl},$$

where

(12) $D_{ijkl}^{ep} = D_{ijkl} - D_{ijtu}\dfrac{\partial f}{\partial \sigma_{tu}}\dfrac{\partial f}{\partial \sigma_{qp}}D_{qpkl}\left(D_{mnrs}\dfrac{\partial f}{\partial \sigma_{mn}}\dfrac{\partial f}{\partial \sigma_{rs}}\right)^{-1}.$

For isotropic materials, $D_{ijkl} = \lambda\delta_{ij}\delta_{kl} + G(\delta_{ik}\delta_{jl} + \delta_{il}\delta_{jk})$, Eq. (11) reduces to

(13) $d\sigma_{ij} = \lambda de_{\alpha\alpha}\delta_{ij} + 2G de_{ij}$
$$- \frac{(\lambda n_{kk}\delta_{ij} + 2G n_{ij})(\lambda n_{\beta\beta}de_{\alpha\alpha} + 2G n_{kl}de_{kl})}{\lambda n_{mm}n_{ss} + 2G},$$

where λ is the Lamé constant, and

(14) $n_{ij} = \dfrac{\partial f}{\partial \sigma_{ij}}\left(\dfrac{\partial f}{\partial \sigma_{kl}}\dfrac{\partial f}{\partial \sigma_{kl}}\right)^{-1/2},$

is a unit outward normal to the yield surface. For materials with plastic deformation insensitive to hydrostatic pressure, we have $n_{kk} = 0$ and Eq. (13) reduces to

(15) $d\sigma_{ij} = \lambda de_{\alpha\alpha}\delta_{ij} + 2G de_{ij} - 2G n_{ij}n_{kl}de_{kl}.$

W. T. Koiter[6.3] (1953) generalized this theory of plasticity by allowing the yield limit to be specified by a set of yield functions,

$$f_1(\sigma_{ij}),\ f_2(\sigma_{ij}),\ldots,\ f_n(\sigma_{ij}).$$

A state of stress is said to be below the yield limit if all these functions are negative. For a state of stress at the *yield limit*, at least one yield function vanishes, while none has a value greater than zero.

If the functions $f_h = \cdots = f_m = 0$, $(1 \le h \le m \le n)$ whereas all other f's are negative, then the Koiter generalization of the flow rule given in Eq. (5) is

(16) $\dot{e}_{ij}^{(p)} = d\Lambda_h\dfrac{\partial f_h}{\partial \sigma_{ij}} + \cdots + d\Lambda_m\dfrac{\partial f_m}{\partial \sigma_{ij}},$

where $d\Lambda_h, \ldots, d\Lambda_m$ are nonnegative proportional factors. Thus, in the case of ideal plasticity, the basic concept leads at once to a general incremental stress-strain relationship.

Example 1. Under the von Mises yield condition, Eq. (6.5:1) and the assumption that the constant k is independent of plastic deformation, we can derive the flow rule below according to Eq. (5):

(17) $\dot{e}_{ij}^{(p)} = \dot{\Lambda}\sigma_{ij}',$

which is the rule presented in Sec. 6.3.

Example 2. Tresca's yield condition can be expressed in terms of Koiter's generalized plastic potential by defining f's as the followings:

$$
\begin{array}{ll}
f_1 = \sigma_2 - \sigma_3 - 2k, & f_2 = \sigma_3 - \sigma_1 - 2k, \\
f_3 = \sigma_1 - \sigma_2 - 2k, & f_4 = -(\sigma_2 - \sigma_3) - 2k, \\
f_5 = -(\sigma_3 - \sigma_1) - 2k, & f_6 = -(\sigma_1 - \sigma_2) - 2k,
\end{array}
\tag{18}
$$

where σ_1, σ_2, σ_3 are the principal stresses. A state of stress is below the yield limit if $f_1, \ldots, f_6$ are all negative. Yielding occurs when one or more of the f's are equal to zero. None of the f's can have a positive value. When yielding occurs, the flow rule is given by Eq. (7). For example, if $f_1 = 0$, while all the other f's are negative, then

$$
\dot{e}_2^{(p)} = \dot{\Lambda}_1, \quad \dot{e}_3^{(p)} = -\dot{\Lambda}_1, \quad \dot{e}_1^{(p)} = 0,
\tag{19}
$$

where $\dot{\Lambda}_1 > 0$, and e_1, e_2, e_3 are the principal strains corresponding to $\sigma_1, \sigma_2, \sigma_3$. The principal axes of the strain tensor coincide with those of the stress tensor under Tresca's condition.

Remark: Formal application of the method of derivation above to a simple load-deflection experiment may appear difficult. If we twist a tube of ideal plastic material in torsion, the yield condition according to von Mises' criterion is reached when the shearing stress $\tau = \kappa$ the yield stress. Plastic flow will continue with no possibility of increasing τ and κ. Hence, if we limit ourselves to torsion and apply Eqs. (1) and (2), we would have obtained $d\tau = 0$, which yields no useful information. To deduce anything significant we would have to consider adding other loads, for example, tension or internal pressure in the tube. These additional varieties of loads will alter the plastic flow, thus providing nontrivial changes $d\sigma_{ij}$ and de_{ij}^e to which the derivation above applies.

    Problem 6.7. A plane strain condition is defined as $e_{33} = e_{31} = e_{32} = 0$. This condition requires $de_{3i}^p = -de_{3i}^e$, for $i = 1, 2, 3$. Consider the von Mises criterion with the associated flow rule $de_{ij}^p = d\Lambda \sigma_{ij}'$. Show that

$$
\sigma_3 = \nu(\sigma_1 + \sigma_2)
$$

where σ_i are the principal stresses.

6.11. FLOW RULE FOR WORK HARDENING MATERIALS

    In this section, the Drucker's hypothesis discussed in Sec. 6.9 is taken to define work-hardening, and von Mises's plastic potential theory is taken as the framework to derive the flow rule. Von Mises[6.3] (1928) suggested that there exists a *plastic potential function* $h(\sigma_{ij})$ so that the plastic strain rate $\dot{e}_{ij}^{(p)}$ could be derived from

(1) $$\dot{e}_{ij}^{(p)} = \lambda \frac{\partial h}{\partial \sigma_{ij}},$$

where λ is a positive scalar factor. If the plastic potential is the same as the yield function: $h = f$, Eq. (1) is called the *associated flow rule*. On the other hand, if $h \neq f$, the flow rule is called *non-associated*. In the preceding section we have seen that $h = f$ for an ideal plasticity body. Now, we shall consider the more general case of work hardening materials by allowing the plastic potential to be a function of not only stresses but also temperature T and internal variables ξ_i. We shall show that under *Drucker's hypothesis* and the assumption that the elastic moduli D_{ijkl} are independent of the plastic deformation, the *yield function f itself is the plastic potential*. Experimental observations show that the associated flow rule characterizes the plastic deformation of metals quite well, but the nonassociated flow rule provides a better representation for the plastic deformation of porous materials such as rocks, concrete, and soils.

Consequences of Drucker's Hypothesis.   We shall prove the following consequences of Drucker's hypothesis from which the term work-hardening in defined:

    A. The yield surface and all subsequent loading surfaces must be *convex*.

    B. The plastic strain increment vector must be *normal to the loading surface* at a regular point, and it must lie between the adjacent normals to the loading surface at a corner of the surface.

    C. The rate of change of plastic strain must be a *linear function of the rate of change of the stress*.

Proof of A.   To facilitate the proof,[3] let us think of an increment of stress $d\sigma_{ij}$ as the components of a vector $d\mathbf{S}$ in the nine-dimensional stress space, and the corresponding plastic strains $de_{ij}^{(p)}$ as the components of a vector $d\mathbf{E}$ in the same (stress) space. Then, by Eq. (6.9:3), we have

(2a) $$d\mathbf{S} \cdot d\mathbf{E} = |d\mathbf{S}||d\mathbf{E}| \cos \psi > 0,$$

which implies that

(2b) $$-\frac{\pi}{2} \leq \psi \leq \frac{\pi}{2},$$

i.e., the angle between $d\mathbf{S}$ and $d\mathbf{E}$ must be acute.

---

[3]The following explanation follows that of P. M. Naghdi[6.3] (1960).

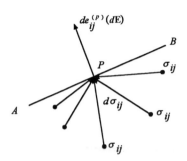

Fig. 6.11:1. Possible stress increments corresponding to a plastic strain increment $d\mathbf{E}$.

Now let P be a regular (smooth) point on the yield surface and $d\mathbf{E}$ be a plastic strain vector at P. According to Eqs. (2a) and (2b), all stress increments $d\mathbf{S}$ that will produce $d\mathbf{E}$, i.e., vectors that represent loading in the stress space and end at P, and corresponding to $d\mathbf{E}$, must form an acute angle with $d\mathbf{E}$. If we represent a hyper-plane normal to $d\mathbf{E}$ by AB in Fig. 6.11:1, then the vectors $d\mathbf{S}(= d\sigma_{ij})$ must all originate in one side of the hyper-plane AB. The initial points of the $d\mathbf{S}$ vectors, represented by stress states inside or on the yield surface, must thus all lie on one side of AB. However, $d\mathbf{S}$ are loading vectors (since $d\mathbf{E}$ exists); they are outward vectors whose directions are bounded by the tangent plane of the yield surface (see Sec. 6.6). Hence, the hyperplane AB must be tangent to the loading surface. Since $d\mathbf{E}$ is normal to AB, it is also normal to the yield surface at the point of tangency. Furthermore, since the yield surface lies on one side of the tangent plane, the yield surface is convex at P. Finally, since P is an arbitrary point on the yield surface, the convexity of the entire yield surface is established.

Proof of B. Since at a regular point on a surface there is a unique tangent plane, the hyper-plane is unique at a regular point of the yield surface. Thus, the direction of $d\mathbf{E}$ normal to the hyper-plane is also unique. In other words, at a smooth point of the yield surface the direction of $d\mathbf{E}$ (i.e. $de_{ij}^{(p)}$) is independent of the direction of $d\sigma_{ij}$. At a corner on the yield surface, there can have been more than one limiting tangent plane; convexity must still hold, but the direction of strain increment $de_{ij}^{(p)}$ may depend on the direction of the loading vector $d\sigma_{ij}$.

Proof of C. The truth of C is exhibited by Eqs. (14), (18) and (19) below. To derive these equations, we need the following formulas for normality and consistency.

Normality Condition. Under Drucker's hypothesis, the *normality of the plastic strain rate vector to the loading surface* at a smooth point of the loading surface requires that

(3) ▲
$$\dot{e}_{ij}^{(p)} = \dot{\Lambda} n_{ij}$$

where $\dot{\Lambda}$ is a proportional scalar factor, which can be a function of stress, strain, strain history, temperature and internal variables, and

(4)
$$n_{ij} = \frac{\partial f}{\partial \sigma_{ij}} \left(\frac{\partial f}{\partial \sigma_{kl}} \frac{\partial f}{\partial \sigma_{kl}} \right)^{-1/2}$$

is a unit normal to the yield surface at the loading point. Since the work done by an external agency during loading must be positive, it is easy to show that $\dot{\Lambda}$ must be nonnegative. Equation (1) is similar to Eq. (6.10:5); thus, the loading function h plays the role of *plastic potential*. We transform Eq. (1) to the form of Eq. (3) so that the *flow equation* takes on the form

(5)
$$\dot{e}_{ij}^{(p)} = \dot{\Lambda} h_{ij} \,,$$

where h_{ij} is a unit normal to a plastic potential $h(\sigma_{ij}, T, \xi_i)$ at a smooth point of the surface,

(6)
$$h_{ij} = \frac{\partial h}{\partial \sigma_{ij}} \left(\frac{\partial h}{\partial \sigma_{kl}} \frac{\partial h}{\partial \sigma_{kl}} \right)^{-1/2} \,.$$

Comparing Eqs. (3) and (5), we obtian

(7)
$$h_{ij} = n_{ij} \,,$$

(8)
$$h = f(\sigma_{ij}, T, \xi_i) \,.$$

In other words, Eq. (5) is an associated flow rule.

Consistency Condition.   As plastic flow proceeds, the subsequent loading surfaces pass through the loading point, we must have

(9)
$$f(\sigma_{ij}, T, \xi_i) = 0 \,.$$

Expressing the *internal variable-rates* in the form

(10)
$$\dot{\xi}_i = \dot{\Lambda} h_i \,,$$

in which h_i is a known function of σ_{ij}, T, and ξ_i, we must have, during loading,

(11) ▲
$$\dot{f} = \frac{\partial f}{\partial \sigma_{ij}} \dot{\sigma}_{ij} + \dot{\Lambda} \frac{\partial f}{\partial \xi_i} h_i = 0 \,,$$

or

(12)
$$\dot{\Lambda} = \frac{n_{ij} \dot{\sigma}_{ij}}{K^p} \,,$$

(13)
$$K^p = -h_i \frac{\partial f}{\partial \xi_i} \left(\frac{\partial f}{\partial \sigma_{kl}} \frac{\partial f}{\partial \sigma_{kl}} \right)^{-1/2} \,,$$

where $K^P(> 0)$ is called the *plastic modulus*. Prager named Eq. (11) the *consistency condition*, which means that loading from a plastic state must lead to another plastic state. Combining Eq. (12) with Eq. (5), we find the *flow rule* in the form

(14) ▲ $$\dot{e}_{ij}^{(p)} = \frac{h_{ij} n_{kl} \dot{\sigma}_{kl}}{K^P} \, .$$

Since $\dot{e}_{ij}^{(p)} = 0$ during unloading, $[n_{ij}\dot{\sigma}_{ij} < 0$, see Eq. (6.6:3)], one may write the *flow rule* in a slightly different form

(15) $$\dot{e}_{ij}^{(p)} = \frac{h_{ij} n_{kl} \dot{\sigma}_{kl}}{K^P} H(n_{mn}\dot{\sigma}_{mn}) \, ,$$

where H is the *Heaviside function* defined as

$$H(x) = 1, \quad \text{if} \quad x \geq 0 \, ,$$
$$= 0, \quad \text{if} \quad x < 0 \, .$$

In other words, it is the unit-step function denoted by $\mathbf{1}(x)$ elsewhere in this book.

Incremental Strain-Stress Relation. Consider the yield surface

(16) $$f(\sigma_{ij}, e_{ij}^{(p)}, \kappa) = 0 \, .$$

The internal variables are $e_{ij}^{(p)}$ and κ, with the latter being a function of $e_{ij}^{(p)}$ also. The flow rule is given by Eq. (5) and the equations for the internal variable-rates are

$$\dot{\kappa} = \frac{\partial \kappa}{\partial e_{kl}^{(p)}} \dot{e}_{kl}^{(p)} = \dot{\Lambda} \frac{\partial \kappa}{\partial e_{kl}}^{(p)} h_{kl} \, .$$

The consistency equation is

$$\dot{f} = \frac{\partial f}{\partial \sigma_{ij}} \dot{\sigma}_{ij} + \left(\frac{\partial f}{\partial e_{rs}^{(p)}} + \frac{\partial f}{\partial \kappa} \frac{\partial \kappa}{\partial e_{rs}^{(p)}} \right) h_{rs} \dot{\Lambda} = 0 \, .$$

We obtain $\dot{\Lambda}$ in the form of Eq. (12) with

(17) $$K^P = - \left(\frac{\partial f}{\partial e_{rs}^{(p)}} + \frac{\partial f}{\partial \kappa} \frac{\partial \kappa}{\partial e_{rs}^{(p)}} \right) h_{rs} \left(\frac{\partial f}{\partial \sigma_{kl}} \frac{\partial f}{\partial \sigma_{kl}} \right)^{-1/2} \, ,$$

for the plastic modulus. These results were first given by Prager[6.3] (1948) and Drucker[6.3] (1959). Equation (14) proves the linearity statement (item C) given in Drucker's hypothesis. Using

$$de_{ij} = D_{ijkl}^{-1} d\sigma_{kl} + de_{ij}^{(p)}$$

and Eq. (14) leads to the *incremental strain-stress relation* (the constitutive equations) for plastic deformation

(18) $$de_{ij} = [D_{ijkl}^{-1} + h_{ij}n_{kl}/K^p]d\sigma_{kl}, \quad \text{or}$$

$$d\sigma_{kl} = [D_{klrs}^{-1} + n_{kl}h_{rs}/K^p]^{-1}de_{rs} = D_{klrs}^{ep}de_{rs}$$

where D_{ijkl}^{-1}, often denoted by C_{ijkl} and called the *elastic flexibility tensor*, are the components of the inverse of the fourth-order elastic modulus tensor D_{ijkl}. Equation (18) breaks down for ideal plasticity because

$$\frac{\partial f}{\partial \xi_m} = 0, \quad \text{i.e.,} \quad \frac{\partial f}{\partial e_{mn}^{(p)}} = \frac{\partial \kappa}{\partial e_{mn}^{(p)}} = 0,$$

and K^p becomes zero. In this case we have to use the incremental stress-strain relation given in Eqs. (6.10:11) and (6.10:12).

Koiter's Generalization. For a loading surface with corners, one can use *Koiter's generalization* to define the flow rule. Such a surface is composed of a number of individual smooth loading surfaces f_r, which meet to form corners. Koiter[6.3] (1953) has shown that if the loading surfaces described by $f_r = 0$ act independently, the total plastic deformation can be written as the sum of contributions from certain of the f_r's, as follows:

(19) $$\dot{e}_{ij}^{(p)} = \sum_{r=1}^{n} C_r \dot{\Lambda}_r (n_{ij})_r (n_{kl})_r \dot{\sigma}_{kl},$$

where $\dot{\Lambda}_r$ are positive functions associated with f_r as defined in Eq. (12) and

(20) $$\begin{aligned} C_r = 0 \quad &\text{if} \quad f_r \leq 0, \quad \text{or} \quad (n_{kl})_r\dot{\sigma}_{kl} < 0, \\ C_r = 1 \quad &\text{if} \quad f_r = 0, \quad \text{and} \quad (n_{kl})_r\dot{\sigma}_{kl} \geq 0, \end{aligned}$$

in which $(n_{kl})_r$ are the components of the unit outward normal to f_r at the loading point. Equations (19) and (20) specify, of course, the condition of yielding and loading.

It should be remarked that the properties deduced above, namely, the convexity, normality, and linearity, follow Drucker's hypothesis — which is often interpreted as a statement that the material is stable. Hence, these properties holds only for the class of stable materials. Unstable materials do exist. Mild steel at its upper yield point (see the point A* in Fig. 6.4:1) is a well-known example. Other engineering materials such as rocks, concrete, and soils exhibit softening phenomena for which

$$d\sigma_{ij}de_{ij} < 0 \quad \text{and} \quad d\sigma_{ij}de_{ij}^{(p)} < 0.$$

A formulation of the plasticity theory in the strain space is needed to describe the softening behavior.

The normality of $de_{ij}^{(p)}$ and the convexity of the yield surface do not hold under Drucker's postulate, if D_{ijkl} is a function of $e_{ij}^{(p)}$, i.e., there is elastic-plastic coupling during the process of plastic deformation. This is also the case for softening materials. Il'iushin (1960) showed that Drucker's postulate results in

$$d\sigma_{ij}[dD_{ijkl}^{-1}\sigma_{kl} + de_{kl}^{(p)}] > 0$$

$$de_{ij}^{(p)} = d\Lambda\frac{\partial f}{\partial\sigma_{ij}} - \sigma_{kl}dD_{ijkl}^{-1}(e_{mn}^{(p)}).$$

Example. If we choose $f = J_2 - k$, where J_2 is the second invariant of the stress deviation, and k is a hardening parameter, which depends on plastic deformation, then the flow rule at yielding is

$$\dot{e}_{ij}^{(p)} = \hat{G}\frac{\partial f}{\partial\sigma_{ij}}\frac{\partial f}{\partial\sigma_{kl}}\dot{\sigma}_{kl} = \hat{G}\left(\frac{\partial f}{\partial J_2}\frac{\partial J_2}{\partial\sigma'_{ij}}\right)\left(\frac{\partial f}{\partial\sigma'_{kl}}\frac{d\sigma'_{kl}}{dt}\right)$$

$$= \hat{G}\sigma'_{kl}\sigma'_{rs}\dot{\sigma}'_{rs} = \hat{G}\sigma'_{kl}\dot{J}_2, \qquad \dot{J}_2 \geq 0$$

where from Eqs. (5), (12) and (17),

$$\hat{G} = \frac{1}{K^p}\left(\frac{\partial f}{\partial\sigma_{mn}}\frac{\partial f}{\partial\sigma_{mn}}\right)^{-1} = \left(\frac{\partial\kappa}{\partial e_{rs}^{(p)}}\sigma'_{rs}\right)^{-1}.$$

If we set $f = F(J_2, J_3) - \kappa$, and assume $\partial\kappa/\partial e_{mn}^{(p)} \neq 0$ for some m and n, then the flow rule is

$$\dot{e}_{ij}^{(p)} = \hat{G}\left[\frac{\partial F}{\partial J_2}\sigma'_{ij} + \frac{\partial F}{\partial J_3}t_{ij}\right]\dot{F}, \qquad \dot{F} \geq 0,$$

where $t_{ij} = \sigma'_{ik}\sigma'_{kj} - \frac{2}{3}J_2\delta_{ij}$. Note that $\partial\sigma_{kk}/\partial\sigma_{ij} = \delta_{ij}$.

Note that the yield-function of a work-hardening material depends on the plastic strain $e_{ij}^{(p)}$ in a significant manner, in contrast to an ideal plastic material, whose yield function is independent of the plastic strain.

6.12. SUBSEQUENT LOADING SURFACES — ISOTROPIC AND KINEMATIC HARDENING RULES

We have discussed yield surfaces and flow rules in previous sections. Now we must consider the third aspect: the determination of subsequent loading surfaces as plastic flow proceeds, i.e., to determine how the

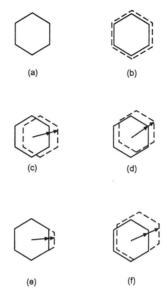

Fig. 6.12:1. Several hardening rules. (a) Initial yield condition (Tresca), (b) isotropic hardening, (c) kinematic hardening (Prager), (d) Kinematic hardening (Ziegler's modification), (e) independently acting plane loading surfaces, (f) interdependent plane loading surfaces. (From Naghdi,[6.3] 1960.)

internal variables including the plastic deformation $e_{ij}^{(p)}$ enter into the loading surface

$$(1) \qquad\qquad f(\sigma_{ij}, T, \xi_i) = 0 \,.$$

Laws governing this aspect are called *hardening rules*.

If we assume that plastic deformation is independent of hydrostatic pressure and that the plastic flow is incompressible, then the yield surfaces in the principal stress space $(\sigma_1, \sigma_2, \sigma_3)$ are cylinders (not necessary with circular cross-section) of infinite length with axis $\sigma_1 = \sigma_2 = \sigma_3$ as illustrated in Fig. 6.12:1. The plane $\sigma_1 + \sigma_2 + \sigma_3 = 0$ is called the π-*plane*, which is perpendicular to the axis. Hence, the yield surfaces can be represented by their cross sections on this plane (see Sec. 6.7). The cross-sectional curves are closed, convex, and piecewise smooth; but they change in size and shape during plastic deformation. For materials involving volume change in the plastic deformation, the yield surface is not parallel to the $\sigma_1 = \sigma_2 = \sigma_3$ axis.

We illustrate several proposed hardening rules in Fig. 6.12:1:

Isotropic Hardening. Isotropic hardening assumes that the material remains isotropic during the process of plastic loading and that the subsequent yield surface is a uniform expansion of the initial yield surface. The initial and subsequent yield surfaces have the same center. Figure 6.12:1 shows an example of isotropic hardening with a Tresca's initial yield surface of a regular hexagon on the π-plane $\sigma_1 + \sigma_2 + \sigma_3 = 0$ [Fig. 6.12:1(a)] and its uniform expansion as a subsequent yield surface [Fig. 6.12:1(b)].

In the general case, if the effect of hydrostatic pressure on yield can be neglected, the subsequent yield surfaces can be expressed in the form

$$(2) \qquad f = f^*(J_2, J_3) - \kappa = 0 \,,$$

where κ is an internal variable that characterizes the hardening of the material. The *strain-hardening hypothesis* assumes that κ is a monotonically increasing function, which depends only on the *effective plastic strain* but *not* the strain path. An effective or equivalent plastic strain, which is an internal variable, can be defined as

$$(3) \qquad \epsilon_e^p = \int d\epsilon_e^p \,,$$

where

$$(4) \quad \blacktriangle \qquad d\epsilon_e^p = \sqrt{\frac{2}{3} de_{ij}^{(p)} de_{ij}^{(p)}} \,.$$

Then κ is a monotonically increasing function of ϵ_e^p. For uniaxial loading, we have

$$(5) \qquad \epsilon_e^p = e^p$$

where e^p is the total uniaxial plastic strain.

Use of Eq. (6.11:12) for $\dot{\Lambda}$ and the flow rule Eq. (6.11:14) for $\dot{e}_{ij}^{(p)}$ in Eq. (4) leads to

$$(6) \qquad \dot{\epsilon}_e^p = \sqrt{\frac{2}{3} \frac{n_{kl}\dot{\sigma}_{kl}}{K^p}} = \dot{\Lambda}\sqrt{\frac{2}{3}} \,.$$

The *internal variable-rate* defined in Eq. (6.11:10) becomes

$$(6a) \qquad \dot{\kappa} = \dot{\xi}_i = \dot{\Lambda} h_i = \dot{\Lambda}\sqrt{\frac{2}{3}} \frac{\partial \kappa}{\partial \epsilon_e^p} \,.$$

For the yield function as defined in Eq. (2), Eq. (6.11:17) gives

$$(7) \qquad K^p = -\sqrt{\frac{2}{3}} \frac{\partial f}{\partial \kappa} \frac{\partial \kappa}{\partial \epsilon_e^p} \left(\frac{\partial f}{\partial \sigma_{kl}} \frac{\partial f}{\partial \sigma_{kl}} \right)^{-1/2} \,,$$

where $\partial f/\partial \kappa = -1$. Once K^p is determined, we can calculate $\dot{\epsilon}_e^p$ and $\dot{e}_{ij}^{(p)}$ from Eqs. (6) and (6.11:14). The calculation of K^p will be discussed later.

Another commonly used parameter for characterizing isotropic hardening is the *total plastic work* defined by

$$(8) \qquad W_p = \int \sigma_{ij} de_{ij}^{(p)} = \int \sigma'_{ij} de_{ij}^{(p)} \, .$$

Assuming κ as a function of W_p is called the *work-hardening hypothesis*. The function $\kappa(W_p)$ or $\sigma_Y(W_p)$ can also be determined from a uniaxal tension test, in which case we have,

$$(9) \qquad K^P = -h_{mn}\sigma'_{mn}\frac{\partial f}{\partial \kappa}\frac{\partial \kappa}{\partial W_p}\left(\frac{\partial f}{\partial \sigma_{kl}}\frac{\partial f}{\partial \sigma_{kl}}\right)^{-1/2} \, .$$

For the von Mises yield function $f = \sqrt{J_2} - \kappa$, we have $\kappa(\epsilon_e^p) = \sigma_Y(\epsilon_e^p)/\sqrt{3}$, where $\sigma_Y(\epsilon_e^p)$ is the tensile yield stress. The yield stress can be determined from the uniaxial curve of σ versus $e^p(=\epsilon_e^p)$ in simple tension and compression. For *linear hardening materials*,

$$(10) \qquad \sigma_Y = \sigma_Y^0 + E^p \epsilon_e^p$$

where E^p is a material constant, from Eq. (8) we find

$$(11) \qquad W_p = \left(\sigma_Y^0 + \frac{1}{2}E^p\epsilon_e^p\right)\epsilon_e^p \, .$$

Then

$$(12) \qquad \sigma_Y^2 = (\sigma_Y^0)^2 + 2E^p\left(\sigma_Y^0 + \frac{1}{2}E^p\epsilon_e^p\right)\epsilon_e^p = (\sigma_Y^0)^2 + 2E^pW^p \, .$$

In this case we have, from Eqs. (7) and (9),

$$\frac{\partial f}{\partial \kappa}\frac{\partial \kappa}{\partial e_e^p} = -\frac{1}{\sqrt{3}}\frac{\partial \sigma_Y}{\partial \epsilon_e^p} = -\frac{E^p}{\sqrt{3}} ,$$

$$\frac{\partial f}{\partial \kappa}\frac{\partial \kappa}{\partial W_p} = -\frac{1}{\sqrt{3}}\frac{\partial \sigma_Y}{\partial W_p} = -\frac{E^p}{\sqrt{3}\sigma_Y} \, .$$

Example of Isotropic Hardening. Consider a thin-walled tube subjected to stretching and torsion, with tensile stress $\sigma(\geq 0)$ and shear stress τ. The Tresca and von Mises yield functions can be expressed in the form $f = \bar{\sigma} - \sigma_Y$, with

$$(13a) \qquad \bar{\sigma} = \sigma_1 - \sigma_3 = \sqrt{(\sigma_{zz} - \sigma_{\theta\theta})^2 + 4\sigma_{z\theta}^2} \qquad \text{(Tresca)},$$

and (13b):

$$\bar{\sigma} = \sqrt{\frac{(\sigma_{zz} - \sigma_{\theta\theta})^2 + (\sigma_{\theta\theta} - \sigma_{rr})^2 + (\sigma_{rr} - \sigma_{zz})^2}{2} + 3(\sigma_{z\theta}^2 + \sigma_{r\theta}^2 + \sigma_{rz}^2)}$$

$$\text{(von Mises)}.$$

For $\sigma_{zz} = \sigma$, $\sigma_{z\theta} = \tau$, $\sigma_{rr} = \sigma_{\theta\theta} = \sigma_{rz} = \sigma_{r\theta} = 0$, Eqs. (13a) and (13b) reduces to Eqs. (6.5:8-9). The yield surface becomes

$$f = \bar{\sigma} - \sigma_Y = \sqrt{\sigma^2 + \alpha\tau^2} - \sigma_Y = 0,$$

where $\alpha = 4$ for the case of Tresca, and $\alpha = 3$ for the case of von Mises. As to the flow rule, we obtain

(14)
$$\dot{e}_{zz}^p = a\dot{\Lambda}\frac{\partial f}{\partial \sigma_{zz}} = \dot{\Lambda}\frac{\sigma}{b},$$

$$\dot{e}_{z\theta}^p = a\frac{\dot{\Lambda}}{2}\frac{\partial f}{\partial \sigma_{z\vartheta}} = \frac{\dot{\Lambda}}{2}\frac{\alpha\tau}{b},$$

$$\dot{e}_{rz}^p = \dot{e}_{r\theta}^p = 0,$$

where a is a normalization factor, $b = \sqrt{\sigma^2 + \alpha^2\tau^2/2}$ and

$$\dot{e}_{\theta\theta}^p = -\dot{e}_{zz}^p, \quad \dot{e}_{rr}^p = 0 \qquad \text{for the Tresca criterion},$$

$$\dot{e}_{\theta\theta}^p = \dot{e}_{rr}^p = -\frac{1}{2}\dot{e}_{zz}^p, \qquad \text{for the von Mises criterion}.$$

Note that the factor $1/2$ for $\dot{e}_{z\theta}^p$ in Eq. (14) is due to the fact that we have treated $\sigma_{z\theta}$ and $\sigma_{\theta z}$ as one rather than two independent variables in the yield function.

From Eq. (7), we have

$$K^p = \sqrt{\frac{2}{3}}\frac{d\sigma_Y}{de_e^p}\frac{\bar{\sigma}}{b}.$$

Then, $\dot{\Lambda}$ can be derived from Eqs. (6) and (7)

$$\dot{\Lambda} = \sqrt{\frac{3}{2}}\left(\frac{d\sigma_Y}{de_e^p}\right)^{-1}\frac{\sigma\dot{\sigma} + \alpha\tau\dot{\tau}}{\bar{\sigma}} = \sqrt{\frac{3}{2}}\left(\frac{d\sigma_Y}{de_e^p}\right)^{-1}\dot{\bar{\sigma}} = \sqrt{\frac{3}{2}}\dot{e}_e^p$$

where $d\sigma_Y/de_e^p$ is the slope of the uniaxial tension σ-e^p (stress-plastic strain) curve evaluating at $e^p = \epsilon_e^p$. The equivalent plastic strain ϵ_e^p is obtained from Eqs. (3) and (4),

$$\epsilon_e^p = \int de_e^p = \int \sqrt{\frac{2}{3}}\, d\Lambda = \int \left(\frac{d\sigma_Y}{de_e^p}\right)^{-1}\frac{\sigma d\sigma + \alpha\tau d\tau}{\bar{\sigma}} = \int \left(\frac{d\sigma_Y}{de_e^p}\right)^{-1} d\bar{\sigma}.$$

For linear hardening materials, Eq. (10) gives

$$\frac{d\sigma_Y}{de^p} = E^p .$$

Finally, for isotropic materials, we have

$$\dot{e}_{zz} = \frac{\dot{\sigma}}{E} + \dot{e}_{zz}^{(p)} = \frac{\dot{\sigma}}{E} + \dot{\Lambda}\frac{\sigma}{b} ,$$

$$2\dot{e}_{z\theta} = \frac{\dot{\tau}}{G} + 2\dot{e}_{z\theta}^{(p)} = \frac{\dot{\tau}}{G} + \dot{\Lambda}\frac{\alpha\tau}{b} .$$

When the stress path is given, we can obtain the strain history by integrating these equations. This solution is complete because the system is statically determinated that the stress field can be determined directly from the applied load.

Lee and Zavenl[6.5] (1978) and Chaboche[6.5] (1977) introduced a nonlinear evolution equation for the isotropic hardening parameter κ:

$$d\kappa = b(\kappa_s - \kappa)d\epsilon_e^p$$

where b and κ_s are material constants. This equation can be integrated to obtain

$$\kappa(\epsilon_e^p) = \kappa_s + (\kappa_0 - \kappa_s)e^{-b\epsilon_e^p} .$$

For cyclic loading, ϵ_e^p increases monotonically and κ approaches κ_s after a number of cycles. The number of cycles required for κ to reach the steady state value depends on the cyclic strain magnitude.

Kinematic Hardening. The kinematic hardening model assumes that during plastic loading the yield surface translates in the stress space without rotation and without change in size and shape. Figure 6.12:1(c) illustrates *Prager's kinematic hardening* showing that the initial yield surface translates in the π-plane. To explain this rule, Prager[6.3] (1954) used a mechanical model, which can be represented as in Fig. 6.12:2. The initial yield surface is regarded as a planar rigid frame lying on the π-plane. The loading point on the π-plane is represented by a small, frictionless pin. If the pin engages the frame, it may push the frame around. Under the assumption of frictionlessness, any motion imparted by the pin to the frame must be normal to the edge in contact. However, when a corner of the frame is caught by the pin, the pin may carry the frame if the direction of motion lies within a certain angle. Rotation of the frame is supposed to be prevented by some mechanism. If the pin disengages and moves away from the frame, the frame stays put, and the change represents an unloading. It is obvious that none of the flow rules deduced from Drucker's hypothesis is violated, and no theoretical objection can be raised against interpreting the motion of the rigid frame as a hardening rule. In fact, the Bauschinger effect is represented very simply, and the development of anisotropy due to plastic deformation appears most naturally.

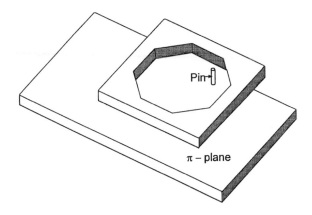

Fig. 6.12:2. A mechanical model used in explaining the kinematic hardening rule.

With some variation in the model, Prager[6.3] (1954) was able to represent various models of plasticity: rigid, perfectly plastic, rigid work-hardening elastic, etc. Almost simultaneously, a similar concept was introduced independently by Ishlinski[6.3] (1954). Further developments were made by Prager,[6.3] Boyce,[6.3] Hodge,[6.3] Novozhilov, and others. Hodge[6.3] (1956) points out that the concept of kinematic hardening can be applied in the nine-dimensional stress space. An example of Prager's kinematic hardening is given in Eq. (6.8:28).

Suppose that the initial yield surface is $f(\sigma_{ij}) = 0$. The subsequent yield surface can be expressed as

$$(15) \qquad f(\sigma_{ij} - \alpha_{ij}) = 0\,,$$

where α_{ij}, usually called the *back stress*, represents the translation of the center of the initial yield surface during the process of plastic loading. The formulation of a kinematic hardening model is to specify the evolution of α_{ij} in terms e_{ij}^p, σ_{ij}, and/or α_{ij} itself. *Prager's linear kinematic hardening model* assumes that

$$(16) \qquad \dot{\alpha}_{ij} = c\dot{e}_{ij}^{(p)} = c\dot{\Lambda}h_{ij}$$

where c is a constant (Shield and Zielger,[6.3] 1958). According to this model the yield surface moves in the direction of the plastic strain rate, i.e., the direction normal to the yield surface at the loading point. From Eqs. (15), (16), (6.11:6) and (6.11:13) with $\dot{\Lambda} h_i \frac{\partial f}{\partial \xi_i} = \frac{\partial f}{\partial \alpha_{ij}} \dot{\alpha}_{ij}$, we obtain,

$$(17) \qquad\qquad K^P = c h_{ij} n_{ij} \,.$$

If the material obeys the associated flow rule, we have

$$K^P = c \,.$$

Prager's linear kinematic model does not give consistent results for three-dimensional and two-dimensional cases. It also introduces a transverse softening in simple tension or compression as shown below. In a uniaxial tension test with load in the 1-direction, the plastic strain increments are

$$(18) \qquad \begin{aligned} de_1^{(p)} &= de^p \,, \\ de_2^{(p)} &= de_3^{(p)} = -\frac{1}{2} de^p \,. \end{aligned}$$

From Eq. (16), we obtain

$$(19) \qquad \begin{aligned} d\alpha_1 &= c \, de^p \,, \\ d\alpha_2 &= d\alpha_3 = -\frac{1}{2} d\alpha_1 = -\frac{1}{2} c \, de^p \,. \end{aligned}$$

The yield surface moves toward the negative 2- and 3-directions causing transverse softening, although no load is applied along these directions.

Ziegler[6.3] (1959) modified Prager's rule Eq. (16) by replacing it with

$$(20) \qquad\qquad \dot{\alpha}_{ij} = \dot{\mu}(\sigma_{ij} - \alpha_{ij})$$

where $\dot{\mu} > 0$. Geometrically, this means the yield surface moves along the direction of $\sigma_{ij} - \alpha_{ij}$, which is the radial vector joining the instantaneous center α_{ij} and the loading point σ_{ij}. *Ziegler's modified kinematic hardening* is illustrated in Fig. 6.12:1(d). The determination of $\dot{\mu}$ will involve the yield criterion.

Armstrong and Frederick (1996) introduced a nonlinear term in Prager's model Eq. (16),

$$(21) \qquad\qquad \dot{\alpha}_{ij} = c \dot{e}_{ij}^{(p)} - \gamma \alpha_{ij} \dot{\epsilon}_e^p = \dot{\Lambda} \left(c h_{ij} - \sqrt{\frac{2}{3}} \gamma \alpha_{ij} \right) \,.$$

Using Eq. (21) as the internal variable-rate and

$$\frac{\partial f}{\partial \alpha_{ij}} = -\frac{\partial f}{\partial \sigma_{ij}} \,,$$

Eq. (6.11:13) reduces to

(22)
$$K^P = \left(ch_{ij} - \sqrt{\frac{2}{3}}\gamma\alpha_{ij} \right) n_{ij}\,.$$

This model was further developed by Chaboche[6.6] (1977, 1986).

Combined Isotropic and Kinematic Hardening. The yield surface of a combined isotropic and kinematic hardening model can be expressed in the form

(23)
$$f(\sigma_{ij} - \alpha_{ij}, \kappa) = 0\,,$$

where α_{ij} and κ are the internal variables. For $\dot{\alpha}_{ij}$ in the form of Eq. (21) and κ as a function of the equivalent plastic strain, Eq. (6.11:13) gives

(24) $$K^P = \left(ch_{ij} - \sqrt{\frac{2}{3}}\alpha_{ij}\gamma \right) n_{ij} - \sqrt{\frac{2}{3}}\frac{\partial f}{\partial \kappa}\frac{\partial \kappa}{\partial \epsilon_e^{(p)}} \left(\frac{\partial f}{\partial \sigma_{ij}}\frac{\partial f}{\partial \sigma_{ij}} \right)^{-1/2}.$$

If the material obeys the associated flow rule ($h_{ij} = n_{ij}$), Eq. (24) becomes

(25) $$K^P = c - \sqrt{\frac{2}{3}}\gamma\alpha_{ij}n_{ij} - \sqrt{\frac{2}{3}}\frac{\partial f}{\partial \kappa}\frac{\partial \kappa}{\partial \epsilon_e^{(p)}} \left(\frac{\partial f}{\partial \sigma_{ij}}\frac{\partial f}{\partial \sigma_{ij}} \right)^{-1/2}.$$

From Eqs. (6.11:4, 14, 18), reproduced here for clarity,

$$n_{ij} = \frac{\partial f}{\partial \sigma_{ij}} \left(\frac{\partial f}{\partial \sigma_{kl}}\frac{\partial f}{\partial \sigma_{kl}} \right)^{-1/2},$$

$$\dot{e}_{ij}^{(p)} = \frac{n_{ij}n_{kl}\dot{\sigma}_{kl}}{K^P},$$

$$de_{rq} = \left[D_{klrq}^{-1} + \frac{n_{kl}n_{rq}}{K^P} \right] d\sigma_{kl}$$

we can determine the plastic strain rate and incremental stress strain relation.

More complicated hardening rules involve translation, expansion and distortion of the yield surface simultaneously. In Fig. 6.12:1(e), the plastic deformation causes a linear segment to move. Figure 6.12:1(f) shows that the loading surface changes with plastic loading in some interdependent manner. Hodge[6.3] (1957) extended the kinematic hardening to include simultaneous expansion of the yield surface. Budiansky[6.3] and

Kliushnikov[6.3] in 1959 considered the possibility of creating corners in subsequent yield surfaces at the point of loading. They achieved a compromise with the Hencky-Nadai "deformation" theory or the "total strain" theory. See Hill[6.2], Nadai[6.2] and papers by Budiansky, Kliushnikov, Naghdi, Sanders, Phillips, etc., in Bibliography 6.2 and 6.4–6.6.

Example of Combined Isotropic and Kinematic Hardening. For the von Mises criterion, the yield surface is

$$f = \frac{1}{2}(\sigma'_{ij} - \alpha'_{ij})(\sigma'_{ij} - \alpha'_{ij}) - \frac{1}{3}\sigma_Y^2 = 0\,,$$

where σ'_{ij} and α'_{ij} are the deviators of σ_{ij} and α_{ij}, respectively. Then

$$\frac{\partial f}{\partial \kappa}\frac{\partial \kappa}{\partial \epsilon_e^p} = -\frac{2\sigma_Y}{3}\frac{d\sigma_Y}{d\epsilon_e^p}\,,$$

$$\left(\frac{\partial f}{\partial \sigma_{ij}}\frac{\partial f}{\partial \sigma_{ij}}\right)^{1/2} = \sqrt{(\sigma'_{ij} - \alpha'_{ij})(\sigma'_{ij} - \alpha'_{ij})} = \sqrt{\frac{2}{3}}\sigma_Y\,,$$

$$n_{ij} = \sqrt{\frac{3}{2}}\frac{\sigma_{ij} - \alpha'_{ij}}{\sigma_Y}\,,$$

and

(26)
$$d\epsilon_e^p = \sqrt{\frac{2}{3}}\frac{n_{ij}d\sigma_{ij}}{K^p} = \frac{(\sigma'_{ij} - \alpha'_{ij})d\sigma_{ij}}{K^p\sigma_Y}\,,$$

Eq. (25) becomes

(27)
$$K^p = c - \gamma\frac{(\sigma'_{ij} - \alpha'_{ij})\alpha'_{ij}}{\sigma_Y} + \frac{2}{3}\frac{d\sigma_Y}{d\epsilon_e^p}\,.$$

Finally we obtain the strain-stress relation for the von Mises criterion with combination of isotropic and kinematic hardening,

$$de_{ij} = \left[D_{ijkl}^{-1} + \frac{3}{2}\frac{(\sigma'_{ij} - \alpha'_{ij})(\sigma'_{kl} - \alpha'_{kl})}{K^p\sigma_Y^2}\right]d\sigma_{kl}\,.$$

We shall illustrate the determination of the yield stress σ_Y (as functions of ϵ_e^p) and the material constants (c, γ) from data measured from uniaxial tests for the von Mises criterion. For uniaxial tension in the 1-direction, on the yield surface, we have

$$de_{11}^{(p)} = de^p, \quad de_{22}^{(p)} = de_{33}^{(p)} = -\frac{1}{2}de^p, \quad d\epsilon_e^p = |de^p|,$$

(28) $\quad \alpha_{11}' = \frac{2}{3}\alpha, \quad \alpha_{22}' = \alpha_{33}' = -\frac{1}{3}\alpha, \quad \sigma_{11} = \sigma, \quad \sigma_{22} = \sigma_{33} = 0,$

$$\sigma_{11}' - \alpha_{11}' = \frac{2}{3}(\sigma - \alpha) = \pm\frac{2}{3}\sigma_Y,$$

$$\sigma_{22}' - \alpha_{22}' = \sigma_{33}' - \alpha_{33}' = -\frac{\sigma - \alpha}{3} = \mp\frac{\sigma_Y}{3}.$$

All other components of e_{ij}, σ_{ij} and α_{ij} are zero. The upper and lower signs in the last two equations of Eq. (28) correspond to loading (tensile plastic flow) and reverse loading (compressive plastic flow), respectively. The yield function can be simplified to become

$$|\sigma - \alpha| = \sigma_Y.$$

From Eqs. (21), (26) and (27), it follows that

(29) $$d\alpha = \frac{3}{2}cde^p - \gamma\alpha|de^p|,$$

(30) $$K^p d\epsilon_e^p = \frac{2}{3}d\sigma,$$

(31) $$\frac{d\sigma}{d\epsilon_e^p} = \frac{3}{2}K^p = \frac{3}{2}c \mp \gamma\alpha + \frac{d\sigma_Y}{d\epsilon_e^p}.$$

Equation (31) shows the different effect of the nonlinear term $\gamma\alpha$ on the tensile and compressive plastic flow. Since $\gamma > 0$, the reverse plastic flow has a higher hardening modulus.

Equation (31) reduces to Prager's linear kinematic hardening if $\gamma = (d\sigma_Y/d\epsilon_e^{(p)}) = 0$. In this case, the yield function is simply

(32) $$\sigma = \alpha \pm \sigma_Y = ce^p \pm \sigma_Y.$$

In general, Eq. (29) can be integrated to give

(33) $$\alpha(e^p) = \pm\frac{3c}{2\gamma} + \left(\alpha_0 \mp \frac{3c}{2\gamma}\right)e^{\mp\gamma(e^p - e_0^p)}.$$

The yield condition becomes

(34) $$\sigma = \alpha(e^p) \pm \sigma_Y = \pm\frac{3c}{2\gamma} + \left(\alpha_0 \mp \frac{3c}{2\gamma}\right)e^{\mp\gamma(e^p - e_0^p)} \pm \sigma_Y,$$

where the upper and lower signs are, respectively, for tensile loading that e^p increases from e_0^p and for compressive loading that e^p decreases from e_0^p.

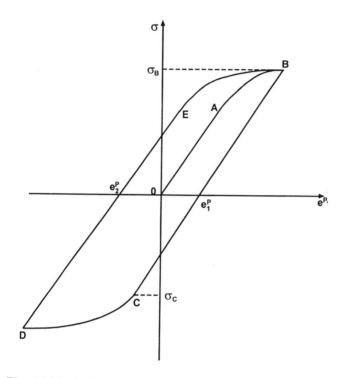

Fig. 6.12:3. A schematic loading path for evaluation of c, γ and σ_Y.

Equations (32) or (34) can be used to determine the material constants (c, γ) and the yield stress σ_Y as function of ϵ_e^p. Consider a load path OABCDE shown in Fig. 6.12:3, where unloading occurs at points B and D, and yielding starts at points A, C and E. Equation (34) gives

$$(35) \qquad \sigma_B = \alpha(e_1^p) + \sigma_Y(e_1^p), \qquad \sigma_E = \alpha(e_2^p) + \sigma_Y(e_2^p),$$

for the tensile loading curve EB,

$$(36) \qquad \sigma_C = \alpha(e_1^p) - \sigma_Y(e_1^p), \qquad \sigma_D = \alpha(e_2^p) - \sigma_Y(e_2^p),$$

for the compressive loading curve CD, where e_1^p and e_2^p are the values of the plastic strain along BC and DE, respectively, and the subscripts B, C, D and E denote the value of σ at those points. Similarly from Eq. (33), we find

(37) $$\alpha(e_1^p) = \frac{3c}{2\gamma}[1 - e^{-\gamma e_1^p}],$$

(38) $$\alpha(e_2^p) = -\frac{3c}{2\gamma} + \left(\alpha(e_1^p) + \frac{3c}{2\gamma}\right) e^{\gamma(e_2^p - e_1^p)}$$

$$= -\frac{3c}{2\gamma} + \frac{3c}{2\gamma}(2 - e^{-\gamma e_1^p})e^{\gamma(e_2^p - e_1^p)}.$$

The following relations are derived for c, γ from Eqs. (35)–(38),

(39) $$\frac{3c}{2\gamma}[1 - e^{-\gamma e_1^p}] = \frac{1}{2}(\sigma_B + \sigma_C),$$

(40) $$-\frac{3c}{2\gamma} + \frac{3c}{2\gamma}(2 - e^{-\gamma e_1^p})e^{\gamma(e_2^p - e_1^p)} = \frac{1}{2}(\sigma_D + \sigma_E).$$

One then determines, from Eqs. (31) and (33) with $\alpha_0 = e_0^p = 0$,

(41) $$\frac{d\sigma_Y}{d\epsilon_e^p} = \frac{d\sigma}{de^p}\bigg|_{e^p = \epsilon_e^p} - \frac{3c}{2}e^{-\gamma \epsilon_e^p}$$

as a function of ϵ_e^p where $\frac{d\sigma}{de^p}|_{e^p = \epsilon_e^p}$ is the measured curve AB. Recall that e^p is the plastic strain measured in the uniaxial tension test. Equation (41) is used to evaluate K^p in Eq. (25) or Eq. (27) for multi-axial loading analyses.

In the derivation c and γ are assumed to be constant. In the more general formulation, c and γ can be functions of plastic deformation.

6.13. MROZ'S, DAFALIAS AND POPOV'S, AND VALANIS' PLASTICITY THEORIES

Hardening is one of the most important mechanical properties in plasticity. The theories in Secs. 6.3–6.12 characterize such behavior by the evolution of *hardening parameters* (e.g., κ and α_{ij}). The models discussed include the isotropic hardening and three types of kinematic hardening (Prager, Ziegler, and Armstrong and Frederick) which are among the simplest plasticity laws regarding this behavior. These models offer a reasonable description of hardening properties for monotonically proportional loading without unloading. The material behaviors under cyclic loading are much too complex to be modeled by these hardening rules. Based on experimental evidence, Drucker and Palgen (1981) identified five basic features of cyclic plasticity:

(1) Plastic strain accumulates, the *cycle creep or ratcheting effect*, in the direction of mean stress for stress cycles.

(2) The mean stress progressively relaxes to zero for strain cycles with nonzero mean.

(3) The elastic-plastic transition is usually smooth.

(4) Material hardens or softens toward a stabilized state with kinematic hardening only.

(5) Extensive plastic loading eliminates most, if not all, of the past effects.

It can be shown, under strain cycles, that isotropic hardening will lead to an elastic state eventually and thus is not capable of simulating the hysteresis loop. Prager's and Ziegler's kinematic hardening models cannot simulate the ratcheting effect.

Recent developments in plasticity theory have been focused on modeling of cyclic loading reversal. We shall discuss some of such developments in the following three sub-sections.

Mroz's Multisurface Model. Mroz[6.6] (1969) introduced a multi-surface model to describe the cyclic effect and the nonlinearity of stress-strain loops. The model uses a number of yield surfaces

$$(1) \qquad f_i(\sigma_{kl} - \alpha_{kl}, \kappa_i) = 0 \quad i = 0, 1, 2, \ldots, n$$

to simulate nonlinear hardening and smooth transition from elastic to plastic deformation. Each of the yield surfaces has its unique plastic modulus κ_i and back stress α_{ij} to represent a combination of linear isotropic hardening and linear kinematic hardening. For a virgin material, the yield surfaces are concentric and similar but of different size. Mroz's model approximates the nonlinear stress-strain curve of plastic deformation by piecewise linear approximation. During loading f_0 yields first. As loading continues, f_0 expands and translates while other surfaces remain stationary. When f_0 touches f_1 at a point A, f_0 and f_1 will move and deform in unison for further loading if the load point stays at point A. The plastic deformation continues until f_1 touches f_2 at the same contact point or the load point moves away from the contact point. When f_0, f_1, and f_2 are in contact, they will move together under further loading if the load point remains at A. This process continues. If the load point moves away from A but remain on the surface $f_0 = 0$, then f_0 may move away from f_1 for further loading. In this case $f_1, f_2, \ldots$ will remain stationary until f_0 touches f_1 again. Then the process will proceed as just described. It is noted that upon unloading, the loading vector moves away or is already away from f_0, and all yield surfaces remain stationary. This means that unloading occurs simultaneously for all yield surfaces and that the contacted surfaces will remain in contact. During reloading f_0 will yield first again and the process will continue as described before.

With a sufficient number of yield surfaces, Mroz's model can describe nonlinear stress-strain loops, the Bauschinger effect, and cyclic hardening or softening of the material. The model requires extensive computational effort in implementation. For more details interested readers are referred to Mroz (1969, 1976), Prevost (1978), Mroz *et al.* (1979).

Dafalias and Popov's Two-Surface Model. Mroz's model cannot simulate ratcheting and mean stress relaxation in strain cyclic loading with nonzero mean. Uniaxial experimental data indicate that, even under complex loading, the stress strain curves tend to converge with definite bounding lines. This suggests the existence of *bounding surface*(s). The plastic modulus $E^p(= \frac{d\sigma}{de^p})$ depends on the distance from the stress-strain curve to the bounding line as well as the recent loading history. Dafalias and Popov[6.6] (1975) proposed a two-surface model to account for these effects. One surface is similar to the conventional yield surface and the other is a bounding surface. For uniaxial loading, the *yield function* f and *bounding function* $\bar{f}$ are in the following forms

$$(2) \qquad f = (\sigma - \alpha)^2 - \sigma_Y^2 = 0 \,, \qquad \bar{f} = (\bar{\sigma} - \bar{\alpha})^2 - \bar{\sigma}_Y^2 = 0 \,,$$

where σ and $\bar{\sigma}$ are the stresses on f and $\bar{f}$, α and $\bar{\alpha}$ their centers, while σ_Y and $\bar{\sigma}_Y$ their respective sizes. Their incremental relations are

$$(3) \qquad de^p = \frac{d\sigma}{K^p} \,, \qquad \text{(flow rule)}$$

$$(4) \qquad d\bar{\sigma} = \bar{K}^p de^p = \frac{\bar{K}^p}{K^p} d\sigma$$

$$(5a) \qquad d\alpha = K^\alpha de^p = \frac{K^\alpha}{K^p} d\sigma \qquad \text{(kinematic hardening)} \,.$$

$$(5b) \qquad d\bar{\alpha} = \bar{K}^\alpha de^p = \frac{\bar{K}^\alpha}{K^p} d\sigma$$

Using the *consistency equation* $df = 0$ and $d\bar{f} = 0$, we obtain

$$(6) \qquad K^\alpha = K^p - \frac{d\sigma_Y}{de^p} \,, \qquad \bar{K}^\alpha = K^p - \frac{d\bar{\sigma}_Y}{de^p} \,.$$

Dafalias and Popov[6.6] assumed that K^p varies continuously, as opposed to the piecewise-constant plastic moduli in Mroz's model, in the form

$$(7) \qquad K^p = K^p(\delta_{in}, \delta)$$

where $\delta = [(\bar{\sigma} - \sigma)(\bar{\sigma} - \sigma)]^{1/2} = |\bar{\sigma} - \sigma|$ is the distance between the present state (on the yield surface), σ, and the stress state, $\bar{\sigma}$, on the bounding surface, with the same outward normal. The quantity δ_{in} is the value of δ at the initiation of a new loading process and measures how far the material state is from the state represented by the bounds. This is used to reflect the effect of loading history. To specify K^p, Dafalias and Popov[6.6] (1976) proposed that

$$(8) \qquad K^p(\delta_{in}, \delta, \epsilon_e^p) = \bar{K}^p(\epsilon_e^p) + \frac{a}{1 + b\delta_{in}} \frac{\delta}{\delta_{in} - \delta},$$

where a and b are constant and $\bar{K}^p$ is the limiting values of the bounds defined as $\frac{\partial \bar{\sigma}_\gamma}{\partial e^p}$. These quantities are determined experimentally. Note that $K^p(\delta_{in}, \delta_{in}, \epsilon_e^p) = \infty$ is specified to give a smooth elastic plastic transition. We can now calculate de^p, $d\alpha$ and $d\bar{\alpha}$ for given $d\sigma$ if the isotropic hardening characteristics $\frac{d\sigma_\gamma}{de^p}$ and $\frac{d\bar{\sigma}_\gamma}{de^p}$ are known.

To generalize into the multiaxial case, the yield surface f and the bounding surface $\bar{f}$ are written in the following forms:

$$(9) \qquad f(\sigma_{ij} - \alpha_{ij}, \kappa) = F(\sigma_{ij} - \alpha_{ij}) - \kappa = 0,$$

$$(10) \qquad \bar{f}(\bar{\sigma}_{ij} - \bar{\alpha}_{ij}, \bar{\kappa}) = \bar{F}(\bar{\sigma}_{ij} - \bar{\alpha}_{ij}) - \bar{\kappa} = 0,$$

where $\bar{\sigma}_{ij}$ is a stress on the bounding surface corresponding to the stress state σ_{ij} on the yield surface. When unloading occurs, σ_{ij} and $\bar{\sigma}_{ij}$ move away from the yield and bounding surfaces in unison. If f and $\bar{f}$ are similar, then $\bar{\sigma}_{ij} - \bar{\alpha}_{ij} = \eta(\sigma_{ij} - \alpha_{ij})$. The flow rules are

$$(11) \qquad de_{ij}^p = \frac{n_{ij}n_{kl}d\sigma_{kl}}{K^p} = \frac{n_{ij}n_{kl}d\bar{\sigma}_{kl}}{\bar{K}^p}$$

which gives the relation between $d\sigma_{ij}$ and $d\bar{\sigma}_{ij}$. The model assumes that

$$d\bar{\alpha}_{ij} - d\alpha_{ij} = d\mu(\bar{\sigma}_{ij} - \sigma_{ij}), \qquad \Lambda \frac{\partial f}{\partial \sigma_{ij}} = \bar{\Lambda} \frac{\partial \bar{f}}{\partial \sigma_{ij}}.$$

Using Eq. (8), the flow rules Eq. (11), and the consistency equations derived from Eqs. (9) and (10), we can express $d\bar{\sigma}_{ij}, d\mu, d\alpha_{ij}, d\bar{\alpha}_{ij}$ in terms of $d\sigma_{ij}, \bar{K}^p, d\bar{\kappa}/d\epsilon_e^p$ and $d\kappa/d\epsilon_e^p$. The last three quantities are usually determined by experiments. In Eq. (8), δ is defined as

$$\delta = \sqrt{(\bar{\sigma}_{ij} - \sigma_{ij})(\bar{\sigma}_{ij} - \sigma_{ij})}.$$

The two-surface model is smooth at the elastic-plastic transition point, but produces an inconsistency in the uniaxial load-unload-reload situation if the unloading step is very small (Chaboche 1986). This model, like the multiple-surface model, requires extensive computational effort in implementation. Interested readers are referred to papers by Dafalias and Popov (1975, 1976), Dafalias (1984), and Chaboche (1986).

Valanis's Endochronic Theory. The plasticity theories discussed so far are based on the concept of yield surface, which divides the elastic and plastic domains. It implies a sharp demarcation between the two domains. As pointed out before, many materials do not exhibit a sharp yield point. Some materials such as aluminum and stainless steel have a nonlinear uniaxial stress-strain curve almost from the start. Valanis[6.6] (1971, 1980) proposed a plastic theory in a convolution integral form for infinitesimal deformation. The concept is that the present state of the material depends on the present values and the past history of observable variables, giving rise to hereditary theories. The theory is called the *endochronic theory*, which assumes

$$(12) \qquad \sigma'_{ij} = 2G \int_0^z \rho(z - z') \frac{de^p_{ij}}{dz'}\, dz'\,,$$

in which yield surface is not *a prior* defined rather a derivable result of the theory. The kernel function $\rho(z)$ is a material function and z is the intrinsic time defined as

$$(13) \qquad dz = \frac{d\varsigma}{f(\varsigma)}\,,$$

where $f(\varsigma)$ is a nonnegative function, called the *intrinsic time scale*, with $f(0) = 1$ and ς defined as

$$(14) \qquad d\varsigma = \sqrt{de^p_{ij} de^p_{ij}}\,.$$

Let
$$(15) \qquad \rho(z) = \rho_0 \delta(z) + \rho_1(z)\,,$$

where $\delta(z)$ is the Dirac delta function, ρ_0 is a material constant and $\rho_1(z)$ is a nonsingular function. A substitution of Eqs. (13) and (15) into Eq. (12) gives

$$(16) \qquad \sigma'_{ij} = S^0_Y \frac{de^p_{ij}}{d\varsigma} f(\varsigma) + \alpha_{ij}\,,$$

where

$$(17) \qquad S^0_Y = 2G\rho_0\,,$$

$$(18) \qquad \alpha_{ij} = 2G \int_0^z \rho_1(z - z') \frac{de^p_{ij}}{dz'}\, dz'\,.$$

The assumption of the kernel function $\rho(z)$ in the form of Eq. (15) effectively divides the constitutive equation, Eq. (16) in two parts: the part represented by the first term on the right hand side of the equation depends on the present values of observable variables, such as S_Y^0, and a set of internal-state variables, which characterize both S_Y^0 and de_{ij}^p/dz. The second term α_{ij} is equivalent to the back stress introduced previously, which depends on the past histories of these variables.

From Eqs. (14) and (16), we obtain

(19) $$(\sigma_{ij}' - \alpha_{ij})(\sigma_{ij}' - \alpha_{ij}) - [S_Y^0 f(\varsigma)]^2 = 0\,,$$

(20) $$de_{ij}^p = \frac{1}{S_Y^0 f(\varsigma)}(\sigma_{ij}' - \alpha_{ij})d\varsigma\,.$$

Note that

(21) $$(\sigma_{ij}' - \alpha_{ij})de_{ij}^p = S_Y^0 f(\zeta)\,d\varsigma > 0\,.$$

Equations (19) and (20) correspond to the yield function and the flow rule, respectively. Equation (19) shows the characteristics of combined isotropic and kinematic hardening. If $f(\varsigma) = 1$, Eq. (19) corresponds to the von Mises yield surface. In this case, the plastic strain increment de_{ij}^p is normal to the yield surface.

Different plasticity theories can be derived through different choices of the kernel function $\rho_1(z)$ and the intrinsic time scale $f(\varsigma)$. Valanis (1980) proved that general form for $\rho_1(z)$ is given by

(22) $$\rho_1(z) = \sum_{j=1}^{\infty} \rho_{1j} e^{-\xi_j z}\,,$$

where ρ_{1j}, ξ_j are material constants to be determined experimentally. Two commonly used forms of $f(\zeta)$ for isotropic hardening are

(23) $$f(\zeta) = 1 + \gamma\zeta\,,$$

and

(24) $$f(\zeta) = \alpha + (1 - \alpha)e^{-\beta\zeta}\,,$$

where α, β, γ, are material constants. The second form reaches a limit asymptotically as ζ increases and is therefore called the *saturated form*. Many features of the endochronic model are described in detail in Watanabe and Atluri[6.6] (1986).

6.14. STRAIN SPACE FORMULATIONS

We have thus far discussed the plasticity theories formulated in the stress space only. This approach has two inherent disadvantages: for ideal plastic materials, a different approach from that for hardening materials is needed to derive the incremental stress-strain relation; and for softening materials, confusion can arise between further plastic loading and elastic unloading (Naghdi and Trapp[6.7] 1975a, 1975b). Parallel to the stress space approach discussed in Sec. 6.11, Il'iushin[6.7] (1961) introduced a strain space formulation to eliminate these drawbacks. Similar to Drucker's postulate, Il'iushin hypothesized that the work done by external forces on a material over a closed strain cycle is nonnegative. In other words, plastic deformation occurs during a closed strain cycle if the work of external forces over the strain cycle is positive. A closed strain cycle is defined as a closed path in the strain space that the strain starts from an equilibrium and compatible state ϵ_0, experiences a change due to external loading, and then reverts to the original ϵ_0. Under Il'iushin's hypothesis, it can be shown that the work done by external forces over a closed strain cycle (the area of ABCDE) is larger than or equal to that done over a closed stress cycle (the area of ABCD). See Fig. 6.14:1.

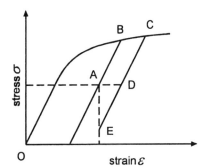

Fig. 6.14:1. Closed strain and stress cycles.

Let σ^p denote the *plastic stress* defined in

(1) $$\sigma_{ij} = D_{ijkl}e_{kl} - \sigma_{ij}^p = D_{ijkl}(e_{kl} - e_{kl}^p),$$

where $\sigma_{ij}, e_{kl}, e_{kl}^p$ are the stress, total strain, and plastic strain tensors, respectively, and D_{ijkl} is the fourth-ordered *elastic modulus tensor*. Then the *plastic stress tensor* is defined in terms of the plastic strain tensor e_{kl}^p in the form

$$\sigma_{ij}^p = D_{ijkl}e_{kl}^p$$

or in an incremental form

(2)
$$d\sigma_{ij}^p = D_{ijkl}de_{kl}^p \,,$$

if the elastic modulus tensor remains constant in the course of plastic deformation. Based on his hypothesis, Il'iushin showed that

(3) $$d\sigma_{ij}^p = \eta \frac{\partial \hat{f}}{\partial e_{ij}} = \hat{\Lambda}\hat{n}_{ij} \,, \qquad \text{and} \qquad \eta = \hat{\Lambda} \left(\frac{\partial \hat{f}}{\partial e_{ij}} \frac{\partial \hat{f}}{\partial e_{ij}} \right)^{-1/2}$$

where $\hat{f}(\mathbf{e}, \boldsymbol{\sigma}^p, \boldsymbol{\alpha}_e, \kappa) = 0$ is the *yield surface*, $\eta, \hat{\Lambda}$ are positive *scalar factors* and $\hat{n}_{ij}$ is a unit normal to the yield surface defined as

(4) $$\hat{n}_{ij} = \frac{\partial \hat{f}}{\partial e_{ij}} \left(\frac{\partial \hat{f}}{\partial e_{ij}} \frac{\partial \hat{f}}{\partial e_{ij}} \right)^{-1/2} ,$$

$\boldsymbol{\alpha}_e$ is called the *back strain tensor* governing kinematic hardening, while κ is a parameter associated with isotropic hardening. Equation (3) is the *flow rule in the strain space formulation* equivalent to Eq. (6.11:5) in the stress space formulation.

The yield function forms a closed surface in strain space enclosing an elastic region with the following features.

- No change in plastic deformation occurs as long as $\hat{f} < 0$.
- Change in plastic deformation occurs when $\hat{f} = 0$.
- No meaning is associated with $\hat{f} > 0$.

Yield surfaces in stress space are more readily available. We can obtain the yield surfaces in the strain space by substituting Eq. (1) into the yield functions in the stress space:

(5) $$f(\boldsymbol{\sigma}, \boldsymbol{\alpha}, \kappa) = f(D_{ijkl}e_{kl} - \sigma_{ij}^p - \alpha_{ij}, \kappa) = \hat{f}(\mathbf{e}, \boldsymbol{\sigma}^p, \boldsymbol{\alpha}_e, \kappa) = 0$$

where $(\alpha_e)_{ij} = D_{ijkl}^{-1}\alpha_{kl}$ and $\boldsymbol{\alpha}_e$ and κ can be functions of $\boldsymbol{\sigma}^p$. For linear kinematic hardening with g_{ijkl} being a 4th-order tensor,

(6) $$d(\alpha_e)_{ij} = g_{ijkl}d\sigma_{kl}^p \,.$$

If κ is a function of the *equivalent plastic stress* σ_e^p with increment

(7) $$d\sigma_e^p = \sqrt{\frac{2}{3} d\sigma_{ij}^p d\sigma_{ij}^p} = \hat{\Lambda}\sqrt{\frac{2}{3}} \,,$$

then

(8) $$d\kappa = \frac{d\kappa}{d\sigma_e^p} d\sigma_e^p = \sqrt{\frac{2}{3}} \frac{d\kappa}{d\sigma_e^p} \hat{\Lambda} \,.$$

If $g_{ijkl} = cD_{ijpq}^{-1}D_{pqkl}^{-1}$, where $D_{ijrs}^{-1}D_{rskl} = (\delta_{ik}\delta_{jl} + \delta_{il}\delta_{jk})/2$ that D_{ijrs}^{-1} is the inverse of **D**, Eq. (6) is Prager's model (6.12:16). For isotropic materials

$$D_{ijkl} = 2G\left(\frac{\delta_{ik}\delta_{jl} + \delta_{il}\delta_{jk}}{2} + \frac{\nu\delta_{ij}\delta_{kl}}{1-2\nu}\right), \quad D_{ijkl}^{-1} = \frac{1}{2G}\left(\frac{\delta_{ik}\delta_{jl} + \delta_{il}\delta_{jk}}{2} - \frac{\nu\delta_{ij}\delta_{kl}}{1+\nu}\right),$$

$$g_{ijkl} = c\left[\frac{\delta_{ik}\delta_{jl} + \delta_{il}\delta_{jk}}{8G^2} + \frac{\nu(2-\nu)}{(1+\nu)^2}\frac{\delta_{ij}\delta_{kl}}{4G^2}\right], \quad (\alpha_e)_{ij} = \frac{c\sigma_{ij}^p}{4G^2}, \quad \sigma_{ij}^p = 2Ge_{ij}^p.$$

The *consistency equation* can be derived from Eq. (5), i.e.,

(9) ▲ $\qquad \hat{f} = \dfrac{\partial \hat{f}}{\partial e_{ij}}de_{ij} + \dfrac{\partial \hat{f}}{\partial \sigma_{ij}^p}d\sigma_{ij}^p + \dfrac{\partial \hat{f}}{\partial (\alpha_e)_{ij}}d(\alpha_e)_{ij} + \dfrac{\partial \hat{f}}{\partial \kappa}d\kappa = 0.$

A substitution of Eqs. (6)–(8) into Eq. (9) yields

(10) $$\Lambda = \frac{\hat{n}_{ij}de_{ij}}{\hat{L}^p},$$

where

(11) $\hat{L}^p\sqrt{\dfrac{\partial \hat{f}}{\partial e_{kl}}\dfrac{\partial \hat{f}}{\partial e_{kl}}} = -\left(\dfrac{\partial \hat{f}}{\partial \sigma_{ij}^p}\hat{n}_{ij} + g_{ijkl}\dfrac{\partial \hat{f}}{\partial (\alpha_e)_{kl}}\hat{n}_{ij} + \dfrac{\partial \hat{f}}{\partial \kappa}\dfrac{d\kappa}{d\sigma_e^p}\sqrt{\dfrac{2}{3}}\right)$

The *flow rule* Eq. (3), together with Eq. (2) can now be written as

(12) ▲ $\qquad d\sigma_{ij}^p = \dfrac{\hat{n}_{kl}de_{kl}}{\hat{L}^p}\hat{n}_{ij} \qquad$ or $\qquad de_{ij}^p = D_{ijrs}^{-1}\hat{n}_{rs}\dfrac{\hat{n}_{kl}de_{kl}}{\hat{L}^p}.$

The *incremental stress-strain relation* becomes

(13) ▲ $\qquad d\sigma_{ij} = \left(D_{ijkl} - \dfrac{\hat{n}_{ij}\hat{n}_{kl}}{\hat{L}^p}\right)de_{kl}.$

Using the yield surface Eq. (5) and noting that

$$\frac{\partial \hat{f}}{\partial e_{ij}} = D_{ijkl}\frac{\partial f}{\partial \sigma_{kl}}, \qquad \frac{\partial \hat{f}}{\partial \sigma_{ij}^p} = -\frac{\partial f}{\partial \sigma_{ij}},$$

$$\frac{\partial \hat{f}}{\partial (\alpha_e)_{ij}} = D_{ijkl}\frac{\partial f}{\partial \alpha_{kl}} = -D_{ijkl}\frac{\partial f}{\partial \sigma_{kl}}$$

we can express quantities in *strain space* in terms of quantities in *stress space*:

(14) $\hat{n}_{ij} = \dfrac{D_{ijkl}n_{kl}}{(D_{rspq}n_{pq}D_{rsuv}n_{uv})^{1/2}}$ (unit normal to yield surface)

(15) $\hat{\Lambda} = \dfrac{D_{ijrs}n_{rs}de_{ij}\sqrt{D_{abkl}n_{kl}D_{abuv}n_{uv}}}{L^{p}}$ (scalar factor)

(16) $d\sigma_{ij}^{p} = \dfrac{D_{ijrs}n_{rs}D_{kluv}n_{uv}de_{kl}}{L^{p}}$

(incremental plastic stress: flow rule)

(17) $d\sigma_{ij} = \left(D_{ijkl} - \dfrac{D_{ijuv}n_{uv}D_{klrs}n_{rs}}{L^{p}}\right)de_{kl} = D_{ijkl}^{ep}de_{kl}$

(incremental stress-strain relation)

where

(18) $L^{p} = D_{ijkl}n_{kl}D_{ijuv}n_{uv}\hat{L}^{p} = D_{ijkl}n_{ij}n_{kl} + K^{p}$,

(19) $K^{p} = n_{ij}D_{ijkl}g_{klrs}D_{rsuv}n_{uv} - \dfrac{\partial f}{\partial \kappa}\dfrac{d\kappa}{de_{e}^{p}}\sqrt{\dfrac{2}{3}}\left(\dfrac{\partial f}{\partial \sigma_{kl}}\dfrac{\partial f}{\partial \sigma_{kl}}\right)^{-1/2}$,

where n_{ij} is a unit normal to the yield surface in stress space defined in Eq. (6.11:4) and $d\sigma_{e}^{p} = de_{e}^{p}\sqrt{D_{abkl}n_{kl}D_{abuv}n_{uv}}$. For isotropic materials,

$\hat{n}_{ij} = n_{ij}, \quad \hat{\Lambda} = \dfrac{4G^{2}n_{ij}de_{ij}}{L^{p}}, \quad d\sigma_{ij} = \lambda e_{kk}\delta_{ij} + 2G\left(de_{ij} - \dfrac{2Gn_{ij}de_{kl}}{L^{p}}\right),$

$L^{p} = 2G + K^{p}, \quad K^{p} = c - \dfrac{\partial f}{\partial \kappa}\dfrac{d\kappa}{de_{e}^{p}}\sqrt{\dfrac{2}{3}}\left(\dfrac{\partial f}{\partial \sigma_{kl}}\dfrac{\partial f}{\partial \sigma_{kl}}\right)^{-1/2}.$

It can be shown that K^{p} has the same definition as Eq. (6.11:13) in the stress space formulation. For ideal plasticity, $K^{p} = 0$ causes the breakdown of the flow rule in the stress space formulation. Since L^{p} is nonzero, there is no such problem in the strain space formulation.

Example. For isotropic von Mises materials, the yield surface in stress space is

(20) $f = \dfrac{1}{2}(\sigma_{kl}' - \alpha_{kl})(\sigma_{kl}' - \alpha_{kl}) - \dfrac{1}{3}\sigma_{Y}^{2} = 0$

with σ_{ij}' being the deviatoric stress and $\alpha_{ij} = ce_{ij}^{p}$ for Prager's linear kinematic hardening model defined by Eq. (6.12:16). For infinitesimal deformation

$\sigma_{ij}' = 2G(e_{ij}' - e_{ij}^{p}) = 2Ge_{ij}' - \sigma_{ij}^{p}$

where e'_{ij} is the deviatoric strain, the corresponding yield surface in strain space can be written as

(21) $\hat{f}(e_{ij} - \hat{\alpha}_{ij}, \kappa) = f(\sigma'_{ij} - \alpha_{ij}, \kappa)$

$$= 2G^2(e'_{kl} - \hat{\alpha}_{kl})(e'_{kl} - \hat{\alpha}_{kl}) - \frac{1}{3}\sigma_Y^2 = 0,$$

where $\hat{\alpha}_{ii} = \sigma_{ij}^p = 0$ and

$$\hat{\alpha}_{ij} = \frac{\sigma_{ij}^p}{2G} + (\alpha_e)_{ij} = \frac{\sigma_{ij}^p}{2G} + \frac{\alpha_{ij}}{2G} = \frac{\sigma_{ij}^p}{2G} + \frac{ce_{ij}^p}{2G} = \left(1 + \frac{c}{2G}\right)\frac{1}{2G}\sigma_{ij}^p.$$

Note that the material exhibits linear kinematic hardening in strain space even if there is no kinematic hardening in the stress space, i.e., $\alpha_{kl} = 0$. Then

$$\frac{\partial \hat{f}}{\partial e_{ij}} = 4G^2(e'_{ij} - \alpha'_{ij}), \quad \hat{n}_{ij} = \frac{\sqrt{6}\,G}{\sigma_Y}(e_{ij} - \hat{\alpha}_{ij}),$$

$$\frac{\partial f}{\partial \sigma_{ij}}\frac{\partial f}{\partial \sigma_{ij}} = \frac{2}{3}\sigma_Y^2, \quad \frac{\partial f}{\partial e'_{ij}}(de'_{ij} - d\hat{\alpha}_{ij}) - \frac{2}{3}\sigma_Y\frac{d\sigma_Y}{d\sigma_e^p}d\sigma_e^p = 0,$$

$$d\sigma_{ij} = 2Gde_{ij} + \lambda de_{kk}\delta_{ij} - \frac{6G^2}{\hat{L}^p\sigma_Y^2}(e'_{ij} - \hat{\alpha}_{ij})(e'_{kl} - \hat{\alpha}_{kl})de_{kl},$$

$$d\hat{\alpha}_{ij} = \left(1 + \frac{c}{2G}\right)\frac{\hat{n}_{ij}\hat{\Lambda}}{2G}, \quad \hat{L}^p = \left(1 + \frac{c}{2G} + \frac{2}{3}\frac{\partial\sigma_Y}{\partial\sigma_e^p}\right)\frac{1}{2G}.$$

These equations are valid for both hardening and ideal plastic materials. For further details, readers are referred to the references such as Il'iushin[6.7] 1960, 1961, Naghdi and Trapp[6.7] 1975(a), 1975(b), 1981, Yoder and Iwan[6.7] 1981, Yoder 1981.

6.15. FINITE DEFORMATION

Considerable attention has been given to the study of finite elastic-plastic deformation in recent years. For finite deformation, it is necessary to distinguish between stress and strain measures defined with reference to the deformed and the undeformed states of a continuum. In formulating a plasticity model, stress and strain rates are required. Unfortunately, depending on the definition, a rate may or may not be objective (independence of coordinate transformation) in large deformation. A key element in finite plasticity formulation is the choice of appropriate objective rates that will correctly reflect the underlying physics of the plastic flow. The choice of rates also affects strongly the form of constitutive equations.

In general, finite deformation will also produce elastic-plastic coupling, thus may make the yield surface nonconvex. The subject of the best choice of rates is still a topic of research. Detailed discussion of finite plasticity theory is beyond the scope of this book.

6.16. PLASTIC DEFORMATION OF CRYSTALS

The underlying physics of plastic flow for metals is the micro-slip of crystallographic planes. Micro-slip in crystals or grains is highly directional and nonuniform. Dislocation theories and the plasticity theories for single- and poly-crystals have provided the basis to quantify plastic deformation at the microscopic level and explain how and why plastic deformation occurs. The approach is based on the movements of atoms and the deformation of crystals and grains along slip planes. The macroscopic movements of metals are the aggregates of single- and poly-crystal responses to applied loads. This physical theory is a very important aspect of plasticity theory. Interested readers are referred to the references listed in Bibliography 6.9 at the end of the book for details.

PROBLEMS

6.8. A solid is assumed to obey the von Mises criterion with isotropic hardening. If the virgin curve in uniaxial tension can be described in the infinitesimal deformation range by $\sigma_Y = F(e^p)$, determine

$$de_{ij}^p = \Lambda \frac{\partial f}{\partial \sigma_{ij}}$$

when κ is assumed to depend on either e^p, or W_p.

6.9. If the yield function is

$$f = \sqrt{J_2} - \kappa = 0 \,,$$

with

$$J_2 = J_2(\sigma_{ij}' - \alpha_{ij}) \,, \quad \alpha_{ij} = c e_{ij}^p \,,$$

where κ and c are constant, show that

$$\dot{\alpha}_{ij} = c(\sigma_{ij}' - \alpha_{ij}) \frac{(\sigma_{kl}' - \alpha_{kl})\dot{\sigma}_{kl}'}{2\kappa^2} \left/ \left(c + \frac{2}{\sqrt{3}} \frac{d\kappa}{de_e^p} \right) \right. \,.$$

6.10. Generalize the preceding result to the case where κ depends on ϵ_e^p, determine the rate equations for both α_{ij} and $\dot{e}_{ij}^p$.

6.11. Consider testing a certain metalic material in two ways: by torsion of a hollow cylindrical specimen as shown in Fig. P6.11(a), and by uniaxial stretching of a tension specimen as shown in Fig. P6.11(b). The torsional test shows that the load-deflection curve is linear for a shearing stress below 25,000 lb/sq-in and that yielding occurs at the stress 25,000 lb/sq-in. If the criterion of yielding for this material is von Mises' $J_2 = k^2$, what is the expected value of the tensile stress at which yielding will occur in the tension test?

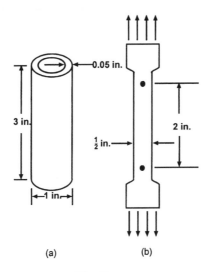

(a)   (b)

Fig. P6.11.

6.12. Consider an isotropic, incompressible, but nonlinearly elastic material. Let e_{ij}', σ_{ij}' be the strain deviation and stress deviation, respectively. If the material were elastic, then

$$e_{ij}' = \frac{1}{2G}\sigma_{ij}'.$$

This can be written, of course, as a matrix equation,

$$\begin{bmatrix} e_{11}' & e_{12}' & e_{13}' \\ e_{21}' & e_{22}' & e_{23}' \\ e_{31}' & e_{32}' & e_{33}' \end{bmatrix} = \frac{1}{2G}\begin{bmatrix} \sigma_{11}' & \sigma_{12}' & \sigma_{13}' \\ \sigma_{21}' & \sigma_{22}' & \sigma_{23}' \\ \sigma_{31}' & \sigma_{32}' & \sigma_{33}' \end{bmatrix},$$

or, in short

$$\mathbf{e}' = \frac{1}{2G}\boldsymbol{\sigma}'.$$

To generalize such a relation to a nonlinear elastic material, we assume that the matrix $\mathbf{e}'$ can be represented as a power series in $\boldsymbol{\sigma}'$,

$$\mathbf{e}' = C_1\boldsymbol{\sigma}' + C_2\boldsymbol{\sigma}'^2 + C_3\boldsymbol{\sigma}'^3 + \cdots.$$

6.13. Prove, by an application of the Cayley–Hamilton theorem, that the most general nonlinear elasticity law of this type can be reduced to the form

$$\mathbf{e}' = \mathbf{P}\boldsymbol{\sigma}' + \mathbf{Q}\boldsymbol{\sigma}'^2 . \qquad \text{(Prager)}$$

6.14. In soil mechanics, Coulomb's yield condition is widely used. Consider a two-dimensional problem. If σ_1 and σ_2 are the principal stresses in the plane of concern, Coulomb's yield condition is specified by the yield functions

$$f_1 = \sigma_1(1 + \sin\phi) - \sigma_2(1 - \sin\phi) - 2c\cos\phi ,$$

$$f_2 = -\sigma_1(1 - \sin\phi) + \sigma_2(1 + \sin\phi) - 2c\cos\phi ,$$

where c = cohesion and ϕ = angle of internal friction. The soil is elastic if both f_1 and f_2 are negative, and it is plastic when f_1 or f_2, or both vanish.

(a) Show that the flow rule derived from this condition by means of Koiter's theory of generalized plastic potential requires that the rate of the planar dilatation $\dot{e}_1^{(p)} + \dot{e}_2^{(p)}$ and the maximum shear rate $|\dot{e}_1^{(p)} - \dot{e}_2^{(p)}|$ have the constant ratio $\sin\phi$ for all states of stress at the yield limit, except the state $\sigma_1 = \sigma_2 = c\cos\phi$.

(b) Generalize the results above to obtain the generalized plastic potentials for three-dimensional problems.

(c) According to your generalization of (b), is there any general relationship between the three-dimensional dilatation and the maximum shear rate for stress states at the yield limit?

6.15. For the yield surface as defined in Eq. (6.14:5), show that

$$\sqrt{D_{ijkl}n_{kl}D_{ijpq}n_{pq}} = 1/\sqrt{D_{ijkl}^{-1}\hat{n}_{kl}D_{ijpq}^{-1}\hat{n}_{pq}} .$$

6.16. For the equivalent plastic stress $(d\sigma_e^p)$ and strain $(d\epsilon_e^p)$ as defined in Eqs. (6.14:7) and (6.12:4), respectively, show that

$$d\sigma_e^p = d\epsilon_e^p \sqrt{D_{ijkl}n_{kl}D_{ijpq}n_{pq}} .$$

6.17. For $\alpha_{ij} = D_{ijkl}(\alpha_e)_{kl}$, $d(\alpha_e)_{ij} = g_{ijkl}d\sigma_{kl}^p, d\epsilon_e^p$ as defined in Eq. (6.12:4), and the yield function as defined in Eq. (6.14:5), show that

(a) $d\alpha_{ij} = D_{ijkl}g_{klrs}D_{rspq}de_{pq}^p$; and

(b) Equation (6.14:19) becomes

$$K^p = -\left(\frac{\partial f}{\partial \alpha_{ij}}D_{ijkl}g_{klrs}D_{rspq}n_{pq} + \frac{\partial f}{\partial \kappa}\frac{d\kappa}{d\epsilon_e^p}\sqrt{\frac{2}{3}} \right)\left(\frac{\partial f}{\partial \sigma_{kl}}\frac{\partial f}{\partial \sigma_{kl}} \right)^{-1/2} ,$$

which is the same as that given in Eq. (6.11:13) in the stress space formulation.

7

LINEARIZED THEORY OF ELASTICITY

In this chapter we shall discuss the classical theory of elasticity. We shall discuss the general structure of the theory and illustrate the applications of the linearized theory by a few examples.

Rectangular Cartesian coordinates of reference will be used throughout. The coordinates will be denoted by x_1, x_2, x_3 or x, y, z unless stated otherwise.

7.1. BASIC EQUATIONS OF ELASTICITY FOR HOMOGENEOUS ISOTROPIC BODIES

An elastic body has a unique *zero-stress* state, to which the body returns when all stress vanish. All stresses, strains, and particle displacements are measured from this *zero-stress* state: their values are counted as zero in that state.

There are two ways to describe a deformed body: the *material* and the *spatial* (see Sec. 5.2). Consider the spatial description. The motion of a continuum is described by the instantaneous velocity field $v_i(x_1, x_2, x_3, t)$. To describe the strain in the body, a displacement field $u_i(x_1, x_2, x_3, t)$ is specified which describes the displacement of a particle located at x_1, x_2, x_3 at time t from its position in the natural state. Various strain measures can be defined for the displacement field. The Almansi strain tensor is expressed in terms of $u_i(x_1, x_2, x_3, t)$ according to Eq. (4.2:4),

$$(1) \qquad e_{ij} = \frac{1}{2}\left[\partial u_j \partial x_i + \frac{\partial u_i}{\partial x_j} - \frac{\partial u_k}{\partial x_i}\frac{\partial u_k}{\partial x_j} \right] .$$

The particle displacements u_i are functions of time and position. The particle velocity is given by the material derivative of the displacement,

$$(2) \qquad v_i = \frac{\partial u_i}{\partial t} + v_j \frac{\partial u_i}{\partial x_j} .$$

The particle acceleration is given by the material derivative of the velocity (5.2:7),

$$(3) \qquad \alpha_i = \frac{\partial v_i}{\partial t} + v_j \frac{\partial v_i}{\partial x_j}$$

The motion of the body must obey the equation of continuity (5.4:3)

$$(4) \qquad \frac{\partial \rho}{\partial t} + \frac{\partial (\rho v_i)}{\partial x_i} = 0 \,,$$

and the equation of motion (5.5:7)

$$(5) \qquad \rho \alpha_i = \frac{\partial \sigma_{ij}}{\partial x_j} + X_i \,.$$

In addition to the field Eqs. (4) and (5), the theory of linear elasticity is based on Hooke's law. For a homogeneous isotropic material, this is Eq. (6.2:7).

$$(6) \qquad \sigma_{ij} = \lambda e_{kk} \delta_{ij} + 2G e_{ij} \,,$$

where λ and G are constants independent of the spatial coordinates.[†]

The famous nonlinear terms in Eqs. (1)–(3) are sources of major difficulty in the theory of elasticity. To make some progress, we are forced to *linearize* by considering small displacements and small velocities, i.e., by restricting ourselves to values of u_i, v_i so small that the nonlinear terms in Eqs. (1)–(3) may be neglected. In such a linearized theory, we have

$$(7) \qquad e_{ij} = \frac{1}{2}(u_{i,j} + u_{j,i}) \,,$$

$$(8) \qquad v_i = \frac{\partial u_i}{\partial t} \,, \qquad \alpha_i = \frac{\partial v_i}{\partial t} \,.$$

Unless stated otherwise, all that is discussed below is subjected to this restriction of linearization. Fortunately, many useful results can be obtained from this linearized theory.

Equations (1)–(6) or (4)–(8) together are 22 equations for the 22 unknowns ρ, u_i, v_i, σ_i, e_{ij}, σ_{ij}. In the infinitesimal displacement theory we may eliminate σ_{ij} by substituting Eq. (6) into Eq. (5) and using Eq. (7) to obtain the well-known *Navier's equation*,

$$(9) \ \blacktriangle \qquad G u_{i,jj} + (\lambda + G) u_{j,ji} + X_i = \rho \frac{\partial^2 u_i}{\partial t^2} \,.$$

This can be written in the form

$$(10) \ \blacktriangle \qquad G \nabla^2 u_i + (\lambda + G) e_{,i} + X_i = \rho \frac{\partial^2 u_i}{\partial t^2} \,,$$

[†]The corresponding equations based on the material description are the following: velocity and acceleration, Eqs. (5.2.3) and (5.2:6); strain measure, the Green's strain tensor, Eq. (4.2:5); the equation of continuity, Eq. (5.2:3); stress tensors, Sec. 16.7; the equations of motion, Eqs. (16.10:1–8); the stress-strain laws, see Sec. 16.11; in particular, Eqs. (16.11:6) and (16.11:7). It can be seen that the kinematical relations appear simpler in the material description, but the equations of motion appear more complicated.

where

(11)
$$e = u_{j,j}$$

(12)
$$\nabla^2 u_i = u_{i,jj} \,.$$

The quantity e is the *divergence* of the displacement vector u_i. ∇^2 is the *Laplace operator.* If we write x, y, z instead of x_1, x_2, x_3, we have

(13)
$$e = \frac{\partial u}{\partial x} + \frac{\partial v}{\partial y} + \frac{\partial w}{\partial z} \,,$$

(14)
$$\nabla^2 = \frac{\partial^2}{\partial x^2} + \frac{\partial^2}{\partial y^2} + \frac{\partial^2}{\partial z^2} \,.$$

Love[1,2] writes Eq. (10) in the form,

(15)
$$G\nabla^2(u,v,w) + (\lambda + G)\left(\frac{\partial}{\partial x}, \frac{\partial}{\partial y}, \frac{\partial}{\partial z}\right)e + (X,Y,Z)$$

$$= \rho\frac{\partial^2}{\partial t^2}(u,v,w) \,,$$

which is a shorthand for three equations of the type

(16) ▲
$$G\nabla^2 u + (\lambda + G)\frac{\partial e}{\partial x} + X = \rho\frac{\partial^2 u}{\partial t^2} \,.$$

This can also be written as

(17) ▲
$$G\left(\nabla^2 u + \frac{1}{1-2\nu}\frac{\partial e}{\partial x}\right) + X = \rho\frac{\partial^2 u}{\partial t^2} \,.$$

If we introduce the rotation vector

(18)
$$(\omega_x, \omega_y, \omega_z) \equiv \frac{1}{2}\mathrm{curl}(u,v,w)$$

$$\equiv \frac{1}{2}\left(\frac{\partial w}{\partial y} - \frac{\partial v}{\partial z}, \frac{\partial u}{\partial z} - \frac{\partial w}{\partial x}, \frac{\partial v}{\partial x} - \frac{\partial u}{\partial y}\right)$$

and use the identity

(19)
$$\nabla^2(u,v,w) = \left(\frac{\partial}{\partial x}, \frac{\partial}{\partial y}, \frac{\partial}{\partial z}\right)e - 2\mathrm{curl}(\omega_x, \omega_y, \omega_z) \,,$$

then Eq. (10) may be written as

(20)
$$(\lambda + 2G)\left(\frac{\partial}{\partial x}, \frac{\partial}{\partial y}, \frac{\partial}{\partial z}\right)e - 2G\,\mathrm{curl}(\omega_x, \omega_y, \omega_z) + (X,Y,Z)$$

$$= \rho\frac{\partial^2}{\partial t^2}(u,v,w) \,.$$

7.2. EQUILIBRIUM OF AN ELASTIC BODY UNDER ZERO BODY FORCE

Consider the conditions of static equilibrium of an elastic body. If the body force vanishes, $X_i = 0$, then by taking divergence of Eq. (7.1:9), we have

$$Gu_{i,jji} + (\lambda + G)u_{j,jii} = 0,$$

or

$$(1) \qquad\qquad (u_{j,j})_{,ii} = 0.$$

In unabridged notations for rectangular Cartesian coordinates, this is

$$(2) \qquad \left(\frac{\partial^2}{\partial x^2} + \frac{\partial^2}{\partial y^2} + \frac{\partial^2}{\partial z^2}\right)e = 0, \qquad \text{or} \qquad \nabla^2 e = 0.$$

Equation (2) is a Laplace equation. A function satisfying Eq. (2) is called a *harmonic function*. Thus, *the dilation e is a harmonic function when the body force vanishes*.

But

$$(3\lambda + 2G)e = \sigma_{xx} + \sigma_{yy} + \sigma_{zz} = 3\sigma,$$

where σ is the mean stress. Hence, *the mean stress is also a harmonic function*:

$$(3) \qquad\qquad \nabla^2 \sigma = 0.$$

If we put $X = 0$, $\partial^2 u / \partial t^2 = 0$, and operate on Eq. (7.1:16) with the Laplacian ∇^2, we have

$$(\lambda + G)\frac{\partial}{\partial x}\nabla^2 e + G\nabla^2\nabla^2 u = 0.$$

With Eq. (2), this implies that

$$(4) \qquad\qquad \nabla^4 u = 0,$$

where in rectangular Cartesian coordinates,

$$(5) \qquad \nabla^4 = \frac{\partial^4}{\partial x^4} + \frac{\partial^4}{\partial y^4} + \frac{\partial^4}{\partial z^4} + 2\frac{\partial^4}{\partial x^2 \partial y^2} + 2\frac{\partial^4}{\partial y^2 \partial z^2} + 2\frac{\partial^4}{\partial z^2 \partial x^2}.$$

Equation (4) is called a *biharmonic* equation, and its solution is called a *biharmonic function*. Hence, *the displacement component u is biharmonic. Similarly, the components v, w are biharmonic*. It follows that *when the* *body force is zero, each of the strain components and each of the stress components, being linear combination of the first derivatives of u, v, w, are all biharmonic functions:*

(6) $$\nabla^4 \sigma_{ij} = 0 \,,$$

(7) $$\nabla^4 e_{ij} = 0 \,.$$

7.3. BOUNDARY VALUE PROBLEMS

Navier's Eq. (7.1:9) combines Hooke's law and the equation of motion. It is to be solved for appropriate boundary and initial conditions. The boundary conditions that occur are usually one of two kinds:

A. *Specified displacements.* The components of displacement u_i are prescribed on the boundary.

B. *Specified surface tractions.* The components of surface traction $\overset{\nu}{T_i}$ are assigned on the boundary.

In most problems of elasticity, the boundary conditions are such that over part of the boundary displacements are specified, whereas over another part the surface tractions are specified. Let the region occupied by an elastic body be denoted by V. Let the boundary surface of V be denoted by S. We separate S into two parts: S_u, where displacements are specified; and S_σ, where surface tractions are specified. Therefore, on S_σ,

$$\overset{\nu}{T_i} = \sigma_{ij}\nu_j = \text{a prescribed function}\,,$$

where ν_j is a unit vector along the outer normal to the surface S_σ. By Hooke's law, this may be written as

$$[\lambda u_{k,k}\delta_{ij} + G(u_{i,j} + u_{j,i})]\nu_j = \text{a prescribed function}\,.$$

Hence, over the entire surface, the boundary conditions are that either u_i, or a combination of the first derivatives of u_i, are prescribed.

In dynamic problems, a set of initial conditions on u_i or σ_{ij} must be specified in the region V and on the boundary S.

The question arises whether a boundary-value problem posed in this way has a solution, and whether the solution is unique or not. The question has two parts. First, do we expect a unique solution on physical grounds? Second, does the specific mathematical problem have a unique solution? In continuum mechanics, there are many occasions in which we do not expect a unique solution to exist. For example, when a thin-walled spherical shell is subjected to a uniform external pressure, the phenomenon of buckling may occur when the pressure exceeds certain specific value. At the buckling load, the shell may assume several different forms of deformation, some stable, some unstable. On the other hand, our everyday experience about the physical world tells us that the vast majority of mechanical cause-and-effect relationships are unique. Theoretically, the physical question is partly answered by thermodynamics. But the mathematical question must be answered by the theory of partial differential equations. A satisfactory theory must bring harmony between the mathematical formulation and the physical world.

In the preceding discussions we have taken the displacements u_i as the basic unknown variables. In problems of static equilibrium, however, it is customary to use an alternate procedure. The equations of equilibrium are first solved for the stresses σ_{ij}. We then use Hooke's law to obtain the strain e_{ij}. This solution will not be unique. In fact, an infinite set of solutions will be found. The correct one is then singled out by the conditions of compatibility. Only the one solution that satisfies both the equations of equilibrium and the equations of compatibility corresponds to a continuous displacement field. This procedure becomes very attractive when stress functions are introduced, which yield general solutions of the equations of equilibrium (see Sec. 9.2).

By means of Hooke's law, the compatibility equation

$$(1) \qquad e_{ij,kl} + e_{kl,ij} - e_{ik,jl} - e_{jl,ik} = 0$$

can be expressed directly in terms of stress components. On substituting

$$e_{ij} = \frac{1+\nu}{E}\sigma_{ij} - \frac{\nu}{E}\theta\delta_{ij}\,, \qquad\qquad \theta = \sigma_{kk}\,,$$

into Eq. (1), we obtain

$$(2) \qquad \sigma_{ij,kl} + \sigma_{kl,ij} - \sigma_{ik,jl} - \sigma_{jl,ik}$$

$$= \frac{\nu}{1+\nu}(\delta_{ij}\theta_{,kl} + \delta_{kl}\theta_{,ij} - \delta_{ik}\theta_{,jl} - \delta_{jl}\theta_{,ik})\,.$$

Since only six of the 81 equations represented by Eq. (1) are linearly independent, the same must be true for Eq. (2). If we combine Eqs. (2) linearly by setting $k = l$ and summing, we obtain

$$(3) \qquad \sigma_{ij,kk} + \sigma_{kk,ij} - \sigma_{ik,jk} - \sigma_{jk,ik}$$

$$= \frac{\nu}{1+\nu}(\delta_{ij}\theta_{,kk} + \delta_{kk}\theta_{,ij} - \delta_{ik}\theta_{,jk} - \delta_{jk}\theta_{,ik})\,,$$

which is a set of nine equations of which six are independent because of the symmetry in i and j. Hence, the number of independent equations is not reduced, and Eqs. (2) and (3) are equivalent. Since $\sigma_{kk} = \theta$ and $\sigma_{ij,kk} = \nabla^2 \sigma_{ij}$, if we use the equation of equilibrium to replace, say, $\sigma_{ik,kj}$ by $-X_{i,j}$, we can write Eq. (3) as

$$(4) \qquad \nabla^2 \sigma_{ij} + \frac{1}{1+\nu}\theta_{,ij} - \frac{\nu}{1+\nu}\delta_{ij}\nabla^2\theta = -(X_{i,j} + X_{j,i}),$$

where X_i is the body force per unit volume. In dynamic problems the inertia force should be included in X_i. With a contraction $i = j$, Eq. (4) furnishes a relation between $\nabla^2\theta$ and $X_{i,i}$. If this is used to transform the third term in Eq. (4), we obtain

$$(5) \blacktriangle \qquad \nabla^2 \sigma_{ij} + \frac{1}{1+\nu}\theta_{,ij} = -\frac{\nu}{1-\nu}\delta_{ij}X_{k,k} - (X_{i,j} + X_{j.i}).$$

Written out *in extenso*, these are

$$\nabla^2\sigma_{xx} + \frac{1}{1+\nu}\frac{\partial^2\theta}{\partial x^2} = -\frac{\nu}{1-\nu}\left(\frac{\partial X}{\partial x} + \frac{\partial Y}{\partial y} + \frac{\partial Z}{\partial z}\right) - 2\frac{\partial X}{\partial x},$$

$$\nabla^2\sigma_{yy} + \frac{1}{1+\nu}\frac{\partial^2\theta}{\partial y^2} = -\frac{\nu}{1-\nu}\left(\frac{\partial X}{\partial x} + \frac{\partial Y}{\partial y} + \frac{\partial Z}{\partial z}\right) - 2\frac{\partial Y}{\partial y},$$

$$\nabla^2\sigma_{zz} + \frac{1}{1+\nu}\frac{\partial^2\theta}{\partial z^2} = -\frac{\nu}{1-\nu}\left(\frac{\partial X}{\partial x} + \frac{\partial Y}{\partial y} + \frac{\partial Z}{\partial z}\right) - 2\frac{\partial Z}{\partial z},$$

$$(7) \qquad \nabla^2\sigma_{yz} + \frac{1}{1+\nu}\frac{\partial^2\theta}{\partial y\partial z} = -\left(\frac{\partial Y}{\partial z} + \frac{\partial Z}{\partial y}\right),$$

$$\nabla^2\sigma_{zx} + \frac{1}{1+\nu}\frac{\partial^2\theta}{\partial z\partial x} = -\left(\frac{\partial Z}{\partial x} + \frac{\partial X}{\partial z}\right),$$

$$\nabla^2\sigma_{xy} + \frac{1}{1+\nu}\frac{\partial^2\theta}{\partial x\partial y} = -\left(\frac{\partial X}{\partial y} + \frac{\partial Y}{\partial x}\right).$$

These equations were obtained by Michell in 1900, and, for the case in which body forces are absent, by Beltrami in 1892. They are known as the *Beltrami–Michell compatibility equations*.

For a simply connected region, satisfaction of the Beltrami–Michell equations implies that the stress system σ_{ij} is derivable from a continuous displacement field. If the region concerned is multiply connected, additional conditions in the form of certain line integrals must be satisfied (see Sec. 4.7).

7.4. EQUILIBRIUM AND UNIQUENESS OF SOLUTIONS

Consider the problem of determining the state of stress and strain in a body of a given shape which is held strained by body forces X_i and surface tractions $\overset{\nu}{T}_i$. Let us assume that a function $W(e_{11}, e_{12}, \ldots, e_{33})$, called *the strain energy function of the elastic material*, exists and has the property

(1)
$$\frac{\partial W}{\partial e_{ij}} = \sigma_{ij}\,.$$

For example, if the material obeys Hooke's law [Eq. (6.1:1)], then

(2)
$$W = \frac{1}{2}C_{ijkl}e_{ij}e_{kl}\,.$$

The existence and the positive definiteness of the strain energy function for an elastic body are discussed in Chapter 12. In Sec. 12.4, it is shown that W is positive definite in the neighborhood of the natural state. A natural state of a material is a stable state in which the material can exist by itself in thermodynamic equilibrium. Limiting ourselves to consider materials which have a unique natural state in the theory of elasticity, we are assured of a positive definite strain energy function in the neighborhood of the natural state.

The equations of equilibrium $\sigma_{ij,j} + X_i = 0$ may be written in terms of W as

(3)
$$\left(\frac{\partial W}{\partial e_{ij}}\right)_{,j} + X_i = 0\,.$$

The boundary conditions over the boundary surface $S_u + S_\sigma$ are:

(4a)   Over S_u, the values of u_i are given,

(4b)   Over S_σ, the tractions $\overset{\nu}{T}_i = \dfrac{\partial W}{\partial e_{ij}}\nu_j$ are specified

(see Sec. 7.3). If S_σ constitutes the entire surface of the body, it is obvious that equilibrium would be impossible unless the system of body forces and surface tractions satisfy the conditions of static equilibrium for the body as a whole. In the same case, the displacement will be indeterminate to the extent of a possible rigid-body motion.

We shall prove the following theorem due to Kirchhoff. *If either the surface displacements or the surface tractions are given, the solution of the problem of equilibrium of an elastic body as specified by Eqs. (1)–(4) is unique in the sense that the state of stress (and strain) is determinate*

*without ambiguity, provided that the magnitude of the stress (and strain) is
so small that the strain energy function exists and remains positive definite.*

Proof. Since a strained state must be either unique or nonunique, a proof
can be constructed by showing that an assumption of nonuniqueness leads
to contradiction. Assume that there exist two sets of displacements u_i'
and u_i'' which define two states of strain, both satisfying Eq. (3) and the
boundary conditions (4a) and (4b). Then the difference $u_i \equiv u_i' - u_i''$ satisfies
the equation

(5)
$$\left(\frac{\partial W}{\partial e_{ij}}\right)_{,j} = 0$$

and the boundary conditions that

(6) $u_i = 0 \quad$ on $\quad S_u \quad$ and $\quad \dfrac{\partial W}{\partial e_{ij}} \nu_j = 0 \quad$ on $\quad S_\sigma$.

From Eq. (5) we have

$$\int_V u_i \left(\frac{\partial W}{\partial e_{ij}}\right)_{,j} dV = 0 ,$$

which, on integrating by parts, becomes

(7) $$\int_S u_i \frac{\partial W}{\partial e_{ij}} \nu_j dS - \int_V \frac{\partial W}{\partial e_{ij}} u_{i,j} dV = 0 .$$

The first surface integral vanishes because of the boundary conditions (6).
The second volume integral may be written as

$$\int_v \frac{\partial W}{\partial e_{ij}} e_{ij} dV .$$

Now, when W is a homogeneous quadratic function of e_{ij} [Eq. (2)], the
integral above is equal to $\int 2W dV$. Since W is assumed to be positive
definite, the integral $\int W dV$ cannot vanish unless W vanishes, which in
turn implies that $e_{ij} = 0$ everywhere. Hence, $u_i = u_i' - u_i''$ corresponds to
the natural, unstrained state of the body. Therefore, the states of strain
(and stress) defined by u_i' and u_i'' are the same, contrary to the assumption.
Hence, the state of strain (and stress) is unique.   Q.E.D.

We must note that the uniqueness theorem is proved only in the neighborhood of the natural state. In fact when the strain energy function fails

to remain positive definite, a multi-valued solution or several solutions may be possible.

Problem 7.1. Prove Neumann's theorem that the solution $u_i(x,t)$, for x in $V + S_u + S_\sigma$ and $t \geq 0$, of the following system of equations, is unique.

$$(8) \qquad \frac{\partial}{\partial x_j}\left(\frac{\partial W}{\partial e_{ij}}\right) + X_i - \rho\frac{\partial^2 u_i}{\partial t^2} = 0, \qquad \text{for } x \text{ in } V,\ t \geq 0,$$

$$(9) \qquad u_i = f_i(x,t), \qquad \text{for } x \text{ on } S_u,\ t \geq 0,$$

$$(10) \qquad \frac{\partial W}{\partial e_{ij}}\nu_j = g_i(x,t), \qquad \text{for } x \text{ on } S_\sigma,\ t \geq 0,$$

$$(11) \qquad u_i = h_i(x), \qquad \dot{u}_i = k_i(x), \qquad \text{when } t = 0, x \text{ in } V + S,$$

$$(12) \qquad e_{ij} = \frac{1}{2}(u_{i,j} + u_{j,i}),$$

where X_i, f_i, g_i, h_i, k_i are preassigned functions and $W(e_{ij})$ is a positive definite quadratic form. *Hint:* Multiply Eq. (8) by $\partial u_i/\partial t$ and integrate over V and $(0, t)$. Note that the kinetic energy is positive definite. (Reference, Love, *Elasticity*,[1.2] p. 176.)

Notes on Possible Loss of Uniqueness

The uniqueness theorem of Kirchhoff is the foundation for the method of potentials (Chapters 8 and 9). For, when uniqueness of solution is established, one needs to find only *a* solution of a given boundary-value problem: that solution is *the* solution.

But it is essential for our theory to be able to violate the uniqueness of solution in one way or another. We know that elastic columns can buckle, thin shells may collapse, airplane wings can flutter, machinery can become unstable in one sense or other. The word "stability" has many meanings; to define a stability problem we must define the sense of the word stability. A large class of practical stability problems is connected with the loss of uniqueness of solution. Under certain circumstances two or more solutions may become possible; some of these may be dangerous from engineering point of view or undesirable for the function of the machinery; and the circumstances are said to cause instability.

The uniqueness theorem can be violated by violating any one of its assumptions. Referring to Kirchhoff's theorem, we have two possibilities:

(a) Loss of positive-definiteness of the strain energy function, $W(e_{ij})$.
(b) Basic changes in the equations of equilibrium (or of motion).

The first possibility arises when the material becomes unstable, as yielding or flow occurs (cf. Sec. 12.5). It is relevant to the plastic buckling of columns, plates, and shells.

The second possibility may arise in a variety of forms. The most important are those due to

(1) finite deformation,
(2) nonconservative forces, and
(3) forces that are functionals of the deformation or history of deformation.

Most buckling problems can be understood only if we realize the basic changes in the equation of equilibrium introduced by finite deformation. For finite deformations the equation of equilibrium (or of motion) is given by Eq. (16.10:7) or (16.10:8). These equations are basically nonlinear because the strains and rotations depend on the stresses. The corresponding equations of equilibrium or motion for a plate are given in Sec. 16.16, e.g., Eqs. (16.16:10) to (16.16:14). The situation with a column is similar. The linearized versions of these equations retain the basic features of these large-deflection equations, so that these problems generally become eigenvalue problems or bifurcationa problems.

Nonconservative forces generally depend on the deformation of the structure in a certain specific manner. For example, consider an axial load applied to the end of a cantilever column. If this load is fixed in direction, it is conservative. If this load is not fixed in direction, but may rotate in the process of buckling, then it is nonconservative. The case (3) listed above is a special but broad class of nonconservative forces. It occurs commonly if a solid body is placed in a flowing fluid. The aerodynamic or hydrodynamic pressure acting on the body depends on the deformation of the body, and on the local and whole body velocity and acceleration. If the wake and vorticity are important, the aerodynamic pressure will depend also on the history of deformation. This is commonly the case of an aircraft lifting surface. Under these loadings the terms X_i and $\overset{\nu}{T_i}$ in Eqs. (7.4:3), (7.4:4), and (7.4:8), (7.4:10) are functions or functionals of $u_i(x_1, x_2, x_3, t)$.

In any of the cases mentioned above the basic equations differ from those assumed in the Kirchhoff theorem. Loss of uniqueness does not necessarily follow, but it becomes a possibility and must be investigated.

7.5. SAINT-VENANT'S THEORY OF TORSION

To illustrate the applications of the theory of elasticity, we shall consider the problem of torsion of a cylindrical body. A cylindrical shaft, with an axis z, is acted on at its ends by a distribution of shearing stresses whose resultant force is zero but whose resultant moment is a torque T. The lateral surface of the shaft is stress free. See Fig. 7.5:1.

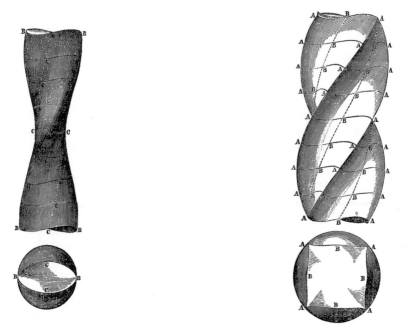

Fig. 7.5:1. Torsion of bars of elliptic and square cross section, as drawn by Saint-Venant.

If the shaft is a circular cylinder, it is very easy to show that all plane cross sections normal to the z-axis remain plane, and the deformation consists of relative rotation θ of the cross sections. The rate of rotation per unit axial length $d\theta/dz$ is proportional to the torque T, with a proportionality constant equal to the product of the shear modulus G of the shaft material, and the polar moment of inertia J of the shaft cross-sectional area:

$$(1) \qquad\qquad GJ\frac{d\theta}{dz} = T\,.$$

The only nonvanishing component of stress is the shear in cross sections perpendicular to z, whose magnitude is

$$(2) \qquad\qquad \tau = \frac{Tr}{J}\,,$$

where r is the radius vector from the central axis z.

If the cross section of the shaft is not circular, a plane cross section does not remain plane; it warps, as is shown in Fig. 7.5:1. The problem is to calculate the stress distribution and the deformation of the shaft.

This is an important problem in engineering; for shafts are used to transmit torques and they are seen everywhere. The celebrated solution to the problem is due to Barre de Saint-Venant (1855), who used the so-called *semi-inverse method*, i.e., a method in which one guesses at part of the solution, tries to determine the rest rationally so that all the differential equations and boundary conditions are satisfied. The torsion problem is not simple. Saint-Venant, guided by the solution of the circular shaft, made a brilliant guess and showed that an exact solution to a well-defined problem can be obtained.

We shall consider, then, a cylindrical shaft with an axis z, with the ends at $z = 0$ and $z = L$. A set of rectangular Cartesian coordinates x, y, z will be used, with the x, y-plane perpendicular to the axis of the shaft. The displacement components in the x, y, z-direction will be written as u, v, w, respectively. Saint-Venant assumed that, as the shaft twists, the plane cross sections are warped but the *projections* on the x, y-plane rotate as a rigid body; i.e.,

$$(3) \qquad u = -\alpha z y, \qquad v = \alpha z x, \qquad w = \alpha \varphi(x, y),$$

where $\varphi(x, y)$ is some function of x and y, called the *warping function*, and α is the angle of twist per unit length of the bar and is assumed to be very small ($\ll 1$). We rely on the function $\varphi(x, y)$ to satisfy the differential equations of equilibrium (without body force)

$$\frac{\partial \sigma_{xx}}{\partial x} + \frac{\partial \sigma_{xy}}{\partial y} + \frac{\partial \sigma_{xz}}{\partial z} = 0,$$

$$(4) \qquad \frac{\partial \sigma_{yx}}{\partial x} + \frac{\partial \sigma_{yy}}{\partial y} \frac{\partial \sigma_{yz}}{\partial z} = 0,$$

$$\frac{\partial \sigma_{zx}}{\partial x} + \frac{\partial \sigma_{zy}}{\partial y} \frac{\partial \sigma_{zz}}{\partial z} = 0,$$

the boundary conditions on the lateral surface of the cylinder

$$\sigma_{xx} \nu_x + \sigma_{xy} \nu_y = 0,$$

$$(5) \qquad \sigma_{yx} \nu_x + \sigma_{yy} \nu_y = 0,$$

$$\sigma_{zx} \nu_x + \sigma_{zy} \nu_y = 0,$$

and the boundary conditions at the ends $z = 0$ and $z = L$:

(6)
$$\sigma_{zz} = 0$$

σ_{zx}, σ_{zy} equipollent to a torque T.

The constants ν_x, ν_y are the direction cosines of the exterior normal to the lateral surface ($\nu_z = 0$).

From Eq. (3), we obtain the stresses according to Hooke's law.

(7)
$$\sigma_{yz} = \alpha G \left(\frac{\partial \varphi}{\partial y} + x \right), \qquad \sigma_{zx} = \alpha G \left(\frac{\partial \varphi}{\partial x} - y \right),$$

$$\sigma_{xy} = \sigma_{xx} = \sigma_{yy} = \sigma_{zz} = 0.$$

A substitution of these values into Eqs. (4) shows that the equilibrium equations will be satisfied if $\varphi(x, y)$ satisfies the equation

(8)
$$\frac{\partial^2 \varphi}{\partial x^2} + \frac{\partial^2 \varphi}{\partial y^2} = 0$$

throughout the cross section of the cylinder. To satisfy the boundary conditions (5), we must have

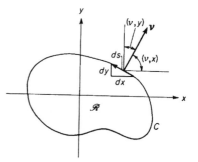

Fig. 7.5:2. Notations.

(9)
$$\left(\frac{\partial \varphi}{\partial x} - y \right) \cos (\nu, x)$$

$$+ \left(\frac{\partial \varphi}{\partial y} + x \right) \cos (\nu, y) = 0,$$

on C,

where C is the boundary of the cross section R (Fig. 7.5:2). But

$$\frac{\partial \varphi}{\partial x} \cos (x, \nu) + \frac{\partial \varphi}{\partial y} \cos (y, \nu) \equiv \frac{\partial \varphi}{\partial \nu};$$

hence, the boundary condition (9) can be written as

(10)
$$\frac{\partial \varphi}{\partial \nu} = y \cos (x, \nu) - x \cos (y, \nu) \qquad \text{on } C.$$

The boundary conditions (6) are satisfied at the ends $z = 0$ and $z = L$ if

(11) $$\iint_R \sigma_{zx}\, dx dy = 0\,, \qquad \iint_R \sigma_{zy}\, dx dy = 0\,,$$

(12) $$\iint_R (x\sigma_{zy} - y\sigma_{zx})\, dx dy = T\,.$$

We can show that Eqs. (11) are readily satisfied if φ satisfies Eqs. (8) and (10); because

$$\iint_R \sigma_{zx}\, dx dy = \alpha G \iint_R \left(\frac{\partial \varphi}{\partial x} - y \right) dx dy$$

$$= \alpha G \iint_R \left\{ \frac{\partial}{\partial x} \left[x \left(\frac{\partial \varphi}{\partial x} - y \right) \right] + \frac{\partial}{\partial y} \left[x \left(\frac{\partial \varphi}{\partial y} + x \right) \right] \right\} dx dy\,,$$

since φ satisfies Eq. (8). On applying Gauss' theorem to the last integral, it becomes a line integral on the boundary C of the region R:

$$\alpha G \int_C x \left[\frac{\partial \varphi}{\partial \nu} - y \cos(x, \nu) + x \cos(y, \nu) \right] ds\,,$$

which vanishes on account of Eq. (10). Similarly, the second of Eqs. (11) is satisfied. Finally, the last condition (12) requires that

(13) $$T = \alpha G \iint_R \left(x^2 + y^2 + x\frac{\partial \varphi}{\partial y} - y\frac{\partial \varphi}{\partial x} \right) dx dy\,.$$

Writing J for the integral

(14) $$J \equiv \iint_R \left(x^2 + y^2 + x\frac{\partial \varphi}{\partial y} - y\frac{\partial \varphi}{\partial x} \right) dx dy\,,$$

we have

(15) $$T = \alpha G J\,.$$

This merely shows that the torque T is proportional to the angle of twist per unit length α, with a proportionality constant GJ, which is usually called the *torsional rigidity* of the shaft. The J represents the polar moment of inertia of the section when the cross section is circular and the warping function φ is zero. However, it is conventional to retain the notation GJ for torsional rigidity, even for noncircular cylinders.

Thus, we see that the problem of torsion is reduced to the solution of Eqs. (8) and (10). The solution will yield a stress system σ_{zx}, σ_{zy}. If the

end sections of the shaft are free to warp, and if the stresses prescribed on
the end sections are exactly the same as those given by the solution, then
an exact solution is obtained, and the solution is unique (see Sec. 7.4). If
the distribution of stresses acting on the end sections, while equipollent to
a torque T, does not agree exactly with that given by Eq. (7), then only
an approximate solution is obtained. According to a principle proposed
by Saint-Venant, the error in the approximation is significant only in the
neighborhood of the end section (see Secs. 10.11–10.13).

Equation (8) is a *potential* equation; its solutions are called *harmonic
functions.* The same equation appears in hydrodynamics. The boundary
condition (10) is similar to that for the velocity potential in hydrodynamics
with prescribed velocity efflux over the boundary. In the hydrodynamics
problem, the condition for the existence of a solution φ is that the total
flux of fluid across the boundary must vanish. Translated to our problem,
the condition for the existence of a solution φ is that the integral of the
normal derivative of the function φ, calculated over the entire boundary C,
must vanish. This follows from the identity

$$(16) \qquad \int_C \frac{\partial \varphi}{\partial \nu} \, ds = \iint_R \operatorname{div}(\operatorname{grad} \varphi) \, dx dy = \iint_R \nabla^2 \varphi \, dx dy$$

and from the fact that $\nabla^2 \varphi = 0$. This condition is satisfied in our case by
Eq. (10), as can be easily shown. Therefore, our problem is reduced to the
solution of a potential problem (called Neumann's problem) subjected to
the boundary condition Eq. (10).

An alternate approach was proposed by Prandtl, who takes the stress
components as the principal unknowns. If we assume that only σ_{xz}, σ_{yz}
differ from zero, then all the equations of equilibrium (4) are satisfied if

$$(17) \qquad \frac{\partial \sigma_{xz}}{\partial x} + \frac{\partial \sigma_{yz}}{\partial y} = 0 \,.$$

Prandtl observes that this equation is identically satisfied if σ_{xz} and σ_{yz}
are derived from a *stress function* $\psi(x, y)$ so that

$$(18) \qquad \sigma_{xz} = \frac{\partial \psi}{\partial y}, \qquad \sigma_{yz} = -\frac{\partial \psi}{\partial x} \,.$$

This corresponds to the stream function in hydrodynamics, if σ_{xz} and σ_{yz}
were identified with velocity components. Although ψ can be arbitrary as
far as equilibrium conditions are concerned, the stress system (18) must
satisfy the boundary conditions (5) and (6), and the compatibility conditions. From Eq. (7.3:6), we see that compatibility requires that (in the

absence of body force),

$$\nabla^2 \sigma_{yz} = 0 ; \qquad \nabla^2 \sigma_{zx} = 0 ,$$

where ∇^2 denotes

$$\left(\frac{\partial^2}{\partial x^2} + \frac{\partial^2}{\partial y^2} \right) .$$

Hence

(19)
$$\frac{\partial}{\partial x} \nabla^2 \psi = 0 , \qquad \frac{\partial}{\partial y} \nabla^2 \psi = 0 .$$

It follows that

(20)
$$\nabla^2 \psi = \text{const} .$$

Of the boundary conditions (5), only the last equation is not identically satisfied. If we note from Fig. 7.5:2, that

(21)
$$\nu_x = \cos (\nu, x) = \frac{dy}{ds} , \qquad \nu_y = \cos (\nu, y) = -\frac{dx}{ds} ,$$

we can write the last of Eq. (5) as

(22)
$$\frac{\partial \psi}{\partial y} \frac{dy}{ds} + \frac{\partial \psi}{\partial x} \frac{dx}{ds} = \frac{d\psi}{ds} = 0 , \qquad \text{on } C .$$

Hence ψ must be a constant along the boundary curve C. For a simply connected region, no loss of generality is involved in setting

(23)
$$\psi = 0 , \qquad \text{on } C .$$

If the cross section occupies a region R that is multi-connected, additional conditions of compatibility must be imposed (see Sec. 4.7).

It remains to examine the boundary conditions (6). The first, $\sigma_{zz} = 0$, follows the starting assumption. The other conditions are stated in Eqs. (11) and (12). Now,

$$\iint_r \sigma_{zx} \, dx dy = \iint_R \frac{\partial \psi}{\partial y} \, dx dy .$$

By Gauss' theorem, this is $\int_C \psi \nu_y ds$, and it vanishes on account of Eq. (23). Similarly, the resultant force in the y-direction vanishes. Thus, Eqs. (11) are satisfied. Finally, Eq. (12) requires that

$$T = - \iint_R \left(x \frac{\partial \psi}{\partial x} + y \frac{\partial \psi}{\partial y} \right) dx dy ,$$

which can be transformed by Gauss' theorem as follows:

$$(24) \qquad T = -\iint_R \left\{ \frac{\partial}{\partial x}(x\psi) + \frac{\partial}{\partial y}(y\psi) - 2\psi \right\} dxdy$$

$$= -\int_C \left\{ x\psi \cos(\nu, x) + y\psi \cos(\nu, y) \right\} ds + \iint_R 2\psi \, dxdy.$$

If R is a simply connected region, the line integral vanishes by the boundary condition (23). Hence,

$$(24a) \qquad\qquad\qquad T = 2\iint_R \psi \, dxdy.$$

Thus, all differential equations and boundary conditions concerning stresses are satisfied if ψ obeys Eqs. (20), (23), and (24). But there remains an indeterminate constant in Eq. (20). This constant has to be determined by boundary conditions on displacements. We have, from Eqs. (3) and (7),

$$(25) \qquad\qquad \frac{\partial w}{\partial x} = \frac{\sigma_{zx}}{G} + \alpha y, \qquad \frac{\partial w}{\partial y} = \frac{\sigma_{zy}}{G} - \alpha x.$$

Differentiating with respect to y and x, respectively, and subtracting, we get

$$(26) \qquad\qquad \frac{1}{G}\left(\frac{\partial \sigma_{zx}}{\partial y} - \frac{\partial \sigma_{zy}}{\partial x} \right) = -2\alpha.$$

Hence, a substitution from Eq. (18) gives

$$(27) \qquad\qquad \frac{\partial^2 \psi}{\partial x^2} + \frac{\partial^2 \psi}{\partial y^2} = -2G\alpha.$$

In this way, the problem of torsion is reduced to the solution of the Poisson Eq. (27) with boundary condition (23).

With either of the two approaches outlined above, the problem of torsion is reduced to standard problems in the theory of potentials in two dimensions. Such potential problems occur also in the theory of hydrodynamics, gravitation, static electricity, steady flow of heat, etc. A great deal is known about these potential problems; many special solutions have been worked out in detail and general methods of solution are available. The most powerful tool for potential theory in two dimensions comes from the theory of functions of a complex variable. Since this branch of applied mathematics is probably well-known to the readers in their study of other branches of physics, we shall not elaborate further. In any case, a detailed account of this beautiful field will require a book in its own name, and indeed many excellent books exist (see, for example, Courant,[5.1] Courant and Hilbert,[10.1]

Kellogg,[5.1] etc.) The complex variable method and the associated singular integral equations approach are developed in the monumental works of Muskhelishvilli;[1.2] a shorter account is given by Sokolnikoff.[1.2] Other methods of solution and detailed examples can be found in the classical books[1.2] of Love, Sokolnikoff, Southwell, Timoshenko and Goodier, Trefftz, Sechler, Green and Zerna. Goodier's article in Flügge's[1.4] *Handbook of Engineering Mechanics* contains results for various cross sections commonly used in engineering.

Example. *Bars with Elliptical Cross Section*
    Let the boundary of the cross section (Fig. 7.5:3) be given by the equation

$$\frac{x^2}{a^2} + \frac{y^2}{b^2} - 1 = 0 \, .$$

Then Eqs. (27) and (23) are satisfied by

$$\psi = -\frac{a^2 b^2 G\alpha}{(a^2 + b^2)} \left(\frac{x^2}{a^2} + \frac{y^2}{b^2} - 1 \right) \, .$$

Equation (24) gives the relation between the torque and the rate of twist $\alpha = d\theta/dz$.

$$T = \frac{\pi a^3 b^3}{(a^2 + b^2)} G \frac{d\theta}{dz} \, .$$

The stresses are given by Eq. (18). Note that the curves $\psi(x, y) =$ const. have an interesting meaning. The slope dy/dx of the tangent to such a curve is determined by the formula

$$\frac{\partial \psi}{\partial x} + \frac{\partial \psi}{\partial y} \frac{dy}{dx} = 0 \, .$$

Hence, according to Eq. (18), we have

(28)
$$\frac{dy}{dx} = \frac{\sigma_{zy}}{\sigma_{zx}} \, .$$

Thus, at each point of the curve $\psi(x, y) = $ const., the stress vector $(\sigma_{zx}, \sigma_{zy})$ is directed along the tangent to the curve. The curves $\psi(x, y) = $ const. are called the *lines of shearing stress*. The magnitude of the tangential stress is

(29)
$$\tau = \sqrt{\sigma_{zx}^2 + \sigma_{zy}^2} = \sqrt{\left(\frac{\partial \psi}{\partial x} \right)^2 + \left(\frac{\partial \psi}{\partial y} \right)^2} \, .$$

Hence, τ is equal to the absolute value of the gradient of the surface $z = \psi(x, y)$. The maximum stress occurs where the gradient is the largest.

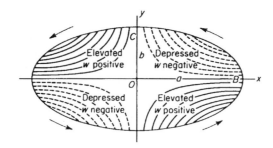

Fig. 7.5:3. Lines of constant warping, according to Saint-Venant.

In the present example, the lines of shearing stress are concentric ellipses. It is easy to see that the spacing of the $\psi = $ const. lines are closest at the end of the minor axis. The maximum shearing stress occurs there and is given by Eq. (29) to be

$$\tau_{\max} = 2G\alpha \frac{a^2 b}{a^2 + b^2}.$$

A general theorem can be proved that the points at which the maximum shearing stress occurs lie on the boundary curve of the cross section.

The warping function $\varphi(x,y)$ is easily shown to be

$$\varphi = -\frac{a^2 - b^2}{a^2 + b^2} xy.$$

Contour lines of constant displacement along the z-axis, $w = \alpha\varphi(x,y) = $ const., are hyperbolas as shown in Figs. 7.5:3, which were taken from Saint-Venant's original publication. The solid lines in the figure indicate where w is positive, the dotted lines where w is negative.

7.6. SOAP FILM ANALOGY

As remarked before, Eqs. (7.5:8) and (7.5:27) occur in many other physical theories entirely unrelated to the torsion problem. For example, if we consider a thin film of liquid, such as that of a soap bubble, we see that the predominant force acting in the film is the surface tension, which may be considered to be constant. The equation of equilibrium of an element of soap film is

$$\frac{T}{R_1} + \frac{T}{R_2} = p,$$

where R_1, R_2 are the principal radii of curvature and p is the pressure per unit area normal to the film. The derivation of this equation is simple and can be understood from elementary considerations as illustrated in Fig. 7.6:1. Now if we take a tube whose cross-sectional shape is the same as

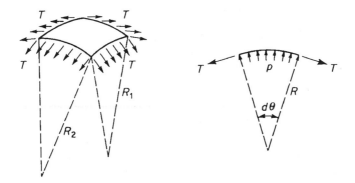

Fig. 7.6:1. Equilibrium of a soap film.

that of the shaft whose torsional property is questioned, cut a plane section, spread a soap film over it under a small pressure p. If the film deflection is sufficiently small, the mean curvature of the film is given by the sum of the second derivatives of the deflection surface. Thus,

$$T\left(\frac{\partial^2 w}{\partial x^2} + \frac{\partial^2 w}{\partial y^2}\right) = p, \qquad w = 0 \qquad \text{on boundary},$$

where w denotes the deflection of the film measured from the plane of the cross section and x, y are a set of rectangular coordinates. These equations are identical with Eqs. (7.5:23) and (7.5:27). Thus, we obtain Prandtl's *soap film analogy*. The gradient of the soap film is proportional to the shear stress in torsion. The volume under the film and above the cross section is proportional to the total torque.

The value of an analogy lies in its power of suggestion. Most people can visualize the shape of a soap film, perhaps because of their long experience with it. Thus, with the soap film analogy it is very easy to explain the stress concentration at an re-entrant corner in a cross section, such as the points marked by P in Fig. 7.6:2.

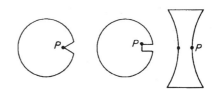

Fig. 7.6:2. Re-entrant corners suggest stress concentration.

Economical and efficient use of materials to transmit forces is an important objective in engineering; and the problem of avoiding the weakest links — points where stress concentrations occur — is obviously of great interest.

7.7. BENDING OF BEAMS

When a shaft is used to transmit bending moments and transverse shear, it is called a *beam*. Since beams are used in every engineering structure, the theory of beams is of great importance. The long history of the development of man's understanding of the action of the beam is a fascinating subject well-recorded in Timoshenko's book.[1.1] Modern investigation began with Galileo, but it is again to the credit of Saint-Venant that the problem is solved within the general theory of elasticity.

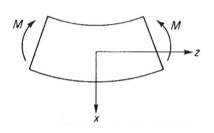

Fig. 7.7:1. Pure bending of a prismatic beam.

Consider first the pure bending of a prismatic beam (Fig. 7.7:1). Let the beam be subjected to two equal and opposite couples M acting in one of its principal planes.[†] Let the origin of the coordinates be taken at the centroid of a cross section, and let the x, z-plane be the principal plane of bending. The usual elementary theory of bending assumes that the stress components are

$$(1) \qquad \sigma_{zz} = \frac{Ex}{R}, \qquad \sigma_{xx} = \sigma_{yy} = \sigma_{xy} = \sigma_{xz} = \sigma_{yz} = 0,$$

in which R is the radius of curvature of the beam after bending. It is easily verified that the stress system (1) satisfies all the equations of equilibrium (7.5:4) and compatibility (7.3:6). The boundary conditions on the lateral surface of the beam are also satisfied. If the normal stress at the ends of the beam is linearly distributed as in Eq. (1), and if the bending moment is

$$(2) \qquad M = \iint \sigma_{zz}\, x dx dy = \iint \frac{1}{R} Ex^2 \, dx dy = \frac{EI}{R},$$

where I is the moment of inertia of the cross section of the beam with respect to the neutral axis parallel to the y-axis, then every condition is satisfied, and Eq. (1) gives an exact solution.

[†] A principal plane is one that contains the principal axes of the moment of area of the cross sections of the beam.

Consider next the same prismatic beam loaded by a lateral force P at the end $z = L$, and clamped at the end $z = 0$ (Fig. 7.7:2). For a beam so loaded, the resultant shear P has to be resisted by the shearing stresses σ_{zx}, σ_{zy}. Hence, the stress system (1) will not suffice.

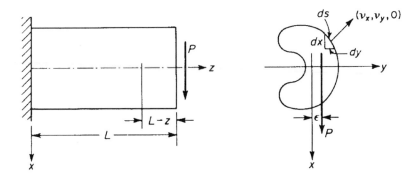

Fig. 7.7:2. Cantilever beam loaded at one end.

Let the force P be parallel to one of the principal axes of the cross section of the beam. (An arbitrary force can be resolved into components parallel to the principal axes, and the action of each component may be considered separately.) Let P be parallel to the x-axis and let the x, z-plane be a principal plane. Saint-Venant, using his semi-inverse method, assumes that

$$(3) \qquad \sigma_{zz} = -\frac{P(L-z)x}{I}, \qquad \sigma_{xx} = \sigma_{yy} = \sigma_{xy} = 0,$$

leaving σ_{zx}, σ_{zy} undetermined. The first term is given by the elementary theory, $P(L-z)$ being the bending moment at section z. Now we must see how the equilibrium, compatibility, and boundary conditions can be satisfied. In the absence of body force, the equilibrium equations (7.5:4) require that

$$(4) \qquad \frac{\partial \sigma_{zx}}{\partial z} = 0, \qquad \frac{\partial \sigma_{yz}}{\partial z} = 0,$$

$$(5) \qquad \frac{\partial \sigma_{xz}}{\partial x} + \frac{\partial \sigma_{yz}}{\partial y} + \frac{Px}{I} = 0.$$

The Beltrami–Michell compatibility conditions (7.3:6) require that

$$(6) \qquad \nabla^2 \sigma_{yz} = 0, \qquad \nabla^2 \sigma_{xz} + \frac{1}{1+\nu}\frac{P}{I} = 0,$$

where $\nabla^2 = \partial^2/\partial x^2 + \partial^2/\partial y^2$. The stress-free boundary condition on the lateral surface requires that [Eq. (7.5:5)]

(7) $$\sigma_{zx} \cos{(\nu, x)} + \sigma_{zy} \cos{(\nu, y)} = 0, \qquad \text{on } C.$$

The end condition at $z = L$ requires that

(8) $$\sigma_{zz} = 0, \qquad \iint \sigma_{zy} \, dx dy = 0,$$

(9) $$\iint \sigma_{zx} \, dx dy = P.$$

The end condition at $z = 0$ is concerned with the conditions of clamping, and it is usually stated in the form

(10) $$u = \frac{\partial u}{\partial z} = 0 \qquad \text{at} \qquad x = y = z = 0.$$

The method of solving Eqs. (4) through (9) is similar to that of Sec. 7.5. Equations (4) imply that both σ_{xz} and σ_{yz} are independent of z. Equation (5) may be written as

(11) $$\frac{\partial}{\partial x}\left(\sigma_{xz} + \frac{Px^2}{2I} - f(y)\right) + \frac{\partial \sigma_{yz}}{\partial y} = 0,$$

where $f(y)$ is a function of y only. Equation (11) can be satisfied identically if the stresses σ_{xz}, σ_{yz} are derived from a stress function $\psi(x,y)$ such that

(12) $$\sigma_{xz} = \frac{\partial \psi}{\partial y} - \frac{Px^2}{2I} + f(y), \qquad \sigma_{yz} = -\frac{\partial \psi}{\partial x}.$$

Equations (6) imply that

(13) $$\frac{\partial}{\partial x}(\nabla^2 \psi) = 0, \qquad \frac{\partial}{\partial y}(\nabla^2 \psi) = \frac{\nu}{1+\nu}\frac{P}{I} - \frac{d^2 f}{dy^2}.$$

Hence,

(14) $$\frac{\partial^2 \psi}{\partial x^2} + \frac{\partial^2 \psi}{\partial y^2} = \frac{\nu}{1+\nu}\frac{Py}{I} - \frac{df}{dy} + c.$$

The integration constant c has a very simple physical meaning. Consider the rotation ω of an element of area in the plane of a cross section.

(15) $$\omega = \frac{1}{2}\left(\frac{\partial v}{\partial x} - \frac{\partial u}{\partial y}\right).$$

The rate of change of this rotation in the z-axis direction is

$$\frac{\partial \omega}{\partial z} = \frac{1}{2} \frac{\partial}{\partial z} \left(\frac{\partial v}{\partial x} - \frac{\partial u}{\partial y} \right)$$

(16)
$$= \frac{1}{2} \frac{\partial}{\partial x} \left(\frac{\partial v}{\partial z} + \frac{\partial w}{\partial y} \right) - \frac{1}{2} \frac{\partial}{\partial y} \left(\frac{\partial u}{\partial z} + \frac{\partial w}{\partial x} \right) = \frac{\partial e_{yz}}{\partial x} - \frac{\partial e_{xz}}{\partial y}$$

$$= \frac{1}{2G} \left(\frac{\partial \sigma_{yz}}{\partial x} - \frac{\partial \sigma_{xz}}{\partial y} \right) = -\frac{1}{2G} \left(\frac{\partial^2 \psi}{\partial x^2} + \frac{\partial^2 \psi}{\partial y^2} + \frac{df}{dy} \right) .$$

In deriving the last line, Hooke's law and Eq. (12) are used. Hence, Eq. (14) leads to

(17)
$$-2G \frac{\partial \omega}{\partial z} = \frac{\nu}{1+\nu} \frac{Py}{I} + c .$$

This shows that c represents a constant rate of rotation, i.e., it corresponds to a rigid-body rotation of a cross section (the same kind as in the torsion problem). It can be shown that by a proper shifting of the load P parallel to itself in the plane $z = L$, the torsional deformation can be eliminated so that $c = 0$. (This leads to the concept of *shear center*, the point through which P must act so that $c = 0$.) In the following discussion we shall assume that P acts through the shear center. The more general problem can be solved obviously by a linear superposition of the solutions of bending and of torsion.

On setting $c = 0$, Eq. (14) becomes

(18)
$$\frac{\partial^2 \psi}{\partial x^2} + \frac{\partial^2 \psi}{\partial y^2} = \frac{\nu}{1+\nu} \frac{Py}{I} - \frac{df}{dy} .$$

The boundary condition (7) now requires that [see Fig. 7.5:2 and Eqs. (7.5:21)]

(19)
$$\frac{\partial \psi}{\partial y} \frac{dy}{ds} + \frac{\partial \psi}{\partial x} \frac{dx}{ds} = \frac{\partial \psi}{\partial s} = \left[\frac{Px^2}{2I} - f(y) \right] \frac{dy}{ds} .$$

From these equations, the value of ψ can be determined up to an integration constant which does not contribute anything to the stress system. The function $f(y)$ is arbitrary; it was introduced by Timoshenko to simplify the solution in case the boundary curve of the cross section can be written in the form

(20)
$$C : \frac{Px^2}{2I} - f(y) = 0 .$$

This would be the case, for example, if C is a circle or an ellipse. In such a case, we choose $f(y)$ according to Eq. (20). Then the boundary condition may be written as

(21) $\psi = 0$ on C.

It remains to show that the load at the end $z = L$ is equipollent to a shear P, i.e., that Eqs. (8) and (9) are satisfied. This is easily done. For example, using Eqs. (5) and (7), we have

$$\iint \sigma_{xz}\, dxdy = \iint \left[\sigma_{xz} + x \left(\frac{\partial \sigma_{xz}}{\partial x} + \frac{\partial \sigma_{yz}}{\partial y} \right) + \frac{Px^2}{I} \right] dxdy$$

$$= P + \int_C x[\sigma_{xz} \cos(\nu, x) + \sigma_{yz} \cos(\nu, y)]ds = P,$$

$$\iint \sigma_{yz}\, dxdy = \iint \left[\sigma_{yz} + y \left(\frac{\partial \sigma_{xz}}{\partial x} + \frac{\partial \sigma_{yz}}{\partial y} \right) + \frac{Pxy}{I} \right] dxdy$$

$$= \iint y[\sigma_{xz} \cos(\nu, x) + \sigma_{yz} \cos(\nu, y)]ds = 0,$$

since $\iint xydxdy$ vanishes because the x-axis is assumed to be a principal axis.

Thus, all the equations are satisfied if a solution $\psi(x, y)$ is found from Eqs. (18) and (19) or (21). This reduces the beam problem to a standard problem in two-dimensional potential theory.

If a solution is obtained and the moment of the shearing stresses is computed and set equal to $P\epsilon$

(22) $$\iint (x\sigma_{yz} - y\sigma_{xz})\, dxdy = P\epsilon,$$

the constant ϵ will give the location of the shear center. The applied load must have an eccentricity ϵ in order to obtain a bending without twisting (see discussion about the constant c above).

Example. *Beam of Circular Cross Section of Radius r*
    The boundary curve C is given by the equation

$$x^2 + y^2 = r^2.$$

Hence, we take

$$f(y) = \frac{P}{2I} (r^2 - y^2).$$

Equation (18) becomes

$$\frac{\partial^2 \psi}{\partial x^2} + \frac{\partial^2 \psi}{\partial y^2} = \frac{1 + 2\nu}{1 + \nu} \frac{Py}{I}.$$

It is easily verified that the solution that satisfies the boundary condition (21) is

$$\psi = -\frac{(1 + 2\nu)}{8(1 + \nu)} \frac{P}{I}(x^2 + y^2 - r^2)y \,.$$

The stress components are

$$\sigma_{xz} = \frac{(3 + 2\nu)}{8(1 + \nu)} \frac{P}{I} \left(r^2 - x^2 - \frac{1 - 2\nu}{3 + 2\nu} y^2 \right) \,,$$

$$\sigma_{yz} = -\frac{(1 + 2\nu)}{4(1 + \nu)} \frac{Pxy}{I} \,.$$

A large number of examples can be found in the books of Love,[1.1] Sokolnikoff,[1.2] Timoshenko,[1.2] etc. An extensive discussion of the shear center can be found in Sechler's and Sokolnikoff's books.[1.2] There are several possible definitions of shear center and center of twist [Trefftz (1935), Goodier (1944) and Weinstein (1947)]; an outline and comparison of various definitions can be found in Fung's *Aeroelasticity*[10.5] (1957) pp. 471–475. See Bibliography 7.2, p. 881.

7.8. PLANE ELASTIC WAVES

As a further illustration of the theory of elasticity, let us consider some simple types of waves in an elastic medium. The displacement components u_1, u_2, u_3 (or in unabridged notations, u, v, w) will be assumed to be infinitesimal so that all equations are linearized. The basic field equation, in the absence of body force, is Navier's Eq. (7.1:9)

$$(1) \qquad\qquad \rho \frac{\partial^2 u_i}{\partial t^2} = G u_{i,jj} + (\lambda + G) u_{j,ji} \,.$$

We shall first verify that the motion

$$(2) \qquad\qquad u = A \sin \frac{2\pi}{l}(x \pm ct) \,, \qquad v = w = 0 \,,$$

where A, l, c are constants, is possible if c assumes the special value c_L,

$$(3) \;\blacktriangle \qquad\qquad c_L = \sqrt{\frac{\lambda + 2G}{\rho}} = \sqrt{\frac{E(1 - \nu)}{(1 + \nu)(1 - 2\nu)\rho}} \,.$$

This can be verified at once by substituting Eqs. (2) into (1). The pattern of motion expressed by Eqs. (2) is unchanged when $x \pm c_L t$ remains constant. Hence, if the negative sign were taken, the pattern would move to the right with a velocity c_L as the time t increases. The constant c_L is called the *phase velocity* of the wave motion. In Eqs. (2), l is the *wave length*, as can be seen from the sinusoidal pattern of u as a function of x, at any instant of time. The particle velocity represented by Eqs. (2) is in the same direction as that of the wave propagation (namely, the x-axis). Such a motion is said to constitute a train of *longitudinal waves*. Since at any instant of time the wave crests lie in parallel planes, the motion represented by Eq. (2) is called a train of *plane waves*.

Next, let us consider the motion

$$(4) \qquad u = 0, \qquad v = A \sin \frac{2\pi}{l} (x \pm ct), \qquad w = 0,$$

which represents a train of plane waves of wave length l propagating in the x-axis direction with a phase velocity c. When Eqs. (4) are substituted into Eqs. (1), it is seen that c must assume the value c_T,

$$(5) \;\blacktriangle \qquad\qquad\qquad c_T = \sqrt{\frac{G}{\rho}} \, .$$

The particle velocity (in the y-direction) represented by Eqs. (4) is perpendicular to the direction of wave propagation (x-direction). Hence, it is said to be a *transverse wave*. The speeds c_L and c_T are called the characteristic *longitudinal wave speed* and *transverse wave speed*, respectively. They depend on the elastic constants and the density of the material. The ratio c_T/c_L depends on Poisson's ratio only,

$$(6) \;\blacktriangle \qquad\qquad\qquad c_T = c_L \sqrt{\frac{1 - 2\nu}{2(1 - \nu)}} \, .$$

If $\nu = 0.25$, then $c_L = \sqrt{3} c_T$.

Similar to Eqs. (4), the following example represents a transverse wave in which the particles move in the z-axis direction.

$$(7) \qquad u = 0, \qquad v = 0, \qquad w = A \sin \frac{2\pi}{l} (x \pm c_T t).$$

The plane parallel to which the particles move [such as the x, y-plane in Eqs. (4), or the x, z-plane in Eqs. (7)], is called the *plane of polarization*.

Table 6.2:1, Sec. 6.2, gives a brief list of the longitudinal wave velocities of some common media. It is very interesting to see that most metals and alloys have approximately the same wave velocities.

Plane waves as described above may exist only in an unbounded elastic continuum. In a finite body, a plane wave will be reflected when it hits a boundary. If there is another elastic medium beyond the boundary, refracted waves occur in the second medium. The features of reflection and refraction are similar to those in acoustics and optics; the main difference is that, in general, an incident longitudinal wave will be reflected and refracted in a combination of longitudinal and transverse waves, and an incident transverse wave will also be reflected in a combination of both types of waves. The details can be worked out by a proper combination of these waves so that the boundary conditions are satisfied. See Sec. 8.14.

7.9. RAYLEIGH SURFACE WAVE

In an elastic body, it is possible to have another type of wave, which is propagated over the surface and which penetrates only a little into the interior of the body. These waves are similar to waves produced on a smooth surface of water when a stone is thrown into it. They are called *surface waves*. The simplest is the *Rayleigh wave* that occurs on the free surface of a homogeneous, isotropic, semi-infinite solid. It is an important type of wave because the largest disturbances caused by an earthquake recorded on a distant seismogram are usually those of Rayleigh waves.

The criterion for surface waves is that the amplitude of the displacement in the medium diminishes exponentially with increasing distance from the boundary.

Let us demonstrate the existence of Rayleigh waves in the simple two-dimensional case. Consider an elastic half-space $y \geq 0$. The surface $y = 0$ is stress-free. Let us consider displacements represented by the real part of the following expressions:

$$u = Ae^{-by} \exp[ik(x - ct)],$$

(1)
$$v = Be^{-by} \exp[ik(x - ct)],$$

$$w = 0,$$

where i is the imaginary unit $\sqrt{-1}$, k is the wave number (a constant), and A and B are complex constants. The coefficient b is supposed to be real and positive so that the amplitude of the waves decreases exponentially with increasing y, and tends to zero as $y \to \infty$.

We would first see if the displacements given by Eq. (1) can satisfy the equations of motion, which, in view of the definitions of c_L and c_T by

Eqs. (7.8:3) and (7.8:5), can be written as

(2) $$\frac{\partial^2 u_i}{\partial t^2} = c_T^2 u_{i,jj} + (c_L^2 - c_T^2) u_{j,ji}.$$

Substituting Eqs. (1) into (2), cancelling the common exponential factor and rearranging terms, we obtain the equations

(3) $$[c_T^2 b^2 + (c^2 - c_L^2)k^2]A - i(c_L^2 - c_T^2)bkB = 0,$$
$$-i(c_L^2 - c_T^2)bkA + [c_L^2 b^2 + (c^2 - c_T^2)k^2]B = 0.$$

The condition for the existence of a nontrivial solution is the vanishing of the determinant of the coefficients, which may be written in the form

(4) $$[c_L^2 b^2 - (c_L^2 - c^2)k^2][c_T^2 b^2 - (c_T^2 - c^2)k^2] = 0.$$

This gives the following roots for b.

(5) $$b' = k\left(1 - \frac{c^2}{c_L^2}\right)^{1/2}, \qquad b'' = k\left(1 - \frac{c^2}{c_T^2}\right)^{1/2}.$$

The assumption that b is real requires that $c < c_T < c_L$. Corresponding to b' and b'', respectively, the ratio B/A can be solved from Eq. (3).

(6) $$\left(\frac{B}{A}\right)' = -\frac{b'}{ik}, \qquad \left(\frac{B}{A}\right)'' = \frac{ik}{b''}.$$

Hence, a general solution of the type (1), satisfying the equations of motion, may be written as

(7) $$u = A'e^{-b'y}\exp[ik(x - ct)] + A''e^{-b''y}\exp[ik(x - ct)],$$
$$v = -\frac{b'}{ik}A'e^{-b'y}\exp[ik(x - ct)] + \frac{ik}{b''}A''e^{-b'y}\exp[ik(x - ct)],$$
$$w = 0.$$

Now, we would like to show that the constants A', A'', k, and c can be so chosen that the boundary conditions on the free surface $y = 0$ can be satisfied:

(8) $$\sigma_{yx} = \sigma_{yy} = \sigma_{yz} = 0, \qquad\qquad \text{on } y = 0.$$

By Hooke's law, and in view of Eq. (7), conditions (8) are equivalent to

(9)
$$\frac{\partial u}{\partial y} + \frac{\partial v}{\partial x} = 0, \qquad \text{on } y = 0,$$

$$\lambda\left(\frac{\partial u}{\partial x} + \frac{\partial v}{\partial y}\right) + 2G\frac{\partial v}{\partial y} = 0, \qquad \text{on } y = 0.$$

On substituting Eqs. (7) into Eq. (9), setting $y = 0$, omitting the common factor $\exp[ik(x - ct)]$, and writing

$$G = \rho c_T^2, \qquad \lambda = \rho(c_L^2 - 2c_T^2),$$

we obtain the results

(10)
$$-2b'A' - \left(b'' + \frac{k^2}{b''}\right)A'' = 0,$$

$$\left((c_L^2 - 2c_T^2) - c_L^2\frac{b'^2}{k^2}\right)A' - 2c_T^2 A'' = 0.$$

This can be written more symmetrically, by Eqs. (5), as

(11)
$$2b'A' + \left(2 - \frac{c^2}{c_T^2}\right)k^2\frac{A''}{b''} = 0,$$

$$\left(2 - \frac{c^2}{c_T^2}\right)A' + 2b''\frac{A''}{b''} = 0.$$

For a nontrivial solution, the determinant of the coefficients of A', A'' must vanish, yielding the characteristic equation for c

(12)
$$\left(2 - \frac{c^2}{c_T^2}\right)^2 = 4\left(1 - \frac{c^2}{c_L^2}\right)^{1/2}\left(1 - \frac{c^2}{c_T^2}\right)^{1/2}.$$

The quantity c^2/c_T^2 can be factored out after rationalization, and Eq. (12), called the *Rayleigh equation*, takes the form

(13)
$$\frac{c^2}{c_T^2}\left[\frac{c^6}{c_T^6} - 8\frac{c^4}{c_T^4} + c^2\left(\frac{24}{c_T^2} - \frac{16}{c_L^2}\right) - 16\left(1 - \frac{c_T^2}{c_L^2}\right)\right] = 0.$$

If $c = 0$, Eqs. (7) are independent of time, and from Eq. (11) we have $A'' = -A'$ and $u = v = 0$. Hence, this solution is of no interest. The second factor in Eq. (13) is negative for $c = 0$, $c_T < c_L$, and is positive for $c = c_T$. There is always a root c of Eq. (13) in the range $(0, c_T)$. Hence, surface waves can exist with a speed less than c_T.

For an incompressible solid $c_L \to \infty$, Eq. (13) becomes

$$(14) \qquad \frac{c^6}{c_T^6} - 8\frac{c^4}{c_T^4} + 24\frac{c^2}{c_T^2} - 16 = 0 \,.$$

This cubic equation in c^2 has a real root at $c^2 = 0.91275c_T^2$, corresponding to surface waves with speed $c \simeq 0.95538c_T$. The other two roots for this case are complex and do not represent surface waves.

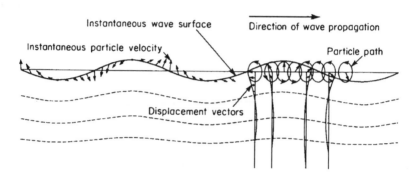

Fig. 7.9:1. Schematic drawing for Rayleigh surface waves.

If the Poisson ratio is $\frac{1}{4}$, so that $\lambda = G$ and $c_L = \sqrt{3}c_T$, Eq. (13) becomes

$$(15) \qquad \frac{c^6}{c_T^6} - 8\frac{c^4}{c_T^4} + \frac{56}{3}\frac{c^2}{c_T^2} - \frac{32}{3} = 0 \,.$$

This equation has three real roots: $c^2/c_T^2 = 4, \, 2+2/\sqrt{3}, \, 2-2/\sqrt{3}$. The last root alone can satisfy the condition that b' and b'' be real for surface waves. The other two roots correspond to complex b' and b'' and do not represent surface waves. In fact, these extraneous roots do not satisfy Eq. (12); they arise from the rationalization process of squaring. The last root corresponds to the velocity

$$(16) \qquad c_R = 0.9194c_T \,,$$

which corresponds, in turn, to the displacements

$$(17) \quad \begin{aligned} u &= A'(e^{-0.8475ky} - 0.5773e^{-0.3933ky}) \cos k(x - c_R t) \,, \\ v &= A'(-0.8475e^{-0.8475ky} + 1.4679e^{-0.3933ky}) \sin k(x - c_R t) \,, \end{aligned}$$

where A' is taken to be a real number, which may depend on k. From Eq. (17), it is seen the particle motion for Rayleigh waves is elliptical retrograde in contrast to the elliptical direct orbit for surface waves on water (see Fig. 7.9:1). The vertical displacement is about 1.5 times the horizontal displacement at the surface. Horizontal motion vanishes at a depth of 0.192 of a wavelength and reverses sign below this.

Figure 7.9:2 shows Knopoff's calculated results for the ratios c_R/c_T, c_R/c_L for Rayleigh waves as functions of Poisson's ratio ν.

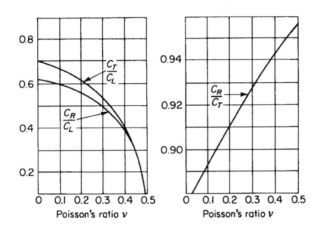

Fig. 7.9:2. Ratios c_R/c_L, c_R/c_T, and c_T/c_L, where $c_R =$ Rayleigh wave speed, $c_L =$ the longitudinal wave speed, $c_T =$ the transverse wave speed.

7.10. LOVE WAVE

In the Rayleigh waves examined in the previous section the material particles move in the plane of propagation. Thus, in Rayleigh waves over

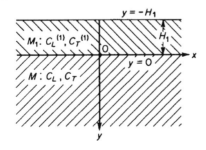

Fig. 7.10:1. A layered half-space.

the half-space $y \geq 0$ along the surface $y = 0$, propagating in the x-direction, the z-component of displacement w vanishes. It may be shown that surface wave with displacements perpendicular to the direction of propagation (the so-called *SH waves*) is impossible in a homogeneous half-space. However, *SH* surface waves are observed as prominently on the Earth's surface as other surface waves. Love showed that a theory sufficient to include SH surface waves can be constructed by having a homogeneous layer of a medium M_1 of uniform thickness H_1, overlying a homogeneous half-space of another medium M.

Using axes as in Fig. 7.10:1, we take $u = v = 0$, and

$$(1) \qquad w = A \exp\left[-k\sqrt{1 - \frac{c^2}{c_T^2}}\, y\right] \exp[ik(x - ct)]$$

in M, and

$$(2) \quad w = \left\{ A_1 \exp\left[-k\sqrt{1 - \left(\frac{c}{c_T^{(1)}}\right)^2}\, y\right] + A_1' \exp\left[k\sqrt{1 - \left(\frac{c}{c_T^{(1)}}\right)^2}\, y\right] \right\}$$
$$\times \exp[ik(x - ct)]$$

in M_1. It is easily verified that these equations satisfy the Navier's equations. If $c < c_T$, then $w \to 0$ as $y \to \infty$, as desired.

The boundary conditions are that w and σ_{xy} must be continuous across the surface $y = 0$, and σ_{zy} zero at $y = -H_1$. On applying these conditions to Eqs. (1) and (2), we obtain

$$(3) \qquad\qquad A = A_1 + A_1',$$

$$(4) \qquad GA[1 - (c/c_T)^2]^{1/2} = G_1(A_1 - A_1')[1 - (c/c_T^{(1)})^2]^{1/2},$$

$$(5) \quad A_1 \exp\{kH_1[1 - (c/c_T^{(1)})^2]^{1/2}\} = A_1' \exp\{-kH_1[1 - (c/c_T^{(1)})^2]^{1/2}\}.$$

Eliminating A from Eqs. (3) and (4), and then using Eq. (5) to eliminate A_1 and A_1', we have

$$\frac{G[1 - (c/c_T)^2]^{1/2}}{G_1[1 - (c/c_T^{(1)})^2]^{1/2}} = \frac{A_1 - A_1'}{A_1 + A_1'} = \tanh\{kH_1[1 - (c/c_T^{(1)})^2]^{1/2}\}.$$

Hence, we have

$$(6) \quad G\left(1 - \frac{c^2}{c_T^2}\right)^{1/2} - G_1\left[\left(\frac{c}{c_T^{(1)}}\right)^2 - 1\right]^{1/2} \tanh\left\{kH_1\left[\left(\frac{c}{c_T^{(1)}}\right)^2 - 1\right]^{1/2}\right\} = 0$$

as the equation to give the SH surface wave velocity c in the present conditions.

If $c_T^{(1)} < c_T$, Eq. (6) yields a real value of c which lies in the range $c_T^{(1)} < c < c_T$ and depends on k and H_1 (as well as on G, G_1, c_T, and $c_T^{(1)}$), because for c in this range the values of the left-hand-side terms in Eq. (6) are real and opposite in sign. Thus, SH surface waves can occur under the stated boundary conditions, provided the shear velocity $c_T^{(1)}$ in the upper layer is less than that in the medium M. These waves are called *Love waves.*

Love waves of general shape may be derived by superposing harmonic Love waves of the type (2) with different k. The dependence of the wave speed c on the wave number k introduces a dispersion phenomenon which will be considered later.

PROBLEMS

7.2.   Derive Navier's equation in spherical polar coordinates.

7.3.   From data given in various handbooks, determine the longitudinal and shear wave speeds in the following materials:

(a) Gases: air at sea level, and at 100,000 ft altitude.
(b) Metals: iron, a carbon steel, a stainless steel, copper, bronze, brass, nickle, aluminium, an aluminium alloy, titanium, titanium carbide, berylium, berylium oxide.
(c) Rocks and soils: a granite, a sandy loam.
(d) Wood: spruce, mahogany, balsa.
(e) Plastics: lucite, a foam rubber.

7.4.   Sketch the instantaneous wave surface, particle velocities, and particle paths of a Love wave.

7.5.   Investigate plane wave propagations in an anisotropic elastic material. Apply the results to a cubic crystal. *Note*:

$$\rho \frac{\partial^2 u_i}{\partial t^2} = C_{ijkl} \frac{\partial^2 u_l}{\partial x_j \partial x_k}, \qquad u_l = A_l e^{-i(\omega t - k_j x_j)}$$

where $\mathbf{k}(k_1, k_2, k_2)$ is the wave number vector normal to the wave front.

7.6.   Determine the stress field in a rotating, gravitating sphere of uniform density.

8

SOLUTIONS OF PROBLEMS IN LINEARIZED THEORY
OF ELASTICITY BY POTENTIALS

In this chapter we shall consider the use of the method of potentials in treating static and dynamic problems of an isotropic elastic body subjected to forces acting on the boundary surface of the body, the forces being independent of the deformation of the body. The main problem is to determine the stresses and displacements at every point in the body under appropriate boundary conditions. The displacements will be assumed to be infinitesimal and Hooke's law is assumed to hold, so that the basic equations are those listed in Sec. 7.1.

In both static and dynamic problems, we may start with Navier's equation and try to determine a continuous and twice differentiable solution u_i under appropriate boundary conditions. In the case of static equilibrium, we may also start with a general solution of the equation of equilibrium and use the compatibility and boundary conditions to determine the unique answer.

To simplify the solution, a number of potentials have been introduced. The most powerful ones are presented below. Potentials related to displacements are the scalar and vector potentials, the Galerkin vectors, and the Papkovich–Neuber functions. These will be discussed in this chapter. Potentials that generate systems of equilibrating stresses are the Airy stress-functions, the Maxwell–Morera stress functions, etc. (see Chapter 9). Some applications will be illustrated in this and subsequent chapters.

8.1. SCALAR AND VECTOR POTENTIALS FOR
DISPLACEMENT VECTOR FIELDS

It is well-known (sometimes referred to as Helmholtz' theorem) that *any analytic vector field* $\mathbf{u}(u_1, u_2, u_3)$ *can be expressed in the form*

(1) ▲
$$u_i = \phi_{,i} + e_{ijk}\psi_{k,j}$$

i.e.,

(2a) ▲ $$\mathbf{u} = \operatorname{grad} \phi + \operatorname{curl} \boldsymbol{\psi}$$

involving three equations of the type

(2b) ▲ $$u_1 = \frac{\partial \phi}{\partial x_1} + \frac{\partial \psi_3}{\partial x_2} - \frac{\partial \psi_2}{\partial x_3},$$

where ϕ is a scalar function and $\boldsymbol{\psi}$ is a vector field with three components.
ψ_1, ψ_2, ψ_3. The requirement (1) leaves ψ_i indeterminate to the extent that
its divergence is arbitrary. For definiteness, we may impose the condition

(3) $$\psi_{i,i} = 0.$$

Then ϕ is called the *scalar potential* and $\boldsymbol{\psi}$ the *vector potential* of the vector
field $\mathbf{u}$. We shall prove this theorem presently by showing that the functions
ϕ and ψ_1, ψ_2, ψ_3 so proposed can be found.

Applying Eq. (1) to the elastic displacement vector u_i, we shall see
that *the dilatation e can be derived from the scalar potential ϕ and that
the rotation can be derived from the vector potential $\boldsymbol{\psi}$*. Note first that if
Eq. (1) is effected, then

(4) $$u_{i,i} = \phi_{,ii} + e_{ijk}\psi_{k,ji}.$$

The last term vanishes when summed in pairs since $\psi_{k,ji} = \psi_{k,ij}$, but
$e_{ijk} = -e_{jik}$. Hence,

(5) $$e = u_{i,i} = \phi_{,ii}.$$

(6) $e_{ijk}u_{k,j} = e_{ijk}\phi_{,kj} + e_{ijk}e_{klm}\psi_{m,lj}$

$\qquad = e_{kij}e_{klm}\psi_{m,lj}$ \qquad [by symmetry of $\phi_{,kj}$]

$\qquad = (\delta_{il}\delta_{jm} - \delta_{im}\delta_{jl})\psi_{m,lj}$ \qquad [by the e-δ identity, Eq. (2.1:11)]

$\qquad = -\psi_{i,jj} + \psi_{j,ij}$

$\qquad = -\psi_{i,jj}$ \qquad [by virtue of Eq. (3)].

The curl of u_i, given by the left-hand side of Eq. (6), is twice the rotation
vector ω_i:

$$\omega_i = \frac{1}{2} e_{ijk} u_{k,j}.$$

Hence, we may write

(7) $$\psi_{i,jj} = -2\omega_i.$$

This shows that each component of the vector ψ_i is related to a component of the rotation vector field.

Now, particular solutions of Eqs. (5) and (7) are known in the theory of Newtonian potentials. They are

$$(8) \qquad \phi(x_1, x_2, x_3) = -\frac{1}{4\pi} \iiint \frac{e(x_1', x_2', x_3')}{r} dx_1' dx_2' dx_3',$$

$$(9) \qquad \bar{\psi}_i(x_1, x_2, x_3) = \frac{1}{2\pi} \iiint \frac{\omega_i(x_1', x_2', x_3')}{r} dx_1' dx_2' dx_3',$$

where

$$r^2 = (x_1 - x_1')^2 + (x_2 - x_2')^2 + (x_3 - x_3')^2$$

and the integration extends over the entire body. But $\bar{\psi}_i$ given in Eq. (9) does not always satisfy Eq. (3). To overcome this difficulty, we note that the curl of the gradient of a scalar function vanishes, and a general vector potential is

$$(10) \qquad \psi_i = \bar{\psi}_i + \theta_{,i},$$

where θ is a scalar function. The added term $\theta_{,i}$ contributes nothing to u_i. However, θ can be determined in such a manner that

$$(11) \qquad \theta_{,ii} = -\bar{\psi}_{i,i}$$

by an integral analogous to Eq. (8). With θ so determined, the function ψ_i given by Eq. (10) will satisfy the requirement Eq. (3).

The general solution of Eq. (5) is the sum of Eq. (8) and an arbitrary harmonic function ϕ^*. The general solution of Eq. (7) is the sum of Eq. (10) and a set of harmonic functions ψ_i^*. Thus, Eq. (1), considered as a set of partial differential equations for the unknown functions ϕ and ψ, can be solved.   Q.E.D.

Love shows (*Mathematical Theory of Elasticity*,[1.1] p. 48), through integration by parts of Eqs. (8) and (9), that the following resolution meets all the potentials requirements:

$$(12) \qquad \phi = -\frac{1}{4\pi} \iiint u_i(x_1', x_2', x_3') \frac{\partial}{\partial x_i}\left(\frac{1}{r}\right) dx_1' dx_2' dx_3',$$

$$(13) \qquad \psi_i = \frac{1}{4\pi} \iiint e_{ijk} u_k(x_1', x_2', x_3') \frac{\partial}{\partial x_j}\left(\frac{1}{r}\right) dx_1' dx_2' dx_3'.$$

Pendse (Phil. Mag. Ser. 7, 39: 862–867, 1948) pointed out, however, that the use of scalar and vector potentials is not always free from ambiguity.

8.2. EQUATIONS OF MOTION IN TERMS OF
DISPLACEMENT POTENTIALS

Let the displacement vector field u_1, u_2, u_3 be represented by a scalar potential $\phi(x_1, x_2, x_3, t)$ and a triple of vector potentials $\psi_i(x_1, x_2, x_3, t)$, $i = 1, 2, 3$, so that

$$(1) \qquad u_i = \frac{\partial \phi}{\partial x_i} + e_{ijk} \frac{\partial \psi_k}{\partial x_j}, \qquad \psi_{i,i} = 0 \,.$$

We shall assume that the body force is zero. [If this is not the case, the body force term can be removed by a particular integral such as Kelvin's, see Eq. (8.8:9). See also Sec. 8.18.] We can write the Navier's equation of linear elasticity for a homogeneous, isotropic body,

$$(2a) \qquad \rho \frac{\partial^2 u_i}{\partial t^2} = G u_{i,jj} + (\lambda + G) u_{j,ji} \,,$$

in the form

$$(2) \qquad \rho \frac{\partial}{\partial x_i} \left(\frac{\partial^2 \phi}{\partial t^2} \right) + \rho e_{ijk} \frac{\partial}{\partial x_j} \left(\frac{\partial^2 \psi_k}{\partial t^2} \right)$$

$$= (\lambda + G) \frac{\partial}{\partial x_i} \nabla^2 \phi + G \frac{\partial}{\partial x_i} \nabla^2 \phi + G e_{ijk} \frac{\partial}{\partial x_j} \nabla^2 \psi_k \,,$$

where ∇^2 is the Laplacian operator

$$\nabla^2 = \frac{\partial^2}{\partial x_1^2} + \frac{\partial^2}{\partial x_2^2} + \frac{\partial^2}{\partial x_3^2} \,.$$

The fact that $\psi_{i,i} = 0$, [Eq. (8.1:3)] is used here. Furthermore, since the displacements are assumed to be infinitesimal, any change in the density ρ is an infinitesimal of the first order, as can be seen easily from the equation of the continuity [Eq. (5.4:4)]. Since in Eq. (2a), ρ is multiplied by small quantities of the first order, u_i, it can be treated as a constant. Now, it is easy to see that *Eq. (2) will be satisfied if the functions ϕ and ψ_i, are solutions of the equations*

$$(3) \; \blacktriangle \qquad \nabla^2 \phi = \frac{1}{c_L^2} \frac{\partial^2 \phi}{\partial t^2} \,, \qquad \nabla^2 \psi_k = \frac{1}{c_T^2} \frac{\partial^2 \psi_k}{\partial t^2} \,,$$

where

$$(4) \qquad c_L = \sqrt{\frac{\lambda + 2G}{\rho}} \,, \qquad c_T = \sqrt{\frac{G}{\rho}} \,.$$

These are wave equations. They indicate that two types of disturbances with velocities c_L and c_T may be propagated through an elastic
solid. A comparison with Eqs. (7.8:3) and (7.8:5) shows that c_L and c_T are
the speeds of plane longitudinal and transverse waves, respectively. These
waves are also referred to as *dilatational* and *distorsional* waves. The latter
are also known as *shear, equivoluminal*, or *rotational* waves. If we write e for
the dilatation $u_{i,i}$ and $\boldsymbol{\omega}(\omega_1, \omega_2, \omega_3)$ for the rotation, then, by Eqs. (8.1:5)
and (8.1:7), we have, from Eq. (3),

$$(5) \qquad \nabla^2 e = \frac{1}{c_L^2} \frac{\partial^2 e}{\partial t^2}, \qquad \nabla^2 \omega_i = \frac{1}{c_T^2} \frac{\partial^2 \omega_i}{\partial t^2}.$$

Thus, by introducing the potentials ϕ and ψ_1, ψ_2, ψ_3, we have reduced the
problems of linear elasticity to that of solving the wave equations.

If static equilibrium is considered, then all derivatives with respect to
time vanish, and we see that Eq. (2) is satisfied if

$$(6) \qquad \nabla^2 \phi = \text{const.}, \qquad \nabla^2 \psi_i = \text{const.}$$

In the following sections we shall consider some applications to statics based
on Eqs. (6).

Of course, not all solutions of Eq. (2) are given by Eqs. (3) and (6).
We shall now consider a more general solution. Such a generalization often
becomes important when we wish to examine the limiting case of a dynamic
problem as the loading tends to a steady state. For example, if we consider a
load suddenly applied at $t = 0$ on a body and then kept constant afterwards,
we may question whether the steady-state solution is given by Eqs. (3) with
$\partial\phi/\partial t$, $\partial\psi_k/\partial t$ terms set to zero. A quick comparison with Eq. (6) shows
that this may not be the case if the constants in Eq. (6) do not vanish.

Let us differentiate Eq. (2) with respect to x_m.

$$(7) \qquad \frac{\partial^2}{\partial x_m \partial x_i}\left(\rho \frac{\partial^2 \phi}{\partial t^2}\right) + e_{ijk}\frac{\partial^2}{\partial x_m \partial x_j}\left(\rho \frac{\partial^2 \psi_k}{\partial t^2}\right)$$

$$= (\lambda + G)\frac{\partial^2}{\partial x_i \partial x_m}\nabla^2\phi + G\frac{\partial^2}{\partial x_i \partial x_m}\nabla^2\phi + Ge_{ijk}\frac{\partial^2}{\partial x_m \partial x_j}\nabla^2\psi_k.$$

A contraction of m with i yields

$$(8) \qquad \nabla^2\left(\nabla^2\phi - \frac{1}{c_L^2}\frac{\partial^2\phi}{\partial t^2}\right) = 0.$$

A multiplication of (7) with e_{iml}, summing over m and summing over i, and using Eq. (8.1:3), yields, on the other hand,

$$(9) \qquad \nabla^2 \left(\nabla^2 \psi_k - \frac{1}{c_t^2} \frac{\partial^2 \psi_k}{\partial t^2} \right) = 0 \, .$$

Hence,

$$(10) \; \blacktriangle \qquad \nabla^2 \phi - \frac{1}{c_L^2} \frac{\partial^2 \phi}{\partial t^2} = \Phi \, , \qquad \nabla^2 \Phi = 0 \, ,$$

$$(11) \; \blacktriangle \qquad \nabla^2 \psi_k - \frac{1}{c_T^2} \frac{\partial^2 \psi_k}{\partial t^2} = \Psi_k \, , \qquad \nabla^2 \Psi_k = 0 \, ,$$

where Φ, Ψ_1, Ψ_2, Ψ_3, are harmonic functions. The Φ and Ψ's are not entirely independent, but are connected through Eq. (2)

$$(12) \; \blacktriangle \qquad \frac{c_L^2}{c_T^2} \frac{\partial \Phi}{\partial x_i} + e_{ijk} \frac{\partial}{\partial x_j} \Psi_k = 0 \, .$$

Equations (10)–(12) give the general solution of (2). Equations (3) are obtained by setting $\Phi = \Psi_k = 0$; Eqs. (6), by taking Φ and Ψ_k to be constants and by letting time derivatives vanish.

8.3. STRAIN POTENTIAL

In this section we shall consider the static equilibrium of elastic bodies in the restricted case in which the components of displacements can be derived from a scalar function $\phi(x_1, x_2, x_3)$ so that

$$(1) \qquad 2Gu_i = \frac{\partial \phi}{\partial x_i} \, .$$

The function ϕ is called Lamé's *strain potential.* With Eq. (1), the dilatation e, the strain tensor e_{ij}, and the stress tensor σ_{ij} are

$$(2) \qquad 2Ge = 2Gu_{i,i} = \phi_{,ii} \, ,$$

$$(3) \qquad e_{ij} = \frac{1}{2}(u_{i,j} + u_{j,i}) = \frac{1}{2G} \phi_{,ij} \, ,$$

$$(4) \qquad \sigma_{ij} = \lambda \delta_{ij} e + 2G e_{ij} = \lambda \delta_{ij} e + \phi_{,ij} \, .$$

In the absence of body force, the function ϕ must satisfy Eqs. (8.2:6). Since our objective is to obtain some solution, not necessarily general, the constant in (8.2:6) will be chosen as zero. Then ϕ satisfies the Laplace equation

$$(6) \qquad \nabla^2 \phi = 0 \, .$$

Hence, it is a *harmonic function* by definition. Under this choice, we have

$$(7) \qquad \sigma_{ij} = \phi_{,ij}.$$

In unabridged notations with respect to rectangular Cartesian coordinates x, y, z, Eqs. (2)–(7) are

$$(8) \qquad 2Gu = \frac{\partial \phi}{\partial x}, \qquad 2Gv = \frac{\partial \phi}{\partial y}, \qquad 2Gw = \frac{\partial \phi}{\partial z},$$

$$(9) \qquad 2Ge = \nabla^2 \phi = \left(\frac{\partial^2}{\partial x^2} + \frac{\partial^2}{\partial y^2} + \frac{\partial^2}{\partial z^2} \right) \phi = 0,$$

$$(10) \qquad \sigma_{xx} = \frac{\partial^2 \phi}{\partial x^2}, \qquad \sigma_{xy} = \frac{\partial^2 \phi}{\partial x \partial y}, \qquad \text{etc.}$$

Corresponding formulas for curvilinear coordinates can be obtained by converting Eqs. (1)–(7) into general tensor equations and then specializing. The results for cylindrical coordinates r, θ, z are as follows, where $(\xi_r, \xi_\theta, \xi_z)$ and $(\sigma_{rr}, \sigma_{r\theta}, \text{etc.})$ denote the physical components of displacement and stress, respectively.

$$(11) \qquad 2G\xi_r = \frac{\partial \phi}{\partial r}, \qquad 2G\xi_\theta = \frac{1}{r}\frac{\partial \phi}{\partial \theta}, \qquad 2G\xi_z = \frac{\partial \phi}{\partial z},$$

$$(12) \qquad \nabla^2 \phi \equiv \left(\frac{\partial^2}{\partial r^2} + \frac{1}{r}\frac{\partial}{\partial r} + \frac{1}{r^2}\frac{\partial^2}{\partial \theta^2} + \frac{\partial^2}{\partial z^2} \right) \phi = 0,$$

$$(13) \qquad \sigma_{rr} = \frac{\partial^2 \phi}{\partial r^2}, \qquad \sigma_{\theta\theta} = \frac{1}{r}\frac{\partial \phi}{\partial r} + \frac{1}{r^2}\frac{\partial^2 \phi}{\partial \theta^2}, \qquad \sigma_{zz} = \frac{\partial^2 \phi}{\partial z^2},$$

$$(14) \qquad \sigma_{r\theta} = \frac{\partial}{\partial r}\left(\frac{1}{r}\frac{\partial \phi}{\partial \theta} \right) = \frac{\partial^2}{\partial r \partial \theta}\left(\frac{\phi}{r} \right),$$

$$(15) \qquad \sigma_{\theta z} = \frac{1}{r}\frac{\partial^2 \phi}{\partial \theta \partial z}, \qquad \sigma_{zr} = \frac{\partial^2 \phi}{\partial z \partial r}.$$

Harmonic functions are well-known. For example, it can easily verified that the functions

$$(16) \qquad A(x^2 - y^2) + 2Bxy,$$

$$(17) \qquad Cr^n \cos n\theta, \qquad r^2 = x^2 + y^2,$$

$$(18) \qquad C \log(r/a), \qquad r^2 = x^2 + y^2,$$

(19) $\qquad\qquad C\theta,$ $\qquad\qquad\qquad \theta = \tan^{-1}(y/x),$

(20) $\qquad\qquad \dfrac{C}{R},$ $\qquad\qquad\qquad R^2 = x^2 + y^2 + z^2,$

(21) $\qquad\qquad C\log(R+z),$

(22) $\qquad C\log\left\{\dfrac{(R_1+z-c)(R_2-z-c)}{r^2}\right\},$ $\qquad \begin{aligned} R_1^2 &= r^2 + (z-c)^2, \\ R_2^2 &= r^2 + (z+c)^2, \end{aligned}$

are harmonic functions, and combinations of them can solve some important practical problems in elasticity. In Eqs. (16)–(19) the coordinate z does not appear. Hence, the component of displacement in the z-direction, $\partial\phi/\partial z$, vanishes, and any combination of these functions can only correspond to a plane strain condition. It is obvious that the solutions (17)–(19) may be useful for cylindrical bodies, (20) may be useful for spheres, and (21) and (22) may be useful for bodies of revolution about the z-axis.

To the stress systems derived from these potentials, we may add the uniform stress distributions given by

(23) $\qquad\qquad\qquad \phi = Cr^2 = C(x^2+y^2),$

(24) $\qquad\qquad\qquad \phi = CR^2 = C(x^2+y^2+z^2),$

which are special solutions of Eq. (8.2:6). With Eq. (23) or (24), Eqs. (9)–(10) and (12)–(15) must be modified because $e \neq 0$.

Example. *Hollow Spheres Subjected to Internal and External Pressure*

Let a hollow sphere of inner radius a and outer radius b be subjected to an internal pressure p and an external pressure q. Because of the spherical symmetry of the problem, it will be advantageous to use spherical coordinates R, θ, φ. All shearing stresses $\sigma_{R\theta}$, $\sigma_{\varphi\theta}$, etc., vanish on account of symmetry.

The solution is furnished by the strain potential (20) and (24),

$$\phi = \frac{C}{R} + DR^2, \qquad R^2 = x^2 + y^2 + z^2,$$

from which

$$\frac{\partial^2\phi}{\partial x^2} = \frac{3Cx^2}{R^5} - \frac{C}{R^3} + 2D, \qquad \frac{\partial^2\phi}{\partial y^2} = \frac{3Cy^2}{R^5} - \frac{C}{R^3} + 2D.$$

Hence, at the point $(R, 0, 0)$

$$\sigma_{RR} = \sigma_{xx} = \frac{2C}{R^3} + 2D,$$

$$\sigma_{\theta\theta} = \sigma_{yy} = -\frac{C}{R^3} + 2D.$$

The constants C and D are easily determined from the boundary conditions,

$$\sigma_{RR} = -p \quad \text{when} \quad R = a, \qquad \sigma_{RR} = -q \quad \text{when} \quad R = b.$$

The results are

$$\sigma_{RR} = -p \frac{(b/R)^3 - 1}{(b/a)^3 - 1} - q \frac{1 - (a/R)^3}{1 - (a/b)^3},$$

$$\sigma_{\theta\theta} = \frac{p}{2} \frac{(b/R)^3 + 2}{(b/a)^3 - 1} - \frac{q}{2} \frac{(a/R)^3 + 2}{1 - (a/b)^3}.$$

Problem 8.1. Find the solution for a solid sphere subjected to a uniform external pressure q.

Problem 8.2. Let the inner wall of the hollow spherical shell, at $R = a$, be rigid (a condition realized by filling the hole with some incompressible material). When the outer surface $R = b$ is subjected to a uniform pressure q, what are the stresses at the inner wall?

Problem 8.3. Show that the function $\phi = C\theta$ solves the problem of a circular disk subjected to uniformly distributed tangential shear on the circumference.

8.4. GALERKIN VECTOR

We have shown in Sec. 8.2 that when the displacement field u_i is expressed in terms of scalar and vector potentials, ϕ and ψ_1, ψ_2, ψ_3, respectively,

(1) $$2Gu_i = \phi_{,i} + e_{ijk}\psi_{k,j}, \qquad \psi_{i,i} = 0$$

the equations of static equilibrium for a homogeneous isotropic linear elastic medium are satisfied if

(2) $$\nabla^2 \phi = \text{const.}, \qquad \nabla^2 \psi_k = \text{const.}$$

Thus, a broad class of problems in elastic equilibrium is solved by determining the four functions ϕ, ψ_1, ψ_2, ψ_3. However, this may not be the general solution. The general solution (8.2:10)–(8.2:12) is more complex.

In searching for other solutions of equal generality, Galerkin introduced, in papers published in 1930, displacement potential functions which satisfy biharmonic equations. Papkovich noted in 1932 that Galerkin's functions are components of a vector. It is perhaps simpler to introduce the Galerkin vector by considering the vector potential ψ_k itself as being generated by another vector field $\tilde{F}_i$,

$$(3) \qquad\qquad \psi_k = -e_{klm}c\tilde{F}_{m,l}\,,$$

where c is a constant. Then Eq. (1) becomes

$$(4) \qquad\qquad 2Gu_i = \phi_{,i} - e_{ijk}e_{klm}c\tilde{F}_{m,lj}\,.$$

This can be simplified by means of the e-δ identity [Eq. (2.1:11)] into

$$(5) \qquad\qquad 2Gu_i = \phi_{,i} - (\delta_{il}\delta_{jm} - \delta_{im}\delta_{jl})c\tilde{F}_{m,lj}$$

$$= \phi_{,i} + c\tilde{F}_{i,jj} - c\tilde{F}_{j,ji}\,.$$

But $c\tilde{F}_{j,j}$ is a scalar function, and can be specified arbitrarily without disturbing the definition (3). This suggests the representation[†]

$$(6) \qquad\qquad 2Gu_i = cF_{i,jj} - F_{j,ji}\,.$$

The hope now arises that, by a judicious choice of the constant c, the Navier Eq. (7.1:9), namely,

$$(\lambda + G)u_{j,jk} + Gu_{k,mm} + X_k = 0\,,$$

where X_k represents the body force per unit volume, can be simplified. Indeed, on substituting Eq. (6) into the Navier equation, and remembering that $\lambda + G = G/(1 - 2\nu)$, we obtain

$$\frac{1}{2(1-2\nu)}\,[cF_{i,jji} - F_{i,jii}]_{,k} + \frac{1}{2}\,[cF_{k,jj} - F_{j,jk}]_{,mm} + X_k = 0\,.$$

With a proper change of dummy indices, this is

$$\left[\frac{c-1}{2(1-2\nu)} - \frac{1}{2}\right]F_{j,jiik} + \frac{c}{2}F_{k,jjii} + X_k = 0\,.$$

The coefficient of the first term vanishes if

$$c = 2(1 - \nu)\,.$$

[†]In order that Eqs. (5) and (6) both represent the same displacement field, we may choose $F_i = \tilde{F}_i + H_i$, where H_i are arbitrary harmonic functions ($H_{i,jj} = 0$) and demand that $\tilde{F}_{j,j} = (\phi + H_{j,j})/(c-1)$.

Therefore, we conclude that *the basic equation of elasticity is satisfied when*

(7) ▲ $$2Gu_i = 2(1-\nu)F_{i,jj} - F_{j,ji} \, ,$$

if F_i satisfies the equation

(8) ▲ $$F_{i,jjmm} = -\frac{X_i}{1-\nu} \, .$$

The vector F_i so defined is called the Galerkin vector. If $X_i = 0$, Eq. (8) is said to be biharmonic and its solutions are called biharmonic functions. Thus, to solve a problem in static equilibrium is to determine the three functions F_1, F_2, F_3.

From Eq. (7), we can derive expressions for the stresses in terms of F_i. In unabridged notation, we have, with

$$\nabla^2 \equiv \frac{\partial^2}{\partial x^2} + \frac{\partial^2}{\partial y^2} + \frac{\partial^2}{\partial z^2} \, ,$$

the following results:

(9) $$\nabla^4 F_1 = -\frac{X_1}{1-\nu} \, ,$$

(10) $$2Gu_1 = 2(1-\nu)\nabla^2 F_1 - \frac{\partial}{\partial x_1}\left(\frac{\partial F_1}{\partial x_1} + \frac{\partial F_2}{\partial x_2} + \frac{\partial F_3}{\partial x_3}\right),$$

(11) $$2G\left(\frac{\partial u_1}{\partial x_1} + \frac{\partial u_2}{\partial x_2} + \frac{\partial u_3}{\partial x_3}\right) = (1-2\nu)\nabla^2\left(\frac{\partial F_1}{\partial x_1} + \frac{\partial F_2}{\partial x_2} + \frac{\partial F_3}{\partial x_3}\right),$$

(12) $$\sigma_{11} = 2(1-\nu)\frac{\partial}{\partial x_1}\nabla^2 F_1$$
$$+\left(\nu\nabla^2 - \frac{\partial^2}{\partial x_1^2}\right)\left(\frac{\partial F_1}{\partial x_1} + \frac{\partial F_2}{\partial x_2} + \frac{\partial F_3}{\partial x_3}\right),$$

(13) $$\sigma_{12} = (1-\nu)\left(\frac{\partial}{\partial x_2}\nabla^2 F_1 + \frac{\partial}{\partial x_1}\nabla^2 F_2\right)$$
$$-\frac{\partial^2}{\partial x_1 \partial x_2}\left(\frac{\partial F_1}{\partial x_1} + \frac{\partial F_2}{\partial x_2} + \frac{\partial F_3}{\partial x_3}\right),$$

(14) $$(\sigma_{11} + \sigma_{22} + \sigma_{33}) = (1+\nu)\nabla^2\left(\frac{\partial F_1}{\partial x_1} + \frac{\partial F_2}{\partial x_2} + \frac{\partial F_3}{\partial x_3}\right).$$

Other components u_2, u_3, σ_{22}, σ_{33}, etc., are obtained by cyclic permutation of the subscripts 1, 2, 3.

Problem 8.4. Determine the body forces, stresses, and displacements defined by the following Galerkin vectors in rectangular Cartesian coordinates:

(a) $F_1 = F_2 = 0$, $F_3 = R^2$,
(b) $F_1 = yR^2$, $F_2 = -xR^2$, $F_3 = 0$,
(c) $F_1 = F_2 = 0$, $F_3 = z^4$,

where $R^2 = x^2 + y^2 + z^2$.

Problem 8.5. Show that the Lamé strain potential ϕ can be identified with the divergence $-F_{i,i}$, if the body force vanishes and if F_i is harmonic.

8.5. EQUIVALENT GALERKIN VECTORS

If two Galerkin vectors $\bar{F}_i$ and $\bar{\bar{F}}_i$ define the same displacements, they are equivalent. Let $\bar{F}_i$ and $\bar{\bar{F}}_i$ be two equivalent Galerkin vectors. Then their difference

$$(1) \qquad\qquad F_i = \bar{F}_i - \bar{\bar{F}}_i$$

is a biharmonic vector and corresponds to zero displacement, $u_i = 0$. A general form of F_i can be obtained if we set, as in Sec. 8.4,

$$(2) \qquad\qquad F_i = cf_{i,jj} + f_{j,ji}$$

and determine c such that $u_i = 0$, where

$$2Gu_i = 2(1 - \nu)F_{i,jj} - F_{j,ji}\,.$$

The result is

$$(3) \qquad\qquad F_i = (1 - 2\nu)f_{i,jj} + f_{j,ji}\,,$$

provided that

$$(4) \qquad\qquad f_{i,jjkk} = 0\,.$$

As an application, let us consider a Galerkin vector the third component of which does not vanish,

$$(5) \qquad\qquad \bar{F}_1, \bar{F}_2, \bar{F}_3 \neq 0\,.$$

It will be shown that an equivalent $\bar{\bar{F}}_i$ with $\bar{\bar{F}}_3 = 0$ can be formed if the body force $X_i = 0$. A particular solution is obtained by taking a special vector f_i

$$(6) \qquad\qquad f_1 = f\,, \qquad f_2 = f_3 = 0\,.$$

Then, from Eqs. (3) and (5),

(7)
$$F_1 = \bar{F}_1 - \bar{\bar{F}}_1 = \left[(1 - 2\nu)\nabla^2 + \frac{\partial^2}{\partial x^2} \right] f ,$$

(8)
$$F_2 = \bar{F}_2 - \bar{\bar{F}}_2 = \frac{\partial^2 f}{\partial x \partial y} ,$$

(9)
$$F_3 = \bar{F}_3 - \bar{\bar{F}}_3 = \bar{F}_3 = \frac{\partial^2 f}{\partial x \partial z} , \qquad \bar{\bar{F}}_3 = 0 .$$

Now, since $X_i = 0$ implies that $\bar{F}_3$ is biharmonic, it will be possible to determine a biharmonic function f that satisfies Eq. (9). When f is so determined, we could compute $\bar{\bar{F}}_1$ and $\bar{\bar{F}}_2$ according to Eqs. (7) and (8). Thus, *when the body forces are zero, any Galerkin vector has an equivalent Galerkin vector with a zero component in the direction of z.*

8.6. EXAMPLE–VERTICAL LOAD ON THE HORIZONTAL SURFACE OF A SEMI-INFINITE SOLID

Consider a semi-infinite homogeneous, isotropic, linear elastic solid which occupies the space $z \geq 0$ as shown in Fig. 8.6:1, and is subjected to a sinusoidally distributed vertical load on the boundary surface $z = 0$, so that the boundary conditions are

(1)
    (a) $\sigma_{zz} = \mathcal{A} \cos \dfrac{\pi x}{l} \cos \dfrac{\pi y}{L} ,$
        at $z = 0 ;$
    (b) $\sigma_{zx} = \sigma_{zy} = 0 .$

The boundary condition at $z = \infty$ is the vanishing of all stress components.

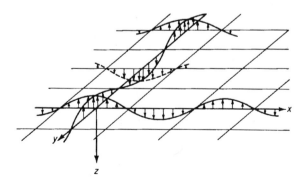

Fig. 8.6:1. Sinusoidally distributed load acting on a semi-infinte solid.

The problem can be solved by taking a Galerkin vector with one non-vanishing component,

(2) $$F_1 = F_2 = 0, \qquad F_3 = Z.$$

The function Z must be biharmonic. It can be easily verified that, if ψ is a harmonic function, then the following function is biharmonic (cf. Sec. 8.11):

(3) $$Z = (A + Bcz)\psi.$$

The form of the boundary condition (1a) suggests that we should try

(4) $$\psi = \cos\frac{\pi x}{l} \cos\frac{\pi y}{L} f(z).$$

On substituting (4) into the Laplace equation $\nabla^2\psi = 0$, one finds that

$$f(z) = e^{-cz},$$

where

(5) $$c = \sqrt{\left(\frac{\pi}{l}\right)^2 + \left(\frac{\pi}{L}\right)^2}.$$

The other solution e^{cz} must be rejected by the boundary condition at infinity.

The stresses corresponding to the Galerkin vector above can be easily derived. Note that

(6) $$\nabla^2 Z = 2Bc\frac{\partial\psi}{\partial z} = -2Bc^2\psi,$$

$$\frac{\partial Z}{\partial z} = -cA\psi + Bc(1 - cz)\psi,$$

$$\left(\frac{\partial^2}{\partial x^2} + \frac{\partial^2}{\partial y^2}\right) Z = -(A + Bcz)c^2\psi.$$

Hence, by Eqs. (8.4:13),

(7) $$\sigma_{zx} = \frac{\partial}{\partial x}[2\nu Bc^2 - (A + Bcz)c^2]\psi.$$

The boundary conditions (1b), that σ_{zx} and σ_{zy} vanish at the surface $z = 0$, are satisfied if

(8) $$A = 2\nu B.$$

Then, Eq. (8.4:12) yields

(9) $$\sigma_{zz} = Bc^3(1 + cz)\psi = Bc^3(1 + cz)\cos\frac{\pi x}{l}\cos\frac{\pi y}{L}e^{-cz}.$$

A comparison with the boundary condition (1a) shows that all boundary conditions are satisfied by taking

$$(10) \qquad B = \frac{A}{c^3}, \qquad A = \frac{2\nu A}{c^3}.$$

The resulting displacements and stresses are as follows, where we write, for convenience,

$$(11) \qquad \alpha = \frac{\pi}{l}, \qquad \beta = \frac{\pi}{L},$$

$$u_x = \frac{A\alpha}{2Gc^2}(-1 + 2\nu + cz)\sin\alpha x \cos\beta y e^{-cz},$$

$$u_y = \frac{A\beta}{2Gc^2}(-1 + 2\nu + cz)\cos\alpha x \sin\beta y e^{-cz},$$

$$u_z = \frac{A}{2Gc}[2(1 - \nu) + cz]\cos\alpha x \cos\beta y e^{-cz},$$

$$\sigma_{xx} = \frac{A}{c^2}(-\alpha^2 - 2\nu\beta^2 + \alpha^2 cz)\psi,$$

$$(12) \qquad \sigma_{yy} = \frac{A}{c^2}(-\beta^2 - 2\nu\alpha^2 + \beta^2 cz)\psi,$$

$$\sigma_{zz} = -A(1 + cz)\psi,$$

$$\sigma_{xy} = \frac{A\alpha\beta}{c^2}(1 - 2\nu - cz)\sin\alpha x \sin\beta y e^{-cz},$$

$$\sigma_{yz} = -A\beta z \cos\alpha x \sin\beta y e^{-cz},$$

$$\sigma_{zx} = -A\alpha z \sin\alpha x \cos\beta y e^{-cz}.$$

Thus, stresses are attenuated exponentially as the depth z is increased; the rate of attenuation c depends on the wave lengths l and L.

Solutions of this form may be superposed together to produce further solutions. The method of Fourier series may be used to obtain periodic loadings on the surface $z = 0$, and the method of Fourier integral may be used to obtain more general loadings.

8.7. LOVE'S STRAIN FUNCTION

A Galerkin vector that has only one nonvanishing component F_3 is called *Love's strain function* and shall be denoted by

$$(1) \qquad F_3 = Z.$$

In many applications it may be desired to express Z both in rectangular coordinates x, y, z and in cylindrical coordinates r, θ, z. In both cases,

(2)
$$\nabla^4 Z = -\frac{X_z}{1-\nu},$$

where

$$\nabla^2 \equiv \frac{\partial^2}{\partial x^2} + \frac{\partial^2}{\partial y^2} + \frac{\partial^2}{\partial z^2} = \frac{\partial^2}{\partial r^2} + \frac{1}{r}\frac{\partial}{\partial r} + \frac{1}{r^2}\frac{\partial^2}{\partial \theta^2} + \frac{\partial^2}{\partial z^2}$$

and X_z is the body force per unit volume in the z-direction — the only nonvanishing component that can be treated by such a strain function.

On putting $F_1 = F_2 = 0$ and $F_3 = Z$ in Eqs. (8.4:10)–(8.4:14), we obtain the following expressions for the physical components of displacements and stresses:

In rectangular coordinates:

$$2Gu_x = -\frac{\partial^2 Z}{\partial x \partial z}, \qquad 2Gu_y = -\frac{\partial^2 Z}{\partial y \partial z},$$

$$2Gu_z = \left[2(1-\nu)\nabla^2 - \frac{\partial^2}{\partial z^2}\right] Z,$$

$$\theta = \sigma_{xx} + \sigma_{yy} + \sigma_{zz} = (1+\nu)\frac{\partial \nabla^2 Z}{\partial z},$$

$$\sigma_{xx} = \frac{\partial}{\partial z}\left(\nu\nabla^2 - \frac{\partial^2}{\partial x^2}\right) Z,$$

(3)
$$\sigma_{yy} = \frac{\partial}{\partial z}\left(\nu\nabla^2 - \frac{\partial^2}{\partial y^2}\right) Z,$$

$$\sigma_{zz} = \frac{\partial}{\partial z}\left[(2-\nu)\nabla^2 - \frac{\partial^2}{\partial z^2}\right] Z,$$

$$\sigma_{zx} = \frac{\partial}{\partial x}\left[(1-\nu)\nabla^2 - \frac{\partial^2}{\partial z^2}\right] Z,$$

$$\sigma_{xy} = -\frac{\partial^3 Z}{\partial x \partial y \partial z},$$

$$\sigma_{zy} = \frac{\partial}{\partial y}\left[(1-\nu)\nabla^2 - \frac{\partial^2}{\partial z^2}\right] Z.$$

In cylindrical coordinates:

$$2G\xi_r = -\frac{\partial^2 Z}{\partial r \partial z}, \qquad 2G\xi_\theta = -\frac{1}{r}\frac{\partial^2 Z}{\partial \theta \partial z},$$

$$2G\xi_z = \left[2(1-\nu)\nabla^2 - \frac{\partial^2}{\partial z^2}\right]Z,$$

$$\theta = \sigma_{rr} + \sigma_{\theta\theta} + \sigma_{zz} = (1+\nu)\frac{\partial \nabla^2 Z}{\partial z},$$

$$\sigma_{rr} = \frac{\partial}{\partial z}\left(\nu\nabla^2 - \frac{\partial^2}{\partial r^2}\right)Z,$$

(4)

$$\sigma_{\theta\theta} = \frac{\partial}{\partial z}\left(\nu\nabla^2 - \frac{1}{r}\frac{\partial}{\partial r} - \frac{1}{r^2}\frac{\partial^2}{\partial \theta^2}\right)Z,$$

$$\sigma_{zz} = \frac{\partial}{\partial z}\left[(2-\nu)\nabla^2 - \frac{\partial^2}{\partial z^2}\right]Z,$$

$$\sigma_{r\theta} = -\frac{\partial^3}{\partial r \partial \theta \partial z}\left(\frac{Z}{r}\right),$$

$$\sigma_{\theta z} = \frac{1}{r}\frac{\partial}{\partial \theta}\left[(1-\nu)\nabla^2 - \frac{\partial^2}{\partial z^2}\right]Z,$$

$$\sigma_{zr} = \frac{\partial}{\partial r}\left[(1-\nu)\nabla^2 - \frac{\partial^2}{\partial z^2}\right]Z.$$

Love introduced, in 1906, the strain function Z as a function of r and z only in treating solids of revolution under axis-symmetric loading.

8.8. KELVIN'S PROBLEM — A SINGLE FORCE ACTING IN THE INTERIOR OF AN INFINITE SOLID

Let a force $2P$ be applied at the origin in the direction of z (Fig. 8.8:1). This concentrated force may be regarded as the limit of a system of loads applied on the surface of a small cavity at the origin. The boundary conditions of the problem are: (1) At infinity, all stresses vanish. (2) At the origin, the stress singularity is equivalent to a concentrated force of magnitude $2P$ in the z-direction.

The symmetry of the problem suggests the use of cylindrical coordinates, and Love's strain function $Z(r, z)$ suggests

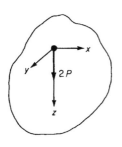

Fig. 8.8:1. Kelvin's problem.

itself. Since body force is absent, Z must be a biharmonic function whose third partial derivatives should define stresses that vanish at infinity, but have a singularity at the origin. One such function is

(1) $$Z = BR = B(z^2 + r^2)^{1/2}$$

(the function $\frac{1}{R}$ is harmonic, hence $R = R^2 \frac{1}{R}$ is biharmonic, see Theorem 2, Sec. 8.11), for which

(2) $$\frac{\partial Z}{\partial r} = \frac{Br}{R}, \qquad \frac{\partial^2 Z}{\partial r^2} = B\left(\frac{1}{R} - \frac{r^2}{R^3}\right) = \frac{Bz^2}{R^3},$$

$$\frac{\partial Z}{\partial z} = \frac{Bz}{R}, \qquad \frac{\partial^2 Z}{\partial z^2} = B\left(\frac{1}{R} - \frac{z^2}{R^3}\right) = \frac{Br^2}{R^3},$$

$$\nabla^2 Z = \frac{2B}{R}.$$

Therefore,

(3) $$2G\xi_r = \frac{Brz}{R^3}, \qquad 2G\xi_\theta = 0,$$

$$2G\xi_z = B\left[\frac{2(1-2\nu)}{R} + \frac{1}{R} + \frac{z^2}{R^3}\right],$$

$$\sigma_{rr} = B\left[\frac{(1-2\nu)z}{R^3} - \frac{3r^2z}{R^5}\right],$$

$$\sigma_{\theta\theta} = \frac{(1-2\nu)Bz}{R^3},$$

$$\sigma_{zz} = -B\left[\frac{(1-2\nu)z}{R^3} + \frac{3z^3}{R^5}\right],$$

$$\sigma_{rz} = -B\left[\frac{(1-2\nu)r}{R^3} + \frac{3rz^3}{R^5}\right],$$

$$\sigma_{z\theta} = \sigma_{r\theta} = 0.$$

These stresses are singular at the origin and vanish at infinity, and they have the correct symmetry. Therefore, the stress singularity can only be equivalent to a vertical force. Thus, Eq. (1) is indeed the desired solution if this force can be made equal to $2P$ by a proper choice of the constant B. To determine B, we consider a cylinder with a cavity at its center at the origin and with bases at $z = \pm a$ (Fig. 8.8:2). Since this cylinder is in equilibrium,

the resultant of surface tractions on the
outer surface must balance the load $2P$
acting on the surface of the cavity at
the origin. Therefore, we must have,
when the radius of this cylinder tends to
infinity,

$$2P = \int_0^\infty 2\pi r\, dr (-\sigma_{zz})_{z=a}$$

$$+ \int_0^\infty 2\pi r\, dr (\sigma_{zz})_{z=-a}$$

$$+ \lim_{r\to\infty} \int_{-a}^a 2\pi r\, dz (\sigma_{rz}).$$

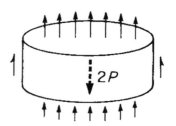

Fig. 8.8:2.   Boundary used in
evaluating the integration constant.

The values of the first and the second integrals are seen to be the same,
and the third integral vanishes in the limit. Noting that $r\,dr = R\,dR$,
we have

$$2P = 2 \int_0^\infty 2\pi R\, dR (-\sigma_{zz})_{z=a}$$

$$= 4\pi B \left[(1 - 2\nu)a \int_a^\infty \frac{R\,dR}{R^3} + 3a^3 \int_a^\infty \frac{R\,dR}{R^5} \right]$$

$$= 8\pi (1 - \nu)B.$$

Hence,

(4) $$B = \frac{P}{4\pi(1 - \nu)}$$

and the solution is completed.

An especially simple situation results when Poisson's ratio is $\frac{1}{2}$. Then
the factor $(1 - 2\nu)$ vanishes, and the stresses become

(5)
$$\sigma_{rr} = -\frac{3Pr^2 z}{2\pi R^5}, \qquad \sigma_{\theta\theta} = 0, \qquad \sigma_{r\theta} = \sigma_{z\theta} = 0,$$

$$\sigma_{zz} = -\frac{3Pz^3}{2\pi R^5}, \qquad \sigma_{rz} = -\frac{3Prz^2}{2\pi R^5}.$$

These results appear even simpler if the stresses are resolved in directions
of spherical coordinates. In any meridional plane (see Fig. 8.8:3), at a
point whose coordinates are (r, z) in cylindrical and (R, φ) in spherical

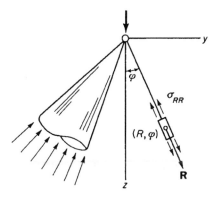

Fig. 8.8:3. Simple interpretation of the solution to Kelvin's problem when Poisson's ratio is 1/2.

coordinates, we find

(6) $$\sigma_{RR} = \sigma_{rr} \sin^2 \varphi + \sigma_{zz} \cos^2 \varphi + 2\sigma_{rz} \sin \varphi \cos \varphi$$

$$= \sigma_{rr} \frac{r^2}{R^2} + \sigma_{zz} \frac{z^2}{R^2} + 2\sigma_{rz} \frac{rz}{R^2} = -\frac{3Pz}{2a\pi R^3} \, .$$

Similarly, we have

(7) $$\sigma_{\varphi\varphi} = \sigma_{R\varphi} = \sigma_{\theta\theta} = 0 \, ,$$

and, by symmetry,

(8) $$\sigma_{R\theta} = \sigma_{\varphi\theta} = 0 \, .$$

Hence, σ_{RR}, $\sigma_{\varphi\varphi}$, $\sigma_{\theta\theta}$ are principal stresses, and the only nonvanishing component is σ_{RR}. If the solid were divided into many cones extending from a common vertex at the origin, each of these cones would transmit its own radial force without reaction from the adjacent cones.

    In the special case of Poisson's ratio $\nu = \frac{1}{2}$, it is evident that the solution of the Kelvin problem also furnishes the solution to Boussinesq's problem of a normal force and Cerruti's problem of a tangential force acting on the boundary of a semi-infinite solid (see Fig. 8.8:4). When $\nu = \frac{1}{2}$, Eqs. (5)–(8) hold for Boussinesq's problem when the solid occupies the space $z \geq 0$, and for Cerruti's problem when the solid occupies the space $x \geq 0$.

    The observation that the solutions of Boussinesq's and Cerruti's problems take on such simple form when Poisson's ratio is $\frac{1}{2}$ led Westergaard

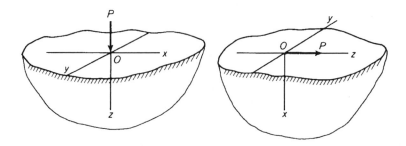

Fig. 8.8:4. Boussinesq and Cerruti's problems.

to consider a method of solution for these problems by perturbation of Poisson's ratio, a method to be described in the next section.

Originally, Lord Kelvin, Boussinesq, and Cerruti obtained solutions to the problems which now are adorned by their names (in 1848, 1878, 1882, respectively) by an extension of the method of singularities in the theory of Newtonian potentials. Lord Kelvin (1848) discovered that a particular solution of Navier's Eq. (7.1:9) is

$$(9) \qquad u_i(x) = A \iiint \left[B \frac{X_i(\xi)}{r} - (x_j - \xi_j)X_j(\xi) \frac{\partial}{\partial x_i} \left(\frac{1}{r} \right) \right] d\xi_1 d\xi_2 d\xi_3 \,,$$

where

$$r = [(x_1 - \xi_1)^2 + (x_2 - \xi_2)^2 + (x_3 - \xi_3)^2]^{1/2} \,,$$

$$A = \frac{\lambda + G}{8\pi G(\lambda + 2G)} \,, \qquad B = \frac{\lambda + 3G}{\lambda + G} \,.$$

In this formula x stands for (x_1, x_2, x_3), and ξ for (ξ_1, ξ_2, ξ_3). The quantity r is the distance between the field point x to the variable point ξ. The functions $X_i(\xi)$ are the components of the body force X_i expressed in terms of the variables of integration ξ_i. If an idealized case is considered in which a small sphere is isolated in the body, in which the body force concentrates, the limiting case will provide a solution of the Kelvin problem named above. A combination of such a solution with some other singular solutions of Navier's equation yields solutions to the Boussinesq and Cerruti problems. A general theory for this adaptation was given by Betti (1872), who showed how to express the dilatation and rotation at every point in a solid body by surface integrals containing explicitly the surface tractions and surface displacements. A lucid exposition of this approach can be found in Love, *Elasticity*,[1.1] Chapter X.

The corresponding problems of specific normal and tangential displacements at the origin on a semi-infinite solid were also solved by Boussinesq (1888). There are numerous ways of arriving at these results. For references, see Love,[1.1] p. 243.

> **Problem 8.6.** Investigate the system of stresses corresponding to
>
> (a) Strain potential $\phi = \log(R + z)$, $R^2 = x^2 + y^2 + z^2$.
> (b) Love's strain functions $Z = \log(R + z)$,
> (c) $Z = x \log(R + z)$.

Find a strain potential that is equivalent to (b).

8.9. PERTURBATION OF ELASTICITY SOLUTIONS BY A CHANGE OF POISSON'S RATIO

In many problems of elasticity, the effect of Poisson's ratio is relatively unimportant. In two-dimensional problems without body force, Poisson's ratio has no effect on the stress distribution at all. In fact, this is the basic justification of most photoelasticity studies, in which the selection of materials for the models can be made regardless of elastic constants.

In three-dimensional problems, the effect of Poisson's ratio generally cannot be disregarded, but the examples of Boussinesq's and Cerruti's problems in Sec. 8.8 show that particularly simple solutions can be obtained when Poisson's ratio assumes a certain particular value. This leads one to consider the effect of Poisson's ratio. Westergaard proposes to obtain a solution by first solving a problem for a specific value of Poisson's ratio m, and then determining the necessary corrections when the actual value of ν is used. The value of the shear modulus G is considered fixed.

Consider a body bounded by a surface S. Let u_i be the solution of a certain problem, i.e., it satisfies the Navier equation (in rectangular Cartesian coordinates).

$$(1) \qquad \frac{1}{1 - 2\nu} e_{,i} + u_{i,\alpha\alpha} + \frac{X_i}{G} = 0, \qquad e = u_{\alpha,\alpha},$$

and assumes on the boundary S surface traction $\overset{\nu}{T_i}$ or surface displacement u_i.

Let u_i^* be a solution of another problem, satisfying, for a different value of Poisson's ratio m, but the same value of G, the equation

$$(2) \qquad \frac{1}{1 - 2m} e_{,i}^* + u_{i,\alpha\alpha}^* + \frac{X_i^*}{G} = 0, \qquad e^* = u_{\alpha,\alpha}^*,$$

$$\overset{\nu}{T_i^*}, \quad \text{or} \quad u_i^* \text{ on } S.$$

This problem shall be assumed to have been solved, and is the starting point for further investigation on the effect of variation of Poisson's ratio.

Subtracting Eq. (2) from Eq. (1), and writing

(3) $\qquad u_i^{**} = u_i - u_i^*, \qquad e_{,i}^{**} = e_{,i} - e_{,i}^*, \qquad X_i^{**} = X_i - X_i^*,$

etc., we obtain

(4) $\qquad \left(\dfrac{1}{1-2\nu} - \dfrac{1}{1-2m} \right) e_{,i}^* + \dfrac{1}{1-2\nu} e_{,i}^{**} + u_{i,\alpha\alpha}^{**} + \dfrac{X_i^{**}}{G} = 0$

and

(5) $\qquad \overset{\nu}{T_i^{**}} = \overset{\nu}{T_i} - \overset{\nu}{T_i^*}, \quad \text{or} \quad u_i^{**} = u_i - u_i^* \text{ on } S.$

Equations (4) and (5) define another problem in the linear theory of elastic equilibrium. Any solution of Eqs. (4) and (5), added to a solution of Eq. (2), which corresponds to Poisson's ratio m, will yield a solution of Eq. (1) with the value of Poisson's ratio equal to ν.

To make this concept useful, let us consider a class of problems such as Boussinesq's, in which the boundary conditions take the form of specified surface tractions; e.g.,

(6) $\qquad \sigma_{33}^*, \sigma_{31}^*, \sigma_{32}^* \quad$ are specified on S: $\quad x_3 = \text{const.}$

Then it is expedient to consider a problem to solve Eq. (4) with

(7) $\qquad \sigma_{33}^{**} = \sigma_{31}^{**} = \sigma_{32}^{**} = 0 \quad \text{on } S: \quad x_3 = \text{const.}$

When such a particular solution is found, it can be added to the solution of Eq. (2) to form a solution of Eq. (1) with the same specified boundary values of σ_{33}, σ_{31}, σ_{32} as listed in Eq. (6).

Westergaard[1.2] shows how to construct a general solution in which σ_{33}^{**}, σ_{31}^{**}, and σ_{32}^{**} vanish identically throughout the elastic body. The method is based on the assumption that the components of displacement are derived from a scalar function φ according to the rule

(8) $\qquad 2Gu_1^{**} = \varphi_{,1}, \quad 2Gu_2^{**} = \varphi_{,2}, \quad 2Gu_3^{**} = -\varphi_{,3}.$

Westergaard calls this a "twinned gradient," the third component $-\varphi_{,3}$ being regarded as the "twin" of the ordinary gardient $\varphi_{,3}$. Now

(9)
$$\sigma_{ij} = \frac{2\nu}{1-2\nu} Ge_{\alpha\alpha}\delta_{ij} + 2Ge_{ij},$$

$$\sigma_{ij}^* = \frac{2m}{1-2m} Ge_{\alpha\alpha}^*\delta_{ij} + 2Ge_{ij}^*.$$

Hence,

(10) $\sigma_{ij}^{**} = \sigma_{ij} - \sigma_{ij}^{*}$

$$= \frac{2G(\nu - m)}{(1 - 2\nu)(1 - 2m)} e_{\alpha\alpha}^{*} \delta_{ij} + \frac{2G\nu}{1 - 2\nu} e_{\alpha\alpha}^{**} \delta_{ij} + 2G e_{ij}^{**} .$$

From Eq. (8), we have

(11) $e_{23}^{**} = 0 , \qquad e_{31}^{**} = 0 .$

Hence,

(12) $\sigma_{23}^{**} = 0 , \qquad \sigma_{31}^{**} = 0 .$

Furthermore,

(13) $2G e_{\alpha\alpha}^{**} = \varphi_{,11} + \varphi_{,22} - \varphi_{,33} = \nabla^2 \varphi - 2\varphi_{,33} ,$

and, from Eq. (10),

(14) $\sigma_{33}^{**} = \dfrac{2G(\nu - m)}{(1 - 2\nu)(1 - 2m)} e_{\alpha\alpha}^{*} + \dfrac{\nu}{1 - 2\nu} (\nabla^2 \varphi - 2\varphi_{,33}) - \varphi_{,33} .$

Hence, if we choose to have

(15) $\sigma_{33}^{**} \equiv 0 ,$

then φ must satisfy the equation

(16) $\nu \nabla^2 \varphi - \varphi_{,33} = -\dfrac{2G(\nu - m)}{(1 - 2m)} e_{\alpha\alpha}^{*} .$

With φ so chosen, Eq. (4) will have to yield the required value of the body force X_i^{**} in order to keep the body in equilibrium. However, it is simpler to proceed as follows. Compute first the stress σ_{11}^{**},

(17) $\sigma_{11}^{**} = \sigma_{11}^{**} - \sigma_{33}^{**} = 2G(e_{11}^{**} - e_{33}^{**}) = \varphi_{,11} + \varphi_{,33} .$

Then the only nonvanishing components of stress are

(18) $\sigma_{11}^{**} = \nabla^2 \varphi - \varphi_{,22} , \quad \sigma_{22}^{**} = \nabla^2 \varphi - \varphi_{,11} , \quad \sigma_{12}^{**} = \varphi_{,12} .$

The equations of equilibrium expressed in stresses show at once that the components of the required body force are

(19) $-X_1^{**} = (\nabla^2 \varphi)_{,1} , \quad -X_2^{**} = (\nabla^2 \varphi)_{,2} .$

The solution is established if a function φ can be found that satisfies Eqs. (16) and (19) simultaneously.

In a majority of significant problems, both the original and the final body forces are zero; i.e.,

$$(20) \qquad X_i = X_i^* = X_i^{**} = 0.$$

Then (16) and (19) can be satisfied if

$$(21) \qquad \nabla^2 \varphi = 0 \quad \text{and} \quad \varphi_{,33} = \frac{2G(\nu - m)}{(1 - 2m)} e_{\alpha\alpha}^* = \frac{\nu - m}{1 + m} \sigma_{\alpha\alpha}^*.$$

But $e_{\alpha\alpha}^*$ is a harmonic function. Hence, these two equations can be satisfied simultaneously, and the method is established.

The formulas given above are valid as long as ν and m are not exactly equal to $\frac{1}{2}$. In case $m \to \frac{1}{2}$, the dilation e^* tends to zero, but the bulk modulus tends to infinity. The sum of normal stresses

$$(22) \qquad \sigma_{ii}^* = \frac{2G(1 + m)}{(1 - 2m)} e_{ii}^*$$

remains finite as $m \to \frac{1}{2}$. It is easy to show that all the formulas in this section remain valid when $m = \frac{1}{2}$, provided that the e_{ii}^* term is replaced by the σ_{ii}^* term according to Eq. (22), as in the last equation of Eq. (21).

8.10. BOUSSINESQ'S PROBLEM

A load P acts at the origin of coordinates and perpendicular to the plane surface of a semi-infinite solid occupying the space $z \geq 0$ (Fig. 8.8:4). When Poisson's ratio is $\frac{1}{2}$, the problem has the simple solution as stated in Eqs. (8.8:5) which gives

$$(1) \qquad \theta^* = \sigma_{11} + \sigma_{22} + \sigma_{33} = -\frac{3Pz}{2\pi R^3}, \qquad R^2 = z^2 + r^2 = x^2 + y^2 + z^2.$$

If Poisson's ratio is ν, and $m = 1/2$, Westergaard's Eq. (8.9:21) becomes, in this case,

$$(2) \qquad \nabla^2 \varphi = 0, \qquad \frac{\partial^2 \varphi}{\partial z^2} = \frac{(1 - 2\nu)P}{2\pi} \frac{z}{R^3}.$$

Integrating, we have

$$(3) \qquad \frac{\partial \varphi}{\partial z} = -\frac{(1 - 2\nu)P}{2\pi R}, \qquad \varphi = -\frac{(1 - 2\nu)P}{2\pi} \log(R + z).$$

It can be shown that φ is harmonic; therefore the problem is solved.

On substituting Eq. (3) into Eqs. (8.9:8)–(8.9:14), and adding to the solution for $m = \frac{1}{2}$, we obtain the final results, in which the only nonvanishing components are

$$(4) \qquad \xi_r = \frac{P}{4\pi GR}\left[\frac{rz}{R^2} - \frac{(1-2\nu)r}{R+z}\right],$$

$$(5) \qquad \xi_z = \frac{P}{4\pi GR}\left[2(1-\nu) + \frac{z^2}{R^2}\right],$$

$$(6) \qquad \theta = \sigma_{rr} + \sigma_{\theta\theta} + \sigma_{zz} = -\frac{(1+\nu)}{\pi}P\frac{z}{R^3},$$

$$(7) \qquad \sigma_{rr} = \frac{P}{2\pi R^2}\left[-\frac{3r^2z}{R^3} + \frac{(1-2\nu)R}{R+z}\right],$$

$$(8) \qquad \sigma_{\theta\theta} = \frac{(1-2\nu)P}{2\pi R^2}\left[\frac{z}{R} - \frac{R}{R+z}\right],$$

$$(9) \qquad \sigma_{zz} = -\frac{3Pz^3}{2\pi R^5},$$

$$(10) \qquad \sigma_{rz} = -\frac{3Prz^2}{2\pi R^5},$$

where $r^2 = x^2 + y^2$, $R^2 = r^2 + z^2 = x^2 + y^2 + z^2$.

Problem 8.7. Solve the Boussinesq problem by a combination of a Galerkin vector

$$F_1 = F_2 = 0, \qquad F_3 = BR, \qquad R = (r^2 + z^2)^{1/2}$$

and a Lamé strain potential $\Phi = c\log(R+z)$. Show that $c = -(1-2\nu)B$, $B = P/2\pi$.

Problem 8.8. Solve Cerruti's problem by the method of "twinned gradient" (Reference, Westergaard,[1.2] p. 142).

8.11. ON BIHARMONIC FUNCTIONS

We have seen that many problems in elasticity are reduced to the solution of biharmonic equations with appropriate boundary conditions. It will be useful to consider the mathematical problem in some detail.

We shall consider the equation

$$(1) \qquad \nabla^2\nabla^2 u = 0,$$

where the operator ∇^2 is, in rectangular Cartesian coordinates (x, y, z),

$$\text{(2)} \qquad \nabla^2 \equiv \frac{\partial^2}{\partial x^2} + \frac{\partial^2}{\partial y^2} + \frac{\partial^2}{\partial z^2} \,,$$

in cylindrical polar coordinates (r, θ, z),

$$x = r \cos \theta, \qquad y = r \sin \theta \,, \qquad z = z \,,$$

$$\text{(3)} \qquad \nabla^2 \equiv \frac{\partial^2}{\partial r^2} + \frac{1}{r} \frac{\partial}{\partial r} + \frac{1}{r^2} \frac{\partial^2}{\partial \theta^2} + \frac{\partial^2}{\partial z^2}$$

and, in spherical polar coordinates (R, φ, θ),

$$x = R \sin \varphi \cos \theta \,, \qquad y = R \sin \varphi \sin \theta \,, \qquad z = R \cos \varphi \,,$$

$$\text{(4)} \qquad \nabla^2 \equiv \frac{1}{R^2} \frac{\partial}{\partial R} \left(R^2 \frac{\partial}{\partial R} \right) + \frac{1}{R^2 \sin \varphi} \frac{\partial}{\partial \varphi} \left(\sin \varphi \frac{\partial}{\partial \varphi} \right) + \frac{1}{R^2 \sin^2 \varphi} \frac{\partial^2}{\partial \theta^2}$$

$$= \frac{\partial^2}{\partial R^2} + \frac{2}{R} \frac{\partial}{\partial R} + \frac{1}{R^2} \frac{\partial^2}{\partial \varphi^2} + \frac{\cot \varphi}{R^2} \frac{\partial}{\partial \varphi} + \frac{1}{R^2 \sin^2 \varphi} \frac{\partial^2}{\partial \theta^2} \,.$$

The ∇^4 operator is obtained by repeated operation of the above. Thus,

$$\text{(5)} \qquad \nabla^4 \equiv \left(\frac{\partial^2}{\partial x^2} + \frac{\partial^2}{\partial y^2} + \frac{\partial^2}{\partial z^2} \right) \left(\frac{\partial^2}{\partial x^2} + \frac{\partial^2}{\partial y^2} + \frac{\partial^2}{\partial z^2} \right)$$

$$= \frac{\partial^4}{\partial x^4} + \frac{\partial^4}{\partial y^4} + \frac{\partial^4}{\partial z^4} + 2 \frac{\partial^4}{\partial x^2 \partial y^2} + 2 \frac{\partial^4}{\partial x^2 \partial z^2} + 2 \frac{\partial^4}{\partial y^2 \partial z^2} \,.$$

A regular solution of Eq. (1) in a region $\mathcal{R}$ is one that is four times continuously differentiable in $\mathcal{R}$. A regular solution of Eq. (1) is called a *biharmonic function*. Since Eq. (1) is obtained by repeated operation of the *Laplace operator* (2), and the regular solution of the equation $\nabla^2 u = 0$ is called a *harmonic function*, it is expected that biharmonic functions are closely connected with harmonic functions. In fact, we have the following theorems due to Almansi.[8.1]

Theorem 1. *If u_1, u_2 are two functions, harmonic in a region $\mathcal{R}(x, y, z)$, then*

$$\text{(6)} \qquad u = x u_1 + u_2$$

is biharmonic in $\mathcal{R}$. Conversely, if u is a given biharmonic function in a region $\mathcal{R}$, and if every line parallel to the x-axis intersects the boundary of

$\mathcal{R}$ in at most two points, then there exist two harmonic functions u_1 and u_2 in $\mathcal{R}$, so that u can be represented in the form of Eq. (6).

Proof. The first part of the theorem can be verified directly according to the identity

$$(7) \qquad \nabla^2(\phi\psi) = \phi\nabla^2\psi + \psi\nabla^2\phi + 2\left(\frac{\partial\phi}{\partial x}\frac{\partial\psi}{\partial x} + \frac{\partial\phi}{\partial y}\frac{\partial\psi}{\partial y} + \frac{\partial\phi}{\partial z}\frac{\partial\psi}{\partial z}\right).$$

To prove the converse theorem, we note that the theorem is established if we can show that there exist a function u_1 such that

$$(8) \qquad \text{(a)} \quad \nabla^2 u_1 = 0, \qquad \text{(b)} \quad \nabla^2(x u_1 - u) = 0.$$

By virtue of Eq. (8a), the second equation can be written as

$$(9) \qquad \nabla^2 u = \nabla^2(x u_1) = 2\frac{\partial u_1}{\partial x},$$

which has a particular solution

$$(10) \qquad \bar{u}_1(x, y, z) = \int_{x_0}^{x} \frac{1}{2}\nabla^2 u(\xi, y, z)\, d\xi,$$

where x_0 is an arbitrary point in the region $\mathcal{R}$. This particular solution does not necessarily satisfy Eq. (8a). However, since u is biharmonic, we have

$$(11) \qquad \frac{\partial}{\partial x}\nabla^2\bar{u}_1 = \nabla^2\frac{\partial\bar{u}_1}{\partial x} = \frac{1}{2}\nabla^4 u = 0.$$

Hence $\nabla^2\bar{u}_1$ is a function $v(y, z)$ of the variables y, z only. Now let us determine a function $\bar{\bar{u}}_1(y, z)$ so that

$$(12) \qquad \left(\frac{\partial^2}{\partial y^2} + \frac{\partial^2}{\partial z^2}\right)\bar{\bar{u}}_1 = -v(y, z);$$

for example, by

$$(13) \qquad \bar{\bar{u}}_1(y, z) = -\frac{1}{2\pi}\iint (\log r)\cdot v(\eta, \zeta)\, d\eta\, d\zeta,$$

where $r^2 = (y - \eta)^2 + (z - \zeta)^2$ and the integral extends through the region $\mathcal{R}$. Then the function $u_1 = \bar{u}_1 + \bar{\bar{u}}_1$ satisfies both conditions (9) and (8a) and the theorem is proved.

By a slight change in the proof it can be shown that Theorem 1 holds as well in the two-dimensional case.

Similarly, we have the following

Theorem 2. *If u_1, u_2 are two harmonic functions in a three-dimensional region $\mathcal{R}$, then*

$$(14) \qquad u = (R^2 - R_0^2)u_1 + u_2\,,$$

is biharmonic in $\mathcal{R}$, where $R^2 = x^2 + y^2 + z^2$ and R_0 is an arbitrary constant. Conversely, if u is a given biharmonic function in a region $\mathcal{R}$, and if $\mathcal{R}$ is such that, with an origin inside $\mathcal{R}$, each radius vector intersects the boundary of $\mathcal{R}$ in at most one point, then two harmonic functions u_1, u_2 can be determined so that Eq. (14) holds.

Proof. The proof of the first part again follows by direct calculation. Since u_1, u_2 are harmonic, an application of the identity (7) yields

$$
(15)
\begin{aligned}
&\nabla^2 u = u_1 \nabla^2 R^2 + 4\left(x\frac{\partial u_1}{\partial x} + y\frac{\partial u_1}{\partial y} + z\frac{\partial u_1}{\partial z}\right) = 6u_1 + 4R\frac{\partial u_1}{\partial R}\,, \\
&\nabla^2 \nabla^2 u = 6\nabla^2 u_1 + 8\left(\frac{\partial^2 u_1}{\partial x^2} + \frac{\partial^2 u_1}{\partial y^2} + \frac{\partial^2 u_1}{\partial z^2}\right) = 0\,.
\end{aligned}
$$

To prove the converse theorem, we note that the theorem is established if we can determine a function u_1 with the properties

$$(16) \qquad \text{(a)} \quad \nabla^2 u_1 = 0\,, \qquad \text{(b)} \quad \nabla^2[u - (R^2 - R_0^2)u_1] = 0\,.$$

Equation (16b) can be simplified, by virtue of Eq. (16a), into

$$(17) \qquad \nabla^2 u = 6u_1 + 4R\frac{\partial u_1}{\partial R}\,.$$

An integral of this differential equation is

$$(18) \qquad u_1 = R^{-3/2} \int_0^R \frac{1}{4}\rho^{1/2}\nabla^2 u\, d\rho\,.$$

It will now be shown that this integral indeed satisfies the condition (16a) and, hence, is the desired function. The demonstration will be simpler if the spherical coordinates are used. From Eq. (18),

$$
\begin{aligned}
(19) \quad \nabla^2 u_1 = &\left\{\frac{1}{R^2}\frac{\partial}{\partial R}\left(R^2\frac{\partial}{\partial R}\right) + \left[\frac{1}{R^2 \sin\varphi}\frac{\partial}{\partial\varphi}\left(\sin\varphi\frac{\partial}{\partial\varphi}\right)\right.\right. \\
&\left.\left. + \frac{1}{R^2 \sin^2\varphi}\frac{\partial^2}{\partial\theta^2}\right]\right\} R^{-3/2}\int_0^R \frac{1}{4}\rho^{1/2}\nabla^2 u\, d\rho\,.
\end{aligned}
$$

The operator in the square bracket can be taken under the sign of integration. But, since u is biharmonic $\nabla^2\nabla^2 u = 0$, we have

$$\left[\frac{1}{\sin\varphi}\frac{\partial}{\partial\varphi}\left(\sin\varphi\frac{\partial}{\partial\varphi}\right) + \frac{1}{\sin^2\varphi}\frac{\partial^2}{\partial\theta^2}\right]\nabla^2 u = -\frac{\partial}{\partial\rho}\left(\rho^2\frac{\partial}{\partial\rho}\right)\nabla^2 u.$$

Therefore, (19) may be written as

$$\nabla^2 u_1 = \frac{1}{R^2}\frac{\partial}{\partial R}\left(R^2\frac{\partial}{\partial R}R^{-3/2}\int_0^R\frac{1}{4}\rho^{1/2}\nabla^2 u\, d\rho\right)$$

$$- R^{-7/2}\int_0^R\frac{1}{4}\rho^{1/2}\frac{\partial}{\partial\rho}\left(\rho^2\frac{\partial}{\partial\rho}\right)\nabla^2 u\, d\rho.$$

On carrying out the indicated differentiation in the first term and integrating the second term twice by parts, we obtain finally

$$\nabla^2 u_1 = \frac{1}{R^2}\frac{\partial}{\partial R}\left\{-\frac{3}{8}R^{-1/2}\int_0^R\rho^{1/2}\nabla^2 u\, d\rho + \frac{R}{4}\nabla^2 u\right\}$$

$$-\frac{1}{4}\frac{1}{R}\frac{\partial\nabla^2 u}{\partial R} + \frac{1}{8}R^{-7/2}\int_0^R\rho^{3/2}\frac{\partial\nabla^2 u}{\partial\rho}\, d\rho$$

$$= \frac{1}{R^2}\left\{\frac{3}{8}\frac{1}{2}R^{-3/2}\int_0^R\rho^{1/2}\nabla^2 u\, d\rho - \frac{3}{8}\nabla^2 u + \frac{1}{4}\nabla^2 u + \frac{R}{4}\frac{\partial\nabla^2 u}{\partial R}\right\}$$

$$-\frac{1}{4}\frac{1}{R}\frac{\partial\nabla^2 u}{\partial R} + \frac{1}{8}\frac{1}{R^2}\nabla^2 u - \frac{1}{8}\frac{3}{2}R^{-7/2}\int_0^R\rho^{1/2}\nabla^2 u\, d\rho$$

$$= 0.$$

Hence, u_1 given by Eq. (18) satisfies all the requirements, and the theorem is proved.

Theorem 2 holds also in the two-dimensional case when R^2 is replaced by $r^2 = x^2 + y^2$. The proof is analogous to the above. It is also evident that the choice of x in Theorem 1 is incidental. The theorem holds when x is replaced by y or z.

Special representation of biharmonic functions in two-dimensions by means of analytic functions of a complex variable will be discussed in Sec. 9.5.

Problem 8.9. *Yih's solution of multiple Bessel equations.* Harmonic functions in cylindrical polar coordinates may assume the form $Z(r)\, e^{i\alpha x}e^{i\beta\theta}$, where $Z(r)$ is a Bessel function. Now consider the hyper-Bessel equation:

$$\left(\frac{d^2}{dr^2} + \frac{1}{r}\frac{d}{dr} - \frac{p^2}{r^2} + k^2\right)^n f = 0, \qquad n,\ \text{positive integer}.$$

With the help of Almansi's theorems discussed above, show that if p (taken to be positive for convenience) is not an integer, the solutions are $r^m J_{\pm(p+m)}(kr)$, in which $m = 0, 1, 2, \ldots, n-1$; otherwise they are $r^m J_{p+m}(kr)$ and $r^m N_{p+m}(kr)$, with m ranging over the same integers. The symbols J and N stand for the Bessel function and the Neumann function, respectively. (Chia Shun Yih, *Quart. Appl. Math.*, **13**, 4, 462–463, 1956.)

Problem 8.10. *Generation of biharmonic functions in cylindrical coordinates.* If $u_1(z, x)$ is a harmonic function in a space $\mathcal{R}(x, y, z)$, then

$$v_1 = \frac{1}{2\pi} \int_0^{2\pi} u_1(z, r\cos\theta) \, d\theta \,,$$

obtained by turning the space around the z-axis, is also harmonic. By selecting $u(z, x)$ as the real part of $(z + ix)^n$, in which n is a positive integer, show that the following functions are harmonic in cylindrical coordinates (r, θ, z), where $r^2 = x^2 + y^2$.

$$\psi_2 = z^2 - \frac{1}{2} r^2 \,,$$

$$\psi_3 = z^3 - \frac{3}{2} z r^2 \,,$$

$$\psi_4 = z^4 - 3 z^2 r^2 + \frac{3}{8} r^4 \,,$$

$$\psi_5 = z^5 - 5 z^3 r^2 + \frac{15}{8} z r^4 \,,$$

$$\psi_6 = z^6 - \frac{15}{2} z^4 r^2 + \frac{45}{8} z^2 r^4 - \frac{5}{16} r^6 \,.$$

Show that $z \psi_n$ and $(z^2 + r^2) \psi_n$ are biharmonic ($n = 2, 3, \ldots,$). *Note:* If $u(z, x)$ is a biharmonic function, then the process indicated above generates a biharmonic function $v(z, r)$.

8.12. NEUBER–PAPKOVICH REPRESENTATION

In Sec. 8.4, the displacement field u_i is represented by the Galerkin vector (F_1, F_2, F_3), in the form

(1) $\qquad\qquad 2Gu_i = 2(1-\nu)\nabla^2 F_i - F_{j,ji} \qquad i = 1, 2, 3 \,.$

If we set

(2) $\qquad\qquad \nabla^2 F_i = \frac{1}{2(1-\nu)} \Phi_i \,, \qquad F_{j,j} = \Psi \,,$

Eq. (1) becomes

(3) $\qquad\qquad 2Gu_i = \Phi_i - \Psi_{,i} \,.$

Consider first the case in which the body force is absent. Since in the absence of body force the Galerkin vector must satisfy the biharmonic equation, we see that Φ_i and Ψ satisfy the equations

$$(4) \qquad \nabla^2 \Phi_i = 0, \qquad \nabla^4 \Psi = 0.$$

Hence the Φ's are the harmonic and the Ψ is biharmonic. They are related through Eq. (2) by

$$(5) \qquad \nabla^2 \Psi = \frac{1}{2(1-\nu)} \Phi_{j,j}.$$

On noting that $\nabla^2(x_j \Phi_j) = 2\Phi_{j,j}$, we see that a particular solution of this equation is $\frac{1}{4(1-\nu)} x_j \Phi_j$. Hence, the general solution can be written as

$$(6) \qquad \Psi = \frac{1}{(1-\nu)} x_j \Phi_j + \phi_0,$$

where ϕ_0 is an arbitrary harmonic function. On substituting into Eq. (3), we obtain

$$(7) \qquad 2Gu_i = \frac{3-4\nu}{4(1-\nu)} \Phi_i - \frac{1}{4(1-\nu)} x_j \Phi_{j,i} - \phi_{0,i}.$$

If we define

$$(8) \qquad \phi_i = \frac{\Phi_i}{4(1-\nu)},$$

$$(9) \qquad \kappa = 3 - 4\nu,$$

we get

$$(10) \; \blacktriangle \qquad 2Gu_i = \kappa \phi_i - x_j \phi_{j,i} - \phi_{0,i}.$$

This formula expresses u_i in terms of four harmonic functions ϕ_0, ϕ_1, ϕ_2, ϕ_3. It was given independently by P. F. Papkovich (1932) and H. Neuber (1934) by different methods. The connection with the Galerkin vector was pointed out by Mindlin (1936). *The special importance of the Neuber–Papkovich solution lies in its strict similarity to a general solution in two dimensions (Sec. 9.6), in which case a well-known procedure exists for the determination of the harmonic functions involved from specified boundary conditions.*

The question of whether all four of the harmonic functions are independent, or whether one of them may be eliminated so that the general solution of the three-dimensional Navier's equation involves only three independent harmonic functions, has been a subject of much discussion. See Sokolnikoff, *Elasticity*,[1.2] 2nd ed. (1956), p. 331, and Naghdi[8.1] (1960).

If the body force does not vanish, a general solution of Navier's equation can be obtained by adding a particular integral to the Neuber–Papkovich solution [see Eq. (8.8:9)].

8.13. OTHER METHODS OF SOLUTION OF ELASTOSTATIC PROBLEMS

Two other classical methods of solving Navier's equations for elastostatic problems must be mentioned. The first is Betti's method, which was referred to in Sec. 8.8. The second is the method of integral transformation (Fourier, Laplace, Hankel, Mellin, Stieltjes, etc.) For the former, see Love, *Mathematical Theory*,[1.2] Chap. 10, and for the latter, Sneddon, *Fourier Transforms*[9.2] (1951), Chap. 10, and Flügge, *Encyclopedia of Physics*,[1.4] Vol. 6 (1962).

8.14. REFLECTION AND REFRACTION OF PLANE P AND S WAVES

So far we have considered only static problems. We shall now consider some dynamic problems in order to illustrate the use of displacement potentials in dynamics.

According to Sec. 8.2, *when the displacements are represented by a scalar potential ϕ and vector potentials ψ_1, ψ_2, ψ_3 through the expression*

$$u_i = \frac{\partial \phi}{\partial x_i} + e_{ijk} \frac{\partial \psi_k}{\partial x_j},$$

a broad class of solution is obtained if ϕ and ψ_k satisfy the wave equations

$$\frac{\partial^2 \phi}{\partial x^2} + \frac{\partial^2 \phi}{\partial y^2} + \frac{\partial^2 \phi}{\partial z^2} = \frac{1}{c_L^2} \frac{\partial^2 \phi}{\partial t^2},$$

$$\frac{\partial^2 \psi_k}{\partial x^2} + \frac{\partial^2 \psi_k}{\partial y^2} + \frac{\partial^2 \psi_k}{\partial z^2} = \frac{1}{c_T^2} \frac{\partial^2 \psi_k}{\partial t^2}.$$

The functions ϕ and ψ_1, ψ_2, ψ_3 define dilatational and distorsional waves, which are called, in seismology, the *primary*, or P (or "push"), waves, and the secondary, or S (or "shake"), waves. If we consider plane waves, as in Sec. 7.8, we see that the S waves are polarized. If an S-wave train propagates along the x-axis in the x, z-plane, (z "vertical", x "horizontal"), and the material particles move in the z-direction (vertical), then we speak of SV waves. If the S waves propagate along the x, z-plane but the particles move in the y-direction ("horizontal"), then we speak of SH waves.

Consider a homogeneous isotropic elastic medium occupying, a half-space $z \geq 0$. Since an elastic medium has two characteristic wave speeds, plane P waves hitting the free boundary $z = 0$ are reflected into plane P waves and plane S waves. Similarly, incident SV waves are reflected as both P and SV waves. If two elastic media are in contact with a "welded" interface, then incident P waves will be reflected in the first medium into P and S waves, and also refracted in the second medium in P and S waves. A similar statement holds for incident SV waves. The SH waves behave simpler. A train of incident SH waves will not generate P waves at the interface, so it is reflected and refracted in SH waves.

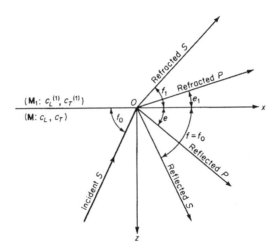

Fig. 8.14:1. Reflection of a SV ray incident against a plane boundary.

We shall show that *the laws of reflection and refraction are the same Snell's law as in optics*. Thus, if we have two homogeneous isotropic elastic media M and M_1, of infinite extent and in "welded" contact at the plane $z = 0$, as shown in Fig. 8.14:1, and if the directions of advance of the waves are all parallel to the x, z-plane as illustrated by rays in the figure, then for incident SV waves,

$$(1) \qquad \frac{c_T}{\cos f_0} = \frac{c_T}{\cos f} = \frac{c_L}{\cos e} = \frac{c_T^{(1)}}{\cos f_1} = \frac{c_L^{(1)}}{\cos e_1}, \qquad (\because f = f_0).$$

In this equation, c_L, c_T are, respectively, the longitudinal and transverse wave speeds of the medium M, and $c_L^{(1)}$, $c_T^{(1)}$ are the corresponding speeds of the medium M_1. The angle f_0 between the ray of the incident waves and the plane boundary is called the *angle of emergence* of the waves. Its complement is called the *angle of incidence*. Similarly, for incident SH waves, we have

$$
(2) \qquad \frac{c_T}{\cos f_0} = \frac{c_T}{\cos f} = \frac{c_T^{(1)}}{\cos f_1} , \qquad (\because f = f_0) ,
$$

and for incident P waves,

$$
(3) \qquad \frac{c_L}{\cos e_0} = \frac{c_L}{\cos e} = \frac{c_T}{\cos f} = \frac{c_L^{(1)}}{\cos e_1} = \frac{c_T^{(1)}}{\cos f_1}
$$

(so that $e = e_0$). If we consider a half-space M with a free surface $z = 0$, we have these same equations, with the $c_L^{(1)}$, $c_T^{(1)}$ terms, which are now irrelevant, omitted, of course.

These results are easily proved. Let us work out the case of SV waves emerging against a free boundary. Other cases are similar.

In SV waves emerging at an angle f_0 from a free plane boundary, the wave front has a normal in the direction of a unit vector whose direction cosines are $(\cos f_0, 0, \sin f_0)$; but in the incident SV waves the normal to the wave front has a different direction, with direction cosines $(\cos f_0, 0, -\sin f_0)$. This change of direction excites a reflected P wave. We assume, therefore, that

$$
(4) \qquad \begin{aligned}
&\phi = \Phi(x \cos e + z \sin e - c_L t) , \quad \psi_1 = \psi_3 = 0 , \\
&\psi_2 = \psi = \Psi_0(x \cos f_0 - z \sin f_0 - c_T t) + \Psi(x \cos f + z \sin f - c_T t) .
\end{aligned}
$$

The displacements are

$$
(5) \qquad u = \frac{\partial \phi}{\partial x} - \frac{\partial \psi}{\partial z} , \qquad w = \frac{\partial \phi}{\partial z} + \frac{\partial \psi}{\partial x} ,
$$

and the stresses are given by

$$
(6) \qquad \begin{aligned}
&\sigma_{zz} = \lambda \left(\frac{\partial^2 \phi}{\partial x^2} + \frac{\partial^2 \phi}{\partial z^2} \right) + 2G \left(\frac{\partial^2 \phi}{\partial z^2} + \frac{\partial^2 \psi}{\partial x \partial z} \right) , \\
&\sigma_{zx} = G \left(2 \frac{\partial^2 \phi}{\partial x \partial z} + \frac{\partial^2 \psi}{\partial x^2} - \frac{\partial^2 \psi}{\partial z^2} \right) .
\end{aligned}
$$

The boundary conditions are

$$
(7) \qquad \text{at} \quad z = 0 , \qquad \sigma_{zz} = \sigma_{zx} = 0 .
$$

On substituting Eqs. (4) and (6) into Eq. (7), we have

(8) $\quad (\lambda + 2G\sin^2 e)\Phi''(x\cos e - c_L t) - 2G[\cos f_0 \sin f_0 \Psi_0''(x\cos f_0 - c_T t)$

$\qquad - \cos f \sin f \Psi''(x\cos f - c_T t)] = 0\,,$

(9) $\quad 2\cos e \sin e\, \Phi''(x\cos e - c_L t) + (\cos^2 f_0 - \sin^2 f_0)\Psi_0''(x\cos f_0 - c_T t)$

$\qquad + (\cos^2 f - \sin^2 f)\Psi''(x\cos f - c_T t) = 0\,.$

These equations can be satisfied for all values of x and t only if the arguments of the various Φ and Ψ functions are in a constant ratio. Hence,

(10) $$\frac{c_T}{\cos f_0} = \frac{c_L}{\cos e} = \frac{c_T}{\cos f} \qquad\qquad \text{Q.E.D.}$$

When Eqs. (10) are satisfied, the functional relationships between Ψ_0, Ψ, and Φ are given by Eqs. (8) and (9). A detailed study of such functional relationships yields information about the partitioning of energy among the various components of reflected and refracted waves — an important subject whose details can be found in Ewing, Jardetzky, and Press.[7.4]

8.15. LAMB'S PROBLEM—LINE LOAD SUDDENLY APPLIED ON ELASTIC HALF-SPACE

Lamb,[7.4] in a classic paper published in 1904, considered the disturbance generated in a semi-infinite medium by an impulsive force applied along a line or at a point on the surface or inside the medium. Lamb's solution, as well as the extensions thereof, has been studied by Nakano, Lapwood, Pekeris, Cagniard, Garvin, Chao, and others. In this section we shall consider only the problem of a line load suddenly applied on the surface. (Fig. 8.15:1.)

Consider a semi-infinite body of homogeneous isotropic linear elastic material occupying the space $z \geq 0$. For time $t < 0$, the medium is stationary. At $t = 0$, a concentrated load is suddenly applied normal to the free surface $z = 0$, along a line coincident with the y-axis. The boundary conditions are therefore two-dimensional. We may assume the deformation state to be plane strain. The displacement v vanishes, and u and w are independent of y. Under this assumption, only one component of the vector potential is required. According to Sec. 8.2, the displacements are represented by

(1) $$u = \frac{\partial\phi}{\partial x} - \frac{\partial\psi}{\partial z}\,, \qquad w = \frac{\partial\phi}{\partial z} + \frac{\partial\psi}{\partial x}\,.$$

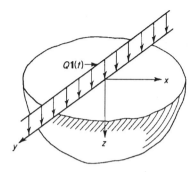

Fig. 8.15:1. A suddenly applied line load on an elastic half-space.

Navier's equations of motion are satisfied if ϕ and ψ satisfy the wave equations

(2) $$\frac{\partial^2 \phi}{\partial x^2} + \frac{\partial^2 \phi}{\partial z^2} = \frac{1}{c_L^2} \frac{\partial^2 \phi}{\partial t^2}, \quad \frac{\partial^2 \psi}{\partial x^2} + \frac{\partial^2 \psi}{\partial z^2} = \frac{1}{c_T^2} \frac{\partial^2 \psi}{\partial t^2}.$$

The stress components are

$$\sigma_{yz} = \sigma_{xy} = 0,$$

(3) $$\sigma_{yy} = \nu(\sigma_{xx} + \sigma_{zz}).$$

$$\sigma_{zx} = G\left(\frac{\partial w}{\partial x} + \frac{\partial u}{\partial z}\right) = G\left(2\frac{\partial^2 \phi}{\partial x \partial z} + \frac{\partial^2 \psi}{\partial x^2} - \frac{\partial^2 \psi}{\partial z^2}\right),$$

(4) $$\sigma_{zz} = \lambda e + 2G\frac{\partial w}{\partial z} = \lambda\left(\frac{\partial^2 \phi}{\partial x^2} + \frac{\partial^2 \phi}{\partial z^2}\right) + 2G\left(\frac{\partial^2 \phi}{\partial z^2} + \frac{\partial^2 \psi}{\partial x \partial z}\right),$$

$$\sigma_{xx} = \lambda e + 2G\frac{\partial u}{\partial x} = \lambda\left(\frac{\partial^2 \phi}{\partial x^2} + \frac{\partial^2 \phi}{\partial z^2}\right) + 2G\left(\frac{\partial^2 \phi}{\partial x^2} - \frac{\partial^2 \psi}{\partial x \partial z}\right).$$

The boundary conditions on the surface $z = 0$ are

(5) $$(\sigma_{zx})_{z=0} = (\sigma_{zy})_{z=0} = 0, \quad (\sigma_{zz})_{z=0} = -Q\delta(x)\mathbf{1}(t),$$

where $\mathbf{1}(t)$ is the unit-step function

(6) $$\mathbf{1}(t) = 0 \quad \text{for} \quad t < 0, \quad \mathbf{1}(t) = 1 \quad \text{for} \quad t > 0,$$

while $\delta(x)$ is the Dirac delta function, i.e., one whose value is zero everywhere except in the neighborhood of $x = 0$ where it becomes infinitely large in such a way that $\int_{-\infty}^{\infty} \delta(x)dx = 1$. It is seen that the surface stress given by Eq. (5) is equivalent to a concentrated load Q per unit length suddenly applied on the line $x = 0$, $z = 0$ and maintained constant afterwards.

For the conditions at infinity, it is reasonable to demand that (a) all displacements and stresses remain finite at infinity, and (b) at large distances from the point of application of the load the disturbance consists of outgoing waves. These are called the *finiteness* and *radiation conditions*, respectively.

In the present problem, the disturbance is propagated outward at a finite velocity, so conditions (a) and (b) are equivalent to the statement that there exists an outgoing wave front, beyond which the medium is undisturbed. The question arises whether the boundary conditions (5) and the finiteness and radiation conditions, in the absence of any other disturbances in the medium, will determine a unique solution of our problem. For a *point* load, the answer is obviously affirmative, because for a suddenly applied point load, the wave front will be at a finite distance from the point of application of the load at any finite time. Hence, if we take a volume sufficiently large so that it includes the wave front in its interior, we have a finite body over whose entire surface the surface tractions are specified. Neumann's uniqueness theorem (Problem 7.1, Sec. 7.4) then guarantees a unique solution. For a *line* load, we do not have such a simple and general proof. In fact, difficulty may arise in two-dimensional problems. (For example, in the corresponding static problem — a static line load acting on the surface — the displacement at infinity is logarithmically divergent. See Sec. 9.4, Example 2.) However, for the present problem, a unique solution can be determined if we assume the deformation to be truly two-dimensional.

The significance of the last assumption is made clear by the following remarks. Note that a cylindrical body subjected to surface forces uniform along the generators *may* have an internal stress state that is not uniform along the axis. For example, transient axial waves may be superposed without disturbing the lateral boundary conditions. In other words, a seemingly two-dimensional problem may actually be three-dimensional. Such an occasion also occurs in fluid mechanics. A nontrivial example in hydrodynamics is the flow around a circular cylinder, with a velocity field uniform at infinity and normal to the cylinder axis. At supercritical Reynolds numbers, the three-dimensionality of the flow in the wake, i.e., variation along the cylinder axis, is very pronounced and becomes a predominant feature.

The boundary conditions (5) can be written in a different form. The unitstep function $\mathbf{1}(t)$ has no Fourier transform. But if we consider it to be the limiting case of the function $e^{-\beta t}\mathbf{1}(t)$, which has a Fourier transform, then the Fourier integral theorem

$$(7) \qquad f(x) = \frac{1}{2\pi} \int_{-\infty}^{\infty} e^{ikx} dk \int_{-\infty}^{\infty} f(\xi) e^{-ik\xi} d\xi ,$$

which is valid for an arbitrary function $f(x)$ that is square-integrable in the Lebesgue sense, yields the representation

$$(8) \qquad \mathbf{1}(t) = \frac{1}{2\pi} \lim_{\beta \to 0} \int_{-\infty}^{\infty} \frac{e^{i\omega t}}{\beta + i\omega} d\omega .$$

Equation (8) is obtained on substituting $e^{-\beta \xi}\mathbf{1}(\xi)$ for $f(\xi)$ in Eq. (7) and changing k to ω. Similarly, $\delta(x)$ has no Fourier transform. But, considering the delta function as the limit of a square wave

$$\frac{1}{\epsilon}\left[\mathbf{1}\left(x + \frac{\epsilon}{2}\right) - \mathbf{1}\left(x - \frac{\epsilon}{2}\right)\right]$$

as $\epsilon \to 0$, we can use Eq. (7) to obtain the representation

$$(9) \qquad \delta(x) = \lim_{\epsilon \to 0} \frac{1}{2\pi\epsilon} \int_{-\infty}^{\infty} \frac{\sin k\epsilon}{k} e^{ikx} dk .$$

Therefore, the second condition in Eq. (5) may be written

$$(10) \quad [\sigma_{zz}]_{z=0} = \frac{-Q}{4\pi^2} \lim_{\substack{\epsilon \to 0 \\ \beta \to 0}} \int_{-\infty}^{\infty}\int_{-\infty}^{\infty} \frac{\sin k\epsilon}{k\epsilon} \frac{1}{\beta + i\omega} e^{i(\omega t + kx)} d\omega dk .$$

From this, we see that if we can obtain an elementary solution of Eqs. (2) satisfying the boundary conditions

$$(11) \qquad [\sigma_{xz}]_{z=0} = 0 , \qquad [\sigma_{zz}]_{z=0} = Ze^{i(\omega t + kx)} ,$$

then by the principle of superposition the solution to the original problem with boundary conditions (5) can be obtained by setting

$$(12) \qquad Z(k, \omega) = \frac{-Q}{4\pi^2} \frac{\sin k\epsilon}{k\epsilon} \frac{1}{\beta + i\omega} ,$$

integrating with respect to k and ω both from $-\infty$ to ∞, and then passing to the limit $\beta \to 0$, $\epsilon \to 0$.

The solution of the elementary problem is obtained by assuming

$$(13) \qquad \phi = Ae^{-\nu z + ikx + i\omega t} , \qquad \psi = Be^{-\nu' z + ikx + i\omega t} ,$$

which satisfy the wave Eqs. (2) if

(14) $\qquad \nu^2 = k^2 - k_\alpha^2 , \quad \nu'^2 = k^2 - k_\beta^2 , \quad k_\alpha = \dfrac{\omega}{c_L} , \quad k_\beta = \dfrac{\omega}{c_T} .$

On substituting Eq. (13) into Eqs. (4) and (11), we obtain

(15) $\qquad \begin{aligned} &-2Ai\nu k - (2k^2 - k_\beta^2)B = 0 , \\ &(2k^2 - k_\beta^2)A - 2Bik\nu' = \frac{1}{G} Z(k,\omega) . \end{aligned}$

Solving these equations for A and B, we obtain

(16) $\qquad A = \dfrac{2k^2 - k_\beta^2}{F(k)} \dfrac{Z(k,\omega)}{G} , \qquad B = \dfrac{-2ik\nu}{F(k)} \dfrac{Z(k,\omega)}{G} ,$

where

(17) $\qquad F(k) \equiv (2k^2 - k_\beta^2)^2 - 4k^2 \nu \nu'$

is Rayleigh's function.

A formal solution of our problem is obtained by substituting Eqs. (16) and (12) into Eq. (13) and integrating with respect to ω and k from $-\infty$ to ∞ as indicated before. In so doing, an appropriate branch of the multi-valued functions ν and ν' must be chosen so that the conditions at infinity are satisfied.

The evaluation of these integrals is a formidable task. Direct integration or numerical integration are exceedingly difficult. Lamb uses the method of contour integration. The variables of integration k and ω are replaced by complex variables, and the contours of integration are deformed in such a way that some explicit results are obtained. Details can be found in Lamb's paper. For alternative approaches, some very elegant, see book by Ewing, Jardetzky and Press, and papers by Cagniard, Pekeris and Lifson listed in the Bibliography 7.4 at the end of this book. See also papers listed in Bibliography 9.2.

PROBLEMS

8.11. Derive potentials to solve the following equations:

Example: $\dfrac{\partial u}{\partial y} - \dfrac{\partial v}{\partial x} = 0$ is solved by taking $u = \dfrac{\partial \phi}{\partial x}, \; v = \dfrac{\partial \phi}{\partial y} .$

(a) $\dfrac{\partial u}{\partial x} + \dfrac{\partial v}{\partial y} = 0.$

(b) Plane stress:

$$\frac{\partial \sigma_x}{\partial x} + \frac{\partial \tau_{xy}}{\partial y} = 0, \qquad \frac{\partial \tau_{xy}}{\partial x} + \frac{\partial \sigma_y}{\partial y} = 0.$$

(c) In the theory of membrane stresses in a flat plate:

$$\frac{\partial N_x}{\partial x} + \frac{\partial N_{xy}}{\partial y} = 0, \qquad \frac{\partial N_{xy}}{\partial x} + \frac{\partial N_y}{\partial y} = 0.$$

(d) In theory of bending of plates:

$$\frac{\partial Q_x}{\partial x} + \frac{\partial Q_y}{\partial y} = 0, \qquad \frac{\partial M_x}{\partial x} + \frac{\partial M_{xy}}{\partial y} = Q_x, \qquad \frac{\partial M_{xy}}{\partial x} + \frac{\partial M_y}{\partial y} = Q_y.$$

8.12. A train of plane wave of wave length L and phase velocity c can be represented as

$$\phi = A \exp\left[i\frac{2\pi}{L}(\nu_1 x + \nu_2 y + \nu_3 z \pm ct)\right]$$

where A is a constant and ν_1, ν_2, ν_3 are the direction cosines of a vector normal to the wave front. Consider suitable superposition of these waves, derive expressions for trains of:

(a) cylindrical waves, and

(b) spherical waves, such as those generated by a point source.

Ans. (a) $BJ_0(kr)e^{-\nu|z|}e^{i\omega t}$, $\qquad\qquad \nu^2 = k^2 - k_\alpha^2, \qquad r^2 = x^2 + y^2.$

(b) $C\dfrac{1}{R}\exp[\pm i(k_\alpha R - \omega t)]$, $\qquad\qquad R^2 = x^2 + y^2 + z^2.$

Here B, C are constants, $k_\alpha = \omega/c$, and k is a parameter. The analysis may be simplified by means of contour integrations, regarding some variable of integration as complex numbers.

8.13. High pressure vessels of steel will be designed on the basis of von Mises' yield criterion. Consider spherical and cylindrical tanks of outer radius b and inner radius a. Compare the maximum internal pressure p at which yielding occurs.

Ans. $J_2 - k^2 = 0$, k = yield stress in simple shear.

$$\text{Sphere: } p_{\text{yield}} = \sqrt{3}\left[\left(\frac{b}{a}\right)^3 - 1\right]\left[\frac{5}{2}\left(\frac{b}{a}\right)^6 + \left(\frac{b}{a}\right)^3 + 1\right]^{-1/2} k.$$

$$\text{Cylinder: } p_{\text{yield}} = \left(1 - \frac{a^2}{b^2}\right)k, \qquad\qquad \text{(far away from the ends)}.$$

If $b/a = 1.1$, $p_{\text{yield}} = 0.523k$ for sphere, $= 0.222k$ for cylinder.

8.14. Consider a spherical fluid gyroscope which consists of a hollow metallic sphere filled with a dense fluid. This sphere is rotated at an angular velocity

ω about its polar axis. In a steady-state rotation (ω = constant), what are the stresses and displacements in the shell due to the fluid pressure and the centrifugal forces acting on the shell?

8.15.   An infinitely long circular cylindrical hole of radius a is drilled in an infinite elastic medium. A pressure load is suddenly applied in the hole and starts to travel at a constant speed, so that the boundary conditions on the surface of the hole are, at $r = a$,

$$\sigma_{rr} = p1\left(t - \frac{|z|}{c}\right) , \quad \sigma_{rz} = \sigma_{r\theta} = 0 .$$

The medium is initially quiescent. Determine the response of the medium.

8.16.   Shock tubes are common tools used in aerodynamic research. A shock tube consists of a long cylindrical shell, closed at both ends. Near one end a thin diaphragm is inserted normal to the tube axis. On one side of the diaphragm the air is evacuated; on the other side, gas at high pressure is stored. In operation, the diaphragm is suddenly split, the onrushing gas from the high pressure side creates a shock front that travels down the evacuated tube.

Elastic waves are generated in the tube wall by the bursting of the diaphragm and by the moving shock wave. These elastic waves have some effect on the instrumentation and measurements. Discuss the transient elastic response of the shock tube wall.

8.17.   An infinite elastic medium contains a spherical cavity of radius a. A sinusoidally fluctuating pressure acts on the surface of the cavity. Determine the displacement field in the medium.

9

TWO-DIMENSIONAL PROBLEMS IN LINEARIZED
THEORY OF ELASTICITY

The application of the Airy stress function reduces elastostatic problems in plane stress and plane strain for isotropic materials to boundary-value problems of a biharmonic equation. A general method of solution using the theory of functions of a complex variable is available. We shall discuss this method briefly and illustrate its utility in solving a few important problems.

Throughout this chapter x, y, z represent a set of rectangular Cartesian coordinates, with respect to which the displacement components are written as u, v, w, the strain components are e_{xx}, e_{xy}, etc., and the stress components are σ_{xx}, σ_{xy}, etc. We recall the factor $\frac{1}{2}$ in the definition of the strain components:

$$e_{ij} = \frac{1}{2} \left(\frac{\partial u_i}{\partial x_j} + \frac{\partial u_i}{\partial x_i} \right).$$

When curvilinear coordinates are used, we retain the notations of Chapter 4, in which u_i and e_{ij} denote the tensor components of the displacement and the strain, respectively; whereas ξ_i, ε_{ij} denote the *physical* components of these tensors. See Secs. 4.10–4.12.

The methods presented in this chapter, as those discussed in the two preceding chapters, are for the linearized theory of elasticity. We assume that the deformation gradient and velocities are small that the nonlinear terms in Eqs. (1)–(3) of Sec. 7.1 are negligible. In other words, the convective acceleration, the convective velocity, and the products of deformation gradients are negligibly small compared with the retained linear terms in the acceleration, velocity, and strain tensor, respectively. There are beautiful methods and results in the linearized theory. There are spectacular panorama in the nonlinear theory. We believe, however, that the best way to learn the nonlinear theory is to master the linearized theory first. This is what this book tries to do.

9.1. PLANE STATE OF STRESS OR STRAIN

If the stress components σ_{zz}, σ_{zx}, σ_{zy} vanish everywhere,

(1)
$$\sigma_{zz} = \sigma_{zx} = \sigma_{zy} = 0,$$

the state of stress is said to be *plane stress* parallel to the x, y-plane. In this case, for isotropic materials,

(2)
$$e_{xx} = \frac{1}{E}\left(\sigma_{xx} - \nu\sigma_{yy}\right), \qquad e_{yy} = \frac{1}{E}\left(\sigma_{yy} - \nu\sigma_{xx}\right),$$

$$e_{zz} = -\frac{\nu}{E}\left(\sigma_{xx} + \sigma_{yy}\right), \qquad e_{xy} = \frac{1}{2G}\sigma_{xy}, \qquad e_{xz} = e_{yz} = 0,$$

(3)
$$\sigma_{xx} = \frac{E}{1-\nu^2}\left(e_{xx} + \nu e_{yy}\right), \qquad \sigma_{yy} = \frac{E}{1-\nu^2}\left(e_{yy} + \nu e_{xx}\right),$$

$$\sigma_{xy} = \frac{E}{1+\nu} e_{xy},$$

(4)
$$\sigma_{xx} + \sigma_{yy} = \frac{E}{1-\nu}\left(e_{xx} + e_{yy}\right),$$

(5)
$$e_{xx} + e_{yy} = \frac{\partial u}{\partial x} + \frac{\partial v}{\partial y}.$$

Substituting Eq. (3) into the equation of equilibrium (7.1:5), we obtain the basic equations for plane stress,

(6) ▲
$$G\left(\frac{\partial^2 u}{\partial x^2} + \frac{\partial^2 u}{\partial y^2}\right) + G\frac{1+\nu}{1-\nu}\frac{\partial}{\partial x}\left(\frac{\partial u}{\partial x} + \frac{\partial v}{\partial y}\right) + X = \rho\frac{\partial^2 u}{\partial t^2},$$

$$G\left(\frac{\partial^2 v}{\partial x^2} + \frac{\partial^2 v}{\partial y^2}\right) + G\frac{1+\nu}{1-\nu}\frac{\partial}{\partial y}\left(\frac{\partial u}{\partial x} + \frac{\partial v}{\partial y}\right) + Y = \rho\frac{\partial^2 v}{\partial t^2}.$$

If the z-component of displacment w vanishes everywhere, and if the displacements u, v are functions of x, y only, and not of z, the body is said to be in *plane strain* state parallel to the x, y-plane. In plane strain we must have

(7)
$$\frac{\partial u}{\partial z} = \frac{\partial v}{\partial z} = w = 0, \quad \text{and} \quad \sigma_{zz} = \nu(\sigma_{xx} + \sigma_{yy}), \quad (\text{since } e_{zz} = 0).$$

The basic equation (7.1:9) becomes, in plane strain,

(8) ▲
$$G\left(\frac{\partial^2 u}{\partial x^2} + \frac{\partial^2 u}{\partial y^2}\right) + \frac{1}{1-2\nu}G\frac{\partial}{\partial x}\left(\frac{\partial u}{\partial x} + \frac{\partial v}{\partial y}\right) + X = \rho\frac{\partial^2 u}{\partial t^2},$$

$$G\left(\frac{\partial^2 v}{\partial x^2} + \frac{\partial^2 v}{\partial y^2}\right) + \frac{1}{1-2\nu}G\frac{\partial}{\partial y}\left(\frac{\partial u}{\partial x} + \frac{\partial v}{\partial y}\right) + Y = \rho\frac{\partial^2 v}{\partial t^2}.$$

If ν is replaced by $\nu/(1+\nu)$ in Eq. (8), then it assumes the form (6). Hence any problem of a plane state of strain may be solved as a problem of a plane state of stress after replacing the true value of ν by the "apparent value" $\nu/(1+\nu)$.[†] Conversely, any plane stress problem may be solved as

---

[†]This substitution refers only to the field Eqs. (6) and (8). The boundary conditions, the stress-strain relationship, and the shear modulus G are not to be changed

a problem of plane strain by replacing the true value of ν by an apparent value $\nu/(1-\nu)$.[†]

The strain state in a long cylindrical body acted on by loads that are normal to the axis of the cylinder and uniform in the axial direction often can be approximated by a plane strain state. A constant axial strain e_{zz} may be imposed on a plane strain state without any change in stresses in the x,y-plane. Hence a minor extension of the definition of plane strain can be formulated by requiring that e_{zz} be a constant, that u and v be functions of x,y only, and that w be a linear function of z only.

The state of stress in a thin flat plate acted on by forces parallel to the midplane of the plate is approximately plane stress. However, since in general e_{zz} does not vanish, the displacements u, v, w are functions of z, and the problem is not truly two-dimensional. In fact, it can be shown that the general state of plane stress, satisfying Eq. (1), the equations of equilibrium, and the Beltrami–Michell compatibility conditions (7.3:6), is one in which the stresses σ_{xx}, σ_{yy}, σ_{xy} are parabolically distributed throughout the thickness of the plate, i.e., of the form $f(x,y) + g(x,y)z^2$. (See Timoshenko and Goodier,[1.2] p. 241.) However, the part proportional to z^2 can be made as small as we please compared with the first term, by restricting ourselves to plates which are sufficiently thin (with the ratio $h/L \to 0$, where h is the plate thickness and L is a characteristic dimension of the plate).

9.2. AIRY STRESS FUNCTIONS FOR TWO-DIMENSIONAL PROBLEMS

For plane stress or plane strain problems, we may try to find general stress systems that satisfy the equations of equilibrium and compatibility and then determine the solution to a particular problem by the boundary conditions.

Let x, y be a set of rectangular Cartesian coordinates. For plane stress and plane strain problems in the x,y-plane, the equations of equilibrium (3.4:2) are specialized into

$$(1) \qquad \frac{\partial \sigma_{xx}}{\partial x} + \frac{\partial \sigma_{xy}}{\partial y} = -X,$$

$$(2) \qquad \frac{\partial \sigma_{yy}}{\partial y} + \frac{\partial \sigma_{yx}}{\partial x} = -Y,$$

with the boundary conditions

$$(3) \qquad \begin{aligned} l\sigma_{xx} + m\sigma_{xy} &= p, \\ m\sigma_{yy} + l\sigma_{xy} &= q, \end{aligned}$$

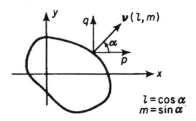

$$l = \cos\alpha$$
$$m = \sin\alpha$$

Fig. 9.2:1. Notations.

where l, m are the direction cosines of the outer normal to the boundary curve and where p, q are surface tractions acting on the boundary surface.

The strain components are,

(a) *In the plane stress case:*

$$e_{xx} = \frac{1}{E}(\sigma_{xx} - \nu\sigma_{yy}), \qquad e_{yy} = \frac{1}{E}(\sigma_{yy} - \nu\sigma_{xx}),$$

(4)

$$e_{xy} = \frac{1}{2G}\sigma_{xy} = \frac{(1+\nu)}{E}\sigma_{xy}.$$

(b) *In the plane strain case:*

$$e_{xx} = \frac{1}{E}[(1-\nu^2)\sigma_{xx} - \nu(1+\nu)\sigma_{yy}],$$

(5)

$$e_{yy} = \frac{1}{E}[(1-\nu^2)\sigma_{yy} - \nu(1+\nu)\sigma_{xx}],$$

$$e_{xy} = \frac{1}{E}(1+\nu)\sigma_{xy}.$$

In view of what was discussed in the preceding section, for very thin plates we may assume σ_{xx}, σ_{yy}, σ_{xy} to be independent of z. Then the plane stress problem becomes truly two-dimensional, as well as the plane strain problem.

The compatibility conditions are as follows (see Sec. 4.6):

$$\frac{\partial^2 e_{xx}}{\partial y^2} + \frac{\partial^2 e_{yy}}{\partial x^2} = 2\frac{\partial^2 e_{xy}}{\partial x \partial y}, \quad \frac{\partial^2 e_{xx}}{\partial y \partial z} = \frac{\partial}{\partial x}\left(-\frac{\partial e_{yz}}{\partial x} + \frac{\partial e_{xz}}{\partial y} + \frac{\partial e_{xy}}{\partial z}\right),$$

(6) $$\frac{\partial^2 e_{yy}}{\partial z^2} + \frac{\partial^2 e_{zz}}{\partial y^2} = 2\frac{\partial^2 e_{yz}}{\partial y \partial z}, \quad \frac{\partial^2 e_{yy}}{\partial x \partial z} = \frac{\partial}{\partial y}\left(\frac{\partial e_{yz}}{\partial x} - \frac{\partial e_{xz}}{\partial y} + \frac{\partial e_{xy}}{\partial z}\right),$$

$$\frac{\partial^2 e_{zz}}{\partial x^2} + \frac{\partial^2 e_{xx}}{\partial z^2} = 2\frac{\partial^2 e_{xx}}{\partial z \partial x}, \quad \frac{\partial^2 e_{zz}}{\partial x \partial y} = \frac{\partial}{\partial z}\left(\frac{\partial e_{yz}}{\partial x} + \frac{\partial e_{xz}}{\partial y} - \frac{\partial e_{xy}}{\partial z}\right).$$

On substituting Eq. (4) into the first equation of Eqs. (6), we obtain, in the plane stress case,

(7) $$\frac{\partial^2}{\partial y^2}(\sigma_{xx} - \nu\sigma_{yy}) + \frac{\partial^2}{\partial x^2}(\sigma_{yy} - \nu\sigma_{xx}) = 2(1+\nu)\frac{\partial^2 \tau_{xy}}{\partial x \partial y}.$$

Differentiating Eq. (1) with respect to x and Eq. (2) with respect to y and adding we obtain

(8) $$\frac{\partial^2 \sigma_{xx}}{\partial x^2} + \frac{\partial^2 \sigma_{yy}}{\partial y^2} + \frac{\partial X}{\partial x} + \frac{\partial Y}{\partial y} = -2\frac{\partial^2 \tau_{xy}}{\partial x \partial y}.$$

Eliminating τ_{xy} between Eqs. (7) and (8), we obtain

$$(9) \qquad \left(\frac{\partial^2}{\partial x^2} + \frac{\partial^2}{\partial y^2}\right)(\sigma_{xx} + \sigma_{yy}) = -(1 + \nu)\left(\frac{\partial X}{\partial x} + \frac{\partial Y}{\partial y}\right).$$

Similarly, in the plane strain case, we have

$$(10) \qquad \left(\frac{\partial^2}{\partial x^2} + \frac{\partial^2}{\partial y^2}\right)(\sigma_{xx} + \sigma_{yy}) = -\frac{1}{(1 - \nu)}\left(\frac{\partial X}{\partial x} + \frac{\partial Y}{\partial y}\right).$$

Equations (1), (2), (3), and (9) or (10) define the plane problems in terms of the stress components σ_{xx}, σ_{yy}, σ_{xy}. If the boundary conditions of a problem are such that surface tractions are all known, then the problem can be solved in terms of stresses, with no need to mention displacements unless they are desired. Even in a mixed boundary-value problem in which part of the boundary has prescribed displacements, it still may be advantageous to solve for the stress state first. These practical considerations lead to the method of Airy stress function.[†]

Airy's method is based on the observation that the left hand side of Eqs. (1) and (2) appears as the divergence of a vector. In hydrodynamics we are familiar with the fact that the conservation of mass, expressed in the equation of continuity

$$(11) \qquad \frac{\partial u}{\partial x} + \frac{\partial v}{\partial y} = 0,$$

where u, v are components of the velocity vector, can be derived from an arbitrary stream function $\psi(x, y)$:

$$(12) \qquad u = \frac{\partial \psi}{\partial y}, \qquad v = -\frac{\partial \psi}{\partial x}.$$

In other words, if u, v are derived from an arbitrary $\psi(x, y)$ according to Eq. (12), then Eq. (11) is satisfied identically.

Let us use the same technique for Eqs. (1) and (2). These equations can be put into the form of Eq. (11) if we assume that the body forces can be derived from a potential V, so that

$$(13) \qquad X = -\frac{\partial V}{\partial x}, \qquad Y = -\frac{\partial V}{\partial y}.$$

A substitution of Eq. (13) into Eqs. (1) and (2) results in

$$(14) \qquad \frac{\partial}{\partial x}(\sigma_{xx} - V) + \frac{\partial \sigma_{xy}}{\partial y} = 0, \qquad \frac{\partial \sigma_{xy}}{\partial x} + \frac{\partial}{\partial y}(\sigma_{yy} - V) = 0.$$

[†]For problems in which displacements are prescribed over the entire boundary, the displacement potential or other devices of the preceding chapter should be tried first.

Now, as in Eq. (11), these equations are identically satisfied if we introduce two stream functions Ψ and χ in such a way that

(15)
$$\sigma_{xx} - V = \frac{\partial\Psi}{\partial y}, \qquad \sigma_{xy} = -\frac{\partial\Psi}{\partial x},$$
$$\sigma_{xy} = -\frac{\partial\chi}{\partial y}, \qquad \sigma_{yy} - V = \frac{\partial\chi}{\partial x}.$$

In other words, a substitution of Eq. (15) into Eq. (14) reduces Eq. (14) into an identity in Ψ and χ. Now, Eqs. (15) can be combined if we let

(16)
$$\chi = \frac{\partial\Phi}{\partial x}, \quad \Psi = \frac{\partial\Phi}{\partial y},$$

i.e.,

(17) ▲ $\quad \sigma_{xx} - V = \dfrac{\partial^2\Phi}{\partial y^2}, \quad \sigma_{xy} = -\dfrac{\partial^2\Phi}{\partial x\partial y}, \quad \sigma_{yy} - V = \dfrac{\partial^2\Phi}{\partial x^2}.$

It is readily verified that if σ_{xx}, σ_{xy}, σ_{yy} are derived from an arbitrary function $\Phi(x, y)$ according to Eq. (17), then Eqs. (14) are identically satisfied. The function $\Phi(x, y)$ is called the *Airy stress function*, in deference to its inventor, the famous astronomer.

An arbitrary function $\Phi(x, y)$ generates stresses that satisfy the equations of equilibrium, but Φ is not entirely arbitrary: it is required to generate only those stress fields that satisfy the condition of compatibility. Since the compatibility condition is given by Eqs. (9) or (10), a substitution gives the requirement that, in the plane stress case,

(18) ▲ $\quad \dfrac{\partial^4\Phi}{\partial x^4} + 2\dfrac{\partial^4\Phi}{\partial x^2\partial y^2} + \dfrac{\partial^4\Phi}{\partial y^4} = -(1-\nu)\left(\dfrac{\partial^2 V}{\partial x^2} + \dfrac{\partial^2 V}{\partial y^2}\right),$

and that, in the plane strain case,

(19) ▲ $\quad \dfrac{\partial^4\Phi}{\partial x^4} + 2\dfrac{\partial^4\Phi}{\partial x^2\partial y^2} + \dfrac{\partial^4\Phi}{\partial y^4} = -\dfrac{(1-2\nu)}{(1-\nu)}\left(\dfrac{\partial^2 V}{\partial x^2} + \dfrac{\partial^2 V}{\partial y^2}\right).$

If the body forces vanish, then, in both plane stress and plane strain, Φ is governed by the equation

(20) ▲ $\quad \dfrac{\partial^4\Phi}{\partial x^4} + 2\dfrac{\partial^4\Phi}{\partial x^2\partial y^2} + \dfrac{\partial^4\Phi}{\partial y^4} = 0.$

A regular solution of Eq. (20) is called a *biharmonic function*. Solution of plane elasticity problems by biharmonic functions will be discussed in the following sections.

How about the other five compatibility conditions in Eqs. (6) left alone so far? In the case of plane strain it is clear that they are identically satisfied. In the case of plane stress, however, they cannot be satisfied in general if we assume σ_{xx}, σ_{yy}, σ_{xy} to be independent of z. For, under such an assumption these compatibility conditions imply that

$$(21) \qquad \frac{\partial^2 e_{zz}}{\partial x^2} = \frac{\partial^2 e_{zz}}{\partial y^2} = \frac{\partial^2 e_{zz}}{\partial x \partial y} = 0 \,.$$

Hence, e_{zz}, and hence $\sigma_{xx} + \sigma_{yy}$, must be a linear function of x and y $[e_{zz} = -\nu(\sigma_{xx} + \sigma_{yy})/E]$, which is the exception rather than the rule in the solution of plane stress problems. Hence, in general, the assumption that plane stress state is two-dimensional, so that $\sigma_{xx}, \sigma_{yy}, \sigma_{xy}$ are functions of x, y only, cannot be true; and the solutions obtained under this assumption cannot be exact. However, as we have discussed previously (Sec. 9.1), they are close approximations for thin plates.

The stress-function method can be extended to three dimensions. The crucial observation is simply that the equation of equilibrium represents a vector divergence of the stress tensor. We are familiar with the stream function in hydrodynamics. In three-dimensions we need a triple of stream functions. Similarly, a generalization of Airy's procedure to equations of equilibrium in three dimensions requires a *tensor* of stress functions. Finzi[8.1] (1934) showed that a general solution to the equations

$$(22) \qquad \sigma_{ij,j} = 0 \,, \qquad \sigma_{ij} = \sigma_{ji}$$

is

$$(23) \qquad \sigma_{ij} = e_{imr} e_{jns} \phi_{rs,mn} \,,$$

where ϕ_{rs} stands for the components of a symmetric second-order tensor of stress functions, while e_{imr} is the usual permutation symbol (Sec. 2.1). Specialization by taking $\phi_{rs} = 0$ $(r \neq s)$, yields Maxwell's stress functions, while taking $\phi_{rr} = 0$ (no sum), yields Morera's stress function, see Sec. 10.9, Eq. (10.9:19), *et seq.* If all elements ϕ_{rs}, except ϕ_{33}, are assumed to vanish, then (23) degenerates into Airy's solution of the two-dimensional equilibrium equations. An elegant proof of Finzi's result that is applicable to n-dimensional Euclidean space was given by Dorn and Schild.[8.1]

Finzi[8.1] (1934) obtained further a beautiful extension to the equations of *motion* of a continuum, with an arbitrary density field, by introducing a fourth dimension. Finzi's arbitrary tensor yields by differentiation a motion and a stress field that satisfy the equation of motion.

For a curved space (non-Euclidean), Truesdell[16.1] obtained related results by methods of calculus of variations. The curved space problem arises naturally in the intrinsic theory of thin shells or membranes. The two-dimensional surface is, to a two-dimensional observer who is not allowed to leave the surface, a non-Euclidean space (imbedded, of course, in a three-dimensional Euclidean space).

Example 1. The following polynomials of second and third degree are obviously biharmonic.

$$\Phi_2 = a_2 x^2 + b_2 xy + c_2 y^2\,,$$

$$\Phi_3 = a_3 x^3 + b_3 x^2 y + c_3 xy^2 + d_3 y^3\,.$$

By adjusting the constants a_2, a_3, etc., many problems in which the stresses are linearly distributed on rectangular boundaries can be solved. Examples of such problems are shown in Fig. 9.2:2.

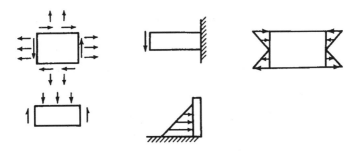

Fig. 9.2:2. Examples of problems solvable by simple polynomials.

Example 2. Consider a rectangular beam (Fig. 9.2:3) supported at the ends and subjected to the surface tractions

$$\text{on}\quad y = c \quad \sigma_{xy} = 0\,, \quad \sigma_{yy} = -B\sin\alpha x\,,$$

$$\text{on}\quad y = -c \quad \sigma_{xy} = 0\,, \quad \sigma_{yy} = -A\sin\alpha x\,.$$

Other edge conditions are unspecified at the beginning, and the body forces are absent.

Solution. Let

$$\Phi = \sin\alpha x f(y)\,,$$

where $f(y)$ is a function of y only. A substitution into Eq. (20) yields

$$\alpha^4 f(y) - 2\alpha^2 f''(y) + f^{iv}(y) = 0\,.$$

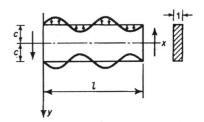

Fig. 9.2:3. A beam subjected to a sinusoidally distributed loading.

The general solution is

$$f(y) = C_1 \cosh \alpha y + C_2 \sinh \alpha y + C_3 y \cosh \alpha y + C_4 y \sinh \alpha y\,.$$

Hence,

$$\Phi = \sin \alpha x (C_1 \cosh \alpha y + C_2 \sinh \alpha y + C_3 y \cosh \alpha y + C_4 y \sinh \alpha y)\,,$$

$$\sigma_{xx} = \sin \alpha x [C_1 \alpha^2 \cosh \alpha y + C_2 \alpha^2 \sinh \alpha y$$

$$+ C_3 (y\alpha^2 \cosh \alpha y, +2\alpha \sinh \alpha y) + C_4 (y\alpha^2 \sinh \alpha y + 2\alpha \cosh \alpha y)]\,,$$

$$\sigma_{yy} = -\alpha^2 \sin \alpha x [C_1 \cosh \alpha y + C_2 \sinh \alpha y + \cdots]\,,$$

$$\sigma_{xy} = -\alpha \cos \alpha x [C_1 \alpha \sinh \alpha y + C_2 \alpha \cosh \alpha y$$

$$+ C_3 (\cosh \alpha y + y\alpha \sinh \alpha y) + C_4 (\sinh \alpha y + y\alpha \cosh \alpha y)]\,.$$

On application of the boundary conditions $\sigma_{xy} = 0$ on $y = \pm c$, we can express C_3 and C_4 in terms of C_1 and C_2. The other boundary conditions then yield the constants

$$C_1 = \frac{(A+B)}{\alpha^2} \frac{\sinh \alpha c + \alpha c \cosh \alpha c}{\sinh 2\alpha c + 2\alpha c}\,,$$

$$C_2 = -\frac{(A-B)}{\alpha^2} \frac{\cosh \alpha c + \alpha c \sinh \alpha c}{\sinh 2\alpha c - 2\alpha c}\,,$$

$$C_3 = \frac{(A-B)}{\alpha^2} \frac{\alpha \cosh \alpha c}{\sinh 2\alpha c - 2\alpha c}\,,$$

$$C_4 = -\frac{(A+B)}{\alpha^2} \frac{\alpha \sinh \alpha c}{\sinh 2\alpha c + 2\alpha c}\,.$$

The details can be found in Timoshenko and Goodier,[1.2] p. 48.

9.3. AIRY STRESS FUNCTION IN POLAR COORDINATES

For two-dimensional problems with circular boundaries, polar coordinates can be used to advantage. Let ξ_r, ξ_θ, ξ_z denote physical components of the displacement, and let $\epsilon_{rr}, \epsilon_{r\theta}, \ldots, \sigma_{rr}, \sigma_{r\theta}, \ldots$, etc., be the physical components of the strain and stress, respectively. The general equations in cylindrical polar coordinates are given in Sec. 4.12. For *plane stress* problems, we assume that

$$(1) \qquad\qquad \sigma_{zz} = \sigma_{zr} = \sigma_{z\theta} = 0\,.$$

For *plane strain* problems, we assume that

$$(2) \qquad\qquad\qquad \xi_z = 0,$$

and that all derivatives with respect to z vanish. In both cases, the strain components are defined as follows:

$$
\epsilon_{rr} = \frac{\partial \xi_r}{\partial r}, \qquad \epsilon_{\theta\theta} = \frac{1}{r}\frac{\partial \xi_\theta}{\partial \theta} + \frac{\xi_r}{r},
$$

(3)

$$
\epsilon_{r\theta} = \frac{1}{2}\left(\frac{1}{r}\frac{\partial \xi_r}{\partial \theta} + \frac{\partial \xi_\theta}{\partial r} - \frac{\xi_\theta}{r}\right).
$$

Furthermore, in plane stress,

$$(4) \quad \epsilon_{rr} = \frac{1}{E}\left(\sigma_{rr} - \nu\sigma_{\theta\theta}\right), \qquad \epsilon_{\theta\theta} = \frac{1}{E}\left(\sigma_{\theta\theta} - \nu\sigma_{rr}\right), \qquad \epsilon_{r\theta} = \frac{1+\nu}{E}\sigma_{r\theta},$$

and in plane strain,

$$
\epsilon_{rr} = \frac{1}{E}\left[(1-\nu^2)\sigma_{rr} - \nu(1+\nu)\sigma_{\theta\theta}\right],
$$

(5)

$$
\epsilon_{\theta\theta} = \frac{1}{E}\left[(1-\nu^2)\sigma_{\theta\theta} - \nu(1+\nu)\sigma_{rr}\right],
$$

$$
\epsilon_{r\theta} = \frac{1+\nu}{E}\sigma_{r\theta},
$$

and the equations of equilibrium become

$$
\frac{1}{r}\frac{\partial(r\sigma_{rr})}{\partial r} + \frac{1}{r}\frac{\partial \sigma_{r\theta}}{\partial \theta} - \frac{\sigma_{\theta\theta}}{r} + F_r = 0,
$$

(6)

$$
\frac{1}{r^2}\frac{\partial(r^2\sigma_{\theta r})}{\partial r} + \frac{1}{r}\frac{\partial \sigma_{\theta\theta}}{\partial \theta} + F_\theta = 0.
$$

If the body forces F_r, F_θ are zero Eqs. (6) are satisfied identically if the stresses are derived from a function $\Phi(r,\theta)$:

$$
\sigma_{rr} = \frac{1}{r}\frac{\partial \Phi}{\partial r} + \frac{1}{r^2}\frac{\partial^2 \Phi}{\partial \theta^2},
$$

(7) ▲

$$
\sigma_{\theta\theta} = \frac{\partial^2 \Phi}{\partial r^2},
$$

$$
\sigma_{r\theta} = -\frac{\partial}{\partial r}\left(\frac{1}{r}\frac{\partial \Phi}{\partial \theta}\right),
$$

as can be verified by direct substitution. The function $\Phi(r, \theta)$ is the *Airy stress function.*

The compatibility conditions (9.2:9) and (9.2:10) are, if body forces are zero,

$$(8) \qquad \left(\frac{\partial^2}{\partial x^2} + \frac{\partial^2}{\partial y^2} \right) (\sigma_{xx} + \sigma_{yy}) = 0 \, .$$

The sum $\sigma_{xx} + \sigma_{yy}$ is an invariant with respect to rotation of coordinates. Hence,

$$(9) \qquad \sigma_{xx} + \sigma_{yy} = \sigma_{rr} + \sigma_{\theta\theta} \, .$$

The Laplace operator is transformed as

$$(10) \qquad \frac{\partial^2}{\partial x^2} + \frac{\partial^2}{\partial y^2} = \frac{\partial^2}{\partial r^2} + \frac{1}{r} \frac{\partial}{\partial r} + \frac{1}{r^2} \frac{\partial^2}{\partial \theta^2} \, .$$

Hence, on substituting Eq. (7) into Eq. (9) and using Eq. (10), Eq. (8) is transformed into

$$(11) \; \blacktriangle \qquad \left(\frac{\partial^2}{\partial r^2} + \frac{1}{r} \frac{\partial}{\partial r} + \frac{1}{r^2} \frac{\partial^2}{\partial \theta^2} \right) \left(\frac{\partial^2 \Phi}{\partial r^2} + \frac{1}{r} \frac{\partial \Phi}{\partial r} + \frac{1}{r^2} \frac{\partial^2 \Phi}{\partial \theta^2} \right) = 0 \, .$$

This is the compatibility equation to be satisfied by the Airy stress function $\Phi(r, \theta)$. If one can find *a* solution of Eq. (11) that also satisfies the boundary conditions, then the problem is solved; since by Kirchhoff's uniqueness theorem, such a solution is unique.

Axially Symmetric Problems. If Φ is a function of r alone and is independent of θ, all derivatives with respect to θ vanish and Eq. (11) becomes

$$(12) \qquad \frac{d^4 \Phi}{dr^4} + \frac{2}{r} \frac{d^3 \Phi}{dr^3} - \frac{1}{r^2} \frac{d^2 \Phi}{dr^2} + \frac{1}{r^3} \frac{d\Phi}{dr} = 0 \, .$$

This is the homogeneous differential equation which can be reduced to a linear differential equation with constant coefficients by introducing a new variable t such that $r = e^t$. The general solution is

$$(13) \qquad \Phi = A \log r + B r^2 \log r + C r^2 + D \, ,$$

which corresponds to

$$\sigma_{rr} = \frac{A}{r^2} + B(1 + 2 \log r) + 2C \, ,$$

$$(14) \qquad \sigma_{\theta\theta} = -\frac{A}{r^2} + B(3 + 2 \log r) + 2C \, ,$$

$$\sigma_{r\theta} = 0 \, .$$

The solutions of all problems of symmetrical stress distribution and no body forces can be obtained from this.

The displacement components corresponding to stresses given by Eq. (14) may be obtained as follows. Consider the *plane stress* case. Substituting Eq. (14) into the first of the Eqs. (4) and (3), we obtain

$$\frac{\partial \xi_r}{\partial r} = \frac{1}{E}\left[\frac{(1+\nu)A}{r^2} + 2(1-\nu)B\log r + (1-3\nu)B + 2(1-\nu)C\right],$$

from which, by integration,

(15)
$$\xi_r = \frac{1}{E}\left[-\frac{(1+\nu)A}{r} + 2(1-\nu)Br\log r\right.$$
$$\left. - B(1+\nu)r + 2C(1-\nu)r\right] + f(\theta),$$

where $f(\theta)$ is an arbitrary function of θ only. From Eq. (14) and the second of the Eqs. (4) and (3), we obtain

$$\frac{\partial \xi_\theta}{\partial \theta} = \frac{4Br}{E} - f(\theta).$$

Hence, by integration,

(16)
$$\xi_\theta = \frac{4Br\theta}{E} - \int_0^\theta f(\theta)d\theta + f_1(r),$$

where $f_1(r)$ is a function of r only. Finally, from the last of Eqs. (14), Eqs. (4) and (3), we find, since $\sigma_{r\theta} = \epsilon_{r\theta} = 0$,

$$\frac{1}{r}\frac{df(\theta)}{d\theta} + \frac{df_1(r)}{dr} + \frac{1}{r}\int_0^\theta f(\theta)d\theta - \frac{1}{r}f_1(r) = 0.$$

Multiplying throughout by r, we find that the first and the third term are functions of θ only and the other two terms are functions of r only. Hence, the only possibility for the last equation to be satisfied is

$$\frac{df(\theta)}{d\theta} + \int_0^\theta f(\theta)d\theta = \alpha, \qquad r\frac{df_1(r)}{dr} - f_1(r) = -\alpha,$$

where α is an arbitrary constant. The solutions are

$$f(\theta) = \alpha \sin \theta + \gamma \cos \theta, \quad f_1(r) = \beta r + \alpha,$$

where α, β, γ are arbitrary constants. Substituting back into Eqs. (15) and (16), we obtain, for the *plane stress case*,

$$\xi_r = \frac{1}{E}\left[-\frac{(1+\nu)A}{r} + 2(1-\nu)Br \log r \right.$$

(17)
$$\left. - B(1+\nu)r + 2C(1-\nu)r \right] + \alpha \sin \theta + \gamma \cos \theta,$$

$$\xi_\theta = \frac{4Br\theta}{E} + \alpha \cos \theta - \gamma \sin \theta + \beta r.$$

The arbitrary constants $A, B, C, \alpha, \beta, \gamma$ are to be determined from the boundary conditions of each special problem. The corresponding expressions for the *plane strain case* are

$$\xi_r = \frac{1}{E}\left[-\frac{(1+\nu)A}{r} - B(1+\nu)r + 2B(1-\nu-2\nu^2)r \log r \right.$$

(18)
$$\left. + 2C(1-\nu-2\nu^2)r \right] + \alpha \sin \theta + \gamma \cos \theta,$$

$$\xi_\theta = \frac{4Br\theta}{E}(1-\nu^2) + \alpha \cos \theta - \gamma \sin \theta + \beta r.$$

Example 1. Uniform Pressure Acting on a Solid Cylinder

The boundary conditions are, at $r = a$,

$$\sigma_{rr} = -p_0, \qquad \sigma_{r\theta} = 0.$$

There should be no singularity in the solid. A glance at Eq. (14) shows that the problem can be solved by taking

$$A = B = 0, \qquad C = -\frac{p_0}{2}.$$

Example 2. Circular Cylindrical Tube Subjected to Uniform Internal and External Pressure

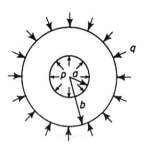

Fig. 9.3:1. Circular tube.

For this problem (Fig. 9.3:1), the boundary conditions are

$$\sigma_{rr} = -p \quad \text{at} \quad r = a\,,$$
(19)
$$\sigma_{rr} = -q \quad \text{at} \quad r = b\,.$$

The form of σ_{rr} in Eq. (14) suggests that the boundary conditions can be satisfied by fixing two of the constants, say A and C. From Eq. (19) alone, however, there is no reason to reject the term involving B; but inspection of the displacements given by Eq. (18) shows that if $B \neq 0$ the circumferential displacement ξ_θ will have a nonvanishing term

$$(20) \qquad\qquad \xi_\theta = \frac{4Br\theta}{E}\,(1 - \nu^2)$$

which is zero when $\theta = 0$ but becomes $\xi_\theta = (1 - \nu^2)8\pi Br/E$ when one traces a circuit around the axis of symmetry and returns to the same point after turning around an angle 2π. Thus the displacement given by (20) is not *single-valued*. Such a *multi-valued* expression for a displacement is physically impossible in a full cylindrical tube. Hence, $B = 0$.

It is simple to derive the constants A and C in the expression

$$\sigma_{rr} = \frac{A}{r^2} + 2C$$

such that Eq. (19) is satisfied. Thus we obtain Lamé's formulas for the stresses,

$$\sigma_{rr} = -p\frac{(b^2/r^2) - 1}{(b^2/a^2) - 1} - q\frac{1 - (a^2/r^2)}{1 - (a^2/b^2)}\,,$$
(21)
$$\sigma_{\theta\theta} = p\frac{(b^2/r^2) + 1}{(b^2/a^2) - 1} - q\frac{1 + (a^2/r^2)}{1 - (a^2/b^2)}\,.$$

It is interesting to note that $\sigma_{rr} + \sigma_{\theta\theta}$ is constant throughout the cylinder. If $q = 0$ and $p > 0$, σ_{rr} is always a compressive stress and $\sigma_{\theta\theta}$ is always a tensile stress, the maximum value of which occurs at the inner radius and is always numerically greater than the internal pressure p.

Example 3. *Pure Bending of a Curved Bar*

The coefficients A, B, C may be chosen to satisfy the conditions $\sigma_{rr} = 0$ for $r = a$ and $r = b$, and

$$\int_a^b \sigma_{\theta\theta} dr = 0,$$

$$\int_a^b \sigma_{\theta\theta} r\, dr = -M, \quad \text{and} \quad \sigma_{r\theta} = 0$$

at the boundary $\theta = \text{const.}$ (see Fig. 9.3:2). See Timoshenko and Goodier,[1.2] p. 61; Sechler,[1.2] p. 143.

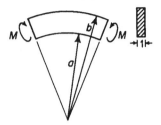

Fig. 9.3:2. Pure bending of a curved bar.

Example 4. *Initial Stress or Residual Stress in a Welded Ring*

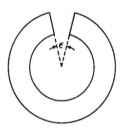

Fig. 9.3:3. Welding of a ring.

If the ends of an opening in a ring, as shown in Fig. 9.3:3, are joined together and welded, the initial stresses can be obtained from the expressions in Eq. (14) by taking

$$(22) \qquad B = \frac{\epsilon E}{8\pi}.$$

See Eq. (17) for circumferential displacements. The constants α, β, γ are all zero.

Problem 9.1. Determine the constant C in the stress function

$$\Phi = C[r^2(\alpha - \theta) + r^2 \sin\theta \cos\theta - r^2 \cos^2\theta \tan\alpha]$$

required to satisfy the conditions on the upper and lower edges of the triangular plate shown in Fig. P9.1. Determine the components of displacement for points on the upper edge.

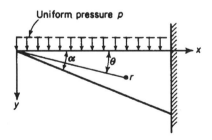

P9.1.

9.4. GENERAL CASE

By direct substitution, it can be verified that a solution of the equation

$$\left(\frac{\partial^2}{\partial r^2} + \frac{1}{r}\frac{\partial}{\partial r} + \frac{1}{r^2}\frac{\partial^2}{\partial \theta^2}\right)\left(\frac{\partial^2 \Phi}{\partial r^2} + \frac{1}{r}\frac{\partial \Phi}{\partial r} + \frac{1}{r^2}\frac{\partial^2 \Phi}{\partial \theta^2}\right) = 0$$

is (J. H. Michell, 1899)

(1)
$$\Phi = a_0 \log r + b_0 r^2 + c_0 r^2 \log r + d_0 r^2 \theta + a_0' \theta$$

$$+ \frac{a_1''}{2} r\theta \sin\theta + \left(\frac{}{a_1} r + b_1 r^3 + a_1' r^{-1} + b_1' r \log r\right)\cos\theta$$

$$+ \frac{c_1''}{2} r\theta \cos\theta + \left(\frac{}{c_1} r + d_1 r^3 + c_1' r^{-1} + d_1' r \log r\right)\sin\theta$$

$$+ \sum_{n=2}^{\infty}(a_n r^n + b_n r^{n+2} + a_n' r^{-n} + b_n' r^{-n+2})\cos n\theta$$

$$+ \sum_{n=2}^{\infty}(c_n r^n + d_n r^{n+2} + c_n' r^{-n} + d_n' r^{-n+2})\sin n\theta .$$

By adjusting the coefficients $a_0, b_0, a_1, \ldots$, etc., a number of important problems can be solved. The general form of Eq. (1) as a Fourier series in θ and power series in r provides a powerful means for solving problems involving circular and radial boundaries. The individual terms of the series provide beautiful solutions to several important engineering problems. Many examples are given in Timoshenko and Goodier,[1.2] pp. 73–130 and Sechler,[1.2] pp. 149–171.

Example 1. *Bending of a Curved Bar By a Force at the End*

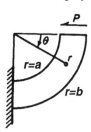

Consider a bar of a narrow rectangular cross section and with a circular axis, as shown in Fig. 9.4:1, loaded by a force P in the radial direction. The bending moment at any cross section is proportional to $\sin\theta$. Since the elementary beam theory suggests that the normal stress $\sigma_{\theta\theta}$ is proportional to the bending moment, it is

Fig. 9.4:1. Bending of a curved bar.

reasonable to try a solution for which $\sigma_{\theta\theta}$, and hence Φ, is proportional to $\sin\theta$. Such a solution can be obtained from the term involving $\sin\theta$ in

Eq. (1). It can be verified that the solution is

(2) $$\Phi = \left(d_1 r^3 + \frac{c_1'}{r} + d_1' r \log r \right) \sin \theta ,$$

with

$$d_1 = \frac{P}{2N} , \quad c_1' = -\frac{P a^2 b^2}{2N} , \quad d_1' = -\frac{P}{N} (a^2 + b^2) ,$$

where $N = a^2 - b^2 + (a^2 + b^2) \log (b/a)$. Note that the term $c_1 r \sin \theta$ in Eq. (1) does not contribute to the stresses and thus is not used here. The stresses are

$$\sigma_{rr} = \frac{P}{N} \left(r + \frac{a^2 b^2}{r^3} - \frac{a^2 + b^2}{r} \right) \sin \theta ,$$

(3) $$\sigma_{\theta\theta} = \frac{P}{N} \left(3r - \frac{a^2 b^2}{r^3} - \frac{a^2 + b^2}{r} \right) \sin \theta ,$$

$$\sigma_{r\theta} = -\frac{P}{N} \left(r + \frac{a^2 b^2}{r^3} - \frac{a^2 + b^2}{r} \right) \cos \theta .$$

If the boundary stress distribution were exactly as prescribed by the equations above, namely,

$$\sigma_{rr} = \sigma_{r\theta} = 0 \quad \text{for} \quad r = a \quad \text{and} \quad r = b ,$$

(4) $$\sigma_{\theta\theta} = 0 , \quad \sigma_{r\theta} = -\frac{P}{N} \left[r + \frac{a^2 b^2}{r^3} - \frac{1}{r} (a^2 + b^2) \right] \quad \text{for} \quad \theta = 0 ,$$

$$\sigma_{r\theta} = 0 , \quad \sigma_{\theta\theta} = \frac{P}{N} \left[3r - \frac{a^2 b^2}{r^3} - (a^2 + b^2) \frac{1}{r} \right] \quad \text{for} \quad \theta = \frac{\pi}{2} ,$$

then an exact solution is obtained.

A detailed examination of the exact solution shows that a commonly used engineering approximation in the elementary beam theory, that plane cross sections of a beam remain plane during bending, gives very satisfactory results.

Example 2. *Concentrated Force at a Point on the Edge of a Semi-Infinite Plate*

Consider a concentrated vertical load P acting on a horizontal straight boundary AB of an infinitely large plate (Fig. 9.4:2). The distribution of the load along the thickness of the plate is uniform. The plate thickness is assumed to be unity, so that P is the load per unit thickness.

This is the static counterpart of Lamb's problem (Secs. 8.15 and 9.7). Its solution was obtained by Flamant (1892) from Boussinesq's (1885) three-dimensional solution (Sec. 8.10). In contrast to the dynamic or the three-dimensional cases, the solution to the present case is given simply by the Airy stress function

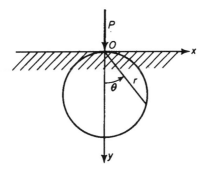

Fig. 9.4:2. Boussinesq–Flamant problem.

$$\Phi = -\frac{P}{\pi} r\,\theta \sin\theta\,,$$
(5)

which gives

$$\sigma_{rr} = -\frac{2P}{\pi}\frac{\cos\theta}{r}\,,\qquad \sigma_{\theta\theta} = 0\,,\qquad \sigma_{r\theta} = 0\,.$$
(6)

From Eq. (6), it is seen that an element at a distance r from the point of application of the load is subjected to a simple compression in the radial direction. Equation (6) shows also that the locus of points where the radial stress is a constant $-\sigma_{rr}$ is a circle $r = (-2P/\pi\sigma_{rr})\cos\theta$, which is tangent to the x-axis, as shown in Fig. 9.4:2.

Using the relations (9.3:3) and (9.3:4) and Eq. (6), we can determine the displacement field ξ_r, ξ_θ. If the constraint is such that the points on the y-axis have no lateral displacement ($\xi_\theta = 0$ when $\theta = 0$), then it can be shown that the elastic displacements are

$$\xi_r = -\frac{2P}{\pi E}\cos\theta \log r - \frac{(1-\nu)P}{\pi E}\theta\sin\theta + B\cos\theta\,,$$

$$\xi_\theta = \frac{2\nu P}{\pi E}\sin\theta + \frac{2P}{\pi E}\log r \sin\theta - \frac{(1-\nu)P}{\pi E}\theta\cos\theta$$
(7)
$$+ \frac{(1-\nu)P}{\pi E}\sin\theta - B\sin\theta\,.$$

The constant B can be fixed by fixing a point, say, $\xi_r = 0$ at $\theta = 0$ and $r = a$. But the characteristic logarithmic singularity at ∞ cannot be removed. This is a peculiarity of the two-dimensional problem. The corresponding three-dimensional or dynamic cases do not have such logarithmically infinite displacements at infinity.

P R O B L E M S

9.2. Verify Eqs. (6) and (7) and show that Eq. (5) yields the exact solution of the problem posed in Example 2.

9.3. Find the stresses in a semi-infinite plate $(-\infty < x < \infty, 0 \le y < \infty)$ due to a shear load of intensity $\tau \cos \alpha x$ acting on the edge $y = 0$, where τ and α are given constants. *Hint:* Consider $\Phi = (Ae^{\alpha y} + Be^{-\alpha y} + Cye^{\alpha y} + Dye^{-\alpha y}) \sin \alpha x$.

9.4. Show that the function $(M_t/2\pi)\theta$, where M_t is a constant, is a stress function. Consider some boundary value problems which may be solved by such a stress function and give a physical meaning to the constant M_t.

9.5. Consider a two-dimensional wedge of perfectly elastic material as shown in Fig. P9.5. If one side of the wedge $(\theta = 0)$ is loaded by a normal pressure distribution $p(r) = Pr^m$, where P and m are constants, while the other side $(\theta = \alpha)$ is stress-free, show that the problem can be solved by an Airy stress function $\Phi(r,\theta)$ expressed in the following form. If $m \ne 0$ (m may be > 0 or < 0),

(1) $\Phi = r^{m+2}[a \cos (m + 2)\theta + b \sin (m + 2)\theta + c \cos m\theta + d \sin m\theta]$.

If $m = 0$,

(2) $$\Phi = Kr^2 \left[-\tan \alpha \cos^2 \theta + \frac{1}{2} \sin 2\theta + \alpha - \theta \right].$$

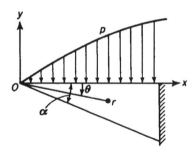

P9.5.

Determine the constants a, b, c, d, K. Discuss the *boundedness* (i.e., whether they are zero, finite, or infinite) of the stresses σ_{rr}, $\sigma_{\theta\theta}$, $\sigma_{r\theta}$; the slope $\partial v/\partial r$, and the second derivative $\partial^2 v/\partial r^2$, in the neighborhood of the tip of the wedge $(r \to 0)$. The symbol v denotes the component of displacement in the direction of increasing θ.

When the wedge angle α is small, v is approximately equal to the vertical displacement. This problem is of interest to the question of curling up of the sharp leading edge of a supersonic wing. [Reference: Fung, *J. Aeronautical Sciences*, **20**, 9 (1953).]

9.6. A semi-infinite plate $(-\infty < x < \infty, 0 \leq y < \infty)$ is subjected to the edge conditions

$$\sigma_{yy} = \frac{q_0}{2} - \frac{4q_0}{\pi^2} \left(\cos \frac{\pi x}{L} + \frac{1}{3^2} \cos \frac{3\pi x}{L} + \frac{1}{5^2} \cos \frac{5\pi x}{L} + \cdots \right),$$

$$\sigma_{xy} = 0,$$

on the edge $y = 0$ (Fig. P9.6). Find the Airy stress function that solves the problem.

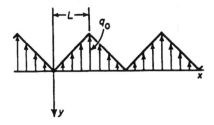

P9.6.

Obtain expression for σ_{xx} and σ_{yy}. Sketch σ_{yy} as a function of x on the line $y = d$, where $d > L$.

9.5. REPRESENTATION OF TWO-DIMENSIONAL BIHARMONIC FUNCTIONS BY ANALYTIC FUNCTIONS OF A COMPLEX VARIABLE

It is well-known that the real and imaginary parts of any analytic function of a complex variable $z = x + iy$ is harmonic, where $i = \sqrt{-1}$. Thus, if $f(z)$ is an analytic function of z, and u and v are the real and imaginary parts of $f(z)$, we have

(1) $f(z) = u + iv,$

(2) $\dfrac{df}{dz} = \dfrac{\partial f}{\partial x} = \dfrac{\partial u}{\partial x} + i\dfrac{\partial v}{\partial x} = -i\dfrac{\partial f}{\partial y} = -i\dfrac{\partial u}{\partial y} + \dfrac{\partial v}{\partial y},$

(3) $\dfrac{\partial u}{\partial x} = \dfrac{\partial v}{\partial y}, \qquad \dfrac{\partial u}{\partial y} = -\dfrac{\partial v}{\partial x},$

(4) $\nabla^2 u = 0, \quad \nabla^2 v = 0, \quad \nabla^2 f = 0,$

where

$$\nabla^2 = \frac{\partial^2}{\partial x^2} + \frac{\partial^2}{\partial y^2},$$

is the two-dimensional Laplace operator. Equation (3) is the well-known Cauchy-Riemann relation.

An alternative proof that an analytic function is harmonic is as follows. Let $\bar{z} = x - iy$ be the complex conjugate of z, then the differential operators in the x, y coordinates are related to those in the z, $\bar{z}$ coordinates by

(5)
$$\frac{\partial}{\partial x} = \frac{\partial}{\partial z} + \frac{\partial}{\partial \bar{z}}, \qquad \frac{\partial}{\partial y} = i\left(\frac{\partial}{\partial z} - \frac{\partial}{\partial \bar{z}}\right),$$

or

(6)
$$2\frac{\partial}{\partial z} = \frac{\partial}{\partial x} - i\frac{\partial}{\partial y}, \qquad 2\frac{\partial}{\partial \bar{z}} = \frac{\partial}{\partial x} + i\frac{\partial}{\partial y}.$$

Equation (6) gives the Laplace operator in the z, $\bar{z}$ coordinates as

(7)
$$4\frac{\partial^2}{\partial z \partial \bar{z}} = \left(\frac{\partial}{\partial x} - i\frac{\partial}{\partial y}\right)\left(\frac{\partial}{\partial x} + i\frac{\partial}{\partial y}\right) = \frac{\partial^2}{\partial x^2} + \frac{\partial^2}{\partial y^2} = \nabla^2,$$

Obviously any analytic function ψ of z or $\bar{z}$ satisfies Eq. (7), i.e., $\nabla^2\psi(z) = 0$ and $\nabla^2\psi(\bar{z}) = 0$; so do the real and imaginary parts of $\psi(z)$ and $\psi(\bar{z})$.

Repeating the operation of Eq. (7) gives the biharmonic operator

(8)
$$\nabla^2\nabla^2 = \left(\frac{\partial^2}{\partial x^2} + \frac{\partial^2}{\partial y^2}\right)\left(\frac{\partial^2}{\partial x^2} + \frac{\partial^2}{\partial y^2}\right) = 16\frac{\partial^4}{\partial z^2 \partial \bar{z}^2}.$$

Thus any bihamornic function can be written in the form of the real or imaginary part of $\bar{z}\psi(z) + \chi(z)$ or $z\psi(\bar{z}) + \chi(\bar{z})$, where $\psi(z)$, $\chi(z)$ are arbitrary analytic functions, because

(9)
$$\nabla^2\nabla^2[\bar{z}\psi(z) + \chi(z)] = \nabla^2\nabla^2[z\psi(\bar{z}) + \chi(\bar{z})] = 0$$

Problem 9.7.

(a) Show that $x - iy$ is not an analytic function of $z = x + iy$.

(b) Determine the real functions of x and y which are real and imaginary parts of the complex functions z^n and $\tanh z$.

(c) Determine the real functions of r and θ which are the real and imaginary parts of the complex function $z \ln z$. Note that $z = re^{i\theta}$.

(d) $z = x + iy$, $\bar{z} = x - iy$, $a = \alpha + i\beta$, $\bar{a} = \alpha - i\beta$, where x, y, α, β are real numbers. Express the real and imaginary parts of the following functions explicitly in terms of x, y, α, β to get acquainted with the notations $f(z), \bar{f}(\bar{z}), f(\bar{z})$ and $\bar{f}(z)$:

1. $f(z) = az$, $\bar{f}(\bar{z}) = \bar{a}\bar{z}$, $f(\bar{z}) = a\bar{z}$, $\bar{f}(z) = \bar{a}z$.
2. $f(z) = e^{iaz}$, $\bar{f}(\bar{z}) = e^{-i\bar{a}\bar{z}}$, $f(\bar{z}) = e^{ia\bar{z}}$, $\bar{f}(z) = e^{-i\bar{a}z}$.
3. $f(z) = az^n$, $\bar{f}(\bar{z}) = \bar{a}\bar{z}^n$, $f(\bar{z}) = a\bar{z}^n$, $\bar{f}(z) = \bar{a}z^n$.

Show that the complex conjugate of $f(z)$ is $\bar{f}(\bar{z})$ in these examples.

(e) Show that the complex conjugate of $f(z) = \sum_{n=0}^{\infty} a_n z^n$ is

$$\bar{f}(\bar{z}) = \sum_{n=0}^{\infty} \bar{a}_n \bar{z}^n .$$

(f) Show that the derivative of $\bar{f}(\bar{z})$ with respect to $\bar{z}$ is equal to the complex conjugate of $df(z)/dz$.

(g) Find a function $v(x, y)$ of two real variables x, y, such that $\ln(x^2 + y^2) + iv(x, y)$ is an analytic function of a complex variable $x + iy$. (Use the Cauchy–Riemann differential equations).

9.6. KOLOSOFF–MUSKHELISHVILI METHOD

The representation of biharmonic functions by analytic functions leads to a general method of solving problems in plane stress and plane strain.

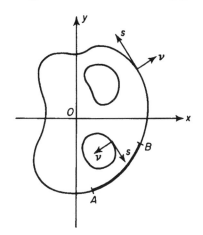

We consider a region — simply or multiply connected — on the x, y-plane bounded by a number of contours (Fig. 9.6:1). The interior of the region is considered to represent a disk of unit thickness. Surface traction is applied on the boundaries of this region. Body forces will be assumed to be zero to simplify the general solution. If body forces actually exist they can be represented by a particular solution so that the general problem is reduced to one without body force.

On introduction of the Airy stress function $\Phi(x, y)$ (Sec. 9.2), the plane stress and plane strain problems are

Fig. 9.6:1. Notations.

reduced to the solution of a biharmonic equation subjected to appropriate boundary conditions. According to the results of Sec. 9.5, the Airy stress function $\Phi(x, y)$ can be written as

(1) ▲ $2\Phi(x, y) = 2\,\mathrm{Rl}[\bar{z}\psi(z) + \chi(z)] = \bar{z}\psi(z) + z\bar{\psi}(\bar{z}) + \chi(z) + \bar{\chi}(\bar{z})$

where $\mathrm{Rl}(.)$ denotes the real part of the quantity in the parenthesis, and $\psi(z)$ and $\chi(z)$ are analytic functions, called the *complex stress functions*. The formula (1) is due to the French mathematician Goursat.

The problem in plane elasticity is now simply to determine the complex stress functions $\psi(z)$ and $\chi(z)$ that satisfy the boundary conditions. For this purpose, we must express all the stresses and displacements in terms of $\psi(z)$ and $\chi(z)$. We shall show that

$$(2) \quad \blacktriangle \qquad \sigma_{xx} + \sigma_{yy} = 2[\psi'(z) + \bar{\psi}'(\bar{z})],$$

$$(3) \quad \blacktriangle \qquad \sigma_{yy} - \sigma_{xx} + 2i\sigma_{xy} = 2[\bar{z}\psi''(z) + \chi''(z)],$$

$$(4) \quad \blacktriangle \qquad 2G(u + iv) = \kappa\psi(z) - z\bar{\psi}'(\bar{z}) - \bar{\chi}'(\bar{z}),$$

where a prime denotes differentiation with respect to the independent variable, i.e., $\psi'(z) = d\psi(z)/dz$ and $\bar{\psi}'(\bar{z}) = d\bar{\psi}(\bar{z})/d\bar{z}$,

$$(5) \qquad \kappa = 3 - 4\nu \qquad \text{for plane strain},$$

$$(6) \qquad \kappa = \frac{3 - \nu}{1 + \nu} \qquad \text{for plane stress},$$

and ν is Poisson's ratio. In addition, we have the following expressions for the resultant forces F_x, F_y and the moment M about the origin of the surface traction on an arc AB of the boundary from a point A to a point B:

$$(7) \quad \blacktriangle \qquad F_x + iF_y = \int_A^B \left(\overset{\nu}{T}_x + i \overset{\nu}{T}_y \right) ds = -i[\psi(z) + z\bar{\psi}'(\bar{z}) + \bar{\chi}'(\bar{z})]_A^B$$

$$(8) \qquad M = \mathrm{Rl}[\chi(z) - \bar{z}\bar{\chi}'(\bar{z}) - z\bar{z}\bar{\psi}'(\bar{z})]_A^B,$$

The physical quantities σ_{xx}, σ_{yy}, σ_{xy}, u, v, etc., are of course real-valued. If the complex-valued functions on the right-hand side of Eqs. (2)–(4) are known, a separation into real and imaginary parts determines all stresses and displacements.

The derivation of these relations is as follows. According to Eqs. (1) and (9.2:17), for vanishing body force, we have

$$(9) \qquad \sigma_{xx} + \sigma_{yy} = \frac{\partial^2\Phi}{\partial x^2} + \frac{\partial^2\Phi}{\partial y^2} = 4\frac{\partial^2\Phi}{\partial z\partial\bar{z}} = 2[\psi'(z) + \bar{\psi}'(\bar{z})].$$

$$(10) \qquad \sigma_{xx} - i\sigma_{xy} = i\frac{\partial}{\partial y}\left(\frac{\partial\Phi}{\partial x} - i\frac{\partial\Phi}{\partial y} \right) = 2i\frac{\partial}{\partial y}\frac{\partial\Phi}{\partial z}.$$

$$(11) \qquad \sigma_{yy} + i\sigma_{xy} = \frac{\partial}{\partial x}\left(\frac{\partial\Phi}{\partial x} - i\frac{\partial\Phi}{\partial y} \right) = 2\frac{\partial}{\partial x}\frac{\partial\Phi}{\partial z}.$$

Taking the difference of Eqs. (10) and (11) and using Eq. (1) lead to Eq. (3).

To derive Eq. (4), we have to consider plane stress and plane strain separately. For plane stress, Hooke's law states

(12) $\qquad e_{xx} = \dfrac{\partial u}{\partial x} = \dfrac{1}{E}\left(\sigma_{xx} - \nu\sigma_{yy}\right) = \dfrac{1}{E}\left(\dfrac{\partial^2\Phi}{\partial y^2} - \nu\dfrac{\partial^2\Phi}{\partial x^2}\right),$

(13) $\qquad e_{yy} = \dfrac{\partial v}{\partial y} = \dfrac{1}{E}\left(\sigma_{yy} - \nu\sigma_{xx}\right) = \dfrac{1}{E}\left(\dfrac{\partial^2\Phi}{\partial x^2} - \nu\dfrac{\partial^2\Phi}{\partial y^2}\right),$

(14) $\qquad e_{xy} = \dfrac{1}{2}\left(\dfrac{\partial u}{\partial y} + \dfrac{\partial v}{\partial x}\right) = \dfrac{1}{2G}\sigma_{xy} = -\dfrac{1}{2G}\dfrac{\partial^2\Phi}{\partial x\partial y}.$

Note that

(15) $\qquad \dfrac{d\psi(z)}{dz} + \dfrac{d\bar\psi(\bar z)}{d\bar z} = \dfrac{\partial}{\partial x}\left[\psi(x+iy) + \bar\psi(x-iy)\right] = 2\dfrac{\partial}{\partial x}\,\mathrm{Rl}(\psi)\,,$

(16) $\qquad \dfrac{d\psi(z)}{dz} + \dfrac{d\bar\psi(\bar z)}{d\bar z} = \dfrac{1}{i}\dfrac{\partial}{\partial y}\left[\psi(x+iy) - \bar\psi(x-iy)\right] = 2\dfrac{\partial}{\partial y}\,\mathrm{Im}(\psi)\,.$

where $\mathrm{Rl}(\,.\,)$ and $\mathrm{Im}(\,.\,)$ denote the real and imaginary parts of the quantity in the parenthesis. From Eqs. (9) and (15) and from Eqs. (9) and (16), we have

(17) $\qquad \dfrac{\partial^2\Phi}{\partial y^2} = 4\dfrac{\partial}{\partial x}\,\mathrm{Rl}(\psi) - \dfrac{\partial^2\Phi}{\partial x^2}\,, \qquad \dfrac{\partial^2\Phi}{\partial x^2} = 4\dfrac{\partial}{\partial y}\,\mathrm{Im}(\psi) - \dfrac{\partial^2\Phi}{\partial y^2}\,,$

respectively. A substitution of Eq. (17) into Eq. (12) yields

$$\frac{\partial u}{\partial x} = \frac{1}{E}\left[4\frac{\partial}{\partial x}\,\mathrm{Rl}(\psi) - (1+v)\frac{\partial^2\Phi}{\partial x^2}\right]$$

An integration with respect to x gives

(18) $\qquad u = \dfrac{1}{E}\left[4\,\mathrm{Rl}(\psi) - (1+\nu)\dfrac{\partial\Phi}{\partial x}\right] + f(y)\,,$

where $f(y)$ is an arbitrary function of y. Similarly, from Eq. (13), we obtain

(19) $\qquad v = \dfrac{1}{E}\left[4\,\mathrm{Im}(\psi) - (1+\nu)\dfrac{\partial\Phi}{\partial y}\right] + g(x)\,,$

where $g(x)$ is an arbitrary function of x. Substituting Eqs. (18) and (19) into Eq. (14), and noticing that $\partial\,\mathrm{Rl}(\psi)/\partial y = -\partial\,\mathrm{Im}(\psi)/\partial x$ we obtain

$$f'(y) + g'(x) = 0$$

Hence, $f(y) = \alpha y + \beta$, $g(x) = -\alpha x + \gamma$, where α, β, γ are constants. The forms of f and g indicate that they represent a rigid motion and can thus be disregarded in the analysis of deformation. If we set $f = g = 0$ in Eqs. (18) and (19), using

$$\frac{\partial \Phi}{\partial x} + i \frac{\partial \Phi}{\partial y} = 2 \frac{\partial \Phi}{\partial \bar{z}} = \psi(z) + z\bar{\psi}'(\bar{z}) + \bar{\chi}'(\bar{z})$$

and combine them properly, we obtain Eqs. (4) and (5). The plane strain case is similar, and Eq. (6) can be obtained without further analysis by applying a rule derived in Sec. 9.1, namely, replacing ν in Eq. (5) by an apparent value $\nu/(1 - \nu)$.

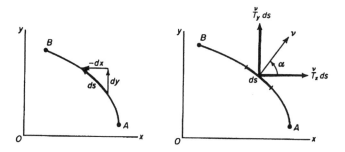

Fig. 9.6:2. Notations.

It remains to prove formulas (7) and (8). Refer to Fig. 9.6:2. Let AB be a continuous arc, ds be an arc length, ν be the unit vector normal to the arc, and $\overset{\nu}{T}_x, \overset{\nu}{T}_y$ be the x's and y's components of the traction acting on the arc. We note that

(20) $$\cos \alpha = \frac{dy}{ds}, \qquad \sin \alpha = -\frac{dx}{ds},$$

(21) $$\overset{\nu}{T}_x = \sigma_{xx} \cos \alpha + \sigma_{xy} \sin \alpha, \qquad \overset{\nu}{T}_y = \sigma_{yy} \sin \alpha + \sigma_{xy} \cos \alpha.$$

Hence by Eq. (9.2:17),

(22) $$\overset{\nu}{T}_x = \frac{\partial^2 \Phi}{\partial y^2} \frac{dy}{ds} + \frac{\partial^2 \Phi}{\partial x \partial y} \frac{dx}{ds} = \frac{d}{ds}\left(\frac{\partial \Phi}{\partial y}\right), \qquad \overset{\nu}{T}_y = -\frac{d}{ds}\left(\frac{\partial \Phi}{\partial x}\right).$$

Therefore, the resultant of forces acting on the arc AB can be written as:

$$F_x + iF_y = \int_A^B (\overset{\nu}{T}_x + i \overset{\nu}{T}_y) \, ds = \left[\frac{\partial \Phi}{\partial y} - i \frac{\partial \Phi}{\partial x}\right]_A^B$$

(23) $$= -i \left[\frac{\partial \Phi}{\partial x} + i \frac{\partial \Phi}{\partial y}\right]_A^B = -2i \left[\frac{\partial \Phi}{\partial \bar{z}}\right]_A^B,$$

which is Eq. (7). The moment about the origin O of the forces acting on the arc AB is

$$
M = \int_A^B (x \overset{\nu}{T_y} + y \overset{\nu}{T_x})\, ds = -\int_A^B \left(x \frac{d}{ds}\frac{\partial \Phi}{\partial x} + y \frac{d}{ds}\frac{\partial \Phi}{\partial y} \right) ds
$$

$$
(24) \qquad = \int_A^B \left(\frac{\partial \Phi}{\partial x}\frac{dx}{ds} + \frac{\partial \Phi}{\partial y}\frac{dy}{ds} \right) ds - \left[x\frac{\partial \Phi}{\partial x} + y\frac{\partial \Phi}{\partial y} \right]_A^B
$$

$$
= \Phi|_A^B - \mathrm{Rl} \left[2\bar{z}\frac{\partial \Phi}{\partial \bar{z}} \right]_A^B .
$$

A substitution of Eq. (1) into Eqs. (23) and (24) verifies Eqs. (7) and (8).

Equations (7) and (8), applied to a closed contour show that if $\psi(z)$ and $\chi(z)$ are single-valued, the resultant forces and moment of the surface traction acting on the contour vanish, since the functions in the brackets return to their initial values when the circuit is completed. If the resultant force of traction acting on a closed boundary does not vanish, the value of the functions in the brackets must not return to their initial values when the circuit is completed. For instance, the function $\ln z = \ln r + i\theta$ does not return to its original value on completing a circuit around the origin, since θ increases by 2π. Thus, if $\psi(z) = C \ln z$, or $\chi(z) = Dz \ln z$, where C and D are complex constants, Eq. (7) will yield a nonzero value of $F_x + iF_y$. Similarly, $\chi(z) = D \ln z$ yields a nonzero value of M according to Eq. (8) if D is imaginary, but a zero value if D is real.

The examples below will show that a number of problems in plane stress and plane strain can be solved by simple stress functions. By reducing the problem into the determination of two analytic functions of a complex variable, Muskhelishvili and his school have devised several general methods of approach. The principal tools for these general approaches are the conformal mapping and Cauchy integral equations. The details are well recorded in Muskhelishvili's books.

Example 1. *Circular hole in a large plate*

Consider a large plate with a circular hole of radius R at the origin. The plate is subjected to uniform remote stresses σ_{xx0}, σ_{yy0}, σ_{xy0}. The traction on the surface of the hole is zero. From Eq. (7), the traction free condition can be written as

$$
(25) \qquad -i(F_x - iF_y) = \bar{\psi}(\bar{z}) + \bar{z}\frac{d\psi(z)}{dz} + \varphi(z) = 0 ,
$$

at $z = Re^{i\theta}$, where φ is a complex stress function defined as

(26)
$$\varphi(z) = \frac{d\chi(z)}{dz},$$

which is introduced to save writing. If we define

(27)
$$\varphi(z) = -\bar{\psi}\left(\frac{R^2}{z}\right) - \frac{R^2}{z}\frac{d\psi(z)}{dz},$$

then the resultant force on a circular arc connecting points P and Q is

$$-i(F_x - iF_y) = \left[\bar{\psi}(\bar{z}) - \bar{\psi}\left(\frac{R^2}{z}\right) + \left(\bar{z} - \frac{R^2}{z}\right)\frac{d\psi(z)}{dz}\right]\bigg|_P^Q$$

which is zero at the hole $(z\bar{z} = R^2)$ for arbitrary $\psi(z)$.

We can use the remote stress conditions to determine $\psi(z)$. Assume that

(28)
$$\psi = Az + \frac{B}{z} + \frac{C}{z^2} + \cdots,$$

where $A, B, C, \ldots$ are complex constants. From Eqs. (2) and (3), we have

$$\frac{\sigma_{xx} + \sigma_{yy}}{2} = A - \frac{B}{z^2} - \frac{2C}{z^3} + \bar{A} - \frac{\bar{B}}{\bar{z}^2} - \frac{2\bar{C}}{\bar{z}^3} + \cdots$$

$$\frac{\sigma_{yy} - \sigma_{xx} + 2i\sigma_{xy}}{2} = \left(\bar{z} - \frac{R^2}{z}\right)\psi''(z) + \left[\bar{\psi}'\left(\frac{R^2}{z}\right) + \psi'(z)\right]\frac{R^2}{z^2}$$

$$= \left(\bar{z} - \frac{R^2}{z}\right)\left(\frac{2B}{z^3} - \frac{6C}{z^4}\right)$$

$$+ \left(\bar{A} - \frac{\bar{B}z^2}{R^4} - \frac{2\bar{C}z^3}{R^6} + A - \frac{B}{z^2} - \frac{2C}{z^3}\right)\frac{R^2}{z^2} + \cdots$$

For $\sigma_{xx} + \sigma_{yy} = \sigma_{xx0} + \sigma_{yy0}$ and $\sigma_{yy} - \sigma_{xx} + 2i\sigma_{xy} = \sigma_{yy0} - \sigma_{xx0} + 2i\sigma_{xy0}$ as z approaches infinite, we obtain

(29) $A + \bar{A} = \dfrac{\sigma_{xx0} + \sigma_{yy0}}{2}$, $B = -\left(\dfrac{\sigma_{yy0} - \sigma_{xx0}}{2} - i\sigma_{xy0}\right)R^2$, $C = \cdots = 0$.

The imaginary part of A is indeterminate since $\psi = \text{Im}(A)z$ represents a rigid body motion, and can be set to zero in the analysis of deformation. Thus, we have

$$\psi = \frac{\sigma_{xx0} + \sigma_{yy0}}{4}z - \left(\frac{\sigma_{yy0} - \sigma_{xx0}}{2} - i\sigma_{xy0}\right)\frac{R^2}{z},$$

$\psi(z)$ can be calculated from Eq. (27) and then

$$\frac{\sigma_{xx} + \sigma_{yy}}{2} = 2A - \frac{B}{z^2} - \frac{\bar{B}}{\bar{z}^2},$$

$$\frac{\sigma_{yy} - \sigma_{xx} + 2i\sigma_{xy}}{2} = \bar{z}\frac{2B}{z^3} + \left(2A - \frac{\bar{B}z^2}{R^4} - \frac{3B}{z^2}\right)\frac{R^2}{z^2}.$$

At $z = \pm R$, one can show that $\sigma_{yy} = 3\sigma_{yy0} - \sigma_{xx0}$, which is the well-known result of stress concentration at a hole in a large plate for $\sigma_{xx0} = 0$.

Example 2. *Interaction of an elliptical hole and a concentrated load*
                    (Tong 1984)

Consider a large plate with an elliptical hole of major and minor axes a, b along the x, y axes at the origin. A concentrated force $F_0 (= F_{x0} + iF_{y0})$ is applied at $z = z_0$. We introduce the transformation

(30) $\qquad z = R\left(\zeta + \frac{m}{\zeta}\right),$ $\qquad$ where $\quad R = \frac{a+b}{2},$ $\quad m = \frac{a-b}{a+b}$

which maps the unit circle at the origin of the ζ-plane to the ellipse in the z-plane. Similar to Eq. (25), the traction free condition on the ellipse can be written as

(31) $\qquad -i(F_x - iF_y) = \bar{\psi}(\bar{\zeta}) + \bar{z}(\bar{\zeta})\frac{d\zeta}{dz}\frac{d\psi(z)}{d\zeta} + \varphi(\zeta) = 0$

on the unit circle in the ζ-plane where

$$\frac{d\zeta}{dz} = 1\Big/\frac{dz}{d\zeta} = 1\left/\left[R\left(1 - \frac{m}{\zeta^2}\right)\right]\right..$$

Let

(32) $\qquad \varphi(\zeta) = -\bar{\psi}\left(\frac{1}{\zeta}\right) - \bar{z}\left(\frac{1}{\zeta}\right)\frac{d\zeta}{dz}\frac{d\psi(z)}{d\zeta},$

then, according to Eq. (7), the resultant force on an arc AB is

(33) $\qquad -i(F_x - iF_y) = \bar{\psi}(\bar{\zeta}) - \bar{\psi}\left(\frac{1}{\zeta}\right) + \left[\bar{z}(\bar{\zeta}) - \bar{z}\left(\frac{1}{\zeta}\right)\right]\frac{d\zeta}{dz}\frac{d\psi(z)}{d\zeta}\Bigg|_A^B,$

which equals to zero on the unit circle for arbitrary $\psi(\zeta)$.

We shall choose $\psi(\zeta)$ to give a net force F_0 at $z = z_0$ (or $\zeta = \zeta_0$) and zero stress at infinite. Assume $\psi(\zeta)$ in the form

(34) $\qquad\qquad\qquad \psi(\zeta) = A\ln(\zeta - \zeta_0) + \psi_1(\zeta),$

where $\psi_1(\zeta)$ is analytic and bounded outside the unit circle. From Eq. (4),

$$2G(u - iv) = \kappa \bar{\psi}(\bar{\zeta}) + \bar{\psi}\left(\frac{1}{\zeta}\right) - \left[\bar{z}(\bar{\zeta}) - \bar{z}\left(\frac{1}{\zeta}\right)\right] \frac{d\zeta}{dz} \frac{d\psi(z)}{d\zeta}$$

$$= \kappa[\bar{A} \ln(\bar{\zeta} - \bar{\zeta}_0) + \bar{\psi}_1(\bar{\zeta})] + \bar{A} \ln\left(\frac{1}{\zeta} - \bar{\zeta}_0\right) + \bar{\psi}_1\left(\frac{1}{\zeta}\right)$$

$$- \left[\bar{z}(\bar{\zeta}) - \bar{z}\left(\frac{1}{\zeta}\right)\right] \frac{d\zeta}{dz}\left[\frac{A}{\zeta - \zeta_0} + \psi_1'(\zeta)\right]$$

To assure that $u - iv$ is single-valued and has no singularity other than $\ln|\zeta - \zeta_0|$ outside the unit circle, we require that

$$\bar{\psi}_1\left(\frac{1}{\zeta}\right) = \kappa \bar{A} \ln(\zeta - \zeta_0) + \frac{A}{\zeta - \zeta_0}\left[\bar{z}(\bar{\zeta}_0) - \bar{z}\left(\frac{1}{\zeta_0}\right)\right] \Big/ R\left(1 - \frac{m}{\zeta_0^2}\right)$$

or

$$\psi_1(\zeta) = \kappa A \ln\left(\frac{1}{\zeta} - \bar{\zeta}_0\right) + \frac{\bar{A}\zeta}{1 - \zeta\bar{\zeta}_0}\left(\zeta_0 + \frac{m}{\zeta_0} - \frac{1}{\bar{\zeta}_0} - m\bar{\zeta}_0\right) \Big/ \left(1 - \frac{m}{\bar{\zeta}_0^2}\right).$$

Clearly $\psi_1(\zeta)$ is analytic and bounded outside the unit circle. To determine A, we examine Eq. (33) in a small neighborhood around ζ_0, i.e., $\zeta = \zeta_0 + \delta e^{i\theta}$,

$$F_{x0} + iF_{y0} = -i[(A \ln \delta + i\theta) - \kappa(A \ln \delta - i\theta) + f(\delta, \theta)]\big|_0^{2\pi} = 2\pi(\kappa + 1)A$$

where $f(\delta, \theta)$ is a single-valued function related to ψ_1. Therefore

$$A = \frac{F_0}{2\pi(\kappa + 1)}.$$

One can then evaluate all the stresses and displacements according to Eqs. (2)–(4).

For $b = 0$, m equals 1 and the ellipse becomes a crack of length $2a$. In this case

$$\frac{d\zeta}{dz} = \frac{2\zeta^2}{a(\zeta^2 - 1)} = \frac{1}{\sqrt{2 a r e^{i\theta}}} + O(1),$$

as $r \to 0$, where $re^{i\theta} = z \mp a$, which gives rise to a $1/\sqrt{r}$ stress singularity at the crack tips $z = \pm a$ or $\zeta = \pm 1$. The stress intensity factors, a measure of the intensity of the singularity, are defined as

$$K_I - iK_{II} = \lim_{|z| \to a^+} \sqrt{\frac{\pi}{a}(z^2 - a^2)}\,(\sigma_{yy} - i\sigma_{xy})$$

(35)

$$= \lim_{|z| \to a^+} 2\sqrt{\frac{\pi}{a}(z^2 - a^2)}\,\frac{d\psi}{dz} = 2\sqrt{\frac{\pi}{a}} \lim_{|\zeta| \to 1^+} \frac{d\psi}{d\zeta}$$

which is proportional to the coefficients of the terms of singular stress. In the equation above, z and ζ approach the limit along the real axis. The $1/\sqrt{r}$ stress singularity at crack tip is the fundamental characteristics of linear fracture mechanics. Interested readers are referred to books on the subject. From Eqs. (34) and (35), we obtain

$$K_1 + iK_{II} = \frac{1}{(1+\kappa)\sqrt{a\pi}} \left[\left(\frac{1}{1-\bar{\zeta}_0} - \frac{\kappa}{1-\zeta_0} \right) \bar{F}_0 + \frac{\zeta_0(\zeta_0\bar{\zeta}_0-1)(\bar{\zeta}_0-\zeta_0)}{\bar{\zeta}_0(1-\zeta_0)^2(\zeta_0^2-1)} F_0 \right] .$$

PROBLEMS

9.8. Discuss the conformal mapping specified by $\zeta = \frac{1}{2}(z + \frac{1}{z})$. Obtain the inverse transformation. This transformation is the basis of airplane wing theory.

9.9. Consider a multiply connected region in the x, y-plane. Express the condition for the single-valuedness of displacements in terms of stresses.

9.10. If $\partial\Phi/\partial x + i\partial\Phi/\partial y = \psi(z) + z\bar{\psi}'(\bar{z}) + \bar{\chi}'(\bar{z})$ is to be single-valued, how arbitrary are the functions $\psi(z)$, $\chi(z)$?

9.11. Consider a doubly connected region R bounded by two concentric circles. If the stress is single-valued in R, how arbitrary are the functions $\psi(z), \chi(z)$ in the Airy stress function? If both stresses and displacements are single-valued in R, how arbitrary are $\psi(z)$, $\chi(z)$?   [Muskhelishvili,[1.2] p. 116–128.]

9.12. Show that, if all stress and displacement components σ_{ij} and u_i vanish on an arc AB, then the stresses and displacements vanish identically in the entire region R containing the arc AB.   [Muskhelishvili,[1.2] p. 132.]

9.13. Show that the transformation $z = e^\zeta$ defines a transformation from rectangular coordinates to polar coordinates.

9.14. A ring $(a < r < b)$ is subjected to uniformly distributed shearing stress in the circumferential direction on the inner and outer surfaces as in Fig. P9.14. The couple of the shearing forces over the circumference is of magnitude M. Determine the stress and displacement components by means of complex potentials

$$\psi(z) = 0, \qquad \chi(z) = A \log z ,$$

where A is a constant which can be complex-valued.

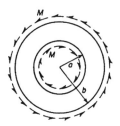

P9.14. Twisting of a disk.

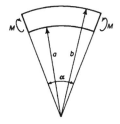

P9.15. Pure bending of a curved bar.

9.15. A curved bar is subjected to a bending moment M at each end. The bar is defined by circular arcs of radius a and b and radial lines with an opening angle α, $(\alpha < 2\pi)$ as in Fig. P9.15. Show that the problem can be solved by complex potentials of the form

$$\psi(z) = Az \log z + Bz,$$

$$\chi(z) = C \log z.$$

Determine the constants A, B, and C.

9.16. Consider a rectangular plate of sides $2l$ and $2L$, and small thickness $2h$. Various loads are applied on the edges, Fig. P9.16. There is no body force. Determine the stresses and displacements in the plate with the suggested stress functions $\psi(z)$ and $\chi(z)$. Express the constants a, b in terms of the external loads.

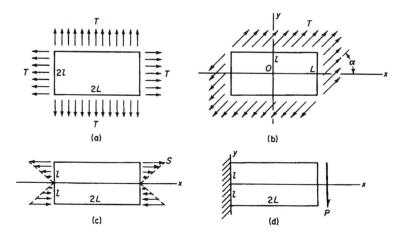

Fig. P9.16.

(a) All-round tension, Fig. P9.16(a):

$$\psi(z) = az, \quad \chi(z) = 0.$$

(b) Uniaxial tension at an angle α to the x-axis, Fig. P9.16(b):

$$\psi(z) = az, \quad \chi'(z) = -2aze^{-2i\alpha}.$$

(c) Pure bending, Fig. P9.16(c):

$$\psi(z) = aiz^2, \quad \chi(z) = -\frac{1}{3}aiz^3.$$

(d) Bending with shear, Fig. P9.16(d):

$$\psi(z) = 2\,aiz^3, \quad \chi'(z) = -4ai(z^3 + 6zb^2).$$

Note: In comparison with the bending of a beam in the three-dimensional case discussed in Sec. 7.7, the example in (d) shows how much simpler is the two-dimensional problem.

9.17. Consider plane-stress or plane-strain states, parallel to the x, y-plane, of elastic bodies occupying regions bounded by circles. Determine the stresses and displacements for the following physical problems with the suggested stress functions (Fig. P9.17):

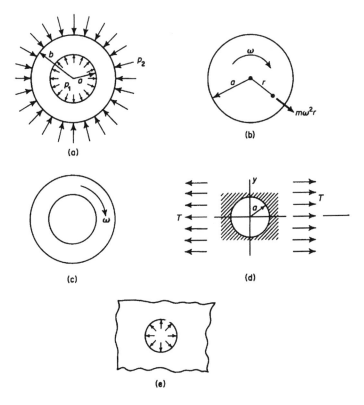

Fig. P9.17.

(a) Cylinder under internal pressure p_1 and external pressure p_2
   [Fig. P9.17(a)]

$$\psi(z) = Az, \quad \chi'(z) = \frac{B}{z}.$$

(b) A disk of radius a rotating about its axis at a constant angular speed ω [Fig. P9.17(b)]: the centrifugal force constitutes a body forces per unit mass $= \omega^2 r$, which has a potential $V = -\frac{1}{2}\omega^2 r^2$. Find a particular integral to Eqs. (9.2:18) and (9.2:19).

(c) Rotating hollow disk [Fig. P9.17(c)]. Solve by combining (a) and (b).

(d) Large plate under tension and containing an unstressed circular hole [Fig. P9.17(d)]:

$$\psi(z) = \frac{1}{4} Tz + \frac{A}{z},$$

$$\chi'(z) = -\frac{1}{2} Tz + \frac{B}{z} + \frac{C}{z^3}.$$

Show that $A = \frac{1}{2} a^2 T$, $B = -\frac{1}{2} Ta^2$, $C = \frac{1}{2} Ta^4$.

(e) Large plate containing a circular hole under uniform pressure [Fig. P9.17(e)]:

$$\psi(z) = 0, \qquad \chi'(z) = \frac{A}{z}.$$

10

VARIATIONAL CALCULUS,
ENERGY THEOREMS,
SAINT-VENANT'S PRINCIPLE

There are at least three important reasons for taking up the calculus of variations in the study of continuum mechanics.

1. Because basic minimum principles exist, which are among the most beautiful of theoretical physics.
2. The field equations (ordinary or partial differential equations) and the associated boundary conditions of many problems can be derived from variational principles. In formulating an approximate theory, the shortest and clearest derivation is usually obtained through variational calculus.
3. The computational methods of solution of variational problems are the most powerful tools for obtaining numerical results in practical problems of engineering importance.

In this chapter we shall discuss several variational principles and their applications.

A brief introduction to the calculus of variations is furnished below. Those readers who are familiar with the mathematical techniques of the calculus of variations may skip over the first six sections.

10.1. MINIMIZATION OF FUNCTIONALS

The calculus of variations is concerned with the minimization of functionals. If $u(x)$ is a function of x, defined for x in the interval (a, b), and if I is a quantity defined by the integral

$$I = \int_a^b [u(x)]^2 \, dx \,,$$

then the value of I depends on the function $u(x)$ as a whole. We may indicate this dependence by writing $I[u(x)]$. Such a quantity I is said to be

313

a *functional* of $u(x)$. Physical examples of functionals are the total kinetic energy of a flow field and the strain energy of an elastic body.

The basic problem of the calculus of variations may be illustrated by the following example. Let us consider a functional $J[u]$ defined by the integral

$$(1) \qquad\qquad J[u] = \int_a^b F(x, u, u')\, dx\,.$$

We shall give our attention to *all* functions $u(x)$ which are continuous and differentiable, with continuous derivatives $u'(x)$ and $u''(x)$ in the interval $a \le x \le b$, and satisfying the boundary conditions

$$(2) \qquad\qquad u(a) = u_0\,, \qquad u(b) = u_1\,,$$

where u_0 and u_1 are given numbers. We assume that the function $F(x, u, u')$ in Eq. (1) is continuous and differentiable with respect to x, and all such u, and u', up to all second-order partial derivatives, which are themselves continuous.

Among all functions $u(x)$ satisfying these continuity conditions and boundary values, we try to find a special one $u(x) = y(x)$, with the property that $J[u]$ attains a minimum when $u(x) = y(x)$, with respect to a sufficiently small neighborhood of $y(x)$. The neighborhood (h) of $y(x)$ is defined as follows. If h is a positive quantity, a function $u(x)$ is said to lie in the neighborhood (h) of $y(x)$ if the inequality

$$(3) \qquad\qquad |y(x) - u(x)| < h$$

holds for all x in (a, b). The situation is illustrated in Fig. 10.1:1.

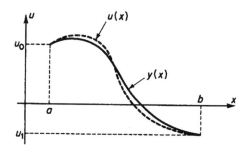

Fig. 10.1:1. Functions $u(x)$ and $y(x)$.

Let us assume that the problem posed above has a solution, which will be designated by $y(x)$; i.e., there exists a function $y(x)$ such that the inequality

(4) $$J[y] \leq J[u]$$

holds for all functions $u(x)$ in a sufficiently small neighborhood (h) of $y(x)$. Let us exploit the necessary consequences of this assumption.

Let $\eta(x)$ be an arbitrary function with the properties that $\eta(x)$ and its derivatives $\eta'(x)$, $\eta''(x)$ are continuous in the interval $a \leq x \leq b$, and that

(5) $$\eta(a) = \eta(b) = 0.$$

Then the function

(6) $$u(x) = y(x) + \epsilon\eta(x)$$

satisfies all the continuity conditions and boundary values specified at the beginning of this section. In fact, any function $u(x)$ satisfying these conditions can be represented with some function $\eta(x)$ in this manner. For a sufficiently small $\delta > 0$, this function $u(x)$ belongs, for all ϵ with $|\epsilon| < \delta$, to a prescribed neighborhood (h) of $y(x)$. Now we introduce the function

(7) $$\phi(\epsilon) = J[y + \epsilon\eta] = \int_a^b F[x, y(x) + \epsilon\eta(x), y'(x) + \epsilon\eta'(x)]\, dx.$$

Since $y(x)$ is assumed to be known, $\phi(\epsilon)$ is a function of ϵ for any specific $\eta(x)$. According to (4), the inequality

(8) $$\phi(0) \leq \phi(\epsilon)$$

must hold for all ϵ with $|\epsilon| < \delta$. In other words, $\phi(\epsilon)$ attains a minimum at $\epsilon = 0$. The function $\phi(\epsilon)$ is differentiable with respect to ϵ. Therefore, the necessary condition for $\phi(\epsilon)$ to attain a minimum at $\epsilon = 0$ must follow,

(9) $$\phi'(0) = 0,$$

where a prime indicates a differentiation with respect to ϵ. Now, a differentiation under the sign of integration yields

(10) $$\phi'(\epsilon) = \int_a^b [F_u(x, y+\epsilon\eta, y'+\epsilon\eta')\eta(x) + F_{u'}(x, y+\epsilon\eta, y'+\epsilon\eta')\eta'(x)]\, dx.$$

where F_u, $F_{u'}$ indicates $\partial F/\partial u$, $\partial F/\partial u'$, respectively. Integrating the last integral by parts, we obtain

(11) $$\phi'(\epsilon) = \int_a^b \left[F_u(x, y + \epsilon\eta, y' + \epsilon\eta') - \frac{d}{dx} F_{u'}(x, y+\epsilon\eta, y'+\epsilon\eta') \right] \eta(x)\, dx$$

$$+ F_{u'}(x, y + \epsilon\eta, y' + \epsilon\eta')\eta(x) \Big|_a^b$$

The last term vanishes according to Eq. (5). $\phi'(\epsilon)$ must vanish at $\epsilon = 0$, at which u equals y. Hence, we obtain the equation

$$(12) \qquad 0 = \phi'(0) = \int_a^b \left[F_y(x, y, y') - \frac{d}{dx} F_{y'}(x, y, y') \right] \eta(x)\, dx\,,$$

which must be valid for an arbitrary function $\eta(x)$.

Equation (12) leads immediately to Euler's differential equation, by virtue of the following "fundamental lemma of the calculus of variations:"

Lemma. *Let $\psi(x)$ be a continuous function in $a \le x \le b$. If the relation*

$$(13) \qquad \int_a^b \psi(x)\eta(x)\, dx = 0$$

holds for all functions $\eta(x)$ which vanish at $x = a$ and b and are continuous together with their first $2n$ derivatives, where n is a positive integer, then $\psi(x) \equiv 0$.

Proof. This lemma is easily proved indirectly. We shall show first that $\psi(x) \equiv 0$ in the open interval $a < x < b$. Let us suppose that this statement is not true, that $\psi(x)$ is different from zero, say positive, at $x = \xi$, where ξ lies in the open interval. Then, according to the continuity of $\psi(x)$, there must exist an interval $\xi - \delta \le x \le \xi + \delta$ (with $a < \xi - \delta, \xi + \delta < b, \delta > 0$), in which $\psi(x)$ is positive. Now we take the function

$$(14) \qquad \eta(x) = \begin{cases} (x - \xi + \delta)^{4n}(x - \xi - \delta)^{4n} \text{ in } \xi - \delta \le x \le \xi + \delta\,, \\ 0 \text{ elsewhere}\,, \end{cases}$$

where n is a positive integer. (See Fig. 10.1:2.) This function $\eta(x)$ satisfies the continuity and boundary conditions specified above. But the choice of $\eta(x)$ as defined by Eq. (14) will make $\int_a^b \psi(x)\,\eta(x)\,dx > 0$, in contradiction to the hypothesis. Thus, the hypothesis is untenable, and $\psi(x) \equiv 0$ in in $a < x < b$.

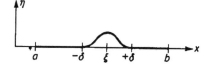

Fig. 10.1:2. The function $\eta(x)$.

According to the continuity of $\psi(x)$, we get $\psi(x) = 0$ also in $a \le x \le b$.

Q.E.D.

This lemma can be extended to hold equally well for multiple integrals.

From Eq. (12), it follows immediately from the lemma that the necessary condition for the functional $J[y + \epsilon\eta]$ in Eq. (7) to have an extremum is that the factor in front of $\eta(x)$ must vanish:

$$(15a) \quad \blacktriangle \qquad F_y(x,y,y') - \frac{d}{dx} F_{y'}(x,y,y') = 0 , \qquad\qquad a \leq x \leq b.$$

This is a differential equation that $y(x)$ must satisfy and is known as Euler's equation. Written out *in extenso*, we have

$$(15b) \quad \blacktriangle \qquad \frac{d^2y}{dx^2} \frac{\partial^2 F}{\partial y' \partial y'} (x,y,y') + \frac{dy}{dx} \frac{\partial^2 F}{\partial y' \partial y} (x,y,y')$$

$$+ \frac{\partial^2 F}{\partial y' \partial x} (x,y,y') - \frac{\partial F}{\partial y} (x,y,y') = 0 .$$

Should the problem be changed to finding the necessary condition for $J[u]$ to attain a maximum with respect to a sufficiently small neighborhood (h) of $y(x)$, the same result would be obtained. Hence, the result: *The validity of Euler's differential Eq. (15) is a necessary condition for a function $y(x)$ to furnish an extremum of the functional $J[u]$ with respect to a sufficiently small neighborhood (h) of $y(x)$.*

The satisfaction of Euler's equation is a necessary condition for $J[u]$ to attain an extremum; but it is not a sufficient condition. The question of sufficiency is rather involved; an interested reader must refer to treatises on calculus of variations such as those listed in Bibliography 10.1.[†]

Now a point of notation. It is customary to call $\epsilon\eta(x)$ the *variation* of $u(x)$ and write

$$(16) \qquad\qquad \epsilon\eta(x) = \delta u(x) .$$

It is also customary to define the first variation of the functional $J[u]$ as

$$(17) \qquad\qquad \delta J = \epsilon\phi'(\epsilon) .$$

On multiplying both sides of Eq. (11) by ϵ, it is evident that it can be

---

[†] Any consideration of the sufficient conditions requires the concept of the second variations and the examination of its positive or negative definiteness. Several conditions are known; they are similar to, but more complex than, the corresponding conditions for maxima or minima of ordinary functions. It is useful to remember that many subtleties exist in the calculus of variations. A physicist or an engineer rarely worries about the mathematical details. His minimum principles are established on physical grounds and the existence of a solution is usually taken for granted. Mathematically, however, many examples can be constructed to show that a functional may not have a maximum or a minimum, or that a solution of the Euler equation may not minimize the functional. An engineer should be aware of these possibilities.

written as

(18)
$$\delta J = \int_a^b \left[F_u(x, u, u') - \frac{d}{dx} F_{u'}(x, u, u') \right] \delta u(x)\, dx$$

$$+ F_{u'}(x, u, u') \delta u(x) \Big|_a^b .$$

This is analogous to the notation of the differential calculus, in which the expression $df = \epsilon f'(x)$ for an arbitrary small parameter ϵ is called the "differential of the function $f(x)$." It is obvious that δJ depends on the function $u(x)$ and its variation $\delta u(x)$. Thus, a necessary condition for $J[u]$ to attain an extremum when $u(x) = y(x)$ is the vanishing of the first variation δJ for all variations δu with $\delta u(a) = \delta u(b) = 0$.

Example 1.

$$J[u] = \int_a^b (1 + u'^2)\, dx = \min, \qquad u(a) = 0, \qquad u(b) = 1.$$

The Euler equation is

$$\frac{d}{dx} F_{y'} = \frac{d}{dx} 2y' = 2y'' = 0.$$

Hence $y(x)$ is a straight line passing through the points $(a, 0)$ and $(b, 1)$.

Example 2.
Which curve minimizes the following functional?

$$I[u] = \int_0^{\pi/2} [(u')^2 - u^2]\, dx, \qquad u(0) = 0, \qquad u\left(\frac{1}{2}\pi\right) = 1.$$

Ans. $u = y = \sin x$.

Example 3. *Minimum surface of revolution.*
Find $y(x)$ that minimizes the functional

$$I[u] = 2\pi \int_a^b u\sqrt{1 + u'^2}\, dx, \qquad u(a) = A, \qquad u(b) = B.$$

Ans. The Euler equation can be integrated to give

$$y\sqrt{1 + y'^2} - \frac{yy'^2}{\sqrt{1 + y'^2}} = c_1.$$

Let $y' = \sinh t$, then $y = c_1 \cosh t$ and

$$dx = \frac{dy}{y'} = \frac{c_1 \sinh t\, dt}{\sinh t} = c_1\, dt, \qquad x = c_1 t + c_2.$$

Hence, the minimum surface is obtained by revolving a curve with the following parametric equations about the x-axis,

$$x = c_1 t + c_2, \qquad y = c_1 \cosh t, \quad \text{or} \quad y = c_1 \cosh \frac{x - c_2}{c_1},$$

which is a family of catenaries. The constants c_1, c_2 can be determined from the values of y at the end points.

10.2. FUNCTIONAL INVOLVING HIGHER DERIVATIVES OF THE DEPENDENT VARIABLE

In an analogous manner one can treat the variational problem connected with the functional

$$(1) \qquad J[u] = \int_a^b F(x, u, u^{(1)}, \ldots, u^{(n)})\, dx$$

involving $u(x)$ and its successive derivatives $u^{(1)}(x), u^{(2)}(x), \ldots, u^{(n)}(x)$, for $a \le x \le b$. To state the problem concisely we denote by D the set of all real functions $u(x)$ with the following properties:

$$(2a) \qquad u(x), \ldots, u^{(2n)}(x) \text{ continuous in } a \le x \le b,$$

$$(2b) \qquad u(a) = \alpha_0, \qquad u^{(\nu)}(a) = \alpha_\nu, \qquad\qquad \nu = 1, \ldots, n-1,$$

$$(2c) \qquad u(b) = \beta_0, \qquad u^{(\nu)}(b) = \beta_\nu, \qquad\qquad \nu = 1, \ldots, n-1,$$

where $\alpha_0, \beta_0, \alpha_\nu$ and β_ν are given numbers. A function $u(x)$ which possesses these continuity properties and boundary values is said to be *admissible* or in the set D. We assume that the function $F(x, u, u^{(1)}, \ldots, u^{(n)})$ has continuous partial derivatives up to the order n with respect to the $n+2$ arguments $x, u, u^{(1)}, \ldots, u^{(n)}$ for u in the set D. We seek a function $u(x) = y(x)$ in D so that $J[u]$ is a minimum (or a maximum) for $u(x) = y(x)$, with respect to a sufficiently small neighborhood (h) of $y(x)$.

Let $\eta(x)$ be an arbitrary function with the properties

$$(3) \qquad \begin{aligned} &\eta(x), \eta^{(1)}(x), \ldots, \eta^{(2n)}(x) \text{ continuous in } a \le x \le b, \\ &\eta(a) = \cdots = \eta^{(n-1)}(a) = \eta(b) = \cdots = \eta^{(n-1)}(b) = 0. \end{aligned}$$

Then the function $u(x) = y(x) + \epsilon\eta(x)$ belongs to the set D for all ϵ. For sufficiently small ϵ's this function belongs to a prescribed neighborhood (h) of $y(x)$. Once again, we introduce the functional

$$(4) \quad \phi(\epsilon) = J[y + \epsilon\eta] = \int_a^b F(x, y + \epsilon\eta, y^{(1)} + \epsilon\eta^{(1)}, \ldots, y^{(n)} + \epsilon\eta^{(n)}) \, dx \,.$$

The assumption that $y(x)$ minimizes $J[u]$ leads to the necessary condition $\phi'(0) = 0$. So we obtain by differentiating under the sign of integration, integrating by parts, and using Eqs. (3), the result

$$
\begin{aligned}
0 = \phi'(0) &= \int_a^b \left\{ \sum_{\nu=0}^n \left[\frac{\partial}{\partial y^{(\nu)}} F(x, y, y^{(1)}, \ldots, y^{(n)}) \right] \eta^{(\nu)}(x) \right\} dx \\
(5) & \\
&= \int_a^b \left\{ \sum_{\nu=0}^n (-1)^\nu \frac{d^\nu}{dx^\nu} \left[\frac{\partial}{\partial y^{(\nu)}} F(x, y, y^{(1)}, \ldots, y^{(n)}) \right] \right\} \eta(x) \, dx \,,
\end{aligned}
$$

which holds for an arbitrary function $\eta(x)$ satisfying Eq. (3). According to the lemma proved in Sec. 10.1, we obtain the Euler equation

$$(6) \quad \blacktriangle \qquad \sum_{\nu=0}^n (-1)^\nu \frac{d^\nu}{dx^\nu} \left[\frac{\partial}{\partial y^{(\nu)}} F(x, y, y^{(1)}, \ldots, y^{(n)}) \right] = 0 \,,$$

which is a necessary condition for the function $u = y(x)$ to minimize (or maximize) the functional $J[u]$ given in Eq. (1).

The notations δu, δJ, etc., introduced in Sec. 10.1, can be extended to this case by obvious changes.

Example
Find the extremal of $J[u] = \int_0^1 (1 + u''^2) \, dx$ satisfying the boundary conditions

$$u(0) = 0 \,, \quad u'(0) = 1 \,, \quad u(1) = 1 \,, \quad u'(1) = 1 \,.$$

Ans. $u = y(x) = x$.

10.3. SEVERAL UNKNOWN FUNCTIONS

The method used in the previous sections can be extended to more complicated functionals. For example, let

$$(1) \qquad J[u_1, u_2, \ldots, u_m] = \int_a^b F(x, u_1, \ldots, u_m; u_1', \ldots, u_m') \, dx$$

be a functional depending on m functions $u_1(x), \ldots, u_m(x)$. We assume that the functions $F(x, u_1, \ldots, u_m')$, $u_1(x), \ldots, u_m(x)$ are twice differentiable, and that the boundary values of $u_1, \ldots, u_m$ are given at $x = a$ and b. We seek a special set of functions $u_\mu(x) = y_\mu(x)$, $\mu = 1, \ldots, m$, in order that $J[u_1, \ldots, u_m]$ attains a minimum (or a maximum) when $u_\mu(x) = y_\mu(x)$, with respect to a sufficiently small neighborhood of the $y_\mu(x)$; i.e., for all $u_\mu(x)$ satisfying the relation

$$|y_\mu(x) - u_\mu(x)| < h_\mu, \qquad h_\mu > 0, \qquad \mu = 1, 2, \ldots, m.$$

Again it is easy to obtain the necessary conditions. Let the set of functions $y_1(x), \ldots, y_m(x)$ be a solution of the variational problem. Let $\eta_1(x), \ldots, \eta_m(x)$ be an arbitrary set of functions with the properties

$$\eta_\mu(x), \eta_\mu'(x), \eta_\mu''(x) \text{ continuous in } a \le x \le b,$$

(2)
$$\eta_\mu(a) = \eta_\mu(b) = 0, \qquad \mu = 1, \ldots, m.$$

Then consider the function

(3) $\phi(\epsilon_1, \ldots, \epsilon_m) = J[y_1 + \epsilon_1 \eta_1, \ldots, y_m + \epsilon_m \eta_m]$

$$= \int_a^b F(x, y_1 + \epsilon_1 \eta_1, \ldots, y_m + \epsilon_m \eta_m; y_1' + \epsilon_1 \eta_1', \ldots, y_m' + \epsilon_m \eta_m') \, dx.$$

Since $y_1(x), \ldots, y_m(x)$ minimize (or maximize) $J[y_1 + \epsilon_1 \eta_1, \ldots, y_m + \epsilon_m \eta_m]$, the following inequality must hold for sufficiently small $\epsilon_1, \ldots, \epsilon_m$:

(4) $\qquad \phi(\epsilon_1, \ldots, \epsilon_m) \ge \phi(0, \ldots, 0), \qquad [\text{or } \phi(\epsilon_1, \ldots, \epsilon_m) \le \phi(0, \ldots, 0)].$

The corresponding necessary conditions are

(5) $\qquad \dfrac{\partial \phi}{\partial \epsilon_\mu} = 0, \qquad \text{for } \epsilon_1 = \epsilon_2 = \cdots = \epsilon_m = 0, \qquad \mu = 1, \ldots, m,$

which, again with F_{y_u}, $F_{y_u'}$ denoting $\partial F/\partial y_u$, $\partial F/\partial y_u'$, lead to

(6) $\qquad 0 = \int_a^b [F_{y_\mu}(x, y_1, \ldots, y_m, y_1', \ldots, y_m') \eta_\mu(x)$

$$+ F_{y_\mu'}(x, y_1, \ldots, y_m, y_1', \ldots, y_m') \eta_\mu'(x)] \, dx, \qquad \mu = 1, \ldots, m.$$

Using integration by parts and the conditions (2), we obtain

(7) $$0 = \int_a^b \left[F_{y_\mu}(x, y_1, \ldots, y_m, y_1', \ldots, y_m') \right.$$

$$\left. - \frac{d}{dx} F_{y_\mu'}(x, y_1, \ldots, y_m, y_1', \ldots, y_m') \right] \eta_\mu(x)\, dx$$

$$+ F_{y_\mu'}(x, y_1, \ldots, y_m, y_1', \ldots, y_m')\eta_\mu(x) \Big|_a^b, \quad \mu = 1, \ldots, m.$$

Equation (7) must hold for any set of functions $\eta_\mu(x)$ satisfying (2). By the lemma of Sec. 10.1 we have the following Euler equations which must be satisfied by $y_1(x), \ldots, y_m(x)$ minimizing (or maximizing) the functional (1):

(8) ▲ $F_{y_\mu}(x, y_1, \ldots, y_m; y_1', \ldots, y_m')$

$$- \frac{d}{dx} F_{y_\mu'}(x, y_1, \ldots, y_m; y_1', \ldots, y_m') = 0, \quad \mu = 1, \ldots, m.$$

As a generalization of the variational notations given in Sec. 10.1, the expression

(9) $$\delta J = \sum_{\mu=1}^{m} \epsilon_\mu \left(\frac{\partial \phi}{\partial \epsilon_\mu} \right)$$

is called the *first variation of the functional.*

Variational problems for functionals involving higher derivatives of $u_1, \ldots, u_m$ can be treated in the same way.

Example

Find the extremal of the functional

$$J[y, z] = \int_0^{\pi/2} (y'^2 + z'^2 + 2yz)dx,$$

$$y(0) = 0, \quad y\left(\frac{\pi}{2}\right) = 1, \quad z(0) = 0, \quad z\left(\frac{\pi}{2}\right) = -1.$$

The Euler equations are

$$y'' - z = 0, \qquad z'' - y = 0.$$

Eliminating z, we have $y^{\text{iv}} - y = 0$. Hence,

$$y = c_1 e^x + c_2 e^{-x} + c_3 \cos x + c_4 \sin x,$$

$$z = y'' = c_1 e^x + c_2 e^{-x} - c_3 \cos x - c_4 \sin x .$$

From the boundary conditions we obtain the solution

$$y = \sin x , \qquad z = -\sin x .$$

10.4. SEVERAL INDEPENDENT VARIABLES

Consider the functional

(1)
$$J[u] = \iint_G F(x, y, u, u_x, u_y) \, dx dy ,$$

where G is a finite, closed domain of the x, y-plane with a boundary curve C which has a piecewise continuously turning tangent, $u(x, y)$ is a continuous twice differentiable function of x, y in G, and u_x, u_y denote partial derivatives $\partial u/\partial x$, $\partial u/\partial y$ respectively. The function $F(x, y, u, u_x, u_y)$ is assumed to be twice continuously differentiable with respect to its five arguments. Let D be the set of all functions $u(x, y)$ with the following properties:

(2)
$$D : \begin{cases} \text{(a) } u(x,y), u_x(x,y), u_y(x,y), u_{xx}(x,y), \\ \qquad u_{xy}(x,y), u_{yy}(x,y) \text{ continuous in } G , \\ \text{(b) } u(x,y) \text{ prescribed on } C . \end{cases}$$

These are the admissible conditions for u. We now seek a special function $u(x, y) = v(x, y)$ in the set D, which minimizes (or maximizes) the functional $J[u]$.

Let $v(x, y)$ be a solution of this variational problem. Let $\eta(x, y)$ denote an arbitrary function with the properties

(3)
$$\begin{cases} \text{(a) } \eta(x,y) \text{ has continuous derivatives up to the} \\ \qquad \text{second order in } G , \\ \text{(b) } \eta(x,y) = 0 \text{ for } (x,y) \text{ on the boundary } C . \end{cases}$$

A consideration of the function

(4)
$$\phi(\epsilon) = J[v + \epsilon \eta] ,$$

which attains an extremum when $\epsilon = 0$, again leads to the necessary condition

(5)
$$\frac{d\phi}{d\epsilon}(0) = 0 ,$$

i.e., explicitly, with subscripts to F denoting partial derivatives of F,

$$(6) \qquad 0 = \iint_G \{F_v(x, y, v, v_x, v_y)\eta + F_{v_x}(x, y, v, v_x, v_y)\eta_x$$

$$+ F_{v_y}(x, y, v, v_x, v_y)\eta_y\}\, dxdy\,.$$

The last two terms can be simplified by Gauss' theorem after rewriting Eq. (6) as[†]

$$0 = \iint_G \left\{ F_v \cdot \eta + \frac{\hat{\partial}}{\partial x}(F_{v_x} \cdot \eta) - \eta\frac{\hat{\partial}}{\partial x}F_{v_x} + \frac{\hat{\partial}}{\partial y}(F_{v_y} \cdot \eta) - \eta\frac{\hat{\partial}}{\partial y}F_{v_y} \right\} dxdy\,.$$

An application of Gauss' theorem to the sum of the second and fourth terms in the integrand gives

$$(7) \qquad 0 = \iint_G \left\{ F_v - \frac{\hat{\partial}}{\partial x}F_{v_x} - \frac{\hat{\partial}}{\partial y}F_{v_y} \right\} \eta(x, y)\, dxdy$$

$$+ \int_C \{F_{v_x}n_1(s) + F_{v_y}n_2(s)\}\eta ds\,.$$

Here $n_1(s)$ and $n_2(s)$ are the components of the unit outward-normal vector $\mathbf{n}(s)$ of C.

The line integral in Eq. (7) vanishes according to Eq. (3). The function $\eta(x, y)$ in the surface integral is arbitrary. The generalized lemma of Sec. 10.1 then leads to the Euler equations

$$F_v - \frac{\hat{\partial}}{\partial x}F_{v_x} - \frac{\hat{\partial}}{\partial y}F_{v_y} = 0\,,$$

i.e.,

$$(9) \quad \blacktriangle \qquad \frac{\partial F}{\partial v} - \frac{\partial^2 F}{\partial v_x \partial x} - \frac{\partial^2 F}{\partial v_y \partial y} - \frac{\partial^2 F}{\partial v_x \partial v}\frac{\partial v}{\partial x} - \frac{\partial^2 F}{\partial v_y \partial v}\frac{\partial v}{\partial y}$$

$$- \frac{\partial^2 F}{\partial v_x^2}\frac{\partial^2 v}{\partial x^2} - 2\frac{\partial^2 F}{\partial v_x \partial v_y}\frac{\partial^2 v}{\partial x \partial y} - \frac{\partial^2 F}{\partial v_y^2}\frac{\partial^2 v}{\partial y^2} = 0\,.$$

Equation (9) is a necessary condition for a function $v(x, y)$ in the set D minimizing (or maximizing) the functional (1).

[†]The symbol $(\hat{\partial}/\partial x)F_{v_x}$ means that F_{v_x} should be considered as a function of x and y, e.g.,

$$\frac{\hat{\partial}}{\partial x}F_{v_x} = \frac{\partial}{\partial x}F_{v_x} + \frac{\partial F_{v_x}}{\partial v}\frac{\partial v}{\partial x} + \frac{\partial F_{v_x}}{\partial v_x}\frac{\partial v_x}{\partial x} + \frac{\partial F_{v_x}}{\partial v_y}\frac{\partial v_y}{\partial x}\,.$$

In the same way one may treat variational problems connected with functionals which involve more than two independent variables, higher derivatives, and several unknown functions.

Example 1

$$J[u] = \iint_D (u_x^2 + u_y^2)\, dxdy \,,$$

with a boundary condition that u is equal to an assigned function $f(x,y)$ on the boundary C of the domain D. The Euler equation is the Laplace equation $v_{xx} + v_{yy} = 0$.

Example 2

$$J[u] = \iint_D (u_{xx}^2 + u_{yy}^2 + 2u_{xy}^2)^2\, dxdy = \min \,.$$

Then v must satisfy the biharmonic equation

$$\frac{\partial^4 v}{\partial x^4} + 2\frac{\partial^4 v}{\partial x^2 \partial y^2} + \frac{\partial^4 v}{\partial y^4} = 0 \,.$$

10.5. SUBSIDIARY CONDITIONS — LAGRANGIAN MULTIPLIERS

In many problems, we are interested in the extremum of a function or functional under certain subsidiary conditions. As an elementary example, let us consider a function $f(x,y)$ defined for all (x,y) in a certain domain G. Suppose that we are interested in the extremum of $f(x,y)$, not for all points in G, but only for those points (x,y) in G which satisfy the relation

(1) $$\phi(x,y) = 0 \,.$$

Thus, if the domain G and the curve $\phi(x,y) = 0$ are as shown in Fig. 10.5:1, then we are interested in finding the extremum of $f(x,y)$ among all points that lie on the segment of the curve AB. For conciseness of expression, we shall designate such a subdomain, the segment AB, by a symbol G_ϕ.

Let us find the necessary conditions for $f(x,y)$ to attain an extreme value at a point $(\bar{x}, \bar{y})$ in G_ϕ, with respect to all points of a sufficiently small neighborhood of $(\bar{x}, \bar{y})$ belonging to G_ϕ.

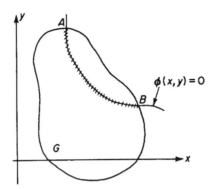

Fig. 10.5:1. Illustrating a subsidiary condition.

Let us assume that the function $\phi(x, y)$ has continuous partial derivatives with respect to x and y, $\phi_x = \partial\phi/\partial x$, $\phi_y = \partial\phi/\partial y$, and that at the point $(\bar{x}, \bar{y})$ not both derivatives are zero; say,

(2) $$\phi_y(\bar{x}, \bar{y}) \neq 0 .$$

Then, according to a fundamental theorem on implicit functions, there exists a neighborhood of $\bar{x}$, say $\bar{x} - \delta \leq x \leq \bar{x} + \delta$ $(\delta > 0)$, where the equation $\phi(x, y) = 0$ can be solved uniquely in the form

(3) $$y = g(x) .$$

The function $g(x)$ so defined is single-valued and differentiable, and $\phi[x, g(x)] = 0$ is an identity in x. Hence

(4) $$0 = \phi_x[x, g(x)] + \phi_y[x, g(x)]\frac{dg(x)}{dx} .$$

Now, let us consider the problem of the extremum value of $f(x, y)$ in G_ϕ. According to (3), in a sufficiently small neighborhood of $(\bar{x}, \bar{y})$, y is an implicit function of x, and $f(x, y)$ becomes

(5) $$\mathcal{F}(x) = f[x, g(x)] .$$

If $\mathcal{F}(x)$ attains an extreme value at $\bar{x}$, then the first derivative of $\mathcal{F}(x)$ vanishes at $\bar{x}$, i.e.,

(6) $$0 = \frac{d\mathcal{F}}{dx}(\bar{x}) = f_x(\bar{x}, \bar{y}) + f_y(\bar{x}, \bar{y})\frac{dg}{dx}(\bar{x}) .$$

But the derivative dg/dx can be eliminated between Eqs. (6) and (4). The result, together with Eq. (1), constitutes the necessary condition for an extremum of $f(x, y)$ in G_ϕ.

The formalism will be more elegant by introducing a number λ, called *Lagrange's multiplier*, defined by

$$(7) \qquad \lambda = -\frac{f_y(\bar{x}, \bar{y})}{\phi_y(\bar{x}, \bar{y})} .$$

A combination of Eqs. (4), (6), and (7) gives

$$(8) \qquad f_x(\bar{x}, \bar{y}) + \lambda \phi_x(\bar{x}, \bar{y}) = 0 ,$$

while Eq. (7) may be written as

$$(9) \qquad f_y(\bar{x}, \bar{y}) + \lambda \phi_y(\bar{x}, \bar{y}) = 0 .$$

Equations (1), (8), and (9) are necessary conditions for the function $f(x, y)$ to attain an extreme at a point $(\bar{x}, \bar{y})$, where $\phi_x^2 + \phi_y^2 > 0$. These conditions constitute three equations for the three "unknowns" x, y, and λ.

These results can be summarized in the following manner. *We introduce a new function*

$$(10) \qquad F(x, y; \lambda) = f(x, y) + \lambda \phi(x, y) .$$

If the function $f(x, y)$ has an extreme value at the point $(\bar{x}, \bar{y})$ with respect to G_ϕ, and if

$$(11) \qquad [\phi_x(\bar{x}, \bar{y})]^2 + [\phi_y(\bar{x}, \bar{y})]^2 > 0 ,$$

then there exists a certain number $\bar{\lambda}$ so that the three partial derivatives of $F(x, y; \lambda)$ with respect to x, y, and λ are zero at $(\bar{x}, \bar{y}, \bar{\lambda})$:

$$(12) \qquad \begin{aligned} \frac{\partial F}{\partial x}(\bar{x}, \bar{y}, \bar{\lambda}) &= f_x(\bar{x}, \bar{y}) + \bar{\lambda} \phi_x(\bar{x}, \bar{y}) = 0 , \\[1mm] \frac{\partial F}{\partial y}(\bar{x}, \bar{y}, \bar{\lambda}) &= f_y(\bar{x}, \bar{y}) + \bar{\lambda} \phi_y(\bar{x}, \bar{y}) = 0 , \\[1mm] \frac{\partial F}{\partial \lambda}(\bar{x}, \bar{y}, \bar{\lambda}) &= \phi(\bar{x}, \bar{y}) = 0 . \end{aligned}$$

If there exist points (x', y') in G with $\phi(x', y') = 0$ and $\phi_x(x', y') = \phi_y(x', y') = 0$, additional considerations are necessary.

In the formulation (10) and (12), the theorem can be generalized to the case of n variables $x_1, \ldots, x_n$ and several subsidiary conditions $0 = \phi_1(x_1, \ldots, x_n) = \cdots = \phi_m(x_1, \ldots, x_n)$, $m < n$.

The application of the Lagrange multiplier method to the minimization of functionals follows a similar reasoning.

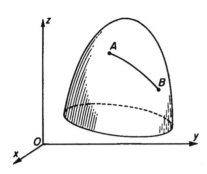

Fig. 10.5:2. A problem of geodesics.

As an example, let us consider the classical problem of geodesics: to find the line of minimal length lying on a given surface $\phi(x, y, z) = 0$ and joining two given points on this surface (Fig. 10.5:2). Here we must minimize the functional

$$(13) \qquad l = \int_{x_0}^{x_1} \sqrt{1 + y'^2 + z'^2}\, dx,$$

with $y(x)$, $z(x)$ satisfying the condition $\phi(x, y, z) = 0$. This problem was solved in 1697 by Johann Bernoulli, but a general method of solution was given by L. Euler and J. Lagrange.

We consider a new functional

$$(14) \qquad I^*[y, z, \lambda] = \int_{x_0}^{x_1} \left[\sqrt{1 + y'^2 + z'^2} + \lambda(x)\phi(x, y, z) \right] dx$$

and use the method of Sec. 10.3 to determine the functions $y(x)$, $z(x)$, and $\lambda(x)$ that minimizes I^*. The necessary conditions are

$$(15) \qquad \lambda \frac{\partial \phi}{\partial y} - \frac{d}{dx} \frac{y'}{\sqrt{1 + y'^2 + z'^2}} = 0,$$

$$(16) \qquad \lambda \frac{\partial \phi}{\partial z} - \frac{d}{dx} \frac{z'}{\sqrt{1 + y'^2 + z'^2}} = 0,$$

$$(17) \qquad \phi(x, y, z) = 0.$$

This system of equations determines the functions $y(x)$, $z(x)$, and $\lambda(x)$.

10.6. NATURAL BOUNDARY CONDITIONS

In previous sections we considered variational problems in which the admissible functions have prescribed values on the boundary. We shall now consider problems in which no boundary values are prescribed for the admissible functions. Such problems lead to the *natural boundary conditions*.

Consider again the functional (10.1:1), but now omit the boundary condition (10.1:2). Following the arguments of Sec. 10.1, we obtain the necessary condition (10.1:12),

$$(1) \qquad 0 = \int_a^b \left\{ F_y - \frac{d}{dx} F_{y'} \right\} \eta(x) dx + F_{y'} \cdot \eta(x) \Big|_a^b.$$

This equation must hold for all arbitrary functions $\eta(x)$ and, in particular, for arbitrary functions $\eta(x)$ with $\eta(a) = \eta(b) = 0$. This leads at once to the Euler equation (10.1:15)

$$(2) \qquad F_y(x, y, y') - \frac{d}{dx} F_{y'}(x, y, y') = 0, \qquad \text{in } a \le x \le b.$$

In contrast to Sec. 10.1, however, the last term $F_{y'} \cdot \eta(x)|_a^b$, does not vanish by prescription. Hence, by Eqs. (1) and (2), we must have

$$(3) \qquad (F_{y'} \cdot \eta)_{x=b} - (F_{y'} \cdot \eta)_{x=a} = 0$$

for all functions $\eta(x)$. But now $\eta(a)$ and $\eta(b)$ are arbitrary. Taking two functions $\eta_1(x)$ and $\eta_2(x)$ with

$$(4) \qquad \eta_1(a) = 1, \qquad \eta_1(b) = 0; \qquad \eta_2(a) = 0, \qquad \eta_2(b) = 1;$$

we get, from (3),

$$(5) \qquad F_{y'}[a, y(a), y'(a)] = 0, \qquad F_{y'}[b, y(b), y'(b)] = 0.$$

The conditions (5) are called the *natural boundary conditions* of our problem. They are the boundary conditions which must be satisfied by the function $y(x)$ if the functional $J[u]$ reaches an extremum at $u(x) = y(x)$, provided that $y(a)$ and $y(b)$ are entirely arbitrary.

Thus, *if the first variation of a functional $J[u]$ vanishes at $u(x) = y(x)$, and if the boundary values of $u(x)$ at $x = a$ and $x = b$ are arbitrary, then $y(x)$ must satisfy not only the Euler equation but also the natural boundary conditions which, in general, involve the derivatives of $y(x)$.* In contrast to the *natural* boundary conditions, the conditions $u(a) = \alpha$, $u(b) = \beta$, which specify the boundary values of $u(x)$ at $x = a$ and b, are called *rigid boundary conditions*.

The concept of natural boundary conditions is also important in a more general type of variational problem in which boundary values occur explicitly in the functionals. It can be generalized also to functionals involving several dependent and independent variables and higher derivatives of the dependent variables.

The idea of deriving natural boundary conditions for a physical problem is of great importance and will be illustrated again and again later. See, for example, Secs. 10.8 and 11.2.

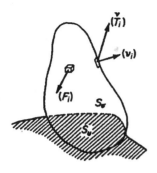

Fig. 10.7:1. Notations.

10.7. THEOREM OF MINIMUM POTENTIAL ENERGY UNDER SMALL VARIATIONS OF DISPLACEMENTS

Let a body be in *static equilibrium* under the action of specified body and surface forces. (Fig. 10.7:1.) The boundary surface S shall be assumed to consist of two parts, S_σ and S_u, with the following boundary conditions.

Over S_σ : The surface traction $\overset{\nu}{T_i}$ is prescribed.

Over S_u : The displacement u_i is prescribed.

We assume that there exists a system of displacements u_1, u_2, u_3 satisfying the Navier's equations of equilibrium and the given boundary conditions. Let us consider a class of arbitrary displacements $u_i + \delta u_i$ consistent with the constraints imposed on the body. Thus δu_i must vanish over S_u, but it is arbitrary over S_σ. We further restrict δu_i to be triply differentiable and to be of such an order of magnitude that the material remains elastic. Such arbitrary displacements δu_i are called *virtual displacements*.

Let us assume that static equilibrium prevails and compute the *virtual work* done by the body force F_i per unit volume and the surface force $\overset{\nu}{T_i^*}$ per unit area:

$$\int_V F_i \delta u_i dv + \int_S \overset{\nu}{T_i^*} \delta u_i dS .$$

On substituting $\overset{\nu}{T_i^*} = \sigma_{ij} \nu_j$ and transforming according to Gauss' theorem, we have

$$(1) \qquad \int_S \overset{\nu}{T_i^*} \delta u_i dS = \int_S \sigma_{ij} \delta u_i \nu_j dS$$

$$= \int_V (\sigma_{ij} \delta u_i)_{,j} \, dv$$

$$= \int_V \sigma_{ij,j} \delta u_i \, dv + \int_V \sigma_{ij} \delta u_{i,j} \, dv .$$

According to the equation of equilibrium the first integral on the right-hand side is equal to $-\int F_i \delta u_i \, dv$. On account of the symmetry of σ_{ij}, the second integral may be written as

$$\int_V \sigma_{ij} \frac{1}{2} \left(\delta u_{i,j} + \delta u_{j,i} \right) dv = \int_V \sigma_{ij} \delta e_{ij} \, dv \,.$$

Therefore, Eq. (1) becomes

(2) ▲ $$\int_V F_i \delta u_i \, dv + \int_{S_\sigma} \overset{\nu}{T_i^*} \, \delta u_i dS = \int_V \sigma_{ij} \delta e_{ij} \, dv \,.$$

This equation expresses the *principle of virtual work*. The surface integral needs only be integrated over S_σ, since δu_i vanishes over the surface S_u where boundary displacements are given.

If the *strain-energy function* $W(e_{11}, e_{12}, \dots)$ exists, so that $\sigma_{ij} = \partial W/\partial e_{ij}$, (Sec. 6.1), then it can be introduced into the right-hand side of Eq. (2). Since

(3) $$\int_V \sigma_{ij} \delta e_{ij} \, dv = \int_V \frac{\partial W}{\partial e_{ij}} \delta e_{ij} \, dv = \delta \int_V W \, dv \,,$$

the principle of virtual work can be stated as

(4) ▲ $$\delta \int_V W \, dv - \int_V F_i \delta u_i \, dv - \int_{S_\sigma} \overset{\nu}{T_i^*} \, \delta u_i dS = 0 \,.$$

Further simplification is possible if the body force F_i and the surface tractions $\overset{\nu}{T_i^*}$ are *conservative* so that

(5) $$F_i = -\frac{\partial G}{\partial u_i} \,, \qquad \overset{\nu}{T_i^*} = -\frac{\partial g}{\partial u_i} \,.$$

The functions $G(u_1, u_2, u_3)$ and $g(u_l, u_2, u_3)$ are called the potential of F_i and $\overset{\nu}{T_i^*}$, respectively. In this case,

(6) $$-\int_V F_i \delta u_i \, dv - \int_{S_\sigma} \overset{\nu}{T_i^*} \, \delta u_i dS = \delta \int_V G \, dv + \delta \int_S g dS \,.$$

Then Eq. (4) may be written as

(7) ▲ $$\delta \mathcal{V} = 0 \,,$$

where

(8) ▲ $$\mathcal{V} \equiv \int_V (W + G) \, dv + \int_{S_\sigma} g dS \,.$$

The function $\mathcal{V}$ is called the *potential energy of the system*. This equation states that the potential energy has a stationary value in a class of admissible variations δu_i of the displacements u_i in the equilibrium state. Formulated in another way, it states that, *of all displacements satisfying the given boundary conditions, those which satisfy the equations of equilibrium are distinguished by a stationary (extreme) value of the potential energy.* For a rigid-body, W vanishes and the familiar form is recognized. We emphasize that the linearity of the stress-strain relationship has not been invoked in the above derivation, so that *this principle is valid for nonlinear, as well as linear, stress-strain law, as long as the body remains elastic.*

That this stationary value is a minimum in the neighborhood of the natural, unstrained state follows the assumption that the strain energy function is positive definite in such a neighborhood (see Secs. 12.4 and 12.5). This can be shown by comparing the potential energy $\mathcal{V}$ of the actual displacements u_i with the energy $\mathcal{V}'$ of another system of displacements $u_i + \delta u_i$ satisfying the condition $\delta u_i = 0$ over S_u. We have

$$(9) \qquad \mathcal{V}' - \mathcal{V} = \int_V [W(e_{11} + \delta e_{11}, \ldots,) - W(e_{11}, \ldots,)]\, dv$$

$$- \int_V F_i \delta u_i\, dv - \int_{S_\sigma} \overset{\nu}{T_i^*}\, \delta u_i\, dS.$$

Expanding $W(e_{11} + \delta e_{11}, \ldots,)$ into a power series, we have

$$(10) \qquad W(e_{11} + \delta e_{11}, \ldots,) = W(e_{11}, \ldots,) + \frac{\partial W}{\partial e_{ij}} \delta e_{ij}$$

$$+ \frac{1}{2} \frac{\partial^2 W}{\partial e_{ij} \partial e_{kl}} \delta e_{ij} \delta e_{kl} + \cdots.$$

A substitution into Eq. (9) yields, up to the second order in δe_{ij},

$$(11) \qquad \mathcal{V}' - \mathcal{V} = \int_V \frac{\partial W}{\partial e_{ij}} \delta e_{ij}\, dv - \int_V F_i \delta u_i\, dv - \int_S \overset{\nu}{T_i^*}\, \delta u_i\, dS$$

$$+ \int_V \frac{1}{2} \frac{\partial^2 W}{\partial e_{ij} \partial e_{kl}} \delta e_{ij} \delta e_{kl}\, dv.$$

The sum of the terms in the first line on the right-hand side vanishes on account of Eq. (4). The sum in the second line is positive for sufficiently small values of strain δe_{ij} as can be seen as follows. Let us set $e_{ij} = 0$ in Eq. (10). The constant term $W(0)$ is immaterial. The linear term must vanish because $\partial W/\partial e_{ij} = \sigma_{ij}$, which must vanish as $e_{ij} \to 0$. Hence, up to the second order,

$$(12) \qquad \frac{1}{2}\frac{\partial^2 W}{\partial e_{ij}\partial e_{kl}}\,\delta e_{ij}\delta e_{kl} = W(\delta e_{ij})\,.$$

Therefore, Eq. (11) becomes

$$(13) \qquad \mathcal{V}' - \mathcal{V} = \int W(\delta e_{ij})\,dv\,.$$

If $W(\delta e_{ij})$ is positive definite, then the last line in Eq. (11) is positive, and

$$(14) \qquad \mathcal{V}' - \mathcal{V} \geq 0$$

and that $\mathcal{V}$ is a minimum is proved. Accordingly, our principle is called the *principle of minimum potential energy*. The equality sign holds only if all δe_{ij} vanish, i.e., if the virtual displacements consist of a virtual rigid-body motion. If there were three or more points of the body fixed in space, such a rigid-body motion would be excluded, and $\mathcal{V}$ is a strong minimum; otherwise it is a weak minimum.

To recapitulate, we remark again that the variational principle (2) is generally valid; Eq. (4) is established whenever the strain energy function $W(e_{11}, e_{12}, \ldots,)$ exists; and Eq. (7) is established when the potential energy $\mathcal{V}$ can be meaningfully defined, but the fact that $\mathcal{V}$ is a minimum for "actual" displacements is established only in the neighborhood of the stable natural state, where W is positive definite.

Conversely, we may show that the variational principle gives the equations of elasticity. In fact, starting from Eq. (7) and varying u_i, we have

$$(15) \qquad \delta\mathcal{V} = \int_V \frac{\partial W}{\partial e_{ij}}\delta e_{ij}\,dv - \int_V F_i\delta u_i\,dv - \int_S \overset{\nu}{T_i^*}\,\delta u_i\,dS = 0\,.$$

But

$$\int_V \frac{\partial W}{\partial e_{ij}}\delta e_{ij}\,dv = \frac{1}{2}\int_V \sigma_{ij}(\delta u_{i,j} + \delta u_{j,i})\,dv$$

$$= -\int_V \sigma_{ij,j}\delta u_i\,dv + \int_S \sigma_{ij}\nu_j\delta u_i\,dS\,.$$

Hence,

$$(16) \qquad \int_V (\sigma_{ij,j} + F_i)\delta u_i\,dv + \int_S (\sigma_{ij}\nu_j - \overset{\nu}{T_i^*})\delta u_i\,dS = 0\,,$$

which can be satisfied for arbitrary δu_i if

(17) $$\sigma_{ij,j} + F_i = 0 \qquad \text{in } V$$

(18) $$\delta u_i = 0 \qquad \text{on } S_u \text{ (rigid boundary condition)},$$

(19) $$\overset{\nu}{T_i^*} = \sigma_{ij}\nu_j \qquad \text{on } S_\sigma \text{ (natural boundary condition)}.$$

The one-to-one correspondence between the differential equations of equilibrium and the variational equation is thus demonstrated; for we have first derived Eq. (7) from the equation of equilibrium and then have shown that, conversely, Eqs. (17)–(19) necessarily follow Eq. (7).

The commonly encountered external force systems in elasticity are conservative systems in which the body force F_i and surface tractions $\overset{\nu}{T_i}$ are independent of the elastic deformation of the body. In this case, $\mathcal{V}$ is more commonly written as

(20) $$\mathcal{V} \equiv \int_V W \, dv - \int_V F_i u_i \, dv - \int_{S_\sigma} \overset{\nu}{T_i^*} u_i \, dS \,.$$

A branch of mechanics in which the external forces are in general nonconservative is the theory of aeroelasticity. In aeroelasticity one is concerned with the interaction of aerodynamic forces and elastic deformation. The aerodynamic forces depend on the flow and the deformation of the entire body, not just the local deformation; thus in general it cannot be derived from a potential function. The principle of virtual work, in the form of Eq. (4), is still applicable to aeroelasticity.

Problem 10.1. The minimum potential energy principle states that elastic equilibrium is equivalent to the condition $J = \min$, where

(a) $$J = \int_V [W(e_{ij}) - F_i u_i] \, dV - \int_{S_\sigma} \overset{\nu}{T_i} \, u_i dS \,, \qquad \text{(varying } u_i \text{)};$$

(b) $$e_{ij} = \frac{1}{2}(u_{i,j} + u_{j,i}) \,.$$

By the method of Lagrange multipliers, the subsidiary condition (b) can be incorporated into the functional J, and we are led to consider the variational equation

(c) $$\delta J' = 0 \,, \qquad \text{(varying } e_{ij}, u_i, \lambda_{ij} \text{ independently)},$$

where

(d) $$J' = J + \int_V \lambda_{ij} \left[e_{ij} - \frac{1}{2}(u_{i,j} + u_{j,i}) \right] dV \,.$$

Since the quantity in [ ] is symmetric in i, j, we may restrict the Lagrange multipliers λ_{ij} to be symmetric, $\lambda_{ij} = \lambda_{ji}$, so that only six independent multipliers are needed. Derive the Euler equations for (c) and show that, $-\lambda_{ij}$ should be interpreted as the stress tensor.

Note: Once Lagrange's multipliers are employed, the phrase "minimum conditions" used in the principle of minimum potential energy has to be replaced by "stationary conditions."

10.8. EXAMPLE OF APPLICATION: STATIC LOADING ON A BEAM—NATURAL AND RIGID END CONDITIONS

As an illustration of the application of the minimum potential energy principle in formulating approximate theories of elasticity, let us consider the approximate theory of bending of a slender beam under static loading. Let the beam be perfectly straight and lying along the x-axis before the application of external loading, which consists of a distributed lateral load $p(x)$ per unit length, a bending moment M_0, and a shearing force Q_0 at the end $x = 0$, and a moment M_l and a shear Q_l at the end $x = l$ (Fig. 10.8:1). We assume that the principal axes of inertia of every cross section of the beam lie in two mutually orthogonal principal planes and that the loading

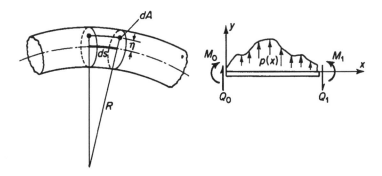

Fig. 10.8:1. Applications to a simple beam.

p, M, Q are applied in one of the principal planes. In accordance with the approximate beam theory, we assume that every plane cross section of the beam remains plane during bending. If the deformation is small, then the analysis given in Sec. 7.7 applies.

Let ds be the arc length along the longitudinal neutral axis. Under these assumptions, we see that when the neutral axis of the beam is bent from the initial straight line into a curve with radius of curvature R, the length of a filament, initially ds and parallel to the neutral axis, is altered by the bending in the ratio $1 : (1 + \eta/R)$, where η is the distance between the filament and the neutral axis. The strain is η/R, and the force acting on the filament is $E\eta dA/R$, where dA is the cross section of the filament. The resultant moment of these forces about the neutral axis is denoted by M, which is

$$M = \int_A \eta \cdot E\frac{\eta}{R} dA = \frac{E}{R} \int_A \eta^2 dA = \frac{EI}{R},$$

where I is the moment of inertia of the area of the beam cross section.

The bending moment acting on the cross sections at the ends of a beam segment of length ds does work if the cross sections rotate relative to each other. Let θ be the angle between a tangent to the neutral axis and the x-axis. In a segment of length ds, the tangent rotates by an amount $d\theta$. The rate of change $d\theta/ds$ is, by definition, the curvature of the neutral axis. R being the radius of curvature, (see Fig. 10.8:1), $d\theta/ds = 1/R$. Therefore, the angle through which the cross sections rotate relative to each other is ds/R. Hence, the work to bend a segment ds of the beam from a curvature of zero to a curvature $1/R$ is

$$\frac{1}{2} M d\theta = \frac{1}{2} EI \frac{ds}{R^2},$$

where the factor $\frac{1}{2}$ is added since the mean work is half the value of the product of the final moment and angle of rotation. Hence, integrating throughout the beam, we have the strain energy

(1)
$$U = \frac{1}{2} \int_0^l EI \frac{ds}{R^2}.$$

If we now assume that the deflection of the beam is *infinitesimal* so that if y denotes the lateral deflection, the curvature $1/R$ is approximated by d^2y/dx^2, and ds is approximated by dx, then

(1a)
$$U = \frac{1}{2} \int_0^l EI \left(\frac{d^2y}{dx^2}\right)^2 dx.$$

The potential energy of the external loading is, with the sign convention specified in Fig. 10.8:1 and with y_0, y_l, $(dy/dx)_0$, (dy/dx), denoting the value of y and dy/dx at $x = 0, l$, respectively,

(2) $$-\int p(x)y(x)\, dx + M_0 \left(\frac{dy}{dx}\right)_0 - M_l \left(\frac{dy}{dx}\right)_l - Q_0 y_0 + Q_l y_l.$$

Hence, the total potential energy is

(3) $$\mathcal{V} = \frac{1}{2} \int_0^l EI \left(\frac{d^2y}{dx^2}\right)^2 dx - \int_0^l py \, dx + M_0 \left(\frac{dy}{dx}\right)_0$$

$$- M_l \left(\frac{dy}{dx}\right)_l - Q_0 y_0 + Q_l y_l \, .$$

At equilibrium, the variation of $\mathcal{V}$ with respect to the virtual displacement δy must vanish. Hence,

(4) $$\delta\mathcal{V} = \int_0^l EI \frac{d^2y}{dx^2} \, \delta \left(\frac{d^2y}{dx^2}\right) dx - \int_0^l p \delta y \, dx$$

(5) $$+ M_0 \delta \left(\frac{dy}{dx}\right)_0 - M_l \delta \left(\frac{dy}{dx}\right)_l - Q_0 \delta y_0 + Q_l \delta y_l = 0 \, .$$

Integrating the first term by parts twice and collecting terms, we obtain

(6) $$\delta\mathcal{V} = \int_0^l \left[\frac{d^2}{dx^2} \left(EI \frac{d^2y}{dx^2}\right) - p \right] \delta y \, dx$$

$$+ \left[EI \left(\frac{d^2y}{dx^2}\right)_l - M_l \right] \delta \left(\frac{dy}{dx}\right)_l - \left[EI \left(\frac{d^2y}{dx^2}\right)_0 - M_0 \right] \delta \left(\frac{dy}{dx}\right)_0$$

$$- \left[\frac{d}{dx} EI \left(\frac{d^2y}{dx^2}\right)_l - Q_l \right] \delta y_l + \left[\frac{d}{dx} EI \left(\frac{d^2y}{dx^2}\right)_0 - Q_0 \right] \delta y_0$$

$$= 0$$

Since δy is arbitrary in the interval $(0, l)$, we obtain the differential equation of the beam

(5) $$\frac{d^2}{dx^2} \left(EI \frac{d^2y}{dx^2}\right) - p = 0 \, , \qquad\qquad 0 \le x \le l \, .$$

In order that the remaining terms in (4) may vanish, it is sufficient to have the end conditions

(6a) Either $\qquad EI \left(\dfrac{d^2y}{dx^2}\right)_l - M_l = 0 \quad$ or $\quad \delta \left(\dfrac{dy}{dx}\right)_l = 0 \, .$

(6b) Either $\qquad EI \left(\dfrac{d^2y}{dx^2}\right)_0 - M_0 = 0 \quad$ or $\quad \delta \left(\dfrac{dy}{dx}\right)_0 = 0 \, .$

(6c)   Either
$$\frac{d}{dx}\left(EI\frac{d^2y}{dx^2}\right)_l - Q_l = 0 \quad \text{or} \quad \delta y_l = 0.$$

(6d)   Either
$$\frac{d}{dx}\left(EI\frac{d^2y}{dx^2}\right)_0 - Q_0 = 0 \quad \text{or} \quad \delta y_0 = 0.$$

If the deflection y_0 is prescribed at the end $x = 0$, then $\delta y_0 = 0$. If the slope $(dy/dx)_0$ is prescribed at the end $x = 0$, then $\delta(dy/dx)_0 = 0$. These are called *rigid boundary conditions*. On the other hand, if the value of y_0 is unspecified and perfectly free, then δy_0 is arbitrary and we must have

(7)
$$\frac{d}{dx}\left(EI\frac{d^2y}{dx^2}\right)_0 - Q_0 = 0$$

as an end condition for a free end subjected to a shear load Q_0; otherwise δV cannot vanish for arbitrary variations δy_0. Equation (7) is called a *natural boundary condition*. Similarly, all the left-hand equations in Eqs. (6a)–(6d) are natural boundary conditions, and all the right-hand equations are rigid boundary conditions.

The distinction between natural and rigid boundary conditions assumes great importance in the application of the direct methods of solution of variational problems; the assumed functions in the direct methods must satisfy the rigid boundary conditions. See Sec. 11.8.

It is worthwhile to consider the following question. The reader must be familiar with the fact that in the engineering beam theory, the end conditions often considered are:

(8a)   *Clamped end* :
$$y = 0, \quad \frac{dy}{dx} = 0,$$

(8b)   *Free end* :
$$EI\frac{d^2y}{dx^2} = 0, \quad \frac{d}{dx}\left(EI\frac{d^2y}{dx^2}\right) = 0,$$

(8c)   *Simply supported end* :
$$y = 0, \quad EI\frac{d^2y}{dx^2} = 0,$$

where $y(x)$ is the deflection function of the beam. May we ask why are the other two combinations, namely

(9a)
$$y = 0, \quad \frac{d}{dx}\left(EI\frac{d^2y}{dx^2}\right) = 0,$$

(9b)
$$\frac{dy}{dx} = 0, \quad EI\frac{d^2y}{dx^2} = 0,$$

never considered?

An acceptable answer is perhaps that the boundary conditions (9a) and (9b) cannot be realized easily in the laboratory. But a more satisfying answer is that they are not proper sets of boundary conditions. If the conditions (9a) or (9b) were imposed, then, according to Eq. (4), it cannot at all be assured that the equation $\delta V = 0$ will be satisfied. Thus, a basic physical law might be violated. These boundary conditions are, therefore, inadmissible.

From the point of view of the theory of differential equations, one may feel that the end conditions (9a) or (9b) are legitimate for Eq. (5). Nevertheless, they are ruled out by the minimum potential energy principle on physical grounds. In fact, in the theory of differential equations the Eq. (5) and the end conditions (8) are known to form a so-called *self-adjoint differential system*, whereas Eqs. (5) and (9) would form a *nonself-adjoint differential system*. Very great difference in mathematical character exists between these two categories. For example, a free vibration problem of a nonself-adjoint system may not have an eigenvector, or it may have complex eigenvalues or complex eigenvectors.

There are other conceivable admissible boundary conditions, such as to require

$$(10) \qquad \frac{dy}{dx} = 0, \qquad \frac{d}{dx}\left(EI\frac{d^2y}{dx^2}\right) = 0, \qquad \text{at } x = 0.$$

Such an end, with zero slope and zero shear, cannot be easily established in the laboratory. Similarly, it is conceivable that one may require that at the end $x = 0$, the following ratios hold:

$$\delta\left(\frac{\partial y}{\partial x}\right) : \delta y = c, \qquad \text{a constant},$$

$$(11)$$

$$\left[\frac{d}{dx}\left(EI\frac{d^2y}{dx^2}\right) - Q_0\right] : \left[EI\frac{d^2y}{dx^2} - M_0\right] = c, \qquad \text{the same constant}.$$

This pair of conditions are also admissible, but are unlikely to be encountered in practice.

10.9. THE COMPLEMENTARY ENERGY THEOREM UNDER SMALL VARIATIONS OF STRESSES

In contrast to the previous sections let us now consider the variation of stresses in order to investigate whether the "actual" stresses satisfy a minimum principle. We pose the problem as in Sec. 10.7 with a body held in equilibrium under the body force per unit volume, F_i, and surface tractions per unit area, $\overset{\nu}{T_i^*}$, over the boundary S_σ, whereas over the boundary S_u the displacements are prescribed. Let σ_{ij} be the "actual" stress field which satisfies the equations of equilibrium and boundary conditions

(1)
$$\sigma_{ij,j} + F_i = 0 \qquad \text{in } V,$$

$$\sigma_{ij}\nu_j = \overset{\nu}{T_i^*} \qquad \text{on } S_\sigma.$$

Let us now consider a system of variations of stresses which also satisfy the equations of equilibrium and the stress boundary conditions

$$(\delta\sigma_{ij})_{,j} + \delta F_i = 0 \qquad \text{in } V,$$

(2)
$$(\delta\sigma_{ij})\nu_j = \delta\overset{\nu}{T_i} \qquad \text{on } S_\sigma,$$

$$\delta\sigma_{ij} \text{ are arbitrary on } S_u.$$

In contrast to the previous sections, we shall now consider the *complementary virtual work*,

$$\int_V u_i \delta F_i \, dv + \int_S u_i \delta\overset{\nu}{T_i} \, dS,$$

which, by virtue of Eq. (2) and through integration by parts,

$$= -\int_V u_i(\delta\sigma_{ij})_{,j} \, dv + \int_S u_i(\delta\sigma_{ij})\nu_j \, dS$$

$$= \int_V (\delta\sigma_{ij})u_{i,j} \, dv - \int_S u_i\nu_j(\delta\sigma_{ij})dS + \int_S u_i(\delta\sigma_{ij})\nu_j \, dS$$

$$= \frac{1}{2}\int_V (\delta\sigma_{ij})(u_{i,j} + u_{j,i}) \, dv$$

$$= \int_V e_{ij}\delta\sigma_{ij} \, dv.$$

Hence,

(3) ▲
$$\int_V e_{ij}\delta\sigma_{ij} \, dv = \int_V u_i \delta F_i \, dv + \int_S u_i \delta\overset{\nu}{T_i} \, dS.$$

This equation may be called the *principle of virtual complementary work*. Now, if we introduce *the complementary strain energy* W_c,[†] which is a

---

[†]The Gibbs' thermodynamic potential (Sec. 12.3) per unit volume, $\rho\Phi$, is equal to the negative of the complementary strain energy function. If the stress-strain law were linear, then $W_c(\sigma_{ij})$ and $W(e_{ij})$ are equal: $-\rho\Phi = W_c = W$ (for linear stress-strain law).

function of the stress components $\sigma_{11}, \sigma_{12}, \ldots,$ and which has the property that,

$$(4) \qquad \frac{\partial W_c}{\partial \sigma_{ij}} = e_{ij}$$

then the complementary virtual work may be written as

$$(5) \qquad \int_V u_i \delta F_i \, dv + \int_S u_i \delta \overset{\nu}{T}_i \, dS = \int_V \frac{\partial W_c}{\partial \sigma_{ij}} \delta \sigma_{ij} dv = \delta \int_V W_c dv \,.$$

Since the volume is fixed and u_i are not varied, the result above can be written as

$$(6) \quad \blacktriangle \qquad\qquad\qquad \delta \mathcal{V}^* = 0 \,,$$

where $\mathcal{V}^*$ as a function of the stresses $\sigma_{11}, \sigma_{12}, \ldots,$ the surface traction $\overset{\nu}{T}_i$ and the body force per unit volume F_i, is defined as the *complementary energy*

$$(7) \quad \blacktriangle \qquad \mathcal{V}^*(\sigma_{11}, \ldots, F_i) \equiv \int_V W_c \, dv - \int_V u_i F_i \, dv - \int_S u_i \overset{\nu}{T}_i \, dS \,.$$

In practice, we would like to compare stress fields which all satisfy the equations of equilibrium, but not necessarily the conditions of compatibility. In other words, we would have $\delta F_i = 0$ in V and $\delta \overset{\nu}{T}_i = 0$ on S_σ. In this case $\delta\sigma_{ij}$ and, hence, $\delta \overset{\nu}{T}_i$ are arbitrary only on that portion of the boundary where displacements are prescribed, S_u. Therefore, only a surface integral over S_u is left in the left-hand side of Eq. (5) and we have

$$(8) \quad \blacktriangle \qquad \mathcal{V}^*(\sigma_{11}, \ldots, \sigma_{33}) \equiv \int_V W_c \, dv - \int_{S_u} u_i \overset{\nu}{T}_i \, dS \,.$$

Therefore, we have the following theorem

Theorem. *Of all stress tensor fields σ_{ij} that satisfy the equation of equilibrium and the boundary conditions where stresses are prescribed, the "actual" one is distinguished by a stationary (extreme) value of the complementary energy $\mathcal{V}^*(\sigma_{11}, \ldots, \sigma_{33})$ as given by Eq. (8).*

In this formulation, the linearity of the stress-strain relationship is *not* required, only the existence of the complementary strain energy function is assumed. However, if the stress-strain law were *linear*, and the material is isotropic and obeying Hooke's law, then

$$(9) \qquad W_c(\sigma_{11}, \ldots, \sigma_{33}) = -\frac{\nu}{2E} (\sigma_{\alpha\alpha})^2 + \frac{1+\nu}{2E} \sigma_{ij}\sigma_{ij} \,.$$

We must remark that the variational Eqs. (10.7:2) and Eq. (3) of the present section are applicable even if the body is *not elastic*, for which the energy functional cannot be defined. These variational equations are used in the analysis of inelastic bodies.

Before we proceed further, it may be well worthwhile to consider the concept of *complementary work* and *complementary strain energy*. Consider a simple, perfectly elastic bar subjected to a tensile load. Let the relationship between the load P and the elongation of the bar u be given by a unique curve as shown in Fig. 10.9:1. Then the work W is the area between the displacement axis and the curve, while the complementary work W_c is that included between the force axis and the curve. Thus, the two areas complement each other in the rectangular area (force) · (displacement), which would be the work if the force were acting with its full intensity from the beginning of the displacement. Naturally, W and W_c are equal if the material follows Hooke's law.

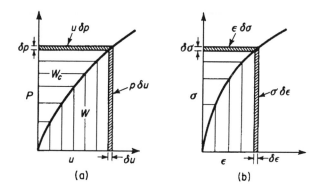

Fig. 10.9:1. Complementary work and strain energy.

The principle of minimum potential energy was formulated by Willard Gibbs; and many beautiful applications were shown by Lord Rayleigh. The complementary energy concept was introduced by F. Z. Engesser; its applications were developed by H. M. Westergaard. In the hands of Kirchhoff, the minimum potential energy theorem becomes the foundation of the approximate theories of plates and shells. Argyris has made the complementary energy theorem the starting point for practical methods of analysis of complex elastic structures using modern digital computers. See Bibliography 10.2, p. 494.

Let us now return to the complementary energy theorem. We shall show that *in the neighborhood of the natural state, the extreme value of the complementary energy V is actually a minimum.* By a natural state is meant a state of stable thermodynamic existence (see Sec. 12.4). In the neighborhood of a natural state, the thermodynamic potential per unit volume $\rho\Phi$ can be approximated by a homogeneous quadratic form of the stresses. For an isotropic material, $-\rho\Phi$ is given by Eq. (9), in which W_c is positive definite.

The proof that $\mathcal{V}^*$ is a minimum is analogous to that in the previous section, and it can be sketched as follows. Comparing $\mathcal{V}^*(\sigma_{ij} + \delta\sigma_{ij})$ with $\mathcal{V}^*(\sigma_{ij})$, we have

$$(10) \quad \mathcal{V}^*(\sigma_{11} + \delta\sigma_{11}, \ldots,) - \mathcal{V}^*(\sigma_{11}, \ldots,)$$

$$= \int_V [W_c(\sigma_{11} + \delta\sigma_{11}, \ldots,) - W_c(\sigma_{11}, \ldots,)] \, dv$$

$$- \int_{S_u} [\overset{\nu}{T_i}(\sigma_{11} + \delta\sigma_{11}, \ldots,) - \overset{\nu}{T_i}(\sigma_{11}, \ldots,)] u_i \, dS$$

$$= \int_V \left[\left(\frac{1+\nu}{E} \sigma_{ij} - \frac{\nu}{E} \sigma_{\alpha\alpha}\delta_{ij} \right) \delta\sigma_{ij} + W_c(\delta\sigma_{11}, \ldots,) \right] dv$$

$$- \int_{S_u} (\delta\sigma_{ij}) \nu_j u_i \, dS$$

$$= \int_V e_{ij}\delta\sigma_{ij} \, dv - \int_{S_u} (\delta\sigma_{ij}) \nu_j u_i \, dS + \int W_c(\delta\sigma_{11}, \ldots,) \, dv$$

$$= \delta\mathcal{V}^* + \int_V W_c(\delta\sigma_{11}, \delta\sigma_{12}, \ldots,) \, dv,$$

where

$$(11) \quad W_c(\delta\sigma_{11}, \ldots,) = \frac{1+\nu}{2E} (\delta\sigma_{ij})(\delta\sigma_{ij}) - \frac{\nu}{2E} (\delta\sigma_{ii})^2 \geq 0.$$

$W_c(\delta\sigma_{11}, \ldots,)$ is positive definite for infinitesimal variations $\delta\sigma_{ij}$. Hence, when $\delta\mathcal{V}^* = 0$,

$$(12) \quad \mathcal{V}^*(\sigma_{11} + \delta\sigma_{11}, \ldots,) - \mathcal{V}^*(\sigma_{11}, \ldots,) \geq 0,$$

and that $\mathcal{V}^*(\sigma_{11}, \ldots,)$ is a minimum is proved.

The converse theorem reads:

Let $\mathcal{V}^*$ be the complementary energy defined by Eq. (8). If the stress tensor field σ_{ij} is such that $\delta\mathcal{V}^* = 0$ for all variations of stresses $\delta\sigma_{ij}$ which satisfy the equations of equilibrium in the body and on the boundary where surface tractions are prescribed, then σ_{ij} also satisfies the equations *of compatibility.* In other words, the conditions of compatibility are the Euler equation for the variational equation $\delta \mathcal{V}^* = 0$.

The proof was given by Richard V. Southwell[1,2] (1936) through the application of Maxwell and Morera stress functions. We begin with the variational equation

(13)
$$\delta \mathcal{V}^* = \int_V e_{ij} \delta \sigma_{ij} \, dv - \int_{S_u} u_i \delta \overset{\nu}{T_i} \, dS = 0 \,.$$

The variations $\delta \sigma_{ij}$ are subjected to the restrictions

(14)
$$(\delta \sigma_{ij})_{,j} = 0 \qquad\qquad \text{in } V \,,$$

(15)
$$(\delta \sigma_{ij}) \nu_j = 0 \qquad\qquad \text{on } S_\sigma \,.$$

To accommodate the restrictions (14) into Eq. (13), we make use of the celebrated result that the equations of equilibrium (14) are satisfied formally, as can be easily verified, by taking

(16)
$$\delta \sigma_{11} = \phi_{22,33} + \phi_{33,22} - 2\phi_{23,23} \,,$$
$$\delta \sigma_{22} = \phi_{33,11} + \phi_{11,33} - 2\phi_{31,31} \,,$$
$$\delta \sigma_{33} = \phi_{11,12} + \phi_{22,11} - 2\phi_{12,12} \,,$$
$$\delta \sigma_{23} = \phi_{31,12} + \phi_{12,13} - \phi_{11,23} - \phi_{23,11} \,,$$
$$\delta \sigma_{31} = \phi_{12,23} + \phi_{23,21} - \phi_{22,31} - \phi_{31,22} \,,$$
$$\delta \sigma_{12} = \phi_{23,31} + \phi_{31,32} - \phi_{33,12} - \phi_{12,33} \,,$$

where $\phi_{ij} = \phi_{ji}$ are arbitrary stress functions. On setting $\phi_{12} = \phi_{23} = \phi_{31} = 0$, we obtain the solutions proposed by James Clerk Maxwell. On taking $\phi_{11} = \phi_{22} = \phi_{33} = 0$, we obtain the solutions proposed by G. Morera.

Let us use the Maxwell system of arbitrary stress functions for the variations $\delta \sigma_{ij}$. Equation (13) may be written as

(17)
$$\delta \mathcal{V}^* = \int_V [e_{11}(\phi_{22,33} + \phi_{33,22}) + e_{22}(\phi_{33,11} + \phi_{11,33})$$
$$+ e_{33}(\phi_{11,22} + \phi_{22,11}) - 2e_{23}\phi_{11,23}$$
$$- 2e_{31}\phi_{22,31} - 2e_{12}\phi_{33,12}] \, dv - \int_{S_u} u_i \delta \overset{\nu}{T_i} \, dS = 0 \,.$$

Integrating by parts twice, we obtain

$$(18) \quad \delta V^* = \int_V [(e_{22,33} + e_{33,22} - 2e_{23,23})\phi_{11}$$

$$+ (e_{33,11} + e_{11,33} - 2e_{31,31})\phi_{22}$$

$$+ (e_{11,22} + e_{22,11} - 2e_{12,12})\phi_{33}] \, dv + \text{a surface integral} = 0 \,.$$

Inasmuch as the stress functions ϕ_{11}, ϕ_{22}, ϕ_{33} are arbitrary in the volume V, the Euler's equations are

$$(19) \qquad\qquad e_{22,33} + e_{33,22} - 2e_{23,23} = 0 \,,$$

etc., which are Saint-Venant's compatibility equations (see Sec. 4.6). The treatment of the surface integral is cumbersome, but it says only that over S_u the values of u_i are prescribed and the stresses are arbitrary; the derivation concerns a certain relationship between ϕ_{ij} and their derivatives to be satisfied on the boundary.

Similarly, the use of Morera system of arbitrary stress functions leads to the other set of Saint-Venant's compatibility equations

$$(20) \qquad\qquad e_{11,23} = -e_{23,11} + e_{31,21} + e_{12,31} \,,$$

etc. [Eq. (4.6:4)].

If we start with δV^* in the form

$$(21) \quad \delta V^* = 0 = \int_V \left(\frac{1+\nu}{E}\sigma_{ij} - \frac{\nu}{E}\sigma_{\alpha\alpha}\delta_{ij} \right) \delta\sigma_{ij} \, dv + \int_{S_u} u_i \delta \overset{\nu}{T_i} \, dS$$

and introduce the stress functions, the Beltrami–Michell compatibility equations

$$(22) \qquad \nabla^2 \sigma_{ij} + \frac{1}{1+\nu}\theta_{,ij} + \frac{\nu}{1-\nu}\delta_{ij}F_{k,k} + F_{i,j} + F_{j,i} \equiv 0 \,, \qquad \text{in } V$$

where $\theta = \sigma_{\alpha\alpha}$, can be obtained directly.

In concluding this section, we remark once more that when we consider the variation of the stress field of a body in equilibrium, the principle of virtual complementary work has broad applicability. The introduction of the complementary energy functional V^*, however, limits the principle to elastic bodies. That V^* actually is a minimum with respect to all admissible variations of stress field is established only if the complementary strain energy function is positive definite.

There are many fascinating applications of the minimum complementary energy principle. In Sec. 10.11, we shall consider its application in proving Saint-Venant's principle in Zanaboni's formulation.

Problem 10.2. The principle of virtual complementary work states that

$$(23) \qquad \int_V e_{ij}\delta\sigma_{ij}dv - \int_{S_u} u_i^*\delta\, \overset{\nu}{T}_i \; dS = 0$$

under the restrictions that

$$(24) \qquad\qquad \delta\sigma_{ij,j} = 0 \quad \text{in } V\,,$$

$$(25) \qquad\qquad \delta\sigma_{ij}\nu_j = 0 \quad \text{on } S_\sigma\,,$$

(26) $u_i = u_i^*$ prescribed on S_u, but $\delta\, \overset{\nu}{T}_i = \delta\sigma_{ij}\nu_j$ arbitrary on S_u.

Using Lagrange multipliers, we may restate this principle as

$$(27) \quad \int_V e_{ij}\delta\sigma_{ij}\, dv - \int_{S_u} u_i^*\delta\, \overset{\nu}{T}_i \; dS + \int_V \lambda_i\delta\sigma_{ij,j}\, dv - \int_{S_\sigma} \mu_i\delta\sigma_{ij}\nu_j dS = 0\,.$$

The six equations (24) and (25), $i = 1, 2, 3$, require six Lagrange multipliers λ_i, μ_i which are functions of (x_1, x_2, x_3). Show that the Euler equations for Eq. (27) yield the physical interpretation

$$(28) \qquad\qquad \lambda_i = u_i\,, \qquad \mu_i = u_i\,.$$

10.10. VARIATIONAL FUNCTIONALS FREQUENTLY USED IN COMPUTATIONAL MECHANICS

We shall introduce hybrid and mixed variational principles, which are used extensively in finite element applications. The finite element method is one of the most powerful numerical methods for solving linear and non-linear problems. The method approximates the field variables in terms of unknown parameters in sub-regions called "*elements*" and then arrives at an approximate solution for the whole domain by enforcing certain relations among the unknowns. Variational principles are often used to establish the relations. The details of finite element applications will be discussed in Chapters 18–21.

Variational principles are classified into *irreducible*, *hybrid*, or *mixed variational principle* on the basis of the nature of the set of unknown functions in the functionals. If the *set* of unknown functions (called *field variables*) is irreducible in some sense or cannot be eliminated from the functional until the solution of the problem has been obtained, then the functional is said to be *irreducible*. For example, in three dimensions, a displacement has components $u_i, i = 1, 2, 3$. In most cases, none of the components can be expressed in terms of others and eliminated from the functional *a priori*. Thus the minimum potential energy principle based on u_i is irreducible. So are the stress components σ_{ij} and thus the complementary energy theorem based on σ_{ij} is also irreducible. On the other hand, a functional of both u_i and σ_{ij} is reducible because stresses can be espressed in terms of strains through the constitutive law and thus be eliminated from the functional. Such a functional is said to be *mixed*. Further, in finite element analysis a continuum is divided into *elements*. One could choose the field variables on the boundary of the element different from those over the domain of the element. Then the functional and the associated variational principle are said to be *hybrid*.

In the subsequent discussion of mixed and hybrid functionals we shall use interchangibly F_i, b_i for the body force; ∂V, S for the boundaries of V; ν_i, n_i for unit normal to a boundary surface; ∂V_u, S_u for the boundaries where displacements are prescribed; ∂V_σ, S_σ for the boundaries where tractions are prescribed; $\bar{u}_i$, u_i^* for the prescribed boundary displacements; and $\overset{\nu}{T}, \overset{\nu}{T}^*$ for the prescribed traction.

Reissner-Hellinger Principle. For materials satisfying Hooke's law, the Reissner-Hellinger principle uses the functional

$$(1) \quad \Pi_R(\sigma_{ij}, u_i) = \int_V \left(\sigma_{ij} e_{ij} - \frac{1}{2} C_{ijkl} \sigma_{ij} \sigma_{kl} - b_i u_i \right) dV - \int_{\partial V_\sigma} \overset{\nu}{T}_i\, u_i\, dS,$$

where V is the volume of the domain, C_{ijkl} are the *elastic flexibility tensor* of rank 4, σ_{ij} and e_{ij} are Cartesian stress and strain tensors of rank 2, u_i are the displacements, b_i are the body force components, and ∂V_σ is the portion of the boundaries of V over which traction $\overset{\nu}{T}_i$ are prescribed. The strains are in terms of the displacements

$$(2) \qquad\qquad e_{ij} = (u_{i,j} + u_{j,i})/2,$$

and the displacements are prescribed over $\partial V_u (= \partial V - \partial V_\sigma)$ so that

$$(3) \qquad\qquad u_i = \bar{u}_i$$

in which $\bar{u}_i$ are known functions. The independent field variables σ_{ij} and u_i are limited by the *admissibility conditions* that either the displacements or the traction $\overset{\nu}{T}_i \, (= n_j \sigma_{ji})$ are C^0 continuous[†] over any internal surfaces and that u_i satisfy Eq. (3). Since Eq. (3) is part of the admissibility requirements, which must be satisfied *a priori* and is called a *rigid condition*. The first variation of Π_R with respect to the field variables is

$$\delta\Pi_R(\sigma_{ij}, u_i) = \int_V \sigma_{ij}\delta e_{ij}\, dV - \int_V b_i \delta u_i\, dV$$

$$+ \int_V (e_{ij} - C_{ijkl}\sigma_{kl})\delta\sigma_{ij}\, dV - \int_{\partial V_\sigma} \overset{\nu}{T}_i \, \delta u_i\, dS \,.$$

The necessary condition for Π_R to be stationary is that $\delta\Pi_R$ vanishes for all admissible variations of the independent field variables. Using the Gauss theorem for the first integral yields the stationarity condition

$$(4) \quad \delta\Pi_R(\sigma_{ij}, u_i) = -\int_V (\sigma_{ji,j} + b_i)\delta u_i\, dV + \int_V (e_{ij} - C_{ijkl}\sigma_{kl})\delta\sigma_{ij} dV$$

$$+ \int_{\partial V_\sigma} (n_j\sigma_{ji} - \overset{\nu}{T}_i)\delta u_i\, dS = 0 \,,$$

where the n's are the components of a unit normal to the element boundaries. Since $\delta\Pi_R$ must be zero for arbitrary admissible δu_i and $\delta\sigma_{ij}$, the integrands of all three integrals must vanish. Vanishing of the three integrals implies, respectively,

$$(5) \quad \sigma_{ji,j} + b_i = 0 \qquad \text{equations of equilibrium in } V,$$

$$(6) \quad e_{ij} = C_{ijkl}\sigma_{kl} \qquad \text{strain-stress relations in } V,$$

$$(7) \quad n_j\sigma_{ji} = \overset{\nu}{T}_i \qquad \text{traction boundary conditions on } \partial V_\sigma.$$

The functional Π_R has both stresses and displacements as independent fields. Since the stresses can be expressed readily in terms of the displacements from Eq. (6), the functional is not irreducible. The variational formulation based on such a functional is a *mixed principle*.

If the domain V is divided into subdomains, called *elements*, for admissible σ_{ij}, u_i we can write Π_R in the form (Pian and Tong 1969)

$$(8) \quad \Pi_R = \sum_{\text{all elements}} \Pi_{Re}(\sigma_{ij}, u_i) \,,$$

[†] A function $u(x, y, z)$, which is continuous in a domain V, is said to be continuous of the order of zero, and is denoted by C^0. If u and its partial derivatives are all continuous in V, then u is said to be continuous of the order of 1, and denoted by C^1. If u and all of its partial derivatives up to order n are continuous in V, then u is said to be C^n continuous.

where Π_{Re} is the corresponding functional for an element

(9a) $\Pi_{\text{Re}}(\sigma_{ij}, u_i) = \displaystyle\int_{V_e} \left(\sigma_{ij} e_{ij} - \frac{1}{2} C_{ijkl} \sigma_{ij} \sigma_{kl} - b_i u_i \right) dV - \int_{\partial V_{\sigma e}} \overset{\nu}{T}_i \, u_i dS$

with V_e and $\partial V_{\sigma e}$ being, respectively, the element volume and the portion of element boundaries over which traction is prescribed. For isotropic materials, Eq. (9a) reduces to

(9b) $\Pi_{\text{Re}}(\sigma_{ij}, u_i) = \displaystyle\int_{V_e} \left[\sigma_{ij} e_{ij} - \frac{1}{2} \left(\frac{1}{2G} \sigma_{ij} \sigma_{ij} + \frac{2G - 3K}{18GK} \sigma_{ii} \sigma_{jj} \right) \right.$

$\left. - b_i u_i \right] dV - \displaystyle\int_{\partial V_{\sigma e}} \overset{\nu}{T}_i \, u_i \, dS$

In finite element applications to be discussed in later chapters, one assumes admissible u's and σ's in terms of unknown parameters within each element and uses the stationarity condition Eq. (4) to establish the equations for the unknown parameters. Note that $(2G - 3K)/(9K) = -\nu/(1 + \nu)$.

A Four-Field Hybrid Variational Principle. One can relax the admissibility requirements at the element boundaries ∂V_e by introducing Lagrangian multipliers $\hat{u}_i$, $\hat{T}_i$ along ∂V_e of each element (Tong and Pian 1969, Pian and Tong 1972) to modify Π_{Re} in Eq. (9) such that

(10) $\Pi_H = \displaystyle\sum_{\text{all elements}} \Pi_{He}(\sigma_{ij}, u_i, \hat{u}_i, \hat{T}_i),$

where

$\Pi_{He}(\sigma_{ij}, u_i, \hat{u}_i, \hat{T}_i) = \displaystyle\int_{\partial V_e} \hat{T}_i(\hat{u}_i - u_i) \, dS$

$+ \displaystyle\int_{V_e} \left(\sigma_{ij} e_{ij} - \frac{1}{2} C_{ijkl} \sigma_{ij} \sigma_{kl} - b_i u_i \right) dV - \int_{\partial V_{\sigma e}} \overset{\nu}{T}_i \, \hat{u}_i \, dS.$

The admissibility conditions for Lagrangian multipliers are: $\hat{u}_i$ are the same for any two adjacent elements along their common boundaries and satisfy the prescribed condition Eq. (3), i.e., $\hat{u}_i = \bar{u}_i$. The functional Π_H has four independent fields σ_{ij}, u_i, $\hat{u}_i$, $\hat{T}_i$, of which σ_{ij}, u_i are defined over V_e, and $\hat{u}_i$, $\hat{T}_i$ are defined on ∂V_e only. As the field variables over ∂V_e differ from those inside V_e, Π_H is a *hybrid functional*. The variational formulation based on such a functional is called a *hybrid principle*.

The stationarity condition of Π_H is

(11) $\delta \Pi_H = \displaystyle\sum_{\text{all elements}} \delta \Pi_{He} = 0$

where

$$\delta\Pi_{He} = -\int_{V_e} (\sigma_{ji,j} + b_i)\delta u_i\, dV + \int_{V_e} (e_{ij} - C_{ijkl}\sigma_{kl})\delta\sigma_{ij}\, dV$$

(11a)
$$+ \int_{\partial V_e} (\hat{u}_i - u_i)\delta\hat{T}_i\, dS + \int_{\partial V_e} (n_j\sigma_{ji} - \hat{T}_i)\delta u_i\, dS$$

$$+ \int_{\partial V_{\sigma e}} (\hat{T}_i - \overset{\nu}{T}_i)\delta\hat{u}_i\, dS + \int_{\partial V_e - \partial V_{\sigma e}} \hat{T}_i\delta\hat{u}_i\, dS\,.$$

Since σ_{ij}, u_i over V_e, $\hat{T}_i$ on $\partial V_{\sigma e}$, and $\hat{u}_i$ on $\partial V_{\sigma e}$ of all elements are independent of each other, Eq. (11) requires that the first five integrals above are zero for each element. The vanish of the first two integrals implies the equations of equilibrium and the strain stress relations within each element. The third through fifth integrals give, respectively,

(12) $$\qquad\qquad \hat{u}_i = u_i \qquad \text{on } \partial V_e\,,$$

(13) $$\qquad\qquad \hat{T}_i = n_j\sigma_{ji} \qquad \text{on } \partial V_e\,,$$

(14) $$\qquad\qquad \hat{T}_i = \overset{\nu}{T}_i \qquad \text{on } \partial V_{\sigma e}\,.$$

Since $\partial\hat{u}_i$ are common for elements along their common boundaries, the last integral of Eq. (11a) require that

(14a) $$\sum_{\text{all elements}} \int_{\partial V_e - \partial V_{\sigma e}} \hat{T}_i\partial\hat{u}_i\, dS = 0\,, \quad \text{or} \quad \int_{S_e} [(\hat{T}_i)_1 - (\hat{T}_i)_2]\partial\hat{u}_i dS = 0\,,$$

where S_e denotes the common boundary of two adjacent elements and $(\hat{T}_i)_1$, $(\hat{T}_i)_2$ are the tractions of the two elements. Equation (14a) implies the traction $\hat{T}_i$ is continuous across inter-element boundaries and is called the *inter-element equilibrium condition*.

One may write Eq. (10) in a slightly different form

$$\Pi_{H1} = \sum_{\text{all elememts}} \Pi_{H1e}(e_{ij}, u_i, \hat{u}_i, \hat{T}_i)\,,$$

where

$$\Pi_{H1e}(e_{ij}, u_i, \hat{u}_i, \hat{T}_i) = \int_{\partial V_e} \hat{T}_i(\hat{u}_i - u_i)\, dS$$

$$+ \int_{V_e} \left[D_{ijkl}e_{kl}\frac{u_{i,j} + u_{j,i}}{2} - D_{ijkl}\frac{e_{ij}e_{kl}}{2} - b_i u_i \right] dV$$

$$- \int_{\partial V_{\sigma e}} \overset{\nu}{T}_i\, \hat{u}_i\, dS\,,$$

in which e_{ij}, u_i, $\hat{u}_i$, $\hat{T}_i$ are the four independent fields and $D_{ijkl}(= C_{ijkl}^{-1})$ is the *elastic coefficient* or *modulus tensor*, the inverse of the elastic flexibility tensor C_{ijkl}. The variational formulation associated with Π_{H1} is called the *hybrid strain principle*.

Hybrid Stress Principle. If the stresses satisfy the equilibrium equations Eq. (5) and the tractions $\hat{T}_i$ satisfy Eq. (13) *a priori*, one can eliminate u's and $\hat{T}$'s from Eq. (10) and derive a new functional (Tong and Pian 1969)

$$(15) \qquad \Pi_{HS} = \sum_{\text{all elements}} \Pi_{HSe}(\sigma_{ij}, \hat{u}_i),$$

where

$$\Pi_{HSe}(\sigma_{ij}, \hat{u}_i) = \int_{\partial V_e} n_j \sigma_{ji} \hat{u}_i dS - \frac{1}{2}\int_{V_e} C_{ijkl}\sigma_{ij}\sigma_{kl} dV - \int_{\partial V_{\sigma e}} \overset{\nu}{\hat{T}}_i \hat{u}_i dS.$$

Π_{HS} is a two-field hybrid functional, which is the basis for the hybrid stress finite element model proposed by Pian (1964). The admissibility conditions require that: $\hat{u}_i$ are common along the common boundaries of any two adjacent elements and satisfy the prescribed displacement condition Eq. (3); and the stresses satisfy the equilibrium equations Eq. (5). The first variation of Π_{HSe} is

$$\delta\Pi_{HSe} = -\int_{V_e} C_{ijkl}\sigma_{kl}\delta\sigma_{ij} dV + \int_{\partial V_e} n_j\hat{u}_i\delta\sigma_{ji} dS$$

$$+ \int_{\partial V_{\sigma e}} (n_j\sigma_{ji} - \overset{\nu}{\hat{T}}_i)\delta\hat{u}_i dS + \int_{\partial V_e - \partial V_{\sigma e}} n_j\sigma_{ji}\delta\hat{u}_i dS.$$

Thus the stationarity condition $\delta\Pi_{HS} = 0$ requires

$$-\int_{V_e} C_{ijkl}\sigma_{kl}\delta\sigma_{ij} dV + \int_{\partial V_e} n_j\hat{u}_i\delta\sigma_{ji} dS = 0,$$

$$\int_{\partial V_{\sigma e}} (n_j\sigma_{ji} - \overset{\nu}{\hat{T}}_i)\delta\hat{u}_i dS = 0,$$

$$\sum_{\text{all elements}} \int_{\partial V_e - \partial V_{\sigma e}} n_j\sigma_{ji}\partial\hat{u}_i dS = 0.$$

The second and third equalities give the traction boundary condition (7) on $\partial V_{\sigma e}$ and the inter-element equilibrium condition (14a), respectively. From Sec. 10.9, the first equality gives the Betrami-Michell compatibility equations as the Euler equation with $\hat{u}_i$ as the element boundary displacements for every element. The compatibility equations for isotropic materials are given in Eq. (10.9:22). An alternative proof of stress compatibility is as follows. For given $\hat{u}_i$, let u_i be the elastic solution in V_e with $u_i = \hat{u}_i$ on ∂V_e. Then the first equality above can be written as

$$\int_{V_e} \left[\frac{1}{2}(u_{i.j} + u_{j,i}) - C_{ijkl}\sigma_{kl} \right] \delta\sigma_{ij} \, dV = 0$$

For arbitrary equilibrium $\delta\sigma_{ij}$, the stresses σ_{kl} must satisfy the strain-stress relations

$$e_{ij} = \frac{1}{2}(u_{i,j} + u_{j,i}) = C_{ijkl}\sigma_{kl} \, .$$

Since e_{ij} are derived from displacement u's, it implies that σ_{kl} satisfy the Betrami-Michell compatibility equations. The variational formulation associated with Π_{HS} is called the *hybrid stress principle*, which forms the basis for the *hybrid stress finite element model*.

Hybrid Displacement Principle. If the stresses satisfy the strain stress relations Eq. (6), one can express the stresses in terms of the strains or displacements and obtain the stress strain relations

(16) $$\sigma_{ij} = D_{ijkl}e_{kl} = D_{ijkl}(u_{k,l} + u_{l,k})/2 \, .$$

One can then eliminate σ's from Eq. (10) and derive a three-field hybrid functional (Tong 1970)

(17) $$\Pi_{HD} = \sum_{\text{all elements}} \Pi_{HDe}(u_i, \hat{u}_i, \hat{T}_i) \, ,$$

where

$$\Pi_{HDe}(u_i, \hat{u}_i, \hat{T}_i) = \int_{\partial V_e} \hat{T}_i(\hat{u}_i - u_i) \, dS$$

$$+ \int_{V_e} \left(\frac{1}{2} D_{ijkl}e_{ij}e_{kl} - b_i u_i \right) dV - \int_{\partial V_{\sigma e}} \overset{\nu}{\overline{T}}_i \, \hat{u}_i \, dS \, ,$$

in which e's are in terms of u's as given in Eq. (2). The admissibility conditions require that $\hat{u}_i$ are common along the common boundary of any two adjacent elements and satisfy the prescribed displacement condition

Eq. (3). The stationarity condition of Π_{HD} is

$$
\begin{aligned}
\delta\Pi_{HD}(u_i, \hat{u}_i, \hat{T}_i) = \sum_{\text{all elements}} \Bigg\{ &-\int_{V_e} [(D_{ijkl}e_{kl}),_j + b_i]\delta u_i \, dV \\
&+ \int_{\partial V_e} (n_j D_{ijkl}e_{kl} - \hat{T}_i)\delta u_i \, dS + \int_{\partial V_e} (\hat{u}_i - u_i)\delta\hat{T}_i \, dS \\
&+ \int_{\partial V_{\sigma e}} (\hat{T}_i - \overset{\nu}{\bar{T}}_i)\delta\hat{u}_i \, dS + \int_{\partial V_e - \partial V_{\sigma e}} \hat{T}_i \delta\hat{u}_i \, dS \Bigg\} = 0.
\end{aligned}
$$

The vanish of the first two integrals gives

(18) $(D_{ijkl}e_{kl}),_j + b_i = 0$ in V_e,

(19) $\hat{T}_i = n_j D_{ijkl}e_{kl}$ on ∂V_e,

which are the equilibrium equations and the stress traction relations Eqs. (5) and (13), respectively, with the stresses being in terms of strains as given by Eq. (16). Equation (18) is the *Navier equation*. The zero conditions of the last three integrals of $\delta\Pi_{HD}$ imply the matching conditions Eq. (12), the prescribed boundary traction conditions Eq. (14), and the continuity of $\hat{T}_i$ across inter-element boundaries. The variational formulation associated with the functional above is called the *hybrid displacement principle*, which forms the basis for the *hybrid displacement finite element model*.

The functional Π_{HD} can be simplified if one requires that u's satisfy the Navier equation and that $\hat{T}$'s satisfy Eq. (19) *a priori*. Applying the Gauss theorem, one obtains a two-field hybrid functional

(20) $$\Pi_{HD1} = \sum_{\text{all elements}} \Pi_{HD1e}(u_i, \hat{u}_i),$$

where

$$
\Pi_{HD1e}(u_i, \hat{u}_i) = \int_{\partial V_e} n_j D_{ijkl}e_{kl}\hat{u}_i \, dS - \frac{1}{2}\int_{V_e} D_{ijkl}e_{ij}e_{kl} \, dV - \int_{\partial V_{\sigma e}} \overset{\nu}{\bar{T}}_i \, \hat{u}_i \, dS,
$$

in which e's are strains in terms of the displacements. If the body force is zero, Π_{HD1e} reduces to

$$
\Pi_{HD1e}(u_i, \hat{u}_i) = \int_{\partial V_e} n_j D_{ijkl}e_{kl}\left(\hat{u}_i - \frac{1}{2}u_i\right) dS - \int_{\partial V_{\sigma e}} \overset{\nu}{\bar{T}}_i \, \hat{u}_i \, dS,
$$

which involves integration over ∂V_e only.   One can show that the stationarity condition of Π_{HD1} gives the matching conditions Eq. (12), the prescribed boundary traction conditions Eq. (14), and the continuity of $n_j D_{ijkl} e_{kl} (= \overset{\nu}{T}_i)$ across inter-element boundaries. The last formulation is very useful for problems involving singularities in finite element applications.

Pian (1995, 1998) reviewed various variational principles of mixed/hybrid formulations. Interested readers are referred to published literature.

Incompressible or Nearly Incompressible Materials. If the material is incompressible, one cannot use the principle of minimum potential energy directly because certain components of the elastic modulus tensor D_{ijkl} becomes infinite. However, the Reissner or hybrid stress functional is still valid which can be seen from Eq. (9b) for isotropic materials as the bulk modulus $K \to \infty$.

We can establish a modified potential energy principle valid for incompressible or nearly incompressible materials. Consider an isotropic material with the elastic modulus tensor being

$$D_{ijkl} = \left(K - \frac{2}{3} G \right) \delta_{ij} \delta_{kl} + G(\delta_{ik} \delta_{jl} + \delta_{il} \delta_{jk})$$

$$= 2G \left(\frac{\nu \delta_{ij} \delta_{kl}}{1 - 2\nu} + \frac{\delta_{ik} \delta_{jl} + \delta_{il} \delta_{jk}}{2} \right).$$

Here K is the bulk modulus, which becomes infinite if the material is incompressible, i.e., when $\nu = 0.5$. Using the *contact transformation*,

$$\frac{1}{2} D_{ijkl} e_{ij} e_{kl} + p e_{kk} = G e'_{ij} e'_{ij} - \frac{p^2}{2K} \,,$$

where p is the hydrostatic pressure and $e'_{ij} (= e_{ij} - \delta_{ij} e_{kk}/3)$ is the deviatoric strain, we can write the strain energy density as

(21) $$\frac{1}{2} D_{ijkl} e_{ij} e_{kl} = G e'_{ij} e'_{ij} - p e_{kk} - \frac{p^2}{2K} \,,$$

which is a function of both u's and p. The *deviatoric energy* $(= G e'_{ij} e'_{kl})$ is bounded for all real materials.

We can now construct a mixed functional with both u's and p as independent field variables. The functional is suitable for applications to incompressible or nearly incompressible materials. (Tong 1969)

(22) $$\Pi_{Me}(u_i, p) = \int_{V_e} \left(G e'_{ij} e'_{ij} - p e_{kk} - \frac{p^2}{2K} - b_i u_i \right) dV - \int_{\partial V_{\sigma e}} \overset{\nu}{T}_i \, u_i dS \,.$$

Penalty Functional. A simple way to relax constraints in the integral formulation is to introduce penalty function in the functional. For instance, if a material is incompressible, one can require that

$$(23) \qquad\qquad\qquad\qquad e_{kk} = 0$$

as a constraint equation. Equation (22) becomes

$$(24) \qquad \Pi_{Me}(u_i) = \int_{V_e} (Ge'_{ij}e'_{ij} - b_i u_i)\, dV - \int_{\partial V_{\sigma e}} \overset{\nu}{T}_i\, u_i\, dS$$

with Eq. (23) as the constraint condition. One can relax the constraint by using the functional

$$(25) \qquad \Pi_{Me}(u_i, p) = \int_{V_e} (Ge'_{ij}e'_{ij} + pe_{kk} - b_i u_i)\, dV - \int_{\partial V_{\sigma e}} \overset{\nu}{T}_i\, u_i\, dS,$$

or introducing a *penalty function* and expressing the functional in the form

$$(26) \quad \Pi_{Me}(u_i, p) = \int_{V_e} \left(Ge'_{ij}e'_{ij} + \frac{c}{2}\, e_{kk}e_{jj} - b_i u_i\right) dV - \int_{\partial V_{\sigma e}} \overset{\nu}{T}_i\, u_i\, dS,$$

where c is an assumed positive constant of the order G or larger. The term associated with c is the penalty function, which enforces incompressibility when $c \to \infty$.

10.11. SAINT-VENANT'S PRINCIPLE

In 1855, Barre de Saint-Venant enunciated the "principle of the elastic equivalence of statically equipollent systems of loads." According, to this principle, the strains that are produced in a body by the application, to a small part of its surface, of a system of forces statically equivalent to zero force and zero couple, are of negligible magnitude at distances which are large compared with the linear dimensions of the part. When this principle is applied to the problem of torsion of a long shaft due to a couple applied at its ends, it states that the shear stress distribution in the shaft at a distance from the ends large compared with the cross-sectional dimension of the shaft will be practically independent of the exact distribution of the surface tractions of which the couple is the resultant. Such a principle is nearly always applied, consciously or unconsciously, when we try to simplify or idealize a problem in mathematical physics. It is used, for example, in devising a simple tension test for a material, when we clamp the ends of a test specimen in the jaws of a testing machine and assume that the action on the central part of the bar is nearly the same as if the forces were uniformly applied at the ends.

The justification of the principle is largely empirical and, as such, its interpretation is not entirely clear.

One possible way to formulate Saint-Venant's principle with mathematical precision is to state the principle in certain sense of average as follows (Zanaboni,[10.3] 1937, Locatelli,[10.3] 1940, 1941). Consider a body as shown in Fig. 10.12:1. A system of forces P in static equilibrium (with zero resultant

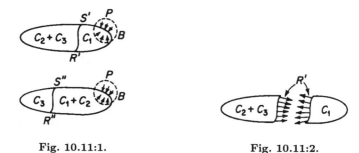

Fig. 10.11:1.   Fig. 10.11:2.

force and zero resultant couple) is applied to a region of the body enclosed in a small sphere B. Otherwise the body is free. Let S' and S'' be two arbitrary nonintersecting cross sections, both outside of B, with S'' farther away from B than S'. Due to the system of loads P, stresses are induced in the body. If we know these stresses, we can calculate the tractions acting on the surfaces S' and S''. Let the body be considered as severed into two parts at S', and let the system of surface tractions acting on the surface S' be denoted by R' which is surely a system of forces in equilibrium (see Fig. 10.12:2). Then a convenient measure of the magnitude of the tractions R' is the total strain energy that would be induced in the two parts should they be loaded by R' alone. Let this strain energy be denoted by $U_{R'}$. We have,

$$U_{R'} = \int_V W(\sigma_{ij}^{(R)})\, dv\,,$$

where W is the strain energy density function, and the stresses $\sigma_{ij}^{(R)}$ correspond to the loading system R'. Similarly, let the magnitude of the stresses at the section S'' be measured by the strain energy $U_{R''}$, which would have been induced by the tractions R'' acting on the surface S'' over the two parts of the body severed at S''. Both $U_{R'}$ and $U_{R''}$ are positive quantities and they vanish only if R', R'' vanish identically. Now we shall formulate Saint-Venant's principle in the following form (Zanaboni,[10.3] 1937).

Let S' and S'' be two nonintersecting sections both outside a sphere B. If the section S'' lies at a greater distance than the section S' from the sphere B in which a system of self-equilibrating forces P acts on the body, then

(1) $$U_{R''} < U_{R'} .$$

In this form, the diminishing influence of the self-equilibrating system of loading P as the distance from B increases is expressed by the functional U_R, which is a special measure of the stresses induced at any section outside B. The reason for the choice of U_R as a measure is its positive definiteness character and the simplicity with which the theorem can be proved. Further sharpening of the principle will be discussed later.

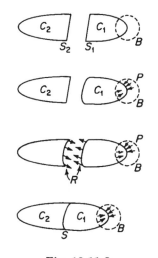

In order to prove the Theorem (1), we first derive an auxiliary principle. Let a self-equilibrating system of forces P be applied to a limited region B at the surface of an otherwise free elastic body C_1 (Fig. 10.11:3). Let U_1 be the strain energy produced by P in C_1. Let us now consider an enlarged body $C_1 + C_2$ by affixing to C_1 an additional body C_2 across a surface S which does not intersect the region B. When P is applied to the enlarged body $C_1 + C_2$, the strain energy induced is denoted by U_{1+2}. Then the lemma states that

Fig. 10.11:3.

(2) $$U_{1+2} < U_1 .$$

Proof of the Lemma. To compute U_{1+2}, we imagine that the stresses in the enlarged body $C_1 + C_2$ are built up by the following steps (see Fig. 10.11:3). First the load P is applied to C_1. The face S_1 of C_1 is deformed. Next a system of surface tractions R is applied to C_1 and C_2 on the surfaces of separation, S_1 and S_2. R will be so chosen that the deformed surfaces S_1 and S_2 fit each other exactly, so that displacements of material points in C_1 and C_2 are continuous as well as the stresses. Now C_1 and C_2 can be brought together and welded, and S becomes merely an interface. The result is the same if C_1 and C_2 were joined in the unloaded state and the combined body $C_1 + C_2$ is loaded by P.

The strain energy U_{1+2} is the sum of the work done by the forces in the above stages. In the first stage, the work done by P is U_1. In the second stage, the work done by R on C_2 is U_{R2}^*; the work done by R on C_1 consists of two parts, U_{R1}^* if C_1 were free, and U_{PR}^*, the work done by the system of loads P due to the deformation caused by R. Hence,

$$(3) \qquad U_{1+2} = U_1 + U_{R1}^* + U_{R2}^* + U_{PR}^* \,.$$

Now the system of forces R represents the internal normal and shear stresses acting on the interface S of the body $C_1 + C_2$. It is therefore determined by the minimum complementary energy theorem. Consider a special variation of stresses in which all the actual forces R are varied in the ratio $1 : (1 + \epsilon)$, where ϵ may be positive or negative. The work U_{R1}^* will be changed to $(1 + \epsilon)^2 U_{R1}^*$, because the load and deformation will both be changed by a factor $(1 + \epsilon)$. Similarly, U_{R2}^* is changed to $(1 + \epsilon)^2 U_{R2}^*$. But U_{PR}^* is only changed to $(1 + \epsilon)U_{PR}^*$, because the load P is fixed, while the deformation is varied by a factor $(1 + \epsilon)$. Hence, U_{1+2} is changed to

$$(4) \qquad U_{1+2}' = U_1 + (1 + \epsilon)^2 U_{R1}^* + (1 + \epsilon)^2 U_{R2}^* + (1 + \epsilon)U_{PR}^* \,.$$

The difference between Eqs. (4) and (3) is

$$\Delta U_{1+2} = \epsilon(2U_{R1}^* + 2U_{R2}^* + U_{PR}^*) + \epsilon^2(U_{R1}^* + U_{R2}^*) \,.$$

For U_{1+2} to be a minimum, ΔU_{1+2} must be positive regardless of the sign of ϵ. This is satisfied if

$$(5) \qquad 2U_{R1}^* + 2U_{R2}^* + U_{PR}^* = 0 \,.$$

On substituting Eqs. (5) into (3), we obtain

$$(6) \qquad U_{1+2} = U_1 - (U_{R1}^* + U_{R2}^*) \,.$$

Since U_{R1}^* and U_{R2}^* are positive definite, we see that Lemma (2) is proved.

Proof of Zanaboni theorem. Now we shall prove Saint-Venant's principle embodied in Eq. (1). Consider an elastic body consisting of three parts $C_1 + C_2 + C_3$ loaded by P in B, as shown in Fig. 10.12:1. Let this body be regarded first as a result of adjoining $C_2 + C_3$ to C_1 with an interface force system R', and then as a result of adjoining C_3 to $C_1 + C_2$ with an interface force R''. We have, by repeated use of Eq. (6),

$$U_{1+(2+3)} = U_1 - (U_{R'1}^* + U_{R'(2+3)}^*) \,,$$

$$U_{(1+2)+3} = U_{1+2} - (U_{R''(1+2)}^* + U_{R''3}^*)$$

$$= U_1 - (U_{R1}^* + U_{R2}^*) - (U_{R''(1+2)}^* + U_{R''3}^*) \,.$$

Equating these expressions, we obtain

$$U^*_{R'1} + U^*_{R'(2+3)} = U^*_{R1} + U^*_{R2} + U^*_{R''(1+2)} + U^*_{R''3},$$

or, since U^*_{R1} and U^*_{R2} are essentially positive quantities,

(7) $$U^*_{R'1} + U^*_{R'(2+3)} > U^*_{R''(1+2)} + U^*_{R''3}.$$

This is Eq. (1), on writing $U_{R'}$ for $U^*_{R'1} + U^*_{R'(2+3)}$, etc. Hence, the principle is proved.

10.12. SAINT-VENANT'S PRINCIPLE — BOUSSINESQ–VON MISES–STERNBERG FORMULATION

The Saint-Venant principle, as enunciated in terms of the strain energy functional, does not yield any detailed information about individual stress components at any specific point in an elastic body. However, such information is clearly desired. To sharpen the principle, it may be stated as follows (von Mises, 1945). "*If the forces acting upon a body are restriced to several small parts of the surface, each included in a sphere of radius ϵ, then the strains and stresses produced in the interior of the body at a finite distance from all those parts are smaller in order of magnitude when the forces for each single part are in equilibrium than when they are not.*"

The classical demonstration of this principle is due to Boussinesq (1885), who considered an infinite body filling the half-space $z \geq 0$ and subjected to several concentrated forces, each of magnitude F, normal to the boundary $z = 0$. If these normal forces are applied to points in a small circle B with diameter ϵ, Boussinesq proved that the largest stress component at a point P which lies at a distance R from B is

(1) of order F/R^2 if the resultant of the forces is of order F,
(2) of order $(\epsilon/R)(F/R^2)$ if the resultant of the forces is zero,
(3) of order $(\epsilon/R)^2(F/R^2)$ if both the resultant force and the resultant moment vanish.

(Cf. Secs. 8.8, 8.10.)

These relative orders of magnitude were believed to have general validity until von Mises (1945) showed that a modification is necessary.

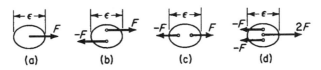

Fig. 10.12:1. von Mises' examples in which forces tangential to the surface of an elastic half-space are applied in a small area.

Consider the half-space $z \geq 0$ again. Let forces of magnitude F *tangent* to the boundary $z = 0$ be applied to points in a small circle B of diameter ϵ. Making use of the well-known Cerruti solution (Sec. 8.8), von Mises obtained the following results for the four cases illustrated in Fig. 10.12:1. The order of magnitude of the largest stress component at a point P which lies at a distance R from B is

(1) of order $\sigma_0 = F/R^2$ in case (a),
(2) of order $(\epsilon/R)\sigma_0$ in case (b),
(3) of order $(\epsilon/R)\sigma_0$ in case (c),
(4) of order $(\epsilon/R)^2\sigma_0$ in case (d).

The noteworthy case is (c), which is drastically different from what one would expect from an indiscriminating generalization of Boussinesq's result, for in this case the forces are in static equilibrium, with zero moment about any axes, but all one could expect is a stress magnitude of order no greater than $(\epsilon/R)\sigma_0$, not $(\epsilon/R)^2\sigma_0$. von Mises found that in this case the order of magnitude of the largest stress component is reduced to $(\epsilon/R)^2\sigma_0$ if and only if the external forces acting upon a small part of the surface are such as to remain in equilibrium when all the forces are turned through an arbitrary angle. (Such a case is called *astatic equilibrium.*)

von Mises examined next the stresses in a finite circular disk due to loads on the circumference, and a similar conclusion was reached. (See Fig. 10.12:2.)

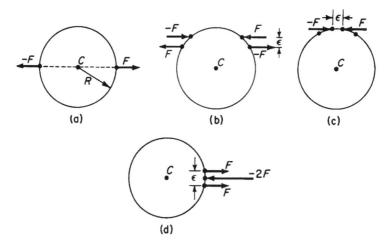

Fig. 10.12:2.   von Mises' examples of a circular disk subjected to loads on the circumference.

These examples show that Saint-Venant's principle, as stated in the traditional form at the beginning of this section, does not hold true.

Accordingly, von Mises[10.3] proposed, and later Sternberg[10.3] proved (1954), the following mathematical statement of the Saint-Venant principle:

Let a body be acted on by surface tractions which are finite and are distributed over a number of regions all no greater than a sphere of diameter ϵ. Consider an interior point x whose distance to any of these loading areas is no less than a characteristic length which will be taken as unity. ϵ is nondimensionalized with respect to this characteristic length. Then, as $\epsilon \to 0$, the order of magnitude of the strain components at x is as follows:

$$e(x, \epsilon) = O(\epsilon^p)$$

where

(a) If the tractions have nonvanishing vector sums in at least one area, then in general, $\rho \geq 2$. (Note that the surface traction is assumed to be finite, so the resultant force $\to 0$ as ϵ^2, since the area on which the surface tractions act $\to 0$ as ϵ^2.)

(b) If the resultant of the surface tractions in every loading area vanishes, then $\rho \geq 3$.

(c) If, in addition, the resultant moment of the surface tractions in every loading area also vanishes, then still we can be assured only of $\rho \geq 3$.

(d) $\rho \geq 4$ in case of astatic equilibrium in every loading area, which may be described by the 12 scalar conditions

$$\int_{S(\epsilon)} T_i^* \, dS = 0, \qquad \int_{S(\epsilon)} T_i^* x_j \, dS = 0, \qquad (i, j = 1, 2, 3)$$

for each loading area $S(\epsilon)$, where T_i^* represents the specified surface traction over $S(\epsilon)$, and x_j is the coordinate of the point of application of T_i^*.

If the tractions applied to $S(\epsilon)$ are parallel to each other and not tangential to the surface, then if they are in equilibrium they are also in astatic equilibrium, and the condition $\rho \geq 4$ prevails.

If, instead of prescribing finite surface traction, we consider finite forces being applied which remain finite as $\epsilon \to 0$, then the exponent ρ should be replaced by 0, 1, 2 in the cases named above, as was illustrated in Figs. 10.12:1 and 10.12:2.

The theorem enunciated above does not preclude the validity of a stronger Saint-Venant principle for special classes of bodies, such as thin plates or shells or long rods. With respect to perturbations that occur at the edges of a thin plate or thin shell, a significant result was obtained by K. O. Friedrichs[10.3] (1950) in the form of a so-called boundary-layer theory. With respect to lateral loads on shells, Naghdi[10.3] (1960) obtained similar results.

10.13. PRACTICAL APPLICATIONS OF
SAINT-VENANT'S PRINCIPLE

It is well-known that Saint-Venant's principle has its analogy in hydrodynamics and that these features are associated with the elliptic nature of the partial differential equations. If the differential equations were hyperbolic and two-dimensional, local disturbances may be propagated far along the characteristics without attenuation. Then the concept of the Saint-Venant's principle will not apply. For example, in the problem of the response of an elastic half-space to a line load traveling at supersonic speeds over the free surface, the governing equations are hyperbolic, and we know that any fine structure of the surface pressure distribution is propagated all the way to infinity.

On the other hand, one feels intuitively that the validity of Saint-Venant's principle is not limited to linear elastic solid or infinitesimal displacements. One expects it to apply in the case of rubber for finite strain or to steel even when yielding occurs. Although no precise proof is available, Goodier[10.3] (1937) has argued on the basis of energy as follows:

Let a solid body be loaded in a small area whose linear dimensions are of order ϵ, with tractions which combine to give zero resultant force and couple. Such a system of tractions imparts energy to the solid through the *relative* displacements of the points in the small loaded area, because no work is done by the tractions in any translation or rotation of the area as a rigid body. Let the tractions be of order p. Let the slope of the stress strain curve of the material be of order E. (The stress-strain relationship does not have to be linear.) Let one element of the loaded area be regarded as fixed in position and orientation. Then since the strain must be of order p/E, the displacements of points within the area are of order $(p\epsilon/E)$. The work done by the traction acting on an element dS is of order $(p\epsilon p \cdot dS/E)$. The order of magnitude of total work is, therefore, $p^2\epsilon^3/E$. Since a stress of order p implies a strain energy of order p^2/E *per unit volume*, the region in which the stress is of order p must have a volume comparable with ϵ^3. Hence, the influence of tractions cannot be appreciable at a distance from the loaded area which are large compared with ϵ.

Goodier's argument can be extended to bodies subjected to limited plastic deformation. In fact, the argument provides an insight to practical judgement of how local self-equilibrating tractions should influence the strain and stress in the interior of a body.

An engineer needs to know not only the order of magnitude comparison such as stated in von Mises–Sternberg theorem; he needs to know also how numerically trustworthy Saint-Venant's principle is to his particular problem. Hoff[10.3] (1945) has considered several interesting examples, two of which are given in Figs. 10.13:1 and 10.13:2 and will be explained below.

In the first example, the torsion of beams with different cross-sections is considered. One end of the beam is clamped, where cross-sectional warping is prevented. The other end is free, where a torque is applied by means of shear stresses distributed according to the requirements of the theory of pure torsion. The difference between the prescribed end conditions at the clamped end from those assumed in Saint-Venant's torsion theory (Sec. 7.5) may be stated in terms of a system of self-equilibrating tractions that act

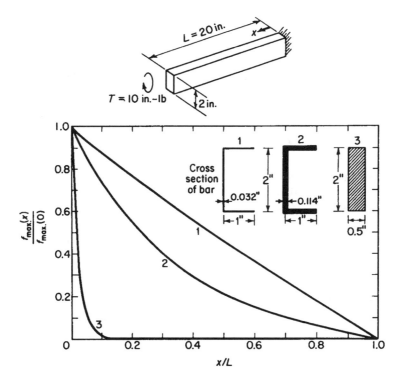

Fig. 10.13:1. Hoff's illustration of Saint-Venant's principle.

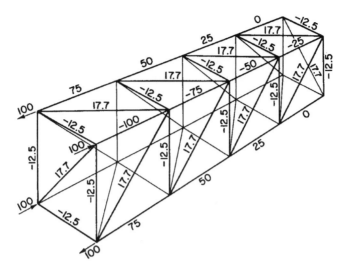

Fig. 10.13:2. Hoff's example illustrating the slow decay of self-equilibrating forces in a space framework.

at the clamped end. Timoshenko has given approximate solutions to these problems. Hoff's example refers to beams with dimensions as shown in the figure and subjected to a torque of 10 in. lb. Due to the restrictions of warping, normal stresses are introduced in the bar in addition to the shearing stresses of Saint-Venant's torsion. For the rectangular beam, the maximum normal stress at the fixed end is equal to 157 lb/sq in. For the other two thin-walled channel sections, the maximum normal stresses at the fixed end are 1230 and 10,900 lb/sq in., respectively, for the thicker and thinner sections. In Fig. 10.13:1 curves are shown for the ratio of maximum normal stress $f_{\max}(x)$ in any section x divided by the maximum normal stress $f_{\max}(0)$ in the fixed-end section, plotted against the ratio distance x of the section from the fixed end of the bar, divided by the total length L of the bar. Inspection of the curves reveals that, while in the case of the solid rectangular section the normal stress caused by the restriction to warping at the fixed end is highly localized, it has appreciable values over the entire length of the channel section bars. Consequently, reliance on Saint-Venant's principle in the calculation of stresses caused by torsion is entirely justified with the bar of rectangular section. In contrast, stresses in the thin-walled section bars depend largely upon the end conditions.

Hoff's second example refers to pin-jointed space frameworks. In one case a statically determinate framework is considered, to one end of which a set of four self-equilibrating concentrated loads are applied, as shown in

Fig. 10.13:2. The figures written on the elements of the framework represent the forces acting in the bars measured in the same units as the applied loads. Negative sign indicates compression. It can be seen that the effect of the forces at one end of the structure is still noticeable at the other end.

Hoff's examples show that Saint-Venant's principle works only if there is a possibility for it to work; in other words, only if there exist paths for the internal forces to follow in order to balance one another within a short distance of the region at which a group of self-equilibrating external forces is applied. This point of view is in agreement with Goodier's reasoning.

Hoff's examples are not in conflict with the von Mises–Sternberg theorem, for the latter merely asserts a certain order of magnitude comparison for the stress and strain as the size of the region of self-equilibrating loading shrinks to zero, and does not state how the stresses are propagated. On the other hand, although Goodier's reasoning does not provide a definitive theorem, it is very suggestive in pointing out the basic reason for Saint-Venant's principle and can be used in estimating the practical efficiency of the principle.

10.14. EXTREMUM PRINCIPLES FOR PLASTICITY

In this section we shall derive the principles of minimum potential energy and minimum complementary energy for displacement and stress increments in elastic-plastic bodies (Hill 1950). We consider a body V subjected to prescribed body forces in V, traction $\overset{\nu}{T}_i$ on ∂V_σ, and displacements on ∂V_u. Suppose that u_i and σ_{ij} are the elastic-plastic solution of the static problem and that du_j and $d\sigma_{ij}$ are the increment solution for *infinitesimal increments* db_i in V, $d\overset{\nu}{T}_i$ on ∂V_σ and $d\bar{u}_i$ on ∂V_u. Then $d\sigma_{ij}$ satisfy the equilibrium equations

$$(1) \qquad\qquad d\sigma_{ij,j} + db_i = 0, \qquad \text{in } V,$$

the traction conditions

$$n_j d\sigma_{ji} = d\overset{\nu}{T}_i, \qquad \text{on } \partial V_\sigma,$$

and can be expressed in terms of displacement increments du_j through the strain increments de_{kl} [Eq. (6.10:11), (6.11:18) or (6.14:17)]

$$(2) \qquad \begin{aligned} d\sigma_{ij} &= D_{ijkl} de_{kl} \qquad \text{for } f < 0 \\ &= D^{ep}_{ijkl} de_{kl} \qquad \text{for } f > 0, \end{aligned}$$

where D^{ep}_{ijkl} is given in Eq. (6.11:11), (6.11:18) or (6.14:17) and f is the yield function.

Extremum Principles for Displacement Increment. Let du_j be the incremental solution over a volume V. Let $d\hat{u}_j$ be a kinematically admissible displacement field that it obeys the internal constraints, if any, and that it satisfies $d\hat{u}_i = d\bar{u}_i$ on ∂V_u. The corresponding strain and stress increments are $(d\hat{e}_{ij}, d\hat{\sigma}_{ij})$. Consider the following functional for displacement increments du_j

$$(3) \quad \Pi_p(du_i) = \frac{1}{2}\int_V d\sigma_{ij} de_{ij}\, dV - \int_V db_i du_i\, dV - \int_{\partial V_\sigma} d\overset{\nu}{T}_i\, du_i dS\,.$$

We can show that

$$\Pi_p(d\hat{u}) - \Pi_p(du) = \frac{1}{2}\int_V (d\hat{\sigma}_{ij}d\hat{e}_{ij} - d\sigma_{ij}de_{ij})\, dV - \int_V db_i(d\hat{u}_i - du_i)\, dV$$

$$- \int_{\partial V_\sigma} d\overset{\nu}{T}_i\, (d\hat{u}_i - du_i)\, dS\,.$$

By taking into account of the relation

$$\int_{\partial V_\sigma} d\overset{\nu}{T}_i\, (d\hat{u}_i - du_i)dS = \int_V d\sigma_{ij}(d\hat{e}_{ij} - de_{ij})\, dV$$

$$+ \int_V d\sigma_{ij,j}(d\hat{u}_i - du_i)\, dV\,,$$

the difference can be written as

$$(4) \quad \Pi_p(d\hat{u}_i) - \Pi_p(du_i) = \frac{1}{2}\int_V (d\hat{\sigma}_{ij}d\hat{e}_{ij} + d\sigma_{ij}de_{ij} - 2d\sigma_{ij}d\hat{e}_{ij})\, dV\,.$$

With the substitution of $de_{ij} = D_{ijkl}^{-1}d\sigma_{kl} + de_{ij}^p$ and $d\hat{e}_{ij} = D_{ijkl}^{-1}d\hat{\sigma}_{kl} + d\hat{e}_{ij}^p$, the integrand becomes

$$D_{ijkl}^{-1}(d\hat{\sigma}_{ij} - d\sigma_{ij})(d\hat{\sigma}_{kl} - d\sigma_{kl}) + d\hat{\sigma}_{ij}d\hat{e}_{ij}^p + d\sigma_{ij}de_{ij}^p - 2d\sigma_{ij}d\hat{e}_{ij}^p\,.$$

The first term is obviously non-negative due to the positive definiteness of D_{ijkl}. For perfectly plastic materials,

$$d\hat{\sigma}_{ij}d\hat{e}_{ij}^p = d\sigma_{ij}de_{ij}^p = 0\,,$$

$$d\sigma_{ij}d\hat{e}_{ij}^p = d\hat{\Lambda}n_{ij}d\sigma_{ij} \le 0\,,$$

[see Eqs. (6.10.3,4)] because $d\hat{\Lambda} \ge 0$ and $n_{ij}d\sigma_{ij} \le 0$. Consequently, the remaining terms of the integrand are never negative, so that

$$(5) \quad \Pi_p(d\hat{u}_i) \ge \Pi_p(du_i)\,.$$

The equality holds only if $d\hat{\sigma}_{ij} = d\sigma_{ij}$.

For work-hardening materials, let

$$(6) \qquad\qquad df = n_{kl}d\sigma_{kl}\,.$$

with $df < 0$ denoting elastic unloading. From Eq. (6.11:15), the flow equation can be written as

$$(7) \qquad\qquad \begin{aligned} K^p de_{ij}^p &= H(df)n_{ij}\,df\,, \\ K^p d\hat{e}_{ij}^p &= H(d\hat{f})n_{ij}d\hat{f}\,. \end{aligned}$$

The sum of the remaining terms of the integrand becomes

$$(8) \qquad R = \frac{1}{K^p}\left[(d\hat{f})^2 H(d\hat{f}) + (df)^2 H(df) - 2(df)(d\hat{f})H(d\hat{f})\right].$$

We can show that

$$(9) \qquad \begin{aligned} R &= 0 & \text{if } df \le 0 \text{ and } d\hat{f} \le 0\,, \\ &= \frac{1}{K^p}(d\hat{f} - df)^2 \ge 0 & \text{if } df > 0 \text{ and } d\hat{f} > 0\,, \\ &= \frac{1}{K^p}d\hat{f}(d\hat{f} - 2df) > 0 & \text{if } df \le 0 \text{ and } d\hat{f} > 0\,, \\ &= \frac{1}{K^p}(df)^2 > 0 & \text{if } df > 0 \text{ and } d\hat{f} \le 0\,. \end{aligned}$$

Consequently, the integrand is never negative except when $d\hat{\sigma}_{ij} = d\sigma_{ij}$, then $\Pi_p(d\hat{u}_i) = \Pi_p(du_i)$. In other words, Eq. (5) also holds for work-hardening materials. $d\hat{\sigma}_{ij} = d\sigma_{ij}$ implies that $d\hat{e}_{ij} = de_{ij}$ and therefore $d\hat{u}_i = du_i$.

The extremum principle for the functional Eq. (3) is also called the *principle of minimum potential energy for the displacement increment*.

Extremum Principle for Stress. The complementary extremum principle concerns a statically and plastically admissible stress increment field. A stress field is statically admissible if it satisfies the equilibrium equations in V and the traction condition on ∂V_σ. A stress field is plastically admissible if it obeys the yield criterion everywhere in V. The *complementary functional for the stress increments* $d\sigma_{ij}$ is

$$(10) \qquad \Pi_c(d\sigma_{ij}) = \frac{1}{2}\int_V d\sigma_{ij}de_{ij}\,dV - \int_{\partial V_u} n_j d\sigma_{ij}d\bar{u}_i dS\,.$$

Expressing de_{ij} in terms of $d\sigma_{ij}$ and de_{ij}^p, we can rewrite Eq. (10) as

$$\Pi_c(d\sigma_{ij}) = \frac{1}{2}\int_V (D_{ijkl}^{-1}d\sigma_{ij}d\sigma_{kl} + d\sigma_{ij}de_{ij}^p)\,dV - \int_{\partial V_u} n_j d\sigma_{ij}d\bar{u}_i dS\,,$$

where

$$
(11) \quad
\begin{aligned}
d\sigma_{ij} de_{ij}^p &= 0\,, & \text{for perfectly plastic materials,} \\
&= \frac{1}{K^p} (df)^2 H(df)\,, & \text{for work-hardening materials.}
\end{aligned}
$$

and $df(= n_{ij} d\sigma_{ij})$ is defined in Eq. (6). Clearly the functional is a function of $d\sigma_{ij}$ only. Let $d\sigma_{ij}$ be the plastic solution of the increment and $d\hat{\sigma}_{ij}$ be a statically admissible plastic stress field. Then

$$
(12) \quad \Pi_c(d\hat{\sigma}_{ij}) - \Pi_c(d\sigma_{ij}) = \frac{1}{2} \int_V \left(d\hat{\sigma}_{ij} \, d\hat{e}_{ij} + d\sigma_{ij} de_{ij} - 2d\hat{\sigma}_{ij} de_{ij}^p \right) dV\,.
$$

In deriving Eq. (12), we have used the relation that

$$
\int_V (d\hat{\sigma}_{ij} - d\sigma_{ij}) de_{ij} \, dV = \int_{\partial V_u} n_i (d\hat{\sigma}_{ij} - d\sigma_{ij}) \, d\bar{u}_j \, dS
$$

for any two admissible stress fields with a continuous displacement $du_j = d\bar{u}_j$ on ∂V_u. The integrand is similar to that in Eq. (4) and may be shown to be positive definite by the same method. Therefore, we have

$$
(13) \quad \Pi_c(d\hat{\sigma}_{ij}) \geq \Pi_c(d\sigma_{ij})\,.
$$

The equality holds only if $d\hat{\sigma}_{ij} = d\sigma_{ij}$.

It can be shown that

$$
\Pi_c(d\sigma_{ij}) = -\Pi_p(du_i)\,,
$$

which implies

$$
(14) \quad \Pi_c(d\hat{\sigma}_{ij}) \geq \Pi_c(d\sigma_{ij}) = -\Pi_p(du_i) \geq -\hat{\Pi}_p(d\hat{u}_{ij})\,.
$$

The extremum principle for Eq. (10) is also called the *principle of minimum complementary energy for $d\sigma_{ij}$*.

The stationary condition of $\partial \Pi_p = 0$ or $\partial \Pi_c = 0$ for functionals in Eqs. (3) and (10) are useful for constructing "weak-form" increment solutions.

Rigid-Plastic Materials. For rigid-plastic materials, the elastic strain-increment is zero, and therefore,

$$
(15) \quad de_{ij} = \frac{1}{2} (du_{i,j} + du_{j,i}) = de_{ij}^p\,.
$$

The principle of extremum plastic work becomes

(16) $$\sigma_{ij}de_{ij} = D_p(de_{ij}) \geq \hat{\sigma}_{ij}de_{ij} ,$$

where the strain increment de_{ij} is normal to the yield surface at the stress point σ_{ij} in the stress space and $\hat{\sigma}_{ij}$ is any stress laying inside or at another point on the yield surface. It is emphasized that $D_p(de_{ij})$ is a function of de_{ij} only.

Let us define another functional

(17) $$\Pi(du_i) = \int_V D_p(de_{ij})\, dV - \int_V b_i du_i dV - \int_{\partial V_\sigma} \overset{\nu}{T}_i\, du_i dS$$

as a function of the velocity field solution du_i. Let $d\hat{u}_i$ be a kinematically admissible with its associated stress $\hat{\sigma}_{ij}$ satisfying the yield criterion in part or all of V [i.e., $f(\hat{\sigma}_{ij}, T, \hat{\xi}_i) < 0$ or $f = 0, \hat{e}_{ij} = \hat{n}_{ij}d\Lambda$]. Then

(18) $$\Pi(d\hat{u}_i) - \Pi(du_i) = \int_V [D_p(d\hat{e}_{ij}) - D_p(de_{ij}) - \sigma_{ij}(d\hat{e}_{ij} - de_{ij})]\, dV$$

$$= \int_V [D_p(d\hat{e}_{ij}) - \sigma_{ij}d\hat{e}_{ij}]\, dV .$$

By the maximum principle of plastic dissipation the integrand is nonnegative. Therefore,

(19) $$\Pi(d\hat{u}_i) \geq \Pi(du_i) .$$

10.15. LIMIT ANALYSIS

We define the *critical state of a body* as a large increase in plastic deformation, much larger than the elastic deformation, with little or no increase in load. In other words, it is a state of impending plastic collapse or incipient plastic flow in which $de_{ij} \neq 0$ under constant load (i.e., $db_i = d\overset{\nu}{T}_i = d\bar{u}_i = 0$). Therefore,

(1) $$\int_V db_i du_i dV + \int_{\partial V_\sigma} d\overset{\nu}{T}_i\, du_i dS = \int_V d\sigma_{ij}de_{ij}\, dV$$

$$= \int_V (D_{ijkl}^{-1}d\sigma_{ij}d\sigma_{kl} + d\sigma_{ij}de_{ij}^p)\, dV = 0 .$$

The positive definiteness of $D_{ijkl}^{-1}d\sigma_{ij}d\sigma_{kl}$ combined with Drucker's inequality implies $d\sigma_{ij} = 0$. This in turns implies $de_{ij}^e = 0$ and $de_{ij} = de_{ij}^p$, i.e., at impending plastic collapse the plastic flow is rigid-plastic. This result was first noted by Drucker *et al.* (1951). It makes possible to use rigid-plastic formulation to establish the upper and lower bounds in limit analysis.

Lower Bound Theorem.   Consider the rigid-plastic deformation with known body force $\bar{b}_i$ in V, traction $\overset{\nu}{\bar{T}}_i$ on ∂V_σ and increment $d\bar{u}_i = 0$ on ∂V_u. Let the applied loads and prescribed displacements be respectively written in the form $P\bar{b}_i$, $P\bar{T}_i$ and $Pd\bar{u}_i$, where P is a load parameter. We further let $(du_i,\ de_{ij}$ and $\sigma_{ij})$ be the actual solution at plastic collapse associated with the load parameter P_l. We have

$$\int_V D_p(de_{ij})\,dV = \int_V \sigma_{ij}de_{ij}\,dV = P_l\left(\int_V \bar{b}_i du_i\,dV + \int_{\partial V_\sigma} \overset{\nu}{\bar{T}}_i\,du_i\,dS\right).$$

If a statically admissible stress $\bar{\sigma}_{ij}$ is in equilibrium with $P_{LB}\bar{b}_i$ and $P_{LB}\overset{\nu}{\bar{T}}_i$, then

(2) $$\int_V \bar{\sigma}_{ij}de_{ij}\,dV = P_{LB}\left(\int_V \bar{b}_i du_i\,dV + \int_{\partial V_\sigma} \overset{\nu}{\bar{T}}_i\,du_i\,dS\right).$$

Using Eq. (2) and the principle of maximum plastic work for rigid-plastic deformation, $D_p(de_{ij}) \geq \bar{\sigma}_{ij}de_{ij}$, we obtain a lower bound of the collapse load

(3) $$P_{LB} = \frac{\displaystyle\int_V \bar{\sigma}_{ij}de_{ij}\,dV}{\displaystyle\int_V \bar{b}_i du_i\,dV + \int_{\partial V_\sigma} \overset{\nu}{\bar{T}}_i\,du_i\,dS}$$

$$\leq \frac{\displaystyle\int_V D_p(de_{ij})\,dV}{\displaystyle\int_V \bar{b}_i du_i\,dV + \int_{\partial V_\sigma} \overset{\nu}{\bar{T}}_i\,du_i\,dS} = P_l\,.$$

Since $d\bar{u}_i$ and de_{ij} are generally not known, we cannot use Eq. (2) directly to evaluate P_{LB}.

Upper Bound Theorem.   For a kinematically admissible increment $d\bar{u}_i$ with $d\bar{u}_i = 0$ on ∂V_u, we have

(4) $$\int_V \sigma_{ij}d\bar{e}_{ij}\,dV = P_l\left(\int_V \bar{b}_i d\bar{u}_i\,dV + \int_{\partial V_\sigma} \overset{\nu}{\bar{T}}_i\,d\bar{u}_i\,dS\right).$$

where σ_{ij} is the solution at impending collapse and P_l is the limit load. Let us define

(5)
$$P_{UB} = \frac{\int_V D_p(d\bar{e}_{ij})dV}{\int_V \bar{b}_i d\bar{u}_i dV + \int_{\partial V_\sigma} \overset{\nu}{\bar{T}}_i \, d\bar{u}_i dS}.$$

Then using Eq. (4) and $D_p(d\bar{e}_{ij}) \geq \sigma_{ij} d\bar{e}_{ij}$ for rigid-plastic deformation, we conclude that

(6)
$$P_{UB} = P_l \frac{\int_V D_p(d\bar{e}_{ij})\, dV}{\int_V \sigma_{ij} d\bar{e}_{ij}\, dV} \geq P_l \,.$$

If one has found a kinematically admissible velocity field having its strain rate associated with a statically admissible stress field everywhere[†] in the domain, then one has a complete solution. This solution predicts the correct collapse load using limit analysis. However, this may not be the exact solution, because the solution may not be unique.

Lower and Upper Bound Loci for Multiple Load Parameters. The results given in Eqs. (3) and (4) can be generalized to include several parameters P_I that

(7)
$$F_i = \sum_I P_I (\bar{b}_i)_I \,, \qquad \text{prescribed body force in } V \,,$$
$$\overset{\nu}{\bar{T}}_i = \sum_I P_I (\overset{\nu}{\bar{T}}_i)_I \,, \qquad \text{prescribed traction on } \partial V_\sigma \,,$$
$$d\bar{u}_i = \sum_I P_I (d\bar{u}_i)_I \,, \qquad \text{prescribed displacement increment in } \partial V_u \,.$$

For a kinematically admissible velocity field $d\bar{u}_i$, we define the generalized velocities

(8)
$$d\bar{v}_I = \int_V (\bar{b}_i)_I d\bar{u}_i \, dV + \int_{\partial V_\sigma} (\overset{\nu}{\bar{T}}_i)_I d\bar{u}_i \, dV \,.$$

Let $(du_i, de_{ij}, \sigma_{ij})$ be the solution at collapse with loading parameters $(P_l)_I$, then

$$\int_V \sigma_{ij} de_{ij} \, dV = \sum_I (P_l)_I d\bar{v}_I \,.$$

[†]In the rigid region where $d\bar{e}_{ij} = 0$, the question of association does not arise.

For a statically admissible plastic stress field $\bar{\sigma}_{ij}$, we have

$$(9) \quad \int_V \bar{\sigma}_{ij} de_{ij}\, dV = \sum_I (P_{LB})_I \left[\int_V (\bar{b}_i)_I\, du_i\, dV + \int_{\partial V_\sigma} (\overset{\nu}{\bar{T}}_i)_I\, du_i\, dV \right]$$

$$= \sum_I (P_{LB})_I\, dv_I .$$

The extremum of plastic dissipation implies that

$$(10) \qquad \sum_I (P_l)_I\, dv_I \geq \sum_I (P_{LB})_I\, dv_I$$

The right hand side of the inequality gives a lower bound locus of the collapse loads.

For a kinematically admissible increment du_i^* and its associated strain-increment de_{ij}^* and stresses σ_{ij}^*, we have

$$(11) \quad \int_V D_p(de_{ij}^*)\, dV = \sum_I (P_{UB})_I \left[\int_V (\bar{b}_i)_I d\bar{u}_i dV + \int_{\partial V_\sigma} (\overset{\nu}{\bar{T}}_i)_I d\bar{u}_i\, dV \right]$$

$$= \sum_I (P_{UB})_I\, dv_I ,$$

$$\int_V \sigma_{ij} de_{ij}^*\, dV = \sum_I (P_l)_I\, dv_I^* .$$

Then

$$(12) \qquad \sum_I (P_l)_I\, dv_I^* \leq \sum_I (P_{UB})_I dv_I^*$$

gives an upper locus of the collapse loads, which can be computed explicitly from the admissible velocity field du_i^*. More general theorems of limit analysis have introduced by Salencon (1977). Interested readers are referred to the literature.

We shall consider a multi-parameter example of a sandwich beam consist of two equal thin flanges of cross-sectional area A and length L. A web of negligible longitudinal strength separates the flanges at a distance d apart. The beam is under an axial force P and a bending moment M, which are treated as two independent loading parameters.

For lower bound analysis, the statically admissible stresses are

$$(13) \qquad \qquad \frac{\sigma_u}{\sigma_l} = \frac{P}{2A} \pm \frac{M}{Ad} ,$$

where σ_u and σ_l are the stress in the upper and lower flanges, respectively. The yield criterion requires that both $|\sigma_u|$ and $|\sigma_l|$ are less or equal to σ_Y, that is,

$$\text{(14)} \qquad \left| \frac{P}{P_l} \pm \frac{M}{M_l} \right| \le 1,$$

which is the lower bound locus with $P_l = 2A\sigma_Y$ and $M_l = \sigma_Y Ad$. Since the problem is statically determinant, the lower bound locus is the actual limit locus with P_l and M_l being the limit loads for simple tensile and pure bending, respectively.

For upper bound analysis, we use the elongation increments of the upper and lower flanges $d\Delta u_u$ and $d\Delta u_l$ as the kinematically admissible increment. Then the plastic dissipation is

$$\int_V D_p(de_{ij}^*)\, dV = A\sigma_Y \left(|d\Delta u_u| + |d\Delta u_l| \right),$$

and the external work is

$$\sum_I (P_{UB})_I \int_{\partial V} (\overset{\nu}{\overline{T}})_I du_i^*\, dS = A\left[\left(\frac{P}{2A} + \frac{M}{Ad} \right) d\Delta u_u + \left(\frac{P}{2A} - \frac{M}{Ad} \right) d\Delta u_u \right].$$

By equating the plastic dissipation and the external work, we obtain the upper bound locus

$$\left(\frac{P}{P_l} + \frac{M}{M_l} \right) d\Delta u_u + \left(\frac{P}{P_l} - \frac{M}{M_l} \right) d\Delta u_u = |d\Delta u_u| + |d\Delta u_l|.$$

PROBLEMS

10.3.   Derive the differential equation and permissible boundary conditions for a membrane stretched over a simply connected regular region A by minimizing the functional I with respect to the displacement w; $p(x,y)$ being a given function

$$I = \frac{1}{2} \iint_A (w_x^2 + w_y^2)\, dx\, dy - \iint_A pw\, dx\, dy.$$

10.4.   Obtain Euler's equation for the functional

$$I' = \frac{1}{2} \iint_A [w_x^2 + w_y^2 + (\nabla^2 w + p)^2 - 2pw]\, dx\, dy.$$

Ans. $\nabla^4 w - \nabla^2 w + \nabla^2 p - p = 0.$

Note: Compare I' with I of Problem 10.3. Since $\nabla^2 w + p = 0$ for a membrane, the solution of Problem 10.3 also satisfies the present problem. But the integrand of I' contains higher derivatives of w and the equation $\delta I' = 0$ is a sharper condition than the equation $\delta I = 0$. These examples illustrate Courant's method of sharpening a variational problem and accelerating the convergence of an approximating sequence in the direct method of solution of variational problems.

10.5. Bateman's principle in fluid mechanics states that in a flow of an ideal, nonviscous fluid (compressible or incompressible) in the absence of external body forces, the "pressure integral"

$$- \iiint p\, dx dy dz \,,$$

which is the integral of pressure over the entire fluid domain, is an extremum. Consider the special case of a steady, irrotational flow of an incompressible fluid, for which the Bernoulli's equation gives

$$p = \text{const.} - \frac{\rho V^2}{2} = \text{const.} - \frac{p}{2}\left[\left(\frac{\partial \phi}{\partial x}\right)^2 + \left(\frac{\partial \phi}{\partial y}\right)^2\right],$$

where ϕ is the velocity potential. Derive the field equation governing ϕ and the corresponding natural boundary conditions according to Bateman's principle. [H. Bateman, *Proc. Roy. Soc. London*, **A 125** (1929) 598–618.]

10.6. Let

$$T = \sum_{\mu,\nu=1}^{m} P_{\mu\nu}(t, q_1, \ldots, q_m)\dot{q}_\mu(t)\dot{q}_\nu(t) \,,$$

$$U = U(t, q_1, \ldots, q_m) \,,$$

and

$$L = T - U \,,$$

where $\dot{q}_\mu(t) = dq_\mu/dt$, and $q_\mu(t_0) = a_\mu$, $q_\mu(t_1) = b_\mu$, $(\mu = 1, \ldots, m)$; a_μ, b_μ are given numbers. Derive the Euler differential equations for the variational problem connected with the functional

$$J[q_1, \ldots, q_m] = \int_{t_0}^{t_1} L\, dt \,.$$

Prove that

$$\frac{d(T + U)}{dt} = 0 \,,$$

if $\partial U/\partial t = 0$, $\partial P_{\mu\nu}/\partial t = 0$, $\mu, \nu = 1, \ldots, m$. *Hint*: Use the Euler's relation for homogeneous functions.

10.7. Find the curves in the x, y-plane such that

$$\int_a^b \sqrt{2E - n^2 y^2}\, ds \,, \quad ds^2 = dx^2 + dy^2$$

is stationary, where E and n are constants and the integral is taken between fixed end points.

10.8.   Let D be the set of all functions $u(x)$ with the following properties:

$$D : \begin{cases} (1) & u(x) = a_1 \sin \pi x + a_2 \sin 2\pi x , \\ (2) & a_1, a_2 \text{ are real, arbitrary numbers}, \\ (3) & a_1^2 + a_2^2 > 0 . \end{cases}$$

Consider the functional

$$J[u] = \frac{\int_0^1 [u'(x)]^2 \, dx}{\int_0^1 [u(x)]^2 \, dx}$$

for all functions $u(x)$ in D.

Questions:

(a) Give necessary conditions for a function $u^*(x)$ in D so that $u^*(x)$ furnishes an extremum of the functional J.

(b) Show that there exist exactly two functions

$$u_1(x) = a_{11} \sin \pi x + a_{12} \sin 2\pi x ,$$

$$u_2(x) = a_{21} \sin \pi x + a_{22} \sin 2\pi x ,$$

notwithstanding a constant factor, which satisfy these necessary conditions mentioned in (a).

(c) Show that for one of the functions mentioned in (b), say, $u_1(x)$, the inequality

$$J[u_1] \le J[u]$$

holds for all functions $u(x)$ in D. Then show that for the other solution, $u_2(x)$, the inequality

$$J[u] \le J[u_2]$$

holds for all functions $u(x)$ in D.

(d) Does there exist any relation between $J[u_1]$, $J[u_2]$ and the eigenvalues of the eigenvalue problem

$$- u''(x) = \lambda u(x) , \qquad\qquad 0 \le x \le 1 ,$$

with

$$u(0) = u(1) = 0 .$$

10.9.   Clapeyron's theorem states that, if a linear elastic body is in equilibrium under a given system of body forces F_i and surface forces T_i, then the strain energy of deformation is equal to one-half the work that would be done by the external forces (of the equilibrium state) acting through the displacements u_i from the unstressed state to the state of equilibrium, i.e.,

$$\int_V F_i u_i \, dv + \int_s \overset{\nu}{T_i} u_i \, dS = 2 \int_V W \, dv .$$

Demonstrate Clapeyron's Theorem for:

(a) A simply supported beam under its own weight [Fig. P10.9(a)].

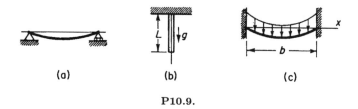

P10.9.

(b) A rod under its own weight [Fig. P10.9(b)],

$$u(x) = \frac{g\rho L^2}{E}\left[\frac{1}{2}\left(\frac{x}{L}\right)^2 - \frac{x}{L}\right].$$

(c) A strip of membrane of infinite length and width b [Fig. P10.9(c)], under a constant pressure p_0, for which

$$w = w_0(1 - 4x^2/b^2), \qquad \frac{w_0}{h} = \sqrt[3]{3(1-\nu^2)64\frac{p_0 b^4}{Eh^4}}.$$

10.10. The *"poker chip" problem (Max Williams).* To obtain a nearly triaxial tension test environment for polymer materials, a "poker chip" test specimen (a short circular cylinder) is glued between two circular cylinders (Fig. P10.10). When the cylinder is subjected to simple tension, the center of the poker chip is subjected to a triaxial tension stress field. Let the elastic constants of the media be E_1, G_1, ν_1 and E_2, G_2, ν_2, as indicated in Fig. P10.10, with $E_1 \ll E_2$, $\nu_1 \cong 0.5$. Assume cylindrical symmetry, and obtain approximate expressions of the stress field by the following methods.

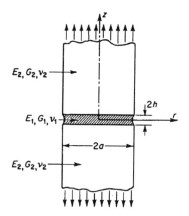

P10.10. Poker chip specimen.

(a) Use the complementary energy theorem and a stress field satisfying stress boundary conditions and the equations of equilibrium.
(b) Use the potential energy theorem and assumed displacements satisfying displacement boundary conditions.
(c) Experimental results suggest that the following displacement field is reasonable:

$$w = w_0 \frac{z}{h}, \qquad\qquad -h \leq z \leq h,$$

$$u = \left(1 - \frac{z^2}{h^2}\right) g(r),$$

where $g(r)$ is as yet an unknown function of r. Obtain the governing equation for $g(r)$ and its appropriate solution for this problem by attempting to satisfy the following equilibrium equations in some average sense with respect to the z-direction:

$$\frac{\partial \sigma_{rr}}{\partial r} + \frac{\partial \sigma_{rz}}{\partial z} + \frac{\sigma_{rr} - \sigma_{\theta\theta}}{r} = 0,$$

$$\frac{\partial \sigma_{rz}}{\partial r} + \frac{\partial \sigma_{zz}}{\partial z} + \frac{\sigma_{rz}}{r} = 0.$$

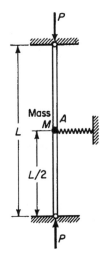

P10.11.

10.11. Consider a pin-ended column of length L and bending stiffness EI subjected to an end thrust P. A spring is attached at the middle of the column as shown in Fig. P10.11. When the column is straight the spring tension is zero. If the column deflects by an amount Δ, the spring exerts a force R on the column:

$$R = K\Delta - \alpha\Delta^3, \qquad (K > 0, \alpha > 0).$$

Derive the equation of equilibrium of the system. It is permissible to use the Euler–Bernoulli approximation for a beam, for which the strain energy per unit length is

$$\frac{1}{2}EI \cdot (\text{curvature})^2$$

and

$$\text{bending moment} = EI \cdot (\text{curvature}).$$

10.12. For the column of Problem 10.11 under the axial load P, is the solution unique? Under what situation is the solution nonunique? What are the possible solutions when the uniqueness is lost?

10.13. A linear elastic beam of bending stiffness EI is supported at four equidistant points $ABCD$, Fig. P10.13. The supports at A and D are pin-ended. At B and C, the beam rests on two identical nonlinear pillars. The characteristic of the pillars may be described as "hardening" and is expressed by the equation

$$K\Delta = R - \beta R^3 \qquad (K > 0, \beta > 0),$$

where R is the reaction of the pillar, Δ is the downward deflection of the beam at the point of attachment of the beam to the pillar, K and β are constants.

A load P is applied to the beam at the midspan point. Find the reactions at the supports B and C. One of the two minimum principles (of potential energy and of complementary energy) is easier to apply to this problem. Solve the problem by a variational method with the appropriate minimum principle.

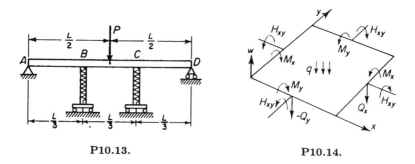

P10.13.   P10.14.

10.14. Consider a square plate loaded in the manner shown in Fig. P10.14. Derive Euler's equation and the natural boundary conditions for V to be a minimum when $w(x, y)$ is varied.

$$V = U + A,$$

$$U = \frac{D}{2} \int_0^1 \int_0^1 [(w_{xx} + w_{yy})^2 - 2(1 - \nu)(w_{xx}w_{yy} - w_{xy}^2)] \, dx \, dy,$$

$$A = \int_0^1 \int_0^1 qw \, dx \, dy - \int_0^1 \left(M_x \frac{\partial w}{\partial x} \right) \Big|_{x=0}^{x=1} dy + \int_0^1 (Q_x w) \Big|_{x=0}^{x=1} dy$$

$$+ \int_0^1 \left(H_{xy} \frac{\partial w}{\partial y} \right) \Big|_{x=0}^{x=1} dy - \int_0^1 \left(M_y \frac{\partial w}{\partial y} \right) \Big|_{y=0}^{y=1} dx$$

$$+ \int_0^1 (Q_y w) \Big|_{y=0}^{y=1} dx + \int_0^1 \left(H_{xy} \frac{\partial w}{\partial x} \right) \Big|_{y=0}^{y=1} dx.$$

10.15. Consider a material with the elastic modulus tensor satisfying

$$D_{ijkk} = D_{kkij} = 3K\delta_{ij} = E\delta_{ij}/(1 - 2\nu).$$

The bulk modulus K is infinite while $D_{ijkl} - \delta_{ij}\delta_{kl}K$ is bounded if the material is incompressible. Show that the *contact transformation* gives

$$D_{ijkl}e_{ij}e_{kl}/2 - pe_{kk} = D'_{ijkl}e'_{ij}e'_{kl}/2 - p^2/(2K),$$

where $D'_{ijkl} = D_{ijkl} - K\delta_{ij}\delta_{kl}$, p is the hydrostatic pressure and e'_{ij} is the deviatoric strain and that the potential energy of the element is

$$\Pi_{Me}(u_i, p) = \int_{V_e} \left(\frac{1}{2} D'_{ijkl}e'_{ij}e'_{kl} + pe_{kk} - \frac{p^2}{2K} - b_i u_i \right) dV - \int_{\partial V_{\sigma e}} \overset{\nu}{T}_i u_i \, dS.$$

The *deviatoric energy* $D'_{ijkl}e'_{ij}e'_{kl}/2$ is bounded.

11

HAMILTON'S PRINCIPLE,
WAVE PROPAGATION,
APPLICATIONS OF GENERALIZED
COORDINATES

In dynamics, the counterpart of the minimum potential energy theorem is Hamilton's principle. In this chapter we shall discuss this important principle and its applications to vibrations and wave propagations in beams.

Toward the end of the chapter a brief discussion of computational methods of solving variational problems is given. The basic idea is to apply the concept of generalized coordinates to obtain approximate solutions for a continuous system by reducing it to one with a finite number of degrees of freedom. Several important methods — those of Euler, Rayleigh-Ritz-Galerkin, and Kantrovich — will be outlined. Much more about computational mechanics is given later in Chapters 17–21.

11.1. HAMILTON'S PRINCIPLE

For an oscillating body, with displacements u_i so small that the acceleration is given by $\partial^2 u_i / \partial t^2$ in Eulerian coordinates (Sec. 5.2), the equation of small motion is

$$(1) \qquad \sigma_{ji \cdot j} + F_i = \rho \frac{\partial^2 u_i}{\partial t^2} ,$$

where ρ is the density of the material and F_i is the body force per unit volume. Let us again consider virtual displacements δu_i as specified in Sec. 10.7, but instead of a body in static equilibrium we now consider a vibrating body. The variations δu_i must vanish over the boundary surface S_u, where values of displacements are prescribed; but are arbitrary, triply differentiable over the domain V; and are arbitrary also over the boundary surface S_σ, where surface tractions are prescribed. The total boundary surface of the volume V is S, and $S = S_u + S_\sigma$.

The virtual work done by the body and surface forces is, as in Sec. 10.7,

$$\int_V F_i \delta u_i dv + \int_S \overset{\nu}{T_i} \, \delta u_i dS .$$

The last integral can be transformed on introducing $\overset{\nu}{T}_i = \sigma_{ij}\nu_j$ and using Gauss' theorem, as in Sec. 10.7:

$$\int_S \overset{\nu}{T}_i \, \delta u_i dS = \int_S \sigma_{ij}\nu_j \delta u_i dS = \int_V (\sigma_{ij}\delta u_i)_{,j} dv$$

$$= \int_V \sigma_{ij,j}\delta u_i dv + \int_V \sigma_{ij}\delta u_{i,j} dv \, .$$

By Eq. (1), and by the symmetry of σ_{ij}, the right hand side is equal to

$$\int_V \left(\rho\frac{\partial^2 u_i}{\partial t^2} - F_i \right)\delta u_i dv + \int_V \sigma_{ij}\delta e_{ij} dv \, .$$

Therefore, we obtain the *variational equation of motion*

(2) ▲ $$\int_V \sigma_{ij}\delta e_{ij} dv = \int_V \left(F_i - \rho\frac{\partial^2 u_i}{\partial t^2} \right)\delta u_i dv + \int_S \overset{\nu}{T}_i \, \delta u_i dS \, .$$

As before, this general equation can be stated more concisely if we introduce various levels of restrictions. Thus, if the body is perfectly elastic and a strain energy function W exists, then the variational equation of motion can be written as

(3) ▲ $$\delta \int_V W dv = \int_V \left(F_i - \rho\frac{\partial^2 u_i}{\partial t^2} \right)\delta u_i dv + \int_S \overset{\nu}{T}_i \, \delta u_i dS \, .$$

The variations δu_i are assumed to vanish over the part of the boundary S_u where surface displacements are prescribed. Hence, the limit for the surface integral can be replaced by S_σ. If the variations δu_i were identified with the actual displacements $(\partial u_i/\partial t)dt$, then the result above states that, in an arbitrary time interval, the sum of the energy of deformation and the kinetic energy increases by an amount that is equal to the work done by the external forces during the same time interval.

If the virtual displacements δu_i are regarded as functions of time and space, not to be identified with the actual displacements, and the variational equation of motion (3) is integrated with respect to time between two arbitrary instants t_0 and t_1, an important variational principle for the moving body can be derived. We have

(4) $$\int_{t_0}^{t_1}\int_V \delta W dv dt = \int_{t_0}^{t_1} dt \int_V F_i \delta u_i dv + \int_{t_0}^{t_1} dt \int_{S_\sigma} \overset{\nu}{T}_i \, \delta u_i dS$$

$$- \int_{t_0}^{t_1} dt \int_V \rho\frac{\partial^2 u_i}{\partial t^2}\delta u_i dv \, .$$

Calling the last term J, inverting the order of integration, and integrating by parts, we obtain

(5) $\quad J = \int_V \rho \frac{\partial u_i}{\partial t} \delta u_i dv \Big|_{t_0}^{t_1} - \int_V dv \int_{t_0}^{t_1} \frac{\partial u_i}{\partial t} \left(\rho \frac{\partial \delta u_i}{\partial t} + \frac{\partial \rho}{\partial t} \delta u_i \right) dt$.

The $\partial \rho / \partial t$ term can be ignored because $\partial \rho / \partial t = -\partial \rho \dot{u}_i / \partial x_i$ according to the equation of continuity, and thus the term $\delta u_i \partial \rho / \partial t$ is an order of magnitude smaller than other terms in this equation. Let us now impose the restriction that at the time t_0 and t_1, the variations δu_i are zero at all points of the body; i.e.,

(6) $\qquad\qquad\qquad \delta u_i(t_0) = \delta u_i(t_1) = 0 \qquad\qquad\qquad \text{in } V$.

Then

(7) $\quad J = -\int_{t_0}^{t_1} \int_V \rho \frac{\partial u_i}{\partial t} \frac{\partial \delta u_i}{\partial t} dv dt$

$\qquad = -\int_{t_0}^{t_1} \int_V \rho \frac{\partial u_i}{\partial t} \delta \frac{\partial u_i}{\partial t} dv dt = -\int_{t_0}^{t_1} \delta \int_V \frac{1}{2} \rho \frac{\partial u_i}{\partial t} \frac{\partial u_i}{\partial t} dv dt$

$\qquad = -\int_{t_0}^{t_1} \delta K dt$,

where

(8) $\qquad\qquad\qquad K = \frac{1}{2} \int_V \rho \frac{\partial u_i}{\partial t} \frac{\partial u_i}{\partial t} dv$

is the kinetic energy of the moving body. Therefore, under the assumption (6), Eq. (4) becomes,

(9) ▲ $\quad \int_{t_0}^{t_1} \delta(U - K) dt = \int_{t_0}^{t_1} \int_V F_i \delta u_i dv dt + \int_{t_0}^{t_1} \int_{S_\sigma} \overset{\nu}{T_i} \delta u_i dS dt$,

where U represents the total strain energy of the body,

$$U = \int_V W dv.$$

If the external forces acting on the body are such that the sum of the integrals on the right-hand side of Eq. (9) represents the variation of a single function — *the potential energy of the loading* — A,

(10) $\qquad\qquad \int_V F_i \delta u_i dv + \int_{S_\sigma} \overset{\nu}{T_i} \delta u_i dS = -\delta A$,

then Eq. (9) can be written as

(11) ▲ $$\delta \int_{t_0}^{t_1} (U - K + A)dt = 0.$$

The term

(12) $$L \equiv U - K + A$$

(or sometimes $-L$) is called the *Lagrangian function* and Eq. (11) represents *Hamilton's principle*, which states that:

   The time integral of the Lagrangian function over a time interval t_0 to t_1 is an extremum for the "actual" motion with respect to all admissable virtual displacements which vanish, first, at instants of time t_0 and t_1 at all points of the body, and, second, over S_u, where the displacements are prescribed, throughout the entire time interval.

   To formulate this principle in another way, let us call $u_i(x_1, x_2, x_3; t)$ a dynamic path. Then Hamilton's principle states that *among all dynamic paths that satisfy the boundary conditions over S_u at all times and that start and end with the actual values at two arbitrary instants of time t_0 and t_1 at every point of the body, the "actual" dynamic path is distinguished by making the Lagrangian function an extremum.*

   In rigid body dynamics the term U drops out, and we obtain Hamilton's principle in the familiar form. The symbol A replaces the usual symbol V in books on dynamics because we have used V for something else.

   Note that the potential energy — A of the external loads exists and is a linear function of the displacements if the loads are independent of the elastic displacements, as is commonly the case. In aeroelastic problems, however, the aerodynamic loading is sensitive to the small surface displacements u_i; moreover, it depends on the time history of the displacements and cannot be derived from a potential. Hence, in aeroelasticity we are generally forced to use the variational form (9) of Hamilton's principle.

   In some applications of the direct method of calculation, it is even desirable to liberalize the variations δu_i at the instants t_0 and t_1 and use Hamilton's principle in the variational form (4) which cannot be expressed elegantly as the minimum of a well-defined functional. On the other hand, such a formulation will be accessible to the direct methods of solution. On introducing Eqs. (5), (7), and (10), we may rewrite Eq. (4) in the form:

(13) $$\int_{t_0}^{t_1} \delta(U - K + A)\, dt$$

$$= \int_{t_0}^{t_1} \int_V F_i \delta u_i \, dv dt + \int_{t_0}^{t_1} \int_S \overset{\nu}{T}_i \, \delta u_i \, dS - \int_V \rho \frac{\partial u_i}{\partial t} \delta u_i \, dv \Big|_{t_0}^{t_1}.$$

Here U is the total strain energy, K is the total kinetic energy, A is the potential energy for the conservative external forces, F_i and $\overset{\nu}{T_i}$ are, respectively, those external body and surface forces that are not included in A, and δu_i are the virtual displacements.

Problem 11.1.   Prove the converse theorem that, for a conservative system, the variational Eq. (11) leads to the equation of motion

$$\rho \frac{\partial^2 u_i}{\partial t^2} = F_i + \frac{\partial}{\partial x_j} \frac{\partial W}{\partial e_{ij}}$$

and the boundary conditions

$$\text{either} \qquad \delta u_i = 0 \quad \text{or} \quad \frac{\partial W}{\partial e_{ij}} \nu_j = \overset{\nu}{T_i} \ .$$

11.2. EXAMPLE OF APPLICATION — EQUATION OF VIBRATION OF A BEAM

As an example of the application of Hamilton's principle in the formulation of approximate theories in elasticity, let us consider the free, lateral vibration of a straight simple beam. We assume that the beam possesses principal planes and that the vibration takes place in one of the principal planes, and let y denote the small deflection of the neutral axis of the beam from its initial, straight configuration. In Sec. 10.8, it is shown that the strain energy of the beam is, for small deflections,

$$(1) \qquad\qquad U = \frac{1}{2} \int_0^l EI \left(\frac{\partial^2 y}{\partial x^2} \right)^2 dx \, ,$$

where E is the Young's modulus of the beam material, I is the cross-sectional moment of inertia, and l is the length of the beam.

The kinetic energy of the beam is derived partly from the translation, parallel to y, of the elements composing it, and partly from the rotation of the same elements about an axis perpendicular to the neutral axis and the plane of vibration. The former part is

$$\frac{1}{2} \int_0^l m \left(\frac{\partial y}{\partial t} \right)^2 dx \, ,$$

where m is the mass per unit length of the beam. The latter part is, for each element dx, the product of moment of inertia times one-half of the square of the angular velocity. Let I_ρ denote the mass moment of inertia about the neutral axis per unit length of the beam. The angular velocity being $\partial^2 y/\partial t\partial x$, the kinetic energy of the beam is

(2)
$$K = \frac{1}{2} \int_0^l m \left(\frac{\partial y}{\partial t} \right)^2 dx + \frac{1}{2} \int_0^l I_\rho \left(\frac{\partial^2 y}{\partial x\partial t} \right)^2 dx.$$

If the beam is loaded by a distributed lateral load of intensity $p(x,t)$ per unit length and moment and shear M and Q, respectively, at the ends as shown in Fig. 11.2:1, then the potential energy of the external loading is

(3) $A = -\displaystyle\int_0^l p(x,t)y(x)\, dx - M_l \left(\frac{\partial y}{\partial x} \right)_l + M_0 \left(\frac{\partial y}{\partial x} \right)_0 + Q_l y_l - Q_0 y_0.$

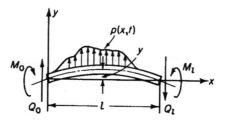

Fig. 11.2:1. Application to a beam.

The equation of motion is given by Hamilton's principle:

(4)
$$\delta \int_{t_0}^{t_1} (U - K + A)dt = 0;$$

i.e.,

(5) $\delta \displaystyle\int_{t_0}^{t_1} \left\{ \int_0^l \left[\frac{1}{2}EI \left(\frac{\partial^2 y}{\partial x^2} \right)^2 - \frac{1}{2}m \left(\frac{\partial y}{\partial t} \right)^2 - \frac{1}{2}I_\rho \left(\frac{\partial^2 y}{\partial x\partial t} \right)^2 - py \right] dx$

$$- M_l \left(\frac{\partial y}{\partial x} \right)_l + M_0 \left(\frac{\partial y}{\partial x} \right)_0 + Q_l y_l - Q_0 y_0 \right\} dt = 0.$$

Following the usual procedure of the calculus of variations, noting that the virtual displacement must be so specified that $\delta y \equiv 0$ at t_0 and t_1, and, hence, $\partial(\delta y)/\partial x = \delta(\partial y/\partial x) \equiv 0$ at t_0 and t_1, we obtain

$$\int_{t_0}^{t_1} \left\{ \int_0^l \left[EI \frac{\partial^2 y}{\partial x^2} \frac{\partial^2 \delta y}{\partial x^2} - m \frac{\partial y}{\partial t} \frac{\partial \delta y}{\partial t} - I_\rho \frac{\partial^2 y}{\partial x\partial t} \frac{\partial^2 \delta y}{\partial x\partial t} - p\delta y \right] dx \right.$$

$$\left. - M_l \delta \left(\frac{\partial y}{\partial x} \right)_l + M_0 \delta \left(\frac{\partial y}{\partial x} \right)_0 + Q_l \delta y_l - Q_0 \delta y_0 \right\} dt = 0.$$

Integrating by parts, we obtain

(6)
$$\int_{t_0}^{t_1} \int_0^l \left[\frac{\partial^2}{\partial x^2} \left(EI \frac{\partial^2 y}{\partial x^2} \right) + m \frac{\partial^2 y}{\partial t^2} \right.$$

$$\left. - \frac{\partial^2}{\partial x \partial t} \left(I_\rho \frac{\partial^2 y}{\partial x \partial t} \right) - p(x,t) \right] \delta y \, dx \, dt$$

$$+ \int_{t_0}^t \left[EI \frac{\partial^2 y}{\partial x^2} - M \right] \delta \left(\frac{\partial y}{\partial x} \right) \Big|_0^l \, dt$$

$$- \int_{t_0}^t \left[\frac{\partial}{\partial x} \left(EI \frac{\partial^2 y}{\partial x^2} \right) - \frac{\partial}{\partial t} \left(I_\rho \frac{\partial^2 y}{\partial x \partial t} \right) - Q \right] \delta y \Big|_0^l \, dt = 0 \,.$$

Hence, the Euler equation of motion is

(7)
$$\frac{\partial^2}{\partial x^2} \left(EI \frac{\partial^2 y}{\partial x^2} \right) + m \frac{\partial^2 y}{\partial t^2} - \frac{\partial^2}{\partial x \partial t} \left(I_\rho \frac{\partial^2 y}{\partial x \partial t} \right) = p(x,t) \,,$$

and a proper set of boundary conditions at each end is

(8a)
$$\text{either} \quad EI \frac{\partial^2 y}{\partial x^2} = M \quad \text{or} \quad \delta \left(\frac{\partial y}{\partial x} \right) = 0$$

and

(8b)
$$\text{either} \quad \frac{\partial}{\partial x} \left(EI \frac{\partial^2 y}{\partial x^2} \right) - \frac{\partial}{\partial t} \left(I_\rho \frac{\partial^2 y}{\partial x \partial t} \right) = Q \quad \text{or} \quad \delta y = 0 \,.$$

These are equations governing the motion of a beam including the effect of the rotary inertia, due to Lord Rayleigh, and known as Rayleigh's equations. If the rotary inertia is neglected and if the beam were uniform, then the governing equation is simplified into:

(9)
$$\frac{\partial^2 y}{\partial t^2} + c_0^2 R^2 \frac{\partial^4 y}{\partial x^4} = \frac{1}{m} p \,,$$

where

(10)
$$c_0^2 = \frac{E}{\rho} \,, \qquad R^2 = \frac{I}{A} \,.$$

The constant c_0 has the dimension of speed and can be identified as the phase velocity of longitudinal waves in a uniform bar.[†] R is the radius of gyration of the cross section. A is the cross-sectional area, so that $m = \rho A$.

[†] See Problem 11.2, p. 390.

In the special case of a uniform beam of infinite length free from lateral loading, $p = 0$, Eq. (9) becomes

(11)
$$\frac{\partial^2 y}{\partial t^2} + c_0^2 R^2 \frac{\partial^4 y}{\partial x^4} = 0 \,.$$

It admits a solution in the form

(12)
$$y = a \sin \frac{2\pi}{\lambda}(x - ct) \,,$$

which represents a progressive wave of phase velocity c and wave length λ. On substituting Eq. (12) into Eq. (11), we obtain the relation

(13)
$$c = \pm c_0 R \frac{2\pi}{\lambda} \,,$$

which states that the phase velocity depends on the wave length and that it tends to infinity for very short wave lengths. Somewhat disconcerting is the fact that, according to Eq. (13), the group velocity (see Sec. 11.3) also tends to infinity as the wave length tends to zero. Since group velocity is the velocity at which energy is transmitted, this result is physically unreasonable. If Eq. (13) were correct, then the effect of a suddenly applied concentrated load will be felt at once everywhere in the beam, as the Fourier representation for a concentrated load contains harmonic components with infinitesimal wave length, and hence infinite wave speed. Thus, Eq. (11) cannot be very accurate in describing the effect of impact loads on a beam.

This difficulty of infinite wave speed is removed by the inclusion of the rotary inertia. However, the speed versus wave length relationship obtained from Rayleigh's Eq. (7) for a uniform beam of circular cross section with radius a, as is shown in Fig. 11.2:2, still deviates appreciably from Pochhammer and Chree's results, which were derived from the exact three-dimensional linear elasticity theory. A much better approximation is obtained by including the shear deflection of the beam, as was first shown by Timoshenko.

To incorporate the shear deformation, we note that the slope of the deflection curve depends not only on the rotation of cross sections of the beam but also on the shear. Let ψ denote the slope of the deflection curve when the shearing force is neglected and β the angle of shear at the neutral axis in the same cross section. Then the total slope is

(14)
$$\frac{\partial y}{\partial x} = \psi + \beta \,.$$

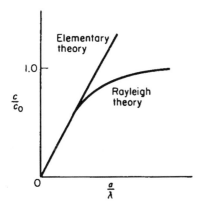

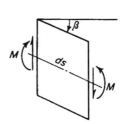

Fig. 11.2:2. Phase velocity curves for flexural elastic waves in a circular cylinder of radius a.

Fig. 11.2:3. A Timoshenko beam element.

The strain energy due to bending, Eq. (1), must be replaced by

$$(15) \qquad \frac{1}{2} \int_0^l EI \left(\frac{\partial \psi}{\partial x} \right)^2 dx \,,$$

because the internal bending moment does no work when shear deformation takes place (see Fig. 11.2:3). The strain energy due to shearing strain β must be a quadratic function of β if linear elasticity is assumed. We shall write

$$(16) \qquad \frac{1}{2} \int_0^l k\beta^2 dx = \frac{1}{2} \int_0^l k \left(\frac{\partial y}{\partial x} - \psi \right)^2 dx$$

for the strain energy for shear. The kinetic energy is

$$(17) \qquad K = \frac{1}{2} \int_0^l m \left(\frac{\partial y}{\partial t} \right)^2 dx + \frac{1}{2} \int_0^l I_\rho \left(\frac{\partial \psi}{\partial t} \right)^2 dx \,,$$

because the translational velocity is $\partial y/\partial t$, but the angular velocity is $\partial \psi/\partial t$. Hence, Hamilton's principle states that

$$(18) \qquad \delta \int_{t_0}^{t_1} \int_0^l \frac{1}{2} \left[EI \left(\frac{\partial \psi}{\partial x} \right)^2 + k \left(\frac{\partial y}{\partial x} - \psi \right)^2 - m \left(\frac{\partial y}{\partial t} \right)^2 \right.$$
$$\left. - I_\rho \left(\frac{\partial \psi}{\partial t} \right)^2 \right] dx dt + \delta A = 0 \,,$$

where A is given by Eq. (3) except that $\partial y/\partial x$ at the ends is to be replaced by ψ. The virtual displacements now consist of δy and $\delta \psi$, which must vanish at t_0 and t_1 and also where displacements are prescribed. On carrying out the calculations, the following two Euler equations are obtained:

(19a)
$$\frac{\partial}{\partial x}\left(EI\frac{\partial \psi}{\partial x}\right) + k\left(\frac{\partial y}{\partial x} - \psi\right) - I_\rho\frac{\partial^2 \psi}{\partial t^2} = 0,$$

(19b)
$$m\frac{\partial^2 y}{\partial t^2} - \frac{\partial}{\partial x}\left[k\left(\frac{\partial y}{\partial x} - \psi\right)\right] - p = 0.$$

The appropriate boundary conditions are, at each end of the beam,

(20a)
$$\text{Either} \quad EI\frac{\partial \psi}{\partial x} = M \quad \text{or} \quad \delta\psi = 0,$$

and

(20b)
$$\text{either} \quad k\left(\frac{\partial y}{\partial x} - \psi\right) = -Q \quad \text{or} \quad \delta y = 0.$$

These are the differential equation and boundary conditions of the so-called *Timoshenko beam theory*.

For a uniform beam, EI, k, m, etc., are constants, and the function ψ can be eliminated from the equations above to obtain the well-known *Timoshenko equation for lateral vibration of prismatic beams*,

(21)
$$EI\frac{\partial^4 y}{\partial x^4} + m\frac{\partial^2 y}{\partial t^2} - \left(I_\rho + \frac{EIm}{k}\right)\frac{\partial^4 y}{\partial x^2 \partial t^2} + I_\rho\frac{m}{k}\frac{\partial^4 y}{\partial t^4}$$
$$= p + \frac{I_\rho}{k}\frac{\partial^2 p}{\partial t^2} - \frac{EI}{k}\frac{\partial^2 p}{\partial x^2}.$$

So far we have not discussed the constants m, I_ρ, and k. For a beam of uniform material, $m = \rho A$, $I_\rho = \rho AR^2$, where ρ is the mass density of the beam material, A is the cross-sectional area, and R is the radius of gyration of the cross section about an axis perpendicular to the plane of motion and through the neutral axis. But k depends on the distribution of shearing stress in the beam cross section. Timoshenko writes

(22)
$$k = k'AG,$$

where G is the shear modulus of elasticity and k' is a numerical factor depending on the shape of the cross section, and ascertains that according to the elementary beam theory, $k' = \frac{2}{3}$ for a rectangular cross section. The use of such a value of k is, however, a subject of controversy in the literature. Mindlin[11.1] suggests that the value of k can be so selected that the solution of Eq. (21) be made to agree with certain solution of the exact three-dimensional equations of Pochhammer (1876) and Chree (1889) (see Love,[1.1] *Elasticity*, 4th edition, pp. 287–92). Indeed, I_ρ, which arises in the assumption of plane sections remain plane in bending, may also be regarded, when such an assumption is relaxed, as an empirical factor to be determined by comparison with exact solutions.

For a uniform beam free from lateral loadings, Eq. (21) can be written as

$$(23) \qquad \frac{\partial^4 y}{\partial x^4} - \left(\frac{1}{c_0^2} + \frac{1}{c_Q^2} \right) \frac{\partial^4 y}{\partial x^2 \partial t^2} + \frac{1}{c_0^2 c_Q^2} \frac{\partial^4 y}{\partial t^4} + \frac{1}{c_0^2 R^2} \frac{\partial^2 y}{\partial t^2} = 0,$$

where

$$(24) \qquad c_0^2 = \frac{EI}{I_\rho} = \frac{E}{\rho}, \qquad c_Q^2 = \frac{k'G}{\rho}, \qquad R^2 = \frac{I}{A}.$$

If the beam is of infinite length, a solution of the form (12) may be substituted into (23), and we see that the wave speed c must satisfy the equation

$$(25) \qquad 1 - \left(\frac{c^2}{c_0^2} + \frac{c^2}{c_Q^2} \right) + \frac{c^4}{c_0^2 c_Q^2} - \frac{c^2}{c_0^2 R^2} \left(\frac{\lambda}{2\pi} \right)^2 = 0.$$

The solution of this equation for c/c_0 versus λ yields two branches, corresponding to two *modes* of motion (two different shear-to-bending deflection ratios for the same wavelength). They are plotted in Fig 11.2:4 for the special case of a beam of circular cross section with radius a. The results of the exact solution of Pochhammer and Chree for Poisson's ratio $\nu = 0.29$ are also plotted there for comparison. It is seen that the Timoshenko theory agrees reasonably well with the exact theory in the first mode, but wide discrepancy occurs in the second mode. The approximate theory gives no information about higher modes: an infinite number of which exist in the exact theory.

The equations derived above are, of course, appropriate for the determination of the free-vibration modes and frequencies of a beam. The effects of rotary inertia and shear are unimportant if the wavelength of the vibration mode is large compared with the cross-sectional dimensions of the beam; but these effects become more and more important with a decrease of wavelength, i.e., with an increase in the frequency of vibration. In the example of a uniform beam with rectangular cross section and simply supported at both ends, with $E = \frac{8}{3}G$ and $k' = \frac{2}{3}$, we find that the shear deflection and rotary inertia reduce the natural frequencies. If the wavelength is ten times larger than the depth of the beam, the correction on the frequency due to rotary inertia alone is about 0.4 per cent, and the correction due to rotary inertia and shear together will be about 2 per cent. For a survey of literature, see Abramson, Plass, and Ripperger.[11.1]

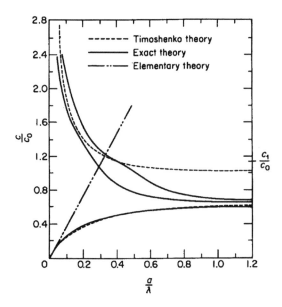

Fig. 11.2:4. Phase velocity curves for flexural elastic waves in a solid circular cylinder of radius a. (From Abramson,[11.1] *J. Acoust. Soc. Am.*, 1957.)

PROBLEMS

11.2.   Consider the free longitudinal vibration of a rod of uniform cross section and length L, as shown in Fig. P11.2. Let us assume that plane cross sections remain plane, that only axial stresses are present, being uniformly distributed over the cross section, and that radial displacements are negligible (i.e., the displacements consist of only one nonvanishing component u in the x-direction). Derive expressions for the potential and kinetic energies and show that the equation of motion is

$$(26) \qquad \frac{\partial^2 u}{\partial x^2} - \frac{1}{c_0^2} \frac{\partial^2 u}{\partial t^2} = 0, \qquad c_0^2 = \frac{E}{\rho}.$$

Show that the general solution is of the form

$$(27) \qquad u = f(x - c_0 t) + F(x + c_0 t),$$

where f and F are two arbitrary functions.

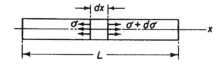

P11.2. Longitudinal vibration of a rod.

11.3. Consider the same problem as above, but now incorporate approximately the transverse inertia associated with the lateral expansion or contraction connected with axial compression and extension, respectively. Let the (Love's) assumption be made that the displacement in the radial direction v is proportional to the radial coordinate r, measured from a centroidal axis, and to the axial strain $\partial u/\partial x$; i.e.,

$$(28) \qquad v = -\nu r \frac{\partial u}{\partial x},$$

where ν is Poisson's ratio. Derive expressions of the kinetic and potential energies and obtain the equation of motion according to Hamilton's principle,

$$(29) \qquad \rho \left[\frac{\partial^2 u}{\partial t^2} - (\nu R)^2 \frac{\partial^4 u}{\partial x^2 \partial t^2} \right] - E \frac{\partial^2 u}{\partial x^2} = 0,$$

where R is the polar radius of gyration of the cross section. The natural boundary condition at the end $x = 0$, if that end is subjected to a stress $\sigma_0(t)$, is

$$(30) \qquad \rho \nu^2 R^2 \frac{\partial^3 u}{\partial x \partial t^2} + E \frac{\partial u}{\partial x} = \sigma_0(t) \qquad\qquad \text{at } x = 0.$$

Note: It is important to note that, according to the last equation, the familiar proportionality between axial stress σ and axial strain $\partial u/\partial x$ does not exist in this theory.

Comparison of the dispersion curves obtained from the elementary theory (Problem 11.2), the Love theory, the Pochhammer-Chree "exact" theory, and another approximate theory due to Mindlin and Herrmann,[11.1] are shown in

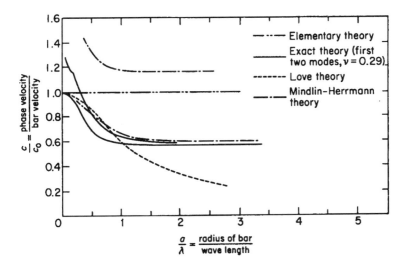

P11.3. Phase velocity curves for longitudinal elastic waves in a solid circular cylinder of radius a. (After Abramson *et al.*, *Adv. Applied Mech.*, 5, 1958.)

Fig. P11.3. The last-mentioned theory accounts for the strain energy associated with the transverse displacement v, of which the most important contribution comes from the shearing strain caused by the lateral expansion of the cross section near a wave front.

11.4. The method of derivation of the various forms of equations of motion of beams as presented above has the advantage of being straightforward, but it does not convey the physical concepts as clearly as in an elementary derivation. Hence, rederive the basic equations by considering the forces that act on an element of length dx, as shown in Fig. P11.2 and Fig. P11.4. Obtain the following equations, and then derive the wave equations by proper reductions.

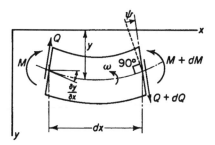

P11.4. Element of a beam in bending.

Longitudinal waves, elementary theory (Fig. P11.2):

$$(31) \qquad \frac{\partial \sigma_{xx}}{\partial x} = \rho \frac{\partial^2 u}{\partial t^2} \qquad \text{(equation of motion)},$$

$$(32) \qquad \frac{\partial^2 u}{\partial x \partial t} = \frac{\partial \epsilon_{xx}}{\partial t} \qquad \text{(equation of strain)},$$

$$(33) \qquad \sigma_{xx} = E \epsilon_{xx} \qquad \text{(equation of material behavior)},$$

where σ_{xx} = axial stress, ϵ_{xx} = axial strain, $\partial u/\partial t$ = axial particle velocity, x = axial coordinate, t = time, E = modulus of elasticity, and ρ = mass density.

Flexural waves, Timoshenko theory (Fig. P11.4):

$$(34) \qquad \frac{\partial M}{\partial x} - Q = \rho I \frac{\partial \omega}{\partial t} \quad \text{(rotational)} \left.\vphantom{\begin{array}{c}1\\1\\1\end{array}}\right\}$$

$$\text{(equations of motion)},$$

$$(35) \qquad \frac{\partial Q}{\partial x} = \rho A \frac{\partial v}{\partial t} \quad \text{(transverse)}$$

$$(36) \qquad \frac{\partial K}{\partial t} = \frac{\partial \omega}{\partial x} \qquad \text{(bending)} \left.\vphantom{\begin{array}{c}1\\1\\1\end{array}}\right\}$$

$$(37) \qquad \frac{\partial \beta}{\partial t} = \frac{\partial v}{\partial x} + \omega \quad \text{(shear)}$$

(38) $M = EIK$ (bending)

$$\left.\phantom{\begin{matrix}a\\a\end{matrix}}\right\} \text{(equations of material behavior)},$$

(39) $Q = GA_s\beta$ (shear)

where M = moment, Q = shear force, K = axial rate of change of section angle $= -\partial\psi/\partial x$, β = shear strain $= \partial y/\partial x - \psi$, ω = angular velocity of section $= -\partial\psi/\partial t$, v = transverse velocity $= \partial y/\partial t$, I = section moment of inertia, A = section area, and A_s = area parameter defined by $\int\int \gamma(z)dA = \beta A_s$ where $\gamma(z)$ is the shear strain at a point z in the cross section. G = shear modulus.

11.3. GROUP VELOCITY

Since we have been concerned in the preceding sections about wave propagations in beams, it seems appropriate to make a digression to explain the concept of *group velocity* as distinguished from the *phase velocity*. We have seen that for certain equations a solution of the following form exists:

$$(1) \qquad\qquad u = a\sin(\mu x - \nu t).$$

If x is increased by $2\pi/\mu$, or t by $2\pi/\nu$, the sine function takes the same value as before, so that $\lambda = 2\pi/\mu$ is the wavelength and $T = 2\pi/\nu$ is the period of oscillation. If $\mu x - \nu t = $ constant, i.e. $x = $ const. $+ \nu t/\mu$, the argument of the sine function remains constant in time; which means that the whole waveform is displaced towards the right with a velocity $c = \nu/\mu$. The quantity c is called the phase velocity, in terms of which Eq. (1) may be exhibited as

$$(2) \qquad\qquad u = a\sin\frac{2\pi}{\lambda}(x - ct).$$

If the phase velocity c depends on the wavelength λ, the wave is said to exhibit *dispersion*. Our examples in the previous section show that dispersion exists in both longitudinal and flexural waves in rods and beams.

What happens when two sine waves of the same amplitude but slightly different wavelengths and frequencies are superposed? Let these two waves be characterized by two sets of slightly different values μ, ν and μ', ν'. The resultant of the superposed waves is

$$u + u' = A[\sin(\mu x - \nu t) + \sin(\mu' x - \nu' t)].$$

Using the well-known formula

$$\sin\alpha + \sin\beta = 2\sin\frac{1}{2}(\alpha+\beta)\cos\frac{1}{2}(\alpha-\beta),$$

we have

(3)
$$u + u' = 2A \sin \left[\frac{1}{2}(\mu + \mu')x - \frac{1}{2}(\nu + \nu')t \right]$$

$$\times \cos \left[\frac{1}{2}(\mu - \mu')x - \frac{1}{2}(\nu - \nu')t \right].$$

This expression represents the well-known phenomenon of "beats." The sine factor represents a wave whose wave number and frequency are equal to the mean of μ, μ' and ν, ν', respectively. The cosine factor, which varies very slowly when $\mu - \mu'$, $\nu - \nu'$ are small, may be regarded as a varying amplitude, as shown in Fig. 11.3:1. The "wave group" ends wherever the

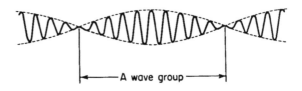

Fig. 11.3:1. An illustration of a wave group.

cosine factor becomes zero. The velocity of advance of these points is called the *group velocity*; its value U is equal to $(\nu - \nu')/(\mu - \mu')$. For long groups (or slow beats), the group velocity may be written with sufficient accuracy as

(4)
$$U = \frac{d\nu}{d\mu}.$$

In terms of the wavelength $\lambda(= 2\pi/\mu)$, we have

(5)
$$U = \frac{d(\mu c)}{d\mu} = c - \lambda \frac{dc}{d\lambda},$$

where c is the phase velocity.

From the fact that no energy can travel past the nodes, one can infer that the rate of transfer of energy is identical with the group velocity. This fact is capable of rigorous proof for single trains of waves.

The most familiar examples of propagation of wave groups are perhaps the water waves. It has often been noticed that when an isolated group of waves, of sensibly the same length, advancing over relatively deep water, the velocity of the group as a whole is less than that of the individual waves composing it. If attention is fixed on a particular wave, it is seen to advance through the group, gradually dying out as it approaches the front, while its

former place in the group is occupied in succession by other waves which
have come forward from the rear. Another familiar example is the wave
train set up by ships. The explanation as presented above seems to have
been first given by Stokes (1876). Other derivations and interpretations of
the group velocity concept can be found in Lamb, *Hydrodynamics*, (1879,
New York: Dover Pub.) Secs. 236 and 237.

If a concentrated lateral load is suddenly and impulsively applied on
an infinitely long beam, the disturbance is propagated out along the beam
by flexural waves. The initial loading may be regarded as composed of
an infinite number of sine-wave components of all wavelengths but of the
same amplitude, with proper phase relationship, so that they reinforce each
other in the limited region where the force is applied, but cancel each other
everywhere outside the region of load application. As time increases, these
sine waves propagate with their own phase velocities, and the pattern of
interference changes with time. Thus, at time t and at a point which
is at a distance x from the initial loading, only a group of waves in the
neighborhood of a specific wavelength can be seen, and the energy of this
wave group is propagated at the group velocity U.

From the dispersion curves of Fig. 11.2:4 for the phase velocity of flexural
elastic waves in a circular cylinder of radius a, the group wave velocity

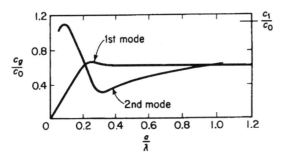

Fig. 11.3:2. Group velocity curves for flexural elastic waves in a solid circular cylinder
of radius a. (After Abramson, *J. Acoust. Soc. Am.*, 1957.)

curves of Fig. 11.3:2 for the two lowest modes can be constructed. These
curves are obtained by Abramson[11.1] for the Pochhammer-Chree theory
and are important in physical considerations. The number c_1 in Fig. 11.3:2
is the dilatational wave speed (see Sec. 7.8) given by the formula

$$c_1 = \sqrt{\frac{(1-\nu)E}{(1+\nu)(1-2\nu)\rho}}\,.$$

The number c_0 is $\sqrt{E/\rho}$. The group velocity is denoted by c_g. The ratio c_1/c_0 depends on the Poisson's ratio ν:

$$\nu = 0.25, \qquad \frac{c_1}{c_0} = 1.095,$$

$$\nu = 0.30, \qquad \frac{c_1}{c_0} = 1.16.$$

Note that the phase velocities in the higher branches exceed the dilatational wave velocity (at long wavelengths), but the group velocities do not. Similar group velocity curves can be constructed for longitudinal waves from Fig. P11.3. Note that the greatest possible velocity of energy transmission is the dilatational velocity, for both flexural and longitudinal waves.

From Fig. 11.3:2 for a circular cylinder of radius a, it can be seen that for the first branch the group velocity reaches a maximum when a/λ is between 0.25 and 0.30, where λ is the wavelength. Hence, when a pulse is propagated along a beam in accordance with the first branch, Fourier components of wavelengths about 0.25 or 0.30 times the radius will be found at the head of the pulse. A detailed study of the transmission of a pulse along the beam can be found in R. M. Davies' paper.[11.2]

We must remark that elastic substances are not inherently dispersive, but the dispersion effect appears in waves traveling through rods and beams because of wave reflections from the boundaries. Therefore, the dispersion in the case of an elastic medium is simply an interface phenomenon and not a physical property of the material. The phenomenon is analogous to that obtained in electrical wave guides and can be discussed in similar terms. In this light, we can answer the question of the existence of phase velocities greater than the dilatational velocity, although no disturbance can be propagated with a velocity greater than that. The point is that energy is propagated at the group velocity and that group velocities greater than the dilatational velocity will not occur. The phase velocities are merely velocities of propagation in the axial direction of the loci of constant phase in specific mode patterns and their values are thus not limited by physical considerations.

11.4. HOPKINSON'S EXPERIMENT

In Section 1.6, we mentioned that John Hopkinson published in 1872 an account of an interesting experiment, the explanation of which will help us understand the nature of elastic wave propagation and its significance

in engineering. Hopkinson wanted to measure
the strength of steel wires when they were suddenly stretched by a falling weight. A ball-shaped
weight pierced by a hole was threaded on the wire
and was dropped from a known height so that
it struck a clamp attached to the bottom of the
wire (Fig. 11.4:1). For a given weight he expected to find a critical height beyond which the
falling weight would break the wire. Using different weights dropped from different heights, however, Hopkinson found that the minimum height
from which a weight had to be dropped to break
the wire was nearly *independent* of the size of the
weight!

Fig. 11.4:1. A sketch of
Hopkinson's experiment.

Now, when different weights are dropped from a given height, the velocity reached at the end is independent of the size of the weight. Hopkinson's
result suggests that in breaking the wire it is the velocity of the loading end
that counts. Following this lead, Hopkinson explains his result on the basis
of elastic wave propagation (ignoring the effects of plasticity). Assuming
that the stress state in the wire is approximately one-dimensional, we see
from Eqs. (11.2:27), or (11.2:31)–(11.2:33), that the stress in the wire is
proportional to the particle velocity:

$$(1) \qquad\qquad \sigma_{xx} = \rho c_0 \frac{\partial u}{\partial t} \,,$$

where $c_0 = \sqrt{E/\rho}$ is the speed of sound of longitudinal waves in the wire.
For steel, c_0 is about 16,000 ft/sec. The largest particle velocity in the wire,
however, is not reached at the instant of the blow, but is reached at the
top after the elastic waves propagate up and down the wire several times.
When this largest particle velocity induces a stress equal to the ultimate
stress of the wire, the wire breaks.

When the weight hits the clamp in Hopkinson's test, the end of the wire
acquires a particle velocity V_0 equal to that of the weight and clamp. A
steep-fronted tension wave is generated and propagated up the wire. In the
meantime, the weight and the clamp are slowed down exponentially by the
tensile load imposed by the wire:

$$(2) \qquad\qquad M\frac{dV}{dt} = A\sigma_{xx} = A\rho c_0 V \,,$$

where M is the mass of the weight and clamp, V is its velocity, and A is the
cross-sectional area of the wire. The elastic wave, on reaching the fixed end
at the top, is reflected as a tension wave of twice the intensity of the incident

wave. The reflected wave, going through the tail of the incident pulse, is reflected again at the lower end as a compression wave, and so on. If the stress at the first reflection (equal to $2\rho c_0 V_0$), is sufficient to break the wire, then fracture will be expected to take place near the top. In a systematic test of gradually increasing the height of drop h (i.e., gradually increasing V_0), this breaking at the first reflection does not happen. In J. Hopkinson's experiments, the head of the stress waves was able to travel the length of wire several times before the weight M was decelerated sufficiently, and the stress wave pattern in the wire was very complicated. Bertram Hopkinson (1905), in repeating his father's experiment, used smaller weights so that the rate of decay was rapid. Nevertheless, as shown by G. I. Taylor[11.2] (1946), the maximum tensile stress in B. Hopkinson's experiment did not occur at the first reflection, when the stress was $2\rho c_0 V_0$, but at the third reflection, i.e., the second reflection at the top of the wire, when the tensile stress reached $2.15\rho c_0 V_0$.

To work out the details, we have to determine the function $f(x-c_0 t)$ and $F(x+c_0 t)$ of Eq. (11.2:27) to satisfy the initial conditions $u = \partial u/\partial t = 0$ at $t = 0$, and the boundary conditions $u = 0$ at the top; $\partial u/\partial t = V$, the velocity of the weight and clamp, at the bottom. The velocity V is determined by Eq. (2). The full solution is given in Taylor's paper.[11.2]

Incidentally, Hopkinson found that the tensile strength of steel wires under rapid loading is much greater than that under a static load.

11.5. GENERALIZED COORDINATES

We shall now turn our attention to the actual construction of the minimizing sequence in solving a variational problem. Our aim is to obtain efficient approximate solutions. For this purpose, we shall discuss first the idea of generalized coordinates and the definitions of the "best approximations."

Let us consider a function $u(x)$ of a real variable x, defined over an interval (a, b). If $u(x)$ can be represented by a power series

$$(1) \qquad u(x) = a_0 + a_1 x + a_2 x^2 + \cdots + a_n x^n + \cdots$$

or by a Fourier series

$$(2) \qquad u(x) = \sum_{n=1}^{\infty} b_n \sin nx$$

or, more generally, by a series of the form

$$(3) \qquad u(x) = \sum_{n=1}^{\infty} c_n \phi_n(x),$$

where $\{\phi_n(x)\}$ is a given set of functions, then $u(x)$ can be specified by the sets of coefficients $\{a_n\}$, $\{b_n\}$, or $\{c_n\}$. These coefficients may be regarded as the generalized coordinates of $u(x)$ referred to the bases $\{x^n\}$, $\{\sin nx\}$, $\{\phi_n\}$, respectively. In Eq. (3), each term $\phi_n(x)$ represents a degree-of-freedom, and the coefficient c_n specifies the extent to which $\phi_n(x)$ participates in the function $u(x)$. For example, if $u(x)$ represents a disturbed state of an elastic string which is anchored at the points $x = 0$ and $x = \pi$, then Eq. (3) states that $u(x)$ can be obtained by a superposition of successive normal modes $\phi_n = \sin nx$, and that the nth mode participates with an amplitude c_n.

This terminology is conventional when c_n is a function of time t. For example, in the dynamics of a vibrating string the displacement $u(x,t)$ may be written as

$$
(4) \qquad u(x,t) = \sum_{n=1}^{\infty} q_n(t) \sin nx \,.
$$

The quantity $q_n(t)$ is known as the generalized coordinate with respect to the mode $\sin nx$. The possibility of generalizing this terminology to cases in which $\mathbf{x}$ and $\mathbf{u}$ represent vectors in vector spaces of dimensions m and k, respectively, is obvious.

The fundamental idea of approximate methods of solution is to reduce a continuum problem (with infinite number of degrees-of-freedom) into one of a finite number of degrees-of-freedom. The reduced problem is solved by usual methods. Then a limiting process is used to extend the restricted solution to the solution of the original problem. In the following discussion we shall describe the ideas about the first step — reducing a given problem to one of finite degrees-of-freedom. The second step — the proof of the validity of the limiting process — is the main mathematical problem. Examples of the mathematical theories which deal with the second step are the theories of Taylor's series, Fourier series, eigenfunction expansions, etc. The theories are beyond the scope of our discussion, and the reader is referred to the Bibliography 11.3.

11.6. APPROXIMATE REPRESENTATION OF FUNCTIONS

Let a function $f(x_1, x_2, \ldots, x_n)$ of n variables $x_1, \ldots, x_n$ be defined in a domain $\mathcal{R}(x_1, x_2, \ldots, x_n)$; or, in brief notation, $f(x_j)$ in $\mathcal{R}(x_j)$. To represent $f(x_j)$ "as well as possible" by a linear combination of given functions $\phi_n(x_j), n = 1, 2, \ldots, N$, say, $Q_N(x_j)$, where

$$
(1) \qquad Q_N(x_j) = \sum_{n=1}^{N} a_n \phi_n(x_j) \,,
$$

the question arises as to the meaning of the term "as well as possible." As soon as the "best approximation" is defined, a definite procedure for calculating a_n can be devised.

The usual definitions of the "best approximation" may be classified into two classes.[†] In the first class the concept of *norm* is introduced. For any function $g(x_j)$ in the domain $\mathcal{R}(x_j)$, a real positive number, called the *norm of $g(x_j)$*, and denoted by $||g(x_j)||$, is defined, with the stipulation that $||g(x_j)|| = 0$ only for $g(x_j) \equiv 0$ in $\mathcal{R}(x_j)$. Then the best approximation is taken to mean that

$$(2) \qquad \left\| f(x_j) - \sum_{n=1}^{N} a_n \phi_n(x_j) \right\| = \min.$$

In the second class, a set of linear homogeneous expressions L_k, $k = 1, 2, \ldots, N$, defined for functions in the domain $\mathcal{R}(x_j)$, is selected, and the best approximation is taken to mean that

$$(3) \qquad L_k(\varepsilon_N) = 0, \qquad\qquad k = 1, 2, \ldots, N,$$

where ε_N is the "error"

$$(4) \qquad \varepsilon_N \equiv \varepsilon_N(x_j) \equiv f(x_j) - \sum_{n=1}^{N} a_n \phi_n(x_j).$$

Examples of the first class are

(A) *Absolute error method*: In this method the norm for any function $g(x_j)$ in $\mathcal{R}(x_j)$ is defined to be the maximum absolute value of $g(x_j)$ in $\mathcal{R}(x_j)$:

$$(5) \qquad ||g(x_j)|| = \max_{x \text{ in } \mathcal{R}} |g(x_j)|.$$

Hence, the best approximation is defined by requiring that

$$(6) \qquad ||\epsilon_N|| = \max_{x \text{ in } \mathcal{R}} |\epsilon_N(x_j)| = \min.$$

This method was used in the early nineteenth century, but is not in favor now because it is difficult to apply.

(B) *Least squares method*: In this method we take a positive integrable function $W(x_j)$ and define the norm

$$(7) \qquad ||g(x_j)|| = \sqrt{\int_{\mathcal{R}} W g^2 dv},$$

[†]We follow Collatz[11.3] (1955) in the subsequent exposition.

where $W(x_j)$ is called the *weighting function*. The volume integral, which may be defined in Riemann or Lebesgue sense, is assumed to exist, dv being a volume element. In this method, the constants a_n are to be determined from the equations

(8)
$$\frac{1}{2}\frac{\partial\|\epsilon_N\|^2}{\partial a_k} = -\int_{\mathcal{R}}\epsilon_N W\phi_k dv = 0, \qquad k = 1, 2, \ldots, N.$$

Examples of the second class are

(C) *Collocation method:* We take N points P_k in the domain $\mathcal{R}(x_j)$ and let

(9)
$$L_k(\epsilon_N) = \epsilon_N(P_k).$$

Hence Eq. (3) implies that the constants a_n are so chosen that $f(x_j)$ is represented exactly at N points. A substituting leads to the equations

(10)
$$\sum a_N \phi_N(P_k) = f(P_k), \qquad k = 1, 2, \ldots, N.$$

If the determinant Δ of coefficients of a_n does not vanish,

$$\Delta = \begin{vmatrix} \phi_1(P_1) & \cdots & \phi_1(P_N) \\ . & \cdots & . \\ . & \cdots & . \\ . & \cdots & . \\ \phi_N(P_1) & \cdots & \phi_N(P_N) \end{vmatrix} \neq 0,$$

the constants a_n can be uniquely determined.

(D) *Orthogonality method:* We choose a set of N linearly independent functions $g_k(x_j)$ and let

(11)
$$L_k(\epsilon_N) = \int_{\mathcal{R}}\epsilon_N g_k dv, \qquad k = 1, 2, \ldots, N.$$

If we choose $g_k = W\phi_k$, then we get the conditions

(12)
$$\int_{\mathcal{R}}\epsilon_N W\phi_k dv = 0$$

identical with the least squares method.

(E) *Subregion method:* As a generalization of the collocation method, we subdivide the region $\mathcal{R}(x_j)$ into N subregions $\mathcal{R}_1, \mathcal{R}_2, \ldots, \mathcal{R}_N$ and define

(13)
$$L_k(\epsilon) = \int_{\mathcal{R}_k}\epsilon dv, \qquad k = 1, \ldots, N.$$

then follow the steps of the collocation method.

11.7. APPROXIMATE SOLUTION OF DIFFERENTIAL EQUATIONS

If we try to approximate the solution $u(x_j)$ of a differential equation defined over a domain $\mathcal{R}(x_1, \ldots, x_N)$ with certain boundary conditions by an expression

$$(1) \qquad Q_N(x_j) = \sum_{n=1}^{N} a_n \phi_n(x_j) ,$$

we may take one of the following approaches: (A) each $\phi_n(x_j)$ satisfies the boundary conditions, (B) each $\phi_n(x_j)$ satisfies the differential equation. In either case an error function can be defined, and it is required to be "as small as possible."

For example, consider the case (A). Let the differential equation be

$$(2) \qquad \mathcal{L}\{u(x_j)\} = 0 .$$

A substitution of Eq. (1) into Eq. (2) gives the error

$$(3) \qquad \epsilon_N(x_j) \equiv \mathcal{L}\left\{ \sum a_n \phi_n(x_j) \right\} ,$$

which must be minimized. The definition of the "best approximation" can be varied, but all methods of the previous section can be used here.

A possible generalization of Eq. (1) to the form

$$Q_N(x_j) = w(x_1, \ldots, x_N; a_1, \ldots, a_N) ,$$

which depends on the N parameters a_n, seems evident, although a mathematical theory of such a generalization is by no means simple.

11.8. DIRECT METHODS OF VARIATIONAL CALCULUS

Ideas similar to those presented in the preceding sections can be applied to variational problems. Let $\mathcal{D}$ be the class of admissible functions $u(x_j)$ of n variables $x_1, x_2, \ldots, x_n$, or x_j for short, defined over a domain $\mathcal{R}(x_j)$ and satisfying all the "rigid" boundary conditions (cf. Sec. 10.8). Let $y(x_j)$, belonging to $\mathcal{D}$, minimize the functional $I[u]$. The following approximate representations of $y(x_j)$ are suggested:

Method of Finite Differences. We choose a set of points P_k, $k = 1, 2, \ldots, N$, in $\mathcal{R}$ and let

$$(1) \qquad y(P_k) = y_k , \qquad\qquad P_k \text{ in } \mathcal{R}, k = 1, 2, \ldots, N .$$

The derivatives of $y(x_j)$ are then replaced by expressions involving successive finite differences of $y_1, y_2, \ldots, y_N$, according to the calculus of finite differences, and the integrations are replaced by a finite summation. Then the functional $I[y]$ becomes a function $\Psi(y_1, y_2, \ldots, y_N)$. We choose the values $y_1, \ldots, y_N$ so that $\Psi(y_j)$ has an extremum; i.e.,

$$(2) \qquad \frac{\partial \Psi}{\partial y_k} = 0, \qquad\qquad k = 1, \ldots, N.$$

This method was used by Euler.

Rayleigh-Ritz Method. We set

$$(3) \qquad y_N(x_j) = \sum_{n=1}^{N} a_n \phi_n(x_j),$$

where $\phi_n(x_j)$ are known functions which are so chosen that $y_N(x_j)$ are admissible, i.e., they belong to the class $\mathcal{D}$. When Eq. (3) is substituted into $I[u]$, the latter becomes a function $\Psi(a_1, a_2, \ldots, a_N)$ of the coefficients a_j. These coefficients $a_1, \ldots, a_N$ are then so chosen that $\Psi(a_j)$ has an extremum; i.e.,

$$(4) \qquad \frac{\partial \Psi}{\partial a_j} = 0, \qquad\qquad j = 1, \ldots, N.$$

Kantorovich's Method. We choose a set of coordinate functions

$$\phi_1(x_1, x_2, \ldots, x_n), \phi_2(x_1, x_2, \ldots, x_n), \ldots, \phi_N(x_1, x_2, \ldots, x_n)$$

and try an approximate solution in the form

$$(5) \qquad y_N = \sum_{k=1}^{N} a_k(x_1) \phi_k(x_1, x_2, \ldots, x_n),$$

where the coefficients $a_k(x_1)$ are no longer constants, but are unknown functions of one of the independent variables. The functional $I[u]$ is reduced to a new functional $\Psi[a_1(x_1), a_2(x_1), \ldots, a_N(x_1)]$, which depends on N functions of one independent variable,

$$(6) \qquad a_1(x_1), a_2(x_1), \ldots, a_N(x_1).$$

The functions (6) must be chosen to minimize the functional Ψ (Sec. 10.3).

It should be recognized that Kantorovich's method is a generalization of a method commonly used in the classical treatment of small oscillations of an elastic body, in which the displacements are represented in the form

$$(7) \qquad u(x,t) = \sum_{k=1}^{N} q_k(t)\phi_k(x) ,$$

where $\phi_k(x)$, depending solely on the spatial coordinates, are the bases of the generalized coordinates. The Euler equations of the energy integral, according to Hamilton's principle, are the Lagrangian equations of motion.

Galerkin's Method. Galerkin's idea of minimization of errors by orthogonalizing with respect to a set of given functions is best illustrated by an example. Consider the simple beam discussed in Sec. 10.8. The minimization of the functional V, Eq. (10.8:3), leads to the variational equation

$$(8) \qquad 0 = \delta V = \int_0^l \left[\frac{d^2}{dx^2}\left(EI\frac{d^2y}{dx^2}\right) - p \right]\delta y \, dx$$

$$+ \left[EI\frac{d^2y}{dx^2} - M\right]\delta\left(\frac{dy}{dx}\right)\Bigg|_0^l - \left[\frac{d}{dx}\left(EI\frac{d^2y}{dx^2}\right) - Q\right]\delta y\Bigg|_0^l .$$

Here EI, and p are known functions of x; M, Q are unspecified if an end is "rigid" [where $\delta y = \delta(dy/dx) = 0$]; or are given numbers if an end is "natural" [where δy and $\delta(dy/dx)$ are arbitrary].

Let us consider a clamped-clamped beam (both ends "rigid"). If we assume

$$(9) \qquad y_N = \sum_{k=1}^{N} a_k\phi_k(x) ,$$

where

$$(10) \qquad \phi_k = \frac{d\phi_k}{dx} = 0 \quad \text{at } x = 0 \text{ and } x = l , \qquad\qquad k = 1, \ldots, N ,$$

then, in general, Eq. (8) cannot be satisfied. The quantity

$$(11) \qquad \frac{d^2}{dx^2}\left(EI\frac{d^2y_n}{dx^2}\right) - p = \epsilon_n(x)$$

may be called an *error*. Now we demand that the coefficients a_k be so chosen that Eq. (8) be satisfied when δy is identified with any of the functions $\phi_1(x), \phi_2(x), \ldots, \phi_n(x)$:

$$(12) \qquad \int_0^l \epsilon_n(x)\phi_k(x)dx = 0 , \qquad\qquad k = 1, \ldots, N .$$

In other words, we demand that the error be orthogonal to $\phi_k(x)$.

If some of the end conditions are "natural," we still represent the approximate solution in the form (9) and require $y_n(x)$ to satisfy only the "rigid" boundary conditions wherever they apply. In this case, Eq. (8) leads to the following in place of Eq. (12).

(13)
$$\int_0^l \epsilon_n(x)\phi_k(x)dx + \left[EI\frac{d^2y_n}{dx^2} - M\right]\frac{d\phi_k}{dx}\Big|_0^l$$

$$- \left[\frac{d}{dx}\left(EI\frac{d^2y_n}{dx^2}\right) - Q\right]\phi_k\Big|_0^l = 0, \qquad k = 1, 2, \ldots, N.$$

Equations (12) or (13) are sets of N linear equations to solve for the N unknowns constants $a_1, \ldots, a_n$.

Trefftz' Method. Consider again Eq. (8). In the Trefftz method, the aproximate solution (9) is so chosen that $\phi_k(x)$ satisfies the Euler equation

$$\frac{d^2}{dx^2}\left(EI\frac{d^2\phi_k}{dx^2}\right) - p = 0$$

but not necessarily the boundary conditions. We now select the coefficients a_k in such a way that Eq. (8) is satisfied if the variations δy were limited to the set of functions ϕ_k:

$$\left(EI\frac{d^2y_n}{dx^2} - M\right)\frac{d\phi_k}{dx}\Big|_0^l - \left[\frac{d}{dx}\left(EI\frac{d^2y_n}{dx^2}\right) - Q\right]\phi_k\Big|_0^l = 0.$$

PROBLEMS

11.5. Saint-Venant's problem of torsion of a shaft with a cross section occupying a region R on the x, y plane bounded by a curve (or curves) C, is discussed in Sec. 7.5. Show that the Saint-Venant's theory is equivalent to finding a function $\psi(x,y)$ (Prandtl's stress function, p. 167) which minimizes the functional

$$I = \iint_R \left[\left(\frac{\partial\psi}{\partial x}\right)^2 + \left(\frac{\partial\psi}{\partial y}\right)^2 - 4G\alpha\psi\right]dxdy,$$

under the boundary condition that $d\psi/ds = 0$ on C. If the region R is simply connected, we may take $\psi = 0$ on C. If R is multiply connected, we take ψ equal to a different constant on each boundary. The torque is then given by Eq. (7.5:24).

Consider a shaft of rectangular cross section. Obtain approximate solutions to the torsion problem by two different direct methods outlined in Sec. 11.8. Compare the values of the torsional rigidity of the shaft obtained in each method.

(Reference, Sokolnikoff[1.2], 2nd edition, Chapter 7, pp. 400, 416, 423, 430, 437, 442.)

11.6. Consider the pin-ended column described in Problem 10.11, p. 313. Assume that the column and the spring are massless but that a lumped mass M is situated on the column at the point A. At the instant of time $t = 0$, the column is plucked so that the deflection at A is $\Delta_0 \neq 0$. By Hamilton's principle, or otherwise, find the equation of motion of the mass. Integrate the equation of motion to determine the motion of the mass in the case in which the lateral spring is not there ($K = 0, \alpha = 0$).

11.7. Let the load P of Problems 10.11 and 11.6 be an oscillatory load,

$$P = P_0 + P_1 \cos \omega t.$$

Could the column become unstable (i.e, with the amplitude of motion unbounded as $t \to \infty$)? In particular, if $P_0 = P_{cr}$, the buckling load, is there a chance that by properly selecting P_1 and ω, the column remains stable? Discuss the case of no lateral spring ($K = 0$, $\alpha = 0$) in greater detail.

11.8. Consider longitudinal wave propagation in a slender rod which in various segments is made of different materials. It is desired to transmit the wave through the rod with as little distortion as possible. How should the material constants be matched? (This is a practical problem which often occurs in instrumentation, such as piezoelectric sensing, etc.) *Hint*: Consider transmission of harmonic waves without reflection at the interfaces between segments. If quantities pertaining to the nth segment are indicated by a subscript n, show that we must have $\rho_n c_n = $ const. for all n or, in another form, $E_n \rho_n = $ const.

11.9. Consider a beam on an elastic foundation. A traveling load moves at a constant speed over the beam. Determine the steady-state elastic responses of the beam.

11.10. Consider the longitudinal vibrations of a slender elastic rod of variable cross-sectional area $A(x)$. Determine the speed of propagation of longitudinal waves, assuming that the rod is so slender that the variation of the elastic displacement over the cross section is negligible. If one end of the beam is subjected to a harmonic longitudinal oscillation with the boundary condition $(u)_{x=0} = ae^{i\omega t}$, whereas the other end is free, determine the beam response for the following cases:

(1) $$A(x) = \text{const}.$$

(2) $$A(x) = A_0 x^2,\qquad\qquad (0 < a \leq x \leq b),$$

(3) $$A(x) = A_0 e^{2x/L},\qquad\qquad (0 \leq x \leq L).$$

12

ELASTICITY AND THERMODYNAMICS

In this chapter we shall consider what restrictions classical thermodynamics imposes on the theory of solids and what information concerning solid continua can be deduced from thermodynamics. We shall discuss also the question of existence and positive definiteness of the strain energy function which is of central importance in the theory of elasticity.

12.1. THE LAWS OF THERMODYNAMICS

A brief summary of the basic structure of the classical theory of thermodynamics is given in this section. We call a particular collection of matter which is being studied, a *system*. Only *closed* systems, i.e., systems which do not exchange matter with their surroundings, are considered here. Occasionally, a further restriction is made that no interactions between the system and its surroundings occur; the system is then said to be *isolated*.

Consider a given system. When all the information required for a complete characterization of the system for a purpose at hand is available, it will be said that the *state* of the system is known. For example, for a certain homogeneous elastic body at rest, a complete description of its thermodynamic state requires a specification of its material content, i.e., the quantity of each chemical substance contained; its geometry in the natural or unstrained state; its deviation from the natural state, or strain field; its stress field; and, if some physical properties depend on whether the body is hot or cold, one extra independent quantity which fixes the degree of hotness or coldness. These quantities are called *state variables*. If a certain state variable can be expressed as a single-valued function of a set of other state variables, then the functional relationship is said to be an *equation of state*, and the variable so described is called a *state function*. The selection of a particular set of independent state variables is important in each problem, but the choice is to a certain extent arbitrary.

If, for a given system, the values of the state variables are independent of time, the system is said to be in *thermodynamic equilibrium*. If the state variables vary with time, then the system is said to undergo a *process*. The number of state variables required to describe a process may be larger than that required to describe the system at thermodynamic equilibrium. For example, in describing the flow of a fluid we may need to know the viscosity.

The boundary or *wall* separating two systems is said to be *insulating* if it has the following property: if any system in complete internal equilibrium is completely surrounded by an insulating wall, then no change can be produced in the system by an external agency except by movement of the wall or by long-range forces such as gravitation. A system surrounded by an insulating wall is said to be *thermally insulated*, and any process taking place in the system is called *adiabatic.*

A system is said to be *homogeneous* if the state variables does not depend on space coordinates. The classical thermodynamics is concerned with the conditions of equilibrium within a homogeneous system or in a heterogeneous system which is composed of homogeneous parts ("phases"). A generalization to nonhomogeneous systems requires certain additional hypotheses which will be considered in Sec. 13.1. *In this chapter we shall restrict our attention to homogeneous systems.*

The first step in the formulation of thermodynamics is to introduce the concept of temperature. It is postulated that *if two systems are each in thermal equilibrium with a third system, they are in thermal equilibrium with each other.* From this it can be shown to follow that the condition of thermal equilibrium between several systems is the equality of a certain single-valued function of the thermodynamic states of the systems, which may be called the temperature $\mathscr{T}$: any one of the systems being used as a "thermometer" reading the temperature $\mathscr{T}$ on a suitable scale. The temperature whose existence is thus postulated is measured on a scale which is determined by the arbitrary choice of the thermometer and is called the *empirical temperature.*

The *first law of thermodynamics* can be formulated as follows. *If a thermally insulated system can be taken from a state I to a state II by alternative paths, the work done on the system has the same value for every such (adiabatic) path.* From this we can deduce that there exists a single-valued function of the state of a system, called its *energy,* such that for an adiabatic process the increase of the energy is equal to the work done on the system. Thus,

(1) Δ energy = work done   (adiabatic process) .

It is to be noticed that for this definition of energy it is necessary and sufficient that it be possible by an adiabatic process to change the system either from state I to state II or from state II to state I.

We now define the heat Q absorbed by a system as the increase in energy of the system, less the work done on the system. Thus,

$$(2) \qquad Q = \Delta \text{ energy} - \text{work done} \qquad (\text{all processes}),$$

or

$$(3) \qquad \Delta \text{ energy} = Q + \text{work done}.$$

If this is regarded as a statement of the conservation of energy and is compared with (1), we observe that the energy of a system can be increased either by work done on it or by absorption of heat.

It is customary to identify several types of energy which make up the total: the kinetic energy K, the gravitational energy G, and the internal energy E. Thus, the equation

$$(4) \qquad \text{energy} = K + G + E$$

may be regarded as the definition of the internal energy.

We now formulate the *second law of thermodynamics for a homogeneous system* as follows.

There exist two single-valued functions of state, T, called the absolute temperature, and S, called the entropy, with the following properties:

I. *T is a positive number which is a function of the empirical temperature $\mathscr{T}$ only.*

II. *The entropy of a system is equal to the sum of entropies of its parts.*

III. *The entropy of a system can change in two distinct ways, namely, by interaction with the surroundings and by changes taking place inside the system. Symbolically, we may write*

$$(5) \qquad dS = d_e S + d_i S,$$

where dS denotes the increase of entropy of the system, $d_e S$ denotes the part of this increase due to interaction with the surroundings, and $d_i S$ denotes the part of this increase due to changes taking place inside the system. Then, if dQ denotes the heat absorbed by the system from its surroundings, we have

$$(6) \qquad d_e S = \frac{dQ}{T}.$$

The change $d_i S$ is never negative. If $d_i S$ is zero the process is said to be reversible. If $d_i S$ is positive the process is said to be irreversible:

$$d_i S > 0 \qquad \text{(irreversible process)},$$

(7)

$$d_i S = 0 \qquad \text{(reversible process)}.$$

The remaining case, $d_i S < 0$, *never occurs in nature.*

The absolute temperature T and the entropy S are two fundamental quantities. We will not attempt to define them in terms of other quantities regarded as simpler. They are defined merely by their properties expressed in the second law. Lord Kelvin has shown how it is possible to calibrate any thermometer into absolute temperature scale. (See Problem 12.3.) The scale of the absolute temperature is fixed by defining the temperature at equilibrium between liquid water and ice at a pressure of 1 atm at 273.16°K, i.e., 273.16 degrees on Kelvin's scale.

These postulates form the basis of the classical thermodynamics. The justification of these postulates is the empirical fact that all conclusions derived from these assumptions are without exception in agreement with the experimentally observed behavior of systems in nature at the macroscopic scale. The form in which these principles are enunciated above is essentially that used by Max Born[12.1] (1921) and I. Prigogine[13.1] (1947).

It is important to familiarize ourselves with the concept that entropy is an attribute to a material body, just as its mass or its electric charges are. A pound of oxygen has a definite amount of entropy, which can be changed by changing the temperature and specific volume of the gas. A pound of steel at a given temperature and in a given state of strain has a definite amount of entropy. It is instructive to consider Problems 12.5 and 12.6 to obtain some intuitive feeling about entropy.

PROBLEMS

12.1. Calculate the work performed by 10 g of hydrogen expanding isothermally at 20°C from 1 to 0.5 atm of pressure. *Note:* The equation of state for a system composed of m grams of a gas whose molecular weight is M is given approximately by $pv = (m/M)RT$, where R is a universal constant (same for all gases); $R = 8.314 \times 10^7$ ergs/degree, or 1.986 cal/degrees.)

12.2. Calculate the energy variation of a system which performs 8×10^8 ergs of work and absorbs 60 cal of heat. (*Note:* 1 cal $= 4.185 \times 10^7$ ergs.)

12.3. To calibrate an empirical temperature scale against the absolute temperature scale, let us consider a heat engine. A working substance is assumed which undergoes a cycle of operations involving changes of temperature, volume, and pressure and which is eventually brought back to its original condition. In a Carnot cycle, an ideal gas is first brought to contact with a large heat source which is maintained at a constant empirical temperature $\mathscr{T}_1$. The gas at the equilibrium temperature $\mathcal{T}_1$, is allowed to expand (isothermally) and absorbs an amount of heat Q_1, from the heat source. Then it is removed and insulated from the heat source and is allowed to expand adiabatically to perform work, lowering its empirical temperature to a value $\mathcal{T}_2$. Next it is put in contact with a heat sink of temperature $\mathcal{T}_2$ and is compressed isothermally to release an amount of heat Q_2. Finally, it is recompressed adiabatically back to its original state. If every process in the Carnot cycle is reversible, the engine is said to be reversible, otherwise it is irreversible.

From the second law of thermodynamics prove that:

(a) When the temperatures of the source and sink are given, a reversible engine is the most efficient type of engine, i.e., it yields the maximum amount of mechanical work for a given abstraction of heat from the source. *Note*: Efficiency $= (Q_1 - Q_2)/Q_1$.

(b) The efficiency of a reversible engine is independent of the construction or material of the engine and depends only on the temperatures of the source and sink.

(c) In a reversible Carnot engine, we must have

$$\frac{Q_2}{Q_1} = f(\mathcal{T}_1, \mathcal{T}_2),$$

where $f(\mathcal{T}_1, \mathcal{T}_2)$ is a universal function of the two temperatures $\mathcal{T}_1$ and $\mathcal{T}_2$. Show that the function $f(\mathcal{T}_1, \mathcal{T}_2)$ has the property

$$f(\mathcal{T}_1, \mathcal{T}_2) = \frac{f(\mathcal{T}_0, \mathcal{T}_2)}{f(\mathcal{T}_0, \mathcal{T}_1)},$$

where $\mathcal{T}_0, \mathcal{T}_1$, and $\mathcal{T}_2$ are three arbitrary temperatures. If $\mathcal{T}_0$ is kept constant, we may consider $f(\mathcal{T}_0, \mathcal{T})$ as being a function of the temperature $\mathcal{T}$ only. Hence, we may write

$$K f(\mathcal{T}_0, \mathcal{T}) = \theta(\mathcal{T}),$$

where K is a constant. Thus,

$$\frac{Q_1}{\theta(\mathcal{T}_1)} = \frac{Q_2}{\theta(\mathcal{T}_2)}.$$

Lord Kelvin perceived (1848) that the numbers $\theta(\mathcal{T}_1)$, $\theta(\mathcal{T}_2)$, may be taken to define an absolute temperature scale. Identify this with the T introduced above.

(d) It is suggested that the reader review the theory of the Joule-Thomson experiment, which is used in practice to calibrate a gas thermometer to absolute temperature scale.

12.4. By considering an ideal gas as the working fluid of a Carnot cycle, show that the gas thermometer reads exactly the absolute temperature.

12.5. Derive the following formula for the entropy of one mole of an ideal gas.

$$\mathcal{S} = C_v \log T + R \log V + a,$$

where C_v is the specific heat at constant volume, R is the universal gas constant, and a is a constant of integration.

12.6. Prove the following analogy. The work that one can derive from a waterfall is proportional to the product of the height of the waterfall and the quantity of the water flow. The work that can be derived from a heat engine performing a Carnot cycle is proportional to the product of the temperature difference between the heat source and the heat sink, and the quantity of entropy flow.

12.2. ENERGY EQUATION

We shall now express the equation of balance of energy, Eqs. (12.1:3) and (12.1:4), in a form more convenient for continuum mechanics. As an exception in the present chapter, this section is not limited to homogeneous systems.

For a body of particles in a configuration occupying a region V bounded by a surface S, the *kinetic energy* is

$$(1) \qquad\qquad K = \int_V \frac{1}{2} \rho v_i v_i dv \,,$$

where v_i are the components of the velocity vector of a particle occupying an element of volume dv, and ρ is the density of the material. The *internal energy* is written in the form

$$(2) \qquad\qquad E = \int_V \rho \mathscr{E} dv \,,$$

where $\mathscr{E}$ is the *internal energy per unit mass.*[†] The heat input into the body must be imparted through the boundary. A new vector, the *heat flux* **h**, with components h_1, h_2, h_3, is defined as follows. Let dS be a surface element in the body, with unit outer normal ν_i. Then the rate at which heat is transmitted across the surface dS in the direction of ν_i is assumed to be representable as $h_i \nu_i dS$. To be specific, we insist on defining the heat flux in the case of a moving medium that the surface element dS be composed of the same particles. The rate of heat input is therefore

$$(3) \qquad\qquad \dot{Q} = -\int_S h_i \nu_i dS = -\int_V h_{j,j} dv$$

[†] It is interesting to verify that if one chooses to use the internal energy *per unit volume* $\mathscr{E}_v$, instead of $\mathscr{E}$ per unit mass, so that $E = \int \mathscr{E}_v dv$ then the energy equation, corresponding to Eq. (7) infra, will be somewhat more complicated.

The rate at which work is done on the body by the body force per unit volume F_i in V and surface traction $\overset{\nu}{T}_i$ in S is the *power*

$$(4) \qquad P = \int F_i v_i dv + \int \overset{\nu}{T}_i\, v_i dS$$

$$= \int F_i v_i dv + \int \sigma_{ij}\nu_j v_i dS$$

$$= \int F_i v_i dv + \int (\sigma_{ij} v_i)_{,j} dv\,.$$

The first law states that

$$(5) \qquad \dot{K} + \dot{E} = \dot{Q} + P\,,$$

where the dot denotes the material derivative D/Dt.

Formula for computing the material derivative of an integral has been given in Eq. (5.3:4). A little calculation yields

$$(6) \qquad \frac{1}{2}\rho\frac{Dv^2}{Dt} + \frac{v^2}{2}\frac{D\rho}{Dt} + \frac{v^2}{2}\rho\operatorname{div}\mathbf{v} + \rho\frac{D\mathscr{E}}{Dt} + \mathscr{E}\frac{D\rho}{Dt} + \mathscr{E}\rho\operatorname{div}\mathbf{v}$$

$$= -h_{j,j} + F_i v_i + \sigma_{ij,j} v_i + \sigma_{ij} v_{i,j}\,.$$

On substituting the equation of continuity and the equations of motion

$$\frac{D\rho}{Dt} + \rho\operatorname{div}\mathbf{v} = 0\,, \qquad \rho\frac{Dv_i}{Dt} = F_i + \sigma_{ij,j}\,,$$

respectively, into Eq. (6), we obtain

$$(7) \quad \blacktriangle \qquad \rho\frac{D\mathscr{E}}{Dt} = -\frac{\partial h_j}{\partial x_j} + \sigma_{ij}V_{ij}$$

where

$$(8) \qquad V_{ij} = \frac{1}{2}\left(\frac{\partial v_i}{\partial x_j} + \frac{\partial v_j}{\partial x_i}\right)$$

is the symmetric part of the tensor $v_{i,j}$, called the *rate of deformation tensor*. The antisymmetric part of $v_{i,j}$ contributes nothing to the sum $\sigma_{ij}v_{i,j}$ since σ_{ij} is symmetric.

In classical thermodynamics we are concerned with the small neighborhood of thermodynamic equilibrium. It suffices in the present chapter to consider infinitesimal strain, which is imposed very slowly. In this case we may write Eq. (7) in a form commonly used in thermodynamics books:

$$(9) \qquad \rho d\mathscr{E} = dQ + \sigma_{ij}de_{ij}$$

Our Eq. (7) gives precise meaning to Eq. (9). If there is no internal entropy production in the process, so that $d_i\mathscr{S} = 0$, then the second law gives $dQ = T\rho d\mathscr{S}$, where $\mathscr{S}$ denotes the *specific entropy*, or the *entropy per unit mass*. Hence,

$$(10) \quad \blacktriangle \qquad d\mathscr{E} = Td\mathscr{S} + \frac{1}{\rho}\sigma_{ij}de_{ij}$$

Problem 12.7. In the case of a fluid under hydrostatic pressure, so that

$$\sigma_{11} = \sigma_{22} = \sigma_{33} = -p\,, \qquad \sigma_{12} = \sigma_{23} = \sigma_{31} = 0\,,$$

show that

$$(11) \qquad d\mathscr{E} = Td\mathscr{S} - pd\mathscr{V}\,,$$

where $\mathscr{V} = 1/\rho$ is the specific volume, i.e., the volume per unit mass of the fluid.

12.3. THE STRAIN ENERGY FUNCTION

In the theory of elasticity, it is often desired to define a function $W(e_{11}, e_{12}, \ldots)$ of the strain components e_{ij}, with the property that[†]

$$(1) \qquad \frac{\partial W}{\partial e_{ij}} = \sigma_{ij} \qquad\qquad (i,j = 1,2,3)\,.$$

Such a function W is called the *strain energy function*. We shall now identify W with the internal energy in an isentropic process and the free energy in all isothermal process.

By definition, the state of stress in an elastic body is a unique function of the strain, and vice versa. Hence, it is sufficient to choose one of these two tensors, e_{ij} and σ_{ij}, as an independent state variable. Now, Eq. (12.2:10),

$$(2a) \qquad d\mathscr{E} = \frac{1}{\rho}\sigma_{ij}de_{ij} + Td\mathscr{S}\,,$$

shows that $\mathscr{E}$ is a state function, $\mathscr{E}(e_{ij}, \mathscr{S})$, of the strain e_{ij} and the entropy per unit mass $\mathscr{S}$. By ordinary rules of differentiation, we have

$$(2b) \qquad d\mathscr{E} = \left(\frac{\partial\mathscr{E}}{\partial\mathscr{S}}\right)_{e_{ij}} d\mathscr{S} + \left(\frac{\partial\mathscr{E}}{\partial e_{ij}}\right)_{\mathscr{S}} de_{ij}\,.$$

[†] We limit our attention to infinitesimal deformations in this chapter. Then e_{ij} is the infinitesimal strain tensor. For finite deformations, see Chapter 16, especially Section 16.11.

On comparing (2a) and (2b), we see that

$$(3) \qquad \rho\left(\frac{\partial \mathscr{E}}{\partial e_{ij}}\right)_{\mathscr{S}} = \sigma_{ij}, \qquad \left(\frac{\partial \mathscr{E}}{\partial \mathscr{S}}\right)_{e_{ij}} = T.$$

Since in infinitesimal deformations the density ρ remains constant, Eq. (3) states that in a reversible adiabatic process there exists a scalar function $\rho\mathscr{E}$ whose partial derivative with respect to a strain component gives the corresponding stress components.[†]

On the other hand, if the process is isothermal ($T = $ const.), we introduce the *Helmholtz' free energy function* per unit mass (also called *Gibbs' work function*) $\mathscr{F}$:

$$(4) \qquad \mathscr{F} \equiv \mathscr{E} - T\mathscr{S}.$$

Then, from Eqs. (4) and (2a), we obtain

$$(5) \qquad d\mathscr{F} = -\mathscr{S}dT + \frac{1}{\rho}\sigma_{ij}de_{ij},$$

whence

$$(6) \qquad \rho\left(\frac{\partial \mathscr{F}}{\partial e_{ij}}\right)_{T} = \sigma_{ij}, \qquad \left(\frac{\partial \mathscr{F}}{\partial T}\right)_{e_{ij}} = -\mathscr{S}.$$

Hence, in an isothermal process there also exists a scalar function, $\rho\mathscr{F}$, whose partial derivative with respect to a strain component gives the corresponding stress component in an elastic body.

Comparing Eq. (1) with Eq. (3) or (6), we see that W can be identified with $\rho\mathscr{E}$ or $\rho\mathscr{F}$, depending on whether an isentropic or an isothermal condition is being considered.

A counterpart to Eq. (6), namely,

$$(7) \qquad \rho\left(\frac{\partial \Phi}{\partial \sigma_{ij}}\right)_{T} = -e_{ij},$$

is obtained from the *Gibbs thermodynamic potential* Φ, which is defined as

$$(8) \qquad \Phi = \mathscr{E} - T\mathscr{S} - \frac{1}{\rho}\sigma_{ij}e_{ij} = \mathscr{F} - \frac{1}{\rho}\sigma_{ij}e_{ij}.$$

For, on differentiating Eq. (8) and using Eq. (5), we have

$$(9) \qquad d\Phi = -\mathscr{S}dT - \frac{1}{\rho}e_{ij}d\sigma_{ij},$$

[†]In the corresponding Eq. (16.7:1) for finite deformation, the density ρ_0 is the uniform density at the natural state of the material.

from which Eq. (7) follows. Thus, $-\rho\Phi$ may be identified with the *complementary energy function* W_c, which has the property that

$$(10) \qquad \frac{\partial W_c}{\partial \sigma_{ij}} = e_{ij} .$$

We have identified the strain energy function for two well-known thermodynamic processes — adiabatic and isothermal. For these two processes, no explicit display of temperature in the function W is necessary. In other words, the stress-strain relation can be written without reference to temperature. In other thermomechanical processes the situation is not so simple; explicit display of temperature is required. In general, W is a function of e_{ij} and T. A simple example is discussed in Sec. 12.7.

12.4. CONDITIONS OF THERMODYNAMIC EQUILIBRIUM

The second law of thermodynamics connects the theory with nature in the statement that the case $d_i\mathscr{S} < 0$ never occurs in nature. It states a direction of motion of events in the world. According to the proverb, "Time is an arrow;" the second law endows the arrow to time.

Thus, the second law indicates a trend in nature. What is the end point of this trend? It is thermodynamic equilibrium. In Sec. 12.1, we have defined thermodynamic equilibrium as a situation in which no state variable varies with time. For a system to be in thermodynamic equilibrium no change in boundary conditions is permissible and no spontaneous process which is consistent with the boundary conditions will occur in it.

To derive the *necessary and sufficient conditions for thermodynamic equilibrium*, we compare the system in equilibrium with those neighboring systems whose state variables are slightly different from those of the equilibrium state, but all subjected to the same boundary conditions. This comparison is to be made in the sense of variational calculus. We shall write δT, $\delta\mathscr{S}$, etc., to denote the first-order infinitesimals, and ΔT, $\Delta\mathscr{S}$, etc., to denote the variations including the first-, second-, and higher-order infinitesimal terms. Thus, if we consider a homogeneous system A and a neighboring state B, we have the following variables:

$$\begin{array}{cl} \textit{System A} & \textit{System B} \\[1em] \textit{In equilibrium} & \textit{Neighboring to A} \\[1em] \mathscr{E}, e_{ij}, \sigma_{ij}, \mathscr{S}, T & \mathscr{E} + \Delta\mathscr{E}, e_{ij} + \Delta e_{ij}, \sigma_{ij} + \Delta\sigma_{ij} \\[1em] & \mathscr{S} + \Delta\mathscr{S},\; T + \Delta T . \end{array}$$

We shall require that the first law be satisfied so that all variations must be subjected to the restriction

$$(1) \qquad \Delta \mathscr{E} = \Delta Q + \frac{1}{\rho}(\sigma_{ij} + \Delta \sigma_{ij})\Delta e_{ij}\,, \qquad \delta \mathscr{E} = \delta Q + \frac{1}{\rho}\sigma_{ij}\delta e_{ij}\,.$$

Let us consider the following particular case. We require the system B to have the same energy as A and that the boundaries are rigid so that no work can be done; thus $\Delta \mathscr{E} = 0$, $\Delta e_{ij} = 0$. It follows from (1) that $\Delta Q = 0$ too. Now system B has the entropy $\mathscr{S} + \Delta \mathscr{S}$. If any spontaneous changes can occur in B at all, it would occur in the direction of increasing entropy. Therefore, if $\Delta \mathscr{S} < 0$, we must say that the second law would permit system B to change itself into the state of A. If $\Delta \mathscr{S} > 0$, no such change is permitted. The second law does not say that if $\Delta \mathscr{S} < 0$, the change must occur. But if *we make an additional assumption — as an addition to the second law — that which is permitted by the second law will actually occur in nature,* then $\Delta \mathscr{S} < 0$ necessarily implies that B will change itself into A. On the other hand, if $\Delta \mathscr{S} < 0$, the second law prohibits A from changing into B. Therefore, *if energy variation is prohibited the necessary and sufficient condition for the system A to be in the state of thermodynamic equilibrium is*

$$(2) \quad \blacktriangle \qquad\qquad (\Delta \mathscr{S})_{\mathscr{E}} < 0\,.$$

In other words, the entropy $\mathscr{S}$ of the system A is the maximum with respect to all the neighboring states which have the same energy. Hence, the variational equation

$$(3) \quad \blacktriangle \qquad\qquad \mathscr{S} = \max. \qquad\qquad (\mathscr{E} = \text{const.}, e_{ij} = \text{const.})$$

for thermodynamic equilibrium. If we restrict ourselves to the first-order infinitesimals, the condition for thermodynamic equilibrium may be written

$$(4) \quad \blacktriangle \qquad\qquad (\delta \mathscr{S})_{\mathscr{E}} \leq 0\,.$$

This is the celebrated condition of equilibrium of Willard Gibbs. Gibbs gave an alternative statement, which is more directly applicable to continuum mechanics: *for the equilibrium of any isolated system it is necessary and sufficient that in all possible variations in the state of the system which do not alter its entropy, the variation of its energy shall either vanish or be positive:*

$$(5) \quad \blacktriangle \qquad\qquad (\Delta \mathscr{E})_{\mathscr{S}} > 0\,, \qquad (\delta \mathscr{E})_{\mathscr{S}} \geq 0\,.$$

In other words, *the internal energy shall be a minimum with respect to all neighboring states which have the same entropy.*

Gibbs proves the equivalence of conditions (2) and (5) from the consideration that it is always possible to increase both the energy and the entropy of a system, or to decrease both together, viz., by imparting heat to any part of the system or by taking it away. Now, if the condition (2) is not satisfied, there must be some neighboring systems for which

$$\Delta \mathscr{S} > 0 \quad \text{and} \quad \Delta \mathscr{E} = 0 \,.$$

Then, by diminishing both the energy and the entropy of the system in this varied state, we shall obtain a state for which

$$\Delta \mathscr{S} = 0 \quad \text{and} \quad \Delta \mathscr{E} < 0 \,;$$

therefore, (5) is not satisfied. Conversely, if condition (5) is not satisfied, there must be a variation in the state of the system for which

$$\Delta \mathscr{E} < 0 \quad \text{and} \quad \Delta \mathscr{S} = 0 \,.$$

Hence, there must also be one for which

$$\Delta \mathscr{E} = 0 \quad \text{and} \quad \Delta \mathscr{S} > 0 \,.$$

Therefore, condition (2) is not satisfied. Taken together, the equivalence of Eqs. (2) and (5) is established.

The derivation of Gibbs' conditions for thermodynamic equilibrium requires a little more than the first and second laws of thermodynamics. The conditions also guarantee more than equilibrium in the ordinary sense — they guarantee *stable* equilibrium, stable in the sense that a neighboring disturbed state will actually tend to return to the equilibrium state.

12.5. THE POSITIVE DEFINITENESS OF THE STRAIN ENERGY FUNCTION

A piece of steel can exist for a long time without any spontaneous changes taking place in it. Thus, it is in a stable thermodynamic equilibrium condition, or said to be in a *zero-stress state*. When we strain a piece of steel with an application of forces, the strained body is in a state disturbed from the zero-stress one. If the metal remains elastic, the strained body will return to its zero-stress state upon removal of the applied forces. This tendency to return to the zero-stress state endows a property to a strain energy function which we shall now investigate.

Since the zero-stress state of a solid body is a stable equilibrium state, it follows from Eq. (12.4:5) that at constant entropy the strain energy must be a minimum. Consider now a strained solid. Let the internal energy of a strained body be $\mathscr{E}$ and that of an unstrained body be $\mathscr{E}_0$. According to Gibbs' theorem, therefore, the stability of the unstrained state implies that the difference $\Delta\mathscr{E} = \mathscr{E} - \mathscr{E}_0$ is positive, and it vanishes only in the trivial case of the zero-stress state itself. Therefore, in isentropic processes

$$(1) \qquad\qquad \Delta\mathscr{E} = \mathscr{E} - \mathscr{E}_0 \geq 0$$

in the neighborhood of the zero-stress state; the equality sign holds only in the zero-stress state. In other words, $\mathscr{E} - \mathscr{E}_0$, is *positive definite*.

In the case of an isothermal process, it can be deduced from the Helmholtz free energy function that $\mathscr{F}$ attains its minimum $\mathscr{F}_0$, at thermodynamic equilibrium. A similar consideration as above shows that $\Delta\mathscr{F} = \mathscr{F} - \mathscr{F}_0$ is positive definite in the neighborhood of the zero-stress state in the isothermal condition.

In either the adiabatic or the isothermal condition, the strain energy function W can be identified as $\rho(\mathscr{E} - \mathscr{E}_0)$ or $\rho(\mathscr{F} - \mathscr{F}_0)$, respectively. Therefore, since $\rho > 0$, the strain energy function W is positive definite in the neighborhood of the zero-stress state.

From the positive definiteness of the strain energy function, the following important theorems are deduced.

1. The uniqueness of the solution in elastostatics and dynamics. (Sec. 7.4.)
2. The minimum potential energy theorem. (Sec. 10.7.)
3. The minimum complementary energy theorem. (Sec. 10.9.)
4. Saint-Venant's principle (in a certain sense). (Sec. 10.11.)

12.6. THERMODYNAMIC RESTRICTIONS ON THE STRESS-STRAIN LAW OF AN ISOTROPIC ELASTIC MATERIAL

We shall consider those restrictions imposed by thermodynamics on the stress-strain law proposed in Sec. 6.2, which is supposed to hold in isentropic or isothermal conditions.

By definition of the strain energy function, $dW = \sigma_{ij}de_{ij}$. If we substitute the stress-strain law Eq. (6.2:7) into this equation, we have

$$(1) \qquad\qquad dW = (\lambda e\delta_{ij} + 2Ge_{ij})de_{ij}$$

$$= \lambda e\,de + 2Ge_{ij}de_{ij}, \qquad\qquad e = e_{\alpha\alpha}.$$

The elastic constants λ and G differ slightly in adiabatic and isothermal conditions (Sec. 14.3), but this difference is of no concern here. On integrating Eq. (1) from the zero-stress to the strained state, we have

(2) $$W = \int_0^e \lambda e \, de + \int_0^{e_{ij}} 2Ge_{ij} \, de_{ij} \, ;$$

i.e.,

(3) $$W = \frac{\lambda e^2}{2} + Ge_{ij}e_{ij} \, .$$

This function is required to be positive definite from thermodynamic considerations. Since all terms appear as squares, an obvious set of sufficient conditions are

(4) $$\lambda > 0 \, , \qquad G > 0 \, .$$

But this condition is not necessary because the term $e_{ij}e_{ij}$ is not independent of e. Therefore, let us introduce the strain deviation $e'_{ij} = e_{ij} - \frac{1}{3}e\delta_{ij}$, whose first invariant vanishes, and rewrite Eq. (3) in the form

(5) $$W = \frac{1}{2}\left(\lambda + \frac{2}{3}G\right)e^2 + G\left(e_{ij} - \frac{1}{3}e\delta_{ij}\right)\left(e_{ij} - \frac{1}{3}e\delta_{ij}\right),$$

On recognizing that $\lambda + \frac{2}{3}G$ is the bulk modulus K and $\frac{1}{2}(e_{ij} - \frac{1}{3}e\delta_{ij})^2$ is the second invariant of the strain deviation e'_{ij},

(6) $$J_2 = \frac{1}{2}e'_{ij}e'_{ij}$$

we see that Eq. (5) may be written as

(7) $$W = \frac{1}{2}Ke^2 + 2GJ_2 \, .$$

The two groups of square terms are now independent of each other. Hence, the necessary and sufficient condition for W to be positive definite is

(8) $$K = \lambda + \frac{2}{3}G > 0, \quad G > 0 \, .$$

Since K and G are related to Young's modulus E and Poisson's ratio ν by

(9) $$K = \frac{E}{3(1 - 2\nu)}, \qquad G = \frac{E}{2(1 + \nu)},$$

an equivalent set of conditions are

(10) $$E > 0, \qquad -1 < \nu < \frac{1}{2} \, .$$

This result is obtained under the assumption that the stress-strain relationship is linear and isotropic, that the strain energy function is positive definite, and, tacitly, that the material is homogeneous and without hierarchical internal structure. The values of E and ν of some common materials are given in Sec. 6.2, Table 6.2:1. Common steels have ν about 0.25, aluminium alloys have ν about 0.32; but the value of ν for beryllium is near zero $(0.01 - 0.06)$, and for lead is near 0.45. For some polymer foams ν can be greater than 1 or smaller than -1, see papers by Lakes and Lee and Lakes listed in Bibliography 12.1.

Smart materials have fine ultrastructures whose states influence the overall mechanical properties of the materials. Hence their state variables are not limited to stress, strain, and temperature. Therefore, they are not bounded by the Eq. (10) which is derived under the restricted set of state variables. This raises the question: under what condition can the ultrastructure be ignored? There is no simple answer, especially in a field like biomechanics, in which organs, tissues, cells, subcellular units, and molecules form successive hierarchies. In fact, investigations of questions at the borderlines of neighboring hierarchies could be most rewarding.

12.7. GENERALIZED HOOKE'S LAW, INCLUDING THE EFFECT OF THERMAL EXPANSION

We shall now consider the stress-strain law in a general thermal environment, not limited to isothermal or isentropic conditions. Let us assume that the strain energy function W can be expanded in a power series

$$(1) \qquad 2W = C_0 + C_{ij}e_{ij} + C'_{ijkl}e_{ij}e_{kl} + \cdots ,$$

where the coefficients C_0, C_{ij}, etc., are functions of the entropy $\mathscr{S}$ and the temperature T. Since $W = 0$ when $e_{ij} = 0$ by definition, the constant C_0 must vanish. As to C_{ij}, we note that since

$$\sigma_{ij} = \frac{\partial W}{\partial e_{ij}} = \frac{C_{ij}}{2} + 0(e_{ij}),$$

C_{ij} means twice the value of the stress component σ_{ij} at zero strain. Such stresses at zero strain can be brought about by a linear expansion if there are temperature changes from a standard state. The simplest assumption is to let C_{ij} be linearly proportional to the temperature change:

$$C_{ij} = -2\beta_{ij}(T - T_0),$$

where β_{ij} are numerical constants. Hence, we have, when e_{ij} are small and terms higher than the second order are neglected,

$$(2) \quad \blacktriangle \qquad W = -\beta_{ij}(T - T_0)e_{ij} + \frac{1}{2}C'_{ijkl}e_{ij}e_{kl} \,.$$

The stress components are, therefore,

$$(3) \qquad \sigma_{ij} = \frac{\partial W}{\partial e_{ij}} = \frac{1}{2}(C'_{ijkl} + C'_{klij})e_{kl} - \beta_{ij}(T - T_0) \,.$$

This is the general *Duhamel–Neumann form of Hooke's law*. If the quadratic form in Eq. (2) is symmetrized in advance, we can write Eq. (3) in the form

$$(4) \quad \blacktriangle \qquad \sigma_{ij} = C_{ijkl}e_{kl} - \beta_{ij}(T - T_0) \,,$$

where

$$(5) \qquad C_{ijkl} = C_{klij}$$

is a tensor of rank 4. Since e_{ij} and σ_{ij} are symmetric tensors, we also have

$$(6) \qquad C_{ijkl} = C_{jikl} \,, \qquad C_{ijkl} = C_{ijlk} \,, \qquad \beta_{ij} = \beta_{ji} \,.$$

Thus, for an anisotropic medium there are 21 independent elastic constants C_{ijkl}, and six independent constants β_{ij}. This number is reduced to two C's and one β if the medium is isotropic. The relationship for an isotropic material is

$$(7) \quad \blacktriangle \qquad \sigma_{ij} = \lambda e\delta_{ij} + 2Ge_{ij} - \beta(T - T_0)\delta_{ij} \,,$$

or

$$(8) \quad \blacktriangle \qquad e_{ij} = -\frac{\nu}{E}s\delta_{ij} + \frac{1+\nu}{E}\sigma_{ij} + \alpha(T - T_0)\delta_{ij} \,,$$

where

$$(9) \qquad e = e_{11} + e_{22} + e_{33} \,, \qquad s = \sigma_{11} + \sigma_{22} + \sigma_{33} \,,$$

$$(10) \qquad \lambda = \frac{E\nu}{(1+\nu)(1-2\nu)} \,, \qquad \beta = \frac{E\alpha}{1-2\nu} \,.$$

The corresponding strain energy function is

$$(11) \quad \blacktriangle \qquad W = Ge_{ij}e_{ij} + \frac{\lambda}{2}e^2 - \beta(T - T_0)e$$

$$= G\left[e_{ij}e_{ij} + \frac{\nu}{1-2\nu}e^2 - \frac{2(1+\nu)}{1-2\nu}\alpha(T - T_0)e\right] \,.$$

Whereas the complementary energy function is

$$W_c = \frac{1}{4G}\left[\sigma_{ij}\sigma_{ij} - \frac{\nu}{1+\nu}s^2\right] + \alpha(T - T_0)s\,.$$

Here α is the linear coefficient of thermal expansion, T_0 the temperature of a reference state of the body, $T - T_0$, the rise of temperature above the reference state.

The strain energy function must be a relative minimum at the zero-stress state (see Sec. 12.4), which implies that, when $T = T_0$, the quadratic form

(12) $$W = \frac{1}{2}C_{ijkl}e_{ij}e_{kl}$$

must be positive definite.

P R O B L E M S

12.8. State the conditions which must be satisfied by the constants C_{ijkl} in order that the quadratic form (12) be positive definite.

12.9. Let $T = T_0$, and derive Clapeyron's formula for a linear elastic material: $W = \frac{1}{2}\sigma_{ij}e_{ij}$.

12.10. In a conventional structural analysis, honeycomb sandwich material is often treated as a homogeneous orthotropic material, provided that the dimensions of the structure are much larger than the dimensions of the individual honeycomb cell. Suggest a stress-strain law for an orthotropic material. Determine explicitly the limitations imposed on the elastic constants that appear in the stress-strain law by the first and second laws of thermodynamics.

12.8. THERMODYNAMIC FUNCTIONS FOR ISOTROPIC HOOKEAN MATERIALS

It will be interesting to record the thermodynamic functions for an isotropic elastic material. Restricted to infinitesimal strains, the derivation of the formulas is quite simple. Consider a perfectly elastic, isotropic body which obeys the Duhamel–Neumann generalization of Hooke's law, Eq. (12.7:7). In a reversible, infinitesimal deformation, we have

(1) $$d\mathscr{F} = -\mathscr{S}dT + \frac{1}{\rho}\sigma_{ij}de_{ij}\,,$$

where $T, \mathscr{E}, \mathscr{S}$, and $\mathscr{F} = \mathscr{E} - T\mathscr{S}$ are, respectively, the absolute temperature, internal energy, entropy, and Helmholtz free energy *per unit mass*. If we substitute σ_{ij} from Eq. (12.7:7) into Eq. (1), we obtain

(2) $$\rho d\mathscr{F} = -\rho\mathscr{S}dT + \lambda ede - \beta(T-T_0)de + 2Ge_{ij}de_{ij}\,, \qquad e = e_{\alpha\alpha}\,,$$

which can be integrated with respect to the strain to yield

(3) $$\rho\mathscr{F} = \frac{\lambda e^2}{2} - \beta(T-T_0)e + Ge_{ij}e_{ij} + C_1(T)\,,$$

where $C_1(T)$ is a function of T. To evaluate C_1, we first note that Eq. (1) implies that $(\partial\mathscr{F}/\partial T)_{e_{ij}} = -\mathscr{S}$ [see Eq. (12.3:6)]. Hence, from Eq. (3),

(4) $$\rho\mathscr{S} = -\frac{e^2}{2}\frac{\partial\lambda}{\partial T} - e_{ij}e_{ij}\frac{\partial G}{\partial T} + e\frac{\partial}{\partial T}[\beta(T-T_0)] - \frac{dC_1}{dT}\,.$$

Now, the function C_1, can be expressed in terms of the heat capacity per unit mass at constant strain C_v. For, on the one hand, since $\mathscr{S}$ is a function of T and e_{ij}, we have that, if $de_{ij} = 0$,

(5) $$dQ = T\rho d\mathscr{S} = T\rho\left(\frac{\partial\mathscr{S}}{\partial T}\right)_{e_{ij}} dT\,.$$

On the other hand, we have, by definition, that at zero strain

(6) $$dQ = \rho C_v dT\,.$$

Hence

(7) $$C_v = T\left(\frac{\partial\mathscr{S}}{\partial T}\right)_{e_{ij}},$$

which relates C_v, with C_1, when $\mathscr{S}$ is substituted from Eq. (4). Simple relationships are obtained if C_v, is evaluated at zero strain. Then

(8) $$(C_v)_{e_{ij}=0} = T\left(\frac{\partial\mathscr{S}}{\partial T}\right)_{e_{ij}=0} = -\frac{T}{\rho}\frac{d^2C_1}{dT^2}\,,$$

or, if we assume that $\mathscr{S}$ and $\mathscr{F}$ vanish for zero strain at $T = T_0$,

(9) $$C_1 = -\int_{T_0}^{T} dT \int_{T_0}^{T} \frac{\rho}{T'}(C_v)_{e_{ij}=0} dT'\,.$$

Now, the internal energy $\rho\mathscr{E}$ per unit volume can be obtained. By Eqs. (3), (4) and (12.6:3)–(12.6:9), we have

(10) $$\rho\mathscr{E} = \rho(\mathscr{F} + T\mathscr{S})$$

$$= \frac{1}{2}e^2\left[K - T\frac{\partial K}{\partial T}\right] + 2J_2\left(G - T\frac{\partial G}{\partial T}\right)$$

$$+ e\left[\beta T_0 + T(T-T_0)\frac{\partial\beta}{\partial T}\right] + C_1 - T\frac{dC_1}{dT}\,.$$

If λ, G, β, and C_v, are independent of temperature, Eqs. (9) and (10) are simplified to

$$(11) \qquad C_1 = -\rho C_v T_0 \left(\frac{T}{T_0} \log \frac{T}{T_0} - \frac{T}{T_0} + 1 \right),$$

$$(12) \qquad \rho \mathcal{E} = \frac{1}{2} K e^2 + 2G J_2 + \beta T_0 e + \rho (C_v)_{e_{ij}=0}(T - T_0),$$

where J_2 is the second strain invariant:

$$(13) \qquad J_2 = \frac{1}{2} \left(e_{ij} - \frac{1}{3} e \delta_{ij} \right) \left(e_{ij} - \frac{1}{3} e \delta_{ij} \right).$$

Problem 12.11.   Starting with the equation

$$(14) \qquad d\Phi = -\mathscr{S} dT - \frac{1}{\rho} e_{ij} d\sigma_{ij},$$

where Φ is Gibbs' thermodynamic potential per unit volume, show that

$$(15) \qquad \rho \Phi = \frac{\nu}{2E} s^2 - \frac{1+\nu}{2E} \sigma_{ij} \sigma_{ij} - \alpha(T - T_0)s + C_2(T),$$

$$(16) \qquad C_2 = \rho \int_{T_0}^{T} dT \int_{T_0}^{T} (C_p)_{\sigma_{ij}=0} \frac{1}{T'} dT',$$

where $s = \sigma_{ii}$ and C_p, is the heat capacity per unit mass at constant stresses.

12.9. EQUATIONS CONNECTING THERMAL AND MECHANICAL PROPERTIES OF A SOLID

It is possible to derive relations which connect the specific heats, the moduli of elasticity, the latent heats of change of strain or stress at constant temperature, etc. These results were given by Lord Kelvin and by Gibbs.

The specific heat at constant strain and the various latent heats of change of strain are defined by the following equation.

$$(1) \qquad dQ = C_v dT + \ell_{e_{ij}} de_{ij},$$

where dQ is the heat required to change the temperature of a unit mass by dT and the strain by de_{ij}, C_v is the specific heat per unit mass measured in the state of constant strain, and $\ell_{e_{ij}}$ are the six latent heats per unit mass due to a change in a component of strain (in each case the temperature and the remaining five strain components are unchanged). Now, under reversible conditions, the change of entropy is

$$(2) \qquad d\mathscr{S} = \frac{dQ}{T} = \frac{C_v}{T} dT + \frac{\ell_{e_{ij}}}{T} de_{ij}.$$

Hence,

$$(3) \qquad \ell_{e_{ij}} = T\frac{\partial \mathscr{S}}{\partial e_{ij}}, \qquad C_v = T\frac{\partial \mathscr{S}}{\partial T}.$$

But from Eq. (12.3:6) we have, by further differentiations,

$$\frac{\partial \mathscr{S}}{\partial e_{ij}} = -\frac{\partial^2 \mathscr{F}}{\partial T \partial e_{ij}} = -\frac{1}{\rho}\frac{\partial \sigma_{ij}}{\partial T}, \qquad \frac{\partial^2 \mathscr{S}}{\partial e_{ij}\partial T} = -\frac{1}{\rho}\frac{\partial^2 \sigma_{ij}}{\partial T^2}.$$

It follows that

$$(4) \qquad \ell_{e_{ij}} = -\frac{T}{\rho}\frac{\partial \sigma_{ij}}{\partial T}, \qquad \frac{\partial C_v}{\partial e_{ij}} = -\frac{T}{\rho}\frac{\partial^2 \sigma_{ij}}{\partial T^2}.$$

In a similar manner we may choose the temperature and the six components of stress as thermodynamic variables and define the specific heat per unit mass at constant stress and the latent heats per unit mass due to change of stress by an equation analogous to Eq. (2),

$$(5) \qquad d\mathscr{S} = \frac{C_p}{T}dT + \frac{1}{T}L_{\sigma_{ij}}d\sigma_{ij}.$$

The specific heat $C_p(T,\sigma)$ is more practically obtainable under usual conditions of measurement where the external forces on the solid are unchanged. The latent heats $L_{\sigma_{11}}$, etc., are defined each at a constant temperature and with five of the stress-components unchanged.

A similar analysis with Gibbs' potential Φ leads to the results

$$(6) \qquad L_{\sigma_{ij}} = \frac{T}{\rho}\frac{\partial e_{ij}}{\partial T}, \qquad \frac{\partial C_p}{\partial \sigma_{ij}} = \frac{T}{\rho}\frac{\partial^2 e_{ij}}{\partial T^2}.$$

The difference between the two specific heats can be expressed as follows. Let an infinitesimal change take place at constant stress. The change of entropy can be expressed in two ways. By Eq. (2), it is

$$d\mathscr{S} = \frac{C_v}{T}dT + \frac{1}{T}\ell_{e_{ij}}\frac{\partial e_{ij}}{\partial T}dT;$$

by Eq. (5), it is

$$d\mathscr{S} = \frac{C_p}{T}dT.$$

Equating these two expressions, we obtain

$$(7) \qquad C_p = C_v + \ell_{e_{ij}}\frac{\partial e_{ij}}{\partial T}.$$

By Eq. (4), this becomes

$$(8) \qquad C_p - C_v = -\frac{T}{\rho}\frac{\partial e_{ij}(T,\sigma)}{\partial T}\frac{\partial \sigma_{ij}(T,e)}{\partial T}$$

Problem 12.12.   Prove the following reciprocal relations.

$$\frac{\partial \ell_{e_{ij}}}{\partial e_{kl}} = \frac{\partial \ell_{e_{kl}}}{\partial e_{ij}}, \quad \frac{\partial \sigma_{ij}}{\partial e_{kl}} = \frac{\partial \sigma_{kl}}{\partial e_{ij}},$$

$$\frac{\partial L_{\sigma_{ij}}}{\partial \sigma_{kl}} = \frac{\partial L_{\sigma_{kl}}}{\partial \sigma_{ij}}, \quad \frac{\partial e_{ij}}{\partial \sigma_{kl}} = \frac{\partial e_{kl}}{\partial \sigma_{ij}}.$$

13

IRREVERSIBLE THERMODYNAMICS
AND VISCOELASTICITY

Problems involving heat conduction, viscosity, and plasticity belong to the realm of irreversible thermodynamics. In this chapter we shall outline the methods of irreversible thermodynamics, limiting ourselves to linear processes. Biot's treatment of relaxation modes and hidden variables will be presented and applied to the stress-strain relationship of viscoelastic materials.

13.1. BASIC ASSUMPTIONS

The classical thermodynamics of Chapter 12 deals essentially with equilibrium conditions of a uniform system (or a heterogeneous system with uniform phases). As far as irreversible processes are concerned, it contains little more than a statement about the trend toward thermodynamic equilibrium. To bring the irreversibility to a sharper focus, we would like to be able to write down definite "equations of evolution" (or equations of "motion"), which describe in precise terms the way an irreversible process evolves. To make the theory useful to continuum mechanics, we must extend the formulation to include nonuniform systems. These two items are the main objectives of the theory of irreversible thermodynamics. Since these objectives are beyond the scope of classical thermodynamics, some new hypotheses must be introduced, whose justification can only be sought by comparing any theoretical deductions with experiments.

The *first assumption* is that the *entropy is a function of state in irreversible processes* as well as in reversible processes. In Chapter 12, we considered reversible processes and showed that the value of the entropy of a system can be computed when the values of the state variables are known (except for an integration constant which is, in general, of no importance). (Near zero absolute temperature, the additive constant of entropy does become important, but in that case its value can be fixed according to Nernst's theorem — the third law of thermodynamics.) We now assume that this same function of state defines entropy even if the system is not in equilibrium. The motivation for this assumption is analogous to our attitude toward the mass distribution of any mechanical system: once the mass of a particle in a continuum is determined, we assume that it remains the same no matter how fast the particle moves.

Just as the constancy of mass must be subjected to relativistic restrictions, the validity of the assumption of entropy as a state function must be restricted to relatively mild processes. Prigogine has investigated this question by comparing our assumption with results of statistical mechanics for some particular models, such as the Chapman–Enskog theory of nonuniform gases. He shows that the domain of validity of this assumption extends throughout the domain of validity of linear phenomenological laws (Fourier's law of heat conduction, Fick's law of mass diffusion, etc.). In the case of chemical reactions, he shows that the reaction rates must be sufficiently slow so as not to perturb Maxwell–Boltzmann equilibrium distribution of velocities of each component to an appreciable extent. This means, for example, that the temperature changes over the length of one mean free path must be much smaller than the absolute temperature itself; and the like for other properties.

The second assumption consists of an extension of the second law of thermodynamics locally to every portion of a continuum, whether it is uniform or nonuniform, and may be explained as follows. We notice first that entropy is an extensive quantity, so it must be subjected to a conservation law: for a given set of particles occupying a domain V the total change of entropy must be equal to the total amount of entropy transferred to these particles through the boundary, plus the entropy produced inside this domain. Let the boundary of this domain be denoted by B. Then we may write

$$(1) \qquad \frac{D}{Dt} \int_V \rho \mathscr{S} dV = - \int_B \dot{\phi} \cdot d\mathbf{A} + \frac{D}{Dt} \int_V \rho({}_i \mathscr{S}) dV ,$$

where $\mathscr{S}$ is the *specific entropy* per unit mass, ${}_i\mathscr{S}$ is the *entropy source* (internal entropy production per unit mass), $\dot{\phi}$ is the *entropy flow vector* on the boundary, $d\mathbf{A}$ is a surface element whose vector direction coincides with the outer normal, so that $\dot{\phi} \cdot d\mathbf{A}$ is a scalar product representing the *outflow*, which is equal to the normal component of $\dot{\phi}$ times the area $d\mathbf{A}$. The material derivative D/Dt is taken with respect to a given set of particles. On transforming the second term into a volume integral and using Eq. (5.3:4) to reduce the material derivative of an integral, and realizing that the domain V is arbitrary, we obtain

$$\rho \frac{D\mathscr{S}}{Dt} + \mathscr{S} \left(\frac{D\rho}{Dt} + \rho \operatorname{div} \mathbf{v} \right) = -\operatorname{div} \dot{\phi} + \rho \frac{D_i \mathscr{S}}{Dt} + {}_i \mathscr{S} \left(\frac{D\rho}{Dt} + \rho \operatorname{div} \mathbf{v} \right)$$

where $\mathbf{v}$ is the particle velocity vector of the flow field. The sum in the parentheses vanishes by the equation of continuity, Eq. (5.4:4). Hence, we have

$$(2) \quad \blacktriangle \qquad \rho\frac{D\mathscr{S}}{Dt} = -\operatorname{div}\dot{\phi} + \rho\frac{D_i\mathscr{S}}{Dt}\,.$$

We can now state the *second hypothesis* (*an extension of the second law of thermodynamics*) as follows.

$$(3) \quad \blacktriangle \qquad \dot{\phi} = \frac{\mathbf{h}}{T}\,, \qquad \mathbf{h} = \text{heat flux vector}\,,$$

$$(4) \quad \blacktriangle \qquad \frac{D_i\mathscr{S}}{Dt} = 0\,, \qquad \text{reversible processes}\,,$$

$$(5) \quad \blacktriangle \qquad \frac{D_i\mathscr{S}}{Dt} > 0\,, \qquad \text{irreversible processes}\,,$$

$$(6) \qquad \frac{D_i\mathscr{S}}{Dt} < 0\,, \qquad \text{never occurs in nature}\,.$$

In Eq. (3), $\mathbf{h}$ is the heat flux vector representing the heat flow per unit area and T is the local absolute temperature. In a moving medium, $\mathbf{h}$ is defined convectively, i.e., with respect to elements of surfaces that are composed always of the same sets of particles.

When Eqs. (3)–(5) are compared with Eqs. (12.1:6) and (12.1:7), we see that the entropy production is now required to be nonnegative everywhere in the system. Such a formulation may be called a *local formulation of the second law* in contrast to the *global formulation* of the classical thermodynamics.

Corresponding to $\dot{\phi}$ we have a vector field ϕ which is the *entropy displacement field* introduced by Biot[13.1] (1956). If we require that

$$(7) \qquad \phi = 0 \qquad \text{at} \qquad t = t_0\,,$$

then the points of the body at time $t > t_0$ are assigned an entropy displacement vector ϕ, just as they are assigned a material displacement vector $\mathbf{u}$. The two vector fields $\mathbf{u}$ and ϕ connect the state of the body at time $t > t_0$ with the natural state at t_0. The material displacement field defines the reversible change of mass distribution by $-\operatorname{div}(\rho\mathbf{u})$; the entropy displacement field ϕ defines a reversible change of entropy by $-\operatorname{div}\phi$. The material displacement field can be analyzed through the displacement gradient tensor $u_{i,j}$; the entropy displacement field can be analyzed through the second order tensor $\phi_{i,j}$.

If we introduce a notation analogous to that of Eq. (12.1:5), we may write the rate of reversible change of entropy as

(8)
$$\rho\frac{D_e\mathscr{S}}{Dt} = -\operatorname{div}\dot{\phi}.$$

Further development of irreversible thermodynamics requires a detailed description of the entropy production $D_i\mathscr{S}/Dt$. To this end we shall first consider a simple example in the following section.

13.2. ONE-DIMENSIONAL HEAT CONDUCTION

Consider the heat transfer in a slender solid bar with continuous temperature gradient in the direction of the lengthwise axis of the bar, x; Fig. 13.2:1. The temperature is assumed to be uniform in each cross section of the bar, and the walls (except the ends) are thermally insulated. It will be assumed that the heat flow takes place in the direction of the temperature gradient, so that the problem is mathematically one-dimensional. Furthermore, we assume that the bar is free of stresses and that the cross section is of unit area.

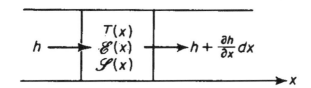

Fig. **13.2:1.** Heat conduction.

Let $T(x)$, $\mathscr{E}(x)$, $\mathscr{S}(x)$ denote, respectively, the temperature, the energy per unit mass, and the entropy per unit mass of the solid. Let h denote the heat flux per unit area per unit time in the x-direction; i.e., if Q represents the heat transported to the right across a unit cross-sectional area, then

(1)
$$h = \frac{\partial Q}{\partial t}.$$

Now consider the changes occurring in a small element of length dx in a small time interval dt. The net increment of heat in this element is evidently

$$dQ = h\,dt - \left(h + \frac{\partial h}{\partial x}dx\right)dt = -\frac{\partial h}{\partial x}dx\,dt.$$

The increase in energy in this element is $(\rho D\mathscr{E}/Dt)dt\,dx$. The stresses being zero, the first law states that the quantities above must be equal. Cancelling the factor $dx\,dt$ we obtain

$$(2) \qquad \rho \frac{D\mathscr{E}}{Dt} = -\frac{\partial h}{\partial x}.$$

The change of entropy in this element is caused by the addition of heat dQ alone. Hence,

$$(3) \qquad \rho \frac{D\mathscr{S}}{Dt}\, dx\, dt = \frac{dQ}{T} = -\frac{1}{T}\frac{\partial h}{\partial x}\, dx\, dt.$$

Hence,

$$(4) \qquad \rho \frac{D\mathscr{S}}{Dt} = -\frac{1}{T}\frac{\partial h}{\partial x} = -\frac{\partial}{\partial x}\left(\frac{h}{T}\right) - h\frac{1}{T^2}\frac{\partial T}{\partial x}.$$

This equation is of the form of Eq. (13.1:2). The entropy flow $\dot{\phi}$ may be identified with h/T. The entropy production per unit time per unit volume is

$$(5) \qquad \rho \frac{D_i\mathscr{S}}{Dt} = -\frac{h}{T^2}\frac{\partial T}{\partial x}.$$

The postulate that $D_i\mathscr{S}/Dt$ must be positive implies that h has an oppositive sign to that of $\partial T/\partial x$; i.e., the heat flows in a direction against the temperature gradient.

Problem 13.1. Generalize the results above to the three-dimensional case: the entropy production per unit volume is

$$(6) \quad \blacktriangle \qquad \rho \frac{D_i\mathscr{S}}{Dt} = -\left(\frac{h_x}{T^2}\frac{\partial T}{\partial x} + \frac{h_y}{T^2}\frac{\partial T}{\partial y} + \frac{h_z}{T^2}\frac{\partial T}{\partial z}\right) = -\frac{\mathbf{h}}{T}\cdot\frac{\operatorname{grad} T}{T},$$

where h_x, h_y, h_z, are the three components of the heat flux vector $\mathbf{h}$, i.e., h_x, is the heat flux per unit area across a surface element normal to the x-axis, etc.

13.3. PHENOMENOLOGICAL RELATIONS — ONSAGER PRINCIPLE

In the preceding section it was shown that in heat conduction the entropy production can be written as the product of a "generalized force" $-(\operatorname{grad} T)/T$ and the corresponding flux $\mathbf{h}/T$. This is true also for other types of irreversible processes such as diffusion, chemical reactions. In chemical reactions the "forces" and "fluxes" are often termed affinities and reaction rates. In general, then, we may write, for an irreversible process,

$$(1) \qquad \frac{D_i\mathscr{S}}{Dt} = \sum_k J_k X_k > 0,$$

where J_k, denotes the kth flux and X_k the corresponding generalized force. For example, in heat conduction

$$
(2) \qquad J_k = \frac{h_k}{T}, \qquad X_k = -\frac{1}{T}\frac{\partial T}{\partial x_k}, \qquad\qquad k = 1, 2, 3.
$$

At thermodynamic equilibrium all processes stop and we have simultaneously for all irreversible processes

$$
(3) \qquad\qquad J_k = 0, \qquad X_k = 0, \qquad\qquad \text{at equilibrium}.
$$

It is quite natural to assume, at least in the neighborhood of an equilibrium condition, that the relations between generalized forces and generalized fluxes are linear. Fourier's law for heat conduction is such an example. Linear laws of this kind are called *phenomenological relations*. For such phenomenological relations, an important symmetry is known, among the coefficients that appear in these relations. This symmetry is embodied in the so-called *Onsager principle*.

Let us consider heat conduction again. It has been known for a long time that in anisotropic crystals a symmetry exists in the heat conduction coefficient matrix, which could not be explained by crystallographic symmetry properties. It was found that when we write

$$
\begin{aligned}
h_x &= L_{11}\frac{\partial T}{\partial x} + L_{12}\frac{\partial T}{\partial y} + L_{13}\frac{\partial T}{\partial z}, \\[4pt]
(4) \qquad h_y &= L_{21}\frac{\partial T}{\partial x} + L_{22}\frac{\partial T}{\partial y} + L_{23}\frac{\partial T}{\partial z}, \\[4pt]
h_z &= L_{31}\frac{\partial T}{\partial x} + L_{32}\frac{\partial T}{\partial y} + L_{33}\frac{\partial T}{\partial z},
\end{aligned}
$$

the matrix L_{kl}, is always symmetric:

$$
(5) \qquad\qquad L_{kl} = L_{lk}, \qquad\qquad l, k = 1, 2, 3,
$$

for arbitrary orientations of the coordinate axes, irrespective of the crystallographic axes. Considerable effort has been spent on accurate experiments to detect any deviation from relation (5). The most ingenious experiments are those of Soret[13.1] and Voigt.[13.1] The symmetry relation is always verified with great accuracy.

A similar symmetry property occurs in other phenomenological relations describing interferences between several simultaneous irreversible processes.

For example, consider the case of one-dimensional thermodiffusion for which two phenomenological relations may be written as

(6)
$$J_1 = L_{11}X_1 + L_{12}X_2,$$
$$J_2 = L_{21}X_1 + L_{22}X_2,$$

where J_1 represents heat flux, J_2 represents the mass flux of a particular component in a mixture, X_1 represents the temperature gradient, and X_2 represents the concentration gradient of that particular component. Then L_{11}, is the heat conductivity, L_{22} is the diffusion coefficient, and L_{12} and L_{21} are coefficients describing the *interference* of the two irreversible processes of heat conduction and diffusion. The coefficient L_{21} is connected with the appearance of a concentration gradient when a temperature gradient is imposed, called the *Soret effect*. The coefficient L_{12} is connected with the inverse phenomenon, the *Dufour effect*, which describes the appearance of temperature difference when a concentration gradient exists. The first equation in Eq. (6) is a modification of Fourier's law of heat conduction to include the Dufour effect. The second equation in Eq. (6) is a modification of the Fick's law of diffusion to include the Soret effect. In this case also, we have

$$L_{12} = L_{21}.$$

Other linear phenomenological laws verified by experiments include Ohm's law between electrical current and potential gradient, Newton's law between shearing force and velocity gradient, the chemical reaction law between reaction rate and chemical potentials, etc. The reciprocal phenomena of thermoelectricity arising from the interference of heat conduction and electrical conduction include the *Peltier effect* (discovered in 1834), which refers to the evolution or absorption of heat at junctions of different metals resulting from the flow of an electric current, the *Seebeck effect* (discovered in 1822), which relates to the electromotive force developed in a circuit made up of different conducting elements, when not all of the junctions are at the same temperature, and the *Thomson effect* (discovered by Lord Kelvin in 1856), which refers to the reversible heat absorption which occurs when an electric current flows in a homogeneous conductor in which there is a temperature gradient.

The simultaneous action of several irreversible processes may cause interferences as typified by the examples named above. As a general form of phenomenological laws, let $J_k(k = 1, 2, \ldots, m)$ represent the generalized *fluxes* (heat flow, electric current, chemical reaction rate, etc.), and let $X_k(k = 1, 2, \ldots, m)$ represent the generalized *forces* (temperature gradient, electric potential gradient, chemical affinity, etc.). Then a linear phenomenological law is

$$(7) \qquad J_k = \sum_{l=1}^{m} L_{kl} X_l , \qquad\qquad k = 1, 2, \ldots, m .$$

To this general form, Onsager principle now states a symmetry property of the coefficients L_{kl} as follows.

When a proper choice of the fluxes J_k and the forces X_l, is made, so that the entropy production per unit time may be written as

$$(8) \qquad \frac{D_i \mathscr{S}}{Dt} = \sum_{k=1}^{m} J_x X_k ,$$

and if the flux and forces are related by linear phenomenological relations,

$$J_k = \sum_{l=1}^{m} L_{kl} X_l ,$$

then the matrix of the coefficients L_{kl} is symmetric; i.e.,

$$(9) \qquad L_{kl} = L_{lk} , \qquad\qquad k, l = 1, 2, \ldots, m .$$

The identities (9) are called the *Onsager reciprocal relations*.

On substituting Eq. (9) into Eq. (8), we obtain a quadratic form

$$(10) \qquad \frac{D_i \mathscr{S}}{Dt} = \sum_{k=1}^{m} \sum_{l=1}^{m} L_{kl} X_l X_k > 0 .$$

This quadratic form has to be positive for all positive or negative values of the variables X_k except when $X_1 = X_2 = \cdots = 0$, in which case the entropy production vanishes. The necessary and sufficient conditions which must be satisfied by the coefficients L_{kl} in order that the quadratic form (10) be positive definite are well known in algebra and matrix theory.

Onsager derived (1931) this principle from statistical mechanics considerations under the assumption of microscopic reversibility; i.e., the symmetry of all mechanical equations of motion of individual particles with respect to time. It furnishes an explanation of the observed relation (5) of the heat conduction for anisotropic crystals. The statistical derivation may be found in Onsager's papers, or in Prigogine[13.1] and De Groot's[13.1] books.

Casimir[13.1] and Tellegen classified irreversible processes into two classes according to whether

$$(11) \qquad J_k(t) = J_k(-t) \qquad\qquad J_k \text{ an even function of time} ,$$

$$(12) \qquad J_k(t) = -J_k(-t) \qquad\qquad J_k \text{ an odd function of time} .$$

Heat flow, mass flow, and so on, defined by expressions such as $\delta Q/\delta t$, are odd functions of time. From the point of view of the kinetic theory of matter, J is even or odd according to whether J does or does not change sign as particle velocity is reversed. So far, we have considered cases in which generalized flows are so chosen that they are all odd. In the more general case, Onsager's principle should be generalized to state: if the irreversible processes i and k are both even or both odd, then

$$(13) \qquad\qquad L_{ik} = L_{ki}\,.$$

If one is even and the other is odd, then

$$(14) \qquad\qquad L_{ik} = -L_{ki}\,.$$

A further modification is necessary if there exist forces depending explicitly on the velocity, such as Lorentz forces, Coriolis forces, and so on. For example, in the presence of a magnetic field, $\mathbf{H}$, two odd or two even processes are related by:

$$(15) \qquad\qquad L_{ik}(\mathbf{H}) = L_{ki}(-\mathbf{H})\,.$$

The generalized theorem is sometimes called the *Onsager–Casimir theorem*.

For simultaneous action of several irreversible processes a general phenomenological law may be written in the form discussed in Sec. 13.3. The Onsager–Casimir theorem simplifies these laws. A further simplification exists in the form of certain symmetry requirements which may dictate whether certain interference coefficients should vanish or not. This is *Curie's symmetry principle* [P. Curie, *Oeuvres* (Paris: Société Français de Physique 1908)], which states that macroscopic causes always have fewer elements of symmetry than the effects they produce. An interference or the irreversible processes is possible only if this general principle is satisfied. For example, chemical affinity (a scalar) cannot produce a directed heat flow (a vector) and the interference coefficient between the heat flow and chemical affinity must vanish.

13.4. BASIC EQUATIONS OF THERMOMECHANICS

Thermomechanical systems are subjected to the same general conservation laws with regard to mass and momentum as were discussed in Chapter 5. However, the law of conservation of energy contains both mechanical and thermal energy. The change of thermal energy is related to the change of entropy. Thus, a complete description of the evolution of a system requires a knowledge of the entropy production. Phenomenological relations, together with the Onsager principle, provide sufficient details about the entropy production. These laws, taken together, determine the evolution of the system.

To demonstrate the procedure let us consider a solid body occupying a regular domain V with boundary B. A set of rectangular Cartesian coordinates will be used for reference; the instantaneous position of a particle will be denoted by (x_1, x_2, x_3). Equations will be written with respect to the instantaneous configuration of the body, i.e., in Eulerian coordinates.

The conservation of mass is expressed by the equation of continuity (5.4:3)

$$(1) \qquad \frac{\partial \rho}{\partial t} + \frac{\partial \rho v_i}{\partial x_i} = 0 \,.$$

The conservation of momentum is expressed by the Eulerian equation of motion (5.5:7), and Cauchy's formula (3.3:2)

$$(2) \qquad \rho \frac{D v_i}{Dt} = \sigma_{ij,j} + \rho F_i \qquad \qquad \text{in } V \,,$$

$$(2a) \qquad \overset{\nu}{T}_i = \sigma_{ij} \nu_j \qquad \qquad \text{on } B \,,$$

$$(2b) \qquad \sigma_{ij} = \sigma_{ji} \qquad \qquad \text{in } V + B \,.$$

The conservation of energy is given by Eq. (12.2:7)

$$(3) \qquad \rho \frac{D\mathscr{E}}{Dt} = \sigma_{ij} v_{i,j} - h_{i,i} \,.$$

In these equations, ρ is the mass density, (v_1, v_2, v_3) is the velocity vector of a particle, D/Dt is the material derivative

$$(4) \qquad \frac{D}{Dt} = \frac{\partial}{\partial t} + v_j \frac{\partial}{\partial x_j} \,,$$

F_i is the body force per unit mass, $\overset{\nu}{T}_i$ is the surface traction per unit area, $\mathscr{E}$ is the internal energy per unit mass, σ_{ij} is the stress tensor, h_i is the heat flux vector, and a comma followed by i indicates partial differentiation with respect to x_i. The indices i, j range over 1, 2, 3; and the summation convention is used.

Next we need the entropy balance equation. According to the first assumption named in Sec. 13.1, we shall assume, *for a solid body*, that the specific entropy $\mathscr{S}$ per unit mass is a function of the internal energy $\mathscr{E}$ and the strain tensor e_{ij},

$$(5) \qquad \mathscr{S} = \mathscr{S}(\mathscr{E}, e_{ij}) \,.$$

This is also expressed by the fact that, in equilibrium, the total differential of $\mathscr{S}$ is given by Gibbs' relation (12.2:10)

$$(6) \qquad \rho T d\mathscr{S} = \rho d\mathscr{E} - \sigma_{ij} de_{ij} \,.$$

Since our hypothesis asserts that $\mathscr{S}$ is related to $\mathscr{E}$ and e_{ij} in the same way even if the system is not in equilibrium, it follows that along the path of motion

$$(7) \qquad \rho T \frac{D\mathscr{S}}{Dt} = \rho \frac{D\mathscr{E}}{Dt} - \sigma_{ij} V_{ij} \,,$$

where $V_{ij} = \frac{1}{2}(v_{i,j} + v_{j,i})$ is the symmetric part of $v_{i,j}$, called the *rate of deformation tensor*. On substituting Eq. (3) into Eq. (7), we obtain

$$(8) \qquad \rho T \frac{D\mathscr{S}}{Dt} = -h_{i,i} \,,$$

or

$$(9) \qquad \rho \frac{D\mathscr{S}}{Dt} = -\frac{h_{i,i}}{T} = -\left(\frac{h_i}{T}\right)_{,i} - h_i \frac{T_{,i}}{T^2} \,.$$

The first term on the right-hand side is the divergence of the entropy flow, the second term is the entropy production which must be positive. This result is in agreement with what was discussed in Sec. 13.2.

The entropy production $-h_i T_{,i}/T^2$ is a product of the flux h_i/T, and the "force," $-T_{,i}/T$. A phenomenological law relates the flux to the force in the form

$$(10) \qquad h_i = -k_{ij} T_{,j}$$

where

$$(11) \qquad k_{ij} = k_{ji} \,.$$

This is all we can get from the general considerations of mass, momentum, energy, and entropy. Equations (1) through (11), taken together, still do not define the strain field uniquely. To complete the formulation, a constitutive law must be added, which relates the stress tensor σ_{ij} to the strain tensor e_{ij}. It can be verified that with the addition of such a constitutive law, a sufficient number of differential equations are obtained for which boundary value problems can be formulated.

As another illustration let us consider a body of *fluid* instead of a solid. A fluid is distinguished from a solid by the fact that it cannot sustain shear stress without motion. In an equilibrium condition, the internal energy must depend only on the volumetric change. Hence, instead of Eq. (5), we now make the assumption

$$(12) \qquad \mathscr{S} = \mathscr{S}(\mathscr{E}, e) \,,$$

where

(13) $$e = e_{ii} = u_{i,i}$$

is the first invariant of the strain tensor. Correspondingly, the stress tensor is separated into a pressure and a stress deviator:

(14) $$\sigma_{ij} = -p\delta_{ij} + \sigma'_{ij} ,$$

(15) $$-p = \frac{\sigma_{ii}}{3} .$$

The Gibbs' relation in thermostatics (12.2:11) is now generalized to read, according to Eq. (12) and our basic assumptions listed in Sec. 13.1,

(16) $$\rho T \frac{D\mathscr{S}}{Dt} = \rho \frac{D\mathscr{E}}{Dt} + p \frac{De}{Dt} .$$

A substitution of Eq. (3) yields

(17) $$\rho T \frac{D\mathscr{S}}{Dt} = -h_{i,i} + \sigma'_{ij} v_{i,j} - p v_{i,i} + p v_{i,i} .$$

Hence,

(18) $$\rho \frac{D\mathscr{S}}{Dt} = -\left(\frac{h_i}{T}\right)_{,i} - h_i \frac{T_{,i}}{T^2} + \frac{1}{T} \sigma'_{ij} v_{i,j} .$$

The first term on the right-hand side of (18) is the divergence of the entropy flow. The last two terms have the significance of entropy production,

(19) $$\rho \frac{D_i \mathscr{S}}{Dt} = -h_i \frac{T_{,i}}{T^2} + \frac{1}{T} \sigma'_{ij} v_{i,j} .$$

Now $v_{i,j}$ can be split into a symmetric part V_{ij} and an antisymmetric part ω_{ij}, called the *rate of deformation* tensor and the *vorticity* tensor respectively,

(20) $$v_{i,j} = V_{ij} + \omega_{ij} ,$$

(21) $$V_{ij} = V_{ji}, \qquad\qquad \omega_{ij} = -\omega_{ji} .$$

(22) $$V_{ij} = \frac{1}{2}(v_{i,j} + v_{j,i}), \qquad \omega_{ij} = \frac{1}{2}(v_{i,j} - v_{j,i})$$

The contraction of the symmetric stress tensor σ'_{ij} with the antisymmetric tensor ω_{ij} vanishes. Hence, $\sigma'_{ij} v_{i,j} = \sigma'_{ij} V_{ij}$ and Eq. (19) becomes

(23) $$\rho \frac{D_i \mathscr{S}}{Dt} = -h_i \frac{1}{T^2} \frac{\partial T}{\partial x_i} + \frac{1}{T} \sigma'_{ij} V_{ij}$$

The linear phenomenological laws assume the form

(24)
$$\frac{h_i}{T} = -C_{ij}^{(1)}\frac{1}{T}\frac{\partial T}{\partial x_j} + C_{ijk}^{(2)}V_{jk}\,,$$

(25)
$$\sigma_{ij}' = C_{ijk}^{(3)}\frac{1}{T}\frac{\partial T}{\partial x_k} + C_{ijkl}^{(4)}V_{kl}\,,$$

where, according to Onsager's principle, the symmetry relation

(26)
$$C_{ijk}^{(2)} = C_{ijk}^{(3)}$$

prevails. Equation (25) is a law for viscous flow. The coupling terms involving $C_{ijk}^{(2)}$, $C_{ijk}^{(3)}$ express the possible interference between viscous flow and heat conductions.[†]

In the case of a fluid, the Eqs. (1)–(3) and (12)–(26) provide the right number of field equations for the determination of the flow field.

13.5. EQUATIONS OF EVOLUTION FOR A LINEAR HEREDITARY MATERIAL

The remainder of this chapter will be devoted to the question of stress-strain law for a linear hereditary material, following a treatment first given by Maurice A. Biot[13.1] (1954). We have in mind such materials as polycrystalline metals, high polymers, etc. When such a material is subjected to a variable strain, many things happen inside, which, however, are not explicitly observed in formulating the stress-strain law. For example, when a polycrystalline metal is uniformly strained in a macroscopic sense, the individual anisotropic crystals are strained differently, and thermal currents that circulate among the crystals are generated. The interstitial atoms move in the crystals or among the crystals. These processes can be accounted for explicitly. However, often we are not interested in them. Our interest may be limited to the extent of their interference with the deformation. They are the "hidden variables" in the stress-strain relationship.

We follow Biot to formulate a problem as follows. Consider a system I with n degrees of freedom defined by n state variables $q_1, q_2, \ldots, q_n$. These q's may represent strain components, local temperature, piezoelectric charges, concentrations such as induced by chemical or solubility processes, etc. It is assumed that the system is under the action of *generalized external forces* denoted by Q_j, the units and senses of which are such that for each $j = 1, 2, \ldots, n$, $Q_j dq_j$ (j not summed) represents the energy furnished to

---

[†]It follows further from Curie's principle that $C_{ijk}^{(3)} = C_{ijk}^{(2)} = 0$. See p. 436.

the system when q_j is changed by an amount dq_j. These forces may be externally applied forces, electromotive forces, or may result from deviations of the Gibbs and chemical potential from the equilibrium state. Inertia forces, if any, are considered as external forces according to D'Alembert's principle.

Let the system I be adjoined to a system II which is a large reservoir at constant temperature T_0, and *let the total system $I + II$ be insulated.* The variables q_j will be defined as the departure from the state of thermodynamic equilibrium; i.e., $q_j = 0$ at the condition of equilibrium at which the temperature is uniform and equal to T_0.

Let the system I be given an initial disturbance from equilibrium. The forces Q_j as functions of time are given. If all coordinates $q_i(t)$ are determined for $t \geq 0$, we say that the *evolution* of the system is known. The time history of $q_j(t)$ is said to be a *trajectory*. The differential equations that describe the evolution of $q_j(t)$ from their initial values are called the *equations of evolution.*

A little reflection will show that if we consider the body represented by the system I to be macroscopically homogeneous and subjected to macroscopically homogeneous stresses, we may let $q_1, q_2, \ldots, q_6$, represent the six independent strain components $e_{11}, e_{12} = e_{21}, \ldots, e_{33}$, and Q_i the corresponding stress components. The relations between $Q_1, \ldots, Q_6$ and $q_1, \ldots, q_6$ are influenced by the other q's and Q's. However, if all other Q's vanish and if we can eliminate all q's other than the first six, then the stress-strain relationship of the material is obtained. It will be seen that the influence of the hidden variables is revealed in the hereditary character of the material. In the present section we shall derive the equations of evolution. The solution of free evolution following an initial disturbance will be discussed in Sec. 13.6, and the forced motion and the elimination of hidden variables will be treated in Secs. 13.7 and 13.8. Biot's results about viscoelastic material will be presented in Sec. 13.9.

We shall now consider what happens in a small time interval. Let the change of the coordinate q_j in a time interval dt be denoted by dq_j. According to the first law of thermodynamics, the heat, $\mathscr{Q}(\mathrm{I})$, required by the system I to bring about a change of state dq_j is the difference between the change in internal energy $\mathscr{E}_\mathrm{I}$ and the work done. Without loss of generality, we shall write the following equations under the assumption that the system I is of unit mass:

$$(1) \qquad d\mathscr{Q}^{(\mathrm{I})} = d\mathscr{E}_\mathrm{I} - \sum_{j=1}^{n} Q_j \, dq_j \,.$$

Since the total system I + II is insulated, we have

$$d\mathcal{Q}^{(\mathrm{I})} + d\mathcal{Q}^{(\mathrm{II})} = 0,$$

where $d\mathcal{Q}^{(\mathrm{II})}$ denotes the heat received by system II. According to our basic assumption stated in Sec. 13.1, Gibbs' relation for change of entropy holds. Hence, the change of entropy of system II is

$$d\mathscr{S}_{\mathrm{II}} = \frac{d\mathcal{Q}^{(\mathrm{II})}}{T_0} = -\frac{d\mathcal{Q}^{(\mathrm{I})}}{T_0} = -\frac{d\mathscr{E}_{\mathrm{I}}}{T_0} + \sum_{j=1}^{n} \frac{Q_j}{T_0} dq_j .$$

The change of entropy of the isolated total system I + II is

$$(2) \qquad d_i \mathscr{S}_{\mathrm{I+II}} = d\mathscr{S}_{\mathrm{I}} + d\mathscr{S}_{\mathrm{II}} = d\mathscr{S}_{\mathrm{I}} - \frac{d\mathscr{E}_{\mathrm{I}}}{T_0} + \sum_{j=1}^{n} \frac{Q_j}{T_0} dq_j .$$

This is the entropy production in the time interval dt. Since $\mathscr{S}_{\mathrm{I}}$ and $\mathscr{E}_{I}$, are functions of state, the total differentials of $d\mathscr{S}_{\mathrm{I}}$ and $d\mathscr{E}_{\mathrm{I}}$ are

$$d\mathscr{S}_{\mathrm{I}} = \sum_j \frac{\partial \mathscr{S}_{\mathrm{I}}}{\partial q_j} dq_j , \qquad d\mathscr{E}_{\mathrm{I}} = \sum_j \frac{\partial \mathscr{E}_{\mathrm{I}}}{\partial q_j} dq_j .$$

Hence,

$$(3) \qquad d_i \mathscr{S}_{\mathrm{I+II}} = \sum_{j=1}^{n} \left(\frac{\partial \mathscr{S}_{\mathrm{I}}}{\partial q_j} - \frac{1}{T_0} \frac{\partial \mathscr{E}_{\mathrm{I}}}{\partial q_j} + \frac{Q_j}{T_0} \right) dq_j .$$

According to the second law of thermodynamics, $d_i \mathscr{S}_{\mathrm{I+II}}$ is nonnegative. If we write dq_j / dt for the generalized flux J_j of the previous section, we have the following general expression for the entropy production:

$$(4) \quad \blacktriangle \qquad d_i \mathscr{S}_{\mathrm{I+II}} = \sum_{j=1}^{n} X_j dq_j ,$$

where X_j are the generalized force conjugate to q_j in the entropy production. To distinguish X_j from the "generalized external forces" Q_i, we shall call X_j the *generalized entropy-production forces*. We remark that X_j, may be identically zero for some index j. Hence, the number of significant terms in the sum on the right-hand side of Eq. (3) may be less than n.

Identifying the general expression (4) with Eq. (3), we see that in our problem

$$(5) \quad \blacktriangle \qquad X_i = \frac{\partial}{\partial q_j} \left(\mathscr{S}_{\mathrm{I}} - \frac{1}{T_0} \mathscr{E}_{\mathrm{I}} \right) + \frac{Q_j}{T_0} .$$

Now, if the phenomenological laws connecting X_j, and $\dot{q}_j$, are linear so that

(6) ▲
$$X_j = \frac{1}{T_0} \sum_{k=1}^{n} b_{jk} \dot{q}_k,$$

then Eq. (5) becomes

(7) ▲
$$\sum_k b_{jk} \dot{q}_k - \frac{\partial}{\partial q_j}(T_0 \mathscr{S}_{\mathrm{I}} - \mathscr{E}_{\mathrm{I}}) = Q_j, \qquad j = 1, 2, \ldots, n,$$

which is the equation of evolution.

Onsager's principle assures that the coefficients b_{ij} in Eq. (6) are symmetric:

(8) ▲
$$b_{ij} = b_{ji}.$$

Let us introduce the quadratic form

(9) ▲
$$\mathscr{D} = \frac{1}{2} \sum_{i,j} b_{ij} \dot{q}_i \dot{q}_j.$$

Then Eq. (6) may be written as

(10) ▲
$$X_i = \frac{1}{T_0} \frac{\partial \mathscr{D}}{\partial \dot{q}_i}.$$

The quadratic form $\mathscr{D}$ is nonnegative because it is proportional to the entropy production, $d_i \mathscr{S}_{\mathrm{I+II}}/dt$, and because T_0 is positive by definition, as can be seen from the following equation:

(11) ▲
$$\frac{d_i \mathscr{S}_{\mathrm{I+II}}}{dt} = \sum_i X_i \dot{q}_i = \sum_i \frac{1}{T_0} \frac{\partial \mathscr{D}}{\partial \dot{q}_i} \dot{q}_i = \frac{2\mathscr{D}}{T_0} \geq 0.$$

In addition, let us define

(12) ▲
$$\mathscr{V} = \mathscr{E}_{\mathrm{I}} - T_0 \mathscr{S}_{\mathrm{I}}.$$

Then the equation of evolution (7) can be written in the familiar Lagrangian form

(13) ▲
$$\frac{\partial \mathscr{D}}{\partial \dot{q}_i} + \frac{\partial \mathscr{V}}{\partial q_i} = Q_i.$$

It is easy to see that in the neighborhood of an equilibrium state (in which $q_i = 0$, $\dot{q}_i = 0$, $Q_i = 0$, $X_i = 0$), $\mathscr{V}$ is a quadratic form in q_i.

For, on expanding $\mathscr{V}$ into a power series in q_i, the constant term has no significance and can be removed, whereas the linear terms must all vanish, because otherwise Eq. (13) cannot be satisfied at the equilibrium condition. Hence, if sufficiently small values of q_i are considered, a quadratic form is obtained if higher powers of q_i are neglected:

(14)   ▲
$$\mathscr{V} = \frac{1}{2} \sum_{i,j} a_{ij} q_i q_j \, .$$

Since only the sum is of interest, we may assume, without loss of generality, that this quadratic form is symmetric,

(15)   ▲
$$a_{ij} = a_{ji} \, .$$

Hence, Eq. (7) can be put in the form

(16)   ▲
$$\sum_j b_{ij} \dot{q}_j + \sum_j a_{ij} q_j = Q_i \, .$$

Equation (16) assumes a form which occurs frequently in the theory of vibration of discrete masses. Evidently, $\mathscr{V}$ plays the role of a potential energy and $\mathscr{D}$ that of the dissipation function which occurs in the usual vibration theory.

13.6. RELAXATION MODES

Let us now consider the solutions of the equation of evolution

(1)
$$\sum_{j=1}^{n} a_{ij} q_j + \sum_{j=1}^{n} b_{ij} \dot{q}_j = Q_i, \qquad\qquad i = 1, 2, \ldots, n \, ,$$

where the coefficients a_{ij} and b_{ij} are real-valued and symmetric,

$$a_{ij} = a_{ji}, \qquad b_{ij} = b_{ji} \, .$$

We shall ignore the inertia forces for the moment, so that Q_i do not involve the acceleration $\ddot{q}_i$. The nature of the solution depends very much on the nature of the quadratic form

(2)
$$\mathscr{D} = \frac{1}{2} \sum_{i=1}^{n} \sum_{j=1}^{n} b_{ij} \dot{q}_i \dot{q}_j \, ,$$

which has been shown above to be proportional to the entropy production $d_i \mathscr{S}_{\text{I+II}}/dt$ and is thus nonnegative. However, $\mathscr{D}$ can be identically zero if the process is reversible. In any case, a coefficient $b_{ij}(i \neq j)$ will be zero if there is no interference between the processes $\dot{q}_i$ and $\dot{q}_j$. For a certain i, the diagonal coefficient b_{ii} will be zero if $\dot{q}_i$ do not participate in the irreversible entropy production (i.e., if q_i is a reversible variable), in which case all coefficients $b_{i1}, \ldots, b_{in}$ vanish. Hence, in general, the quadratic form $\mathscr{D}$ is not positive definite. However, it can be reduced into a reduced positive definite form if all the reversible variables are eliminated (unless $\mathscr{D} \equiv 0$, in which case there is no problem). Let us assume that this reduction has been done, so that the variables $\dot{q}_1, \dot{q}_2, \ldots, \dot{q}_m (m \leq n)$ really participate in the irreversible entropy production. Then we can assert that the quadratic form

$$(3) \qquad \mathscr{D} = \frac{1}{2} \sum_{i=1}^{m} \sum_{j=1}^{m} b_{ij} \dot{q}_i \dot{q}_j$$

is positive definite in the m variables $\dot{q}_i (i = 1, \ldots, m)$.

Equations (1) can then be separated into a group of m equations involving $\dot{q}_1, \ldots, \dot{q}_m$, and another group of $(n - m)$ equations which does not contain time derivatives. The latter group of equations can be solved for $q_{m+1}, \ldots, q_n$, which can be substituted in turn into the first m equations. In this way we obtain a sharper restatement of the equations of evolution:

$$(4) \qquad \sum_{j=1}^{m} a'_{ij} q_j + \sum_{j=1}^{m} b_{ij} \dot{q}_j = Q'_i, \qquad\qquad i = 1, 2, \ldots, m,$$

where the matrix $[b_{ij}]$ is real-valued, symmetric, and positive definite. It is easy to see that the new coefficients a'_{ij} are again real and symmetric.

Let us now consider the solutions of Eqs. (4) in the homogeneous case in which $Q'_i = 0$. The equations

$$(5) \qquad \sum_{j=1}^{m} a'_{ij} q_j + \sum_{j=1}^{m} b_{ij} \dot{q}_j = 0, \qquad\qquad i = 1, \ldots, m,$$

admit a solution of the form

$$(6) \qquad q_j = \psi_j e^{-\lambda t}, \qquad\qquad j = 1, 2, \ldots, m,$$

where ψ_j are constants. On substituting Eq. (6) into Eq. (5), we have

$$(7) \qquad \sum_{j=1}^{m} (a'_{ij} - \lambda b_{ij}) \psi_j = 0, \qquad\qquad i = 1, \ldots, m.$$

To find a set of nontrivial solutions ψ_j poses an eigenvalue problem. The eigenvalues satisfy the determinantal equation

(8) $$\det |a_{ij} - \lambda b_{ij}| = 0.$$

Since the square matrices $[a'_{ij}]$, $[b_{ij}]$ are real, symmetric, and $[b_{ij}]$ is, positive definite, all eigenvalues $\lambda_1, \ldots, \lambda_m$ are real-valued. Whether they are all positive or not depends on the nature of the quadratic form

(9) $$\mathscr{V} = \frac{1}{2} \sum_{i=1}^{m} \sum_{j=1}^{m} a'_{ij} q_i q_j.$$

The irreversible process is considered as a disturbed motion about an equilibrium state. If the equilibrium state is *stable*, then $\mathscr{V}$ is positive definite and all roots $\lambda_1, \ldots, \lambda_m$ are positive. In that case, every disturbance tends to zero with increasing time [see Eq. (6)]. If the equilibrium is *unstable*, then $\mathscr{V}$ is indefinite and some of the roots λ will be negative. In that case, the disturbed motion will increase exponentially with time. If the equilibrium is neutrally stable with respect to some coordinates, then $\mathscr{V}$ is positive semi-definite; some of the roots λ may be zero, while the others are positive.

It is known that corresponding to the m eigenvalues $\lambda_1, \ldots, \lambda_m$ (some of them may be zero or may be multiple roots), there exist m eigenvectors $\{\psi_j^{(1)}\}, \ldots, \{\psi_j^{(m)}\}$ which are linearly independent and orthonormal:

(10)
$$\sum_{i=1}^{m} \sum_{j=1}^{m} b_{ij} \psi_i^{(k)} \psi_j^{(l)} = \begin{cases} 0 & \text{if } k \neq l, \\ 1 & \text{if } k = l, \end{cases}$$

$$\sum_{i=1}^{m} \sum_{j=1}^{m} a'_{ij} \psi_i^{(k)} \psi_j^{(l)} = \begin{cases} 0 & \text{if } k \neq l, \\ \lambda_k & \text{if } k = l. \end{cases}$$

The general solution of Eqs. (5) is

(11) $$q_j = \sum_{k=1}^{m} \psi_j^{(k)} e^{-\lambda_k t}, \qquad j = 1, 2, \ldots, m.$$

Each solution corresponding to a $\lambda_k (k = 1, \ldots, m)$,

(12) $$q_j^{(k)} = \psi_j^{(k)} e^{-\lambda_k t}, \qquad j = 1, 2, \ldots, m,$$

is called a *relaxation mode*.

Thus, for an irreversible process with m fluxes $\dot{q}_1, \ldots, \dot{q}_m$ participating, there exist m linearly independent orthonormal relaxation modes.

13.7. NORMAL COORDINATES

Now we shall consider the nonhomogeneous case, so that we may find the evolution of an irreversible process under a given set of forcing functions. As in the theory of mechanical vibrations, an introduction of the relaxation modes as the basis of generalized coordinates will decouple the equations of motion and thus lead to simple solutions.

We introduce the linear transformation from the state variables $\{q_i\}$ to the normal coordinates $\{\xi_i\}$ by the equations

$$(1) \qquad q_i = \sum_{k=1}^{m} \psi_i^{(k)} \xi_k , \qquad i = 1, \ldots, m ,$$

where $\{\psi_i^{(k)}\}$ is the modal column of the kth relaxation mode. Let the relaxation modes be normalized as in Eq. (13.6:10). Then the potential and dissipation functions assume the form

$$(2) \qquad \begin{aligned} \mathscr{V} &= \frac{1}{2} \sum_{k=1}^{m} \lambda_k \xi_k^2 , \\[2mm] \mathscr{D} &= \frac{1}{2} \sum_{k=1}^{m} \dot{\xi}_k^2 . \end{aligned}$$

The Lagrangian equations of motion, Eq. (13.6:4), now become decoupled,

$$(3) \qquad \frac{\partial \mathscr{V}}{\partial \xi_k} + \frac{\partial \mathscr{D}}{\partial \dot{\xi}_k} = \Xi_k ,$$

where Ξ_k is the generalized force corresponding to ξ_k,

$$(4) \qquad \Xi_k = \sum_{j=1}^{m} \psi_j^{(k)} Q_j' .$$

From Eqs. (2) and (3) we have

$$(5) \qquad \dot{\xi}_k + \lambda_k \xi_k = \Xi_k , \qquad\qquad k = 1, 2, \ldots, m .$$

The Laplace transform of Eq. (5) is, for zero initial condition ($\xi_k = 0$ when $t = 0$),

$$(6) \qquad (s + \lambda_k)\bar{\xi}_k = \bar{\Xi}_k , \qquad\qquad k = 1, 2, \ldots, m ,$$

where

$$(7) \qquad \bar{\xi}_k = \int_0^{\infty} e^{-st} \xi_k(t)\, dt, \qquad \bar{\Xi}_k = \int_0^{\infty} \Xi_k(t) e^{-st}\, dt .$$

The solution of Eq. (6) as algebraic equations in s is

$$(8) \qquad \bar{\xi}_k = \frac{\bar{\Xi}_k}{s + \lambda_k}, \qquad k = 1, 2, \ldots, m.$$

Substituting back into Eq. (1), we obtain the Laplace transform of $q_i(t)$,

$$(9) \qquad \bar{q}_i = \int_0^\infty e^{-st} q_i(t)\, dt = \sum_{k=1}^m \psi_i^{(k)} \frac{\bar{\Xi}_k}{s + \lambda_k}$$

$$= \sum_{k=1}^m \sum_{j=1}^m \frac{\psi_i^{(k)} \psi_j^{(k)} \bar{Q}_j'}{s + \lambda_k}.$$

Equation (9) provides the solution to the response problem in Laplace transformation language. On transforming back to the physical plane, the time history of the evolution of an irreversible process in response to external forcing functions is obtained:

$$(10) \quad \blacktriangle \qquad q_i(t) = \sum_{j=1}^m \sum_{k=1}^m \psi_i^{(k)} \psi_j^{(k)} \int_0^t e^{-\lambda_k \tau} Q_j'(t - \tau)\, d\tau,$$

or

$$(10a) \qquad q_i(t) = \sum_{j=1}^m \sum_{k=1}^m \psi_i^{(k)} \psi_j^{(k)} \int_0^t e^{-\lambda_k (t - \tau)} Q_j'(\tau)\, d\tau.$$

This gives the complete solution if $m = n$. The case $m < n$ [cf. discussion in Sec. 13.6 following Eq. (13.6:2)] is treated in the exercise below (Problem 13.2).

It is seen from Eq. (17) below, that, in general, if some coordinates are reversible, q_i are related to Q_i by equations of the form

$$(11) \quad \blacktriangle \qquad \bar{q}_i(s) = \sum_{j=1}^m \left(\sum_{k=1}^m \frac{A_{ij}^{(k)}}{s + \lambda_k} + A_{ij} \right) \bar{Q}_j(s),$$

$$(12) \quad \blacktriangle \qquad q_i(t) = \sum_{j=1}^m \left[\sum_{k=1}^m A_{ij}^{(k)} \int_0^t e^{-\lambda_k (t - \tau)} Q_j(\tau)\, d\tau + A_{ij} Q_j(t) \right],$$

where

$$(13) \quad \blacktriangle \qquad A_{ij}^{(k)} = A_{ji}^{(k)}, \qquad A_{ij} = A_{ji}.$$

Problem 13.2. Write Eq. (13.6:1) in the form of partitioned matrices

(14)
$$
\begin{bmatrix} \mathbf{A} & \vdots & \mathbf{F} \\ \cdots & & \cdots \\ \mathbf{C} & \vdots & \mathbf{H} \end{bmatrix}
\begin{bmatrix} \mathbf{q}^{(1)} \\ \cdots \\ \mathbf{q}^{(2)} \end{bmatrix}
+
\begin{bmatrix} \mathbf{B} & \vdots & 0 \\ \cdots & & \cdots \\ 0 & \vdots & 0 \end{bmatrix}
\begin{bmatrix} \dot{\mathbf{q}}^{(1)} \\ \cdots \\ 0 \end{bmatrix}
=
\begin{bmatrix} \mathbf{Q}^{(1)} \\ \cdots \\ \mathbf{Q}^{(2)} \end{bmatrix} ,
$$

where

$$
\mathbf{q}^{(1)} = \begin{bmatrix} q_1 \\ \cdot \\ \cdot \\ \cdot \\ q_m \end{bmatrix} , \quad
\mathbf{Q}^{(1)} = \begin{bmatrix} Q_1 \\ \cdot \\ \cdot \\ \cdot \\ Q_m \end{bmatrix} ,
$$

$$
\mathbf{q}^{(2)} = \begin{bmatrix} q_{m+1} \\ \cdot \\ \cdot \\ q_n \end{bmatrix} , \quad
\mathbf{Q}^{(2)} = \begin{bmatrix} Q_{m+1} \\ \cdot \\ \cdot \\ Q_n \end{bmatrix} ,
$$

$$
\mathbf{A} = \begin{bmatrix} a_{11} & a_{12} & \cdots & a_{1m} \\ \cdot & & \cdot & \cdot \\ \cdot & & \cdot & \cdot \\ \cdot & & \cdot & \cdot \\ a_{m1} & a_{m2} & \cdots & a_{mm} \end{bmatrix} , \quad
\mathbf{B} = \begin{bmatrix} b_{11} & b_{12} & \cdots & b_{1m} \\ \cdot & & \cdot & \cdot \\ \cdot & & \cdot & \cdot \\ \cdot & & \cdot & \cdot \\ b_{m1} & b_{m2} & \cdots & b_{mm} \end{bmatrix} ,
$$

$$
\mathbf{F} = \begin{bmatrix} a_{1,m+1} & \cdots & a_{1n} \\ \cdot & & \cdot \\ \cdot & & \cdot \\ \cdot & & \cdot \\ a_{m1} & \cdots & a_{mn} \end{bmatrix} , \quad
\mathbf{C} = \begin{bmatrix} a_{m+1,1} & \cdots & a_{m+1,m} \\ \cdot & & \cdot \\ \cdot & & \cdot \\ \cdot & & \cdot \\ a_{n1} & \cdots & a_{nm} \end{bmatrix} ,
$$

$$
\mathbf{H} = \begin{bmatrix} a_{m+1,m+1} & \cdots & a_{m+1,n} \\ \cdot & & \cdot \\ \cdot & & \cdot \\ \cdot & & \cdot \\ a_{n,m+1} & \cdots & a_{nn} \end{bmatrix} .
$$

Assume that det $|\mathbf{H}| \neq 0$ so that $\mathbf{H}^{-1}$ exists and that at $t = 0$, $\mathbf{q}^{(1)} = 0$, $\mathbf{q}^{(2)} = 0$. Show that if Eq. (13.6:4) is written as

(15)
$$
\mathbf{A}' \cdot \mathbf{q}^{(1)} + \mathbf{B} \cdot \dot{\mathbf{q}}^{(1)} = \mathbf{Q}' ,
$$

then

(16)
$$
\mathbf{A}' = \mathbf{A} - \mathbf{F}\mathbf{H}^{-1}\mathbf{C} ,
$$
$$
\mathbf{Q}' = \mathbf{Q}^{(1)} - \mathbf{F}\mathbf{H}^{-1}\mathbf{Q}^{(2)} .
$$

Let $\mathbf{\Psi}$ be a square matrix formed by the columns of relaxation modes, $\mathbf{\Psi}^T$ be the transpose of $\mathbf{\Psi}$, and $\mathbf{\Lambda}$ be a diagonal matrix of the relaxation spectrum:

$$\Psi = \begin{bmatrix} \psi_1^{(1)} & \psi_1^{(2)} & \cdots & \psi_1^{(m)} \\ & \cdot & & \cdot \\ \cdot & \cdot & & \cdot \\ \cdot & \cdot & & \cdot \\ \psi_n^{(1)} & \psi_n^{(2)} & \cdots & \psi_n^{(m)} \end{bmatrix}, \qquad \Lambda = \begin{bmatrix} \lambda_1 & 0 & \cdots & 0 \\ 0 & \lambda_2 & \cdots & 0 \\ \cdot & \cdot & & \cdot \\ \cdot & \cdot & & \cdot \\ \cdot & \cdot & & \cdot \\ 0 & 0 & \cdots & \lambda_m \end{bmatrix}.$$

Then the Laplace transforms of $\mathbf{q}^{(1)}$, $\mathbf{q}^{(2)}$, $\mathbf{Q}^{(1)}$, $\mathbf{Q}^{(2)}$ are related by

(17)
$$\bar{\mathbf{q}}^{(1)}(s) = \Psi(\Lambda + s\mathbf{I})^{-1}\Psi^T(\bar{\mathbf{Q}}^{(1)} - \mathbf{FH}^{-1}\bar{\mathbf{Q}}^{(2)}),$$
$$\bar{\mathbf{q}}^{(2)}(s) = \mathbf{H}^{-1}\bar{\mathbf{Q}}^{(2)} - \mathbf{H}^{-1}\mathbf{C}\Psi(\Lambda + s\mathbf{I})^{-1}\Psi^T(\bar{\mathbf{Q}}^{(1)} - \mathbf{FH}^{-1}\bar{\mathbf{Q}}^{(2)}),$$

where

$$(\Lambda + s\mathbf{I})^{-1} = \begin{bmatrix} \dfrac{1}{s + \lambda_1} & 0 & \cdots & & 0 \\ & \cdot & & & \\ \cdot & & \cdot & & \cdot \\ \cdot & & & \cdot & \cdot \\ \cdot & & & & \cdot \\ 0 & \cdots & 0 & & \dfrac{1}{s + \lambda_m} \end{bmatrix}.$$

13.8. HIDDEN VARIABLES AND THE FORCE-DISPLACEMENT RELATIONSHIP

In many physical problems a great many variables are "hidden," that is, we do not observe them. For example, we may have a system with n variables in which only k internal forces $Q_1, \ldots, Q_k$, corresponding to the displacements $q_1, \ldots, q_k$, are examined, while the remaining coordinates constitute an "internal system." An example of this occurs in the stress-strain relationship of a polycrystalline metal with stresses $Q_1 = \sigma_{11}$, $Q_2 = \sigma_{12} = \sigma_{21}, \ldots, Q_6 = \sigma_{33}$, strains $q_1 = e_{11}, q_2 = e_{12} = e_{21}, \ldots, q_6 = e_{33}$; whereas the intercrystalline thermal currents, movement of interstitial atoms, twining, etc., are represented by $q_7, q_8, \ldots, q_n$, for which no corresponding external forces are applied. We wish to know the influence of these hidden variables on the stress-strain relationship.

In the general case, some of the hidden variables $q_{k+1}, q_{k+2}, \ldots, q_n$ may not participate in the entropy production: they may be reversible. These reversible hidden variables are related to the other variables by linear algebraic equations, and can be eliminated easily. To simplify our expressions we shall assume that this elimination has been done so that every flux $\dot{q}_{k+1}, \dot{q}_{k+2}, \ldots, \dot{q}_n$ participates in the entropy production.

We now divide all the variables into two groups:

(1)
$$\mathbf{q}^{(1)} = \begin{bmatrix} q_1 \\ q_2 \\ \vdots \\ q_k \end{bmatrix}, \qquad \mathbf{q}^{(2)} = \begin{bmatrix} q_{k+1} \\ q_{k+2} \\ \vdots \\ q_n \end{bmatrix},$$

where $\mathbf{q}^{(1)}$ is the vector to be observed, the elements of which correspond to the generalized forces $Q_1, \ldots, Q_k$ written as a vector

(2)
$$\mathbf{Q} = \begin{bmatrix} Q_1 \\ Q_2 \\ \vdots \\ Q_k \end{bmatrix}.$$

We may now write the equations of evolution in the form of partitioned matrices:

(3)
$$\begin{bmatrix} \mathbf{A}_{11} & \vdots & \mathbf{A}_{12} \\ \cdots & & \cdots \\ \mathbf{A}_{12}^T & \vdots & \mathbf{A}_{22} \end{bmatrix} \begin{bmatrix} \mathbf{q}^{(1)} \\ \cdots \\ \mathbf{q}^{(2)} \end{bmatrix} + \begin{bmatrix} \mathbf{B}_{11} & \vdots & \mathbf{B}_{12} \\ \cdots & & \cdots \\ \mathbf{B}_{12}^T & \vdots & \mathbf{B}_{22} \end{bmatrix} \begin{bmatrix} \dot{\mathbf{q}}^{(1)} \\ \cdots \\ \dot{\mathbf{q}}^{(2)} \end{bmatrix} = \begin{bmatrix} \mathbf{Q} \\ \cdots \\ 0 \end{bmatrix},$$

where a superscript T indicates the transpose of the matrix. The submatrices $\mathbf{A}_{11}$, $\mathbf{A}_{22}$, $\mathbf{B}_{11}$, $\mathbf{B}_{22}$ are symmetric. The matrix $\mathbf{B}_{22}$ is positive definite because we have assumed that every element of $\dot{\mathbf{q}}^{(2)}$ participates in the entropy production.

Our problem is to eliminate $\mathbf{q}^{(2)}$ and relate $\mathbf{Q}$ to $\mathbf{q}^{(1)}$ directly. Let us first consider the subsystem

(4)
$$\mathbf{A}_{22}\mathbf{q}^{(2)} + \mathbf{B}_{22}\dot{\mathbf{q}}^{(2)} = 0 .$$

Since $\mathbf{B}_{22}$ is a positive definite square matrix of order $n-k$, there exists $n-k$ linearly independent orthonormal relaxation modes $\boldsymbol{\psi}^{(1)}$, $\boldsymbol{\psi}^{(2)}, \ldots, \boldsymbol{\psi}^{(n-k)}$, corresponding to the real-valued eigenvalues $\lambda_1, \lambda_2, \ldots, \lambda_{n-k}$. Let $\boldsymbol{\Psi}$ be a square matrix composed of the columns of relaxation modes,

(5)
$$\boldsymbol{\Psi} = \begin{bmatrix} \boldsymbol{\psi}^{(1)} & \boldsymbol{\psi}^{(2)} & \cdots & \boldsymbol{\psi}^{(n-k)} \end{bmatrix}, \qquad \boldsymbol{\psi}^{(i)} = \begin{bmatrix} \psi_1^{(i)} \\ \vdots \\ \psi_{n-k}^{(i)} \end{bmatrix}$$

and let the relaxation modes be so normalized that

(6)
$$\boldsymbol{\Psi}^T \mathbf{B}_{22} \boldsymbol{\Psi} = \mathbf{I}, \qquad \boldsymbol{\Psi}^T \mathbf{A}_{22} \boldsymbol{\Psi} = \boldsymbol{\Lambda},$$

where $\mathbf{I}$ is a unit matrix and $\mathbf{\Lambda}$ is diagonal:

$$(7) \qquad \mathbf{I} = \begin{bmatrix} 1 & \cdots & 0 \\ & \cdot & \cdot & \cdot \\ \cdot & & \ddots & \cdot \\ \cdot & & & \cdot \\ 0 & \cdots & 1 \end{bmatrix}, \qquad \mathbf{\Lambda} = \begin{bmatrix} \lambda_1 & \cdots & & 0 \\ & \cdot & \cdot & \cdot \\ \cdot & & \ddots & \cdot \\ \cdot & & & \cdot \\ 0 & \cdots & & \lambda_{n-k} \end{bmatrix}.$$

By introducing the normal coordinates $\boldsymbol{\xi}$ defined by the relation

$$(8) \qquad \mathbf{q}^{(2)} = \mathbf{\Psi}\,\boldsymbol{\xi},$$

we are ready to simplify Eq. (3). Let us introduce Laplace transformation with respect to time for the $\mathbf{q}$'s and $\mathbf{Q}$'s:

$$(9) \qquad \bar{q}_i = \int_0^\infty e^{-st} q_i(t)\,dt,$$

etc., and write $\bar{\mathbf{q}}^{(1)}$ for the column matrix whose elements are $\bar{q}_1, \bar{q}_2, \ldots, \bar{q}_k$, etc. For our purpose it is sufficient to consider zero initial condition, $\mathbf{q}^{(1)} = 0$, $\mathbf{q}^{(2)} = 0$, when $t = 0$. Then, since the Laplace transformation of $\dot{q}_i$ is equal to $s\bar{q}_i$, we can reduce the differential Eq. (3) into the algebraic equation

$$(10) \qquad \left\{ \begin{bmatrix} \mathbf{A}_{11} & \mathbf{A}_{12} \\ \mathbf{A}_{12}^T & \mathbf{A}_{22} \end{bmatrix} + s \begin{bmatrix} \mathbf{B}_{11} & \mathbf{B}_{12} \\ \mathbf{B}_{12}^T & \mathbf{B}_{22} \end{bmatrix} \right\} \left\{ \begin{matrix} \bar{\mathbf{q}}^{(1)} \\ \bar{\mathbf{q}}^{(2)} \end{matrix} \right\} = \left\{ \begin{matrix} \bar{\mathbf{Q}} \\ 0 \end{matrix} \right\}.$$

Finally, let us define an $n \times n$ matrix $\mathbf{\Phi}$

$$(11) \qquad \mathbf{\Phi} = \begin{bmatrix} \mathbf{I}^{(1)} & 0 \\ 0 & \mathbf{\Psi} \end{bmatrix}.$$

Then it is easy to verify, on account of Eqs. (6) and (7), that

$$(12) \qquad \mathbf{\Phi} \begin{bmatrix} \bar{\mathbf{q}}^{(1)} \\ \bar{\boldsymbol{\xi}} \end{bmatrix} = \begin{bmatrix} \bar{\mathbf{q}}^{(1)} \\ \bar{\mathbf{q}}^{(2)} \end{bmatrix},$$

$$(13) \qquad \mathbf{\Phi}^T \begin{pmatrix} \mathbf{A}_{11} & \mathbf{A}_{12} \\ \mathbf{A}_{12}^T & \mathbf{A}_{22} \end{pmatrix} \mathbf{\Phi} = \begin{pmatrix} \mathbf{A}_{11} & \mathbf{A}_{12}\mathbf{\Psi} \\ \mathbf{\Psi}^T \mathbf{A}_{12}^T & \mathbf{\Lambda} \end{pmatrix},$$

$$(14) \qquad \mathbf{\Phi}^T \begin{pmatrix} \mathbf{B}_{11} & \mathbf{B}_{12} \\ \mathbf{B}_{12}^T & \mathbf{B}_{22} \end{pmatrix} \mathbf{\Phi} = \begin{bmatrix} \mathbf{B}_{11} & \mathbf{B}_{12}\mathbf{\Psi} \\ \mathbf{\Psi}^T \mathbf{B}_{12}^T & \mathbf{I} \end{bmatrix}.$$

Premultiply Eq. (10) by $\boldsymbol{\Phi}^T$, substituting Eq. (12), and using Eqs. (13) and (14), we obtain

(15)
$$\begin{bmatrix} \mathbf{A}_{11} + s\mathbf{B}_{11} & \mathbf{A}_{12}\boldsymbol{\Psi} + s\mathbf{B}_{12}\boldsymbol{\Psi} \\ \boldsymbol{\Psi}^T\mathbf{A}_{12}^T + s\boldsymbol{\Psi}^T\mathbf{B}_{12}^T & \boldsymbol{\Lambda} + s\mathbf{I} \end{bmatrix} \begin{bmatrix} \bar{\mathbf{q}}^{(1)} \\ \bar{\boldsymbol{\xi}} \end{bmatrix} = \begin{bmatrix} \bar{\mathbf{Q}} \\ 0 \end{bmatrix} .$$

Expanding by rows, we see that this is equivalent to two equations:

(16)
$$(\mathbf{A}_{11} + s\mathbf{B}_{11})\bar{\mathbf{q}}^{(1)} + (\mathbf{A}_{12} + s\mathbf{B}_{12})\boldsymbol{\Psi}\bar{\boldsymbol{\xi}} = \bar{\mathbf{Q}} .$$

(17)
$$\boldsymbol{\Psi}^T(\mathbf{A}_{12}^T + s\mathbf{B}_{12}^T)\bar{\mathbf{q}}^{(1)} + (\boldsymbol{\Lambda} + s\mathbf{I})\bar{\boldsymbol{\xi}} = 0 .$$

From Eq. (17), we have

(18)
$$\bar{\boldsymbol{\xi}} = -(\boldsymbol{\Lambda} + s\mathbf{I})^{-1}\boldsymbol{\Psi}^T(\mathbf{A}_{12}^T + s\mathbf{B}_{12}^T)\bar{\mathbf{q}}^{(1)} .$$

A substitution back into Eq. (16) then gives the final result:

(19) $\bar{\mathbf{Q}} = [\mathbf{A}_{11} + s\mathbf{B}_{11} - (\mathbf{A}_{12} + s\mathbf{B}_{12})\boldsymbol{\Psi}(\boldsymbol{\Lambda} + s\mathbf{I})^{-1}\boldsymbol{\Psi}^T(\mathbf{A}_{12}^T + s\mathbf{B}_{12}^T)]\bar{\mathbf{q}}^{(1)} .$

It is clear that the matrix in the brackets is symmetric. The inverse matrix $(\boldsymbol{\Lambda} + s\mathbf{I})^{-1}$ is simply

(20)
$$(\boldsymbol{\Lambda} + s\mathbf{I})^{-1} = \begin{Bmatrix} \dfrac{1}{s+\lambda_1} & 0 & \cdots & 0 \\ 0 & \dfrac{1}{s+\lambda_2} & \cdots & 0 \\ \vdots & \vdots & & \vdots \\ 0 & 0 & \cdots & \dfrac{1}{s+\lambda_{n-k}} \end{Bmatrix} .$$

The general term in Eq. (19) may be written in the form

(21) ▲
$$\bar{Q}_i = \sum_{j=1}^{k}\left(c_{ij} + sc'_{ij} + \sum_{\alpha=1}^{n-k}\frac{c_{ij}^{(\alpha)}}{s+\lambda_\alpha}\right)\bar{q}_j ,$$

where

(22) ▲
$$c_{ij} = c_{ji} , \quad c'_{ij} = c'_{ji}, \quad c_{ij}^{(\alpha)} = c_{ji}^{(\alpha)} .$$

By taking the inverse transform, we obtain the general form of the generalized force-generalized displacement relationship,

(23) ▲ $Q_i(t) = \displaystyle\sum_{j=1}^{k}\left[c_{ij}q_j(t) + c'_{ij}\dot{q}_j(t) + \sum_{\alpha=1}^{n-k}c_{ij}^{(\alpha)}\int_0^t e^{-\lambda_\alpha(t-\tau)}q_j(\tau)d\tau\right] .$

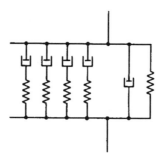

Fig. 13.8:1. Model suggested by Eq. (13.8:23).

If this equation is interpreted in terms of mechanical models, we see that any relaxation phenomena may be represented by a spring, a dashpot, and a sum of a great many elements made up of a Maxwell type material (Fig. 13.8:1).

13.9. ANISOTROPIC LINEAR VISCOELASTIC MATERIALS

Let the six independent components of the stress tensor σ_{ij} play the role of the generalized forces $Q_1, \ldots, Q_6$, and let the six independent components of the strain tensor e_{ij} play the role of the generalized coordinates $q_1, \ldots, q_6$. According to the results of Sec. 13.8, the effect of relaxation modes renders the stress-strain relationship into the form (with summation convention used)

(1) ▲
$$\bar{\sigma}_{ij} = \bar{C}_{ij}^{kl}(s)\bar{e}_{kl},$$

with

(2) ▲
$$\bar{C}_{ij}^{kl}(s) = \sum_{\alpha=1}^{n} \frac{D_{ij}^{kl(\alpha)}}{s + \lambda_\alpha} + D_{ij}^{kl} + sD_{ij}'^{kl}.$$

The restrictions imposed by thermodynamics are

(3) ▲
$$\bar{C}_{ij}^{kl} = \bar{C}_{ji}^{kl} = \bar{C}_{ij}^{lk}, \qquad \bar{C}_{ij}^{kl} = \bar{C}_{kl}^{ij}.$$

Strictly speaking, the last two Laplace transform terms in Eq. (2) have no inverse in the time domain. But the inverse of $\bar{C}_{ij}^{kl}/s^2$ exists. Hence, if $s^2\bar{e}_{kl}$ has a Laplace transform, the inverse of $\bar{\sigma}_{ij}$ given by Eq. (1) can still be defined. However, as the delta function and higher order delta functions are often used in mathematical physics as generalized functions, we can formally write

(4)
$$C_{ij}^{kl}(t) = \sum_\alpha D_{ij}^{kl(\alpha)} e^{-\lambda_\alpha t} 1(t) + D_{ij}^{kl} \delta(t) + D_{ij}'^{kl} \delta_1(t),$$

where $\mathbf{1}(t)$, $\delta(t)$, $\delta_1(t)$ are, respectively, the unit-step function, the delta function, and the first-order delta function. The inverse of (1) then gives, formally,

$$(5) \quad \blacktriangle \quad \sigma_{ij}(t) = \sum_\alpha \int_0^t e^{-\lambda_\alpha(t-\tau)} D_{ij}^{kl(\alpha)} e_{kl}(\tau) d\tau + D_{ij}^{kl} e_{kl}(t) + D_{ij}^{\prime kl} \frac{\partial e_{kl}}{\partial t}.$$

A further generalization is possible if the hidden variables are so numerous so that the summation over α is replaced by an integration over λ_α. See M. A. Biot[3.1] (1956).

On the other hand, in analogy to the Neumann-Duhamel generalization of Hooke's law in thermoelasticity, we may wish to include the deviation of local temperature from the uniform reference temperature, $T - T_0$, as an observable variable in the stress-strain relationship. This can be done by adding another coordinate q_7, to be the local temperature deviation, while the corresponding force Q_7, is taken to mean the local value of entropy per unit volume above that at the reference state.

PROBLEMS

13.3. Specialize the results of Sec. 13.9 to the case of an isotropic material. (See M. A. Biot,[13.1] 1954.)

13.4. From Eq. (13.7:9) or (13.7:11), derive the form of the coefficients A_{ij}^{kl} that relates strain to stress:

$$\bar{e}_{ij} = \bar{A}_{ij}^{kl}(s)\bar{\sigma}_{kl}.$$

13.5. If, for purely mathematical generality, we write the most general linear relation between the Laplace transforms of stress and strain as

$$\left(\sum_\alpha a_{\mu\nu}^{ij(\alpha)} s^\alpha\right) \bar{\sigma}_{\mu\nu} = \left(\sum_\alpha b_{\mu\nu}^{ij(\alpha)} s^\alpha\right) \bar{e}_{\mu\nu},$$

where s is the Laplace transformation variable, and

$$\bar{\sigma}_{\mu\nu}(s) = \int_0^\infty \sigma_{\mu\nu}(t) e^{-st} dt,$$

etc., enumerate the ways that this general expression may not satisfy thermodynamic requirements. (See Biot,[13.1] 1958.)

14

THERMOELASTICITY

Basic equations of thermoelasticity will be discussed in this chapter, and their applications will be illustrated in some typical problems.

14.1. BASIC EQUATIONS

We shall consider a solid body subjected to external forces and heating. We assume that the material is linear elastic, and that it is stress-free at a uniform temperature T_0 when all external forces are removed. The stress-free state will be referred to as the *reference state*, and the temperature T_0 as the *reference temperature*. A system of rectangular Cartesian coordinates x_i will be chosen. The displacement u_i of every particle in the instantaneous state from its position in the reference state will be assumed to be small, so that the infinitesimal strain components are

$$(1) \qquad e_{ij} = \frac{1}{2}(u_{i,j} + u_{j,i}), \qquad\qquad i,j = 1,2,3.$$

The instantaneous absolute temperature will be denoted by T, and the difference $T - T_0$ by θ:

$$(2) \qquad \theta = T - T_0.$$

Under these conditions the basic equations of thermoelasticity are

The Constitutive Equation. Duhamel–Neumann law. Eq. (12.7:4.)

$$(3) \qquad \sigma_{ij} = C_{ijkl}e_{kl} - \beta_{ij}(T - T_0).$$

Conservation of Mass. Continuity equation.

$$(4) \qquad \frac{\partial \rho}{\partial t} + \frac{\partial \rho v_i}{\partial x_i} = 0, \qquad\qquad v_i = \frac{\partial u_i}{\partial t}.$$

Conservation of Momentum. Newton's law.

$$(5) \qquad \rho \dot{v}_i = \frac{\partial \sigma_{ij}}{\partial x_j} + X_i.$$

Conservation of Energy. Eq. (12.2:7).

$$
(6) \qquad \dot{\mathscr{E}} = T\dot{\mathscr{S}} + \frac{1}{\rho}\sigma_{ij}V_{ij}\,, \qquad V_{ij} = \frac{1}{2}\left(\frac{\partial v_i}{\partial x_j} + \frac{\partial v_j}{\partial x_i}\right).
$$

Rate of Change of Entropy. Eqs. (13.1:2) and (13.2:6).

$$
(7) \qquad \rho\dot{\mathscr{S}} = -\frac{1}{T}\frac{\partial h_i}{\partial x_i} = -\frac{\partial}{\partial x_i}\left(\frac{h_i}{T}\right) - \frac{h_i}{T^2}\frac{\partial T}{\partial x_i}\,.
$$

Heat Conduction. Fourier's law. Eqs. (13.3:4) and (13.4:10).

$$
(8) \qquad h_i = -k_{ij}\frac{\partial T}{\partial x_j}\,.
$$

Definition of Specific Heat. If $\dot{e}_{ij} = 0$ $(i, j = 1, 2, 3)$, then

$$
(9) \qquad -\frac{\partial h_i}{\partial x_i} = \rho C_v \dot{T}\,.
$$

In these equations, all the indices range over 1, 2, 3; ρ is the mass density, σ_{ij} are components of the stress tensor, C_{ijkl} are the elastic moduli; β_{ij} are the thermal moduli; v_i are components of the velocity vector; X_i are components of the body force per unit volume; $\mathscr{E}$ is the internal energy per unit mass; $\mathscr{S}$ is the entropy per unit mass; h_i are the components of the heat flux vector; k_{ij} are the heat conduction coefficients; and C_v the heat capacity per unit mass at constant volume of the solid. A dot above a variable denotes the material derivative of that variable:

$$
(\dot{\,}) = \frac{D(\,)}{Dt} = \frac{\partial(\,)}{\partial t} + v_j\frac{\partial(\,)}{\partial x_j}\,,
$$

which, under the approximation of small velocities, is the same as the partial derivative with respect to time. In this chapter, we assume

$$
(\dot{\,}) = \frac{\partial(\,)}{\partial t}\,.
$$

A comma indicates partial differentiation with respect to a spatial coordinate, thus $\theta_{,i}$ means $\partial\theta/\partial x_i$. Summation convention for repeated indices is used. Equation (3) is discussed in Sec. 12.7; Eqs. (4) and (5) are discussed in Sec. 5.4; Eq. (6) is discussed in Sec. 12.2; Eq. (7) is discussed in Secs. 13.1 and 13.2, where h_i/T is identified as the rate of change of entropy displacement and $-(h_i/T^2)\partial T/\partial x_i$ as the entropy production. We have assumed in Eqs. (7) and (9) that there is no heat source in the material; otherwise, to the right-hand side of thses equations we should add a term representing the strength of the heat source (heat generation per unit time per unit volume). Equation (8) is discussed in Sec. 13.3. The coefficients C_{ijkl}, β_{ij}, and k_{ij} have the following symmetry properties as discussed in Secs. 12.7 and 13.3.

$$(10) \qquad C_{ijkl} = C_{klij} = C_{ijlk} = C_{jikl} ,$$

$$(11) \qquad \beta_{ij} = \beta_{ji} ,$$

$$(12) \qquad k_{ij} = k_{ji} .$$

The energy equation may be put in a more convenient form as follows. On introducing the free energy $\mathscr{F}$

$$\mathscr{F} = \mathscr{E} - T\mathscr{S} ,$$

we obtain, according to Eq. (12.3:6),

$$(13) \qquad \rho \left(\frac{\partial \mathscr{F}}{\partial e_{ij}} \right)_T = \sigma_{ij} , \qquad \left(\frac{\partial \mathscr{F}}{\partial T} \right)_{e_{ij}} = -\mathscr{S} .$$

With Eq. (13), Eq. (7) can be written as

$$(14) \qquad -\frac{1}{T} \frac{\partial h_i}{\partial x_i} = \rho \dot{\mathscr{S}} = \rho \frac{\partial \mathscr{S}}{\partial e_{ij}} \dot{e}_{ij} + \rho \frac{\partial \mathscr{S}}{\partial T} \dot{T}$$

$$= -\rho \frac{\partial^2 \mathscr{F}}{\partial e_{ij} \partial T} \dot{e}_{ij} - \rho \frac{\partial^2 \mathscr{F}}{\partial T^2} \dot{T} .$$

A multiplication by T and comparison with Eq. (9) when $\dot{e}_{ij} = 0$ yields

$$(15) \qquad C_v = -T \frac{\partial^2 \mathscr{F}}{\partial T^2} .$$

Also, from Eqs. (13) and (3), it is seen that

$$(16) \qquad \rho \frac{\partial^2 \mathscr{F}}{\partial e_{ij} \partial T} = \frac{\partial \sigma_{ij}}{\partial T} = -\beta_{ij} .$$

Equation (14) therefore takes the form

$$(17) \quad \blacktriangle \qquad -\frac{\partial h_i}{\partial x_i} = \rho C_v \dot{T} + T \beta_{ij} \dot{e}_{ij} ,$$

which reduces to Eq. (9) if $\dot{e}_{ij} = 0$. Finally, on substituting Eq. (8) into Eq. (17), we obtain

$$(18) \quad \blacktriangle \qquad \frac{\partial}{\partial x_i} \left(k_{ij} \frac{\partial T}{\partial x_j} \right) = \rho C_v \frac{\partial T}{\partial t} + T \beta_{ij} \frac{\partial e_{ij}}{\partial t} .$$

If the material is isotropic, these equations are simplified as follows.

$$(19) \qquad \sigma_{ij} = \lambda e_{\mu\mu}\delta_{ij} + 2Ge_{ij} - \beta\delta_{ij}\theta\,,$$

$$(20) \qquad \beta_{ij} = \beta\delta_{ij}\,, \qquad \beta = \frac{\alpha E}{1 - 2\nu} = (3\lambda + 2G)\alpha\,,$$

$$(21) \qquad k_{ij} = k\delta_{ij}\,.$$

The constant α is the thermal coefficient of linear expansion and k is the heat conductivity.

A problem in thermoelasticity is formulated when appropriate boundary conditions and initial conditions are specified. Several simple examples will be discussed below. A proof for the existence and uniqueness of solution under suitable continuity conditions can be constructed in a way analogous to that discussed in Sec. 7.4; an explicit proof in the case of isotropic materials can be found in Boley and Weiner[14.1], pp. 38–40.

14.2. THERMAL EFFECTS DUE TO A CHANGE OF STRAIN; KELVIN'S FORMULA

It is a familiar fact that an adiabatic expansion of a gas is accompanied by a drop in its temperature. Similarly, a solid body changes its temperature when the state of strain of the body is altered adiabatically. For a material like steel, there will be a fall of temperature when the body is strained to expand adiabatically.

Equation (14.1:18) gives the relationship between the rate of change of the temperature and the strain with the heat conduction. If heat conduction is prevented $(\partial h_i/\partial x_i = 0)$, the left-hand side of Eq. (14.1:18) should vanish, and we obtain at once

$$(1) \qquad \frac{\partial T}{\partial t} = -\frac{T}{\rho C_v}\beta_{ij}\frac{\partial e_{ij}}{\partial t}\,,$$

which is Kelvin's formula for the change of temperature of an insulated body due to strain rate.

14.3. RATIO OF ADIABATIC TO ISOTHERMAL ELASTIC MODULI

If an elastic modulus is measured on a sample that is completely insulated, or if the change in strain takes place so rapidly that the heat does **not** have time to escape, the measured value may be called the *adiabatic modulus*. On the other hand, it is called the *isothermal* modulus if it is measured in such a way that the temperature is kept uniform and constant throughout the process.

The coefficients C_{ijkl} in the Duhamel–Neumann law (14.1:3) is the isothermal modulus of elasticity at $\theta = 0$:

(1) $$\sigma_{ij} = C_{ijkl}e_{kl} - \beta_{ij}\theta, \qquad\qquad \theta = T - T_0.$$

If a deformation is adiabatic, and no heat conduction takes place, then Eq. (14.1:17) gives

(2) $$\rho C_v \dot\theta + T_0\beta_{ij}\dot e_{ij} = 0,$$

or

(3) $$\rho C_v\theta + T_0\beta_{kl}e_{kl} = \text{const}.$$

The constant of integration is zero if $e_{ij} = 0$ when $\theta = 0$. On solving Eq. (3) for θ and substituting into Eq. (1), we obtain in an adiabatic process

(4) $$\sigma_{ij} = \left(C_{ijkl} + \frac{T_0}{\rho C_v}\beta_{ij}\beta_{kl}\right)e_{kl} = C'_{ijkl}e_{kl}.$$

Thus, the adiabatic modulus of elasticity C'_{ijkl} is related to the isothermal modulus C_{ijkl} by the equation

(5) $$C'_{ijkl} = C_{ijkl} + \frac{T_0}{\rho C_v}\beta_{ij}\beta_{kl}.$$

PROBLEMS

14.1. Derive the relationship of the temperature and the elastic constant $(\partial p/\partial V)_T$ for an ideal gas, and compare qualitatively the temperature change in an adiabatic expansion of a gas with the temperature change of a similar process in a solid.

14.2. Show that the Young's modulus of steel determined by an adiabatic process such as the propagation of elastic waves or the longitudinal vibration of a rod is higher than that determined by a slow static test in which the strains are maintained constant for a sufficiently long time for the temperature to become uniform.

14.3. For steel at $T = 274.7°$ K, or nearly $1.6°$C, $\rho = 7.0$ gm/cm^3, $\alpha = 1.23 \times 10^{-5}$, $C_v = 0.102$ cal/(gm-$°$C), show that $dT = -0.125°$C if a wire is subjected adiabatically to a sudden increment of tensile stress of 1.09×10^9 dyne/sq cm. Over a century ago, Joule [*Phil. Trans. Roy. Soc. London,* **149** (1859), p. 91] gave an observed value $dT = -0.1620°$C for the above case by experiments on cylindrical bars.

14.4. Consider simple tension and derive the following relationship between the adiabatic and isothermal Young's modulus, E' and E, respectively:

$$\frac{1}{E'} = \frac{1}{E} - \frac{\alpha^2 T_0}{JC_v\rho},$$

which shows that the adiabatic modulus E' is always greater than the isothermal modulus E. In the equation above α is the linear coefficient of expansion, T_0 is the equilibrium temperature, J is the mechanical equivalent of heat, C_v is the specific heat per unit mass at zero strain rate, ρ is the density of the material.

14.5.   Show that the theoretical value of the ratio E'/E is of order 1.003 for steel and copper at room temperature.

14.4. UNCOUPLED, QUASI-STATIC THERMOELASTIC THEORY

The basic equations given in Sec. 14.1 combine the theory of elasticity with heat conduction under transient conditions. Boundary-value problems involving these equations are rather difficult to solve. Fortunately, in most engineering applications it is possible to omit the mechanical coupling term in the energy Eqs. (14.1:6) or in the heat conduction Eq. (14.1:18) and the inertia term in the equation of motion (14.1:5) without significant error. When these simplifying assumptions are introduced, the theory is referred to as an *uncoupled, quasi-static theory*; it degenerates into heat conduction and thermoelasticity as two separate problems.

A plausible argument for the smallness of the thermoelastic coupling is as follows. We have seen in Sec. 14.2 that the change of temperature of an elastic body due to adiabatic straining is, in general, very small. If this interaction between strain and temperature is ignored, then the only effects of elasticity on the temperature distribution are effects of change in dimensions of the body under investigation. The change in dimension of a body is of the order of the product of the linear dimension of the body L, the temperature rise T, and the coefficient of thermal expansion α. If $L = 1$ in., $\theta = 1000°F$, $\alpha = 10^{-5}$ per °F, the change in dimension is 10^{-2} in., which is negligible in problems of heat conduction.

Again, if the temperature rise from 0 to 1000°F were achieved in a time interval of 0.1 sec, then the acceleration is of the order $10^{-2} \div (0.1)^2 = 1$ in./sec^2. The change of stress due to this acceleration may be estimated from the equation of equilibrium

$$\Delta\sigma_{xx} \cong \Delta x \rho \frac{d^2 u}{dt^2}.$$

If the specific gravity of the material is ten and the material is 1 in. thick, we have

$$\Delta\sigma_{xx} = \frac{1}{12} \times \frac{1}{32.2} \times 10 \times 62.4 \times \frac{1}{12} \times \frac{1}{144} \cong 0.001 \text{ lb/sq in}.$$

This stress is negligible in most structural problems in which the magnitude of the stresses concerned are of the order of the yielding stress or ultimate stress of the material.

Some examples of the coupled theory are presented in Chapter 2 of Boley and Weiner's book.[14.1] It is pointed out that the thermomechanical coupling is important in the problem of internal friction of metals.

14.5. TEMPERATURE DISTRIBUTION

We shall denote the temperature above a reference temperature by θ, the space coordinates by (x, y, z), and time by t. In an uncoupled theory for an isotropic material, Eq. (14.1:18) becomes

$$
(1) \qquad \frac{\partial}{\partial x}\left(k\frac{\partial \theta}{\partial x}\right) + \frac{\partial}{\partial y}\left(k\frac{\partial \theta}{\partial y}\right) + \frac{\partial}{\partial z}\left(k\frac{\partial \theta}{\partial z}\right) = \rho C_v \frac{\partial \theta}{\partial t} .
$$

In the following discussion we shall assume k and C_v to be constants. Then,

$$
(2) \qquad k\left[\frac{\partial^2 \theta}{\partial x^2} + \frac{\partial^2 \theta}{\partial y^2} + \frac{\partial^2 \theta}{\partial z^2}\right] = \rho C_v \frac{\partial \theta}{\partial t} .
$$

In cylindrical coordinates (x, r, ψ), we have

$$
(3) \qquad k\left[\frac{\partial^2 \theta}{\partial x^2} + \frac{\partial^2 \theta}{\partial r^2} + \frac{1}{r}\frac{\partial \theta}{\partial r} + \frac{1}{r^2}\frac{\partial^2 \theta}{\partial \psi^2}\right] = \rho C_v \frac{\partial \theta}{\partial t} .
$$

In spherical polar coordinates (r, ψ, φ), we have

$$
(4) \qquad k\left[\frac{1}{r^2}\frac{\partial}{\partial r}\left(r^2\frac{\partial \theta}{\partial r}\right) + \frac{1}{r^2 \sin\psi}\frac{\partial}{\partial \psi}\left(\sin\psi\frac{\partial \theta}{\partial \psi}\right) + \frac{1}{r^2 \sin^2\psi}\frac{\partial^2 \theta}{\partial \varphi^2}\right]
$$
$$
= \rho C_v \frac{\partial \theta}{\partial t} .
$$

For the special case of steady heat flow, $\partial \theta / \partial t$ vanishes, then

$$
(5) \qquad \nabla^2 \theta = 0 .
$$

Numerous examples of solution of these equations can be found in books on the conduction of heat in solids, such as Carslaw and Jaeger, McAdams, etc. (See Bibliographies 14.1 and 14.2.)

Example. *Steady Temperature Distribution in a Disk Cooled by Air*

If there is a heat source in the material, of Q *Btu* per unit volume per unit time, the equation of heat conduction is modified into

$$(6) \qquad k\nabla^2\theta + Q = \rho C_v \frac{\partial\theta}{\partial t}.$$

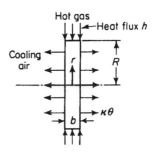

Fig. 14.5:1. A turbine disk.

An atomic reactor with a radiation source may present problems of this kind. A turbine disk may also be described approximately by an equation of this type. Consider a circular disk as shown in Fig. 14.5:1. The heat flux from the rim represents the heat from the turbine blades operating in hot combustion gas. The heat loss on the disk surface represents the effect of cooling air. Let $\theta(r)$ be the average temperature (above the cooling air) across the disk at any radius r. The heat loss per unit area on each face is $-\kappa\theta$. The total heat loss per unit volume is $Q = -2\kappa\theta/b$. Hence, for the axially symmetric temperature distribution, the heat conduction equation at the steady state is

$$(7) \qquad \frac{d^2\theta}{dr^2} + \frac{1}{r}\frac{d\theta}{dr} - \frac{2\kappa}{kb}\theta = 0.$$

This is the differential equation for the modified Bessel functions. The appropriate solution is

$$(8) \qquad \theta = AI_0\left(\sqrt{\frac{2\kappa}{kb}}\, r\right),$$

where $I_0(z)$ is the modified Bessel function of the first kind and zeroth order.

The constant A is determined by the boundary condition at the rim $r = R$, where a heat flux h enters the disk from the hot gas:

$$(9) \qquad k\frac{d\theta}{dr} = h \qquad\qquad\qquad \text{at } r = R.$$

Therefore,

$$(10) \qquad kA\sqrt{\frac{2\kappa}{kb}}I_0'\left(\sqrt{\frac{2\kappa}{kb}}R\right) = h,$$

where I_0' is the derivative of I_0 and is equal to I_1. Thus, finally,

$$(11) \qquad \theta = \frac{h}{\sqrt{2\kappa k/b}}\frac{I_0(\sqrt{2\kappa/kb}\,r)}{I_1(\sqrt{2\kappa/kb}R)}.$$

14.6. THERMAL STRESSES

In the uncoupled, quasi-static theory, the stress and strain fields are computed for each instantaneous temperature distribution $\theta(x_1, x_2, x_3)$ according to Eqs. (14.1:1) to (14.1:5), with appropriate boundary conditions. For an isotropic material, we have

(1) $$\sigma_{ij} = \lambda e_{\mu\mu}\delta_{ij} + 2Ge_{ij} - \beta\delta_{ij}\theta\,,$$

(2) $$\sigma_{ij,j} + X_i = 0\,.$$

The strain field

(3) $$e_{ij} = \frac{1}{2}\left(u_{i,j} + u_{j,i}\right)$$

must satisfy the compatibility conditions

(4) $$e_{ij,kl} + e_{kl,ij} - e_{ik,jl} - e_{jl,ik} = 0\,.$$

As a conjugate to Eq. (1), we have

(5) $$e_{ij} = \frac{1+\nu}{E}\sigma_{ij} - \frac{\nu}{E}\sigma_{\mu\mu}\delta_{ij} + \alpha\theta\delta_{ij}\,, \qquad\qquad \beta = \frac{E\alpha}{1 - 2\nu}\,.$$

Let us first observe that if $\sigma_{ij} = 0$, then

(6) $$e_{ij} = \alpha\theta\delta_{ij}$$

and Eq. (4) is reduced to

(7) $$\theta_{,kl}\delta_{ij} + \theta_{,ij}\delta_{kl} - \theta_{,jl}\delta_{ik} - \theta_{,ik}\delta_{jl} = 0\,,$$

which is satisfied if

(8) $$\theta_{,ij} = 0\,, \qquad\qquad i, j = 1, 2, 3\,.$$

The solution is

(9) $$\theta = a_0 + a_1 x_1 + a_2 x_2 + a_3 x_3\,,$$

with arbitrary coefficients a_i. Hence, if θ is a linear function of spatial coordinates x_1, x_2, x_3, and if the displacements on the boundary are unrestrained, then it is possible to satisfy the compatibility condition without calling stresses into play. Generally, for an arbitrary temperature field, the strain field corresponding to thermal expansion alone, Eq. (6), will not be compatible, in which case thermal stresses must be called into action.

If we substitute Eqs. (1) and (3) into Eq. (2), we obtain the generalized Navier's equation

(10) $$Gu_{i,\mu\mu} + (\lambda + G)u_{\mu,\mu i} + X_i - \beta\theta_{,i} = 0\,.$$

This is particularly convenient to use if the boundary condition is specified in terms of displacements:

$$(11) \qquad u_i = f_i(x_1, x_2, x_3) \qquad \text{on the boundary}.$$

On the other hand, if tractions are specified on the boundary, i.e., if

$$(12) \qquad \overset{\nu}{T_i} = \sigma_{ij}\nu_j = g_i(x_1, x_2, x_3) \qquad \text{on the boundary},$$

where ν_j is the outer normal vector to the boundary surface, and g_i's are specified, we must have

$$(13) \qquad \nu_j[\lambda u_{\mu,\mu}\delta_{ij} + Gu_{i,j} + Gu_{j,i} - \beta\delta_{ij}\theta] = g_i \qquad \text{on the boundary}.$$

By comparing Eqs. (10)–(13) with the corresponding equations in linear elasticity (Secs. 7.1 and 7.3), we see that the effect of the temperature change θ is equivalent to replacing the body force X_i in Navier's equations by $X_i - \beta\theta_{,i}$ and to substituting for the surface tractions g_i by $g_i + \beta\nu_i\theta$. Thus the displacements u_1, u_2, u_3 produced by a temperature change θ are the same as those produced by the body forces $-\beta\theta_{,i}$ and the normal tractions $\beta\theta$ acting on the surface of a body of the same shape but throughout which the temperature is uniform. These facts can be stated in a theorem.[†]

Theorem: Duhamel–Neumann Analogy. *Consider two bodies of exactly the same shape but with conditions prescribed as shown in Fig. 14.6 : 1. Then*

$$u_i^{(I)}(x_1, x_2, x_3; t) = u_i^{(II)}(x_1, x_2, x_3; t),$$

$$\sigma_{ij}^{(I)} = \sigma_{ij}^{(II)} - \beta\theta^{(I)}\delta_{ij}.$$

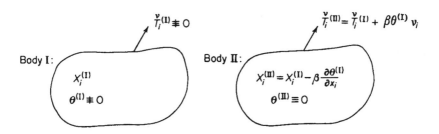

Fig. 14.6:1. Duhamel–Neumann analogy.

[†]The Duhamel–Neumann theorem was stated in this form by A. J. A. Morgan.

Problem 14.6.   *Thermal Stresses in Plates with Clamped Edges.* Consider a flat plate of arbitrary shape clamped at all edges. Let the midplane of the plate in the unloaded position be the x, y-plane and let the temperature be uniform in the plane of the plate but variable throughout the thickness, i.e., $\theta = \theta(z)$. Show that under these conditions the plate will remain flat, i.e., no bending will be introduced except possibly near the edges. Determine the thermal stresses in the plate. This example shows the extreme importance of the edge constraints on the thermal stress problem.

14.7. PARTICULAR INTEGRAL-GOODIER'S METHOD

By introducing suitable particular integrals, the problem of thermal stresses can be reduced to the solution of the homogeneous Navier's equations. Let us assume a solution of Eq. (14.6:10) in the form

$$(1) \qquad\qquad u_i = \frac{\partial \phi}{\partial x_i}$$

where ϕ is a *displacement potential.* Let us assume also that the body forces are conservative so that

$$(2) \qquad\qquad X_i = -\frac{\partial P}{\partial x_i},$$

then the generalized Navier's Eq. (14.6:10) may be written as

$$(3) \qquad\qquad (\lambda + 2G)\phi_{,i\mu\mu} = P_{,i} + \beta\theta_{,i}.$$

An integration yields

$$(4) \qquad\qquad \phi_{,\mu\mu} = \frac{1}{\lambda + 2G}(P + \beta\theta).$$

The constant of integration may be absorbed in the potential P. Since Eq. (4) is linear its solution may be written

$$(5) \qquad\qquad \phi = \phi^{(p)} + \phi^{(c)},$$

where $\phi^{(p)}$ is a particular integral satisfying Eq. (4), and $\phi^{(c)}$ is the complementary solution satisfying the equation

$$(6) \qquad\qquad \nabla^2 \phi^{(c)} = 0.$$

A particular integral can be taken in the form of the gravitational potential due to a distribution of matter of density $(P + \beta\theta)/(\lambda + 2G)$, i.e.,

$$(7) \qquad \phi^{(p)}(x_i) = \frac{-1}{4\pi(\lambda + 2G)} \iiint \frac{1}{r}[P(x_i') + \beta\theta(x_i')]\, dx_1' dx_2' dx_3',$$

where

(8) $$r = [(x_1 - x_1')^2 + (x_2 - x_2')^2 + (x_3 - x_3')^2]^{1/2}.$$

Once $\phi^{(p)}$ is determined, the boundary conditions for the complementary function $\phi^{(c)}$ can be derived.

There are problems which cannot be solved by the method of scalar potential. However, the particular solution $\phi^{(p)}$ can still be used. From $\phi^{(p)}$, the displacements $u_i^{(p)}$ are computed according to Eq. (1). Then we put

(9) $$u_i = u_i^{(c)} + u_i^{(p)}.$$

The equations governing $u_i^{(c)}$ are the homogeneous equations

(10) $$(\lambda + G)\frac{\partial e^{(c)}}{\partial x_i} + G\nabla^2 u_i^{(c)} = 0, \qquad e^{(c)} = u_{\alpha,\alpha}^{(c)}, \qquad i = 1, 2, 3.$$

The boundary conditions for $u_i^{(c)}$ must be derived, of course, from the original boundary conditions by subtracting the contributions of the particular integral on the boundary.

Problem 14.7. Consider the special case in which $X_i = 0$. Show that, in view of the heat conduction equation

$$k\nabla^2 \theta = \rho C_v \frac{\partial \theta}{\partial t},$$

a particular solution $\phi^{(p)}$ of Eq. (4) is

(11) $$\phi^{(p)}(x_1, x_2, x_3; t) = \frac{\alpha k}{\rho C_v}\frac{1+\nu}{1-\nu} \int_t^\infty \theta(x_1, x_2, x_3; \xi)d\xi,$$

provided that $\theta \to 0$ as $t \to \infty$.

14.8. PLANE STRAIN

A long cylindrical body is said to be in the state of plane strain parallel to the x, y-plane if the displacement component w vanishes and the components u and v are functions of x, y, but not of z; and the temperature distribution is independent of z also. Let the body occupy a domain $V(x, y)$ with boundary B in the x, y-plane (Fig. 14.8:1). Then, in $V + B$,

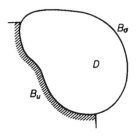

Fig. 14.8:1. Notations.

(1) $u = u(x, y)$, $v = v(x, y)$, $w = 0$, $\theta = \theta(x, y)$.

It is easily verified that

(2) $e_{xz} = e_{yz} = e_{zz} = 0$,

(3) $\sigma_{xz} = \sigma_{yz} = 0$,

(4) $\sigma_{zz} = \lambda(e_{xx} + e_{yy}) - \beta\theta = \nu(\sigma_{xx} + \sigma_{yy}) - \alpha E\theta$.

We may derive Duhamel's analogy explicitly in this two-dimensional case. For a body in $V + B$, subjected to a temperature rise $\theta(x, y)$, but with zero body force and zero surface traction over B_σ, and assigned displacements u_i on B_u, where $B = B_\sigma + B_u$, the stress field is the same as that given by a superposition of a hydrostatic pressure

(5) $\sigma_{xx} = \sigma_{yy} = -\beta\theta$, $\sigma_{xy} = 0$,

with that given by another problem in which the same body in $V + B$ is kept at the uniform reference temperature, but subjected to a body force $X = -\beta(\partial\theta/\partial x)$; $Y = -\beta(\partial\theta/\partial y)$, and a tension $\beta\theta$ over B_σ.

Let us see how the last-mentioned problem can be solved. The equation of equilibrium is

(6) $\dfrac{\partial\sigma_{xx}}{\partial x} + \dfrac{\partial\sigma_{xy}}{\partial y} - \beta\dfrac{\partial\theta}{\partial x} = 0$, $\dfrac{\partial\sigma_{xy}}{\partial x} + \dfrac{\partial\sigma_{yy}}{\partial y} - \beta\dfrac{\partial\theta}{\partial y} = 0$,

and the compatibility condition is

(7) $\dfrac{\partial^2 e_{xx}}{\partial y^2} + \dfrac{\partial^2 e_{yy}}{\partial x^2} = 2\dfrac{\partial^2 e_{xy}}{\partial x \partial y}$,

which can be reduced to

(8) $\nabla^2(\sigma_{xx} + \sigma_{yy}) = \dfrac{\beta}{1-\nu}\nabla^2\theta$.

This equation can be obtained by substituting $-\beta\partial\theta/\partial x$ and $-\beta\partial\theta/\partial y$ for X and Y in Eq. (9.2:10). If we introduce the Airy stress function Φ so that

(9) $\sigma_{xx} = \dfrac{\partial^2\Phi}{\partial y^2} + \beta\theta$, $\sigma_{yy} = \dfrac{\partial^2\Phi}{\partial x^2} + \beta\theta$, $\sigma_{xy} = -\dfrac{\partial^2\Phi}{\partial x \partial y}$,

then Eq. (6) is identically satisfied and Eq. (7) becomes

(10) $$\nabla^4\Phi = -\frac{\alpha E}{1-\nu}\nabla^2\theta, \qquad\qquad \nabla^2 = \frac{\partial^2}{\partial x^2} + \frac{\partial^2}{\partial y^2},$$

where $\alpha = (1-2\nu)\beta/E$ is the linear coefficient of expansion. But

(11) $$k\nabla^2\theta = \rho C_v \frac{\partial\theta}{\partial t}.$$

Hence, finally,

(12) $$\nabla^4\Phi = -\frac{\alpha E \rho C_v}{k(1-\nu)}\frac{\partial\theta}{\partial t}.$$

The details are left to the reader.

Example

If a long cylinder with axis z is subjected to *steady heat flow* with resulting temperature distribution $\theta(x,y)$ independent of t, then $\nabla^2\theta = 0$ and Eq. (12) implies that

(13) $$\nabla^4\Phi = 0.$$

Let the boundary surface of the cylinder be unrestrained. According to Duhamel analogy, the boundary condition is a tensile traction $\beta\theta$. Hence, according to Eq. (9), the boundary conditions can be satisfied by taking

(14) $$\frac{\partial^2\Phi}{\partial y^2} = 0, \quad \frac{\partial^2\Phi}{\partial x^2} = 0, \quad \frac{\partial^2\Phi}{\partial x\partial y} = 0 \qquad \text{on the boundary}.$$

A solution is evidently $\Phi \equiv 0$. For this solution, a superposition of Eqs. (5) and (9) yields

(15) $$\sigma_{xx} = \sigma_{yy} = \sigma_{xy} = 0,$$

whereas, from Eq. (4),

(16) $$\sigma_{zz} = -\alpha E\theta.$$

If the region $V(x,y)$ occupied by the cylinder is simply connected, and if plane strain condition can be assumed, we can quote the uniqueness theorem to say that this is *the* solution. Hence, we conclude that *if a cylinder is simply connected and unrestrained on the surface, then under steady two-dimensional heat flow and plane strain condition, there will be no thermal stress in any surface element parallel to the cylinder axis.*

PROBLEMS

14.8. *Plane Stress.* A body is said to be in a state of plane stress parallel to the x_1, x_2-plane if

$$\sigma_{13} = \sigma_{23} = \sigma_{33} = 0$$

so that

$$e_{33} = -\frac{\nu}{1-\nu}\frac{\partial u_i}{\partial x_i} + \frac{1+\nu}{1-\nu}\alpha\theta, \qquad e_{13} = e_{23} = 0,$$

where $i = 1, 2$. Prove the following Duhamel analogy (Fig. 14.8:2):

Problem I: Heating   **Problem II: Nonheating**

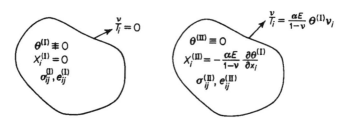

Fig. 14.8:2. Duhamel analogy in the case of plane stress.

Then

$$e_{ij}^{(I)} = e_{ij}^{(II)}$$

$$\sigma_{ij}^{(I)} = \sigma_{ij}^{(II)} - \frac{\alpha E\theta^{(I)}}{1-\nu}\delta_{ij}.$$

It is seen that the role of the parameter $\beta = \alpha E/(1-2\nu)$ is now played by the parameter $\alpha E/(1-\nu)$. If Problem II in plane stress is posed in terms of stresses and Airy stress functions so that

$$\sigma_{11} = \frac{\partial^2\Phi}{\partial x_2^2} + \frac{\alpha E\theta}{1-\nu}, \qquad \sigma_{22} = \frac{\partial^2\Phi}{\partial x_1^2} + \frac{\alpha E\theta}{1-\nu}, \qquad \sigma_{12} = -\frac{\partial^2\Phi}{\partial x_1\partial x_2},$$

then

$$\nabla^4\Phi = -\alpha E\nabla^2\theta, \qquad\qquad \nabla^2 = \frac{\partial^2}{\partial x_1^2} + \frac{\partial^2}{\partial x_2^2}.$$

14.9. Consider a thin plate parallel to the x, y-plane subjected to a temperature distribution $T(x,y)$ which is independent of z. The Duhamel analogy can be applied either in the three-dimensional form or in the plane stress form with the vanishing of σ_{zz}, σ_{zx}, σ_{zy}, introduced explicitly at the outset. Demonstrate the similarity and differences of these two analogies.

14.9. AN EXAMPLE — STRESSES IN A TURBINE DISK

Let the cooled turbine disk which is treated in Sec. 14.5 (see Fig. 14.5:1) be rotating at an angular velocity ω. The body force per unit volume is then radial and equal to $\omega^2\rho r$, where ρ is the density of the disk material. If the thickness of the disk is small compared with its radius R, the axial stress is negligible. The disk is then in a state of plane stress.

On account of the fact that all stresses are considered to be functions of r only and the boundary condition $\tau_{rz} = 0$ at $r = R$, the shearing stress τ_{rz} vanishes and only the tension strains e_{rr}, $e_{\theta\theta}$ and tension stress σ_{rr}, $\sigma_{\theta\theta}$ are to be determined. The elastic displacement has only the radial component u, which is a function of r. Then

(1) $$e_{rr} = \frac{du}{dr}, \qquad e_{\theta\theta} = \frac{u}{r},$$

(2) $$\frac{du}{dr} = \frac{1}{E}(\sigma_{rr} - \nu\sigma_{\theta\theta}) + \alpha\theta,$$

(3) $$\frac{u}{r} = \frac{1}{E}(\sigma_{\theta\theta} - \nu\sigma_{rr}) + \alpha\theta.$$

The equation for the equilibrium of stresses and body forces is

(4) $$\frac{d\sigma_{rr}}{dr} + \frac{\sigma_{rr} - \sigma_{\theta\theta}}{r} + \omega^2\rho r = 0.$$

To solve for stresses directly, we eliminate u from Eqs. (2) and (3). Thus,

(5) $$\frac{1}{1+\nu}\left(\frac{d\sigma_{\theta\theta}}{dr} - \nu\frac{d\sigma_{rr}}{dr}\right) = \frac{\sigma_{rr} - \sigma_{\theta\theta}}{r} - \frac{\alpha E}{1+\nu}\frac{d\theta}{dr}.$$

An elimination of $(\sigma_{rr} - \sigma_{\theta\theta})/r$ from Eqs. (4) and (5) yields

(6) $$\frac{d\sigma_{rr}}{dr} + \frac{d\sigma_{\theta\theta}}{dr} + \alpha E\frac{d\theta}{dr} + (1+\nu)\rho\omega^2 r = 0.$$

But Eq. (4) gives

(7) $$\sigma_{\theta\theta} = \frac{d}{dr}(r\sigma_{rr}) + \omega^2\rho r^2.$$

Therefore, from Eqs. (6) and (7), the equation for σ_{rr} is

(8) $$\frac{1}{r^2}\frac{d}{dr}\left(r^3\frac{d\sigma_{rr}}{dr}\right) + \alpha E\frac{d\theta}{dr} + (3+\nu)\omega^2\rho r = 0.$$

On integrating twice, we obtain

(9) $$\sigma_{rr} = -\frac{1}{2}\frac{C_1}{r^2} + C_2 - \alpha E\int_0^r \frac{1}{\eta^3}d\eta\left[\int_0^\eta \xi^2\frac{d\theta(\xi)}{d\xi}d\xi\right] - \frac{3+\nu}{8}\omega^2\rho r^2.$$

The integral involving θ may be simplified by interchanging the order of integration. Thus

$$I \equiv \int_0^r \xi^2\frac{d\theta(\xi)}{d\xi}d\xi\int_\xi^r \frac{1}{\eta^3}d\eta = \int_0^r \xi^2\frac{d\theta(\xi)}{d\xi}\left(\frac{1}{2\xi^2} - \frac{1}{2r^2}\right)d\xi$$

$$= -\frac{\theta(0)}{2} + \frac{1}{r^2}\int_0^r \xi\theta(\xi)d\xi.$$

Hence, Eq. (9) may be written as

$$(10) \qquad \sigma_{rr} = -\frac{1}{2}\frac{C_1}{r^2} + C_2 + \frac{\alpha E\theta(0)}{2} - \frac{\alpha E}{r^2}\int_0^r \xi\theta(\xi)\,d\xi - \frac{3+\nu}{8}w^2\rho r^2 .$$

The boundary conditions are that $\sigma_{rr} = 0$ at $r = R$ and σ_{rr} remain finite at $r = 0$. Hence,

$$(11) \quad \sigma_{rr} = \alpha E\left[\frac{1}{R^2}\int_0^R \xi\theta(\xi)d\xi - \frac{1}{r^2}\int_0^r \xi\theta(\xi)\,d\xi\right] + \frac{3+\nu}{8}w^2\rho(R^2 - r^2) .$$

Equation (7) then gives

$$(12) \qquad \sigma_{\theta\theta} = \alpha E\left[\frac{1}{R^2}\int_0^R \xi\theta(\xi)\,d\xi + \frac{1}{r^2}\int_0^r \xi\theta(\xi)\,d\xi - \theta(r)\right]$$
$$+ w^2\rho\frac{(3+\nu)R^2 - (1+3\nu)r^2}{8} .$$

It is interesting to note that the first integral in Eq. (11) is one-half the average temperature throughout the whole disk. The second integral is one-half the average temperature in the inner portion of the disk (between 0 and r). Therefore, the thermal stress in the radial direction, i.e., the part of σ_{rr} caused by $\theta(r)$, is proportional to the difference between the average temperature throughout the whole disk and the average temperature for the inside portion between 0 and r. The circumferential thermal stress differs from that in the radial direction by

$$(13) \qquad \sigma_{\theta\theta} - \sigma_{rr} = \alpha E[\theta_{\text{avg}(0\text{ to }r)} - \theta(r)] + \frac{(1-\nu)\rho r^2 w^2}{4} .$$

For the particular problem of the turbine disk treated in Sec. 14.5, the temperature is given by Eq. (14.5:11). Since

$$\int_0^z zI_0(z)dz = zI_1(z) ,$$

we have

$$\frac{1}{r^2}\int_0^r \xi\theta(\xi)d\xi = \frac{h}{2\kappa}\frac{b}{r}\frac{I_1(\sqrt{2\kappa/kb}\,r)}{I_1(\sqrt{2\kappa/kb}\,R)}$$

and

$$\frac{1}{R^2}\int_0^R \xi\theta(\xi)d\xi = \frac{h}{2\kappa}\frac{b}{R} .$$

With these relations, the thermal stresses can be easily calculated.

14.10. VARIATIONAL PRINCIPLE FOR UNCOUPLED THERMOELASTICITY

In uncoupled thermoelasticity theory, the temperature field is determined by heat conduction and the influence of the latent heat due to change of strain is ignored. Variational principles analogous to those of Chapter 10 can be derived. In fact, the only difference between the linear elasticity of Chapters 7–9, which is valid in isothermal or isentropic conditions, and the thermoelasticity of the present chapter lies in a difference in the stress-strain law. Furthermore, in linear thermoelasticity concerning infinitesimal displacement, even this difference in stress-strain is hidden in the following expressions which apply to both elasticity and thermoelasticity:

$$(1) \qquad \frac{\partial W}{\partial e_{ij}} = \sigma_{ij} \,, \qquad \frac{\partial W_c}{\partial \sigma_{ij}} = e_{ij} \,.$$

Here $W(e_{ij}, T)$ and $W_c(\sigma_{ij}, T)$ are the strain energy function and the complementary strain energy function, respectively. $W(e_{ij}, T)$ is expressed in strain components; $W_c(\sigma_{ij}, T)$ is expressed in stress components; in linear theory they are equal: $W = W_c$. For isothermal or isentropic elasticity (Chapter 7) W is given by Eq. (12.6:3). For general thermoelasticity (Chapter 14), W is given by Eq. (12.7:2) or (12.7:11).

With these remarks we may derive the following theorems.

Let $\mathscr{U}$ be the total strain energy of a body which occupies a region V with boundary B,

$$(2) \qquad \mathscr{U} = \int_V W(e_{ij}, T) \, dv \,,$$

where $W(e_{ij}, T)$ is given by Eq. (12.7:2).

Let the body be subjected to a body force X_i per unit volume in V, and surface tractions $\overset{\nu}{T}_i$ on B_σ (B_σ in B). Both X_i and $\overset{\nu}{T}_i$ are assumed to be functions of time and space not influenced by the small elastic displacements. Let $\mathscr{K}$ be the kinetic energy of the body and $\mathscr{A}$ be the potential of the external loading:

$$(3) \qquad \mathscr{K} = \frac{1}{2} \int_V \frac{\partial u_i}{\partial t} \frac{\partial u_i}{\partial t} \rho \, dv$$

$$(4) \qquad \mathscr{A} = - \int_V X_i u_i \, dv - \int_{B_\sigma} \overset{\nu}{T}_i \, u_i \, dS \,.$$

Then *the Hamilton's principle states that of all displacements u_i that satisfy the boundary condition that*

$$(5) \qquad u_i = g_i(x_1, x_2, x_3) \quad \text{over} \quad B_u = B - B_\sigma \,,$$

the one that also satisfies the equations of motion minimizes the integral

(6)
$$\int_{t_0}^{t_1} (\mathscr{U} - \mathscr{K} + \mathscr{A})\,dt = \text{minimum}\,,$$

where t_0, t_1 are two arbitrary instants, and where the admissible variations δu_i are triply differentiable and satisfy the conditions

(7)
$$\delta u_i = 0 \quad \text{on } B_u \text{ in } t_0 \leq t \leq t_1\,,$$

(8)
$$\delta u_i = 0 \quad \text{at } t = t_0 \text{ and } t = t_1 \text{ in } V + B\,.$$

If we consider static equilibrium, then *the principle of minimum potential energy for thermoelasticity states that of all displacements that are continuous and triply differentiable in a body $V + B$, and satisfying the specified boundary displacements u_i over the surface B_u, the one that also satisfies the condition of equilibrium minimizes the potential*

(9)
$$\int_V W(e_{ij}, T)\,dv - \int_V X_i u_i\,dv - \int_{B_\sigma} \overset{\nu}{T}_i\, u_i\,dS = \text{minimum}\,.$$

Similarly, on varying the stresses σ_{ij}, *the complementary energy theorem in thermoelasticity states that the stress system σ_{ij} that satisfies the equations of elasticity minimizes the complementary energy with respect to variations of σ_{ij},*

(10)
$$\int_V W_c(\sigma_{ij}, T)\,dv - \int_V X_i u_i\,dv - \int_{B_u} \overset{\nu}{T}_i\, u_i\,dS = \text{minimum}\,.$$

$W_c = -\rho\Phi$ is given by Eq. (12.8:15). The variations $\delta\sigma_{ij}$ are subjected to the conditions $(\delta\sigma_{ij})_{,j} = 0$ in V and $\delta\sigma_{ij}\nu_j = 0$ over B_σ, where surface tractions are specified.

Problem 14.10. Derive Euler equations for the functionals in Eqs. (6), (9), and (10) and compare the results with the basic equations of Sec. 14.1.

14.11. VARIATIONAL PRINCIPLE FOR HEAT CONDUCTION

The basic equations for heat conduction, Eqs. (14.1:8) and (14.1:9), are

(1)
$$h_i = -k_{ij}\frac{\partial\theta}{\partial x_j}\,, \qquad\qquad i = 1, 2, 3\,,$$

(2)
$$-\frac{\partial h_i}{\partial x_i} = \rho C_v \frac{\partial\theta}{\partial t}\,,$$

where C_v is the specific heat per unit mass at constant volume, ρ is the density, k_{ij} are the coefficients of heat conduction, $\theta = T - T_0$ is the difference of local temperature T from a uniform reference temperature T_0, and $\mathbf{h}$ (h_1, h_2, h_3) is the heat flux vector per unit area. A problem in heat conduction for a body occupying a volume V with boundary B is to find a continuously differentiable function $\theta(x_1, x_2, x_3; t)$ which satisfies Eqs. (1) and (2), the boundary conditions

$$\text{(3)} \qquad\qquad \theta = \theta_0(x, t) \qquad\qquad\qquad \text{on } B_1,$$

$$\text{(4)} \qquad\quad h_n = \kappa(x, t)[\theta_0(x, t) - \theta(x, t)] \qquad\qquad \text{on } B - B_1,$$

and the initial condition at $t = 0$

$$\text{(5)} \qquad\qquad\qquad \theta = \theta_0(x, 0) \qquad\qquad\qquad \text{in } V \text{ and } B,$$

where h_n is the component of the heat flux vector $\mathbf{h}$ in the direction of the *outer normal* to the boundary, i.e., $h_n = h_i \nu_i$ and θ_0 is an assigned temperature. Equation (4) represents Newton's law of surface heat transfer; $\theta_0(x, t)$ is the temperature just outside the boundary layer, and κ is the heat transfer coefficient which usually depends on the space coordinate $\mathbf{x}$.

We shall now prove *Biot's variational principle which states that if C_v is a constant, then Eqs. (1)–(4) are the necessary conditions for the following variational equation to hold for an arbitrary variation of the vector H_i*

$$\text{(6)} \qquad \delta \int_V \frac{\rho C_v \theta^2}{2} dv + \int_V \lambda_{ij} \dot{H}_j \delta H_i dv + \int_{B_1} \theta_0 \delta H_i \nu_i dS = 0,$$

under the stipulations that

$$\text{(7)} \qquad\qquad\qquad \rho C_v \theta = -H_{i,i}$$

and that $\delta H_i = 0$ over the boundary $B - B_1$, where heat flux is so specified that $\partial H_i / \partial t = h_i$ satisfies Eq. (4). The matrix (λ_{ij}), called the thermal resistivity matrix, is the inverse of the conductivity matrix (k_{ij}):

$$\text{(8)} \qquad\qquad\qquad (\lambda_{ij}) = (k_{ij})^{-1}.$$

Proof. First, we have

$$\delta \int_V \rho C_v \frac{\theta^2}{2} dv = \delta \int_V \frac{1}{2\rho C_v} H_{i,i} H_{j,j} dv \qquad \text{(by substitution)}$$

$$= \int_V \frac{1}{\rho C_v} H_{i,i} \delta H_{j,j} dv \qquad \text{(i, j being dummy indices)}$$

$$= \int_V \left[\left(\frac{1}{\rho C_v} H_{i,i} \delta H_j \right)_{,j} - \left(\frac{1}{\rho C_v} H_{i,i} \right)_{,j} \delta H_j \right] dv$$

$$= \int_B \frac{1}{\rho C_v} H_{i,i} \delta H_j \nu_j dS - \int_V \left(\frac{1}{\rho C_v} H_{i,i} \right)_{,j} \delta H_j dv$$

$$\text{(by Gauss' theorem)}$$

$$= - \int_B \theta \delta H_j \nu_j dS + \int_V \theta_{,j} \delta H_j dv .$$

Hence, Eq. (6) is

$$0 = \int_V (\theta_{,j} + \lambda_{ij} \dot{H}_j) \delta H_i dv - \int_{B_1} (\theta - \theta_0) \delta H_i \nu_i dS - \int_{B-B_1} \theta \delta H_i \nu_i dS .$$

To satisfy this equation by arbitrary δH_i in V and on B_1, the necessary conditions are

$$\theta_{,j} + \lambda_{ij} \dot{H}_j = 0 , \quad \text{or} \quad \dot{H}_j = -k_{ij} \theta_{,j} \qquad \text{in } V ,$$

$$\theta = \theta_0 \qquad \text{on } B_1 ,$$

$$\delta H_j \nu_j = 0 \qquad \text{on } B - B_1 .$$

The first equation, together with Eq. (7), is exactly Eq. (1). The second equation is the same as Eq. (3), the last equation is in accordance with the stipulation stated.   Q.E.D.

Applications of this variational principle will be shown later in Secs. 14.13. A few remarks will be made here. First, the vector H_i is related to the heat flux h_i:

(9) $$h_i = \dot{H}_i .$$

If we impose the condition that

(10) $$H_i = 0 \quad \text{when} \quad \theta = 0 ,$$

then Eqs. (2) and (7) are consistent since C_v is a constant. Comparing Eq. (9) with Eq. (13.1:3), it is seen that H_i is proportional to the entropy displacement introduced by Biot.

Second, the special way in which the second term in Eq. (6) is posed should be noted. Biot denotes this term by δD and calls it *the variation of a dissipation function.* The justification of this terminology becomes clear when generalized coordinates are introduced. Let $q_1, \ldots, q_n$ be the generalized coordinates so that

(11) $$H_j = H_j(q_1, q_2, \ldots, q_n; x_1, x_2, x_3) .$$

Then

$$\dot{H}_j = \frac{\partial H_j}{\partial q_k} \dot{q}_k .$$

Hence,

(12) $$\frac{\partial \dot{H}_j}{\partial \dot{q}_k} = \frac{\partial H_j}{\partial q_k} .$$

Therefore,

$$\int_V \lambda_{ij} \dot{H}_j \delta H_i \, dv = \int_V \lambda_{ij} \dot{H}_j \frac{\partial H_i}{\partial q_k} \delta q_k \, dv$$

$$= \delta q_k \int_V \lambda_{ij} \dot{H}_j \frac{\partial \dot{H}_i}{\partial \dot{q}_k} \, dv$$

$$= \delta q_k \frac{\partial}{\partial \dot{q}_k} \int_V \frac{1}{2} \lambda_{ij} \dot{H}_j \dot{H}_i \, dv .$$

It is then clear that if we define a dissipation function

(13) $$D = \frac{1}{2} \int_V \lambda_{ij} \dot{H}_j \dot{H}_i \, dv ,$$

then

(14) $$\int_V \lambda_{ij} \dot{H}_j \delta H_i \, dv = \frac{\partial D}{\partial \dot{q}_k} \delta q_k .$$

The application of generalized coordinates will be illustrated in Sec. 14.13.

It may appear unnatural to introduce the vector field H_i rather than vary the temperature θ directly. A little reflection will show, however, that the thermal evolution is determined by the heat flow. The three components of the heat flow vector are capable of independent variations. Hence, as a proper choice of variables, we use the components of the heat flow vector. Equivalently, the three components of the temperature gradient may be used also, but then the analysis will be purely formal; the functionals do not have as simple an interpretation as those encountered above.

14.12. COUPLED THERMOELASTICITY

A variational equation corresponding to the basic equations of Sec. 14.1 is given by Biot under the assumptions that the elastic constants C_{ijkl}, the thermal stress constants β_{ij}, the specific heat per unit mass at constant strain C_v, and the heat conductivity matrix (k_{ij}) or its inverse $(\lambda_{ij}) = (k_{ij})^{-1}$, are independent of the temperature and time, and that the temperature change θ is small compared with T_0, the absolute temperature of the reference state. Under these assumptions, Eq. (14.1:17) can be integrated to give

(1) $$-H_{i,i} = \rho C_v \theta + T_0 \beta_{ij} e_{ij} \,,$$

where

$$\frac{\partial H_i}{\partial t} \cong h_i \,,$$

$$H_i = 0 \quad \text{when} \quad \theta = e_{ij} = 0 \,.$$

The vector H_i so defined is proportional to the entropy displacement. Biot's principle now states that *the thermoelastic equations are necessary conditions for the following variational equation to hold for arbitrary δu_i and δH_i, except that $\delta u_i = 0$ over the portion of the boundary where the displacement u_i is prescribed, and $\delta H_i = 0$ where the heat flow is prescribed:*

(2) $$\delta \mathscr{V} + \delta \mathscr{D} = \int_V (X_i - \rho \ddot{u}_i) \delta u_i \, dv + \int_B \left(\overset{\nu}{T_i} \, \delta u_i - \theta_0 \frac{\delta H_i}{T_0} \nu_i \right) dS \,,$$

where

(3) $$\mathscr{V} = \int_V \left(\frac{1}{2} C_{ijkl} e_{ij} e_{kl} - \beta_{ij} e_{ij} \theta + \frac{1}{2} \frac{\rho C_v \theta^2}{T_0} \right) dv \,,$$

(4) $$\delta \mathscr{D} = \int_V \frac{1}{T_0} \lambda_{ij} \dot{H}_j \delta H_i \, dv \,,$$

θ_0 is the boundary value of θ, ν_i are the direction cosines of the outward normal to the surface B (see Sec. 14.11), and $\mathscr{V}$ is the *thermoelastic potential* of Biot. On solving Eq. (1) for θ, substituting θ and $e_{ij} = \frac{1}{2}(u_{i,j} + u_{j,i})$ into Eq. (3), and varying u_i and H_i, the statement above can be verified readily.

We may also use the method of Lagrangian multipliers as in Sec. 14.10. The details are left to the reader.

Note that if Eq. (2) is integrated with respect to time between an arbitrary interval (t_0, t_1), under the stipulation that $\delta u_i = 0$ at $t = t_0, t_1$, the term involving the inertia force $-\rho \ddot{u}_i$ can be expressed as $-\delta \mathcal{K}$, where $\mathcal{K}$ is the kinetic energy. Then a formula similar to the Hamilton's principle, now including a dissipation function, is obtained.

Biot[14.4] derived the expressions above for $\mathcal{V}$ and $\mathcal{D}$ from a reasoning which has been discussed in Sec. 13.5. For small deviations from equilibrium, the equation of evolution (13.5:13) is a necessary condition for the variational equation

$$(5) \qquad \left(\frac{\partial \mathcal{V}}{\partial q_i} + \frac{\partial \mathcal{D}}{\partial \dot{q}_i} - Q_i \right) \delta q_i = 0$$

to hold for arbitrary variations δq_i. It was shown in Sec. 13.5 that $\mathcal{V}$ is the sum of $\mathcal{E} - T_0 \mathcal{S}$ over the entire system, and $2\mathcal{D}/T_0$ is equal to the rate of entropy production. Although no heat input on the boundary was considered in Sec. 13.5, the generalization of Q_i to include heat conduction on the boundary is justified by the theorem just proved above.

Let us now evaluate $\mathcal{V} = \int \rho(\mathcal{E} - T_0 \mathcal{S}) dv$, when a temperature increment $\theta(x_1, x_2, x_3; t)$ is imposed on the body without changing any other state variables. The absolute temperature distribution is $T_0 + \theta$. To change the temperature by $d\theta$ at a particular point requires an amount of heat $\rho C_v d\theta$ per unit mass, where C_v is the specific heat per unit mass at constant strain. Hence, the change of internal energy per unit volume, $\mathcal{E}$, is $\rho C_v d\theta$, the change in completementary work $-\beta_{ij} e_{ij} d\theta$ and the change of entropy per unit volume is $-\rho C_v d\theta/(T_0 + \theta)$. Thus, the imposition of a temperature increment θ changes $\mathcal{V}$ by

$$(6) \qquad \mathcal{V}_c = \int_V dv \left(\int_0^\theta \rho C_v \, d\theta - T_0 \int_0^\theta \frac{\rho C_v d\theta}{T_0 + \theta} - \int_0^\theta \beta_{ij} e_{ij} \, d\theta \right)$$

$$= \int_V dv \left(\int_0^\theta \frac{\rho C_v \theta}{T_0 + \theta} d\theta - \int_0^\theta \beta_{ij} e_{ij} \, d\theta \right).$$

If $\theta \ll T_0$, then

$$(7) \qquad \mathcal{V}_c \doteq \frac{1}{2} \int_V \frac{\rho C_v \theta^2}{T_0} \, dv - \beta_{ij} e_{ij} \theta.$$

The total value of $\mathcal{V}$ is obtained by adding to $\mathcal{V}_c$ the value of $\mathcal{V}$ at the constant temperature T_0. The latter is the classical Helmholtz free energy at T_0. The term $\mathcal{V}_c$ is familiar to power engineering: it represents the heat that may be transformed into useful work. Indeed, $\mathcal{V}_c$ is an integral of the heat $\rho C_v d\theta$ multiplied by the Carnot efficiency $\theta/(T_0 + \theta) \doteq \theta/T_0$, integrated first with respect to θ from 0 to θ, then over the region V.

The special treatment of the dissipation function $\mathscr{D}$ in the variational equations should be noted. The variational invariant $\delta\mathscr{D}$ was defined directly. When the equations of motion are written in the Lagrangian form, Eq. (5), $\mathscr{D}$ can be identified as a functional which is equal to $T_0/2$ times the rate of entropy production of the entire system [see Eq. (13.5:11), assuming $\theta \ll T_0$]. From Eq. (13.2:6), the entropy production per unit volume is $-h_j T_{,j}/T_0^2$, where h_j is the heat flux and $T_{,j}$ is the temperature gradient. On introducing Fourier's law of heat conduction in the form $T_{,j} = -\lambda_{jl} h_l$, the rate or entropy production per unit volume is $\lambda_{jl} h_j h_l / T_0^2$. Hence,

$$(8) \qquad \mathscr{D} = \frac{1}{2T_0} \int_v \lambda_{jl} h_j h_l \, dv \,.$$

In Sec. 14.11, for the boundary condition over $B - B_1$, where the wall temperature is unspecified, it is required that the heat flow be specified exactly. For a variational approach, this boundary condition can be relaxed by including in the dissipation function a term corresponding to entropy production at the boundary. Let there be a heat flux per unit area $h_j \nu_j$ leaving the body at a boundary, where ν_j is the direction cosine of the outer normal to the surface. Let T_a be the temperature *outside* the heat transfer layer at the wall and T be that of the body inside. (For example, T_a represents the adiabatic wall temperature in the boundary-layer flow in aerodynamic heating.) If $T_a \neq T$, the entropy production per unit wall area is $-h_j \nu_j (T^{-1} - T_a^{-1})$. A phenomenological law may be posed as

$$(9) \qquad h_n = h_j \nu_j = -K\left(\frac{1}{T} - \frac{1}{T_a}\right).$$

Therefore, the contribution of heat transfer at the boundary to the dissipation function is

$$(10) \qquad \mathscr{D}_B = \frac{T_0}{2} \int_B K\left(\frac{1}{T} - \frac{1}{T_a}\right)^2 dS = \frac{T_0}{2} \int_B \frac{1}{K} h_n^2 \, dS \,.$$

If $\mathscr{D}_B$ is added to $\mathscr{D}$ from (8), we have

$$(11) \qquad \mathscr{D} = \frac{1}{2T_0} \int_V \lambda_{jl} h_j h_l \, dv + \frac{T_0}{2} \int_B \frac{1}{K} h_n^2 \, dS \,,$$

which must be used if the heat flow at the wall is to be considered as a natural boundary condition. When the expression (11) is used, the variation δH_i can be regarded as arbitrary over the entire boundary.

14.13. LAGRANGIAN EQUATIONS FOR HEAT CONDUCTION AND THERMOELASTICITY

When n generalized coordinates $q_1, q_2, \ldots, q_n$ are introduced to represent the displacement vector and heat flow vector by expressions

(1)
$$u_i = u_i(q_1, q_2, \ldots, q_n; x_1, x_2, x_3) \, , \qquad i = 1, 2, 3 \, ,$$
$$H_i = H_i(q_1, q_2, \ldots, q_n; x_1, x_2, x_3) \, , \qquad i = 1, 2, 3 \, .$$

the variational principles of the preceding section may be written as

(2)
$$\frac{\partial \mathscr{V}}{\partial q_j} + \frac{\partial \mathscr{D}}{\partial \dot{q}_j} = Q_j \, , \qquad j = 1, 2, \ldots, n \, ,$$

where $\mathscr{V}$ and $\mathscr{D}$ are given by Eqs. (14.12:3) and (14.12:8) respectively, and

(3)
$$Q_j = \int_V (X_i - \rho \ddot{u}_i) \frac{\partial u_i}{\partial q_j} dv + \int_B \left(\overset{\nu}{T_i} \frac{\partial u_i}{\partial q_j} + \frac{\theta_0}{T_0} \frac{\partial H_n}{\partial q_j} \right) dS \, .$$

The applications of these Lagrangian equations are well-known for the stress problems (Chapters 10 and 11). In what follows we shall give illustration to heat conduction problems.

Example 1. *Prescribed Wall Temperature*

Consider a semi-infinite homogeneous, isotropic solid, with constant parameters k and C_v, initially at a uniform temperature $\theta = 0$ (Fig. 14.13:1). The boundary at $x = 0$ is heated to a variable temperature $\theta_0(t)$. On ignoring the inertia force and thermoelastic coupling, it is desired to find the temperature distribution in the half-space.

We propose to solve this problem approximately. Following Biot[14.4] (1957), we assume the temperature distribution to be parabolic and represented by

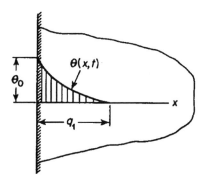

Fig. 14.13:1. Penetration of heat in one-dimensional flow.

(4)
$$\theta(x, t) = \theta_0(t) \left[1 - \frac{x}{q_1(t)} \right]^2 \, , \qquad x < q_1 \, ,$$
$$\theta(x, t) = 0 \, , \qquad x > q_1 \, .$$

The function $q_1(t)$ can be interpreted physically as a "penetration depth," and will be taken as our generalized coordinate (see Fig. 14.13:1).

The heat flow vector H_i is defined by Eqs. (14.11:7), (14.11:9), and (14.11:10)

(5) $$\dot{H}_i = h_i, \qquad -H_{i,i} = \rho C_v \theta.$$

In the one-dimensional problem posed here, we may assume that only one component H_1 is different from zero and that H_1 is a function of x alone. Hence, from Eq. (5),

(6) $$\frac{\partial H_1}{\partial x} = -\rho C_v \theta.$$

For $x > q_1$, thermal equilibrium is undisturbed so that $H_1 = 0$. Hence, Eqs. (6) and (4) can be integrated to give

(7) $$H_1(x,t) = \rho C_v \int_x^{q_1(t)} \theta(\xi,t)\, d\xi = \rho C_v \theta_0 \left(\frac{q_1}{3} - x + \frac{x^2}{q_1} - \frac{x^3}{3q_1^2} \right).$$

To evaluate $\mathscr{V}$ and $\mathscr{D}$, it is sufficient to consider a semi-infinite cylinder of unit cross section with axis parallel to x. Hence,

(8) $$\mathscr{V} = \frac{\rho C_v}{2T_0} \int_0^{q_1} \theta^2\, dx = \frac{\rho C_v \theta_0^2}{10T_0}\, q_1,$$

(9) $$\mathscr{D} = \frac{1}{2kT_0} \int_0^{q_1} \left(\frac{\partial H_1}{\partial t} \right)^2 dx,$$

which, from Eq. (7), is

$$\mathscr{D} = \frac{\rho^2 C_v^2}{2kT_0} \int_0^{q_1} \left[\dot{\theta}_0 \left(\frac{q_1}{3} - x + \frac{x^2}{q_1} - \frac{x^3}{3q_1^2} \right) + \theta_0 \dot{q}_1 \left(\frac{1}{3} - \frac{x^2}{q_1^2} + \frac{2x^3}{3q_1^3} \right) \right]^2 dx$$

$$= \frac{\rho^2 C_v^2}{2kT_0}\, q_1 \left(\frac{13}{315} \dot{q}_1^2 \theta_0^2 + \frac{1}{21} q_1 \theta_0 \dot{q}_1 \dot{\theta}_0 + \frac{1}{63} \dot{\theta}_0^2 q_1^2 \right).$$

The generalized force Q_1 is, according to Eq. (3),

(10) $$Q_1(t) = \frac{\theta_0}{T_0} \left(\frac{\partial H_1}{\partial q_1} \right)_{x=0} = \frac{\rho C_v \theta_0^2}{3T_0}.$$

Hence, the Lagrangian equation for heat conduction (2) is

(11) $$\frac{\rho C_v \theta_0^2}{10T_0} + \frac{\rho^2 C_v^2}{2kT_0}\, q_1 \left(\frac{26}{315} \dot{q}_1 \theta_0^2 + \frac{1}{21} q_1 \theta_0 \dot{\theta}_0 \right) = \frac{\rho C_v \theta_0^2}{3T_0}.$$

Let

(12) $$z = q_1^2.$$

Then Eq. (11) reduces to

(13)
$$\dot{z} + \frac{15}{13}\frac{\dot{\theta}_0}{\theta_0} z = \frac{147}{13}\frac{k}{\rho C_v} .$$

This is a standard differential equation with an integration factor

$$e^{\int (15/13)(\dot{\theta}_0/\theta_0)\, dt} = [\theta_0(t)]^{15/13} ;$$

i.e.,

$$\frac{d}{dt}[z\theta_0^{15/13}] = \frac{147}{13}\frac{k}{\rho C_v}\theta_0^{15/13} .$$

Hence, on noting the initial condition $q_1 = z = 0$ when $t = 0$, we obtain the solution

(14)
$$z = \frac{147}{13}\frac{k}{\rho C_v}[\theta_0(t)]^{-15/13} \int_0^t [\theta_0(\tau)]^{15/13}\, d\tau .$$

If it is assumed that the wall temperature follows a power law,

(15)
$$\theta_0(t) = \alpha t^n , \qquad\qquad n \geq 0 ,$$

we find

(16)
$$q_1^2 = z = \frac{147}{13}\frac{k}{\rho C_v}\frac{t}{\frac{15}{13}n + 1} .$$

This shows that the penetration depth q_1 varies with $\sqrt{t}$. It is independent of α and depends on the exponent n only through a constant factor $[\frac{15}{13}n + 1]^{-1/2}$. The case where $\theta_0 = $ const. corresponds to $n = 0$ and yields the result

(17)
$$q_1 = 3.36\sqrt{\frac{kt}{\rho C_v}} .$$

Comparison of this simple solution, with only one generalized coordinate, with the exact solution of the problem was made by Biot[14.4] (1957) who shows that the approximation (4) is valid if the temperature increases or decreases monotonically. Biot also points out that if this is not the case, one may split up the time interval into segments for which θ_0 varies monotonically, and then apply Eq. (13) to each segment, using the principle of superposition. Use can also be made of the power law solution (16) by dividing the time history of the temperature into segments, each of which may be approximated by a power law, or by an additive combination of such terms, including the constant value.

Example 2. *Prescribed Heat Flux*

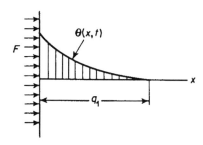

Fig. 14.13:2. Heating of a half-space with prescribed heat flux.

The semi-infinite solid described in Example 1 is heated at the wall $x = 0$ with a uniform, constant heat flux $h_x = F$ at the wall (see Fig. 14.13:2) (Lardner, T. J., AIAA J. *1*, 1, 1963, p. 196–206).

The problem is one-dimensional with one component of the vector **H**, which will be written as H. Then

(18) $$\frac{\partial H}{\partial x} = -\rho C_v \theta.$$

Again, we assume a parabolic temperature distribution

(19) $$\theta(x,t) = A(t)\left[1 - \frac{x}{q_1(t)}\right]^2,$$

with q_1 representing a "penetration depth." The function $A(t)$ must be so chosen to satisfy the boundary condition at $x = 0$. An integration of Eq. (18) gives

(20) $$H = \frac{A\rho C_v q_1}{3}\left(1 - \frac{x}{q_1}\right)^3,$$

whence

(21) $$h_x = \dot{H} = \dot{A}\frac{\rho C_v}{3}q_1\left(1 - \frac{x}{q_1}\right)^3 + \frac{A}{3}\rho C_v \dot{q}_1\left(1 - \frac{x}{q_1}\right)^3$$
$$+ \frac{xA\rho C_v \dot{q}_1}{q_1}\left(1 - \frac{x}{q_1}\right)^2.$$

At $x = 0$, the boundary condition $h_x = F$ requires that

$$(\dot{A}q_1 + \dot{q}_1 A)\frac{\rho C_v}{3} = F,$$

or

(22) $$A = \frac{3Ft}{\rho C_v q_1}.$$

On substituting this expression for A into Eqs. (19) and (20) and then computing the thermal potential $\mathscr{V}$ and dissipation function $\mathscr{D}$ according to Eqs. (14.12:3) and (14.12:11), we obtain

$$\frac{\partial \mathscr{V}}{\partial q_1} = -\frac{9F^2t^2}{10\rho C_v q_1^2},$$

$$\frac{\partial \mathscr{D}}{\partial \dot{q}_1} = \frac{F^2t}{k}\left(\frac{3\dot{q}_1 t}{35q_1} + \frac{3}{42}\right),$$

$$Q_1 = \frac{9}{10}t.$$

The governing differential equation (2) becomes

(23)
$$\frac{k}{\rho C_v}\left[\frac{3}{35}q_1\dot{q}_1 t + \frac{3}{42}q_1^2\right] = \frac{9}{10}t.$$

The solution for the penetration depth is

(24)
$$q_1 = 2.81\left(\frac{k}{\rho C_v}t\right)^{1/2}.$$

The surface temperature is

(25)
$$\theta_0 = \frac{3Ft}{\rho C_v q_1} = 1.065\frac{F}{k}\left(\frac{kt}{\rho C_v}\right)^{1/2}.$$

The exact solution is known to be

(26)
$$\theta_0 = 1.128\frac{F}{k}\left(\frac{kt}{\rho C_v}\right)^{1/2}.$$

PROBLEMS

14.11. Consider a slab of thickness b, heated on the surface $x = 0$. Initial condition at $t = 0$ is $\theta = 0$ throughout the plate. The boundary condition at the surface $x = 0$ is $\theta = \theta_0(t)$ for $t > 0$ (same as in Example 1), and that at $x = b$ is $h_x = \partial\theta/\partial x = 0$. At certain time t_0, the penetration depth q_1 will be equal to b. For time $t > t_0$, the temperature profile will be as shown in Fig. P14.11. Obtain an approximate solution by taking q_3 as the generalized coordinate.

14.12. Consider the same problem as in Problem 14.11 except that the boundary condition at $x = 0$ is, as shown in Fig. P14.12, $h_x = F$, a const. (same as in Example 2).

14.13. A long, circular, cylindrical, solid-fuel rocket, upon curing, releases its heat of polymerization throughout its mass at a steady rate, say, Q_0. The external surfaces may be assumed held at constant temperature T_0. Assuming

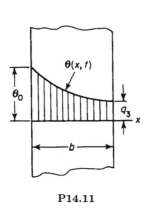

P14.11

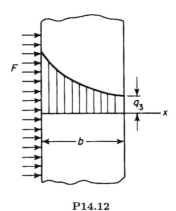

P14.12

that the cylinder is so long that over the body of the cylinder the heat flow is essentially two-dimensional, determine approximately the temperature distribution in the cross section of the cylinder.

Consider other cylindrical rockets whose cross-sectional shapes are (1) an isosceles right triangle, (2) an equilateral triangle, (3) a square. Let all of these rockets be made of the same material and have the same cross-sectional area. Compare the steady-state temperatures at the centroids of these rockets.

15

VISCOELASTICITY

In this chapter we shall generalize the ideas discussed in Sec. 1.2 to deal with a three-dimensional continuum. It is popular to call a material that obeys a linear hereditary law *viscoelastic*, although the logic of this terminology is by no means certain. The origin of the term lies in the simple models such as those of Maxwell, Voigt, or the standard linear solid, which are built of springs and dashpots. The springs are elastic, the dashpots are viscous; hence the name. Such an etymology is entirely different from that of viscoplasticity, hyperelasticity, hypoelasticity, elastic-ideally plastic materials, etc. However, the terminology of identifying a material obeying a linear hereditary law as viscoelastic is popular and well-accepted.

15.1. VISCOELASTIC MATERIAL

Let a rectangular Cartesian frame of reference be chosen, and let the position vector of a point be denoted by (x_1, x_2, x_3). A function of position $f(x_1, x_2, x_3)$ will be written as $f(x)$ for short. Let σ_{ij} and e_{ij} be the stress and strain tensors defined at every point x of a body and in the time interval $(-\infty < t < \infty)$. The strain field $e_{ij}(x, t)$ and the displacement field $u_i(x, t)$, as well as the velocity field $v_i(x, t)$, will be assumed to be infinitesimal, and

$$(1) \qquad e_{ij} = \frac{1}{2}(u_{i,j} + u_{j,i}),$$

where a comma indicates a partial differentiation. Under the assumption of infinitesimal strain and velocity, the partial derivative with respect to time $\partial e_{ij}/\partial t$ is equal to the material derivative $\dot{e}_{ij}$ within the first order. Now we define a *linear viscoelastic* material to be one for which $\sigma_{ij}(x, t)$ is related to $e_{ij}(x, t)$ by a convolution integral as follows.

$$(2) \quad \blacktriangle \qquad \sigma_{ij}(x, t) = \int_{-\infty}^{t} G_{ijkl}(x, t - \tau)\frac{\partial e_{kl}}{\partial \tau}(x, \tau)d\tau\,,$$

where G_{ijkl} is a tensor field of order 4 and is called the *tensorial relaxation function* of the material. Equation (2) is called the stress-strain law of the *relaxation type.* Its inverse,

$$(3) \quad \blacktriangle \qquad e_{ij}(x,t) = \int_{-\infty}^{t} J_{ijkl}(x, t - \tau) \frac{\partial \sigma_{kl}}{\partial \tau}(x, \tau) d\tau \,,$$

if it exists, is called the stress-strain law of the *creep type*. The fourth-order tensor J_{ijkl} is called the *tensorial creep function*. It can be shown (Sternberg and Gurtin[15.1] 1962, Theorem 3.3), that the inverse (3) of Eq. (2) exists if $G_{ijkl}(x,t)$ is twice differentiable and if the initial value of $G_{ijkl}(x,t)$ at $t = 0$ is not zero.

The lower limits of integration in Eqs. (2) and (3) are taken as $-\infty$, which is to mean that the integration is to be taken before the very beginning of motion. If the motion starts at time $t = 0$, and $\sigma_{ij} = e_{ij} = 0$ for $t < 0$, Eq. (2) reduces to

$$(4) \quad \sigma_{ij}(x,t) = e_{kl}(x, 0+) G_{ijkl}(x,t) + \int_{0}^{t} G_{ijkl}(x, t - \tau) \frac{\partial e_{kl}}{\partial \tau}(x, \tau)\, d\tau \,,$$

where $e_{ij}(x, 0+)$ is the limiting value of $e_{ij}(x,t)$ when $t \to 0$ from the positive side. The first term in Eq. (4) gives the effect of initial disturbance; it arises from the jump of $e_{ij}(x,t)$ at $t = 0$. In fact, it was tacitly assumed that $e_{ij}(x,t)$ is continuous and differentiable when Eq. (2) was written. Any discontinuity of $e_{ij}(x,t)$ in the form of a jump will contribute a term similar to the first term in Eq. (4). For example, if $e_{ij}(x,t)$ has another jump $\Delta e_{ij}(x, \xi)$ at $t = \xi$, as shown in Fig. 15.1:1, while G_{ijkl} and $\partial e_{ij}/\partial t$ are continuous elsewhere, then we have

$$(5) \quad \sigma_{ij}(x,t) = e_{kl}(x, 0+) G_{ijkl}(x,t) + \Delta e_{kl}(x, \xi) G_{ijkl}(x, t - \xi) 1(t - \xi)$$

$$+ \int_{0}^{t} G_{ijkl}(x, t - \tau) \frac{\partial e_{kl}}{\partial \tau}(x, \tau)\, d\tau \,,$$

where $1(t)$ is the unit-step function. In such cases the following forms, which are equivalent to Eq. (4) when $\partial e_{ij}/\partial t$, $\partial G_{ijkl}/\partial t$ exist and are continuous

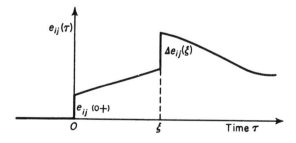

Fig. 15.1:1. Illustration for a loading history with jumps.

in $0 \leq t < \infty$, may be used to advantage:

$$(6) \qquad \sigma_{ij}(x,t) = G_{ijkl}(x,0)e_{kl}(x,t) + \int_0^t e_{kl}(x,t-\tau)\frac{\partial G_{ijkl}}{\partial \tau}(x,\tau)\,d\tau$$

$$= \frac{\partial}{\partial t}\int_0^t e_{kl}(x,t-\tau)G_{ijkl}(x,\tau)\,d\tau.$$

These constitutive equations are appropriate for *isothermal conditions*. For the influence of temperature, see references listed in Bibliography 15.2, p. 505.

A viscoelastic material is defined by a specific relaxation function or creep function. In the following discussions we shall regard G_{ijkl} or J_{ijkl} as experimentally determined functions. Restrictions imposed on these functions by thermodynamics, as well as their origin in hidden coordinates, are discussed in Chapter 13.

In the treatment of viscoelasticity, we would have to write many convolution integrals. A shorthand is therefore desired. We shall introduce the notation of *composition products*. Let ϕ and ψ be functions defined on the intervals $0 \leq t < \infty$ and $-\infty < t < \infty$, respectively, and let the integral

$$(7) \qquad I(t) = \int_0^t \phi(t-\tau)\frac{d\psi}{d\tau}(\tau)d\tau + \phi(t)\psi(0)$$

exist for all t in $(0,\infty)$. Then the function $I(t)$ is called the *convolution of* ϕ *and* ψ and is denoted by a "composition" product

$$(8) \qquad I(t) = \phi * d\psi.$$

The integral in Eq. (7) may be understood in Riemannian or Stieltjes' sense, the latter being more general and better suited to our purpose. Sternberg and Gurtin[15.1] developed the theory of viscoelasticity on the basis of Stieltjes' convolution.

The following properties of the convolution of ϕ with ψ and θ (also defined over $-\infty < t < \infty$) can be verified.

$$(9) \qquad \phi * d\psi = \psi * d\phi \qquad \text{(commutativity)},$$

$$(10) \qquad \phi * d(\psi * d\theta) = (\phi * d\psi) * d\theta = \phi * d\psi * d\theta \qquad \text{(associativity)},$$

$$(11) \qquad \phi * d(\psi + \theta) = \phi * d\psi + \phi * d\theta \qquad \text{(distributivity)},$$

$$(12) \qquad \phi * d\psi \equiv 0 \quad \text{implies}$$

$$\phi \equiv 0 \quad \text{or} \quad \psi \equiv 0 \qquad \text{(Titchmarsh theorem)}.$$

With these notations, Eqs. (2) and (3) may be written as

(2a)   ▲ $\sigma_{ij} = G_{ijkl} * de_{kl} = e_{kl} * dG_{ijkl}$,

(3a)   ▲ $e_{ij} = J_{ijkl} * d\sigma_{kl} = \sigma_{kl} * dJ_{ijkl}$.

The symmetry of the stress and strain tensors either requires or permits that

(13) $G_{ijkl} = G_{jikl} = G_{ijlk}$,

(14) $J_{ijkl} = J_{jikl} = J_{ijlk}$.

Furthermore, it is natural to require that the action of a loading at a time t_0 will produce a response only for $t \geq t_0$. Hence, we must have

(15) $G_{ijkl} = 0$, $J_{ijkl} = 0$,   for $-\infty < t < 0$.

This requirement is sometimes called *the axiom of nonretroactivity.*

If G_{ijkl} is invariant with respect to rotation of Cartesian coordinates, the material is said to be *isotropic*. It can be shown that a material is isotropic if and only if G_{ijkl} is an isotropic tensor. A fourth-order isotropic tensor with the symmetry properties of Eq. (13) can be written as

(16) $$G_{ijkl} = \frac{G_2 - G_1}{3} \delta_{ij}\delta_{kl} + \frac{G_1}{2}\left(\delta_{ik}\delta_{jl} + \delta_{il}\delta_{jk}\right),$$

where G_1, G_2 are scalar functions satisfying Eq. (15). If Eq. (16) is substituted into Eq. (2a), we obtain the stress-strain law which can be put in the form

$$\sigma'_{ij} = e'_{ij} * dG_1 = G_1 * de'_{ij} ,$$

(17)   ▲

$$\sigma_{kk} = e_{kk} * dG_2 = G_2 * de_{kk} ,$$

where σ'_{ij} and e'_{ij} are the components of the stress and strain deviators:

(18) $\sigma'_{ij} = \sigma_{ij} - \dfrac{1}{3}\delta_{ij}\sigma_{kk}$, $e'_{ij} = e_{ij} - \dfrac{1}{3}\delta_{ij}e_{kk}$.

The functions G_1 and G_2, are referred to as the relaxation functions in shear and in isotropic compression, respectively.

The corresponding stress-strain laws of the creep type for an isotropic material is

$$e'_{ij} = \sigma'_{ij} * dJ_1 = J_1 * d\sigma'_{ij} ,$$

(19)   ▲

$$e_{kk} = \sigma_{kk} * dJ_2 = J_2 * d\sigma_{kk} ,$$

where J_1, J_2 are called the creep functions in shear and in isotropic compression, respectively.

If G_{ijkl}, J_{ijkl} or G_1, G_2, J_1, J_2 are step functions in time, then the stress-strain laws (2), (3), (17), and (19) reduce to those of linear elastic solids.

The Fourier transforms of G and J can be interpreted as the responses to appropriate harmonic forcing functions. Within a certain range of frequencies, measurements of harmonic responses of many materials are feasible. The bulk of our information about the viscoelastic behavior of high polymers is presented in the form of *frequency responses*.

15.2. STRESS-STRAIN RELATIONS IN DIFFERENTIAL EQUATION FORM

In Secs. 13.8 and 13.9, it was shown that when the relaxation function consists of a finite discrete spectrum, the stress-strain relation may be put in the form of a differential equation. Simple examples are the Maxwell, Voigt, or the standard linear models discussed in Sec. 1.2. A more general expression may be given as follows. Let D denote the time-derivative operator, or *differential operator* defined by

$$(1) \qquad Df = \frac{\partial f(t)}{\partial t}, \qquad D^2 f = \frac{\partial^2 f}{\partial t^2}, \quad \text{etc.,}$$

where f is a function of time. Let us consider the polynomials

$$(2) \qquad P_1(D) = \sum_{k=0}^{n_1} a_k D^k, \qquad Q_1(D) = \sum_{k=0}^{m_1} b_k D^k,$$

$$P_2(D) = \sum_{k=0}^{n_2} c_k D^k, \qquad Q_2(D) = \sum_{k=0}^{m_2} d_k D^k,$$

where a_k, b_k, c_k, d_k are real-valued functions of the spatial coordinates x_1, x_2, x_3. We assume that the coefficients a_{n_1}, b_{m_1}, c_{n_2}, d_{m_2} are different from zero, so that P_1, P_2, Q_1, Q_2 are polynomials of degree n_1, n_2, m_1, m_2, respectively. Then the stress-strain relations

$$(3) \quad \blacktriangle \qquad P_1(D)\sigma'_{ij} = Q_1(D)e'_{ij}, \qquad P_2(D)\sigma_{kk} = Q_2(D)e_{kk}$$

specify an isotropic linear viscoelastic material. Here σ_{ij}, e_{ij}, σ'_{ij}, e'_{ij} [see Eq. (15.1:17)] shall be understood to be functions of x_1, x_2, x_3 and t. In Eq. (3), the dependence on x is not explicitly shown, but shall be understood.

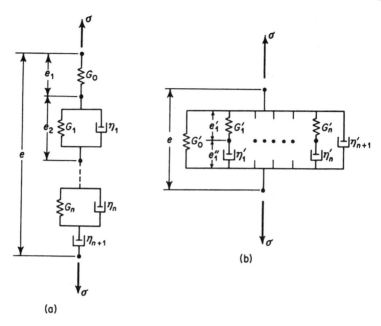

Fig. 15.2:1. Generalized Kelvin and Maxwell models.

If we consider the load-deflection relationship of a network of springs and dashpots such as those pictured in Fig. 15.2:1, it can be shown that the load σ and the deflection e are related by an equation of the form of Eq. (3) (see Problem 15.1). For this reason, a viscoelastic body is often represented by a mechanical model. Naturally, not all conceivable polynomials (2) may represent physically realizable systems. Some restrictions are imposed by thermodynamic considerations, as discussed in Sec. 13.9 and Problem 13.5.

If the stress-strain law for a given material can be expressed in both the integral form and the differential form, then the relationship between the relaxation (or creep) function and the differential operators can be determined. The result is particularly simple if we assume that

$$(4) \qquad\qquad \sigma_{ij}(t) = e_{ij}(t) = 0 \qquad\qquad \text{for} \quad t < 0,$$

and that the Laplace transformation (with respect to time) of all the functions concerned exist. Let the Laplace transformation of a function $f(t)$ be indicated by a bar, thus

$$\bar{f}(s) = \int_0^\infty e^{-st} f(t)\,dt, \qquad\qquad \text{Rl}\, s > s_0,$$

where s_0 is the abscissa of convergence of the Laplace integral.

The Laplace transformations of Eqs. (15.1:17) and (15.1:19) are

(5) $\bar{\sigma}'_{ij}(s) = s\bar{G}_1(s)\bar{e}'_{ij}(s)\,, \qquad \bar{\sigma}_{kk}(s) = s\bar{G}_2(s)\bar{e}_{kk}(s)\,,$

(6) $\bar{e}'_{ij}(s) = s\bar{J}_1(s)\bar{\sigma}'_{ij}(s)\,, \qquad \bar{e}_{kk}(s) = s\bar{J}_2(s)\bar{\sigma}_{kk}(s)\,.$

The Laplace transformations of Eqs. (3) are

(7a) $\bar{P}_1(s)\bar{\sigma}'_{ij}(s) - \sum\limits_{k=1}^{n_1} a_k \left[s^{k-1}\sigma_{ij}(0) + s^{k-2}\dfrac{\partial\sigma'_{ij}}{\partial t}(0) + \cdots + \dfrac{\partial^{k-1}\sigma'_{ij}}{\partial t^{k-1}}(0) \right]$

$= \bar{Q}_1(s)\bar{e}'_{ij}(s)$

$- \sum\limits_{k=1}^{m_1} b_k \left[s^{k-1}e'_{ij}(0) + s^{k-2}\dfrac{\partial e'_{ij}}{\partial t}(0) + \cdots + \dfrac{\partial^{k-1}e'_{ij}}{\partial t^{k-1}}(0) \right],$

(7b) $\bar{P}_2(s)\bar{\sigma}_{jj}(s) - \sum\limits_{k=1}^{n_2} c_k \left[s^{k-1}\sigma_{jj}(0) + \cdots + \dfrac{\partial^{k-1}\sigma_{jj}}{\partial t^{k-1}}(0) \right]$

$= \bar{Q}_2(s)\bar{e}_{jj}(s) - \sum\limits_{k=1}^{m_2} d_k \left[s^{k-1}e_{jj}(0) + \cdots + \dfrac{\partial^{k-1}e_{jj}}{\partial t^{k-1}}(0) \right],$

where

(8)
$$\bar{P}_1(s) = \sum_{k=0}^{n_1} a_k s^k\,, \qquad \bar{Q}_1(s) = \sum_{k=0}^{m_1} b_k s^k\,,$$

$$\bar{P}_2(s) = \sum_{k=0}^{n_2} c_k s^k\,, \qquad \bar{Q}_2(s) = \sum_{k=0}^{m_2} d_k s^k\,,$$

and $\sigma'_{ij}(0)$, $(\partial\sigma'_{ij}/\partial t)(0)$ etc., are the inital values of σ'_{ij}, $\partial\sigma'_{ij}/\partial t$, etc., i.e., the value of $\sigma'_{ij}(x,t)$ and its time derivatives as $t \to 0$ from the positive side.

Whether Eqs. (7) should be identified with the relaxation law (5) or with the creep law (6) depends on whether $n_1 \geq m_1$, $n_2 \geq m_2$ or not. If $n_1 \geq m_1$, then Eq. (7a) may be written as*

(9) ▲
$$\bar{\sigma}'_{ij}(s) = \frac{\bar{Q}_1(s)}{\bar{P}_1(s)}\bar{e}'_{ij}(s)\,,$$

---

*Note that a polynomial of s has no continuous inverse function. The inverse of s is $\delta(t)$; those of s^2, s^3 are higher order singularities.

provided that the following initial condition holds:

$$(10) \quad \sum_{k=1}^{n_1} a_k \left[s^{k-1} \sigma'_{ij}(0) + \cdots + \frac{\partial^{k-1} \sigma'_{ij}}{\partial t^{k-1}}(0) \right]$$

$$- \sum_{k=1}^{m_1} b_k \left[s^{k-1} e'_{ij}(0) + \cdots + \frac{\partial^{k-1} e'_{ij}}{\partial t^{k-1}}(0) \right] \equiv 0.$$

In this case we may identify Eq. (9) with the relaxation law (5), with

$$(11) \quad \blacktriangle \qquad \bar{G}_1(s) = \frac{\bar{Q}_1(s)}{s\bar{P}_1(s)}.$$

The initial condition (10) is an identity as a polynomial in s; every coefficient of the polynomial must vanish. Thus, if $m_1 = n_1$,

$$a_{n_1} \sigma'_{ij}(0) = b_{n_1} e'_{ij}(0),$$

$$(12) \quad \blacktriangle \qquad \cdots\cdots\cdots\cdots\cdots\cdots\cdots\cdots\cdots\cdots\cdots\cdots\cdots,$$

$$a_{n_1} \frac{\partial^{n_1-1} \sigma'_{ij}}{\partial t^{n_1-1}}(0) + a_{n_1-1} \frac{\partial^{n_1-2} \sigma'_{ij}}{\partial t^{n_1-2}}(0) + \cdots + a_1 \sigma'_{ij}(0)$$

$$= b_{n_1} \frac{\partial^{n_1-1} e'_{ij}}{\partial t^{n_1-1}}(0) + \cdots + b_1 e'_{ij}(0).$$

If $m_1 < n_1$, then those coefficients b_k in Eqs. (12) with subscript $k > m_1$ must be replaced by zero.

In the alternative case $n_1 \le m_1$, Eq. (7a) may be written as

$$(13) \quad \bar{e}'_{ij}(s) = \frac{\bar{P}_1(s)}{\bar{Q}_1(s)} \bar{\sigma}'_{ij}(s),$$

provided that the initial condition (10) holds. We may identify Eq. (13) with the creep law (6), with

$$(14) \quad \bar{J}_1(s) = \frac{\bar{P}_1(s)}{s\bar{Q}_1(s)}.$$

If $n_1 = m_1$, then the material may be represented by stress-strain laws of both the relaxation and the creep type.

An analogous situation exists for the laws governing the mean stress and mean strain, σ_{kk} and e_{kk}.

The initial conditions (12) are interesting and important. They represent the proper initial conditions that must be imposed when Eqs. (3) are regarded as differential equations to solve for σ'_{ij}, with e'_{ij} regarded as given forcing functions; or vice versa.

What is the physical significance of the initial conditions (12)? We recognized their necessity through an identification of Eq. (15.1:17) in the preceding Section, which is valid for t in $(-\infty, \infty)$, with Eq. (3) of this Section, which applies for t in $(0, \infty)$. For the purpose of identification we have assumed $e'_{ij} = \sigma'_{ij} = 0$ for t in $(-\infty, 0)$. If

$$(15) \qquad e'_{ij} = \frac{\partial e'_{ij}}{\partial t} = \cdots = \frac{\partial^{n_1-1} e'_{ij}}{\partial t^{n_1-1}} = 0 \qquad\qquad \text{at} \quad t = 0 \,,$$

then Eqs. (12) imply

$$(16) \qquad \sigma'_{ij} = \frac{\partial \sigma'_{ij}}{\partial t} = \cdots = \frac{\partial^{n_1-1} \sigma'_{ij}}{\partial t^{n_1-1}} = 0 \qquad\qquad \text{at} \quad t = 0 \,.$$

Thus the transition is smooth, as expected. The surprising aspect of Eqs. (12) arises only when $e'_{ij}(t)$ is continuous in $(0, \infty)$ but the limiting

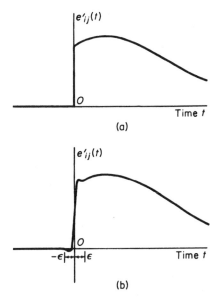

values $e'_{ij}(0+)$, $(\partial e'_{ij}/\partial t)(0+)$, etc., are nonzero as $t \to 0$ from the positive side. In this case there is a jump in the value of e'_{ij} or its derivatives in the neighborhood of $t = 0$. The initial conditions (12) must stem from the fact that the differential operator in Eq. (3) is expected to hold, in a certain sense, during the jump.

To see that this is indeed the case, let us smooth off the jump in an interval $(-\epsilon, \epsilon)$ (see Fig. 15.2:2). Let $e'_{ij}(t, \epsilon)$ be a smooth function (bounded, continuous, and n_1 times differentiable) in the range $(-\epsilon, \epsilon)$, which has the property that $e'_{ij}(t, \epsilon)$ and its derivatives vanish at the left end, $t = -\epsilon$; whereas assume the values $e'_{ij}(0+)$, $(\partial e'_{ij}/\partial t)(0+)$, etc., at the right

Fig. 15.2:2. Smoothing of a jump.

end, $t = +\epsilon$. Such a function serves to describe the jump if we let $\epsilon \to 0$. Let $\sigma'_{ij}(t, \epsilon)$ be the corresponding smoothing function describing the jump in the stress deviation. We now assume that $e'_{ij}(t, \epsilon)$ and $\sigma'_{ij}(t, \epsilon)$ are connected by the viscoelasticity law (3); i.e.,

$$(17) \qquad \sum_{k=0}^{N} a_k \frac{\partial^k \sigma'_{ij}(t,\epsilon)}{\partial t^k} = \sum_{k=0}^{N} b_k \frac{\partial^k e'_{ij}(t,\epsilon)}{\partial t^k},$$

where N is equal to the larger of m_1, and n_1, with some of the coefficients a_k, b_k set equal to zero to eliminate any terms which may be absent from Eq. (3).

An integration of Eq. (17) with respect to t gives

$$(18) \qquad \sum_{k=1}^{N} a_k \frac{\partial^{k-1} \sigma'_{ij}(t,\epsilon)}{\partial t^{k-1}} + a_0 \int_{-\epsilon}^{t} \sigma'_{ij}(\tau,\epsilon)\, d\tau$$

$$= \sum_{k=1}^{N} b_k \frac{\partial^{k-1} e'_{ij}(t,\epsilon)}{\partial t^{k-1}} + b_0 \int_{-\epsilon}^{t} e'_{ij}(\tau,\epsilon)\, d\tau.$$

Now let $t = \epsilon$, and take the limit $\epsilon \to 0$. Since $\sigma'_{ij}(t,\epsilon), e_{ij}(t,\epsilon)$ are bounded in $(-\epsilon, \epsilon)$, the integrals in Eq. (18) are of order ϵ and vanish as $\epsilon \to 0$. Therefore, at $t = \epsilon$ and in the limit $\epsilon \to 0$, we obtain the necessary initial condition

$$\sum_{k=1}^{N} a_k \frac{\partial^{k-1} \sigma'_{ij}}{\partial t^{k-1}}(0+) = \sum_{k=1}^{N} b_k \frac{\partial^{k-1} e'_{ij}}{\partial t^{k-1}}(0+),$$

which is one of the equations in Eq. (12). Integrating Eq. (18) again from $-\epsilon$ to ϵ and repeating the arguments as above, we obtain the second initial condition

$$\sum_{k=2}^{N} a_k \frac{\partial^{k-2} \sigma'_{ij}}{\partial t^{k-1}}(0+) = \sum_{k=2}^{N} b_k \frac{\partial^{k-2} e'_{ij}}{\partial t^{k-2}}(0+).$$

All Eqs. (12) can be obtained in this way by repeated applications of the same process.

Problem 15.1. Let σ denote loads and e denote displacements. Show that, for the generalized Kelvin model shown in Fig. 15.2:1(a),

$$G_0 e_1 = \sigma, \qquad G_1 e_2 + \eta_1 D e_2 = \sigma, \qquad \dots, \qquad e = e_1 + e_2 + \cdots + e_n,$$

so that

$$\left(\frac{1}{G_0} + \frac{1}{G_1 + \eta_1 D} + \cdots + \frac{1}{G_n + \eta_n D} + \frac{1}{\eta_{n+1} D} \right) \sigma = e.$$

For the generalized Maxwell model shown in Fig. 15.2:1(b), we have

$$G_1' e_1' = \sigma_1, \qquad \eta_1' D e_1'' = \sigma_1,$$

$$e = e_1' + e_1'' = \left(\frac{1}{G_1'} + \frac{1}{\eta_1' D} \right) \sigma_1,$$

$$\sigma = \sigma_0 + \sigma_1 + \cdots + \sigma_{n+1}$$

$$= \left(G_0' + \frac{G_1 \eta_1 D}{G_1 + \eta_1 D} + \cdots + \frac{G_n \eta_n D}{G_n + \eta_n D} + \eta_{n+1} D \right) e.$$

Reduce these relations to the form of Eq. (3).

Problem 15.2. Let a special case of Eq. (3) be specified by

$$P_1 \sigma_{ij} = a_2 \ddot{\sigma}_{ij} + a_1 \dot{\sigma}_{ij} + a_0 \sigma_{ij},$$

$$Q_1 e_{ij} = b_2 \ddot{e}_{ij} + b_1 \dot{e}_{ij} + b_0 e_{ij},$$

where a dot denotes a differentiation with respect to time. Deduce proper initial conditions. Interpret the usual statement that "the initial response of a viscoelastic body is purely elastic."

15.3. BOUNDARY-VALUE PROBLEMS AND INTEGRAL TRANSFORMATIONS

The motion of a viscoelastic body is governed by the laws of conservation of mass and momentum, the stress-strain relations, and the boundary conditions and initial conditions. Let u_i, e_{ij}, σ_{ij}, and X_i denote the Cartesian components of the displacement, strain, stress, and body force per unit volume, respectively, and ρ be the mass density. Let us restrict our consideration to homogeneous and isotropic viscoelastic solids, and infinitesimal displacements. Under these limitations the field equations are:

(1) *Definition of strain:* $e_{ij} = \frac{1}{2}(u_{i,j} + u_{j,i})$,

(2) *Equation of continuity:* $\frac{\partial \rho}{\partial t} + \frac{\partial}{\partial x_i}(\rho \frac{\partial u_i}{\partial t}) = 0$,

(3) *Equations of motion:* $\sigma_{ij,j} + X_i = \rho \frac{\partial^2 u_i}{\partial t^2}$, $\qquad\qquad$ $\sigma_{ij} = \sigma_{ji}$.

The stress-strain relation may assume any of the following forms (see Sec. 15.2):

(4a) *Relaxation law:* $\sigma_{ij}' = e_{ij}' * dG_1$, $\qquad$ $\sigma_{kk} = e_{kk} * dG_2$,

(4b) *Creep law:* $e_{ij}' = \sigma_{ij}' * dJ_1$, $\qquad$ $e_{kk} = \sigma_{kk} * dJ_2$,

(4c) *Differential operator law:* $P_1(D)\sigma_{ij}' = Q_1(D)e_{ij}'$, $\qquad$ $P_2(D)\sigma_{kk} = Q_2(D)e_{kk}$, where e'_{ij}, σ'_{ij} are the stress and strain deviators defined by

(5) $$\sigma'_{ij} = \sigma_{ij} - \frac{1}{3}\delta_{ij}\sigma_{kk}, \qquad e'_{ij} = e_{ij} - \frac{1}{3}\delta_{ij}e_{kk},$$

G_1, G_2 are the relaxation functions, J_1, J_2 are the creep functions, and P_1, P_2, Q_1, Q_2 are polynomials of the time-derivative operator D as discussed in Sec. 15.2. In view of the assumed homogeneity of the material, these functions and operators are independent of position.

If the body is originally undisturbed, then the initial conditions are

(6) $$u_i = e_{ij} = \sigma_{ij} = 0 \qquad\qquad \text{for } -\infty < t < 0.$$

If the differential operator law (4c) is used and there is a jump condition at $t = 0$, then the initial conditions may assume the form of specific assigned values of $e'_{ij}(0+)$, $\frac{\partial e'_{ij}}{\partial t}(0+), \ldots, \frac{\partial^n e'_{ij}}{\partial t^n}(0+)$, and of $\sigma'_{ij}(0+)$, $\frac{\partial \sigma'_{ij}}{\partial t}(0+), \ldots, \frac{\partial^n \sigma'_{ij}}{\partial t^n}(0+)$, which must be connected by the necessary conditions [Eq. (15.2:12)]:

(7) $$\sum_{k=r}^{n} a_k \frac{\partial^{k-r}\sigma'_{ij}}{\partial t^{k-r}}(0+) = \sum_{k=r}^{n} b_k \frac{\partial^{k-r}e'_{ij}}{\partial t^{k-r}}(0+), \qquad r = 1, 2, \ldots, n.$$

where n is the larger of the degrees of $P_1(D)$ and $Q_1(D)$. Similar statements hold for initial values of mean stress and mean strain.

The boundary conditions may take the form of either assigned traction over a surface S_σ with an outward-pointing unit normal vector ν_i,

(8) $$\overset{\nu}{T}_i = \sigma_{ij}\nu_j = f_i \qquad\qquad \text{over } S_\sigma,$$

or of assigned displacement over a surface S_u,

(9) $$u_i = g_i \qquad\qquad \text{over } S_u,$$

where f_i, g_i are prescribed functions of position and time, and $S_\sigma + S_u = S$, the total surface of the body.

The problems in the theory of linear viscoelasticity usually consist of determining u_i, e_{ij}, and σ_{ij} for prescribed X_i, f_i, g_i and initial conditions. Except for the stress-strain law, these same equations occur in the linearized theory of elasticity.

In the lineared theory of elasticity we have applied the Laplace transformation to solve dynamic problems. The transformation translates an original problem involving derivatives and convolution integrals with respect to the time t into an algebraic problem with respect to a parameter s. After the algebraic problem is solved, the solution is translated back to the time domain. The last step, of course, may not be easy.

It appears natural to apply the Laplace transformation also to problems in linear viscoelasticity. In fact, the transformed problem, involving the parameter s, can be put into the same form as that in the theory of linear elasticity. If the latter problem can be solved, the viscoelasticity solution can be obtained by an inverse transformation from s back to the time domain. The last step, again, may not be easy.

The Fourier transform may be used in place of the Laplace transform. The selection of the appropriate transform depends on the nature of the functions involved. A function which has a Laplace transform may not necessarily have a Fourier transform, and vice versa. For example, the unit-step function $\mathbf{1}(t)$, or a power t^n, has no Fourier transform in the interval $(0, \infty)$.

The identification of a problem in linear elasticity with one in viscoelasticity in the transformed plane, is called the *correspondence principle*. Applications of the correspondence principle will become apparent by examining some examples in Secs. 15.4 and 15.5.

Problem 15.3. Assuming that the functions u_i, G_1, G_2 are continuous and can be differentiated as many times as may be desired, show that the equation of motion may be written (in the form of Navier's equation) as

$$(u_i)_{,jj} * dG_1 + (u_{j,j})_{,i} * dK + 2X_i = 2\rho \frac{\partial^2 u_i}{\partial t^2}, \qquad i = 1, 2, 3,$$

where $3K = G_1 + 2G_2$.   (Gurtin and Sternberg 1962[15.1])

Problem 15.4.   From the strain equations of compatibility

$$e_{ij,kk} + e_{kk,ij} - e_{ik,jk} - e_{jk,ik} = 0,$$

the stress-strain relations

$$e_{ij} = \sigma_{ij} * dJ_1 + \frac{1}{3}\delta_{ij}\sigma_{kk} * d(J_2 - J_1),$$

and the equations of equilibrium

$$\sigma_{ik,k} = -X_i,$$

deduce the stress equations of compatibility

$$(\sigma_{ij})_{,kk} * dJ_1 + \sigma_{kk,ij} * d\Lambda = \Theta_{ij},$$

where

$$\Theta_{ij} = \delta_{ij} X_{k,k} * d\Omega - (X_{i,j} + X_{j,i}) * dJ_1 \,,$$

$$3\Lambda = 2J_1 + J_2 \,,$$

$$9\Omega = J_1 * d(J_2 - J_1) * d(J_1 + 2J_2)^{-1} \,.$$

(Gurtin and Sternberg 1962[15.1])

Problem 15.5.   Let $\mathbf{u}(u_i), \boldsymbol{\epsilon}(e_{ij}), \boldsymbol{\sigma}(\sigma_{ij})$ satisfy the equations of viscoelasticity, and $G_1 \not\equiv 0, 2G_1 + G_2 \not\equiv 0$. Show that, when the body force X_i is absent,

$$\nabla^2(\nabla \cdot \mathbf{u}) = 0, \quad \nabla^2(\nabla \times \mathbf{u}) = 0, \quad \nabla^2 e_{kk} = 0, \quad \nabla^2 \sigma_{kk} = 0,$$

$$\nabla^4 \mathbf{u} = 0, \quad \nabla^4 \boldsymbol{\epsilon} = 0, \quad \nabla^4 \boldsymbol{\sigma} = 0.$$

(Gurtin and Sternberg 1962[15.1])

15.4. WAVES IN AN INFINITE MEDIUM

Let us consider a viscoelastic medium of infinite extent and look for solutions of Eqs. (15.3:1)–(15.3:6), in which all dependent variables, including body forces, vary sinusoidally with time. We shall use the complex representation for sinusoidal oscillations. Since $\cos \omega t$ is equal to the real part of $e^{i\omega t}$, a real-valued function $f(x,t)$ that varies like $\cos \omega t$ at every point x may be represented as $\mathrm{Rl}\,[F(x)e^{i\omega t}]$, where $F(x)$ is real. A real-valued function $f(x,t)$ that varies like $\cos(\omega t + \phi)$ may be represented as $\mathrm{Rl}\,[F(x)e^{i(\omega t + \phi)}]$ or as $\mathrm{Rl}\,[\bar{F}(x)e^{i\omega t}]$, where $\bar{F}(x)$ now stands for a complex number $F(x)e^{i\phi}$. It is elementary to show that

$$\mathrm{Rl}\,\bar{F}(x) = F(x)\cos\phi, \quad \mathrm{Im}\,\bar{F}(x) = F(x)\sin\phi\,,$$

$$\mathrm{Rl}\,[\bar{F}(x)e^{i\omega t}] = \mathrm{Rl}\,\bar{F}\,\mathrm{Rl}\,e^{i\omega t} - \mathrm{Im}\,\bar{F}\,\mathrm{Im}\,e^{i\omega t}$$

$$= F\cos\phi\cos\omega t - F\sin\phi\sin\omega t$$

$$= F\cos(\omega t + \phi)\,.$$

Thus it is clear that the real part of $\bar{F}(x)$ is *in phase* with $\cos\omega t$ and that the imaginary part of $\bar{F}(x)$ *leads* $\cos\omega t$ by a phase angle $\phi = \pi/2$. In general, ϕ is a function of x. In this way a sinusoidal oscillation of a solid medium whose amplitude and phase angle vary from point to point may be represented by a complex function $\bar{F}(x)$. A multiplication of $\bar{F}(x)$ by the imaginary number i means an advance of phase angle by $\pi/2$; a multiplication by $-i$ means a lag by $\pi/2$.

With these interpretations of complex representation we let

(1) $\qquad e_{ij} = \text{Rl}\,(\bar{e}_{ij} e^{i\omega t}), \qquad \sigma_{ij} = \text{Rl}\,(\bar{\sigma}_{ij} e^{i\omega t}), \quad \text{etc.},$

where $\bar{e}_{ij}$, $\bar{\sigma}_{ij}$, etc., are complex-valued functions of the spatial coordinates only. The basic Eqs. (15.3:1)–(15.3:4) now read

(2) $\qquad\qquad\qquad\qquad \bar{e}_{ij} = \dfrac{1}{2}(\bar{u}_{i,j} + \bar{u}_{j,i}),$

(3) $\qquad\qquad\qquad\qquad \bar{\sigma}_{ij,j} + \bar{X}_i + \rho\omega^2 \bar{u}_i = 0,$

(4) $\qquad\qquad\qquad\qquad \bar{\sigma}'_{ij} = i\omega \bar{G}_1(\omega) \bar{e}'_{ij},$

$\qquad\qquad\qquad\qquad \bar{\sigma}_{kk} = i\omega \bar{G}_2(\omega) \bar{e}_{kk},$

where $\bar{G}_1(\omega)$, $\bar{G}_2(\omega)$ are the deviatoric and the dilatational complex moduli, and are the Fourier transforms of the relaxation functions $G_1(t)$ and $G_2(t)$, respectively. If we write

(5) $\qquad \bar{\lambda}(\omega) = \dfrac{1}{3}i\omega[\bar{G}_2(\omega) - \bar{G}_1(\omega)], \qquad \bar{G}(\omega) = \dfrac{1}{2}i\omega \bar{G}_1(\omega),$

and call $\bar{\lambda}(\omega)$ and $\bar{G}(\omega)$ the complex Lamé constants for a viscoelastic material, we see that Eqs. (2)–(5) are identical with those governing linear elasticity theory (Sec. 7.1), except that the Lamé constants are replaced by complex moduli. Hence, without much ado, we can write down the Navier equation [cf. Eq. (7.1:9)]

(6) $\qquad \bar{G}(\omega)\bar{u}_{i,jj} + [\bar{\lambda}(\omega) + \bar{G}(\omega)]\bar{u}_{j,ji} + \bar{X}_i + \rho\omega^2 \bar{u}_i = 0$

and the solutions, in case $\bar{X}_i \equiv 0$,

(7) $\qquad u_j = \text{Rl}\left\{ An_j \exp\left[i\omega\left(t \pm \sqrt{\dfrac{\rho}{3\bar{\lambda}(\omega) + 2\bar{G}(\omega)}}\, n_k x_k \right)\right]\right\}$

and

(8) $\qquad\qquad u_j = \text{Rl}\left\{ C_j \exp\left[i\omega\left(t \pm \sqrt{\dfrac{\rho}{\bar{G}(\omega)}}\, n_k x_k \right)\right]\right\},$

where A is a complex constant, n_j is an arbitrary vector, and C_j are components of a vector perpendicular to n_i, i.e., $C_j n_j = 0$.

Equation (7) represents a plane dilatational wave. Equation (8) represents a plane shear wave. The exponential factors in Eqs. (7) and (8) are complex. A little reflection shows that the real parts of the factors in front of n_k, x_k, are the inverse of wave velocities, v_D, v_R; whereas the imaginary parts are attenuation factors α_D, α_R.

(9) *Dilatational waves:*

$$v_D = \left\{ \mathrm{Rl} \left[\sqrt{\left(\frac{\rho}{3\bar{\lambda}(\omega) + 2\bar{G}(\omega)} \right)} \right] \right\}^{-1},$$

$$\alpha_D = -\omega \, \mathrm{Im} \left[\sqrt{\left(\frac{\rho}{3\bar{\lambda}(\omega) + 2\bar{G}(\omega)} \right)} \right],$$

(10) *Rotational waves:*

$$v_R = \left\{ \mathrm{Rl} \left[\sqrt{\left(\frac{\rho}{\bar{G}(\omega)} \right)} \right] \right\}^{-1}, \qquad \alpha_R = -\omega \, \mathrm{Im} \left[\sqrt{\left(\frac{\rho}{\bar{G}(\omega)} \right)} \right].$$

If the material is elastic, $\bar{\lambda}(\omega)$, $\bar{G}(\omega)$ become real numbers, independent of ω, in which case the attenuation factors α_D and α_R vanish and the wave speeds v_D, v_R are independent of frequency.

Any formal solution of Navier's equation in the classical theory of linear elasticity, of the form $f = \mathrm{Rl}\,[\bar{f}(x_1, x_2, x_3)e^{i\omega t}]$, offers a corresponding solution for a linear viscoelastic body, if the elastic moduli that occur in $\bar{f}(x_1, x_2, x_3)$ are replaced by the corresponding complex moduli of the material. If the boundary conditions for the elastic and viscoelastic problems are identical, then a solution in viscoelasticity can be obtained by this correspondence principle. Note, however, that the operation of separating a complex modulus into its real and imaginary parts is one that has no counterpart in the theory of linear elasticity. Hence, such an operation is excluded from the correspondence principle. For example, if we wish to find the maximum of $|f|$ with respect to ω, we must separate the real and imaginary parts, and no direct analogy will be available.

Problem 15.6.   Work out the details of the derivation of Eqs. (7) and (8) by reference to Sec. 7.8.

Problem 15.7.   Find the speed and attenuation factor for Rayleigh surface waves propagating over a viscoelastic half-space (see Sec. 7.9). Reference Bland, *Linear Viscoelasticity*[15.1] pp. 73–5.

Problem 15.8.   Consider a uniform cantilever beam the clamped end of which is forced to oscillate at a constant amplitude ω_0 (the clamping wall moves, while the other end is free). Fig. P15.8.

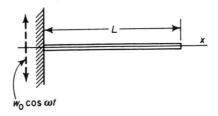

Fig. P15.8.

For an elastic beam, the equation of motion of the beam is

$$EI\frac{\partial^4 w}{\partial x^4} + m\frac{\partial^2 w}{\partial t^2} = 0,$$

where w is the beam deflection, m is the mass per unit length, E is Young's modulus, and I the moment of inertia of the cross section. The boundary conditions are at $x = 0$:

$$w = w_o \cos \omega t, \qquad \frac{\partial w}{\partial x} = 0;$$

at $x = L$:

$$\frac{\partial^2 w}{\partial x^2} = \frac{\partial^3 w}{\partial x^3} = 0.$$

Determine a solution $w(x,t)$ in the form $\bar{w}(x)e^{i\omega t}$.

If the beam material is viscoelastic, determine (a) the corresponding complex bending rigidity $EI(\omega)$ from the stress-strain relationship of the material and (b) the solution $w(x,t)$.

The amplitude ratio and phase lag of the oscillations at the free and the clamped ends can be used for experimental determination of the complex modulus. Reference Bland and Lee, *J. Appl. Phys.*, *26* (1955), p. 1497.

15.5. QUASI-STATIC PROBLEMS

If all functions of concern vanish for $t < 0$ and have Laplace transformations, the transforms of the basic Eqs. (15.3:1)–(15.3:9) assume exactly the same form as in the theory of linear elasticity. In quasi-static problems, the loading is assumed to be so slow that inertia forces may be neglected. In this case, the corresponding elastic problems are static.

Let the Laplace transform of a variable be denoted by a bar over it. The basic equations for quasi-static problems of a viscoelastic body occupying a region V with boundary $B = B_\sigma + B_u$ are

(1) $$\bar{e}_{ij} = \frac{1}{2}(\bar{u}_{i,j} + \bar{u}_{j,i}) \qquad \text{in} V,$$

(2) $$\bar{\sigma}_{ij,j} + \bar{X}_i = 0 \qquad \text{in } V,$$

(3a) $$\bar{\sigma}'_{ij} = s\bar{G}_1(s)\bar{e}'_{ij}, \qquad \bar{\sigma}_{kk} = s\bar{G}_2(s)\bar{e}_{kk} \qquad \text{in } V,$$

or

(3b) $$\bar{e}'_{ij} = s\bar{J}_1(s)\bar{\sigma}'_{ij}, \qquad \bar{e}_{kk} = s\bar{J}_2(s)\bar{\sigma}_{kk} \qquad \text{in } V,$$

or

(3c) $$P_1(s)\bar{\sigma}'_{ij} = Q_1(s)\bar{e}'_{ij}, \qquad P_2(s)\bar{\sigma}_{kk} = Q_2(s)\bar{e}_{kk} \qquad \text{in } V,$$

where

(4) $$\bar{\sigma}_{ij}\nu_j = \bar{f}_i \qquad \text{on } B_\sigma,$$

(5) $$\bar{u}_i = \bar{g}_i \qquad \text{on } B_u.$$

The boundary surface enclosing V is $B = B_\sigma + B_u$. The initial conditions (15.3:6) and (15.3:7) are assumed to be satisfied both in V and on B. The material properties $\bar{G}_1(s)$, $\bar{G}_2(s)$; or $\bar{J}_1(s)$, $\bar{J}_2(s)$; or $P_1(s)$, $P_2(s)$, $Q_1(s)$, $Q_2(s)$; are given. The problem is to determine $\bar{u}_i$, $\bar{e}_{ij}$, $\bar{\sigma}_{ij}$ for assigned forcing functions $\bar{X}_i$ in V and $\bar{f}_i$, $\bar{g}_i$ on the boundary. If the bars were removed, these would be the same equations that govern the static equilibrium of an elastic body of the same geometry. If the solution of the latter were known, the transformed solution would be obtained, and the solution of the original problem could be obtained by an inversion.

The counterpart of Young's modulus E, the shear modulus G, the bulk modulus K, Poisson's ratio ν, and one of Lamé constants λ. (the other Lamé constant is $\mu = G$) can be derived from Eqs. (3a), (3b), and (3c), and are

(6) $$\bar{E}(s) = \frac{3Q_1(s)Q_2(s)}{Q_1(s)P_2(s) + 2P_1(s)Q_2(s)} = \frac{3s\bar{G}_1(s)\bar{G}_2(s)}{2\bar{G}_2(s) + \bar{G}_1(s)}$$

$$= \frac{3}{s[2\bar{J}_1(s) + \bar{J}_2(s)]},$$

(7) $$\bar{G}(s) = \frac{1}{2}\frac{Q_1(s)}{P_1(s)} = \frac{1}{2}s\bar{G}_1(s) = \frac{1}{2}\frac{1}{s\bar{J}_1(s)},$$

(8) $$\bar{K}(s) = \frac{1}{3}\frac{Q_2(s)}{P_2(s)} = \frac{1}{3}s\bar{G}_2(s) = \frac{1}{3}\frac{1}{s\bar{J}_2(s)},$$

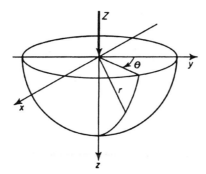

Fig. 15.5:1. Boussinesq problem over a viscoelastic material.

(9)
$$\bar{\nu}(s) = \frac{P_1(s)Q_2(s) - Q_1(s)P_2(s)}{Q_1(s)P_2(s) + 2P_1(s)Q_2(s)} = \frac{\bar{G}_2(s) - \bar{G}_1(s)}{2\bar{G}_2(s) + \bar{G}_1(s)}$$

$$= \frac{\bar{J}_1(s) - \bar{J}_2(s)}{2\bar{J}_1(s) + \bar{J}_2(s)},$$

(10)
$$\bar{\lambda}(s) = \bar{K}(s) - \frac{2}{3}\bar{G}(s) = \frac{1}{3}s[\bar{G}_2(s) - \bar{G}_1(s)].$$

Example 1. *Boussinesq Problem*

Consider the problem of a concentrated load Z on a half-space (Fig. 15.5:1). The elastic solution is given in Sec. 8.10. For example, the stress component σ_{rr} is given by Eq. (8.10:7) as

(11)
$$\sigma_{rr}(r, z) = \frac{Z}{2\pi}\left\{(1 - 2\nu)\left[\frac{1}{r^2} - \frac{z}{r^2}\frac{1}{(r^2 + z^2)^{1/2}}\right]\right.$$

$$\left. - 3r^2 z\frac{1}{(r^2 + z^2)^{5/2}}\right\}.$$

On applying the correspondence principle, the solution of $\bar{\sigma}_{rr}$ is

(12)
$$\bar{\sigma}_{rr}(r, z; s) = \frac{\bar{Z}(s)}{2\pi}\left\{[1 - 2\bar{\nu}(s)]\left[\frac{1}{r^2} - \frac{z}{r^2}\frac{1}{(r^2 + z^2)^{1/2}}\right]\right.$$

$$\left. - 3r^2 z\frac{1}{(r^2 + z^2)^{5/2}}\right\},$$

where $\bar{\nu}(s)$ is given by Eq. (9). An inverse Laplace transformation gives, therefore,

(13) $\quad \sigma_{rr}(r, z; t) = \dfrac{1}{2\pi}\Bigg\{ \left[\dfrac{1}{r^2} - \dfrac{z}{r^2}\dfrac{1}{(r^2 + z^2)^{1/2}} \right]$

$$\times \int_0^t Z(t - \tau)\Phi(\tau)\,d\tau - 3r^2 z\dfrac{1}{(r^2 + z^2)^{5/2}}\, Z(t) \Bigg\},$$

where

(14) $$\Phi(t) = \int_{c-i\infty}^{c+i\infty} [1 - 2\bar{\nu}(s)]e^{st}\,ds$$

is the inverse Laplace transformation of $1 - 2\bar{\nu}(s)$, the constant c being any number greater than the abscissa of convergence.

Problem 15.9. A concentrated load $Z(t) = Z_0 1(t)$ (Z_0 a constant, $1(t)$ a unit-step function) acts normal to the free surface of a half-space of Voigt material, for which

$$\sigma'_{ij} = (2\eta D + 2G)e'_{ij}, \qquad \sigma_{jj} = 3Ke_{jj}.$$

Show that

$$1 - 2\bar{\nu}(s) = \dfrac{3(\eta s + G)}{3K + \eta s + G}$$

and

$$\sigma_{rr} = \dfrac{Z_0}{2\pi}\Bigg\{ \dfrac{3}{3K + G}\left[G + 3K\exp\left(-\dfrac{3K + G}{\eta}t \right) \right]$$

$$\times \left[\dfrac{1}{r^2} - \dfrac{z}{r^2}\dfrac{1}{(r^2 + z^2)^{1/2}} \right] - 3r^2 z\dfrac{1}{(r^2 + z^2)^{5/2}} \Bigg\}1(t).$$

Problem 15.10. Consider a long, thick-walled circular cylinder subjected to a uniform internal pressure p_i and uniform external pressure p_0. Fig. P15.10. Regarding this as a plane strain problem, show that the stress

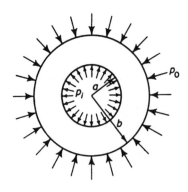

Fig. P15.10.

distribution in the cylinder is the same whether the cylinder material is linear elastic or linear viscoelastic. Derive the history of deformation $u_i(t)$ if the cylinder is viscoelastic.

Note: In the theory of elasticity, the biharmonic equation governing the Airy stress function in plane stress or plane strain problems is independent of the elastic constants if the body force vanishes. Note, however, that if the boundary conditions of a two-dimensional problem involve displacements, then the associated elastic problem will involve elastic moduli and the correspondence principle applies.

15.6. RECIPROCITY RELATIONS

In the beginning of Sec. 1.1, we derived the Maxwell and Betti–Rayleigh reciprocity relations. These relations can be generalized to an elastic or viscoelastic continuum and, with proper interpretations, they are valid not only for statics but also for dynamics.

The most general reciprocal theorem in dynamics was asserted by Horace Lamb[15.3] to be derivable from a remarkable formula established by Lagrange in the *Mécaniqlie Analytique* (1809) by way of a prelude to Lagrange's theory of the variation of arbitrary constants. Lamb showed how the reciprocal theorems of von Helmholtz, in the theory of least action in acoustics and optics, and of Lord Rayleigh, in acoustics, can be derived from Lagrange's formula. Rayleigh[15.3] extended the reciprocal theorem to include the action of dissipative forces, and Lamb[15.3] showed the complete reciprocity relationships in a moving fluid with reversed-flow conditions.

In the theory of elasticity, a generalization of the reciprocity theorem to dynamic problems was given by Graffi,[15.3] and certain applications of Graffi's results to the problem of elastic wave propagation were pointed out by Di Maggio and Bleich.[15.3]

We shall derive the dynamic reciprocity relationship for a viscoelastic solid, which includes the linear elastic solid as a special case. We assume the stress-strain law to be given by one of the equations (15.3:4). The equations of motion and boundary conditions are given by Eqs. (15.3:3) and (15.3:8) or (15.3:9), respectively. Let us consider the problems in which the body force $X_i(x_1, x_2, x_3; t)$, the specified surface tractions $f_i(x_1, x_2, x_3; t)$ and the specified displacements $g_i(x_1, x_2, x_3; t)$ are given functions of space and time, which starts its action at $t > 0$, under the initial conditions

$$(1) \qquad u_i = \frac{\partial u_i}{\partial t} = \cdots = \frac{\partial^N u_i}{\partial t^N} = 0 \quad \text{when} \quad t \leq 0,$$

where N is either unity or equal to the highest derivative that occurs in the stress-strain law (4c). We apply the Laplace transform with respect to the time t to every dependent variable, assuming that the transforms exist. Let the Laplace transform of u be written as $\bar{u}$,

$$(2) \qquad \bar{u} = \int_0^\infty e^{-st} u(t) dt \,.$$

Similarly, a bar over other variable denotes the Laplace transform of that variable. Then, we have, due to the initial conditions indicated above,

$$
\begin{aligned}
\bar{\sigma}_{ij} &= \bar{\lambda}(s)\bar{u}_{k,k}\delta_{ij} + \bar{G}(s)(\bar{u}_{i,j} + \bar{u}_{j,i})\,, \\
s^2 \rho \bar{u}_i &= \bar{X}_i + \bar{\sigma}_{ij,j} \quad \text{in} \quad V \,, \\
\bar{\sigma}_{ij}\nu_j &= \bar{f}_i \quad \text{on} \quad S_\sigma \\
\bar{u}_i &= \bar{g}_i \quad \text{on} \quad S_u
\end{aligned}
$$

(3)

where $\bar{\lambda}(s)$ and $\bar{G}(s)$ are given by Eqs. (15.5:10) and (15.5:7), respectively.

Now consider two problems where the applied body force and the surface tractions and displacements are specified differently. Let the variables involved in these two problems be distinguished by superscripts in parentheses. Then,

$$(4a) \qquad \text{In } V: \quad s^2 \rho \bar{u}_i^{(1)} = \bar{X}_i^{(1)} + \bar{\sigma}_{ij,j}^{(1)}$$

$$(4b) \qquad \qquad \quad s^2 \rho \bar{u}_i^{(2)} = \bar{X}_i^{(2)} + \bar{\sigma}_{ij,j}^{(2)}$$

$$(5a) \qquad \text{On } S_\sigma: \quad \bar{\sigma}_{ij}^{(1)} \nu_j = \bar{f}_i^{(1)}$$

$$(5b) \qquad \qquad \quad \bar{\sigma}_{ij}^{(2)} \nu_j = \bar{f}_i^{(2)}$$

$$(6a) \qquad \text{On } S_u: \quad \bar{u}_i^{(1)} = \bar{g}_i^{(1)}$$

$$(6b) \qquad \qquad \quad \bar{u}_j^{(2)} = \bar{g}_i^{(2)} \,.$$

Multiplying Eq. (4a) by $\bar{u}_i^{(2)}$ and Eq. (4b) by $\bar{u}_i^{(1)}$, subtracting, and integrating over the region V, we obtain

$$(7) \quad \int_V \bar{X}_i^{(1)} \bar{u}_i^{(2)} dv + \int_V \bar{\sigma}_{ij,j}^{(1)} \bar{u}_i^{(2)} dv = \int_V \bar{X}_i^{(2)} \bar{u}_i^{(1)} dv + \int_V \bar{\sigma}_{ij,j}^{(2)} \bar{u}_i^{(1)} dv \,.$$

Now

(8)
$$\int_V \bar{\sigma}_{ij,j}^{(1)} \bar{u}_i^{(2)} dv = \int_V [(\bar{\lambda}\bar{u}_{k,k}^{(1)} \delta_{ij})_{,j} \bar{u}_i^{(2)} + (\bar{G}\bar{u}_{i,j}^{(1)} + \bar{G}\bar{u}_{j,i}^{(1)})_{,j} \bar{u}_i^{(2)}] dv$$

$$= \int_s \bar{\lambda}\bar{u}_{k,k}^{(1)} \bar{u}_i^{(2)} \nu_i dS - \int_V \bar{\lambda}\bar{u}_{k,k}^{(1)} \bar{u}_{j,j}^{(2)} dv$$

$$+ \int_S \bar{G}\bar{u}_{i,j}^{(1)} \bar{u}_i^{(2)} \nu_j dS + \int_S \bar{G}\bar{u}_{j,i}^{(1)} \bar{u}_i^{(2)} \nu_j dS$$

$$- \int_V \bar{G}\bar{u}_{i,j}^{(1)} \bar{u}_{i,j}^{(2)} dv - \int_V \bar{G}\bar{u}_{j,i}^{(1)} \bar{u}_{i,j}^{(2)} dv .$$

A similar expression is obtained for the integral

(9)
$$\int_V \bar{\sigma}_{ij,j}^{(2)} \bar{u}_i^{(1)} dv .$$

When these expressions are substituted into Eq. (7), we see that a number of volume integrals cancel each other. The surface integrals contributed by the integrals (8) and (9) to Eq. (7) are

$$\int_S \bar{\lambda}\bar{u}_{k,k}^{(1)} \bar{u}_i^{(2)} \nu_i dS + \int_S \bar{G}\bar{u}_{i,j}^{(1)} \bar{u}_i^{(2)} \nu_j dS + \int_S \bar{G}\bar{u}_{j,i}^{(1)} \bar{u}_i^{(2)} \nu_j dS$$

(10)
$$- \left(\int_S \bar{\lambda}\bar{u}_{k,k}^{(2)} \bar{u}_i^{(1)} \nu_i dS + \int_S \bar{G}\bar{u}_{i,j}^{(2)} \bar{u}_i^{(1)} \nu_j dS + \int_S \bar{G}\bar{u}_{j,i}^{(2)} \bar{u}_i^{(1)} \nu_j dS \right)$$

$$= \int_S \bar{\sigma}_{ij}^{(1)} \bar{u}_i^{(2)} \nu_j dS - \int_S \bar{\sigma}_{ij}^{(2)} \bar{u}_i^{(1)} \nu_j dS .$$

Recall that S_σ is the part of S over which surface tractions are specified, S_u is the part of S over which displacements are specified, $S = S_\sigma + S_u$. Now, substituting Eqs. (8), (9), (10) and the boundary conditions (5a), (5b), (6a), (6b) into Eq. (7), we obtain finally:

(11)
$$\int_V \bar{X}_i^{(1)} \bar{u}_i^{(2)} dv + \int_{S_\sigma} \bar{f}_i^{(1)} \bar{u}_i^{(2)} dS + \int_{S_u} \bar{\sigma}_{ij}^{(1)} \bar{g}_i^{(2)} \nu_j dS$$

$$= \int_V \bar{X}_i^{(2)} \bar{u}_i^{(1)} dv + \int_{S_\sigma} \bar{f}_i^{(2)} \bar{u}_i^{(1)} dS + \int_{S_u} \bar{\sigma}_{ij}^{(2)} \bar{g}_i^{(1)} \nu_j dS .$$

This is the *general reciprocal relation in the Laplace transformation form.* If we remove the bars and consider the variables to be in the real-time domain, then it becomes Betti's reciprocal relation in elastostatics.

Since the inverse transform of the product of two functions is the convolution of the inverses, we obtain:

(12)
$$\int_V \int_0^t X_i^{(1)}(x, t - \tau)u_i^{(2)}(x, \tau)d\tau dv$$

$$+ \int_{S_\sigma} \int_0^t f_i^{(1)}(x, t - \tau)u_i^{(2)}(x, \tau)d\tau \, dS$$

$$+ \int_{S_u} \int_0^t \sigma_{ij}^{(1)}(x, t - \tau)g_i^{(2)}(x, \tau)\nu_j d\tau \, dS$$

$$= \int_V \int_0^t X_i^{(2)}(x, t - \tau)u_i^{(1)}(x, \tau)d\tau \, dv$$

$$+ \int_{S_\sigma} \int_0^t f_i^{(2)}(x, t - \tau)u_i^{(1)}(x, \tau)d\tau \, dS$$

$$+ \int_{S_u} \int_0^t \sigma_{ij}^{(2)}(x, t - \tau)g_i^{(1)}(x, \tau)\nu_j d\tau \, dS.$$

This is the *general reciprocal relation for elasto-kinetics.* Whether the material is viscoelastic or purely elastic makes no difference in the final result. Note that this result holds for variable density $\rho(x_1, x_2, x_3)$ and nonhomogeneous material properties.

Generalization to Infinite Region. A generalization of the above result to an infinite or semi-infinite region is possible. Since with a finite wave speed there always exists a finite boundary surface, at any finite $t > 0$, which is yet uninfluenced by the loading initiated at $t = 0$. Let S_u be such a surface, Then $g_i = 0$ on S_u, and the remainder of the equation holds without question.

Examples of Applications

(a) *Space-time-separable body forces, surface tractions and displacements.* If

$$X_i^{(1)} = \Xi_i^{(1)}(x)h(t), \qquad X_i^{(2)} = \Xi_i^{(2)}(x)h(t),$$
$$f_i^{(1)} = P_i^{(1)}(x)h(t), \qquad f_i^{(2)} = P_i^{(2)}(x)h(t),$$
$$g_i^{(1)} = W_i^{(1)}(x)h(t), \qquad g_i^{(2)} = W_i^{(2)}(x)h(t),$$

then Eq. (11) can be written, on cancelling $h(s)$ from every term, as

$$\int_V \Xi_i^{(1)}\bar{u}_i^{(2)}dv + \int_{S_\sigma} P_i^{(1)}\bar{u}_i^{(2)}dS + \int_{S_u} W_i^{(2)}\bar{\sigma}_{ij}^{(1)}\nu_j dS$$

$$= \int_V \Xi_i^{(2)}\bar{u}_i^{(1)}dv + \int_{S_\sigma} P_i^{(2)}\bar{u}_i^{(1)}dS + \int_{S_u} W_i^{(1)}\bar{\sigma}_{ij}^{(2)}\nu_j dS.$$

The inverse transformation gives

$$\int_V \Xi_i^{(1)} u_i^{(2)}(x,t)\,dv + \int_{S_\sigma} P_i^{(1)} u_i^{(2)}(x,t)\,dS$$

$$+ \int_{S_u} \sigma_{ij}^{(1)}(x,t) W_i^{(2)} \nu_j\,dS\,,$$

$$= \int_V \Xi_i^{(2)} u_i^{(1)}(x,t)\,dv + \int_{S_\sigma} P_i^{(2)} u_i^{(1)}(x,t)\,dS$$

$$+ \int_{S_u} \sigma_{ij}^{(2)}(x,t) W_i^{(1)} \nu_j\,dS$$

Graffi's well-known formula results if $g_i^{(1)} = g_i^{(2)} = 0$ on S_u.

(b) *Forces applied at different times.* If

$$X_i^{(1)} = \Xi_i^{(1)}(x)h(t - T_1)\,, \qquad X_i^{(2)} = \Xi_i^{(2)}(x)h(t - T_2)$$

$$f_i^{(1)} = P_i^{(1)}(x)h(t - T_1)\,, \qquad f_i^{(2)} = P_i^{(2)}(x)h(t - T_2)\,,$$

$$g_i^{(1)} = 0\,, \qquad\qquad\qquad g_i^{(2)} = 0\,,$$

where $h(t) = 0$ for $t \leq 0$, then on substituting the above into Eq. (11) and taking the inverse Laplace transform, one obtains

$$\int_V \Xi_i^{(1)}(x) u_i^{(2)}(x, t - T_1)\,dv + \int_{S_\sigma} P_i^{(1)}(x) u_i^{(2)}(x, t - T_1)\,dS$$

$$= \int_V \Xi_i^{(2)}(x) u_i^{(1)}(x, t - T_2)\,dv + \int_{S_\sigma} P_i^{(2)}(x) u_i^{(1)}(x, t - T_2)\,dS\,.$$

(c) *Concentrated forces.* If the loading consists of concentrated loads $F_i^{(1)}$ and $F_i^{(2)}$ acting at points p_1, p_2 respectively, we may consider Ξ_i or P_i of the Cases (a) and (b) above as suitable delta functions and obtain at once

$$F_i^{(1)}(p_1) u_i^{(2)}(p_1, t - T_1) = F_i^{(2)}(p_2) u_i^{(1)}(p_2, t - T_2)\,,$$

or, if $T_1 = T_2$,

$$F_i^{(1)}(p_1) u_i^{(2)}(p_1, t) = F_i^{(2)}(p_2) u_i^{(1)}(p_2, t)\,.$$

This is the extension of the conventional elastostatic Betti–Rayleigh reciprocal relation to kinetics.

(d) *Impulsive and traveling concentrated forces.* Let an impulsive concentrated force act at a point p_1,

$$X_i^{(1)} = F_i^{(1)}\delta(p_1)\delta(t),$$

where $\delta(t)$ is a unit-impulse or delta function, and let a concentrated force $F_i^{(2)}$ be applied at the origin at $t = 0$, and thereafter moved along the x_1 axis at uniform speed U:

$$X_i^{(2)} = F_i^{(2)}\delta\left(t - \frac{x_1}{U}\right)\delta(x_2)\delta(x_3).$$

No other surface loading or displacement is imposed. Then Eq. (12) gives

$$F_i^{(1)}u_i^{(2)}(p_1, t) = F_i^{(2)}\iiint \delta(x_2)\delta(x_3)dx_1dx_2dx_3$$

$$\times \int_0^t \delta\left(\tau - \frac{x_1}{U}\right)u_i^{(1)}(x_1, x_2, x_3, t - \tau)d\tau$$

and therefore,

$$F_i^{(1)}u_i^{(2)}(p_1, t) = F_i^{(2)}\int_{-\infty}^{x_1/U} u_i^{(1)}\left(x_1, 0, 0, t - \frac{x_1}{U}\right)dx_1.$$

If $u_i^{(1)}(x_1, 0, 0, t - x_1/U)$ is known, then $u_i^{(2)}(p_1, t)$ can be found from the above equation.

(e) *Suddenly started line load over an elastic half-space* Ang[9.2] considered the problem of suddenly started line load acting on the surface of a half-space. Now, according to the reciprocal theorem Ang's problem can be solved by one integration of the solution of Lamb's problem: the impulsive loading at one point (not traveling) inside a two-dimensional half-space. Only the surface displacement due to the point loading needs to be known.

Problem 15.11. The equation of transverse motion of a string stretched between two points is

$$c^2\frac{\partial^2 w}{\partial x^2} - \frac{\partial^2 w}{\partial t^2} = F(x, t), \qquad 0 \le x \le L.$$

Let $F^{(1)}(x, t)$ be a dynamic loading corresponding to a solution $w^{(1)}(x, t)$; $F^{(2)}(x, t)$ be a dynamic loading corresponding to a solution $w^{(2)}(x, t)$. Generalize the dynamic reciprocity relationship to the present problem by integrating over the length $0 \le x \le L$ as well as over the time $(0, t)$.

Next consider the specific case in which $F^{(2)}(x,t)$ is an impulsive concentrated load acting at x : $F^{(2)}(x,t) = \delta(x)\delta(t)$; and $F^{(1)}(x,t)$ is a traveling load $F^{(1)}(x,t)$ $\delta(x - Vt)$, where V is a constant. Show that $w^{(1)}(x,t)$ can be derived by the reciprocity relation from the solution $w^{(2)}(x,t)$ corresponding to $F^{(2)}(x,t)$.

16

LARGE DEFORMATION

In Chapter 4, we discussed briefly the different strain measures for large deformation. In this chapter we shall examine further the consequences of the finiteness of strain fields. The generalization from infinitesimal strains to the nonlinear field theory of finite strains opens up a tremendous vista of the theory. The nonlinear field theory associated with finite strain is difficult and extensive. In this book, we cover only the theory needed for the development of numerical solutions of three-dimensional elastic solid bodies including plates and shells. In this chapter, the strain measures are developed further, the stresses conjugate to various finite strain measures are defined, the equations of motion are derived, some constitutive equations are discussed, and the variational principles for large deformation are presented.

16.1. COORDINATE SYSTEMS AND TENSOR NOTATION

In this chapter, *rectangular Cartesian coordinates in three-dimension will be employed* to formulate the large deformation theory unless otherwise specified. Occasionally we use orthogonal curvilinear coordinates such as cylindrical and spherical coordinates in examples. In that case, it will be clearly specified. *We shall use a combination of the matrix, and indicial notations.* Normally, the Roman indices have a range from 1 to 3, and the Greek indices from 1 to 2. *Bold face characters represent vectors, tensors or matrices. To simplify the discussion in this section only, a lower case bold face character denotes a vector, a tensor of rank 1, and an upper case for a tensor of rank 2,* unless specifically noted otherwise. Vectors and tensors may be expressed in component form:

$$(1) \qquad \mathbf{u} = u_I \varepsilon_I, \qquad \mathbf{A} = A_{IJ} \varepsilon_I \varepsilon_J \,,$$

in which the *vector* $\mathbf{u}$ *and the second rank tensor* $\mathbf{A}$ *in 3-dimensions are decomposed into components measured with respect to the unit base vectors* ε_I of a reference Cartesian system of coordinates. If the reference frame is transformed to another Cartesian coordinate system with unit base vectors $\mathbf{e}_i$, the vector and the tensor may be written as

(2) $\qquad\qquad \mathbf{v} = v_i \mathbf{e}_i , \qquad \mathbf{B} = B_{ij} \mathbf{e}_i \mathbf{e}_j ,$

to show their decomposition with respect to the new frame of reference. One may view $\mathbf{e}_i \mathbf{e}_j$ as a tensor product of two vectors resulting in a tensor of rank 2. In general, $\mathbf{e}_i \mathbf{e}_j \neq \mathbf{e}_j \mathbf{e}_i$ if $i \neq j$, i.e., tensor product is not commutative. The unit base vectors of the two systems are related by

(3) $\qquad\qquad\qquad \varepsilon_I = R_{Ij} \mathbf{e}_j ,$

in which the determinant of the matrix (R_{Ij}), denoted as $\det(\mathbf{R})$, is equal to 1.

In *cylindrical coordinates* (r, θ, z), a vector $\mathbf{v}$ can be referred to a set of orthogonal unit base vectors $(\mathbf{e}_r, \mathbf{e}_\theta, \mathbf{e}_z)$ in the form

(4) $\qquad\qquad\qquad \mathbf{v} = v_r \mathbf{e}_r + v_\theta \mathbf{e}_\theta + v_z \mathbf{e}_z ,$

Here (v_r, v_θ, v_z) are the *physical, not tensorial*, components of $\mathbf{v}$ (see Sec. 2.14) because $(\mathbf{e}_r, \mathbf{e}_\theta, \mathbf{e}_z)$ are unit vectors. The base vectors of cylindrical coordinates are related to the base vectors $(\mathbf{e}_1, \mathbf{e}_2, \mathbf{e}_3)$ of the Cartesian frame of reference through

$$\mathbf{e}_r = \mathbf{e}_1 \cos\theta + \mathbf{e}_2 \sin\theta , \qquad \mathbf{e}_\theta = -\mathbf{e}_1 \sin\theta + \mathbf{e}_2 \cos\theta , \qquad \mathbf{e}_z = \mathbf{e}_3 .$$

According to the notation of Eq. (3), we have

$$R_{r1} = R_{\theta 2} = \cos\theta , \quad R_{r2} = -R_{\theta 1} = \sin\theta ,$$

$$R_{r3} = R_{\theta 3} = R_{z1} = R_{z2} = 0 , \quad R_{z3} = 1 .$$

The coordinates of the two systems are related by

$$x_1 = x = r \cos\theta , \quad x_2 = y = r \sin\theta , \quad x_3 = z .$$

In *spherical coordinates*, the unit base vectors $(\mathbf{e}_r, \mathbf{e}_\phi, \mathbf{e}_\theta)$ are related to the base vectors $(\mathbf{e}_1, \mathbf{e}_2, \mathbf{e}_3)$ of the Cartesian frame by:

$$\mathbf{e}_r = (\mathbf{e}_1 \cos\theta + \mathbf{e}_2 \sin\theta) \sin\phi + \mathbf{e}_3 \cos\phi ,$$

$$\mathbf{e}_\phi = (\mathbf{e}_1 \cos\theta + \mathbf{e}_2 \sin\theta) \cos\phi - \mathbf{e}_3 \sin\phi ,$$

$$\mathbf{e}_\theta = -\mathbf{e}_1 \sin\theta + \mathbf{e}_2 \cos\theta .$$

The coordinates (r, ϕ, θ) and (x_1, x_2, x_3) are related by

$$x_1 = x = r \sin\phi \cos\theta , \quad x_2 = y = r \sin\phi \sin\theta , \quad x_3 = z = r \cos\phi .$$

In Cartesian coordinates, the physical and tensorial components are the same. Similarly one can find the decomposition of tensors in cylindrical and spherical coordinates. The components will be physical rather than tensorial, since the base vectors are unit vectors.

The two indices of $\mathbf{A}$ may be referred to different coordinate systems, e.g., $\mathbf{A} = A_{iI}\mathbf{e}_i\boldsymbol{\varepsilon}_I$. In this case, $\mathbf{A}$ is called a *mixed or two-point tensor of rank 2* as opposed to a *uniform tensor* for which all indices refer to the same coordinate frame. One may similarly define *mixed tensors of higher rank*, for which each index of a tensor associates with a different coordinate system.

We summarize the basic vector and tensor notations and operations as follows:

- If $\mathbf{B}(= B_{ij}\mathbf{e}_i\mathbf{e}_j)$ is a tensor of rank 2, then $\mathbf{B}^T(= B_{ji}\mathbf{e}_i\mathbf{e}_j)$ is the transpose of $\mathbf{B}$. If $B_{ij} = B_{ji}$, $\mathbf{B}^T = \mathbf{B}$ and $\mathbf{B}$ is symmetric. If $\mathbf{B}(= B_{iJ}\mathbf{e}_i\boldsymbol{\varepsilon}_J)$ is a two-point tensor, then $\mathbf{B}^T(= B_{iJ}\boldsymbol{\varepsilon}_J\mathbf{e}_i)$.

- If $\mathbf{A}(= A_{IJ}\boldsymbol{\varepsilon}_I\boldsymbol{\varepsilon}_J)$ and $\mathbf{B}(= B_{ij}\mathbf{e}_i\mathbf{e}_j)$ are tensors of rank 2, then $\mathbf{AB}(= A_{IJ}B_{ij}\boldsymbol{\varepsilon}_I\boldsymbol{\varepsilon}_J\mathbf{e}_i\mathbf{e}_j)$ denotes a tensor of rank 4.

- $\mathbf{e}_i \cdot \mathbf{e}_j = \delta_{ij}$ is the dot product of two orthogonal unit base vectors, where δ_{ij} is the Kronecker delta.

- With $\mathbf{u} = u_I\boldsymbol{\varepsilon}_I$, $\mathbf{v} = v_i\mathbf{e}_i$, and $\boldsymbol{\varepsilon}_I = R_{Ij}\mathbf{e}_j$, then the *dot product of the two vectors* $\mathbf{u}$ and $\mathbf{v}$, which involves the dot products of base vectors, is $\mathbf{u} \cdot \mathbf{v} = u_I v_j(\boldsymbol{\varepsilon}_I \cdot \mathbf{e}_j) = u_I R_{Ij} v_j$. Note that $\mathbf{u} \cdot \mathbf{v} = \mathbf{v} \cdot \mathbf{u}$. If $\boldsymbol{\varepsilon}_I = \mathbf{e}_i$ for $I = i$, one has $R_{Ij} = \delta_{ij}$, then $\mathbf{u}\cdot\mathbf{v} = u_i v_i$. For $\mathbf{u} = \mathbf{e}_i$, one has $\mathbf{e}_i\cdot\mathbf{v} = v_i$, which is called the *projection or component* of $\mathbf{v}$ along $\mathbf{e}_i$.

- If $\mathbf{B} = B_{ij}\mathbf{e}_i\mathbf{e}_j$ and $\mathbf{v} = v_i\mathbf{e}_i$, then the *dot product of the vector and the second-rank tensor* is a vector defined by $\mathbf{B} \cdot \mathbf{v} = B_{ik}v_k\mathbf{e}_i$ and $\mathbf{v} \cdot \mathbf{B} = v_k B_{ki}\mathbf{e}_i$. Thus, $\mathbf{B} \cdot \mathbf{v}$ and $\mathbf{v} \cdot \mathbf{B}$ are vectors with components $B_{ik}v_k$ and $v_k B_{ki}$, respectively. In general, $B_{ik}v_k \neq v_k B_{ki}$, i.e. $\mathbf{B} \cdot \mathbf{v} \neq \mathbf{v} \cdot \mathbf{B}$. One can similarly obtain the dot products of vectors and tensors whose components refer to different coordinate systems.

- If $\mathbf{A} = A_{IJ}\boldsymbol{\varepsilon}_I\boldsymbol{\varepsilon}_J$, $\mathbf{B} = B_{ij}\mathbf{e}_i\mathbf{e}_j$, and $\boldsymbol{\varepsilon}_I = R_{Ij}\mathbf{e}_j$, then the *dot product of the two second rank tensors* is defined as $\mathbf{A} \cdot \mathbf{B} = A_{IJ}B_{ij}\boldsymbol{\varepsilon}_I(\boldsymbol{\varepsilon}_J \cdot \mathbf{e}_i)\mathbf{e}_j = A_{IJ}R_{Ji}B_{ij}\boldsymbol{\varepsilon}_I\mathbf{e}_j$. $\mathbf{A} \cdot \mathbf{B}$ is a tensor of rank 2. In general, $\mathbf{A} \cdot \mathbf{B} \neq \mathbf{B} \cdot \mathbf{A}$. Let $\mathbf{C} = \mathbf{A} \cdot \mathbf{B} = C_{Ij}\boldsymbol{\varepsilon}_I\mathbf{e}_j$, where $C_{Ij} = A_{IJ}R_{Ji}B_{ij}$. $\mathbf{C}$ is a *two-point tensor* whose first index refers to a coordinate frame with base vectors $\boldsymbol{\varepsilon}_I$ and the second index refers to $\mathbf{e}_j$. If $\boldsymbol{\varepsilon}_J = \mathbf{e}_i$ for $J = i$, one has $R_{Ji} = \delta_{ij}$ and $\mathbf{C} = \mathbf{A}\cdot\mathbf{B} = A_{ik}B_{kj}\mathbf{e}_i\mathbf{e}_j$. Then $\mathbf{C}$ is a uniform tensor of rank 2. One can generalize the dot product for two tensors of arbitrary rank: $\mathbf{A} \cdot \mathbf{B}$ involves the contraction of the last index of $\mathbf{A}$ and the first index of $\mathbf{B}$.

The resulting product $\mathbf{A} \cdot \mathbf{B}$ is a tensor of rank $r_A + r_B - 2$, where r_A, r_B are the ranks of $\mathbf{A}$ and $\mathbf{B}$, respectively. Note that a vector is a tensor of rank 1 and a scalar is a zero rank tensor.

- If $\mathbf{A}$ is an arbitrary tensor of rank 2, then $\mathbf{A}^2 = \mathbf{A} \cdot \mathbf{A}$.
- The *trace of a uniform tensor* $\mathbf{A}$ of rank 2 is a scalar defined as $\mathrm{tr}(\mathbf{A}) = A_{II}$.
- The *inner product of two tensors* $\mathbf{A}$, $\mathbf{B}$ of rank 2 is a scalar defined as $\mathbf{A} : \mathbf{B} = \mathrm{tr}(\mathbf{A} \cdot \mathbf{B}^T)$. For example, if $\mathbf{A} = A_{ij}\mathbf{e}_i\mathbf{e}_j$, $\mathbf{B} = B_{ij}\mathbf{e}_i\mathbf{e}_j$, are two uniform tensors of the same reference frame, then $\mathbf{A} : \mathbf{B} = A_{ij}B_{ij}$. If $\mathbf{A} = A_{IJ}\varepsilon_I\varepsilon_J$, $\mathbf{B} = B_{ij}\mathbf{e}_i\mathbf{e}_j$, then $\mathbf{A} : \mathbf{B} = A_{IJ}B_{ij}(\varepsilon_I \cdot \mathbf{e}_i)(\varepsilon_J \cdot \mathbf{e}_j) = A_{IJ}R_{Ii}R_{Jj}B_{ij}$. If $\mathbf{A} = A_{iJ}\mathbf{e}_i\varepsilon_J$ and $\mathbf{B} = B_{iJ}\mathbf{e}_i\varepsilon_J$, then $\mathbf{A} : \mathbf{B} = A_{iJ}B_{iJ}$. Since $\mathrm{tr}(\mathbf{A} \cdot \mathbf{B}^T) = \mathrm{tr}(\mathbf{A}^T \cdot \mathbf{B})$, one has $\mathbf{A} : \mathbf{B} = \mathbf{B} : \mathbf{A}$ for second rank tensors.
- If $\mathbf{A}$ and $\mathbf{B}$ are tensors of unequal rank, then the inner product $\mathbf{A} : \mathbf{B}$ involves the contraction of the last index pair of $\mathbf{A}$ and the first index pair of $\mathbf{B}$, which results in a tensor of rank $r_A + r_B - 4$. For example, $\mathbf{A} : \mathbf{B} = A_{IJ}B_{IJkl}\mathbf{e}_k\mathbf{e}_l = \mathbf{C}$ for $\mathbf{A} = A_{IJ}\varepsilon_I\varepsilon_J$, $\mathbf{B} = B_{IJkl}\varepsilon_I\varepsilon_J\mathbf{e}_k\mathbf{e}_l$, where $\mathbf{C} = C_{kl}\mathbf{e}_k\mathbf{e}_l$, $C_{kl} = A_{IJ}B_{IJkl}$, $r_A = r_C = 2$ and $r_B = 4$. Note that $\mathbf{B} : \mathbf{A} = B_{IJij}A_{KL}R_{iK}R_{jL}\varepsilon_I\varepsilon_J$. In general $\mathbf{A} : \mathbf{B} \neq \mathbf{B} : \mathbf{A}$.
- If $\mathbf{C} = C_{IJij}\varepsilon_I\varepsilon_J\mathbf{e}_i\mathbf{e}_j$ is a *mixed tensor* of rank 4, and $\mathbf{A}(= A_{IJ}\varepsilon_I\varepsilon_J)$ and $\mathbf{B}(= B_{ij}\mathbf{e}_i\mathbf{e}_j)$ are tensors of rank 2, then the *inner product* $\mathbf{A} : \mathbf{C} : \mathbf{B} = A_{IJ}C_{IJij}B_{ij}$ is a scalar.
- For base vectors $\mathbf{e}_i$ of a rectangular Cartesian frame, the *cross product* $\mathbf{e}_j \times \mathbf{e}_k = e_{ijk}\mathbf{e}_i$ is a vector, where e_{ijk} is the *permutation tensor* of rank 3. If Eq. (3) holds, then $\varepsilon_J \times \mathbf{e}_k = R_{Ji}\mathbf{e}_i \times \mathbf{e}_k = R_{Jj}e_{ijk}\mathbf{e}_i$. One can easily determine $\mathbf{u} \times \mathbf{v}$ for $\mathbf{u}$ and $\mathbf{v}$ defined in Eqs. (1) and (2). If $\varepsilon_I = \mathbf{e}_i$ for $I = i$, then $\mathbf{u} \times \mathbf{v} = e_{ijk}u_jv_k\mathbf{e}_i$. Note that $(\mathbf{u} \times \mathbf{v}) \times \mathbf{r} = (\mathbf{r} \cdot \mathbf{u})\mathbf{v} - (\mathbf{r} \cdot \mathbf{v})\mathbf{u}$.
- If $\mathbf{u} = u_I\varepsilon_I$, $\mathbf{v} = v_i\mathbf{e}_i$, then $\frac{\partial \mathbf{u}}{\partial \mathbf{v}} = \frac{\partial u_I}{\partial v_i}\varepsilon_I\mathbf{e}_i$ is a second rank tensor.
- If $\mathbf{A} = A_{IJ}\varepsilon_I\varepsilon_J$, $\mathbf{B} = B_{ij}\mathbf{e}_i\mathbf{e}_j$, then $\frac{\partial \mathbf{A}}{\partial \mathbf{B}} = \frac{\partial A_{IJ}}{\partial B_{ij}}\varepsilon_I\varepsilon_J\mathbf{e}_i\mathbf{e}_j$ is a fourth rank tensor.
- The *del operator* is

$$\nabla = \mathbf{e}_i \frac{\partial}{\partial x_i} \qquad \text{(Cartesian coordinates)}$$

$$= \mathbf{e}_r \frac{\partial}{\partial r} + \frac{\mathbf{e}_\theta}{r} \frac{\partial}{\partial \theta} + \mathbf{e}_z \frac{\partial}{\partial z} \qquad \text{(cylindrical coordinates)}$$

$$= \mathbf{e}_r \frac{\partial}{\partial r} + \frac{\mathbf{e}_\theta}{r \sin \phi} \frac{\partial}{\partial \theta} + \frac{\mathbf{e}_\phi}{r} \frac{\partial}{\partial \phi} \qquad \text{(spherical coordinates)}.$$

- The *gradient* of a vector $\mathbf{u}$ is defined as

$$\nabla \mathbf{u} = \frac{\partial \mathbf{u}}{\partial x_j}\mathbf{e}_j = \frac{\partial u_I}{\partial x_j}\varepsilon_I\mathbf{e}_j = \frac{\partial \mathbf{u}}{\partial \mathbf{x}},$$

which is a tensor of rank 2. In cylindrical coordinates, the gradient of $\mathbf{u}$ becomes

$$\nabla \mathbf{u} = \frac{\partial \mathbf{u}}{\partial r}\mathbf{e}_r + \frac{1}{r}\frac{\partial \mathbf{u}}{\partial \theta}\mathbf{e}_\theta + \frac{\partial \mathbf{u}}{\partial z}\mathbf{e}_z.$$

In computing the partial derivatives of $\mathbf{u}$, one must take into account that the base vectors of the reference frame for $\mathbf{u}$ can be functions of r, θ and z. For example, expressing $\mathbf{u}$ in the form $\mathbf{u} = u_r\mathbf{e}_r + u_\theta\mathbf{e}_\theta + u_z\mathbf{e}_z$ where $\mathbf{e}_r$ and $\mathbf{e}_\theta$ are functions of θ, one has

$$\frac{\partial \mathbf{u}}{\partial \theta} = \frac{\partial u_r}{\partial \theta}\mathbf{e}_r + u_r\frac{\partial \mathbf{e}_r}{\partial \theta} + \frac{\partial u_\theta}{\partial \theta}\mathbf{e}_\theta + u_\theta\frac{\partial \mathbf{e}_\theta}{\partial \theta} + \frac{\partial u_z}{\partial \theta}\mathbf{e}_z$$

$$= \left(\frac{\partial u_r}{\partial \theta} - u_\theta\right)\mathbf{e}_r + \left(\frac{\partial u_\theta}{\partial \theta} + u_r\right)\mathbf{e}_\theta + \frac{\partial u_z}{\partial \theta}\mathbf{e}_z.$$

In spherical coordinates, the gradient of $\mathbf{u}$ is

$$\nabla \mathbf{u} = \frac{\partial \mathbf{u}}{\partial r}\mathbf{e}_r + \frac{1}{r\sin\phi}\frac{\partial \mathbf{u}}{\partial \theta}\mathbf{e}_\theta + \frac{1}{r}\frac{\partial \mathbf{u}}{\partial \phi}\mathbf{e}_\phi.$$

- The gradient of a tensor is similarly defined as $\nabla \mathbf{A} = \frac{\partial \mathbf{A}}{\partial x_i}\mathbf{e}_i$, which is a tensor of rank $n+1$ and n is the rank of $\mathbf{A}$. For cylindrical and spherical coordinates, one simply replaces $\frac{\partial}{\partial x_i}\mathbf{e}_i$ with the appropriate *del operator*. In Cartesian coordinates, $\nabla \mathbf{A} = \frac{\partial A_{ij}}{\partial x_k}\mathbf{e}_i\mathbf{e}_j\mathbf{e}_k$ for a second rank tensor $\mathbf{A}$.
- The *divergence* of a vector $\mathbf{u}$ is defined as $\nabla \cdot \mathbf{u} = \mathrm{tr}(\nabla\mathbf{u})$ and is given by

$$\nabla \cdot \mathbf{u} = \mathbf{e}_i \cdot \frac{\partial \mathbf{u}}{\partial x_i} = \frac{\partial u_i}{\partial x_i}$$

in Cartesian coordinates,

$$\nabla \cdot \mathbf{u} = \mathbf{e}_r \cdot \frac{\partial \mathbf{u}}{\partial r} + \frac{\mathbf{e}_\theta}{r} \cdot \frac{\partial \mathbf{u}}{\partial \theta} + \mathbf{e}_z \cdot \frac{\partial \mathbf{u}}{\partial z} = \frac{1}{r}\frac{\partial r u_r}{\partial r} + \frac{1}{r}\frac{\partial u_\theta}{\partial \theta} + \frac{\partial u_z}{\partial z}$$

in cylindrical coordinates, and

$$\nabla \cdot \mathbf{u} = \mathbf{e}_r \cdot \frac{\partial \mathbf{u}}{\partial r} + \frac{\mathbf{e}_\theta}{r\sin\phi} \cdot \frac{\partial \mathbf{u}}{\partial \theta} + \frac{\mathbf{e}_\phi}{r} \cdot \frac{\partial \mathbf{u}}{\partial \phi}$$

$$= \frac{1}{r^2}\frac{\partial r^2 u_r}{\partial r} + \frac{1}{r\sin\phi}\frac{\partial u_\theta}{\partial \theta} + \frac{1}{r\sin\phi}\frac{\partial \sin\phi u_\phi}{\partial \phi}$$

in spherical coordinates. The divergence of a vector is a scalar.

- If **B** is a tensor of rank 2 or higher, the divergence of **B** is

$$\nabla \cdot \mathbf{B} = \mathbf{e}_i \cdot \frac{\partial \mathbf{B}}{\partial x_i}.$$

The dot operation by the base vectors must be executed from the left as defined. Note that $\frac{\partial \mathbf{B}}{\partial x_i} \cdot \mathbf{e}_i \neq \mathbf{e}_i \cdot \frac{\partial \mathbf{B}}{\partial x_i}$. However, if **B** is a vector, no distinction is needed. For **B** of rank 2, $\nabla \cdot \mathbf{B} = \frac{\partial B_{ij}}{\partial x_i} \mathbf{e}_j$.

- The *curl* of a vector is defined as

$$\nabla \times \mathbf{u} = \mathbf{e}_i \times \frac{\partial \mathbf{u}}{\partial x_i} \qquad \text{(Cartesian coordinates)}$$

$$= \mathbf{e}_r \times \frac{\partial \mathbf{u}}{\partial r} + \mathbf{e}_\theta \times \frac{1}{r} \frac{\partial \mathbf{u}}{\partial \theta} + \mathbf{e}_z \times \frac{\partial \mathbf{u}}{\partial z} \qquad \text{(cylindrical coordinates)}$$

$$= \mathbf{e}_r \times \frac{\partial \mathbf{u}}{\partial r} + \mathbf{e}_\theta \times \frac{1}{r \sin \phi} \frac{\partial \mathbf{u}}{\partial \theta} + \mathbf{e}_\phi \times \frac{1}{r} \frac{\partial \mathbf{u}}{\partial \phi} \qquad \text{(spherical coordinates)}.$$

Note that $\mathbf{e}_i \times \mathbf{e}_j = e_{ijk} \mathbf{e}_k$.

- The *Laplace* operation is defined as $\nabla^2 \mathbf{v} = \nabla \cdot (\nabla \mathbf{v}) = \mathbf{e}_i \cdot \frac{\partial \nabla \mathbf{v}}{\partial x_i}$. On has $\nabla^2 \mathbf{v} = \frac{\partial^2 v_j}{\partial x_j \partial x_i} \mathbf{e}_i$ in the Cartesian coordinates. One can similarly determine the Laplace operation in other coordinates.

We shall close the section by identifying some frequently used relations of the inner product and properties of matrix operations. The inner product of stress and strain rate is the rate at which work is done by the forces acting on the surface of a unit cube as the body deforms. It is related to the rate at which the internal energy of the material changes by the first law of thermodynamics. Hence the inner product defined earlier is a frequently used operation. The followings are true:

$$\mathbf{A} : \mathbf{B} = \mathbf{B} : \mathbf{A} = \mathbf{A}^T : \mathbf{B}^T,$$

$$(\mathbf{A} \cdot \mathbf{B}) : \mathbf{C} = \mathbf{C} : (\mathbf{A} \cdot \mathbf{B}) = (\mathbf{B} \cdot \mathbf{C}^T) : \mathbf{A}^T = (\mathbf{C} \cdot \mathbf{B}^T) : \mathbf{A}$$

(5)

$$= (\mathbf{A}^T \cdot \mathbf{C}) : \mathbf{B} = (\mathbf{C}^T \cdot \mathbf{A}) : \mathbf{B}^T = \mathbf{B} : (\mathbf{A}^T \cdot \mathbf{C})$$

$$= \mathbf{A} : (\mathbf{C} \cdot \mathbf{B}^T),$$

for any second rank tensors **A**, **B** and **C**. If **S** is a symmetric tensor, then

$$\mathbf{S} : \mathbf{B} = \mathbf{S} : \mathbf{B}^T = \mathbf{S} : (\mathbf{B} + \mathbf{B}^T)/2 = \mathbf{S} : (\mathbf{B})_s, \qquad \mathbf{S} : (\mathbf{S})_a = 0,$$

where $(\mathbf{B})_s = (\mathbf{B} + \mathbf{B}^T)/2$ and $(\mathbf{B})_a = (\mathbf{B} - \mathbf{B}^T)/2$ are the symmetric and skew-symmetric parts of **B**, respectively. If A is skew-symmetric, then

$$\mathbf{A} : \mathbf{B} = -\mathbf{A}^T : \mathbf{B} = -\mathbf{A} : \mathbf{B}^T = \mathbf{A} : (\mathbf{B} - \mathbf{B}^T)/2 = \mathbf{A} : (\mathbf{B})_a,$$

$$\mathbf{A} : (\mathbf{B})_s = 0.$$

From time to time, we will use matrix notations and follow the rules of matrix operations for scalar, vectors, and tensors or rank 2. In this case, the vectors and tensors are referred to the same orthogonal coordinate system. The vector (*column matrix*) $\mathbf{u}$ and tensor $\mathbf{A}$ defined in Eq. (1) can be written in matrix notation as

$$(6) \qquad \mathbf{u} = \begin{bmatrix} u_1 \\ u_2 \\ u_3 \end{bmatrix} = [u_i], \qquad \mathbf{A} = \begin{bmatrix} A_{11} & A_{12} & A_{13} \\ A_{21} & A_{22} & A_{23} \\ A_{31} & A_{32} & A_{33} \end{bmatrix} = [A_{ij}].$$

The vector $\mathbf{u}$ as defined in the equation above is called a *column vector*. In matrix notation, the two subscript indices of A's can have different ranges. For example, if the range of first index is from 1 to 3, and the second from 1 to 4, we write

$$\mathbf{A} = \underset{3 \times 4}{\mathbf{A}} = \begin{bmatrix} A_{11} & A_{12} & A_{13} & A_{14} \\ A_{21} & A_{22} & A_{23} & A_{24} \\ A_{31} & A_{32} & A_{33} & A_{34} \end{bmatrix},$$

where the first and second numbers in the underscore of $\mathbf{A}$ denote the ranges of the first and second indices, respectively. In this case, $\mathbf{A}$ is a 3×4 matrix, i.e, a matrix with 3 rows and 4 columns. Note that a column vector of rank p is a "$p \times 1$" matrix and a row vector of rank p is a "$1 \times p$" matrix (*row matrix*). When the two ranges of the indices are equal, $\mathbf{A}$ is called a *square matrix*. For simplicity, we drop the underscore unless it is needed for clarity. We implicitly assume the ranges of indices are known. The transpose of a column vector $\mathbf{u}$ is a row vector, i.e.,

$$\mathbf{u}^T = \begin{bmatrix} u_1 & u_2 & u_3 \end{bmatrix},$$

which is different from the tensor notation where the transpose of a vector is the vector itself. The transpose of $\mathbf{A}$ is $\mathbf{A}^T$ and is defined as follows: if $\mathbf{C} = \mathbf{A}^T$, then the components $C_{ij} = A_{ji}$. If $A_{ij} = A_{ji}$, $\mathbf{A}$ is symmetric.

If $\mathbf{u}$ is a column vector and $\mathbf{A}$ is matrix, their products are defined as follows:

$$\mathbf{Au} = \lfloor A_{ij}u_j \rfloor, \qquad \mathbf{u}^T\mathbf{A} = [u_i A_{ij}],$$

where repeated indices denote summation over appropriate ranges. In the first case, the number of columns of $\mathbf{A}$ must be the same as the number of rows of $\mathbf{u}$. In the second case, the number of rows of $\mathbf{A}$ must be the same as the number of rows of $\mathbf{u}$. Note that the components of $\mathbf{Au}$ (matrix multiplication) equal those of $\mathbf{A} \cdot \mathbf{u}$ (inner product) and those of $\mathbf{u}^T \mathbf{A}$ equal $\mathbf{u} \cdot \mathbf{A}$ in the tensor expression. If $\underset{m \times n}{\mathbf{A}}$, $\underset{n \times p}{\mathbf{B}}$ are two nonsquare matrices with their ranges of indices indicated by the underscores, their product is

$$\underset{m \times p}{\mathbf{C}} = \underset{m \times n}{\mathbf{A}} \underset{n \times p}{\mathbf{B}} = \lfloor A_{ij} B_{jk} \rfloor,$$

which is the same as $\mathbf{A} \cdot \mathbf{B}$ in tensor notation. Note that the number of columns $\mathbf{A}$ must be the same as the number of rows of $\mathbf{B}$. Finally, if $\mathbf{A}$ is a square matrix, the inverse of $\mathbf{A}$ is $\mathbf{A}^{-1}$ such that

$$\mathbf{A}\mathbf{A}^{-1} = \mathbf{A}^{-1}\mathbf{A} = \mathbf{I},$$

where $\mathbf{I}$ is an identity matrix whose diagonal terms are all 1 and the off-diagonal terms are zero.

16.2. DEFORMATION GRADIENT

Consider a body which deforms from a volume V in the original or undeformed configuration R_0 into a volume v in the current or deformed configuration R. Let $\mathbf{x}$ denote the position of a point in R associated with a material point originally located at $\mathbf{X}$ in R_0. Then we have

$$\mathbf{x} = x_i \mathbf{e}_i, \qquad \mathbf{X} = X_I \varepsilon_I,$$

where $\mathbf{e}_i$ and ε_I are the unit base-vectors in R and R_0, respectively. The former, $\mathbf{e}_i$, is called the *Eulerian base-vector* and the latter, ε_I, the *Lagrangian base-vector*. The base vectors may translate in the Cartesian frame, but are not allowed to rotate. As a rule, we use capital indices referring to the undeformed coordinates and the lower case indices for the deformed coordinates. We assume the mechanics to be Newtonian. The motion of the body is described by the relation

(1) $$\mathbf{x} = \mathbf{x}(\mathbf{X}, t),$$

where t is time and $\mathbf{x}$ is continuously differentiable with respect to $\mathbf{X}$ and t. To find the transformation between line elements in R and R_0, we consider a material line element $d\mathbf{X}$ at $\mathbf{X}$, which becomes $d\mathbf{x}$ after deformation. The relation between the two elements is then

(2) $$d\mathbf{x} = \frac{\partial \mathbf{x}}{\partial \mathbf{X}} \cdot d\mathbf{X} = \nabla_{\mathbf{X}} \mathbf{x} \cdot d\mathbf{X} = \mathbf{F} \cdot d\mathbf{X},$$

where ∇_X denotes the gradient operation with respect to the X coordinates, and F is called the *deformation gradient tensor* defined as

$$(3) \ \blacktriangle \qquad F = \frac{\partial x}{\partial X} = F_{iI} e_i \varepsilon_I , \qquad \text{or} \qquad F_{iI} = \partial x_i / \partial X_I ,$$

The first index of F_{iI} refers to an *Eulerian* and the second index to a *Lagrangian* basis. F is a second-rank tensor related to the *Eulerian* and the *Lagrangian* coordinates, and is therefore a *two-point tensor*. Equation (2) can be inverted to give

$$(4) \qquad\qquad dX = F^{-1} \cdot dx$$

providing that the *Jacobian* $J \equiv \det[F(X,t)]$ is nonsingular, i.e., $J \neq 0$. The Jacobian is denoted also commonly in alternative forms

$$(5) \qquad J = \det[F] = \det\left[\frac{\partial x_i}{\partial X_I}\right] = \det\left|\frac{\partial x_i}{\partial X_I}\right| = e_{ijk} \frac{\partial x_i}{\partial X_1} \frac{\partial x_j}{\partial X_2} \frac{\partial x_k}{\partial X_3} .$$

The inverse of F can be written as

$$(6) \qquad\qquad F^{-1} = \frac{\partial X}{\partial x} , \qquad \text{or} \qquad (F_{iI})^{-1} = \frac{\partial X_I}{\partial x_i} .$$

Stretches of Line, Area, and Volume. We shall show how line, area, and volume elements in R are related to their counterparts in R_0 through F. First, the well-known Cauchy and Almansi tensors C and B^{-1}, respectively, are defined as

$$(7) \ \blacktriangle \qquad C = F^T \cdot F , \qquad \text{or} \qquad C_{IJ} = \frac{\partial x_i}{\partial X_I} \frac{\partial x_i}{\partial X_J} ,$$

$$(8) \ \blacktriangle \qquad B^{-1} = (F^{-1})^T \cdot F^{-1} , \qquad \text{or} \qquad B_{ij}^{-1} = \frac{\partial X_K}{\partial x_i} \frac{\partial X_K}{\partial x_j} .$$

They both are symmetric and positive definite. Clearly C is a Lagrangian based tensor because the Eulerian indices i have been contracted; whereas B^{-1} is an Eulerian based tensor as a result of the contraction of the Lagrangian indices K.

Now we can establish the stretching of line elements under deformation. Consider an infinitesimal line element $dx(= s\,ds)$ at (x, t) in R and its corresponding line element $dX(= S\,dS)$ at (X, t) before deformation, where ds and dS are the lengths of the line elements before and after the deformation, and s and S are unit vectors along the directions of dx and dX. ds/dS is called the *stretch ratio* of the element. Denoting it by λ_s, we can show that

$$(9) \ \blacktriangle \qquad \lambda_s = \frac{ds}{dS} = \frac{\sqrt{dX \cdot F^T \cdot F \cdot dX}}{|dX|} = \sqrt{S \cdot C \cdot S} ,$$

for $dS \to 0$. We can also write

(10) $$\lambda_s = \frac{ds}{dS} = \frac{|d\mathbf{x}|}{\sqrt{d\mathbf{x} \cdot \mathbf{B}^{-1} \cdot d\mathbf{x}}} = \frac{1}{\sqrt{\mathbf{s} \cdot \mathbf{B}^{-1} \cdot \mathbf{s}}}.$$

Next, consider the differential geometry of a surface in the region R. Let $d\mathbf{x}^{(1)}$ and $d\mathbf{x}^{(2)}$ be two infinitesimal noncollinear line elements lying in the deformed surface and let $d\mathbf{X}^{(1)}$ and $d\mathbf{X}^{(2)}$ be the corresponding line segments in the undeformed configuration. The cross product $d\mathbf{x}^{(1)} \times d\mathbf{x}^{(2)} / |d\mathbf{x}^{(1)} \times d\mathbf{x}^{(2)}|$ is a unit vector normal to $d\mathbf{x}^{(1)}$ and $d\mathbf{x}^{(2)}$. We define

$$d\mathbf{a} = d\mathbf{x}^{(1)} \times d\mathbf{x}^{(2)} = \mathbf{n} da, \qquad \text{or} \qquad n_i da = e_{ijk} dx_j^{(1)} dx_k^{(2)},$$

that $d\mathbf{a}$ is the area of an element in the deformed configuration with $\mathbf{n}$ being a unit normal to the plane of $d\mathbf{x}^{(1)}$ and $d\mathbf{x}^{(2)}$. The corresponding area element in the undeformed configuration is

(11) $$d\mathbf{A} = d\mathbf{X}^{(1)} \times d\mathbf{X}^{(2)} = \mathbf{N} dA, \qquad \text{or} \qquad N_I dA = e_{IJK} dX_J^{(1)} dX_K^{(2)},$$

where $\mathbf{N}$ is a unit normal to $d\mathbf{X}^{(1)}$ and $d\mathbf{X}^{(2)}$.

We shall establish the relation between $d\mathbf{a}$ and $d\mathbf{A}$. Expressing dX's in terms dx's yields

$$N_I dA = e_{IJK} \frac{\partial X_J}{\partial x_j} \frac{\partial X_K}{\partial x_k} dx_j^{(1)} dx_k^{(2)}.$$

Multiplying both sides of Eq. (11) by $J\mathbf{F}^{-1}$ (or $J \partial X_I / \partial x_i$) and simplifying the result by making use of Eq. (5) and the definition of determinant

$$e_{ijk} \det \left| \frac{\partial X_L}{\partial x_m} \right| = e_{IJK} \frac{\partial X_I}{\partial x_i} \frac{\partial X_J}{\partial x_j} \frac{\partial X_K}{\partial x_k},$$

we obtain

(12) $$J(\mathbf{F}^{-1})^T \cdot d\mathbf{A} = d\mathbf{a} \qquad \text{or}$$

$$J \frac{\partial X_M}{\partial x_i} N_M dA = J e_{LMN} \frac{\partial X_L}{\partial x_i} \frac{\partial X_M}{\partial x_j} \frac{\partial X_N}{\partial x_k} dx_j^{(1)} dx_k^{(2)} = n_i da.$$

The details of the proof are left as an exercise. Using Eq. (7), yields

$$|d\mathbf{a}| = J\sqrt{d\mathbf{A} \cdot \mathbf{F}^{-1} \cdot (\mathbf{F}^{-1})^T \cdot d\mathbf{A}} = J\sqrt{d\mathbf{A} \cdot \mathbf{C}^{-1} \cdot d\mathbf{A}}.$$

The *area stretch ratio* is then

(13) ▲ $\Lambda_N = \dfrac{|d\mathbf{a}|}{|d\mathbf{A}|} = J\dfrac{\sqrt{d\mathbf{A}\cdot\mathbf{C}^{-1}\cdot d\mathbf{A}}}{|d\mathbf{A}|} = J\sqrt{\mathbf{N}\cdot\mathbf{C}^{-1}\cdot\mathbf{N}}\,.$

Similarly

(14) ▲ $\Lambda_n = \dfrac{|d\mathbf{a}|}{|d\mathbf{A}|} = J\dfrac{|d\mathbf{a}|}{\sqrt{d\mathbf{a}\cdot\mathbf{B}\cdot d\mathbf{a}}} = \dfrac{J}{\sqrt{\mathbf{n}\cdot\mathbf{B}\cdot\mathbf{n}}}\,.$

Note that $\Lambda_N = \Lambda_n$. From Eqs. (12)–(14), one obtains

$$\mathbf{n} = \dfrac{|d\mathbf{A}|}{|d\mathbf{a}|}\dfrac{J(\mathbf{F}^{-1})^T\cdot d\mathbf{A}}{|d\mathbf{A}|} = (\mathbf{F}^{-1})^T\cdot\mathbf{N}\cdot\sqrt{\mathbf{n}\cdot\mathbf{B}\cdot\mathbf{n}} = \dfrac{(\mathbf{F}^{-1})^T\cdot\mathbf{N}}{\sqrt{\mathbf{N}\cdot\mathbf{C}^{-1}\cdot\mathbf{N}}}\,,$$

which is the relation between the normal to $d\mathbf{a}$ and that to $d\mathbf{A}$.

Finally, we shall determine the relation between the deformed and undeformed volumes. The volume of the element tetrahedron dv formed by the infinitesimal line elements $d\mathbf{x}^{(1)}$, $d\mathbf{x}^{(2)}$ and $d\mathbf{x}^{((3)}$ in R is

(15) $$dv = \dfrac{1}{6}\left(d\mathbf{x}^{(1)}\times d\mathbf{x}^{(2)}\right)\cdot d\mathbf{x}^{(3)}\,.$$

It is related to its corresponding undeformed volume dV by

(16) ▲ $$\dfrac{dv}{dV} = J = \det(\mathbf{F})\,.$$

Details of the proof are left as an exercise. Let ρ and ρ_0 be the mass density per unit deformed and undeformed volumes, respectively. From conservation of mass, we have

(17) $$J = \rho_0/\rho\,.$$

Problem 16.1. Consider an area element vector $d\mathbf{a}[= d\mathbf{x}^{(1)}\times d\mathbf{x}^{(2)}]$ in the deformed configuration R. The corresponding area vector in the undeformed configuration is $d\mathbf{A}[= d\mathbf{X}^{(1)}\times d\mathbf{X}^{(2)}]$. Show that

(18) $$d\mathbf{a} = Jd\mathbf{A}\cdot\mathbf{F}^{-1} = J(\mathbf{F}^{-1})^T\cdot d\mathbf{A}\,.$$

Problem 16.2. Show that Eq. (16.2:16) holds for the volume ratio of the deformed to the undeformed configuration. *Hint*: To determine the relationship between area elements and between volume elements, one needs to use the permutation symbols and identities defined in Chapter 2.

$$e_{ijk} = (\mathbf{e}_i\times\mathbf{e}_j)\cdot\mathbf{e}_k\,,\qquad e_{IJK} = (\mathbf{e}_I\times\mathbf{e}_J)\cdot\mathbf{e}_K\,,\qquad \mathbf{e}_i\times\mathbf{e}_j = e_{ijm}\mathbf{e}_m\,,$$

$$e_{KLN}J = e_{KLN}\det(\mathbf{F}) = e_{ijm}F_{iK}F_{jL}F_{mN}\,,$$

$$e_{KLN}(F_{lN})^{-1}dX_K^{(1)}dX_L^{(2)} = (d\mathbf{X}^{(1)}\times d\mathbf{X}^{(2)})\cdot\mathbf{F}^{-1}\,.$$

16.3. STRAINS

As discussed in Chapter 4, strain is a measure of deformation. In this section, we shall discuss various strain measures and the relations among them. Let $ds^2 (= dx \cdot dx)$ be the length square of the line element dx in R and $dS^2 (= dX \cdot dX)$ be that of the corresponding undeformed element dX in R_0. They can be expressed in terms of the deformation gradient:

$$(1) \quad ds^2 = dx \cdot dx = dX \cdot \mathbf{F}^T \cdot \mathbf{F} \cdot dX = dX \cdot \mathbf{C} \cdot dX = C_{IJ} dX_I dX_J,$$

$$(2) \quad dS^2 = dX \cdot dX = dx \cdot (\mathbf{F}^{-1})^T \cdot \mathbf{F}^{-1} \cdot dx = dx \cdot \mathbf{B}^{-1} \cdot dx = B_{ij}^{-1} dx_i dx_j.$$

If the Lagrangian coordinates are used to describe positions of material particles in the undeformed body, then on allowing the metric tensor to become $\mathbf{C}$, the deformed body is obtained. Hence, the deformation is like a transformation from an Euclidean geometry into a non-Euclidean geometry with metric tensor $\mathbf{C}$. On the other hand, if the Eulerian coordinates are used to describe positions in R_0, then $\mathbf{B}^{-1}$ is the metric tensor that turns the deformed body into the undeformed body of non-Euclidean geometry. A necessary condition (also locally sufficient) for a symmetric positive definite metric $\mathbf{C}$ to be derivable from a configuration x is the vanishing of a fourth-rank tensor field called the *Riemann–Christoffel tensor* [See Problems 2.24, 2.31 and 4.9; and Green and Zerna (1968), Eringen (1962)]. The condition is called the compatibility condition, (see Sec. 4.6).

The *Green strain tensor* $\mathbf{E}$ and the *Almansi strain tensor* $\mathbf{e}$ are defined by

$$(3) \quad \blacktriangle \qquad ds^2 - dS^2 = 2d\mathbf{X} \cdot \mathbf{E} \cdot d\mathbf{X} = 2E_{IJ} dX_I dX_J,$$

$$(4) \quad \blacktriangle \qquad ds^2 - dS^2 = 2d\mathbf{x} \cdot \mathbf{e} \cdot d\mathbf{x} = 2e_{ij} dx_i dx_j,$$

where

$$(5) \quad \blacktriangle \qquad \mathbf{E} = \frac{1}{2}(\mathbf{C} - \mathbf{I}), \qquad E_{IJ} = \frac{1}{2}(C_{IJ} - \delta_{IJ}),$$

like $\mathbf{C}$, is a Lagrangian tensor referring to the undeformed coordinates; whereas

$$(6) \quad \blacktriangle \qquad \mathbf{e} = \frac{1}{2}(\mathbf{I} - \mathbf{B}^{-1}), \qquad e_{ij} = \frac{1}{2}(\delta_{ij} - B_{ij}^{-1}),$$

like $\mathbf{B}$, is an Eulerian tensor referring the deformed coordinates.

We shall examine the characteristics of the various deformation measures. We first consider $\mathbf{C}$ of Eq. (16.2:7). Since $\mathbf{C}$ is positive definite, its eigenvalues can be written in the form λ_i^2, $i = 1, 2, 3$. We call the directions of the eigenvectors the *principal directions*. Consider a point $\mathbf{X} + d\mathbf{X}$ in R_0 which is displaced to $\mathbf{x}(\mathbf{X} + d\mathbf{X}, t)[= \mathbf{x}(\mathbf{X}, t) + d\mathbf{x}]$ in R. Using Eqs. (16.2:2) and (16.2:7), we obtain

$$dx \cdot dx = d\mathbf{X} \cdot \mathbf{C} \cdot d\mathbf{X} + O(|d\mathbf{X}|^3).$$

If $\mathbf{X} + d\mathbf{X}$ is a point on a small ellipsoid centered at $\mathbf{X}$ with the semi-axes r/λ_i oriented along the principal directions of $\mathbf{C}$. Then

$$dx \cdot dx = d\mathbf{X} \cdot \mathbf{C} \cdot d\mathbf{X} = r^2,$$

which indicates that the image of an ellipsoid in R_o is approximately a sphere of radius r in R. Expressing $d\mathbf{X}$ in terms of dx leads to

$$d\mathbf{X} \cdot d\mathbf{X} = dx \cdot \mathbf{B}^{-1} \cdot dx.$$

If $d\mathbf{X} \cdot d\mathbf{X} = r^2$, the equation above is an ellipsoid of the principal semi-axes $r\lambda_i$ for dx as the eigenvalues of $\mathbf{B}^{-1}$ are $1/\lambda_i^2$. Thus, the image of a sphere of radius r in R_o is approximately an ellipsoid of the principal semi-axes $r\lambda_i$ in R. For line element $d\mathbf{X}$ along the principal direction associated with eigenvalue λ_i, we have

$$dx = \lambda_i d\mathbf{X}.$$

Therefore, λ_i, $i = 1, 2, 3$, are called the *principal stretches* and the ellipsoid in R is called the *strain ellipsoid*.

16.4. RIGHT AND LEFT STRETCH AND ROTATION TENSORS

Another useful strain measure is $\mathbf{U}$ defined as

(1) $$\mathbf{U} \equiv \mathbf{C}^{1/2}.$$

If $\mathbf{N}^i$ is an eigenvectors of $\mathbf{C}$ with λ_i^2 being the corresponding eigenvalue, then $\mathbf{N}^i$ is also the eigenvector of $\mathbf{U}$ corresponding to the eigenvalue λ_i. Hence $\mathbf{C}$ and $\mathbf{U}$ are coaxial. Clearly, like $\mathbf{C}$, $\mathbf{U}$ is also symmetric and positive definite that satisfies

(2) ▲ $$\mathbf{U} \cdot \mathbf{U} = \mathbf{C} = \mathbf{F}^T \cdot \mathbf{F}.$$

Let

(3) $$\mathbf{R} = \mathbf{F} \cdot \mathbf{U}^{-1}.$$

We can easily show that

$$\mathbf{R}^T \cdot \mathbf{R} = (\mathbf{U}^{-1})^T \cdot \mathbf{F}^T \cdot \mathbf{F} \cdot \mathbf{U}^{-1} = (\mathbf{U}^{-1}) \cdot \mathbf{U} \cdot \mathbf{U} \cdot \mathbf{U}^{-1} = \mathbf{I},$$

therefore, $\mathbf{R}$ is a *rotation tensor*. Equation (3) gives

(4) ▲ $$\mathbf{F} = \mathbf{R} \cdot \mathbf{U},$$

which is called the *right polar decomposition* of $\mathbf{F}$, and $\mathbf{R}$ and $\mathbf{U}$ are called the *rotation tensor* and the *right stretch tensor*, respectively.

From Eq. (4), we can interpret the deformation from $d\mathbf{X}$ to $d\mathbf{x}$ as a two-stage transformation and write Eq. (16.2:2) as

(5) $$d\mathbf{X}^* = \mathbf{U} \cdot d\mathbf{X},$$

(6) $$d\mathbf{x} = \mathbf{R} \cdot d\mathbf{X}^*.$$

In other words, the line element $d\mathbf{X}$ is first stretched and rotated by $\mathbf{U}$ to become $d\mathbf{X}^*$. (For line elements along the direction of an eigenvector of $\mathbf{U}$, the deformation involves stretching only.) Then the line element $d\mathbf{X}^*$ is rotated by $\mathbf{R}$ to become $d\mathbf{x}$. The second stage deformation involves no stretching. Such a description of the deformation of $d\mathbf{X}$ is unique only to a rigid body translation.

The deformation gradient $\mathbf{F}$ can also be decomposed into

(7) ▲ $$\mathbf{F} = \mathbf{V} \cdot \mathbf{R},$$

where $\mathbf{V}$ is a symmetric tensor and satisfies

(8) $$\mathbf{V} \cdot \mathbf{V} = \mathbf{V} \cdot \mathbf{R} \cdot \mathbf{R}^T \cdot \mathbf{V}^T = \mathbf{F} \cdot \mathbf{F}^T \equiv \mathbf{B}.$$

Equation (7) is called the *left polar decomposition* and $\mathbf{V}$ is called the *left stretch tensor*. Note that $\mathbf{B}$ and $\mathbf{V}$ have the same eigenvectors, and therefore are coaxial. Clearly $\mathbf{V}$ is positive definite and is linked to $\mathbf{U}$ of the right polar decomposition by

$$\mathbf{V} = \mathbf{F} \cdot \mathbf{R}^T = \mathbf{R} \cdot \mathbf{U} \cdot \mathbf{R}^T.$$

Writing Eq. (16.2:2) in the form

(9) $$d\mathbf{x}' = \mathbf{R} \cdot d\mathbf{X},$$

(10) $$d\mathbf{x} = \mathbf{V} \cdot d\mathbf{x}',$$

we can interpret that the deformation first rotates $d\mathbf{X}$ as a rigid body by $\mathbf{R}$ to form $d\mathbf{x}'$, then stretches and rotates it by $\mathbf{V}$ to form $d\mathbf{x}$. If $d\mathbf{x}'$ is along a principal direction of $\mathbf{V}$, it is only stretched to become $d\mathbf{x}$.

If λ_i is an *eigenvalue* of $\mathbf{U}$ associated with the *eigenvector* $\mathbf{N}^i$, then

$$\mathbf{U} \cdot \mathbf{N}^i = \lambda_i \mathbf{N}^i \qquad (i \text{ not summed}),$$

and

$$\lambda_i \mathbf{R} \cdot \mathbf{N}^i \cdot \mathbf{R}^T = \mathbf{R} \cdot \mathbf{U} \cdot \mathbf{R}^T \cdot \mathbf{R} \cdot \mathbf{N}^i \cdot \mathbf{R}^T = \mathbf{V} \cdot \mathbf{R} \cdot \mathbf{N}^i \cdot \mathbf{R}^T, \qquad (i \text{ not summed}),$$

which shows that λ_i is also an *eigenvalue* of $\mathbf{V}$ associated with the *eigenvector*

(11) $$\mathbf{n}^i = \mathbf{R} \cdot \mathbf{N}^i \cdot \mathbf{R}^T.$$

In other words, $\mathbf{V}$ and $\mathbf{U}$ have same eigenvalues, and their eigenvectors are linked by Eq. (11). Note that $\mathbf{N}^i$ and $\mathbf{n}^i$ are, respectively, also the eigenvectors of $\mathbf{C}$ and $\mathbf{B}$ associated with the eigenvalue λ_i^2.

16.5. STRAIN RATES

We shall now define the time rates of the strain measures discussed in the previous section. The Eulerian descriptions of the *velocity field* $\mathbf{v} = \mathbf{v}(\mathbf{x}, t)$ and the *velocity gradient* $\mathbf{L}$ at time t are defined as

(1) $$\mathbf{v}(\mathbf{x}, t) \equiv \dot{\mathbf{x}}(\mathbf{X}, t) \equiv \frac{\partial \mathbf{x}(\mathbf{X}, t)}{\partial t},$$

(2) $$\mathbf{L} \equiv \frac{\partial \mathbf{v}}{\partial \mathbf{x}}, \quad \text{or} \quad L_{ij} = \frac{\partial v_i}{\partial x_j}.$$

Then the *material time derivative* of deformation gradient is

(3) ▲ $$\dot{\mathbf{F}} = \frac{D\mathbf{F}}{Dt} = \frac{\partial \dot{\mathbf{x}}}{\partial \mathbf{X}} = \frac{\partial \dot{\mathbf{x}}}{\partial \mathbf{x}} \cdot \frac{\partial \mathbf{x}}{\partial \mathbf{X}} = \mathbf{L} \cdot \mathbf{F}, \qquad \text{or}$$

$$\dot{F}_{iJ} = \frac{D}{Dt}\left(\frac{\partial x_i}{\partial X_J}\right) = \frac{\partial \dot{x}_i}{\partial X_J} = \frac{\partial v_i}{\partial x_j}\frac{\partial x_j}{\partial X_J} = L_{ij}F_{jJ}.$$

Again $\dot{\mathbf{F}}$ *is a two-point tensor*. Using $\mathbf{F} \cdot \mathbf{F}^{-1} = \mathbf{I}$ and $\dot{\mathbf{F}} \cdot \mathbf{F}^{-1} + \mathbf{F} \cdot \dot{\mathbf{F}}^{-1} = 0$, we establish that

(3a) $$\frac{D\mathbf{F}^{-1}}{Dt} = -\mathbf{F}^{-1} \cdot \dot{\mathbf{F}} \cdot \mathbf{F}^{-1} = -\mathbf{F}^{-1} \cdot \mathbf{L}.$$

In general, $(\dot{\mathbf{F}})^{-1}(=\mathbf{F}^{-1} \cdot \mathbf{L}^{-1}) \neq \frac{D\mathbf{F}^{-1}}{Dt}$. From Eq. (16.2:7), we find

(4) $$\dot{\mathbf{C}} = \frac{D\mathbf{C}}{Dt} = \mathbf{F}^T \cdot \dot{\mathbf{F}} + \dot{\mathbf{F}}^T \cdot \mathbf{F} = \mathbf{F}^T \cdot (\mathbf{L} + \mathbf{L}^T) \cdot \mathbf{F} = 2\mathbf{F}^T \cdot \mathbf{D} \cdot \mathbf{F},$$

where

(5) ▲ $$\mathbf{D} \equiv (\mathbf{L} + \mathbf{L}^T)/2$$

is the symmetric part of $\mathbf{L}$ called the (*Eulerian*) *deformation* or *stretching rate tensor*. Another useful rate tensor derived from $\mathbf{L}$ is

(6) ▲ $$\boldsymbol{\Omega} = (\mathbf{L} - \mathbf{L}^T)/2\,,$$

which is skew-symmetric and called the *spin tensor*.

From Eqs. (5) and (16.3:5), we find the *Green's strain rate*

(7) ▲ $$\dot{\mathbf{E}} = \mathbf{F}^T \cdot \mathbf{D} \cdot \mathbf{F}\,.$$

The rate of the right stretch tensor can be derived from Eqs. (7) and (16.4:2)

(8) $$\dot{\mathbf{U}} \cdot \mathbf{U} + \mathbf{U} \cdot \dot{\mathbf{U}} = 2\dot{\mathbf{E}} = 2\mathbf{R}^T \cdot \mathbf{D} \cdot \mathbf{F}\,.$$

We then obtain

(9) $$\frac{D\mathbf{B}^{-1}}{Dt} = -(\mathbf{L}^T \cdot \mathbf{B}^{-1} + \mathbf{B}^{-1} \cdot \mathbf{L})\,,$$

(9a) ▲ $$\dot{\mathbf{B}} = -\mathbf{B} \cdot \frac{D\mathbf{B}^{-1}}{Dt} \cdot \mathbf{B} = \mathbf{B} \cdot \mathbf{L}^T + \mathbf{L} \cdot \mathbf{B}\,.$$

Again $\frac{D\mathbf{B}^{-1}}{Dt} \neq (\dot{\mathbf{B}})^{-1}$. The time derivative of Eq. (16.3:6) gives the rate of the Almansi strain tensor

(10) ▲ $$\dot{\mathbf{e}} = \mathbf{D} - (\mathbf{e} \cdot \mathbf{L} + \mathbf{L}^T \cdot \mathbf{e})\,.$$

From Eqs. (3) and (8), $\dot{\mathbf{R}} \cdot \mathbf{R}^T + \mathbf{R} \cdot \mathbf{R}^T = 0$ and $\dot{\mathbf{F}} = \dot{\mathbf{R}} \cdot \mathbf{U} + \mathbf{R} \cdot \dot{\mathbf{U}}$ we have

(11) $$\dot{\mathbf{R}} = \boldsymbol{\omega} \cdot \mathbf{R}\,. \quad \dot{\mathbf{U}} = \mathbf{R}^T \cdot (\mathbf{L} - \boldsymbol{\omega}) \cdot \mathbf{F}\,,$$

where $\boldsymbol{\omega}$ is a skew-symmetric tensor.

16.6. MATERIAL DERIVATIVES OF LINE, AREA AND VOLUME ELEMENTS

In Sec. 16.2, it was shown that $d\mathbf{x}$, $d\mathbf{a}$, and dv in R are related to their undeformed counterparts $d\mathbf{X}$, $d\mathbf{A}$ and dV in R_0 through $\mathbf{F}$. In the following, we shall show that the stretching rate tensor $\mathbf{D}$ defined in Eq. (16.5:5) relates the material rate derivatives of $d\mathbf{x}$, $d\mathbf{a}$, dv to those of $d\mathbf{X}$, $d\mathbf{A}$, dV. The time rate of a line element $d\mathbf{x}$ is

$$d\dot{\mathbf{x}} \equiv \frac{D}{Dt}d\mathbf{x} = \dot{\mathbf{F}} \cdot d\mathbf{X} = \mathbf{L} \cdot d\mathbf{x}\,,$$

in which the first equality is the definition of the dot, the second equality is true because $d\mathbf{X}$ is independent of t, whereas the third equality introduces the velocity gradient tensor defined in Eq. (16.5:2). The stretch rate of the element is

$$(1) \qquad l_s \equiv \frac{1}{|d\mathbf{x}|} \frac{D}{Dt} |d\mathbf{x}| = \frac{1}{|d\mathbf{x}|} \frac{D}{Dt} \sqrt{d\mathbf{x} \cdot d\mathbf{x}}$$

$$= \frac{d\dot{\mathbf{x}} \cdot d\mathbf{x} + d\mathbf{x} \cdot d\dot{\mathbf{x}}}{2|d\mathbf{x}|^2} = \frac{d\mathbf{x} \cdot \mathbf{D} \cdot d\mathbf{x}}{|d\mathbf{x}|^2} = \bar{\mathbf{s}} \cdot \mathbf{D} \cdot \bar{\mathbf{s}},$$

where $\bar{\mathbf{s}} = d\mathbf{x}/|d\mathbf{x}|$ is the unit vector along the direction of $d\mathbf{x}$. Since $ds = \sqrt{d\mathbf{x} \cdot d\mathbf{x}} = |d\mathbf{x}|$ and $ds/dS = \lambda_s$ [Eq. (16.2:10)], it follows that

$$(2) \qquad l_s = \frac{D}{Dt} \ln(ds) = \frac{D}{Dt} \ln \left(\frac{ds}{dS} \right) = \frac{D}{Dt} \ln(\lambda_s),$$

where dS is the undeformed length of ds and is independent of t. The equation above gives

$$(3) \qquad \dot{\lambda}_s = \lambda_s l_s = \lambda_s \bar{\mathbf{s}} \cdot \mathbf{D} \cdot \bar{\mathbf{s}} = \lambda_s \bar{\mathbf{s}} \cdot \mathbf{L} \cdot \bar{\mathbf{s}} = \lambda_s \bar{\mathbf{s}} \cdot \mathbf{L}^T \cdot \bar{\mathbf{s}}.$$

Using Eq. (1) we can show that

$$(4) \qquad \dot{\bar{\mathbf{s}}} = \frac{d\dot{\mathbf{x}}}{|d\mathbf{x}|} - d\mathbf{x} \frac{d\mathbf{x} \cdot d\dot{\mathbf{x}}}{|d\mathbf{x}|^3} = \mathbf{L} \cdot \bar{\mathbf{s}} - (\bar{\mathbf{s}} \cdot \mathbf{D} \cdot \bar{\mathbf{s}})\bar{\mathbf{s}} = (\mathbf{L} - l_s \mathbf{I}) \cdot \bar{\mathbf{s}}.$$

The quantity $\ln \lambda_s$ is often called the *natural* or *logarithmic strain*. Equation (2) establishes that the stretch rate of a deformed element is equal to the rate of logarithmic strain. In Eq. (1), taking $\bar{\mathbf{s}} = \mathbf{e}_i$ leads to $l_s = D_{ii}$ (i not summed). Thus, with respect to Cartesian coordinates, the diagonal components D_{11}, D_{22} and D_{33} are the logarithmic strain rates along the directions of base vectors $\mathbf{e}_1$, $\mathbf{e}_2$ and $\mathbf{e}_3$, respectively.

To determine the meaning of the off-diagonal components of $\mathbf{D}$, we consider two infinitesimal line elements $d\mathbf{x}^{(1)}$ and $d\mathbf{x}^{(2)}$ along the directions $\bar{\mathbf{s}}^{(1)}$ and $\bar{\mathbf{s}}^{(2)}$, respectively. The dot product of the two line elements gives

$$d\mathbf{x}^{(1)} \cdot d\mathbf{x}^{(2)} = |d\mathbf{x}^{(1)}| |d\mathbf{x}^{(2)}| \cos \theta,$$

where θ is the angle between the two vector differentials. Then

$$\frac{D}{Dt} [d\mathbf{x}^{(1)} \cdot d\mathbf{x}^{(2)}] = 2d\mathbf{x}^{(1)} \cdot \mathbf{D} \cdot d\mathbf{x}^{(2)},$$

$$\frac{D}{Dt} [|d\mathbf{x}^{(1)}| |d\mathbf{x}^{(2)}| \cos \theta] = \cos \theta \frac{D}{Dt} [|d\mathbf{x}^{(1)}| |d\mathbf{x}^{(2)}|] - |d\mathbf{x}^{(1)}| |d\mathbf{x}^{(2)}| \sin \theta \frac{D\theta}{Dt}.$$

If $dx^{(1)}$ is perpendicular to $dx^{(2)}$, $\cos\theta = 0$ and $\sin\theta = 1$. Equating the two equations above gives

(5)
$$-\frac{D\theta}{Dt} = 2\frac{dx^{(1)} \cdot \dot{\mathbf{D}} \cdot dx^{(2)}}{|dx^{(1)}||dx^{(2)}|} = 2\bar{\mathbf{s}}^{(1)} \cdot \mathbf{D} \cdot \bar{\mathbf{s}}^{(2)}.$$

But $-D\theta/Dt$ is the decreasing rate of a right angle, which is precisely the shearing rate. If $\bar{\mathbf{s}}^{(1)}$ and $\bar{\mathbf{s}}^{(2)}$ are in the directions of two base-vectors $\mathbf{e}_i$ and $\mathbf{e}_j$ where $i \neq j$, then $\bar{\mathbf{s}}^{(1)} \cdot \mathbf{D} \cdot \bar{\mathbf{s}}^{(2)} = D_{ij}$.

The rate of the volume ratio is determined as follows. Notice that

$$\frac{\partial J}{\partial F_{iI}} = \text{cofactor} \quad (F_{iI}) = J(F_{iI}^T)^{-1}, \quad \text{or} \quad \frac{\partial J}{\partial \mathbf{F}} = J(\mathbf{F}^T)^{-1}.$$

It is straightforward to find

(6)
$$\frac{D}{Dt}\left(\frac{dv}{dV}\right) = \frac{DJ}{Dt} = \frac{\partial J}{\partial \mathbf{F}} : \dot{\mathbf{F}} = J(\mathbf{F}^T)^{-1} : (\mathbf{L} \cdot \mathbf{F})$$

$$= JX_{K,i}L_{ij}x_{j,K} = \text{tr}(\mathbf{L})J = \text{tr}(\mathbf{D})J.$$

Then

(7) ▲
$$\frac{D}{Dt}\ln(J) = \text{tr}(\mathbf{D}) = \frac{\partial v_i}{\partial x_i} = \nabla \cdot \mathbf{v},$$

where $\nabla (= \mathbf{e}_i \frac{\partial}{\partial x_i})$ is the Eulerian gradient operator. The divergence of the velocity vector is the logarithmic rate of the volume ratio.

To compute the time rate of the area element $d\mathbf{a}$, whose magnitude is da, we derive from Eqs. (16.2:12), (16.1:15), (3) and (6) that

(8)
$$d\dot{\mathbf{a}} = \left\{\frac{DJ}{Dt}(\mathbf{F}^{-1})^T + J\frac{D}{Dt}[(\mathbf{F}^{-1})^T]\right\} \cdot d\mathbf{A} = [\text{tr}(\mathbf{L})\mathbf{I} - \mathbf{L}^T] \cdot d\mathbf{a}.$$

The use of Eq. (16.2:14) and $|d\mathbf{a}|^2 = d\mathbf{a} \cdot d\mathbf{a}$ leads to

$$\dot{\Lambda}_n \equiv \frac{D}{Dt}\frac{|d\mathbf{a}|}{|d\mathbf{A}|} = \frac{|d\mathbf{a}|}{|d\mathbf{A}|}\frac{1}{|d\mathbf{a}|}\frac{D|d\mathbf{a}|}{Dt} = \frac{\Lambda_n d\mathbf{a} \cdot d\dot{\mathbf{a}}}{|d\mathbf{a}|^2}.$$

From Eqs. (1) and (8), the equation above becomes

(9)
$$\dot{\Lambda}_n = \Lambda_n[\text{tr}(\mathbf{D}) - \mathbf{n} \cdot \mathbf{L}^T \cdot \mathbf{n}] = \Lambda_n[\text{tr}(\mathbf{D}) - l_s],$$

where $\mathbf{n}$ is the unit vector normal to the surface and l_s is the logarithmic stretch rate of the line element along the normal direction. Then

(10)
$$\frac{D}{Dt}\ln(\Lambda_n) = \text{tr}(\mathbf{D}) - l_s.$$

Introducing the *Jacobian* $J \equiv \det[\mathbf{F}(\mathbf{X},t)]$, and using Eqs. (2), (6) and (10), we obtain

$$\frac{D}{Dt}\ln(J) = \frac{D}{Dt}\ln(\Lambda_n) + \frac{D}{Dt}\ln(\lambda_s) = \frac{D}{Dt}\ln(\lambda_s \Lambda_n),$$

which implies

(11) $$J = \lambda_s \Lambda_n.$$

Equation (11) indicates that the volume ratio equals the product of the stretch ratio and the area ratio provided that the direction of the stretch is normal to the area element.

In summary, the rates of the logarithmic line, area and volume ratios of finite deformation are

$$\frac{D}{Dt}\ln(\lambda_s) = \bar{\mathbf{s}}\cdot\mathbf{D}\cdot\bar{\mathbf{s}}, \qquad \frac{D}{Dt}\ln(\Lambda_n) = \mathrm{tr}(\mathbf{D}) - \mathbf{n}\cdot\mathbf{D}\cdot\mathbf{n}, \qquad \frac{D}{Dt}\ln(J) = \mathrm{tr}(\mathbf{D}).$$

These results reduce to those of the infinitesimal strain in Chapter 4.

Problem 16.3. Show that

$$\frac{\partial J}{\partial \mathbf{F}} = J(\mathbf{F}^T)^{-1}, \qquad \text{and} \qquad \frac{\partial J}{\partial \mathbf{E}} = \mathbf{C}^{-1},$$

where $J = \det(\mathbf{F})$.

16.7. STRESSES

The stress tensor introduced in Chapter 3 is valid under large deformation, provided that all quantities are Eulerian referred to the current configuration. This stress tensor is called the *Cauchy stress* and denoted by $\boldsymbol{\tau}$, which is an Eulerian tensor with components τ_{ij}. The rate of work done per unit volume by the Cauchy stress acting on the surface of an infinitesimal volume element is $\tau_{ij}D_{ij}$, which is the inner product of the stress tensor and the deformation rate tensor [see Sec. 12.3, Eqs. (12.3:7,8)]. Let ρ be the density of the material at the deformed state. The rate of work done per unit mass by the stress is $\dot{W} = \tau_{ij}D_{ij}/\rho$. Then

(1) $$\rho\dot{W} = \boldsymbol{\tau}:\mathbf{D} = \tau_{ij}D_{ij} = \tau_{ij}V_{ij},$$

where $\mathbf{D}$ and V_{ij} are given in Eq. (16.5:5) and (12.2:8), respectively. We define that a stress is *conjugate to* (or the *work conjugate variable* of) a strain rate, and vice versa, if their inner product is the work rate per unit deformed or undeformed volume. Thus the Cauchy stress $\boldsymbol{\tau}$ is the stress measure conjugate to, or the work conjugate variable of, the deformation rate tensor $\mathbf{D}$ when the work rate is referred to the deformed volume.

Since the mass does not change during deformation, we see that the work rate referred to a unit volume of the material in the undeformed state is

$$(2) \qquad \rho_0 \dot{W} = \frac{\rho_0}{\rho}\, \rho \dot{W} = J\boldsymbol{\tau} : \mathbf{D} = J\tau_{ij} D_{ij} \,,$$

where $J [= \det(\mathbf{F})]$ is the *Jacobian* of the deformation gradient tensor $\mathbf{F}$.

Chapter 4 and Secs. 16.3 and 16.4 shows that there are other strain measures, each has a claim to advantage in certain circumstances. Corresponding to each strain measure is a stress measure whose inner product with the time rate of the strain measure gives the rate of work done per unit deformed or undeformed volume. Then the strain rate and the corresponding stress measure are said to be *conjugate* to each other. The stress measures are defined as follows.

Kirchhoff Stress. Kirchhoff defined

$$(3) \qquad\qquad\qquad \boldsymbol{\sigma} \equiv J\boldsymbol{\tau} \,,$$

now known as the *Kirchhoff stress tensor*. Obviously $\boldsymbol{\sigma}$ is symmetric and $\boldsymbol{\sigma} : \mathbf{D}$ equals to $\rho_0 \dot{W}$, the time rate of strain energy density per unit undeformed volume. Therefore, $\boldsymbol{\sigma}$ is conjugate to $\mathbf{D}$ when the work rate is referred to the undeformed configuration.

First Piola–Kirchhoff Stress. The *first Piola-Kirchhoff stress tensor*, also called the *Lagrangian stress tensor*, is defined as

$$(4) \qquad\qquad \mathbf{T} \equiv \mathbf{F}^{-1} \cdot \boldsymbol{\sigma} = J\mathbf{F}^{-1} \cdot \boldsymbol{\tau} \,,$$

which is a nonsymmetric two-point tensor with $\mathbf{F}^{-1}$ based on the undeformed state and $\boldsymbol{\tau}$ referred to the deformed state. Since $\boldsymbol{\sigma}$ is symmetric, using Eq. (16.5:3), one can show that

$$(5) \qquad \mathbf{T} : \dot{\mathbf{F}}^T = (\mathbf{F}^{-1} \cdot \boldsymbol{\sigma}) : \dot{\mathbf{F}}^T = (\dot{\mathbf{F}} \cdot \mathbf{F}^{-1})^T : \boldsymbol{\sigma} = \boldsymbol{\sigma} : \mathbf{L} = \boldsymbol{\sigma} : \mathbf{D} \,.$$

Thus $\mathbf{T} : \dot{\mathbf{F}}^T$ equals to $\rho_0 \dot{W}$ and, therefore, $\mathbf{T}$ is conjugate to $\dot{\mathbf{F}}^T$.

Second Piola–Kirchhoff Stress. A widely used stress measure is

$$(6) \qquad \mathbf{S} \equiv \mathbf{T} \cdot (\mathbf{F}^{-1})^T = \mathbf{F}^{-1} \cdot \boldsymbol{\sigma} \cdot (\mathbf{F}^{-1})^T = J\mathbf{F}^{-1} \cdot \boldsymbol{\tau} \cdot (\mathbf{F}^{-1})^T \,,$$

known as the *second Piola–Kirchhoff stress tensor*. Clearly $\mathbf{S}$ is symmetric and is defined in Lagrangian description. It can be shown that

(7) $\mathbf{S} : \dot{\mathbf{E}} = [\mathbf{F}^{-1} \cdot \boldsymbol{\sigma} \cdot (\mathbf{F}^{-1})^T] : \dot{\mathbf{E}} = (\mathbf{F}^{-1} \cdot \boldsymbol{\sigma}) : (\dot{\mathbf{E}} \cdot \mathbf{F}^{-1})$

$= \boldsymbol{\sigma} : [(\mathbf{F}^{-1})^T \cdot \dot{\mathbf{E}} \cdot \mathbf{F}^{-1}] = \boldsymbol{\sigma} : \mathbf{D} = \rho_0 \dot{W} \,,$

which establishes $\mathbf{S}$ being conjugate to Green's strain rate $\dot{\mathbf{E}}$ defined in Eq. (16.5:7).

Biot-Lur'e Stress. Another stress measure is the *Biot–Lur'e stress tensor* which is defined as

(8) $\mathbf{r}^* = \mathbf{F}^{-1} \cdot \boldsymbol{\sigma} \cdot \mathbf{R} = \mathbf{T} \cdot \mathbf{R} = \mathbf{S} \cdot \mathbf{U} \,.$

One can show that

(8a) $\mathbf{S} : \dot{\mathbf{E}} = \mathbf{S} \cdot \mathbf{U} : \dot{\mathbf{U}} = \mathbf{r}^* : \dot{\mathbf{U}} = \mathbf{r} : \dot{\mathbf{U}} \,,$

where

$$\mathbf{r} = (\mathbf{r}^* + \mathbf{r}^{*T})/2$$

is the symmetric part of $\mathbf{r}^*$. Then the conjugate stress measure of $\dot{\mathbf{U}}$ is $\mathbf{r}^*$ since $\mathbf{r}^* : \dot{\mathbf{U}} = \rho_0 \dot{W}$.

Corotational Cauchy Stress. Green and Naghdi (1965) introduced

(9) $\boldsymbol{\sigma}_r = \mathbf{R}^T \cdot \boldsymbol{\tau} \cdot \mathbf{R} J = \mathbf{U} \cdot \mathbf{S} \cdot \mathbf{U}$

as the symmetric corotational Cauchy stress. This stress is obtained by rotating the Cauchy stress tensor back to the undeformed body. The conjugate strain measure is $\mathbf{R}^T \cdot \mathbf{D} \cdot \mathbf{R}$ that the strain energy per unit undeformed volume is

(9a) $\boldsymbol{\sigma}_r : (\mathbf{R}^T \cdot \mathbf{D} \cdot \mathbf{R}) = \boldsymbol{\sigma} : \mathbf{D} = \mathbf{S} : \dot{\mathbf{E}} \,.$

A summary of the relations among the different stress measures is as follows:

(10) ▲ $\boldsymbol{\sigma} = J\boldsymbol{\tau} = \mathbf{F} \cdot \mathbf{T} = \mathbf{F} \cdot \mathbf{S} \cdot \mathbf{F}^T = \mathbf{F} \cdot \mathbf{r}^* \cdot \mathbf{R}^T = \mathbf{R} \cdot \boldsymbol{\sigma}_r \cdot \mathbf{R}^T \,.$

The hydrostatic pressure is usually defined as the mean Cauchy stress

$$p = -(\boldsymbol{\tau} : \mathbf{I})/3 \,.$$

It follows from Eq. (10)

(11) ▲ $$-p = \frac{\tau:\mathbf{I}}{3} = \frac{\sigma_r:\mathbf{I}}{3J} = \frac{\sigma:\mathbf{I}}{3J} = \frac{\mathbf{T}:\mathbf{F}^T}{3J} = \frac{\mathbf{S}:\mathbf{C}}{3J} = \frac{\mathbf{r}^*:\mathbf{U}}{3J} = \frac{\mathbf{r}:\mathbf{U}}{3J},$$

which gives the consistent definition of hydrostatic pressure based on different stress measures. Similarly one can consistently define the deviatoric stresses for different stress measures, namely

$$\tau'' = \tau + p\mathbf{I} = \tau - \frac{1}{3}(\tau:\mathbf{I})\mathbf{I},$$

$$\sigma'' = \sigma + pJ\mathbf{I} = \sigma - \frac{1}{3}(\sigma:\mathbf{I})\mathbf{I},$$

$$\mathbf{T}'' = \mathbf{T} + pJ(\mathbf{F}^{-1})^T = \mathbf{T} - \frac{1}{3}(\mathbf{T}:\mathbf{F}^T)(\mathbf{F}^{-1})^T,$$

(12) $$\mathbf{S}'' = \mathbf{S} + pJ\mathbf{C}^{-1} = \mathbf{S} - \frac{1}{3}(\mathbf{S}:\mathbf{C})\mathbf{C}^{-1},$$

$$\mathbf{r}^{*\prime\prime} = \mathbf{r}^* + pJ\mathbf{U}^{-1} = \mathbf{r}^* - \frac{1}{3}(\mathbf{r}^*:\mathbf{U})\mathbf{U}^{-1},$$

$$\mathbf{r}'' = \mathbf{r} + pJ\mathbf{U}^{-1} = \mathbf{r} - \frac{1}{3}(\mathbf{r}^*:\mathbf{U})\mathbf{U}^{-1},$$

$$\sigma_r'' = \sigma_r + Jp\mathbf{I} = \sigma_r - \frac{1}{3}(\sigma_r:\mathbf{I})\mathbf{I}.$$

Note that $\mathbf{F}^{-1}:\mathbf{F}^T = 3$. We have

$$\tau'':\mathbf{I} = \sigma'':\mathbf{I} = \mathbf{T}'':\mathbf{F}^T = \mathbf{S}'':\mathbf{C} = \mathbf{r}^{*\prime\prime}:\mathbf{U} = \mathbf{r}'':\mathbf{U} = \sigma_r'':\mathbf{I} = 0.$$

(13)

We have established in Eqs. (5), (7), (8a) and (9a) that the work conjugate variables of σ, $\mathbf{T}$, $\mathbf{S}$, $\mathbf{r}^*$ and σ_r are $\dot{\mathbf{D}}$, $\dot{\mathbf{F}}^T$, $\dot{\mathbf{E}}$, $\dot{\mathbf{U}}$ and $\mathbf{R}\cdot\mathbf{D}\cdot\mathbf{R}^T$, respectively. Let us define

$$P_d \equiv \int_{R_0} \rho_0 \dot{W} dV = \int_R \rho \dot{W} dv$$

as the *deformation power*, where R_0 is the undeformed body. We can express P_d in alternative forms based on different stress measures, that is,

(14) ▲ $$P_d = \int_R \tau:\mathbf{D}dv = \int_{R_0} \sigma:\mathbf{D}dV = \int_{R_0} \mathbf{T}:\dot{\mathbf{F}}^T dV = \int_{R_0} \mathbf{S}:\dot{\mathbf{E}}dV$$

$$= \int_{R_0} \mathbf{r}^*:\dot{\mathbf{U}}dV = \int_{R_0} \mathbf{r}:\dot{\mathbf{U}}dV = \int_{R_0} \sigma_r:(\mathbf{R}^T\cdot\mathbf{D}\cdot\mathbf{R})dV.$$

Using Eqs. (10) and (16.2:12), one can derive the following relations among the stress measures:

$$\boldsymbol{\tau}{\cdot}d\mathbf{a} = \mathbf{T}^T{\cdot}d\mathbf{A} = \mathbf{F}{\cdot}\mathbf{S}{\cdot}d\mathbf{A} = \mathbf{R}{\cdot}\mathbf{r}^{*T}{\cdot}d\mathbf{A} = \mathbf{R}{\cdot}\boldsymbol{\sigma}_r{\cdot}\mathbf{U}^{-1}{\cdot}d\mathbf{A} = \mathbf{R}{\cdot}\boldsymbol{\sigma}_r{\cdot}\mathbf{R}^T{\cdot}d\mathbf{a}/J,$$

which means that the transformation of the force vector $\boldsymbol{\tau} \cdot d\mathbf{a}$ by $\mathbf{I}$, $\mathbf{F}^{-1}$, $\mathbf{R}^T$ gives the force vectors $\mathbf{T}^T \cdot d\mathbf{A}$, $\mathbf{S} \cdot d\mathbf{A}$, $\mathbf{r}^{*T} \cdot d\mathbf{A}$, respectively on the undeformed surface. We also have the rotated force vector $\mathbf{R}^T \cdot (\boldsymbol{\sigma} \cdot d\mathbf{a}) \cdot \mathbf{R} = \boldsymbol{\sigma}_r \cdot d\mathbf{a}^*$ where $d\mathbf{a}^* = \mathbf{R}^T \cdot d\mathbf{a} \cdot \mathbf{R}$ is a rotated surface of $d\mathbf{a}$. The geometric interpretation of the different force vectors is explained in more details below.

Graphic Description of Stress Measures. Because of the importance of the stress tensors, we would like to offer a graphic description of the Lagrangian, the second Piola–Krichhoff and Biot–Lur'e stresses below. This is to supplement the graphic details presented in Chapter 3 for the Cauchy stress, and to do the derivation once more in the indicial notation, which was used in Chapters 3–15. Consider an element of a strained solid as shown on the right-hand side of Fig. 16.7:1. Assume that in the original (undeformed) state this element has the configuration as shown on the left side of Fig. 16.7:1. A force vector $d\overset{\nu}{\mathbf{T}}$ acts on the surface $PQRS$. A corresponding force vector $d\overset{\nu}{\mathbf{T}}_0$ acts on the surface $P_0Q_0R_0S_0$. If we assign a rule of correspondence between $d\overset{\nu}{\mathbf{T}}$ and $d\overset{\nu}{\mathbf{T}}_0$ for every corresponding pair of surfaces, and define stress vectors in each case as the limiting ratios $d\overset{\nu}{\mathbf{T}}/da$, $d\overset{\nu}{\mathbf{T}}_0/dA$, where da and dA are the areas of $PQRS$, $P_0Q_0R_0S_0$, respectively, then by the method of Chapter 3 we can define stress tensors

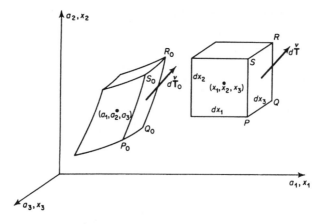

Fig. 16.7:1. The corresponding tractions in the original and deformed state of body.

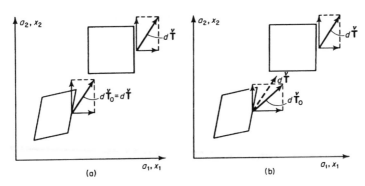

Fig. 16.7:2. The correspondence of force vectors in defining (a) Lagrange's and (b) the second Piola-Kirchhoff stress, illustrated in a two-dimensional case.

in both configurations. The assignment of a correspondence rule between $d\overset{\nu}{\mathbf{T}}$ and $d\overset{\nu}{\mathbf{T}}_0$ is based on the equations derived above.

The following alternative rules are known as the Lagrangian, the second Piola-Kirchhoff and the Biot-Lur'e rules, respectively (see Fig. 16.7:2)

$$(15) \quad d\overset{\nu}{T}{}_{0i}^{(L)} = d\overset{\nu}{T}_i \qquad \text{or} \quad d\overset{\nu}{\mathbf{T}}{}_0^{(L)} = d\overset{\nu}{\mathbf{T}} \qquad \text{(Lagrangian rule)},$$

$$(16) \quad d\overset{\nu}{T}{}_{0i}^{(K)} = \frac{\partial X_i}{\partial x_j} d\overset{\nu}{T}_j \qquad \text{or} \quad d\overset{\nu}{\mathbf{T}}{}_0^{(K)} = \mathbf{F}^{-1} \cdot d\overset{\nu}{\mathbf{T}} \qquad \text{(Piola-Kirchhoff rule)},$$

$$(17) \quad d\overset{\nu}{T}{}_{0i}^{(B)} = R_{ij} d\overset{\nu}{T}_j \qquad \text{or} \quad d\overset{\nu}{\mathbf{T}}{}_0^{(B)} = \mathbf{R} \cdot d\overset{\nu}{\mathbf{T}} \qquad \text{(Biot-Lur'e rule)}.$$

In other words, the Lagrangian rule assigns the force vector $d\overset{\nu}{\mathbf{T}}{}_0^{(L)}$ acting on the surface element dA in the undeformed configuration to be equal to the force vector $d\overset{\nu}{\mathbf{T}}$ acting on the corresponding surface element da in the current configuration; the second Piola-Kirchhoff rule specifies $d\overset{\nu}{\mathbf{T}}{}_0^{(K)}$, the force vector acting on the surface element dA, to be related to $d\overset{\nu}{\mathbf{T}}$ by the same rule as the transformation $dX_i = \frac{\partial X_i}{\partial x_j} dx_j$; whereas the second Biot-Lur'e rule specifies $d\overset{\nu}{\mathbf{T}}{}_0^{(B)}$ to be $d\overset{\nu}{\mathbf{T}}$ rotated by $\mathbf{R}$. Note that, in the present case, both the deformed and undeformed configurations have the same base vectors. There is no need to use capital subscripts to distinguish indices referred to the undeformed configuration.

Let $\mathbf{n}$ and $\mathbf{N}$ be the unit vectors normal to da and dA, respectively. If τ_{ij} is the Cauchy stress tensor referred to the strained state, from *Cauchy's relation*, the vector $d\overset{\nu}{\mathbf{T}}$, with components $d\overset{\nu}{T}_i$, can be written as

$$(18) \qquad d\overset{\nu}{T}_i = \tau_{ji} n_j da, \qquad \text{or} \qquad d\overset{\nu}{\mathbf{T}} = \boldsymbol{\tau} \cdot d\mathbf{a}.$$

We now define stress components referred to the original state by a law similar to Eq. (18). If Eq. (15) is used, we write

$$(19) \qquad d\overset{\nu}{T}{}^{(L)}_{0i} = T_{ji}N_j dA = d\overset{\nu}{T}_i , \qquad d\overset{\nu}{\mathbf{T}}{}^{(L)}_0 = \mathbf{T}^T \cdot d\mathbf{A} = d\overset{\nu}{\mathbf{T}} ,$$

according to the Lagrangian rule. Note that $T_{ij} \neq T_{ji}$. If Eq. (16) is used, we write

$$(20) \quad d\overset{\nu}{T}{}^{(K)}_{0i} = S_{ji}N_j dA = \frac{\partial X_i}{\partial x_j} d\overset{\nu}{T}_j , \quad d\overset{\nu}{\mathbf{T}}{}^{(K)}_0 = \mathbf{S} \cdot d\mathbf{A} = \mathbf{F}^{-1} \cdot d\overset{\nu}{\mathbf{T}} ,$$

according to the second Piola–Kirchhoff rule. Note that $S_{ij} = S_{ji}$ and that $\tau_{ij}, T_{ij}, S_{ij}$ are the Cauchy, the Lagrangian, and the second Piola-Kirchhoff stress tensors, respectively.

The relationship among τ_{ij}, T_{ij} and S_{ij} can now be derived. Using the relation between $n_i da$ and $N_i dA$ given in Eq. (16.2:12), from Eqs. (18) and (19), we obtain

$$T_{ji}N_j dA = \tau_{ji}n_j da = J\tau_{mi}\frac{\partial X_j}{\partial x_m} N_j dA , \qquad \text{or}$$

$$\mathbf{N} \cdot \mathbf{T} dA = \boldsymbol{\tau} \cdot \mathbf{n} da = J\boldsymbol{\tau} \cdot \frac{\partial \mathbf{X}}{\partial \mathbf{x}} \cdot \mathbf{N} dA ,$$

which implies

$$(21) \quad T_{ji} = J\frac{\partial X_j}{\partial x_m}\tau_{mi} = \frac{\partial X_j}{\partial x_m}\sigma_{mi} , \qquad \text{or} \qquad \mathbf{T} = J\mathbf{F}^{-1} \cdot \boldsymbol{\tau} = \mathbf{F}^{-1} \cdot \boldsymbol{\sigma} ,$$

which is the same as that defined in Eq. (4), where J is the Jacobian of the transformation from the undeformed to the deformed configuration defined in Eqs. (16.2:5), (16.2:16), or (16.2:17). Similarly, from Eqs. (18) and (20),

$$(22) \qquad S_{ji} = J\frac{\partial X_i}{\partial x_m}\frac{\partial X_j}{\partial x_n} \tau_{nm} = \frac{\partial X_i}{\partial x_m}\frac{\partial X_j}{\partial x_n}\sigma_{nm} , \qquad \text{or}$$

$$\mathbf{S} = J\mathbf{F}^{-1} \cdot \boldsymbol{\tau} \cdot (\mathbf{F}^{-1})^T = \mathbf{F}^{-1} \cdot \boldsymbol{\sigma} \cdot (\mathbf{F}^{-1})^T$$

as given in Eq. (6). Hence, from the two equations above, we find the relation between $\mathbf{S}$ and $\mathbf{T}$

$$S_{ji} = T_{in}\frac{\partial X_j}{\partial x_n} , \qquad \text{or} \qquad \mathbf{S} = \mathbf{T} \cdot (\mathbf{F}^{-1})^T .$$

Similarly, we can show the relation between $\boldsymbol{\tau}$ and σ_r using Eq. (17).

The Eulerian (Cauchy) stress tensor τ_{ij} is symmetric as shown in Chapter 3. From Eq. (21), we see that, in general, the Lagrangian stress tensor T_{ji} is not symmetric, whereas from Eq. (22), we see that the second Piola–Kirchhoff's stress tensor S_{ji} is. The Lagrangian tensor is inconvenient to use in a stress-strain law in which the strain tensor is always symmetric, but is often convenient in laboratory work. The second Piola-Kirchhoff stress tensor is more suitable for the purpose.

From the identities

$$\delta_{ij} = \frac{\partial X_i}{\partial x_m}\frac{\partial x_m}{\partial X_j}, \qquad \delta_{ij} = \frac{\partial x_i}{\partial X_m}\frac{\partial X_m}{\partial x_j},$$

we find at once the relations

(23)
$$\tau_{ji} = \frac{1}{J}\frac{\partial x_i}{\partial X_m}T_{mj} = \frac{1}{J}\frac{\partial x_i}{\partial X_m}\frac{\partial x_j}{\partial X_n}S_{nm}, \quad \text{or}$$

$$\boldsymbol{\tau} = \frac{\mathbf{F}\cdot\mathbf{T}}{J} = \frac{\mathbf{F}\cdot\mathbf{S}\cdot\mathbf{F}^T}{J},$$

(24)
$$T_{ij} = S_{im}\frac{\partial x_j}{\partial X_m}, \quad \text{or } \mathbf{T} = \mathbf{S}\cdot\mathbf{F}^T.$$

16.8. EXAMPLE: COMBINED TENSION AND TORSION LOADS

Consider a thin-walled circular cylinder subjected to combined tension and torsion loads (Khan and Huaung 1995). The geometry and deformation are described in terms of polar coordinates. The geometry of the tube is

$$R_m - h/2 \le R \le R_m + h/2, \qquad 0 \le \theta \le 2\pi, \quad 0 \le Z \le L,$$

where h/R_m, $h/L \ll 1$. The point at (R, Θ, Z) before deformation moves to (r, θ, z). In polar coordinates, the deformation can be written as

$$r = \alpha R, \qquad \theta = \Theta + \omega Z, \qquad z = \lambda Z.$$

Here ω and λ are the angle of twist ratio and the axial stretch ratio per unit undeformed length, respectively. The base-vectors for the deformed and undeformed coordinates are, respectively, $(\mathbf{e}_r, \mathbf{e}_\theta, \mathbf{e}_z)$ and $(\mathbf{e}_R, \mathbf{e}_\Theta, \mathbf{e}_Z)$. They are related to the fixed Cartesian base vectors $(\mathbf{e}_X, \mathbf{e}_Y, \mathbf{e}_Z)$ by

(1) $\mathbf{e}_r = \cos\theta\mathbf{e}_X + \sin\theta\mathbf{e}_Y = \cos(\Theta + \omega Z)\mathbf{e}_X + \sin(\Theta + \omega Z)\mathbf{e}_Y,$

(2) $\mathbf{e}_\theta = -\sin\theta\mathbf{e}_X + \cos\theta\mathbf{e}_Y = -\sin(\Theta + \omega Z)\mathbf{e}_X + \cos(\Theta + \omega Z)\mathbf{e}_Y,$

and

$$\mathbf{e}_z = \mathbf{e}_Z, \qquad \mathbf{e}_R = \cos\Theta\mathbf{e}_X + \sin\Theta\mathbf{e}_Y, \qquad \mathbf{e}_\Theta = -\sin\Theta\mathbf{e}_X + \cos\Theta\mathbf{e}_Y.$$

The base vectors $\mathbf{e}_r$, $\mathbf{e}_\theta$ are functions of time and the undeformed coordinates. The deformed position and the velocity vector are, respectively,

$$(3) \qquad \mathbf{x} = r\mathbf{e}_r + z\mathbf{e}_z = Ra\mathbf{e}_r + Z\lambda\mathbf{e}_z\,,$$

$$(4) \qquad \dot{\mathbf{x}} = R\dot{a}\mathbf{e}_r + RZa\dot{\omega}\mathbf{e}_\theta + Z\dot{\lambda}\mathbf{e}_z\,.$$

The second term on the right hand side of the last equation arises from the time derivative of $\mathbf{e}_r$. The deformation gradient tensor $\mathbf{F}$ and its material time derivative $\dot{\mathbf{F}}$ can be derived by taking the spatial derivatives of Eqs. (3) and (4) with respect to the undeformed coordinates. In doing so, one must account for the fact also that both $\mathbf{e}_r$ and $\mathbf{e}_\theta$ are functions of Θ and Z.

As an example, consider the derivation of $\dot{\mathbf{F}}$ from $\dot{\mathbf{x}}$ through the spatial derivative with respect to the undeformed coordinates:

$$\dot{\mathbf{F}} = \frac{\partial \dot{\mathbf{x}}}{\partial R}\mathbf{e}_R + \frac{1}{R}\frac{\partial \dot{\mathbf{x}}}{\partial \Theta}\mathbf{e}_\Theta + \frac{\partial \dot{\mathbf{x}}}{\partial Z}\mathbf{e}_Z\,,$$

where

$$(5) \qquad \begin{aligned} \frac{\partial \dot{\mathbf{x}}}{\partial R} &= \mathbf{e}_r \dot{F}_{rR} + \mathbf{e}_\theta \dot{F}_{\theta R} + \mathbf{e}_z \dot{F}_{zR} = \mathbf{e}_r \dot{a} + \mathbf{e}_\theta Za\dot{\omega}\,, \\[2mm] \frac{1}{R}\frac{\partial \dot{\mathbf{x}}}{\partial \Theta} &= \mathbf{e}_r \dot{F}_{r\Theta} + \mathbf{e}_\theta \dot{F}_{\theta\Theta} + \mathbf{e}_z \dot{F}_{z\Theta} = -\mathbf{e}_r Za\dot{\omega} + \mathbf{e}_\theta \dot{a}\,, \\[2mm] \frac{\partial \dot{\mathbf{x}}}{\partial Z} &= \mathbf{e}_r \dot{F}_{rZ} + \mathbf{e}_\theta \dot{F}_{\theta Z} + \mathbf{e}_z \dot{F}_{zZ} \\[2mm] &= -\mathbf{e}_r RZa\omega\dot{\omega} + \mathbf{e}_\theta R(a\dot{\omega} + \dot{a}\omega) + \dot{\lambda}\mathbf{e}_z\,, \end{aligned}$$

where $\dot{F}_{rR}, \dot{F}_{r\Theta}, \ldots$ are the components of the two-point tensor $\dot{\mathbf{F}}$. The lower-case subscripts denote the association with the current coordinate frame and the upper-case subscripts denote the association with the reference frame. A comparison of the components on both sides of Eq. (5) yields $\dot{F}_{rR}, \dot{F}_{r\Theta}, \ldots$, etc. Following this procedure, one obtains

$$(6) \qquad \mathbf{F} = \begin{bmatrix} \dfrac{\partial r}{\partial R} & \dfrac{1}{R}\dfrac{\partial r}{\partial \Theta} & \dfrac{\partial r}{\partial Z} \\[3mm] r\dfrac{\partial \theta}{\partial R} & \dfrac{r}{R}\dfrac{\partial \theta}{\partial \Theta} & r\dfrac{\partial \theta}{\partial Z} \\[3mm] \dfrac{\partial z}{\partial R} & \dfrac{1}{R}\dfrac{\partial z}{\partial \Theta} & \dfrac{\partial z}{\partial Z} \end{bmatrix} = \begin{bmatrix} \alpha & 0 & 0 \\[2mm] 0 & \alpha & \omega\alpha R_m \\[2mm] 0 & 0 & \lambda \end{bmatrix},$$

$$(7) \qquad \dot{\mathbf{F}} = \begin{bmatrix} \dot{\alpha} & -Za\dot{\omega} & -R_m Za\omega\dot{\omega} \\[2mm] Za\dot{\omega} & \dot{\alpha} & R_m(a\dot{\omega} + \dot{a}\omega) \\[2mm] 0 & 0 & \dot{\lambda} \end{bmatrix},$$

in which the mean undeformed radius R_m has been used to replace R. From Eq. (6), it follows that

$$\mathbf{F}^{-1} = \begin{bmatrix} 1/\alpha & 0 & 0 \\ 0 & 1/\alpha & -\omega R_m/\lambda \\ 0 & 0 & 1/\lambda \end{bmatrix}, \qquad J = \det(\mathbf{F}) = \alpha^2 \lambda.$$

To obtain $\mathbf{R}$, $\mathbf{U}$ and $\mathbf{V}$ in the polar decomposition, one writes the rotational tensor in the form

$$\mathbf{R} = \begin{bmatrix} 1 & 0 & 0 \\ 0 & \cos\phi & \sin\phi \\ 0 & -\sin\phi & \cos\phi \end{bmatrix}.$$

The solution of the equations $\mathbf{F} = \mathbf{R} \cdot \mathbf{U} = \mathbf{V} \cdot \mathbf{R}$ gives the right and left stretch tensors

$$\mathbf{U} = \alpha \begin{bmatrix} 1 & 0 & 0 \\ 0 & \cos\phi & \sin\phi \\ 0 & \sin\phi & \omega R_m \sin\phi + \dfrac{\lambda}{\alpha}\cos\phi \end{bmatrix},$$

$$\mathbf{V} = \begin{bmatrix} \alpha & 0 & 0 \\ 0 & \alpha\omega R_m \sin\phi + \lambda\cos\phi & \lambda\sin\phi \\ 0 & \lambda\sin\phi & \lambda\cos\phi \end{bmatrix},$$

where

$$\cos\phi = \frac{\lambda + \alpha}{d}, \qquad \sin\phi = \frac{\alpha\omega R_m}{d}, \qquad d = \sqrt{(\lambda + \alpha)^2 + (\alpha\omega R_m)^2}.$$

The other quantities can be determined in a straightforward manner:

$$\mathbf{C} = \mathbf{F}^T \cdot \mathbf{F} = \begin{bmatrix} \alpha^2 & 0 & 0 \\ 0 & \alpha^2 & \omega\alpha^2 R_m \\ 0 & \omega\alpha^2 R_m & \lambda^2 + \omega^2\alpha^2 R_m^2 \end{bmatrix},$$

$$\mathbf{B}^{-1} = (\mathbf{F}^{-1})^T \cdot \mathbf{F}^{-1} = \begin{bmatrix} \dfrac{1}{\alpha^2} & 0 & 0 \\[2ex] 0 & \dfrac{1}{\alpha^2} & -\dfrac{\omega R_m}{\alpha \lambda} \\[2ex] 0 & -\dfrac{\omega R_m}{\alpha \lambda} & \dfrac{1}{\lambda^2} + \dfrac{\omega^2 R_m^2}{\lambda^2} \end{bmatrix} .$$

The Lagrangian and the Eulerian strains can be obtained by using $\mathbf{E} = (\mathbf{C} - \mathbf{I})/2$ and $\mathbf{e} = (\mathbf{I} - \mathbf{B}^{-1})/2$. The velocity gradient tensor is simply

$$\mathbf{L} = \dot{\mathbf{F}} \cdot \mathbf{F}^{-1} = \begin{bmatrix} \dot{\alpha}/\alpha & -\dot{\omega} Z & 0 \\[1ex] \dot{\omega} Z & \dot{\alpha}/\alpha & \dot{\omega}\alpha R_m/\lambda \\[1ex] 0 & 0 & \dot{\lambda}/\lambda \end{bmatrix} .$$

Hence the stretch and spin rates are

$$\mathbf{D} = \begin{bmatrix} \dot{\alpha}/\alpha & 0 & 0 \\[1ex] 0 & \dot{\alpha}/\alpha & \dot{\omega}\alpha R_m/(2\lambda) \\[1ex] 0 & \dot{\omega}\alpha R_m/(2\lambda) & \dot{\lambda}/\lambda \end{bmatrix} ,$$

$$\mathbf{\Omega} = \begin{bmatrix} 0 & -\dot{\omega} Z & 0 \\[1ex] \dot{\omega} Z & 0 & \dot{\omega}\alpha R_m/(2\lambda) \\[1ex] 0 & -\dot{\omega}\alpha R_m/(2\lambda) & 0 \end{bmatrix} .$$

One can obtain $\mathbf{L}$ directly by expressing $\dot{\mathbf{x}}$ in terms of the base vectors of the current configuration

$$\dot{\mathbf{x}} = \frac{r\dot{\alpha}}{\alpha} \mathbf{e}_r + \frac{rz\dot{\omega}}{\lambda} \mathbf{e}_\theta + \frac{z\dot{\lambda}}{\lambda} \mathbf{e}_z ,$$

and taking the spatial derivatives with respect to the deformed coordinates, i.e., $\mathbf{L} = \frac{\partial \dot{\mathbf{x}}}{\partial \mathbf{x}}$.

The Cauchy, first Piola–Kirchhoff and second Piola-Kirchhoff stresses are given by:

$$\boldsymbol{\tau} = \begin{bmatrix} 0 & 0 & 0 \\ 0 & 0 & \tau \\ 0 & \tau & \sigma \end{bmatrix} ,$$

$$\mathbf{T} = J\mathbf{F}^{-1} \cdot \tau = \begin{bmatrix} 0 & 0 & 0 \\ 0 & -\tau\alpha^2\omega R_m & \tau\alpha\lambda - \sigma\alpha^2\omega R_m \\ 0 & \tau\alpha^2 & \sigma\alpha^2 \end{bmatrix},$$

$$\mathbf{S} = \mathbf{T} \cdot (\mathbf{F}^{-1})^T = \begin{bmatrix} 0 & 0 & 0 \\ 0 & -2\tau\alpha\omega R_m + \dfrac{\sigma\alpha^2\omega^2 R_m^2}{\lambda} & \tau\alpha - \dfrac{\sigma\alpha^2\omega R_m}{\lambda} \\ 0 & \tau\alpha - \dfrac{\sigma\alpha^2\omega R_m}{\lambda} & \sigma\dfrac{\alpha^2}{\lambda} \end{bmatrix}.$$

Here τ is deduced from the applied axial and torsion loads. A constitutive law is needed to determine the relation between τ, σ and λ, ω.

Problem 16.4. Show that $\mathbf{E} = \mathbf{F}^T \cdot \mathbf{e} \cdot \mathbf{F}$ where $\mathbf{E}$ is the Green strain and $\mathbf{e}$ is the Almansi strain.

Problem 16.5. In a simple shear experiment, the motion is described by

$$x = X + \gamma Y, \qquad y = Y, \qquad z = Z.$$

Find $\mathbf{F}$, $\mathbf{C}$, $\mathbf{B}$, $\mathbf{R}$, $\mathbf{U}$, $\mathbf{V}$, the principal stretches and the eigenvectors of $\mathbf{C}$ and $\mathbf{B}$.

16.9. OBJECTIVITY

Truesdell and Noll (1965) discussed the concept of *frame indifference* of world events and the significance of the concept in the formulation of mechanics. Eringen (1967) introduced the term *axiom of objectivity*. Simply stated, the constitutive equation must be invariant under changes of reference frame. A quantity or an equation is *frame indifference* or *objective* if it is invariant under changes of reference frame. We give below the mathematical definition of the term and examine what quantities are objective and what are not. To appreciate the beauty and significance of these concepts, one must read the original papers.

Consider two motions $\mathbf{x}(\mathbf{X}, t)$ and $\bar{\mathbf{x}}(\mathbf{X}, t)$ of a body that differ by a rigid-body motion, that is,

(1) $$\bar{\mathbf{x}}(\mathbf{X}, t) = \mathbf{Q}(t) \cdot \mathbf{x}(\mathbf{X}, t) + \mathbf{c}(t),$$

where $\mathbf{Q}(t)$ is a rotational tensor and $\mathbf{c}(t)$ is a vector representing a rigid body translation. All quantities associated with the two deformations will have different values in general. In the subsequent discussion, we shall use the notation $\overline{(\cdot)}$ and $(\cdot)$ to denote the same relevant quantity in the barred ($\bar{\mathbf{x}}$) and the unbarred ($\mathbf{x}$) configurations. For instance, $\bar{\mathbf{F}}$ and $\mathbf{F}$ will denote the deformation gradients in the $\bar{\mathbf{x}}$ and $\mathbf{x}$ configurations, respectively. Truesdell and Noll, and Eringen defined that a vector or a tensor of rank 2 is *frame indifferent* or *objective* if its corresponding quantities in the two configurations are related by the following transformation rules:

(2) ▲
$$\bar{\mathbf{v}} = \mathbf{Q} \cdot \mathbf{v} ,$$

if $\mathbf{v}$ is a vector, and

(3) ▲
$$\bar{\mathbf{G}} = \mathbf{Q} \cdot \mathbf{G} \cdot \mathbf{Q}^T ,$$

if $\mathbf{G}$ is a tensor of rank 2.

We shall now examine the objectivity of various quantities associated with the two deformations. First consider a line element $d\bar{\mathbf{x}}$ in the barred configuration and its corresponding line element $d\mathbf{x}$ in the unbarred configuration. From Eq. (1), one finds

(4)
$$d\bar{\mathbf{x}} = \mathbf{Q} \cdot d\mathbf{x} ,$$

which satisfies the transformation rule Eq. (2) for a vector. Hence the line element $d\mathbf{x}$ *is an objective vector*. One can also show that all area elements and all volume elements are objective. In fact a volume element is always objective because it is a scalar, which is always so.

Now consider the deformation gradients $\bar{\mathbf{F}}$, $\mathbf{F}$ in the two configurations. Since

$$\bar{\mathbf{F}} \cdot d\mathbf{X} = d\bar{\mathbf{x}} = \mathbf{Q} \cdot d\mathbf{x} = \mathbf{Q} \cdot \mathbf{F} \cdot d\mathbf{X} ,$$

it implies

(5)
$$\bar{\mathbf{F}} = \mathbf{Q} \cdot \mathbf{F} .$$

This shows that $\mathbf{F}$ *is not objective* because $\mathbf{F}$, being a tensor of rank 2, does not follow the transformation rule Eq. (3) for tensor of rank 2. Note that $\mathbf{F}$ is a two-point tensor. The components associated with the deformed coordinates change with the reference frame while those associated with the undeformed coordinates do not. This is because the undeformed coordinates are fixed whereas the reference frame for the deformed coordinates changes. Following Eq. (5), one finds

(6)
$$\bar{\mathbf{C}} = \bar{\mathbf{F}}^T \cdot \bar{\mathbf{F}} = \bar{\mathbf{F}}^T \cdot \mathbf{Q} \cdot \mathbf{Q}^T \cdot \bar{\mathbf{F}} = \mathbf{F}^T \cdot \mathbf{F} = \mathbf{C} ,$$

which indicates that $\mathbf{C}$ is also *not objective* because the transformation does not follow the rule given in Eq. (3). We should expect $\mathbf{C}$ not to change as shown in Eq. (6) for different reference deformed coordinates because $\mathbf{C}$ is a Lagrangian tensor referred to the original frame, which does not depend on the current reference frame. In the mean time,

$$\bar{\mathbf{B}} = \bar{\mathbf{F}} \cdot \bar{\mathbf{F}}^T = \mathbf{Q} \cdot \mathbf{B} \cdot \mathbf{Q}^T \,,$$

which transforms like Eq. (3). Therefore, $\mathbf{B}$ *is an objective tensor*. This is also expected because $\mathbf{B}$ is an Eulerian tensor referred to the reference frame. Similarly, the *Lagrangian strain tensor* $\mathbf{E}$ *is not objective*, because

$$\bar{\mathbf{E}} = \mathbf{E} \,,$$

which does not satisfy Eq. (3). On the other hand the Almansi strain tensor e, an Eulerian tensor, transforms as

$$\bar{\mathbf{e}} = \mathbf{Q} \cdot \mathbf{e} \cdot \mathbf{Q}^T \,,$$

and is, therefore, objective. Using the polar decomposition and Eq. (5), one derives

$$\bar{\mathbf{F}} = \bar{\mathbf{R}} \cdot \bar{\mathbf{U}} = \mathbf{Q} \cdot \mathbf{F} = \mathbf{Q} \cdot \mathbf{R} \cdot \mathbf{U} \,.$$

Since $\mathbf{Q} \cdot \mathbf{R}$ is still a rotational tensor, and $\bar{\mathbf{U}}$ and $\mathbf{U}$ are symmetric, and since the polar decomposition is unique, the equation above implies

(7) $$\bar{\mathbf{R}} = \mathbf{Q} \cdot \mathbf{R} \,,$$

(8) $$\bar{\mathbf{U}} = \mathbf{U} \,,$$

which shows that neither $\mathbf{R}$ nor $\mathbf{U}$ is objective. However, for the left polar decomposition,

$$\bar{\mathbf{F}} = \bar{\mathbf{V}} \cdot \bar{\mathbf{R}} = \mathbf{Q} \cdot \mathbf{V} \cdot \mathbf{R} \,,$$

one obtains

(9) $$\bar{\mathbf{V}} = \mathbf{Q} \cdot \mathbf{V} \cdot \mathbf{R} \cdot \bar{\mathbf{R}}^T = \mathbf{Q} \cdot \mathbf{V} \cdot \mathbf{Q}^T \,,$$

which means the left stretch tensor $\mathbf{V}$ is objective. Thus $\mathbf{B}$, $\mathbf{e}$, and $\mathbf{V}$, which are all Eulerian tensors, are objective while $\mathbf{C}$, $\mathbf{E}$, and $\mathbf{U}$, all Lagrangian, are not.

Vectors and tensors related to the velocity field are in general not objective. To show this result, we first establish the properties of the rotation rate tensor. Let us define $\mathbf{\Psi} \equiv \mathbf{Q}^T \cdot \dot{\mathbf{Q}}$. Since $\mathbf{Q}^T \cdot \mathbf{Q} = \mathbf{I}$, it follows that

$$\frac{D}{Dt} (\mathbf{Q}^T \cdot \mathbf{Q}) = \dot{\mathbf{Q}}^T \cdot \mathbf{Q} + \mathbf{Q}^T \cdot \dot{\mathbf{Q}} = \mathbf{\Psi}^T + \mathbf{\Psi} = 0 \,,$$

which indicates that $\mathbf{\Psi}(= -\mathbf{\Psi}^T)$ is antisymmetric and that

$$\dot{\mathbf{Q}} = -\mathbf{Q} \cdot \dot{\mathbf{Q}}^T \cdot \mathbf{Q} \,.$$

For an objective vector $\mathbf{v}$, i.e., $\bar{\mathbf{v}} = \mathbf{Q} \cdot \mathbf{v}$, the following is true

$$\dot{\bar{\mathbf{v}}} = (\mathbf{Q} \cdot \mathbf{v})^{\displaystyle\cdot} = \mathbf{Q} \cdot (\dot{\mathbf{v}} + \mathbf{\Psi} \cdot \mathbf{v}), \tag{10}$$

which means that $\dot{\mathbf{v}}$ is in general not objective. Similarly, for an objective tensor $\mathbf{G}$ of rank 2, i.e., $\bar{\mathbf{G}} = \mathbf{Q} \cdot \mathbf{G} \cdot \mathbf{Q}^T$, one can show that

$$\dot{\bar{\mathbf{G}}} = \mathbf{Q} \cdot (\dot{\mathbf{G}} + \mathbf{\Psi} \cdot \mathbf{G} - \mathbf{G} \cdot \mathbf{\Psi}) \cdot \mathbf{Q}^T. \tag{11}$$

Now consider the deformation rate. It can be shown that

$$\dot{\bar{\mathbf{F}}} = \mathbf{Q} \cdot (\dot{\mathbf{F}} + \mathbf{\Psi} \cdot \mathbf{F}) = \bar{\mathbf{L}} \cdot \bar{\mathbf{F}},$$

which leads to

$$\bar{\mathbf{L}} = \mathbf{Q} \cdot (\dot{\mathbf{F}} + \mathbf{\Psi} \cdot \mathbf{F}) \cdot \mathbf{F}^{-1} \cdot \mathbf{Q}^T = \mathbf{Q} \cdot (\mathbf{L} + \mathbf{\Psi}) \cdot \mathbf{Q}^T = \mathbf{Q} \cdot (\mathbf{D} + \mathbf{\Omega} + \mathbf{\Psi}) \cdot \mathbf{Q}^T, \tag{12}$$

where $\mathbf{L} = \mathbf{D} + \mathbf{\Omega}$, and $\mathbf{D}$ and $\mathbf{\Omega}$ are the deformation and spin rate tensors. Splitting $\bar{\mathbf{L}}$ into the symmetric part $\bar{\mathbf{D}}$ and the antisymmetric parts $\bar{\mathbf{\Omega}}$, i.e., $\bar{\mathbf{L}} \equiv \bar{\mathbf{D}} + \bar{\mathbf{\Omega}}$, we obtain from Eq. (12)

$$\bar{\mathbf{D}} = \mathbf{Q} \cdot \mathbf{D} \cdot \mathbf{Q}^T, \tag{13}$$

$$\bar{\mathbf{\Omega}} = \mathbf{Q} \cdot (\mathbf{\Omega} + \mathbf{\Psi}) \cdot \mathbf{Q}^T = \mathbf{Q} \cdot \mathbf{\Omega} \cdot \mathbf{Q}^T + \dot{\mathbf{Q}} \cdot \mathbf{Q}^T. \tag{14}$$

One sees that only $\mathbf{D}$ satisfies the transformation law of objectivity while both the velocity gradient tensor $\mathbf{L}$ and the spin tensor $\mathbf{\Omega}$ do not.

We shall now examine the objectivity of stress measures. The objectivity of the surface traction $\overset{\nu}{\mathbf{T}}$ is based on physical grounds. Then

$$\bar{\boldsymbol{\tau}} \cdot \bar{\mathbf{n}} = \overset{\nu}{\bar{\mathbf{T}}} = \mathbf{Q} \cdot \overset{\nu}{\mathbf{T}} = \mathbf{Q} \cdot \boldsymbol{\tau} \cdot \mathbf{Q}^T \cdot \mathbf{Q} \cdot \mathbf{n} = \mathbf{Q} \cdot \boldsymbol{\tau} \cdot \mathbf{Q}^T \cdot \bar{\mathbf{n}},$$

where $\mathbf{n}$ is the unit normal to the element. Therefore

$$\bar{\boldsymbol{\tau}} = \mathbf{Q} \cdot \boldsymbol{\tau} \cdot \mathbf{Q}^T, \tag{15}$$

which indicates that the Cauchy stress is objective. This is expected because, similar to the case of strain measures, the Cauchy stress is an Eulerian tensor and, therefore, is objective. For the Lagrangian stress $\mathbf{T}$ and the second Piola–Kirchhoff stress $\mathbf{S}$, it is easy to show that

$$\bar{\mathbf{T}} = \mathbf{T} \cdot \mathbf{Q}^T, \qquad \bar{\mathbf{S}} = \mathbf{S}, \qquad \bar{\mathbf{r}}^* = \mathbf{r}^* \cdot \mathbf{Q}^T, \qquad \bar{\sigma}_r = \sigma_r, \tag{16}$$

all of which are not objective. This can be expected because $\mathbf{T}$ and $\mathbf{r}^*$ are two-point tensor and $\mathbf{S}$ and σ_r are Lagrangian tensor.

Finally, we shall discuss the transformation of stress rates. Stress rates are especially needed in plasticity theory where the constitutive equations are often expressed in terms of stress increments (rates). They are also needed in the incremental approach for solving large deformation problems. From Eq. (16.7:10), using Eqs. (16.5:3a) and (16.6:6) for the rates of deformation, one find the rates of various stress measures:

(17) $\quad \dot{\mathbf{T}} = J\mathbf{F}^{-1} \cdot [\dot{\boldsymbol{\tau}} + \mathrm{tr}(\mathbf{D})\boldsymbol{\tau} - \mathbf{L} \cdot \boldsymbol{\tau}] \,,$

(18) $\quad \dot{\mathbf{S}} = J\mathbf{F}^{-1} \cdot [\dot{\boldsymbol{\tau}} + \mathrm{tr}(\mathbf{D})\boldsymbol{\tau} - \mathbf{L} \cdot \boldsymbol{\tau} - \boldsymbol{\tau} \cdot \mathbf{L}^T] \cdot (\mathbf{F}^{-1})^T \,,$

(19) $\quad \dot{\mathbf{r}}^* = J\mathbf{F}^{-1} \cdot [\dot{\boldsymbol{\tau}} + \mathrm{tr}(\mathbf{D})\boldsymbol{\tau} - \mathbf{L} \cdot \boldsymbol{\tau} + \boldsymbol{\tau} \cdot \boldsymbol{\Omega}] \cdot \mathbf{R} \,,$

(20) $\quad \dot{\boldsymbol{\sigma}}_r = \mathbf{R}^T \cdot [\dot{\boldsymbol{\tau}} - \boldsymbol{\Omega} \cdot \boldsymbol{\tau} + \boldsymbol{\tau} \cdot \boldsymbol{\Omega}\,\mathrm{tr}(\mathbf{D})\boldsymbol{\tau}] \cdot \mathbf{R} \,.$

Using Eq. (14) that $\dot{\mathbf{Q}} = \bar{\boldsymbol{\Omega}} \cdot \mathbf{Q} - \mathbf{Q} \cdot \boldsymbol{\Omega}$ and differentiating Eqs. (15) and (16), one can establish that

(21) $\quad \dot{\bar{\boldsymbol{\tau}}} - \bar{\boldsymbol{\Omega}} \cdot \bar{\boldsymbol{\tau}} + \bar{\boldsymbol{\tau}} \cdot \bar{\boldsymbol{\Omega}} = \mathbf{Q} \cdot (\dot{\boldsymbol{\tau}} - \boldsymbol{\Omega} \cdot \boldsymbol{\tau} + \boldsymbol{\tau} \cdot \boldsymbol{\Omega}) \cdot \mathbf{Q}^T \,,$

(22) $\quad \dot{\bar{\mathbf{T}}} + \bar{\mathbf{T}} \cdot \bar{\boldsymbol{\Omega}} = (\dot{\mathbf{T}} + \mathbf{T} \cdot \boldsymbol{\Omega}) \cdot \mathbf{Q}^T \,,$

(23) $\quad \dot{\bar{\mathbf{S}}} = \dot{\mathbf{S}} \,,$

(24) $\quad \dot{\bar{\mathbf{r}}}^* + \bar{\mathbf{r}}^* \cdot \bar{\boldsymbol{\Omega}} = (\dot{\mathbf{r}}^* + \mathbf{r}^* \cdot \boldsymbol{\Omega}) \cdot \mathbf{Q}^T \,,$

(25) $\quad \dot{\bar{\boldsymbol{\sigma}}}_r = \dot{\boldsymbol{\sigma}}_r \,.$

None of the material rates $\dot{\boldsymbol{\tau}}$, $\dot{\mathbf{T}}$, $\dot{\mathbf{S}}$, $\dot{\mathbf{r}}^*$ and $\dot{\boldsymbol{\sigma}}_r$ is objective. However, from Eq. (21) we see that

(26) $\qquad\qquad\qquad \boldsymbol{\tau}^\partial = \dot{\boldsymbol{\tau}} - \boldsymbol{\Omega} \cdot \boldsymbol{\tau} + \boldsymbol{\tau} \cdot \boldsymbol{\Omega}$

is objective and called the *Jaumann (corotational) stress rate*.

In general, for an objective tensor $\mathbf{G}$, rates such as

$$\mathbf{G}^\partial = \dot{\mathbf{G}} - \boldsymbol{\Omega} \cdot \mathbf{G} + \mathbf{G} \cdot \boldsymbol{\Omega} \,, \qquad \text{(Jaumann rate)}$$

$$\mathbf{G}^\nabla = \dot{\mathbf{G}} - \mathbf{L} \cdot \mathbf{G} - \mathbf{G} \cdot \mathbf{L}^T \,, \qquad \text{(Oldroyd rate)}$$

$$\mathbf{G}^\circ = \dot{\mathbf{G}} + \mathbf{L}^T \cdot \mathbf{G} + \mathbf{G} \cdot \mathbf{L} \qquad \text{(Cotter–Rivilin rate)}$$

are all objective. They are respectively, called the *Jaumann, Oldroyd*, and *Cotter-Rivlin rates*. It can also be shown that the rate $\dot{\mathbf{G}} + \mathrm{tr}(\mathbf{D})\mathbf{G} + \mathbf{L}^T \cdot \mathbf{G} + \mathbf{G} \cdot \mathbf{L}$ introduced by Truesdell is objective if $\mathbf{G}$ is so.

To conclude this section, we mention two useful properties of rates: if $\mathbf{G}$ is a symmetric tensor, then

$$\frac{D}{Dt}(\mathbf{G} : \mathbf{G}) = 2\mathbf{G} : \mathbf{G}^\partial \,;$$

and if $\mathbf{G}$ is a deviatoric tensor, so is $\mathbf{G}^\partial$.

Problem 16.6. Let $\mathbf{G}^*$ be a Lagrangian tensor and $\mathbf{G}$ an Eulerain tensor. If $\mathbf{G}$ is related to $\mathbf{G}^*$ by $\mathbf{G} = \mathbf{F}^T \cdot \mathbf{G}^* \cdot \mathbf{F}^{-1}$, the Cotter-Rivlin rate of $\mathbf{G}$ is linearly related to $\dot{\mathbf{G}}^*$; and if $\mathbf{G} = \mathbf{F} \cdot \mathbf{G}^* \cdot \mathbf{F}^T / J$, then the Truesdell rate is linearly related to $\dot{\mathbf{G}}^*$.

16.10. EQUATIONS OF MOTION

Consider the body of a continuum occupying a region v with a boundary surface ∂v in the deformed state. The corresponding region in the original (natural or zero-stress) state is V with a boundary surface ∂V. The body is subjected to external loads, which consist of the inertia force, the body force $\mathbf{b}$ (with components b_1, b_2, b_3) per unit mass acting on a volume element dv, and the surface traction $\overset{\nu}{\mathbf{T}}$ acting on a surface element $d\mathbf{a}$ whose unit outer normal is $\mathbf{n}$. All these variables are in the Eulerian description. The corresponding variables for the body force and surface traction are $\mathbf{b}_0$ in dV and $\overset{\nu}{\mathbf{T}}_0$ on $d\mathbf{A}$ with outward unit normal $\mathbf{N}$ in the Lagrangian description. The original density is ρ_0 and the corresponding density in the deformed state is ρ. The notations to be used are given in the accompanying table below.

	In Deformed Configuration	In Undeformed Configuration
Region	v	V
Boundary surface	∂v	∂V
Volume element	$dv = dx_1 dx_2 dx_3$	$dV = dX_1 dX_2 dX_3$
Surface element	$d\mathbf{a}$	$d\mathbf{A}$
Unit outer normal	$\mathbf{n}, (n_1, n_2, n_3)$	$\mathbf{N}, (N_1, N_2, N_3)$
Particle coordinates	$\mathbf{x}, (x_1, x_2, x_3)$	$\mathbf{X}, (X_1, X_2, X_3)$
Body force per unit mass	$\mathbf{b}, (b_1, b_2, b_3)$	$\mathbf{b}_0, (b_{0_1}, b_{0_2}, b_{0_3})$
Surface traction	$\overset{\nu}{\mathbf{T}}, (\overset{\nu}{T}_1, \overset{\nu}{T}_2, \overset{\nu}{T}_3)$	$\overset{\nu}{\mathbf{T}}_0, (\overset{\nu}{T}_{0_1}, \overset{\nu}{T}_{0_2}, \overset{\nu}{T}_{0_3})$
Density	ρ	ρ_0
Stresses	$\boldsymbol{\tau}, (\tau_{ij})$ (Cauchy)	$\mathbf{T}, (T_{Ji})$ (1st Piola-Kirchhoff)
	$\boldsymbol{\sigma}, (\sigma_{ij})$ (Kirchhoff)	$\mathbf{S}, (S_{IJ})$ (2nd Piola-Kirchhoff)

Note that the traction $\overset{\nu}{\mathbf{T}}$ is a vector (tensor of rank 1) and the first Piola-Kirchhoff stress $\mathbf{T}$ is a tensor of rank 2.

Consider the motion of the body. Conservation of mass requires that $\rho \, dv = \rho_0 \, dV$ and $\mathbf{b} = \mathbf{b}_0$, i.e., the mass of a volume element dV and the body force per unit mass do not change as the body deforms. The conservation of momentum gives the Eulerian equation of motion of the continuum

$$(1) \qquad \nabla_{\mathbf{x}} \cdot \boldsymbol{\tau} + \rho \mathbf{b} = \rho \frac{D\mathbf{v}}{Dt}$$

[Eq. (5.5:7) in the present notation]. This is the equation of motion in the current configuration. We can write the equation of motion in a different reference frame and express the variables in terms of coordinates in that configuration. One way of doing this is to write the equation of motion in the integral form as follows: the sum of the total body force and inertia force acting on the region $v(V)$ is

$$(2) \qquad \int_v \left(\mathbf{b} - \frac{D\mathbf{v}}{Dt} \right) \rho dv = \int_V \left(\mathbf{b}_0 - \frac{D\mathbf{v}}{Dt} \right) \rho_0\, dV, \qquad \text{or}$$

$$\int_v \left(b_i - \frac{Dv_i}{Dt} \right) \rho\, dv = \int_V \left(b_{0i} - \frac{Dv_i}{Dt} \right) \rho_0\, dV.$$

The resultant of the surface traction $\overset{\nu}{\mathbf{T}}$ acting on the surface ∂v (or ∂V) is

$$(3) \qquad \int_{\partial v} \overset{\nu}{\mathbf{T}} da = \int_{\partial v} \mathbf{n} \cdot \boldsymbol{\tau}\, da = \int_{\partial V} \mathbf{N} \cdot \mathbf{T}\, dA, \qquad \text{or}$$

$$\int_{\partial v} \overset{\nu}{T}_i\, da = \int_{\partial v} \tau_{ji} n_j\, da = \int_{\partial V} N_I T_{Ii}\, dA,$$

in which Eq. (16.2:12) has been used and $\mathbf{T}$ is the *Lagrangian* or the *first Piola–Kichhoff stress tensor*. The use of Gauss' theorem leads to

$$(4) \qquad \int_{\partial v} \overset{\nu}{\mathbf{T}} da = \int_V \nabla \cdot \mathbf{T}\, dV, \qquad \text{or} \qquad \int_{\partial v} \overset{\nu}{T}_i\, da = \int_V \frac{\partial T_{Ii}}{\partial X_I}\, dV,$$

where ∇ is the gradient operator with respect to the $\mathbf{X}$-coordinates. Hence, by Eqs. (2) and (4), the equation of motion can be written as

$$(5) \qquad \int_V \left(\rho_0 \mathbf{b}_0 - \rho_0 \frac{D\mathbf{v}}{Dt} + \nabla \cdot \mathbf{T} \right) dV = 0.$$

Since this equation must be valid for an arbitrary region V, the integrand must vanish. Hence

$$(6) \ \blacktriangle \qquad \nabla \cdot \mathbf{T} + \rho_0 \mathbf{b}_0 = \rho_0 \frac{D\mathbf{v}}{Dt}, \qquad T_{Ki,K} + \rho_0 b_{0i} = \rho_0 \frac{Dv_i}{Dt}$$

is the equation of motion in terms of the *first Piola–Kirchhoff stress*. The divergent operation is with respect to the undeformed coordinates. A substitution of Eq. (16.7:6) into Eq. (6) yields the equation of motion in terms of the *second Piola–Kirchhoff stress*:

(7) ▲
$$\boldsymbol{\nabla} \cdot (\mathbf{S} \cdot \mathbf{F}^T) + \rho_0 \mathbf{b}_0 = \rho_0 \frac{D\mathbf{v}}{Dt},$$
$$\frac{\partial}{\partial X_J} \left(S_{JK} \frac{\partial x_i}{\partial X_K} \right) + \rho_0 b_{0_i} = \rho_0 \frac{Dv_i}{Dt}.$$

Since $\mathbf{x} = \mathbf{X} + \mathbf{u}$, where $\mathbf{u}$ is the displacement vector, Eq. (7) can be written as

(8) ▲
$$\boldsymbol{\nabla} \cdot \left\{ \mathbf{S} \cdot \left[\mathbf{I} + \left(\frac{\partial \mathbf{u}}{\partial \mathbf{X}} \right)^T \right] \right\} + \rho_0 \mathbf{b}_0 = \rho_0 \frac{D\mathbf{v}}{Dt},$$

or in indicial notation

(8a) ▲
$$\frac{\partial}{\partial X_K} \left[S_{KJ} \left(\delta_{IJ} + \frac{\partial u_I}{\partial X_J} \right) \right] + \rho_0 b_{0_I} = \rho_0 \frac{Dv_I}{Dt},$$

if the deformed and undeformed configurations have the same base vectors. In this case, there is no distinction between the upper and lower case indices. Equation (8) exhibits the geometric effect of finite deformation through the displacement gradient $\partial u_I / \partial X_J$.

16.11. CONSTITUTIVE EQUATIONS OF THERMOELASTIC BODIES

Equations that describe the mechanical properties of materials are called the *constitutive equations*. Materials may be classified according to their constitutive equations. The constitutive equations do have to be restricted by the laws of thermodynamics. The case of small deformation is discussed in Chapters 12 and 13. Large deformation does make the mathematical theory more complex. In this section, the *thermo-mechanically-simple* case of an elastic material will be discussed. An *elastic material* is defined as one in which the Cauchy stress $\boldsymbol{\tau}$ depends only on the strain, the internal energy, the temperature and the entropy; it does not depend on the strain rate, stress rate, electromagnetism, chemical reactions, and biological processes. The material has no energy dissipation or energy creation mechanism and is subjected to reversible processes. There is an extensive literature on the subject (Jones and Treloar 1975, Ling *et al.* 1993, Ogden 1986, Valanis and Landel 1967, Wang *et al.* 1995). A review of hyperelasticity of rubber, elastomers and biological tissues is given by Beatty (1987).

The constitutive equations must be restricted by the laws of thermodynamics. The *first law of thermodynamics* states that, for a closed system,

the rate of work done on the systems by all external agencies must equal
to the rate of increase of the energy of the system. The rate of mechanical
work done on a volume v by the external forces is

$$(1) \qquad \dot{W}_m = \int_{\partial v} \overset{\nu}{\mathbf{T}} \cdot \mathbf{v} \, da + \int_v \rho \mathbf{b} \cdot \mathbf{v} \, dv \,,$$

where $\overset{\nu}{\mathbf{T}}$ is the surface traction and $\mathbf{b}$ is the body force per unit mass. The
use of the equation $\overset{\nu}{\mathbf{T}} = \mathbf{n} \cdot \boldsymbol{\tau}$, Newton's law and Gauss' theorem in Eq. (1)
leads to

$$(2) \qquad \dot{W}_m = \frac{D}{Dt} \int_v \frac{1}{2} \rho \mathbf{v} \cdot \mathbf{v} \, dv + \int_v \boldsymbol{\tau} : \mathbf{D} \, dv$$

$$= \frac{D}{Dt} \int_v \frac{1}{2} \rho v_i v_i dv + \int_v \tau_{ij} \dot{e}_{ij} \, dv \,,$$

where $\dot{e}_{ij}$ are the components of $\mathbf{D}$. The heat flow into the body is

$$Q = - \int_{\partial v} \mathbf{h} \cdot \mathbf{n} \, da + \int_v \rho q \, dv \,,$$

where $\mathbf{h}$ is the heat flux vector, $\mathbf{n}$ is a unit normal to ∂v, and q is the
heat source per unit mass. The total energy of the system consists of the
kinetic energy and the internal energy. The *first law of thermodynamics* or
principle of energy balance requires that the rate of increase of the total
energy of the system is equal to the sum of the mechanical work rate and
the heat flow. Hence

$$(3) \qquad \frac{D}{Dt} \int_v \frac{1}{2} \rho v_i v_i dv + \frac{D}{Dt} \int_v \rho \mathcal{E} \, dv = \dot{W}_m + Q \,,$$

where $\mathcal{E}$ is the internal energy per unit mass. Since the volume v is arbitrary,
Eq. (3) implies

$$(4) \qquad \rho \dot{\mathcal{E}} = \boldsymbol{\tau} : \mathbf{D} + \rho q - \nabla \cdot \mathbf{h} \,, \qquad \text{or} \qquad \rho \dot{\mathcal{E}} = \tau_{ij} \dot{e}_{ij} + \rho q - h_{i,i} \,.$$

The physical implication of reversibility of processes in an elastic body has
been discussed in Chapter 12. It is shown in Sec. 12.3 that in isotropic
process, the internal energy can serve as the strain energy function so that

$$(5) \qquad \boldsymbol{\tau} = \rho \frac{\partial \mathcal{E}}{\partial \mathbf{e}} \,, \qquad \text{or} \qquad \tau_{ij} = \rho \frac{\partial \mathcal{E}}{\partial e_{ij}} \,.$$

In isothermal processes, the *Helmholtz free energy per unit undeformed
volume* W_0 can serve as the *strain energy function*, so that

$$(6) \qquad \boldsymbol{\tau} = \frac{1}{J} \frac{\partial W_0}{\partial \mathbf{e}} \,, \qquad \text{or} \qquad \tau_{ij} = \frac{1}{J} \frac{\partial W_0}{\partial e_{ij}} \,.$$

Alternatively

(7)
$$\mathbf{S} = \frac{\partial W_0}{\partial \mathbf{E}}.$$

Hence, from Eq. (16.7:7) the Cauchy stress can be expressed as

(8)
$$\boldsymbol{\tau} = \frac{1}{J}\mathbf{F} \cdot \frac{\partial W_0}{\partial \mathbf{E}} \cdot \mathbf{F}^T.$$

If the material is *isotropic*, W_0 must be a function of invariant of $\mathbf{E}$ or $\mathbf{C}(=\mathbf{I}+2\mathbf{E})$. A set of such invariant is

$$I_1 = \mathrm{tr}(\mathbf{C}) = 3 + 2E_{KK}, \quad I_2 = (I_1^2 - \mathbf{C}:\mathbf{C})/2, \quad I_3 = \det(\mathbf{C}) = J^2.$$

One can then write W_0 in the form $W_0(T, I_1, I_2, I_3)$. It can be shown that

$$\frac{\partial I_1}{\partial \mathbf{C}} = \mathbf{I}, \quad \frac{\partial I_2}{\partial \mathbf{C}} = I_1\mathbf{I} - \mathbf{C}, \quad \frac{\partial I_3}{\partial \mathbf{C}} = I_3\mathbf{C}^{-1}.$$

Then the *constitutive law* becomes

(9) ▲
$$\mathbf{S} = 2\frac{\partial W_0}{\partial \mathbf{C}} = C_1\mathbf{I} + C_2(I_1\mathbf{I} - \mathbf{C}) + C_3 I_3 \mathbf{C}^{-1},$$

where

(10)
$$C_1 = 2\frac{\partial W_0}{\partial I_1}, \quad C_2 = 2\frac{\partial W_0}{\partial I_2}, \quad C_3 = 2\frac{\partial W_0}{\partial I_3},$$

Since $\mathbf{S} = 0$ for zero deformation, we require that

(11) $C_1 + 2C_2 + C_3 = 0$ or $\dfrac{\partial W_0}{\partial I_1} + 2\dfrac{\partial W_0}{\partial I_2} + \dfrac{\partial W_0}{\partial I_3} = 0$

as $I_1 = 3$, $I_2 = 3$, $I_3 = 1$ and $\mathbf{C} = \mathbf{I}$. From Eqs. (9) and (16.7:6), one obtains the constitutive law for the Cauchy stress

(12) ▲
$$\boldsymbol{\tau} = [C_1\mathbf{B} + C_2(I_1\mathbf{B} - \mathbf{B}^2) + C_3 I_3 \mathbf{I}]/J.$$

From Eq. (16.7:11), $p = -(\boldsymbol{\tau}:\mathbf{I})/3 = -(\mathbf{S}:\mathbf{C})/3J$, where p is the hydrostatic pressure, it follows

(13)
$$p = -\frac{2}{3J}\left[\frac{\partial W_0}{\partial I_1}I_1 + 2\frac{\partial W_0}{\partial I_2}I_2 + 3\frac{\partial W_0}{\partial I_3}I_3\right]$$

$$= -\frac{1}{3J}[C_1 I_1 + 2C_2 I_2 + 3C_3 I_3].$$

One can split as follows the stresses into a distortional part, which involves the hydrostatic pressure p, and a deviatoric part, which has a zero mean as defined in Eq. (16.7:13):

$$(14) \quad \mathbf{S} = -pJ\mathbf{C}^{-1} + C_1\left(\mathbf{I} - \frac{1}{3}I_1\mathbf{C}^{-1}\right) + C_2\left(I_1\mathbf{I} - \mathbf{C} - \frac{2}{3}I_2\mathbf{C}^{-1}\right),$$

$$(15) \quad \tau = -p\mathbf{I} + C_1\left(\mathbf{B} - \frac{1}{3}I_1\mathbf{I}\right)\frac{1}{J} + C_2\left(I_1\mathbf{B} - \mathbf{B}^2 - \frac{2}{3}I_2\mathbf{I}\right)\frac{1}{J}.$$

A special form of the energy density is

$$(16a) \quad W_0 = [C_1(I_1 - 3) + C_2(I_2 - 3) + b_3(I_3 - 1) + b_4(I_3 - 1)^2]/2,$$

with C_1, C_2, b_3, b_4 being constant and $C_3 = b_3 + 2b_4(I_3 - 1)$. Equation (16a) represents an expansion of W_0 in a power series of $I_1 - 3$, $I_2 - 3$, and $I_3 - 1$. If the last two terms of Eq. (16a) are dropped, it reduces to the *Mooney-Rivlin* model for certain incompressible rubber-like materials.

For infinitesimal deformation, we can approximate

$$I_2 = 3 + 4E_{KK}, \quad I_3 = 1 + 2E_{KK}, \quad J = 1 + E_{KK}, \quad \mathbf{C}^{-1} = \mathbf{I} - 2\mathbf{E}.$$

where E's are strains. The energy density becomes

$$W_0 = (C_1 + 2C_2 + b_3)E_{KK} - (b_3 + C_2)E_{IJ}E_{IJ} + (b_3 + 2b_4 + C_2)(E_{KK})^2.$$

To assure that W_0 is consistent with Hooke's law, we require

$$2C_1 + 4C_2 + 2b_3 = 0, \quad -b_3 - C_2 = G, \quad 2b_3 + 4b_4 + 2C_2 = \lambda = \frac{2G\nu}{1 - 2\nu},$$

i.e.,

$$C_2 = G - C_1, \quad b_3 = C_1 - 2G, \quad b_4 = \frac{(1 - \nu)G}{2(1 - 2\nu)},$$

where G is the shear modulus, λ is the Lamé constant and ν is Poisson's ratio. From Eqs. (9) and (12) or Eqs. (14) and (15), we obtain Hooke's law

$$\tau = \mathbf{S} = 2G\mathbf{E} + \lambda E_{KK}\mathbf{I} = 2G\left(\mathbf{E} - \frac{1}{3}E_{KK}\mathbf{I}\right) - p\mathbf{I}.$$

Another commonly used energy density is

$$(16b) \quad W_0 = [C_1(I_1 - 3) + C_2(I_2 - 3) + b_3(1/I_3^2 - 1) + b_4(I_3 - 1)^2]/2.$$

For infinitesimal deformation, the energy density becomes

$$W_0 = (C_1 + 2C_2 - 2b_3)E_{KK} + (2b_3 - C_2)E_{IJ}E_{IJ} + (4b_3 + 2b_4 + C_2)(E_{KK})^2.$$

with

$$2C_1 + 4C_2 - 4b_3 = 0, \qquad 2b_3 - C_2 = G, \qquad 8b_3 + 4b_4 + 2C_2 = \lambda = \frac{2G\nu}{1 - 2\nu},$$

i.e.,

$$C_1 + C_2 = G, \qquad b_3 = \frac{1}{2}C_1 + C_2, \qquad b_4 = \frac{(5\nu - 2)C_1 + (11\nu - 5)C_2}{2(1 - 2\nu)}.$$

For *incompressible materials*, $J = 1$, Eqs. (7) and (8) no longer hold and so are Eqs. (9) and (12). Since the rate of volume change is

$$\frac{DJ}{Dt} = \frac{DJ}{D\mathbf{E}} : \dot{\mathbf{E}} = J\mathbf{C}^{-1} : \dot{\mathbf{E}},$$

the *incompressibility condition* requires

(17) ▲ $$\mathbf{C}^{-1} : \dot{\mathbf{E}} = 0,$$

which represents an internal constraint to **E**. We can modify Eq. (7) to reflect the constraint of Eq. (17) and obtain the constitutive equation

(18) ▲ $$\mathbf{S} = -p'\mathbf{C}^{-1} + \frac{\partial W_0}{\partial \mathbf{E}},$$

where p' is a Lagrangian multiplier relating to the hydrostatic pressure. The expression Eq. (18) is not unique for incompressible materials when $J = 1$ is taken into account. For example, if W_0 is replaced by $W_0 + f(\mathbf{C})(J - 1)$, the value of W_0 remains the same, but the derivative of W_0 with respect to **E** changes. The constitutive equation becomes

$$\mathbf{S} = -p'\mathbf{C}^{-1} + \frac{\partial W_0}{\partial \mathbf{E}} + f(\mathbf{C})J\mathbf{C}^{-1}.$$

However no ambiguity arises in the determination of the deviatoric stress as defined in Eq. (16.7:12)

(19) $$\mathbf{S}'' = \mathbf{S} - \frac{1}{3}(\mathbf{S} : \mathbf{C})\mathbf{C}^{-1} = \frac{\partial W_0}{\partial \mathbf{E}} - \frac{1}{3}\left(\frac{\partial W_0}{\partial \mathbf{E}} : \mathbf{C}\right)\mathbf{C}^{-1}.$$

The additional term in **S** gives no contribution to the deviatoric stress. We can then rewrite the constitutive equation in the form

(20) $$\mathbf{S} = -p\mathbf{C}^{-1} + \frac{\partial W_0}{\partial \mathbf{E}} - \frac{1}{3}\left(\frac{\partial W_0}{\partial \mathbf{E}} : \mathbf{C}\right)\mathbf{C}^{-1},$$

where

(21) $$p = p' - \frac{1}{3}\left(\frac{\partial W_0}{\partial \mathbf{E}} : \mathbf{C}\right)$$

is the hydrostatic pressure as defined in Eq. (16.7:11) and has to be determined directly from the field solution. Accordingly the constitutive equation for the Cauchy stress becomes

$$(22) \quad \blacktriangle \qquad \tau = -p\mathbf{I} + \frac{1}{J}\left[\mathbf{F}\cdot\frac{\partial W_0}{\partial \mathbf{E}}\cdot\mathbf{F}^T - \frac{1}{3}\left(\frac{\partial W_0}{\partial \mathbf{E}}:\mathbf{C}\right)\mathbf{I}\right].$$

If the material is isotropic and incompressible, one can write Eqs. (20) and (22) in the form of Eqs. (14) and (15). In other words, the constitutive laws Eqs. (14) and (15) are valid for incompressible materials, except that the hydrostatic pressure must be determined directly from the field equations instead of Eq. (13). For infinitesimal deformation, Eqs. (14) and (15) reduce to

$$\tau = \mathbf{S} = -p\mathbf{I} + 2G(\mathbf{E} - E_{KK}\mathbf{I}/3).$$

We can establish a formulation valid for compressible as well as incompressible materials. Following the division of the stresses into the deviatoric stress and hydrostatic pressure, we split the deformation into a dilatational part, representing the volume change of a differential element of the body, and a distortion of its shape at constant volume. The procedure is known as *kinematic split*, which involves multiplicative decomposition in the case of finite strain. This formulation is particularly useful for treating nearly incompressible materials. Let us define

$$(23) \qquad\qquad \mathbf{U}' = \mathbf{U}J^{-1/3},$$

then

$$\mathbf{U} = J^{1/3}\mathbf{U}' = \mathbf{U}'\cdot\mathbf{I}J^{1/3}, \qquad \det(\mathbf{U}') = \det(\mathbf{U})J^{-1} = 1.$$

We thus obtain the multiplicative decomposition with $\mathbf{U}'$ *representing the distortion of the right stretch tensor* $\mathbf{U}$, whereas $J^{1/3}\mathbf{I}$ is the *dilatation*. As we shall see later the decomposition of the deformation rate becomes additive and is in the same form as that in the infinitesimal strain case.

Accordingly, one can define

$$\mathbf{C}' = \mathbf{C}J^{-2/3},$$

and write the strain energy density in the form

$$(24) \qquad\qquad W_0^*(\mathbf{C}', J) = W_0(\mathbf{C})$$

such that

$$(25) \quad \blacktriangle \qquad \mathbf{S}' = \frac{\partial W_0^*(\mathbf{C}', J)}{\partial \mathbf{C}'}, \qquad p = -\frac{\partial W_0^*(\mathbf{C}', J)}{\partial J}.$$

The relationship between the second Piola-Kirchhoff stress $\mathbf{S}$ and $\mathbf{S}'$, and p can be derived as follows:

$$(26) \; \blacktriangle \qquad \mathbf{S} : \delta \mathbf{E} = \delta W_0 = \frac{\partial W_0^*(\mathbf{C}', J)}{\partial \mathbf{C}'} : \delta \mathbf{C}' + \frac{\partial W_0^*(\mathbf{U}', J)}{\partial J} \delta J$$

$$= \mathbf{S}' : \delta \mathbf{C}' - p \delta J \,.$$

From the property of determinant $[J = \sqrt{\det(\mathbf{C})}]$, it can be shown that

$$\delta J = \frac{\partial J}{\partial \mathbf{C}} : \delta \mathbf{C} = \frac{J}{2} \mathbf{C}^{-1} : \delta \mathbf{C} \,.$$

Using the result above, one derives

$$\delta \mathbf{C}' = J^{-2/3} \delta \mathbf{C} - \frac{2}{3} J^{-5/3} \mathbf{C} \delta J = J^{-2/3} \delta \mathbf{C} - \frac{1}{3} J^{-2/3} \mathbf{C}(\mathbf{C}^{-1} : \delta \mathbf{C}) \,.$$

A substitution of the last two equations into Eq. (26) leads to

$$(27) \; \blacktriangle \qquad \mathbf{S} = 2J^{-2/3} \mathbf{S}' - 2J^{-2/3}(\mathbf{S}' : \mathbf{C})\mathbf{C}^{-1}/3 - pJ\mathbf{C}^{-1} \,.$$

Taking the inner product of the equation above with $\mathbf{C}$, one can show that

$$(28) \qquad\qquad p = -\frac{1}{3J}(\mathbf{S} : \mathbf{C}) \,,$$

in which the result $\mathbf{C}^{-1} : \mathbf{C} = 3$ has been used. Equation (28) establishes that p as defined in Eq. (25) is indeed the hydrostatic pressure, the mean of the Cauchy stress. Note that $\mathbf{S}'$ is not deviatoric, since $\mathbf{S}' : \mathbf{C} \neq 0$.

Problem 16.7. Let I_1, I_2 and I_3 be the principal invariance of $\mathbf{C}$. Show that

$$I_2 = (I_1^2 - \mathbf{C} : \mathbf{C})/2 = I_3 \operatorname{tr}(\mathbf{C}^{-1}) \,.$$

For incompressible materials $I_3 = 1$, then $I_2 = \operatorname{tr}(\mathbf{C}^{-1})$. (*Hint*: Express the invariance in terms of the eigenvalues of $\mathbf{C}$.)

Problem 16.8. Show that

$$\mathbf{C} = I_1 \mathbf{I} - I_2 \mathbf{C}^{-1} + I_3 \mathbf{C}^{-2} \,,$$

without using the Cayley-Hamilton theorem. (Hint: Use the same approach as that for Problem 16.7 or differentiate the result of Problem 16.7 with respect to $\mathbf{C}$.)

Problem 16.9. Using $\mathbf{C} = I_1 \mathbf{I} - I_2 \mathbf{C}^{-1} + I_3 \mathbf{C}^{-2}$ and $I_2 = I_3 \operatorname{tr}(\mathbf{C}^{-1})$, show that $\mathbf{B}^2 = I_1 \mathbf{B} - I_2 \mathbf{I} + I_3 \mathbf{B}^{-1}$ and that

$$\mathbf{S} = -pJ\mathbf{C}^{-1} + C_1 \left(\mathbf{I} - \frac{1}{3} I_1 \mathbf{C}^{-1} \right) + C_2 \left(\frac{1}{3} I_2 \mathbf{C}^{-1} - J^2 \mathbf{C}^{-2} \right) \,,$$

$$\boldsymbol{\tau} = -p\mathbf{I} + C_1 \left(\mathbf{B} - \frac{1}{3} I_1 \mathbf{I} \right) \frac{1}{J} + C_2 \left(\frac{1}{3J} I_2 \mathbf{I} - J\mathbf{B}^{-1} \right) \,.$$

16.12. MORE EXAMPLES

Uniform Tension. We consider a unit cube with sides parallel to the Cartesian axes (X, Y, Z) and is deformed into a cuboid of dimensions $\lambda_1, \lambda_2, \lambda_3$ parallel to the X, Y, Z axes, respectively. The material is compressible and the energy density is in the form as Eq. (16.11:16a). The deformed coordinates are referred to the Cartesian axes x, y, z, such that

(1) $$x = \lambda_1 X, \quad y = \lambda_2 Y, \quad z = \lambda_3 Z.$$

The components of the deformation gradient $\mathbf{F}$, the metric tensors $\mathbf{C}$, $\mathbf{B}$ and the Green strain tensor $\mathbf{E}$ are

$$\mathbf{F} = \begin{bmatrix} \lambda_1 & 0 & 0 \\ 0 & \lambda_2 & 0 \\ 0 & 0 & \lambda_3 \end{bmatrix}, \quad \mathbf{C} = \mathbf{B} = \begin{bmatrix} \lambda_1^2 & 0 & 0 \\ 0 & \lambda_2^2 & 0 \\ 0 & 0 & \lambda_3^2 \end{bmatrix},$$

$$\mathbf{E} = \frac{1}{2} \begin{bmatrix} \lambda_1^2 - 1 & 0 & 0 \\ 0 & \lambda_2^2 - 1 & 0 \\ 0 & 0 & \lambda_3^2 - 1 \end{bmatrix}.$$

in which $\mathbf{B}$ is objective and $\mathbf{F}$, $\mathbf{C}$ and $\mathbf{E}$ are normally not. The invariants are

$$I_1 = \mathrm{tr}(\mathbf{C}) = \lambda_1^2 + \lambda_2^2 + \lambda_3^2,$$

$$I_2 = (I_1^2 - \mathbf{C} : \mathbf{C})/2 = \lambda_1^2 \lambda_2^2 + \lambda_2^2 \lambda_3^2 + \lambda_2^2 \lambda_1^2,$$

$$I_3 = \det(\mathbf{C}) = J^2 = \lambda_1^2 \lambda_2^2 \lambda_3^2.$$

The inverse of $\mathbf{C}$ is simply

$$\mathbf{C}^{-1} = \begin{bmatrix} 1/\lambda_1^2 & 0 & 0 \\ 0 & 1/\lambda_2^2 & 0 \\ 0 & 0 & 1/\lambda_3^2 \end{bmatrix}.$$

From Eqs. (16.11:9), (16.11:12), it follows

$$\mathbf{S} = C_1 \begin{bmatrix} 1 & 0 & 0 \\ 0 & 1 & 0 \\ 0 & 0 & 1 \end{bmatrix} + C_2 \begin{bmatrix} \lambda_2^2 + \lambda_3^2 & 0 & 0 \\ 0 & \lambda_3^2 + \lambda_1^2 & 0 \\ 0 & 0 & \lambda_1^2 + \lambda_2^2 \end{bmatrix}$$

$$+ C_3 \lambda_1^2 \lambda_2^2 \lambda_3^2 \begin{bmatrix} 1/\lambda_1^2 & 0 & 0 \\ 0 & 1/\lambda_2^2 & 0 \\ 0 & 0 & 1/\lambda_3^2 \end{bmatrix},$$

$$\tau = \frac{C_1}{J} \begin{bmatrix} \lambda_1^2 & 0 & 0 \\ 0 & \lambda_2^2 & 0 \\ 0 & 0 & \lambda_3^2 \end{bmatrix}$$

$$+ \frac{C_2}{J} \begin{bmatrix} (\lambda_2^2 + \lambda_3^2)\lambda_1^2 & 0 & 0 \\ 0 & (\lambda_3^2 + \lambda_1^2)\lambda_2^2 & 0 \\ 0 & 0 & (\lambda_1^2 + \lambda_2^2)\lambda_3^2 \end{bmatrix}$$

$$+ \frac{C_3}{J} \lambda_1^2\lambda_2^2\lambda_3^2 \begin{bmatrix} 1 & 0 & 0 \\ 0 & 1 & 0 \\ 0 & 0 & 1 \end{bmatrix},$$

where, from Eq. (16.11:16a)

$$C_3 = b_3 + 2b_4(\lambda_1^2\lambda_2^2\lambda_3^2 - 1).$$

Note that τ is objective and $\mathbf{S}$ is not. All stress components are constant and therefore the equilibrium equations are satisfied. Since the deformation involves uniform stretching only, all shear stresses are zero and the components of $\mathbf{S}$ are proportional to those of τ, i.e., $\tau_{ii} = S_{ii}\lambda_i^2/J$, $i = 1, 2, 3$ and are not summed. These results are also true locally for nonuniform deformation if the referenced coordinates coincide with the principal stress directions where λ_i is the stretch in the ith principal direction.

For the particular case of simple extension under a force parallel to the x-axis, $\lambda_2 = \lambda_3$, $\tau_{22} = \tau_{33} = 0$, hence

$$\lambda_1\tau_{22} = C_1 + C_2(\lambda_1^2 + \lambda_2^2) + [b_3 + 2b_4(\lambda_1^2\lambda_2^4 - 1)]\lambda_1^2\lambda_2^2 = 0.$$

One can solve for λ_2^2 in terms of λ_1^2. The above equation gives only one real solution for λ_2^2. Replacing λ_K^2 with $1 + 2E_{KK}$, $(K = 1, 2,$ not summed) and solving for E_{22}, one finds

$$\frac{E_{22}}{E_{11}} = -\nu(E_{11}),$$

in which $\nu(E_{11})$ is a complicated function of E_{11}. For infinitesimal deformation, it reduces to the well-known Poisson's ratio

$$\nu(E_{11}) = \frac{C_2 + b_3 + 2b_4}{C_2 + b_3 + 4b_4} = \frac{\lambda}{2(\lambda + G)}.$$

For incompressible materials, $I_3 = 1$, the Cauchy stress for the Mooney-Rivlin material becomes

$$\tau = -\left(p + \frac{C_1 I_1}{3} + \frac{2C_2 I_2}{3}\right)\mathbf{I} + C_1 \begin{bmatrix} \lambda_1^2 & 0 & 0 \\ 0 & \lambda_2^2 & 0 \\ 0 & 0 & \lambda_3^2 \end{bmatrix}$$

$$+ C_2 \begin{bmatrix} (\lambda_2^2 + \lambda_3^2)\lambda_1^2 & 0 & 0 \\ 0 & (\lambda_3^2 + \lambda_1^2)\lambda_2^2 & 0 \\ 0 & 0 & (\lambda_1^2 + \lambda_2^2)\lambda_3^2 \end{bmatrix}.$$

Problem 16.10. Find the Cauchy stress for the uniform extension problem with the strain energy density

$$W_0 = [b_1(I_1 - 3) + b_2(I_2 - 3) + b_3(J - 1) + b_4(J - 1)^2]/2,$$

where b's are constants. Show that (a) we must have

$$2b_1 + 4b_2 + b_3 = 0,$$

to assure that the stresses are zero for zero deformation; (b) for simple extension in the X-direction

$$\frac{E_{22}}{E_{11}} = -\nu(E_{11})$$

$$= -\frac{2(b_2 + b_4)E_{11} + (b_1 + 2b_2 + b_4)(1 - \sqrt{1 + 2E_{11}})}{4[b_2 + b_4(1 + 2E_{11})]E_{11}}.$$

Simple Shear. Consider the simple shear deformation

(2) $$x = X + \gamma Y, \qquad y = Y, \qquad z = Z,$$

where (x, y, z) and (X, Y, Z) are, respectively, the deformed and undeformed Cartesian coordinates. The matrices for deformation and deformation rates are

$$\mathbf{F} = \begin{bmatrix} 1 & \gamma & 0 \\ 0 & 1 & 0 \\ 0 & 0 & 1 \end{bmatrix}, \quad \mathbf{R} = \begin{bmatrix} \cos\phi & \sin\phi & 0 \\ -\sin\phi & \cos\phi & 0 \\ 0 & 0 & 1 \end{bmatrix},$$

$$\mathbf{U} = \begin{bmatrix} \cos\phi & \sin\phi & 0 \\ \sin\phi & (1 + \sin^2\phi)/\cos\phi & 0 \\ 0 & 0 & 1 \end{bmatrix},$$

$$C = \begin{bmatrix} 1 & \gamma & 0 \\ \gamma & 1+\gamma^2 & 0 \\ 0 & 0 & 1 \end{bmatrix}, \quad C^{-1} = \begin{bmatrix} 1+\gamma^2 & -\gamma & 0 \\ -\gamma & 1 & 0 \\ 0 & 0 & 1 \end{bmatrix},$$

$$B = \begin{bmatrix} 1+\gamma^2 & \gamma & 0 \\ \gamma & 0 & 0 \\ 0 & 0 & 1 \end{bmatrix}, \quad E = \frac{1}{2}\begin{bmatrix} 0 & \gamma & 0 \\ \gamma & \gamma^2 & 0 \\ 0 & 0 & 0 \end{bmatrix},$$

$$D = \frac{1}{2}\begin{bmatrix} 0 & \dot{\gamma} & 0 \\ \dot{\gamma} & 0 & 0 \\ 0 & 0 & 0 \end{bmatrix}, \quad \Omega = \frac{1}{2}\begin{bmatrix} 0 & \dot{\gamma} & 0 \\ -\dot{\gamma} & 0 & 0 \\ 0 & 0 & 0 \end{bmatrix},$$

where $\phi = \tan^{-1}(\gamma/2)$. The strain invariants are

$$I_1 = I_2 = 3 + \gamma^2, \qquad I_3 = J^2 = 1.$$

Note that the simple shear deformation does not involve any dilatation. It follows from Eqs. (16.11:9) and (16.11:12) that

$$S = C_1 \begin{bmatrix} 1 & 0 & 0 \\ 0 & 1 & 0 \\ 0 & 0 & 1 \end{bmatrix} + C_2 \begin{bmatrix} 2+\gamma^2 & -\gamma & 0 \\ -\gamma & 2 & 0 \\ 0 & 0 & 2+\gamma^2 \end{bmatrix}$$

$$+ C_3 \begin{bmatrix} 1+\gamma^2 & -\gamma & 0 \\ -\gamma & 1 & 0 \\ 0 & 0 & 1 \end{bmatrix},$$

$$\tau = C_1 \begin{bmatrix} 1+\gamma^2 & \gamma & 0 \\ \gamma & 1 & 0 \\ 0 & 0 & 1 \end{bmatrix} + C_2 \begin{bmatrix} 2+\gamma^2 & \gamma & 0 \\ \gamma & 2 & 0 \\ 0 & 0 & 2+\gamma^2 \end{bmatrix}$$

$$+ C_3 \begin{bmatrix} 1 & 0 & 0 \\ 0 & 1 & 0 \\ 0 & 0 & 1 \end{bmatrix}.$$

The stresses are uniform and therefore satisfy the equilibrium equations automatically. It is interesting to note that $\tau_{33} = S_{33} \neq 0$ unless $\gamma = 0$. In other words, if the body is compressible, we cannot have a traction free plane for simple shear deformation.

This would not be true if the body is incompressible. In this case, the Cauchy stresses are in the form

$$\tau = -p\mathbf{I} + C_1 \begin{bmatrix} 2\gamma^2/3 & \gamma & 0 \\ \gamma & -\gamma^2/3 & 0 \\ 0 & 0 & -\gamma^2/3 \end{bmatrix}$$

$$+ C_2 \begin{bmatrix} \gamma^2/3 & \gamma & 0 \\ \gamma & -2\gamma^2/3 & 0 \\ 0 & 0 & \gamma^2/3 \end{bmatrix},$$

where p is to be evaluated from the field equations. If there is no stress ($\tau_{i3} = S_{I3} = 0$, $i, I = 1, 2, 3$) across the planes where $z = $ constant, it requires that $p = (C_2 - C_1)\gamma^2/3$. Then the stresses become

$$\tau = \begin{bmatrix} C_1\gamma^2 & (C_1 + C_2)\gamma & 0 \\ (C_1 + C_2)\gamma & -C_2\gamma^2 & 0 \\ 0 & 0 & 0 \end{bmatrix}.$$

We can easily compute the surface forces on any surface. For example, let us consider a surface originally at $X = $ constant. The unit vector normal to the surface at the deformed state [Eq. (16.2:12)] is

$$\mathbf{n} = \frac{1}{\sqrt{1+\gamma^2}} \begin{bmatrix} 1 \\ -\gamma \\ 0 \end{bmatrix}.$$

The surface traction acting over this surface is given by the vector

$$\overset{\nu}{\mathbf{T}} = \tau \cdot \mathbf{n}^T = \frac{1}{\sqrt{1+\gamma^2}} \begin{bmatrix} -C_2\gamma^2 \\ (C_1 + C_2)\gamma + C_2\gamma^3 \\ 0 \end{bmatrix}.$$

The normal and tangential components of $\overset{\nu}{\mathbf{T}}$ are, respectively,

$$[-C_2\gamma^4 - (C_1 + 2C_2)\gamma^2]/(1+\gamma^2), \quad (C_1 + C_2)\gamma/(1+\gamma^2).$$

For the surface originally at Y = constant, the unit normal to its deformed surface is still in the Y-direction. The surface traction is

$$\overset{\nu}{\mathbf{T}} = \begin{bmatrix} (C_1 + C_2)\gamma \\ -C_2\gamma^2 \\ 0 \end{bmatrix}.$$

The normal and tangential components of the surface traction are, respectively, $[-C_2\gamma^2, -(C_1 + C_2)\gamma]$. The normal component is proportional to γ^2 as compared to γ^4 for the formal case.

More examples including simultaneous extension, inflation and torsion of a cylindrical tube, flexure of a circular cylinder, zero stress state of artery can be found in Green and Zenna (1968), Green and Adkins (1960), and Fung (1980).

Anisotropic Materials. Isotropic materials are a special group in nature. Almost all biological tissues are anisotropic. Most crystalline materials are also anisotropic. An extensive and complete discussion of the various kinds of symmetry that can be expected in crystals and their reflection on constitutive equations is given by Green and Alkins (1960). An account of the constitutive equations of biological tissues is given in Fung (1993). Current literature can be found in Applied Mechanics Reviews.

16.13. VARIATIONAL PRINCIPLES FOR FINITE ELASTICITY: COMPRESSIBLE MATERIALS

In this section, we shall discuss variational principles for finite strain analysis of hyperelastic materials. An easly work is by Koiter (1973) on the complementary energy principle. A general variational principle can involve displacements, and the different strain and stress measures defined in Secs. 16.2–7 as independent fields. The following are frequently used relations from Sec. 16.11. Denoting the *strain energy per unit undeformed volume* as $W_0(\mathbf{C})$, one obtains the following constitutive equation

(1) $$\frac{\partial W_0}{\partial \mathbf{C}} = \frac{\partial W_0}{\partial \mathbf{F}^T \cdot \mathbf{F}} = \frac{1}{2}\mathbf{S} = \frac{1}{2}S_{IJ}\epsilon_I\epsilon_J.$$

where $\mathbf{F}$ is the *deformation gradient tensor*, $\mathbf{C}(= \mathbf{F}^T \cdot \mathbf{F})$ is the *metric tensor of the deformed body* referred to the undeformed configuration [See Eq. (16.3:1)], and $\mathbf{S}$ is the *second Piola-Kirchhoff stress tensor*. These quantities are defined in Eqs. (16.2:3), (16.2:7) and (16.7:6). Since

$$(2) \quad \frac{\partial \mathbf{F}^T \cdot \mathbf{F}}{\partial \mathbf{F}} = \frac{\partial (F_{jI} F_{jJ} \varepsilon_I \varepsilon_J)}{\partial F_{iK}} \mathbf{e}_i \varepsilon_K = (F_{iI} \delta_{JK} + F_{iJ} \delta_{IK}) \varepsilon_I \varepsilon_J \mathbf{e}_i \varepsilon_K \,,$$

one can show that

$$(3) \quad \frac{\partial W_0}{\partial \mathbf{F}} = \frac{\partial W_0}{\partial \mathbf{F}^T \cdot \mathbf{F}} : \frac{\partial \mathbf{F}^T \cdot \mathbf{F}}{\partial \mathbf{F}} = \frac{1}{2} S_{IJ} (F_{iI} \delta_{JK} + F_{iJ} \delta_{IK}) \mathbf{e}_i \varepsilon_K$$

$$= F_{iJ} S_{JK} \mathbf{e}_i \varepsilon_K = \mathbf{F} \cdot \mathbf{S} = \mathbf{T}^T \,,$$

which is the constitutive equation for $\mathbf{T}$, the *Lagrangian* or *first Piola-Kirchhoff stress tensor*. In Eqs. (1) and (2), the base vectors must be in the order consistent with that of the subscripts of F's as shown.

We shall use Q_0 to denote the strain energy per unit undeformed volume when it is a function of $\mathbf{U}$, the *right stretch tensor* defined in Eq. (16.4:4). This is to distinguish Q_0 from W_0, the strain energy density in terms of $\mathbf{F}$ or $\mathbf{C}$. We define

$$Q_0(\mathbf{U}) \equiv W_0(\mathbf{C})$$

for the same deformation. Then, it can be shown that

$$(4) \quad \frac{\partial Q_0}{\partial \mathbf{U}} = (\mathbf{T} \cdot \mathbf{R})_s = \mathbf{r}$$

where the subscript $(.)_s$ denotes the symmetric part of the relevant quantity and $\mathbf{R}$ is the *rotation tensor*. Equation (4) is an alternative expression of the constitutive equation for $\mathbf{T}$.

Following Atluri and Cazzani (1995), we can derive a *mixed four-field variational principle*, which uses the displacement vector $\mathbf{u}(= \mathbf{x} - \mathbf{X})$, the rotational tensor $\mathbf{R}$, the right stretch tensor $\mathbf{U}$, and the Lagrangian stress tensor $\mathbf{T}$ as independent fields. The functional is in the form

$$(5) \quad \Pi_1(\mathbf{u}, \mathbf{R}, \mathbf{U}, \mathbf{T}) = \int_{V_0} [Q_0(\mathbf{U}) + \mathbf{T}^T : (\mathbf{I} + \nabla_0 \mathbf{u} - \mathbf{R} \cdot \mathbf{U})$$

$$- \rho_0 \mathbf{b} \cdot \mathbf{u}] \, dV - \int_{S_{u0}} \mathbf{N} \cdot \mathbf{T} \cdot (\mathbf{u} - \bar{\mathbf{u}}) dA - \int_{S_{\sigma 0}} \overset{\nu}{\mathbf{T}} \cdot \mathbf{u} dA \,,$$

where $\mathbf{b}$ is the body force per unit mass, V_0 is the undeformed volume, $\mathbf{N}$ is a unit outward normal to the boundary surface $\partial V_0 (= S_0)$, and $S_{\sigma 0} + S_{u0} = \partial V_0 = S_0$ with $S_{\sigma 0}$ and S_{u0} being the parts of the surface of ∂V_0 where the traction $\overset{\nu}{\mathbf{T}}$ and the displacement $\bar{\mathbf{u}}$ are prescribed, respectively. The *admissible requirements* for the four field variables are: $\mathbf{u}$ *must be continuous,* **R** *orthogonal,* **U** *symmetric and* **T** *a general tensor of rank* 2. In Eq. (5), the gradient operator refers to the undeformed coordinates, i.e., $\mathbf{\nabla}_0 = \boldsymbol{\varepsilon}_I \frac{\partial}{\partial X_I}$.

A four-field mixed principle having **u**, **R**, **U** and **T** as independent fields is stated as the stationarity conditions of the functional Π_1. *To derive a solution of the problem based on the stationarity conditions of the functional is called an integral formulation.* Following the derivation in Secs. 10.1-3, the *first variation* of Π_1 with respect to the independent field variables is

$$(6) \quad \delta\Pi_1 = \int_{V_0} \left\{ \frac{\partial Q_0}{\partial \mathbf{U}} : \delta\mathbf{U} + \delta\mathbf{T}^T : [\mathbf{I} + \mathbf{\nabla}_0\mathbf{u} - \mathbf{R}\cdot\mathbf{U}] - \mathbf{T}^T : (\delta\mathbf{R}\cdot\mathbf{U}) \right.$$

$$- \mathbf{T}^T : (\mathbf{R}\cdot\delta\mathbf{U}) + \mathbf{T}^T : \mathbf{\nabla}_0\delta\mathbf{u} - \rho_0\mathbf{b}\cdot\delta\mathbf{u}]\, dV$$

$$- \int_{S_{u0}} [\mathbf{N}\cdot\delta\mathbf{T}\cdot(\mathbf{u}-\bar{\mathbf{u}}) + \mathbf{N}\cdot\mathbf{T}\cdot\delta\mathbf{u}]\, dA - \int_{S_{\sigma0}} \overset{\nu}{\mathbf{T}}\cdot\delta\mathbf{u}\, dA,$$

where $\delta(.)$ denotes the variation of the corresponding field variable, e.g., $\delta\mathbf{u}$ is the variation of **u**. Taking into account the properties of (:) as given in Sec. 16.1, one can show that

$$(7) \qquad \mathbf{T}^T : \mathbf{\nabla}_0\delta\mathbf{u} = \mathbf{\nabla}\cdot(\mathbf{T}\cdot\delta\mathbf{u}) - (\mathbf{\nabla}_0\cdot\mathbf{T})\cdot\delta\mathbf{u}, \quad \text{or}$$

$$T_{Ji}\frac{\partial\delta u_i}{\partial X_J} = \frac{\partial(T_{Ji}\delta u_i)}{\partial X_J} - \frac{\partial T_{Ji}}{\partial X_J}\delta u_i,$$

$$(8) \qquad \mathbf{T}^T : (\delta\mathbf{R}\cdot\mathbf{U}) = (\mathbf{T}^T\cdot\mathbf{U}\cdot\mathbf{R}^T) : (\delta\mathbf{R}\cdot\mathbf{R}^T)$$

$$= (\mathbf{T}^T\cdot\mathbf{U}\cdot\mathbf{R}^T)_a : (\delta\mathbf{R}\cdot\mathbf{R}^T),$$

$$(9) \qquad \mathbf{T}^T : (\mathbf{R}\cdot\delta\mathbf{U}) = (\mathbf{R}^T\cdot\mathbf{T}^T) : \delta\mathbf{U}$$

$$= \frac{1}{2}(\mathbf{R}^T\cdot\mathbf{T}^T + \mathbf{T}\cdot\mathbf{R}) : \delta\mathbf{U}.$$

The proof is left to the reader. A substitution of Eqs. (7)–(9) into Eq. (6) and the use of Gauss' theorem yield

$$\delta\Pi_1 = \int_{V_0} \left\{ \left[\frac{\partial Q_0}{\partial \mathbf{U}} - (\mathbf{T}\cdot\mathbf{R})_s\right] : \delta\mathbf{U} + \delta\mathbf{T}^T : (\mathbf{I} + \mathbf{\nabla}_0\mathbf{u} - \mathbf{R}\cdot\mathbf{U}) \right.$$

$$- (\mathbf{T}^T\cdot\mathbf{U}\cdot\mathbf{R}^T)_a : (\delta\mathbf{R}\cdot\mathbf{R}^T) - (\mathbf{\nabla}_0\cdot\mathbf{T} + \rho_0\mathbf{b})\cdot\delta\mathbf{u}\Big\}\, dV$$

$$- \int_{S_{u0}} \mathbf{N}\cdot\delta\mathbf{T}\cdot(\mathbf{u}-\bar{\mathbf{u}})\, dA - \int_{S_{u0}} (\overset{\nu}{\mathbf{T}} - \mathbf{N}\cdot\mathbf{T})\cdot\delta\mathbf{u}\, dA.$$

The vanish of $\delta\Pi_1$ for arbitrary continuous $\delta\mathbf{u}$, antisymmetric $\delta\mathbf{R} \cdot \mathbf{R}^T$, symmetric $\delta\mathbf{U}$ and general tensor $\delta\mathbf{T}$ gives

(10) $\quad \dfrac{\partial Q_0}{\partial \mathbf{U}} = (\mathbf{T} \cdot \mathbf{R})_s \qquad\qquad$ (Constitutive law in V_0),

(11) $\quad \mathbf{I} + \boldsymbol{\nabla}_0 \mathbf{u} = \mathbf{R} \cdot \mathbf{U} \qquad\qquad$ (Compatibility condition in V_0),

(12) $\quad (\mathbf{T}^T \cdot \mathbf{U} \cdot \mathbf{R}^T)_a$

$\qquad\qquad = (\mathbf{R} \cdot \mathbf{U} \cdot \mathbf{T})_a = 0 \qquad$ (Angular momentum balance in V_0),

(13) $\quad \boldsymbol{\nabla}_0 \cdot \mathbf{T} + \rho_0 \mathbf{b} = 0 \qquad\qquad$ (Equilibrium equation in V_0),

(14) $\quad \mathbf{N} \cdot \mathbf{T} = \overset{\nu}{\mathbf{T}} \qquad\qquad$ (Traction boundary condition on $S_{\sigma 0}$),

(15) $\quad \mathbf{u} = \bar{\mathbf{u}} \qquad\qquad$ (Displacement boundary condition on S_{u0}).

This variational principle is valid for general elastic materials. In this formulation both the displacement and traction conditions are *natural boundary conditions* (See Sec. 10.6). Note that the angular momentum equation requires $\mathbf{R} \cdot \mathbf{U} \cdot \mathbf{T}$ to be symmetric. As a matter of fact, $\mathbf{R} \cdot \mathbf{U} \cdot \mathbf{T}(= \boldsymbol{\sigma})$ is the *Kirchhoff stress tensor*, which of course should be symmetric.

One can obtain other variational principles by selectively enforcing Eqs. (10)–(15) *a priori*. For example, setting

$$\mathbf{U} = [\mathbf{R}^T \cdot (\mathbf{I} + \boldsymbol{\nabla}_0 \mathbf{u})]_s$$

in Eq. (5) will eliminate $\mathbf{U}$ from the functional and result in a variational principle with only three dependent variables $\mathbf{u}$, $\mathbf{R}$, and $\mathbf{T}$.

One can also establish a variational principle *involving only the displacements as the field variable* by requiring *a priori*

$$\mathbf{I} + \boldsymbol{\nabla}_0 \mathbf{u} = \mathbf{R} \cdot \mathbf{U} \qquad \text{in} \quad V_0$$

and

$$\mathbf{u} = \bar{\mathbf{u}} \qquad \text{on} \quad S_{u0}.$$

Equation (5) then becomes

(16) $\quad \Pi_3(\mathbf{u}) = \displaystyle\int_{V_0} \{W_0[\mathbf{F}^T \cdot \mathbf{F}] - \rho_0 \mathbf{b} \cdot \mathbf{u}\}\, dV - \int_{S_{\sigma 0}} \overset{\nu}{\mathbf{T}} \cdot \mathbf{u}\, dA,$

wherein

$$\mathbf{F} = \mathbf{I} + \boldsymbol{\nabla}_0 \mathbf{u}.$$

Equation (16) is the functional for the commonly known *principle of minimum potential energy*, which involves only one field variable **u**. The associated field equations are

$$\nabla_0 \cdot \frac{\partial W_0}{\partial \mathbf{F}^T} + \rho_0 \mathbf{b} = \nabla_0 \cdot (\mathbf{S} \cdot \mathbf{F}^T) + \rho_0 \mathbf{b} = 0$$

in V_0, and the traction condition Eq. (14) on $S_{\sigma 0}$ as a *natural boundary condition*. The displacement boundary condition Eq. (15) is now a *rigid boundary condition* (see Sec. 10.6). One enforces the rigid condition in the integral formulation and uses $\delta \mathbf{u} = 0$ on S_{u0} in deriving the stationarity condition for Π_3.

Atluri included **R** as an additional field variable by modifying Π_3 in the form

(17) $$\Pi_3^*(\mathbf{u}, \mathbf{R}) = \int_{V_0} \left\{ W_0(\mathbf{F}^T \cdot \mathbf{F}) + \frac{\gamma}{4}(\mathbf{R}^T \cdot \mathbf{F})_a^2 - \rho_0 \mathbf{b} \cdot \mathbf{u} \right\} dV$$

$$- \int_{S_{\sigma 0}} \overset{\nu}{\mathbf{T}} \cdot \mathbf{u} \, dA \,,$$

wherein

$$(\mathbf{R}^T \cdot \mathbf{F})_a^2 = (\mathbf{R}^T \cdot \mathbf{F} - \mathbf{F}^T \cdot \mathbf{R}) : (\mathbf{R}^T \cdot \mathbf{F} - \mathbf{F}^T \cdot \mathbf{R}) \,,$$

γ is a constant, and $(.)_a$ denotes the *skew symmetric part* of the corresponding tensor of rank 2. The constant is of the magnitude of the elastic modulus of the material. One can show that the variation of $(\mathbf{R}^T \cdot \mathbf{F})_a^2$ with respect to **R** and **u** gives

$$\delta(\mathbf{R}^T \cdot \mathbf{F})_a^2 = 4(\mathbf{F} - \mathbf{R} \cdot \mathbf{F}^T \cdot \mathbf{R}) : \nabla_0 \delta \mathbf{u}$$

$$+ 4[(\mathbf{R}^T \cdot \mathbf{F} - \mathbf{F}^T \cdot \mathbf{R}) \cdot \mathbf{F}^T \cdot \mathbf{R}] : (\delta \mathbf{R} \cdot \mathbf{R})$$

by using the properties of (:) operation described in Sec. 16.1. Applying Eq. (3), one can derive the field equations

(18) $$\nabla_0 \cdot [\mathbf{S} \cdot \mathbf{F}^T + \gamma(\mathbf{F} - \mathbf{R} \cdot \mathbf{F}^T \cdot \mathbf{R})] + \rho_0 \mathbf{b} = 0 \,,$$

(19) $$(\mathbf{F}^T \cdot \mathbf{R})^2 = (\mathbf{R}^T \cdot \mathbf{F})^2 \,,$$

in V_0 and the natural boundary condition Eq. (14) on $S_{\sigma 0}$. The boundary condition Eq. (15) is a rigid condition. Equation (19) implies that $\mathbf{R}^T \cdot \mathbf{F}$ is symmetric, i.e.,

(20) $$\mathbf{R}^T \cdot \mathbf{F} = \mathbf{F}^T \cdot \mathbf{R} \quad \text{or} \quad \mathbf{F} = \mathbf{R} \cdot \mathbf{F}^T \cdot \mathbf{R} \,.$$

One sees that γ is the constant factor in the penalty function approach. Atluri and Cazzani (1995) showed that the finite element solution based on Π_3^* provided better accuracy than that based on Π_3.

One can also establish different three-field variational principles through contact transformation. For example, using the following contact transformation

(21) $$-Q_c(\mathbf{r}) = Q_0(\mathbf{U}) - (\mathbf{T} \cdot \mathbf{R})_s : \mathbf{U}\,,$$

where $\mathbf{r} = (\mathbf{T} \cdot \mathbf{R})_s$ is the *symmetrized Biot-Lur's stress tensor* and Q_c is called the *complementary energy density* in terms of $\mathbf{r}$ (see Sec. 10.9). One can show that Q_c does not explicitly depend on $\mathbf{U}$, i.e.,

$$-\frac{\partial Q_c}{\partial \mathbf{U}} = \frac{\partial Q_0}{\partial \mathbf{U}} - \mathbf{r} = 0\,.$$

A substitution of Eq. (21) into Eq. (5) yields

(22) $$\Pi_4(\mathbf{u}, \mathbf{R}, \mathbf{T}) = \int_{V_0} \{-Q_c[(\mathbf{T} \cdot \mathbf{R})_s] + \mathbf{T}^T : (\mathbf{I} + \nabla_0 \mathbf{u}) - \rho_0 \mathbf{b} \cdot \mathbf{u}\} dV$$
$$- \int_{S_{u0}} \mathbf{N} \cdot \mathbf{T} \cdot (\mathbf{u} - \bar{\mathbf{u}})\, dA - \int_{S_{\sigma 0}} \overset{\nu}{\mathbf{T}} \cdot \mathbf{u}\, dA\,.$$

The admissible requirements for $\mathbf{u}$, $\mathbf{R}$, and $\mathbf{T}$ are the same as those of Eq. (5). The corresponding field equations of the functional are

(23) $$\frac{\partial Q_c}{\partial \mathbf{r}} = \mathbf{R}^T \cdot (\mathbf{I} + \nabla_0 \mathbf{u}) = \mathbf{U} \quad \text{in} \quad V_0\,,$$

and Eqs. (12)–(15). Apparently, Eq. (23) is a compatibility condition for the new functional replacing Eqs. (10) and (11).

One can establish variational principles using $\mathbf{u}$, $\mathbf{R}$, $\mathbf{U}$, $\mathbf{r}^*$ as independent fields, where $\mathbf{r}^*$ is the *Biot-Lur'e stress tensor* defined in Eq. (16.7:8). For example, by substituting $\mathbf{T}$ by $\mathbf{r}^* \cdot \mathbf{R}^T$ into Eq. (5), one obtains

(24) $$\Pi_5(\mathbf{u}, \mathbf{R}, \mathbf{U}, \mathbf{r}^*) = \int_{V_0} \{Q_0(\mathbf{U}) + \mathbf{r}^{*T} : [\mathbf{R}^T \cdot (\mathbf{I} + \nabla_0 \mathbf{u}) - \mathbf{U}]$$
$$-\rho_0 \mathbf{b} \cdot \mathbf{u}\}\, dV - \int_{S_{u0}} \mathbf{N} \cdot \mathbf{r}^* \cdot \mathbf{R}^T \cdot (\mathbf{u} - \bar{\mathbf{u}})\, dA - \int_{S_{\sigma 0}} \overset{\nu}{\mathbf{T}} \cdot \mathbf{u} dA\,.$$

One can derive other variational principles by imposing different conditions *a priori* as before.

PROBLEMS

16.11.   Show Eq. (16.13:7) using the indicial notation.

$$\mathbf{T}^T : \boldsymbol{\nabla}_0 \delta\mathbf{u} = T_{Ii}\frac{\partial \delta u_i}{\partial X_I} = \frac{\partial T_{Ii}\delta u_i}{\partial X_I} - \frac{\partial T_{Ii}}{\partial X_I}\delta u_i$$

$$= \boldsymbol{\nabla}_0 \cdot (\mathbf{T}\cdot\delta\mathbf{u}) - (\boldsymbol{\nabla}_0\cdot\mathbf{T})\cdot\delta\mathbf{u}\,.$$

16.12.   Show Eq. (16.13:8) using the properties of the (:) operation. [*Hint*: $\mathbf{R}\cdot\mathbf{R}^T = \mathbf{I}$ and $\partial\mathbf{R}\cdot\mathbf{R}^T$ is antisymmetric.]

16.13.   Show Eq. (16.13:9) using the properties of the (:) operation.

16.14.   Derive the field equations including the natural boundary conditions for the functional given in Eq. (16.13:5) with $\mathbf{U} = [\mathbf{R}^T\cdot(\mathbf{I}+\boldsymbol{\nabla}_0\mathbf{u})]_s$ satisfied *a priori*. Identify the admissible conditions for the independent field variables.

16.15.   Show that Eq. (16.13:23) is a field equation for the functional Eq. (16.13:22).

16.14. VARIATIONAL PRINCIPLES FOR FINITE ELASTICITY: NEARLY INCOMPRESSIBLE OR INCOMPRESSIBLE MATERIALS

For materials that are nearly incompressible, the energy density function W_0 or Q_0 given in the previous section becomes ill defined and is not suitable for evaluating the mean stress. When the material is incompressible, one must use the constitutive law as defined in Eq. (16.11:18) or Eqs. (16.11:20) and (16.11:22). In this section, we shall discuss the variational formulation that is valid for compressible as well as incompressible materials by using the kinematic split formulation discussed at the end of Sec. 16.11. This is to multiplicatively decompose the finite strain into a dilatational part and a distortion one and to divide the stress into the deviatoric stress and the hydrostatic pressure. As an illustration we shall derive the variation principles involving the *Biot-Lur's stress tensor* $\mathbf{r}^*$.

Following Eq. (16.11:23), we split the right stretch tensor as $\mathbf{U} = \mathbf{U}'J^{1/3}$ to give $\det(\mathbf{U}') = 1$, where J is the Jacobian, $\mathbf{U}'$ represents the *distortion of the right stretch tensor* $\mathbf{U}$ and $J^{1/3}$ the *dilatation*. If the strain is small, this decomposition becomes additive as that for the infinitesimal strain case. Similar to the derivation in Sec. 16.11, we define

$$(\mathbf{r'}^*)_s = \frac{\partial Q_0^*(\mathbf{U'}, J)}{\partial \mathbf{U'}}, \qquad -p = \frac{\partial Q_0^*(\mathbf{U'}, J)}{\partial J},$$

where $\mathbf{Q}_0^*(\mathbf{U'}, J) \equiv Q_0(\mathbf{U})$ is the strain energy density function in terms of $\mathbf{U}$. The relations among $\mathbf{r}^*$, $\mathbf{r'}^*$, p are

(1) $\qquad (\mathbf{r}^*)_s = J^{-1/3}(\mathbf{r'}^*)_s - \frac{1}{3} J^{-1/3}[(\mathbf{r'}^*)_s : \mathbf{U}]\mathbf{U}^{-1} - pJ\mathbf{U}^{-1},$

(2) $\qquad -p = \frac{1}{3J}[(\mathbf{r}^*)_s : \mathbf{U}] = \frac{1}{3}(\tau : \mathbf{I}).$

Note that p is the hydrostatic pressure and $\mathbf{r'}^*$ is not deviatoric since $\mathbf{r'}^* : \mathbf{I} \neq 0$.

We can establish variational principles having $\mathbf{u}$, $\mathbf{R}$, $\mathbf{U'}$, J, $\mathbf{r'}^*$, p, $\overset{\nu}{\mathbf{T}}$ as independent field variables where $\overset{\nu}{\mathbf{T}}$ is a traction vector defined only on S_{u0} (Atluri and Cazzani 1995). For example,

(3) $\Pi(\mathbf{u}, \mathbf{R}, \mathbf{U'}, J, \mathbf{r'}^*, p, \overset{\nu}{\mathbf{T}})$

$$= \int_{V_0} \{Q_0^*(\mathbf{U'}, J) + \mathbf{r'}^{*T} : [\mathbf{R}^T \cdot (\mathbf{I} + \nabla_0\mathbf{u})J_u^{-1/3} - \mathbf{U'}]$$

$$- (J_u - J)p - rho_0\mathbf{b} \cdot \mathbf{u}\} \, dV - \int_{S_{\sigma0}} \overset{\nu}{\mathbf{T}} \cdot \mathbf{u} \, dA - \int_{S_{u0}} \overset{\nu}{\mathbf{T}} \cdot (\mathbf{u} - \bar{\mathbf{u}}) \, dA$$

is the functional for a variational principle of the seven field variables, in which

(4) $\qquad J_u \equiv \det(\hat{\mathbf{U}}), \qquad \hat{\mathbf{U}} = [\mathbf{R}^T \cdot (\mathbf{I} + \nabla_0\mathbf{u})]_s.$

The admissible requirements are: $\mathbf{u}$ is continuous, $\mathbf{R}$ orthogonal, $\mathbf{U'}$ symmetric, J positive, $\mathbf{r'}^*$ a second rank tensor, p piecewise continuous and $\overset{\nu}{\mathbf{T}}$ piecewise continuous vector.

To derive the field equations associated with the functional given in Eq. (3), we needs to take into account that

$$\mathbf{r'}^{*T} : [\delta\mathbf{R}^T \cdot (\mathbf{I} + \nabla_0\mathbf{u})] = [(\mathbf{I} + \nabla_0\mathbf{u}) \cdot \mathbf{r'}^*] : \delta\mathbf{R}$$

$$= [\mathbf{R}^T \cdot (\mathbf{I} + \nabla_0\mathbf{u}) \cdot \mathbf{r'}^*] : (\mathbf{R}^T \cdot \delta\mathbf{R}),$$

$$\mathbf{r'^{*T}} : (\mathbf{R}^T \cdot \boldsymbol{\nabla}_0 \delta \mathbf{u}) = (\mathbf{R} \cdot \mathbf{r'^{*T}}) : \boldsymbol{\nabla}_0 \delta \mathbf{u} \,,$$

$$\delta J_u = \frac{\partial J_u}{\partial \hat{\mathbf{U}}} : \delta \hat{\mathbf{U}} = J_u \hat{\mathbf{U}}^{-1} : \delta \hat{\mathbf{U}}$$

$$= J_u \hat{\mathbf{U}}^{-1} : [\delta \mathbf{R}^T \cdot (\mathbf{I} + \boldsymbol{\nabla}_0 \mathbf{u}) + \mathbf{R}^T \cdot \boldsymbol{\nabla}_0 \delta \mathbf{u}]$$

$$= J_u [\mathbf{R}^T \cdot (\mathbf{I} + \boldsymbol{\nabla}_0 \mathbf{u}) \cdot \hat{\mathbf{U}}^{-1}] : (\mathbf{R}^T \cdot \delta \mathbf{R}) + J_u (\hat{\mathbf{U}}^{-1} \cdot \mathbf{R}^T) : (\boldsymbol{\nabla}_0 \delta \mathbf{u})^T \,.$$

From the last equation above, one derives

$$J_u^{1/3} \mathbf{r'^{*T}} : [\mathbf{R}^T \cdot (\mathbf{I} + \boldsymbol{\nabla}_0 \mathbf{u}) \delta J_u^{-1/3}]$$

$$= -[\mathbf{r'^{*T}} : \mathbf{R}^T \cdot (\mathbf{I} + \boldsymbol{\nabla}_0 \mathbf{u})](\hat{\mathbf{U}}^{-1} \cdot \mathbf{R}^T) : (\boldsymbol{\nabla}_0 \delta \mathbf{u})^T / 3$$

$$- [\mathbf{r'^{*T}} : \mathbf{R}^T \cdot (\mathbf{I} + \boldsymbol{\nabla}_0 \mathbf{u})][\mathbf{R}^T \cdot (\mathbf{I} + \boldsymbol{\nabla}_0 \mathbf{u}) \cdot \hat{\mathbf{U}}^{-1}] : (\mathbf{R}^T \cdot \delta \mathbf{R}) / 3 \,.$$

The stationary condition of Π for arbitrary variation of the admissible field variables gives the following field equations: the constitutive laws

$$(5) \qquad\qquad (\mathbf{r'^{*}})_s = \frac{\partial Q_0^*(\mathbf{U'}, J)}{\partial \mathbf{U'}} \,,$$

$$(6) \qquad\qquad -p = \frac{\partial Q_0^*(\mathbf{U'}, J)}{\partial J} \,;$$

the compatibility conditions

$$(7) \qquad\qquad \mathbf{U'} = \mathbf{R}^T \cdot (\mathbf{I} + \boldsymbol{\nabla}_0 \mathbf{u}) J_u^{-1/3} \,,$$

$$(8) \qquad\qquad J = J_u \,;$$

the angular momentum balance condition

$$(9) \qquad \{\mathbf{R}^T \cdot (\mathbf{I} + \boldsymbol{\nabla}_0 \mathbf{u}) \cdot [3\mathbf{r'^{*}} - (\mathbf{r'^{*}} : (\mathbf{I} + \boldsymbol{\nabla}_0 \mathbf{u})^T \cdot \mathbf{R}) \cdot \hat{\mathbf{U}}^{-1}$$

$$+ 3p J_u^{4/3} \hat{\mathbf{U}}^{-1}]\}_a = 0 \,,$$

[note that $\mathbf{r'^{*}} : (\mathbf{I} + \boldsymbol{\nabla}_0 \mathbf{u})^T \cdot \mathbf{R} = \mathbf{r'^{*T}} : \mathbf{R}^T \cdot (\mathbf{I} + \boldsymbol{\nabla}_0 \mathbf{u})$]; the linear momentum balance condition

$$(10) \quad \boldsymbol{\nabla}_0 \cdot \{[J_u^{-1/3} \mathbf{r'^{*}} - (J_u^{-1/3} \mathbf{r'^{*}} : (\mathbf{I} + \boldsymbol{\nabla}_0 \mathbf{u})^T \cdot \mathbf{R}/3 + p J_u) \hat{\mathbf{U}}^{-1}] \cdot \mathbf{R}^T\}$$

$$+ \rho_0 \mathbf{b} = 0 \,,$$

in V_0; the traction boundary condition

$$(11) \quad \mathbf{N} \cdot \left\{ \frac{\mathbf{r'^{*}}}{J_u^{1/3}} - \left[\frac{\mathbf{r'^{*}} : (\mathbf{I} + \boldsymbol{\nabla}_0 \mathbf{u})^T \cdot \mathbf{R}}{3 J_u^{1/3}} + p J_u \right] \hat{\mathbf{U}}^{-1} \right\} \cdot \mathbf{R}^T = \overset{\nu}{\mathbf{T}} \,,$$

on $S_{\sigma 0}$, and the displacement boundary condition and the traction compatibility condition

(12) $$\mathbf{u} = \bar{\mathbf{u}},$$

(13) $$\mathbf{N} \cdot \left\{ \frac{\mathbf{r'^*}}{J_u^{1/3}} - \left[\frac{\mathbf{r'^*} : (\mathbf{I} + \boldsymbol{\nabla}_0 \mathbf{u})^T \cdot \mathbf{R}}{3 J_u^{1/3}} + p J_u \right] \hat{\mathbf{U}}^{-1} \right\} \cdot \mathbf{R}^T = \overset{\nu}{\mathbf{T}},$$

on S_{u0}. Note that since $\mathbf{U}'$ is symmetric, the compatibility condition assures that $\mathbf{U} = \mathbf{R}^T \cdot (\mathbf{I} + \boldsymbol{\nabla}_0 \mathbf{u})$ or $\mathbf{U} = (\mathbf{I} + \boldsymbol{\nabla}_0 \mathbf{u})^T \cdot \mathbf{R}$ is symmetric and, therefore, equals to $\hat{\mathbf{U}}(= [\mathbf{R}^T \cdot (\mathbf{I} + \boldsymbol{\nabla}_0 \mathbf{u})]_s)$. Consequently

$$\frac{\mathbf{r'^*} : (\mathbf{I} + \boldsymbol{\nabla}_0 \mathbf{u})^T \cdot \mathbf{R}}{3 J_u^{1/3}} \hat{\mathbf{U}}^{-1} + p J_u \hat{\mathbf{U}}^{-1} - \frac{\mathbf{r'^*}}{J_u^{1/3}}$$

$$= \frac{\mathbf{r'^*} : \mathbf{U}}{3 J^{1/3}} \mathbf{U}^{-1} + p J \mathbf{U}^{-1} - \frac{\mathbf{r'^*}}{J^{1/3}}$$

and Eqs. (11) and (13) become, respectively,

$$\mathbf{N} \cdot \left\{ \left[\mathbf{r'^*} - \frac{1}{3}(\mathbf{r'^*} : \mathbf{U})\mathbf{U}^{-1} \right] J^{-1/3} - p J \mathbf{U}^{-1} \right\} \cdot \mathbf{R}^T = \overset{\nu}{\mathbf{T}} \qquad \text{on } S_{\sigma 0},$$

$$\mathbf{N} \cdot \left\{ \left[\mathbf{r'^*} - \frac{1}{3}(\mathbf{r'^*} : \mathbf{U})\mathbf{U}^{-1} \right] J^{-1/3} - p J \mathbf{U}^{-1} \right\} \cdot \mathbf{R}^T = \overset{\nu}{\mathbf{T}} \qquad \text{on } S_{u0}.$$

One can obtain a *five-field variational principle through the contact transformation*

(14) $$-Q_c^*[(\mathbf{r'^*})_s, p] = Q_0^*(\mathbf{U}', J) - \mathbf{U}' : (\mathbf{r'^*})_s + p J.$$

A substitution of Eq. (14) into Eq. (3) yields

(15) $$\Pi_8(\mathbf{u}, \mathbf{R}, \mathbf{r'^*}, p, \overset{\nu}{\mathbf{T}}) = \int_{V_0} \left\{ -Q_c^*[(\mathbf{r'^*})_s, p] + \frac{\mathbf{r'^{*T}} : \mathbf{R}^T \cdot (\mathbf{I} + \boldsymbol{\nabla}_0 \mathbf{u})}{J_u^{1/3}} \right.$$

$$\left. - J_u p - \rho_0 \mathbf{b} \cdot \mathbf{u} \right\} dV - \int_{S_{\sigma 0}} \overset{\nu}{\mathbf{T}} \cdot \mathbf{u} \, dA - \int_{S_{u0}} \overset{\nu}{\mathbf{T}} \cdot (\mathbf{u} - \bar{\mathbf{u}}) \, dA,$$

which is the functional for a five-field principle. The admissible requirements for the independent field $\mathbf{u}, \mathbf{R}, \mathbf{r'^*}, p, \overset{\nu}{\mathbf{T}}$ are the same as those in Π.

We can similarly establish the variational principles involving the first Piola-Kirchhoff stress $\mathbf{T}$. Let

(16) $$\mathbf{F}' = \mathbf{F} J^{-1/3},$$

(17) $$W_0^*(\mathbf{F}', J) \equiv W_0(\mathbf{F}^T \cdot \mathbf{F}).$$

From the last two equations, we can show that

(18) $\displaystyle (\mathbf{T}')_s = \frac{\partial W_0^*(\mathbf{F}', J)}{\partial \mathbf{F}'}$, $\displaystyle -p = \frac{\partial W_0^*(\mathbf{F}', J)}{\partial J}$,

(19) $\displaystyle \mathbf{T}^T = J^{-1/3}\mathbf{T}'^T - \frac{1}{3}J^{-1/3}[\mathbf{T}'^T : \mathbf{F}]\mathbf{F}^{-1} - pJ\mathbf{F}^{-1}$.

One can then establish the variational principle using $\mathbf{u}$, $\mathbf{F}'$, J, $\mathbf{T}'$, p $\overset{\nu}{\mathbf{T}}$ as independent field variables.

One can also derive a *two-field principle* through the contact transformation

(20) $\displaystyle -W_m(\mathbf{F}', p) = W_0^*(\mathbf{F}', J) + pJ$,

The enforcement of compatibility conditions Eqs. (7) and (8) *a priori* results in

$$\mathbf{F}' = (\mathbf{I} + \boldsymbol{\nabla}_0\mathbf{u})J^{-1/3} ,$$

$$J = \det(\mathbf{I} + \boldsymbol{\nabla}_0\mathbf{u}) .$$

Then in Eq. (3), replacing $Q_0^*(\mathbf{U}', J)$ with $W_0^*(\mathbf{F}', J)$, enforcing the displacement boundary condition Eq. (12) on S_{u0}, and introducing the penalty function $\gamma[\mathbf{R}^T \cdot (\mathbf{I} + \boldsymbol{\nabla}_0\mathbf{u})]_a^2$, one obtains

(21) $\displaystyle \Pi_9(\mathbf{u}, \mathbf{R}, p) = \int_{V_0} \left\{ -W_m(\mathbf{F}', p) - p\det(\mathbf{I} + \boldsymbol{\nabla}_0\mathbf{u}) \right.$

$$\left. + \frac{\gamma}{4}[\mathbf{R}^T \cdot (\mathbf{I} + \boldsymbol{\nabla}_0\mathbf{u})]_a^2 - \rho_0\mathbf{b} \cdot \mathbf{u} \right\}dV - \int_{S_{\sigma0}} \overset{\nu}{\mathbf{T}} \cdot \mathbf{u}\, dA .$$

The penalty function is introduced to improve the accuracy of the weak solution (Atluri and Cazzani 1995). The functional Π_9 for infinitesimal strain with $\gamma = 0$ was derived by Tong (1969).

Note that the function Π of Eq. (3) cannot be used for incompressible materials because $Q_0^*(\mathbf{U}', J)[\equiv Q_0(\mathbf{U})]$ is not defined when $J = 1$. However, both Π_8 and Π_9 of Eqs. (15) and (21) are suitable for incompressible and nearly incompressible materials.

Before leaving this section, we summarize below the definitions of the various energy density functions and the associated constitutive laws developed in the last two sections:

Energy density functions	constitutive laws
$Q_0(\mathbf{U}) \equiv W_0(\mathbf{C}) \equiv W_0(\mathbf{F}^T \cdot \mathbf{F})$	$\dfrac{\partial W_0}{\partial \mathbf{C}} = \dfrac{1}{2}\,\mathbf{S}, \; \dfrac{\partial W_0}{\partial \mathbf{F}} = \mathbf{T}^T,$ $\dfrac{\partial Q_0}{\partial \mathbf{U}} = (\mathbf{T}\cdot\mathbf{R})_s$
$-Q_c(\mathbf{r}) \equiv Q_0(\mathbf{U}) - (\mathbf{T}\cdot\mathbf{R})_s : \mathbf{U}$	$\dfrac{\partial Q_c}{\partial \mathbf{r}} = \mathbf{U}$
$Q_0^*(\mathbf{U}', J) \equiv Q_0(\mathbf{U}', J)$	$(\mathbf{r}'^*)_s = \dfrac{\partial Q_0^*(\mathbf{U}', J)}{\partial \mathbf{U}'},$ $p = -\dfrac{\partial Q_0^*(\mathbf{U}', J)}{\partial J}$
$-Q_c^*[(\mathbf{r}'^*)_s, p] \equiv Q_0^*(\mathbf{U}', J) - \mathbf{U}' : (\mathbf{r}'^*)_s + pJ$	$\mathbf{U}' = \dfrac{\partial Q_c}{\partial (\mathbf{r}'^*)_s},$ $p = -\dfrac{\partial Q_c}{\partial J}$
$W_0^*(\mathbf{F}', J) \equiv W_0(\mathbf{F}'^T \cdot \mathbf{F}' J^{2/3})$	$(\mathbf{T}')_s = \dfrac{\partial W_0^*(\mathbf{F}', J)}{\partial \mathbf{F}'},$ $p = -\dfrac{\partial W_0^*(\mathbf{F}', J)}{\partial J}$
$-W_m(\mathbf{F}', p) \equiv W_0^*(\mathbf{F}', J) + pJ$	$(\mathbf{T}')_s = \dfrac{\partial W_m(\mathbf{F}', p)}{\partial \mathbf{F}'},$ $J = -\dfrac{\partial W_m(\mathbf{F}', p)}{\partial p}$

16.15. SMALL DEFLECTION OF THIN PLATES

In the following two sections we shall consider the basic equations in the theory of isotropic elastic plates. Section 16.15 serves as an introduction by considering small deflection only. Section 16.16 treats the main subject of large deflection of plates. From the standpoint of engineering applications, the theory of plates is one of the most important and most interesting topics in the theory of elasticity.

A *plate* is a body bounded by two surfaces of small curvature, the distance between these surfaces, called the *thickness*, being small in comparison with the dimensions of the surface. The thickness of the plate is denoted by h. The surface equidistant to the bounding surfaces is called the *middle surface or midplane.* When h is constant, the plate is said to be of *uniform thickness;* when the middle surface is a plane in the underformed configuration, the plate is said to be *flat.* We shall consider only flat plates of uniform thickness and of homogeneous linear elastic material. Our purpose is to illustrate the application of the concepts developed in Secs. 16.1 to 16.11 to the derivation of basic equations for plates.

A principal feature in straining a plate or a shell is the relative smallness of the traction acting on surfaces parallel to the middle surface as compared with the maximum bending or stretching stresses in the body. For example, the aerodynamic pressure acting on the wings of an airplane in flight is of the order of 1 to 10 pounds per square inch (7 to 70 KPa), whereas the bending stress in the wing skin is likely to range from 10,000 to 200,000 pounds per square inch (7×10^4 to 1.4×10^6 KPa). In many other applications, plates and shells are used to transmit forces and moments acting on their edges, and no load acts on the faces at all. This practical situation is important in simplifying the theory. When a plate is very thin, small traction on the external faces imply that the traction on any surface parallel to the middle surface is also small.

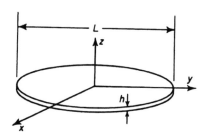

Fig. 16.15:1. A thin plate and the coordinate system.

A flat thin plate is depicted in Fig. 16.15:1. The x, y-plane coincides with the middle surface of the plate in its initial, unloaded state, and the z-axis is normal to it. The thickness of the plate is h, which is much smaller than the dimensions of the plate in the plane of x, y. Under load, the plate deforms, causing elastic displacements u_i (u_x, u_y, u_z).

In particular, $|u_i| \ll h$ is considered. The strains are small. The constitutive equation is Hooke's law.

A general feature of the theory of plates and shells is that it is concerned with applications in which the stress components $\sigma_{zx}, \sigma_{zy}, \sigma_{zz}$ in this coordinate system are very small throughout the plate. Hence, in the determination of the deformation of the plate, it is a good approximation to assume that

(1) $$\sigma_{zx} = 0, \quad \sigma_{zy} = 0, \quad \sigma_{zz} = 0, \quad -h/2 \leq z \leq h/2.$$

The plate deformation is mainly stretching and bending, hence one may assume that σ_{xx} and σ_{yy} vary linearly throughout the thickness, i.e.

(2) $\sigma_{xx} = a_1(x,y) + b_1(x,y)z, \qquad \sigma_{yy} = a_2(x,y) + b_2(x,y)z .$

These assumptions are to be verified a *posteriori* for each boundary-value problem. With the assumptions listed in Eqs. (1) and (2), we have, by Hooke's law,

(3) $e_{zx} = \dfrac{1}{2} \left(\dfrac{\partial u_x}{\partial z} + \dfrac{\partial u_z}{\partial x} \right) = 0, \qquad e_{zy} = \dfrac{1}{2} \left(\dfrac{\partial u_y}{\partial z} + \dfrac{\partial u_z}{\partial y} \right) = 0,$

(4) $e_{zz} = \dfrac{\partial u_z}{\partial z} = -\dfrac{\nu}{E}(\sigma_{xx} + \sigma_{yy}).$

Equations (2) and (4) together yield

(5) $u_z = w(x,y) - \dfrac{\nu}{E}[a_1(x,y) + a_2(x,y)]z - \dfrac{\nu}{E}[b_1(x,y) + b_2(x,y)]\dfrac{z^2}{2} .$

The function $w(x,y)$ represents the vertical deflection of the middle surface of the plate. In view of the thinness of the plate, the last two terms in Eq. (5) are in general small and will be neglected in comparison with $w(x,y)$.[1] Substituting $w(x,y)$ for u_z in Eq. (3) and integrating it, we obtain

(6) $u_x = u(x,y) - z\dfrac{\partial w(x,y)}{\partial x} \quad u_y = v(x,y) - z\dfrac{\partial w(x,y)}{\partial y}, \quad u_z = w(x,y)$

where $u(x,y)$ and $v(x,y)$ are the in-plane displacements of the midsurface (Fig. 16.15:2).

To simply the discussion, we shall employ matrix notations and operations in the remaining part of this chapter. From Eq. (6) we deduce the engineering strain matrix:

(7) $\varepsilon = e - z\kappa ,$

where e are the in-plane strain column matrix at the mid-surface and κ is the curvature column matrix whose components are defined below:

(8) $\varepsilon^T = [\varepsilon_x \quad \varepsilon_y \quad 2\varepsilon_{xy}], \qquad e^T = \left[\dfrac{\partial u}{\partial x} \quad \dfrac{\partial v}{\partial y} \quad \dfrac{\partial u}{\partial y} + \dfrac{\partial v}{\partial x} \right],$

$\kappa^T = \left[\dfrac{\partial^2 w}{\partial x^2} \quad \dfrac{\partial^2 w}{\partial y^2} \quad 2\dfrac{\partial^2 w}{\partial x \partial y} \right],$

---

[1]The consistency of this statement has to be verified a *posteriori*. Note, however, that for a successful design using plates as part of a safe and stable structure, the stresses σ_{xx}, σ_{yy} must be bounded by the yielding stress, $\sigma_{Y.P.}$, of the material. Hence, according to Eq. (16.15:4), e_{zz} is bounded by $2\nu\sigma_{Y.P.}/E$. Thus, the last two terms in Eq. (16.15:5) must be smaller than $2\nu(\sigma_{Y.P.}/E)h$. For most structural materials, $(\sigma_{Y.P.}/E)$ is of order 10^{-3}. Hence the last two terms in Eq. (16.15:5) are negligible if $w/h \gg 10^{-3}$.

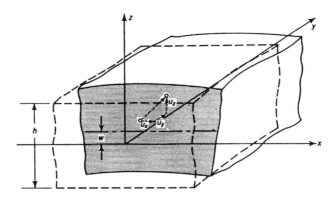

Fig. 16.15:2. Notations for components of displacements.

where the superscript $(\cdot)^T$ denotes the transpose of the relevant quantity. The transpose of a column matrix is a row matrix. In the present definition, ε, $\mathbf{e}$, and $\boldsymbol{\kappa}$ are no longer tensor. By means of Hooke's law, we find the stress matrix

$$(9) \qquad \boldsymbol{\sigma} = \begin{bmatrix} \sigma_{xx} \\ \sigma_{yy} \\ \sigma_{xy} \end{bmatrix} = \mathbf{D}(\mathbf{e} - z\boldsymbol{\kappa}),$$

where

$$(10) \qquad \mathbf{D}_e = \frac{E}{1-\nu^2} \begin{bmatrix} 1 & \nu & 0 \\ \nu & 1 & 0 \\ 0 & 0 & (1-\nu)/2 \end{bmatrix}$$

is the elastic coefficient matrix for a homogeneous isotropic material. In this way, the stresses and deformations in the plate are determined on the basis of the assumptions Eqs. (1) and (2).

The approximate results embodied in Eq. (6) are called *Kirchhoff's hypothesis*, which states that *every straight line in the plate that was originally perpendicular to the plate middle surface remains straight and perpendicular to the deflected middle surface after the strain.* The theory of plates based on this hypothesis is called the *Kirchhoff theory*.

In most cases the *ad hoc* assumptions Eqs. (1) and (2) are approximations. In general, they are not even consistent. Kirchhoff alluded to the theory of simple beams for justification of his hypothesis; but we know from Saint-Venant's beam theory (see Sec. 7.7) that the assumption "plane cross sections remain plane" is incorrect if the resultant shear does not vanish.

For these reasons many authors dislike these *ad hoc* assumptions, and have tried other formulations. A systematic scheme of successive approximations has been developed, of which the "first-order" approximation coincides with those of the Kirchhoff theory. The criticisms provide impetus for further developments of plate theory. However, in retrospect, the truth of the matter is that the recognition of Kirchhoff's hypothesis Eq. (6) was the most important discovery in the theory of plates.

The Kirchhoff hypothesis reduces the equations of three-dimensional elastic continuum to those of a two-dimensional non-Euclidean continuum of the middle surface, describing the behavior of the plate in terms of the displacements (and their derivatives) of points on the middle surface.

Let us now consider the equations of equilibrium. Since the approximate stress distribution throughout the thickness is already known [see Eq. (9)], we shall derive the approximate equations of equilibrium by integrating Eqs. (7.1:5) over the thickness. For small strain, the Cauchy and Kirchhoff stress tensors [see Eq. (16.7:3)] are approximately the same. We shall use σ's to denote stress in the deformed configuration. We assume that over the faces of the plate ($z = \pm h/2$) there act the normal stress $\sigma_{zz}(\pm h/2)$ and the shear stresses $\sigma_{zx}(\pm h/2)$, $\sigma_{zy}(\pm h/2)$. Figure 16.15:3 shows the directions of stress vectors, if the external loads $\sigma_{zx}(\pm h/2)$, $\sigma_{zy}(\pm h/2)$, $\sigma_{zz}(\pm h/2)$ are positive. By integrating the stresses σ_{xx}, σ_{xy}, σ_{yy} through the thickness, we obtain the stress resultants N_{xx}, N_{xy}, N_{yy} defined by

$$(11) \quad N_x = \int_{-h/2}^{h/2} \sigma_{xx}\, dz\,, \quad N_{xy} = \int_{-h/2}^{h/2} \sigma_{xy}\, dz\,, \quad N_y = \int_{-h/2}^{h/2} \sigma_{yy}\, dz\,.$$

Now integrating the first two equations of Eq. (7.1:5) with respect to z from $-h/2$ to $h/2$, we obtain

$$(12) \quad \frac{\partial N_x}{\partial x} + \frac{\partial N_{xy}}{\partial y} + f_x = 0\,, \qquad \frac{\partial N_{xy}}{\partial x} + \frac{\partial N_y}{\partial y} + f_y = 0\,,$$

where f's are the *external loads tangent to the plate* (forces per unit area)

$$f_x = \sigma_{zx}(h/2) - \sigma_{zx}(-h/2) + \int_{-h/2}^{h/2} b_{0x}\, dz\,,$$

$$f_y = \sigma_{zy}(h/2) - \sigma_{zy}(-h/2) + \int_{-h/2}^{h/2} b_{0y}\, dz\,,$$

where b's are the components of body force per unit volume.

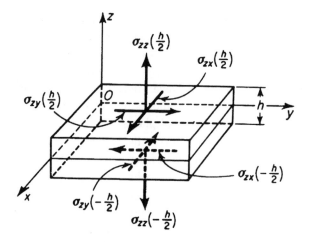

Fig. 16.15:3. External loads acting on the plate surfaces.

Now if we multiply the first two equations of Eq. (7.1:5) by zdz and integrate from $-h/2$ to $h/2$, we will obtain the results

(13) $\dfrac{\partial M_x}{\partial x} + \dfrac{\partial M_{xy}}{\partial y} - Q_x + m_x = 0, \quad \dfrac{\partial M_{xy}}{\partial x} + \dfrac{\partial M_y}{\partial y} - Q_y + m_y = 0,$

where

(14) $M_x = \displaystyle\int_{-h/2}^{h/2} \sigma_{xx}z\,dz, \quad M_{xy} = \int_{-h/2}^{h/2} \sigma_{xy}z\,dz, \quad M_y = \int_{-h/2}^{h/2} \sigma_{yy}z\,dz,$

(15) $Q_x = \displaystyle\int_{-h/2}^{h/2} \sigma_{zx}\,dz, \qquad Q_y = \int_{-h/2}^{h/2} \sigma_{zy}\,dz,$

$$m_x = \frac{h}{2}\left[\sigma_{zx}(h/2) + \sigma_{zx}(-h/2)\right] + \int_{-h/2}^{h/2} b_{0x}\,z\,dz,$$

$$m_y = \frac{h}{2}\left[\sigma_{zy}(h/2) + \sigma_{zy}(-h/2)\right] + \int_{-h/2}^{h/2} b_{0y}z\,dz.$$

The quantities M_x, M_y are called the *bending moment* and M_{xy} the *twisting moment*; m_x, m_y are the *resultant external moment* per unit area about the middle surface; Q_x, Q_y, of the dimensions force per unit length, are called the *transverse shear*. Clearly, the moment arm is $h/2$ for shear on the faces and z in the plate. These stress resultants and moments are illustrated in Figs. 16.15:4 and 16.15:5.

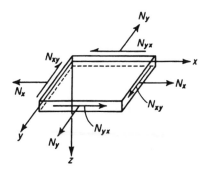

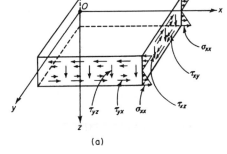

(a)

Fig. 16.15:4. Stress resultants.

If the third equation of Eq. (7.1:5) is multiplied by dz and integrated from $-h/2$ to $h/2$, we obtain

(16) $$\frac{\partial Q_x}{\partial x} + \frac{\partial Q_y}{\partial y} + p = 0$$

where p is the external load per unit area normal to the middle surface defined as

$$p = \sigma_{zz}(h/2) - \sigma_{zz}(-h/2)$$
$$+ \int_{-h/2}^{h/2} b_{0z}\, dz\,.$$

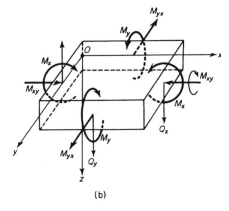

(b)

Fig. 16.15:5. Moments.

Eliminating Q_x, Q_y from Eq. (16), we have the equation of equilibrium in moments,

(17) $$\frac{\partial^2 M_x}{\partial x^2} + 2\frac{\partial^2 M_{xy}}{\partial x \partial y} + \frac{\partial^2 M_y}{\partial y^2} = -\frac{\partial m_x}{\partial x} - \frac{\partial m_y}{\partial y} - p\,.$$

To complete the formulation, we substitute σ_{xx}, σ_{yy}, σ_{xy} from Eq. (9) into Eqs. (11) and (14) to obtain

(18) $$\mathbf{N} = \begin{bmatrix} N_x \\ N_y \\ N_{xy} \end{bmatrix} = \mathbf{D}_n \mathbf{e}\,,$$

(19)
$$\mathbf{M} = \begin{bmatrix} M_x \\ M_y \\ M_{xy} \end{bmatrix} = -\mathbf{D}_b \boldsymbol{\kappa},$$

where

(20)
$$\mathbf{D}_n = \int_{-h/2}^{h/2} \mathbf{D}_e \, dz, \qquad \mathbf{D}_b = \int_{-h/2}^{h/2} z \mathbf{D}_e \, dz,$$

with $\mathbf{D}_e$ given in Eq. (10). For homogeneous isotropic materials,

$$\mathbf{D}_n = \frac{Eh}{1 - \nu^2} \begin{bmatrix} 1 & \nu & 0 \\ \nu & 1 & 0 \\ 0 & 0 & (1-\nu)/2 \end{bmatrix},$$

$$\mathbf{D}_b = D \begin{bmatrix} 1 & \nu & 0 \\ \nu & 1 & 0 \\ 0 & 0 & (1-\nu)/2 \end{bmatrix}, \quad D = \frac{Eh^3}{12(1-\nu^2)}$$

in which D is called the *bending rigidity of the plate*.

Equations (17) and (19), involving the deflection $w(x, y)$, define the so-called *bending problem*. Equations (12) and (18), involving the displacements u, v in the midplane of the plate, define the so-called *stretching problem*. An examination of these equations shows that in the linear theory of plates, under the assumption of infinitesimal displacements, the bending and the stretching of the plate are independent of each other, i.e., no coupling between these two responses. It will be seen later that for large deflection, the bending and stretching become coupled and introduce nonlinearity. A substitution of Eq. (19) into Eq. (17) yields the fundamental equation of the linear theory of bending of a plate, which is a fourth order equation in terms of w.

For plates of uniform thickness, the equation becomes

(21)
$$\frac{\partial^4 w}{\partial x^4} + 2\frac{\partial^4 w}{\partial x^2 \partial y^2} + \frac{\partial^4 w}{\partial y^4} = \frac{1}{D}\left(p + \frac{\partial m_x}{\partial x} + \frac{\partial m_y}{\partial y}\right).$$

Thus the problem of bending is reduced to the solution of a biharmonic equation with appropriate boundary conditions. The mathematical problem is, therefore, the same as that of the Airy stress functions in plane elasticity. The mathematical problem of plate stretching is identical with the plane stress problem, and Airy stress functions can be used.

16.16. LARGE DEFLECTION OF PLATES

The theory of small deflections of plates was derived in the previous section under the assumption of infinitesimal displacements. Unfortunately, the results are valid literally only for very small displacements. When the deflections are as large as the thickness of the plate, the results become quite inaccurate. This is in sharp contrast with the theory of beams, for which the linear equation is valid as long as the slope of the deflection curve is small in comparison to unity.

A well-known theory of large deflections of plates is due to von Kármán. In this theory the following assumptions are made.

(HI) *The plate is thin. The thickness h is much smaller than the characteristic dimension L of the plate in its plane, i.e., $h \ll L$.*

(H2) *The magnitude of the deflection w is of the same order of magnitude as the thickness of the plate, h, but small compared with the typical plate dimension L, i.e.,*

$$|w| = O(h), \qquad |w| \ll L.$$

(H3) *The slope is everywhere small, $|\partial w/\partial x| \ll 1$, $|\partial w/\partial y| \ll 1$.*

(H4) *The tangential displacements, u, v are infinitesimal. In the strain-displacement relations, only those nonlinear terms, which depend on $\partial w/\partial x$ and $\partial w/\partial y$ are retained. All other nonlinear terms are neglected.*

(H5) *All strain components are small. Hooke's law holds.*

(H6) *Kirchhoff's hypotheses hold; i.e., the traction on surfaces parallel to the middle surface are small, strains vary linearly with z, the distance from the midplane, within the plate, and normal remains normal.*

Thus, von Karman's theory of plates differs from the linear theory of plates only in retaining certain powers of the derivatives $\partial w/\partial x$ and $\partial w/\partial y$ in the strain-displacement relationship.

We now have a basic small parameter h. The elastic displacement w is assumed to be of the same order of magnitude as h. In other words, the term *large deflection* refers to the fact that w is no longer small compared with h. Thus, as shown in Fig. 16.16:1, the deformed configuration differs considerably from the original one. We cannot imitate the procedure of Sec. 16.15 without further care. For example, if we retain the rectangular coordinates, with the z-plane fixed as the original middle plane of the plate in the undeformed position, then the limits of integration across the plate thickness for a loaded plate can no longer be from $-h/2$ to $h/2$. If we wish to retain the convenience of fixed coordinates for the surfaces, we have two alternative courses. The first is to introduce "convective" coordinates,

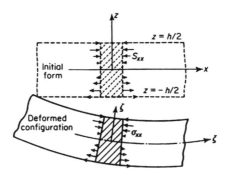

Fig. 16.16:1. Corresponding stresses in the initial and deformed configuration of a plate.

which become curvilinear in the deformed state, i.e., a set of curvilinear coordinates imbedded in the plate so that the middle surface is always the surface $z = 0$, and the faces of the plate $z = \pm h/2$. The other is to adopt the Lagrangian description for the plate. In Lagrangian coordinates the plate surfaces are always designated as $z = \pm h/2$. The second alternative is simpler, and will be pursued below.

In the following we use the *Lagrangian description* and *Green's strain tensor*, referred to the initial configuration. The strain components are [identifying X_1, X_2, X_3 with x, y, z] are given in Eq. (16.3:5). According to the *ad hoc* assumptions (H6), we have the displacement components as given in Eq. (16.15:6). According to (H4) the higher powers of the derivatives of u and v are negligible in comparison with their first powers; and since $|z| \leq h \ll L$, it can be argued that the only nonlinear terms to be retained in the strain-displacement relations Eqs. (16.15:7) and (16.15:8) are the squares and products of $\partial w/\partial x$, $\partial w/\partial y$. Hence,

(1) $$\mathbf{E} = \mathbf{e} + \frac{\mathbf{e}_N}{2} - z\boldsymbol{\kappa},$$

where $\mathbf{e}_N$ are the nonlinear part of the strain at the mid-plane, and

(2) $$\mathbf{e}_N^T = \left[\left(\frac{\partial w}{\partial x}\right)^2 \quad \left(\frac{\partial w}{\partial y}\right)^2 \quad 2\frac{\partial w}{\partial x}\frac{\partial w}{\partial y} \right].$$

Conjugate to Green's strain tensor is the second Piola-Kirchhoff stress tensor $\mathbf{S}$ [see Eq. (16.7:6)], which can be used in the Lagrangian description. We assume, according to (H5), that $\mathbf{S}$ is related to $\mathbf{E}$ through Hooke's law,

$$\mathbf{S} = \begin{bmatrix} S_{xx} \\ S_{yy} \\ S_{xy} \end{bmatrix} = \mathbf{D}\left(\mathbf{e} + \frac{\mathbf{e}_N}{2} - z\boldsymbol{\kappa}\right),$$

or in terms of the stress resultants and moments

$$(3) \quad \mathbf{N}' = \begin{bmatrix} N'_x \\ N'_y \\ N'_{xy} \end{bmatrix} = \mathbf{D}_n(\mathbf{e} + \mathbf{e}_N/2), \qquad \mathbf{M}' = \begin{bmatrix} M'_x \\ M'_y \\ M'_{xy} \end{bmatrix} = -\mathbf{D}_b\boldsymbol{\kappa},$$

where $\mathbf{D}_n$ and $\mathbf{D}_b$ are given in Eq. (16.15:20) and the stress resultants N's and the moments M's are defined as

$$(4) \quad N'_x = \int_{-h/2}^{h/2} S_{xx}dz, \quad N'_{xy} = \int_{-h/2}^{h/2} S_{xy}dz, \quad N'_y = \int_{-h/2}^{h/2} S_{yy}dz,$$

$$(5) \quad M'_x = \int_{-h/2}^{h/2} S_{xx}zdz, \quad M'_{xy} = \int_{-h/2}^{h/2} S_{xy}zdz, \quad M'_y = \int_{-h/2}^{h/2} S_{yy}zdz.$$

The prime indicates that N's and M's are based on the second Piola-Kirchhoff stresses and are referred to the initial, unloaded configuration of the plate.

The stress tensor $\mathbf{S}$, the resultants N'_x, N'_y, N'_{xy}, and the moments M'_x, M'_y, M'_{xy} are defined with respect to the original, unloaded configuration of the plate. Since all partial derivatives u, v, w are small in comparison with unity, the values of $\mathbf{S}$, N'_x, N'_y, N'_{xy}, M'_x, M'_y, M'_{xy} are in the first order approximation *equal*, respectively, to the Cauchy or Kirchhoff stress tensor σ_{ij}, the stress resultants N_x, N_y, N_{xy}, and the moments M_x, M_y, M_{xy} defined in the deformed configuration of the plate. See Eq. (16.7:10) which relates the Kirchhoff stress tensor σ_{ij} to the second Piola-Kirchhoff stress tensor S_{ij}. We will drop the prime in the subsequent discussion.

In the Lagrangian description, the condition of equilibrium is given by Eq. (16.10:7)

$$\frac{\partial}{\partial X_J}\left(S_{JK}\frac{\partial x_i}{\partial X_K}\right) + b_{0i} = 0,$$

in indicial notation, where b's are the body force per unit volume. Identifying X_1, X_2, X_3 with x, y, z and identifying x_1, x_2, x_3 with $x + u_x$, $y + u_y$, $z + u_z$, we obtain, on retaining only the first-order terms (i.e., neglecting any product terms that include u or v), the equations become

$$(6) \quad \frac{\partial S_{xx}}{\partial x} + \frac{\partial S_{xy}}{\partial y} + \frac{\partial S_{zx}}{\partial z} + b_{0x} = 0, \quad \frac{\partial S_{xy}}{\partial x} + \frac{\partial S_{yy}}{\partial y} + \frac{\partial S_{zy}}{\partial z} + b_{0y} = 0,$$

$$(7) \quad \frac{\partial}{\partial x}\left(S_{xx}\frac{\partial w}{\partial x} + S_{xy}\frac{\partial w}{\partial y} + S_{zx}\right) + \frac{\partial}{\partial y}\left(S_{xy}\frac{\partial w}{\partial x} + S_{yy}\frac{\partial w}{\partial y} + S_{zy}\right)$$

$$+ \frac{\partial}{\partial z}\left(S_{zx}\frac{\partial w}{\partial x} + S_{zy}\frac{\partial w}{\partial y} + S_{zz}\right) + b_{0z} = 0.$$

Note that u_x, u_y, u_z are given in Eq. (16.15:6).

The rest of the development of the large deflection theory parallels closely that of the classical linear theory. Hence, our presentation will be brief, and we refer to the preceding section for many details. If Eq. (6) is multiplied successively by dz and zdz and integrated from $-h/2$ to $h/2$ and if only the first order terms are retained, the governing equations for the membrane stress resultants N's and the moments M's are the same as Eqs. (16.15:12) and (16.15:13) with σ_{ij} replaced by S_{ij} in the formulas for the f's, Q's, m's, and M's.

Finally, an integration of Eq. (7) with respect to z from $-h/2$ to $h/2$ gives

(8)
$$\frac{\partial}{\partial x}\left(Q_x + N_x\frac{\partial w}{\partial x} + N_{xy}\frac{\partial w}{\partial y}\right)$$
$$+ \frac{\partial}{\partial y}\left(Q_y + N_{xy}\frac{\partial w}{\partial x} + N_y\frac{\partial w}{\partial y}\right) + p = 0$$

where p' is the lateral load per unit area (of the undeformed middle plane) acting on the plate,

(9)
$$p = \left(S_{zx}\frac{\partial w}{\partial x} + S_{zy}\frac{\partial w}{\partial y} + S_{zz}\right)_{-h/2}^{h/2} + \int_{-h/2}^{h/2} b_{0z}dz\,.$$

The last two terms on the right-hand side of Eq. (9) are the familiar lateral load. The first and second terms represent contributions to the lateral load due to shear acting on the surfaces that are rotated in the deformed position. Equation (8) is substantially different from the corresponding Eq. (16.15:16) of the small-deflection theory. We can eliminate Q_x, Q_y between Eqs. (8) and (16.15:13) to obtain

$$\frac{\partial^2 M_x}{\partial x^2} + 2\frac{\partial^2 M_{xy}}{\partial x\partial y} + \frac{\partial^2 M_y}{\partial y^2} = -p - \frac{\partial m_x}{\partial x} - \frac{\partial m_y}{\partial y}$$
$$- \frac{\partial}{\partial x}\left(N_x\frac{\partial w}{\partial x} + N_{xy}\frac{\partial w}{\partial y}\right) - \frac{\partial}{\partial y}\left(N_{xy}\frac{\partial w}{\partial x} + N_y\frac{\partial w}{\partial y}\right).$$

Finally, a substitution of Eq. (3) into the above gives, for a plate of uniform thickness,

(10) ▲
$$\frac{\partial^4 w}{\partial x^4} + 2\frac{\partial^4 w}{\partial x^2\partial y^2} + \frac{\partial^4 w}{\partial y^4} = \frac{1}{D}\left[p + \frac{\partial m_x}{\partial x} + \frac{\partial m_y}{\partial y}\right.$$
$$\left. + \frac{\partial}{\partial x}\left(N_x\frac{\partial w}{\partial x} + N_{xy}\frac{\partial w}{\partial y}\right) + \frac{\partial}{\partial y}\left(N_{xy}\frac{\partial w}{\partial x} + N_y\frac{\partial w}{\partial y}\right)\right].$$

A substitution of Eq. (3) into Eq. (16.15:12) yields immediately the basic equations for u and v. Alternatively, a stress function $F(x, y)$ can be introduced that

$$N_x = \frac{\partial^2 F}{\partial y^2} - P_x, \quad N_y = \frac{\partial^2 F}{\partial x^2} - P_y, \quad N_{xy} = -\frac{\partial^2 F}{\partial x \partial y}$$

satisfy Eq. (16.15:12) identically, where the P's are the potentials of the tangential forces given by the indefinite integrals

$$(11) \qquad P_x = \int f_x dx, \qquad P_y = \int f_y dy.$$

The compatibility condition can be derived by eliminating u and v from Eq. (3). Using Eq. (11), we obtain, for a plate of uniform thickness,

$$(12) \; \blacktriangle \qquad \frac{\partial^4 F}{\partial x^4} + 2\frac{\partial^4 F}{\partial x^2 \partial y^2} + \frac{\partial^4 F}{\partial y^4} = Eh\left[\left(\frac{\partial^2 w}{\partial x \partial y}\right)^2 - \frac{\partial^2 w}{\partial x^2}\frac{\partial^2 w}{\partial y^2}\right]$$
$$+ \frac{\partial^2(P_y - \nu P_x)}{\partial x^2} + \frac{\partial^2(P_x - \nu P_y)}{\partial y^2}.$$

Equation (10) and (12) are the famous *von Karman equations for large deflection of plates*.

In most problems the external loads m's, and P's are zero. We then have the well-known pair of equations[2]

$$(13) \; \blacktriangle \qquad \nabla^4 w = \frac{1}{D}\left(p + \frac{\partial^2 F}{\partial y^2}\frac{\partial^2 w}{\partial x^2} - 2\frac{\partial^2 F}{\partial x \partial y}\frac{\partial^2 w}{\partial x \partial y} + \frac{\partial^2 F}{\partial x^2}\frac{\partial^2 w}{\partial y^2}\right),$$

$$(14) \; \blacktriangle \qquad \nabla^4 F = Eh\left[\left(\frac{\partial^2 w}{\partial x \partial y}\right)^2 - \frac{\partial^2 w}{\partial x^2}\frac{\partial^2 w}{\partial y^2}\right],$$

where ∇^4 is the two-dimensional biharmonic operator.

The most important features of Eqs. (13) and (14) are that they are coupled and nonlinear. The nonlinear terms are fairly easy to interpret. Equation (8) shows that the membrane stress N_x, N_y, N_{xy} multiplied by the corresponding curvatures and twist, $\partial^2 w/\partial x^2$, $\partial^2 w/\partial y^2$, $\partial^2 w/\partial x \partial y$ respectively, are equivalent to a lateral force. The nonlinear terms in Eq. (10) or (13) represent such feedback of induced membrane stress on bending. On

---

[2]These equations were given without proof by von Karman in 1910. Most books give derivations of these equations without a clear indication as to whether Lagrangian or Eulerian descriptions were used. The explicit use of Lagrangian description introduces a degree of clarity not hitherto achieved.

the other hand, the reader may recall from differential geometry that if a surface is represented by the equation $z = w(x, y)$, the total or Gaussian curvature of the surface is given by the quantity $\frac{\partial^2 w}{\partial x^2} \frac{\partial^2 w}{\partial y^2} - \left(\frac{\partial^2 w}{\partial x \partial y} \right)^2$. The Gaussian curvature of a surface is equal to the product of the principal curvatures at a point. The Gaussian curvature vanishes if the surface is developable. Hence, the right-hand side of Eq. (14) vanishes if the deflection surface is developable. If a flat plate is deformed into a nondevelopable surface, its middle surface will be stretched in some way and the right-hand side of Eq. (14) does not vanish. Thus, the nonlinear term on the right-hand side of Eq. (14) arises from the stretching of the middle surface of a flat plate due to bending into a nondevelopable surface. When the nonlinear terms are neglected, these equations reduce to the corresponding equations of the small-deflection theory.

There are a large number of papers on the theory of plates. Some are devoted to justifying the *ad hoc* assumptions (HI)–(H6), or to their removal or generalization; others are concerned with the solution of boundary-value problems. See Bibliography 16.14. We shall conclude our discussion here, merely pointing out that the mathematical structure of the in-plane deformation of plates in small deflection is the same as that of the plane problems of linear elasticity, whereas the large-deflection equations are essentially nonlinear. The theory of large deflection of plates includes many interesting problems on the stability, responses, vibrations, subharmonic resonances, edge layers, etc. The explicit introduction of both Eulerian and Lagrangian descriptions of stress and strain clarifies the foundation of the large-deflection theory of plates. Readers are referred to literature for the theory of shells, e.g. Atluri (1984).

Problem 16.16. Section 16.9 establishes that the displacement vector $\mathbf{u}$, the Almansi strain tensor $\mathbf{e}$, the metric tensor related the deformed body to the undeformed body $\mathbf{B}^{-1}$, the left stretch tensor $\mathbf{V}$, the symmetric part of the deformation rate tensor $\mathbf{D}$, and the Cauchy and Kirchhoff stress tensors $\boldsymbol{\tau}$ and $\boldsymbol{\sigma}$ are objective and that the velocity vector $\mathbf{v}$, the Green strain tensor $\mathbf{E}$, the metric tensor related the underformed body to the deformed body $\mathbf{C}$, the deformation gradient $\mathbf{F}$, the rotational tensor $\mathbf{R}$, the right stretch tensor $\mathbf{U}$, the Lagrange stress tensor $\mathbf{T}$, the second Piola-Kirchhoff stress tensor $\mathbf{S}$, the Biot stress tensors $\mathbf{r}^*$, $\mathbf{r}$, the corotational stress tensor $\boldsymbol{\sigma}_r$ the velocity gradient $\mathbf{L}$, the spin rate $\boldsymbol{\Omega}$, and all stress rates are not objective. Work out the details to prove the results.

17

INCREMENTAL APPROACH TO
SOLVING SOME NONLINEAR PROBLEMS

The field equations of the mechanics of solids subjected to large deformation and finite strain are nonlinear with respect to displacements and velocities. The constitutive equation is linear or nonlinear depending on the material. The boundary conditions could be nonlinear if the external forces and the displacements of particles on the boundary are coupled in a certain nonlinear way. The external loading may depend on the elastic deformation as in aeroelasticity or in some special structural problems. These features may result in a nonlinear boundary-value or initial-value problem, which is difficult to solve. One observes, however, that large deformation or motion are arrived at through many infinitesimal steps, and for each infinitesimal step it is possible that the linearized equations prevail in some sense. If we can solve the linearized equations and keep track of all the infinitesimal steps, then we can solve the nonlinear problem. This is the basic idea of the incremental approach. The mathematics of the incremental approach will be described in this chapter. The incremental approach is used not only for problems of elasticity. For some problems of plasticity, visco-elasticity, and biomechanics, it is also a natural approach because sometimes the constitutive laws of the materials involved in these fields can be described only incrementally.

In this chapter, we shall derive the linearized equations for the increments of deformations and stresses, the linearized equations of motion in terms of these increments, and the corresponding variational principles for integral formulation. Computational methods for solving these linearized incremental equations will be discussed in the sequent chapters. Finite element modelling of nonlinear elasticity, vicoelasticity, plasticity, viscoplasticity, and creep is discussed in Chapter 21.

17.1. UPDATED LAGRANGIAN DESCRIPTION

The following derivation follows mostly that by Atluri and Cazzani (1995). We describe the deformation as a multi-increment process whose evolution is described by means of a parameter t, which can actually be time as in dynamic or creep problems but often it is just the loading magnitude of a proportional load. Implicit in the assumption is that the solution is independent of the rate of change of the parameter. Let D_0 be the *reference configuration* with region V_0 occupied by the body at $t = t_0$ and $D_1, D_2, \ldots$, be the sequence of configurations occupied by the body at $t = t_1, t_2, \ldots$, respectively. Thus, D_i represent the configurations at different states of deformation. We shall use D_N to denote the current configuration, D_{N+1} the configuration at the subsequent increment, and D_0 the undeformed state. The increment approach seeks the incremental solution at $t = t_{N+1}$ with the solution at t_N known. We shall use **bold** characters such as $\mathbf{A}_I^J(\mathbf{X})$ as a tensor or vector, and non-bold *italic* characters such as $A_I^J(\mathbf{X})$ for scalar at the material point $\mathbf{X}$. The superscript J and the subscript I denote that the concerned quantity is in the D_J configuration (i.e., at $t = t_J$) but referred to the configuration D_I (i.e., the configuration at $t = t_I$). For example, we use $\mathbf{u}_I^J(\mathbf{X})$ to describe the relative displacement vector at the material point at $\mathbf{X}$ between the deformation state t_J and t_I. As another example, the first Piola-Kirchhoff stress tensor $\mathbf{T}$ as defined in Eq. (16.7:4) is a stress measures in the current configuration (i.e., at $t = t_N$) referred to the undeformed corrdinates. Hence, we have $\mathbf{T} = \mathbf{T}_0^N(\mathbf{X})$ as the subscript 0 refers to the undeformed configuration and the superscript N denotes the current configuration. On the other hand, $\mathbf{T}_I^N$ is the first Piola-Kirchhoff stress tensor defined according to the Lagrangian rule Eq. (16.7:15) with reference surface $P_0\,Q_0\,R_0\,S_0$ as shown in Fig. 16.7:1 being the surface at the deformation state $t = t_I$, i.e., the D_I configuration. In subsequent discussion, for simplicity, we shall drop the explicit reference to the material point $\mathbf{X}$ by simply writing $\mathbf{A}_I^J$, A_I^J for $\mathbf{A}_I^J(\mathbf{X})$, $A_I^J(\mathbf{X})$, respectively, except occasionally use $\mathbf{X}$ to emphasis clarity.

In Sec. 5.2, the material and spatial descriptions of changing configurations of matter are discussed, and the associated Lagrangian and Eulerian variables are considered. In solid mechanics, the material or Lagrangian description is used more often. In the incremental approach, one further distinguishes the *total Lagrangian description* and the *updated Lagrangian description*. The former refers all quantities of the latest increment to the reference configuration D_0, denoted as $\mathbf{A}_0^{N+1}$ or A_0^{N+1}, while the latter refers quantities to the current configuration, denoted by $\mathbf{A}_N^{N+1}$ or A_N^{N+1}. From computational point of view, the use of the *updated Lagrangian description* requires the updated configuration at each increment, whereas the *total Lagrangian description* needs no configuration update but leads to a much more cumbersome expression. The updated Lagrangian description is often preferred in practical applications and is used in the subsequent discussion to establish the linearized equations.

Consider a quantity $\mathbf{A}$, which associates with a parameter t. In the updated Lagrangian description, the *material rate* of the quantity corresponding to the parameter at $t = t_N$ is defined as

$$\text{(1)} \quad \blacktriangle \qquad \overset{\circ}{\mathbf{A}}{}^N(\mathbf{X}) \equiv \frac{\mathbf{A}_N^{N+1}(\mathbf{X}) - \mathbf{A}_N^N(\mathbf{X})}{\Delta t^N},$$

where $\mathbf{A}$ can be a tensor, a vector or a scalar (if it is a non-boldfaced italic character) and $\Delta t^N (= t_{N+1} - t_N)$ is assumed to be small. In subsequent discussion, for simplicity, we shall drop the explicit reference to the material point $\mathbf{X}$ and the deformation state N by simply using $\overset{\circ}{\mathbf{A}}$ and Δt except occasionally include $\mathbf{X}$ and N to emphasis clarity. Then

$$\text{(2)} \qquad \mathbf{A}_N^{N+1} = \mathbf{A}_N^N + \overset{\circ}{\mathbf{A}} \Delta t.$$

The incremental approach is to determine $\overset{\circ}{\mathbf{A}}$ approximately by linearizing the governing equations for all material rates at D_N and evaluate the incremental solution according to Eq. (2).

On the other hand, the *material rate in the total Lagrangian description* is defined as

$$\text{(3)} \qquad \frac{D\mathbf{A}_0^N(\mathbf{X})}{Dt} = \frac{\mathbf{A}_0^{N+1}(\mathbf{X}) - \mathbf{A}_0^N(\mathbf{X})}{\Delta t^N}.$$

The material rates in the total and updated Lagrangian descriptions are related. The relationship depends on the definition of $\mathbf{A}$. We shall examine the relationship for specific cases in subsequent sections.

If $\mathbf{A}$ represents a strain measure and t is actually the time, than in the limit of $t_{N+1} \to t_N$, the material rate described in Eq. (3) is the strain rate $\dot{\mathbf{A}}$ defined in Sec. 16.5. Thus, we may approximate the material rate

$$\frac{D\mathbf{A}_0^N(\mathbf{X})}{Dt} \cong \dot{\mathbf{A}},$$

when Δt is small but finite, and write

$$\text{(4)} \qquad \mathbf{A}_0^{N+1} \cong \mathbf{A}_0^N + \dot{\mathbf{A}}\Delta t.$$

In general, the material rate $\overset{\circ}{\mathbf{A}}$ in the update Lagrangian description differs from the material rate $\dot{\mathbf{A}}$. Especially the definition of stress rate in the updated Lagrangain description needs much more care. Since stress is defined by a force vector and a surface, one must consider both the rate of change of the force vector and the rate of change of the associated surface with respect to the parameter when one attempts to define a stress rate. In fact, stress rate can be defined in a number of ways. By introducing the parameter t in the updated and total Lagrangian descriptions, one liberalizes but not necessarily simplifies or settles the matter. The simplified linearized case will be discussed in Sec. 17.3.

17.2. LINEARIZED RATES OF DEFORMATION

We shall use the notations of Sec. 16.2. Let $\mathbf{u}$ denote a relative displacement vector. Then, $\mathbf{u}_N^N$ is the relative displacement vector in the same deformation state N and is therefore zero. Hence

$$(1) \qquad \mathbf{u}_N^N = 0 \,.$$

$\mathbf{F}_N^N$ is the deformation gradient tensor associated with the displacement field $\mathbf{u}_N^N$ and is then an identity,

$$(2) \qquad \mathbf{F}_N^N = \mathbf{I} \,.$$

From the definition in Eqs. (16.2:5) and (16.4:4), we find

$$(3) \qquad J_N^N = \det(\mathbf{F}_N^N) = \det(\mathbf{I}) = 1 \,,$$

$$(4) \qquad \mathbf{U}_N^N = \mathbf{R}_N^N = \mathbf{I} \,.$$

In deriving the approximate material rates in the updated Lagrangain description, we assume that Δt is small. The rate of change of displacement $\mathbf{u}$ is the *velocity-like vector* denoted by

$$(5) \qquad \overset{\circ}{\mathbf{u}} \equiv \frac{\mathbf{u}_N^{N+1}}{\Delta t} \cong \mathbf{v} \,.$$

There is no definition for an acceleration-like vector in the updated Lagrangian description. In subsequent discussion, we shall use "$\equiv$" and "$\cong$" to denote definition and approximation.

In general, since the spatial coordinates of quantities are referred to the current configuration in the updated Lagrangian description, the material rate of a strain measure is equivalent to the corresponding rate in the Eulerian description. Thus, one can determine the rates of deformation in the updated Lagrangian description from the rates given in Secs. 16.5 and 16.6 by replacing the undeformed state with the current one as the reference configuration. From Eqs. (2) and (16.5:3), (16.5:5) and (16.5:6), the updated Lagrangain rate of the deformation gradient tensor is

$$(6) \; \blacktriangle \qquad \overset{\circ}{\mathbf{F}} \equiv \dot{\mathbf{F}}_N^N = (\mathbf{L} \cdot \mathbf{F})_N^N = \mathbf{L}_N^N \cdot \mathbf{F}_N^N = \mathbf{L} = \mathbf{D} + \boldsymbol{\Omega} \,,$$

where $\mathbf{L}$ is the *velocity-like gradient* and $\mathbf{D}$ and Ω are the symmetric and antisymmetric parts of $\mathbf{L}$. Recall that τ, $\mathbf{L}$, $\mathbf{D}$, Ω and all rates ($\overset{\circ}{\ }$) are referred to the current configuration $\mathbf{D}_N$ and that $\mathbf{F}$, $\mathbf{R}$ and J are the current deformation gradient tensor, rotation tensor and Jocobian associated with zero displacement. Equation (6) can be derived also directly from Eq. (17.1:1),

$$\overset{\circ}{\mathbf{F}} \equiv \frac{\mathbf{F}_N^{N+1} - \mathbf{F}_N^N}{\Delta t} = \left[\frac{\partial(\mathbf{X}^N + \mathbf{u}_N^{N+1})}{\partial \mathbf{X}^N} - \mathbf{I} \right] \frac{1}{\Delta t}$$

$$= \frac{\partial}{\partial \mathbf{X}^N} \frac{\mathbf{u}_N^{N+1}}{\Delta t} = \frac{\partial \mathbf{v}}{\partial \mathbf{X}^N} = \mathbf{L}.$$

Since $\mathbf{F} = \mathbf{R} \cdot \mathbf{U}$, one has

(7) $\overset{\circ}{\mathbf{F}} \cong (\mathbf{R} \cdot \dot{\mathbf{U}} + \dot{\mathbf{R}} \cdot \mathbf{U})_N^N = \mathbf{R}_N^N \cdot \overset{\circ}{\mathbf{U}} + \overset{\circ}{\mathbf{R}} \cdot \mathbf{U}_N^N = \overset{\circ}{\mathbf{U}} + \overset{\circ}{\mathbf{R}}$.

Notice that the decomposition of the deformation gradient rate is additive as opposed to the multiplicative polar decomposition for finite deformation given in Eq. (16.5:3). Since $\mathbf{R} \cdot \mathbf{R}^T = \mathbf{I}$, its deformation rate is zero,

$$\overset{\circ}{\mathbf{R}} \cdot (\mathbf{R}_N^N)^T + \mathbf{R}_N^N \cdot \overset{\circ}{\mathbf{R}}^T = \overset{\circ}{\mathbf{R}} + \overset{\circ}{\mathbf{R}}^T = \mathbf{0},$$

which implies that $\overset{\circ}{\mathbf{R}}$ is antisymmetric. Since $\mathbf{U}$ is symmetric, then

$$\overset{\circ}{\mathbf{U}} \equiv \frac{\mathbf{U}_N^{N+1} - \mathbf{U}_N^N}{\Delta t}$$

is symmetric. We call $\overset{\circ}{\mathbf{U}}$ the *updated Lagrangian strain rate tensor* and $\overset{\circ}{\mathbf{R}}$ the *updated Lagrangian spin tensor*. A comparison of Eqs. (6) and (7), and the use of the symmetric and antisymmetric properties of $\overset{\circ}{\mathbf{U}}$ and $\overset{\circ}{\mathbf{R}}$ give

(8) ▲ $\overset{\circ}{\mathbf{U}}{}^N \cong \mathbf{D} = (\mathbf{L})_s$,

(9) ▲ $\overset{\circ}{\mathbf{R}}{}^N \cong \Omega = (\mathbf{L})_a$,

where $(\cdot)_s$ and $(\cdot)_a$ denote the symmetric and antisymmetric parts of the revelant quantity. Thus, the *updated Lagrangian strain rate* and the *updated Lagrangian spin rate tensors* are the symmetric and antisymmetric parts of the *velocity-like gradient tensor*. Similarly, from Eq. (16.6:6), one finds

(10) ▲ $\overset{\circ}{J} \cong \operatorname{tr}(\mathbf{L}) J_N^N = \operatorname{tr}(\mathbf{L}) = \operatorname{tr}(\mathbf{D})$.

One can also show that the *Cauchy* and *Almansi tensor rates* are

$$\overset{\circ}{\mathbf{C}} = \overset{\circ}{\mathbf{B}} = 2\mathbf{D} .$$

The use of the definitions above gives the updated solution of various quantities

(11) $$\mathbf{F}_N^{N+1} = \mathbf{I} + \overset{\circ}{\mathbf{F}}\,\Delta t = \mathbf{I} + \mathbf{L}\Delta t\,,$$

(12) $$\mathbf{U}_N^{N+1} = \mathbf{I} + \overset{\circ}{\mathbf{U}}\,\Delta t \cong \mathbf{I} + \mathbf{D}\Delta t\,,$$

(13) $$\mathbf{R}_N^{N+1} = \mathbf{I} + \overset{\circ}{\mathbf{R}}\,\Delta t \cong \mathbf{I} + \mathbf{\Omega}\Delta t\,,$$

(14) $$\mathbf{u}_N^{N+1} = \overset{\circ}{\mathbf{u}}\,\Delta t = \mathbf{v}\Delta t\,,$$

(15) $$J_N^{N+1} = \det(\mathbf{F}_N^{N+1}) = 1 + \overset{\circ}{J}\,\Delta t \cong 1 + \operatorname{tr}(\mathbf{D})\Delta t$$

in terms of their updated Lagrangian material rate. Since $\mathbf{F} = \mathbf{I} + \nabla\mathbf{u}$, one has

(16) $$\mathbf{F}_N^{N+1} = \mathbf{I} + \nabla_N\mathbf{u}_N^{N+1} = \mathbf{I} + \nabla_N\mathbf{v}\Delta t\,,$$

where ∇_N is the gradient operator defined in the current configuration. Comparing Eq. (16) with Eq. (11), one finds

(17) $$\mathbf{L} = \nabla_N\mathbf{v}\,.$$

Note that $\mathbf{L}$, $\mathbf{D}$, $\mathbf{\Omega}$ and all variables $(\overset{\circ}{\cdot})$ are referred to the current configuration D_N and that $\mathbf{F}(=\mathbf{I})$, $\mathbf{R}(=\mathbf{I})$ and $J(=1)$ are the current deformation gradient tensor, rotation tensor and Jocobian in the updated Lagrangain description.

With $\mathbf{X}^N$ denoting the material point in D_N, its corresponding point in D_{N+1} is

(18) $$\mathbf{X}^{N+1} = \mathbf{X}^N + \mathbf{u}_N^{N+1}\,,$$

which is the *updated coordinates* for the next increment.

The gradient operation ∇_N is calculated as follows. If the current configuration is characterized by generalized coordinates $(\eta^i)^N$ with corresponding contravariant base vectors $(\mathbf{g}^i)^N$, the gradient operator is in the form

$$\nabla_N = \frac{\partial}{\partial(\eta^i)^N}(\mathbf{g}^i)^N \qquad (N \text{ not summed})\,.$$

Then

(19) $$\nabla_N\mathbf{v} = \frac{\partial\mathbf{v}}{\partial(\eta^i)^N}(\mathbf{g}^i)^N \qquad (N \text{ not summed})\,,$$

$$= \mathbf{e}_j\mathbf{e}_i\frac{\partial v_j}{\partial(X^i)^N} \qquad (\text{in Cartesian coordinates})\,,$$

where $\mathbf{e}_i$ are the base vectors of the Cartesian coordinates. Note that in general $\mathbf{e}_j\mathbf{e}_i \neq \mathbf{e}_i\mathbf{e}_j$ for $i \neq j$.

One can easily establish the following relations between the update Lagrangian quantities and those referenced to the undeformed state:

(19) $\quad \mathbf{u}_0^{N+1} = \mathbf{u}_0^N + \mathbf{v}\Delta t$,

(20) $\quad \mathbf{F}_0^{N+1} = \dfrac{\partial \mathbf{X}^N}{\partial \mathbf{X}^0} + \dfrac{\partial \mathbf{u}_N^{N+1}}{\partial \mathbf{X}^N} \cdot \dfrac{\partial \mathbf{X}^N}{\partial \mathbf{X}^0} = (\mathbf{I} + \mathbf{L}\Delta t) \cdot \mathbf{F}_0^N$,

(21) $\quad \mathbf{C}_0^{N+1} = (\mathbf{F}_0^{N+1})^T \cdot \mathbf{F}_0^{N+1} = (\mathbf{F}_0^N)^T \cdot [\mathbf{I} + 2\mathbf{D}\Delta t + \mathbf{L}^T \cdot \mathbf{L}(\Delta t)^2] \cdot \mathbf{F}_0^N$,

(22) $\quad \mathbf{R}_0^{N+1} = \mathbf{R}_0^N + (\mathbf{R}_0^{N+1} - \mathbf{R}_0^N) \cong (\mathbf{I} + \mathbf{\Omega}\Delta t) \cdot \mathbf{R}_0^N$,

(23) $\quad \mathbf{U}_0^{N+1} = \mathbf{R}_0^{N+1} \cdot \mathbf{F}_0^{N+1} \cong [\mathbf{I} + (\mathbf{R}_0^N)^T \cdot \mathbf{D} \cdot \mathbf{R}_0^N \Delta t] \cdot \mathbf{U}_0^N$.

In Eq. (22), the second order term is retained to derive the constitutive laws in Sec. 17.5. Equation (20) is by definition and Eqs. (21) and (22) follow. Equations (23) and (24) are approximate.

17.3. LINEARIZED RATES OF STRESS MEASURES

The Cauchy stress tensor in the current configuration D_N is τ_N^N. For simplicity, we shall drop the super- and subscript N and denote it as τ. The Kirchhoff stress tensor $\boldsymbol{\sigma}$, the first and second Piola-Kirchhoff stress tensors $\mathbf{T}$ and $\mathbf{S}$, and the Biot-Lur'e stress tensor $\mathbf{r}^*$ referring to D_N are

(1) $\qquad\qquad \sigma_N^N = J_N^N \tau_N^N = \tau$,

(2) $\qquad\qquad \mathbf{T}_N^N = J_N^N (\mathbf{F}_N^N)^{-1} \cdot \tau_N^N = \tau$,

(3) $\qquad\qquad \mathbf{S}_N^N = \mathbf{T}_N^N \cdot [(\mathbf{F}^{-1})_N^N]^T = \tau$,

(4) $\qquad\qquad \mathbf{r}_N^{*N} = J_N^N (\mathbf{F}_N^N)^{-1} \cdot \tau_N^N \cdot \mathbf{R}_N^N = \tau$.

In other words, all *updated Lagrangian stress tensors* in the current configuration are equal to the Cauchy stress tensor. We shall determine the *material rates* of these quantities in the *updated Lagrangian description*. First, consider the first Piola–Kirchhoff stress tensor $\mathbf{T}$ whose material rate in the updated Lagrangian description is

$$\overset{\circ}{\mathbf{T}} \equiv \frac{\mathbf{T}_N^{N+1} - \mathbf{T}_N^N}{\Delta t},$$

which gives

(5) $\qquad\qquad \mathbf{T}_N^{N+1} = \tau + \overset{\circ}{\mathbf{T}} \Delta t$.

From $\mathbf{T}_N^{N+1}$, one can evaluate the updated Cauchy stress tensor τ_N^{N+1} in the $(N+1)^{\text{th}}$ configuration. Since the Cauchy stress tensor is independent of the reference configuration, we can simply denote τ_N^{N+1} as τ^{N+1}. Using Eq. (16.7:21), we have

$$\tau^{N+1} = \mathbf{F}_N^{N+1} \cdot \mathbf{T}_N^{N+1} / J_N^{N+1} \, .$$

In the *linearized form*, the Cauchy stress in the incremented configuration becomes

(6a) $$\tau^{N+1} \cong (\mathbf{I} + \mathbf{L}\Delta t) \cdot (\tau + \overset{\circ}{\mathbf{T}} \Delta t)/[1 + \text{tr}(\mathbf{L})\Delta t]$$

$$\cong \tau + [\overset{\circ}{\mathbf{T}} + \mathbf{L} \cdot \tau - \text{tr}(\mathbf{L})\tau]\Delta t \, ,$$

which will be used to determine the solution for the next increment. Since $\overset{\circ}{\tau}$ is the limit of the solution of the following equation as $\Delta t \to 0$,

(6b) $$\tau^{N+1} = \tau + \overset{\circ}{\tau} \Delta t \, ,$$

we can equate Eqs. (6a) and (6b) to determine the *Cauchy stress rate* $\overset{\circ}{\tau}$ in the updated Lagrangian description. We obtain:

(7) $$\overset{\circ}{\tau} \cong \overset{\circ}{\mathbf{T}} + \mathbf{L} \cdot \tau - \text{tr}(\mathbf{L})\tau \, , \qquad \text{or} \qquad \overset{\circ}{\mathbf{T}} \cong \overset{\circ}{\tau} - \mathbf{L} \cdot \tau + \text{tr}(\mathbf{L})\tau \, .$$

By this choice, we adopted a definition of the stress rate $\overset{\circ}{\tau}$ from among many possible definitions mentioned at the end of Sec. 17.1. Equation (7) can be obtained also directly by expressing the material derivative of the Cauchy stress tensor in the current configuration. Since $\tau = \mathbf{F} \cdot \mathbf{T}/J$ [Eq. (16.7:22)], one derives

$$\overset{\circ}{\tau} = \mathbf{F}_N^N \cdot \overset{\circ}{\mathbf{T}}/J_N^N + \overset{\circ}{\mathbf{F}} \cdot \mathbf{T}_N^N/J_N^N - \mathbf{F}_N^N \cdot \mathbf{T}_N^N \overset{\circ}{J}^N /(J_N^N)^2 \cong \overset{\circ}{\mathbf{T}} + \mathbf{L} \cdot \tau - \overset{\circ}{J} \tau \, .$$

Because τ^{N+1}, τ and $\text{tr}(\mathbf{L})\tau$ are symmetric, so is $\overset{\circ}{\mathbf{T}} + \mathbf{L} \cdot \tau$, hence the antisymmetric part of $\overset{\circ}{\mathbf{T}} + \mathbf{L} \cdot \tau$ vanishes:

(8) $$(\overset{\circ}{\mathbf{T}} + \mathbf{L} \cdot \tau)_a = 0 \, .$$

With $\overset{\circ}{\tau}$ as defined in Eq. (7), we can determine the updated Lagrangain rates of the *Kirchhoff stress tensor* $\sigma (= J\tau)$, the *Biot-Lur'e stress tensor* $\mathbf{r}^* (= \mathbf{F}^{-1} \cdot \tau \cdot \mathbf{R}J)$, the *second Piola-Kirchhoff stress tensor* $\mathbf{S}[= \mathbf{T} \cdot (\mathbf{F}^{-1})^T]$

and the Green Naghdi stress tenor $\sigma_r(= J\mathbf{R}^T \cdot \tau \cdot \mathbf{R})$. On examining the equations

$$\sigma_N^{N+1} = \tau + \overset{\circ}{\sigma} \, \Delta t = J_N^{N+1}\tau_N^{N+1}$$

$$\cong (1 + \overset{\circ}{J} \, \Delta t)(\tau + \overset{\circ}{\tau} \, \Delta t) \cong \tau + (\overset{\circ}{J} \, \tau + \overset{\circ}{\tau})\Delta t \,,$$

$$\mathbf{r}_N^{*N+1} = \tau + \overset{\circ}{\mathbf{r}}{}^* \Delta t = (\mathbf{F}^{-1})_N^{N+1} \cdot \tau^{N+1} \cdot \mathbf{R}_N^{N+1} J_N^{N+1}$$

$$\cong (\mathbf{I} - \mathbf{L}\Delta t) \cdot (\tau + \overset{\circ}{\tau} \, \Delta t) \cdot (\mathbf{I} + \Omega\Delta t)[1 + \text{tr}(\mathbf{D})\Delta t]$$

$$\cong \tau + [\overset{\circ}{\tau} + \tau \cdot \text{tr}(\mathbf{D}) - \mathbf{L} \cdot \tau + \tau \cdot \Omega]\Delta t \,,$$

$$\mathbf{S}_N^{N+1} = \tau + \overset{\circ}{\mathbf{S}} \, \Delta t = \mathbf{T}_N^{N+1} \cdot (\mathbf{F}^{-1})^{T_N^{N+1}}$$

$$\cong (\tau + \overset{\circ}{\mathbf{T}} \, \Delta t) \cdot (\mathbf{I} - \mathbf{L}^T \Delta t) = \tau + (\overset{\circ}{\mathbf{T}} - \tau \cdot \mathbf{L}^T)\Delta t \,,$$

$$\sigma_{rN}^{N+1} = (\mathbf{I} + \Omega^T \Delta t) \cdot (\tau + \overset{\circ}{\tau} \, \Delta t) \cdot (\mathbf{I} + \Omega\Delta t)[1 + \text{tr}(\mathbf{D})\Delta t]$$

$$= \tau + (\overset{\circ}{\tau} + \Omega^T \cdot \tau + \tau \cdot \Omega + \text{tr}(\mathbf{D})\tau)\Delta t \,,$$

one can derive the following rates of stress measures:

(9) ▲ $\qquad \overset{\circ}{\sigma} \cong \overset{\circ}{\tau} + \overset{\circ}{J} \, \tau = \overset{\circ}{\tau} + \text{tr}(\mathbf{D})\tau \,,$

(10) ▲ $\qquad \overset{\circ}{\mathbf{r}}{}^* \cong \overset{\circ}{\tau} - \mathbf{L} \cdot \tau + \text{tr}(\mathbf{D})\tau + \tau \cdot \Omega \,,$

(11) ▲ $\qquad \overset{\circ}{\mathbf{S}} \cong \overset{\circ}{\mathbf{T}} - \tau \cdot \mathbf{L}^T = \overset{\circ}{\tau} - \mathbf{L} \cdot \tau - \tau \cdot \mathbf{L}^T + \text{tr}(\mathbf{D})\tau \,.$

(12) ▲ $\qquad \overset{\circ}{\sigma}_r \cong \overset{\circ}{\tau} - \Omega \cdot \tau + \tau \cdot \Omega + \text{tr}(\mathbf{D})\tau \,.$

Note that $\text{tr}(\mathbf{D}) = \text{tr}(\mathbf{L})$. From Eqs. (7) and (10), one establishes the relation between the rates of $\mathbf{T}$ and $\mathbf{r}^*$ in the updated Lagrangian description as

$$\overset{\circ}{\mathbf{r}}{}^* = \overset{\circ}{\mathbf{T}} + \tau \cdot \Omega \,.$$

Also from Eq. (10), since $\overset{\circ}{\tau}$ and τ are symmetric,

$$(\overset{\circ}{\mathbf{r}}{}^* + \mathbf{L} \cdot \tau - \tau \cdot \Omega)_a = (\overset{\circ}{\mathbf{r}}{}^* + \mathbf{D} \cdot \tau)_a = 0 \,.$$

The updated Cauchy stress $\tau^{N+1}(= \tau^N + \overset{\circ}{\tau} \, \Delta t)$, is evaluated from Eq. (6b) and the stress rate $\overset{\circ}{\tau}$ in the updated Lagrangian description is evaluated from Eq. (7), or Eqs. (9)–(12). The choice of the specific expression for $\overset{\circ}{\tau}$ will depend on which of these stress rates $\overset{\circ}{\mathbf{T}}, \overset{\circ}{\mathbf{r}}{}^*$ and $\overset{\circ}{\mathbf{S}}$ is directly available. Depending on the solution approach, one solves for $\overset{\circ}{\mathbf{T}}$, $\overset{\circ}{\mathbf{r}}{}^*$ or $\overset{\circ}{\mathbf{S}}$ directly and then determines $\overset{\circ}{\tau}$. If one uses $\overset{\circ}{\mathbf{T}}$ or $\overset{\circ}{\mathbf{r}}{}^*$, one needs the value of the spin rate $\mathbf{\Omega}$, i.e. the antisymmetric part of $\mathbf{L}$. The accuracy of $\mathbf{\Omega}$ strongly affects the accuracy of the overall solution (Iura and Atluri 1992). Details will be discussed later in the numerical implementation sections.

Note that σ_N^N, $\mathbf{T}_N^N$, $\mathbf{S}_N^N$, $\mathbf{r}_N^{*N}$ all equal $\tau_N^N (= \tau^N = \tau)$, and therefore differ from σ_0^N, $\mathbf{T}_0^N$, $\mathbf{S}_0^N$, $\mathbf{r}_0^{*N}$, respectively, with the latter referring to the undeformed coordinates. However, we have $\tau_N^N = \tau_0^N$, because the Cauchy stress tensor is the stress at the deformed state D_N and is independent of the original configuration. This distinction is important for the determination of the constitutive relations. The rates of the stress measures in the total Lagrangian description are simply those of τ_0^N, σ_0^N, $\mathbf{t}_0^N$, $\mathbf{S}_0^N$, $\mathbf{r}_0^{*N}$. The rates of τ in both descriptions are the same because $\tau_N^N = \tau_0^N$, i.e., $\overset{\circ}{\tau} = \dot{\tau}$. From Eq. (16.7:3), the rate of σ_0^N is

$$\frac{D\sigma}{Dt} = J\frac{D\tau}{Dt} + \frac{DJ}{Dt}\tau = J[\dot{\tau} + \mathrm{tr}(\mathbf{D})\tau]\,,$$

in which the rate for J in Eq. (16.6:6) has been used and the superscript "N" and the subscript "0" have been dropped to make the notation consistent with Sec. 16.6. One can also obtain the material rates of $\mathbf{T}_0^N$, $\mathbf{S}_0^N$, $\mathbf{r}_0^{*N}$ by taking the time derivatives of Eqs. (16.7:4,6,8). The results are given in Eqs. (16.9:17)–(16.9:20). Comparing them with Eqs. (7) and (9)–(12) with $\overset{\circ}{\tau} = \dot{\tau}$, we obtain

$$(13) \qquad \overset{\circ}{\mathbf{T}} = \mathbf{F} \cdot \frac{D\mathbf{T}}{Dt}\frac{1}{J}\,,$$

$$(14) \qquad \overset{\circ}{\sigma} = \frac{D\sigma}{Dt}\frac{1}{J}\,,$$

$$(15) \qquad \overset{\circ}{\mathbf{r}}{}^* = \mathbf{F} \cdot \frac{D\mathbf{r}^*}{Dt} \cdot \mathbf{R}^T \frac{1}{J}\,,$$

$$(16) \qquad \overset{\circ}{\mathbf{S}} = \mathbf{F} \cdot \frac{D\mathbf{S}}{Dt} \cdot \mathbf{F}^T \frac{1}{J}\,,$$

$$(17) \qquad \overset{\circ}{\sigma}_r = \mathbf{R} \cdot \frac{D\sigma_r}{Dt} \cdot \mathbf{R}^T\,,$$

which give the relation between the time rates in the updated Lagrangian description and those in the total Lagrangian description. For materials which have well defined strain energy (called *hyper-elastic materials*), the energy density function is often in terms of strain measures referred to the undeformed state. The material rates of stresses in the total Lagrangian description can be expressed in terms of the derivatives of the energy density function with respect to deformation. In turn, which will be shown later, we can determine the material rates of stresses in the updated Lagrangian description from the derivatives of the energy density function.

Recall that τ, $\mathbf{L}$, $\mathbf{D}$, $\mathbf{\Omega}$, $\mathbf{F}$, $\mathbf{R}$, J and all concerned variables $(\overset{\circ}{}\,)$ refer to the current configuration D_N and that $\mathbf{F}, \mathbf{R}$ and J are the deformation gradient tensor, the rotation tensor and the Jocobian.

Problem 17.1. Prove that $d\mathbf{F}^{-1}/Dt = -\mathbf{F}^{-1} \cdot \mathbf{L}$ where $\mathbf{F}$ is the deformation gradient tensor and $\mathbf{L}$ is the velocity gradient tensor.

17.4. INCREMENTAL EQUATIONS OF MOTION

The equation of motion in terms of the Cauchy stress in D_N is

$$(1) \qquad \mathbf{\nabla}_N \cdot \tau + \rho_N \mathbf{b}^N = 0 \,,$$

where $\mathbf{\nabla}_N \cdot$ is the divergent operator, ρ_N the density and $\mathbf{b}^N$ the sum of the body force and the inertia force per unit mass referring to the deformation state in D_N. The inertia force is mass time acceleration, and is a function of the rigid body motion and the deformation of the body. The acceleration must be updated in real time. Equation (1) is the equation of equilibrium, which, when expressed in terms of $\mathbf{T}_N^{N+1}$ in D_N, becomes

$$(2) \qquad \mathbf{\nabla}_N \cdot \mathbf{T}_N^{N+1} + \rho_N \mathbf{b}_N^{N+1} = 0 \,.$$

Using $\mathbf{b}_N^{N+1} = \mathbf{b}^N + \overset{\circ}{\mathbf{b}}\,\Delta t$ and $\mathbf{T}_N^N = \tau^N$, one gets

$$\mathbf{\nabla}_N \cdot (\tau + \overset{\circ}{\mathbf{T}}\,\Delta t) + \rho_N (\mathbf{b}^N + \overset{\circ}{\mathbf{b}}\,\Delta t) = 0 \,,$$

or

$$(3) \;\blacktriangle \qquad \mathbf{\nabla}_N \cdot \overset{\circ}{\mathbf{T}} + \rho_N \overset{\circ}{\mathbf{b}} = 0 \,,$$

where $\overset{\circ}{\mathbf{b}}$ is the body force in the updated Lagrangian description. Equation (3) is linear in $\overset{\circ}{\mathbf{T}}$ and is similar in form as the equation of motion in terms of $\mathbf{T}_N^{N+1}$ in Eq. (2), but $\overset{\circ}{\mathbf{b}}$ can be a nonlinear function of strain, strain rate, displacement and velocities. When the system is in equilibrium so that the inertia force vanishes, then the effect of large deformation appears only implicitly in the equilibrium equation through the gradient operator $\mathbf{\nabla}_N$. A substitution of $\mathbf{T}_N^{N+1} = \mathbf{r}_N^{*N+1} \cdot (\mathbf{R}_N^{N+1})^T$ into Eq. (2) yields

$$\mathbf{\nabla}_N \cdot [\mathbf{r}_N^{*N+1} \cdot (\mathbf{R}_N^{N+1})^T] + \rho_N \mathbf{b}_N^{N+1} = 0 \,.$$

Using $\mathbf{r}_N^{*N} = \boldsymbol{\tau}$ and $\mathbf{R}_N^{N+1}$ from Eq. (17.2:13), one finds

$$\boldsymbol{\nabla}_N \cdot [(\boldsymbol{\tau} + \overset{\circ}{\mathbf{r}}{}^* \Delta t) \cdot (\mathbf{I} + \boldsymbol{\Omega} \Delta t)^T] + \rho_N (\mathbf{b}^N + \overset{\circ}{\mathbf{b}} \, \Delta t) = 0 \,.$$

Expanding and linearizing the equation above and taking into account that $\boldsymbol{\Omega}^T = -\boldsymbol{\Omega}$, one obtains

(4) $$\boldsymbol{\nabla}_N \cdot (\overset{\circ}{\mathbf{r}}{}^* - \boldsymbol{\tau} \cdot \boldsymbol{\Omega}) + \rho_N \overset{\circ}{\mathbf{b}} = 0 \,,$$

the equilibrium equation for the Biot-Lur'e stress rate. Equation (4) shows explicitly the effect of large deformation through the term $\boldsymbol{\tau} \cdot \boldsymbol{\Omega}$. The large deformation also affects the equilibrium equation implicitly through the gradient operator $\boldsymbol{\nabla}_N\cdot$. Similarly, the replacement of $\mathbf{T}$ by $\mathbf{S} \cdot \mathbf{F}^T$ in Eq. (2)

$$\boldsymbol{\nabla}_N \cdot [(\boldsymbol{\tau} + \overset{\circ}{\mathbf{S}} \, \Delta t) \cdot (\mathbf{I} + \mathbf{L}^T \Delta t)] + \rho_N (\mathbf{b}^N + \overset{\circ}{\mathbf{b}} \, \Delta t) = 0 \,,$$

gives

(5) ▲ $$\boldsymbol{\nabla}_N \cdot (\overset{\circ}{\mathbf{S}} + \boldsymbol{\tau} \cdot \mathbf{L}^T) + \rho_N \overset{\circ}{\mathbf{b}} = 0 \,,$$

the equilibrium equation for the second Piola-Kirchhoff stress rate with $\overset{\circ}{\mathbf{b}}$ representing the rate of inertia and body forces. In this case, the large deformation effects are through the term $\boldsymbol{\tau} \cdot \mathbf{L}$ and the gradient operator, and the rate of inertia force $\overset{\circ}{\mathbf{b}}$. The equilibrium equation for the corrotational stress rate in the updated Lagrangian approach is

(6) ▲ $$\boldsymbol{\nabla}_N \cdot (\overset{\circ}{\boldsymbol{\sigma}}_r - \mathbf{D} \cdot \boldsymbol{\tau} - \boldsymbol{\tau} \cdot \boldsymbol{\Omega}) + \rho_N \overset{\circ}{\mathbf{b}} = 0 \,.$$

17.5. CONSTITUTIVE LAWS

A constitutive law describes the mechanical properties of a material. The law should be independent of the frame of reference. Hence, ideally every term in a constitutive law must be objective, i.e., unaffected by translation and rotation of the frame of reference. A stress rate can be brought about not only by the strain rate $\mathbf{D}$ but also by the spin rate $\boldsymbol{\Omega}$ of the material element. However, if the strain rate is zero or the local material motion is characterized by pure-spin, objectivity requires that there is no change in the stress noted by observer who is spinning along with the material. Therefore, in establishing a consistent rate constitutive law, we must define a stress rate that does not changes in the body fixed coordinates when the body undergoes a rigid spin (Truesdel and Noll 1965, Atluri 1980, 1984, Reed and Atluri 1983, Rubinstein and Atluri 1983). Such a rate is called an *objective rate*. We then assume that the objective stress rate is a function of the strain rate only (not a function of the spin rate) and postulate the relation between the strain rate and the objective stress rate in the spinning coordinate system.

Objective tensors are discussed in Sec. 16.9. The updated stress rates have been mentioned in Sec. 17.3. Stresses and their rate in the updated Lagrangian description are in general not objective. However, for an objective tensor $\mathbf{A}$ and its updated Lagrangian rate $\overset{\circ}{\mathbf{A}}$, the following rate tensors are objective:

(1) $$\overset{\partial}{\mathbf{A}} \equiv \overset{\circ}{\mathbf{A}} - \mathbf{\Omega} \cdot \mathbf{A} + \mathbf{A} \cdot \mathbf{\Omega} \qquad \text{(Jaumann rate)},$$

(2) $$\overset{\triangledown}{\mathbf{A}} \equiv \overset{\circ}{\mathbf{A}} - \mathbf{L} \cdot \mathbf{A} - \mathbf{A} \cdot \mathbf{L}^T \qquad \text{(Oldroyd rate)}.$$

Here $(\overset{\circ}{\ })$, $(\overset{\partial}{\ })$, and $(\overset{\triangledown}{\ })$ denote, respectively, the *updated Lagrangian, Jaumann, and Oldroyd rates* of the relevant quantity in the parenthesis, $\mathbf{\Omega}$ is the spin tensor and $\mathbf{L}$ is the velocity gradient tensor. The Jaumann and Oldroyd rates of an objective tensor $\mathbf{A}$ differ from one another by terms that are bilinear in $\mathbf{A}$ and the deformation rate tensor $\mathbf{L}$ or the spin rate tensor $\mathbf{\Omega}$. There are many objective rates of an objective tensor $\mathbf{A}$, but the Jaumann or the Oldroyd rates are best known.

For the Kirchhoff stress, the Jaumann and Oldroyd rates are, respectively,

(3) $$\overset{\partial}{\sigma} = \overset{\circ}{\sigma} - \mathbf{\Omega} \cdot \sigma_N^N + \sigma_N^N \cdot \mathbf{\Omega}$$

$$= \overset{\circ}{\tau} + \text{tr}(\mathbf{D})\tau - \mathbf{\Omega} \cdot \tau + \tau \cdot \mathbf{\Omega} = \overset{\partial}{\tau} + \text{tr}(\mathbf{D})\tau \qquad \text{(Jaumann rate)},$$

(4) $$\overset{\triangledown}{\sigma} = \overset{\circ}{\sigma} - \mathbf{L} \cdot \sigma_N^N - \sigma_N^N \cdot \mathbf{L}^T$$

$$= \overset{\circ}{\tau} + \text{tr}(\mathbf{D})\tau - \mathbf{L} \cdot \tau - \tau \cdot \mathbf{L}^T = \overset{\triangledown}{\tau} + \text{tr}(\mathbf{D})\tau \qquad \text{(Oldroyd rate)}.$$

Clearly the two rates are linked by the following equation

(5) $$\overset{\partial}{\sigma} = \overset{\triangledown}{\sigma} + \mathbf{D} \cdot \tau + \tau \cdot \mathbf{D}.$$

The relation between the rate of various stress measures and the *Jaumann rate of the Kirchhoff stress tensor* in the updated Lagrangian description can be derived from Eqs. (17.3:7) and (17.3:9)–(17.3:12). One finds

(6) ▲ $\overset{\circ}{\mathbf{T}} = \overset{\circ}{\tau} + \mathrm{tr}(\mathbf{D})\tau - \mathbf{\Omega} \cdot \tau - \mathbf{D} \cdot \tau = \overset{\partial}{\sigma} - \mathbf{D} \cdot \tau - \tau \cdot \mathbf{\Omega}$,

(7) ▲ $\overset{\circ}{\mathbf{r}}^{*} = \overset{\circ}{\tau} + \mathrm{tr}(\mathbf{D})\tau - \mathbf{\Omega} \cdot \tau + \tau \cdot \mathbf{\Omega} - \mathbf{D} \cdot \tau = \overset{\partial}{\sigma} - \mathbf{D} \cdot \tau$,

(8) ▲ $\overset{\circ}{\mathbf{S}} = \overset{\circ}{\mathbf{T}} - \tau \cdot \mathbf{L}^{T} = \overset{\partial}{\sigma} - \mathbf{D} \cdot \tau - \tau \cdot \mathbf{D} = \overset{\triangledown}{\sigma}$.

(8a) ▲ $\overset{\circ}{\sigma}_{r} = \overset{\partial}{\sigma}$.

Clearly $\overset{\circ}{\sigma}_{r}$ is objective. From Eqs. (5) and (8), one can show that $\overset{\circ}{\mathbf{S}}$ is also objective. Note that $\overset{\circ}{\mathbf{T}}$ and $\overset{\circ}{\mathbf{r}}^{*}$ are not objective.

Following Hill (1967), one can establish a *rate potential* $\overset{\circ}{V}_{J}$ in terms of the strain rate tensor $\mathbf{D}$ such that the *constitutive equation for the Jaumann rate of the Kirchhoff stress tensor* in the update Lagrangian description is

(9)
$$\overset{\partial}{\sigma} = \frac{\partial \overset{\circ}{V}_{J}(\mathbf{D})}{\partial \mathbf{D}}.$$

Once $\overset{\circ}{V}_{J}$ is known, the corresponding rate potentials and constitutive equations for stress rates such as $\overset{\circ}{\mathbf{T}}, \overset{\circ}{\mathbf{S}}$ and $\overset{\circ}{\mathbf{r}}^{*}$ can be derived. Let $\overset{\circ}{U}$, $\overset{\circ}{W}$ and $\overset{\circ}{Q}$ denote the *rate potentials* for the first Piola-Kirchhoff stress rate tensor $\overset{\circ}{\mathbf{T}}$, the second Piola-Kirchhoff stress rate tensor $\overset{\circ}{\mathbf{S}}$, the Biot-Lur'e stress rate tensor $\overset{\circ}{\mathbf{r}}^{*}$ and the corotational stress rate tensor $\overset{\circ}{\sigma}_{r}$, respectively. Then

(10) $\overset{\circ}{\mathbf{T}} = \dfrac{\partial \overset{\circ}{U}(\mathbf{L})}{\partial \mathbf{L}^{T}}$, $\quad \overset{\circ}{\mathbf{S}} = \dfrac{\partial \overset{\circ}{W}(\mathbf{D})}{\partial \mathbf{D}}$, $\quad \overset{\circ}{\mathbf{r}} \equiv (\overset{\circ}{\mathbf{r}}^{*})_{s} = \dfrac{\partial \overset{\circ}{Q}(\mathbf{D})}{\partial \mathbf{D}}$, $\quad \overset{\circ}{\sigma}_{r} = \dfrac{\partial \overset{\circ}{P}(\mathbf{D})}{\partial \mathbf{D}}$.

Then, noting that

$$\overset{\circ}{\mathbf{r}} = \overset{\circ}{\mathbf{S}} + \frac{1}{2}(\tau \cdot \mathbf{D} + \mathbf{D} \cdot \tau), \quad \overset{\circ}{\sigma}_{r} = \overset{\circ}{\mathbf{S}} + \mathbf{D} \cdot \tau + \tau \cdot \mathbf{D}$$

and using Eqs. (8)–(10), we can show that the rate potential is

(11)
$$\overset{\circ}{W}(\mathbf{D}) = \overset{\circ}{V}_{J}(\mathbf{D}) - \tau : (\mathbf{D} \cdot \mathbf{D}).$$

Equations (8a), (9), and (10) give

(11a)
$$\overset{\circ}{P}(\mathbf{D}) = \overset{\circ}{V}_{J}(\mathbf{D}) = \overset{\circ}{W}(\mathbf{D}) + \tau : (\mathbf{D} \cdot \mathbf{D}).$$

From Eqs. (7), (9) and (10), we have

(12) $\overset{\circ}{Q}(\mathbf{D}) = \overset{\circ}{V}_{J}(\mathbf{D}) - \dfrac{1}{2}\tau : (\mathbf{D} \cdot \mathbf{D}) = \overset{\circ}{W}(\mathbf{D}) + \dfrac{1}{2}\tau : (\mathbf{D} \cdot \mathbf{D})$.

Finally, from Eqs. (6), (9) and (10), we obtain

$$(13) \qquad \overset{\circ}{U}\,(\mathbf{L}) = \overset{\circ}{V}_J\,(\mathbf{D}) - \boldsymbol{\tau} : (\mathbf{D} \cdot \mathbf{D}) + \frac{1}{2}\boldsymbol{\tau} : (\mathbf{L}^T \cdot \mathbf{L})\,.$$

The relations among the rate potentials can be easily derived

$$(14) \quad \overset{\circ}{E}\,(\mathbf{v}) \equiv \overset{\circ}{W}\,(\mathbf{D}) + \frac{\boldsymbol{\tau} : (\mathbf{L}^T \cdot \mathbf{L})}{2} = \overset{\circ}{Q}\,(\mathbf{D}) + \frac{1}{2}\boldsymbol{\tau} : (\mathbf{L}^T \cdot \mathbf{L}) - \frac{\boldsymbol{\tau} : (\mathbf{D} \cdot \mathbf{D})}{2}$$

$$= \overset{\circ}{P}\,(\mathbf{D}) + \frac{1}{2}\boldsymbol{\tau} : (\mathbf{L}^T \cdot \mathbf{L}) - \boldsymbol{\tau} : (\mathbf{D} \cdot \mathbf{D}) = \overset{\circ}{U}\,(\mathbf{L})\,.$$

Depending on applications, one expresses $\mathbf{L}$ and $\mathbf{D}$ in terms of displacement increments, $\boldsymbol{\Omega}$, etc.

The corresponding constitutive law for the *Oldroyd rate of the Kirchhoff stress tensor* $\boldsymbol{\sigma}$ is:

$$(15) \qquad \overset{\triangledown}{\boldsymbol{\sigma}} = \frac{\partial \overset{\circ}{V}_0\,(\mathbf{D})}{\partial \mathbf{D}}\,,$$

where the *Oldroyd rate potential* $\overset{\circ}{V}_0$ is related to the *Jaumann rate potential* $\overset{\circ}{V}_J$ by

$$(16) \qquad \overset{\circ}{V}_0\,(\mathbf{D}) = \overset{\circ}{V}_J\,(\mathbf{D}) + \boldsymbol{\tau} : (\mathbf{D} \cdot \mathbf{D})\,,$$

We can then express the rate potentials $\overset{\circ}{U}, \overset{\circ}{W}$ and $\overset{\circ}{Q}$ in terms $\overset{\circ}{V}_0$ by the following equations

$$(17) \quad \blacktriangle \qquad\qquad \overset{\circ}{W}\,(\mathbf{D}) = \overset{\circ}{V}_0\,(\mathbf{D})\,,$$

$$(18) \quad \blacktriangle \qquad\qquad \overset{\circ}{Q}\,(\mathbf{D}) = \overset{\circ}{V}_0\,(\mathbf{D}) + \frac{1}{2}\boldsymbol{\tau} : (\mathbf{D} \cdot \mathbf{D})\,,$$

$$(18a) \quad \blacktriangle \qquad\qquad \overset{\circ}{P}\,(\mathbf{D}) = \overset{\circ}{V}_0\,(\mathbf{D}) + \boldsymbol{\tau} : (\mathbf{D} \cdot \mathbf{D})$$

$$(19) \quad \blacktriangle \qquad\qquad \overset{\circ}{U}\,(\mathbf{L}) = \overset{\circ}{V}_0\left(\frac{\mathbf{L} + \mathbf{L}^T}{2}\right) + \frac{1}{2}\boldsymbol{\tau} : (\mathbf{L}^T \cdot \mathbf{L})\,.$$

Note that the *Oldroyd rate potential* $\overset{\circ}{V}_0$ is the rate potential $\overset{\circ}{W}$ for the *rate of the second Piola-Kirchhoff stress* in the updated Lagrangian description. The equations above are similar to Eqs. (11)–(13). Substituting Eq. (15) into Eqs. (6)–(8), one can express the updated Lagrangian stress rates in terms of the Oldroyd rate of the Kirchhoff stress

$$(20) \qquad \overset{\circ}{\mathbf{T}} = \overset{\triangledown}{\sigma} + \tau \cdot \mathbf{L}^T ,$$

$$(21) \qquad \overset{\circ}{\mathbf{r}}^* = \overset{\triangledown}{\sigma} + \tau \cdot \mathbf{D} ,$$

$$(21a) \qquad (21a)\ \overset{\circ}{\sigma}_r = \overset{\triangledown}{\sigma} + \mathbf{D} \cdot \tau + \tau \cdot \mathbf{D}$$

$$(22) \qquad \overset{\circ}{\mathbf{S}} = \overset{\triangledown}{\sigma} .$$

Thus the Oldroyd rate of the Kirchhoff stress is actually the rate of the second Piola–Kirchhoff stress in the updated Lagrangian description.

For hyperelastic materials, there exists a strain energy function $W_0(\mathbf{C})$ with its second variation defined as $\frac{\partial^2 W_0}{\partial C_{KL} \partial C_{IJ}} \delta C_{KL} \delta C_{IJ}$. One can show that the rate potential $\overset{\circ}{W}$ for the second Piola–Kirchhoff stress rate is derivable from the second variation of $W_0(\mathbf{C})$. From Eq. (16.11:7) and (16.11:9), we have

$$\mathbf{S} = \frac{\partial W_0}{\partial \mathbf{E}} = 2\frac{\partial W_0}{\partial \mathbf{C}} = 2\frac{\partial W_0}{\partial C_{IJ}} \varepsilon_I \varepsilon_J ,$$

where $\mathbf{S}$ is the second Piola-Kirchhoff stress referred to the *undeformed coordinates*. Then its material rate in the total Lagrangian description is

$$\frac{D\mathbf{S}}{Dt} = 2\frac{D}{Dt}\left(\frac{\partial W_0}{\partial \mathbf{C}}\right) = 2\frac{\partial^2 W_0}{\partial \mathbf{C}^2} : \frac{D\mathbf{C}}{Dt} = 4\frac{\partial^2 W_0}{\partial \mathbf{C}^2} : (\mathbf{F}^T \cdot \mathbf{D} \cdot \mathbf{F}) ,$$

or, in indicial notation

$$\frac{D\mathbf{S}}{Dt} = 4\frac{\partial^2 W_0}{\partial C_{KL}\partial C_{IJ}} \frac{\partial x_k}{\partial X_K} \frac{\partial x_l}{\partial X_L} D_{kl}\varepsilon_I\varepsilon_J$$

where x_i and X_I are the *deformed and the undeformed coordinates referring to the D_N and D_0 configurations*, respectively. From Eq. (17.3:16), one has

$$(23)\ \blacktriangle \quad \overset{\circ}{\mathbf{S}} = \mathbf{F}\cdot\frac{D\mathbf{S}}{Dt}\cdot\mathbf{F}^T\frac{1}{J} = d_{ijkl}D_{kl}\mathbf{e}_i\mathbf{e}_j = \mathbf{d} : \mathbf{D} , \quad \text{or} \quad \overset{\circ}{S}_{ij} = d_{ijkl}D_{kl} ,$$

where $\mathbf{d} = d_{ijkl}\mathbf{e}_i\mathbf{e}_j\mathbf{e}_k\mathbf{e}_l$ is a fourth rank tensor with $\mathbf{e}_i$ being the unit base vectors in the current configuration D_N and

$$(24) \qquad d_{ijkl} = \frac{4}{J}\frac{\partial^2 W_0}{\partial C_{KL}\partial C_{IJ}} \frac{\partial x_i}{\partial X_I} \frac{\partial x_j}{\partial X_J} \frac{\partial x_k}{\partial X_K} \frac{\partial x_l}{\partial X_L} ,$$

which have the same symmetry properties as those of the elastic constants for anisotropic materials discussed in Sec. 6.1. Note that $\overset{\circ}{\mathbf{S}}$ and $\mathbf{D}$ are objective. From Eq. (23), the rate potential can be written as

$$\overset{\circ}{W}(\mathbf{D}) = d_{ijkl}D_{ij}D_{kl}/2\,,\tag{25}$$

which is a quadratic function of D_{ij} related to the second variation of W_0 (Pian and Tong 1971). With ϵ_I, $\mathbf{e}_i$ being unit base vectors, $\frac{2}{J}\frac{\partial^2 W_0}{\partial C_{KL}\partial C_{IJ}}$ are the physical components of a fourth rank tensor referred to the undeformed coordinates whereas d_{ijkl}'s are the components of the corresponding tensor referred to the deformed coordinates. If ϵ_I, $\mathbf{e}_i$ are contravariant base vectors, then d_{ijkl} are the corresponding contravariant components of the tensor $\frac{2}{J}\frac{\partial^2 W_0}{\partial C_{KL}\partial C_{IJ}}$ in the deformed coordinates. In this case, we generally denote ϵ_I, $\mathbf{e}_i$, d_{ijkl} by $\mathbf{G}^I$, $\mathbf{g}^i$, d^{ijkl}, respectively.

If the hyper-elastic material is isotropic with $W_0(\mathbf{C}) = W_0(I_1, I_2, I_3)$, we have

$$\frac{\partial^2 W_0}{\partial C_{KL}\partial C_{IJ}}\,\epsilon_I\epsilon_J\epsilon_K\epsilon_L = \frac{\partial W_0}{\partial I_2}(\mathbf{II}-\mathbf{H}) + \frac{\partial W_0}{\partial I_3}\mathbf{A} + \mathbf{G}\,,$$

where $\mathbf{H}$, $\mathbf{A}$, $\mathbf{G}$ are fourth rank tensors defined as

$$\mathbf{H} = \frac{\partial \mathbf{C}}{\partial \mathbf{C}} = \delta_{IK}\delta_{JL}\epsilon_I\epsilon_J\epsilon_K\epsilon_L\,,$$

$$\mathbf{A} = \frac{\partial I_3\mathbf{C}^{-1}}{\partial \mathbf{C}} = e_{JKN}e_{ILM}C_{MN}\epsilon_I\epsilon_J\epsilon_K\epsilon_L\,,$$

$$\mathbf{G} = \frac{\partial^2 W_0}{\partial I_1^2}\,\mathbf{II} + \frac{\partial^2 W_0}{\partial I_2^2}\,(I_1\mathbf{I}-\mathbf{C})(I_1\mathbf{I}-\mathbf{C}) + \frac{\partial^2 W_0}{\partial I_3^2}\,I_3^2\mathbf{C}^{-1}\mathbf{C}^{-1}$$

$$+ \frac{\partial^2 W_0}{\partial I_1\partial I_2}\,(2I_1\mathbf{II}-\mathbf{IC}-\mathbf{CI}) + \frac{\partial^2 W_0}{\partial I_1\partial I_3}\,I_3(\mathbf{IC}^{-1}+\mathbf{C}^{-1}\mathbf{I})$$

$$+ \frac{\partial^2 W_0}{\partial I_2\partial I_3}\,I_3[\mathbf{C}^{-1}(I_1\mathbf{I}-\mathbf{C}) + (I_1\mathbf{I}-\mathbf{C})\mathbf{C}^{-1}]\,,$$

where e_{JKN} is the permutation tensor. The derivation is left to the readers. Note that $\mathbf{I}(= \delta_{IJ}\epsilon_I\epsilon_J)$ is the identity tensor. Both $\mathbf{II}(= \delta_{IJ}\delta_{KL}\epsilon_I\epsilon_J\epsilon_K\epsilon_L)$ and $\mathbf{IC}[= (\mathbf{CI})^T]$ are tensor of rank 4.

A substitution of Eq. (25) into Eq. (11), (12) and (14) yields the rate potentials for other stress measures. The substitution of Eq. (25) into Eq. (16), (18) and (19) yields the rate potentials for other stress measures related to the Oldroyd rate [Eq. (15)]. From Eqs. (6)–(8) or (20)–(22), the constitutive laws can be written as

$$(26) \quad \begin{aligned} &\overset{\circ}{\mathbf{S}} = \mathbf{d} : \mathbf{D}, & &\overset{\circ}{\mathbf{T}} = \mathbf{d} : \mathbf{D} + \boldsymbol{\tau} \cdot \mathbf{L}^T, \\ &\overset{\circ}{\boldsymbol{\sigma}} = \mathbf{d} : \mathbf{D} + \mathbf{L} \cdot \boldsymbol{\tau} + \boldsymbol{\tau} \cdot \mathbf{L}^T, & &\overset{\circ}{\boldsymbol{\tau}} = \mathbf{d} : \mathbf{D} + \mathbf{L} \cdot \boldsymbol{\tau} + \boldsymbol{\tau} \cdot \mathbf{L}^T - \mathrm{tr}(\mathbf{L})\boldsymbol{\tau} \\ &\overset{\circ}{\mathbf{r}}^* = \mathbf{d} : \mathbf{D} + \boldsymbol{\tau} \cdot \mathbf{D}, & &\overset{\circ}{\boldsymbol{\sigma}}_r = \mathbf{d} : \mathbf{D} + \mathbf{D} \cdot \boldsymbol{\tau} + \boldsymbol{\tau} \cdot \mathbf{D}. \end{aligned}$$

One sees that the constitutive laws consist of two parts: one part relates to the energy density represented by the first term on the right hand side of the equations above. The second part includes the remaining terms of the equations involving the interaction between the current Cauchy stress tensor $\boldsymbol{\tau}$ and the incremental change of geometry.

Problem 17.2. Show that

$$\frac{\partial \mathbf{C}^{-1}}{\partial \mathbf{C}} = -\mathbf{C}^{-1}\mathbf{C}^{-1} + \frac{e_{JKN}e_{ILM}C_{MN}}{I_3} \boldsymbol{\epsilon}_I \boldsymbol{\epsilon}_J \boldsymbol{\epsilon}_K \boldsymbol{\epsilon}_L.$$

(Hint: Define $\mathbf{C}' = I_3 \mathbf{C}^{-1}$ and show that $C'_{JI} = e_{JKN}e_{ILM}C_{KL}C_{NM}/2$.)

17.6. INCREMENTAL VARIATIONAL PRINCIPLES IN TERMS OF $\overset{\circ}{\mathbf{T}}$

We shall construct variational functionals involving stress and deformation rates in such a way that the conditions for the vanishing of the variation of the functional leads to the Eulerian equations for the rate variables, which are exactly the same as the field equations for these rate variables in the continuum. Thus, the solution of the field equations of the boundary-value problem corresponds to the stationary condition of the functional and vice versa. Pian and Tong (1971) introduced the second variation of the original functional as an incremental functional for finite deformation analyses. One can often derive different variation principles from a given functional (Pian and Tong 1969) by selectively relaxing constraints on the field variables of the functional (see Sec. 10.5). One can also derive different variational principles by enforcing selected constraints, *a priori*, and thus, reducing the number of independent field variables. The derivation in the next two sections follows mostly those given by Atluri (1979, 1980a) and Atluri and Cazzani (1995).

A Four-Field Principle. Atluri (1979, 1980a) constructed the following variational principle which involving four tensors: the displacement rate tensor $\mathbf{v}$, the spin rate tensor $\boldsymbol{\Omega}$, the stretch rate tensor $\mathbf{D}$, and the material rate of the first Piola–Kirchhoff stress tensor $\overset{\circ}{\mathbf{T}}$ in the updated Lagrangian description

$$(1) \quad \overset{\circ}{\Pi}_1 (\mathbf{v}, \mathbf{\Omega}, \mathbf{D}, \overset{\circ}{\mathbf{T}}) = \int_{V_N} \left\{ \overset{\circ}{Q} (\mathbf{D}) + \frac{1}{2} \boldsymbol{\tau} : (\mathbf{\Omega}^T \cdot \mathbf{\Omega}) + \boldsymbol{\tau} : (\mathbf{\Omega}^T \cdot \mathbf{D}) \right.$$

$$\left. + \overset{\circ}{\mathbf{T}}{}^T : (\boldsymbol{\nabla}_N \mathbf{v} - \mathbf{\Omega} - \mathbf{D}) - \rho_N \overset{\circ}{\mathbf{b}} \cdot \mathbf{v} \right\} dV$$

$$- \int_{S_{uN}} \mathbf{n} \cdot \overset{\circ}{\mathbf{T}} \cdot (\mathbf{v} - \bar{\mathbf{v}}) dA - \int_{S_{\sigma N}} \overset{\circ\nu}{\mathbf{T}} \cdot \mathbf{v} dA,$$

where $\overset{\circ\nu}{\mathbf{T}}$, $\bar{\mathbf{v}}$ are the prescribed traction and displacement rates on the boundaries, V_N is the volume of the body in the reference configuration D_N, $\mathbf{n}$ is the unit outward normal to the surface S_N, and $S_{\sigma N}$, S_{uN} are the portions of S_N where the traction and displacement increments are prescribed, respectively. The requirements for the field variables are: $\mathbf{v}$ is continuous, $\mathbf{\Omega}$ skew-symmetric and $\mathbf{D}$ symmetric. The last two requirements are called the rigid constraint, which must be enforced in constructing a weak form solution. In this case, $\mathbf{\Omega}$ has only three independent components and $\mathbf{D}$ has six rather than nine normally for a tensor of rank 2.

We shall derive the stationary condition of $\overset{\circ}{\Pi}_1$ for arbitrary variation of the four field variables subjected to the rigid constraints to obtain the appropriate field equations and natural boundary conditions. To derive the variation of $\overset{\circ}{\Pi}_1$, one must take into account that

$$\overset{\circ}{\mathbf{T}}{}^T : \delta\mathbf{D} = (\overset{\circ}{\mathbf{T}}{}^T)_s : \delta\mathbf{D},$$

$$\overset{\circ}{\mathbf{T}}{}^T : \delta\mathbf{\Omega} = - \overset{\circ}{\mathbf{T}}{}^T : \delta\mathbf{\Omega}^T = - \overset{\circ}{\mathbf{T}} : \delta\mathbf{\Omega},$$

for symmetric $\delta\mathbf{D}$ and skew-symmetric $\delta\mathbf{\Omega}$. One also needs to use the following trace and divergent properties:

$$\boldsymbol{\tau} : (\mathbf{\Omega}^T \cdot \delta\mathbf{D}) = \boldsymbol{\tau} : (\delta\mathbf{D} \cdot \mathbf{\Omega}) = -\boldsymbol{\tau} : (\delta\mathbf{D} \cdot \mathbf{\Omega}^T) = -(\boldsymbol{\tau} \cdot \mathbf{\Omega}) : \delta\mathbf{D},$$

$$\boldsymbol{\tau} : (\delta\mathbf{\Omega}^T \cdot \mathbf{D}) = (\boldsymbol{\tau} \cdot \mathbf{D}) : \delta\mathbf{\Omega}^T = (\mathbf{D} \cdot \boldsymbol{\tau}) : \delta\mathbf{\Omega},$$

$$\boldsymbol{\tau} : \delta(\mathbf{\Omega}^T \cdot \mathbf{\Omega}) = \boldsymbol{\tau} : (\delta\mathbf{\Omega}^T \cdot \mathbf{\Omega} + \mathbf{\Omega}^T \cdot \delta\mathbf{\Omega})$$

$$= (\boldsymbol{\tau} \cdot \mathbf{\Omega}^T) : \delta\mathbf{\Omega}^T + (\mathbf{\Omega} \cdot \boldsymbol{\tau}) : \delta\mathbf{\Omega} = 2(\mathbf{\Omega} \cdot \boldsymbol{\tau}) : \delta\mathbf{\Omega},$$

$$\int_{V_N} \overset{\circ}{\mathbf{T}} : \boldsymbol{\nabla}_N \delta\mathbf{v} dV = \int_{S_N} \mathbf{n} \cdot \overset{\circ}{\mathbf{T}} \cdot \delta\mathbf{v} dA - \int_{V_N} (\boldsymbol{\nabla}_N \cdot \overset{\circ}{\mathbf{T}}) \cdot \delta\mathbf{v} dV.$$

The first variation of $\overset{\circ}{\Pi}_1$ is

$$\delta \overset{\circ}{\Pi}_1 = \int_{V_N} \left\{ \left[\frac{\partial \overset{\circ}{Q}}{\partial \mathbf{D}} - (\overset{\circ}{\mathbf{T}} + \boldsymbol{\tau} \cdot \boldsymbol{\Omega})_s \right] : \delta \mathbf{D} + \partial \overset{\circ}{\mathbf{T}}{}^T : (\boldsymbol{\nabla}_N \mathbf{v} - \boldsymbol{\Omega} - \mathbf{D}) \right.$$

$$+ (\overset{\circ}{\mathbf{T}} + \mathbf{D} \cdot \boldsymbol{\tau} + \boldsymbol{\Omega} \cdot \boldsymbol{\tau}) : \delta \boldsymbol{\Omega} - (\boldsymbol{\nabla}_N \cdot \overset{\circ}{\mathbf{T}} + \rho_N \overset{\circ}{\mathbf{b}}) \cdot \delta \mathbf{v} \left. \right\} dV$$

$$- \int_{S_{\sigma N}} (\overset{o\nu}{\mathbf{T}} - \mathbf{n} \cdot \overset{\circ}{\mathbf{T}}) \cdot \delta \mathbf{v} dA - \int_{S_{uN}} \mathbf{n} \cdot \delta \overset{\circ}{\mathbf{T}} \cdot (\mathbf{v} - \bar{\mathbf{v}}) dA \,,$$

with the rigid constraints $(\delta \mathbf{D})_a = 0$ and $(\delta \boldsymbol{\Omega})_s = 0$. The stationary condition of $\overset{\circ}{\Pi}_1$ gives the following incremental field equations,

$$(2) \qquad \frac{\partial \overset{\circ}{Q}}{\partial \mathbf{D}} = (\overset{\circ}{\mathbf{T}} + \boldsymbol{\tau} \cdot \boldsymbol{\Omega})_s \qquad \text{(constitutive law in } V_N),$$

$$(3) \qquad \boldsymbol{\nabla}_N \mathbf{v} = \mathbf{D} + \boldsymbol{\Omega} \qquad \text{(compatibility condition in } V_N),$$

$$(4) \; (\overset{\circ}{\mathbf{T}} + \mathbf{D} \cdot \boldsymbol{\tau} + \boldsymbol{\Omega} \cdot \boldsymbol{\tau})_a = 0 \qquad \text{(angular momentum balance in } V_N),$$

$$(5) \qquad \boldsymbol{\nabla}_N \cdot \overset{\circ}{\mathbf{T}} + \rho_N \overset{\circ}{\mathbf{b}} = 0 \qquad \text{(linear momentum balance in } V_N),$$

and the following natural boundary conditions,

$$(6) \qquad \mathbf{n} \cdot \overset{\circ}{\mathbf{T}} = \overset{o\nu}{\mathbf{T}} \qquad \text{(traction boundary condition on } S_{\sigma N}),$$

$$(7) \qquad \mathbf{v} = \bar{\mathbf{v}} \qquad \text{(displacement boundary condition on } S_{uN}).$$

One can establish different variational principles by imposing *a priori* conditions to reduce the number of field variables in Eq. (1). One can also derive additional variational principles by using the Lagrangian multiplier to introduce new field variables. We will discuss several of such principles in the following subsections.

A Three-Field Principle. A variational principle with a functional involving three tensors $\mathbf{v}$, $\boldsymbol{\Omega}$ and $\overset{\circ}{\mathbf{T}}$ can be obtained by imposing

$$(8) \qquad \mathbf{D} = (\boldsymbol{\nabla}_N \mathbf{v})_s \,,$$

a priori in Eq. (1) to give

$$(9) \qquad \overset{\circ}{\Pi}_2(\mathbf{v}, \boldsymbol{\Omega}, \overset{\circ}{\mathbf{T}}) = \int_{V_N} \left\{ \overset{\circ}{Q} \left[(\nabla_N \mathbf{v})_s \right] \right.$$

$$+ \frac{1}{2} \boldsymbol{\tau} : (\boldsymbol{\Omega}^T \cdot \boldsymbol{\Omega}) + \boldsymbol{\tau} : [\boldsymbol{\Omega}^T \cdot (\nabla_N \mathbf{v})_s]$$

$$\left. + \overset{\circ}{\mathbf{T}}{}^T : [(\nabla_N \mathbf{v})_a - \boldsymbol{\Omega}] - \rho_N \overset{\circ}{\mathbf{b}} \cdot \mathbf{v} \right\} dV$$

$$- \int_{S_{uN}} (\mathbf{n} \cdot \overset{\circ}{\mathbf{T}}) \cdot (\mathbf{v} - \bar{\mathbf{v}}) dA - \int_{S_{\sigma N}} \overset{o\nu}{\mathbf{T}} \cdot \mathbf{v} dA.$$

The admissible requirements for the field variables are: $\mathbf{v}$ is continuous and $\boldsymbol{\Omega}$ is skew-symmetric. The stationary condition of $\overset{\circ}{\Pi}_2$ gives the same set of equations as Eqs. (2)–(7) except that $\mathbf{D}$ is replaced by $(\nabla_N \mathbf{v})_s$ in Eqs. (1) and (4) and that Eq. (3) becomes

$$(10) \qquad\qquad (\nabla_N \mathbf{v})_a = \boldsymbol{\Omega}.$$

A Purely Kinematic Principle. A variational principle with a functional involving $\mathbf{v}$ only can be derived by requiring, *a priori*,

$$(11) \qquad \nabla_N \mathbf{v} = \mathbf{D} + \boldsymbol{\Omega} = \mathbf{L}, \quad \mathbf{D} = (\nabla_N \mathbf{v})_s, \quad \boldsymbol{\Omega} = (\nabla_N \mathbf{v})_a$$

and $\mathbf{v} = \bar{\mathbf{v}}$ on S_{uN}. Equation (1) reduces to

$$(12) \qquad \overset{\circ}{\Pi}_3(\mathbf{v}) = \int_{V_N} \left\{ \overset{\circ}{E}(\mathbf{v}) - \rho_N \overset{\circ}{\mathbf{b}} \cdot \mathbf{v} \right\} dV - \int_{S_{\sigma N}} \overset{o\nu}{\mathbf{T}} \cdot \mathbf{v} dA,$$

where $\overset{\circ}{E}$ is given in Eq. (17.5:14). The specific expression depends on the choice of rate potential. In Eq. (12), $\mathbf{v}$ should be continuous and equals $\bar{\mathbf{v}}$ on S_{uN}.

If $\overset{\circ}{E}$ is in terms of $\overset{\circ}{W}$, Eq. (12) is the second variation of the functional of Eq. (16.13:16) (Pian and Tong 1971) derived by using Eqs. (17.2:20)–(17.2:22). One can show that

$$W(\mathbf{C}_0^{N+1}) = W(\mathbf{C}_0^N) + \boldsymbol{\tau}^N : \mathbf{D}\Delta t + \left[\overset{\circ}{W}(\mathbf{D}) + \frac{1}{2} \boldsymbol{\tau}^N : (\mathbf{L}^T \cdot \mathbf{L}) \right] \Delta t^2,$$

where $\overset{\circ}{W}(\mathbf{D})$ is given in Eq. (17.5:25). Equation (12) then becomes

$$(12a) \quad \overset{\circ}{\Pi}_3 (\mathbf{v}) = \int_{V_N} \left\{ \overset{\circ}{W} \left[(\boldsymbol{\nabla}_N \mathbf{v})_s \right] + \frac{1}{2} \boldsymbol{\tau} : \left[(\boldsymbol{\nabla}_N \mathbf{v})^T \cdot (\boldsymbol{\nabla}_N \mathbf{v}) \right] \right.$$

$$\left. - \rho_N \overset{\circ}{\mathbf{b}} \cdot \mathbf{v} \right\} dV - \int_{S_{\sigma N}} \overset{\circ\nu}{\bar{\mathbf{T}}} \cdot \mathbf{v} dA \,.$$

One can introduce $\boldsymbol{\Omega}$ as an additional independent field variable by modifying $\overset{\circ}{\Pi}_3$ as

$$(13) \quad \overset{\circ}{\Pi}{}^*_3(\mathbf{v}) = \int_{V_N} \left\{ \overset{\circ}{W} \left[(\boldsymbol{\nabla}_N \mathbf{v})_s \right] + \frac{1}{2} \boldsymbol{\tau} : \left[(\boldsymbol{\nabla}_N \mathbf{v})^T \cdot (\boldsymbol{\nabla}_N \mathbf{v}) \right] \right.$$

$$\left. + \frac{c}{2} \left[(\boldsymbol{\nabla}_N \mathbf{v})_a - \boldsymbol{\Omega} \right]^2 - \rho_N \overset{\circ}{\mathbf{b}} \cdot \mathbf{v} \right\} dV - \int_{S_{\sigma N}} \overset{\circ\nu}{\bar{\mathbf{T}}} \cdot \mathbf{v} dA \,,$$

where c is a given positive constant, and

$$\left[(\boldsymbol{\nabla}_N \mathbf{v})_a - \boldsymbol{\Omega} \right]^2 = \left[(\boldsymbol{\nabla}_N \mathbf{v})_a + \boldsymbol{\Omega} \right] : \left[(\boldsymbol{\nabla}_N \mathbf{v})_a - \boldsymbol{\Omega} \right] \,.$$

The admissibility requirements for the field variables are: $\mathbf{v}$ is continuous and equals $\bar{\mathbf{v}}$ on S_{uN}, and $\boldsymbol{\Omega}$ is skew-symmetric. Equation (13) is equivalent to the *penalty function approach*. One usually chooses c to be of the order of the elastic modulus of the material.

A Complementary Three-Field Principle. We shall derive a *complementary principle* with a functional involving three tensors $\mathbf{v}$, $\boldsymbol{\Omega}$ and $\overset{\circ}{\mathbf{T}}$, which is a variational principle based on the *complementary strain energy density* (a function of stresses). The concept of complementary work and complementary strain energy are explained in Sec. 10.9. We shall use the terminology and call the variational principle complementary when it is based on a rate potential, which is a function of stress-rate tensor. The gradient of the rate potential with respect to the stress-rate is a deformation-rate tensor. We introduce the contact transformation

$$(14) \quad -\overset{\circ}{Q}_c \left[(\overset{\circ}{\mathbf{T}} + \boldsymbol{\tau} : \boldsymbol{\Omega})_s \right] = \overset{\circ}{Q} (\mathbf{D}) + \boldsymbol{\tau} : (\boldsymbol{\Omega}^T \cdot \mathbf{D}) - (\overset{\circ}{\mathbf{T}}{}^T)_s : \mathbf{D} \,,$$

to eliminate the explicit dependence on $\mathbf{D}$ in Eq. (1). Now $\overset{\circ}{Q}_c$ is a function of a stress rate $(\overset{\circ}{\mathbf{T}} + \boldsymbol{\tau} : \boldsymbol{\Omega})_s$ and is called the *complementary rate potential*. We first show that the gradient of $\overset{\circ}{Q}_c$ with respect to $(\overset{\circ}{\mathbf{T}} + \boldsymbol{\tau} : \boldsymbol{\Omega})_s$ is the stretch rate tensor $\mathbf{D}$. Consider the variation of $\overset{\circ}{Q}_c$

$$\delta \overset{\circ}{Q}_c = - \left[\frac{\partial \overset{\circ}{Q}(\mathbf{D})}{\partial \mathbf{D}} - (\tau \cdot \mathbf{\Omega}^T)_s \right] : \delta \mathbf{D} - \tau : (\delta \mathbf{\Omega}^T \cdot \mathbf{D})$$

$$+ (\overset{\circ}{\mathbf{T}}^T)_s : \delta \mathbf{D} + (\delta \overset{\circ}{\mathbf{T}}^T)_s : \mathbf{D} \,.$$

Taking into account that

$$\frac{\partial \overset{\circ}{Q}(\mathbf{D})}{\partial \mathbf{D}} - (\overset{\circ}{\mathbf{T}}^T + \tau \cdot \mathbf{\Omega}^T)_s = 0 \,,$$

$$\tau : (\delta \mathbf{\Omega}^T \cdot \mathbf{D}) = (\mathbf{D} \cdot \tau) : \delta \mathbf{\Omega} \,,$$

one finds

$$\delta \overset{\circ}{Q}_c = \mathbf{D} : (\delta \overset{\circ}{\mathbf{T}})_s + \mathbf{D} : (\delta \tau \cdot \mathbf{\Omega})_s = \mathbf{D} : \delta(\overset{\circ}{\mathbf{T}} + \tau \cdot \mathbf{\Omega})_s \,.$$

In other words, $\overset{\circ}{Q}_c$ is a function of $(\overset{\circ}{\mathbf{T}} + \tau \cdot \mathbf{\Omega})_s$ only and

(15)
$$\frac{\partial \overset{\circ}{Q}_c}{\partial(\overset{\circ}{\mathbf{T}} + \tau \cdot \mathbf{\Omega})_s} = \mathbf{D} \,.$$

A substitution of Eq. (14) into Eq. (1) yields

(16) $$\overset{\circ}{\Pi}_4 (\mathbf{v}, \mathbf{\Omega}, \overset{\circ}{\mathbf{T}}) = \int_{V_N} \left\{ -\overset{\circ}{Q}_c \left[(\overset{\circ}{\mathbf{T}} + \tau \cdot \mathbf{\Omega})_s\right] + \frac{1}{2} \tau : (\mathbf{\Omega}^T \cdot \mathbf{\Omega}) \right.$$

$$\left. - \rho_N \overset{\circ}{\mathbf{b}} \cdot \mathbf{v} + \overset{\circ}{\mathbf{T}}^T : (\nabla_N \mathbf{v} - \mathbf{\Omega}) \right\} dV$$

$$- \int_{S_{uN}} \mathbf{n} \cdot \overset{\circ}{\mathbf{T}} \cdot (\mathbf{v} - \bar{\mathbf{v}}) dA - \int_{S_{\sigma N}} \overset{\circ}{\bar{\mathbf{T}}} \cdot \mathbf{v} dA \,.$$

The admissibility requirements for the field variables are: $\mathbf{v}$ is continuous, $\overset{\circ}{\mathbf{T}}$ is piecewise continuous and $\mathbf{\Omega}$ is skew-symmetric. It can be shown that the field equations associated with $\overset{\circ}{\Pi}_4$ are

(17) $$\nabla_N \mathbf{v} = \frac{\partial \overset{\circ}{Q}_c}{\partial(\overset{\circ}{\mathbf{T}} + \tau \cdot \mathbf{\Omega})_s} + \mathbf{\Omega} \quad \text{(compatibility condition)},$$

(18) $$\left[\overset{\circ}{\mathbf{T}} + \frac{\partial \overset{\circ}{Q}_c}{\partial(\overset{\circ}{\mathbf{T}} + \tau \cdot \mathbf{\Omega})_s} \cdot \tau + \mathbf{\Omega} \cdot \tau \right]_a = 0 \quad \text{(angular moment balance)},$$

and the equilibrium equation Eq. (5) in V_N. The natural boundary conditions are Eqs. (6) and (7).

One can independently enforce the condition Eq. (18) also by adding a term $c[\overset{\circ}{\mathbf{T}} + (\boldsymbol{\nabla}_N \mathbf{v})^T \cdot \tau^N]_a^2/2$ to the volume integral in Eq. (16). The new functional is

$$(19) \qquad \overset{\circ}{\Pi}_4{}^* (\mathbf{v}, \boldsymbol{\Omega}, \overset{\circ}{\mathbf{T}}) = \int_{V_N} \left\{ -\overset{\circ}{Q}_c \left[(\overset{\circ}{\mathbf{T}} + \boldsymbol{\tau} \cdot \boldsymbol{\Omega})_s \right] + \frac{1}{2} \boldsymbol{\tau} : (\boldsymbol{\Omega}^T \cdot \boldsymbol{\Omega}) \right.$$

$$+ \overset{\circ}{\mathbf{T}} : (\boldsymbol{\nabla}_N \mathbf{v} - \boldsymbol{\Omega}) + \frac{c}{2} [\overset{\circ}{\mathbf{T}} + (\boldsymbol{\nabla}_N \mathbf{v}) \cdot \boldsymbol{\tau}]_a^2 - \rho_N \overset{\circ}{\mathbf{b}} \cdot \mathbf{v} \Big\} dV$$

$$- \int_{S_{uN}} \mathbf{n} \cdot \overset{\circ}{\mathbf{T}} \cdot (\mathbf{v} - \bar{\mathbf{v}}) dA - \int_{S_{\sigma N}} \overset{\circ}{\bar{\mathbf{T}}} \cdot \mathbf{v} dA,$$

where c is a positive constant of the penalty function approach. In Eq. (19), c is of order $1/|\tau^N|$ or larger.

17.7. INCREMENTAL VARIATIONAL PRINCIPLES IN TERMS OF $\overset{\circ}{\mathbf{r}}{}^*$

One can derive incremental variational principles in terms $\overset{\circ}{\mathbf{r}}{}^*$ simply by substituting

$$\overset{\circ}{\mathbf{T}} = \overset{\circ}{\mathbf{r}}{}^* - \boldsymbol{\tau} \cdot \boldsymbol{\Omega}$$

into Eq. (17.6:1) and using the following properties

$$-(\boldsymbol{\Omega}^T \cdot \boldsymbol{\tau}) : (\boldsymbol{\nabla}_N \mathbf{v} - \boldsymbol{\Omega} - \mathbf{D})$$

$$= (\boldsymbol{\Omega} \cdot \boldsymbol{\tau}) : (\boldsymbol{\nabla}_N \mathbf{v} - \boldsymbol{\Omega} - \mathbf{D})$$

$$= \boldsymbol{\tau} : (\boldsymbol{\Omega}^T \cdot \boldsymbol{\nabla}_N \mathbf{v}) - \boldsymbol{\tau} : (\boldsymbol{\Omega}^T \cdot \boldsymbol{\Omega}) - \boldsymbol{\tau} : (\boldsymbol{\Omega}^T \cdot \mathbf{D}).$$

One finds

$$(1) \quad \overset{\circ}{\Pi}_5 (\mathbf{v}, \boldsymbol{\Omega}, \mathbf{D}, \overset{\circ}{\mathbf{r}}{}^*) = \int_{V_N} \left\{ \overset{\circ}{Q} (\mathbf{D}) - \frac{1}{2} \boldsymbol{\tau} : (\boldsymbol{\Omega}^T \cdot \boldsymbol{\Omega}) + \boldsymbol{\tau} : (\boldsymbol{\Omega}^T \cdot \boldsymbol{\nabla}_N \mathbf{v}) \right.$$

$$+ \overset{\circ}{\mathbf{r}}{}^{*T} : (\boldsymbol{\nabla}_N \mathbf{v} - \boldsymbol{\Omega} - \mathbf{D}) - \rho_N \overset{\circ}{\mathbf{b}} \cdot \mathbf{v} \Big\} dV$$

$$- \int_{S_{uN}} \mathbf{n} \cdot (\overset{\circ}{\mathbf{r}}{}^{*T} - \boldsymbol{\tau} \cdot \boldsymbol{\Omega}) \cdot (\mathbf{v} - \bar{\mathbf{v}}) dA - \int_{S_{\sigma N}} \overset{\circ\nu}{\bar{\mathbf{T}}} \cdot \mathbf{v} dA.$$

The admissibility requirements for the field variables are: $\mathbf{v}$ is continuous, $\boldsymbol{\Omega}$ skew-symmetric, $\mathbf{D}$ symmetric, and $\overset{\circ}{\mathbf{r}}{}^{*}$ piecewise continuous.

By requiring $\mathbf{D} = (\nabla_N \mathbf{v})_s$ a priori in Eq. (1), one obtains a three-field principle in terms $\mathbf{v}$, $\boldsymbol{\Omega}$ and $\overset{\circ}{\mathbf{r}}{}^{*}$. The functional is

$$(2) \qquad \overset{\circ}{\Pi}_6 (\mathbf{v}, \boldsymbol{\Omega}, \overset{\circ}{\mathbf{r}}{}^{*})$$

$$= \int_{V_N} \left\{ \overset{\circ}{Q} \left[(\nabla_N \mathbf{v})_s\right] - \frac{1}{2}\boldsymbol{\tau} : (\boldsymbol{\Omega}^T \cdot \boldsymbol{\Omega}) \right.$$

$$\left. + \boldsymbol{\tau} : (\boldsymbol{\Omega}^T \cdot \nabla_N \mathbf{v}) + \overset{\circ}{\mathbf{r}}{}^{*T} : [(\nabla_N \mathbf{v})_a - \boldsymbol{\Omega}] - \rho_N \overset{\circ}{\mathbf{b}} \cdot \mathbf{v} \right\} dV$$

$$- \int_{S_{uN}} \mathbf{n} \cdot (\overset{\circ}{\mathbf{r}}{}^{*T} - \boldsymbol{\tau} \cdot \boldsymbol{\Omega}) \cdot (\mathbf{v} - \bar{\mathbf{v}}) dA - \int_{S_{\sigma N}} \overset{o\nu}{\bar{\mathbf{T}}} \cdot \mathbf{v} dA.$$

The admissibility requirements for the field variables are the same as those for the functional $\overset{\circ}{\Pi}_5$, except that $\mathbf{D}$ does not appear here. If we further require that $\boldsymbol{\Omega} = (\nabla_N \mathbf{v})_a$ and $\mathbf{v} = \bar{\mathbf{v}}$ on S_{uN} are satisfied a priori, $\overset{\circ}{\Pi}_6$ reduces to $\overset{\circ}{\Pi}_3$ as given in Eq. (17.6:12).

As in the previous subsection, the use of the contact transformation

$$(3) \qquad \overset{\circ}{Q}(\mathbf{D}) - (\overset{\circ}{\mathbf{r}}{}^{*})_s : \mathbf{D} = -\overset{\circ}{Q}_c \left[(\overset{\circ}{\mathbf{r}}{}^{*})_s\right]$$

in Eq. (1) leads to an alternate three-field principle with the functional

$$(4) \qquad \overset{\circ}{\Pi}_7 (\mathbf{v}, \boldsymbol{\Omega}, \overset{\circ}{\mathbf{r}}{}^{*})$$

$$= \int_{V_N} \left\{ -\overset{\circ}{Q}_c \left[(\overset{\circ}{\mathbf{r}}{}^{*})_s\right] - \frac{1}{2}\boldsymbol{\tau} : (\boldsymbol{\Omega}^T \cdot \boldsymbol{\Omega}) \right.$$

$$\left. + \boldsymbol{\tau} : (\boldsymbol{\Omega}^T \cdot \nabla_N \mathbf{v}) + \overset{\circ}{\mathbf{r}}{}^{*T} : [(\nabla_N \mathbf{v})_a - \boldsymbol{\Omega}] - \rho_N \overset{\circ}{\mathbf{b}} \cdot \mathbf{v} \right\} dV$$

$$- \int_{S_{uN}} \mathbf{n} \cdot (\overset{\circ}{\mathbf{r}}{}^{*T} - \boldsymbol{\tau} \cdot \boldsymbol{\Omega}) \cdot (\mathbf{v} - \bar{\mathbf{v}}) dA - \int_{S_{\sigma N}} \overset{o\nu}{\bar{\mathbf{T}}} \cdot \mathbf{v} dA.$$

The admissibility requirements for the field variables are also the same of those for the functional $\overset{\circ}{\Pi}_5$ without the dependence on $\mathbf{D}$.

Problem 17.3. Show that a four-field principle involving $\mathbf{v}$, $\boldsymbol{\Omega}$, $\mathbf{D}$ and $\overset{\circ}{\mathbf{S}}$ can be written as

$$\mathring{\Pi}_1 (\mathbf{v}, \mathbf{\Omega}, \mathbf{D}, \mathring{\mathbf{S}}) = \int_{V_N} \left\{ \mathring{W} (\mathbf{D}) - \frac{1}{2} \boldsymbol{\tau} : [(\mathbf{D} + \mathbf{\Omega})^T \cdot (\mathbf{D} + \mathbf{\Omega})] \right.$$

$$+ \boldsymbol{\tau} : [(\nabla_N \mathbf{v})^T \cdot (\mathbf{D} + \mathbf{\Omega})]$$

$$\left. + \mathring{\mathbf{S}} : (\nabla_N \mathbf{v} - \mathbf{\Omega} - \mathbf{D}) - \rho_N \mathring{\mathbf{b}} \cdot \mathbf{v} \right\} dV$$

$$- \int_{S_{uN}} \mathbf{n} \cdot (\mathring{\mathbf{S}} + \boldsymbol{\tau} \cdot \mathbf{L}^T) \cdot (\mathbf{v} - \bar{\mathbf{v}}) \, dA$$

$$- \int_{S_{\sigma N}} \mathring{\mathbf{T}}^{\nu} \cdot \mathbf{v} \, dA.$$

17.8. INCOMPRESSIBLE AND NEARLY INCOMPRESSIBLE MATERIALS

For materials approaching incompressibility, the formulation in Secs. 17.6 and 17.7 breaks down because the equivalent bulk modulus tends to infinity. To circumvent the difficulty, we shall modify the variational principles discussed before. We first introduce the deviatoric variables

$$(1) \qquad\qquad \mathbf{L}' = \mathbf{L} - \mathbf{I} \, \mathring{J} \, /3 \, ,$$

$$(2) \qquad\qquad \mathbf{D}' = \mathbf{D} - \mathbf{I} \, \mathring{J} \, /3 \, ,$$

in which $\mathring{J}$ is, from Eq. (17.2:10),

$$(3) \qquad\qquad \mathring{J} = \mathrm{tr}(\mathbf{L}) = \mathrm{tr}(\mathbf{D}) \, .$$

Denoting the rate potentials given in Sec. 17.5 as follows:

$$(4) \quad \mathring{Q}^*(\mathbf{D}', \mathring{J}) \equiv \mathring{Q} (\mathbf{D}), \quad \mathring{W}^*(\mathbf{D}', \mathring{J}) \equiv \mathring{W} (\mathbf{D}), \quad \mathring{U}^*(\mathbf{L}', \mathring{J}) \equiv \mathring{U} (\mathbf{L}),$$

one has the constitutive laws

$$(5) \quad \mathring{\mathbf{S}}' \equiv \frac{\partial \mathring{W}^*(\mathbf{D}', \mathring{J})}{\partial \mathbf{D}'} \, , \quad (\mathring{\mathbf{r}}'^*)_s \equiv \frac{\partial \mathring{Q}^*(\mathbf{D}', \mathring{J})}{\partial \mathbf{D}'} \, , \quad \mathring{\mathbf{T}}'^T \equiv \frac{\partial \mathring{U}^*(\mathbf{L}', \mathring{J})}{\partial \mathbf{L}'} \, ,$$

(6) $\quad -\overset{\circ}{p}_r \equiv \dfrac{\partial \overset{\circ}{Q}{}^*(\mathbf{D}', \overset{\circ}{J})}{\partial \overset{\circ}{J}} = \dfrac{1}{3}\dfrac{\partial \overset{\circ}{Q}(\mathbf{D})}{\partial \mathbf{D}} : \mathbf{I} = \dfrac{1}{3}\,\overset{\circ}{\mathbf{r}}{}^* : \mathbf{I} = \dfrac{1}{3}\,\mathrm{tr}(\overset{\circ}{\mathbf{r}}{}^*),$

$\quad\quad -\overset{\circ}{p}_s \equiv \dfrac{\partial \overset{\circ}{W}{}^*(\mathbf{D}', \overset{\circ}{J})}{\partial \overset{\circ}{J}} = \dfrac{1}{3}\,\mathrm{tr}(\overset{\circ}{\mathbf{S}}),$

$\quad\quad -\overset{\circ}{p}_t \equiv \dfrac{\partial \overset{\circ}{U}{}^*(\mathbf{L}', \overset{\circ}{J})}{\partial \overset{\circ}{J}} = \dfrac{1}{3}\,\mathrm{tr}(\overset{\circ}{\mathbf{T}}),$

which relate deformation rates to different measures of stress rates. One can show that

$$(\overset{\circ}{\mathbf{r}}{}^*)_s : \delta \mathbf{D} = \delta \overset{\circ}{Q} = \dfrac{\partial \overset{\circ}{Q}{}^*(\mathbf{D}', \overset{\circ}{J})}{\partial \mathbf{D}'} : \delta \mathbf{D}' + \dfrac{\partial \overset{\circ}{Q}{}^*(\mathbf{D}', \overset{\circ}{J})}{\partial \overset{\circ}{J}}\,\delta \overset{\circ}{J}$$

$$= (\overset{\circ}{\mathbf{r}}{}'{}^*)_s : (\delta \mathbf{D} - \mathbf{I}\delta \overset{\circ}{J}/3) - \overset{\circ}{p}_r\,\delta \overset{\circ}{J}$$

$$= (\overset{\circ}{\mathbf{r}}{}'{}^*)_s : \delta \mathbf{D} - [(\overset{\circ}{\mathbf{r}}{}'{}^*)_s : \mathbf{I}](\mathbf{I} : \delta \mathbf{D})/3 - \overset{\circ}{p}_r\,\mathbf{I} : \delta \mathbf{D},$$

which implies

(7) $\quad\quad (\overset{\circ}{\mathbf{r}}{}^*)_s = (\overset{\circ}{\mathbf{r}}{}'{}^*)_s - [(\overset{\circ}{\mathbf{r}}{}'{}^*)_s : \mathbf{I}]\mathbf{I}/3 - \overset{\circ}{p}_r\,\mathbf{I}.$

Similarly, one can establish that

(8) $\quad\quad \overset{\circ}{\mathbf{S}} = \overset{\circ}{\mathbf{S}}{}' - (\overset{\circ}{\mathbf{S}}{}' : \mathbf{I})\mathbf{I}/3 - \overset{\circ}{p}_s\,\mathbf{I},$

(9) $\quad\quad \overset{\circ}{\mathbf{T}} = \overset{\circ}{\mathbf{T}}{}' - (\overset{\circ}{\mathbf{T}}{}' : \mathbf{I})\mathbf{I}/3 - \overset{\circ}{p}_t\,\mathbf{I}.$

The relations between various updated Lagrangian pressure rates can be obtained by differentiating Eq. (16.7:11) with respect to t with all quantities referring to the current configuration in D_N:

(10) $\quad -\overset{\circ}{p} = \overset{\circ}{\tau} : \mathbf{I}/3 = -\overset{\circ}{p}_t + \mathrm{tr}(\mathbf{D})p + \tau : \mathbf{D}/3$

$\quad\quad\quad = -\overset{\circ}{p}_s + \mathrm{tr}(\mathbf{D})p + 2\tau : \mathbf{D}/3 = -\overset{\circ}{p}_r + \mathrm{tr}(\mathbf{D})p + \tau : \mathbf{D}/3,$

where $\overset{\circ}{p}$ is the hydrostatic pressure rate and $p[=-(\tau : \mathbf{I})/3]$ is the hydrostatic pressure. In general, $\overset{\circ}{p}_t = \overset{\circ}{p}_r \neq \overset{\circ}{p}_s \neq \overset{\circ}{p}$.

If one replaces

$$\overset{\circ}{Q} \to \overset{\circ}{Q}{}^*, \quad \overset{\circ}{W} \to \overset{\circ}{W}{}^*, \quad \overset{\circ}{U} \to \overset{\circ}{U}{}^*, \quad \mathbf{D} \to \mathbf{D}' + \overset{\circ}{J}\mathbf{I}, \quad \mathbf{L} \to \mathbf{L}' + \overset{\circ}{J}\mathbf{I},$$

by substituting Eqs. (1)–(4) and (7)–(9) into the functional defined in Secs. 17.6 and 17.7, one can obtain various new variational principles in terms of the new field variables. We shall illustrate the process with the following example. A substitution of Eqs. (2)–(4) and (7) into Eq. (17.7:1) yields

(11) $\overset{\circ}{\Pi}\left(\mathbf{v}, \mathbf{\Omega}, \mathbf{D}', \overset{\circ}{J}, \overset{\circ}{\mathbf{r}}{}'^*, \overset{\circ}{p}_r\right)$

$$= \int_{V_N} \left\{ \overset{\circ}{Q}{}^*(\mathbf{D}', \overset{\circ}{J}) - \frac{1}{2}\boldsymbol{\tau} : (\mathbf{\Omega}^T \cdot \mathbf{\Omega}) + \boldsymbol{\tau} : (\mathbf{\Omega}^T \cdot \boldsymbol{\nabla}_N \mathbf{v}) \right.$$

$$- \overset{\circ}{p}_r\left[\mathrm{tr}(\boldsymbol{\nabla}_N \mathbf{v}) - \overset{\circ}{J}\right] + \overset{\circ}{\mathbf{r}}{}'^{*T} : \left[\boldsymbol{\nabla}_N \mathbf{v} - \mathbf{\Omega} - \mathbf{D}' - \mathrm{tr}(\boldsymbol{\nabla}_N \mathbf{v})\mathbf{I}/3\right]$$

$$\left. - \rho_N \overset{\circ}{\mathbf{b}} \cdot \mathbf{v} \right\} dV - \int_{S_{\sigma N}} \overset{\circ\nu}{\mathbf{\bar{T}}} \cdot \mathbf{v} dA$$

$$- \int_{S_{uN}} \mathbf{n} \cdot \left[\overset{\circ}{\mathbf{r}}{}'^* - \frac{1}{3}\mathrm{tr}(\overset{\circ}{\mathbf{r}}{}'^*)\mathbf{I} - \overset{\circ}{p}_r\mathbf{I} - \boldsymbol{\tau} \cdot \mathbf{\Omega}\right](\mathbf{v} - \bar{\mathbf{v}}) dA.$$

The functional holds for all materials for continuous $\mathbf{v}$, skew symmetric $\mathbf{\Omega}$, symmetric deviatoric $\mathbf{D}'$, positive $\overset{\circ}{J}$, piecewise continuous deviatoric $\overset{\circ}{\mathbf{r}}{}'^*$ and piecewise continuous $\overset{\circ}{p}_r$. This functional is slightly different from that derived by Seki and Atluri (1994). In a similar fashion, one can derive the variational functionals using $\mathbf{v}$, $\mathbf{\Omega}$, $\mathbf{L}'$, $\overset{\circ}{J}$, $\overset{\circ}{\mathbf{T}}{}'$, $\overset{\circ}{p}_t$ or $\mathbf{v}$, $\mathbf{\Omega}$, $\mathbf{D}'$, $\overset{\circ}{J}$, $\overset{\circ}{\mathbf{S}}{}'$, $\overset{\circ}{p}_s$ as the independent field variables. The details are left to readers.

One can derive different variational principles from Eq. (11) using the contact transformation

(12) $\qquad -\overset{\circ}{Q}_c\left[(\overset{\circ}{\mathbf{r}}{}'^*)_s, \overset{\circ}{p}_r\right] = \overset{\circ}{Q}{}^*(\mathbf{D}', \overset{\circ}{J}) - (\overset{\circ}{\mathbf{r}}{}'^*)_s : \mathbf{D}' + \overset{\circ}{p}_r\overset{\circ}{J}$.

One can show that

$$\frac{\partial \overset{\circ}{Q}_c\left[(\overset{\circ}{\mathbf{r}}{}'^*)_s, \overset{\circ}{p}_r\right]}{\partial(\overset{\circ}{\mathbf{r}}{}'^*)_s} = \mathbf{D}', \qquad -\frac{\partial \overset{\circ}{Q}_c\left[(\overset{\circ}{\mathbf{r}}{}'^*)_s, \overset{\circ}{p}_r\right]}{\partial \overset{\circ}{p}_r} = \overset{\circ}{J} .$$

A substitution of Eq. (12) into Eq. (11) yields

(13) $\overset{\circ}{\Pi}_8 \, (\mathbf{v}, \boldsymbol{\Omega}, \overset{\circ}{\mathbf{r}}'^*, \overset{\circ}{p}_r)$

$$= \int_{V_N} \left\{ - \overset{\circ}{Q}_c \, (\overset{\circ}{\mathbf{r}}'^*, \overset{\circ}{p}_r) - \frac{1}{2} \boldsymbol{\tau} : (\boldsymbol{\Omega}^T \cdot \boldsymbol{\Omega}) + \boldsymbol{\tau} : (\boldsymbol{\Omega}^T \cdot \boldsymbol{\nabla}_N \mathbf{v}) \right.$$

$$+ \overset{\circ}{\mathbf{r}}'^{*T} : [\boldsymbol{\nabla}_N \mathbf{v} - \boldsymbol{\Omega} - \mathrm{tr}(\boldsymbol{\nabla}_N \mathbf{v})\mathbf{I}/3]$$

$$\left. - \overset{\circ}{p}_r \, \mathrm{tr}(\boldsymbol{\nabla}_N \mathbf{v}) - \rho_N \, \overset{\circ}{\mathbf{b}} \cdot \mathbf{v} \right\} dV - \int_{S_{\sigma N}} \overset{\circ\nu}{\mathbf{T}} \cdot \mathbf{v} dA$$

$$- \int_{S_{uN}} \mathbf{n} \cdot \left[\overset{\circ}{\mathbf{r}}^* - \frac{1}{3} \, \mathrm{tr}(\overset{\circ}{\mathbf{r}}'^*)\mathbf{I} - \overset{\circ}{p}_r \, \mathbf{I} - \boldsymbol{\tau} \cdot \boldsymbol{\Omega} \right] \cdot (\mathbf{v} - \bar{\mathbf{v}}) \, dA,$$

which is a functional with the four-field variables $\mathbf{v}$, $\boldsymbol{\Omega}$, $\overset{\circ}{\mathbf{r}}'^*$, $\overset{\circ}{p}_r$. The stationary condition of $\overset{\circ}{\Pi}_8$ under the variation of the field variables subjected to the rigid condition $(\boldsymbol{\Omega})_s = 0$ gives the appropriate field equations and natural boundary conditions. The derivation is straightforward and is left to the reader.

One can also introduce a mixed contact transformation

(14) $$\overset{\circ}{Q}_m \, (\mathbf{D}', \overset{\circ}{p}_r) = \overset{\circ}{Q}^* (\mathbf{D}', \overset{\circ}{J}) + \overset{\circ}{p}_r \overset{\circ}{J}$$

such that

$$\frac{\partial \overset{\circ}{Q}_m \, (\mathbf{D}', \overset{\circ}{p}_r)}{\partial \mathbf{D}'} = (\overset{\circ}{\mathbf{r}}'^*)_s, \qquad - \frac{\partial \overset{\circ}{Q}_m \, (\mathbf{D}', \overset{\circ}{p}_r)}{\partial \overset{\circ}{p}_r} = \overset{\circ}{J} \, .$$

A substitution of Eq. (14) into Eq. (11) yields

(15) $\overset{\circ}{\Pi}_9 \, (\mathbf{v}, \boldsymbol{\Omega}, \mathbf{D}', \overset{\circ}{\mathbf{r}}'^*, \overset{\circ}{p}_r)$

$$= \int_{V_N} \left\{ \overset{\circ}{Q}_m \, (\mathbf{D}', \overset{\circ}{p}_r) - \boldsymbol{\tau} : (\boldsymbol{\Omega}^T \cdot \boldsymbol{\nabla}_N \mathbf{v}) - \frac{1}{2} \boldsymbol{\tau} : (\boldsymbol{\Omega}^T \cdot \boldsymbol{\Omega}) \right.$$

$$+ \overset{\circ}{\mathbf{r}}'^{*T} : [\boldsymbol{\nabla}_N \mathbf{v} - \boldsymbol{\Omega} - \mathbf{D}' - \mathrm{tr}(\boldsymbol{\nabla}_N \mathbf{v})\mathbf{I}/3]$$

$$\left. - \overset{\circ}{p}_r \, \mathrm{tr}(\boldsymbol{\nabla}_N \mathbf{v}) - \rho_N \, \overset{\circ}{\mathbf{b}} \cdot \mathbf{v} \right\} dV - \int_{S_{\sigma N}} \overset{\circ\nu}{\mathbf{T}} \cdot \mathbf{v} dA$$

$$- \int_{S_{uN}} \mathbf{n} \cdot \left[\overset{\circ}{\mathbf{r}}'^* - \frac{1}{3} \, \mathrm{tr}(\overset{\circ}{\mathbf{r}}'^*) \, \mathbf{I} - \overset{\circ}{p}_r \, \mathbf{I} - \boldsymbol{\tau} \cdot \boldsymbol{\Omega} \right] (\mathbf{v} - \bar{\mathbf{v}}) \, dA,$$

which is a functional with the five-field variables $\mathbf{v}$, $\boldsymbol{\Omega}$, $\mathbf{D}'$, $\overset{\circ}{\mathbf{r}}'^*$, $\overset{\circ}{p}_r$.

In Eqs. (14) and (15), $\overset{\circ}{J}$ does not appear explicitly and both $\overset{\circ}{Q}_c$ and $\overset{\circ}{Q}_m$ are well behaved as the material approaches incompressibility. Thus both functionals $\overset{\circ}{\Pi}_8$ and $\overset{\circ}{\Pi}_9$ are suitable for incompressible or nearly incompressible materials.

One can derive additional functionals by improsing constrain condition *a priori*. For example, if the compatibility condition are satisfied *a priori*

$$(16) \qquad \mathbf{D}' = (\nabla_N \mathbf{v})_s - \mathbf{I} : (\nabla_N \mathbf{v})_s / 3 \qquad \text{in } V_N ,$$

$$(17) \qquad \boldsymbol{\Omega} = (\nabla_N \mathbf{v})_a \qquad \text{in } V_N ,$$

$$(18) \qquad \mathbf{v} = \bar{\mathbf{v}} \qquad \text{on } S_{uN} .$$

Then Eq. (15) is reduced to

$$(19) \quad \overset{\circ}{\Pi}_{10} (\mathbf{v}, \overset{\circ}{p}_r) = \int_{V_N} \left\{ \overset{\circ}{Q}_m \left[(\nabla_N \mathbf{v})_s - \frac{1}{3} \operatorname{tr}(\nabla_N \mathbf{v}), \overset{\circ}{p}_r \right] \right.$$
$$- \frac{1}{2} \boldsymbol{\tau} : [(\nabla_N \mathbf{v})_a^T \cdot (\nabla_N \mathbf{v})_a] + \boldsymbol{\tau} : [(\nabla_N \mathbf{v})_a^T \cdot \nabla_N \mathbf{v}]$$
$$\left. - \overset{\circ}{p}_r \operatorname{tr}(\nabla_N \mathbf{v}) - \rho_N \overset{\circ}{\mathbf{b}} \cdot \mathbf{v} \right\} dV - \int_{S_{\sigma N}} \overset{o\nu}{\mathbf{T}} \cdot \mathbf{v} dA ,$$

which is a functional of the two fields, piecewise continuous $\overset{\circ}{p}_r$ and continuous $\mathbf{v}$. Tong (1969) first derived $\overset{\circ}{\Pi}_{10}$ for infinitesimal deformation.

When $\boldsymbol{\Omega}$ is re-introduced as part of penalty function, we can extend the two-field functional $\overset{\circ}{\Pi}_{10}$ to the following three-field functional:

$$(20) \quad \overset{\circ}{\Pi}{}^*_{10}(\mathbf{v}, \boldsymbol{\Omega}, \overset{\circ}{p}_r) = \int_{V_N} \left\{ \overset{\circ}{Q}_m \left[(\nabla_N \mathbf{v})_s - \frac{1}{3} \operatorname{tr}(\nabla_N \mathbf{v}), \overset{\circ}{p}_r \right] \right.$$
$$- \frac{1}{2} \boldsymbol{\tau} : [(\nabla_N \mathbf{v})_a^T \cdot (\nabla_N \mathbf{v})_a] + \boldsymbol{\tau} : [(\nabla_N \mathbf{v})_a^T \cdot \nabla_N \mathbf{v}]$$
$$\left. - \overset{\circ}{p}_r \operatorname{tr}(\nabla_N \mathbf{v}) + \frac{c}{2} [(\nabla_N \mathbf{v})_a - \boldsymbol{\Omega}]^2 - \rho_N \overset{\circ}{\mathbf{b}} \cdot \mathbf{v} \right\} dV$$
$$- \int_{S_{\sigma N}} \overset{o\nu}{\mathbf{T}} \cdot \mathbf{v} dA ,$$

where c is the constant factor of the penalty function. The functional $\overset{\circ}{\Pi}{}^*_{10}$, with the condition Eq. (16) satisfied *a priori*, is different from $\overset{\circ}{\Pi}_9$ in which $\overset{\circ}{\mathbf{r}}'{}^*$ is an independent field variable.

Having Ω as an independent field rather than computing it from $(\nabla_N \mathbf{v})_a$ is important from the computational point of view. As demonstrated by Cazzani and Atluri (1992), and Seki and Atluri (1994), the solution accuracy for large deformation problems depends strongly on the accuracy of Ω. The spin rate is needed to compute the updated solution. The formulation of Eq. (20) is generally more accurate as it gives Ω directly without having to differentiate the numerical solution of $\mathbf{v}$.

17.9. UPDATED SOLUTION

Depending on the solution approach, during each increment, one solves directly for the stress increment rates $\overset{\circ}{\mathbf{T}}$, $\overset{\circ}{\mathbf{r}}{}^{*}$ or $\overset{\circ}{\mathbf{S}}$, the deformation rates $\mathbf{D}$, Ω and the displacement rate $\mathbf{v}$. One then evaluates the Cauchy stress increment rate $\overset{\circ}{\tau}$ from Eq. (17.3:7), or Eqs. (17.3:9)–(17.3:12). In solving for the stress rates, the information needed includes: the Cauchy stress $\tau(= \tau^N)$, the increment rate of boundary traction $\overset{\circ\nu}{\mathbf{T}}$ on $S_{\sigma N}$, the increment rate of the body force $\overset{\circ}{\mathbf{b}}$, the increment rate of boundary displacement $\bar{\mathbf{v}}$ on S_{uN}, the deformed geometry of the body in D_N, and the constitutive laws that relate the stress rates to deformation rates. The constitutive law generally depends on the total deformation gradient $\mathbf{F}_0^N$ and the Cauchy stress τ. If a constitutive relation is known in the form of the gradient of a rate-potential in terms of a specific strain rate, e.g., Eq. (17.5:9) or (17.5:15), one can derive other rate-potentials through Eq. (17.5:11)–(17.5:13) or Eqs. (17.5:16)–(17.5:19) and determine the corresponding constitutive laws accordingly. Hill (1967) and Atluri (1979, 1980a) show that a rate potential $\overset{\circ}{V}_J(\mathbf{D})$ can be expressed in terms of the strain rate tensor $\mathbf{D}$. Pian and Tong (1971) show that the rate-potential $\overset{\circ}{W}(\mathbf{D})$ can be derived from the strain energy density $W_0(\mathbf{C})$. Equations (17.5:24) and (17.5:25) show the dependence of $\overset{\circ}{W}(\mathbf{D})$ on the total deformation gradient and the related deformation matrices.

The updated Cauchy stress is simply $\tau^{N+1} = \tau + \overset{\circ}{\tau}\,\Delta t$, which is used to determine the solution of the next increment. The current Cauchy stress is simply the sum of all previous increments, i.e.,

$$(1) \qquad \tau^{N+1} = \sum_{i=1}^{N} \overset{\circ}{\tau}_i\, \Delta t_i \,,$$

where $\overset{\circ}{\tau}_i\, \Delta t_i$ is the Cauchy stress increment of the ith step. The direct summation of the increments of other stress measures such as $\overset{\circ}{\mathbf{T}}, \overset{\circ}{\mathbf{r}}{}^{*}$ and

$\overset{\circ}{\mathbf{S}}$ does not have any physical meaning, because each increment refers to a different configuration.

The updated total displacement vector and the updated position vector of material point after each increment are

(2) ▲ $$\mathbf{u}^{N+1} = \mathbf{u}^N + \mathbf{v}\Delta t\,,$$

(3) ▲ $$\mathbf{X}^{N+1} = \mathbf{X}^N + \mathbf{v}\Delta t = \mathbf{X}^0 + \mathbf{u}^{N+1}\,.$$

Based on the updated $\mathbf{u}^{N+1}$ and $\mathbf{X}^{N+1}$, one can evaluate the updated total deformation gradient and the related deformation matrices. Together with the updated $\boldsymbol{\tau}$, one can update the rate-potential and the constitutive equation, e.g., Eqs. (17.5:25) and (17.5:26), for the solution of the next increment.

In Secs. 16.9 and 17.5, we discuss the need of maintaining objectivity in establishing the constitutive laws. There is also a need of maintaining *objectivity in computation*. The rate of change of stress depends not only on the strain rate $\mathbf{D}$ but also on the spin rate $\boldsymbol{\Omega}$ of the material element. An objective constitutive law gives the relation between the strain rate and the objective stress rate in the coordinate frame rigidly spinning with the material element. The relation between the strain rate, the spin rate, and the rates of other stress measures can be obtained by transformation of the objective stress rate tensor and the strain rate from the spinning coordinate system.

How well objectivity is maintained in the computation depends on the accuracy of the computed spin rate and spinning coordinates. The covariant base vector $\mathbf{g}_M$ of a rigid frame spinning at a rate of $\omega(t)$ satisfies the equation

(4) $$\frac{d\mathbf{g}_M}{dt} = \boldsymbol{\omega} \cdot \mathbf{g}_M\,,$$

at any loading stage. The solution is

$$\mathbf{g}_M(t) = \mathbf{R}(t) \cdot \mathbf{g}_M(t_N)$$

where $\mathbf{g}_M(t_N)$ is the initial value at $t = t_N$, $\mathbf{R}(t)$ is the rotation tensor of the deformation if $\omega(t)$ is as defined in Eq. (16.5:11). Thus to maintain objectivity, we should assure that the computed $\mathbf{R}(t)$ is a rotation tensor (Hughes and Winget 1981, Rubinstein and Atluri 1983).

It can be shown that the linearized solution of Eq. (16.5:11) is

$$\mathbf{R}(t_N + \Delta t) = [\mathbf{I} + \boldsymbol{\Omega}(t_N)\Delta t] \cdot \mathbf{R}(t_N)\,,$$

where $\Omega(t_N)$ is the spin rate of the deformation in the updated Lagrange description. Then, obviously

$$
(5) \qquad \mathbf{R}(t_N + \Delta t) \cdot \mathbf{R}(t_N + \Delta t)^T
$$

$$
= \mathbf{R}(t_N)^T \cdot [\mathbf{I} + \Omega(t_N)^T \Delta t] \cdot [\mathbf{I} + \Omega(t_N)\Delta t] \cdot \mathbf{R}(t_N)
$$

$$
= \mathbf{I} + \mathbf{R}(t_N)^T \cdot \Omega(t_N) \cdot \Omega(t_N)^T \cdot \mathbf{R}(t_N)\Delta t^2 \,,
$$

since $\mathbf{R}(t_N)^T \cdot \mathbf{R}(t_N) = \mathbf{I}$ and $\Omega(t_N)^T + \Omega(t_N) = 0$. The error in rotation is of the order Δt^2. Objectivity is maintained only in the limit of $\Delta t \to 0$.

In practical computation, the increment Δt is finite. In order to improve the preservation of objectivity in the incremental solution, we modify the computational algorithm to minimize the error caused by finite incremental steps. The accuracy of integrating Eq. (4) is improved, if an appropriate value of $\Omega^*[= \Omega(t^*)$, where $t_N \le t^* \le t_N + \Delta t]$ is used instead of $\Omega(t_N)$ in the Euler approximation given above. Hughes and Winget (1981), Rubinstein and Atluri (1983), Key (1980), Courant and Hilbert (1953) and others have shown that the *midpoint rule* of integration (i.e., use $t^* = t_N + \Delta t/2$) provides an accuracy of the order $(\Delta t)^2$.

The midpoint rule can be easily implemented by evaluating $\Omega(t)$ at the midpoint of each increment, i.e., $\Omega(t_N + \Delta t/2)$. In other words, in determining the solution in D_{N+1} configuration, instead of using the operator ∇_{N+1}, we use

$$
(6) \qquad \nabla_{N+1/2} = \varepsilon_I \frac{\partial}{\partial X_I^{N+1/2}} \,,
$$

where $\mathbf{X}^{N+1/2} = \mathbf{X}^N + \mathbf{v}\Delta t/2$. Then, we define the rotation and the deformation gradient as

$$
(7) \qquad \mathbf{R}(t_N + \Delta t) = [\mathbf{I} + \Omega(t_N + \Delta t/2)\Delta t] \cdot \mathbf{R}(t_N)
$$

$$
(8) \qquad \mathbf{F} = \mathbf{F}_0^{N+1} = \frac{\partial \mathbf{X}^{N+1/2}}{\partial \mathbf{X}^0} = \mathbf{F}\left(t_N + \frac{\Delta t}{2}\right) \,,
$$

where

$$
(9) \qquad \Omega\left(t_N + \frac{\Delta t}{2}\right) = \frac{1}{2}[\nabla_{N+1/2}\mathbf{v} - (\nabla_{N+1/2}\mathbf{v})^T] \,.
$$

We may also compute the orthogonal rotation by

$$
(10) \qquad \mathbf{R}(t_N + \Delta t) = \exp(\Omega\Delta t)\mathbf{R}(t_N) \,,
$$

which can further be approximated as

$$(11) \qquad \mathbf{R}(t_N + \Delta t) = [\mathbf{I} + \mathbf{\Omega}\Delta t + \mathbf{\Omega}^T \cdot \mathbf{\Omega}(\Delta t)^2/2]\mathbf{R}(t_N) \,.$$

We calculate $\mathbf{C}_0^{N+1}$, $\mathbf{B}_0^{N+1}$ and $\mathbf{U}_0^{N+1}$ from $\mathbf{F}_0^{N+1}$ and $\mathbf{R}_0^{N+1}$. Depending on the approach one has the solution in terms of $\overset{\circ}{\mathbf{T}}$, $\overset{\circ}{\mathbf{r}}{}^*$, $\overset{\circ}{\mathbf{S}}$ or $\overset{\circ}{\sigma}_r$. Then one of the following relations is used to calculate $\overset{\circ}{\tau}$,

$$(12) \qquad \overset{\circ}{\tau} = \overset{\circ}{\sigma}_r + \mathbf{\Omega} \cdot \tau - \tau \cdot \mathbf{\Omega} - \mathrm{tr}(\mathbf{L})\tau \,,$$

$$(13) \qquad \overset{\circ}{\tau} = \overset{\circ}{\mathbf{S}} + \mathbf{L} \cdot \tau + \tau \cdot \mathbf{L}^T - \mathrm{tr}(\mathbf{L})\tau \,,$$

$$(14) \qquad \overset{\circ}{\tau} = \overset{\circ}{\mathbf{r}}{}^* + \mathbf{L} \cdot \tau - \mathrm{tr}(\mathbf{L})\tau - \tau \cdot \mathbf{\Omega} \,,$$

$$(15) \qquad \overset{\circ}{\tau} = \overset{\circ}{\mathbf{T}} + \mathbf{L} \cdot \tau - \mathrm{tr}(\mathbf{L})\tau \,,$$

where $\mathbf{L}$ and $\mathbf{\Omega}$ are the half-step values defined in Eqs. (6) and (9). Incremental numerical solutions are numerous, e.g., Pian and Tong (1971) Murakawa and Atluri (1978, 1979) and Seki and Atluri (1994).

17.10. INCREMENTAL LOADS

The value of load increment at each increment step depends on the nature of applied loads. Normally the applied loads are of either a *deadweight* or *pressure type*. The direction of a deadweight load does not change (e.g., the gravitational force per unit mass) and its magnitude is given in terms of unit undeformed volume or surface. On the other hand, a pressure loads is in terms of unit deformed area and both its direction and magnitude can change at each increment.

We are going to determine the applied traction rate $\overset{o\nu}{\mathbf{T}}$ in the updated Lagrangian description in the N configuration. The rate depends on the nature of the applied loads. From Eqs. (17.6:6) and (17.3:7), we obtain

$$(1) \qquad \overset{o\nu}{\mathbf{T}} = \mathbf{N} \cdot \overset{\circ}{\mathbf{T}} = \mathbf{N} \cdot \overset{\circ}{\tau} + \mathbf{N} \cdot [\mathrm{tr}(\mathbf{L})\mathbf{I} - \mathbf{L}] \cdot \tau \,,$$

where $\mathbf{N}$ is a unit normal to $d\mathbf{A}^N$ that $d\mathbf{A}^N = \mathbf{N}dS^N$, $\mathbf{L}$ is the rate of deformation gradient and $\overset{\circ}{\mathbf{T}}$, $\overset{\circ}{\tau}$ are, respectively, the Lagrangian and the Cauchy stress rates in the updated Lagrangian description. The first term of the right most equation above reflects the change in the loads. The second term $[\mathrm{tr}(\mathbf{L})\mathbf{I} - \mathbf{L}] \cdot \tau$ results from the deformation of the boundary surface.

For applied *pressure* loads (without shear),

$$\bar{p}^{N+1}d\mathbf{A}^{N+1} = (\bar{p}^N + \overset{\circ}{p}\,\Delta t)[1 + \text{tr}(\mathbf{D})\Delta t](\mathbf{I} - \mathbf{L}^T\Delta t)\cdot d\mathbf{A}^N$$

$$= \bar{p}^N d\mathbf{A}^N + \{\overset{\circ}{p}\,\mathbf{I} + [\text{tr}(\mathbf{D})\mathbf{I} - \mathbf{L}^T]\bar{p}^N\}\Delta t\cdot d\mathbf{A}^N\ ,$$

where $\overset{\circ}{p}$ is the *prescribed incremental rate of pressure*. Note that, from Eq. (16.2:18), we have

$$d\mathbf{A}^{N+1} = J_N^{N+1}[(\mathbf{F}^{-1})^T]_N^{N+1}d\mathbf{A}^N\ .$$

Then, from $\overset{o\nu}{\mathbf{T}}\,dA^N = \bar{p}^{N+1}d\mathbf{A}^{N+1} - \bar{p}^N d\mathbf{A}^N$, the applied load increment is

(2) ▲ $$\overset{o\nu}{\mathbf{T}} = \overset{\circ}{p}\,\mathbf{N} + [\text{tr}(\mathbf{D})\mathbf{N} - \mathbf{L}^T\cdot\mathbf{N}]\bar{p}^N\ ,$$

for the *rate of traction increment* on the *deformed traction-prescribed sur-faces* in the updated Lagrangian description.

For a *deadweight type* of applied loads, the forces are normally referred to the undeformed configuration. Let $d\,\overset{\nu}{\mathbf{T}}{}^{N+1}$ be the applied force vector acting on an infinitesimal surface $d\mathbf{A}^{N+1}$ on the boundary in the D_{N+1} configuration. We have

(3) $$d\,\overset{\nu}{\mathbf{T}}{}^{N+1} = \mathbf{T}_0^{N+1}\cdot d\mathbf{A}^0 = (\mathbf{T}_0^N + \Delta\mathbf{T})\cdot d\mathbf{A}^0$$

$$= \tau^N\cdot d\mathbf{A}^N + \Delta\,\overset{\nu}{\mathbf{T}}\,dS^0 = \overset{\nu}{\mathbf{T}}{}^N\,dS^N + \Delta\,\overset{\nu}{\mathbf{T}}\,dS^0\ ,$$

where the superscript "0" refers to the *undeformed state*, $\mathbf{T}_0^{N+1}$, $\mathbf{T}_0^N$ are the first Piola-Kirchhoff stress at the $N+1$ and N deformation states referred to the undeformed state. Thus $\Delta\,\overset{\nu}{\mathbf{T}}$ is the *given traction increment*. The force vector $d\,\overset{\nu}{\mathbf{T}}{}^{N+1}$ also relates to the traction $\overset{\nu}{\mathbf{T}}{}^N$ and the surface area $dS^N = |d\mathbf{A}^N|$ in the D_N configuration by the equation

(4) $$d\,\overset{\nu}{\mathbf{T}}{}^{N+1} = \overset{\nu}{\mathbf{T}}{}^{N+1}\,dS^N = (\overset{\nu}{\mathbf{T}}{}^N + \overset{o\nu}{\mathbf{T}}\Delta t)dS^N\ .$$

Comparing Eqs. (3) and (4) gives

$$\overset{o\nu}{\mathbf{T}} = \frac{\Delta\,\overset{\nu}{\mathbf{T}}}{\Delta t}\frac{dS^0}{dS^N}\ ,$$

Using Eqs. (16.2:14), one finds

(5) ▲ $$\overset{o\nu}{\mathbf{T}} = \frac{\Delta\overset{\nu}{\mathbf{T}}}{\Delta t}\frac{\sqrt{\mathbf{N}\cdot\mathbf{B}_0^N\cdot\mathbf{N}}}{J_0^N}\ ,$$

for the *rate of traction increment* on the *deformed traction-prescribed sur-faces* in the updated Lagrangian description, where $\mathbf{B}_0^N[=\mathbf{F}_0^N(\mathbf{F}_0^N)^T]$ is the Almansi tensor in terms of the deformation gradient $\mathbf{F}_0^N$ given in Eq. (16.2:8), and J_0^N is the Jacobian referred to the undeformed state. Note, lower case n is used as the unit normal to the deformed surface da in Eq. (16.2:8) and presently they are denoted by $\mathbf{N}$ and dA^N, respectively.

For body forces proportional to the mass per unit undeformed volume, the *rate of body force increment* in the updated Lagrangian description is

$$\overset{\circ}{\mathbf{b}} = \frac{\Delta \mathbf{b}}{\Delta t}\frac{dV^0}{dV^N} = \frac{\Delta \mathbf{b}}{\Delta t}\frac{1}{J_0^N}\,,$$

where $\Delta \mathbf{b}$ is the body force increment per unit mass at the Nth step of increment.

We summarize the rate potentials and the constitutive laws developed in this chapter as follows:

Rate potential	Constitutive law
$\overset{\circ}{\mathbf{W}}(\mathbf{D}) = \overset{\circ}{V}_0(\mathbf{D}) = d_{ijkl}D_{ij}D_{kl}/2$	$\overset{\circ}{\mathbf{S}} = \frac{\partial \overset{\circ}{W}(\mathbf{D})}{\partial \mathbf{D}}$
$\overset{\circ}{Q}(\mathbf{D}) = \overset{\circ}{V}_0(\mathbf{D}) + \frac{1}{2}\boldsymbol{\tau}:(\mathbf{D}\cdot\mathbf{D})$	$\overset{\circ}{\mathbf{r}} \equiv (\overset{\circ}{\mathbf{r}}^*)_s = \frac{\partial \overset{\circ}{Q}(\mathbf{D})}{\partial \mathbf{D}}$
$\overset{\circ}{\mathbf{U}}(\mathbf{L}) = \overset{\circ}{V}_0\left(\frac{\mathbf{L}+\mathbf{L}^T}{2}\right) + \frac{1}{2}\boldsymbol{\tau}:(\mathbf{L}^T\cdot\mathbf{L})$	$\overset{\circ}{\mathbf{T}} = \frac{\partial \overset{\circ}{U}(\mathbf{L})}{\partial \mathbf{L}^T} = \frac{\partial \overset{\circ}{U}}{\partial(\nabla_N\mathbf{v})}$
$\overset{\circ}{P}(\mathbf{D}) = \overset{\circ}{V}(\mathbf{D}) + \boldsymbol{\tau}:(\mathbf{D}\cdot\mathbf{D})$	$\overset{\circ}{\sigma}_r = \frac{\partial \overset{\circ}{P}(\mathbf{D})}{\partial \mathbf{D}}$
$\overset{\circ}{Q}_c[(\overset{\circ}{\mathbf{T}} + \boldsymbol{\tau}:\boldsymbol{\Omega})_s] = (\overset{\circ}{\mathbf{T}}^T)_s : \mathbf{D}$	$\mathbf{D} = \frac{\partial \overset{\circ}{Q}_c}{\partial(\overset{\circ}{\mathbf{T}}+\boldsymbol{\tau}\cdot\boldsymbol{\Omega})_s}$
$\quad - \overset{\circ}{Q}(\mathbf{D}) + \boldsymbol{\tau}:(\boldsymbol{\Omega}\cdot\mathbf{D})$	
$\overset{\circ}{W}^*(\mathbf{D}', \overset{\circ}{J}) = \overset{\circ}{W}(\mathbf{D})$	$\overset{\circ}{\mathbf{S}}' = \frac{\partial \overset{\circ}{W}^*(\mathbf{D}', \overset{\circ}{J})}{\partial \mathbf{D}'}, \quad \overset{\circ}{p}_s = \frac{\partial \overset{\circ}{W}^*(\mathbf{D}', \overset{\circ}{J})}{\partial \overset{\circ}{J}}$
$\overset{\circ}{Q}^*(\mathbf{D}', \overset{\circ}{J}) = \overset{\circ}{Q}(\mathbf{D})$	$(\overset{\circ}{\mathbf{r}}^*)_s = \frac{\partial \overset{\circ}{Q}^*(\mathbf{D}', \overset{\circ}{J})}{\partial \mathbf{D}'}, \quad \overset{\circ}{p}_r = \frac{\partial \overset{\circ}{Q}^*(\mathbf{D}', \overset{\circ}{J})}{\partial \overset{\circ}{J}}$
$\overset{\circ}{U}^*(\mathbf{L}', \overset{\circ}{J}) = \overset{\circ}{U}(\mathbf{L})$	$\overset{\circ}{\mathbf{T}}'^T = \frac{\partial \overset{\circ}{U}^*(\mathbf{L}', \overset{\circ}{J})}{\partial \mathbf{L}'}, \quad \overset{\circ}{p}_t = \frac{\partial \overset{\circ}{U}^*(\mathbf{L}', \overset{\circ}{J})}{\partial \overset{\circ}{J}}$

Note: $\overset{\circ}{V}_0(\mathbf{D}) = \overset{\circ}{V}_j(\mathbf{D}) + \boldsymbol{\tau}:(\mathbf{D}\cdot\mathbf{D})$

17.11. INFINITESIMAL STRAIN THEORY

The incremental variational principles previously formulated can be reduced to the case of the infinitesimal strain theory by simply regarding the current configuration D_N as the initial configuration and assuming no initial stress, i.e., $\tau^N = 0$. In this case, all stress measures become the same

$$\tau^{N+1} = \mathbf{t}^{N+1} = \mathbf{r}^{*N+1} = \mathbf{S}^{N+1} = \overset{\circ}{\tau}\,\Delta t = \overset{\circ}{\mathbf{T}}\,\Delta t = \overset{\circ}{\mathbf{r}}{}^*\Delta t = \overset{\circ}{\mathbf{S}}\,\Delta t\,,$$

$$\overset{\circ}{\mathbf{T}}{}' = \overset{\circ}{\mathbf{r}}{}'^* = \overset{\circ}{\mathbf{S}}{}' = \overset{\circ}{\tau}{}' = \overset{\circ}{\tau} - \mathrm{tr}(\overset{\circ}{\tau})/3\,,$$

$$\overset{\circ}{p} = \overset{\circ}{p}_r = \overset{\circ}{p}_t = \overset{\circ}{p}_s\,,$$

The following rate potentials are also equal,

$$\overset{\circ}{V}_J\,(\mathbf{D}) = \overset{\circ}{Q}\,(\mathbf{D}) = \overset{\circ}{W}\,(\mathbf{D}) = \overset{\circ}{U}\,(\mathbf{L}) = \overset{\circ}{P}\,(\mathbf{D})\,.$$

For linear elastic isotropic material, one has

$$\overset{\circ}{Q}\,(\mathbf{D}) = \frac{1}{2}\lambda(D_{ii})^2 + GD_{ij}D_{ij}\,,$$

$$\overset{\circ}{Q}{}^*(\mathbf{D}',\overset{\circ}{J}) = \frac{1}{2}\lambda\,\overset{\circ}{J}{}^2 + G\left(D'_{ij} + \frac{1}{3}\,\overset{\circ}{J}\delta_{ij}\right)\left(D'_{ij} + \frac{1}{3}\,\overset{\circ}{J}\delta_{ij}\right)$$

$$= \frac{1}{2}\left(\lambda + \frac{2G}{3}\right)\overset{\circ}{J}{}^2 + GD'_{ij}D'_{ij}\,,$$

$$\overset{\circ}{Q}_m\,(\mathbf{D}',\dot{p}) = \overset{\circ}{Q}{}^*(\mathbf{D}',\overset{\circ}{J}) - \overset{\circ}{p}\overset{\circ}{J} = GD'_{ij}D'_{ij} - \frac{1}{2K}\,\overset{\circ}{p}{}^2\,,$$

$$-\overset{\circ}{Q}_c\,(\tau',\overset{\circ}{p}) = \overset{\circ}{Q}{}^*(\mathbf{D}',\overset{\circ}{J}) - \tau'_{ij}D'_{ij} - \overset{\circ}{p}\overset{\circ}{J} = -\frac{1}{4G}\,\tau'_{ij}\tau'_{ij} - \frac{1}{2K}\,\overset{\circ}{p}{}^{-2}\,,$$

where λ is the Lamé constant, G the shear modulus and K the bulk modulus

$$K = \lambda + \frac{2G}{3}\,.$$

For materials approaching incompressibility, λ and K tend to infinity while $\overset{\circ}{J}$ approaches zero. Therefore $\overset{\circ}{Q}$ and $\overset{\circ}{Q}{}^*$ are not suitable for numerical computation. However, both $\overset{\circ}{Q}_m$ and $\overset{\circ}{Q}_c$ are well behaved since $\overset{\circ}{p}$ is finite.

PROBLEMS

17.4. Derive a variational principle using $\mathbf{v}$, $\boldsymbol{\Omega}$, $\mathbf{L}'$, $\overset{\circ}{J}$, $\overset{\circ}{\mathbf{T}}{}'$, $\overset{\circ}{p}_t$ as the field variables where $\overset{\circ}{\mathbf{T}}{}'$, $\overset{\circ}{p}_t$ are defined in Eqs. (17.8:5) and (17.8:6). Also, derive the field equations and the natural boundary conditions associated with the stationary condition of the functional.

17.5. Derive the field equations and the natural boundary conditions for the functionals $\overset{\circ}{\Pi}_8, \overset{\circ}{\Pi}_9, \overset{\circ}{\Pi}_{10}, \overset{\circ}{\Pi}{}^*_{10}$ defined in Sec. 17.8.

18

FINITE ELEMENT METHODS

Numerical approach is a "must" to establish satisfactory approximate solutions for most practical problems. There are many numerical methods. A treatise on the entire subject is beyond the scope of this book. In this chapter, we focus on the fundamentals of the *finite element method*. We will explore the finite element concept, the approximations involved, the construction of the finite element equations and the use of incremental approach to nonlinear problems.

The finite element method is one of the most important developments in numerical analysis. It is versatile and powerful for solving linear as well as nonlinear complex problems. The method was developed in the 1950's by engineers as an outgrowth of the so-called matrix method (Argyris 1955, 1966, Argyris and Kelsey 1960) for systematically analyzing complex structures containing a large number of components. Over the years, the finite element method has spread to applications in all fields of engineering, science and medicine.

The finite element method establishes approximate solutions in terms of unknown parameters in sub-regions called "elements" and then deduces an approximate solution for the whole domain by enforcing certain relations among the solutions of all elements. For structural analysis the procedure is to express the relations between the displacements and internal forces at the selected nodal points of individual structural components in the form of a system of algebraic equations. The unknowns for the equations are nodal displacements, nodal internal forces, or both. Depending on the unknowns and dependent variables selected, the method is qualified with word like *displacement, force, hybrid or mixed*. The system of equations is written most conveniently in matrix notation, and the solution of these equations is obtained efficiently by high-speed computation.

The foundations for many finite element methods have been established. Formulations of methods with variational principles or principles of virtual work were developed by Courant (1943), Gallagher *et al.* (1962), Besseling (1963), Jones (1964), Pian (1964), Fraeijs de Veubeke (1964), Herrmann (1965), Prager (1967, 1968), Tong and Pian (1967), Pian and Tong (1969),

Oden (1969), Tong (1970), Zienkiewicz (1971), Atluri (1980, 1995), Hughes (1987), and Pian (1995). Tong (1966) developed the early finite element method for fluid mechanics. The method has become a major tool for computational fluid dynamics (Hughes and Franca 1987, Tezduyar *et al.* 1988, Gunzburger 1989, Shakib *et al.* 1991, Capon and Jimack 1995), magneto-hydrodynamics, radiation analysis, and heat transfer. A major branch of the finite element method is the *boundary element method* (Cruse 1969, Curse and Rizzo, eds. 1975, Brebbia 1984, Chen and Zhou 1992, Camp and Gipson 1992, Hall 1994) in which one transforms the problem to have all the unknown field quantities appear on the boundaries only to avoid dealing with the entire domain. One then establishes the algebraic equations for the unknowns of selected discrete locations. The accuracy of the approximate solution depends on the smoothness of the field variable. When a physical problem is analyzed using the finite element method, it involves approximation in the geometry as well as local solutions. The errors can be estimated from the approximate solutions. If the errors are large, then the finite element model is refined by improving the geometrical approximation and/or the representation of the local solutions. The new model is re-analyzed until the estimated errors fall below the specified limits. This method is called the *adaptive finite element method*. A recent review of the methods for error estimation and the adaptive refinement processes is given by Li and Bettess (1997). The topics of fluid mechanics, magneto-hydrodynamics, etc. (topics other than solid mechanics), the boundary element methods, and adaptive finite element methods are beyond the scope of this book. Interested readers are referred to the published literature.

There are many publications on finite element method in journals, proceedings and books. An up-to-date source of references is the Journal Applied Mechanics Reviews, which publishes periodically review articles in different areas of the finite element method. For example, in its October, 1998 issue, Mackerle gave a bibliographical review of the finite element method applied to biomechanics published from 1987 through 1997.

Numerous finite-element computer programs are commercially available for specific or general applications. To list a few, they include PINTO (by the present junior author and is not commonly available), DYNA3D (Hallquist 1992, Hallquist *et al.* 1993), ANSYS, ABACUS, ADINA, STRUDL and NASTRAN.

The finite element method has been extensively applied to complicated problems. The description of the problem itself can be very involved and prone to errors, e.g., the detailed geometry of the brain, a nuclear reactor, the complete structure of an aircraft. Results of analysis are usually voluminous. A new field has evolved specifically to ease the burden of preparing the complex input information and processing the voluminous output. This includes techniques for automatic mesh generation, contour plots, graphics for input and output. Commercially available software in the area includes INGRID (Stillman 1993), PATRAN, HYPERMESH, etc. All software can be run interactively or in batch mode. Detailed information on the commercially available software can be found in their web-sites on the internet.

To lay down the foundation for all aspects of the finite element method with such extensive applications is of course not easy; the learning of it will demand great attention. But this is the task we set for ourselves for this chapter. In the subsequent discussion, we will use bold letters to denote matrices, unless specified otherwise.

18.1. BASIC APPROACH

Turner *et al.* (1956) in their celebrated paper first applied the displacement method to plane stress problems. They divided a structure into triangular and rectangular subdomains called *"elements"* and designated the vertices of the elements as nodes. The *behavior of each element was represented by the nodal displacements and an element-stiffness matrix that related the forces at the nodal points of the element to the nodal displacements.* They assembled the element-stiffness matrices to form a system of algebraic equations and solved them using high-speed computers. **There are two key aspects in the finite element method. One is the determination of the element stiffness matrices. The other is the assembly of element-stiffness matrices to form a system of algebraic equations, called the global system, for the selected unknowns. The former involves the local approximation for establishing the relations between element nodal forces and nodal displacements. The latter enforces the appropriate relations between the nodal quantities of adjacent elements at the inter-element boundaries.**

Equally important, if not more, in the finite element method is its *implementation.* The **success of the finite element method lies largely in the development of efficient pre- and post-processors, and algorithms for solving large systems of equations.** The pre-processor enables users to describe efficiently and in a relatively error free manner a complex problem in terms of its geometry, configuration, material properties, loading conditions, etc., through the input. The post-processor handles the voluminous output making it easily understandable through interactive graphic, tables, charts, and summaries. The pre- and post-processors also enable users to examine at ease any specific location of the domain or aspect of the solution prior to, during, or after the analysis. The need for efficient solution algorithms for large system goes without saying. The finite element method often deals with hundreds of thousands, even millions, of equations. Even with the advent of computers today, efficient algorithms make it possible to obtain the solution, especially for nonlinear problems, at a reasonable cost and in a reasonable time. The concept of finite element method dates back to 1943 when Courant used a triangular mesh to solve a two-dimensional Laplace equation, or earlier when mathematicians formulated differential calculus with piece-wise smooth functions. But the method did not catch on until the availability of high speed computers to solve the large number of algebraic equations and to handle the tedious input and the voluminous output.

In this book, due to scope limitation, we will not address the issues of pre- and post-processors. We will focus on the finite element formulation and will touch upon lightly the solution algorithms for large systems. We will discuss methods for determining element stiffness matrices, which are normally symmetric, and the process to assemble the element matrices into a global system of algebraic equations for numerical solution.

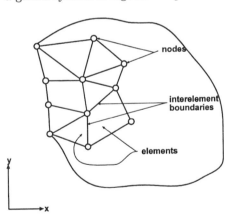

In finite element method, the continuum is divided into a finite number of sub-regions V_e in 3-dimension or A_e in 2-dimension domains, called *elements* (see Fig. 18.1:1). The boundary between two adjacent elements, called the *interelement boundary*, is a point for one-dimensional, a line or a curve for 2-dimensional, and a plane or a curve surface for 3-dimensional continuum. One may visualize that two adjacent elements are connected at a number of discrete points, called *nodes*, on their common boundaries. The nodes are usually the vertices of an element, the midpoint, the one-third and two-third points along an element boundary line, the centroid or some specific points of the element boundary surfaces. Points within an element may also be selected as nodes, which are called interior nodes of the

Fig. 18.1:1. Two dimensional elements in a continuum.

element. *The finite element method approximates the field variables within an element by interpolating their values at the nodes by shape or interpolation functions.* The interpolated functions approximate the behaviors of the element through an *integral formulation*, which will be defined later. The nodal values are parameters, called *degrees-of-freedom* or *generalized coordinates*, to be determined. These two names are used interchangeably throughout the remaining chapters. One may also include the nodal values of the spatial derivatives of the field variables for this purpose, especially when dealing with higher order differential equations. *Each node and each element can have different number of degrees-of-freedom.* The element behavior is characterized by these parameters (such as the nodal values of the field variable) in the form of element matrices such as *element stiffness matrix* and *element applied force matrix*. The local solution of a field variable must satisfy certain continuity requirements within the elements and along the common boundaries of adjacent elements. This is to assure that the finite-element solution will converge to the exact solution as the size of elements reduces to zero as the number of degrees-of-freedom per element increases to infinity. One usually **constructs the element matrices from a functional associated with a variational principle or from an integrally weighted-average of the governing differential equations.** Then, one **derives the equations relating the nodal parameters of all elements based on a stationary condition of the variational functional, or the weighted-average of the governing differential equations.** This process is called an *integral formulation*.

In the finite element method, **the interpolation functions, the assembling process and the solution methods for a large system of equations are generic. They can be applied to almost any types of problem.** In the subsequent sections, we shall discuss the shape functions, the element matrices, the assembly of element matrices, and methods for solving large systems of algebraic equations. For efficient implementation of the finite element method, we employ a *local system to sequentially label the quantities of each element* and a *global scheme to number the quantities of the entire system*. The list identifying the global labels of the generalized coordinates of each element is called the *assembly list* whose utility will be discussed in detail in Sec. 18.4.

In this chapter, we deal extensively with matrix operation. We shall use **bold** letters, both lower and upper cases, to denote *matrices*, unless otherwise specified. Thus, **AB** denotes a product of two matrices **A** and **B**. If **A** and **B** are of the order $m \times n$ and $n \times r$, respectively, the resulting matrix **AB** is of the order $m \times r$. A $m \times 1$ matrix is a column matrix of order m, whereas a $1 \times n$ matrix is a row matrix. They are often called vectors in the literature. Since they do not follow tensor transformation rules, in this book, we will just call them matrix to distinct them from *vector*, a tensor of rank 1. Bold Greek letters such as $\boldsymbol{\alpha}$, $\boldsymbol{\beta}$, etc., and some bold lowercase Latin letters are used to denote column or row matrices for generalized coordinates. Specifically, we use bold Greek letter $\boldsymbol{\sigma}$ and bold **e** to represent the stress and strain column matrices (not tensors), which will be defined later. We shall also use bold capital **T** to denote the traction column matrix and subscripted capital T_j to represent the traction components unless otherwise specified.

In the next section, for clarity, we shall write out explicitly most of the components of matrices in one-dimensional problems. In later sections, we shall resort to matrix notations and operations to simplify the expressions of field variables and the derivation of equations.

18.2. ONE-DIMENSIONAL PROBLEMS GOVERNED BY A SECOND ORDER DIFFERENTIAL EQUATION

In this section, we shall discuss the construction of interpolation functions and element matrices for one-dimensional problems. In the process, we will point out features that are important to the numerical implementation of the finite element process. The following notation is used to describe the order of continuity of interpolation functions.

C^p **Interpolation Function** (p^{th} order continuity). A function is said to have C^p *continuity* if the derivatives of the function up to order p are continuous. Thus, a C^0 function is continuous and a C^1 function not only is continuous but also has continuous first derivatives.

Let $u(X)$ be a function of the *spatial coordinate* X over $0 \leq X \leq L$. The domain is divided into a finite number of small, non-overlapping segments, called *elements* (Fig. 18.2:1). The *inter-element boundary* of two adjacent elements is a point, designated as a *node*. Additional points within the element may also be selected as nodes. The behavior of an element, i.e., the representation of $u(X)$ within the element, is approximated by the interpolation of selected parameters associated with $u(X)$ at the nodes of the element.

We label each node of the whole domain with a unique number, called the *global nodal number*. For instance, for the element over the region $X_2 \leq X \leq X_3$ with nodes at $X = X_2, X_3$ (Fig. 18.2:1a), the subscripts 2 and 3

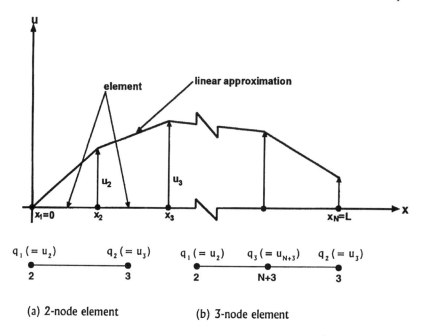

Fig. 18.2:1. Subdivision of a one-dimensional domain.

are the *global nodal labels* denoting the 2nd and 3rd nodes of the domain. Let the values of $u(X)$, a *field variable*, be u_2, u_3 at X_2, X_3, respectively. Thus, subscripts 2 and 3 of the field variable identify the nodal association of relevant quantities. The quantities u_2, u_3, called the *degrees-of-freedom* or *generalized coordinates*, are generally not known *a priori*. In other words u_2, u_3 are 2nd and 3rd degrees-of-freedom of the problem, i.e., subscripts 2 and 3 also denote the *global labels of the degrees-of-freedom* of the system. If there is only one degree-of-freedom at each node, the global degree-of-freedom label and the global nodal label are the same. In case that there are more than one field-variable and/or more than one parameter for the field variable at each node, the global degree-of-freedom label at a node will be different from the global nodal label of the node. Properly labeling all nodes of the whole domain and/or all degrees-of-freedom are essential to the success of the finite element analysis, because most of practical problems involve large number of degrees-of-freedom.

Linear Interpolation (C^0 function). We seek to represent a function u in an element between nodes X_2 and X_3 in terms of its values u_2, u_3 at the nodes (Fig. 18.2:1a). If u is smooth and if $X_3 - X_2$ is much smaller than the overall dimension, u can be approximated by the first two terms of the Taylor series as

$$(1) \qquad u(X) = \alpha_1 + \alpha_2(X - X_2).$$

Expressing α's in terms of nodal values of u gives

$$(2) \qquad u(X) = u_2 \left(1 - \frac{X - X_2}{X_3 - X_2} \right) + u_3 \frac{X - X_2}{X_3 - X_2}.$$

Approximating u in the form above for all elements yields a piecewise linear representation for u over the whole domain. This approximate function is obviously C^0 continuous.

We shall write the approximate function u in a form more suitable for the finite element process. First we introduce a new coordinate ξ defined as

$$(3) \quad \blacktriangle \qquad \xi = (2X - X_2 - X_3)/\varepsilon,$$

$$(4) \quad \blacktriangle \qquad \varepsilon = X_3 - X_2,$$

where ε is the *size of the element* and ξ is called the *normalized local coordinate*, or *natural coordinate* of the element. The origin of ξ is at the midpoint of the element and the range of ξ is between ± 1. We write Eq. (2) as

$$(5) \quad \blacktriangle \qquad u(\xi) = q_1 h_1(\xi) + q_2 h_2(\xi),$$

where

$$(6) \quad \blacktriangle \qquad h_1(\xi) = (1 - \xi)/2, \qquad h_2(\xi) = (1 + \xi)/2,$$

$$(7) \quad \blacktriangle \qquad u_2 = q_1, \qquad u_3 = q_2.$$

The h's are called *shape* or *interpolation functions* satisfying the conditions

$$(8) \qquad h_1(-1) = h_2(1) = 1 \qquad h_1(1) = h_2(-1) = 0.$$

The functions h_i as defined are called the *Lagrange interpolation* where h_i equals to 1 at the ith (a local label) node of the element and is zero at all other nodes. The q's, like u_i, have the simple physical meaning being the values of the field variable at the corresponding nodes and are called the *local generalized coordinates* as contrast to u_i, with the subscript being a global label, called the *global generalized coordinates*. Here, the subscripts of h's and q's refer to the local label, as opposed to the global label for u's. In the present case, the 1st and 2nd nodes (the local label) of the element are, respectively, the 2nd and 3rd nodes (the global label) of the system.

Through Eq. (7), the expression in Eq. (5) represents a *continuous function* over the whole domain.

Equation (1) is the first two terms of the Taylor series expansion. The error in appoximation for a smooth function is obviously of the order $O(\varepsilon^2)$, which is small if the element size ε is so. *Equation (2) is a typical finite element representation for u by linear functions within the element, whereas Eq. (5) is a generic expression* useful in numerical implementation of the finite method. This last expression has the following properties:

1. The range of the *natural coordinate* ξ is always from -1 to 1 for all elements. Equation (3) is the transformation between the global coordinate (the coordinate for the entire system) and the natural coordinate (the coordinate chosen to conveniently describe the element). The use of local coordinate is not essential here but is often necessary for more complex problems.

2. When the interpolation function h_i is expressed in terms of the *natural coordinate* ξ. The geometric characteristics of an element, represented by the element size ε here, do not appear at all.

3. We use q_i to denote the value of u at the ith node of the element. The subscript i (with $i = 1, 2$) is a label referencing to the nodes of the element and is thus a local nodal label.

The representation of a field variable in the generic form of Eq. (5) does not depend on the specific location of the element, or the global labels of the degrees-of-freedom of the element. The field variable in other elements can be represented in exactly the same way, e.g., u in the element $X_n \leq X \leq X_{n+1}$ can be expressed in the same generic form. The element specifics are reflected only through Eqs. (3), (4) and (7). Thus, *the generic expression allows us to represent a field variable in an element without knowing the specifics of the variable itself.* In other words, *we can approximate any continuous function in a small neighborhood in the form of Eq. (5) with the interpolation functions of Eq. (6). This property forms the basis of general and efficient algorithms for computing element matrices.* Equation (5) relates u at different locations through the subscripts in Eq. (7), which describes the relation between the degrees-of-freedom q's identified by the local nodal label and the nodal values of u identified by the global label. The array relates the local label of a degree-of-freedom to the global label of the same degree-of-freedom of every element is called the *assembly list* which is needed for assembling the element matrices into a system of equations. We shall further examine this point in the discussion of assembling.

Using the natural coordinate ξ rather than the global spatial coordinate X can provide advantages other than just representing a function by the generic shape functions. The details will be considered when we discussed two-dimensional problems.

Quadratic Interpolation (C^0 function). One can improve the approximation by shrinking the element, i.e., refining the subdivision, or by increasing the order of the Taylor expansion. The latter approach means constructing *higher order element* in the finite element language. For example, a representation of u as

$$\text{(9)} \quad \blacktriangle \qquad u = \alpha_1 + \alpha_2(X - X_2) + \alpha_3(X - X_2)^2$$

can be expressed in terms of the nodal values of u at the two-end nodes and the mid-point node of the element (Fig. 18.2:1b), i.e.,

$$\text{(10)} \quad \blacktriangle \qquad u(\xi) = q_1 h_1(\xi) + q_2 h_2(\xi) + q_3 h_3(\xi),$$

where

$$\text{(11)} \quad h_1(\xi) = -\frac{(1-\xi)\xi}{2}, \quad h_2(\xi) = \frac{\xi(1+\xi)}{2}, \quad h_3(\xi) = (1-\xi)(1+\xi).$$

The interpolation functions h's are zero at $\xi = \pm 1$ and 0 except for $h_1(-1) = h_2(1) = h_3(0) = 1$. The functions are the *Lagrange interpolation*. The midpoint node is within the element and is called the *interior node*. Nodes 1 and 2 are at the two ends and are *boundary nodes*. The truncation error of Eq. (9) is of the order $O(\varepsilon^3)$ which represents an improvement in the approximation over that of Eq. (1). Again, the generic representation in Eq. (10) does not depend on element specifics. The relationship between the local and global degrees-of-freedom is based on the requirement that

$$\text{(12)} \qquad u_2 = q_1, \qquad u_3 = q_2, \qquad u_{N+3} = q_3.$$

The subscripts 2 and 3 for u are the global labels of the degrees-of-freedom of the end nodes, which are labeled as the 1st and 2nd nodes of the element. The subscript $N + 3$ is the global label of the midpoint node at $X = (X_2 + X_3)/2$, Fig. 18.2:1(b), which is designated as the 3rd node of the element. The representation Eq. (9) with the interpolation function in the form of Eq. (11) can be used for any element. Similarly, one can establish representations of higher order (Tong and Rossettos 1977).

Problem 18.1. Derive the Lagrangian interpolation, which gives unity at $\xi = \xi_k$ and equals zero at other m points.

Element Matrices. How q's relates to other field quantities and how the values of q's will be determined depend on the nature of the problem and the governing differential equations or the variational formulation for the problem. These are the roles of element matrices. We shall discuss the construction of element matrices using the *variational functional of the element*.

Consider a rod subjected to a distributed load $P(X)$ along its axis with boundary conditions

(13) $\qquad\qquad u = \bar{u} \qquad \text{at} \qquad X = 0 \text{ (rigid condition)}$

(14) $\qquad EA(L)\dfrac{du}{dX} = \bar{T} \qquad \text{at} \qquad X = L \text{ (natural condition)}.$

The longitudinal deformation satisfies the second-order differential equation

(15) $\qquad\qquad \dfrac{d}{dX}\left[EA(X)\dfrac{du}{dX}\right] = P(X).$

We subdivide the domain into elements and introduce the natural coordinate $\xi = (2X - X_n - X_{n+1})/\varepsilon$, $\varepsilon = X_{n+1} - X_n$ for the element $X_n \le X \le X_{n+1}$ (Fig. 18.2:1). We *express the elastic modulus EA and the load P in terms of the natural coordinate and denote them by lower case symbols as*

$$ea(\xi) = EA(X), \qquad p(\xi) = P(X).$$

Following the derivation of Sec. 10.1, the functional associated with the *principle of minimum potential energy* is

(16) $\quad \Pi = \displaystyle\int_0^L \left[\dfrac{EA(X)}{2}\dfrac{du}{dX}\dfrac{du}{dX} + P(X)u\right]dX - u(L)\bar{T} = \sum_{n=1}^{N}\Pi_n,$

where

$$\Pi_n = \int_{-1}^{1}\dfrac{\varepsilon}{2}\left[\dfrac{2ea(\xi)}{\varepsilon^2}\dfrac{du}{d\xi}\dfrac{du}{d\xi} + p(\xi)u\right]d\xi - \alpha q_2 \bar{T},$$

in which α is a constant that $\alpha = 0$ for $n < N$ and $\alpha = 1$ for $n = N$ with node $N+1$ at $X = L$. Thus $q_2[= u(L)]$ is the unknown parameter associated with the node at $X = L$ in the last element. The admissibility condition requires that u be C^0 continuous over $0 \le X \le L$ and satisfies the *rigid boundary condition* Eq. (13) at $X = 0$ (see Sec. 10.6). The admissibility condition assures *the sum of the functionals of all elements equal to the functional of the whole domain*, a feature of great importance to the finite element method. We call Π a *functional* and u the *field* or *field variable* of the functional.

The variational principle requires that the *first variation* of the functional with arbitrary variation δu equals to zero, i.e.,

$$(17) \qquad \delta\Pi = \int_0^L \left[EA(X)\frac{du}{dX}\frac{d\delta u}{dX} + P(X)\delta u \right] dX - \bar{T}\delta u(L)$$

$$= \sum_{n=1}^N \delta\Pi_n = 0\,,$$

where

$$(18) \qquad \delta\Pi_n = \frac{\varepsilon}{2}\int_{-1}^1 \left[\frac{4ea(\xi)}{\varepsilon^2}\frac{du}{d\xi}\frac{d\delta u}{d\xi} + p(\xi)\delta u \right] d\xi - \alpha\delta q_2\bar{T}\,.$$

In Eq. (18), u and its variation δu must be C^0 continuous for $0 \le X \le L$, and $u = \bar{u}, \delta u = 0$ at $X = 0$ as required by the admissibility condition. The function with its variation satisfying these conditions are called the *admissible function* associated with the functional. The determination of the solution based on the vanishing of the first variation of a functional as given in Eq. (17) for arbitrary admissible functions is called an *integral formulation*.

The finite element method constructs an approximate solution based on an *integral formulation* using a selected set of admissible functions. The linear and the quadratic interpolation functions discussed before give continuous u in the whole region and are, therefore, admissible to the present problem. The rigid condition is satisfied by requiring $u_1[= u(0)]$ equal to $\bar{u}$ and $\delta u_1 = 0$ which is equivalent to imposing constraint(s) on the system of algebraic equations to be discussed later. The boundary condition Eq. (14) at $X = L$ is accounted for by the term $u(L)\bar{T}$ in Π or $\alpha q_2\bar{T}$ in Π_n of the last element. There is no restriction on u and δu at $X = L$ since the condition is a *natural boundary condition* (see Sec. 10.6). We call the boundary with prescribed natural boundary conditions a *force boundary*.

Substituting Eqs. (5) and (6) or Eqs. (10) and (11) into Eq. (18), and using $u = \sum_j h_j(\xi)q_j$ and $\delta u = \sum_i h_j(\xi)\delta q_j$ give

$$(19) \qquad \delta\Pi_n = \sum_{i,j} \delta q_i q_j \frac{2}{\varepsilon}\int_{-1}^1 ea(\xi)\frac{dh_i}{d\xi}\frac{dh_j}{d\xi}\,d\xi$$

$$+ \sum_i \delta q_i \frac{\varepsilon}{2}\int_{-1}^1 p(\xi)h_i(\xi)\,d\xi - \alpha\delta q_2\bar{T}\,,$$

where the range of summation is from 1 to 2 for linear interpolation and from 1 to 3 for quadratic interpolation. The rigid condition at $X = 0$ is satisfied by requiring

$$(20) \qquad\qquad q_1 = \bar{u}, \qquad \delta q_1 = 0$$

in the first element whose first node is at $X = 0$.

Equation (19) can be written in the matrix form

$$(21) \qquad\qquad \delta\Pi_n = \delta\mathbf{q}^T\mathbf{Q} = \delta\mathbf{q}^T\mathbf{kq} - \delta\mathbf{q}^T\mathbf{f},$$

where

$$\mathbf{q} = [q_1 \ q_2 \ \ldots]^T, \qquad \mathbf{Q} = \mathbf{kg} - \mathbf{f}$$

are, respectively, the *generalized element nodal displacement* and *force matrices* of the element, $\mathbf{k}$ is the *element stiffness matrix* and $\mathbf{f}$ is the *element external* or *applied force matrix* from the applied loads distributed within the element and on the element boundary. The terminology of stiffness and force is borrowed from structural mechanics. Note that $\mathbf{q}$, $\mathbf{Q}$, and $\mathbf{f}$ are column matrices. In general, $\mathbf{q}$ and $\mathbf{Q}$ may not have the dimension of displacement and force, respectively. The components of $\mathbf{k}$ and $\mathbf{f}$ are

$$(22) \ \blacktriangle \qquad\qquad k_{ij} = \frac{2}{\varepsilon}\int_{-1}^{1} ea(\xi)\frac{dh_i}{d\xi}\frac{dh_j}{d\xi}\,d\xi,$$

$$(23) \ \blacktriangle \qquad\qquad f_i = -\frac{\varepsilon}{2}\int_{-1}^{1} p(\xi)h_i(\xi)\,d\xi + \delta_{i2}\alpha\bar{T},$$

in which the range of i, j is from one to the number of degrees-of-freedom of the element and δ_{ij} is the Kronecker delta.

The element stiffness and external force matrices relate the element nodal displacement $\mathbf{q}$ uniquely to an element nodal force $\mathbf{Q}$. The unique relation is an essential requirement for the interpolation functions that it does not allow assuming the displacement u as a constant within an element because it leads to zero stiffness for the element. However, it is legitimate for the solution $\mathbf{q}$ to have all its components be of the same value.

We see from Eqs. (22) and (23) that the element matrices are *independent of the boundary conditions* except for the element on the force boundary where the element applied force matrix involves the applied boundary force. The element stiffness matrix $\mathbf{k}$ is symmetric and positive semi-definite provided that $ea(\xi) > 0$. It has one zero eigenvalue associated with the eigenvector $q_1 = q_2$, called the *zero eigenvector*. Physically, the zero eigenvector(s) associates with rigid body motion and gives zero strain energy.

Thus, the rank of $\mathbf{k}$ equals the number of q's minus the number of zero eigenvectors, which is one in the present case, if the shape function can represent constant du/dX exactly within element.

Linear Element. For linear interpolation, there are only two degrees-of-freedom per element

$$\mathbf{q} = \begin{bmatrix} q_1 \\ q_2 \end{bmatrix}, \qquad \delta\mathbf{q} = \begin{bmatrix} \delta q_1 \\ \delta q_2 \end{bmatrix}.$$

If ea and p are or can be approximated as constant, the integration of Eqs. (22) and (23) gives

$$(24) \qquad \mathbf{k} = \frac{ea}{\varepsilon} \begin{bmatrix} 1 & -1 \\ -1 & 1 \end{bmatrix},$$

$$(25) \qquad \mathbf{f} = \begin{bmatrix} f_1 \\ f_2 \end{bmatrix} = -\frac{p\varepsilon}{2} \begin{bmatrix} 1 \\ 1 \end{bmatrix} + \alpha \begin{bmatrix} 0 \\ \bar{T} \end{bmatrix}.$$

In this case, the rank of $\mathbf{k}$ is 1. The finite element process lumps half of the applied distributed load over the element to each node.

Since all interpolation functions involved are linear in X, we call the element a *linear element*.

Quadratic Element. We use higher order approximations for the field variable to derive higher order elements. Using the quadratic interpolation functions of Eq. (11) and linear approximation for ea and p, i.e.,

$$ea(\xi) = ea_0 + ea_1\varepsilon\,\xi/2,$$

$$p(\xi) = p_0 + p_1\varepsilon\,\xi/2,$$

where ea_0, ea_1, p_0, p_1 are constants, we obtain from of Eqs. (22) and (23)

$$(26) \quad \mathbf{k} = \frac{ea_0}{3\varepsilon} \begin{bmatrix} 7 & & \text{sym} \\ 1 & 7 & \\ -8 & -8 & 16 \end{bmatrix} + \frac{2}{3} ea_1 \begin{bmatrix} -1 & & \text{sym} \\ 0 & 1 & \\ 1 & -1 & 0 \end{bmatrix},$$

$$(27) \quad \mathbf{f} = -\frac{p_0\varepsilon}{6} \begin{bmatrix} 1 \\ 1 \\ 4 \end{bmatrix} - \frac{p_1\varepsilon^2}{12} \begin{bmatrix} -1 \\ 1 \\ 0 \end{bmatrix} + \alpha \begin{bmatrix} 0 \\ \bar{T} \\ 0 \end{bmatrix}.$$

Since the interpolation functions involve quadratic functions of X, we call the element a *quadratic element*. If p_1 is zero, the finite element process lumps 1/6 of distributed load over the element to each of the end nodes and 2/3 to the midpoint node. There are three degrees-of-freedom for the element. The rank of $\mathbf{k}$ is two if ea_0 and ea_1 are both positive. The rank remains to be two for a limited range of negative ea_1. We leave it to the reader to show that such a negative ea_1 outside the range limits is unphysical.

One can similarly derive elements of even higher orders (Tong and Rossettos 1977). Analytical integration of Eqs. (22) and (23) is not esssential because we can always evaluate the integrals numerically. In certain practical applications to be discussed later, one sometime purposely uses lower order approximation in integration to reduce the rigidity of the stiffness matrix. Also, in practice, ea or p may be available only in numerical form that Eqs. (22) and (23) can only be integrated numerically. The formation of the system of algebraic equations for the whole domain from the element matrices and the solution of the system of equations will be discussed in later sections.

Problem 18.2.   Find the condition that the rank of $\mathbf{k}$ in Eq. (18.2:26) is one and explain why such a condition is unphysical.

18.3. SHAPE FUNCTIONS AND ELEMENT MATRICES FOR HIGHER ORDER ORDINARY DIFFERENTIAL EQUATIONS

The interpolation functions and element matrices derived up to this point are not suitable for problems governed by differential equations of the order higher than 2. Consider the bending of a beam

$$(1) \qquad \frac{d^2}{dX^2}\left[EI(X)\frac{d^2u}{dX^2}\right] - P(X) = 0$$

over $0 \leq X \leq L$, with boundary conditions

$$(2) \quad u = \bar{u}, \qquad\qquad du/dX = \bar{u}', \qquad\qquad \text{at } X = 0 \text{ (rigid condition)},$$

$$(3) \quad EI(X)\frac{d^2u}{dX^2} = \bar{M}, \quad \frac{d}{dX}\left[EI(X)\frac{d^2u}{dX^2}\right] = \bar{Q},$$

$$\text{at } X = L \text{ (natural condition)},$$

where $\overline{(\cdot)}$ denotes that the concerned quantity is prescribed. The functional for the principle of minimum potential energy of beams is given in

Eq. (10.8:3). The first variation of the potential energy of an element in terms of the local coordinate is

$$(4) \qquad \delta \Pi_n = \frac{\varepsilon}{2} \int_{-1}^{1} \left[\frac{16 ei(\xi)}{\varepsilon^4} \frac{d^2 u}{d\xi^2} \frac{d^2 \delta u}{d\xi^2} - p(\xi) \delta u \right] d\xi$$

$$- \alpha \left(\frac{2 \bar{M}}{\varepsilon} \frac{du}{d\xi} - \bar{Q} u \right) \Bigg|_{\substack{\xi = 1 \\ (X = L)}},$$

where the lower case $ei(\xi)[= EI(X)]$ is used to denote the bending rigidity expressed in terms of the natural coordinate ξ and α is a constant that $\alpha = 0$ for all elements except for the one containing the node at $X = L$, then $\alpha = 1$. The admissibility condition requires that u be C^1 continuity, i.e., u and du/dx of adjacent elements are equal at the common node and that u satisfies the rigid boundary conditions Eq. (2) at $X = 0$. The boundary conditions at $X = L$ are natural conditions associated with applied boundary moment and shear force. The boundary point at $X = L$ is said to be a *force boundary*.

Obviously, u is not C^1 continuous over the whole domain $0 \leq X \leq L$ if u is defined as in Eq. (18.2:5) or (18.2:10) with the interpolation functions given in Eq. (18.2:6) or (18.2:11) and q's satisfying Eq. (18.2:7) or (18.2:12), respectively. We need to construct new interpolation functions to make u meet the C^1 requirement. Let us define $u_{2n-1} = u(X_n)$ and $u_{2n} = du(X_n)/dX$ at the node $X = X_n$ as unknown parameters for the field variable and call them the *generalized coordinates* or *degrees-of-freedom*. Now there are two degress-of-freedom at each node and they have different physical dimension. The interpolation functions can be constructed as follows. First we assume u in the form

$$u = \alpha_1 + \alpha_2 (X - X_n) + \alpha_3 (X - X_n)^2 + \alpha_4 (X - X_n)^3$$

or in terms of the local coordinate ξ,

$$(5) \qquad u = q_1 h_1(\xi) + q_2 h_2(\xi) + q_3 h_3(\xi) + q_4 h_4(\xi)$$

for the element between X_n and X_{n+1}, where ξ has a range from -1 to 1. The interpolation functions h's have the properties that h_i and $\frac{dh_i}{d\xi}$ are zero at $\xi = -1$ and 1 except for

$$h_1(-1) = h_3(1) = 1, \qquad \frac{dh_2(-1)}{d\xi} = \frac{dh_4(1)}{d\xi} = \frac{\varepsilon}{2}.$$

An interpolation function that satisfies these conditions is known as the

Hermite interpolation. One can show that

$$
(6) \ \blacktriangle \quad
\begin{aligned}
h_1(\xi) &= (2+\xi)(\xi-1)^2/4, & h_2(\xi) &= \varepsilon(\xi+1)(\xi-1)^2/8, \\
h_3(\xi) &= (2-\xi)(\xi+1)^2/4, & h_4(\xi) &= \varepsilon(\xi-1)(\xi+1)^2/8.
\end{aligned}
$$

In this case, q_1, q_3 are the nodal values of u and q_2, q_4 the values of du/dX at $\xi = \pm 1$, respectively. If we require

$$
(7)
\begin{aligned}
u_{2n-1} &= u(X_{n-1}) = q_1, & u_{2n} &= \frac{du(X_{n-1})}{dX} = q_2, \\[2mm]
u_{2n+1} &= u(X_n) = q_3, & u_{2n+2} &= \frac{du(X_n)}{dX} = q_2,
\end{aligned}
$$

then u in the form of Eq. (5) is C^1 continuous over $0 \le X \le L$.

A substitution of Eq. (5) into Eq. (4) yields

$$
\delta\Pi_n = \delta\mathbf{q}^T(\mathbf{kq} - \mathbf{f})
$$

for the element where the components of $\mathbf{k}$ and $\mathbf{f}$ are

$$
(8) \qquad k_{ij} = \frac{8}{\varepsilon^3} \int_{-1}^{1} ei(\xi) \frac{d^2 h_i}{d\xi^2} \frac{d^2 h_j}{d\xi^2}\, d\xi,
$$

$$
(9) \qquad f_i = \frac{\varepsilon}{2} \int_{-1}^{1} p(\xi) h_i\, d\xi + \alpha(\delta_{i4}\bar{M} - \delta_{i3}\bar{Q})
$$

with α and δ's defined as before. The element stiffness matrix $\mathbf{k}$ is symmetric and positive semidefinite provided that $ei(\xi) > 0$. It has two zero eigenvectors associated with the rigid body translation and rotation. Thus, if Eq. (8) is integrated exactly, the rank of $\mathbf{k}$ equals the number of q's minus 2. One often integrates Eq. (8) approximately by numerical means. The approximation may lower the rank of $\mathbf{k}$ and introduce spurious deformation mode(s), i.e., nonrigid body motion with no contribution to energy.

If ei and p are treated as constant within the element, $\mathbf{k}$ and $\mathbf{f}$ can be written as

$$
(10) \qquad \mathbf{k} = \frac{ei}{\varepsilon^3}
\begin{bmatrix}
12 & & & \text{sym} \\
6\varepsilon & 4\varepsilon^2 & & \\
-12 & -6\varepsilon & 12 & \\
6\varepsilon & 2\varepsilon^2 & -6\varepsilon & 4\varepsilon^2
\end{bmatrix},
$$

$$
(11) \qquad \mathbf{f} = \frac{p\varepsilon}{12} \begin{bmatrix} 6 \\ \varepsilon \\ 6 \\ -\varepsilon \end{bmatrix} + \alpha \begin{bmatrix} 0 \\ 0 \\ -\bar{Q} \\ \bar{M} \end{bmatrix} .
$$

Each element involves 4 degrees-of-freedom and each node has two. For an element with its 1st and 2nd nodes being the I_1^{th} and I_2^{th} global nodes, the relationship between the global and the local labels is given in Table 18.3:1 as an array. The arrays for all elements, denoting by I, is called the *assembly list* that each array of the list specifies the global labels of an element. The list is needed in the assembly process to form $\mathbf{K}$ and $\mathbf{F}$.

Table 18.3:1. Global and local degree and nodal label.

Global degree label	Global nodal label	Local degree label	Local nodal label
$2I_1 - 1$	I_1	1	1
$2I_1$	I_1	2	1
$2I_2 - 1$	I_2	3	2
$2I_2$	I_2	4	2

From Eqs. (10) and (18.2:24), we see that $\mathbf{k}$ is independent of loads and boundary conditions. Equations (11) and (18.2:25) show that interior loads contribute to the element applied force matrix $\mathbf{f}$. The boundary loads contribute to $\mathbf{f}$ of the elements on the force boundary only.

One can use Eqs. (8) and (9) to evaluate element matrices for higher order elements provided higher order shape functions are used. For an element with a total of m nodes at $\xi_j, j = 1, \ldots, m$, the interpolated function is in the form

$$
u = \sum_{j=1}^{2m} q_j h_j(\xi),
$$

where

$$
q_{2j-1} = u(\xi_j), \qquad q_{2j} = \frac{du(\xi_j)}{dX},
$$

$$
(12) \qquad h_{2j-1} = \left[1 + \frac{d\psi_j(\xi_j)}{d\xi} \frac{\xi_j - \xi}{\psi_j(\xi_j)}\right] \frac{\psi_j(\xi)}{\psi_j(\xi_j)}, \qquad h_{2j} = \frac{\varepsilon(\xi - \xi_j)\psi_j(\xi)}{2\psi_j(\xi_j)},
$$

in which j's are not summed and

$$\psi_j(\xi) = \frac{1}{(\xi - \xi_j)^2} \prod_{i=1}^{m} (\xi - \xi_i)^2 .$$

Obviously, the interpolation functions given in Eq. (6) or (12) are applicable to the differential equation such as

(13) $$\frac{d^2}{dX^2}\left[EI(X)\frac{d^2u}{dX^2}\right] - \frac{d}{dX}\left[N(X)\frac{du}{dX}\right] + K(X)u - P(X) = 0$$

with the first variation of the potential energy in the form

(14) $$\delta\Pi_n = \frac{\varepsilon}{2}\int_{-1}^{1}\left[\frac{16ei(\xi)}{\varepsilon^4}\frac{d^2u}{d\xi^2}\frac{d^2\delta u}{d\xi^2} + \frac{4n(\xi)}{\varepsilon^2}\frac{du}{d\xi}\frac{d\delta u}{d\xi}\right.$$

$$\left. + k(\xi)u\delta u - p(\xi)\delta u\right]d\xi + B,$$

where $ei(\xi) = EI(X), n(\xi) = N(X), k(\xi) = K(X), p(\xi) = P(X)$ within the element, and B involves terms related to natural boundary conditions if the element has a force boundary. The element matrices can be computed in a similar fashion and assembled into a system of simultaneous equations for numerical solution by computers. **The finite element process reduces the task of solving differential equations to numerically solving algebraic equations. The complexity of constructing analytical solutions is replaced by the routine integration, which can be performed numerically, in evaluating the element matrices.**

Problem 18.3.   Show that the rank of **k** given in Eq. (18.3:10) is two.

Problem 18.4.   Using the interpolation functions defined in Eqs. (18.2:6) and (18.3:6), to determine the matrices associated with $\int_{-1}^{1} k(\xi)u\delta u d\xi$ for $k = k_0 + k_1\varepsilon\xi/2$.

Problem 18.5.   Consider the bending of a beam as defined in Eq. (18.3:1) with prescribed boundary conditions

$$EI(X)\frac{d^2u}{dX^2} = \bar{M}, \qquad \frac{d}{dX}\left[EI(X)\frac{d^2u}{dX^2}\right] = \bar{Q}, \qquad \text{at } X = 0,$$

$$u = \bar{u}, \qquad\qquad du/dX = \bar{u}', \qquad\qquad \text{at } X = L.$$

(a) Show that the first variation of the potential energy of an element is

$$\delta\Pi_e = \frac{\varepsilon}{2}\int_{-1}^{1}\left[\frac{16ei(\xi)}{\varepsilon^4}\frac{d^2u}{d\xi^2}\frac{d^2\delta u}{d\xi^2} - p(\xi)\delta u\right]d\xi - \alpha\left(\bar{Q}u - \frac{2\bar{M}}{\varepsilon}\frac{du}{d\xi}\right)\Bigg|_{\substack{\xi=-1\\(X=0)}},$$

where $\alpha = 0$ for all elements except for the one containing the node $X = 0$, then $\alpha = 1$. The equation above has a same form as Eq. (18.3:4) except for the last term, which has a sign difference. This shows clearly that the element stiffness matrix does not depend on the boundary conditions.

(b) Explain the reason for the sign difference.

(c) Derive the element force matrix for the element with the node at X=0.

Problem 18.6. Repeat the previous problem with prescribed boundary conditions

$$u = \bar{u}, \qquad EI(X)\frac{d^2u}{dX^2} = \bar{M} \qquad \text{at } X = 0,$$

$$\frac{du}{dX} = \bar{u}', \qquad \frac{d}{dX}\left[EI(X)\frac{d^2u}{dX^2}\right] = \bar{Q} \qquad \text{at } X = L.$$

18.4. ASSEMBLING AND CONSTRAINING GLOBAL MATRICES

We need to determine matrices $\mathbf{K}$ and $\mathbf{F}$ to solve for the unknown q's. It is $\mathbf{K}$ and $\mathbf{F}$ that relate the *nodal generalized coordinates* of one node to those of all other nodes. The finite element method divides the solution process in three steps:

1. Assemble the element matrices into the global matrices based on Eq. (18.2:17) with $\delta\Pi_n$ in the form of Eq. (18.2:21). $\delta\Pi_n$ is given in Eq. (18.2:19) for rods and Eq. (18.3:4) for beams.
2. Impose rigid constraints to the assembled matrices, i.e., enforcing the rigid condition such as Eq. (18.2:20) or (18.3:2).
3. Solve the system of algebraic equations by computers.

The process of assembling element matrices to form $\mathbf{K}$ and $\mathbf{F}$ can be symbolized as

$$\mathbf{K} = \sum_m \mathbf{k}_m, \qquad \mathbf{F} = \sum_m \mathbf{f}_m,$$

where the subscript m denotes the mth element. Assembling is actually a numerical summing process to carry out the summation of Eq. (18.2:17) for all elements. Starting with null matrices $\mathbf{K}$ and $\mathbf{F}$, we add to them the contributions from $\mathbf{k}$ and $\mathbf{f}$ of each element. When the last element has been added, the global $\mathbf{K}$ and $\mathbf{F}$ are formed. We have implicitly assume that all elements have the same definition of generalized coordinates at their common node(s) to allow the direct addition of the components of $\mathbf{k}$ and $\mathbf{f}$ from different elements. If the generalized coordinates of an element are different from those of the global system, a transformation of $\mathbf{k}$ and $\mathbf{f}$ is necessary before addition. Such transformation will be discussed in a later section.

Let the components of $\mathbf{K}$ and $\mathbf{F}$ be K_{IJ} and F_I, respectively, while those of $\mathbf{k}_m$ and $\mathbf{f}_m$ of the mth element be $(k_{ij})_m$ and $(f_i)_m$, where I and J denote global degree-of-freedom labels and i and j denote local labels. The key in assembling is to know the global degree-of-freedom labels, I and J, of the ith and j^{th} degrees-of-freedom of the mth element being assembled. This information is contained in the assembly list. Then, we simply add $(k_{ij})_m$ and $(f_i)_m$ to the corresponding K_{IJ} and F_I, respectively.

For example, for the linear element considered in Sec. 2, the relation between the global and local degree-of-freedom labels are described in Eq. (18.2:7), i.e., the global degrees-of-freedom 2 and 3 are, respectively, the 1st and the 2nd degrees-of-freedom of the element. Therefore, assembling the element is simply to

add k_{11} to K_{22} add k_{12} to K_{23}

add k_{21} to K_{32} add k_{22} to K_{33}

add f_1 to F_2 add f_2 to F_3.

Another example is the quadratic element shown in Fig. 18.2:1. The relation between the global and local labels of the element of Eq. (18.2:12) is as follows:

Global label	Local label
2	1
3	2
$N + 3$	3

where N is a known integer depending on how the global nodes are numbered. Assembling the element is simply to

add k_{11} to K_{22}

add k_{21} to K_{32} add k_{22} to K_{33}

add k_{31} to $K_{N+3,2}$ add k_{32} to $K_{N+3,3}$ add k_{33} to $K_{N+3,N+3}$

add f_1 to F_2 add f_2 to F_3 add f_3 to F_{N+3}.

Only the assembling of the lower triangle of $\mathbf{k}$ is illustrated above because $\mathbf{k}$ and $\mathbf{K}$ are symmetric.

In general, the assembly procedure can be described as follows: Let I_r be the global degree-of-freedom label of the rth degree-of-freedom of the element being assembled and p be its number of degrees-of-freedom of the element. Then the assembly of the element is simply

(1)   add k_{rs} to $K_{I_r I_s}$,   add f_r to F_{I_r} ,   for $r, s = 1, 2, \ldots, p$.

The rest of the components of $\mathbf{K}$ and $\mathbf{F}$ do not change in assembling this element. The array I with components $I_r, r = 1, 2, \ldots, p$, represents the transformation from the local degree label r to the global degree label I_r for the element being assembled.

In many practical problems, $\mathbf{k}$ and $\mathbf{K}$ are symmetric, and we therefore only have to assemble the lower (or the upper) triangle of $\mathbf{K}$ using the lower (or upper) triangle of $\mathbf{k}$. In this case the process will have to be modified slightly

(2) add k_{rs} to K_{MN} ,   add f_r to F_{I_r} ,   for $r = 1, 2 \ldots, p; s = 1, \ldots, r$.

where

$$M = \mathrm{Max}(I_r, I_s), \qquad N = \mathrm{Min}(I_r, I_s) .$$

After we proceed through all the elements, we obtain the final global matrices $\mathbf{K}$ and $\mathbf{F}$.

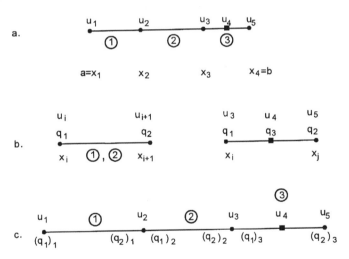

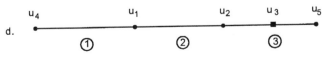

Fig. 18.4:1. Local and global labels for three-finite-element division of a domain. (c) and (d) are two alternate global label schemes.

Consider the problem described by Eqs. (18.2:13)–(18.2:15) with a three-element division. The global labels of the generalized coordinates of the label scheme are shown in Fig. 18.4:1(a). The assembly list, the global labels of the generalized coordinates of all elements, is given in Table 18.4:1 below

Table 18.4:1. Assembly list.

Element 1	1	2	-
Element 2	2	3	-
Element 3	3	5	4

Note that the assembly list is a two-dimensional array. It is important to note that for element 3 the global label of its 2nd degree-of-freedom is 5 and that of its 3rd degree-of-freedom [the one associated with the interior node of the element as defined in Eq. (18.2:10)] is 4. This information is essential for the computer to assemble the element matrices correctly. The first two elements have two degrees-of-freedom each with element matrices given in Eqs. (18.2:24) and (18.2:25). Element 3 has three degrees-of-freedom and its element matrices are given in Eqs. (18.2:26) and (18.2:27).

Consider the problem described by Eqs. (18.2:13)–(18.2:15). Applying the assembly procedure to the three-element division of a domain as shown in Fig. 18.4:1(a) yields

$$
(3) \quad \mathbf{K} = \begin{bmatrix}
\left(\dfrac{ea}{\varepsilon}\right)_1 & & & & \text{sym} \\
-\left(\dfrac{ea}{\varepsilon}\right)_1 & \left(\dfrac{ea}{\varepsilon}\right)_1 + \left(\dfrac{ea}{\varepsilon}\right)_2 & & & \\
0 & -\left(\dfrac{ea}{\varepsilon}\right)_2 & \left(\dfrac{ea}{\varepsilon}\right)_2 + \left(\dfrac{7ea}{3\varepsilon}\right)_3 & & \\
0 & 0 & -\left(\dfrac{8ea}{3\varepsilon}\right)_3 & \left(\dfrac{16ea}{3\varepsilon}\right)_3 & \\
0 & 0 & \left(\dfrac{ea}{3\varepsilon}\right)_3 & -\left(\dfrac{8ea}{3\varepsilon}\right)_3 & \left(\dfrac{7ea}{3\varepsilon}\right)_3
\end{bmatrix},
$$

$$
(4) \quad \mathbf{F} = \begin{bmatrix}
(p\varepsilon/2)_1 \\
-(p\varepsilon/2)_1 - (p\varepsilon/2)_2 \\
-(p\varepsilon/2)_2 - (p\varepsilon/6)_3 \\
-(2p\varepsilon/3)_3 \\
-(p\varepsilon/6)_3 + \bar{T}
\end{bmatrix},
$$

in which the subscript $(\cdot)_j$ signifies that the quantity in the parenthesis is from the jth element. In Eqs. (3) and (4), for simplicity, we have set a_1, p_1 of Eqs. (18.2:26) and (18.2:27) to zero.

At the beginning of the assembly process, all entries of $\mathbf{K}$ and $\mathbf{F}$ are set to zero. In assembling the first element, the quantities with subscript 1, i.e., $(\cdot)_1$, are added to $\mathbf{K}$ and $\mathbf{F}$ in Eqs. (3) and (4). After the second element is assembled, the quantities with subscripts 1 and 2 have been added; and so on. In this case, when the third element is assembled, the term associated with $\bar{T}$ is added to F_5 to account for the contribution from the natural boundary condition at the 2nd node of the last element. The global label for the degree-of-freedom associated with this node is 5.

Up to this point we have not discussed the rigid boundary conditions. The $\mathbf{K}$ as assembled is in general singular,[1] i.e., there exists at least one non-zero column matrix, say $\mathbf{u}_0$, such that $\mathbf{K}\mathbf{u}_0 = 0$. For example, the column matrix with all components being the same is obviously one such column matrix. We must constrain $\mathbf{K}$ before we can solve the equation $\mathbf{K}\mathbf{u} = \mathbf{F}$ for $\mathbf{u}$. Constraints are related to enforcing the rigid boundary conditions. In the finite element method, the rigid boundary condition(s) is enforced as part of the admissibility requirements whereas the natural boundary condition(s) is automatically accounted for through the integral formulation. For the present problem $u = \bar{u}$ at $X = 0$ is a rigid boundary condition, which must be enforced, i.e., one must have $\delta u_1 = 0$ and $u_1 = \bar{u}$. Therefore, of the five nodal parameters $(u_1, u_2, u_3, u_4, u_5)$, only four are actually unknowns. Constraining means using just the last four equations of $\mathbf{K}\mathbf{u} = \mathbf{F}$ together with $u_1 = \bar{u}$ to solve for the four unknowns.

The actual procedure of enforcing the rigid boundary conditions in the finite element method is quite simple. Let us call the components of $\mathbf{u}$ associated with rigid boundary conditions the constrained degrees-of-freedom. The constraining procedure is as follows. We multiply the components of the columns of $\mathbf{K}$ associated with the constrained degrees-of-freedom by the corresponding constrained values, and subtracts them from $\mathbf{F}$. Then we remove the rows of the resulting $\mathbf{F}$ and the rows and columns of $\mathbf{K}$ associated with the constrained degrees-of-freedom to obtain the reduced set of equations denoted as

$$(5) \qquad\qquad \mathbf{K}^*\mathbf{u}^* = \mathbf{F}^*,$$

where $\mathbf{u}^*$ is the column matrix $\mathbf{u}$ excluding the constrained degrees-of-freedom. This is called the *compact process*, which physically reduces the

---

[1]Except for the case of Eq. (18.3:13) with $N(X)$ and $K(X) > 0$.

size of the global matrices. For the present case, u_1 is constrained to be $\bar{u}$. One modifies $\mathbf{F}$ by subtracting it from the first column of $\mathbf{K}$ multiplied by $\bar{u}$, removing the first row of the resulting $\mathbf{F}$ to form $\mathbf{F}^*$. One also removes the first row and the first column of $\mathbf{K}$ to form $\mathbf{K}^*$. The column matrix $\mathbf{u}^*$ involves the four components (u_2, u_3, u_4, u_5). The resulting $\mathbf{K}^*$ and $\mathbf{F}^*$ are

$$
\mathbf{K}^* = \begin{bmatrix}
\left(\dfrac{ea}{\varepsilon}\right)_1 + \left(\dfrac{ea}{\varepsilon}\right)_2 & & & \text{sym} \\[2ex]
-\left(\dfrac{ea}{\varepsilon}\right)_2 & \left(\dfrac{ea}{\varepsilon}\right)_2 + \left(\dfrac{7ea}{3\varepsilon}\right)_3 & & \\[2ex]
0 & -\left(\dfrac{8ea}{3\varepsilon}\right)_3 & \left(\dfrac{16ea}{3\varepsilon}\right)_3 & \\[2ex]
0 & -\left(\dfrac{ea}{3\varepsilon}\right)_3 & -\left(\dfrac{8ea}{3\varepsilon}\right)_3 & \left(\dfrac{7ea}{3\varepsilon}\right)_3
\end{bmatrix},
$$

$$
\mathbf{F}^* = \begin{bmatrix}
-(p\varepsilon/2)_1 - (p\varepsilon/2)_2 + \left(\dfrac{ea}{\varepsilon}\right)_1 \bar{u} \\[2ex]
-(p\varepsilon/2)_2 - (p\varepsilon/6)_3 \\[2ex]
-(p\varepsilon/3)_3 \\[2ex]
-(p\varepsilon/6)_3 + \bar{T}
\end{bmatrix}.
$$

From computational point of view, reducing the number of equations to enforce the rigid constraints may not be convenient for large number of equations. The process involves altering the data structure for storing $\mathbf{K}$ and $\mathbf{F}$, even though it reduces the number of equations to be solved. A practical way to enforce the constraints is to change the numerical value of some components of $\mathbf{K}$ and $\mathbf{F}$ as shown below

$$
(6)\quad \mathbf{K}^* = \begin{bmatrix}
1 & & & & \text{sym} \\[2ex]
0 & \left(\dfrac{ea}{\varepsilon}\right)_1 + \left(\dfrac{ea}{\varepsilon}\right)_2 & & & \\[2ex]
0 & -\left(\dfrac{ea}{\varepsilon}\right)_2 & \left(\dfrac{ea}{\varepsilon}\right)_2 + \left(\dfrac{7ea}{3\varepsilon}\right)_3 & & \\[2ex]
0 & 0 & -\left(\dfrac{8ea}{3\varepsilon}\right)_3 & \left(\dfrac{16ea}{3\varepsilon}\right)_3 & \\[2ex]
0 & 0 & \left(\dfrac{ea}{3\varepsilon}\right)_3 & -\left(\dfrac{8ea}{3\varepsilon}\right)_3 & \left(\dfrac{7ea}{3\varepsilon}\right)_3
\end{bmatrix},
$$

$$\mathbf{F}^* = \begin{bmatrix} \bar{u} \\ -(p\varepsilon/2)_1 - (p\varepsilon/2) + (ea/\varepsilon)_1\bar{u} \\ -(p\varepsilon/2)_2 - (p\varepsilon/6)_3 \\ -(p\varepsilon/3)_3 \\ -(p\varepsilon/6)_3 + \bar{T} \end{bmatrix}.$$

The process involves first multiplying the components of the columns of $\mathbf{K}$ associated with the constrained degrees-of-freedom by the corresponding constrained values, and subtracting them from $\mathbf{F}$. One then sets the rows and the columns of $\mathbf{K}$ associated with rigidly constrained degrees-of-freedom to zero except for the diagonal term, which is set to 1. In the mean time, the rows of $\mathbf{F}$ associated with the constrained degrees-of-freedom are set to the corresponding constrained values. The solution of the modified equations

(7) $$\mathbf{K}^*\mathbf{u} = \mathbf{F}^*$$

automatically gives the correct values for the constrained degrees-of-freedom. This is called the *constraining process*.

In the present case, only the first degree-of-freedom is constrained with the value of $\bar{u}$. We set the first row and the first column of $\mathbf{K}$ to be zero except for K_{11}, which is set to 1. We modify $\mathbf{F}$ and set F_1 to $\bar{u}$. The process may look cumbersome but can be implemented easily and efficiently in high-speed computation because $\mathbf{K}, \mathbf{F}$ and $\mathbf{K}^*, \mathbf{F}^*$ have the same data structure.

Constraints can be imposed as each individual element is being assembled or after the assembling of all elements has been completed. For convenience, we will use the notation

(8) $$\sum \mathbf{kq} = \sum \mathbf{f}$$

to represent symbolically the assembled and constrained finite element equations Eq. (5) or (7).

No special effort is required to enforce the natural boundary conditions. As mentioned before, they are accounted for by the integral formulation and, therefore, are satisfied automatically in an approximate sense in the finite element method. This makes it easy to meet natural boundary conditions. The influence of natural boundary conditions is in fact reflected in the generalized force matrix, e.g., in the element applied force matrix for elements including boundary with prescribed natural boundary conditions. In the present case, $EA(L)du/dX = \bar{T}$ at $X = L$ is a natural boundary condition. The term involving $\bar{T}$ in the last element accounts for the condition in the integral formulation.

The ease of handling boundary conditions, rigid as well as natural, for any boundaries is an important advantage of the finite element method over other approximation methods such as the finite difference method.

The matrix $\mathbf{K}$ shows the *banding or sparseness* characteristics typical in the finite element method. One can determine the bandwidth of any degrees-of-freedom from Eq. (18.2:17). The bandwidth of a row of a matrix is defined as the number of components, named also as *entries*, between the first to the last nonzero components of the row. Another often-used term is the *semi-bandwidth,* which is defined to be the number of components between the first nonzero entry of the row to the diagonal. For the one-dimensional example, since u_j appears at most in two elements, if we sequentially label the nodes and parameters in the global system, the semi-bandwidth is at most two for nodes associated with linear elements as given in Eq. (18.2:6). For nodes associated with quadratic element as given in Eq. (18.2:11), the semi-bandwidth is three. The semi-bandwidth depends strongly on the way the global degrees-of-freedom are labeled. For the label scheme shown in Fig. 18.4:1(c), the semi-bandwidths for u_1, u_2, u_3, u_4, u_5 are 1, 2, 2, 2, 3, respectively, while the corresponding semi-bandwidths are 1, 2, 2, 4, 4 for Fig. 18.4:1(d).

The semi-bandwidth strongly affects the computer storage of the assembled matrices and the computer time needed for solving the system of equations. This is important for large system of equations. For the *variable bandwidth solution algorithm* (Tong and Rossettos 1977), the optimum-labeling scheme is the one that approximately gives the minimum average semi-bandwidth. Obviously for the simple problem considered, the labeling scheme shown in Fig. 18.4:1(d) does not give a minimum semi-bandwidth. However, for more complicated 2- or 3-dimensional problems, determining the optimum-labeling scheme that minimizes the average semi-bandwidth of the global stiffness matrix is non-trivial. Readers are referred to the extensive literature on this subject (e.g., George and Liu 1981).

One can construct $\mathbf{K}$ and $\mathbf{F}$ in a similar fashion for more elements. Thus one can refine the analysis in a systematic manner to improve the accuracy of the solution. One may also construct $\mathbf{K}$ and $\mathbf{F}$ for higher order elements such as the three-node element with the element matrices defined in Eqs. (18.2:26) and (18.2:27). Each element has three parameters, i.e., three degrees-of-freedom. If all elements have three nodes, $\mathbf{K}$ and $\mathbf{F}$ will involve 7 degrees-of-freedom for the 3-element domain. The semi-bandwidth is 3 for the degrees-of-freedom associated the element boundary nodes and is 2 for those associated with the interior nodes of elements if the global degrees-of-freedom are labeled sequentially. For the same number of elements, the system associated with higher order elements has more degrees-of-freedom and higher semi-bandwidth for $\mathbf{K}$. In the mean time, the higher order system generally gives a more accurate solution.

Problem 18.7. Construct the four-node Lagrange interpolation functions over $0 \leq x \leq \varepsilon$. The four nodes are the two-end nodes and two equally spaced interior nodes.

Problem 18.8. Consider the problem defined by Eqs. (18.2:13)–(18.2:15) with EA being constant over the entire domain. (a) Show the finite element solution using linear elements always gives the exact solution at the nodes for any loading distribution $p(X)$ and any number of elements of non-uniform size. (*Hint*: see Tong 1970a) (b) Do you expect the same is true if quadratic elements are used? Prove your result.

18.5. EQUATION SOLVING

For convenience we shall drop the superscript '$*$' of $\mathbf{K}$ and $\mathbf{F}$ in the subsequent discussion. The finite element analysis inevitably ends up solving a large set of linear algebraic equations like Eq. (18.4:7). The feasibility of the application of the finite element method hinges on ability to solve a large system of equations by computers in a reasonable time and with acceptable accuracy. The second point associates with round-off error, which can render the numerical solution meaningless. The first point depends on the operation counts involved in the solution algorithm for a linear system of equations. Usually the resulting $\mathbf{K}$ from the finite element method is *symmetric and sparse* with many of its nonzero entries lying within a narrow band of the diagonal. Sparse is defined as having only few components or entries, of $\mathbf{K}$ being nonzero. Both the sparseness and the semi-bandwidth of each row play important roles affecting the operation counts of a solution algorithm.

There are two kinds of methods for solving a large system of equations: (1) direct methods, and (2) iterative methods. A direct method is one, which after a finite number of operations, if all computations were carried out without roundoff, would lead to the "exact" solution of the algebraic system. An iterative method usually requires an infinite number of iterations to converge to the "exact" solution. Detailed discussion of solution methods is beyond the scope of this book. We shall just briefly outline the concept of direct methods and identify a number of iterative schemes. Readers are referred to literature for details.

The most commonly used direct methods of solution are the Gauss elimination and the triple factoring method. The idea underlying the Gauss elimination is simple. For a system of n algebraic equations, we use the first equation to eliminate the "first" variable u_1 from the last $n - 1$ equations, then use the second equation to eliminate the second variable u_2 from the updated last $n - 2$ equations, and so on. After $n - 1$ such elimination, the resulting matrix equation is triangular. This process is termed *forward elimination*. At this stage the last equation involves only one unknown u_n, which can be determined. Then one can determine u_{n-1} from $(n - 1)^{\text{th}}$ equation, and so on until all unknowns are obtained. The last procedure is called *backward substitution*.

Many time saving techniques can be applied to reduce the operation counts of this method. The main one is to recognize the zero entries of **K** in the elimination procedure. During the elimination of the jth variable from an equation, if the coefficient associated with this variable is zero, all operations to eliminate this variable from the equation can be skipped. For a sparse matrix, this step provides great timesaving. Recognizing zero entries (components of **K**) in **K** efficiently and skipping the unnecessary operations are keys to the success of implementation of the finite element method in practical applications.

The essential information enabling us to recognize the unnecessary operation in both the forward elimination and backward substitution processes is the column number of the first nonzero entry of each row, denoted by $C(i)$ for the ith row. This means $K_{ij} = 0$ for $j < C(i)$. The semi-bandwidth of the ith row is simply $i - C(i) + 1$. All operations to eliminate the variable 1 through $C(i) - 1$ from the ith row can be skipped in forward elimination. In addition, there is no need to store K_{ij} for all $j < C(i)$. This helps greatly in reducing the data storage requirement in practice, as the semi-bandwidth of K usually varies greatly. Therefore, it is important to label the nodes in such way to give a minimum average semi-bandwidth.

For symmetric **K**, the first zero row of the ith column of the resulting matrix after forward elimination is also $C(i)$. This allows us to use the same computer storage for the lower triangle of **K** before forward elimination process and for the resulting upper triangle after the process. This helps further reduce computer storage requirements. The process for determining $C(i)$ and utilizing this information was discussed by Tong and Rossettos (1977).

Another commonly used direct method is the triple factoring method. The basic idea is to express a symmetric **K** in the form

$$\mathbf{K} = \mathbf{LDL}^T,$$

where $\mathbf{L}$ is a lower triangular matrix with all diagonal terms being unity and the entries of the upper triangle being zero and $\mathbf{D}$ is a diagonal matrix. The process of determining $\mathbf{L}$ and $\mathbf{D}$ is named *factorization*. We can then easily solve the equation

$$\mathbf{L}\mathbf{u}^* = \mathbf{F} ,$$

because $\mathbf{L}$ is a lower triangular matrix. The procedure of solving for $\mathbf{u}^*$ is called *forward substitution*. We can solve for $\mathbf{u}$ from

$$\mathbf{L}^T\mathbf{u} = \mathbf{D}^{-1}\mathbf{u}^* ,$$

by *backward substitution*. One can show that $\mathbf{L}$ and $\mathbf{K}$ have the same $C(i)$. Therefore one can use the same computer storage for $\mathbf{L}$ and $\mathbf{D}$. One can also use the same storage for $\mathbf{F}$, $\mathbf{u}^*$, and $\mathbf{u}$ as they are generated sequentially.

Factorization in the triple factoring method or forward elimination in the Gauss elimination method are the most time consuming part of the solution procedure. For many practical problems, 85 to 95 percent of the total computational effort is for factorization or forward elimination. The multiplication count for both the triple factoring and the Gaussian elimination methods are of the order n^3 for a fully populate matrix of order n. The count is reduced to the order nB^2 for a banded matrix where B is the average semi-bandwidth. This count is further reduced if the variable bandwidth of the matrix is accounted for. In the process of forward elimination or fractorization of a highly sparse matrix, only some of the zero entries between the first nonzero entry and the diagonal of a row will be *filled-in* i.e., become nonzero, while most of zero entries remain untouched. Recognizing those unaltered entries and avoiding the operations (i.e., multiplication by zero) to update them can reduce the operation even much further. Therefore, it is important to recognize the number of '*filled-in*' in order to reduce both memory requirements and the number of arithmetic operations. Algorithms and software have been developed to reorder the original matrix $\mathbf{K}$ for this purpose. Unfortunately, an algorithm that is best to minimize the average semi-bandwidth is in general not the best for minimizing 'filling-in'. The reversed *Cuthill–McKee and Gipspoole–Stockmyer reordering algorithms* (George and Liu 1981, Cuthill and McKee 1969, Crane *et al.* 1976) are effective schemes for *variable bandwidth* (also called *skyline*) equation solution algorithms. The *modified minimum degree* and the *nested disection reordering algorithms* (George and Liu 1981, Lewis *et al.* 1989, Liu 1987, Gilbert and Zmijewski 1987, Ng and Peyton 1993) are more suitable for large sparse matrices. The details of efficient computer implementation of the triple factoring method based on the variable band-width strategy can be found in Tong and Rossettos (1977). The development of efficient equation solvers for sparse matrices is an important area of on-going research. Work on the subject includes Chen et al. (1996) and Nguyen et al. (1996).

Many advances in the direct method are in the adaptation for parallel computation. This is to having many central process units (CPUs) working together that each CPU does part of the work of solving the equations. The computation process can be very involved. Since the direct method is inherently a sequential process, many operations require information from the results of earlier steps, which may be computed by different CPUs. Also, often the assembled global stiffness matrix is so huge, that no single computer can store all the information. When a large number of CPU's is involved, communication between different CPU's, and synchronizing the processing[2] can consume both CPU time and clock time. The speed up of computer time is not linearly proportional to the number of CPU used. Balancing the work loads among CPU's and minimize communications among them are essential for efficient parallel computation. Efficient parallel computation algorithms are under intensive development (Noor 1987, Belytscho et al. 1990, John et al. 1992, Law and Mackay 1993, Zheng and Chang 1995, Duff and Reid 1995, Golub and Vanloan 1996, Nyuyen et al. 1996).

There are numerous iterative schemes. The methods proceed from an initial guessed solution $\mathbf{u}^{(0)}$ and define a sequence of successive approximation $\mathbf{u}^{(r)}$, which is required to converge to the exact solution. The procedure can be symbolished in the form

$$\mathbf{N}\mathbf{u}^{(r)} = \mathbf{P}\mathbf{u}^{(r-1)} + \mathbf{F},$$

where $\mathbf{N}$ and $\mathbf{P}$ are matrices derived from $\mathbf{K}$. The matrix $\mathbf{N}$ is usually diagonal or triangular so that $\mathbf{u}^{(r)}$ can be readily solved. In each step, most of the operations required are the multiplication of $\mathbf{P}$ and $\mathbf{u}^{(r-1)}$. One advantage of some of the iterative methods is that the product can be computed at the element level. This makes it unnecessary to physically form the global $\mathbf{P}$ and allows the saving of computer storage and is easily adaptable for parallel computation. The commonly used methods include Jacob's iteration, the Gauss–Seidel iteration, the conjugate gradient method, and the variable metric method. Interested readers are referred to literature (Isaacson and Keller 1966).

The accuracy of the solution by the direct method and the number iterations required by the iterative methods to reached a converged solution

---

[2]Often the needed information is generated by a different CPU. The task to generate the information may not have been completed when it is needed by another CPU.

depend strongly on the *condition number*, the ratio of the highest to the lowest eigenvalue of the constrained $\mathbf{K}$. The condition number of the finite element equations can be estimated from the element stiffness matrices (Tong 1970b, Hughes 1987). Reconditioning and reordering (including pivoting) $\mathbf{K}$ are often necessary in order to obtain numerical solution of acceptable accuracy (George and Liu 1981, Chen *et al.* 1996) for very large systems.

Within a tolerable error, there is no clear-cut answer as to which method is best. Generally, the iterative methods are better for very large and highly sparse systems, usually over one hundred thousand equations. It is also easier to implement an iterative scheme for parallel computation if the scheme does not require the assembled global matrices. The direct methods are often faster, if they can be used at all. The practical considerations of adopting a specific method are which of the two methods is more familiar to the user and what computer programs are available.

The finite element process of assembling, constraining, and solving algebraic equations is very general and applicable to one, two or three-dimensional problems. In assembling an element, one needs to know the assembly list, which identifies the global labels of the generalized coordinates of each element. In constraining the assembled matrices, one needs the information on which degree-of-freedom is to be constrained and its constrained value. Constraining can be performed as an element is being assembled or after assembling has been completed. This is a matter of choice in computer programming. In solving the algebraic method by a direct method, one needs to know the semi-bandwidth of each degree-of-freedom in the assembled stiffness matrix for efficient computation. This is extremely important not only for saving computer time but also for saving the computer storage for the global stiffness matrix.

18.6. TWO-DIMENSIONAL PROBLEMS BY ONE-DIMENSIONAL ELEMENTS

A class of two-dimensional problems solvable by one-dimensional elements is plane truss. A plane truss assembly consists of members describable by one-dimensional elements. Consider a two-member plane truss as shown in Fig. 18.6:1. Each member can be approximated by one or more one-dimensional elements of Eqs. (18.2:24) and (18.2:25) or Eqs. (18.2:26) and (18.2:27). (More than one elements or higher order elements are needed for a truss member only if the trust member has nonuniform cross-section. See Problem 18.2) We use X, Y as the global coordinates for the plane truss and the nodal values of the displacements u, v in the X-, Y-directions as

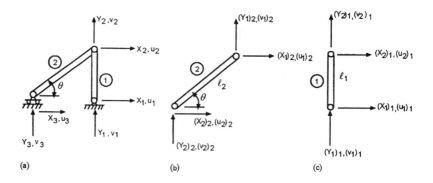

Fig. 18.6:1. Two-member plane truss assembly.

the generalized coordinates. Let x, y be the local coordinates of a truss member with x being along the axis of the member. Then, q's in the one-dimensional element are nodal displacements along the x-direction. In order to assemble an one-dimensional element in the plane truss application, we express the q's of a node in terms of the nodal values of u, v, which are the common generalized coordinates for all elements at the same node. We then derive the transformed element matrices that are associated with the common generalized coordinates. By so doing we can then simply add the transformed element matrices to form the global matrices.

Assuming uniform crosssection for the member, the element stiffness and force matrices are in the form given by Eqs. (18.2:24) and (18.2:25) with q's being the axial displacements at the nodes. From Fig. 18.6:1, q's are related to u, v by rotation. Let $\bar{q}_{2i-1} = u_i$ and $\bar{q}_{2i} = v_i$ for $i = 1, 2$. Then the relation between $\bar{q}$'s and q's is

$$q_i = \bar{q}_{2i-1}c + \bar{q}_{2i}s \,,$$

where $c = \cos\theta$, $s = \sin\theta$ with θ denoting the orientation of the element with respect to the X-axis and i is the local nodal label. In the matrix form

$$\mathbf{q} = \mathbf{R}\bar{\mathbf{q}} \,,$$

where $\mathbf{R}$ is a rotational matrix defined as

$$\mathbf{q} = \begin{bmatrix} q_1 \\ q_2 \end{bmatrix}, \quad \mathbf{R} = \begin{bmatrix} c & s & 0 & 0 \\ 0 & 0 & c & s \end{bmatrix},$$

$$\bar{\mathbf{q}}^T = [\bar{q}_1 \ \bar{q}_2 \ \bar{q}_3 \ \bar{q}_4] \quad \text{for 2-node element} \,,$$

$$\mathbf{q} = \begin{bmatrix} q_1 \\ q_2 \\ q_3 \end{bmatrix}, \quad \mathbf{R} = \begin{bmatrix} c & s & 0 & 0 & 0 & 0 \\ 0 & 0 & c & s & 0 & 0 \\ 0 & 0 & 0 & 0 & c & s \end{bmatrix},$$

$$\bar{\mathbf{q}}^T = [\bar{q}_1, \ldots, \bar{q}_6] \quad \text{for 3-node element},$$

The element matrices associated with $\bar{\mathbf{q}}$ can be derived based on the first variation of the functional for an element

$$\delta \bar{\mathbf{q}}^T (\bar{\mathbf{k}} \bar{\mathbf{q}} - \bar{\mathbf{f}}) = \delta \mathbf{q}^T (\mathbf{k} \mathbf{q} - \mathbf{f}).$$

The element matrices associated the $\bar{\mathbf{q}}$ are

$$\bar{\mathbf{k}} = \mathbf{R}^T \mathbf{k} \mathbf{R}, \qquad \bar{\mathbf{f}} = \mathbf{R}^T \mathbf{f}.$$

This process is called *rotation*.

In terms of $\bar{q}$'s the 2- and 3-node elements have 4 and 6 degrees-of-freedom, respectively. We can determine the assembly list, which specifies the global degree-of-freedom label $I_r, r = 1, 2, \ldots, p$, of the rth degree-of-freedom of each element, where p is the number of degrees-of-freedom of the element. We can then assemble the element matrices as described in Eqs. (18.4:1) and (18.4:2)

Similarly one can generate 6- and 9-degree-of-freedom 3-dimensional elements from the 2- and 3-node one-dimensional elements. This is accomplished by expressing the axial displacement of the one-dimensional element in terms of the 3 displacements in the 3-dimensional space.

18.7. GENERAL FINITE ELEMENT FORMULATION

We have demonstrated the construction of finite element equations for rods (governed by a second order ordinally differential equation) and beams (governed by a fourth order ordinally differential equation). The process involves steps including: *dividing the domain into a finite number of discrete elements, evaluating the element matrices based on the functional of a variational principal, assembling the element matrices based on the condition that the functional of the whole system equals to the sum of the functionals of all elements, imposing constraints to enforce the rigid boundary conditions, and finally solving the constrained algebraic equations numerically by high-speed computers.*

The determination of element stiffness matrices requires the establishment of the relations between the generalized element nodal forces and nodal displacements. In the process, the local approximate solution is expressed in terms of *nodal parameters or generalized coordinates* interpolated by interpolation functions. The nodal parameters and interpolation functions are chosen to satisfy the admissibility requirements of the functional. Such a concept can be generalized to problems involving several unknown functions in two- or three-dimensional domain.

Consider a general problem of seeking the solution $\mathbf{u}(\mathbf{X})$ of the differential equations

$$(1) \qquad \mathbf{A}(\mathbf{u}) \equiv \begin{bmatrix} A_1(\mathbf{u}) \\ A_2(\mathbf{u}) \\ \vdots \end{bmatrix} = 0$$

in a domain V, subjected to the boundary condition

$$(2) \qquad \mathbf{B}(\mathbf{u}) \equiv \begin{bmatrix} B_1(\mathbf{u}) \\ B_2(\mathbf{u}) \\ \vdots \end{bmatrix} = 0$$

on the boundaries ∂V of the domain, where $\mathbf{X}$ represents the spatial coordinates or the independent variables, $\mathbf{u}$ is a vector of several variables and is a function of $\mathbf{X}$, A's are differential operators defined in V and B's are differential operators defined on ∂V. Thus, the differential equation may be one or a set of simultaneous equations.

We seek the solution in the form

$$(3) \qquad \delta\Pi = \int_V W(\mathbf{u}, \delta\mathbf{u})\, dV + \int_{\partial V} w(\mathbf{u}, \delta\mathbf{u})\, dS = 0 \,,$$

for all admissible $\mathbf{u}$ and $\delta\mathbf{u}$, where $W(\mathbf{u}, \delta\mathbf{u}), w(\mathbf{u}, \delta\mathbf{u})$ are the variation of the integrands of a functional, or simply $\mathbf{A}(\mathbf{u})\delta\mathbf{u}$ and $\mathbf{B}(\mathbf{u})\delta\mathbf{u}$, respectively. The admissible requirements are: the functions $\mathbf{u}$ and $\delta\mathbf{u}$ must be sufficient smooth to allow $\delta\Pi$ be defined mathematically, and in addition, $\mathbf{u}$ must satisfy the rigid boundary conditions (see Sec. 10.6) and $\delta\mathbf{u}$ is zero over the boundaries where rigid boundary conditions are prescribed.

The specific form of Eq. (3) depends obviously on the nature of the problem. For instance, for the rod problem discussed in Sec. 18.2.2, $\mathbf{A}, \mathbf{u}, \mathbf{X}$ are scalars. We have

$$\mathbf{X} = X \,,$$

as the spatial coordinate,

$$\mathbf{A}(\mathbf{u}) = \frac{d}{dX}\left[EA(X)\frac{du}{dX} \right] - P(X) = 0 \,,$$

as the differential equation over $0 \leq X \leq L$, and

$$B_1(\mathbf{u}) = u - \bar{u} = 0 \qquad\qquad \text{at } X = 0\,,$$

$$B_2(\mathbf{u}) = EA(L)\frac{du}{dX} - \bar{T} = 0 \qquad \text{at } X = L\,,$$

as the boundary conditions. The functional for the principle of minimum potential energy is given in Eq. (18.2:16) with its first variation being

$$\delta\Pi = \int_0^L \left[EA(X)\frac{du}{dX}\frac{d\delta u}{dX} + P(X)\delta u \right] dX - \bar{T}\delta u(L) = 0\,,$$

shown in Eq. (18.2:17). The admissibility requirements are: u and δu are C^0 continuous over $0 \leq X \leq L$ and $u = \bar{u}$ and $\delta u = 0$ at $X = 0$.

In the finite element method, the domain is divided into elements and Eq. (3) is written in the form

(4) ▲
$$\delta\Pi = \sum_{\text{all elements}} \delta\Pi_e = 0\,,$$

which is similiar to Eq. (18.2:17), where

(5) ▲
$$\delta\Pi_e = \int_{V_e} W(\mathbf{u}, \delta\mathbf{u})\, dV + \int_{\delta\bar{V}_e} w(\mathbf{u}, \delta\mathbf{u})\, dS\,,$$

in which V_e is the domain of an element and $\partial\bar{V}_e$ is the portion of ∂V_e (the boundary of V_e) where natural boundary conditions are prescribed. The *admissibility requirements* are that $\mathbf{u}$ and $\delta\mathbf{u}$ are sufficiently smooth over the whole domain that the sum of $\delta\Pi_e$ of all elements equals $\delta\Pi$ and that $\mathbf{u}$ and $\delta\mathbf{u}$ must also satisfy the *rigid boundary conditions*. Functions that satisfy the admissibility requirements are called *admissible functions*. The continuity requirements at inter-element boundaries are also called the *compatibility conditions*. Generally admissibility requires C^0 continuity for $\mathbf{u}$ and $\delta\mathbf{u}$ if the differential operators in Eq. (1) is second order, and C^1 continuity for a fourth order differential operators.

For the case of rods as described in Sec. 18.2, we have

$$\int_{V_e} W(\mathbf{u}, \delta\mathbf{u})dV = \frac{\varepsilon}{2}\int_{-1}^{1} \left[\frac{4ea(\xi)}{\varepsilon^2}\frac{du}{d\xi}\frac{\delta u}{d\xi} + p(\xi)\delta u \right] d\xi\,,$$

$$\int_{\partial\bar{V}_e} w(\mathbf{u}, \delta\mathbf{u})dS = -\alpha\delta q_2\bar{T}$$

as given in Eq. (18.2:18) with the requirement of C^0 continuity for u and δu. We shall derive the expressions of Π_e or $\delta\Pi_e$ for different finite element models in the subsequent sections.

Establishing the solution based on Eq. (3) or (4) is termed the *integral formulation*. It is also identified as the *'weak' formulation* meaning that the requirements of satisfying the governing differential equations everywhere in the domain and the natural boundary conditions on the force boundaries have been relaxed or weakened. Equations (3) and (4) are the *"weak" or integral forms of the whole domain* and Eq. (5) is the *"weak" or integral form of an element*. The components of u in Eqs. (3) and (4) are often called *fields* or *field variables* of the functional or integral form.

The solution within an element is approximated in the form

$$(6) \qquad\qquad \mathbf{u} = \mathbf{h}(\mathbf{X})\mathbf{q},$$

where $\mathbf{h}(\mathbf{X})$, the *shape or interpolation function matrix*, is a pre-determined matrix function of independent variables $\mathbf{X}$, and $\mathbf{q}$ is an unknown *parameter matrix* representing the *generalized coordinates* or *parameters* associated with nodal values of u which may also include the spatial derivatives of u. The approximation must meet the requirements of *completeness* to be discussed in the next section to assure that u converges to the exact solution, and admissibility to assure the existence of $\delta\Pi$. The admissibility and completeness conditions, due to the continuity requirement, restrict the selection of $\mathbf{h}(\mathbf{X})$ in different applications.

Continuity normally does not cause any problems within the element, where the interpolation functions are generally smooth. However, the combination of completeness and smoothness often poses difficulty along inter-element boundaries for 2- and 3-dimensional problems, especially for problems involving fourth order differential equations. In practical applications, the *admissibility condition is sometime violated along a portion of inter-element boundaries*, especially for problems governed by fourth order differential equations. Elements with all interpolation functions meeting the admissibility requirements are named *compatible elements* and those do not are identified as *incompatible elements*.

The variation $\delta\mathbf{u}$ is usually, but not necessarily, expressed in the form

$$(7) \qquad\qquad \delta\mathbf{u} = \mathbf{h}(\mathbf{X})\delta\mathbf{q},$$

i.e., $\delta\mathbf{u}$ has the same spatial variation as u and its generalized coordinates $\delta\mathbf{q}$ is arbitrary subject to the constraint of rigid boundary conditions. This usually requires certain components of $\mathbf{q}$ at nodes associated with rigid boundary condition equal to the prescribed values and the corresponding components of $\delta\mathbf{q}$ equal to zero. In the subsequent discussion, we will assume that u and $\delta\mathbf{u}$ are similar in form as given in Eqs. (6) and (7), therefore there will be no need to discuss them separately.

Again, for the rod problem considered in Sec. 18.2.2, we have

$$\mathbf{u} = \mathbf{h}(\mathbf{X})\mathbf{q} = u = [h_1(\xi) \ \ h_2(\xi)] \begin{bmatrix} q_1 \\ q_2 \end{bmatrix},$$

for linear interpolation, where h's are interpolation or shape function in terms of the natural coordinates as those defined in Eq. (18.2:6), or

$$\mathbf{u} = u = [h_1(\xi) \ \ h_2(\xi) \ \ h_3(\xi)] \begin{bmatrix} q_1 \\ q_2 \\ q_3 \end{bmatrix}$$

for quadratic interpolation as defined in Eq. (18.2:11). The rigid constraints are equivalent to set $q_1 = \bar{u}$ and $\delta q_1 = 0$ in the first element. For elastic analysis the weak form in Eq. (3) is the first variation of the potential energy or the principle of virtual work.

Equation (3) or (4) is used to derive the algebraic equations for the approximate solution. We write the algebraic equations in the matrix notation

(8) $$\mathbf{Kr} = \mathbf{F},$$

where $\mathbf{K}$ and/or $\mathbf{F}$ depend on $\mathbf{q}$ if the problem is nonlinear, $\mathbf{q}$ is a *generalized coordinate, or parameter matrix* of an element, whereas $\mathbf{r}$ is that of the whole system. The well-known Rayleigh–Ritz process uses the similar approach in which the shape function matrix $\mathbf{h}(\mathbf{X})$ in Eq. (6) is expressed in terms of functions that are valid throughout the whole region and $\mathbf{q}$ is just a parameter matrix, which may or may not associate with nodal values of $\mathbf{u}$. In this case there is usually no distinction between $\mathbf{q}$ and $\mathbf{r}$.

The *basic premise of the finite element method is to use elements that are small and simple in shape.* Over a small region, we can approximate, with good accuracy, any field variable by polynomials (shape or interpolation functions), e.g., Eqs. (18.2:1) and (18.2:9), or special functions that interpolate the values of the field variable at selected points within the small region. From this point of view, $\mathbf{q}$ does not have to be the nodal values of the field variables and/or its spatial derivatives. For instance, in the Galerkin method, the components of $\mathbf{q}$ are simply parameters associated with some base functions for the field variables. However, having the type of $\mathbf{q}$ as chosen in Sec. 18.2 makes it easier to meet the continuity requirements along inter-element boundaries, which we shall see soon.

Continuity requirement generally does not cause any difficulty for one-dimensional problems. *Elements* for 2-dimensional problems are planes or surfaces and those in 3-dimension are solids. The common boundaries of elements are lines or curves for 2-dimensional and planes or curved surfaces for 3-dimensional problems. The element boundaries are defined by a finite number of discrete points, called *nodes*. The elements are connected to each other at a finite number of discrete nodes. Figure 18.1:1 shows plane elements with inter-element boundaries connected by two nodes. The finite element method employs the following strategy to meet the continuity requirements for problems of higher dimensions. First, nodal values of $\mathbf{u}$, and its spatial derivatives if needed, are chosen as parameters. This approach automatically assures the continuity of the field variable at the interconnecting nodes as the nodal parameters of all elements have the same values at their common nodes. This also provides simple rules for assembling element matrices to form the global matrices, i.e., to establish the algebraic equations for the whole domain. Second, the interpolation functions within elements are chosen to

1. allow $\mathbf{q}$ of Eq. (6) easily expressible in terms of the nodal parameters of $\mathbf{u}$;
2. assure that the compatibility of $\mathbf{u}$ at the nodes automatically guarantees compatibility along the entire inter-element boundaries (an important consideration for 2- and 3-dimensional problems), and
3. uniquely determine the value of the integral form or the functional of the element defined in Eq. (5) as a function of $\mathbf{q}$ except for possible rigid body motions. This is to avoid the *zero energy modes* to be discussed later.

The element matrices such as the *element stiffness matrix* and the *force matrix* (to be defined) can then be evaluated from Eq. (5). The unknown parameter or generalized coordinate matrix may include values of the spatial derivatives of the field variables, e.g., Eq. (18.3:5). This is often the case for higher order differential equations such as bending of beams (Sec. 18.3), plates and shells, which involve a fourth order differential equation for the out-of-plane displacement. Generally, higher order of continuity is required for the field variables involving differentials of higher order.

The key difference between the conventional Ritz method and the finite element method is the manner in which the field variables are approximated. The Ritz process uses *"assumed functions"* that is valid for the entire domain leading generally to a fully populated matrix $\mathbf{K}$ in Eq. (8). In the finite element process, interpolation functions are assumed for each element and each nodal parameter associates only with elements connected to the node. This characteristics leads to a sparse and usually banded matrix $\mathbf{K}$, e.g., Eq. (18.4:3) has a semi-band width of 2 or 3. One can also use this property to determine the semi-bandwidth of a particular degree-of-freedom. By its nature the conventional Ritz method is limited to problems with region of relatively simple shape since the assumed functions must be valid for the entire domain and satisfy all rigid boundary conditions. For finite element method, interpolation functions can be constructed easily because elements generally have simple shapes. Assemblies of elements of simple shapes are then used to represent complex configurations.

Equation (4) or Eqs. (18.2:16) and (18.2:17), which represents the functional of the whole domain as the sum of the functional of individual element or sub-domain, is a very powerful and versatile concept. It permits the use of different approximations for each element. We must, of course, make sure that the sum of the functionals of all elements is indeed equal to the functional of the whole system as demanded by the admissibility requirement, e.g., u must be continuous over the entire domain in Eqs. (18.2:16) and (18.2:17). Allowing different approximations for differemt element is particularly useful in dealing with problems with singularity or steep gradient. For instance the hybrid singular element (Tong *et al.* 1973, Tong 1977, 1984) is used for domain involving singularity. The flexibility also allows easy handling of problems with many components such as solid-fluid interactions where part of domain is solid and part is liquid, and structures involving solid, beams and shells where part of the domain is governed by second order differential equations and part by fourth order equations.

Therefore, the finite element method is inherently an *element base method* in which local approximate solutions in terms of unknown parameters are constructed based on the integral form for each element or sub-domain. The local solutions would have to satisfy certain compatibility and completeness requirements to assure convergence. The approximate solution for the whole domain is determined from the integral form as given by Eq. (4).

The finite element method permits the systematic selection of interpolation functions, the construction of element matrices, and the assembly of the element matrices into a system of simultaneous algebraic equations for the final solution. The interpolation functions, the assembled processes, and the solution methods for a large system of equations are generic, i.e., they can be applied to almost any types of problem. The element stiffness matrix depends on problem type such as beams, plane problems, 3-dimensional solids, etc., rather than on the specifics of the problem such as the dimensions, the boundary conditions, and the load distributions. The element applied force matrix depends on the distributed loads, but the process of determining it is again generic. We can use the same type of element stiffness matrix to solve problems of a plane with a square hole or with a circular hole. The nature of the problem is reflected in the rigid and natural constraints on the boundaries, the sizes, shapes, number of elements, and the values of the distributed load. These inherent generic characteristics make it possible to apply the finite element method to almost any problems with ease.

18.8. CONVERGENCE

We can examine the convergence of finite-element solution to the exact solution from two points of view. One is to increase the number of degrees-of-freedom per element to infinity with the element size fixed. The other is, by subdivision of the domain into smaller and smaller elements, to reduce the size of each element to zero with the number of degrees-of-freedom per element (and hence the order of accuracy of the local approximate solution) fixed. If one uses polynomials as the approximation within the element, the former procedure leads to the so-called *p-convergence*, where p denotes the degree of complete polynomial used as shape functions. This procedure is not commonly used in practice because increasing the number of degrees-of-freedom of an element means reformulating the local approximation with higher order interpolation functions, which can be tedious. The latter procedure leads to the so-called *h-convergence* where h denotes the size of the element. In the latter procedure, we only have to continuously sub-divide the domain being considered, which is relatively straightforward and can be used to test for the rate of convergence.

The *p-convergence*, i.e., increasing the order of polynomials to infinite as the approximate function, can be proved from the convergence of the *Taylor series expansion*. The *h-convergence*, i.e., reducing the element size to zero, can be established as follows. The finite element method constructs the solution of the entire domain from the local approximate solutions based on an *integral formulation*. If the solution is to converge to the exact solution, the *functional* or the *integral form* itself must converge as the size of the element diminishes. In proving convergence, then one only has to show that the convergence of the functional or the integral form implies the convergence of the function itself.

In practice, there are two convergence criteria:

1. The functional itself must be mathematically defined over the entire domain. In other words, the assumed functions for the independent fields must meet the admissibility requirements of the functional. The admissibility requirements are: the enforcement of rigid

boundary conditions and C^0 continuity for independent fields of the functional involving first order differentials, C^1 continuity for independent fields involving second order differentials and so on.

2. The highest differentials of the dependent variables (in terms of the generalized coordinates) in the integrand of the functional must be able to represent any constant within an element as the element size approaches zero.

The second criterion is called the *completeness requirements*. Interested readers are referred to the literature for the details of convergence proof (Tong and Pian 1967, Strang and Fix 1973, Babuska 1971, Oden 1969, Szabo *et al.* 1988).

In the subsequent sections, we shall discuss the shape functions that meet the criteria set forth above, the element matrices, and the assembly of element matrices.

18.9. TWO-DIMENSIONAL SHAPE FUNCTIONS

Consider a two-dimensional domain A shown in Fig. 18.1:1, which is divided into a finite number of *triangular and quadrilateral elements*. The common boundaries, *inter-element boundaries*, of elements are lines or curves defined by two or more nodes. The nodes are usually the vertices of the element, the midpoints, the one-third and two-third points, etc. along the element boundaries. Elements are connected to each other at these nodes. Additional points within the element may also be selected as nodes, which are called *interior nodes*. We will illustrate the use of the nodal values of the field variables and their spatial derivatives as parameters to characterize the element behavior. The latter is especially needed for higher order differential equations. These parameters are called *degrees-of-freedom* or *generalized coordinates*. For simplicity, in the subsequent discussion, **we will use the local spatial coordinates only to describe element.**

Consider a family of triangular elements with progressively increasing number of nodes on the element boundaries (Fig. 18.9:1). The nodes are indicated by dots. We seek to approximate a 2-dimensional smooth function in an element by interpolating the values of the function at the nodes. Each nodal value is a parameter called a *degree-of-freedom*. In this case the number of degrees-of-freedom to approximate a function equals the number of nodes in the element. For the time being, only polynomials are used as the *interpolation* or *shape function*. Therefore, the number of element nodes equals the number of terms needed in the polynomial representation.[3]

---

[3]If the derivatives of the function are also used as parameter, more polynomial terms will be needed.

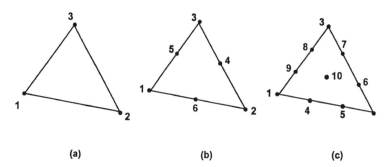

Fig. 18.9:1. (a) Linear, (b) quadratic, (c) cubic triangular elements. Node 10 is an interior node at the centroid.

The polynomial representation is a *truncated Taylor series* expansion. The order of approximation equals one plus the highest order of complete polynomial[4] in the expansion. Therefore, it is desirable to have an expansion which possesses the highest order of complete polynomial possible for a given number of degrees-of-freedom. This will give the highest order of approximation for a given number of parameters. The more the degrees-of-freedom, the higher the order of complete polynomials can be obtained. Thus elements with more nodes generally give higher order approximation and are called *higher order elements*.[5]

Shape Function Variation on Element Boundaries. Other than choosing polynomial terms to give the highest order of complete polynomial for a shape function, we also need to select the type of polynomial terms to ensure full inter-element compatibility, i.e., to have the approximate function meet the admissibility requirement of the integral formulation. This may limit the acceptability of certain higher order polynomial terms and thus the acceptability of a higher order complete polynomial. Let us consider the case of C^0 *continuity*. An interpolated function is smooth and generally satisfies the continuity requirement within each element. Since the interpolated function has the same value for all elements at their common nodes, the interpolated function will also meet the compatibility condition at all nodes. However, we must assure that compatibility at the common nodes of two adjacent elements implies *compatibility* along their entire inter-element boundary. This requirement limits the variation of the

---

[4]A complete polynomial that includes all polynomial terms up to certain order, e.g., a complete linear polynomial includes the constant and all linear terms, a complete quadratic polynomial includes the constant, all linear and quadratic terms, etc.

[5]Elements with more degrees-of-freedom do not give a high order approximation if lower order polynomial terms are left out in the shape functions.

shape function at element boundaries. For plane elements, element bound-arieis are straight lines or curves. If the shape function is a polynomial of order j along a boundary with $j + 1$ (the number of terms in a jth order complete polynomial) nodes, then full compatibility between the two elements sharing this boundary is assured. Thus, shape functions can be at most linear on an edge with two nodes [Fig. 18.9:1(a)] to ensure full compatibility over the entire edge. A six-node triangle [Fig. 18.9:1(b)] has three nodes on each side. Full compatibility is assured, if the shape function does not have any terms higher than quadratic on the side. For a 10-node triangle [Fig. 18.9:1(c)], the shape functions can be cubic at most and so on.

In summary, the number of degree-of-freedom for a field variable is the same as the number of nodes in the element if nodal values of these nodes are used as the generalized coordinates. It is desirable that the shape function has the highest order of complete polynomial for a given number of degree-of-freedom in order to get the highest order of approximation. However, the type of polynomial terms is restricted by the compatibility requirement, which can limit the allowable order of complete polynomials. We shall further examine these points below.

For one-dimensional problems, an inter-element boundary is simply a node. If an interpolated function is compatible at the inter-element nodes, it always satisfies the compatibility conditions. For two and three-dimensional problems, inter-element boundaries are lines and planes, respectively. The finite element method enforces the compatibility condition only at the inter-connecting nodes. Therefore, *the interpolation function must be chosen in such a way that compatibility at the common nodes of an inter-element boundary automatically guarantees compatibility over the entire boundary.*

Shape Functions for 3-Node Triangle. Over the area of a three-node triangle [Fig. 18.9:1(a)] with the nodal coordinates $(x_i, y_i), i = 1, 2, 3$, we can approximate a field variable ϕ by the first 3 terms of the Taylor expansion as

$$(1) \qquad \phi = \alpha_1 + \alpha_2 x + \alpha_3 y \,.$$

The error in the approximation within the triangle is of the order ε^2 where ε is the maximum linear dimension of the triangle. The approximated ϕ can be written in the form

$$(2) \qquad \phi(x, y) = \sum_{j=1}^{3} q_j h_j(x, y) = \mathbf{hq} \,,$$

where

$$(3) \qquad \mathbf{h}^T = \begin{bmatrix} h_1(x,y) \\ h_2(x,y) \\ h_3(x,y) \end{bmatrix}, \qquad \mathbf{q} = \begin{bmatrix} q_1 \\ q_2 \\ q_3 \end{bmatrix} = \begin{bmatrix} \phi(x_1,y_1) \\ \phi(x_2,y_2) \\ \phi(x_3,y_3) \end{bmatrix},$$

with $\mathbf{h}$ being the *shape function matrix* and $\mathbf{q}$ being the *element generalized coordinate matrix* whose components q's are the nodal values of ϕ at the vertices. One can determine the shape functions by evaluating Eq. (1) at the three vertices (nodes) and solving for α's in terms of q's. The solution is

$$(4) \qquad h_j(x,y) = \zeta_j, \qquad j = 1,2,3,$$

$$(5) \qquad \begin{bmatrix} \zeta_1 \\ \zeta_2 \\ \zeta_3 \end{bmatrix} = \frac{1}{2\Delta} \begin{bmatrix} a_1 & b_1 & c_1 \\ a_2 & b_2 & c_2 \\ a_3 & b_3 & c_3 \end{bmatrix} \begin{bmatrix} 1 \\ x \\ y \end{bmatrix},$$

$$(6) \qquad \Delta = (x_2 y_3 - x_3 y_2 + x_3 y_1 - x_1 y_3 + x_1 y_2 - x_2 y_1)/2,$$

$$(7) \qquad a_1 = x_2 y_3 - x_3 y_2, \qquad b_1 = y_2 - y_3, \qquad c_1 = x_3 - x_2,$$

whereas a_2, b_2, c_2 and a_3, b_3, c_3 are obtainable from Eq. (7) by cyclic permutation of the subscripts 1, 2, 3, and Δ is the area of the triangle. Note that ζ's satisfy

$$(8) \qquad \zeta_i(x_j, y_j) = \delta_{ij}$$

i.e., $\zeta_i = h_i = 1$ at node i and equals zero at the other two nodes. This is a *Lagrange interpolation*. The ζ's are commonly referred to as the *triangular coordinates*. One can show that ζ's are the areas of certain normalized triangles that

$$(5a) \qquad \zeta_1 = \frac{\Delta_{P23}}{\Delta}, \qquad \zeta_2 = \frac{\Delta_{P31}}{\Delta}, \qquad \zeta_3 = \frac{\Delta_{P12}}{\Delta},$$

where Δ_{Pjk} denotes the area of the triangle with nodes P, j, k as shown Fig. 18.9:2 with P being a point within the triangle. The ζ's are not independent that they satisfy the equation

$$\zeta_1 + \zeta_2 + \zeta_3 = 1.$$

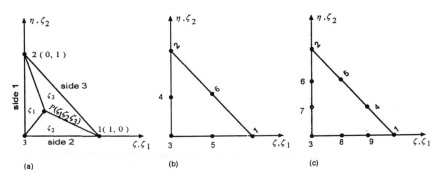

Fig. 18.9:2. Triangular coordinates.

One can also show that $\zeta_j = $ constant is a linear function of x, y and is a straight line parallel to the base of the triangle opposite to node j. The triangular coordinates are useful for constructing higher order shape functions for triangular elements.

If ϕ is defined as in Eq. (2) for all triangles that cover a domain and if ϕ is continuous at all inter-connecting nodes, then ϕ is a piecewise linear and continuous over the whole domain. Thus ϕ meets the C^0 continuity requirement over the whole domain. From Eq. (1), h_i are the shape functions of the complete linear polynomial. In Eq. (2), the subscript of q_j refers to the local label, which means that q_j is the jth generalized coordinate of the element. For assembly of element matrices to form the global matrices, we need the information on the global nodal labels I_j of the vertices j, $j = 1, 2, 3$, of each element in order to establish the assembly list for all elements.

Shape Functions for 6-Node Triangle. Consider a six-node triangle [Fig. 18.9:1(b)] with nodes 1, 2, 3 at the vertices, and nodes 4, 5, 6 at the midsides of the edges. The nodes are located at x_j, y_j, $j = 1, 2, \ldots, 6$. The representation

$$\phi = \alpha_1 + \alpha_2 x + \alpha_3 y + \alpha_4 x^2 + \alpha_5 xy + \alpha_6 y^2$$

gives an error of the order of ε^3 in approximation over the triangle, where ε is the maximum linear dimension of the triangle. Expressing ϕ in terms of the nodal values q_j, one obtains

(9)
$$\phi(x, y) = \sum_{j=1}^{6} q_j h_j(x, y) = \mathbf{hq},$$

which is in the same form as Eq. (2) except now both $\mathbf{h}$ and $\mathbf{q}$ are column matrices of six components. The components of $\mathbf{h}$ are

$$(10) \qquad h_i(x,y) = \zeta_i(2\zeta_i - 1), \qquad h_{i+3}(x,y) = 4\zeta_j\zeta_k,$$

where $i = 1, 2, 3$, and i, j, k are the cyclic permutations of 1, 2, 3. The shape functions satisfy

$$(11) \qquad h_i(x_j, y_j) = \delta_{ij} \qquad i, j = 1, \ldots, 6.$$

In other words, h_i is associated with node i that it equals 1 at node i and zero at all other nodes. Thus, h's are the *Lagrange interpolation*.

The shape functions can be obtained in general by solving for α's in terms of q's. They can be also systematically derived by the use of the triangular coordinates (Zienkiewicz 1971, Gallagher 1974, 1975, Tong and Rossettos 1977, Zienkiewicz and Taylor 1989). The shape function h_j is a normalized product of a minimum number of line functions determined by the rules that all nodes except node j must be passed by at least one line and that each line must pass through at least one node other than node j. The product is normalized to unity at node j. As an example, we see that $\zeta_i = $ constant is linear in x, y and represents a line in the xy plane. The shape function $h_1(x, y) = \zeta_1(2\zeta_1 - 1)$, with node 1 being a corner node, is the product of the two line functions. The line $\zeta_1 = 0$ passes through nodes 2, 4, 3 on the side of the element opposite to node 1 [Fig. 18.9:1(b)]. The other line $2\zeta_1 - 1 = 0$ passing through nodes 5 and 6 is a line parallel to the same side. The function h_1 equals unity at $\zeta_1 = 1$. It requires at least two lines to cover the five of the six nodes. Therefore, the minimum number of lines required is two and the resulting shape function is quadratic in x, y. The shape function $h_4(x, y) = 4\zeta_3\zeta_2$ (with node 4 being a midside node) is a product of the two line-functions with one passing through nodes 1, 5, 3 and the other passing through nodes 1, 6, 2. These two lines are the sides of the triangle opposite to node 4. The factor 4 is to make h_4 equal to unity at $\zeta_3 = \zeta_2 = 1/2$. Similarly, one can derive other shape functions.

The product of two line-functions is a quadratic function. The 6-node triangle has three nodes on each side. With the shape functions being quadratic, the compatibility at the three nodes on the side guarantees compatibility along the whole side. The six quadratic functions given Eq. (10) are all linearly independent, therefore, h_j are the shape functions of complete quadratic polynomial. The function ϕ as defined in Eq. (9) is piecewise quadratic over the domain. The array I_i, specifying the global nodal label of the ith node of the element, has six components i.e., $i = 1, \ldots, 6$.

Using the triangular coordinates, one can derive families of higher order interpolation functions for triangular elements with ease. For cubic elements, one can have four nodes on each side for a total of 9 boundary nodes. A complete cube polynomial has 10 terms that allows one to have an extra node to be placed in the interior of the triangle. The shape function for the tenth node is $27\zeta_1\zeta_2\zeta_3$, which is the normalized product of three line functions representing the sides of the triangle and is unity at the centroid of triangle for $\zeta_1 = \zeta_2 = \zeta_3 = 1/3$. This shape function is also called the *bulb function* because it vanishes on all sides. The other 9 shape functions are products of three line functions of which one passes through the centroid. For example, in $h_1(x, y) = \zeta_1(3\zeta_1 - 1)(3\zeta_1 - 2)/2$ and $h_4(x, y) = 2\zeta_1\zeta_2(3\zeta_1 - 1)/9$, the line function that passes through the centroid is $3\zeta_1 - 1 = 0$. This choice of line functions assures that all the shape functions associated with boundary nodes are zero at the centroid. The complete fourth order polynomial has 15 terms, which leads to a 15-node triangle with 12 nodes on the sides and 3 in the interior.

One can derive an element with shape functions being complete polynomial only if the number of boundary nodes needed by the polynomial expansion to assure compatibility is less or equal to the number of the complete polynomial terms. It can be shown (Problem 18.10) that triangular element allows complete polynomials of all orders in the Cartesian coordinates. For the cubic expansion discussed above, one needs four boundary nodes on each side to assure full compatibility of cubic functions. This makes a total of nine boundary nodes needed for a triangle. The complete cubic polynomial has 10 terms. Thus we can have a triangular element with complete cubic polynomials as shape functions. In Fig. 18.9:3, the complete polynomial terms needed for a triangular element are shown in the rows above and at the same level of the Pascal triangle. The needed terms

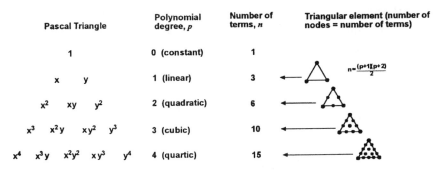

Pascal Triangle	Polynomial degree, p	Number of terms, n	Triangular element (number of nodes = number of terms)
1	0 (constant)	1	
$x \quad y$	1 (linear)	3	
$x^2 \quad xy \quad y^2$	2 (quadratic)	6	
$x^3 \quad x^2y \quad xy^2 \quad y^3$	3 (cubic)	10	
$x^4 \quad x^3y \quad x^2y^2 \quad xy^3 \quad y^4$	4 (quartic)	15	

$n = \dfrac{(p+1)(p+2)}{2}$

Fig. 18.9:3. Relation between type of triangular elements and polynomial terms used for interpolation.

are arranged in the shape of a triangle called the *truncated Pascal triangle*. Thus, terms through the second row are needed for linear elements, through the third row for a quadratic elements, and so on.

The representation of a field variable in the form of Eq. (2) or (9), derived without knowing the specifics of the variable itself, is generic. The use of natural coordinates makes it easy to construct C^0 continuous shape functions. The generic representation can be used to construct finite element solutions for any problems that require C^0 continuity.

Problem 18.9.   Use the triangular coordinates to derive the shape functions that satisfy the C^0 continuity requirement for a 10-node triangular element. The 10 nodes are the three vertices, the 1/3- and 2/3-points of each edge, and the centroid of the triangle.

Problem 18.10.   Show that it is possible to construct the shape functions of complete polynomial of any order higher or equal to one for triangular elements. [Hint: Show that the minimum number of boundary nodes required for complete p^{th} order polynomial expansion is $3p$ while the number of complete polynomial terms is $(p+1)(p+2)/2$.]

18.10. ELEMENT MATRICES FOR A SECOND-ORDER ELLIPTICAL EQUATION

To illustrate the derivation of element matrices, we consider a general second-order elliptical equation

$$(1) \qquad \frac{\partial}{\partial x_\alpha} D_{\alpha\beta} \frac{\partial \phi}{\partial x_\beta} - g\phi = b \qquad \text{on } A\,,$$

where $D_{12} = D_{21}$, the range of α, β is from 1 to 2 with $x_1 = x, x_2 = y$, and D's, g and b can all be functions of x, y. For Eq. (1) to be elliptical, it means that the quadratic form $D_{\alpha\beta}x_\alpha x_\beta$ is positive definite, i.e., $D_{\alpha\beta}x_\alpha x_\beta > 0$ for all x's not equal to zero. Recall that repeated indices denote summation. The boundary conditions are

$$(2) \qquad \phi = \bar{\phi} \qquad \text{on } \partial A_\phi\,,$$

$$(3) \qquad n_\alpha D_{\alpha\beta} \frac{\partial \phi}{\partial x_\beta} = \bar{\phi}_\sigma \qquad \text{on } \partial A_\sigma\,,$$

where ∂A_ϕ and ∂A_σ are the boundaries over which ϕ and $n_\alpha D_{\alpha\beta} \frac{\partial \phi}{\partial x_\beta}$ are specified to be $\bar{\phi}$ and $\bar{\phi}_\sigma$, respectively, and n_α are the components of the unit outward normal to the boundary. The boundary of A is ∂A which equals $\partial A_\phi + \partial A_\sigma$.

The integral formulation of Eq. (1) can be expressed as the first variation of the potential energy Eq. (18.7:4). The variation of the functional for an element with $\phi = \mathbf{h}\mathbf{q}$ and $\delta\phi = \mathbf{h}\delta\mathbf{q}$ can be written in the matrix form

$$(4) \qquad \delta\Pi_e = \delta\mathbf{q}^T(\mathbf{k}\mathbf{q} - \mathbf{f}),$$

where

$$\delta\mathbf{q}^T\mathbf{k}\mathbf{q} = \int_{A_e}\left(D_{\alpha\beta}\frac{\partial\phi}{\partial x_\alpha}\frac{\partial\delta\phi}{\partial x_\beta} + g\phi\delta\phi\right)dA,$$

$$\delta\mathbf{q}^T\mathbf{f} = -\int_{A_e} b\delta\phi dA + \int_{\delta A_{\sigma e}} \bar{\phi}_\sigma\delta\phi ds,$$

in which A_e is the area of the element and $\partial A_{\sigma e}$ is the portion of the boundary ∂A_σ occupied by the element. The admissible requirements are that ϕ be C^0 continuous and satisfies the rigid boundary condition. In the integral formulation, Eq. (2) is a rigid boundary condition, and Eq. (3) is a natural one. The term associated with integration over $\partial A_{\sigma e}$ accounts for the natural boundary condition automatically while it is necessary to enforce the rigid condition. The components of $\mathbf{k}$ and $\mathbf{f}$ are

$$(5) \qquad k_{ij} = \int_{A_e}\left(D_{\alpha\beta}\frac{\partial h_i}{\partial x_\alpha}\frac{\partial h_j}{\partial x_\beta} + gh_ih_j\right)dA,$$

$$(6) \qquad f_i = -\int_{A_e} bh_i dA + \int_{\delta A_{\sigma e}} \bar{\phi}_\sigma h_i ds.$$

The range of i, j is from 1 to 3 for a three-node triangle and is from 1 to 6 for a 6-node triangle. One can also express $\mathbf{k}$ and $\mathbf{f}$ in the matrix form

$$(7) \qquad \mathbf{k} = \int_{A_e}(\mathbf{B}^T\mathbf{D}_e\mathbf{B} + g\mathbf{h}^T\mathbf{h})\,dA,$$

$$(8) \qquad \mathbf{f} = -\int_{A_e} b\mathbf{h}^T dA + \int_{\delta A_{\sigma e}} \bar{\phi}_\sigma\mathbf{h}^T ds,$$

where

$$(9) \ \ \mathbf{h} = [h_1 \ h_2 \ \cdots \ h_p], \quad \mathbf{B} = \mathbf{dh} = \begin{bmatrix} \dfrac{\partial\mathbf{h}}{\partial x} \\[2mm] \dfrac{\partial\mathbf{h}}{\partial y} \end{bmatrix}, \quad \mathbf{D}_e = \begin{bmatrix} D_{11} & D_{12} \\ D_{12} & D_{22} \end{bmatrix},$$

in which $\mathbf{D}_e$ is the coefficient matrix of the elliptical equation or the elastic modulus matrix in elasticity. If $\mathbf{h}$ is an *interpolation (row) matrix* in terms of the triangular coordinates, the differentials with respect to x, y can be written as

$$\frac{\partial h}{\partial x} = \sum_{i=1}^{3} \frac{b_i}{2\Delta} \frac{\partial h}{\partial \varsigma_i}, \qquad \frac{\partial h}{\partial y} = \sum_{i=1}^{3} \frac{c_i}{2\Delta} \frac{\partial h}{\partial \varsigma_i},$$

where b's, c's and Δ are defined in Eqs. (18.9:6) and (18.9:7).

One can in principle integrate Eqs. (5) and (6) analytically by approximating D's, g, b, $\bar{\phi}_\sigma$ as polynomials. In practice, numerical integration is frequently used, which will be discussed later in the chapter.

Clearly $\mathbf{k}$ is a symmetric $p \times p$ matrix. The rank of the matrix is p if $g > 0$ in the element. This is because the first term of the integrand of Eq. (5) is non-negative for elliptical equation and the second term of the integrand is always positive. If $g = 0$, the rank of the matrix is at most $p - 1$ because $\mathbf{k}$ has at least one zero-eigenvector, which corresponds $\phi = $ constant, i.e., all components of $\mathbf{q}$ are the same, resulting in $\partial \phi / \partial x_\alpha = 0$. As pointed out before, the rank of $\mathbf{k}$ can be lower if one uses approximate integration for Eq. (5).

Three-node Triangular Element. For a three-node triangular element, the shape functions are given in Eq. (18.9:4). For the special case that $D_{11}, D_{22}, D_{12}(= D_{21})$ and g are constant, we find

$$\mathbf{k} = \frac{D_{11}}{4\Delta} \begin{bmatrix} b_1^2 & & \text{sym} \\ b_2 b_1 & b_2^2 & \\ b_3 b_1 & b_3 b_2 & b_3^2 \end{bmatrix} + \frac{D_{22}}{4\Delta} \begin{bmatrix} c_1^2 & & \text{sym} \\ c_2 c_1 & c_2^2 & \\ c_3 c_1 & c_3 c_2 & c_3^2 \end{bmatrix}$$

$$+ \frac{D_{12}}{4\Delta} \begin{bmatrix} 2b_1 c_1 & & \text{sym} \\ b_2 c_1 + b_1 c_2 & 2b_2 c_2 & \\ b_3 c_1 + b_1 c_3 & b_3 c_2 + b_2 c_3 & 2b_3 c_3 \end{bmatrix} + \frac{g\Delta}{12} \begin{bmatrix} 2 & & \text{sym} \\ 1 & 2 & \\ 1 & 1 & 2 \end{bmatrix}.$$

If b is constant and if no element boundary is on ∂A_σ, then

$$\mathbf{f} = -\frac{b\Delta}{3} \begin{bmatrix} 1 \\ 1 \\ 1 \end{bmatrix}.$$

The finite element process lumps $1/3$ of the applied load over the element to each of the vertices.

Six-node Triangular Element. For a six-node triangle, the interpolation functions are given in Eq. (18.9:10). In this case, $\mathbf{k}$ is a 6×6 matrix and $\mathbf{f}$ is a 6×1 matrix. The components of $\mathbf{k}$ and $\mathbf{f}$ can be evaluated from Eqs. (5) and (6) or (7) and (8). If D_{11}, D_{22}, $D_{12}(= D_{21})$, g and p are constant, we find

$$\mathbf{k} = [D_{11}\bar{\mathbf{k}}(b_i, b_i) + 2D_{12}\bar{\mathbf{k}}(b_i, c_i) + D_{22}\bar{\mathbf{k}}(c_i, c_i)]/\Delta + g\mathbf{k}_2\Delta .$$

Both $\bar{\mathbf{k}}$ and $\mathbf{k}_2$ are symmetric. The components of the lower triangle of $\bar{\mathbf{k}}(b_i, c_i)$ are

$$\bar{k}_{ii} = b_i c_i/2 \qquad i = 1, 2, 3 \ (i \text{ not summed}) ,$$

$$\bar{k}_{ij} = (b_i c_j + b_j c_i)/12 \qquad i, j = 1, 2, 3 \ (i \neq j) ,$$

$$\bar{k}_{61} = \bar{k}_{62} = (b_2 c_1 + b_1 c_2)/6 ,$$

$$\bar{k}_{42} = \bar{k}_{43} = (b_3 c_2 + b_2 c_3)/6 ,$$

$$\bar{k}_{53} = \bar{k}_{51} = (b_3 c_1 + b_1 c_3)/6 ,$$

$$\bar{k}_{63} = \bar{k}_{41} = \bar{k}_{52} = 0 ,$$

$$\bar{k}_{66} = [(b_1 + 2b_2)c_2 + (2b_1 + b_2)c_1]/3 ,$$

$$\bar{k}_{44} = [(b_2 + 2b_3)c_3 + (2b_2 + b_3)c_2]/3 ,$$

$$\bar{k}_{55} = [(b_3 + 2b_1)c_1 + (2b_3 + b_1)c_3]/3 ,$$

$$\bar{k}_{64} = [(b_2 + 2b_3)c_1 + (b_1 + 2b_2 + b_3)c_2 + (2b_1 + b_2)c_3]/6 ,$$

$$\bar{k}_{54} = [(b_3 + 2b_1)c_2 + (b_2 + 2b_3 + b_1)c_3 + (2b_2 + b_3)c_1]/6 ,$$

$$\bar{k}_{65} = [(b_1 + 2b_2)c_3 + (b_3 + 2b_1 + b_2)c_1 + (2b_3 + b_1)c_2]/6 ,$$

$$\mathbf{k}_2 = \frac{1}{180}
\begin{bmatrix}
6 & & & & & \\
-1 & 6 & & & \text{sym} & \\
-1 & -1 & 6 & & & \\
-4 & 0 & 0 & 32 & & \\
0 & -4 & 0 & 16 & 32 & \\
0 & 0 & -4 & 16 & 16 & 32
\end{bmatrix} , \quad
\mathbf{f} = \frac{b\Delta}{3}
\begin{bmatrix}
0 \\ 0 \\ 0 \\ 1 \\ 1 \\ 1
\end{bmatrix} .$$

We obtain $\bar{\mathbf{k}}(b_i, b_i)$ from $\bar{\mathbf{k}}(b_i, c_i)$ by replacing c_i with b_i and similarly $\bar{\mathbf{k}}(c_i, c_i)$ by replacing b_i with c_i. The expression for $\mathbf{f}$ above is only for elements with $\partial A_{\sigma e} = 0$ or $\bar{\phi}_\sigma = 0$ on $\partial A_{\sigma e}$. Otherwise we must include the contributions from the line integral along $\partial A_{\sigma e}$. We see from $\mathbf{f}$ that the finite element process lumps $1/3$ of the applied load over the element to each of the three midside boundary nodes and nothing to the vertices.

This is different from the linear triangle, which lumps $1/3$ of the applied load to each of the vertices.

For higher order elements, the process of deriving element matrices is identical to that for the lower order elements. One simply uses higher order interpolation functions in Eqs. (5) and (6) or (7) and (8). The evaluation of higher order elements involves the integration of quantities defined in terms of triangular coordinates over a triangular region. The following exact integration formula has been proven to be useful

$$(10) \qquad \int_{\Delta} \zeta_1^{\alpha} \zeta_2^{\beta} \zeta_3^{\gamma} \, dA = 2\Delta \frac{\alpha!\beta!\gamma!}{(\alpha + \beta + \gamma + 2)!} \, .$$

18.11. COORDINATE TRANSFORMATION

In principle, triangular elements can approximate a domain of any shape. For a domain with curved boundaries, it may require many small elements to get a good geometrical approximation. We can get a better geometrical representation by using elements with curved sides but then the difficulty is to find interpolation functions in terms of x, y that can fulfill the compatibility requirements. The strategy is to map curved triangles into an isosceles right triangle and map curved quadrilaterals to a square (to be discussed later), then all the shape functions we have developed will become usable. The key requirements for the transformation are that *there is no gap between mapped elements* and that *compatibility at the connecting nodes of two adjacent elements implies compatibility along the whole curved inter-element boundaries of the two elements.*

The mapping of an element with curved boundaries into one with straight boundary edges involves a proper transformation of geometry. Consider a transformation of variables from ξ, η to x, y

$$(1) \qquad x = x(\xi, \eta), \qquad y = y(\xi, \eta) \, .$$

The specific forms of $x(\xi, \eta)$, $y(\xi, \eta)$ depends on the shape of the element and will be discussed case by case. We shall first identify the relationships between ξ, η and x, y. To shorten the derivation and simplify the final results, we define

$$\mathbf{x} = \begin{bmatrix} x \\ y \end{bmatrix}, \qquad \psi = \begin{bmatrix} \xi \\ \eta \end{bmatrix} \, .$$

For later convenience, we take a digression to define the following transformation matrices:

(2)
$$\frac{\partial \mathbf{\Psi}}{\partial x} = \begin{bmatrix} \dfrac{\partial \xi}{\partial x} & \dfrac{\partial \eta}{\partial x} \end{bmatrix}^{T}, \qquad \frac{\partial \mathbf{\Psi}}{\partial y} = \begin{bmatrix} \dfrac{\partial \xi}{\partial y} & \dfrac{\partial \eta}{\partial y} \end{bmatrix}^{T}$$

(3)
$$\mathbf{J} = \begin{bmatrix} \dfrac{\partial \mathbf{x}}{\partial \xi} & \dfrac{\partial \mathbf{x}}{\partial \eta} \end{bmatrix}^{T} = \begin{bmatrix} \dfrac{\partial x}{\partial \xi} & \dfrac{\partial y}{\partial \xi} \\ \dfrac{\partial x}{\partial \eta} & \dfrac{\partial y}{\partial \eta} \end{bmatrix},$$

$$\tilde{\mathbf{J}} = \begin{bmatrix} \dfrac{\partial \mathbf{\Psi}}{\partial x} & \dfrac{\partial \mathbf{\Psi}}{\partial y} \end{bmatrix}^{T} = \begin{bmatrix} \dfrac{\partial \xi}{\partial x} & \dfrac{\partial \eta}{\partial x} \\ \dfrac{\partial \xi}{\partial y} & \dfrac{\partial \eta}{\partial y} \end{bmatrix},$$

the first order differential operators

(4)
$$\tilde{\mathbf{d}} = \frac{\partial}{\partial \mathbf{\Psi}} = \begin{bmatrix} \dfrac{\partial}{\partial \xi} \\ \dfrac{\partial}{\partial \eta} \end{bmatrix} = \mathbf{J} \begin{bmatrix} \dfrac{\partial}{\partial x} \\ \dfrac{\partial}{\partial y} \end{bmatrix} = \mathbf{J} \frac{\partial}{\partial \mathbf{x}} = \mathbf{J}\mathbf{d},$$

(5)
$$\mathbf{d} = \frac{\partial}{\partial \mathbf{x}} = \begin{bmatrix} \dfrac{\partial}{\partial x} \\ \dfrac{\partial}{\partial y} \end{bmatrix} = \tilde{\mathbf{J}}\tilde{\mathbf{d}},$$

(6)
$$\mathbf{d}_{e} = \begin{bmatrix} \dfrac{\partial}{\partial x} & 0 \\ 0 & \dfrac{\partial}{\partial y} \\ \dfrac{\partial}{\partial y} & \dfrac{\partial}{\partial x} \end{bmatrix}.$$

The inverse transformation $\xi = \xi(x, y)$ and $\eta = \eta(x, y)$ of Eq. (1) is in general not available. We cannot calculate $\tilde{\mathbf{J}}$ or any partial derivatives of ξ, η with respect to x, y directly. We can use the following relations derived from Eqs. (3)–(5) for the purpose:

$$(7) \qquad J = \det(\mathbf{J}) = \frac{\partial x}{\partial \xi}\frac{\partial y}{\partial \eta} - \frac{\partial y}{\partial \xi}\frac{\partial x}{\partial \eta},$$

$$(8) \qquad \tilde{\mathbf{J}} = \mathbf{J}^{-1} = \frac{1}{J}\begin{bmatrix} \dfrac{\partial y}{\partial \eta} & -\dfrac{\partial y}{\partial \xi} \\[2ex] -\dfrac{\partial x}{\partial \eta} & \dfrac{\partial x}{\partial \xi} \end{bmatrix},$$

$$(9) \qquad \frac{\partial \boldsymbol{\Psi}}{\partial x} = \frac{1}{J}\begin{bmatrix} \dfrac{\partial y}{\partial \eta} & -\dfrac{\partial y}{\partial \xi} \end{bmatrix}^{T}, \qquad \frac{\partial \boldsymbol{\Psi}}{\partial y} = \frac{1}{J}\begin{bmatrix} -\dfrac{\partial x}{\partial \eta} & \dfrac{\partial x}{\partial \xi} \end{bmatrix}^{T},$$

where J is the Jacobian of the transformation. For physical reasons, we must limit the transformation to $J > 0$.

The integration of a function with respect to ξ, η, and to x, y are related by

$$\int_{A_e} u(x,y)\,dx\,dy = \int_{A_e(\xi,\eta)} u[x(\xi,\eta), y(\xi,\eta)]J d\xi\,d\eta.$$

For triangular elements, we generally use the triangular coordinates for the transformation. Equation (1) is written in the form

$$x = x(\zeta_1, \zeta_2, \zeta_3), \qquad y = y(\zeta_1, \zeta_2, \zeta_3),$$

with

$$\xi = \zeta_1, \qquad \eta = \zeta_2, \qquad \zeta_3 = 1 - \zeta_1 - \zeta_2.$$

The derivatives with respect to ξ, η can be written as

$$\frac{\partial}{\partial \xi} = \frac{\partial}{\partial \zeta_1} - \frac{\partial}{\partial \zeta_3}, \qquad \frac{\partial}{\partial \eta} = \frac{\partial}{\partial \zeta_2} - \frac{\partial}{\partial \zeta_3}.$$

The differential operator $\tilde{\mathbf{d}}$ and the transformation matrices $\mathbf{J}$ and $\tilde{\mathbf{J}}$ become

$$(10) \quad \tilde{\mathbf{d}} = \begin{bmatrix} \dfrac{\partial}{\partial \xi} \\[2ex] \dfrac{\partial}{\partial \eta} \end{bmatrix} = \begin{bmatrix} \dfrac{\partial}{\partial \zeta_1} - \dfrac{\partial}{\partial \zeta_3} \\[2ex] \dfrac{\partial}{\partial \zeta_2} - \dfrac{\partial}{\partial \zeta_3} \end{bmatrix},$$

$$(11) \quad \mathbf{J} = \begin{bmatrix} \dfrac{\partial x}{\partial \zeta_1} - \dfrac{\partial x}{\partial \zeta_3} & \dfrac{\partial y}{\partial \zeta_1} - \dfrac{\partial y}{\partial \zeta_3} \\[2ex] \dfrac{\partial x}{\partial \zeta_2} - \dfrac{\partial x}{\partial \zeta_3} & \dfrac{\partial y}{\partial \zeta_2} - \dfrac{\partial y}{\partial \zeta_3} \end{bmatrix}, \quad \tilde{\mathbf{J}} = \frac{1}{J}\begin{bmatrix} \dfrac{\partial y}{\partial \zeta_2} - \dfrac{\partial y}{\partial \zeta_3} & \dfrac{\partial y}{\partial \zeta_3} - \dfrac{\partial y}{\partial \zeta_1} \\[2ex] \dfrac{\partial x}{\partial \zeta_3} - \dfrac{\partial x}{\partial \zeta_2} & \dfrac{\partial x}{\partial \zeta_1} - \dfrac{\partial x}{\partial \zeta_3} \end{bmatrix},$$

For triangular elements with all sides straight, the geometry transformation is

$$x = x_i \zeta_i , \qquad y = y_i \zeta_i .$$

We can calculate the transformation matrices and differential operators as defined above. With $\varphi^T = [\zeta_1 \ \zeta_2 \ \zeta_3]$, the differential operators can be expressed also in terms of the *triangular coordinates directly* as shown below:

$$(12) \qquad \mathbf{d} = \frac{\partial}{\partial \mathbf{x}} = \begin{bmatrix} \dfrac{\partial \varphi}{\partial x} \\[2mm] \dfrac{\partial \varphi}{\partial y} \end{bmatrix} \frac{\partial}{\partial \varphi} = \underset{(2 \times 3)(3 \times 1)}{\hat{\mathbf{J}} \ \hat{\mathbf{d}}}, \quad \hat{\mathbf{d}} = \frac{\partial}{\partial \varphi} = \begin{bmatrix} \dfrac{\partial}{\partial \zeta_1} & \dfrac{\partial}{\partial \zeta_2} & \dfrac{\partial}{\partial \zeta_3} \end{bmatrix}^T ,$$

$$(13) \qquad \frac{\partial \varphi}{\partial x} = \frac{1}{2\Delta}[b_1 \ b_2 \ b_3], \qquad \frac{\partial \varphi}{\partial y} = \frac{1}{2\Delta}[c_1 \ c_2 \ c_3],$$

where

$$(14) \qquad \hat{\mathbf{J}} = \frac{1}{2\Delta} \begin{bmatrix} b_1 & b_2 & b_3 \\ c_1 & c_2 & c_3 \end{bmatrix} ,$$

in which b's, c's and Δ are defined in Eqs. (18.9:6) and (18.9:7). One can then calculate $\mathbf{d}_e$ according to Eq. (6).

Recall that, in Chapters 2–4, the components of coordinates are defined as the contravariant components of a tensor, and the components of deformation gradient or strains are the covariant components of a tensor. The transformation of components between different coordinate systems follows the tensor transformation laws. If the coordinate systems are orthogonal and if the physical components for those quantities are used, there is no need to distinguish the covariant and contravariant transformations. The natural coordinates for triangles and those for general quadrilaterals used in the finite element methods are not orthogonal. The transformation described in this section is the covariant transformation of gradient operators in the matrix form.

18.12. TRIANGULAR ELEMENTS WITH CURVED SIDES

A family of such elements is shown in Fig. 18.12:1. There is no need for curved 3-node triangular element since the approximation over the element is linear. Thus only higher order elements are considered. In the ξ, η-plane, the element is an isosceles right triangle A_e located at the origin in the first

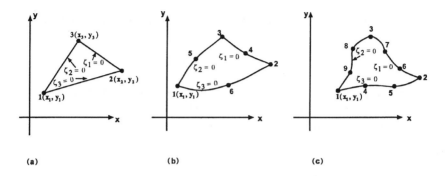

Fig. 18.12:1. Triangular elements with curved sides: (a) linear, (b) quadratic, and (c) cubic.

quadrant with $0 \le \xi,\, \eta \le 1$ (Fig. 18.9:2). The area of triangle is $1/2$. The specific form of $x(\xi, \eta)$, $y(\xi, \eta)$ will depend on the order of the curves of the element. The transformation can be written in the form

(1)
$$x = \sum_{\text{all nodes}} x_j h_j(\zeta_1, \zeta_2, \zeta_3),$$

$$y = \sum_{\text{all nodes}} y_j h_j(\zeta_1, \zeta_2, \zeta_3),$$

where h's are shape function discussed earlier, x's and y's are the coordinates of the nodes and

$$\xi = \zeta_1, \qquad \eta = \zeta_2, \qquad \zeta_3 = 1 - \zeta_1 - \zeta_2.$$

Elements derived by the transformation using shape functions of Eq. (1) are called *isoparametric elements* (Taig 1961, Irons 1966, Coons 1967) as one uses same shape functions for the interpolation of the field variable as well as for the transformation of the spatial coordinates.

The element matrices are in the exact same form as given in Eqs. (18.10:7) and (18.10:8) except that the integrands are now in terms of ζ and

(2) $\mathbf{B} = \tilde{\mathbf{J}} \tilde{\mathbf{d}} \mathbf{h} = \tilde{\mathbf{J}} \begin{bmatrix} \dfrac{\partial h}{\partial \xi} \\[2mm] \dfrac{\partial h}{\partial \eta} \end{bmatrix} = \dfrac{1}{J} \begin{bmatrix} \dfrac{\partial y}{\partial \zeta_2} - \dfrac{\partial y}{\partial \zeta_3} & \dfrac{\partial y}{\partial \zeta_3} - \dfrac{\partial y}{\partial \zeta_1} \\[3mm] \dfrac{\partial x}{\partial \zeta_3} - \dfrac{\partial x}{\partial \zeta_2} & \dfrac{\partial x}{\partial \zeta_1} - \dfrac{\partial x}{\partial \zeta_3} \end{bmatrix} \begin{bmatrix} \dfrac{\partial h}{\partial \zeta_1} - \dfrac{\partial h}{\partial \zeta_3} \\[3mm] \dfrac{\partial h}{\partial \zeta_2} - \dfrac{\partial h}{\partial \zeta_3} \end{bmatrix},$

where J is the Jacobian.

The boundary integral in Eq. (18.10:8) becomes an integral along a straight line. Suppose that we have to integrate along the side of the triangle from corner i to corner j, where $i, j = 1, 2$, or $3 (i \neq j)$. Let corner k be the third vertex. Along the side, we have $\zeta_i = 1 - \zeta_j$ and

$$(3) \quad \int_{\delta A_{\sigma e}} \mathbf{h}^T \bar{\phi}_\sigma \, ds = \int_0^1 \mathbf{h}^T \bar{\phi}_\sigma \sqrt{\left(\frac{\partial x}{\partial \varsigma_j} - \frac{\partial x}{\partial \varsigma_i} \right)^2 + \left(\frac{\partial y}{\partial \varsigma_j} - \frac{\partial y}{\partial \varsigma_i} \right)^2} \Bigg|_{\substack{\varsigma_k = 0 \\ \varsigma_i = 1 - \varsigma_j}} d\varsigma_j$$

(j not summed). The integration with respect s is now with respect to ζ_j ranging from 0 to 1.

One can in principle express the area integrals in Eqs. (18.10:7) and (18.10:8) in the form

$$\int_{A_e} (\cdots) \, dx \, dy = \int_0^1 \int_0^{1-\varsigma_2} (\cdots) J \, d\varsigma_1 \, d\varsigma_2$$

with $\zeta_3 = 1 - \zeta_1 - \zeta_2$. In practice, we carry out the integration either numerically or according to the formula of Eq. (18.10:10). It is not essential to use the expression above.

It can be seen that under the transformation the interpolated function is still continuous over the inter-element boundaries. Note that the transformation of the common boundaries of two adjacent elements is the same, even though the transformation of the two elements involves different transformation functions. This is because, from Eq. (1), the transformation of a side depends only the coordinates of nodes on that side, as shape functions associated with nodes not on the side vanish. Thus, the transformation is independent of the geometry of the element other than the side of interest. Now, we have the same coordinates along the common boundary of the two adjacent elements before and after transformation, which implies that there is no gap between adjacent elements after transformation and that the compatibility conditions for the interpolated functions are no different from those without transformation. Therefore, compatibility at inter-element nodes guarantees compatibility along the entire curved inter-element boundary.

The element matrices are in the same form as given by Eqs. (18.10:7) and (18.10:8). The assembling, constraining and equation solving procedures are the same as described in Secs. 18.4 and 18.5. No further elaboration is needed.

Six-node Curved Triangle. The triangle shown in Fig. 18.12:1(b) has six nodes. Each side of the triangle is described by a quadratic curve. The coordinate transformation is

$$(4) \qquad x = \sum_{j=1}^{6} x_j h_j(\zeta_1, \zeta_2, \zeta_3), \qquad y = \sum_{j=1}^{6} y_j h_j(\zeta_1, \zeta_2, \zeta_3),$$

where x_j, y_j are the coordinates of the nodes in the x, y-plane and h_j are the shape functions for the quadratic element given in Eq. (18.9:10).

18.13. QUADRILATERAL ELEMENTS

Although triangles have the advantage of approximating domain of any boundary shape, the linear triangular elements are not very accurate that a fine finite element mesh is often needed to achieve acceptable results in applications. To gain better accuracy, one has the option of using higher order triangular elements or elements of other shapes. Rectangles or quadrilaterals are commonly used as rectangular and quadrilateral elements are generally more accurate than triangular elements. We shall derive shape functions and construct the matrices for these elements.

Rectangular Element. Shape functions for the rectangle shown in Fig. 18.13:1 can be easily derived. The *natural* or *intrinsic coordinates* are

$$(1) \qquad \begin{aligned} \xi &= 2\frac{x - x_1}{\varepsilon_x} - 1, \\ \eta &= 2\frac{y - y_1}{\varepsilon_y} - 1, \end{aligned}$$

where x_1, y_1 are the coordinates of node 1, ε_x, ε_y are the element size in the x, y plane and

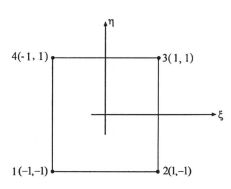

Fig. 18.13:1. A four-node square element.

ξ, η are defined over the domain $-1 \le \xi, \eta \le 1$ regardless the size of the element. A field variable can be expressed in terms of shape functions

$$\phi(x, y) = \sum_{j=1}^{4} q_j h_j(\xi, \eta),$$

where

$$h_1(\xi, \eta) = (1 - \xi)(1 - \eta)/4, \qquad h_2(\xi, \eta) = (1 + \xi)(1 - \eta)/4,$$
$$h_3(\xi, \eta) = (1 + \xi)(1 + \eta)/4, \qquad h_4(\xi, \eta) = (1 - \xi)(1 + \eta)/4.$$

These functions are called bilinear function since they are linear in ξ and in η. The shape function can be written in the more compact form:

$$(2) \qquad h_j(\xi, \eta) = (1 + \xi_j \xi)(1 + \eta_j \eta)/4 \qquad j = 1, \ldots, 4.$$

where $\xi_1 = \xi_4 = \eta_1 = \eta_2 = -1$ and $\xi_2 = \xi_3 = \eta_3 = \eta_4 = 1$. Obviously h_j equals 1 at the jth node and 0 at the other three nodes. Therefore, like the triangular element,

$$q_j = \phi(x_j, y_j) = \phi_j \qquad j = 1, \ldots, 4.$$

Since ϕ varies linearly along each side and is compatible at the two corners, therefore, it is compatible with its adajcent elements along the entire inter-element boundary between the two corners. In other words, ϕ is C^0 continuous over the whole domain. If $D_{11}, D_{22}, D_{12}(= D_{21}), g$ and p are all constant, the element matrices for the problem defined by Eqs. (18.10:1)–(18.10:3) are

$$
\mathbf{k} = \left(\frac{D_{11}}{6} \frac{\varepsilon_y}{\varepsilon_x} + \frac{D_{22}}{6} \frac{\varepsilon_x}{\varepsilon_y} \right)
\begin{bmatrix}
2 & & & \text{sym} \\
-2 & 2 & & \\
-1 & 1 & 2 & \\
1 & -1 & -2 & 2
\end{bmatrix}
$$

$$
+ \frac{D_{12}}{2}
\begin{bmatrix}
1 & & & \text{sym} \\
0 & -1 & & \\
-1 & 0 & 1 & \\
0 & 1 & 0 & 1
\end{bmatrix}
+ g \varepsilon_x \varepsilon_y \mathbf{k}_2 ,
$$

$$
\mathbf{k}_2 = \frac{1}{36}
\begin{bmatrix}
4 & & & \text{sym} \\
2 & 4 & & \\
1 & 2 & 4 & \\
2 & 1 & 2 & 4
\end{bmatrix}
, \qquad
\mathbf{f} = -\frac{b \varepsilon_x \varepsilon_y}{4}
\begin{bmatrix}
1 \\
1 \\
1 \\
1
\end{bmatrix}.
$$

Just like the case of triangular element, the applied force matrix is for elements with $\partial A_{\sigma e} = 0$ or $\bar{\phi}_\sigma = 0$ on $\partial A_{\sigma e}$ only. Otherwise one must include the contributions from the line integral along $\partial A_{\sigma e}$. The finite element process accounts for the distributed load over the element by lumping it equally to the 4 corner nodes.

Figures 18.13:2 and 18.13:3 show families of *rectangular serendipity*[6] and *Lagrange elements* where progressively increasing and equal number of nodes are placed on the element boundaries. Heavy dots denote element boundary nodes, whereas open circles indicate interior nodes. The variation of the function on the edges to ensure continuity is linear, quadratic, cubic, etc. for the increasingly higher order elements. The difference of two families is in the number of interior nodes. One can use the approach similar to that for triangular elements to generate the shape functions for higher order elements by multiplying line functions that passes through nodes of the element.

Alternatively we can generate shape functions by utilizing the bubble function. Consider a quadratic Lagrange element. The bubble function, which is zero on the element boundaries, is simply $h_9 = (1 - \xi^2)(1 - \eta^2)$, the shape function associated with the interior (central) node 9. The shape function for the midside node j involves the bubble function and the

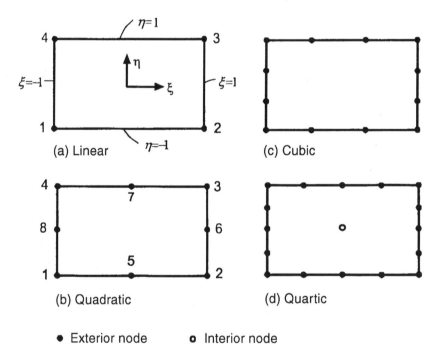

Fig. 18.13:2. A family of rectangular '*Serendipity*' elements: (a) linear, (b) quadratic, (c) cubic and (d) quartic.

---

[6]The name was coined by Zienkiewicz after the famous Prince of Serendip noted for their chance of discoveries.

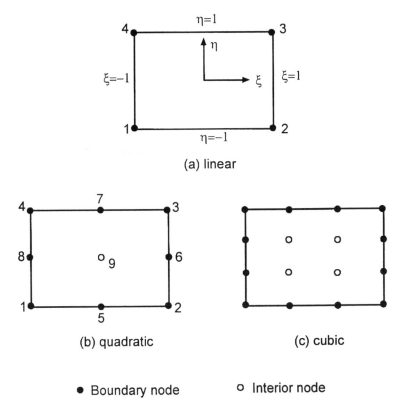

(a) linear

(b) quadratic   (c) cubic

● Boundary node   ○ Interior node

Fig. 18.13:3. A family of rectangular Lagrange elements: (a) linear, (b) quadratic, (c) cubic.

products of functions representing all the sides that do not include node j. For example, the shape function for node 5, the midside node between nodes 1 and 2 (Fig. 18.13:3), is

$$h_5 = (1 - \xi^2)(1 - \eta)/2 - h_9/2.$$

The first term on the right hand side is the product of the line functions for the sides $\xi = \pm 1$ and $\eta = 1$. The factor $1/2$ is to normalize h_5 to unity at node 5 ($\xi = 0, \eta = -1$). The bubble function is introduced to make $h_5 = 0$ at node 9. The shape function for a corner node can be obtained by subtracting from the bilinear shape function associated with the node half of the shape functions of the two adjacent midside nodes and a quarter of the bubble function. For corner node 1, we have

$$h_1 = (1 - \xi)(1 - \eta)/4 - (h_5 + h_8)/2 - h_9/4.$$

Table 18.13:1. Shape functions of a rectangle with missing node(s).

Shape function	Include only if node j is present in the element				
	$j = 5$	$j = 6$	$j = 7$	$j = 8$	$j = 9$
$h_1 = (1 - \xi)(1 - \eta)/4$	$-h_5/2$			$-h_8/2$	$-h_9/4$
$h_2 = (1 + \xi)(1 - \eta)/4$	$-h_5/2$	$-h_6/2$			$-h_9/4$
$h_3 = (1 + \xi)(1 + \eta)/4$		$-h_6/2$	$-h_7/2$		$-h_9/4$
$h_4 = (1 - \xi)(1 + \eta)/4$			$-h_7/2$	$-h_8/2$	$-h_9/4$
$h_5 = (1 - \xi^2)(1 - \eta)/2$					$-h_9/2$
$h_6 = (1 + \xi)(1 - \eta^2)/2$					$-h_9/2$
$h_7 = (1 - \xi^2)(1 + \eta)/2$					$-h_9/2$
$h_8 = (1 - \xi)(1 - \eta^2)/2$					$-h_9/2$
$h_9 = (1 - \xi^2)(1 - \eta^2)$					

The subtracted terms are to make h_1 equal to zero at the midsides and the center.

Table 18.13:1 gives the shape functions for the *9-node quadrilateral Lagrange element*. If h_9 is set to zero in Table 18.13:1, we obtain the 8 shape functions for the *8-node quadrilateral serendipity element*. One can obtain the shape functions for elements with missing midside node(s) from Table 18.13:1. If a midside node is missing, we exclude the shape function of the missing node and remove from the expressions of the shape functions for the corner nodes the terms involved the shape functions of the missing nodes. For example, for an element without node 5 as shown in Fig. 18.13:4, we remove h_5 from h_1 and h_2 to obtain

$$h_1 = (1 - \xi)(1 - \eta)/4 - h_8/2 - h_9/4\,,$$
$$h_2 = (1 - \xi)(1 - \eta)/4 - h_6/2 - h_9/4\,.$$

The other shape functions do not involve h_5 and remain unchanged. On the side with missing midside node, the field variable varies linearly. The side can be connected to a linear element while the other sides are connected to quadratic elements. This arrangement is useful for transition from linear elements to quadratic elements in the discretization of the domain.

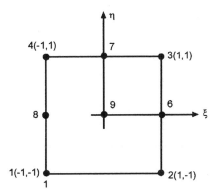

Fig. 18.13:4. Rectangular element with no midside node between nodes 1 and 2.

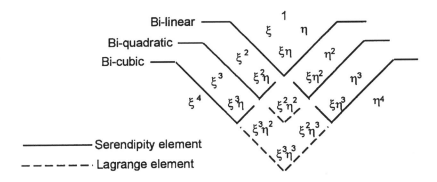

Fig. 18.3:5. Polynomial terms for plane Serendipity and Lagrange elements.

It is instructional to examine the polynomial terms that are included in (or missed from) the shape functions. In Fig. 18.13:5 polynomial terms above the solid lines are those included in the expansions of shape function of the corresponding Serendipity elements while those above dotted lines are included in the corresponding Lagrange elements. The linear element includes one quadratic term $\xi\eta$ and the highest order of complete polynomial is only linear. For the quadratic and cubic elements, the highest complete polynomials are quadratic and cubic, respectively. However, for the quartic Serendipity element, the expansion of the shape functions associated with the 12 boundary nodes misses the $\xi^2\eta^2$ term. The error in approximation would be of the order ε^4 unless an interior node is included. Then the shape functions can form a complete fourth order polynomial and the error in approximation is of the order ε^5. Therefore, it is essential to have the central node added.

The higher order Lagrange elements contain some very high order terms while omitting some lower order ones. For example, a cubic element has 16 nodes. It contains the term $\xi^3\eta^3$ while omits terms like ξ^4 and η^4. The accuracy of the higher order Lagrange elements is limited by the omitted terms. A cubic Serendipity element has the same order of truncation error as the cubic Lagrange element but has only 12 nodes.

Isoparametric Quadrilateral Elements. Rectangular elements are generally more accurate than the triangular elements of the same order. However, rectangular element suffers the shortcomings of not being able to approximate irregular domains. We shall introduce a family of quadrilateral elements (Fig. 18.13:6) that can approximate domains of any shape and in the mean time provides better accuracy than the triangular elements of the same order. Following the approach employed for curved triangular elements, we use shape functions to map the quadrilateral elements into a square and then construct the element matrices in the transformed coordinates. The transformation is in the same form as Eq. (18.12:1) that

$$(3) \qquad x = \sum_{\text{all nodes}} x_j h_j(\xi,\eta), \qquad y = \sum_{\text{all nodes}} y_j h_j(\xi,\eta),$$

where h_j are the shape functions associated with node j of the element. If so desired, one may use the nodes on the element boundaries only for the transformation. For a 4-node quadrilateral element, the shape functions are defined in Eq. (2). For an 8-node *serendipity* or 9-node *Lagrange* quadrilateral elements, the shape functions associated with boundary nodes are

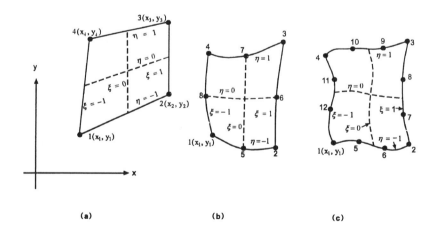

(a)   (b)   (c)

Fig. 18.3:6. Quadrilateral elements.

the first eight given in Table 18.13:1. The mapped element is a square over $-1 \leq \xi, \eta \leq 1$ in the ξ, η-plane (Fig. 18.13:1).

The element matrices given in Eqs. (18.10:7) and (18.10:8) can be now written as

$$(4) \qquad \mathbf{k} = \int_{-1}^{1} \int_{-1}^{1} (\mathbf{B}^T \mathbf{D}_e \mathbf{B} + g\mathbf{h}^T \mathbf{h}) J d\xi d\eta \,,$$

$$(5) \qquad \mathbf{f} = -\int_{-1}^{1} \int_{-1}^{1} b\mathbf{h}^T J d\xi d\eta + \int_{\delta A_{\sigma e}} \bar{\phi}_\sigma \mathbf{h}^T ds \,,$$

where, similar to Eq. (18.12:2),

$$(6) \qquad \mathbf{B} = \tilde{\mathbf{J}} \tilde{\mathbf{d}} \mathbf{h} = \tilde{\mathbf{J}} \begin{bmatrix} \dfrac{\partial \mathbf{h}}{\partial \xi} \\[2mm] \dfrac{\partial \mathbf{h}}{\partial \eta} \end{bmatrix} = \dfrac{1}{J} \begin{bmatrix} \dfrac{\partial y}{\partial \eta} & -\dfrac{\partial y}{\partial \xi} \\[2mm] -\dfrac{\partial x}{\partial \eta} & \dfrac{\partial x}{\partial \xi} \end{bmatrix} \begin{bmatrix} \dfrac{\partial \mathbf{h}}{\partial \xi} \\[2mm] \dfrac{\partial \mathbf{h}}{\partial \eta} \end{bmatrix} \,.$$

The resulting element is called an *isoparametric* element as the shape functions are used for coordinate transformation and for interpolation of the field variable(s).

The boundary integral in Eq. (5) involves integration along the side for ξ or $\eta = \pm 1$ if $\delta A_{\sigma e}$ is not null. If the integration is along the side $\xi = \alpha$ where $\alpha = 1$ or -1, the boundary integral can be written as

$$(7) \qquad \int_{\delta A_{\sigma e}} \bar{\phi}_\sigma \mathbf{h}^T ds = \int_{-1}^{1} \bar{\phi}_\sigma \mathbf{h}(\xi, \eta)^T \sqrt{\left(\dfrac{\partial x}{\partial \eta}\right)^2 + \left(\dfrac{\partial y}{\partial \eta}\right)^2} \Bigg|_{\xi=\alpha} d\eta \,.$$

If the integration is along $\eta = \alpha$, the boundary integral is

$$(8) \qquad \int_{\delta A_{\sigma e}} \bar{\phi}_\sigma \mathbf{h}^T ds = \int_{-1}^{1} \bar{\phi}_\sigma \mathbf{h}(\xi, \eta)^T \sqrt{\left(\dfrac{\partial x}{\partial \xi}\right)^2 + \left(\dfrac{\partial y}{\partial \xi}\right)^2} \Bigg|_{\eta=\alpha} d\xi \,.$$

The integration in Eqs. (4) and (5) is complicated, because of the presence of J in the denominator in Eq. (6). Even for the simplest 4-node isoparametric element with constant $\mathbf{D}_e$, analytical integration, though possible, is not practical. Numerical integration is the only practical means to evaluate the element matrices for isoparametric elements.

Static Condensation. There are boundary and interior nodes for cubic triangle, and quadratic Lagrange and quartic Serendipity quadrilaterals or elements of higher order. The shape functions associated with interior nodes are bubble functions. The generalized coordinates associated with the interior nodes do not appear in domain outside the element and can be eliminated from the rest of equations by the stationary condition of the functional. The process is referred to as *static condensation*.

We partition $\mathbf{q}, \mathbf{k}, \mathbf{f}$ of Eq. (18.10:4) in the corresponding forms

$$\mathbf{q} = \begin{bmatrix} \mathbf{q}_b \\ \mathbf{q}_a \end{bmatrix}, \qquad \mathbf{k} = \begin{bmatrix} \mathbf{k}_{bb} & \mathbf{k}_{ab} \\ \mathbf{k}_{ab}^T & \mathbf{k}_{aa} \end{bmatrix}, \qquad \mathbf{f} = \begin{bmatrix} \mathbf{f}_b \\ \mathbf{f}_a \end{bmatrix},$$

where subscripts a and b denote that the concerned quantities are associated with the generalized coordinates of interior and boundary nodes, respectively. For quadratic Lagrange element $\mathbf{q}_a = (\mathbf{q}_9)$ is 9. For the generalized coordinate of the only interior-node cubic Lagrange element, there are three interior nodes (Fig. 18.13:3). By virtue of the fact that $\mathbf{q}_a$ does not appear in other elements, the vanish of the first variation of the functional with respect to $\mathbf{q}_a$ implies $\delta \Pi_e = 0$, i.e.,

$$\mathbf{k}_{ab} \mathbf{q}_b + \mathbf{k}_{aa} \mathbf{q}_a - \mathbf{f}_a = 0 \qquad \text{or} \qquad \mathbf{q}_a = -\mathbf{k}_{aa}^{-1} \mathbf{k}_{ab} \mathbf{q}_b + \mathbf{k}_{aa}^{-1} \mathbf{f}_a \,,$$

which can be used to eliminate $\mathbf{q}_a$ from Eq. (18.10:4). This process leads to the following reduced element stiffness matrix and force matrix:

$$\mathbf{k}^* = \mathbf{k}_{bb} - \mathbf{k}_{ab}^T \mathbf{k}_{aa}^{-1} \mathbf{k}_{ab} \,, \qquad \mathbf{f}^* = \mathbf{f} + \mathbf{k}_{ab}^T \mathbf{k}_{aa}^{-1} \mathbf{f}_a \,,$$

which associates only with the generalized coordinates $\mathbf{q}_b$ of the boundary nodes of the element. One assembles $\mathbf{k}^*$ and $\mathbf{f}^*$ to form the global matrices.

18.14. PLANE ELASTICITY

We have derived shape functions for a variety of element shapes. Up to now, we have considered primarily problems with one field variable. Generalizing the finite element process to problems of multiple field variables is straightforward. Let us examine problems of plane elasticity, which involves the following field variables:

(1)
$$\mathbf{u} = \begin{bmatrix} u \\ v \end{bmatrix}, \qquad \boldsymbol{\sigma} = \begin{bmatrix} \sigma_x \\ \sigma_y \\ \sigma_{xy} \end{bmatrix},$$

$$\mathbf{e} = \begin{bmatrix} e_x \\ e_y \\ \gamma_{xy} \end{bmatrix} = \begin{bmatrix} \partial u/\partial x \\ \partial v/\partial y \\ \partial u/\partial y + \partial v/\partial x \end{bmatrix} = \mathbf{d}_e \mathbf{u} \,,$$

where $\mathbf{u}$, $\boldsymbol{\sigma}$ and $\mathbf{e}$ are column matrices, whose components are the displacements, the stresses and the engineering strains, respectively. For 3-dimensional problems $\mathbf{u}$ has 3 components and both $\boldsymbol{\sigma}$ and $\mathbf{e}$ have 6 components. They are **not** tensor as defined in previous chapters. The components of $\mathbf{u}$, $\boldsymbol{\sigma}$ and $\mathbf{e}$ are the components of the corresponding quantities referred to a specific coordinate system. Thus they are not equal in different coordinate systems. Consider two coordinate systems x^i and $\bar{x}^i$ which are related by $\bar{x}^i = R_{ij}x^j$. Then, in the matrix form, we have

$$\bar{\mathbf{u}} = \mathbf{R}\mathbf{u}\,,$$

with the components of $\mathbf{R}$ being R_{ij}. The relationship between $\boldsymbol{\sigma}$, $\mathbf{e}$ and $\bar{\boldsymbol{\sigma}}$, $\bar{\mathbf{e}}$ can be derived from the **tensor transformation rules**. For 2-dimensional case, the tensor rule gives

$$\begin{bmatrix} \bar{\sigma}^{11} & \bar{\sigma}^{12} \\ \bar{\sigma}^{21} & \bar{\sigma}^{22} \end{bmatrix} = \begin{bmatrix} R_{11} & R_{21} \\ R_{12} & R_{22} \end{bmatrix} \begin{bmatrix} \sigma^{11} & \sigma^{12} \\ \sigma^{21} & \sigma^{22} \end{bmatrix} \begin{bmatrix} R_{11} & R_{12} \\ R_{21} & R_{22} \end{bmatrix}\,,$$

then

$$\bar{\boldsymbol{\sigma}} = \begin{bmatrix} \bar{\sigma}^{11} \\ \bar{\sigma}^{22} \\ \bar{\sigma}^{12} \end{bmatrix} = \begin{bmatrix} R_{11}^2 & R_{12}^2 & 2R_{11}R_{12} \\ R_{21}^2 & R_{11}^2 & 2R_{22}R_{21} \\ R_{11}R_{21} & R_{22}R_{12} & R_{11}R_{22} + R_{12}R_{21} \end{bmatrix} \begin{bmatrix} \sigma^{11} \\ \sigma^{22} \\ \sigma^{12} \end{bmatrix}\,.$$

In general, $\mathbf{u} \neq \bar{\mathbf{u}}$, $\boldsymbol{\sigma} \neq \bar{\boldsymbol{\sigma}}$ and $\mathbf{e} \neq \bar{\mathbf{e}}$ based on the matrix definition. Based on the tensor definition, $\mathbf{u} = \bar{\mathbf{u}}$, $\boldsymbol{\sigma} = \bar{\boldsymbol{\sigma}}$ and $\mathbf{e} = \bar{\mathbf{e}}$, but their components will be different in different coordinate systems. Note that $\boldsymbol{\sigma}^T\mathbf{e} = \mathbf{e}^T\boldsymbol{\sigma}$ in the present matrix definition and is equal to $\boldsymbol{\sigma}^T : \mathbf{e}(= \mathbf{e}^T : \boldsymbol{\sigma})$ if $\boldsymbol{\sigma}$ and $\mathbf{e}$ are the stress and strain tensors of the previous chapters. The engineering shear strain γ_{xy} rather then the tensor shear strain $e_{xy}[= (\partial u/\partial y + \partial v/\partial x)/2]$ [Eq. (4.2:7)] is used in Eq. (1) to make the elastic modulus matrix $\mathbf{D}_e$ of Eq. (2) below symmetric.

The generalized Hooke's law can be written in the matrix form

$$\boldsymbol{\sigma} = \mathbf{D}_e\mathbf{e}\,,$$

where

(2)
$$\mathbf{D}_e = \begin{bmatrix} D_{11} & & \text{sym} \\ D_{12} & D_{22} & \\ D_{13} & D_{23} & D_{33} \end{bmatrix}$$

is the *elastic constant*, or *modulus matrix*. The matrix $\mathbf{D}_e$ is also **not** a tensor but its components are related to the elastic modulus

tensor C^{ijkl} of Eq. (6.1:1). The strains can be expressed in terms of displacements

$$(3) \qquad\qquad \mathbf{e} = \mathbf{d}_e \mathbf{u},$$

where $\mathbf{d}_e$ is the differential operator defined in Eq. (18.11:6).

Consider a plane elasticity problem over a domain A with prescribed boundary conditions

$$(4) \qquad\qquad \mathbf{u} = \bar{\mathbf{u}} \qquad \text{on} \quad \partial A_u,$$

$$(5) \qquad\qquad \mathbf{T} = \bar{\mathbf{T}} \qquad \text{on} \quad \partial A_\sigma,$$

where $\mathbf{T}$ is the boundary traction that its components are

$$T_1 = n_1 \sigma_{xx} + n_2 \sigma_{xy},$$
$$T_2 = n_1 \sigma_{xy} + n_2 \sigma_{yy},$$

in which $\mathbf{n}$ is a unit normal to ∂A_σ. In this and subsequent chapters, we shall *use the capital $\mathbf{T}$ to denote traction rather than the Lagrangian stress as used in the previous chapters.*

In integral formulation, the functional for the potential energy of an element is

$$(6) \qquad \Pi_e = \int_{A_e} \left(\frac{1}{2} \mathbf{e}^T \mathbf{D}_e \mathbf{e} - \mathbf{u}^T \mathbf{b} \right) dA - \int_{\partial A_{\sigma e}} \mathbf{u}^T \bar{\mathbf{T}} \, ds$$

$$= \int_{A_e} \left[\frac{1}{2} (\mathbf{d}_e \mathbf{u})^T \mathbf{D}_e (\mathbf{d}_e \mathbf{u}) - \mathbf{u}^T \mathbf{b} \right] dA - \int_{\partial A_{\sigma e}} \mathbf{u}^T \bar{\mathbf{T}} ds,$$

where $\mathbf{b}$ is the distributed body force vector. *The integral formulation corresponds to the vanishing of the first variation of the functional for the whole domain with respect to the field variable $\mathbf{u}$ subjected to admissibility constraint.* The prescribed displacement condition Eq. (4), a rigid condition, must be enforced whereas the prescribed traction condition Eq. (5), a natural condition, is accounted for by the integral over ∂A_σ.

Note that the form of Eq. (6) would have to be modified slightly for mixed boundary value problems. If the mixed boundary conditions are

$$u_n = \bar{u}_n, \qquad T_t = \bar{T}_t \qquad \text{on} \quad \partial A'_{\sigma e},$$
$$u_t = \bar{u}_t, \qquad T_n = \bar{T}_n \qquad \text{on} \quad \partial A''_{\sigma e}$$

where n and t denote the normal and tangential direction along the element boundaries. Additional terms must be added to Π_e of Eq. (6) to account for the prescribed traction components. We have

$$(6a) \qquad \Pi_e = \int_{A_e} \left(\frac{1}{2} \mathbf{e}^T \mathbf{D}_e \mathbf{e} - \mathbf{u}^T \mathbf{b} \right) dA - \int_{\partial A_{\sigma e}} \mathbf{u}^T \bar{\mathbf{T}} \, ds$$

$$- \int_{\partial A'_{\sigma e}} u_t \bar{T}_t \, ds - \int_{\partial A''_{\sigma e}} u_n \bar{T}_n ds \, .$$

We can easily establish the finite element equations using the integral formulation. Consider an element of p nodes with nodal coordinate x_j, y_j for the jth node. Let us denote the nodal values of u and v as

$$q_{2j-1} = u(x_j, y_j), \qquad q_{2j} = v(x_j, y_j),$$

where q's are the generalized coordinates with their subscripts being the local label. Each node now has *two degrees-of-freedom* as opposed to one considered earlier. Utilizing the shape functions derived earlier, we can interpolate both field variables in the form

$$(7) \qquad u = \sum_{j=1}^{p} h_j(x, y) q_{2j-1} \, ,$$

$$(8) \qquad v = \sum_{j=1}^{p} h_j(x, y) q_{2j} \, .$$

In matrix form, they can be written as

$$(9) \qquad \mathbf{u} = \mathbf{h} \mathbf{q} \, , \qquad \mathbf{e} = \mathbf{d}_e \mathbf{u} = \mathbf{B} \mathbf{q} \, ,$$

where $\mathbf{d}_e$ a differential operator matrix defined in Eq. (18.11:6),

$$(10) \qquad \underset{2 \times 2p}{\mathbf{h}} = \begin{bmatrix} h_1 & 0 & \cdots & h_p & 0 \\ 0 & h_1 & \cdots & 0 & h_p \end{bmatrix} ,$$

$$(11) \qquad \mathbf{B} = \mathbf{d}_e \mathbf{h} \, ,$$

with $\mathbf{h}$ being a $2 \times 2p$ *interpolation matrix* [$\mathbf{h}$ is a row matrix for problems of one field variable in Eq. (10.10:9)] and $\mathbf{B}$ is the *strain-displacement matrix* relating the nodal displacement matrix

$$\underset{2p \times 1}{\mathbf{q}} = [q_1 \ q_2 \ \cdots \ q_{2p-1} \ q_{2p}]^T$$

to the strain matrix $\mathbf{e}$.

A substitution of Eq. (9) into Eq. (6) yields the standard equation for element

$$(12) \qquad \Pi_e = \frac{1}{2}\mathbf{q}^T\mathbf{k}\mathbf{q} - \mathbf{q}^T\mathbf{f},$$

where

$$(13) \qquad \mathbf{k} = \int_{A_e} \mathbf{B}^T\mathbf{D}_e\mathbf{B}dA,$$

$$(14) \qquad \mathbf{f} = \int_{A_e} \mathbf{h}^T\mathbf{b}dA + \int_{\partial A_{\sigma e}} \mathbf{h}^T\bar{\mathbf{T}}ds.$$

These equations are similar in form as Eqs. (18.10:7) and (18.10:8) except that $\mathbf{D}_e$, $\mathbf{B}$ are now defined in Eqs. (2) and (11), respectively.

The stiffness matrix $\mathbf{k}$ is symmetric and has the rank of $2p-3$, because there are 3 rigid body motions for plane elasticity. However, the rank can be higher if the shape functions are incapable of representing certain rigid body modes. The rank can be lower also if the integration in Eq. (13) is carried out only approximately, which can introduce spurious deformation.

For a three-node triangle, $p = 3, h_j$ are linear functions of x, y [or $\zeta_1, \zeta_2, \zeta_3$, Eq. (18.9:4)] and the strain $\mathbf{e}$ represented by $\mathbf{Bq}$ is constant. Therefore, a three node triangular element is also called the *constant strain triangle*. A 6-node triangular element uses quadratic shape functions Eq. (18.9:10) and has linear strains. It is called the *linear strain triangle*.

For elements with curved sides, the element matrices can be written in the similar form

$$(15) \qquad \mathbf{k} = \int_{A_e} \mathbf{B}^T\mathbf{D}_e\mathbf{B}Jd\xi d\eta,$$

$$(16) \qquad \mathbf{f} = \int_{A_e} \mathbf{h}^T\mathbf{b}Jd\xi d\eta + \int_{\partial A_{\sigma e}} \mathbf{h}^T\bar{\mathbf{T}}ds,$$

where ξ, η are the transformed coordinates, A_e is a right isosceles triangle of area $1/2$ for triangular elements and a square of area 4 for quadrilateral elements, and

$$(17) \qquad \mathbf{B} = \mathbf{d}_e\mathbf{h} = \begin{bmatrix} \dfrac{\partial}{\partial x} & 0 \\[2mm] 0 & \dfrac{\partial}{\partial y} \\[2mm] \dfrac{\partial}{\partial y} & \dfrac{\partial}{\partial x} \end{bmatrix}\mathbf{h},$$

which can be written as

$$
\mathbf{B} = \begin{bmatrix} \dfrac{\partial \xi}{\partial x}\dfrac{\partial}{\partial \xi} + \dfrac{\partial \eta}{\partial x}\dfrac{\partial}{\partial \eta} & 0 \\[3mm] 0 & \dfrac{\partial \xi}{\partial y}\dfrac{\partial}{\partial \xi} + \dfrac{\partial \eta}{\partial y}\dfrac{\partial}{\partial \eta} \\[3mm] \dfrac{\partial \xi}{\partial y}\dfrac{\partial}{\partial \xi} + \dfrac{\partial \eta}{\partial y}\dfrac{\partial}{\partial \eta} & \dfrac{\partial \xi}{\partial x}\dfrac{\partial}{\partial \xi} + \dfrac{\partial \eta}{\partial x}\dfrac{\partial}{\partial \eta} \end{bmatrix} \mathbf{h},
$$

in which $\mathbf{h}$ is defined in Eq. (10). The boundary integral of Eq. (16) is in the form of Eq. (18.12:3) for triangular elements and Eq. (18.13:7) or (18.13:8) for quadrilateral elements with $\mathbf{h}^T \phi_\sigma$ replaced by $\mathbf{h}^T \bar{T}$.

The evaluation of the matrix $\mathbf{B}$ is straightforward but tedious. According to Eqs. (18.11.3), (18.11:8) and (18.13:3),

$$
(18) \qquad \mathbf{J} = \begin{bmatrix} \dfrac{\partial x}{\partial \xi} & \dfrac{\partial y}{\partial \xi} \\[3mm] \dfrac{\partial x}{\partial \eta} & \dfrac{\partial y}{\partial \eta} \end{bmatrix} = \sum_{\text{all nodes}} \begin{bmatrix} x_i \dfrac{\partial h_i}{\partial \xi} & y_i \dfrac{\partial h_i}{\partial \xi} \\[3mm] x_i \dfrac{\partial h_i}{\partial \eta} & y_i \dfrac{\partial h_i}{\partial \eta} \end{bmatrix},
$$

$$
(19) \qquad \mathbf{J}^{-1} = \begin{bmatrix} \dfrac{\partial \xi}{\partial x} & \dfrac{\partial \eta}{\partial x} \\[3mm] \dfrac{\partial \xi}{\partial y} & \dfrac{\partial \eta}{\partial y} \end{bmatrix} = \dfrac{1}{J} \sum_{\text{all nodes}} \begin{bmatrix} y_i \dfrac{\partial h_i}{\partial \eta} & -y_i \dfrac{\partial h_i}{\partial \xi} \\[3mm] -x_i \dfrac{\partial h_i}{\partial \eta} & x_i \dfrac{\partial h_i}{\partial \xi} \end{bmatrix},
$$

$$
(20) \qquad J = \frac{\partial x}{\partial \xi}\frac{\partial y}{\partial \eta} - \frac{\partial y}{\partial \xi}\frac{\partial x}{\partial \eta},
$$

where repeated indices denote summation over all nodes involved. Then from Eqs. (7)–(9), we obtain

$$
(21)\ \mathbf{e} = \mathbf{Bhq} = \begin{bmatrix} \dfrac{\partial u}{\partial x} \\[3mm] \dfrac{\partial v}{\partial y} \\[3mm] \dfrac{\partial u}{\partial y} + \dfrac{\partial v}{\partial x} \end{bmatrix}
$$

$$
= \sum_{i=1}^{p} \begin{bmatrix} \left(\dfrac{\partial h_i}{\partial \xi}\dfrac{\partial y}{\partial \eta} - \dfrac{\partial h_i}{\partial \eta}\dfrac{\partial y}{\partial \xi} \right) \dfrac{q_{2i-1}}{J} \\[4mm] \left(\dfrac{\partial h_i}{\partial \eta}\dfrac{\partial x}{\partial \xi} - \dfrac{\partial h_i}{\partial \xi}\dfrac{\partial x}{\partial \eta} \right) \dfrac{q_{2i}}{J} \\[4mm] \left(\dfrac{\partial h_i}{\partial \eta}\dfrac{\partial x}{\partial \xi} - \dfrac{\partial h_i}{\partial \xi}\dfrac{\partial x}{\partial \eta} \right) \dfrac{q_{2i-1}}{J} + \left(\dfrac{\partial h_i}{\partial \xi}\dfrac{\partial y}{\partial \eta} - \dfrac{\partial h_i}{\partial \eta}\dfrac{\partial y}{\partial \xi} \right) \dfrac{q_{2i}}{J} \end{bmatrix} = \mathbf{Bq}.
$$

The range of summation in Eqs. (18) and (19) is the same as that of Eq. (21), if the transformation Eq. (18.13:3) involves all nodes of the element. Note that we have

$$(22) \qquad B_{1,2i-1} = \frac{\partial h_i}{\partial x} = \left(\frac{\partial h_i}{\partial \xi} \frac{\partial y}{\partial \eta} - \frac{\partial h_i}{\partial \eta} \frac{\partial y}{\partial \xi} \right) \frac{1}{J},$$

$$B_{2,2i} = \frac{\partial h_i}{\partial y} = \left(\frac{\partial h_i}{\partial \eta} \frac{\partial x}{\partial \xi} - \frac{\partial h_i}{\partial \xi} \frac{\partial x}{\partial \eta} \right) \frac{1}{J},$$

$$B_{3,2i-1} = B_{2,2i}, \qquad B_{3,2i} = B_{1,2i-1}, \qquad i = 1, \ldots, p,$$

with unlisted B's being zero. The geometry of the element affects the element matrices through the values J and $\mathbf{J}^{-1}$.

For a 4-node quadrilateral element, from Eq. (18.13:2), we have

$$\frac{\partial h_i}{\partial \xi} = \frac{1}{4} \xi_i (1 + \eta_i \eta), \qquad \frac{\partial h_i}{\partial \eta} = \frac{1}{4} \eta_i (1 + \xi_i \xi) \qquad (i \text{ not summed}).$$

Then

$$\mathbf{J} = \frac{1}{4} \sum_{i=1}^{4} \begin{bmatrix} x_i \xi_i (1 + \eta_i \eta) & y_i \xi_i (1 + \eta_i \eta) \\ x_i \eta_i (1 + \xi_i \xi) & y_i \eta_i (1 + \xi_i \xi) \end{bmatrix},$$

$$(22a) \qquad \mathbf{J}^{-1} = \frac{1}{4J} \sum_{i=1}^{4} \begin{bmatrix} y_i \eta_i (1 + \xi_i \xi) & -y_i \xi_i (1 + \eta_i \eta) \\ -x_i \eta_i (1 + \xi_i \xi) & x_i \xi_i (1 + \eta_i \eta) \end{bmatrix},$$

$$J = \frac{1}{16} \sum_{i,j=1}^{4} (x_i y_j - y_i x_j) \xi_i \eta_j (1 + \eta_i \eta)(1 + \xi_j \xi),$$

The Jacobian J is not a constant for elements other than rectangle and parallelogram. From the expressions above, one can conclude that it is best to numerically integrate Eqs. (15) and (16) for given $\mathbf{D}_e$.

In integral formulation, the sum of the variation $\delta\Pi_e$ of all elements must be zero, i.e.,

$$\sum_{\text{all element}} \delta\Pi_e = \sum_{\text{all element}} \delta\mathbf{q}^T (\mathbf{kq} - \mathbf{f}) = 0$$

with the components of $\delta\mathbf{q}$ corresponding to rigid constraints being zero.

Plane Stress. For plane stress problems, if the material is orthotropic, we have

$$
(23) \qquad \mathbf{D}_e = \frac{1}{1 - \nu_{xy}\nu_{yx}}
\begin{bmatrix}
E_x & & \text{sym} \\
E_x\nu_{xy} & E_y & \\
0 & 0 & G_{xy}(1 - \nu_{xy}\nu_{yx})
\end{bmatrix},
$$

in which $E_x\nu_{xy} = E_y\nu_{yx}$. If the material is *isotropic* then $E_x = E_y = E$, $\nu_{xy} = \nu_{yx} = \nu$, and $G_{xy} = G = E/2(1+\nu)$, where E is Young's modulus and ν is Poisson's ratio.

Plane Strain. For plane strain problems, if the material is isotropic

$$
(24) \qquad \mathbf{D}_e = \frac{E}{(1+\nu)(1-2\nu)}
\begin{bmatrix}
1-\nu & & \text{sym} \\
\nu & 1-\nu & \\
0 & 0 & (1-2\nu)/2
\end{bmatrix}.
$$

Axisymmetric Problems. For axisymmetric problem, there are four components of strain

$$
e_r = \frac{\partial u}{\partial r}, \qquad e_\theta = \frac{u}{r}, \qquad e_z = \frac{\partial w}{\partial z}, \qquad \gamma_{rz} = \frac{\partial u}{\partial z} + \frac{\partial w}{\partial r},
$$

which can be written in the form

$$
\mathbf{e} = \mathbf{d}_a \mathbf{u} = \mathbf{d}_a \begin{bmatrix} u \\ w \end{bmatrix},
$$

where

$$
\mathbf{d}_a^T =
\begin{bmatrix}
\dfrac{\partial}{\partial r} & \dfrac{1}{r} & 0 & \dfrac{\partial}{\partial z} \\[2mm]
0 & 0 & \dfrac{\partial}{\partial z} & \dfrac{\partial}{\partial r}
\end{bmatrix}.
$$

Expressing u, w in the form of Eqs. (7) and (8), we obtain the element matrices

$$
\mathbf{k} = \int_{A_e} (\mathbf{d}_a \mathbf{h})^T \mathbf{D}_a (\mathbf{d}_a \mathbf{h}) r\, dr\, dz,
$$

$$
\mathbf{f} = \int_{A_e} \mathbf{h}^T \mathbf{b} r\, dr\, dz + \int_{\partial A_{\sigma e}} \mathbf{h}^T \bar{\mathbf{T}} ds,
$$

which are similar to Eqs. (13) and (14) except that $\mathbf{D}_e$ is a 4×4 elastic constant matrix and $\mathbf{d}_a$ is a 4×2 differential operator defined above. For isotropic materials

$$(25) \quad \mathbf{D}_e = \frac{E}{(1+\nu)(1-2\nu)} \begin{bmatrix} 1-\nu & & & \text{sym} \\ \nu & 1-\nu & & \\ \nu & \nu & 1-\nu & \\ 0 & 0 & 0 & (1-2\nu)/2 \end{bmatrix} .$$

Problem 18.11. Determine the nominal rank of $\mathbf{k}$ for axial symmetric deformation.

Problem 18.12. Consider the coordinate transform

$$r = r(\xi, \eta), \qquad z = z(\xi, \eta).$$

Find $\mathbf{d}_a$ in terms of differentials $\partial/\partial\xi, \partial/\partial\eta$.

Incompatible Element. The accuracy of triangular and rectangular elements, more so for triangles, deteriorates when the *aspect ratio* (the ratio of one of the dimension of the element to the other) becomes large. In particular, when Poison's ratio approaches $1/2$, i.e., the material is approaching incompressible, the element will be overly stiff. The effect is specially pronounced in simulating bending of slender beams using plane elements. If only one element is used through the thickness, the numerical result is practically worthless. It will be shown later that the high stiff behavior is caused by spurious shear. Wilson *et al.* (1973) introduced quadratic terms in the deformation that

$$(26) \qquad \mathbf{u} = \mathbf{h}\mathbf{q} + (1 - \xi^2)\mathbf{a}_5 + (1 - \eta^2)\mathbf{a}_6 ,$$

where $\mathbf{h}$ is the same as that for a 4-node rectangle given in Eq. (10) and $\mathbf{a}$'s are unknown parameters. The additional terms are zero at the 4 nodes and, therefore, do not affect the physical meaning of $\mathbf{q}$. But these terms result in discontinuous displacements between two elements along the interelement boundaries

The strain matrix can be written as

$$(27) \qquad \mathbf{e} = \mathbf{B}\mathbf{q} + \frac{2\xi}{J}\mathbf{B}_5\mathbf{a}_5 + \frac{2\eta}{J}\mathbf{B}_6\mathbf{a}_6 ,$$

where

$$(28) \qquad \mathbf{B}_5 = \begin{bmatrix} -y_\eta & 0 \\ 0 & x_\eta \\ x_\eta & -y_\eta \end{bmatrix} , \qquad \mathbf{B}_6 = \begin{bmatrix} y_\xi & 0 \\ 0 & -x_\xi \\ -x_\xi & y_\xi \end{bmatrix} ,$$

in which $\mathbf{B}$ is the same as that of Eq. (11), J is given in Eq. (22a), and $(\cdot)_\xi, (\cdot)_\eta$ denote partial differentiation with respect to ξ, η, respectively. Note that both $\mathbf{B}_5$ and $\mathbf{B}_6$ are constant for rectangle and parallelogram. A substitution of Eq. (27) into Eq. (6) yields

$$(29) \qquad \Pi_e = \frac{1}{2}\mathbf{q}^T\mathbf{k}\mathbf{q} + \mathbf{a}^T\mathbf{k}_{aq}\mathbf{q} + \frac{1}{2}\mathbf{a}^T\mathbf{k}_{aa}\mathbf{a} - \mathbf{q}^T\mathbf{f} - \mathbf{a}^T\mathbf{f}_a\,,$$

where $\mathbf{k}$ and $\mathbf{f}$ are the usual 8×8 stiffness and 8×1 force matrices of the 4-node quadrilateral as those given in Eqs. (15) and (16), and

$$\mathbf{k}_{aq} = \int_{A_e}\begin{bmatrix} 2\xi\mathbf{B}_5 & 2\eta\mathbf{B}_6 \end{bmatrix}^T D_e\mathbf{B}\,d\xi d\eta\,,$$

$$\mathbf{k}_{aa} = \int_{Ae}\begin{bmatrix} 2\xi\mathbf{B}_5 & 2\eta\mathbf{B}_6 \end{bmatrix}^T D_e\begin{bmatrix} 2\xi\mathbf{B}_5 & 2\eta\mathbf{B}_6 \end{bmatrix}\frac{d\xi d\eta}{J}\,,$$

$$\mathbf{a} = \begin{bmatrix} \mathbf{a}_5 \\ \mathbf{a}_6 \end{bmatrix}\,,$$

$$\mathbf{f}_a = \int_{A_e}\begin{bmatrix} (1-\xi^2)\mathbf{b} \\ (1-\eta^2)\mathbf{b} \end{bmatrix} J d\xi d\eta + \int_{\partial A_{\sigma e}}\begin{bmatrix} (1-\xi^2)\bar{\mathbf{T}} \\ (1-\eta^2)\bar{\mathbf{T}} \end{bmatrix} ds\,.$$

Defining $\mathbf{q}'^T = \begin{bmatrix} \mathbf{q}^T & \mathbf{a}^T \end{bmatrix}$, we can write Eq. (29) in the standard form with the element stiffness and force matrices associated with $\mathbf{q}'$ being

$$\underset{12\times 12}{\mathbf{k}'} = \begin{bmatrix} \mathbf{k} & \mathbf{k}_{aq} \\ \mathbf{k}_{aq}^T & \mathbf{k}_{aa} \end{bmatrix}\,, \qquad \mathbf{f}' = \begin{bmatrix} \mathbf{f} \\ \mathbf{f}_a \end{bmatrix}\,.$$

By virtue of the fact that $\mathbf{a}$ does not appear in other elements, the vanish of the first variation of the functional with respect to $\mathbf{a}$ implies $\delta\Pi_e = 0$, i.e.,

$$\mathbf{k}_{aq}\mathbf{q} + \mathbf{k}_{aa}\mathbf{a} - \mathbf{f}_a = 0 \qquad \text{or} \qquad \mathbf{a} = -\mathbf{k}_{aa}^{-1}\mathbf{k}_{aq}\mathbf{q} + \mathbf{k}_{aa}^{-1}\mathbf{f}_a\,,$$

which can be used to eliminate $\mathbf{a}$ from Eq. (29). This *static condensation process* leads to the following modified element stiffness and force matrices associated with the column matrix $\mathbf{q}$

$$\mathbf{k}^* = \mathbf{k} - \mathbf{k}_{aq}^T\mathbf{k}_{aq}^{-1}\mathbf{k}_{aq}\,, \qquad \mathbf{f}^* = \mathbf{f} - \mathbf{k}_{aq}^T\mathbf{k}_{aa}^{-1}\mathbf{f}_a\,,$$

for the incompatible element.

The incompatible element provides improved performance in bending applications. However, its behavior is erratic when the element is not a rectangle or parallelogram. Taylor *et al.* (1976) modified the incompatible modes by treating $\mathbf{B}_5, \mathbf{B}_6$ of Eq. (28) as constant with their values evaluated at $\xi = \eta = 0$. The motivation of the modification is to remove the spurious shear induced by the isoparametric transformation when the element is not a rectangle or parallelogram. Approximating $\mathbf{B}_5, \mathbf{B}_6$ as a constant in the evaluation of element matrices is a form of *reduced integration*, which is a technique often used to reduce the stiff behavior of certain elements. Taylor's modification also makes the element satisfy the *patch test*, a condition required for the convergence of incompatible elements. The details of reduced integration and patch test will be discussed in later sections.

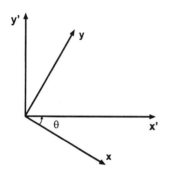

Fig. 18.14:1. Coordinate rotation.

Rotation. In order to use the assembly procedures described in Sec. 18.4, all generalized coordinates of different elements must have the same physical meaning at their common nodes. For example, in plane elasticity just considered, the components of the displacement vectors of different elements at a common node must refer to a common coordinate system. In practice, only one such coordinate system, the *global* system, is used for all the nodes of the whole body. However, in generating the element matrices, a local coordinate system may be more convenient. Modeling a two dimensional truss by one-dimensional rods as discussed in Sec. 18.6 is such an example. For orthotropic materials, it may be desirable to have the local coordinates aligned with the material coordinates of the element in deriving the element matrices. Let x', y' be the global coordinates and x, y be the local coordinates as shown in Fig. 18.14:1. The two systems are related by

$$\mathbf{x} = \beta \mathbf{x}',$$

where

$$\beta = \begin{bmatrix} \cos\theta & -\sin\theta \\ \sin\theta & \cos\theta \end{bmatrix}.$$

The displacements at the jth node is transformed by

$$\begin{bmatrix} q_{2j-1} \\ q_{2j} \end{bmatrix} = \beta_j \begin{bmatrix} q'_{2j-1} \\ q'_{2j} \end{bmatrix},$$

where β_j is the *rotation matrix for node j* which can be different for different nodes. For an element with p nodes, we have

$$(30) \qquad\qquad \mathbf{q} = \underset{2p\times 2p}{\mathbf{R}}\,\mathbf{q}' \,,$$

where

$$\underset{2p\times 2p}{\mathbf{R}} = \begin{bmatrix} \beta_1 & & & 0 \\ & \beta_2 & & \\ & & \cdots & \\ 0 & & & \beta_p \end{bmatrix}$$

is the *rotation matrix for the element*. In general, β's can be different from each other. The number of submatrices on the diagonal of $\mathbf{R}$ is equal to the number of nodes in the element. The element matrices associated with $\mathbf{q}'$ are derived by substituting Eq. (30) into Eq. (12):

$$\delta\Pi_e = \delta\mathbf{q}^T(\mathbf{kq}-\mathbf{f}) = \delta\mathbf{q}'^T(\mathbf{k}'\mathbf{q}'-\mathbf{f}') \,,$$

where

$$(31) \qquad\qquad \mathbf{k}' = \mathbf{R}^T\mathbf{k}\mathbf{R}\,, \qquad \mathbf{f}' = \mathbf{R}^T\mathbf{f}\,.$$

The element matrices $\mathbf{k}'$ and $\mathbf{f}'$ are used for assembly.

There are other circumstances that one must rotates the coordinates. In mixed-boundary problems, for instance, the components of the displacement can be prescribed in certain directions while the components of traction are prescribed in another direction, and these directions are not the same as the global coordinates or the local coordinates of the element. This will require a transformation of the coordinates of those boundary nodes to the directions of prescribed displacements and tractions, and a transformation of other nodes to the global directions, if needed. The element matrices would have to be modified accordingly.

In summary, the finite element method is to derive an approximate solution based on the weak form formulation, the principle of virtual work or variational calculus. The method expresses all field variables in terms of shape functions in the construction of the element matrices. For most cases, the shape functions are simple polynomials independent of the type of problems. Thus the method allows generating element matrices in a generic way to approximate the behavior of elements. Then, before solving for the nodal unknowns on high speed computers, the element matrices are assembled systematically into a system of algebraic equations for the unknowns and constraints are imposed to enforce the rigid boundary conditions. The process of constructing and assembling element matrices, and enforcing rigid constraints is based on the condition that the sum of the first variation of the functionals of all elements is zero, i.e.,

$$\sum_{\text{all element}} \delta\Pi_e = \sum_{\text{all element}} \delta\mathbf{q}^T(\mathbf{kq} - \mathbf{f}) = 0$$

with the components of $\delta\mathbf{q}$ corresponding to rigid constraints being zero. We denote *symbolically* the process as

$$(32) \quad \mathbf{Kq} - \mathbf{F} = \int_A \mathbf{B}^T \mathbf{D}_e \mathbf{Bq} dA - \left(\int_A \mathbf{h}^T \mathbf{b} dA + \int_{\partial A_\sigma} \mathbf{h}^T \bar{T} ds \right) = 0 .$$

Here $\mathbf{q}$ represents the nodal unknown matrix for the whole system. Equation (32) is then solved using high speed computers. Thus in the finite element method, *solving a set of differential equations with prescribed boundary conditions becomes the task of discretizing the domain by elements, constructing the element matrices, labeling the nodes, assembling the element matrices to form a set of algebraic equations, imposing constraints to enforce the rigid boundary conditions, and finally solving the algebraic equations.*

Problem 18.13.   Show that rotation does not change the rank of $\mathbf{k}$.

Problem 18.14.   Let the constitutive law be in the form

$$\sigma = \mathbf{D}_t \begin{bmatrix} e_x \\ e_y \\ e_{xy} \end{bmatrix} ,$$

where

$$e_{xx} = \frac{\partial u}{\partial x}, \qquad e_{yy} = \frac{\partial v}{\partial y}, \qquad e_{xy} = \frac{1}{2}\left(\frac{\partial u}{\partial y} + \frac{\partial v}{\partial x}\right)$$

are components of the strain tensor. Show that $\mathbf{D}_t$ is in general not symmetric.

18.15.   THREE-DIMENSIONAL SHAPE FUNCTIONS

Generalizing the finite element process to three-dimensional continua is conceptually straightforward. It is just a matter of discretizing the three dimensional body into a finite number of *solid elements*, deriving the appropriate *three-dimensional shape functions*, and generating the corresponding element matrices based on an appropriate integral formulation. The procedures of assembling, constraining, and solving algebraic equations for one-, two- or three-dimensional problems are all the same. The complication of three-dimensional problems is more in the implementation such as the description of the geometry, loading, higher semi-bandwidth, many more algebraic equations, etc. Discretizing a complex three-dimensional body into a finite number of solids, especially the hexahedral ones, is not a trivial matter. Labeling all the nodes to give an optimum average semi-bandwidth is much more complicated than that for the two dimensional case.

In this section we shall limit ourselves to the derivation of three dimensional shape functions and the corresponding element matrices.

Shape Function Variation on Element Boundaries. The variation of shape functions is restricted by the requirement of *full compatibility*, i.e., the compatibility of a field variable at the inter-connecting nodes of two adjacent elements guarantees compatibility over the entire inter-element boundary containing these nodes. If the shape functions on an element boundary surface are polynomials of the surface coordinates and if the highest order of the polynomials is j, it will require $(j+1)(j+2)/2$, the number of independent terms in a complete j^{th} order polynomial, matching conditions to assure that the interpolated functions of the adjacent elements are fully matched, i.e., continuous everywhere over the inter-element boundary. In other words, if the shape function involves j^{th} order terms of polynomial on element boundaries, it needs $(j+1)(j+2)/2$ inter-connecting nodes on each inter-element surface to guarantee continuity of two functions over the common boundary of two elements. For example, continuity is assured for a 3-node surface if all shape functions are linear, for a 6-node surface if they are quadratic, for a 10-node surface if they are cubic, and so on.

Shape Functions for Tetrahedron. Consider a family of progressively higher order tetrahedral elements with node j at (x_j, y_j, z_j) as shown Fig. 18.15:1. For convenience, we shall describe the elements using the *volume coordinates*,

$$(1) \qquad \zeta_j = \frac{a_j + b_j x + c_j y + d_j z}{6\Delta_v} \qquad j = 1, \dots, 4$$

where Δ_v is the volume of the tetrahedral element

$$(2) \qquad \Delta_v = \frac{1}{6} \begin{vmatrix} 1 & 1 & 1 & 1 \\ x_1 & x_2 & x_3 & x_4 \\ y_1 & y_2 & y_3 & y_4 \\ z_1 & z_2 & z_3 & z_4 \end{vmatrix},$$

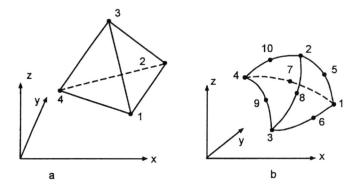

Fig. 18.15:1. Tetrahedral elements.

(3) $\quad a_1 = \begin{vmatrix} x_2 & x_3 & x_4 \\ y_2 & y_3 & y_4 \\ z_2 & z_3 & z_4 \end{vmatrix}, \quad b_1 = - \begin{vmatrix} 1 & 1 & 1 \\ y_2 & y_3 & y_4 \\ z_2 & z_3 & z_4 \end{vmatrix},$

$\quad\quad c_1 = \begin{vmatrix} 1 & 1 & 1 \\ x_2 & x_3 & x_4 \\ z_2 & z_3 & z_4 \end{vmatrix}, \quad d_1 = - \begin{vmatrix} 1 & 1 & 1 \\ x_2 & x_3 & x_4 \\ y_2 & y_3 & y_4 \end{vmatrix},$

with other b's, c's and c's obtainable through the cyclic permutation of the subscripts 1, 2, 3, 4 and Δ_v being the volume of the tetrahedron. The ζ's are the natural coordinates for tetrahedron and, similar to the triangular coordinates for triangle, are normalized volume of certain tetrahedron, e.g.,

(4) $$\zeta_1 = \frac{\Delta_{P234}}{\Delta_v},$$

where Δ_{P234} is the volume of the tetrahedron with nodes P, 2, 3, 4 in which P is a point within the tetrahedron. The other volume coordinates ζ_2, ζ_3, ζ_4 are obtained by the cyclic permutations of the subscripts. Note that ζ's are not independent that they satisfy

$$\zeta_1 + \zeta_2 + \zeta_3 + \zeta_4 = 1.$$

The evaluation of the element matrices of higher order elements will involve integration of quantities defined in terms of the volume coordinates. The following exact integration formula is useful

$$\int_{\Delta_v} \zeta_1^\alpha \zeta_2^\beta \zeta_2^\gamma \zeta_4^\lambda \, dV = 6\Delta_v \frac{\alpha! \beta! \gamma! \lambda!}{(\alpha + \beta + \gamma + \lambda + 3)!}.$$

which is similar to the formula for the triangular coordinates in two-dimension. A field variable ϕ can be represented over an element by a three-dimensional function that interpolates the nodal values q_j of the field variable, i.e.,

$$(5) \qquad\qquad \phi = \sum_{j=1}^{p} h_j(x, y, z) q_j \,,$$

where p is the number of nodes of the element and h_j are shape functions.

Four-node Tetrahedron. A four-node tetrahedral is the simplest solid element [Fig. 18.15:1(a)]. The shape functions are simply

$$(6) \qquad\qquad h_j = \zeta_j \,,$$

which are linear functions of x, y, z. The shape functions are linear on the element boundary and thus requires 3 inter-connecting nodes between adjacent element for full compatibility. Since each side of tetrahedron has 3 nodes, full compatibility can be assured. The strains in the element are constant. Therefore, a 4-node tetrahedron element is also called a constant strain tetrahedron.

Ten-node Tetrahedron [Fig. 18.15:1(b)]. The shape function h_j, which equals 1 at node j and zero at all other nodes, can be generated from the product of two functions, each of which represents a plane passing through nodes other than node j. The shape functions are

$$h_j = \zeta_j(2\zeta_j - 1) \quad j = 1, 2, 3, 4 \,,$$

$$(7) \qquad h_5 = 4\zeta_1\zeta_2 \,, \quad h_6 = 4\zeta_1\zeta_3 \,, \quad h_7 = 4\zeta_1\zeta_4 \,, \quad h_8 = 4\zeta_2\zeta_3 \,,$$

$$h_9 = 4\zeta_3\zeta_4 \,, \quad h_{10} = 4\zeta_4\zeta_2 \,.$$

These shape functions can represent a complete quadratic polynomial. Since a complete quadratic polynomial on a surface has 6 terms, it requires 6 inter-connecting nodes for full compatibility. This condition is met by the 10-node tetrahedron.

Higher order shape functions can be derived from the products of three plane functions, the products of four plane functions, and so on. For a 20-node tetrahedron, there are 16 boundary nodes on the edges and 4 at the centroids of the boundary faces. The shape functions are cubic. The quartic tetrahedron has 35 nodes, 22 of which are on the six edges, 12 on the four faces, and one, an interior node, at the volume centroid of the tetrahedron. One can show that polynomial shape functions can also meet the full compatibility requirement.

Shape Functions for Hexahedron. Two hexahedral elements are shown in Fig. 18.15:2. The shape functions for hexahedron are very much like those of quadrilateral elements except that now we have three independent variables x, y, z. The shape functions can be derived easily by the use of the natural coordinates ξ, η, ζ. In natural coordinates, the element is a cube with faces bounded by $\xi = \pm 1$, $\eta = \pm 1$, $\zeta = \pm 1$. The shape functions are simply the product of a minimum number of functions of planes passing one or more nodes of the element. For the shape function associated with node j, all nodes except node j must be passed through by at least one of these planes and none of these planes should pass through node j. For a 8-node cube, the shape functions in terms of the natural coordinates are

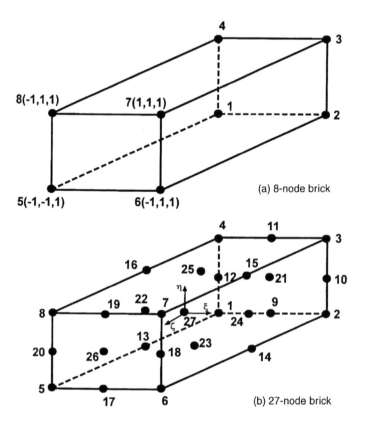

Fig. 18.15:2. 8- and 27-node brick elements.

(8) $$h_j(\xi, \eta) = (1 + \xi_j\xi)(1 + \eta_j\eta)(1 + \zeta_j\zeta)/8,$$

(j not summed), where

$$\xi_1 = \xi_4 = \xi_5 = \xi_8 = \eta_1 = \eta_2 = \eta_5 = \eta_6 = \zeta_1 = \zeta_2 = \zeta_3 = \zeta_4 = -1,$$
$$\xi_2 = \xi_3 = \xi_6 = \xi_7 = \eta_3 = \eta_4 = \eta_7 = \eta_8 = \zeta_5 = \zeta_6 = \zeta_7 = \zeta_8 = 1.$$

Equation (8) is a product of three plane functions $1 + \xi_j\xi = 0$, $1 + \eta_j\eta = 0$ and $1 + \zeta_j\zeta = 0$. The shape functions on an element boundary surface are bilinear and can have at most 4 independent terms. Since each boundary surface has four node and thus assures full compatibility.

The shape functions for a 27-node cube can be derived also. The shape function associated node 27, the centroid node, is a bulb function

$$h_{27} = (1 - \xi^2)(1 - \eta^2)(1 - \zeta^2).$$

The shape functions associated nodes 21, 22 and 23, the midface nodes adjacent to node 1 at $\varsigma = -1$, $\xi = -1$ and $\eta = -1$, respectively, are

$$h_{21} = (1 - \xi^2)(1 - \eta^2)(1 - \zeta)/2 - h_{27}/2,$$
$$h_{22} = (1 - \eta^2)(1 - \varsigma^2)(1 - \xi)/2 - h_{27}/2,$$
$$h_{23} = (1 - \varsigma^2)(1 - \xi^2)(1 - \eta)/2 - h_{27}/2.$$

The shape functions associated nodes 9, 12 and 13, the midedge nodes adjacent to node 1, are

$$h_9 = (1 - \xi)(1 - \eta)(1 - \zeta)(1 + \xi)/4 - (h_{21} + h_{23})/2 - h_{27}/4,$$
$$h_{12} = (1 - \xi)(1 - \eta)(1 - \zeta)(1 + \eta)/4 - (h_{22} + h_{21})/2 - h_{27}/4,$$
$$h_{13} = (1 - \xi)(1 - \eta)(1 - \zeta)(1 + \varsigma)/4 - (h_{23} + h_{22})/2 - h_{27}/4.$$

The shape function h_9 is the product of the functions of four planes normalized to unity at $\xi = 0$, $\eta = \varsigma = -1$ minus a quarter of h_{27} and one half of h_{21} and h_{23}. The shape function associated with node 1, a corner-node, is

(9) $$h_1 = \frac{(1 - \xi)(1 - \eta)(1 - \zeta)}{8} - \frac{h_9 + h_{12} + h_{13}}{2} - \frac{h_{21} + h_{22} + h_{33}}{4} - \frac{h_{27}}{8},$$

which equals the corresponding shape function of the 8-node brick element minus one eighth of the bulb function, one quarter of the sum of the shape functions of the three midedge nodes and one half of the sum of the shape functions of the three midface nodes adjacent to node 1. The shape functions are bi-quadratic on the element boundaries. The bi-quadratic functions have 9 independent terms. The 27-node hexahedron has 9 nodes on each face and thus assures full compatibility. The shape functions for a 20-node element can be obtained from those of the 27-node element by setting $h_{21} = \cdots = h_{27} = 0$.

Problem 18.15. Derive the shape functions for triangular prisms as shown. The heavy dots are nodes of the element. The prism has two triangular and three rectangular surfaces.

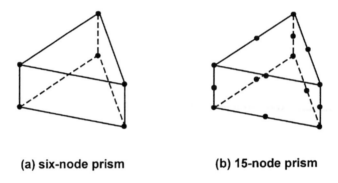

(a) six-node prism   **(b) 15-node prism**

Fig. P18.15. Triangular prism: (a) 6-node prism and (b) 15-node prism.

Problem 18.16. Derive the shape functions for a quartic tetrahedral element. Show that an interior node inside the tetrahedron is needed to achieve complete quartic polynomial for the shape functions.

18.16. THREE-DIMENSIONAL ELASTICITY

We shall express the field variables of three-dimensional elasticity in matrix form

(1) $$\mathbf{u}^T = [u \ v \ w],$$

(2) $$\boldsymbol{\sigma}^T = [\sigma_{xx} \ \sigma_{yy} \ \sigma_{zz} \ \sigma_{xy} \ \sigma_{yz} \ \sigma_{zx}],$$

(3) $$\mathbf{e}^T = [e_{xx} \ e_{yy} \ e_{zz} \ \gamma_{xy} \ \gamma_{yz} \ \gamma_{zx}],$$

where $\mathbf{u}$ is the displacement matrix, $\boldsymbol{\sigma}$ the stress matrix (**not tensor**), and $\mathbf{e}$ the strain matrix (**not tensor**) in which $\gamma_{xy}, \gamma_{yz}, \gamma_{zx}$ are the *engineering shear strains*. The stresses and strains are related by the constitutive law

$$\boldsymbol{\sigma} = \mathbf{D}_e \mathbf{e},$$

where

(4)
$$\mathbf{D}_e = \begin{bmatrix} D_{11} & & & & \text{sym} \\ D_{12} & D_{22} & & & \\ \cdots & \cdots & \cdots & & \\ D_{16} & D_{26} & \cdots & D_{66} \end{bmatrix}$$

is the elastic modulus matrix. The components of $\mathbf{D}_e$ are related to the coefficients of Hooke's law ($\sigma_{ij} = d_{ijkl}e_{kl}$) by the following equations:

(4a) $\qquad D_{ij} = D_{ji} = d_{iijj}, \quad D_{i,j+3} = D_{j+3,i} = d_{iijm(j)}$,

$$D_{i+3,j+3} = D_{j+3,i+3} = d_{im(i),jm(j)},$$

where $i, j = 1, 2, 3$ and repeated indices are not summed, and $m(j) = j + 1$ for $j = 1, 2$; $m(j) = 1$ for $j = 3$. The use of engineering shear strains in $\mathbf{e}$ makes $\mathbf{D}_e$ symmetric. In general, it will not be symmetric if the tensoral shear strains are used. For isotropic materials,

$$D_{11} = D_{22} = D_{33} = \frac{E(1 - \nu)}{(1 + \nu)(1 - 2\nu)},$$

$$D_{44} = D_{55} = D_{66} = \frac{E}{2(1 + \nu)},$$

$$D_{12} = D_{21} = D_{31} = D_{13} = D_{32} = D_{23} = \frac{E\nu}{(1 + \nu)(1 - 2\nu)},$$

while all unlisted D's are zero. The strain can be expressed in terms of the displacement as

(5) $\qquad\qquad\qquad\qquad \mathbf{e} = \mathbf{d}_e \mathbf{u}$,

where $\mathbf{d}_e$ is a matrix differential operator that its transpose is

(6)
$$\mathbf{d}_e^T = \begin{bmatrix} \dfrac{\partial}{\partial x} & 0 & 0 & \dfrac{\partial}{\partial y} & 0 & \dfrac{\partial}{\partial z} \\[2ex] 0 & \dfrac{\partial}{\partial y} & 0 & \dfrac{\partial}{\partial x} & \dfrac{\partial}{\partial z} & 0 \\[2ex] 0 & 0 & \dfrac{\partial}{\partial z} & 0 & \dfrac{\partial}{\partial y} & \dfrac{\partial}{\partial x} \end{bmatrix}.$$

Consider a problem over a domain V with prescribed boundary conditions

(7) $\qquad\qquad\qquad\qquad \mathbf{u} = \bar{\mathbf{u}} \quad \text{on} \quad \partial V_u$,

(8) $\qquad\qquad\qquad\qquad \mathbf{T} = \bar{\mathbf{T}} \quad \text{on} \quad \partial V_\sigma$,

where $\partial V_u + \partial V_\sigma = \partial V$ is the boundary of V and $\mathbf{T}$ is the boundary traction with components

$$T_1 = n_1\sigma_{xx} + n_2\sigma_{xy} + n_3\sigma_{xz},$$
$$T_2 = n_1\sigma_{xy} + n_2\sigma_{yy} + n_3\sigma_{yz},$$
$$T_3 = n_1\sigma_{xz} + n_2\sigma_{yz} + n_3\sigma_{zz},$$

in which $\mathbf{n}$ is a unit normal to ∂V_σ. The potential energy for an element is

$$(9) \qquad \Pi_e = \int_{V_e} \left(\frac{1}{2}\mathbf{e}^T\mathbf{D}_e\mathbf{e} - \mathbf{u}^T\mathbf{b} \right) dV - \int_{\partial V_{\sigma e}} \mathbf{u}^T\bar{\mathbf{T}}\,dS,$$

where V_e is the volume of the element and $\mathbf{b}$ is the distributed body force matrix. Here, $\mathbf{b}$ has three components as opposed to two in Eq. (18.14:6) for plane elasticity. If there are mixed boundary conditions, i.e. a combination of components of the displacement and traction are prescribed, terms similar to those included in Eq. (18.14:6a) must be added to Eq. (9) to account for the prescribed traction component(s). The admissibility conditions are that $\mathbf{u}$ is C° continuous and $\mathbf{u} = \bar{\mathbf{u}}$ on ∂V_{ue}. The first variation of the potential energy with respect to the field variable $\mathbf{u}$ is

$$(10) \qquad \delta\Pi_e = \int_{V_e} (\delta\mathbf{e}^T\mathbf{D}_e\mathbf{e} - \delta\mathbf{u}^T\mathbf{b})\,dV - \int_{\partial V_{\sigma e}} \delta\mathbf{u}^T\bar{\mathbf{T}}\,dS$$
$$= \int_{V_e} [(\mathbf{d}_e\delta\mathbf{u})^T\mathbf{D}_e(\mathbf{d}_e\mathbf{u}) - \delta\mathbf{u}^T\mathbf{b}]\,dV - \int_{\partial V_{\sigma e}} \delta\mathbf{u}^T\bar{\mathbf{T}}\,dS.$$

In integral formulation, the sum of the first variation $\delta\Pi_e$ of all elements is required to be zero

$$\sum_{\text{all element}} \delta\Pi_e = 0$$

for all admissible $\delta\mathbf{u}$. The prescribed displacement Eq. (7) is a rigid condition, which must be enforced, i.e., the components of $\delta\mathbf{u}$ corresponding to rigid constraint are set to zero. The prescribed traction Eq. (8) is a natural condition, which is accounted for automatically by the integral term over $\partial V_{\sigma e}$.

The procedure for establishing the finite element equations using integral formulation is identical to that for plane elasticity with the two-dimensional quantities being replaced by the corresponding three-dimensional ones. Consider an element with p nodes at $x_j, y_j, z_j,\ j = 1, \ldots, p$. Let the nodal values of u, v and w be

$$q_{3j-2} = u(x_j, y_j, z_j), \quad q_{3j-1} = v(x_j, y_j, z_j), \quad q_{3j} = w(x_j, y_j, z_j),$$

where q's are the generalized coordinates. Each node now has *three degrees-of-freedom* as opposed to one or two considered earlier. We interpolate the field variables that

(11)
$$u = \sum_{j=1}^{p} h_j(x, y, z) q_{3j-1}, \quad v = \sum_{j=1}^{p} h_j(x, y, z) q_{3j-2},$$

$$w = \sum_{j=1}^{p} h_j(x, y, z) q_{3j},$$

or in the matrix form

(12)
$$\mathbf{u} = \mathbf{h} \mathbf{q}, \qquad \mathbf{e} = \mathbf{d}_e \mathbf{u} = \mathbf{d}_e \mathbf{h} \mathbf{q} = \mathbf{B} \mathbf{q},$$

where $\mathbf{h}$ is the $3 \times 3p$ interpolation matrix and $\mathbf{q}$ is the nodal displacement matrix of $3p$ components

(13)
$$\underset{3 \times 3p}{\mathbf{h}} = \begin{bmatrix} h_1 & 0 & 0 & \cdots & h_p & 0 & 0 \\ 0 & h_1 & 0 & \cdots & 0 & h_p & 0 \\ 0 & 0 & h_1 & \cdots & 0 & 0 & h_p \end{bmatrix},$$

$$\underset{1 \times 3p}{\mathbf{q}^T} = \begin{bmatrix} q_1 & q_2 & \cdots & q_{3p-1} & q_{3p} \end{bmatrix}.$$

A substitution of Eq. (12) into Eq. (10) yields the standard equation for the element

(14)
$$\delta \Pi_e = \delta \mathbf{q}^T (\mathbf{k} \mathbf{q} - \mathbf{f}),$$

where

(15)
$$\mathbf{k} = \int_{V_e} \mathbf{B}^T \mathbf{D}_e \mathbf{B} \, dV,$$

(16)
$$\mathbf{f} = \int_{V_e} \mathbf{h}^T \mathbf{b} \, dV + \int_{\partial V_{\sigma e}} \mathbf{h}^T \bar{\mathbf{T}} \, dS.$$

These equations are in the similar form as Eqs. (18.14:13) and (18.14:14) except that $\mathbf{D}_e, \mathbf{d}_e, \mathbf{B}$ and $\mathbf{h}$ are defined in Eqs. (4), (6), (12) and (13), respectively, for 3-dimensional body. The h's of a four-node tetrahedron are linear functions of x, y, z, and the corresponding strain $\mathbf{e}(= \mathbf{B} \mathbf{q})$ is constant. Thus, a four-node tetrahedral element is called a *constant strain tetrahedron*. A 10-node tetrahedral element has quadratic displacements and linear strains and is, therefore, called a *linear strain tetrahedron*. The rank of $\mathbf{k}$ is $3p - 6$ if the shape functions are capable of representing all 6 rigid body modes. Approximate integration of Eq. (15) can lower the rank of $\mathbf{k}$ and introduce spurious deformation mode(s).

For an element with curved surfaces, we introduce the transformation

$$x = \sum_{\text{all element nodes}} x_j h_j(\xi, \eta, \zeta),$$

$$y = \sum_{\text{all element nodes}} y_j h_j(\xi, \eta, \zeta),$$

$$z = \sum_{\text{all element nodes}} z_j h_j(\xi, \eta, \zeta)$$

to map a tetrahedral element V_e into a right angle tetrahedron in the first quadrant with $0 \le \xi, \eta, \zeta \le 1$ and a hexahedral element into a cube with $-1 \le \xi, \eta, \zeta \le 1$. The element matrices can be written in the similar form as Eqs. (18.14:15) and (18.14:16)

$$(17) \qquad \mathbf{k} = \int_{V_e} \mathbf{B}^T \mathbf{D} \mathbf{B} J d\xi \, d\eta \, d\zeta \,,$$

$$(18) \qquad \mathbf{f} = \int_{V_e} \mathbf{h}^T \mathbf{b} J d\xi \, d\eta \, d\zeta + \int_{\partial V_{\sigma e}} \mathbf{h}^T \bar{\mathbf{T}} \, dS \,,$$

in terms of the natural coordinates ξ, η, ζ where

$$\mathbf{B} = \mathbf{d}_e \mathbf{h}$$

with the components of $\mathbf{d}_e$ [Eq. (18.6:6)] calculated from

$$\frac{\partial}{\partial \alpha} = \frac{\partial \xi}{\partial \alpha} \frac{\partial}{\partial \xi} + \frac{\partial \eta}{\partial \alpha} \frac{\partial}{\partial \eta} + \frac{\partial \zeta}{\partial \alpha} \frac{\partial}{\partial \zeta} \,, \qquad \alpha = x, y, z \,.$$

Since the transformation $\xi = \xi(x, y, z)$, $\eta = \eta(x, y, z)$ and $\zeta = \zeta(x, y, z)$ is generally not available, we have to evaluate $\partial \xi / \partial x, \dots$, using the following relations:

$$(19) \quad \mathbf{J} = \begin{bmatrix} \dfrac{\partial x}{\partial \xi} & \dfrac{\partial y}{\partial \xi} & \dfrac{\partial z}{\partial \xi} \\[2mm] \dfrac{\partial x}{\partial \eta} & \dfrac{\partial y}{\partial \eta} & \dfrac{\partial z}{\partial \eta} \\[2mm] \dfrac{\partial x}{\partial \zeta} & \dfrac{\partial y}{\partial \zeta} & \dfrac{\partial z}{\partial \zeta} \end{bmatrix} \,, \qquad \mathbf{J}^{-1} = \begin{bmatrix} \dfrac{\partial \xi}{\partial x} & \dfrac{\partial \eta}{\partial x} & \dfrac{\partial \zeta}{\partial x} \\[2mm] \dfrac{\partial \xi}{\partial y} & \dfrac{\partial \eta}{\partial y} & \dfrac{\partial \zeta}{\partial y} \\[2mm] \dfrac{\partial \xi}{\partial z} & \dfrac{\partial \eta}{\partial z} & \dfrac{\partial \zeta}{\partial z} \end{bmatrix} \,.$$

The integration over $\partial V_{\sigma e}$ becomes integration over a plane surface(s) in the transformed coordinates, which is a right isosceles triangle of area of 1/2 for tetrahedral elements or a square of area of 4 for cubic elements. If the surface coordinates of the boundary are ξ_1, ξ_2, which are two of the three natural coordinates ξ, η, ς. Then

$$dS = \sqrt{\frac{\partial x_k}{\partial \xi_1}\frac{\partial x_k}{\partial \xi_1} + \frac{\partial x_k}{\partial \xi_2}\frac{\partial x_k}{\partial \xi_2} - \frac{\partial x_k}{\partial \xi_1}\frac{\partial x_k}{\partial \xi_2}}\, d\xi_1\, d\xi_2,$$

where $x = x_1, y = x_2, z = x_3$ and repeated indices denote summation from 1 to 3. For a boundary surface with the normal in the ς-direction, ξ_1, ξ_2 are ξ, η and so on.

For tetrahedral elements, the shape functions are actually in terms of the volume coordinates $\zeta_1, \zeta_2, \zeta_3, \zeta_4$. We have

$$\xi = \zeta_1, \quad \eta = \zeta_2, \quad \zeta = \zeta_3, \quad \zeta_4 = 1 - \zeta_1 - \zeta_2 - \zeta_3.$$

Differentiations with respect to the coordinates of the two systems are related by

$$\frac{\partial}{\partial \xi} = \frac{\partial}{\partial \zeta_1} - \frac{\partial}{\partial \zeta_4}, \quad \frac{\partial}{\partial \eta} = \frac{\partial}{\partial \zeta_2} - \frac{\partial}{\partial \zeta_4}, \quad \frac{\partial}{\partial \zeta} = \frac{\partial}{\partial \zeta_3} - \frac{\partial}{\partial \zeta_4}.$$

We see the similarity among the element matrices for one-, two- and three-dimensional problems. For a n-dimensional continuum, the evaluation of element stiffness matrix and the portion of element force matrix due to the distributed body force requires n-dimensional integration. The evaluation of the element force matrix associated with the applied boundary loads requires $(n-1)$-dimensional integration. The latter element force is a quantity evaluated at one or more boundary points for one-dimensional problems while it involves line and surface integrals, respectively, for two- and three-dimensional problems. The finite element formulation for elasticity discussed so far involves only displacements as field variables and is thus called the *displacement model*. The number of degrees-of-freedom per node in the displacement model is normally n for n-dimensional elasticity.

The finite element method permits a systematic process for selecting interpolation functions, constructing element matrices, assembling them into a system of simultaneous algebraic equations and determining the solution. Interpolation functions, assembling process, and solution methods for a large system of equations are usually generic, i.e., they can be applied to almost any types of problem. The element stiffness matrix depends on problem type rather than the specifics of the problem. The element applied force matrix depends on the distributed loads, but the process of determining it is again generic. For example, we have derived element stiffness

matrices for beams, plane problems, 3-dimensional solids, etc. We can use
the element stiffness matrices for plane elasticity to solve a plane problem
with a square hole or with a circular hole. The nature of the problem is
reflected in the geometry (element sizes and shapes), the rigid constraints
on the boundaries, and the value of the distributed load. These inherent
generic characteristics of the finite element method make it easy to apply
to almost any problems.

In integral formulation, the sum of the variation $\delta\Pi_e$ of all elements
is zero

$$(20) \quad \sum_{\text{all element}} \delta\Pi_e = \sum_{\text{all element}} \left[\int_{V_e} (\delta e^T D_e e - \delta u^T b) dV - \int_{\partial V_{\sigma e}} \delta u^T \bar{T} dS \right]$$

$$= \sum_{\text{all element}} \delta q^T (kq - f) = 0$$

with the components of δq corresponding to rigid constraints being zero.
We denote *symbolically* the summation as

$$(21) \quad Kq - F = \int_V B^T D_e Bq \, dV - \left(\int_V h^T b \, dV + \int_{\partial V_\sigma} h^T \bar{T} \, dS \right) = 0 \,,$$

which is the same as Eq. (18.14:32) except that the area and line integrals
are replaced by volume and surface integrals, respectively. Since $\sigma = D_e e = D_e Bq$, Eq. (18.16:21) can be written as

$$(22) \qquad\qquad \int_V B^T \sigma \, dV - F = 0 \,.$$

18.17. DYNAMIC PROBLEMS OF ELASTIC SOLIDS

The finite-element formulation given in this section can be generalized
to time-dependent problems involving of dynamic behavior of structures. In
such problems the finite-element idealization leads to a set of simultaneous
dynamic equations of the form

$$(1) \qquad\qquad M\ddot{u} + C\dot{u} + Ku = F \,,$$

where u represents the matrix of generalized coordinates or unknown parameters for the entire structure and dot(s) over a character denotes differentiation with respect to time t. The matrices M and C are the assembled
mass and damping matrices, respectively. For time independent problems
these matrices do not appear because $\ddot{u}, \dot{u} = 0$, and Eq. (1) reduces to the

equation governing the static elastic response of the structure, where $\mathbf{K}$ is the familiar constrained assembled stiffness matrix introduced earlier. As will be seen the matrix $\mathbf{M}$ is symmetric and positive definite, and when linear viscous damping prevails, the matrix $\mathbf{C}$ is also symmetric. The dynamic problem now becomes an initial-valued problem of finding a column matrix $\mathbf{u} = \mathbf{u}(t)$, satisfying Eq. (1) and the given initial conditions:

$$\mathbf{u}(0) = \mathbf{u}_0, \quad \dot{\mathbf{u}}(0) = \mathbf{v}_0.$$

A variational formulation can be used in conjunction with the finite element discretization to derive appropriate the mass and damping matrices. The process is an application of Hamilton's principle to a dynamic system. We consider the integral

$$(2) \qquad\qquad H = \int_{t_1}^{t_2} (U - W - T)\, dt,$$

where T is the kinetic energy; U the strain energy; and W the work of the applied loads. Hamilton's principle may be written as

$$(3) \qquad\qquad \delta H = 0,$$

and stated as follows: among all possible time histories of displacement configurations that satisfy compatibility and rigid boundary conditions and the conditions at time t_1 and t_2, the actual solution makes H a stationary value.

The following simple example can illustrate these ideas further. Consider the spatially one-dimensional dynamic problem of the axial displacement $u(x,t)$ of a rod with prescribed loads $\bar{F}_1$, $\bar{F}_2$ at both ends. The governing equation is

$$(4) \qquad\qquad \frac{\partial}{\partial x}\left(AE\frac{\partial u}{\partial x}\right) + p(x,t) = \rho A\ddot{u}$$

where ρA is the mass per unit length, p is an applied distributed axial load, and AE is the rod axial stiffness. All these quantities can be a function of x, t. Also assume that $u(x, t_1)$ and $u(x, t_2)$ are prescribed values of u at t_1 and t_2. Let the rod be divided into N elements. For the eth element between points x_e and x_{e+1}, the nodal values at these points are denoted by $u(x_e, t) = q_e$. An integral formulation of Eq. (4), which can also be identified as the Hamilton's principle, is expressed by $\delta H = 0$ with respect to u, where

(5)
$$H = \int_{t_1}^{t_2} \left\{ \sum_{e=1}^{N} \int_{x_e}^{x_{e+1}} \left[\frac{1}{2} AE(x,t) \left(\frac{\partial u}{\partial x} \right)^2 - p(x,t)u \right. \right.$$

$$\left. \left. - \frac{1}{2} \rho(x,t) A \dot{u}^2 \right] dx + \bar{F}_1(t)u|_{x_0} - \bar{F}_2(t)u|_{x_{N+1}} \right\} dt \,.$$

In most elasticity applications, the density ρA and the equivalent modulus AE are independent of time. However, this is not true for biological or living materials, where the time variation of the material plays an essential role influencing the behavior of the system.

We assume u in the form

(6)
$$u = \mathbf{h}(x)\mathbf{q}_e(t), \quad \delta u = \mathbf{h}(x)\delta \mathbf{q}_e(t) \,,$$

where $\mathbf{h}(x)$ is the shape function matrix, which is the same as that of the static problems and $\mathbf{q}_e$ is the generalized-coordinate matrix for the eth element. The assumption of a field variable in the form of Eq. (6) is most fundamental in the finite element approach to dynamic problems. It separates the field variables as the product of a space-like function $\mathbf{h}(x)$ and a time-like function $\mathbf{q}_e(t)$ where $\mathbf{h}(x)$ is in the exact same form as in static problems. This allows us to construct the element stiffness and force matrices in the same manner discussed before. We then have to construct, in addition to the stiffness matrix, only the mass and damping matrices for dynamic applications.

We can write H in the form

$$H = \int_{t_1}^{t_2} \left\{ \sum_{e=1}^{N} \left[\frac{1}{2} \mathbf{q}_e^T \mathbf{k}_e(t)\mathbf{q}_e - \mathbf{q}_e^T \mathbf{f}_e(t) - \frac{1}{2} \dot{\mathbf{q}}_e^T \mathbf{m}_e(t)\dot{\mathbf{q}}_e \right] \right.$$

$$\left. + \bar{F}_1(t)q_1 - \bar{F}_2(t)q_{N+1} \right\} dt \,,$$

where $\mathbf{k}_e$, and $\mathbf{f}_e$ are the element stiffness matrix and the element force matrix, which are derived in the same manner as before with the exception that $\mathbf{k}_e$ and $\mathbf{f}_e$ can be a function of time for dynamic problems. For this example, $\mathbf{m}_e$ is the *element-mass matrix* defined by

(7)
$$\int_{x_e}^{x_{e+1}} \frac{1}{2}\rho A(x,t)\dot{u}^2 dx = \frac{1}{2}\dot{\mathbf{q}}_e^T \mathbf{m}_e \dot{\mathbf{q}}_e \,.$$

After all elements are summed, we have

(8)
$$H = \int_{t_1}^{t_2} \left(\frac{1}{2}\mathbf{u}^T \mathbf{K}\mathbf{u} - \mathbf{u}^T \mathbf{F} - \frac{1}{2}\dot{\mathbf{u}}^T \mathbf{M}\dot{\mathbf{u}} + \bar{F}_1 q_1 - \bar{F}_1 q_{N+1} \right) dt \,,$$

where $\mathbf{u}$ is the nodal parameter matrix of u. Setting the first variation δH with respect to u's to zero, integrating by parts, and noting that $\delta\mathbf{u}(t_1) = \delta\mathbf{u}(t_2) = 0$, we obtain the Euler equations,[1]

$$\mathbf{M\ddot{u}} + \mathbf{Ku} = \mathbf{F}\,,$$

for the dynamic system. The matrices $\mathbf{K}$ and $\mathbf{F}$ are from the strain energy and work terms, respectively, in Eq. (5) and $\mathbf{M}$ results from the consideration of the kinetic-energy.

When the interpolation function used in Eq. (7) is the same as those used to obtain $\mathbf{k}_e$, the mass matrix is referred to as a *consistent mass matrix*. Consistent element-mass matrix is usually fully populated. In practice, *diagonal or lumped mass matrices* are often employed due to their general economy especially in the explicit time-integration methods. As will be seen later in this section, each approach has certain advantages and disadvantages with regard to computer implementation.

The lumped mass formulation assumes that a certain amount of structural mass surrounding a given node is concentrated or *lumped* at that node. There are many ways to construct lumped mass matrices (Tong *et al.* 1971, Hughes *et al.* 1976, Fried and Malkus 1976, Hinton *et al.* 1976). However, most techniques are *ad hoc*. One approach is to integrate the element mass matrix using an integration formula with integration points at nodes only. This effectively diagonalizes the mass matrix. Another technique is to sum the quantities of each row of the consistent matrix to the diagonal if all generalized coordinates have the same physical dimension. Hinton *et al.* proposed setting the entries of the lumped mass matrix proportionally to the diagonal entries. Goudreau (1970) showed that averaging the consistent and lumped mass matrices for the one-dimensional second order equation improves the rate of convergence of the finite element method. No general theory of obtaining higher-order accurate mass matrices has yet been established.

To continue with the example of Eq. (4), we can use the same linear interpolation as before at the element level:

$$u = q_1(1 - \xi)/2 + q_2(1 + \xi)/2\,,$$

where $\xi = (2x - x_e - x_{e+1})/\varepsilon$, $\varepsilon = x_{e+1} - x_e$ and q's are the nodal nodal values of u at the first and second nodes. To derive a consistent mass matrix

---

[1]If the displacements are prescribed at the ends, δq's corresponding to the prescribed values of u are zero to fulfill the rigid constrained conditions. One must then use the constrained $\mathbf{M}$ and $\mathbf{K}$ in the Euler equations.

for constant density over an element from Eq. (7), we integrate

$$\rho A \int_{x_e}^{x_{e+1}} \dot{u}^2 dx = \rho A \int_{-1}^{1} [q_1(1-\xi)/2 + q_2(1+\xi)/2]^2 dx = \dot{\mathbf{q}}_e^T \mathbf{m}_e \dot{\mathbf{q}}_e \,,$$

where the subscript e denotes values associated the eth element and $\mathbf{m}_e$ is the element mass matrix

$$(9) \qquad\qquad \mathbf{m}_e = \frac{\rho A \varepsilon}{3} \begin{bmatrix} 1 & 1/2 \\ 1/2 & 1 \end{bmatrix}.$$

In the lumped mass approach, we lump half of the element mass to each node. The matrix becomes

$$(10) \qquad\qquad \mathbf{m}_e = \frac{\rho A \varepsilon}{2} \begin{bmatrix} 1 & 0 \\ 0 & 1 \end{bmatrix}.$$

The consistent mass matrices for beam and triangular plane-stress elements can be derived in similar fashion. For a beam [Eq. (18.3:1)], we approximate the deflection in terms of $u(= q_1)$ and $du/dX(= q_2)$ at $\xi = -1$ and the corresponding values q_3, q_4 at $\xi = 1$. We can then employ the same cubic interpolation Eq. (18.3:6) used to derive the element-stiffness matrix $\mathbf{k}$ to calculate the kinetic energy. The consistent element mass matrix is given by

$$(11) \qquad\qquad \mathbf{m}_e = \frac{\rho A \varepsilon}{420} \begin{bmatrix} 156 & & & \text{sym} \\ 22\varepsilon & 4\varepsilon^2 & & \\ 54 & 13\varepsilon & 156 & \\ -13\varepsilon & -3\varepsilon^2 & -22\varepsilon & 4\varepsilon^2 \end{bmatrix},$$

for an element having a uniform mass distribution where again ρA is mass per unit length. The corresponding lumped mass matrix can be obtained by lumping half the beam element mass and rotary inertia onto each node,

$$(12) \qquad\qquad \mathbf{m}_e = \frac{\rho A \varepsilon}{2} \begin{bmatrix} 1 & 0 & 0 & 0 \\ 0 & \varepsilon^2/12 & 0 & 0 \\ 0 & 0 & 1 & 0 \\ 0 & 0 & 0 & \varepsilon^2/12 \end{bmatrix}.$$

The rotary inertia is associated with the angular deflections du/dx at the nodes. Lumping the rotary inertia at a node as $(\varepsilon^2/12)$ is *ad hoc* at best.

One can even neglect them completely in Eq. (12). In this case, the mass matrix becomes singular. This makes the numerical integration of Eq. (1) by an explicit integration scheme more cumbersome.

For the triangular elements in plane elasticity, we use the same interpolation functions as those in Sec. 18.9 for evaluating the element-stiffness matrix, i.e.,

$$\mathbf{u} = \mathbf{hq},$$

where h's are defined in Eqs. (18.9:4) and (18.9:10). The expression for the kinetic energy of an element of thickness d, area A_e, and mass per unit volume ρ is

$$(13) \qquad \frac{1}{2} \int_{A_e} \rho d\dot{\mathbf{u}}^T \dot{\mathbf{u}} \, dA = \frac{1}{2} \dot{\mathbf{q}}_e^T \left(\int_{A_e} \rho d\mathbf{h}^T \mathbf{h} dA \right) \dot{\mathbf{q}}_e = \frac{1}{2} \dot{\mathbf{q}}_e^T \mathbf{m}_e \dot{\mathbf{q}}_e .$$

Thus, the consistent mass matrix is given by

$$(14) \qquad \mathbf{m}_e = \frac{\rho A_e d}{12} \begin{bmatrix} 2 & & & & & \\ 0 & 2 & & & \text{sym} & \\ 1 & 0 & 2 & & & \\ 0 & 1 & 0 & 2 & & \\ 1 & 0 & 1 & 0 & 2 & \\ 0 & 1 & 0 & 1 & 0 & 2 \end{bmatrix}$$

for an element of uniform mass and thickness. One defines the corresponding lumped mass matrix by lumping one-third of the element mass onto each node

$$(15) \qquad \mathbf{m}_e = \frac{\rho A_e d}{3} \begin{bmatrix} 1 & & & & & \\ 0 & 1 & & & \text{sym} & \\ 0 & 0 & 1 & & & \\ 0 & 0 & 0 & 1 & & \\ 0 & 0 & 0 & 0 & 1 & \\ 0 & 0 & 0 & 0 & 0 & 1 \end{bmatrix} .$$

Note that the total mass of the element is equally divided among the nodes.

In Eqs. (13)–(15), the mass matrix is based on the total mass of the element. In Eqs. (18.14:13) and (18.14:14), the element stiffness and loading matrices are values per unit thickness. For plane stress, before assembling them for dynamic study, if the thickness is uniform, we simply multiply those matrices by the element thickness. If the thickness varies over the element, we must include the thickness as a multiplication factor in the *elastic modulus matrix* $\mathbf{D}_e$ in Eq. (18.14:13) or (18.14:15), the body force $\mathbf{b}$, and the boundary traction $\bar{\mathbf{T}}$ as in Eq. (18.14:14) or (18.14:16). In plane strain, one only has to deal with problems per unit thickness. In this case, one may simply put $d = 1$ in Eqs. (14) and (15).

In order to determine the damping matrix $\mathbf{C}$, we need to consider the virtual work done by damping forces. As an example, consider the case where the damping force is proportional to velocity (i.e., damping force $= \gamma \dot{u} dA$). The virtual work done is then

$$\int_{A_e} \delta \dot{\mathbf{u}}^T \gamma \dot{\mathbf{u}} \, dA = \delta \dot{\mathbf{q}}_e^T \left(\int_{A_e} \gamma \mathbf{h}^T \mathbf{h} dA \right) \dot{\mathbf{q}}_e = \delta \dot{\mathbf{q}}_e^T \mathbf{c}_e \dot{\mathbf{q}}_e \,,$$

which is similar to the element mass matrix. An often-used form of $\mathbf{C}$ is the *Rayleigh damping matrix*

$$\mathbf{C} = a\mathbf{M} + b\mathbf{K} \,,$$

where a and b are constant parameters.

Both matrices $\mathbf{M}$ and $\mathbf{C}$ for the entire structure are assembled from element matrices $\mathbf{m}_e$ and $\mathbf{c}_e$ in a fashion analogous to that for obtaining $\mathbf{K}$. Once the system of equations implied by Eq. (1) is established, its solution can be obtained by time integration or modal analysis. The eigenvalue problems (free vibration) corresponding to homogeneous system of Eq. (1) can also be treated by well-established techniques (Clough and Bathe 1972).

From the point of view of computer implementation and the questions of accuracy, let us examine the pros and cons of employing the consistent and lumped mass matrices. The lumped mass matrix is often a diagonal matrix. This requires less computer storage and less time to generate than the corresponding consistent mass matrix (a consistent mass matrix requires the same amount of storage space as the stiffness matrix). The diagonal form also facilitates calculation when an explicit time-integration scheme is used, where new vectors to be computed in a given time step are functions only of known vectors computed in previous steps (Clough and Bathe 1972; Tong and Rossettos 1974). As an example of the integration of Eq. (18.17:1) with $\mathbf{C} = 0$ by the central difference scheme, we have

$$\mathbf{M}\mathbf{u}_{n+1} = \mathbf{M}(2\mathbf{u}_n - \mathbf{u}_{n-1}) - \mathbf{K}\mathbf{u}_n(\Delta t)^2 + \mathbf{F}(t_n) \,, \quad \mathbf{u}(0) = \mathbf{u}_0 \,,$$

where $\mathbf{u}_n$ denotes the solution for $\mathbf{u}$ at time $t = t_n(= n\Delta t)$ with Δt being the time increment of each integration step. If $\mathbf{M}$ is a diagonal matrix,

$\mathbf{u}_{n+1}$ can be evaluated rapidly from $\mathbf{u}_n$ and $\mathbf{u}_{n-1}$. Otherwise, solving for $\mathbf{u}_{n+1}$ at each step requires considerably more computer time.

A particularly useful feature and the principle advantage of using consistent mass matrix is, when the interpolation is conforming and the evaluation of all element matrices is carried out by full integration, the natural frequencies obtained for vibration problem are upper bounds. The use of lumped mass matrix in compatible model tends to lower the frequencies. There may be instances where more accurate frequencies are obtained by the lumped mass approach.

In regard to the rate of convergence of mode shapes and frequencies by the finite-element method using consistent and lumped mass formulations, the following results have been established (Tong, Pian and Bucciarfelli 1971). In cases where certain lowest-order elements are used, such as the constant-strain triangle or four-node quadrilateral elements, a lumped mass matrix provides the same order of approximation as a consistent mass matrix for second-order differential equations. However, no systematic procedure has been found to lump the mass that guarantees improvement in the convergence of higher order elements or for problems involving higher order differential equations. As a matter of fact, consistent mass matrix often gives higher-order approximation. In general, what can be recommended is that lumped mass matrix be used for low-order elements described above, where it provides the same order of approximation.

In order to treat problems in real structures, however, such as auto or rail vehicles, buildings, or bridges, where the mass distribution itself is often distinctly concentrated in certain areas, one often uses a gross approximation for the stiffness description. In these cases, one should use a lumped-mass approach. An example where this concept has been used to advantage concerns the structural deformation of vehicles in a crash (Rossettos and Weinstock 1974; Tong and Rossettos 1974). The vehicle can be conveniently divided into individual modules, each with its own stiffness and mass characteristics. The modules are connected by the standard finite-element methodology. Mass lumping becomes a clear choice for such physical problems (e.g., engine mass, etc.).

One may solve Eq. (1) by modal analysis or by step-by-step integration in time. Readers are referred to any books on dynamics regarding modal analysis. We will discuss only the time integration methods, also called the direct methods. The accuracy of the direct methods is measured in terms of the truncation error. If the truncation error is of the order Δt^{m+1}, the integration scheme is said to be m^{th}-order accurate.

Newmark β-Method (Newmark 1959). This is one of the most widely used direct integration methods for 2nd order ordinary differential equations. The *Newmark β-method* rewrites Eq. (1) in the approximate form:

(16)
$$\mathbf{M}_{n+1}\mathbf{a}_{n+1} + \mathbf{C}_{n+1}\mathbf{v}_{n+1} + \mathbf{K}_{n+1}\mathbf{q}_{n+1} = \mathbf{F}_{n+1},$$
$$\mathbf{q}_{n+1} = \mathbf{q}_n + \Delta t_n \mathbf{v}_n + \Delta t^2[(1-2\beta)\mathbf{a}_n + 2\beta\mathbf{a}_{n+1}]/2,$$
$$\mathbf{v}_{n+1} = \mathbf{v}_n + \Delta t[(1-\gamma)\mathbf{a}_n + \gamma\mathbf{a}_{n+1}],$$

where Δt is time step size and $(\cdot)_n$ denotes the value of the concerned quantity at $t = t_n$ that $\mathbf{q}_n$, $\mathbf{v}_n$, and $\mathbf{a}_n$ are the approximations of $\mathbf{q}(t_n)$, $\dot{\mathbf{q}}(t_n)$, $\ddot{\mathbf{q}}(t_n)$, respectively. In most of structural problems, $\mathbf{M}$ and $\mathbf{C}$ are independent of time. This is not in the case of biological systems.

The following recursive formula for $\mathbf{a}_{n+1}$ can be derived from Eq. (16)

(17) $(\mathbf{M} + \gamma\Delta t\mathbf{C} + \beta\Delta t^2\mathbf{K})_{n+1}\mathbf{a}_{n+1} = \mathbf{F}_{n+1} - \mathbf{C}_{n+1}\hat{\mathbf{v}}_{n+1} - \mathbf{K}_{n+1}\hat{\mathbf{q}}_{n+1},$

and

(17a) $\mathbf{q}_{n+1} = \hat{\mathbf{q}}_{n+1} + \beta\Delta t^2\mathbf{a}_{n+1}/2,$ $\mathbf{v}_{n+1} = \hat{\mathbf{v}}_{n+1} + \gamma\Delta t\mathbf{a}_{n+1},$

where

(17b)
$$\hat{\mathbf{q}}_{n+1} = \mathbf{q}_n + \Delta t_n \mathbf{v}_n + \Delta t^2(1-2\beta)\mathbf{a}_n/2,$$
$$\hat{\mathbf{v}}_{n+1} = \mathbf{v}_n + \Delta t(1-\gamma)\mathbf{a}_n.$$

To start the integration from the initial condition $\mathbf{q} = \mathbf{q}_0$ and $\mathbf{v} = \mathbf{v}_0$, one first calculates $\mathbf{a}_0$ from the first equation of Eq. (16)

$$\mathbf{M}_0\mathbf{a}_0 = \mathbf{F}_0 - \mathbf{C}_0\mathbf{v}_0 - \mathbf{K}_0\mathbf{q}_0,$$

and evaluates $\hat{\mathbf{q}}_1$ and $\hat{\mathbf{v}}_1$ from Eq. (17b). One then solves for $\mathbf{a}_1$ from Eq. (17), determines $\mathbf{q}_1$, $\mathbf{v}_1$ from Eq. (17a), proceeds to determine $\mathbf{a}_2$, $\mathbf{q}_2$ and $\mathbf{v}_2$ of the next time increment and so on.

An integration scheme is called *explicit* if $\mathbf{a}_{n+1}$ can be obtained without solving the system of algebraic equations Eq. (17). Otherwise the scheme is called *implicit*. Obviously if $\mathbf{M}$ and $\mathbf{C}$ are diagonal and $\beta = 0$ ($\mathbf{K}$ is usually not diagonal), we have an explicit scheme. In practice, we often called the scheme explicit as long as $\beta = 0$ even when $\mathbf{M}$ and $\mathbf{C}$ are not diagonal. Many well-known integration schemes are special cases of the Newmark method. For example, the trapezoidal rule corresponds to $\beta = 1/4$ and $\gamma = 1/2$, which is an unconditionally stable implicit scheme; $\beta = 1/6$ and $\gamma = 1/4$ gives the linear acceleration method, an implicit scheme; and $\beta = 0$ and $\gamma = 1/2$ gives the central difference method which is an explicit scheme. It can be shown that the accuracy is of the order Δt^2, said to be *second order accurate*, for both cases. An integral algorithm is always stable regardless of the size of Δt is said to be unconditionally stable.

For some values of β, γ and Δt, the integrated solution grows exponentially. This phenomenon is called *numerical instability*. The parameters β, γ and Δt also affect the accuracy of the solution. In practice, one must choose proper values for β, γ and Δt to control both the accuracy and stability of the algorithm.

One can determine the condition when instability will occur. For the Newmark method with $\mathbf{C} = a\mathbf{M} + b\mathbf{K}$, the stability conditions (Goudreau and Taylor 1972, Krieg and Key 1973) are

$$\text{Unconditionally stable}: \quad 2\beta \geq \gamma \geq 0.5$$

$$\text{Conditionally stable}: \quad 2\beta < \gamma, \quad \omega\Delta t \leq \Omega_{\text{crit}},$$

where

$$\Omega_{\text{crit}} = \frac{\xi(\gamma - 0.5) + [0.5\gamma - \beta + \xi^2(\gamma - 0.5)^2]^{1/2}}{0.5\gamma - \beta}, \qquad \xi = \frac{1}{2}\left(\frac{a}{\omega} + b\omega\right).$$

The quantity ω is the maximum undamped frequency of the system and ξ is the corresponding damping ratio. For $\gamma > 1/2$, one can use $\Omega_{\text{crit}} = (0.5\gamma - \beta)^{-1/2}$ as a conservative value when a realistic estimate of the damping coefficient is not available. The maximum frequency of the system can be obtained from the maximum eigenvalues of individual elements (Tong 1970b). Hughes (1987) listed the maximum frequencies for a number of elements.

For explicit procedures, the time increment Δt required for stability is inversely proportional to the maximum frequency (Tong *et al.* 1971, Clough and Bathe 1972, Hughes 1987). Since the lumped mass approach usually gives to a lower maximum frequency as compares to that of the consistent mass formulation, a larger time step could be used for integration with lumped mass matrix. This particular advantage is no longer clear when an implicit integration scheme is employed.

Multi-Step Methods for First-Order Equations (Gear 1971). The multi-step methods for first-order equations are used to integrate equations of the form

$$(18) \qquad\qquad \dot{\mathbf{y}} = \mathbf{f}(\mathbf{y}, t).$$

One can use the multi-step method to integrate Eq. (1) by converting it to a set of first-order equations

$$\dot{\mathbf{y}} = \begin{bmatrix} \ddot{\mathbf{q}} \\ \dot{\mathbf{q}} \end{bmatrix} = \begin{bmatrix} \dot{\mathbf{v}} \\ \dot{\mathbf{q}} \end{bmatrix} = \mathbf{f}(\mathbf{y}, t),$$

$$\mathbf{f}(\mathbf{y}, t) = \begin{bmatrix} \mathbf{M} & 0 \\ 0 & \mathbf{I} \end{bmatrix}^{-1} \left(\begin{bmatrix} \mathbf{F} \\ 0 \end{bmatrix} - \begin{bmatrix} \mathbf{C} & \mathbf{K} \\ \mathbf{I} & 0 \end{bmatrix} \mathbf{y} \right).$$

The integration is defined by

(19) $$\sum_{i=0}^{k} [\alpha_i \mathbf{y}_{n+1-i} + \beta_i \Delta t \mathbf{f}(\mathbf{y}_{n+1-i}, t_{n+1-i})] = 0,$$

where α's and β's are constant parameters. The method is called explicit if $\beta_0 = 0$; otherwise it is implicit.

Many algorithms of practical interest take the form of Eq. (19). For example, with $k = 1$, $\alpha_0 = -\alpha_1 = 1$, $\beta_0 = \alpha$ and $\beta_1 = 1 - \alpha$, Eq. (19) becomes a one-step *trapezoidal method* in the form

(20) $$\mathbf{y}_{n+1} - \mathbf{y}_n = \Delta t [\alpha \mathbf{f}(\mathbf{y}_{n+1}, t_{n+1}) + (1 - \alpha)\mathbf{f}(\mathbf{y}_n, t_n)].$$

The forward difference method corresponds to $\alpha = 0$ and is an explicit method; and the backward difference method corresponds to $\alpha = 1$ and is an implicit method.

For linear system, $\mathbf{f}$ can be written in the form

$$\mathbf{f}(\mathbf{y}, t) = \mathbf{H}\mathbf{y} + \mathbf{F},$$

where $\mathbf{H}$ and $\mathbf{F}$ are matrix functions independent of $\mathbf{y}$. The stability of the multi-step method for constant $\mathbf{H}$ can be determined as follows: let

(21) $$\mathbf{y}_j = \mathbf{a}_\lambda \eta^j,$$

where $\mathbf{a}_\lambda$ is the eigenvector of $\mathbf{H}$ associated with the *eigenvalue* λ and η is a parameter. Substituting Eq. (21) into Eq. (19) and setting $\mathbf{F} = 0$ (the particular solution has no effect on stability), we obtain the stability equation

$$\sum_{i=0}^{k} (\alpha_i + \Delta t \lambda \beta_i) \eta^{n+1-i} = 0.$$

If the magnitude of all the roots η of the stability equations is less than one for all λ, the integration scheme is stable. Usually it is the maximum eigenvalue $\mathbf{H}$ that controls stability. Dahlquist (1963) showed that

- No explicit unconditionally stable multi-step method exists.
- No 3rd-order accurate unconditionally stable multi-step method exists.
- The trapezoidal rule has the smallest error constant among the unconditionally stable multi-step methods that are 2nd order accurate.

For one-step method, the stability equation gives

$$\eta = [1 + (1 - \alpha)\lambda\Delta t]/(1 - \alpha\lambda\Delta t).$$

We see that the backward difference method ($\alpha = 1$) is unconditionally stable if $\text{Re}(\lambda) < 0$ for all the eigenvalues λ of $\mathbf{H}$; while the forward difference method ($\alpha = 0$) is stable only if $|1 + \lambda\Delta t| \leq 1$ for all eigenvalues.

Park (1975) proposed a second-order accurate 3-step implicit method, which is unconditionally stable with good accuracy in the low frequencies and strong dissipation in high frequencies. The Park method is defined by

$$k = 3, \quad \alpha_0 = -1, \quad \alpha_1 = 1.5, \quad \alpha_2 = -0.6,$$

$$\alpha_3 = 0.1, \quad \beta_0 = 0.6, \quad \beta_1 = \beta_2 = \beta_3 = 0.$$

Multi-Step Methods for Second-Order Equations. Consider a system of second-order equations in the form

$$(22) \qquad \ddot{\mathbf{y}} = \mathbf{f}_1(\mathbf{y}, t)\dot{\mathbf{y}} + \mathbf{f}_0(\mathbf{y}, t).$$

Most of structural dynamics equations can be put in this form. The multi-step method is defined by (Geradin 1974)

$$(23) \qquad \sum_{i=0}^{k} [\alpha_i \mathbf{y}_{n+1-i} + \beta_i \Delta t \mathbf{f}_1(\mathbf{y}_{n+1-i}, t_{n+1-i}) \mathbf{y}_{n+1-i}$$

$$+ \gamma_i \Delta t^2 \mathbf{f}_0(\mathbf{y}_{n+1-i}, t_{n+1-i})] = 0.$$

One sees that the Newmark method is a two-step method. The stability properties of Eq. (23) can be examined in a similar fashion (Krieg 1973).

Generally, conditionally stable algorithms require a time step that is of the order of the shortest vibration period of the structure (the inverse of the highest frequency). Thus one often is forced to use time steps much smaller than those needed for accuracy. Therefore, unconditionally stable algorithms may be preferable, especially if only low-frequency response is of interested. In this case, it would be desirable if the algorithm can effectively damp out the high-frequency modes. Assessment of numerical dissipation can be found in Hughes (1983).

There are many time integration schemes such as the Wilson θ-method (Wilson 1968), Rungi-Kutta predictor-corrector method and the α-method (Hilber *et al.* 1977). Readers are referred to literature (e.g., Hughes 1987) for the properties of various integration schemes.

18.18. NUMERICAL INTEGRATION

Derivation of element matrices involves line, area, or volume integral. With exception of simple elements exact integration could be very complicated. Numerical integration is essential almost for all practical applications. We shall discuss briefly the principle of numerical integration and give the tables of numerical coefficients for commonly used schemes.

If a function $u(x)$ is approximated by shape functions that $u(x) \approx \sum h_j(x)q_j$ with q's being parameters associated with the values of u at selected points, then the integral of u over a domain $-1 \le x \le 1$ is approximately

$$\int_{-1}^{1} u(x)\,dx \cong \sum W_j q_j ,$$

where

$$W_j = \int_{-1}^{1} h_j(x)\,dx .$$

In the ordinary rules of integration – generally called quadrature formula – q's are the values of u at the selected locations. There are many quadrature rules. The *Gauss quadrature* stands out as being generally more accurate for a fixed number of integration points and is described below.

One-Dimensional Integration. The limits of integration can always be transformed to become from -1 to $+1$. If we approximate u by the mth order *Hermite interpolation*, we have

$$I = \int_{-1}^{1} u(\xi)\,d\xi \approx \sum_{j=1}^{m} W_j u(\xi_j) + \sum_{j=1}^{m} W_j' \frac{du(\xi_j)}{d\xi} ,$$

where

$$W_j = \int_{-1}^{1} h_{2j-1}(\xi)\,d\xi , \qquad W_j' = \int_{-1}^{1} h_{2j}(\xi)\,d\xi ,$$

in which h's are defined in Eq. (18.3:12) and W's are called the *weighting factors*. If the stations $\xi_1, \xi_2, \ldots, \xi_m$ are chosen such that all W_j' vanish, then

(1)
$$I = \int_{-1}^{1} u(\xi)\,d\xi \approx \sum_{j=1}^{m} W_j u(\xi_j)$$

Table 18.18:1. Stations and weighting factors of Gauss quadrature.

No. of stations	ξ_i (Station locations)	weighting factors (W_j)	Order	Error
1	0	2	linear	$O(\varepsilon^2)$
2	$\pm 1/\sqrt{3}$	1	quadratic	$O(\varepsilon^4)$
3	0	8/9	cubic	$O(\varepsilon^6)$
	$\pm\sqrt{3/5}$	5/9		

and the resulting formula is called the mth order Gauss quadrature. Such a quadrature represents a function by a polynomial of degree $2m - 1$. The *degree-of-accuracy* of a quadrature is the degree of the highest order complete polynomial that can be integrated exactly by the formula. Hence $2m - 1$ is the degree-of-accuracy of a m-point quadrature and the error of the approximated integration is thus of the order ε^{2m}. Table 18.18:1 gives the stations and the weighting factors of the Gauss quadrature of different order. Any integration of a one-dimensional function can be approximated now as the summation of the products of the function's value at selected stations and their corresponding weighting factor.

Two- and Three-Dimensional Integration. Multi-dimensional Gauss integration rules are formed by successive application of the one-dimensional Gauss rule. In two-dimension, for a square domain, we have

$$(2) \qquad I = \int_{-1}^{1} \int_{-1}^{1} u(\xi, \eta) \, d\xi \, d\eta \approx \int_{-1}^{1} \sum_i W_i u(\xi_i, \eta) \, d\eta$$

$$\approx \sum_i \sum_j W_i W_j u(\xi_i, \eta_j) \,,$$

where W's are the same as those of one-dimensional integration. For Eq. (18.14:15), the integrand $\mathbf{B}^T \mathbf{D}_e \mathbf{B} J$ is a $n \times n$ symmetric matrix for an element of n degrees-of-freedom. The matrix has $n(n + 1)/2$ distinguished entries which are functions of ξ and η. We would have to integrate all entries individually. A 4-node quadrilateral in plane elasticity has 8 degrees-of-freedom and would require 36 separate integration for the determination of $\mathbf{k}$.

In three-dimension, the integration rule has the similar form

$$(3) \qquad I = \int_{-1}^{1} \int_{-1}^{1} \int_{-1}^{1} u(\xi, \eta, \zeta) \, d\xi \, d\eta \, d\zeta$$

$$\approx \sum_i \sum_j \sum_k W_i W_j W_k u(\xi_i, \eta_j, \zeta_k) \,.$$

A mth order Gauss quadrature can correctly integrate any term $\xi^i \eta^j \varsigma^k$, $i, j, k \leq 2m - 1$. For integral of Eq. (18.16:17), an 8-node hexahedron for 3-dimensional elasticity has 24 degrees-of-freedom. There are 300 distinguished entries of $\mathbf{B}^T \mathbf{DBJ}$ to be integrated individually. A 20-node hexahedron has 60 degrees-of-freedom with 1891 distinguished entries. The required number of integration increases rapidly for 3-dimensional element of high order. In practice, it is not necessary to use the same number of weighting station in all directions of integration. One may use the appropriate weights and number of Gauss points according to needs.

The Gauss quadrature is most popular for the integration of one-dimensional, quadrilateral and brick elements where the domains of integration in natural coordinates are $-1 \leq \xi \leq 1$, $-1 \leq \xi, \eta \leq 1$ and $-1 \leq \xi, \eta, \varsigma \leq 1$, respectively. More efficient integration rules can be designed to integrate complete polynomials (Irons 1971) of order $2m-1$. The integration takes the form

$$(4) \quad \int_{-1}^{1} \int_{-1}^{1} \int_{-1}^{1} u(\xi, \eta, \varsigma) \, d\xi \, d\eta \, d\varsigma = Au(0,0,0) \quad \text{(1 term)}$$

$$+ B[u(-b,0,0) + u(b,0,0) + u(0,-b,0) + \ldots] \quad \text{(6 terms)}$$

$$+ C[u(-c,-c,-c) + u(c,-c,-c) + u(c,c,-c) + \ldots] \quad \text{(8 terms)}$$

$$+ D[u(-d,-d,0) + u(d,-d,0) + \ldots$$

$$+ u(d,0,-d) + u(-d,0,-d) + \ldots] \quad \text{(12 terms)} .$$

The weighting factors A, B, C, D and the corresponding coordinates are listed in Table 18.18:2. From the truncation errors, we see that, to correctly

Table 18.18:2. Weighting factors and coordinates for Eq. (18.18:4).

Number of integration points	Coordinates	Weighting factors*	Order	Error
1		$A = 8$	Linear	$O(\varepsilon^2)$
6	$b = 1$	$B = 8/6$	Cubic	$O(\varepsilon^4)$
14	$b = 0.795822426$	$B = 0.886426593$	Quintic	$O(\varepsilon^6)$
	$c = 0.758786911$	$C = 0.335180055$		
27		$A = 0.788073483$	Hept	$O(\varepsilon^8)$
	$b = 0.848418001$	$B = 0.499362002$		
	$c = 0.652816472$	$C = 0.478508449$		
	$d = 1.106412899$	$D = 0.032303742$		

*unlisted weighting factors are zero

integrate a given complete order polynomial, the *Irons formulae* require less number of integration points than the Gauss quadrature.

Problem 18.17. Derive the equivalence of the Irons formulae for square: Determine (a) the number of points needed for integration; (b) the weight factors: and (c) the corresponding coordinates of the integration points.

Table 18.18:3. Numerical integration stations and weights for triangles.

Figure	Integration Points	Triangular coordinates	Weights	Order	Error
	a	1/3,1/3,1/3	1	Linear	$O(\varepsilon^2)$
	a	1/2,1/2,0	1/3	Quadratic	$O(\varepsilon^3)$
	b	0,1/2,1/2	1/3		
	c	1/2,0,1/2	1/3		
	a	1/3,1/3,1/3	−27/48	Cubic	$O(\varepsilon^4)$
	b	0.6, 0.2, 0.2			
	c	0.2, 0.6, 0.2	25/48		
	d	0.2, 0.2, 0.6			
	a	1/3,1/3,1/3	0.2250000000	Quintic	$O(\varepsilon^6)$
	b	a_1, b_1, b_1			
	c	b_1, a_1, b_1	0.1323941527		
	d	b_1, b_1, a_1			
	e	a_2, b_2, b_2			
	f	b_2, a_2, b_2	0.1259391805		
	g	b_2, b_2, a_2			

$$a_1 = 0.0597158717$$
$$b_1 = 0.4701420641$$
$$a_2 = 0.7974269853$$
$$b_2 = 0.1012865073$$

Numerical Integration for Triangles or Tetrahedrons. A triangle can be treated as a degenerated quadrilateral. Hence, an integral over a triangular area can be obtained by the Gauss quadrature. However, it is not advisable to use the degenerated integration involving many integration stations because a singularity occurs in mapping a triangle to a square, which will lead to loss of accuracy. An alternative is to apply the numerical integration approach described for one-dimensional integration directly to an triangular area

$$I = \int_0^1 \int_0^{1-\eta} u d\xi \, d\eta \approx \sum_j W_j u_j .$$

The limits of integration now involve the independent variables themselves. A series of sampling stations, weights and degrees of precision for different integration formulae (Hammer *et al.* 1956, Felippa 1966, Cowper 1973) is given in Table 18.18:3. Similarly one can derive the formulae for tetrahedrons. The results are presented in Table 18.18:4 (Hammer *et al.* 1956).

Table 18.18:4. Numerical integration stations and weights for tetrahedrons.

Figure	Integration Points	Tetrahedral coordinates	Weights	Order	Error
	a	1/4, 1/4, 1/4, 1/4	1	Linear	$O(\varepsilon^2)$
	a	$\alpha, \beta, \beta, \beta$	1/4	Quadratic	$O(\varepsilon^3)$
	b	$\beta, \alpha, \beta, \beta$	1/4		
	c	$\beta, \beta, \alpha, \beta$	1/4		
	d	$\beta, \beta, \beta, \alpha$	1/4		
		$\alpha = 0.58541021$			
		$\beta = 0.13819660$			
	a	1/4, 1/4, 1/4, 1/4	$-4/5$	Cubic	$O(\varepsilon^4)$
	b	1/3, 1/6, 1/6, 1/6	9/20		
	c	1/6, 1/3, 1/6, 1/6	9/20		
	d	1/6, 1/6, 1/3, 1/6	9/20		
	e	1/6, 1/6, 1/6, 1/3	9/20		

Required Order of Numerical Integration. Numerical integration is a source of errors in the finite element method. The accuracy of integration is proportional (not necessary linearly) to the number of integration points. As numerical integration can consume a significant amount of computer time, it is of interest to determine (a) the minimum integration points required for convergence and (b) the number of integration points necessary to preserve the order of convergence of the method, which would result if exact integration were used.

To establish the minimum number of integration points, we first note that, for convergence, the integration must be able to reproduce the size of the domain as all elements approach zero. This is obviously a necessary condition, otherwise the functional will not converge to a constant when the integrand of the functional is a constant. This means the error in evaluating $\int_{-1}^{1} d\xi$, $\int_{A_e} d\xi \, d\eta$, $\int_{V_e} d\xi \, d\eta \, d\zeta$ by the integration formulae must be of the order $O(\varepsilon)$ or higher. All quadrature formulae cited in the tables meet this requirement. Convergence of the finite element method based on an integral formulation can occur providing arbitrary constant value of the integrand of the functional can be reproduced. Thus, any integration with order of error $O(\varepsilon)$ suffices.

The finite element method approximates the functional to the order $2(p-m)$ where p is the order of complete polynomial of the shape functions and m the order of differentials appearing in the functional. If the error of the integration is at least $O[\varepsilon^{2(p-m)+1}]$, no loss of the order of convergence will occur. Thus, integration with an order of accuracy higher than the approximation of the functional by the shape functions will not change the order of convergence of the overall solution. In fact, it will show later that using integration of the order higher than actually needed under (b) may sometime affect adversely the accuracy of the finite element solution.

18.19. PATCH TESTS

We have presented finite element methods involving displacements that violate the admissibility conditions of the weak form formulation. The general theory can no longer assure the convergence of the weak form solution. Irons (1966, 1966a) first introduced the concept of *patch test* to examine the correctness of the finite element formulation. The patch test has now been generalized to determine whether or not an element employing nonstandard features such as *reduced integration, incompatible interpolations* satisfies the condition for convergence. Even though compatible (i.e., all assumed field variables meet the admissibility requirements) displacement, hybrid and mixed elements automatically pass the patch test, the assumed field variables may contain deformation modes not called for, one still can use the test to check the correctness of the finite element formulation and computer implementation, as Irons originally intended.

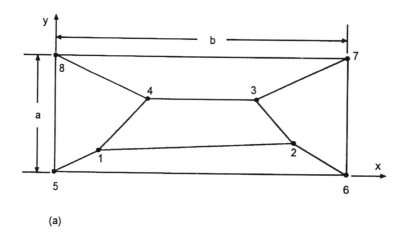

(a)

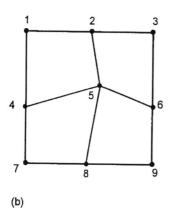

(b)

Fig. 18.19:1. A two-dimensional patch of elements.

A *patch* is a small group of interconnecting irregularly shaped elements. Figure 18.19:1 shows two typical 2-dimensional patches frequently used in benchmark test (Robinson and Blackham 1979, Hughes 1987). An element passes the patch test if its shape functions can represent exactly the linear states of deformation for a second order system. Plane elasticity is a two-dimensional second order system and the linear states of deformation are simply 1, x, and y.

For plane elasticity, there are two versions of patch test. Version 1 specifies the displacements

(1) $$u = a_1 + a_2 x + a_3 y, \qquad v = a_4 + a_5 x + a_6 y$$

at all exterior nodes [nodes 5–8 for Fig. 18.19:1(a) or nodes 1–4 and 6–9 for Fig. 18.19:1(b)] for arbitrary a's. There are six independent choices of these constants. Thus one has to solve six displacement-boundary-value problems in order to examine all deformations of the six linear states. The patch test is passed if, in the limit of the element size approaching zero, (1) the displacement solutions at all interior nodes [nodes 1–4 for Fig. 18.19:1(a) or node 5 for Fig. 18.19:1(b)] are exact, i.e., give the values u, v of Eq. (1) evaluated at these nodes, and (2) the displacement gradients, consequently the strains and stresses, are exact within each element, i.e., $\partial u/\partial x = a_2, \partial u/\partial y = a_3, \partial v/\partial x = a_5, \partial v/\partial y = a_6$.

For strain and stress to be exact, it implies that the interior nodal forces are zero, i.e.,

(2) $$\sum_{\text{all patch elements}} \int_{A_e} \mathbf{q}^T (\mathbf{d}_e \mathbf{h})^T \sigma \, dA = 0 \quad \text{or}$$

$$\sigma^T \left\{ \sum_{\text{all patch elements}} \left[\int_{A_e} \mathbf{d}_e \mathbf{h} \, dA \right] \mathbf{q} \right\} = 0$$

for any arbitrary q's associated with the interior nodes of the patch, where $\sigma^T (= [\sigma_{xx} \; \sigma_{yy} \; \sigma_{xy}])$ is constant. This is equivalent to the requirement of self-equilibrium at all interior nodes. If the generalized coordinates q'_j with their associated shape functions h'_j are defined within a single element only, from Eq. (2), we obtain

(3) $$\left(\int_{A_e} \mathbf{d}_e \mathbf{h}' dA \right) \mathbf{q}' = 0$$

for the element. This is a necessary condition for any arbitrary $\mathbf{q}'$ parameters and their associated shape function to pass the constant strain patch test. Parameters associated with bubble functions or some incompatible shape functions are such type of generalized coordinates.

In version 2, the test involves two parts. First, rigid body modes are tested. For a patch in plane elasticity as shown in Fig. 18.19:2, the rigid body motion is specified

$$u = a_1 + a_2 y, \qquad v = a_3 - a_2 x$$

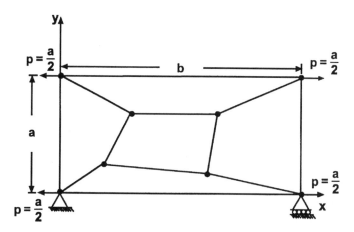

Fig. 18.19:2. Patch test with applied loads.

at all nodes, interior as well as exterior. The resulting nodal forces must be zero at all nodes for arbitrary a's [Eq. (2)]. In the second part of the test, we impose a minimum number of exterior constraints, just enough to restrain the rigid body motion, and apply appropriate loads to the remaining exterior nodes that will produce a constant state of stress over the patch. The patch test is passed if the finite element solution achieves the desired constant stress for all elements.

The second part of the test requires the calculation of the appropriate nodal loads corresponding to the desired state of deformation. These loads are applied to the unconstrained exterior nodes. This is quite easy for linear deformation. For plane elasticity, the linear deformation that satisfies the displacement constraints shown in Fig. 18.19:2 is

$$u = e_{xx}x + \alpha e_{xy}y , \quad v = e_{yy}y + (1 - \alpha)e_{xy}x ,$$

where e's are the three independent constant engineering strains and α is an arbitrary constant. The nodal forces at the exterior nodes are

$$\sum_{\text{all patch elements}} \int_{\partial A_{\sigma e}} \mathbf{h}^T \begin{bmatrix} n_x\sigma_{xx} + n_y\sigma_{xy} \\ n_x\sigma_{xy} + n_y\sigma_{yy} \end{bmatrix} ds ,$$

where σ's are the stresses associated with the three strains e_{xx}, e_{yy}, e_{xy} given above, n's are the components of a unit normal to the exterior boundaries and $\partial A_{\sigma e}$ is the portion of element boundaries coinciding with the exterior boundaries of the patch. The h's involve only the shape functions associated with nodes on $\partial A_{\sigma e}$. For elements with straight edges,

$n_x\sigma_{xx} + n_y\sigma_{xy}$ and $n_x\sigma_{xy} + n_y\sigma_{yy}$ are constant. The integration above can be carried out easily.

A patch test can be applied to a single element. In this case, the test can determine whether strains within the element are correctly evaluated from nodal displacements or forces but it cannot determine whether the element is compatible with its adjacent element. Therefore, a single element patch test is less useful than a multi-element patch test. However, a single element test is good to detect the existence of spurious zero energy modes, which are usually induced by the approximate integration of element matrices. Zero-energy modes will be discussed in more detail later.

We can similarly establish higher order match tests, e.g., requiring that the shape functions exactly represent quadratic states of deformation. This is needed for higher order equations such as bending of thin plates and shells. In this case, the curvatures and twists are proportional to second derivatives of the out-of-plane displacement. Quadratic test is a minimum requirement for fourth order equations since quadratic deformation represents only constant curvature and twist. One generally has to solve many more boundary-value problems for higher order patch test.

If the material is homogeneous, the patch test will be independent of element size for plane elasticity.

18.20. LOCKING-FREE ELEMENTS

The incorrect interpolation of a displacement field and its derivatives from its values at nodes causes the failure of patch tests. There are two sources of errors. Approximation for higher order functions occurs naturally since the finite element method uses mostly polynomials of finite order. Errors are also introduced through the nonlinearity of isoparametric coordinate transformation (Tong and Rossettos 1977, MacNeal 1994). For example, for an 8-node quadrilateral element, the shape functions contain only complete quadratic polynomials in ξ, η. However, the transformation

$$x = \sum_{j=1}^{8} x_j h_j(\xi, \eta), \quad \text{etc.},$$

given in Eq. (18.13:3) includes all quadratic and some higher order terms of ξ, η. Then x^2, xy, y^2 will involves terms like $\xi^2\eta^2, \eta\xi^3$, etc. Thus the shape functions cannot represent x^2, xy, y^2 exactly within a general quadrilateral.

Interpolation failure has other consequences. Among the more serious ones is the phenomenon of locking, a condition of excessive stiffness. Locking often occurs in bending caused by spurious shear. It also occurs in problems involving incompressible or nearly incompressible materials resulted from incorrect representation of dilatation. Discussions of the subject can be found in Hughes (1987), Babuska and Suri (1990), MacNeal (1994). Babuska's work identifies parameters characterizing locking and its effects on convergent rates.

Increasing the order of shape functions generally reduces the disorder due to lower interpolation errors and thus reduces the seriousness of locking or rids it completely for regular elements. Unfortunately, this is usually not true for elements involving nonlinear coordinate transformation. For example, isoparametric transformation introduces quadratic interpolation errors for the 8-node quadrilateral element (Tong and Rossettos 1977). Barlow (1989) gave a comprehensive analysis of errors due to shape distortions for 8-node quadrilaterals and 20-node bricks.

As mentioned before, locking is the result of incorrect representation of some, not all, strain components. However, there are locations within the element where the values of strains are correct. Thus, evaluating the incorrectly represented strain components only at the locations where their values are correct could eliminate locking. This procedure is, therefore, called *selectively reduced integration*. If one reduces the number of integration points for all strain components, the procedure is called *uniformly reduced integration*. Uniformly reduced integration generally introduces *spurious zero energy modes*, i.e., nonrigid body motion that contributes zero energy, which will be discussed in the next section, while selectively reduced integration may retain sufficient number of evaluations of strain components to prevent spurious mode. Reduced integration reduces the number of evaluation in the integration.

Reduced integration for 4-node quadrilateral was first introduced by Doherty *et al.* (1969). More discussion of selectively reduced integration can be found in MacNeal (1978), Malkus and Hughes (1978), Hughes (1980), Belytschko *et al.* (1981, 1984). Hybrid and mixed formulations are alternatives to circumvent locking. The full potential of the hybrid and mixed approaches is yet to be explored. We shall illustrate shear and dilatation locking by examples below and discuss locking-free elements.

Shear Locking. Consider a 4-node orthotropic rectangle ($-L \leq x \leq L$, $-1 \leq y \leq 1$) subjected to bending shown in Fig. 18.20:1 A bending displacement field is

$$(1) \qquad\qquad u = xy, \quad v = -x^2/2,$$

which gives a constant curvature along the x-axis. The corresponding strains, stresses, and strain energy per unit thickness for the *plane stress*

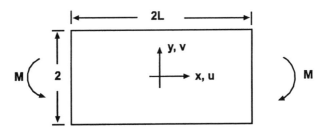

Fig. 18.20:1. Four-node rectangular element subjected to in-plane bending.

element are

$$e_x = u_{,x} = y\,, \quad e_y = v_{,y} = 0\,, \quad \gamma_{xy} = u_{,y} + v_{,x} = 0\,,$$

$$\sigma_x = E_x y/(1 - \nu_{xy}\nu_{yx})\,, \quad \sigma_y = E_x \nu_{xy} y/(1 - \nu_{xy}\nu_{yx})\,, \quad \sigma_{xy} = 0\,,$$

$$W = \frac{1}{2}\int_{-1}^{1}\int_{-L}^{L} \sigma^T \mathbf{e}\,dx\,dy = \frac{2 E_x L}{3(1 - \nu_{xy}\nu_{yx})}\,,$$

where E_x is Young's modulus and ν_{xy}, ν_{yx} are Poisson's ratios. The 4-node finite element representation (bilinear) of the displacement field is

(2) $$u_e = xy\,, \quad v_e = -L^2/2\,.$$

One can verify that Eqs. (1) and (2) have same displacements at the four corner nodes. The finite element representation gives

$$e_x = y\,, \quad e_y = 0\,, \quad \gamma_{xy} = x\,,$$

$$\sigma_x = \frac{E_x y}{1 - \nu_{xy}\nu_{yx}}\,, \quad \sigma_y = \frac{E_x \nu_{xy} y}{1 - \nu_{xy}\nu_{yx}}\,, \quad \sigma_{xy} = G_{xy} x\,,$$

$$W_e = \frac{1}{2}\int_{-1}^{1}\int_{-L}^{L} \sigma^T \mathbf{e}\,dx\,dy = \frac{2 E_x L}{3(1 - \nu_{xy}\nu_{yx})}\left[1 + \frac{G_{xy}}{E_x}(1 - \nu_{xy}\nu_{yx})L^2\right]\,,$$

where G_{xy} is the shear modulus. Obviously, the shear strain, the shear stress and the strain energy per unit thickness are not correct. The ratio of the strain energy of the finite element representation to the correct strain energy is $1 + \frac{G_{xy}}{E_x}(1 - \nu_{xy}\nu_{yx})L^2$, which represents the ratio of finite element bending stiffness to the correct bending stiffness. The ratio is proportional to L^2 and thus is large even for a moderately slander element (moderately large L). It is also large if G_{xy}/E_x is large, even if the slander ratio is of order 1. The high stiffness of the element is mainly from the incorrect shear strain. This phenomenon is called *shear locking*, which precludes the use of regular 3-dimensional displacement elements to represent thin plates or shells by simply reducing the dimension of the 3-dimensional element in the thickness direction. This is due to the characteristics of bending deformation that the maximum displacement ratio, Max (v/u), is the same order as the slander ratio of the element. The situation is further exacerbated if G_{xy}/E_x is large for orthotropic materials. More on the elimination or minimizing shear locking in plate and shell applications will be discussed in the next chapter.

An effective remedy to shear locking is to use the value of γ_{xy} at $x = y = 0$ for all integration points (Doherty *et al.* 1969), since at $x = y = 0$, the finite element representation of shear is correct for such deformation. This technique is called selectively reduced-integration because, in effect, we employ a single point to evaluate the shear strain energy and the 2×2 or higher order Gauss integration rule to evaluate the strain energy contribution from e_x and e_y. Complication arises if the shear strain is elastically coupled to the direct strains, i.e., $D_{13}, D_{23} \neq 0$, where D's are the components of the elastic modulus matrix defined in Eq. (18.14:2). In evaluating $(D_{13}e_x + D_{23}e_y)\gamma_{xy}$, one would have to use the shear strain value at the center for those at the 2×2 Gauss points. Another remedy is to use nonconforming element by adding a deformation mode proportional to x^2 (Bazeley *et al.* 1966, Taylor *et al.* 1976) in the y-direction. One can also use the hybrid formulation to avoid the spurious shear strain modes (Razzaque 1973).

Similarly, one can show that the 8-node rectangle also locks, but to a much lesser degree. For the bending deformation

$$u = x^2 y \,, \qquad v = -x^3/3 \,, \qquad e_x = 2xy \,, \qquad e_y = e_{xy} = 0$$

the finite element representation is

$$u = x^2 y \,, \quad v = -L^2 x/3 \,, \quad e_x = 2xy \,, \quad e_y = 0 \,, \quad \gamma_{xy} = x^2 - L^2/3 \,.$$

The dominant locking mode is a result of the spurious shear strain. Since the finite element representation of the shear strain is correct at the Gauss stations $(\xi_i = \pm 1/\sqrt{3})$ of the 2×2 integration rule, the problem can be corrected by using the 2×2 uniformly reduced integration (Zienkiewicz *et al.* 1971) for all strains. The situation can be corrected also by the 2×2 selectively reduced rule for the shear energy and the 3×3 integration rule for the bending energy. Note that uniformly reduced integration introduces rank deficiency. If the shear strain couples elastically with the tensile strains, one would have to extrapolate the shear strain from the 2×2 Gauss points to the 3×3 Gauss points ($\xi_i = 0, \pm\sqrt{0.6}$). Unfortunately the extrapolation makes the element fail the patch test. However, it generally still provides good results in many practical applications.

The 8-node brick element also locks when its dimension in one direction is much less than the other two dimensions. Shear locking can be avoided by using one-point integration for the shear strain energy. The constant strain triangle experiences interpolation errors for all quadratic or higher order displacements. As a result, it locks for in-plane bending deformation. The constant strain triangle is of little practical use for such situation. The 6-node triangular element does not lock for any quadratic deformation because it can correctly interpolate such displacements. However, for elements with curve sides, locking returns due to errors in the nonlinear isoparametric transformation.

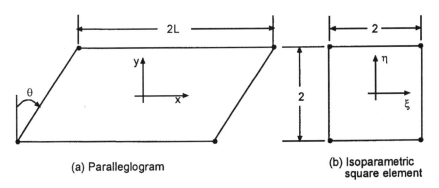

(a) Parallelglogram

(b) Isoparametric square element

Fig. 18.20:2. Parallelogram element.

Reduced integration often fails to fix the locking of general distorted elements. For example, consider the parallelogram shown in Fig. 18.20:2 with the displacement field (MacNeal 1994)

$$(3) \qquad\qquad u = xy, \quad v = -x^2/2 \,.$$

The strains are $e_x = y$ and $e_y = \gamma_{xy} = 0$. The transformation to the natural coordinates gives

$$x = L\xi + \eta \tan\theta \,, \qquad y = \eta \,,$$

$$u = L\xi\eta + \eta^2 \tan\theta \,, \qquad v = -(L^2\xi^2 + 2L\xi\eta \tan\theta + \eta^2 \tan^2\theta)/2 \,.$$

The 4-node isoparametric finite element representation is

$$u = L\xi\eta + \tan\theta, \qquad v = -(L^2 + 2L\xi\eta\tan\theta + \tan^2\theta)/2.$$

with the strain field

$$e_x = \eta, \quad e_y = \tan\theta(-L\xi + \eta\tan\theta), \quad \gamma_{xy} = L\xi - 2\eta\tan\theta.$$

Clearly γ_{xy} is in error and will lead to locking if L is large compare to one. In the mean time the error in e_y will also be significant and contribute to locking if $L\tan\theta$ is large. The error in strains cannot be eleminated by the selectively reduced integration of γ_{xy} at $\xi = \eta = 0$. On the other hand, if one evaluates all strains at the origin to eliminate the error of e_y and γ_{xy}, the element would have no stiffness at all for in-plane bending. Similar results can be shown for a 4-node trapezoid.

Mixed Formulation. An alternative to reduced integration is to use mixed formulation. To simplify the discussion, we shall consider isotropic materials only. We separate the functional of Eq. (18.14:6) into three parts

$$(4) \qquad \Pi_e = \Pi_{de} + \Pi_{se} - \Pi_{fe}$$

where Π_{de}, Π_{se}, Π_{fe} are the distortion energy, the shear energy and the external work, respectively. For plane stress,

$$(4a) \qquad \Pi_{de} = \int_{A_e} \frac{E}{2(1-\nu)^2}(u_{,x}^2 + v_{,y}^2 + 2\nu u_{,x}v_{,y})\,dxdy,$$

$$(4b) \qquad \Pi_{se} = \int_{A_e} \frac{G}{2}(u_{,y} + v_{,x})^2\,dxdy,$$

$$(4c) \qquad \Pi_{fe} = \int_{A_e} (ub_x + vb_y)\,dxdy + \int_{\partial A_{\sigma e}} (u\bar{T}_x + v\bar{T}_y)\,dxdy.$$

Consider a 4-node rectangular element of height $2l$ and length $2h$. The displacements can be written in terms of natural coordinates ($x = h\xi, y = l\eta$) as

$$(5) \qquad u = a_1 + a_2\xi + a_3\eta + a_4\xi\eta, \quad v = b_1 + b_2\xi + b_3\eta + b_4\xi\eta,$$

where

$$(6) \quad \begin{aligned} a_1 &= (u_1 + u_2 + u_3 + u_4)/4, & a_2 &= (-u_1 + u_2 + u_3 - u_4)/4, \\ a_3 &= (-u_1 - u_2 + u_3 + u_4)/4, & a_4 &= (u_1 - u_2 + u_3 - u_4)/4, \end{aligned}$$

in which u's are the nodal values of u. Similarly one can express b's in terms of the nodal values of v. From the strain displacement relation, one finds

$$(7) \qquad \gamma_{xy} = \frac{a_3}{l} + \frac{b_2}{h} + \frac{a_4}{l}\xi + \frac{b_4}{h}\eta.$$

The shear energy within the element is

$$(8) \qquad \Pi_{se} = \frac{G}{2}4hl\left[\left(\frac{a_3}{l} + \frac{b_2}{h}\right)^2 + \frac{1}{3}\left(\frac{a_4}{l}\right)^2 + \frac{1}{6}\left(\frac{b_4}{h}\right)^2\right].$$

At the high aspect ratio $(h/l \to \infty)$ limit, the shear energy approaches zero. This is equivalent to forcing

$$(9a) \qquad a_3/l + b_2/h \to 0,$$

$$(9b) \qquad a_4 \to 0, \quad b_4 \to 0.$$

These three conditions together over constrain the system and lead to shear locking. Prathap (1993) argued that if a constraint, such as Eq. (9a), involves coefficients from all related field variables, it can be correctly enforced in the limit. The representation of the strains by those terms (like a_3, b_2) are said to be *field-consistent*. If a constraint, such as Eq. (9b), has no contribution from one or more related field variables, it may incorrectly constrain the deformation and cause locking. Such terms are called *field-inconsistent representation*.

We can remove the spurious constraints using the mixed or hybrid formulation. Let us replace the shear energy Π_{se} of Eq. (4b) by the Reissner–Hellinger functional [Eq. (10.10:9)]

$$(4d) \qquad \Pi_{se} = \int_{A_e}\left[G\tilde{\gamma}_{xy}(u_{,y} + v_{,x}) - \frac{G}{2}\tilde{\gamma}_{xy}^2\right]dxdy,$$

in which $\tilde{\gamma}_{xy}$ is a new independent field variable for the element. From the stationary condition of Π_{se} with respect to $\tilde{\gamma}_{xy}$,

$$(10) \qquad \delta\Pi_{se} = \int_{A_e} G[u_{,y} + v_{,x} - \tilde{\gamma}_{xy}]\delta\tilde{\gamma}_{xy}\,dxdy = 0,$$

we can establish the relation between $\tilde{\gamma}_{xy}$ and $u_{,y} + v_{,x}$. Judicious selection of the shear strain $\tilde{\gamma}_{xy}$ can remove the spurious constraint(s) and alleviate locking. An element with $\tilde{\gamma}_{xy}$ being field-consistent is called a *field-consistent element*, otherwise a *field-inconsistent element*.

4-Node Field-Consistent Rectangular Element. For a 4-node rectangle, assuming constant $\tilde{\gamma}_{xy}$ within the element, from Eqs. (10) and (6), one can show that

(11) $\tilde{\gamma}_{xy} = \dfrac{a_3}{l} + \dfrac{b_2}{h}$ or

$$\tilde{\gamma}_{xy} = \frac{1}{4l}(-u_1 - u_2 + u_3 + u_4) + \frac{1}{4h}(-v_1 + v_2 + v_3 - v_4),$$

A substitution of $\tilde{\gamma}_{xy}$ into Eq. (4d) yields

(12) $\Pi_{se} = 2Ghl \left(\dfrac{-u_1 - u_2 + u_3 + u_4}{4l} + \dfrac{-v_1 + v_2 + v_3 - v_4}{4h} \right)^2 .$

Using Π_{se} above, together with Π_{de}, Π_{fe} of Eqs. (4a) and (4c), one can drive the element matrices. Equation (12) is equivalent to evaluating the shear strain energy Π_{se} from Eq. (4b) by one point integration. The modified element performs well for $h/l \gg 1$ in modeling bending of thin beams. For stress calculation, *one should use field-consistent $\tilde{\gamma}_{xy}$ since the shear strain derived from the assumed displacements [Eq. (7)] often shows rapid linear variation within an element.* Note that Wilson's 4-node incompatible rectangular element discussed in Sec. 18.14 is field-consistent.

Generally field-inconsistent quadrilateral elements lock badly in modeling flexure of thin beams. In this case, optimum field-consistency is yet to be found. The modified version of Wilson's 4-node incompatible element (Sec. 18.14) provides satisfactory solution. *Field-consistency may not guarantee locking-free* as the shear strains of constant triangle and tetrahedron are field-consistent and yet they still lock in thin beams, plate and shell applications due to insufficient deformation modes.

8-Node Field-Consistent Rectangular Element. For 8-node elements, the displacements are

(13)
$$u = a_1 + a_2\xi + a_3\eta + a_4\xi\eta + a_5\xi^2 + a_6\eta^2 + a_7\xi^2\eta + a_8\xi\eta^2,$$
$$v = b_1 + b_2\xi + b_3\eta + b_4\xi\eta + b_5\xi^2 + b_6\eta^2 + b_7\xi^2\eta + b_8\xi\eta^2,$$

where a's and b's can be expressed in terms of the nodal values of u and v, respectively. The shear strain derived from the displacements is

(14) $\gamma_{xy} = \dfrac{a_3}{l} + \dfrac{b_2}{h} + \left(\dfrac{a_4}{l} + \dfrac{2b_5}{h} \right)\xi + \left(\dfrac{2a_6}{l} + \dfrac{b_4}{h} \right)\eta + \dfrac{a_7}{l}\xi^2$

$$+ 2\left(\dfrac{a_8}{l} + \dfrac{b_7}{h} \right)\xi\eta + \dfrac{b_8}{h}\eta^2 .$$

The spurious constraints are

$$a_7 \to 0, \quad b_8 \to 0.$$

Expanding Eq. (14) in terms of the Legrendre polynomial

$$(15) \qquad \gamma_{xy} = \frac{a_3}{l} + \frac{b_2}{h} + \frac{1}{3}\frac{a_7}{l} + \frac{1}{3}\frac{b_8}{h} + \left(\frac{a_4}{l} + \frac{2b_5}{h}\right)\xi$$

$$+ \left(\frac{2a_6}{l} + \frac{b_4}{h}\right)\eta + 2\left(\frac{a_8}{l} + \frac{b_7}{h}\right)\xi\eta$$

$$+ \frac{a_7}{l}\left(\xi^2 - \frac{1}{3}\right) + \frac{b_8}{h}\left(\eta^2 - \frac{1}{3}\right),$$

and using its orthogonal properties, one can show that the field-consistent shear strain is

$$(16) \qquad \tilde{\gamma}_{xy} = \frac{a_3}{l} + \frac{b_2}{h} + 3\frac{a_7}{l} + 3\frac{b_8}{h} + \left(\frac{a_4}{l} + \frac{2b_5}{h}\right)\xi$$

$$+ \left(\frac{2a_6}{l} + \frac{b_4}{h}\right)\eta + 2\left(\frac{a_8}{l} + \frac{b_7}{h}\right)\xi\eta.$$

Equation (16) is also the field-consistent shear strain of 9-node rectangular elements.

The shear stress should be calculated from $\tilde{\gamma}_{xy}$. Field-consistency usually gives smoother stresses in modeling flexure of thin beams. The shear stress calculated from the assumed displacements often shows rapid quadratic oscillation within an element (Prathap 1993).

8-Node Field-Consistent Brick Element (Fig. 18.15:2). In 3-dimensional linear theory of isotropic materials, the shear strain functionals are

$$(17a) \qquad \Pi_{se} = \int_{V_e} \frac{G}{2}(\gamma_{xy}^2 + \gamma_{yz}^2 + \gamma_{zx}^2)\,dxdydz,$$

$$(17b) \qquad \Pi_{se} = \int_{V_e} G[(\tilde{\gamma}_{xy}\gamma_{xy} + \tilde{\gamma}_{yz}\gamma_{yz} + \tilde{\gamma}_{zx}\gamma_{zx})$$

$$- \frac{1}{2}(\tilde{\gamma}_{xy}^2 + \tilde{\gamma}_{yz}^2 + \tilde{\gamma}_{zx}^2)]\,dxdydz,$$

for displacement and mixed formulations, respectively, where γ's are shear strains derived from the assumed displacements. For an 8-node brick

$(2h \times 2l \times 2d)$, the trilinear displacements are:

(18) $u = a_1 + a_2\xi + a_3\eta + a_4\varsigma + a_5\xi\eta + a_6\eta\varsigma + a_7\varsigma\xi + a_8\xi\eta\varsigma$,

$v = b_1 + b_2\xi + b_3\eta + \ldots$, $w = c_1 + c_2\xi + c_3\eta + \ldots$,

where a', b' and c's are related to the nodal values of u, v and w, respectively. The shear strain γ_{xy} derived from the displacements is

$$\gamma_{xy} = \frac{a_3}{l} + \frac{b_2}{h} + \left(\frac{a_6}{l} + \frac{b_7}{h}\right)\varsigma + \frac{a_5}{l}\xi + \frac{b_5}{h}\eta + \frac{a_8}{l}\xi\varsigma + \frac{b_8}{h}\eta\varsigma.$$

In the limit of $\gamma_{xy} \to 0$ in bending applications to slender beams, all polynomial coefficients vanish. Of all the constraints,

$$a_5 \to 0, \quad a_8 \to 0, \quad b_5 \to 0, \quad b_8 \to 0,$$

are spurious and can cause locking. One can similarly obtain other spurious constraints by considering γ_{yz} and γ_{zx}. It can be shown that the field-consistent shear strains are

(19) $\tilde{\gamma}_{xy} = \dfrac{a_3}{l} + \dfrac{b_2}{h} + \left(\dfrac{a_6}{l} + \dfrac{b_7}{h}\right)\varsigma$, $\quad \tilde{\gamma}_{yz} = \dfrac{b_4}{d} + \dfrac{c_3}{h} + \left(\dfrac{b_7}{d} + \dfrac{c_5}{h}\right)\xi$,

$\tilde{\gamma}_{zx} = \dfrac{c_2}{l} + \dfrac{a_4}{d} + \left(\dfrac{c_5}{l} + \dfrac{a_6}{d}\right)\eta$.

Another alternative to reduce locking is to use incompatible elements. Introducing bubble functions in the displacements, one can write

(20) $u = a_1 + a_2\xi + a_3\eta + a_4\varsigma + a_5\xi\eta + a_6\eta\varsigma + a_7z\xi + a_8\xi\eta\varsigma$

$$+ a_9(1 - \eta^2) + a_{10}(1 - \varsigma^2) + a_{11}(1 - \xi^2),$$

and similar v and w. The last three terms are added for mitigating dilatation locking to be discussed later. It can be shown that the displacement in Eq. (20) is incompatible but the derived shear and dilatation strains are field-consistent.

27-Node Brick Element. The 20-node brick element performs poorly in slender beam, thin plate and thin shell applications. It is difficult to find the field-consistent strains that will make the element robust in those applications. For a 27-node brick as shown in Fig. 18.15:2, the displacements are triquadratic functions of the natural coordinates ξ, η, ς. The derived shear strains γ's are also triquadratic. It can be shown that the field-consistent shear strains are bilinear in ξ, η and quadratic in ζ for $\tilde{\gamma}_{xy}$, bilinear in η, ζ and quadratic in ξ for $\tilde{\gamma}_{yz}$, and so on. Similar to Eq. (15), if γ_{xy} is expanded in terms of the Legendre polynomial of ξ, η, from orthogonal properties of the polynomial, then $\tilde{\gamma}_{xy}$ is in the form similar to Eq. (16) which involves, the zeroth and first order Legendre polynomials in ξ, η. Similarly one can derive $\tilde{\gamma}_{yz}$ and $\tilde{\gamma}_{zx}$. The proof is left for the readers.

Using the field-consistent $\tilde{\gamma}$'s in Eq. (17b) is equivalent to evaluating the shear strain energy from Eq. (17a) with the $2 \times 2 \times 3$ integration rule for γ_{xy}, the $2 \times 3 \times 2$ rule for γ_{yz} and so on. Again stresses derived from the assumed displacements shown rapid quadratic oscillations within the element. Thus stresses should be calculated from the field-consistent $\tilde{\gamma}$'s at the nodes.

If the edge and surface nodes of the brick element (nodes 9 to 26, Fig. 18.15:2) are not located at the middle of the edges or the center of the sides, the isoparametric transformation of the coordinates involves nonlinear mapping. The $\tilde{\gamma}$'s or the integration rules derived in the last paragraph are no longer field-consistent. Establishing proper $\tilde{\gamma}$'s become formidable. Using Eq. (17a) with the $2 \times 2 \times 2$ integration rule for all γ's is simple and practical even at the risk of introducing zero energy modes, which in some rare occasions may lead to poor results. Stresses should be evaluated at the integration points and extrapolated to nodes.

Locking is a result of interpolation error. Higher order elements usually do not lock or locks to a lesser extent. Selectively reduced integration does wonder in removing shear locking for rectangular elements but is ineffective for general quadrilaterals as isoparametric mapping introduces spurious strain modes other than shear (MacNeal 1990). One may have to resort to the field-consistent approach or incompatible elements to reduce locking. The effective of reduced integration being limited to rectangles may not be serious in practice. Normally it is the spurious out-of-plane shears that cause locking in bending of thin beams, plates and shells. In practical applications, the sides normal to the midplane are generally rectangular or nearly rectangular for which shear locking can be effectively eliminated by selectively reduced integration. We shall consider the general distorted brick elements for thin shell applications in the next chapter.

Problem 18.18. Consider a trapezoid as shown in the figure with the displacements given in Eq. (18.20:1). The transformation of Cartesian coordinates to the natural coordinates is

$$x = L\xi(1 - \alpha\eta)\,, \quad y = \eta\,, \quad \text{where } 0 < \alpha < 1\,.$$

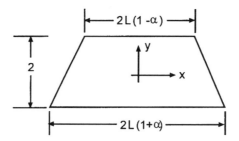

Fig. P18.17. Trapezoid.

(a) Determine the finite element representations of the displacement and strain fields in terms of the natural coordinates. From the result, show that locking cannot be eliminated by reduced integration. (b) Compute the ratio of the strain energy of the finite element representation to the correct strain energy of the element.

Dilatation Locking (MacNeal 1994). The problem is commonly associated with nearly incompressible materials, i.e. $\nu \to 1/2$. Consider the in-plane bending of a rectangle with

$$u = xy, \quad v = -x^2/2 - \nu y^2/2(1-\nu)$$

$$e_x = y, \quad e_y = -\nu y/(1-\nu), \quad \gamma_{xy} = 0,$$

(21)

$$\sigma_x = Ey/(1-\nu^2), \quad \sigma_y = 0, \quad \sigma_{xy} = 0,$$

$$W = \frac{1}{2}\int_{-1}^{1}\int_{-L}^{L}\sigma^T e\, dx\, dy = \frac{2EL}{3(1-\nu^2)},$$

in *plane strain*. The dilatation is

$$e = e_x + e_y + e_z = (1-2\nu)y/(1-\nu),$$

which tends to zero as ν approaches 0.5.

For 4-node rectangular element the deformation is represented by

$$u = xy, \quad v = -L^2/2 - \nu/2(1-\nu),$$

$$e_x = y, \quad e_y = 0, \quad \gamma_{xy} = x,$$

$$\sigma_x = \frac{E(1-\nu)y}{(1+\nu)(1-2\nu)}, \quad \sigma_y = \frac{E\nu y}{(1+\nu)(1-2\nu)}, \quad \sigma_{xy} = \frac{Ex}{2(1+\nu)},$$

$$e = e_x + e_y + e_z = y,$$

$$W_e = \frac{2EL}{3(1+\nu)}\left(\frac{1-\nu}{1-2\nu}+\frac{L^2}{2}\right),$$

in which the dilatation is independent of Poisson's ratio. This error in dilatation magnifies the strain energy by a factor proportional to $1/(1-2\nu)$, which is large as ν approaches 0.5. The error in the shear strain thus contributes to high stiffness if L is lange.

Nagtegaal *et al.* (1974) introduced a strain-projection scheme to treat the nearly incompressible case. The strain-displacement matrix $\mathbf{B}$ as defined in Eq. (18.16:12) can be written in the form

$$(22) \qquad \mathbf{B} = \lfloor \mathbf{B}_1^{\text{dev}} + \mathbf{B}_1^{\text{dil}} \quad \mathbf{B}_2^{\text{dev}} + \mathbf{B}_2^{\text{dil}} \quad \cdots \quad \mathbf{B}_p^{\text{dev}} + \mathbf{B}_p^{\text{dil}} \rfloor,$$

where p is the number of nodes of the element and $\mathbf{B}_i^{\text{dev}}, \mathbf{B}_i^{\text{dil}}$ denote, respectively, the deviation and dilatation parts of $\mathbf{B}$ associated with node i, and are defined as

$$(23) \qquad \mathbf{B}_i^{\text{dev}} = \frac{1}{3} \begin{bmatrix} 2B_{xi} & -B_{yi} & -B_{zi} \\ -B_{xi} & 2B_{yi} & -B_{zi} \\ -B_{xi} & -B_{yi} & 2B_{zi} \\ B_{yi} & B_{xi} & 0 \\ 0 & B_{zi} & B_{yi} \\ B_{zi} & 0 & B_{xi} \end{bmatrix},$$

$$\mathbf{B}_i^{\text{dil}} = \frac{1}{3} \begin{bmatrix} B_{xi} & B_{yi} & B_{zi} \\ B_{xi} & B_{yi} & B_{zi} \\ B_{xi} & B_{yi} & B_{zi} \\ 0 & 0 & 0 \\ 0 & 0 & 0 \\ 0 & 0 & 0 \end{bmatrix},$$

$$B_{\alpha i} = \frac{\partial h_i}{\partial \alpha}, \qquad \alpha = x, y, z.$$

For nearly incompressible materials, one modifies the values of $\mathbf{B}_i^{\text{dil}}$ by approximating it as a constant equal to its value at the center of the element or its average value over the element. One can also approximate $\mathbf{B}_i^{\text{dil}}$ by the interpolated function that interpolates the values of $\mathbf{B}_i^{\text{dil}}$ at the Gauss points based on a reduced integration rule. Unfortunately, these techniques cannot eliminate dilatation locking completely.

An alternative to remove dilatation locking is to use the mixed or hybrid formulation. Again we shall use isotropic materials to illustrate the idea. Separating Π_{de} of Eq. (4a) into two parts,

$$(24) \qquad\qquad \Pi_{de} = \Pi_{ve} + \Pi_{ae} \,,$$

where Π_{ae}, Π_{ve} are the strain energies associated with the deviatoric strains and the dilatation e, respectively, i.e.,

$$(24a) \quad \Pi_{ae} = \frac{1}{2} \int_{V_e} \frac{E}{1+\nu} \left[\left(u_{,x} - \frac{1}{3} e \right)^2 + \left(v_{,y} - \frac{1}{3} e \right)^2 \right.$$

$$\left. + \left(w_{,z} - \frac{1}{3} e \right)^2 \right] dxdydz \,,$$

$$(24b) \quad \Pi_{ve} = \frac{1}{2} \int_{V_e} \frac{E}{3(1-2\nu)} e^2 \, dxdydz \qquad \text{(displacement model)}\,,$$

$$(24c) \quad \Pi_{ve} = \int_{V_e} \frac{E}{3(1-2\nu)} \left(\tilde{e}e - \frac{1}{2}\tilde{e}^2 \right) dxdydz \qquad \text{(mixed model)}\,,$$

in which $e = u_{,x} + v_{,y} + w_{,z}$. In addition to u, v and w, $\tilde{e}$ is now another independent field variable. From the stationary condition of Eq. (24c) with respect to $\tilde{e}$,

$$(25) \qquad\qquad \delta\Pi_{ve} = \int_{V_e} \frac{1}{3(1-2\nu)} \delta\tilde{e}(e - \tilde{e}) \, dxdydz = 0 \,,$$

we can determine $\tilde{e}$.

8-Node Brick. For an 8-node element, from Eq. (18), we can calculate

$$e = \frac{a_2}{h} + \frac{b_3}{l} + \frac{c_4}{d} + \left(\frac{b_5}{l} + \frac{c_7}{d} \right) \xi + \left(\frac{a_5}{h} + \frac{c_6}{d} \right) \eta + \left(\frac{b_6}{l} + \frac{a_7}{h} \right) \varsigma$$

$$+ \frac{a_8}{h} \eta\varsigma + \frac{b_8}{l} \varsigma\xi + \frac{c_8}{d} \xi\eta \,.$$

As the material approaches incompressible, all coefficients of the equations above tend to zero. Following the field-consistent argument, we conclude that, of all the constraints, only

$$\frac{a_2}{h} + \frac{b_3}{l} + \frac{c_4}{d} \to 0$$

is genuine while all others are spurious. Assuming constant $\tilde{e}$, from Eq. (25), we find the field-consistent dilatation:

$$\tilde{e} = \frac{a_2}{h} + \frac{b_3}{l} + \frac{c_4}{d} \,,$$

which is in terms of the nodal values of u, v and w. A substitution of $\tilde{e}$ into Eq. (24c) yields

$$\Pi_{ve} = \frac{1}{2}\frac{8Elhd}{3(1-2\nu)}\left(\frac{a_2}{h} + \frac{b_3}{l} + \frac{c_4}{d}\right)^2.$$

This is equivalent to evaluating the dilatation energy from Eq. (24b) by one-point integration. Using Π_{ve} above, together with Π_{ae} of Eq. (24a) and Π_{se} of Eq. (17a) or (17b), we can establish the element stiffness matrix in terms of the nodal values of u, v and w.

This element is able to respond to nearly incompressible deformation. However, this simple dilatation is not sufficient to represent flexure action, which involves linear strain variation. A remedy is to introduce in the displacements incompatible bubble functions such as the last three terms in Eq. (20). The dilatation derived from the assumed displacements is field-consistent and can be used directly in Eq. (24b) to construct the element stiffness matrix. Such an element is free from dilatation locking and able to represent bending.

27-Node Brick. The dilatation e of a 27-node brick is a triquadratic function of the natural coordinates. It can be shown that the constraints associated with ξ^2, η^2 and ζ^2 are spurious. Thus the field-consistent dilatation is simply a trilinear function of ξ, η, ζ. Expanding e in terms of the Legendre polynomial in ξ, η, ζ and dropping all the terms proportional to $1 - \xi^2/3$, $1 - \eta^2/3$ and $1 - \zeta^2/3$ give the field-consistent $\tilde{e}$.

More on incompressible and nearly incompressible materials will be discussed in Sec. 19.6. For distorted bricks, field-consistent dilatation is yet to be determined.

Problem 18.19. Use the displacement field in Eq. (18.20:21) to examine the dilatation locking of a 4-node rectangular element in plane stress.

Problem 18.20. For 27-node $2h \times 2l \times 2d$ brick elements, the displacements can be written in the form

$$u = \sum_{i,j,k=0}^{2} a_{ijk}\xi^i\eta^j\zeta^k, \quad v = \sum_{i,j,k=0}^{2} b_{ijk}\xi^i\eta^j\zeta^k, \quad w = \sum_{i,j,k=0}^{2} c_{ijk}\xi^i\eta^j\zeta^k,$$

where a', b', c's are constants in terms of the nodal values of u, v, w. Derived the field-consistent shear strains. Hint:

$$\gamma_{xy} = \sum_{i,j,k=0}^{2} \left(\frac{j}{l} a_{ijk} \xi^{i} \eta^{j-1} + \frac{i}{h} b_{ijk} \xi^{i-1} \eta^{j} \right) \varsigma^{k}$$

$$= \sum_{k=0}^{2} \sum_{j=1}^{2} \frac{j}{l} \varsigma^{k} \eta^{j-1} \left[\left(a_{0jk} + \frac{1}{3} a_{2jk} \right) + a_{1jk} \xi + a_{2jk} \left(\xi^{2} - \frac{1}{3} \right) \right] + \dots .$$

18.21. SPURIOUS MODES IN REDUCED INTEGRATION (MacNeal 1994)

MacNeal defined a working definition of reduced integration. If p is the degree of complete polynomials in the element shape functions, the Gauss integration with $p + 1$ or more points in each direction is full and the integration with p points or less in a direction is reduced. Reduced order of integration has many benefits. Obviously, it reduces computation time as the number of multiplication in the calculation of element matrices is proportional to the number of integration point. In the last section, we showed that reduced integration is a potential remedy for many locking problems resulting from interpolation failure. Barlow (1968, 1976) noticed that strains computed at the reduced-order Gauss points in rectangular elements were more accurate than strains computed at other points. Unfortunately, reduced integration has its shortcomings. For one, it is less accurate and can increase the integration error. Also, more seriously, it may introduce *spurious zero energy deformation modes*, which can destroy the accuracy of the finite element analysis.

The spurious modes are the eigenvectors of the stiffness matrix associated with zero eigenvalue but nonzero strains. They are distinguished from a rigid body mode as the latter is a zero eigenvector associated with a zero strain state. Since strains are evaluated at integration points, a non-zero strain will give zero strain energy if the strain is zero at those points. Thus, the number of strain evaluations at all integration points represents the number of constraints for the occurrence of spurious modes. Let S_e and S_s be the number of strain evaluations and the number of independent strain states provided by the nodal displacements, respectively. Spurious modes occur if $S_e < S_s$. The difference between the two quantities is the maximum number of spurious modes that can exist for the element. The S_e equals the number of strain components times the total number of integration points. In plane elasticity, there are 3 strains. We have

$$S_e = 3p^2,$$

for $p \times p$ integration and

$$S_e = 2(p+1)^2 + p^2 \,,$$

for selectively reduced integration with $(p+1) \times (p+1)$ integration for direct strains and $p \times p$ for shear. For 3-dimensional elasticity, the number of strain components is 6 and $S_e = 6p^3$ for $p \times p \times p$ integration.

In the mean time, S_s equal to the total number of degrees-of-freedom minus the number of rigid body modes. The 8-node serendipity element in plane elasticity has 16 degrees-of-freedom and 3 rigid body modes. Therefore, $S_s = 13$. If we use the 2×2 reduced integration rule, $S_e = 12$. Then, there will be at least 1 spurious zero energy mode.

Spurious Modes in Quadrilateral Elements. For a 4-node plane quadrilateral element, $S_s = 5(= 8 - 3)$. The 2×2 Gauss integration provides 12 strain evaluations so that there will be no spurious modes in the element. On the other hand, $S_e = 3$ for one-point integration. Thus, the element will have at least two spurious modes, which are $u = \xi\eta$, $v = 0$ and $u = 0$, $v = \xi\eta$. Similarly, one can show that, for an 8-node brick with one integration point, there are 12 spurious modes of which one of the displacements u, v, or w equals $\xi\eta$, $\eta\zeta$, $\zeta\xi$, or $\xi\eta\zeta$ while others are zero.

We can find the spurious modes for higher order Lagrange elements under the $p \times p$ Gauss integration rule. Consider a 2-dimensional displacement field in the form

(1) $$u = f_p(\xi)f_p(\eta)\,, \quad v = 0$$

for a $(p+1)^2$-node quadrilateral elements, where $f_p(\xi)$ is a Legendre polynomial of degree p, which vanishes at the p Gauss integration points. The displacements u and v vanish at all the Guass points for $p \times p$ integration. So are the corresponding strains

$$e_x = u_{,x} = f_p'(\xi)f_p(\eta)\xi_{,x} + f_p(\xi)f_p'(\eta)\eta_{,x}\,,$$

$$e_y = v_{,y} = 0\,,$$

$$\gamma_{xy} = u_{,y} + v_{,x} = f_p'(\xi)f_p(\eta)\xi_{,y} + f_p(\xi)f_p'(\eta)\eta_{,y}\,.$$

The displacement field given in Eq. (1) is a spurious mode. The other spurious mode is given by the displacements of Eq. (1) with u and v interchanged.

Another type of spurious mode is given by

(2)
$$u = f_p'(\xi)f_p(\eta)\xi_{,x} - f_p(\xi)f_p'(\eta)\eta_{,x}\,,$$

$$v = f_p'(\xi)f_p(\eta)\xi_{,y} - f_p(\xi)f_p'(\eta)\eta_{,y}\,,$$

provided that the isoparametric mapping is linear, i.e., $\xi_{,x}$, $\xi_{,y}$, $\eta_{,x}$, $\eta_{,y}$ are constant. This is the case of rectangle or parallelogram. For example, for an 8-node ($p = 2$) rectangle, $\xi_{,y} = \eta_{,x} = 0$ and $\xi_{,x}, \eta_{,y}$ are constant. The spurious mode is

$$u = \xi(\eta^2 - 1/3)\xi_{,x}, \quad v = -\eta(\xi^2 - 1/3)\eta_{,y},$$

where the Gauss integration points are $\xi, \eta = \pm 1/\sqrt{3}$.

Spurious Modes in Hexahedral Elements. For $(p+1)^3$-node Lagrange hexahedral elements in 3-dimensional elasticity, one type of spurious modes of $p \times p \times p$ reduced integration is in the form

$$
\begin{aligned}
u &= f_p(\xi)f_p(\eta)(a_1 + \cdots + a_{p+1}\varsigma^p) + f_p(\eta)f_p(\varsigma)(a_{p+2} + \cdots + a_{2p+1}\xi^{p-1}) \\
&\quad + f_p(\varsigma)f_p(\xi)(a_{2p+2} + \cdots + a_{3p+1}\eta^{p-1}) \\
v &= 0, \quad w = 0,
\end{aligned}
$$

where the isoparametric mapping is linear and a's are constants. The other types of spurious mode are given by

$$
\begin{aligned}
u &= [f_p'(\xi)f_p(\eta)\xi_{,x} - f_p(\xi)f_p'(\eta)\eta_{,x}](a_{3p+2} + \cdots + a_{4p+1}\varsigma^{p-1}), \\
v &= [f_p'(\xi)f_p(\eta)\xi_{,y} - f_p(\xi)f_p'(\eta)\eta_{,y}](a_{3p+2} + \cdots + a_{4p+1}\varsigma^{p-1}), \\
w &= 0,
\end{aligned}
$$

One can obtain other spurious modes by interchanging u, ξ with v, η, respectively, or u, ξ with w, ζ. For $p = 2$, the 27-node brick element, there are 21 spurious modes of the first type and 6 of the second type.

The determination of spurious zero energy modes resulting from reduced integration for triangular elements and elements involving nonlinear isoparametric mapping (the Jacobian of transformation is not constant) is not trivial. One can always determined the spurious modes from the zero eigenvectors of the element stiffness matrix numerically.

If a spurious mode of a single element exists also in the assembly of elements, it is called a *communicating spurious mode*. If the same type of a spurious mode exists over the entire domain, then it is a *global spurious mode*. On the other hand if the same mode occurs over only part of the domain, it is called the *local mode*. If the spurious mode exists only in the single element, the mode is *non-communicating*. Communicating spurious mode is a serious flaw in practical applications. If it exists after boundary constraints are imposed, the global stiffness matrix is singular and the finite element solution is not unique or is very sensitive to small changes in loading or boundary condition. It can render the solution useless (Irons and Hellen 1978, Bic'anic' and Hinton 1979). MacNeal (1994) showed that many of type 1 spurious modes above are communicating. Those in plane strain or plane stress are global while those in three dimensions are local. On the other hand, type 2 spurious modes are generally non-communicating. There is at least one exception — a beam modeled by reduced integration of 20-node brick elements with displacements normal to the beam. There are as many spurious modes as the number of elements plus or minus one, depending on end conditions.

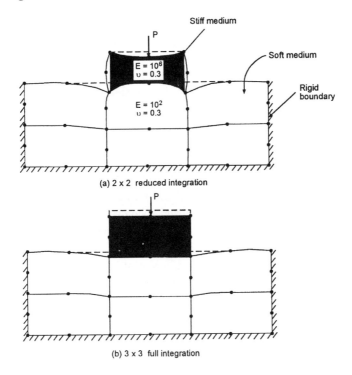

Fig. 18.21:1 Propagation of spurious mode from reduced-integration elements.

Spurious modes can also cause errors when there is a mixture of stiff and soft materials even when, strictly speaking, a global spurious mode does not exist. Consider the example of a stiff block on the top of a soft foundation shown in Fig. 18.21:1 (Zienkiewicz and Taylor, volume 1, 1989). The use of 2×2 quadrature throughout leads to the answer shown in Fig. 18.21:1(a) while the result for 3×3 quadrature given in Fig. 18.21:1(b) is an appropriate solution. Even though no zero-energy mode exists for the assembled system, the low resistance of the material in the foundation allows the penetration of spurious deformation into the soft elements.

We have given a broad coverage of the displacement model of the finite element method. The discussion focuses primarily on problems governed by linear second order differential equations. In the remaining chapters, we will examine mixed and hybrid finite element models, plates and shells (problems involve fourth order differential equations), and non-linear problems.

18.22. PERSPECTIVE

Since its introduction in the early 1960s, the finite element method has become one of the most important numerical method for analysis and design. The method has been widely applied to all fields of engineering, science and medicine. The mathematical foundation of the method has been well established and understood. Three-node triangular and 4-node quadrilateral membrane elements were first introduced because of their simplicity. The problem of locking in applications to bending of thin plates and shells and to incompressible or nearly incompressible materials was soon discovered. Reduced integration was introduced to alleviate the problem. Unfortunately such technique also brings with it zero energy deformation mode(s) which often requires specialized schemes to stabilize the stiffness matrix. Mixed and hybrid methods, such as the field-consistent approach, originally developed for circumventing the difficulty in meeting the C^1 continuity requirement for fourth order differential equations, are alternatives for mitigating locking and require further development.

In spite of the advent of computer technology, computational efficiency is still a consideration in engineering applications. Adaptive methods for remeshing and for the analysis of strain localization are needed. Simple and efficient elements are still essential to computational efficiency and accuracy especially in thin plate and shell applications. The quest for such perfect elements continues.

The greatest future of the finite element method lies in its applications. In the areas of metal forming and bioengineering, nonlinear material behaviors and large deformations are the essence. The material and thin-film technologies in microelectronics necessitate computational modeling at the atomic level. Materials can be tailored to optimize its strength, stiffness, fracture toughness and acoustic characteristics. The meso- (atomic level) and micro-mechanics of composite materials will make the finite element method indispensable in material development. Applications of smart materials in sensors and actuators increasingly require precise knowledge of material performance and reliability as miniaturization of these devices continues. Material modeling to examine slip planes and effects of grain boundaries of multiple crystal systems would be required. Material failure often associates with local material instability or strain localization including material separation or void growth. Finite element method is again indispensable in simulation and prediction of such phenomena.

19

MIXED AND HYBRID FORMULATIONS

So far we have considered mostly the displacement type of finite element formulation. There are other types of formulation that demand less stringent admissible requirements on continuity. Among them are the mixed and hybrid formulations, which provide flexibility to deal with problems involving singularities, infinite domain, and higher order differential equations. The displacement formulation is based on Eq. (18.14:6) or Eq. (18.16:10) and is an *irreducible formulation*, whereas the mixed and hybrid finite element formulations are based on the mixed and hybrid variational principles, which use multiple fields as dependent variables (see Sec. 10.3). An irreducible formulation is one for which none of the field variables can be eliminated without solving the differential equations or the weak form of equations. The hybrid formulation allows some field variables defined only along the element boundaries to be different from those within the element. The hybrid method is therefore most versatile. R. H. Gallagher proposed in 1981 to restrict the definition of the hybrid method as one which is formulated by multi-variable variational principles, yet the resulting matrix equations consist only nodal values of displacements as unknown. Pian (1995, 1998) surveyed and assessed the advances in hybrid/mixed finite element method.

19.1. MIXED FORMULATIONS

We can establish the finite element equations using the *Reissner-Hellinger principle* as given in Eqs. (10.10:8) and (10.10:9) that employs both stress and displacement as independent fields. We shall illustrate the derivation by considering problems of plane state. For plane stress or plane strain, the functional Π_{Re} of an element can be written in the form:

$$(1) \; \blacktriangle \qquad \Pi_{Re} = \int_{A_e} \left(\boldsymbol{\sigma}^T \mathbf{e} - \frac{1}{2} \boldsymbol{\sigma}^T \mathbf{C}_e \boldsymbol{\sigma} - \mathbf{u}^T \mathbf{b} \right) dA - \int_{\partial A_{\sigma e}} \mathbf{u}^T \bar{\mathbf{T}} \, ds,$$

in which $\mathbf{e}$ is the engineering *strain matrix* (**not tensor**, see comments in Sec. 18.14) in terms of the displacement $\mathbf{u}$, $\boldsymbol{\sigma}$ is the *stress matrix* (**not tensor**), $\mathbf{b}$ is the *body force per unit volume*, $\bar{\mathbf{T}}$ is the prescribed *boundary* *traction*, (Like last chapter, we use capital $\mathbf{T}$ to denote traction.) $\mathbf{C}_e$ is the elastic flexibility matrix, the inverse of the elastic modulus matrix $\mathbf{D}_e$ defined in Eq. (18.14:2), A_e is the area of the element, and ∂A_e is the boundary of A_e. For linearized theory of elasticity, $\mathbf{u}$, $\boldsymbol{\sigma}$, $\mathbf{e}$ are given in Eq. (18.14:1) for plane stress and strain, and in Eqs. (18.16:1)–(18.16:3) for 3-dimensional problems.

The functional for 3-dimensional elasticity is similar except that the area (over A_e) and the line (over ∂A_e) integrals in Eq. (1) are replaced by the volume (over V_e) and the surface (over ∂V_e) integrals, respectively. The variational functional for axisymmetric problems is similar. If the material is isotropic, then

$$\mathbf{C}_e = \frac{1}{E} \begin{bmatrix} 1 & & \text{sym} \\ -\nu & 1 & \\ 0 & 0 & 2(1+\nu) \end{bmatrix} \qquad \text{(for plane stress)},$$

$$\mathbf{C}_e = \frac{1+\nu}{E} \begin{bmatrix} 1-\nu & & \text{sym} \\ -\nu & 1-\nu & \\ 0 & 0 & 2 \end{bmatrix} \qquad \text{(for plane strain)},$$

$$\mathbf{C}_e = \frac{1}{E} \begin{bmatrix} 1 & & & & & \\ -\nu & 1 & & & \text{sym} & \\ -\nu & -\nu & 1 & & & \\ 0 & 0 & 0 & 2(1+\nu) & & \\ 0 & 0 & 0 & 0 & 2(1+\nu) & \\ 0 & 0 & 0 & 0 & 0 & 2(1+\nu) \end{bmatrix} \qquad \text{(3-D elasticity)}.$$

The displacement boundary condition given in Eq. (18.14:4) or (10.10:3) is a rigid condition whereas the traction condition given in Eq. (18.14:5) is natural. Using the stationary condition of the variational functional, one can derive the equilibrium equations in terms of stresses, the constitutive equations relating strains to stresses, and the traction conditions on $\partial A_{\sigma e}$ as those given in Eqs. (10.10:5)–(10.10:7) in index notation.

In finite element formulation, both the stresses and the displacements are approximated in the usual manner by interpolating their nodal parameters with appropriate shape functions. The admissibility requirement are that either the stresses or the displacements are C^0 continuous and that the displacements satisfy the rigid condition(s). The use of C^0 continuous stresses may result in higher accuracy than possible with the displacement formulation previously discussed (Zienkiewicz *et al.* 1985, 1985a). Poor choice of shape functions may result in poor results. Babuska (1971) and Brezzi (1974) formulated a complex criterion for the choice of dependent variables, known as the LBB condition. Tong and Pian (1969) argued in favor of requiring that the assumed stresses and strains derived from the assumed displacements have the same level of approximation. Fraeijs de Veubeke (1964, 1965), Fraeijs de Veubeke and Zienkiewicz (1967) showed that if the approximation for stresses is capable of reproducing precisely the same type of variation as that determinable from the displacement model, then the irreducible and mixed formulations will yield identical answers. In many cases, the mixed method can provide superior results for stresses by requiring continuity of traction between elements.

In mixed formulation, we assume the displacement in the usual manner by the matrix equation $\mathbf{u} = \mathbf{h}\mathbf{q}$ and derive the matrix equation for the strain $\mathbf{e} = \mathbf{B}\mathbf{q}$. We also independently assume the stresses in the form

$$\boldsymbol{\sigma} = \mathbf{h}_\sigma \bar{\boldsymbol{\sigma}} \,,$$

where $\bar{\boldsymbol{\sigma}}$ is the unknown parameter column matrix for the stress and $\mathbf{h}_\sigma$ is the corresponding shape function matrix, which is in general different from $\mathbf{h}$. Discontinuous $\mathbf{u}$ between elements is allowed, if the assumed $\boldsymbol{\sigma}$ is C^0 continuous. In terms of $\mathbf{q}$ and $\bar{\boldsymbol{\sigma}}$, Π_{Re} becomes

(2) ▲ $$\Pi_{\mathrm{Re}} = \bar{\boldsymbol{\sigma}}^T \mathbf{k}_{\sigma u} \mathbf{q} - \frac{1}{2}\, \bar{\boldsymbol{\sigma}}^T \mathbf{k}_\sigma \bar{\boldsymbol{\sigma}} - \mathbf{q}^T \mathbf{f} \,,$$

where

(3) ▲ $$\mathbf{k}_\sigma = \int_{A_e} \mathbf{h}_\sigma^T \mathbf{C}_e \mathbf{h}_\sigma \, dA\,, \quad \mathbf{k}_{\sigma u} = \int_{A_e} \mathbf{h}_\sigma^T \mathbf{B} dA \,,$$

$$\mathbf{f} = \int_{A_e} \mathbf{h}^T \mathbf{b} dA + \int_{\partial A_{\sigma e}} \mathbf{h}^T \overset{v}{\mathbf{T}} \, ds \,.$$

In these formulas, all products are matrix products. Note that $\mathbf{k}_\sigma$ is symmetric and positive definite and that $\mathbf{f}$ is in the same form as Eq. (18.14:14). One can assemble $\mathbf{k}_\sigma$, $\mathbf{k}_{\sigma u}$, $\mathbf{f}$ and enforce the displacement constraint(s) at the boundaries to derive the matrices of the entire system for finite element solution.

The system of algebraic equations for all unknown parameters is derived from the stationary condition of Π_R defined in Eq. (10.10:8) with appropriate constraints and symbolically can be written as

$$
\begin{bmatrix} -\sum \mathbf{k}_\sigma & \sum \mathbf{k}_{\sigma u} \\ \sum \mathbf{k}_{\sigma u}^T & 0 \end{bmatrix} \begin{bmatrix} \sum \bar{\sigma} \\ \sum \mathbf{q} \end{bmatrix} = \begin{bmatrix} 0 \\ \sum \mathbf{f} \end{bmatrix},
$$

where the summation denotes assembling of all elements. We borrow the terminology of the displacement model and call $\begin{bmatrix} -\sum \mathbf{k}_\sigma & \sum \mathbf{k}_{\sigma u} \\ \sum \mathbf{k}_{\sigma u}^T & 0 \end{bmatrix}$ the assembled or constrained stiffness matrix.

With the added independent field σ, generally there will be more algebraic equations. Though the assembled matrix is still symmetric, unlike the displacement model, it is no longer positive definite as all diagonal terms associated with displacement degrees-of-freedom are zero. These terms will become nonzero in the process of factorization or Gauss elimination if the unknowns are properly ordered and if the constrained stiffness matrix is nonsingular.

If σ is chosen independently for each element with the displacement continuous, $\bar{\sigma}$ can be eliminated from the element by the stationary condition of Π_e with respect to $\bar{\sigma}$,

$$
\mathbf{k}_\sigma \bar{\sigma} - \mathbf{k}_{\sigma u} \mathbf{q} = 0, \qquad \text{or} \qquad \bar{\sigma} = \mathbf{k}_\sigma^{-1} \mathbf{k}_{\sigma u} \mathbf{q}.
$$

Note that the matrix $\mathbf{k}_\sigma$ is symmetric and positive definite. Substituting $\bar{\sigma}$ above into Eq. (2) leads to the typical element functional,

$$
(4) \qquad\qquad \Pi_{\mathrm{Re}} = \frac{1}{2} \mathbf{q}^T \mathbf{k} \mathbf{q} - \mathbf{q}^T \mathbf{f},
$$

where

$$
(5) \qquad\qquad \mathbf{k} = \mathbf{k}_{\sigma u}^T \mathbf{k}_\sigma^{-1} \mathbf{k}_{\sigma u}
$$

is the *deduced element stiffness matrix*. One can then assembled the element matrices in the usual manner.

Obviously the deduced $\mathbf{k}$ is symmetric. Its rank depends on the form and number of assumed stress and displacement modes. Let n_σ and n_d be the number of unknown stress and displacement parameters, and r the number of rigid body modes associated with the shape functions $\mathbf{h}$. Then

$$
(6) \qquad\qquad \text{rank of } \mathbf{k} \leq \min(n_\sigma, n_d - r).
$$

The rank of $\mathbf{k}$ will be deficient and there will be spurious deformation mode(s) if $n_\sigma < n_d - r$.

19.2. HYBRID FORMULATIONS

A special mixed formulation is the *hybrid finite element method*, which uses different field variables in different parts of the domain. The hybrid concept was initiated by Pain (1964). Tong and Pian (1969) rationalized the approach using the variational formulation with the functional defined in Eq. (10.10:10) or (10.10:15). The concept was later generalized into different finite element models (Tong 1970, Pian and Tong 1969, 1972) and used in numerous applications (Tong and Fung 1971, Tong *et al.* 1973, Chen and Mei 1974, Pian and Chen 1982, Atluri 1980, 1980a, 1995).

Again, using plane elasticity as an example, we can write the hybrid element functional of Eq. (10.10:10) in a form of a matrix products equation:

$$(1) \quad \Pi_{\text{He}} = \int_{\partial A_e} (\hat{\mathbf{u}}^T - \mathbf{u}^T)\widehat{\mathbf{T}}\, ds$$

$$+ \int_{A_e} \left(\boldsymbol{\sigma}^T \mathbf{e} - \frac{1}{2}\boldsymbol{\sigma}^T \mathbf{C}_e \boldsymbol{\sigma} - \mathbf{u}^T \mathbf{b} \right) dA - \int_{\partial A_{\sigma e}} \hat{\mathbf{u}}^T \bar{\mathbf{T}}\, ds,$$

where $\widehat{\mathbf{T}}$ is the boundary traction for each element and $\hat{\mathbf{u}}$ is the inter-element boundary displacement. In other words, $\widehat{\mathbf{T}}$ of different elements are independent of each other while $\hat{\mathbf{u}}$ is the same for adjacent elements along their common boundaries and satisfies the condition

$$(2) \qquad\qquad \hat{\mathbf{u}} = \bar{\mathbf{u}} \qquad \text{on} \quad \partial A_u,$$

where the displacement is prescribed.

The hybrid functional Π_H given in Eq. (10.10:10) for the entire domain involves multiple fields, namely $\mathbf{u}$ and $\boldsymbol{\sigma}$ within each element (the strain is in terms of $\mathbf{u}$) as well as $\widehat{\mathbf{T}}$ and $\hat{\mathbf{u}}$ on the element boundaries. The admissibility requirements are: piecewise continuity of $\hat{\mathbf{u}}$ along inter-element boundaries, $\hat{\mathbf{u}} = \bar{\mathbf{u}}$, $\delta\bar{\mathbf{u}} = 0$ on ∂A_u, and smooth function for other field variables within the element.

The stationary condition of Π_H with respect to the admissible fields gives the Euler equation

$(3) \quad \boldsymbol{\sigma}_{,\mathbf{x}} + \mathbf{b} = 0$ (equilibrium equation in A_e),

$(4) \quad \mathbf{e} = \mathbf{C}_e \boldsymbol{\sigma}$ (constitutive law in A_e),

$(5) \quad \hat{\mathbf{u}} = \mathbf{u}$ (element displacement condition on ∂A_e),

$$(6) \quad \widehat{\mathbf{T}} = \begin{bmatrix} n_x \sigma_{xx} + n_y \sigma_{xy} \\ n_x \sigma_{xy} + n_y \sigma_{yy} \end{bmatrix} \qquad \text{(element traction condition on } \partial A_e),$$

(7) $\widehat{\mathbf{T}} = \bar{\mathbf{T}}$ (traction boundary condition on $\partial A_{\sigma e}$),

(8) $\displaystyle\sum_{\text{all elements}} \int_{\partial A_e - \partial A_{\sigma e}} \delta\hat{\mathbf{u}}^T \widehat{\mathbf{T}}\, ds = 0,$ (interelement equilibrium),

where $\mathbf{n}$ is the unit normal to ∂A_e. Equations (3)–(7) do not involve summation over elements because $\delta\widehat{\mathbf{T}}$, $\delta\mathbf{u}$ of different elements are independent of each other. On the other hand, $\delta\hat{\mathbf{u}}$ are common to adjacent elements and Eq. (8) implies inter-element equilibrium between two adjacent elements

(9) $\widehat{\mathbf{T}}_I + \widehat{\mathbf{T}}_{II} = 0\,,$

where the subscripts I and II signify that the quantities are associated with elements I and II, respectively. In the formulation above, inter-element compatibility for $\mathbf{u}$ is satisfied in an average sense through Eq. (5) and inter-element equilibrium for $\boldsymbol{\sigma}$ is achieved in average through Eqs. (6) and (8). The derivation of the element matrices based on Eq. (1) by independently interpolating all field variables in terms of their unknown parameters is straightforward. Since the inter-element admissibility requirement involves only $\hat{\mathbf{u}}$ on the element boundaries while imposes no restriction on $\mathbf{u}$ and $\boldsymbol{\sigma}$, one has great flexibility to select the shape functions to represent these variables. This alleviates many difficulties encountered in problems governed by fourth order differential equations and provides easy means to treat problems with singularities (Tong and Pian 1972). Also because stress is an independent field, it permits a better stress description to improve the stress approximation.

Through selectively enforcing Eqs. (3)–(9) *a priori*, the number of independent field variables of Eq. (1) can be reduced, which leads to different hybrid models to be discussed below.

Hybrid Stress Model. This is a widely used hybrid model originally proposed by Pian. Choosing an equilibrium stress $\boldsymbol{\sigma}$ that satisfies also the element traction condition Eq. (6) *a priori*, we can simplify Π_{He}. Denoting it as Π_{HSe}, we obtain

(10) ▲ $\displaystyle \Pi_{HSe} = \int_{\partial A_e} \hat{\mathbf{u}}^T \mathbf{T}\, ds - \int_{A_e} \frac{1}{2}\boldsymbol{\sigma}^T \mathbf{C}_e \boldsymbol{\sigma}\, dA - \int_{\partial A_{\sigma e}} \hat{\mathbf{u}}^T \bar{\mathbf{T}}\, ds\,,$

for the hybrid stress principle Eq. (10.10:15). To derive the element matrices, we shall consider only the case of zero body force, i.e., $\mathbf{b} = 0$. We assume $\boldsymbol{\sigma}$ in the form

(11) $\boldsymbol{\sigma} = \mathbf{h}_\beta \boldsymbol{\beta}\,,$

where

$$(12) \qquad \underset{3\times p}{\mathbf{h}_\beta} = \begin{bmatrix} 1 & 0 & 0 & y & 0 & x & 0 & \dots \\ 0 & 1 & 0 & 0 & x & 0 & y & \dots \\ 0 & 0 & 1 & 0 & 0 & -y & -x & \dots \end{bmatrix},$$

$$\boldsymbol{\beta}^T = [\beta_1 \ \beta_2 \ \dots \ \beta_p].$$

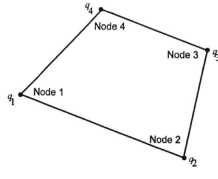

The matrix $\mathbf{h}_\beta$ is so chosen that $\boldsymbol{\sigma}$ satisfies the equilibrium equations for arbitrary $\boldsymbol{\beta}$. We then assume $\hat{\mathbf{u}}$ in terms of its nodal values along the element boundaries. An example of quadrilateral element is shown in Fig. 19.2:1 with (q_1, q_3, q_5, q_7) and (q_2, q_4, q_6, q_8) being the nodal values of $\hat{u}$, $\hat{v}$, respectively. Then $\hat{\mathbf{u}}$ can be written as

Fig. 19.2:1. Quadrilateral element.

$$(13) \qquad \hat{\mathbf{u}} = \begin{bmatrix} \hat{u} \\ \hat{v} \end{bmatrix} = \mathbf{hq},$$

where

$$\mathbf{h} = \begin{bmatrix} 1 - s/s_{1,2} & 0 & s/s_{1,2} & 0 & 0 & 0 & 0 & 0 \\ 0 & 1 - s/s_{1,2} & 0 & s/s_{1,2} & 0 & 0 & 0 & 0 \end{bmatrix}$$

(between nodes 1 and 2),

$$\mathbf{h} = \begin{bmatrix} 0 & 0 & 1 - s/s_{2,3} & 0 & s/s_{2,3} & 0 & 0 & 0 \\ 0 & 0 & 0 & 1 - s/s_{2,3} & 0 & s/s_{2,3} & 0 & 0 \end{bmatrix}$$

(between nodes 2 and 3),

$$\dots$$

$$\mathbf{h} = \begin{bmatrix} s/s_{4,1} & 0 & 0 & 0 & 0 & 0 & 1 - s/s_{4,1} & 0 \\ 0 & s/s_{4,1} & 0 & 0 & 0 & 0 & 0 & 1 - s/s_{4,1} \end{bmatrix}$$

(between nodes 4 and 1).

Here s is the distance along the element boundary and $s_{i,j}$ is the distance between nodes i and j. Note that $\hat{\mathbf{u}}$ has the same displacement along the element boundaries as that of a 4-node quadrilateral displacement element. Also, at the element boundaries, one has

$$\mathbf{T} = \mathbf{n}'^T \boldsymbol{\sigma} = \mathbf{n}'^T \mathbf{h}_\beta \boldsymbol{\beta} = \boldsymbol{\beta}^T \mathbf{h}_\beta \mathbf{n}',$$

where

$$\mathbf{n}'^T = \begin{bmatrix} n_x & 0 & n_y \\ 0 & n_y & n_x \end{bmatrix},$$

in which n's are the components of a unit normal to the boundaries. Recall that in Chapters 16 and 17, we defined traction as $\mathbf{T} = \mathbf{n} \cdot \boldsymbol{\sigma}$ in tensor notation, where $\boldsymbol{\sigma}$ is a second rank tensor and $\mathbf{n}$ a unit vector. Presently $\boldsymbol{\sigma}$ is defined in Eq. (18.14:1) as a column matrix. In matrix operation, we have to define n'^T as above to obtain the correct traction

$$\mathbf{T} = \begin{bmatrix} n_x \sigma_{xx} + n_y \sigma_{xy} \\ n_x \sigma_{xy} + n_y \sigma_{yy} \end{bmatrix}.$$

A substitution of Eqs. (11) and (13) into Eq. (10) yields

(14) $$\Pi_{HSe} = \boldsymbol{\beta}^T \mathbf{k}_{\beta q} \mathbf{q} - \frac{1}{2} \boldsymbol{\beta}^T \mathbf{k}_{\beta\beta} \boldsymbol{\beta} - \mathbf{q}^T \mathbf{f},$$

where

(15) ▲ $$\mathbf{k}_{\beta\beta} = \int_{A_e} \mathbf{h}_\beta^T \mathbf{C}_e \mathbf{h}_\beta dA, \qquad \mathbf{k}_{\beta q} = \int_{\partial A_e} \mathbf{h}_\beta^T \mathbf{n}' \mathbf{h}\, ds,$$

$$\mathbf{f} = \int_{\partial A_{\sigma e}} \mathbf{h}^T \bar{\mathbf{T}} ds.$$

The element force matrix $\mathbf{f}$ is the same as that of the displacement model. For a constant body force $\mathbf{b}^T = [b_x \ b_y]$,

$$\boldsymbol{\sigma} = \mathbf{h}_\beta \boldsymbol{\beta} + \boldsymbol{\sigma}_p = \mathbf{h}_\beta \boldsymbol{\beta} + [b_x x \ \ b_y y]^T.$$

There will be additional terms in $\mathbf{f}$ through $\boldsymbol{\sigma}_p$ (Tong and Rossettos 1977). Then, the element force matrices of the two approaches can be different.

In Eq. (15) $\boldsymbol{\beta}$ of different elements are independent of each other and be eliminated from the stationary condition of Π_{HSe} with respect to $\boldsymbol{\beta}$.

(16) $$\mathbf{k}_{\beta\beta} \boldsymbol{\beta} = \mathbf{k}_{\beta q} \mathbf{q} \qquad \text{or} \qquad \boldsymbol{\beta} = \mathbf{k}_{\beta\beta}^{-1} \mathbf{k}_{\beta q} \mathbf{q}.$$

Substituting Eq. (16) into Eq. (14), we obtain again the quadratic form

(17) $$\Pi_{HSe} = \frac{1}{2} \mathbf{q}^T \mathbf{k} \mathbf{q} - \mathbf{q}^T \mathbf{f},$$

for the element, where

(18) $$\mathbf{k} = \mathbf{k}_{\beta q}^T \mathbf{k}_{\beta\beta}^{-1} \mathbf{k}_{\beta q} .$$

The element matrices can be assembled and constrained in the usual manner to obtain the finite element equations for the entire system.

Similar to Eq. (19.1:6), we have

$$\text{rank of } \mathbf{k} \leq \min(n_\beta, n_q - n_r),$$

in which n_β, n_q, n_r, are, respectively, the number of β, q and rigid body modes. In other words, the rank of $\mathbf{k}$ is no larger than the number of β's. For a quadrilateral element, there are 8 q's and 3 rigid body motions. If the number of β's is less than 5, the rank of $\mathbf{k}$ is deficient and spurious deformation mode(s) which contributes no energy to the element will exist. The five-β hybrid stress element is commonly used. In this case, Eq. (12) becomes

$$\mathbf{h}_\beta_{\atop 3\times 5} = \begin{bmatrix} 1 & 0 & 0 & y & 0 \\ 0 & 1 & 0 & 0 & x \\ 0 & 0 & 1 & 0 & 0 \end{bmatrix}, \qquad \boldsymbol{\beta}^T = [\beta_1 \ \beta_2 \ \beta_3 \ \beta_4 \ \beta_5].$$

For elements of distorted geometry, instead of using $\mathbf{h}_\beta$ as function of the Cartesian coordinates as in Eq. (12), which makes the assumed stresses satisfy equilibrium condition exactly, Pian recommended expressing stresses in terms of the nature coordinates ξ, η of the element. In this case the assumed stresses satisfy equilibrium condition only in an integral or average sense (Pian and Sumihara 1984).

Hybrid Displacement Model (Tong 1970). If the constitutive laws are satisfied, and if the stress is expressed in terms of strains, the functional Eq. (1) becomes the hybrid displacement principle Eq. (10.10:17) and can be written as

(19) ▲ $$\Pi_{HDe} = \int_{\partial A_e} (\hat{\mathbf{u}}^T - \mathbf{u}^T)\widehat{\mathbf{T}} \, ds$$

$$+ \int_{A_e} \left(\frac{1}{2} \mathbf{e}^T \mathbf{D}_e \mathbf{e} - \mathbf{u}^T \mathbf{b} \right) dA - \int_{\partial A_{\sigma e}} \hat{\mathbf{u}}^T \bar{\mathbf{T}} ds,$$

where $\mathbf{D}_e$ is the elastic modulus matrix relating strain $\mathbf{e}$ to stress $\boldsymbol{\sigma}$. Along the traction prescribed boundaries, there is no adjacent element and no restriction on the selection of $\hat{\mathbf{u}}$, one can simply set $\hat{\mathbf{u}} = \mathbf{u}$ on $\partial A_{\sigma e}$. In this case, Eq. (19) becomes

$$(20) \qquad \Pi_{HDe} = \int_{\partial A_e - \partial A_{\sigma e}} (\hat{\mathbf{u}}^T - \mathbf{u}^T)\widehat{\mathbf{T}}\, ds$$

$$+ \int_{A_e} \left(\frac{1}{2} \mathbf{e}^T \mathbf{D}_e \mathbf{e} - \mathbf{u}^T \mathbf{b} \right) dA - \int_{\partial A_{\sigma e}} \mathbf{u}^T \bar{\mathbf{T}}\, ds\,,$$

in which the integration over $\partial A_{\sigma e}$ is removed from the first integral and $\hat{\mathbf{u}}$ is replaced by $\mathbf{u}$ in the last integral.

One can reduce further the number of field variables in the functional by requiring the traction stress relation, the equilibrium equation and the stress displacement relation

$$(21) \qquad \widehat{\mathbf{T}} = \mathbf{T} = \mathbf{n}'^T \boldsymbol{\sigma}\,, \qquad \boldsymbol{\sigma}_{,\mathbf{x}} + \mathbf{b} = 0\,, \qquad \boldsymbol{\sigma} = \mathbf{D}_e \mathbf{e} = \mathbf{D}_e \mathbf{d}_e \mathbf{u}$$

to be satisfied *a priori*, where $\mathbf{d}_e$ is defined in Eq. (18.11:6) and $\mathbf{d}_e \mathbf{u}$ is the strain matrix. Then Eq. (19) is reduced to [Eq. (10.10:20)]

$$(22) \qquad \Pi_{HDe} = \int_{\partial A_e - \partial A_{\sigma e}} \hat{\mathbf{u}}^T \mathbf{T}\, ds$$

$$- \frac{1}{2} \int_{A_e} \mathbf{e}^T \mathbf{D}_e \mathbf{e}\, dA + \int_{\partial A_{\sigma e}} \mathbf{u}^T (\mathbf{T} - \bar{\mathbf{T}})\, ds\,.$$

If the displacement $\mathbf{u}$ is divided into the homogeneous and particular parts

$$(23) \qquad\qquad \mathbf{u} = \mathbf{u}_h + \mathbf{u}_p$$

such that their corresponding stresses $\boldsymbol{\sigma}_h$ and $\boldsymbol{\sigma}_p$ satisfy

$$(24) \qquad\qquad \frac{\partial \boldsymbol{\sigma}_h}{\partial \mathbf{x}} = 0\,, \qquad \frac{\partial \boldsymbol{\sigma}_p}{\partial \mathbf{x}} + \mathbf{b} = 0\,.$$

Then Π_{HDe} becomes

$$(25) \quad \Pi_{HDe} = \int_{\partial A_e - \partial A_{\sigma e}} \hat{\mathbf{u}}^T (\mathbf{T}_h + \mathbf{T}_p) ds - \int_{\partial A_e} \frac{1}{2}(\mathbf{u}_h^T + \mathbf{u}_p^T)\mathbf{T}_h\, ds$$

$$+ \int_{\partial A_{\sigma e}} \mathbf{u}_h^T (\mathbf{T}_h + \mathbf{T}_p - \bar{\mathbf{T}})\, ds + \text{const}\,,$$

which involves integrals along element boundaries only. Note that $\mathbf{u}_p$, $\mathbf{T}_p$, $\bar{\mathbf{T}}$ are known functions. The functional above involves only two independent field variables $\mathbf{u}_h$ and $\hat{\mathbf{u}}$.

We further separate $\mathbf{u}_h$ into two parts

$$(26) \qquad\qquad \mathbf{u}_h = \mathbf{u}_{hh} + \mathbf{u}_{th}$$

in the way that the traction $\mathbf{T}_{hh}$, $\mathbf{T}_{th}$, derived from $\mathbf{u}_{hh}$, $\mathbf{u}_{th}$ respectively, satisfy the condition

$$(27) \qquad \mathbf{T}_{hh} = 0, \qquad \mathbf{T}_{th} = \bar{\mathbf{T}} - \mathbf{T}_p \qquad \text{on} \qquad \partial A_{\sigma e},$$

Let $\mathbf{T}_h$ be the traction derived from $\mathbf{u}_h$, then

$$(28) \qquad\qquad\qquad \mathbf{T}_h = \mathbf{T}_{hh} + \mathbf{T}_{th}.$$

A substitution of Eqs. (26)–(28) into Eq. (25) yields

$$(29) \quad \Pi_{HDe} = \int_{\partial A_e - \partial A_{\sigma e}} \left(\hat{\mathbf{u}}^T \mathbf{T}_{hh} - \frac{1}{2} \mathbf{u}_{hh}^T \mathbf{T}_{hh} \right) ds$$

$$+ \int_{\partial A_e - \partial A_{\sigma e}} \hat{\mathbf{u}}^T (\mathbf{T}_{th} + \mathbf{T}_p) \, ds$$

$$- \int_{\partial A_e - \partial A_{\sigma e}} \left[\frac{1}{2} \mathbf{u}_{hh}^T \mathbf{T}_{th} + \left(\frac{1}{2} \mathbf{u}_{th}^T + \mathbf{u}_p^T \right) \mathbf{T}_{hh} \right] ds$$

$$+ \int_{\partial A_{\sigma e}} \frac{1}{2} \mathbf{u}_{hh}^T (\mathbf{T}_p - \bar{\mathbf{T}}) \, ds + \text{constant}.$$

Note that $\mathbf{T}_{hh}$ and $\mathbf{T}_{th}$ are traction associated with $\mathbf{u}_{hh}$ and $\mathbf{u}_{th}$, respectively, and that the displacement $\mathbf{u}_{hh}$ satisfies both the homogeneous equilibrium equations and zero traction condition on the traction prescribed surfaces. Only the first integral involves quadratic terms of the field variables $\mathbf{u}_{hh}$ since $\mathbf{T}_{hh}$ is a function of $\mathbf{u}_{hh}$, whereas the rest of the integrals involve only linear terms of $\mathbf{u}_{hh}$. If there is no body force. i.e., $\mathbf{b} = 0$, we have $\mathbf{u}_p$, $\mathbf{T}_p = 0$. If the traction on the traction prescribed surface is also zero, then $\mathbf{u}_{th}$, $\mathbf{T}_{th} = 0$ and all last three integrals in Eq. (29) vanish. For 3-dimensional problems, one simply replaces A's by V's and ds by dS.

Problem 19.1.   Show that the element stiffness matrix of the hybrid stress model is independent of the body force.

Problem 19.2.   Derive the element force matrix of the hybrid stress model for plane elasticity with a constant body force.

Problem 19.3.   Show that the hybrid stress stiffness matrix for a triangular element is the same as that of the constant strain triangle of the displacement model regardless of the number of β's in the stress mode.

Problem 19.4.   Derive the hybrid stress element stiffness matrix for a 8-node hexahedral. Determine the rank of **k** for different number of stress modes.

Problem 19.5.   Derive the element stiffness matrix of the hybrid displacement model:

(a) Use all three variables **u**, **û**, $\widehat{\mathbf{T}}$ as independent fields.
(b) Enforce $\widehat{\mathbf{T}} = \mathbf{T}$ *a priori* and just use **u**, **û** as the independent field variables.
(c) Discuss how the rank of **k** is affected by the number of unknown parameters of **u** and $\widehat{\mathbf{T}}$.

Problem 19.6.   Derive the element force matrix of the hybrid displacement model corresponding to the field variable **û**.

Problem 19.7.   If the conditions given in Eq. (19.2:21) are satisfied *a priori*, show that Eq. (19.2:19) can be reduced to Eq. (19.2:22).

Problem 19.8.   Consider the case that $\mathbf{T}(= n_j\sigma_{ji})$ and that σ_{ji} is in terms of displacements through the stress strain relation (i.e., $\boldsymbol{\sigma} = \mathbf{D}_e\mathbf{e} = \mathbf{D}_e\mathbf{d}_e\mathbf{u}$). Show that the first variation of Π_{HDe} in Eq. (19.2:19) with respect to **u** gives the matching condition

$$\hat{\mathbf{u}} = \mathbf{u} \qquad \text{on } \partial A_e\,,$$

$$\mathbf{T} = \bar{\mathbf{T}} \qquad \text{on } \partial A_{\sigma e}\,,$$

and the equilibrium equation in terms of **u** in A_e.

Problem 19.9.   In Problem 19.7, if the displacement is separated into the homogeneous and particular solutions $\mathbf{u}_h$ and $\mathbf{u}_p$ such that Eq. (19.2:24) is satisfied, show that Π_{HDe} can be reduced to Eq. (19.2:25).

Problem 19.10.   In Problem 19.9, if one further separates $\mathbf{u}_h = \mathbf{u}_{hh} + \mathbf{u}_{th}$ as in Eq. (19.2:26) with corresponding $\mathbf{T}_{hh}$, $\mathbf{T}_{th}$ satisfying Eq. (19.2:27). Show that Eq. (19.2:25) reduces to Eq. (19.2:29).

19.3. HYBRID SINGULAR ELEMENTS (SUPER-ELEMENTS)

For problems whose solution is smooth, the order of the finite element approximation is proportional to the order of the complete polynomial used for the shape functions. When a problem involves singularity, the order of approximation is dominated by the order of the singularity (Key 1966, Tong and Pian 1972). For example, in fracture mechanics, the asymptotic solution near a crack tip is (Sih and Liebowitz 1968)

$$
(1) \qquad
\begin{bmatrix} \sigma_{xx} \\ \sigma_{yy} \\ \sigma_{xy} \end{bmatrix}
= \frac{K_I}{\sqrt{2\pi r}} \cos\frac{\theta}{2}
\begin{bmatrix}
1 - \sin\dfrac{\theta}{2}\sin\dfrac{3\theta}{2} \\[2ex]
1 + \sin\dfrac{\theta}{2}\sin\dfrac{3\theta}{2} \\[2ex]
\sin\dfrac{\theta}{2}\cos\dfrac{3\theta}{2}
\end{bmatrix}
$$

$$
+ \frac{K_{II}}{\sqrt{2\pi r}}
\begin{bmatrix}
-\sin\dfrac{\theta}{2}\left(2 + \cos\dfrac{\theta}{2}\cos\dfrac{3\theta}{2}\right) \\[2ex]
\cos\dfrac{\theta}{2}\sin\frac{\theta}{2}\cos\dfrac{3\theta}{2} \\[2ex]
\cos\frac{\theta}{2}\left(1 - \sin\dfrac{\theta}{2}\sin\dfrac{3\theta}{2}\right)
\end{bmatrix} + \cdots,
$$

$$
(2) \qquad
\begin{bmatrix} u \\ v \end{bmatrix}
= \frac{K_I}{2G}\sqrt{\frac{r}{2\pi}}(\kappa - \cos\theta)
\begin{bmatrix} \cos\dfrac{\theta}{2} \\[2ex] \sin\dfrac{\theta}{2} \end{bmatrix}
$$

$$
+ \frac{K_{II}}{2G}\sqrt{\frac{r}{2\pi}}
\begin{bmatrix}
\sin\dfrac{\theta}{2}\left(2 + \kappa + \cos\theta\right) \\[2ex]
\cos\dfrac{\theta}{2}\left(2 - \kappa - \cos\theta\right)
\end{bmatrix} + \cdots,
$$

where G is the shear modulus, r is the distance from the crack tip, θ is the angle measuring from the crack plane, and $\kappa = 3 - 4\nu$ (for plane strain) and $\kappa = (3 - \nu)/(1 + \nu)$ (for plane stress), in which ν is the Poisson's ratio, K_I and K_{II} are constants commonly referred to as the *stress intensity factors associated with symmetric (Mode I) and antisymmetric (Mode II) deformations* with respect to the crack plane. The stresses and strains are proportional to $1/\sqrt{r}$ and the displacements are proportional to $\sqrt{r}$.

Near the singularity, the polynomial-base finite element method performs badly. Henshell and Shaw (1975) and Barsoum (1976, 1977) shifted the midside node to the quarter point along a side in quadratic isoparametric elements. The isoparametric transformation has a singularity at the corner near the shifted node. This generates a singular strain of $1/\sqrt{r}$ along the edge with the shifted nodes. To illustrate this approach, let us consider a 6-node rectangle (Fig. 19.3:1) with the 5th and 6th (identified as node 8)

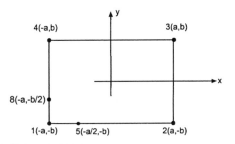

Fig. 19.3:1. A quarter point rectangular element.

nodes being at the quarter point from node 1 along side between nodes 1 and
2 and between nodes 1 and 4, respectively. Isoparametric transformation
is used to map the two quarter point nodes to the corresponding midsides
in the ξ, η plane. The transformation is

$$x = \sum_{i=1}^{5} x_i h_i + x_8 h_8 = a\xi + x_5 h_5 = a\left[\xi - \frac{1}{4}(1 - \xi^5)(1 - \eta)\right],$$

$$y = \sum_{i=1}^{5} y_i h_i + y_8 h_8 = b\eta + y_8 h_8 = b\left[\eta - \frac{1}{4}(1 - \eta^2)(1 - \xi)\right],$$

which maps the midside points $(0, -1)$ and $(-1, 0)$ in the ξ, η plane to the
quarter points $(-a/2, -b)$ and $(-a, -b/2)$ in the x, y plane. The inverse of
the Jacobian

$$\mathbf{J}^{-1} = \begin{bmatrix} \dfrac{\partial\xi}{\partial x} & \dfrac{\partial\eta}{\partial x} \\[3mm] \dfrac{\partial\xi}{\partial y} & \dfrac{\partial\eta}{\partial y} \end{bmatrix}$$

$$= \begin{bmatrix} \dfrac{2 + \eta - \xi\eta}{a} & \dfrac{\eta^2 - 1}{2a} \\[3mm] \dfrac{\xi^2 - 1}{2b} & \dfrac{2 + \xi - \xi\eta}{b} \end{bmatrix} \dfrac{8}{(3\xi\eta - \xi - \eta - 5)(\xi\eta - \xi - \eta - 3)}$$

is singular at $\xi = n = -1$. It can be shown that $\mathbf{J}^{-1}$ has $1/\sqrt{r}$ singularity
at node 1 at $x = -a$ and $y = -b$, where r is the distance from the node.
The strains e_{xx}, e_{yy} and e_{xy} are proportional to $\mathbf{J}^{-1}$ and thus have the
same singularity as $\mathbf{J}^{-1}$.

The hybrid approach allows the incorporation of appropriate singularity in the shape functions of field variables (Tong *et al.*, 1973, Tong 1977, 1984, Lin and Tong 1980, Piltner 1985, Jirousek 1987, Jeong *et al.* 1996). Consider a domain with a crack as shown in Fig. 19.3:2. One can use one element to surround the crack tip and incorporate the appropriate singularity in the shape functions for the element. There is no singularity in the region outside the crack element. Regular polynomial-based elements discussed before can be used. The assumed functions for the crack element and those for the surrounding elements are, in general, not continuous along their common boundaries. The hybrid model enforces continuity along the inter-element boundaries in an average sense in the final solution. Such an element is also called a *super-element*.

The crack tip can be anywhere within the element. Thus, one can use the same element with the crack tip at different locations (Fig. 19.3:3). This feature makes the crack element ideal for crack growth simulation. As

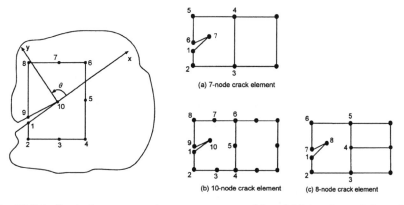

Fig. 19.3:2. Crack element superelement connects to: (a) and (c) 4-node quadrilaterals and (b) 8-node quadrilaterals.

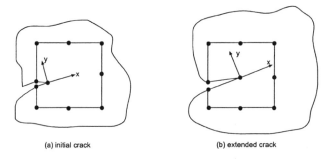

Fig. 19.3:3. Crack growth.

the crack advances within the element, one simply changes the crack tip location to reflect the crack extension, calculates new crack element stiffness matrix and analyzes the new configuration. No remeshing is required. When the crack tip gets too close to the element boundary, the solution will deteriorate because the adjacent element is normally polynomial based. One may then include the elements close to the crack tip to form a new and enlarged crack element. At the same time the part of the crack element far from the crack tip can be converted into regular elements. In this process, one often can retain the same global nodal labels for the whole domain with changes in the assembly list only, which defines the relation between the local and the global labels of each element, for the altered elements. Even if some minor adjustment of nodal locations may be needed for the new elements to avoid having elements with bad aspect ratio or severely acute angle, there will be no complicated remeshing involved.

The shape of crack element, generally a polygon, and the number of nodes along the element boundaries can be arbitrary. The spacing of nodes can be varied to coincide with the nodes of adjacent elements. A crack element with 10 nodes is shown in Fig. 19.3:2(b). The origin of the coordinates is at the crack tip and the crack surfaces are along the negative x-axis (i.e., $y = 0^{\mp}, x < 0$).

The procedure of deriving the element matrices based on the hybrid displacement functional Eq. (19.2:19) is as follows: the asymptotic solution of Eq. (1) is used to guide the selection of the assumed functions for the field variables within the element. The crack element connects to the adjacent elements along the boundaries by nodes from 1 through 9. There is no singularity along these boundaries. We simply assume the element boundary displacements $\hat{\mathbf{u}}$ in the same form as those of the adjacent elements to assure interelement compatibility. The crack surfaces are between nodes 9 and 10 and between nodes 10 and 1 where traction is prescribed. Since no adjacent element is involved along the crack surfaces, there is no restriction on the selection of $\hat{\mathbf{u}}$. We simply assume $\hat{\mathbf{u}} = \mathbf{u}$ along these surfaces. Then, Eq. (19.2:20) is used to evaluate the element matrices.

We shall first discuss the selection of $\hat{\mathbf{u}}$. Figure 19.3:2(a) depicts a 7-node crack element connecting to a 4-node quadrilateral element while Fig. 19.3:2(b) illustrates a 10-node crack element adjacent to an 8-node quadrilateral. Thus, $\hat{\mathbf{u}}$ is needed from nodes 1 to 6 for Fig. 19.3:2(a) and from nodes 1 to 9 for Fig. 19.3:2(b). In the former case, $\hat{\mathbf{u}}$ should vary linearly between nodes in the form of Eq. (19.2:13) in which q_{2j-1}, q_{2j} are nodal values of $\hat{u}$, $\hat{v}$ at node j. For instance, the crack element is adjacent to a 4-node quadrilateral between nodes 3 and 4, all components of $\mathbf{h}$ of

Eq. (19.2:13) for the edge between these two nodes are zero except

(3) $h_{1,5} = h_{2,6} = 1 - s/s_{3,4}$, $h_{1,7} = h_{2,8} = s/s_{3,4}$,

where s is the distance from first node (node 3 in this case) and $s_{i,j}$ denotes the distance between nodes i and j. In Fig. 19.3:2(b), the element adjacent to nodes 4 through 6 is an 8-node quadrilateral and $\hat{\mathbf{u}}$ should be quadratic between these nodes. Then, all components of $\mathbf{h}$ for this edge are zero except

$$h_{1,7} = h_{2,8} = \left(1 - \frac{s}{s_{4,6}}\right)\left(1 - \frac{s}{s_{4,5}}\right),$$

(4) $$h_{1,9} = h_{2,10} = \frac{s}{s_{4,5}}\left(\frac{s_{4,6} - s}{s_{4,6} - s_{4,5}}\right),$$

$$h_{1,11} = h_{2,12} = \left(\frac{s_{4,5} - s}{s_{4,5} - s_{4,6}}\right)\frac{s}{s_{4,6}}.$$

We can similarly define h's of $\hat{\mathbf{u}} = \mathbf{hq}$ for other edges.

To simplify subsequent discussion, we introduce a complex function $\hat{u}_c$ and a complex interpolation function matrix (row vector) $\hat{\mathbf{h}}_c$ such that

(5) $$\hat{\mathbf{u}}_c = \hat{u} + i\hat{v} = \hat{\mathbf{h}}_c \mathbf{q},$$

where $\hat{u}$, $\hat{v}$ are the x, y components of $\hat{\mathbf{u}}$. The components of $\hat{\mathbf{h}}_c$ are

(6) $$\hat{h}_{2j-1} = h_{1,2j-1},   \hat{h}_{2j} = ih_{2,2j}$$

in which $i = \sqrt{-1}$ is the imaginary number and h's are real function. Note that $\hat{h}_{2j-1} = \hat{h}_{2j}/i$. We will use i as $\sqrt{-1}$ throughout the remaining chapter.

Hybrid Displacement Element. We shall discuss a method of assuming $\boldsymbol{\sigma}, \mathbf{u}, \hat{\mathbf{u}}, \mathbf{T}, \hat{\mathbf{T}}$ within the element. Consider the choice of field variables with

(7) $$\hat{\mathbf{T}} = \mathbf{T} = \mathbf{n}'^T \boldsymbol{\sigma},   \boldsymbol{\sigma} = \mathbf{D}_e \mathbf{e} = \mathbf{D}_e \mathbf{d}_e \mathbf{u}.$$

In this case, we only need to assume $\mathbf{u}$ inside the element and $\hat{\mathbf{u}}$ along the element boundaries, and derive the stresses and the traction from equation above. Since there is no compatibility constraint on $\mathbf{u}$, one just pick functions that contain appropriate singularity for $\mathbf{u}$. We introduce the following coordinate transformation:

(8) $$x = \xi^2 - \eta^2,   y = 2\xi\eta,   z = x + iy = (\xi + i\eta)^2,$$

(9) $$\xi = \sqrt{(r+x)/2},   \eta = \text{sgn}(y)\sqrt{(r-x)/2}.$$

where $r = \sqrt{x^2 + y^2}$ and $\mathrm{sgn}(y)$ equals 1 if $y \geq 0$ and equals -1 if $y < 0$.
Note that ξ and η are not the natural coordinates defined in Chapter 18.
We assume $\mathbf{u}$ in the form

$$\mathbf{u} = \mathbf{p}\boldsymbol{\alpha} = \begin{bmatrix} 1 & \xi & \eta & \xi^2 & \cdots & 0 & 0 & 0 & 0 & \cdots \\ 0 & 0 & 0 & 0 & \cdots & 1 & \xi & \eta & \xi^2 & \cdots \end{bmatrix} \boldsymbol{\alpha},$$

where

$$\boldsymbol{\alpha}^T = [\alpha_1 \ \alpha_2 \ \cdots \ \alpha_p \ \alpha_{p+1} \ \cdots \ \alpha_{2p}].$$

In the equations above, p equals 3 for linear and equals 6 for complete
quadratic polynomials.

The derivatives with respect to x, y are calculated as follows:

$$\frac{\partial}{\partial x} = \frac{\partial \xi}{\partial x}\frac{\partial}{\partial \xi} + \frac{\partial \eta}{\partial x}\frac{\partial}{\partial \eta} = \frac{1}{2r}\left(\xi\frac{\partial}{\partial \xi} - \eta\frac{\partial}{\partial \eta}\right)$$

$$\frac{\partial}{\partial y} = \frac{\partial \xi}{\partial y}\frac{\partial}{\partial \xi} + \frac{\partial \eta}{\partial y}\frac{\partial}{\partial \eta} = \frac{1}{2r}\left(\eta\frac{\partial}{\partial \xi} + \xi\frac{\partial}{\partial \eta}\right),$$

which give an $1/\sqrt{r}$ singularity. Any polynomial of ξ, η, is a polynomial of
$\sqrt{r}$ near the crack tip. If $\mathbf{u}$ is assumed to be a polynomial of ξ, η, then $\mathbf{u}$ is
a function of $\sqrt{r}$ and its gradients in the z-plane is proportional to $1/\sqrt{r}$.
The singularity of strains and stresses are $1/\sqrt{r}$ at the crack tip. Thus, the
singularity of the strain energy is $1/r$. Substituting the assumed functions

$$\hat{\mathbf{u}} = \hat{\mathbf{h}}_c \mathbf{q}, \qquad \mathbf{u} = \mathbf{p}\boldsymbol{\alpha}, \qquad \mathbf{T} = \mathbf{n}'^T \mathbf{D}_e \mathbf{d}_e \mathbf{p}\boldsymbol{\alpha}$$

into the hybrid displacement functional Eq. (19.2:20), we obtain

(10) $$\Pi_{HDe} = \boldsymbol{\alpha}^T \mathbf{k}_{u\hat{u}} \mathbf{q} - \frac{1}{2}\boldsymbol{\alpha}^T \mathbf{k}_{uu}\boldsymbol{\alpha} - \boldsymbol{\alpha}^T \mathbf{f}_u,$$

where

$$\boldsymbol{\alpha}^T \mathbf{k}_{u\hat{u}} \mathbf{q} = \int_{\partial A_e - \partial A_{\sigma e}} \hat{\mathbf{u}}^T \mathbf{T} ds = \boldsymbol{\alpha}^T \left[\int_{\partial A_e - \partial A_{\sigma e}} (\mathbf{d}_e \mathbf{p})^T \mathbf{D}_e \mathbf{n}' \mathbf{h} ds \right] \mathbf{q},$$

(11) $$\boldsymbol{\alpha}^T \mathbf{k}_{uu}\boldsymbol{\alpha} = 2 \int_{\partial A_e} \mathbf{u}^T \mathbf{T} ds - \int_{A_e} \mathbf{e}^T \mathbf{D}_e \mathbf{e} dA$$

$$= \boldsymbol{\alpha}^T \left[2 \int_{\partial A_e - \partial A_{\sigma e}} \mathbf{p}^T \mathbf{n}' \mathbf{D}_e \mathbf{d}_e \mathbf{p} ds - \int_{A_e} (\mathbf{d}_e \mathbf{p})^T \mathbf{D}_e \mathbf{d}_e \mathbf{p} dA \right] \boldsymbol{\alpha}$$

$$\boldsymbol{\alpha}^T \mathbf{f}_u = \int_{A_e} \mathbf{u}^T \mathbf{b} dA + \int_{\partial A_{\sigma e}} \mathbf{u}^T \bar{\mathbf{T}} ds$$

$$= \boldsymbol{\alpha}^T \left(\int_{A_e} \mathbf{p}^T \mathbf{b} dA + \int_{\partial A_{\sigma e}} \mathbf{p}^T \bar{\mathbf{T}} ds \right).$$

Note that $\mathbf{d}_e\mathbf{p}$ is equivalent to $\mathbf{B}$ of the displacement model. Since the assumed $\mathbf{u}$ of different elements is independent of each other, we can eliminate $\boldsymbol{\alpha}$ from the stationary condition Π_{HDe} with respect to $\boldsymbol{\alpha}$:

(12) $\mathbf{k}_{uu}\boldsymbol{\alpha} = \mathbf{k}_{u\hat{u}}\mathbf{q} - \mathbf{f}_u ,$ or $\boldsymbol{\alpha} = \mathbf{k}_{uu}^{-1}\mathbf{k}_{u\hat{u}}\mathbf{q} - \mathbf{k}_{uu}^{-1}\mathbf{f}_u .$

A substitution of the result above into Eq. (10) yields

(13) $\Pi_{HDe} = \dfrac{1}{2}\mathbf{q}^T\mathbf{k}\mathbf{q} - \mathbf{q}^T\mathbf{f} ,$

where

(14) $\mathbf{k} = \mathbf{k}_{u\hat{u}}^T\mathbf{k}_{uu}^{-1}\mathbf{k}_{u\hat{u}} ,$ $\mathbf{f} = \mathbf{k}_{u\hat{u}}^T\mathbf{k}_{uu}^{-1}\mathbf{f}_u .$

Care must be taken in carrying out the integration near the crack tip in Eq. (11). The integration along $\partial A_{\sigma e}$ involves terms of order $\sqrt{r}$ and the integrand over the element involves terms of order $1/r$. Using the Gauss quadrature based on the x, y coordinates for numerical integration will result in lose of accuracy (convergent rate) because numerical integration simply approximates the singular integrand as a polynomial. Therefore, we ought to carry out the integration analytically near the crack tip or over the ξ, η-plane. Unfortunately the element shape in the transformed plane is no longer a simple rectangle, which makes the area integration nontrivial. Either analytical integration or integration over the ξ, η-plane makes the implementation of the hybrid approach more complicated.

Alternative Hybrid Formulation.   To circumvent the difficulty discussed above, we choose $\mathbf{u}$ in the form of Eq. (19.2:23) such that the corresponding stresses $\boldsymbol{\sigma}_h$ and $\boldsymbol{\sigma}_p$ are the homogeneous and particular solutions of the equilibrium equation defined in Eq. (19.2:24). We can then evaluate the element matrices using the functional given in Eq. (19.2:25), which involves line integrals only. Using this type of assumed stresses and displacements, the hybrid displacement and hybrid stress models become the same.

For simplicity, we shall consider only the case of *zero body force* in the subsequent discussion. Then $\mathbf{u}_p = \mathbf{T}_p = 0$ and Eq. (19.2:25) reduces to

(15) $\Pi_{HDe} = \displaystyle\int_{\partial A_e - \partial A_{\sigma e}} \hat{\mathbf{u}}^T\mathbf{T}_h ds - \dfrac{1}{2}\int_{\partial A_e} \mathbf{u}_h^T\mathbf{T}_h ds$

$$+ \int_{\partial A_{\sigma e}} \mathbf{u}_h^T(\mathbf{T}_h - \bar{\mathbf{T}})\, ds + \text{constant} .$$

For isotropic materials, the homogeneous solution $\mathbf{u}_h$ and $\boldsymbol{\sigma}_h$ can be derived from the complex potentials ψ, χ as given in Eqs. (9.6:3)–(9.6:5), which are repeated here for convenience

$$\sigma_{xx} + \sigma_{yy} = 2[\psi'(z) + \overline{\psi'(z)}]\,,$$

(16)
$$\sigma_{yy} - \sigma_{xx} + 2i\sigma_{xy} = 2[\bar{z}\psi''(z) + \chi''(z)]\,,$$

$$2G(u + iv) = \kappa\psi(z) - z\overline{\psi'(z)} - \overline{\chi'(z)}\,,$$

where $z = x + iy$ is a complex variable, the over bar $\overline{(.)}$ denotes the complex conjugate of the relevant function, one and two primes denote first and second differentives with respect to z, and

$$\kappa = 3 - 4\nu \qquad \text{for plane strain}$$

$$= (3 - \nu)/(1 + \nu) \qquad \text{for plane stress}\,.$$

The complex potentials ψ, χ are two arbitrary analytic functions of z.

To construct the element matrices, we assume the complex potentials ψ, χ' in terms of unknown parameters and derive the stresses and the displacements according to Eq. (16). (Note that χ does not appear explicitly. Thus, we may assume χ' directly.) The derived stresses and displacements automatically satisfy both the compatibility and equilibrium equations, within each element. We shall choose ψ, χ' containing the appropriate singularity at the crack tip. This can be done easily by introducing the transformation

$$z = \varsigma^2 = (\xi + i\eta)^2\,,$$

which is known as the conforming mapping (Muskhelishvili 1953) that maps the x, y plane to the right half ξ, η plane. It can be shown that ξ and η are related to x and y by Eqs. (8) and (9). The upper $(y = 0^+)$ and lower $(y = 0^-)$ crack surfaces on the negative x-axis are mapped, respectively, to the positive and negative parts of the η-axis. If we assume ψ, χ' as polynomials of ς, the derived stresses and strains will contain automatically the appropriate square root singularity at the crack tip.

The derivatives with respect to z and ζ are related by the equation $\frac{d}{dz} = \frac{1}{2\varsigma}\frac{d}{d\varsigma}$. From Eq. (16), one can write the stresses and displacements as functions of ξ, η

$$\sigma_{xx} + \sigma_{yy} = \frac{1}{\varsigma}\frac{d\psi(\varsigma)}{d\varsigma} + \frac{1}{\bar{\varsigma}}\overline{\frac{d\psi(\varsigma)}{d\varsigma}}\,,$$

(17)
$$\sigma_{yy} - \sigma_{xx} + 2i\sigma_{xy} = \frac{\bar{\varsigma}^2}{2\varsigma}\frac{d}{d\varsigma}\left[\frac{1}{\varsigma}\frac{d\psi(\varsigma)}{d\varsigma}\right] + \frac{1}{\varsigma}\frac{d\chi'(\varsigma)}{d\varsigma}\,,$$

$$2G(u + iv) = \kappa\psi(\varsigma) - \frac{\varsigma^2}{2\bar{\varsigma}}\overline{\frac{d\psi(\varsigma)}{d\varsigma}} - \overline{\chi'(\varsigma)}\,.$$

Clearly, the displacements and stresses are proportional to ζ, $1/\zeta$ (i.e., $\sqrt{r}$, $1/\sqrt{r}$), respectively, at the crack tip.

We can enforce the traction condition *a priori* to further simplify the finite element formulation. Let us assume ψ, χ' in the form

$$\psi = \psi_{th} + \psi_{hh}, \qquad \chi' = \chi'_{th} + \chi'_{hh},$$

where ψ_{th}, χ'_{th} are particular solutions, satisfying the non-zero prescribed boundary traction while ψ_{hh}, χ'_{hh} denote the homogeneous solution satisfying traction free condition on the crack surfaces. For a prescribed traction of the form

$$\bar{\mathbf{T}} = \begin{bmatrix} (\sigma_{xy})_1 + (\sigma_{xy})_2 x \\ (\sigma_{yy})_1 + (\sigma_{yy})_2 x \end{bmatrix} \quad \text{on the lower crack surface } (y = 0^-, x < 0).$$

$$\bar{\mathbf{T}} = -\begin{bmatrix} (\sigma_{xy})_1 + (\sigma_{xy})_2 x \\ (\sigma_{yy})_1 + (\sigma_{yy})_2 x \end{bmatrix} \quad \text{on the upper crack surface } (y = 0^+, x < 0).$$

where the $(.)$'s are known constants, one can show that

(18) $$\psi_{th} = c_2 z^2, \qquad \chi'_{th} = c_1 z,$$

where

$$c_1 = (\sigma_{yy})_1 + i(\sigma_{xy})_1, \qquad c_2 = (\sigma_{yy})_2/6 + i(\sigma_{xy})_2/2.$$

The particular solution for ψ_{th}, χ'_{th} is not unique. It can be shown that the solution for ψ, χ' is independent of the specific form of the particular solution if the assumed homogeneous solution is sufficiently general. For a more general prescribed traction, one can approximate it as a polynomial near the crack tip and derive the particular solution accordingly.

We can derive the homogeneous solution using the analytic continuation technique [Muskhelishvili 1953, Tong 1984]. From Eq. (9.6:7), the traction free condition can be written as

$$\overline{\psi_{hh}(z)} + \bar{z}\psi'_{hh}(z) + \chi'_{hh}(z) = 0$$

on the crack surfaces, $z = x(< 0)$, in the z-plane, or

(19) $$\overline{\psi_{hh}(\varsigma)} + \bar{\varsigma}^2 \frac{1}{2\varsigma} \frac{d\psi_{hh}(\varsigma)}{d\varsigma} + \chi'_{hh}(\varsigma) = 0$$

on the imaginary axis, $\varsigma = i\eta$, in the ς-plane. The analytic continuation in the present case simply replaces $\bar{\varsigma}$ by $-\varsigma$ in Eq. (19) because $\bar{\varsigma} = -\varsigma$ on the imaginary axis. This makes Eq. (19) an analytical equation of ζ. Thus χ'_{hh} can be written as an analytic function in terms of ψ_{hh}:

$$(20) \qquad \chi'_{hh}(\varsigma) = -\bar{\psi}_{hh}(-\varsigma) - \frac{\varsigma}{2}\frac{d\psi_{hh}(\varsigma)}{d\varsigma},$$

where the short bar over ψ_{hh} denotes the complex conjugate of the function while the independent variable itself remains in the form as given in the argument. For instance, if

$$(21) \qquad \psi_{hh}(\varsigma) = a_1\varsigma + a_2\varsigma^2 + a_3\varsigma^3 + \cdots,$$

then

$$\bar{\psi}_{hh}(-\varsigma) = \bar{a}_1(-\varsigma) + \bar{a}_2(-\varsigma)^2 + \bar{a}_3(-\varsigma)^3 + \cdots.$$

With χ'_{hh} in terms of ψ_{hh} as in Eq. (20), the traction free condition Eq. (19) is automatically satisfied, since $\bar{\varsigma} = -\varsigma$ on the imaginary axis in the ζ-plane.

We assume the complex potential ψ_{hh} as a simple polynomial of ζ as in Eq. (21) and derive χ'_{hh} according to Eq. (20)

$$(22) \quad \chi'_{hh}(\varsigma) = -\bar{\psi}_{hh}(-\varsigma) - \frac{\varsigma}{2}\frac{d\psi_{hh}}{d\varsigma}$$

$$= \left(\bar{a}_1 - \frac{1}{2}a_1\right)\varsigma - (\bar{a}_2 + a_2)\varsigma^2 + \left(\bar{a}_3 - \frac{3}{2}a_3\right)\varsigma^3 - \cdots$$

where a's are complex constants. The stresses derived from the assumed ψ_{hh}, χ'_{hh} in Eqs. (21) and (22) satisfy zero traction on the crack surfaces. One can show that the imaginary part of a_2 gives no contribution to stresses and, therefore, has no contribution to the element matrices. It actually represents a rigid body motion and can be set to zero without any loss of generality.

As mentioned before, the assumed complex potentials of Eqs. (21) and (22) give an $1/\sqrt{r}$ *stress singularity* at the crack tip. The coefficient of the singular term at the crack tip is a measure of the intensity of the singularity. Thus, one defines [Eq. (9.6:35) with the coordinates at the center of the crack]

$$(23) \qquad K_I - iK_{II} = 2\sqrt{2\pi}\,\lim_{z\to 0}\sqrt{z}\frac{d\psi}{dz} = \sqrt{2\pi}\frac{d\psi}{d\varsigma}\bigg|_{\varsigma=0} = \sqrt{2\pi}\,a_1,$$

where K_I, K_{II} are real constants often referred to as the *stress intensity factors* (Sih and Liebowitz 1968). They are associated with the symmetric (*mode I*) and antisymmetric (*mode II*) deformations about the x-axis, respectively. Mode I is a crack opening deformation and mode II corresponds to shear along the crack plane at the crack tip. Using the hybrid approach, we can evaluate the stress intensity factors directly from a_1. Stress intensity factors are important parameters in fracture mechanics for structural integrity assessment (Tong 1984).

For problems with non-zero prescribed surface traction, let $\mathbf{u}_{th}$, $\mathbf{T}_{th}$ and $\mathbf{u}_h$, $\mathbf{T}_h$ be the quantities derived from ψ_{th}, χ'_{th} and ψ_{hh}, χ'_{hh}, respectively. Taking into account that $\mathbf{T}_{th} = \bar{\mathbf{T}}$ and $\mathbf{T}_{hh} = 0$ on the crack surface $\partial A_{\sigma e}$, we can show that Eq. (15) reduces to Eq. (19.2:29) with $\mathbf{u}_p = \mathbf{T}_p = 0$ in the form

$$(24) \quad \Pi_{HDe} = \int_{\partial A_e - \partial A_{\sigma e}} \left(\hat{\mathbf{u}}^T \mathbf{T}_{hh} - \frac{1}{2} \mathbf{u}_{hh}^T \mathbf{T}_{hh} \right) ds + \int_{\partial A_e - \partial A_{\sigma e}} \hat{\mathbf{u}}^T \mathbf{T}_{th} \, ds$$

$$- \frac{1}{2} \int_{\partial A_e - \partial A_{\sigma e}} (\mathbf{u}_{hh}^T \mathbf{T}_{th} + \mathbf{u}_{th}^T \mathbf{T}_{hh}) \, ds - \int_{\partial A_{\sigma e}} \frac{1}{2} \mathbf{u}_{hh}^T \bar{\mathbf{T}} \, ds \, ,$$

which involves line integration only. All integrals except the last one do not involve the crack tip. One can integrate them using regular numerical integration. The following identity is used for the integration in Eq. (24)

$$(25) \quad T_c = T_x - iT_y = \frac{\sigma_{xx} + \sigma_{yy}}{2} e^{-i\theta} - \frac{\sigma_{yy} - \sigma_{xx} + 2i\sigma_{xy}}{2} e^{i\theta}$$

$$= \left[\frac{1}{\varsigma} \frac{d\psi(\varsigma)}{d\varsigma} + \frac{1}{\bar{\varsigma}} \frac{\overline{d\psi(\varsigma)}}{d\varsigma} \right] e^{-i\theta} - \left\{ \frac{\bar{\varsigma}^2}{2\varsigma} \frac{d}{d\varsigma} \left[\frac{1}{\varsigma} \frac{d\psi(\varsigma)}{d\varsigma} \right] + \frac{1}{\varsigma} \frac{d\chi'(\varsigma)}{d\varsigma} \right\} e^{i\theta} \, .$$

where θ is the angle between the x-axis and the unit normal to the boundary. For elements with straight edges, θ is a constant along each side. Substituting ψ_{th}, χ'_{th} from Eq. (18) and ψ_{hh}, χ'_{hh} from Eqs. (21) and (22) into Eq. (25), we obtain

$$(26) \qquad\qquad T_{hh} + T_{th} \equiv T_c = T_x - iT_y$$

$$(27) \qquad\qquad u_{hh} + u_{th} \equiv u + iv$$

where the subscripts hh and th denote quantities derived from ψ_{hh}, χ'_{hh} and ψ_{th}, χ'_{th}, respectively. Note that bold $\mathbf{u}$ and $\mathbf{T}$ denote matrices and subscripted italic u and T denote the corresponding complex quantities, e.g.,

$$u_{hh} = \text{Re}\,(u_{hh}) + i\,\text{Im}\,(u_{hh}), \qquad T_{hh} = \text{Re}\,(T_{hh}) + i\,\text{Im}\,(T_{hh}),$$

$$\mathbf{u}_{hh} = \begin{bmatrix} u \\ v \end{bmatrix}_{hh} = \begin{bmatrix} \text{Re}(u_{hh}) \\ \text{Im}(v_{hh}) \end{bmatrix}, \qquad \mathbf{T}_{hh} = \begin{bmatrix} T_x \\ T_y \end{bmatrix}_{hh} = \begin{bmatrix} \text{Re}(T_{hh}) \\ \text{Im}(T_{hh}) \end{bmatrix},$$

where Re(.) and Im(.) are, respectively, the real and the imaginary parts of the corresponding function. Thus $\mathrm{Re}(u_{hh})$ and $\mathrm{Im}(u_{hh})$ are, respectively, the x- and y-components of $\mathbf{u}_{hh}$ and similarly for the traction $\mathbf{T}_{hh}$. All u's and T's are complex functions of ς, $\bar{\varsigma}$ and can be written in the form

(27a)
$$T_{hh} = \mathbf{G}_c \boldsymbol{\alpha}\,, \qquad u_{hh} = \mathbf{h}_c \boldsymbol{\alpha}$$

in which

$$\mathbf{G}_c = [G_1 \ G_2 \ \cdots]\,, \quad \mathbf{h}_c = [h_{c1} \ h_{c2} \ \cdots]\,, \quad \boldsymbol{\alpha}^T = [\alpha_1 \ \alpha_2 \ \cdots]\,,$$

$$a_1 = \alpha_1 + i\alpha_2\,, \quad a_2 = \alpha_3\,, \quad a_p = \alpha_{2p-2} + i\alpha_{2p-1} \quad \text{for } p \geq 3$$

with α's being real unknowns and a's being the coefficients of Eq. (21). Using Eq. (18), (21), (22), (25) and (26), after some algebraic manipulations, we find

$$G_1 = \left(\frac{1}{\varsigma} + \frac{1}{\bar{\varsigma}}\right)\exp(-i\theta) + \frac{\bar{\varsigma}^2 - \varsigma^2}{\varsigma^3}\frac{\exp(i\theta)}{2}\,,$$

$$G_2 = i\left[\left(\frac{1}{\varsigma} - \frac{1}{\bar{\varsigma}}\right)\exp(-i\theta) + \frac{\bar{\varsigma}^2 + 3\varsigma^2}{\varsigma^3}\frac{\exp(i\theta)}{2}\right]\,,$$

$$G_3 = 8\,\cos\,\theta\,,$$

$$G_4 = 3\left[(\varsigma + \bar{\varsigma})\exp(-i\theta) - \frac{\bar{\varsigma}^2 - \varsigma^2}{\varsigma}\frac{\exp(i\theta)}{2}\right]\,,$$

$$G_5 = 3i\left[(\varsigma - \bar{\varsigma})\exp(-i\theta) - \frac{\bar{\varsigma}^2 - 5\varsigma^2}{\varsigma}\frac{\exp(i\theta)}{2}\right]\,;$$

$$\cdots\,$$

$$h_{c1} = (2\kappa\varsigma - \bar{\varsigma} - \varsigma^2/\bar{\varsigma})/(4G)\,, \qquad h_{c2} = i(2\kappa\varsigma - 3\bar{\varsigma} + \varsigma^2/\bar{\varsigma})/(4G)\,,$$

$$h_{c3} = (\kappa\varsigma^2 + 2\bar{\varsigma}^2 - \varsigma^2)/(2G)\,, \qquad h_{c4} = (2\kappa\varsigma^3 + \bar{\varsigma}^3 - 3\varsigma^2\bar{\varsigma})/(4G)\,,$$

$$h_{c5} = i(2\kappa\varsigma^3 - 5\bar{\varsigma}^3 + 3\varsigma^2\bar{\varsigma})/(4G)\,;$$

$$\cdots\,$$

$$T_{th} = 2(\bar{c}_2\bar{\varsigma}^2 + c_2\varsigma^2)\exp(-i\theta) - (2c_2\bar{\varsigma}^2 + c_1)\exp(i\theta)\,,$$

$$u_{th} = (\kappa c_2\varsigma^4 - \bar{c}_1\bar{\varsigma}^2 - 2\bar{c}_2\varsigma^2\bar{\varsigma}^2)/(2G)\,.$$

Note that u_{th} and T_{th} are known functions of the prescribed quantities such as the constants c_1 and c_2. Using Eqs. (5), (26), (27) and (27a), we can express the integrands of Eq. (24) in terms of the complex function matrices $\hat{\mathbf{h}}_c$ and $\mathbf{G}_c$ as

$$\hat{\mathbf{u}}^T \mathbf{T}_{hh} = \mathrm{Re}(\hat{u}T_{hh}) - \mathrm{Im}\,(\hat{v}T_{hh})$$

$$= \mathrm{Re}\,(\hat{u}_c T_{hh}) = \mathbf{q}^T\,\mathrm{Re}\,(\hat{\mathbf{h}}_c^T \mathbf{G}_c)\boldsymbol{\alpha}\,,$$

$$\hat{\mathbf{u}}^T \mathbf{T}_{th} = \mathrm{Re}\,(\hat{u}_c T_{th}) = \mathbf{q}^T\,\mathrm{Re}\,(\hat{\mathbf{h}}_c^T T_{th}),$$

$$\mathbf{u}_{hh}^T \mathbf{T}_{hh} = \mathrm{Re}\,(u_{hh}T_{hh}) = \boldsymbol{\alpha}^T\,\mathrm{Re}\,(\mathbf{h}_c^T \mathbf{G}_c)\boldsymbol{\alpha}\,,$$

$$\mathbf{u}_{hh}^T \mathbf{T}_{th} + \mathbf{u}_{th}^T \mathbf{T}_{hh} = \mathrm{Re}\,(u_{hh}T_{th} + u_{th}T_{hh})$$

$$= \boldsymbol{\alpha}^T\,\mathrm{Re}\,(\mathbf{h}_c^T T_{th} + \mathbf{G}_c^T u_{th})$$

on $\partial A_e - \partial A_{\sigma e}$, and

$$\mathbf{u}_{hh}^T \bar{\mathbf{T}} = \mathrm{Re}\,(u_{hh}T_{th}) = \boldsymbol{\alpha}^T\,\mathrm{Re}\,(\mathbf{h}_c^T T_{th})$$

on $\partial A_{\sigma e}$. A substitution of these complex functions into Eq. (24) yields

(28) $$\Pi_{HDe} = \boldsymbol{\alpha}^T \mathbf{k}_{u\hat{u}} \mathbf{q} - \frac{1}{2}\boldsymbol{\alpha}^T \mathbf{k}_{uu} \boldsymbol{\alpha} - \boldsymbol{\alpha}^T \mathbf{f}_u - \mathbf{q}^T \mathbf{f}_{\hat{u}}\,,$$

where

$$\boldsymbol{\alpha}^T \mathbf{k}_{u\hat{u}} \mathbf{q} = \int_{\partial A_e - \partial A_{\sigma e}} \hat{\mathbf{u}}^T \mathbf{T}_{hh}\,ds = \boldsymbol{\alpha}^T\,\mathrm{Re}\left(\int_{\partial A_e - \partial A_{\sigma e}} \mathbf{G}_c^T \hat{\mathbf{h}}_c\,ds\right)\mathbf{q}\,,$$

$$\boldsymbol{\alpha}^T \mathbf{k}_{uu} \boldsymbol{\alpha} = \int_{\partial A_e - \partial A_{\sigma e}} \mathbf{u}_{hh}^T \mathbf{T}_{hh}\,ds = \boldsymbol{\alpha}^T\,\mathrm{Re}\left(\int_{\partial A_e - \partial A_{\sigma e}} \mathbf{h}_c^T \mathbf{G}_c\,ds\right)\boldsymbol{\alpha}\,,$$

$$\boldsymbol{\alpha}^T \mathbf{f}_u = \int_{\partial A_e - \partial A_{\sigma e}} \frac{\mathbf{u}_{hh}^T \mathbf{T}_{th} + \mathbf{u}_{th}^T \mathbf{T}_{hh}}{2}\,ds + \int_{\partial A_{\sigma e}} \frac{\mathbf{u}_{hh}^T \mathbf{T}_{th}}{2}\,ds$$

$$= \frac{\boldsymbol{\alpha}^T}{2}\,\mathrm{Re}\left[\int_{\partial A_e - \partial A_{\sigma e}} (\mathbf{h}_c^T T_{th} + \mathbf{G}_c^T u_{th})\,ds + \int_{\partial A_{\sigma e}} \mathbf{h}_c^T T_{th}\,ds\right]$$

$$\mathbf{q}^T \mathbf{f}_{\hat{u}} = -\int_{\partial A_e - \partial A_{\sigma e}} \hat{\mathbf{u}}^T \mathbf{T}_{th}\,ds = -\mathbf{q}^T\,\mathrm{Re}\left(\int_{\partial A_e - \partial A_{\sigma e}} \mathbf{h}_c^T T_{th}\,ds\right).$$

It can be shown that $\mathbf{k}_{uu}$ is symmetric and positive definite. If we had allowed $\mathrm{Im}(a_2)$ to be an unknown parameter in Eq. (21), $\mathbf{k}_{uu}$ would be positive semi-definite with $u + iv = \mathrm{Im}(a_2)$ being the zero eigenvector.

Note, that $T_{th} = \bar{T}_x - i\bar{T}_y$ on the crack surfaces $\partial A_{\sigma e}$ where $\bar{T}_x$, $\bar{T}_y$ are the prescribed traction and that u_{th} is a known complex function related to $\bar{T}_x$, $\bar{T}_y$ in terms of the prescribed constants c_1 and c_2.

The integration along $\partial A_e - \partial A_{\sigma e}$ (from nodes 1 to 9, Fig. 19.3:2b) can be performed numerically in the z-plane where ξ, η are related to x, y by Eq. (9). For the integrals over $\partial A_{\sigma e}$, it can be integrated analytically in the ζ-plane if the prescribed traction is a polynomial. Note, along $\partial A_{\sigma e}$, $\varsigma = i\eta$ and $ds = -2\eta d\eta$.

Like before, using the stationary condition of Π_{HDe} with respect to α leads to

$$\mathbf{k}_{u\hat{u}}\mathbf{q} - \mathbf{k}_{uu}\boldsymbol{\alpha} - \mathbf{f}_u = 0 \,.$$

Solving for $\boldsymbol{\alpha}$ and substituting into Eq. (28) yields

$$\Pi_{HDe} = \frac{1}{2}\mathbf{q}^T\mathbf{k}\mathbf{q} - \mathbf{q}^T\mathbf{f} \,,$$

where

$$\mathbf{k} = \mathbf{k}_{u\hat{u}}^T\mathbf{k}_{uu}^{-1}\mathbf{k}_{u\hat{u}} \,, \qquad \mathbf{f} = \mathbf{f}_{\hat{u}} + \mathbf{k}_{u\hat{u}}^T\mathbf{k}_{uu}^{-1}\mathbf{f}_u \,,$$

are the equivalent element stiffness matrices and force matrices, respectively. These element matrices are to be assembled with those from the other part of the domain to solve the crack problem.

When the crack element is too small or when the crack tip is too close to an element boundary, the crack element will lose its effectiveness (Orringer et al. 1977) because the element boundary dispalcement $\hat{\mathbf{u}}$ and the displacements of the adjacent elements are usually polynomials. To improve the accuracy of crack analysis, instead of reducing the size of the crack element, one increases the number of boundary nodes of the crack element and the order of the polynomial for the complex potential ψ. One should place the element boundaries a distance away from the crack tip, if possible.

Problem 19.11. Derive the following expression from Eq. (19.3:17)

$$2(\sigma_{yy} + i\sigma_{xy}) = \frac{1}{\varsigma}\frac{d\psi(\varsigma)}{d\varsigma} + \frac{1}{\bar{\varsigma}}\overline{\frac{d\psi(\varsigma)}{d\varsigma}}$$

$$+ \frac{\bar{\varsigma}^2}{2\varsigma}\frac{d}{d\varsigma}\left[\frac{1}{\varsigma}\frac{d\psi(\varsigma)}{d\varsigma}\right] + \frac{1}{\varsigma}\frac{d\chi'(\varsigma)}{d\varsigma} \,.$$

Find a particular solution ψ_p, χ'_p that satisfies the traction condition

$$\bar{\mathbf{T}} = \pm\begin{bmatrix} (\sigma_{xy})_n x^n \\ (\sigma_{yy})_n x^n \end{bmatrix} \qquad \text{on the crack surfaces}, y = 0^{\mp}, x < 0 \,,$$

where n is an integer and $(\sigma_{xy})_n$, $(\sigma_{yy})_n$ are known constants.

Problem 19.12.   Show that Eq. (19.3:25) is true.

19.4. ELEMENTS FOR HETEROGENEOUS MATERIALS

The super-element concept can be applied also to heterogeneous materials. Tong and Mei (1992) treated composite materials made of periodic fibers imbedded in a matrix material. They used the *method of homogenization* and modeled an individual fiber and its surrounding material as a basic cell of the structure. The cell was modeled as an *super-element of polygons.* Accorsi (1988), Ghosh and his associates (1991, 1994), Zhang and Katsube (1994) modeled the inclusions of a heterogeneous material as two-dimensional elliptical particles of various sizes, aspect ratios, and orientations scattered in the parent material. They used polygonal elements, called them *Voronoi cell elements*, each of which contains one or more inclusion. The hybrid approach is used to determine the stiffness matrices of such elements for inclusions of various shapes and sizes. Such approach can be applied to elastic-plastic and thermo-elastic problems (Ghosh and Moorthy 1995, Ghosh and Liu 1995).

19.5. ELEMENTS FOR INFINITE DOMAIN

The hybrid technique provides a means to treat problems involving an infinite domain. For a problem in which the solution decays rapidly toward infinity, one may approximate an infinite domain with finite boundaries far from the region of interest. One then solves the problem using elements in the finite domain. However, for certain problems (such as those involving wave propagation), the disturbance is often transmitted to infinity. Any truncation to a finite domain will produce a false solution.

To treat the problem using a hybrid approach, we divide the infinite domain into two regions R_1 and R_2 separated by a curve C (Fig. 19.5:1). The R_1 is sufficiently large to enclose the area of interest including the body B and R_2 is the rest of the domain extending to infinity. The region R_1 is divided into a finite number of elements in which the finite element approximation discussed earlier is used. The region R_2 will be considered as a single infinite- or super-element. Within R_2 the asymptotic solution is used as the approximation. The hybrid technique is then used to match the solutions of the two regions. We shall consider two examples to illustrate the idea.

(a) **Plane Elasticity**: Consider a plane elasticity problem with zero body force and zero deformation gradient at infinity. If neither condition is actually so, we can introduce a particular solution so that the general problem is reduced to one having zero body force and decaying deformation gradient at the far field. A variational functional for the entire domain is written as

$$\Pi = \Pi_{R1} + \Pi_{R2} \,,$$

where Π_{R1} is the functional for the finite domain R_1 which can be the sum of the functionals for all the element within the region and Π_{R2} is the hybrid functional for R_2 which is in the same form as Eq. (19.2:19)

$$(1) \qquad \Pi_{R_2} = \int_C (\hat{\mathbf{u}}^T - \mathbf{u}^T)\widehat{\mathbf{T}}ds + \int_{R_2} \frac{1}{2}\mathbf{e}^T\mathbf{D}_e\mathbf{e}dA \,,$$

where C is the curve separates R_1 and R_2. We have made an implicit assumption here that the integration over the infinite domain is bounded. We assume $\hat{\mathbf{u}}$ as those over the region R_1 along the inter-element boundaries. The form of $\hat{\mathbf{u}}$ is the similar to that for the crack element as illustrated in Eqs. (19.3:3) and (19.3:4). Thus, interelement compatibility in the sense of Eq. (19.2:5) is ensure. By assuming $\widehat{\mathbf{T}}$ in terms of $\mathbf{u}$ as in Eq. (19.2:21) and choosing $\mathbf{u}$ in terms of asymptotic solutions (satisfying the equilibrium equations and decaying to zero at infinity), we can avoid the integration over the infinite domain. Equation (1) becomes

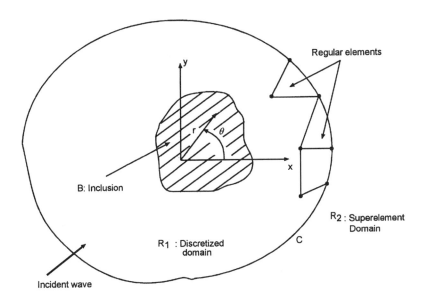

Fig. 19.5:1. An infinite-domain element.

(2) $$\Pi_{R_2} = \int_C (\hat{\mathbf{u}}^T - \mathbf{u}^T)\mathbf{T}ds + \frac{1}{2}\int_C \mathbf{u}^T \mathbf{T}ds + \frac{1}{2}\int_{R\to\infty} \mathbf{u}^T \mathbf{T}ds$$

$$= \int_C \left(\hat{\mathbf{u}}^T - \frac{1}{2}\mathbf{u}^T\right)\mathbf{T}ds\,.$$

This functional is the same as the homogeneous part of Eq. (19.2:29), i.e., the part excludes $\mathbf{u}_{th}$, $\mathbf{u}_p$, $\mathbf{T}_{th}$, $\mathbf{T}_p$, $\bar{\mathbf{T}}$. The integral over the curve with the radius $R \to \infty$ vanishes due to the decaying nature of the asymptotic solution.

An asymptotic solution of $\mathbf{u}$ can be expressed in terms of complex potentials of the following form

(3) $$\psi(z) = (\alpha_1 + i\alpha_2)/z + (\alpha_3 + i\alpha_4)/z^2$$

$$+ \cdots + (\alpha_{2p-1} + i\alpha_{2p})/z^p\,,$$

(4) $$\mathcal{X}'(z) = (\alpha_{2p+1} + i\alpha_{2p+2})/z + (\alpha_{2p+3} + i\alpha_{2p+4})/z^2$$

$$+ \cdots + (\alpha_{4p-1} + i\alpha_{4p})/z^p\,,$$

where p is an integer and α's are real constants. The total number of α's is $4p$. It is obvious that the derived displacements and stresses satisfy the equilibrium equations and decay to zero at infinity. Using Eqs. (19.3:16) and (19.3:25), after some algebraic manipulation, we find

(5) $$T_c = T_x - iT_y = \mathbf{G}_c\alpha\,, \qquad u_c = u + iv = \mathbf{h}_c\alpha\,,$$

where the components of $\mathbf{G}_c$ and $\mathbf{h}_c$ are

$$G_{2j-1} = -j(1/\bar{z}^{j+1} + 1/z^{j+1})e^{-i\theta} - j(j+1)e^{i\theta}\bar{z}/z^{j+2}\,,$$

$$G_{2j} = i[j(1/\bar{z}^{j+1} - 1/z^{j+1})e^{-i\theta} - j(j+1)e^{i\theta}\bar{z}/z^{j+2}]\,,$$

$$G_{2p+2j-1} = G_{2p+2j}/i = j/z^{j+1}\,,$$

$$(h_c)_{2j-1} = (\kappa/z^j + jz/\bar{z}^{j+1})/(2G)\,,$$

$$(h_c)_{2j} = i(\kappa/z^j - jz/\bar{z}^{j+1})/(2G)\,,$$

$$(h_c)_{2p+2j-1} = -(h_c)_{2p+2j}/i = -1/(2\bar{z}^j G)\,,$$

for $j = 1, 2, \ldots, p$. Using $\hat{\mathbf{u}}$ in the form in Eq. (19.3:5) on C, and u_c and T_c of Eq. (5), one can write Eq. (2) in the form

(6) $$\Pi_{R_2} = \alpha^T\mathbf{k}_{u\hat{u}}\mathbf{q} - \frac{1}{2}\alpha^T\mathbf{k}_{uu}\alpha\,,$$

where

$$\alpha^T \mathbf{k}_{u\hat{u}} \mathbf{q} = \int_C \hat{\mathbf{u}}^T \mathbf{T} ds = \alpha^T \text{Re} \left(\int_C \mathbf{G}_c^T \hat{\mathbf{h}}_c ds \right) \mathbf{q},$$

$$\alpha^T \mathbf{k}_{uu} \alpha = \int_C \mathbf{u}^T \mathbf{T} ds = \alpha^T \text{Re} \left(\int_C \mathbf{G}_c^T \mathbf{h}_c ds \right) \alpha,$$

which are similar to the element matrices defined in Eq. (19.3:28). From the zero condition of the first variation of Π_{R_2} with respect to α,

$$\mathbf{k}_{u\hat{u}} \mathbf{q} - \mathbf{k}_{uu} \alpha = 0,$$

one can eliminate α from Eq. (6) to obtain

$$\Pi_{R_2} = \frac{1}{2} \mathbf{q}^T \mathbf{k} \mathbf{q},$$

which is a function of $\mathbf{q}$ only, where

$$\mathbf{k} = \mathbf{k}_{u\hat{u}}^T \mathbf{k}_{uu}^{-1} \mathbf{k}_{u\hat{u}}$$

is the equivalent element stiffness matrix of the infinite element. The element stiffness matrix can be assembled together with the matrices from R_1 in the usual fashion to form the finite element equation.

(b) **Long Wave in Potential Flow**: Figure 19.5:1 shows an incident wave being scattered by a solid body B in the x, y-plane. Under the long wave theory for potential flow (Stoker 1957, Chen and Mei 1974), the velocity potential $\phi \exp(-i\omega t)$ is governed by

$$(7) \qquad \boldsymbol{\nabla} \cdot (h \boldsymbol{\nabla} \phi) + \omega^2 \phi/g = 0 \qquad \text{in} \qquad R_1 + R_2,$$

where $h(x, y)$ is the depth of the fluid, and g is the gravitational acceleration. The boundary condition on ∂B is zero normal flow velocity, i.e., $\partial \phi / \partial n = 0$. Consider the case that the water depth is a constant at the far field, an incident wave can be written in the form

$$(8) \qquad a\phi^I = -igc_0 \exp[ikr\cos(\theta - \alpha)]/\omega,$$

where r, θ are the cylindrical coordinates, $k = \omega/\sqrt{gh}$, c_0 is a constant and α is the angle of the incident wave direction. Interested readers can prove that $\phi^I \exp(-i\omega t)$ is a plane wave satisfying Eq. (7). The scattered wave potential, defined as $\phi^S = \phi - \phi^I$, must satisfy the far field conditions

$$(9) \qquad \lim_{r\to\infty} \sqrt{r} \left(\frac{\partial}{\partial r} - ik \right) \phi^S = 0, \qquad \lim_{r\to\infty} \sqrt{r} \phi^S \text{ is bounded},$$

which represents the behavior of an outgoing wave and is called the *radiation condition*. As before, the hybrid functional can be written as

$$(10) \qquad \Pi(\phi, \hat{\phi}) = \Pi_{R1}(\hat{\phi}) + \Pi_{R2}(\phi, \hat{\phi}),$$

where now

$$\Pi_{R1}(\hat{\phi}) = \frac{1}{2} \int_{R_1} \left[h(\nabla \hat{\phi})^2 - \frac{\omega^2}{g} \hat{\phi}^2 \right] dA,$$

$$\Pi_{R2}(\phi, \hat{\phi}) = \int_C \left[h(\hat{\phi} - \phi) \frac{\partial(\phi - \phi^I)}{\partial n} + h\hat{\phi}\frac{\partial \phi^I}{\partial n} \right] ds - \frac{1}{2} \int_{r \to \infty} ikh(\phi - \phi^I)^2 ds$$

$$+ \frac{1}{2} \int_{R_2} \left\{ h[\nabla(\phi - \phi^I)]^2 - \frac{\omega^2}{g}(\phi - \phi^I)^2 \right\} dA,$$

in which $\partial\phi/\partial n$ is the outward normal derivative along $\partial R_2 (= C)$. In the equations above, Π_{R1} is a regular functional over a finite domain surrounding the body B and Π_{R2} involves infinite domain. It can shown that the vanish of the first variation of Π with respect to $\hat{\phi}$, ϕ gives Eq. (7) as the Euler equations for the two field variables in R_1 and R_2, Eq. (9) as the rigid boundary condition for $\phi^S (= \phi - \phi^I)$, and the matching conditions

$$\hat{\phi} = \phi, \qquad \frac{\partial \hat{\phi}}{\partial n} = \frac{\partial \phi}{\partial n}$$

on curve C. If one makes ϕ satisfy Eq. (7) in R_2 and $\phi - \phi^I$ satisfy Eq. (9) *a priori*, one can use the Gauss theorem to eliminate the integration over the infinite domain in the functional Π_{R2} such that

$$(11) \qquad \Pi_{R2}(\phi, \hat{\phi}) = \int_C h\left(\hat{\phi} - \phi^I - \frac{\phi^S}{2}\right) \frac{\partial \phi^S}{\partial n} ds$$

$$+ \int_C h\hat{\phi}\frac{\partial \phi^I}{\partial n} ds.$$

which involves only line integrals along curve C in a finite domain.

To construct the finite element equations, we shall again divide R_1 into elements as shown in Fig. 19.5:1 and regard R_2 as a superelement. The hybrid element matrices for the infinite domain R_2 can easily be evaluated using Eq. (11). Consider the case where h is constant in R_2. We select the scattered-wave potential ϕ^S, which satisfies the Helmholtz equation Eq. (7)

and the radiation condition Eq. (9), in the form

$$(12) \qquad \phi^S = \phi - \phi^I = \alpha_0 H_0(kr)$$

$$+ \sum_{j=1}^{N} (\alpha_j \cos j\theta + \alpha_{j+N} \sin j\theta) H_j(kr) = \mathbf{G}\boldsymbol{\alpha},$$

in which

$$\mathbf{G} = [H_0(kr) \quad H_1(kr)\cos\theta \quad H_1(kr)\sin\theta \quad H_2(kr)\cos 2\theta \cdots].$$

where H's are the Hankel functions of the first kind and α's are unknown parameters. We assume $\hat{\phi}$ of R_2 to be the same as $\hat{\phi}$ of R_1 along C. The interpolation functions h's for $\hat{\phi}$ are similar to that of $\hat{u}$ in the form of Eq. (19.3:3) or (19.3:4) depending on whether $\hat{\phi}$ is linear or quadratic along the inter-element boundaries on C. We have

$$(13) \qquad\qquad \hat{\phi} = \hat{\mathbf{h}}\mathbf{q},$$

where all components of $\mathbf{h}$ are zero between nodes j and $j+1$, except

$$h_j = 1 - s/s_{j,j+1}, \qquad h_{j+1} = s/s_{j,j+1},$$

for linear $\mathbf{h}$ between nodes j and $j+1$, or

$$h_j = \left(1 - \frac{s}{s_{j,j+2}}\right)\left(1 - \frac{s}{s_{j,j+1}}\right), \qquad h_{j+1} = \left(\frac{s_{j,j+2} - s}{s_{j,j+2} - s_{j,j+1}}\right)\frac{s}{s_{j,j+1}},$$

$$h_{j+2} = \left(\frac{s_{j,j+1} - s}{s_{j,j+1} - s_{j,j+2}}\right)\frac{s}{s_{j,j+2}},$$

for quadratic $\mathbf{h}$ between nodes j and $j+2$. Substituting Eq. (12) and (13) into Eq. (11) yields

$$(14) \qquad \Pi_{R_2} = \boldsymbol{\alpha}^T \mathbf{k}_{\phi\hat{\phi}}\mathbf{q} - \frac{1}{2}\boldsymbol{\alpha}^T \mathbf{k}_{\phi\phi}\boldsymbol{\alpha} - \boldsymbol{\alpha}^T \mathbf{f}_\phi - \mathbf{q}^T \mathbf{f}_{\hat{\phi}},$$

where

$$\boldsymbol{\alpha}^T \mathbf{k}_{\phi\hat{\phi}}\mathbf{q} = \int_C \frac{\partial \phi^S}{\partial n}\hat{\phi}hds, \qquad \boldsymbol{\alpha}^T \mathbf{k}_{\phi\phi}\boldsymbol{\alpha} = \int_C \phi^S \frac{\partial \phi^S}{\partial n}hds,$$

$$\boldsymbol{\alpha}^T \mathbf{f}_\phi = \int_C \frac{\partial \phi}{\partial n}\phi^I hds, \qquad \mathbf{q}^T \mathbf{f}_{\hat{\phi}} = -\int_C \hat{\phi}\frac{\partial \phi^I}{\partial n}hds.$$

In principle, one can eliminate $\boldsymbol{\alpha}$ using stationary condition of Π_{R2} and obtains

$$\Pi_{R_2} = \frac{1}{2}\mathbf{q}^T \mathbf{k}\mathbf{q} - \mathbf{q}^T \mathbf{f} + \text{constant},$$

where

$$\mathbf{k} = \mathbf{k}_{\phi\hat{\phi}}^T \mathbf{k}_{\phi\phi}^{-1} \mathbf{k}_{\phi\hat{\phi}}, \qquad \mathbf{f} = \mathbf{k}_{\phi\hat{\phi}}^T \mathbf{k}_{\phi\phi}^{-1} \mathbf{f}_{\phi} + \mathbf{f}_{\hat{\phi}}.$$

These element matrices can be assembled like regular element matrices. The size of $\mathbf{k}$ can be quite large since it involves the nodes along the entire curve C.

One may also simply assemble the matrices of the infinite element in Eq. (14) with the element matrices of R_1 without first eliminating $\boldsymbol{\alpha}$ and use the assembled equations to solve for $\mathbf{q}$ and $\boldsymbol{\alpha}$ simultaneously.

If C is a circle of radius r_0, $\mathbf{k}_{\phi\phi}$ can be inverted analytically. Note that $\frac{\partial}{\partial n} = -\frac{\partial}{\partial r}$ at $r = r_0$ in Eq. (11). In this case, we have

$$(15) \qquad \boldsymbol{\alpha}^T \mathbf{k}_{\phi\phi} \boldsymbol{\alpha} = \pi r_0 kh \left[2\alpha_0^2 H_0 H_0' + \sum_{j=1}^N (\alpha_j^2 + \alpha_{j+N}^2) H_j H_j' \right],$$

where

$$H_j = H_j(kr_0), \qquad H_j' = \frac{dH_j}{dr}(kr_0).$$

The matrix $\mathbf{k}_{\phi\phi}$ is a diagonal matrix and can be inverted easily. A more general discussion of the waterwave problem and corresponding finite-element solutions can be found in Chen and Mei (1974) and Mei and Chen (1975).

19.6. INCOMPRESSIBLE OR NEARLY INCOMPRESSIBLE ELASTICITY

Many physical problems involve deformation that essentially preserves volume locally. Rubber is often modeled as such a material. In linearized theory of elasticity of isotropic materials obeying Hook's law, the incompressible or nearly incompressible condition is expressed by Poisson's ratio approaching $1/2$. This is seen by the ratio of bulk modulus K to shear modulus G shown in Eq. (6.2:9)

$$\frac{K}{G} = \frac{2(1+\nu)}{3(1-2\nu)}.$$

The limit creates problems in elasticity for isotropic materials since the constitutive equations Eq. (6.2:7) are

$$\sigma_{ij} = \lambda u_{k,k}\delta_{ij} + G(u_{i,j} + u_{j,i}).$$

(recall that $x_1 = x$, $x_2 = y$, $x_3 = z$, $u_1 = u$, $u_2 = v$, $u_3 = w$ in index notation.) The Lamé constant $\lambda[= 2\nu G/(1-2\nu)]$ becomes infinity at $\nu = 1/2$.

An alternative formulation is to include the *hydrostatic pressure p* as an independent field where

$$p/K = -e = -(u_{,x} + v_{,y} + w_{,z}) = -u_{j,j} \,,$$

in which e is the *dilatation* as defined in Eq. (7.1:11). A penalty type of variational functional (see Sec. 10.10) for an element can be written as

(1) ▲ $$\Pi_e = \int_{V_e} \left[\frac{1}{2} \mathbf{e}^T \mathbf{D}_I \mathbf{e} - \alpha \left(\frac{p^2}{2K} + p u_{k,k} \right) - \mathbf{u}^T \mathbf{b} \right] dV$$

$$- \int_{\partial V_{\sigma e}} \mathbf{u}^T \bar{\mathbf{T}}_n \, dS \,,$$

with the penalty function $f = K(u_{k,k})^2 + 2p u_{k,k} + p^2/K$ and

(2) $$\mathbf{D}_I = \mathbf{D}_e - \alpha K \widetilde{\mathbf{D}} \,, \qquad \underset{6 \times 6}{\widetilde{\mathbf{D}}} = \begin{bmatrix} 1 & & & & & \\ 1 & 1 & & & \text{sym} & \\ 1 & 1 & 1 & & & \\ & & & 0 & & 0 \\ & & & \scriptstyle 1 \times 3 & & \scriptstyle 3 \times 3 \end{bmatrix} \,,$$

$\mathbf{D}_e$ is the elastic modulus matrix of the material (see Sec. 18.14 for plane elasticity and Sec. 18.16 for 3-dimensional elasticity), $\mathbf{0}$ is a zero matrix with all entries being zero, K is the bulk modulus and α is a constant of order 1 chosen to ensure that $\mathbf{D}_I$ is bounded and positive definite. The choice of α is not unique except for incompressible materials.

In the case of an isotropic material, we choose α in the form

$$\alpha = (\lambda - 2G\beta) \frac{1}{K} = 1 - \frac{2G}{K} \left(\frac{1}{3} + \beta \right) \,,$$

where β is a selected constant of order 1. Then

(3) $$\mathbf{D}_I = G(\mathbf{D}' + 2\beta \widetilde{\mathbf{D}}) \,, \qquad \mathbf{D}' = G \, \text{diag}(2 \; 2 \; 2 \; 1 \; 1 \; 1) \,,$$

i.e., $\mathbf{D}'$ is a diagonal matrix. When the material is incompressible, α becomes $3\nu/(1 + \nu)(= 1)$. For any non-negative β, $\mathbf{D}_I$ is positive definite. If $\beta = 0$, Eq. (2) reduces to the conventional mixed functional (Herrmann 1965, Tong 1969) involving the two fields $\mathbf{u}, p$.

(4) $$\Pi'_e = \int_{V_e} \left(\frac{1}{2} \mathbf{e}'^T \mathbf{D}' \mathbf{e}' - \frac{p^2}{2K} - p u_{k,k} - \mathbf{u}^T \mathbf{b} \right) dV$$

$$- \int_{\partial V_{\sigma e}} \mathbf{u}^T \bar{\mathbf{T}}_n \, dS \,,$$

where $\mathbf{e}'$ is the *deviatoric engineering strain matrix* and $\mathbf{D}'$ is always bounded and given by Eq. (3) for isotropic materials. In Eq. (1), p plays the role of the Langrange multiplier to enforce the pressure-dilatation strain relation. In the limiting case of incompressible materials, it enforces the incompressible condition by requiring the dilatation to be zero.

The integrand of the functional above involves both strains and hydro-static pressure and therefore Π'_e a mixed functional. Admissibility simply requires that $\mathbf{u}$ be C^0 continuous. Discontinuous p is permitted. The mathematical convergence theory for mixed finite element methods can be found in Tong (1969), Babuska (1971), Brezzi and Pitkaranta (1974), Oden and Carey (1984), and Brezzi (1984).

In finite element formulation, we express

(5) $\mathbf{u} = \mathbf{hq}, \qquad p = \mathbf{gp}, \qquad e = \mathrm{div}(\mathbf{u}) = d_d \mathbf{hq} = \mathbf{B}_d \mathbf{q}$

where $\mathbf{q}$, $\mathbf{p}$ are nodal value matrices of $\mathbf{u}$, p with $\mathbf{h}$ and $\mathbf{g}$ being their interpolation matrices, $\mathbf{B}_d$ relates the nodal displacement to the volumeric strains, and $\mathbf{d}_d$ is a differential operator that

$$
\mathbf{d}_d = \begin{bmatrix} \dfrac{\partial}{\partial x} & 0 & 0 \\[2mm] 0 & \dfrac{\partial}{\partial y} & 0 \\[2mm] 0 & 0 & \dfrac{\partial}{\partial z} \end{bmatrix} .
$$

The interpolation matrix $\mathbf{h}$ is in the form of Eq. (18.14:10) for plane or axisymmetric problems and in the form of Eq. (18.16:13) for three-dimensional elasticity. A substitution of Eq. (5) into (1) yields

(6) $\Pi_e = \dfrac{1}{2} \mathbf{q}^T \mathbf{k}_{qq} \mathbf{q} - \mathbf{q}^T \mathbf{k}_{qp} \mathbf{p} - \dfrac{1}{2} \mathbf{p}^T \mathbf{k}_{pp} \mathbf{p} - \mathbf{q}^T \mathbf{f} ,$

where

(7) $\mathbf{k}_{qq} = \displaystyle\int_{V_e} \mathbf{B}^T \mathbf{D}_I \mathbf{B} J d\xi d\eta d\zeta , \qquad \mathbf{k}_{qp} = \alpha \displaystyle\int_{V_e} \mathbf{B}_d^T \mathbf{g} J d\xi d\eta d\zeta ,$

$\mathbf{k}_{pp} = \alpha \displaystyle\int_{V_e} \dfrac{\mathbf{g}^T \mathbf{g}}{K} J d\xi d\eta d\zeta , \qquad \mathbf{f} = \displaystyle\int_{V_e} \mathbf{h}^T \mathbf{b} \, dV + \displaystyle\int_{\partial V_{\sigma e}} \mathbf{h}^T \bar{\mathbf{T}} \, dS .$

Note that $\mathbf{f}$ is the same as that defined in Eq. (18.16:16).

In general, we use lower order shape functions for the pressure because the continuity requirement for p is lower as compared to that for $\mathbf{u}$. For linear $\mathbf{u}$, we use constant for $p(= p_1)$ in the element; for quadratic $\mathbf{u}$, we use linear $p(= p_1 + p_2 x + p_3 y + p_4 z)$ and so on. If one assumes discontinuous pressure between elements, in the compressible case, one can eliminate the pressure degrees-of-freedom at the element level, i.e.,

$$\mathbf{k}_{pp}\mathbf{p} + \mathbf{k}_{qp}^T\mathbf{q} = 0, \qquad \text{or} \qquad \mathbf{p} = -\mathbf{k}_{pp}^{-1}\mathbf{k}_{qp}^T\mathbf{q}.$$

A substitution of $\mathbf{p}$ into Eq. (6) reduces Π_e to the standard form

$$\Pi_e = \frac{1}{2}\mathbf{q}^T\mathbf{kq} - \mathbf{q}^T\mathbf{f},$$

where

$$\mathbf{k} = \mathbf{k}_{qq} + \mathbf{k}_{qp}\mathbf{k}_{pp}^{-1}\mathbf{k}_{qp}^T.$$

We then can derive the global stiffness and force matrices by assemblying the element matrices in the usual manner. The procedure of first eliminating the pressure degrees-of-freedom fails for incompressible materials because $\mathbf{k}_{pp}^{-1}$ becomes singular (K becomes infinite). The procedure does not offer much advantage for nearly incompressible materials as the terms involving $\mathbf{k}_{pp}^{-1}$ can dominate $\mathbf{k}_{qq}$ when K is large. It is better off to solve the pressure and nodal-displacement equations simultaneously. However, some precautions should be taken. For example, an equation corresponding to a zero or nearly zero diagonal must not appear as the first equation of the global system. When this happens, the direct method of solution cannot proceed.

Displacement and pressure are two independent fields. However, arbitrary combinations of displacement and pressure interpolations may not be effective. Each pressure degree-of-freedom is equivalent to a constraint to the deformation mode. Too few pressure degrees-of-freedom can lead to poor approximation. It can also lead to spurious deformation with zero strain energy for the functional in Eq. (4), which can make the equation unsolvable. For example, pure dilatation with zero pressure ($\gamma_{xy} = \gamma_{yz} = \gamma_{zx} = p = 0, e_{xx} = e_{yy} = e_{zz}$) give zero strain energy in Eq. (4).

Too many pressure modes will overly constrain the deformation and lead to locking. Consider the constant strain triangle with one pressure degree-of-freedom. For the element mesh shown in Fig. 19.6:1, the constant volume (area) condition for element 2 precludes the horizontal motion of node 1. In the mean time, the same condition for element 1 restrains node 1 to move only horizontally. As a result, the displacements at node 1 can only be zero. With displacements at node 1 being zero, it infers that nodes 2

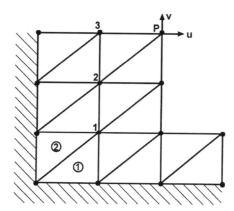

Fig. 19.6:1. Mesh for which incompressibility dictates zero displacement.

and 3 have zero displacements and so on. We conclude that every node in the entire mesh must have zero displacement. The result holds no matter how many elements are in the mesh.

Hughes (1987) introduced a heuristic approach for determining the ability of an element to perform well in incompressible and nearly incompressible applications. Consider a finite element mesh. Let n_{eq} represent the total number of displacement equations after boundary conditions have been imposed and n_c be the total number of incompressibility constraints. For linear problems, n_c is the total number of pressure degrees of freedom. The *constraint ratio* r is defined as

$$r = n_{eq}/n_c.$$

Hughes conjectured that r should mimic the behavior of the number of equilibrium equations divided by the number of incompressibility conditions for the system of the governing partial differential equations, i.e., $r = n$ where n is the number of space dimensions. So in two dimensions, the ideal value of r should be 2. A value of r less than 2 would indicate a tendency to lock. If $r \leq 1$, there are more constraints on $\mathbf{q}$ than the number of q's available. Severe locking will be resulted. If r is much greater than 2, it indicates a lack of incompressibility constraint which can lead to poor solution. The results (Huges 1987) are illustrated in Figs. 19.6:2 and 19.6:3.

Another approach to handle the nearly incompressible issue is to split the elastic coefficient matrix $\mathbf{D}_e$ into the G-part (shear) and λ-parts (Lamé constant) (Hughes 1987)

$$\mathbf{D}_e = \widehat{\mathbf{D}} + \lambda\widetilde{\mathbf{D}},$$

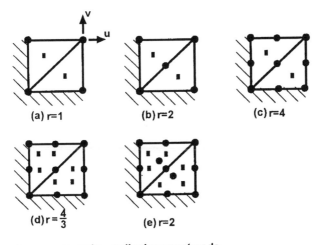

• pressure node   **●** displacement node

Fig. 19.6:2. Triangular elements with discontinuous pressure

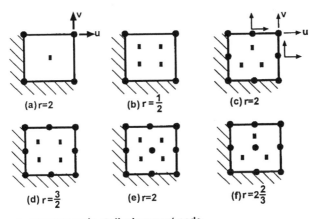

• pressure node   **●** displacement node

Fig. 19.6:3. Rectangular elements with discontinuous pressure

where $\widehat{\mathbf{D}} = \mathbf{D}'$ is given in Eq. (3) and $\widetilde{\mathbf{D}}$ is given in Eq. (2) for isotropic materials. For plane strain,

$$\widehat{\mathbf{D}} = G\,\mathrm{diag}\,(2\ \ 2\ \ 1), \qquad \widetilde{\mathbf{D}}_{3\times3} = \begin{bmatrix} 1 & 1 & 0 \\ 1 & 1 & 0 \\ 0 & 0 & 0 \end{bmatrix}.$$

Using the displacement formulation, from Eq. (18.8:23), the element stiffness matrix can be written as

$$\mathbf{k} = \hat{\mathbf{k}} + \tilde{\mathbf{k}},$$

where

$$\hat{\mathbf{k}} = \int_{V_e} \mathbf{B}^T \hat{\mathbf{D}} \mathbf{B} dV, \qquad \tilde{\mathbf{k}} = \lambda \int_{V_e} \mathbf{B}^T \tilde{\mathbf{D}} \mathbf{B} \, dV \quad \mathbf{B} = \mathbf{d}_e \mathbf{h},$$

in which $\mathbf{d}_e$ is given in Eq. (18.6:6).

If λ/G is very large, the numerical values of $\tilde{\mathbf{k}}$ can much larger than $\hat{\mathbf{k}}$. One simple and important practical way to lower the stiffness of an element is to reduce the order of numerical quadrature by 1 below that 'normally' used in evaluating $\tilde{\mathbf{k}}$. For instance, for the Gauss integration, if we use $p + 1$ integration points to evaluate $\hat{\mathbf{k}}$, we will use only p-point to evaluate $\tilde{\mathbf{k}}$. This is a form of *selectively reduced integration*, which is a remedy to eliminate shear locking as discussed before. Malkus and Hughes (1978) showed that many elements of mixed formulation are equivalent to elements of selectively reduced integration.

20

FINITE ELEMENT METHODS FOR
PLATES AND SHELLS

Plates and shells are particular forms of a three-dimensional solid in which the thickness is small as compared with other dimensions. Conceptually, they do not present difficulties for the finite element method. However, the 3-dimensional elements for such structures generally would have very large aspect ratio (in-plane dimension to thickness), which makes the elements overly stiff in the out-plane direction. This 'overly stiff' phenomenon is known as '*locking*' as defined in Sec. 18.20. In order to obtain a solution of acceptable accuracy in the 3-dimensional model, it often requires elements of size of the order of thickness or less. This would be costly. Approximate theories such as the *Kirchhoff theory* for thin plates and shells and the *Reissner–Mindlin theory* for moderately thick plates and shells have been developed to reduce the plate and shell problems to two dimensions. In this chapter, we shall discuss the finite element method to model the Kirchhoff (Secs. 16.15 and 16.16) and Reissner–Mindlin theories of plates and shells. Other approaches, including the reduced integration and mixed formulations to avoid locking, will be discussed also. Throughout this chapter, we shall use t as the plate or shell thickness.

20.1. LINEARIZED BENDING THEORY OF THIN PLATES*

Let the midplane of the plate be on the x, y plane in a Cartesian system of coordinates. The *Kirchhoff theory* assumes that (1) the deformation in the midplane is like that of the plane stress state discussed in Sec. 9.1,

---

*With focus on the finite element approach, the linearized small deflection theory of plates and shells is disucssed in this chapter. The linearized small deflection theory of plates is presented in Sec. 16.15. To solve the nonlinear problem, a fundamental method of incremental approach is discussed in Chapter 17. Using finite element method for incremental approach, we need the linearized solution at every step. Hence the material discussed herein is important.

(2) strains vary linearly along lines normal to the midplane, and (3) lines normal to the midplane remain normal to the deformed midsurface after deformation. Under the Kichhoff assumptions, the state of deformation off the midplane is determined only by the lateral deflection, w, normal to the midplane. The displacements at any point of the plate are given by

$$ u = u_0 - z\frac{\partial w}{\partial x}, \qquad v = v_0 - z\frac{\partial w}{\partial y}, \qquad w = w, $$

where z is the distance from the midplane and u_0, v_0 are the in-plane displacements determined by plane-stress analysis based on the assumption (1) mentioned above. Thus, u_0, v_0 associate with midplane stretching and shear, w relates to bending, and $\partial w/\partial y$ and $-\partial w/\partial x$ represent the *rotations* along the x-, y-axes of the midplane, respectively. In the linearized theory of plates, stretching and bending are decoupled. Therefore, in Secs. 1 through 4, since only plate bending is considered, we shall drop terms involving u_0, v_0. It is only after Sec. 4, when shells are considered, we will include the midsurface in-plane displacements to account for stretching and bending coupling.

The strains associated with the off-midplane deflections are proportional to the changes of the curvature of the midplane and called the bending strains. These strains forms a column matrix $\boldsymbol{\kappa}$ with components consisting of the second derivatives of w.

$$ (1) \qquad \boldsymbol{\kappa} = [\kappa_x \ \kappa_y \ \kappa_{xy}]^T = \mathbf{d}_e \mathbf{d} w, $$

which represents the *change of curvature* as discussed in Sec. 16.15, and $\mathbf{d}_e$ and $\mathbf{d}$ are matrix differential operators defined in Eqs. (18.11:5) and (18.11:6):

$$ \mathbf{d}_e = \begin{bmatrix} \dfrac{\partial}{\partial x} & 0 & \dfrac{\partial}{\partial y} \\[2mm] 0 & \dfrac{\partial}{\partial y} & \dfrac{\partial}{\partial x} \end{bmatrix}^T, \qquad \mathbf{d} = \begin{bmatrix} \dfrac{\partial}{\partial x} \\[2mm] \dfrac{\partial}{\partial y} \end{bmatrix}. $$

Thus $\mathbf{d}_e\mathbf{d}$ is a *second order matrix differential operator* that

$$ (2) \qquad (\mathbf{d}_e\mathbf{d})^T = \begin{bmatrix} \dfrac{\partial^2}{\partial x^2} & \dfrac{\partial^2}{\partial y^2} & 2\dfrac{\partial^2}{\partial x\partial y} \end{bmatrix}. $$

The stress measures are *bending moments*, $\mathbf{M}$, and *transverse shear forces*, $\mathbf{Q}$, per unit length along lines on the midsurface of the plate. They are, in matrix form,

$$ (3) \qquad \mathbf{M} = [M_x \ M_y \ M_{xy}]^T, \qquad \mathbf{Q} = [Q_x \ Q_y]^T. $$

As discussed in Sec. 16.15, $\mathbf{Q}$ is related to $\mathbf{M}$ by Eq. (16.15:13), and $\mathbf{M}$ is related to the lateral displacement w through the *constitutive law* Eq. (16.15:19), which in matrix form, is

(4) $$\mathbf{M} = -\mathbf{D}_b\boldsymbol{\kappa} = -\mathbf{D}_b\mathbf{d}_e dw \,,$$

where $\mathbf{D}_b$ is the *bending rigidity matrix*. For isotropic elastic material, $\mathbf{D}_b$ is (Sec. 16.15)

(5) $$\mathbf{D}_b = \frac{Et^3}{12(1-\nu^2)}\begin{bmatrix} 1 & \nu & 0 \\ \nu & 1 & 0 \\ 0 & 0 & (1-\nu)/2 \end{bmatrix},$$

where E represents Yong's modulus, ν is Poisson's ratio (see Sec. 6.2), and t is the thickness of the plate. The shear $\mathbf{Q}$ is related to $\mathbf{M}$ through the moment equilibrium equation (16.15:13).

Consider a domain A with the following two sets of boundary conditions:

(6) $$w = \bar{w} \qquad\qquad \text{on } \partial A_w \,,$$

(7) $$n_x Q_x + n_y Q_y - \frac{dM_{ns}}{ds} = \bar{Q} - \frac{d\overline{M}_{ns}}{ds} \qquad \text{on } \partial A_Q \,,$$

and

(8) $$\frac{\partial w}{\partial n} = \bar{w}_n \qquad\qquad \text{on } \partial A_{wn} \,,$$

(9) $$n_x^2 M_x + 2n_x n_y M_{xy} + n_y^2 M_y = \overline{M}_{nn} \qquad \text{on } \partial A_M \,,$$

where

$$M_{ns} = n_x n_y (M_y - M_x) + (n_x^2 - n_y^2) M_{xy} \,.$$

The overbar denotes prescribed quantities with $\overline{M}_{nn}$, $\overline{M}_{ns}$ being the prescribed normal and twist moments, n_x, n_y are the direction cosines of the unit outward normal to the boundary, M's, Q's are in terms of w through the constitutive law, $\partial/\partial n$ denotes the outward normal derivative along the boundary, and the subscripted ∂A's are part of the domain boundaries with the subscripts identifying the prescribed quantities. At any given boundary location, one and only one condition from each of the two sets above can be specified. Equations (6) and (8) are *rigid* and Eqs. (7) and (9) are *natural boundary conditions*.

The functional for the *principle of minimum potential energy theorem* is

$$\Pi = \int_A \left[\frac{1}{2}(\mathbf{d}_e dw)^T \mathbf{D}_b \mathbf{d}_e dw - pw \right] dxdy + \int_{\partial A_M} \overline{M}_{nn}\frac{\partial w}{\partial n} ds - \int_{\partial A_Q} \bar{Q}w ds \,,$$

where $p = p(x, y)$ is the distributed lateral pressure on the plate. The functional can be written as the sum of the functional of all elements provided that w is C^1 continuous

$$\Pi = \sum_{\text{all elements}} \Pi_{Be} \,,$$

where Π_{Be} is the functional of an element in the same form as Π except that the domains of integration are replaced with the corresponding domains of the element. We further divide the element functional into two parts, namely,

(10) $$\Pi_{Be} = \Pi_{be} - \Pi_{bfe}$$

where

$$\Pi_{be} = \int_{A_e} \frac{1}{2} (\mathbf{d}_e \mathrm{d} w)^T \mathbf{D}_b \mathbf{d}_e \mathrm{d} w \, dx \, dy \,,$$

$$\Pi_{bfe} = \int_{A_e} pw \, dxdy - \int_{\partial A_{M_e}} \overline{M}_{nn} \frac{\partial w}{\partial n} \, ds + \int_{\partial A_{Q_e}} \bar{Q} w \, ds \,,$$

in which A_e is the area of the element, and ∂A_{M_e} and ∂A_{Q_e} are the portions of the element boundary coinciding with ∂A_M and ∂A_Q, respectively. Let us introduce additional subscripts for Π's such that Π_{Be} is the potential energy, Π_{be} represents the strain energy, and Π_{bfe} denotes the external work on the element associated with bending deformation. This is to distinguish them from the corresponding quantities associated with membrane action to be discussed later.

For *convergence*, the shape functions must be able to approximate the *state of constant curvature* within an element. This requires that the shape functions have the *complete quadratic terms*. The governing differential equation for w is fourth order. The *condition number* (see Sec. 18.5 and Tong 1970c) of the resulting stiffness matrix grows as $1/\varepsilon^4$ where ε is the minimum dimension of all elements. Thus the condition number grows rapidly as the size of elements decreases which can lead to round-off problem in the finite element solution for fine meshes.

Determination of C^1 shape functions is much harder than the determination of C^0 functions. Elements with the shape functions that preserve continuity but may violate slope continuity between elements are called *non-conforming* or *incompatible elements* (see Sec. 20). For incompatible elements, we have to rely on the *patch test* to ensure the convergence of the approximate solution. Constructing multi-dimensional C^1-interpolations created considerable difficulties in the early development of the finite element method, especially for triangular elements. Various conforming plate elements based upon the displacement model have subsequently been developed. Unfortunately, the resulting compatible shape functions are generally complicated. Hybrid or mixed formulations are alternative approaches to circumvent the difficulty.

The different elements that have evolved are distinguished by the type of shape functions, generalized coordinates at nodes, geometrical shape, and the number of nodes. Clough and Tocher (1966) were among the first who successfully constructed a compatible low-order triangular bending element. The element was very stiff that the solution using the element converges very slowly. Bell (1969), Cowper *et al.* (1969), Holand and Bell (1972) derived conforming higher order triangular elements that use six quantities w, $w_{,x}$, $w_{,y}$, $w_{,xx}$, $w_{,xy}$, and $w_{,yy}$ at the vertices as the generalized nodal coordinates. The subscripts with a comma denote partial differentiation with respect to the subscript variable. This convention will be used throughout the remaining of this book. Clough and Felippa (1968) developed a similar bending element for quadrilaterals. The elements is quite accurate but computationally expensive. For many structural applications, discontinuity is natural that abrupt changes in thickness or composition due to discrete stiffening or members meeting at angles is part of design. At such locations, continuity should *not* be enforced. Conforming elements, especially the lower order ones, may yield inferior accuracy as compared to non-conforming elements. Therefore, non-conforming elements are widely used in practice.

Higher order bending elements can improve the accuracy. Unfortunately the higher order conforming elements are even more complicated. Therefore, if one weighs the expenses against the incremental increase in accuracy, simpler non-compatible elements can be a good choice. It should be emphasized that in specific applications where accurate stress (bending moments, which are proportional to the second derivatives of w) prediction is important, especially in regions of high stress gradients, higher-order elements may be necessary.

In the following, two non-compatible elements based the variational functional given in Eq. (10) will be developed.

Incompatible Rectangular Element. Consider the rectangular plate element in the x, y-plane shown in Fig. 20.1:1 and take the nodal values of w, $w_{,x}$, $w_{,y}$ at the four corners as the generalized coordinates. The element has a total of 12 degrees-of-freedom. The transpose of $\mathbf{q}_B$ is:

$$\mathbf{q}_B^T = [w(\xi_1,\eta_1)\, w_{,x}(\xi_1,\eta_1)\, w_{,y}(\xi_1,\eta_1) \cdots w_{,x}(\xi_1,\eta_2)\, w_{,y}(\xi_1,\eta_2)],$$

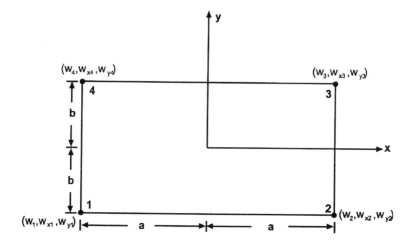

Fig. 20.1:1. A rectangular element.

where q_B is a column matrix, $\xi_1 = \eta_1 = -1$, $\xi_2 = \eta_2 = 1$, and $\xi = x/a$, $\eta = y/b$ are the *natural coordinates* of the element. We approximate w within the element as a polynomial in terms of 12 parameters:

$$(11) \quad w(x,y) = \alpha_1 + \alpha_2 x + \alpha_3 y + \alpha_4 x^2 + \alpha_5 xy + \alpha_6 y^2 + \alpha_7 x^3 + \alpha_8 x^2 y$$

$$+ \alpha_9 xy^2 + \alpha_{10} y^3 + \alpha_{11} x^3 y + \alpha_{12} xy^3 = \mathbf{P}\boldsymbol{\alpha}\,,$$

where $\mathbf{P}$ and $\boldsymbol{\alpha}$ are matrices defined as

$$\underset{(1\times12)}{\mathbf{P}} = \begin{bmatrix} 1 & x & y & x^2 & \cdots & xy^3 \end{bmatrix},$$

$$\boldsymbol{\alpha}^T = \begin{bmatrix} \alpha_1 & \alpha_2 & \alpha_3 & \cdots & \alpha_{12} \end{bmatrix}.$$

Along lines x or y = constant, the displacement w is cubic, which can be uniquely defined by four parameters. Since the inter-element boundaries are lines with either x or y being constant, the two nodal values of w and its *tangential derivatives* at the two ends of a boundary of the element define w along the boundary. The *tangential derivative* of w along the boundary x = constant is $w_{,y}$, whereas that along y = constant is $w_{,x}$. As a result, if two neighboring elements have the same nodal values, continuity of w is assured along their common boundary. From Eq. (11), since both $w_{,x}$ and $w_{,y}$ vary as a cubic function, the derivative of w normal to any boundary is cubic in general. The normal derivative of w along an element boundary is matched with that of the adjacent element only at the two end-points of the boundary line. Therefore, continuity in the *normal derivative* of w

between the two points *cannot* be assured. Thus, the interpolation function given in Eq. (11) is nonconforming.

Using more degrees of freedom for each node can resolve the problem just described. In particular, if one adds the twist $w_{,xy}$ as an additional degree-of-freedom at each node (thereby resulting in an element of 16 degrees-of-freedom with four at each node), the uniqueness of normal slope of w along the inter-element boundaries, therefore compatibility, can be established. In this case, four additional terms $(\alpha_{13}x^2y^2 + \alpha_{14}x^3y^2 + \alpha_{15}x^2y^3 + +\alpha_{16}x^3y^3)$ are added to Eq. (11), thereby allowing higher-order matching between adjacent elements.

Further development of the element may take two alternative approaches. In the first approach, one expresses w in terms of q's and the shape functions:

$$(12) \qquad w(x,y) = \sum_{i=1}^{2} \sum_{j=1}^{2} [h_{ij}(\xi,\eta)w(\xi_i,\eta_j) + h_{xij}(\xi,\eta)w_{,x}(\xi_i,\eta_j)$$

$$+ h_{yij}(\xi,\eta)w_{,y}(\xi_i,\eta_j)] = \mathbf{h}\mathbf{q}_B ,$$

where $x = a\xi$, $y = b\eta$, and

$$\mathbf{h} = [h_{11} \ h_{x11} \ h_{y11} \ h_{21} \ h_{x21} \ h_{y21} \ h_{22} \ h_{x22} \ \cdots \ h_{y12}] ,$$

in which

$$h_{ij} = \frac{\xi_i\eta_j}{8}(2 + \xi\xi_i + \eta\eta_j - \xi^2 - \eta^2)(\xi + \xi_i)(\eta + \eta_j) ,$$

$$(13) \qquad h_{xij} = \frac{\eta_j}{8}(\xi^2 - 1)(\xi + \xi_i)(\eta + \eta_j)a ,$$

$$h_{yij} = \frac{\xi_i}{8}(\eta^2 - 1)(\xi + \xi_i)(\eta + \eta_j)b$$

(i, j not summed). A substitution of Eq. (12) into Eq. (10) yields

$$(14) \qquad \Pi_{Be} = \frac{1}{2}\mathbf{q}_B^T \mathbf{k}_B \mathbf{q}_B - \mathbf{q}_B^T \mathbf{f}_B ,$$

where the matrices $\mathbf{k}_B$, $\mathbf{f}_B$ are

$$(15) \qquad \mathbf{k}_B = \int_{Ae} (\mathbf{d}_e\mathbf{dh})^T \mathbf{D}_b \mathbf{d}_e \mathbf{dh} \, dxdy ,$$

$$(16) \qquad \mathbf{f}_B = \int_{Ae} p\mathbf{h}^T \, dxdy - \int_{\partial A_{Me}} \overline{M}_{nn} \frac{\partial \mathbf{h}^T}{\partial n} \, ds + \int_{\partial A_{Qe}} \bar{Q}\mathbf{h}^T \, ds .$$

In terms of the natural coordinates, Eqs. (14) and (15) can be written as

(17)
$$\mathbf{k}_B = ab \int_{-1}^{1} \int_{-1}^{1} (\widetilde{\mathbf{d}}_e \widetilde{\mathbf{dh}})^T \mathbf{D}_b(x,y) \widetilde{\mathbf{d}}_e \widetilde{\mathbf{dh}} \, d\xi d\eta \, ,$$

(18)
$$\mathbf{f}_B = ab \int_{-1}^{1} \int_{-1}^{1} p(x,y) \mathbf{h}^T \, d\xi d\eta$$

$$- \int_{\partial A_{Me}} \overline{M}_n \frac{\partial \mathbf{h}^T}{\partial n} \, ds + \int_{\partial A_{Qe}} \overline{Q} \mathbf{h}^T \, ds \, ,$$

where
$$(\widetilde{\mathbf{d}}_e \widetilde{\mathbf{dh}})^T = \begin{bmatrix} \dfrac{1}{a^2} \dfrac{\partial^2 \mathbf{h}^T}{\partial \xi^2} & \dfrac{1}{b^2} \dfrac{\partial^2 \mathbf{h}^T}{\partial \eta^2} & \dfrac{2}{ab} \dfrac{\partial^2 \mathbf{h}^T}{\partial \xi \partial \eta} \end{bmatrix}.$$

They are similar in form as Eqs. (18.14:15) and (18.14:16). Equations (16) and (17) can be evaluated analytically if the *bending modulus matrix* $\mathbf{D}_b$ and the pressure load p are polynomials, or numerically otherwise.

In the second approach, we obtain the relation between α's and q's in the matrix form

$$\mathbf{A}\boldsymbol{\alpha} = \mathbf{q} \, ,$$

by evaluating w and its partial derivatives at the nodes. We evaluate the element matrices in terms α's using w as defined in Eq. (11). We then express α's in terms q's by the transformation matrix $\mathbf{A}^{-1}$ above. The components of the element matrices in terms of α's are mostly zero if $\mathbf{D}_b$ and p are constant.

Distorted quadrilateral elements based on the isoparametric transformation of Eq. (18.13:3) perform badly. In this case, the interpolated function cannot represent exactly the deformation of constant curvature except for parallelogram (Dawe 1966, Argyris 1966).

Incompatible Triangular Element. The early development of triangular elements encountered considerable difficulty when attempting to construct the shape functions in terms of x, y directly. For an element with three degrees-of-freedom per node, as shown in Fig. 20.1:2, there are only nine parameters whereas the complete cubic polynomial has ten terms. Several possibilities have been investigated to limit the number of unknowns to nine but various problems arose, which include the lack of symmetric appearance of the ten terms, the non-uniqueness of shape functions, etc.

The asymmetric appearance can be avoided by using the triangular coordinates. Bazeley *et al.* (1966) assumed w in the form

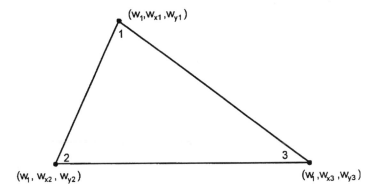

Fig. 20.1:2. Triangular element with three degrees-of-freedom per node.

$$w = \alpha_1\varsigma_1 + \alpha_2\varsigma_2 + \alpha_3\varsigma_3 + \alpha_4\varsigma_1 \left(\varsigma_2^2 + \frac{\varsigma_2\varsigma_3}{2}\right) + \alpha_5\varsigma_1 \left(\varsigma_3^2 + \frac{\varsigma_3\varsigma_2}{2}\right)$$

$$+ \cdots + \alpha_9\varsigma_3 \left(\varsigma_2^2 + \frac{\varsigma_2\varsigma_1}{2}\right) .$$

Expressing α's in terms of the nodal values w_i, $w_{,xi}$, $w_{,yi}$ of the three nodes, one obtains the shape functions associated with the nodal parameters.

The element failed the patch test for arbitrary mesh configurations. The shape functions were subsequently amended to pass the patch test by Specht (1988) who assumed w in the form

$$w = [\varsigma_1, \varsigma_2, \varsigma_3, \varsigma_1\varsigma_2, \varsigma_2\varsigma_3, \varsigma_3\varsigma_1 ,$$

$$\varsigma_1^2\varsigma_2 + \varsigma_1\varsigma_2\varsigma_3\{3(1 - \mu_3)\varsigma_1 - (1 + 3\mu_3)\varsigma_2 + (1 + 3\mu_3)\varsigma_3\}/2 ,$$

$$\varsigma_2^2\varsigma_3 + \varsigma_1\varsigma_2\varsigma_3\{3(1 - \mu_1)\varsigma_2 - (1 + 3\mu_1)\varsigma_3 + (1 + 3\mu_1)\varsigma_1\}/2 ,$$

$$\varsigma_3^2\varsigma_1 + \varsigma_1\varsigma_2\varsigma_3\{3(1 - \mu_2)\varsigma_3 - (1 + 3\mu_2)\varsigma_1 + (1 + 3\mu_2)\varsigma_2\}/2]\alpha = \mathbf{P}\alpha ,$$

where $\mu_i = (l_k^2 - l_j^2)/l_i^2$ with i, j, k being the cyclic permutations of subscripts 1, 2, 3 and l's being the side lengths of the triangle. The parameter μ_i relates to the normal derivative on side i by

$$\frac{\partial}{\partial n} = \frac{l_i}{4\Delta} \left[\frac{\partial}{\partial \varsigma_j} + \frac{\partial}{\partial \varsigma_k} - 2\frac{\partial}{\partial \varsigma_i} + \mu_i \left(\frac{\partial}{\partial \varsigma_k} - \frac{\partial}{\partial \varsigma_j} \right) \right] .$$

The shape functions corresponding to the nodal parameters w, $w_{,x}$, $w_{,y}$ at node i are

$$
\begin{bmatrix} h_{3(i-1)+1} \\ h_{3(i-1)+2} \\ h_{3(i-1)+3} \end{bmatrix} = \begin{bmatrix} \varsigma_i - \varsigma_i\varsigma_j + \varsigma_i\varsigma_k + 2(P_{i+6} - P_{k+6}) \\ -b_j(P_{k+6} - \varsigma_i\varsigma_k) - b_k P_{i+6} \\ -c_j(P_{k+6} - \varsigma_i\varsigma_k) - c_k P_{i+6} \end{bmatrix},
$$

where b's and c's are defined in Eq. (18.9:7) and P's are the components of **P**. It can be shown that the shape functions are C^0 continuous for arbitrary μ's, which are chosen to satisfy the patch test.

The element matrices are given by Eqs. (14) and (15) with A_e being the area of the triangular element. The differential operator for the element stiffness matrix in Eq. (14) is defined in Eq. (20.1:2) as

$$
(19) \qquad \mathbf{d}_e\mathbf{d} = \frac{1}{4\Delta^2} \sum_{i,j=1}^{3} [b_ib_j \; c_ic_j \; 2b_ic_j]^T \frac{\partial^2}{\partial\varsigma_i\partial\varsigma_j} .
$$

The integrals in Eqs. (14) and (15) can be evaluated analytically using the general formula Eq. (18.10:10). However, the explicit integration is lengthy and error prone (Cheung et al. 1968). It is more practical to use numerical integration.

It is possible to construct simpler elements of C^0 continuity. Morley (1971) first proposed an element with w at the triangle vertices and its normal slope at the midsides of the element as nodal parameters. The shape functions form a complete quadratic polynomial and can represent constant curvatures. If the material and the thickness of the element are uniform, this element is identical to the hybrid stress element of the same displacement at the element boundaries because the constant curvatures satisfy the homogeneous moment equilibrium equations. With constant moments, the element satisfies exactly interelement equilibrium conditions (Hinton and Huang 1986, Zienkiewicz et al. 1990).

The discussion up to this point has been focused on the approaches to overcome compatibility difficulties of the displacement model of plates. Alternative is to use the hybrid stress method (Pian 1964) to be discussed later. The success of Pian's approach in resolving the compatibility dilemma lies in its flexibility in the element-formulation, whereby stress and displacement fields are assumed over different portions of the element. Hybrid stress plate elements have yielded excellent results for stress distribution, deflection and vibration (Mau et al. 1973, Pian and Mau 1972, Rossettos et al. 1972, Spilker and Munir 1980).

20.2. REISSNER-MINDLIN PLATES

For moderately thick plates (Fig. 16.15:5) the transverse shear stresses τ_{xz}, τ_{yz} across the thickness of the plate become significant. The *Reissner-Mindlin plate theory* (Reissner 1946) includes transverse shear by allowing the rotations of the fibers normal to the midplane about the x-, y-axes to be different from the slopes of the midplane $w_{,y}$ and $-w_{,x}$. Like thin plates, the in-plane stretching and shear decouple from the bending and transverse shear deformations and can be considered separately. The theory still assumes that the stress normal to the plate is zero, i.e., $\sigma_{zz} = 0$, and that the displacements off the midplane are

$$(1) \qquad\qquad u = -z\theta_x\,, \qquad v = -z\theta_y\,, \qquad w = w(x, y)\,,$$

where z is the distance off the midplane, and θ_y and $-\theta_x$ are the rotations of the cross sectional plane about the x- and y-axes, respectively. The transverse shear strains are then $w_{,x} - \theta_x$ and $w_{,y} - \theta_y$. Strictly speaking the assumption $\sigma_{zz} = 0$ contradicts the assumption $w = w(x, y)$. However, the inconsistency does not affect the usefulness of the resulting theory in practical engineering applications.

Following the procedure for the Kirchhoff plate theory, one can derive the finite element equations for Reissner-Mindlin plates: the bending and transverse shear strains and the rotation matrices are defined as

$$(2) \quad \boldsymbol{\kappa} = \begin{bmatrix} \kappa_x \\ \kappa_y \\ \kappa_{xy} \end{bmatrix} = -\mathbf{d}_e\boldsymbol{\theta}\,, \qquad \boldsymbol{\gamma} = \begin{bmatrix} \gamma_x \\ \gamma_y \end{bmatrix} = \mathbf{d}w - \boldsymbol{\theta}\,, \qquad \boldsymbol{\theta} = \begin{bmatrix} \theta_x \\ \theta_y \end{bmatrix}.$$

The matrix constitutive law is

$$(3) \qquad\qquad \mathbf{M} = -\mathbf{D}_b\mathbf{d}_e\boldsymbol{\theta}\,, \qquad \mathbf{Q} = \mathbf{D}_s(\mathbf{d}w - \boldsymbol{\theta})\,,$$

where $\mathbf{D}_b$ is the bending rigidity matrix and $\mathbf{D}_s$ is the shear rigidity matrix. Both $\mathbf{D}_b$ and $\mathbf{D}_s$ are symmetric and positive definite. For isotropic materials, $\mathbf{D}_b$ is given in Eq. (20.1:5) and

$$(4) \qquad\qquad \mathbf{D}_s = \kappa tG\mathbf{I}\,,$$

where $\kappa(\cong 5/6)$ is a constant to account for the nonuniform shear strain distribution through the thickness. The element functional for *the principle of minimum potential energy* becomes

$$(5) \qquad\qquad \Pi_{Be} = \Pi_{be} + \Pi_{se} - \Pi_{bfe}\,,$$

where

(6)
$$\Pi_{be} = \frac{1}{2} \int_{A_e} (\mathbf{d}_e \boldsymbol{\theta})^T \mathbf{D}_b \mathbf{d}_e \boldsymbol{\theta} \, dxdy \,,$$

(7)
$$\Pi_{se} = \frac{1}{2} \int_{A_e} \boldsymbol{\gamma}^T \mathbf{D}_s \boldsymbol{\gamma} \, dxdy \,,$$

(8)
$$\Pi_{bf_e} = \int_{A_e} pw \, dxdy - \int_{\partial A_{Mne}} \overline{M}_{nn} \theta_n \, ds$$
$$- \int_{\partial A_{Mse}} \overline{M}_{ns} \theta_s \, ds + \int_{\partial A_{Qe}} \overline{Q} w \, ds \,,$$

where Π_{be}, Π_{se}, Π_{bf_e} are associated with the bending energy, the transverse shear energy and the work by external loads, respectively. The boundary conditions are divided into three sets:

Set 1.

(9)
$$w = \bar{w} \qquad \text{on } \partial A_{we} \,,$$

(10)
$$n_x Q_x + n_y Q_y = \bar{Q} \qquad \text{on } \partial A_{Qe} \,;$$

Set 2.

(11)
$$n_x \theta_x + n_y \theta_y = \bar{\theta}_n \qquad \text{on } \partial A_{\theta ne} \,,$$

(12)
$$n_x^2 M_x + 2 n_x n_y M_{xy} + n_y^2 M_y = \overline{M}_{nn} \qquad \text{on } \partial A_{Mne} \,;$$

Set 3.

(13)
$$-n_y \theta_x + n_x \theta_y = \bar{\theta}_s \qquad \text{on } \partial A_{\theta se} \,,$$

(14)
$$n_x n_y (M_y - M_x) + (n_x^2 - n_y^2) M_{xy} = \overline{M}_{ns} \qquad \text{on } \partial A_{Mse} \,,$$

where M's and Q's are in terms of θ_x, θ_y and w through the constitutive laws given in Eq. (3), θ_s, $-\theta_n$ are the rotations about the normal and the tangent to the boundary, and the subscripted ∂A's are part of the element boundaries that are also the domain boundaries with the first subscript denoting the prescribed quantity. At any given boundary location, one and only one condition from each of the three sets above is to be specified. Equations (9), (11) and (13) are *rigid boundary conditions* while Eqs. (10), (12) and (14) are *natural* for the functional Eq. (5). The admissibility requirements are: θ's are C^0 continuous, w is piecewise continuous and both of them satisfy rigid boundary condition(s) if prescribed.

In the Reissner-Mindlin theory, there are three boundary conditions at a given point as compared to two for thin plate theory. The so-called simply supported condition (i.e., prescribed w and M_{nn} in the thin plate theory) refers the specification of w, M_{nn} and M_{ns} at the boundary as *soft support* (a more realistic condition) and the specification of w, M_{nn} and θ_s as *hard support*. A boundary subjected to both rigid and natural conditions is called a *mixed boundary*, e.g., a boundary with prescribed displacement and moments. Simply support is a mixed boundary condition.

The functional in Eq. (5) involves the first order differentials of θ_x, θ_y, and w. The bending strain energy depends on the rotations in a manner analogous to the dependence of membrane strain energy on the in-plane displacements. Thus, the formulation of the bending part of Reissner-Mindlin plate elements is same as the formulation of plane elastic elements. One simply uses the C^0 shape functions derived in Sec. 18.14 [with the substitution of θ for $\mathbf{u}$ in Eq. (18.14:9) and $\mathbf{D}_b$ for $\mathbf{D}_e$ in Eq. (18.14:13)] to construct the element matrices for bending.

To evaluate the strain energy associated with the transverse shear, one can use the same type of shape functions for w. In this case, θ will contain higher degree terms not present in $w_{,x}$ and $w_{,y}$. These higher degree terms cause transverse shear locking in thin plate applications. For example, for a pure bending deformation, the change of curvature is constant and the transverse shear is zero. This implies that θ is linear and w quadratic. Therefore, unless the shape functions can correctly represent a quadratic field, locking will occur. Increasing the order of the approximation of θ and w can eliminate or minimize locking. However, this will be at the expense of increase in the number of degrees-of-freedom per element. One can get the similar result by increasing the order of approximation for w to one order higher than that of θ.

A simple alternative is to use *uniformly or selectively reduced integration*. In uniformly reduced integration, both the bending and shear terms are integrated with same lower-order integration rule. In selectively reduced integration, the bending term is integrated with the normal rule, whereas the transverse shear term with a lower-order rule. Unfortunately, reduced integration introduces spurious zero energy mode(s), which can cause serious errors in applications. A summary of integration rules for quadrilateral elements and their associated zero-energy modes for the case of parallelogram is given in Table 20.2:1.

Table 20.2:1. Reduced integration rules and the associated zero-energy modes for thick parallergram plates.

Element	4-node	8-node (Serendipity)	9-node (Lagrange)
shape function for w, θ_x, θ_y	Bilinear	Incomplete biquadratic	Biquadratic
Selectively reduced integration	1×1 shear 2×2 bending (S1)	2×2 shear 3×3 bending (S2)	2×2 shear 3×3 bending (S2)
Zero-energy modes	1. $w = \xi\eta$, (G) $\theta_x = \theta_y = 0$ 2. $w = 0$, $\theta_x = y$ $\theta_y = -x$ (G)		1. $w = \xi^2\eta^2$ $-\frac{\xi^2+\eta^2}{3}$ $\theta_x = \theta_y = 0$ (G)
Uniformly reduced integration	1×1 (U1) (Gauss station $\xi = \eta = 0$)	2×2 (U2) (Gauss stations $\xi = \eta = \pm 1\sqrt{3}$)	2×2 (U2)
Zero-energy modes for parallelogram	In addition to those of S1 3. $w = \theta_y = 0$ $\theta_x = \xi\eta$ (G) 4. $w = \theta_x = 0$ $\theta_y = \xi\eta(G)$	1. $w = 0$ (N) $\theta_x = \xi\left(\eta^2 - \frac{1}{3}\right)\xi_{,x}$ $-\eta\left(\xi^2 - \frac{1}{3}\right)\eta_{,x}$ $\theta_y = \xi(\eta^2 - \frac{1}{3})\xi_{,y}$ $-\eta\left(\xi^2 - \frac{1}{3}\right)\eta_{,y}$ ($\xi_{,x}, \xi_{,y}, \eta_{,x}, \eta_{,y}$ are constant)	(In addition to that of S2) 2. $w = \theta_x = 0$ (G) $\theta_y = \left(\xi^2 - \frac{1}{3}\right)\left(\eta^2 - \frac{1}{3}\right)$ 3. $w = \theta_y = 0$ (G) $\theta_x = \left(\xi^2 - \frac{1}{3}\right)\left(\eta^2 - \frac{1}{3}\right)$ 4. $w = 0$ (N) $\theta_x = \xi\left(\eta^2 - \frac{1}{3}\right)\xi_{,x}$ $-\eta\left(\xi^2 - \frac{1}{3}\right)\eta_{,x}$ $\theta_y = \xi\left(\eta^2 - \frac{1}{3}\right)\xi_{,y}$ $-\eta\left(\xi^2 - \frac{1}{3}\right)\eta_{,y}$

S – selectively reduced integration; U – uniformly reduced integration; G – global communicating mode; N – non communicating mode.

Problem 20.1. The *Legendre polynomial* $f_n(\xi)$ of degree n satisfies the recursive formula

$$(n + 1)f_{n+1}(\xi) = (2n + 1)\xi f_n(\xi) - nf_{n-1}(\xi).$$

Show that

(a) $f_0(\xi) = 1$, $f_1(\xi) = \xi$, $f_2(\xi) = (3\xi^2 - 1)/2$, $f_3(\xi) = \xi(5\xi^2 - 3)/2, \ldots$.

(b) for parallelogram, the displacement and rotations

$$w = 0, \qquad \theta_x = f_n'(\xi)f_n(\eta)\xi_{,x} - f_n(\xi)f_n'(\eta)\eta_{,x},$$

$$\theta_y = f_n'(\xi)f_n(\eta)\xi_{,y} - f_n(\xi)f_n'(\eta)\eta_{,y}$$

give zero bending and transverse strains at the integration points of the $n \times n$ Gauss quadrature, where $(.)'$ denotes the derivative of the function. Note that $f_n(\xi)$, $f_n(\eta)$ equal to zero at the corresponding Gauss points.

Locking Modes of 8-Node Quadrilateral. An 8-node serendipity element of general quadrilateral shape other than parallelogram cannot interpolate $w = x^2$ correctly, because the shape functions include only the terms $1, \xi, \eta, \xi^2, \xi\eta, \eta^2, \xi^2\eta, \xi\eta^2$, while $x^2\{= [\sum_{i=1}^{8} h_i(\xi, \eta)x_i]^2\}$ involves $\xi^2\eta^2$. The incorrect interpolation will not only introduce spurious shear, causing locking, but also make the element fail the patch test. A remedy is to use reduced (2×2) integration (Zienkiewicz et al. 1971). The reduced integration element has received wide use, but behaves poorly in the thin plate limit (Pugh et al. 1978). Another remedy is to add a term to the shape functions to ensure correct interpolation of quadratic polynomials (MacNeal and Harder 1992). The 9-node Lagrange element (Fig. 18.13:3) includes the term $\xi^2\eta^2$ in the base functions. Unfortunately, the 9-node element has an additional spurious communicating zero energy mode, which if not suppressed can destroy the accuracy of analysis. Hughes and Cohen (1978) introduced the *Heterosis* element, which modifies the 9-node Lagrange element by using just the 8 boundary nodes for w while 9 nodes for $\boldsymbol{\theta}$. They used the 2×2 integration rule for transverse shear and the 3×3 rule for bending curvature.

To illustrate locking, consider the deformation

$$\theta_x = x^2y, \qquad \theta_y = \frac{1}{3}x^3, \qquad w = \frac{1}{3}x^3y,$$

with the curvature and transverse shear strains

(15) $\qquad \kappa_{xx} = 2xy, \qquad \kappa_{xy} = 2x^2, \qquad \gamma_x = \gamma_y = \kappa_{yy} = 0.$

An 8-node rectangular $(2a \times 2b)$ element with bi-quadratic interpolation functions, in general, cannot correctly interpolate x^3. The finite element representation of such deformation is

$$\theta_x = x^2y, \qquad \theta_y = \frac{1}{3}xa^2, \qquad w = \frac{1}{3}a^2xy,$$

(16) $\qquad \kappa_{xx} = 2xy, \qquad \kappa_{xy} = x^2 + a^2/3, \qquad \gamma_y = \kappa_{yy} = 0,$

$$\gamma_x = y(x^2 - a^2/3).$$

The transverse shear strain γ_x is spurious and κ_{xy} is only approximate. The ratio of spurious strain energy from the strains given in Eq. (16) to the correct strain energy from Eq. (15) is

$$\frac{(\Pi_{be})_{\text{finite element}}}{(\Pi_{be})_{\text{exact}}} = 1 + \left(\frac{4}{9}\frac{a^2}{t^2}\kappa - \frac{1}{3}\frac{a^2}{b^2}\right) \Big/ \left[\frac{a^2}{b^2} + \frac{10}{9(1-\nu)}\right],$$

which is proportional to the square of the element size-to-thickness ratio (a^2/t^2), where ν is Poisson's ratio. For a thin plate element of aspect ratio (a/b) about 1, the above *energy ratio* can be very large if the element size is much bigger than the plate thickness, which means locking for such deformation. For 2×2 reduced integration, shears are evaluated at $x = \pm a/\sqrt{3}$. At the Gauss points the shears are correct and thus locking is avoided. Even though the finite element representation of the twist curvature κ_{xy} is incorrect, the influence is not serious. The shear locking mode and the incorrect twist curvature are not the elastic modes required for convergence, so that convergence is slow but not affected when the size of elements progressively reduces.

Problem 20.2. Consider a beam along the x-axis subjected to a concentrated load at $x = 0$, show that the deflection near the origin can be expressed in the form $a_0 + a_1 x + a_2 x^2 + a_3 x^3 + \cdots$, where $a_3 \neq 0$.

Four-Node Element. A four-node quadrilateral cannot correctly represent a quadratic displacement field. However, if there are locations at which the first derivatives of a quadratic field are correct, one can obtain the correct strains by using only the values at those points. For example, the spurious transverse shear strains $\gamma_x = x$ and $\gamma_y = y$ cause locking for pure bending ($\theta_x = x$, $\theta_y = 0$, $w = x^2/2$, or $\theta_x = 0$, $\theta_y = y$, $w = y^2/2$, see Sec. 18.20). But, their values at $x, y = 0$ are correct. One can obtain correct shear strains by using only the value of γ_x at $x = 0$ and the value of γ_y at $y = 0$. Hughes *et al.* (1977) used a single-point quadrature for both transverse shears. MacNeal (1978) used the values of γ_x at $\xi = 0$, $\eta = \pm 1/\sqrt{3}$ and the values of γ_y at $\xi = \pm 1/\sqrt{3}$, $\eta = 0$. Both approaches work well for rectangular elements but fail to eliminate shear locking for elements of other shapes.

Zienkiewicz *et al.* (1971), Pawsey and Clough (1971), Cook (1974), Zienkiewicz and Hinton (1976), and Pugh *et al.* (1978) showed the performance of rectangular elements by analyzing a square plate ($2a \times 2a$) subjected to a concentrated load at the center. As a/t becomes large, the results of the 8-node serendipity element (Fig. 18.13.3) with full integration deteriorates rapidly from the thin plate solution. Reduced integration improves the result drastically for the simply supported case. If the boundaries are clamped, the answers are not satisfactory even with reduced integration when a/t is very large. The 9-node Lagrange element with reduced integration performs well in most circumstances.

Elements with Discrete Constraints. For general quadrilateral elements, reduced integration generally fails to eliminate locking. One alternative is to impose discrete constraints on the shear strains. In evaluating Π_{Be} of Eq. (5), one assume

$$(17) \quad \boldsymbol{\gamma} = \begin{bmatrix} \gamma_x \\ \gamma_y \end{bmatrix} = \mathbf{P}(\xi, \eta)\mathbf{a}, \qquad \boldsymbol{\theta} = \begin{bmatrix} \theta_x \\ \theta_y \end{bmatrix} = \mathbf{h}_\theta \mathbf{q}_\theta, \qquad w = \mathbf{h}_w \mathbf{q}_w,$$

where $\mathbf{P}$ is a polynomial matrix of the natural coordinates, $\mathbf{a}$ is a column matrix of unknown parameters, and $\mathbf{h}_\theta$, $\mathbf{h}_w$ are interpolation matrices associated with nodal parameter matrices $\mathbf{q}_\theta$, $\mathbf{q}_w$ for the field variables $\boldsymbol{\theta}$ and w, respectively. Rather than evaluating the transverse shear strains $\boldsymbol{\gamma}$ according to the strain/displacement relations Eq. (2), one uses the approximation of Eq. (17) for $\boldsymbol{\gamma}$ in Eq. (7) and obtains

$$(18) \qquad \Pi_{se} = \frac{1}{2} \mathbf{a}^T \left(\int_{Ae} \mathbf{P}^T \mathbf{D}_s \mathbf{P} \, dxdy \right) \mathbf{a} = \frac{1}{2} \mathbf{a}^T \mathbf{k}_{sa} \mathbf{a}.$$

The $\mathbf{a}$'s are to be expressed in terms of the nodal values of $\boldsymbol{\theta}$ and w by collocating the tangential component of $\boldsymbol{\gamma}$ at selected points with that derived from the assumed $\boldsymbol{\theta}$ and w according to Eq. (2):

$$\mathbf{a} = \mathbf{A}_\theta \mathbf{q}_\theta + \mathbf{A}_w \mathbf{q}_w.$$

Then Eq. (18) can be written as

$$\Pi_{se} = \frac{1}{2} = \begin{bmatrix} \mathbf{q}_\theta^T & \mathbf{q}_\theta^T \end{bmatrix} \mathbf{k}_s \begin{bmatrix} \mathbf{q}_\theta \\ \mathbf{q}_w \end{bmatrix},$$

where

$$\mathbf{k}_s = \begin{bmatrix} \mathbf{A}_\theta & \mathbf{A}_w \end{bmatrix}^T \mathbf{k}_{sa} \begin{bmatrix} \mathbf{A}_\theta & \mathbf{A}_w \end{bmatrix}.$$

The other element matrices are evaluated according to Eqs. (6) and (8).

Hughes and Tezduyar (1981), and MacNeal (1982) assume

$$\gamma = \sum_{i=1}^{3} \varsigma_i \begin{bmatrix} a_i \\ a_{i+3} \end{bmatrix}$$

for triangles and

$$\gamma = \mathbf{J}^{-1}\tilde{\gamma} = \mathbf{J}^{-1} \begin{bmatrix} 1 & \eta & 0 & 0 \\ 0 & 0 & 1 & \xi \end{bmatrix} \mathbf{a},$$

for quadrilaterals, where $\tilde{\gamma}$ is the strain associated with the natural coordinates and $\mathbf{J}^{-1}$ is the inverse of the coordinate transformation matrix:

$$\mathbf{J}^{-1} = \frac{1}{x_{,\xi}\, y_{,\eta} - y_{,\xi}\, x_{,\eta}} \begin{bmatrix} y_{,\eta} & -y_{,\xi} \\ -x_{,\eta} & x_{,\xi} \end{bmatrix}.$$

The tangential shear strain along each side of the assumed γ is set equal to that derived from the assumed θ and w at the midpoint of each side. Discrete constriant has achieved varied degrees of success in reducing locking.

Problem 20.3. (a) Derive the element matrices of a quadrilateral element with 8 degrees-of-freedom (DOFs) for θ and 8 DOFs for w as shown. (b) Show that the element no longer locks for the deformations

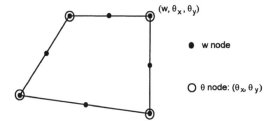

(w, θ_x, θ_y)

● w node

○ θ node: (θ_x, θ_y)

$$\theta_x = x, \ \theta_y = 0, \ w = x^2/2,$$

and

$$\theta_x = 0, \ \theta_y = y, \ w = y^2/2.$$

(c) Program the new element and solve the bending of a cantilever plate using a rectangular mesh and the mesh as shown in Fig. P20.3. Compare the results to the quadrilateral elements with 8 DOFs for θ and 4 DOFs at the vertices for w.

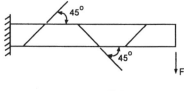

P20.3.

Problem 20.4. Repeat Problem 20.3 with triangular elements of 6 DOFs for θ and 6 DOFs (3 at the vertices and 3 at the midsides) for w and compare with the results of triangles of 6 DOFs for θ and 3 DOFs at the vertices for w. The mesh is the same as before with each quadrilateral replaced by two triangles.

20.3. MIXED FUNCTIONALS FOR REISSNER PLATE THEORY

A mixed Reissner functional (Herrmann 1966, Pian and Tong 1972, 1980, 1987) for an element can be written in the same form as Eq. (20.2:5)

$$(1) \qquad \Pi_{Be} = \Pi_{be} + \Pi_{se} - \Pi_{fe}\,,$$

where Π_{be}, Π_{bfe} are the same as those of the displacement formulation in Eqs. (20.2:6) and (20.2:8) while

$$(2) \qquad \Pi_{se} = \int_{A_e} \left[\tilde{\mathbf{Q}}^T (\mathrm{d}w - \boldsymbol{\theta}) - \frac{1}{2}\, \tilde{\mathbf{Q}}^T \mathbf{D}_s^{-1} \tilde{\mathbf{Q}} \right] dxdy$$

is the Reissner functional for the transverse shear energy where $\tilde{\mathbf{Q}}$ is a new independent field variable. An alternative expression of Π_{se} is

$$(2a) \qquad \Pi_{se} = \int_{Ae} \left[\tilde{\boldsymbol{\gamma}}^T \mathbf{D}_s (\mathrm{d}w - \boldsymbol{\theta}) - \frac{1}{2}\, \tilde{\boldsymbol{\gamma}}^T \mathbf{D}_s \tilde{\boldsymbol{\gamma}} \right] dxdy\,,$$

where $\tilde{\boldsymbol{\gamma}} = [\tilde{\gamma}_{xz}\ \tilde{\gamma}_{yz}]^T$ is an independent field representing the transverse shear strains within the element. Judicious selection of $\tilde{\mathbf{Q}}$ or $\tilde{\boldsymbol{\gamma}}$ can provide flexibility to avoid shear locking.

The admissibility requirements for the independent fields $(\boldsymbol{\theta}, \tilde{\boldsymbol{\gamma}}, w)$ in Eq. (2a) are C^0 continuity for $\boldsymbol{\theta}$ and w or $\tilde{\gamma}_n$, the component of $\tilde{\boldsymbol{\gamma}}$ normal to the interelement boundaries. We assume

$$(3) \qquad \boldsymbol{\theta} = \mathbf{h}_\theta \mathbf{q}_\theta\,, \qquad \tilde{\boldsymbol{\gamma}} = \mathbf{h}_\gamma \boldsymbol{\beta}\,, \qquad w = \mathbf{h}_w \mathbf{q}_w$$

and derive the element matrices from Eq. (1), where $\mathbf{h}$'s are shape function matrices and $\mathbf{q}$'s, $\boldsymbol{\beta}$ are parameter matrices associated with variables denoted by their subscripts. Normally, one chooses $\mathbf{h}$'s to assure that $\boldsymbol{\theta}$ and w are C^0 continuous while $\tilde{\mathbf{Q}}$ or $\tilde{\boldsymbol{\gamma}}$ is independently assumed for each element. In this case, $\boldsymbol{\beta}$ can be eliminated at the element level from Eq. (2a). A possible $\mathbf{h}_\gamma$ is

$$(4) \qquad \mathbf{h}_\gamma = \begin{bmatrix} 1 & 0 & x & 0 & y & 0 & \cdots \\ 0 & 1 & 0 & y & 0 & x & \cdots \end{bmatrix}.$$

One can also assume $\mathbf{h}_\gamma$ in terms of polynomials of the natural coordinates ξ, η. Judicious selection of $\tilde{\mathbf{Q}}$ or $\tilde{\boldsymbol{\gamma}}$ to relieve locking will soon be discussed.

A substitution of Eq. (3) into Eq. (1) yields

(5)
$$\Pi_{Be} = \frac{1}{2}\mathbf{q}_\theta^T\mathbf{k}_b\mathbf{q}_\theta + \boldsymbol{\beta}^T(\mathbf{g}_{\gamma w}\mathbf{q}_w - \mathbf{g}_{\gamma\theta}\mathbf{q}_\theta)$$
$$-\frac{1}{2}\boldsymbol{\beta}^T\mathbf{c}\boldsymbol{\beta} - \mathbf{q}_w^T\mathbf{f}_w - \mathbf{q}_\theta^T\mathbf{f}_\theta\,,$$

where the first term on the right hand side is derived from Π_{be}, the second and third terms from Π_{se}, and the last two terms from Π_{fe}. Thus

$$\mathbf{k}_b = \int_{A_e}(\mathbf{d}_e\mathbf{h}_\theta)^T\mathbf{D}_b\mathbf{d}_e\mathbf{h}_\theta\,dxdy = \int_{A_e}\mathbf{B}_b^T\mathbf{D}_b\mathbf{B}_b\,dxdy\,,$$

$$\mathbf{g}_{\gamma w} = \int_{A_e}\mathbf{h}_\gamma^T\mathbf{D}_s\mathbf{d}\mathbf{h}_w\,dxdy\,, \qquad \mathbf{g}_{\gamma\theta} = \int_{A_e}\mathbf{h}_\gamma^T\mathbf{D}_s\mathbf{h}_\theta\,dxdy\,,$$

$$\mathbf{c} = \int_{A_e}\mathbf{h}_\gamma^T\mathbf{D}_s\mathbf{h}_\gamma\,dxdy\,,$$

$$\mathbf{f}_w = \int_{A_e}\mathbf{h}_w^Tp\,dxdy + \int_{\partial A_{Qe}}\mathbf{h}_w^T\bar{Q}\,ds\,,$$

$$\mathbf{f}_\theta = -\int_{\partial A_{Mne}}\mathbf{h}_\theta^T\begin{bmatrix}\partial y/\partial s \\ -\partial x/\partial s\end{bmatrix}\overline{M}_{nn}ds - \int_{\partial A_{Mse}}\mathbf{h}_\theta^T\begin{bmatrix}\partial x/\partial s \\ \partial y/\partial s\end{bmatrix}\overline{M}_{ns}\,ds\,.$$

Note that
$$\theta_s = \begin{bmatrix}\dfrac{\partial x}{\partial s} & \dfrac{\partial y}{\partial s}\end{bmatrix}\boldsymbol{\theta}\,, \qquad \theta_n = \begin{bmatrix}\dfrac{\partial y}{\partial s} & -\dfrac{\partial x}{\partial s}\end{bmatrix}\boldsymbol{\theta}\,.$$

Since $\boldsymbol{\beta}$ for each element are independently chosen, from the stationary condition of Π_{Be} of Eq. (2a) with respect to $\tilde{\gamma}$, we obtain

(6) $\mathbf{c}\boldsymbol{\beta} = \mathbf{g}_{\gamma w}\mathbf{q}_w - \mathbf{g}_{\gamma\theta}\mathbf{q}_\theta\,,$ or $\boldsymbol{\beta} = \mathbf{c}^{-1}(\mathbf{g}_{\gamma w}\mathbf{q}_w - \mathbf{g}_{\gamma\theta}\mathbf{q}_\theta)\,.$

Using Eq. (6), we can eliminate $\boldsymbol{\beta}$ from Eq. (5) and obtain

(7) $\Pi_{Be} = \dfrac{1}{2}\mathbf{q}_\theta^T\mathbf{k}_b\mathbf{q}_\theta + \dfrac{1}{2}\mathbf{q}_B^T\mathbf{k}_s\mathbf{q}_B - \mathbf{q}_B^T\mathbf{f}_B = \dfrac{1}{2}\mathbf{q}_B^T\mathbf{k}_B\mathbf{q}_B - \mathbf{q}_B^T\mathbf{f}_B\,,$

in which the matrices are defined as follows:

(8) $\mathbf{q}_B = \begin{bmatrix}\mathbf{q}_w \\ \mathbf{q}_\theta\end{bmatrix}, \quad \mathbf{f}_B = \begin{bmatrix}\mathbf{f}_w \\ \mathbf{f}_\theta\end{bmatrix}, \quad \mathbf{k}_s = \begin{bmatrix}\mathbf{g}_{\gamma w}^T \\ -\mathbf{g}_{\gamma\theta}^T\end{bmatrix}\mathbf{c}^{-1}[\mathbf{g}_{\gamma w} \quad -\mathbf{g}_{\gamma\theta}]\,,$

$$\mathbf{k}_B = \begin{bmatrix}\mathbf{g}_{\gamma w}^T\mathbf{c}^{-1}\mathbf{g}_{\gamma w} & -\mathbf{g}_{\gamma w}^T\mathbf{c}^{-1}\mathbf{g}_{\gamma\theta} \\ -\mathbf{g}_{\gamma\theta}^T\mathbf{c}^{-1}\mathbf{g}_{\gamma w} & \mathbf{k}_\theta + \mathbf{g}_{\gamma\theta}^T\mathbf{c}^{-1}\mathbf{g}_{\gamma\theta}\end{bmatrix}\,.$$

Tong and Pian (1969), Babuska (1971) and Brezzi (1974) examined the stability associated with assumed functions for multiple field functional. One issue is the existence of spurious modes in the finite element solution. To avoid spurious modes, the following *stability criterion*

$$(9) \qquad\qquad n_w \leqslant \max(3, n_c)$$

should be satisfied, where 3 is the number of rigid body modes, n_c is the number of constraints on the parameters for w, and n_γ, n_w are the number of parameters for $\widetilde{\gamma}$ and w, respectively. This stability criterion can be applied to an assembly of elements if all elements are similar. However, the criterion is not applicable if some elements of the assembly fail while some pass.

Under the stationary condition

$$(10) \qquad \delta\Pi_{se} = \int_{Ae} \delta\widetilde{\gamma}^T \mathbf{D}_s(\mathbf{d}w - \boldsymbol{\theta} - \widetilde{\gamma})\,dxdy = 0\,,$$

for arbitrary $\delta\widetilde{\gamma}$, the second variation of the functional (1) with respect to its field variables is

$$(11) \qquad \delta^2\Pi_{Be} = \frac{1}{2}\int_{Ae}\left[(\mathbf{d}_e\delta\theta)^T\mathbf{D}_b(\mathbf{d}_e\delta\theta) - \delta\widetilde{\gamma}^T\mathbf{D}_s\delta\widetilde{\gamma}\right]dxdy\,.$$

Thus, the mixed functional given in Eq. (1) is a minimum-maximum principle (Tong and Pian 1969). In other words, for fixed $\boldsymbol{\theta}$ and w (q_θ's and q_w's), the functional is a maximum principle with respect to $\widetilde{\gamma}$ (β's). With $\widetilde{\gamma}$ fixed or in terms of $\boldsymbol{\theta}$ and w, the functional [such as the one given in Eq. (7) for an element] is a minimum principle with respect to $\boldsymbol{\theta}$ and w. One can also show that the mixed functional is bounded from above by the functional for the principle of minimum potential energy defined in Eq. (20.2:5). In finite element formulation, increasing the number of β (n_γ) tends to increase the value of the maximum for given $\boldsymbol{\theta}$ and w and thus, increase the minimum of the maxima. Physically, increasing the number of β makes the model more stiff. In developing a new element, if one starts with the minimum number of n_γ needed by Eq. (9) and if the element is too flexible, then increase n_γ; on the other hand, if the element is already too stiff, one has to reduce the number of β's or increase the order of $\boldsymbol{\theta}$ and/or w at the risk of introducing spurious mode(s).

Field-Consistent Element. We can examine locking relieve from another point of view. For clarity, we shall consider isotropic materials only. Equations (20.2:6), (20.2:7) and (2a) become

(12a) $\quad \Pi_{be} = \displaystyle\int_{Ae} \frac{Et^3}{12(1-\nu^2)} \left[\theta_{x,x}^2 + \theta_{y,y}^2 \right.$

$$+ 2\nu\theta_{x,x}\theta_{y,y} + \frac{1-\nu}{2} \left. (\theta_{x,y} + \theta_{y,x})^2 \right] dxdy ,$$

(12b) $\quad \Pi_{se} = \displaystyle\int_{Ae} \frac{Et\kappa}{2(1+\nu)} \left[(w_{,x} - \theta_x)^2 + (w_{,y} - \theta_y)^2 \right] dxdy$

(displacement model),

(12c) $\quad \Pi_{se} = \displaystyle\int_{Ae} \frac{Et\kappa}{2(1+\nu)} \left[\tilde{\gamma}_{xz}(w_{,x} - \theta_x) + \tilde{\gamma}_{yz}(w_{,y} - \theta_y) \right.$

$$\left. - \frac{\tilde{\gamma}_{xz}^2 + \tilde{\gamma}_{yz}^2}{2} \right] dxdy \qquad\qquad \text{(mixed model)}.$$

4-Node Field-Consistent Rectangle. Consider a 4-node $2l \times 2h$ rectangular element with $x = l\xi$ and $y = h\eta$. We can write

(13) $\quad w = a_1 + a_2\xi + a_3\eta + a_4\xi\eta , \quad \theta_x = b_1 + b_2\xi + b_3\eta + b_4\xi\eta ,$

$$\theta_y = c_1 + c_2\xi + c_3\eta + c_4\xi\eta ,$$

where a', b' and c's are functions of the nodal values of the corresponding variables, e.g. [see Eq. (18.20:6)],

(14) $\quad \begin{aligned} &a_1 = (w_1 + w_2 + w_3 + w_4)/4 , &&a_2 = (-w_1 + w_2 + w_3 - w_4)/4 , \\ &a_3 = (-w_1 - w_2 + w_3 + w_4)/4 , &&a_4 = (w_1 - w_2 + w_3 - w_4)/4 . \end{aligned}$

In displacement formulation, the shear strains γ_{xz}, γ_{yz} derived from the assumed w and θ_x are

(15) $\quad \begin{aligned} \gamma_{xz} &= \frac{a_2}{l} - b_1 + \left(\frac{a_4}{l} - b_3 \right) \eta - b_2\xi - b_4\xi\eta , \\ \gamma_{yz} &= \frac{a_3}{h} - c_1 + \left(\frac{a_4}{h} - c_2 \right) \xi - c_3\eta - c_4\xi\eta . \end{aligned}$

As t approaches zero, shear strains of the finite element solution approaches zero. This is equivalent to forcing

(16a) $\qquad\qquad\qquad b_1 - \dfrac{a_2}{l} \to 0 , \qquad b_3 - \dfrac{a_4}{l} \to 0 ,$

(16b) $\qquad\qquad\qquad b_2 \to 0 , \qquad\qquad b_4 \to 0 .$

Equations (16b) involve the coefficients of one field variable only and thus are considered to be spurious constraints that can lead to locking in the limit of thin plate. There are similar spurious shear constraints for c's.

In the mixed formulation, we assume

$$(17) \qquad \tilde{\gamma}_{xz} = \beta_1 + \beta_2 \eta, \qquad \tilde{\gamma}_{yz} = \beta_3 + \beta_4 \xi.$$

Using Eqs. (10) and (13), one can show that

$$
\begin{aligned}
\beta_1 &= \frac{a_2}{l} - b_1 = \frac{-w_1 + w_2 + w_3 - w_4}{4l} - \frac{\theta_{x1} + \theta_{x2} + \theta_{x3} + \theta_{x4}}{4}, \\[2mm]
(18) \quad \beta_2 &= \frac{a_4}{l} - b_3 = \frac{w_1 - w_2 + w_3 - w_4}{4l} - \frac{-\theta_{x1} - \theta_{x2} + \theta_{x3} + \theta_{x4}}{4}, \\[2mm]
\beta_3 &= \frac{a_3}{h} - c_1 = \frac{-w_1 - w_2 + w_3 + w_4}{4h} - \frac{\theta_{y1} + \theta_{y2} + \theta_{y3} + \theta_{y4}}{4}, \\[2mm]
\beta_4 &= \frac{a_4}{h} - c_2 = \frac{w_1 - w_2 + w_3 - w_4}{4h} - \frac{-\theta_{y1} + \theta_{y2} + \theta_{y3} - \theta_{y4}}{4}.
\end{aligned}
$$

From Eq. (12c), one has

$$(19) \qquad \Pi_{se} = \frac{\kappa E t a b}{1+\nu} \left(\beta_1^2 + \frac{1}{3}\beta_2^2 + \beta_3^2 + \frac{1}{3}\beta_4^2 \right),$$

which can be expressed in terms of the nodal values w's and θ's as given in Eq. (18). One can then derive the element matrices as defined in Eq. (8) in a straightforward matter.

4-Node Field-Consistent Quadrilateral. For general quadrilateral elements, it is difficult to determine the appropriate $\tilde{\gamma}$ precisely. We shall determine an approximate $\tilde{\gamma}$ using the covariant rotations and shear strains defined as

$$
\begin{bmatrix} \theta_\xi \\ \theta_\eta \end{bmatrix} = \mathbf{J} \begin{bmatrix} \theta_x \\ \theta_y \end{bmatrix}, \qquad
\begin{bmatrix} \gamma_{\xi z} \\ \gamma_{\eta z} \end{bmatrix} = \begin{bmatrix} w_{,\xi} - \theta_\xi \\ w_{,\eta} - \theta_\eta \end{bmatrix} = \mathbf{J} \begin{bmatrix} \gamma_{xz} \\ \gamma_{yz} \end{bmatrix},
$$

$$(20) \qquad \mathbf{J} = \begin{bmatrix} x_{,\xi} & y_{,\xi} \\ x_{,\eta} & y_{,\eta} \end{bmatrix}.$$

Assuming w, θ_ξ, θ_η in the form of Eq. (13) and $\tilde{\gamma}_{\xi z}, \tilde{\gamma}_{\eta z}$ in the form of Eq. (17) and, instead of following Eq. (10), using the approximate stationary condition

$$(21) \qquad \int_{-1}^{1} \int_{-1}^{1} [\delta \tilde{\gamma}_{\xi z}(\gamma_{\xi z} - \tilde{\gamma}_{\xi z}) + \delta \tilde{\gamma}_{\eta z}(\gamma_{\eta z} - \tilde{\gamma}_{\eta z})] \, d\xi d\eta = 0,$$

we find that the β's of $\tilde{\gamma}_{\xi z}$ and $\tilde{\gamma}_{\eta z}$ are given by Eq. (18) with θ_x replaced by θ_ξ and θ_y by θ_η. The nodal values of θ_ξ, θ_η can be transformed to those of θ_x, θ_y by Eq. (20). We then defined

$$(22) \qquad \begin{bmatrix} \tilde{\gamma}_{xz} \\ \tilde{\gamma}_{yz} \end{bmatrix} = \mathbf{J}^{-1} \begin{bmatrix} \tilde{\gamma}_{\xi z} \\ \tilde{\gamma}_{\eta z} \end{bmatrix} = \mathbf{J}^{-1} \begin{bmatrix} \beta_1 & \beta_2\eta & 0 & 0 \\ 0 & 0 & \beta_3 & \beta_3\xi \end{bmatrix}$$

with β's in terms of the nodal values of θ_x, θ_y and w. We also approximate Π_{se} of Eq. (12c) as

$$(23) \qquad \Pi_{se} = \frac{1}{2} \int_{Ae} \frac{Et\kappa}{2(1+\nu)} (\tilde{\gamma}_{xz}^2 + \tilde{\gamma}_{yz}^2) \, dxdy \,.$$

Using Eq. (12a) and (23), we can derive the element stiffness matrix. This element passes the patch test.

9-Node Field-Consistent Quadrilateral. For a 9-node quadrilateral, w, θ_ξ, θ_η are biquadratic functions of the natural coordinates in the form

$$(24) \quad w = \sum_{i,j=0}^{2} a_{ij}\xi^i\eta^j \,, \qquad \theta_\xi = \sum_{i,j=0}^{2} b_{ij}\xi^i\eta^j \,, \qquad \theta_\eta = \sum_{i,j=0}^{2} c_{ij}\xi^i\eta^j \,,$$

where a', b' and c's are related to the nodal values of w, θ_ξ, θ_η, respectively. The transverse shear strains can be expressed in terms of Legendre polynomials of ξ, η as follows:

$$(25) \quad \begin{aligned} \gamma_{\xi z} &= \sum_{j=0}^{2} \left[a_{1j} - b_{0j} - \frac{1}{3} b_{2j} + (2a_{2j} - b_{1j})\xi - b_{2j}\left(\xi^2 - \frac{1}{3}\right) \right] \eta^j \,, \\ \gamma_{\eta z} &= \sum_{j=0}^{2} \left[a_{j1} - c_{j0} - \frac{1}{3} c_{j2} + (2a_{j2} - c_{j1})\eta - c_{j2}\left(\eta^2 - \frac{1}{3}\right) \right] \xi^j \,, \end{aligned}$$

The coefficients of the second order Legendre polynomials $\xi^2 - 1/3$ and $\eta^2 - 1/3$ depend on the parameters of one of the rotations only and, therefore, are field-inconsistent. The element derived from such field-inconsistent strains converges slowly and will have strong quadratic oscillation in the shear stresses within an element.

Using the orthogonal properties of the Legendre polynomial, from the approximate stationary condition Eq. (21), we find the field-consistent shear strains

$$\tilde{\gamma}_{\xi z} = \sum_{j=0}^{2} \left[a_{1j} - b_{0j} - \frac{1}{3} b_{2j} + (2a_{2j} - b_{1j})\xi \right] \eta^j ,$$

(26)

$$\tilde{\gamma}_{\eta z} = \sum_{j=0}^{2} \left[a_{j1} - c_{j0} - \frac{1}{3} c_{j2} + (2a_{j2} - c_{j1})\eta \right] \xi^j .$$

The approximate field-consistent Cartesian strains are

(27)
$$\begin{bmatrix} \tilde{\gamma}_{xz} \\ \tilde{\gamma}_{yz} \end{bmatrix} = \mathbf{J}^{-1} \begin{bmatrix} \tilde{\gamma}_{\xi z} \\ \tilde{\gamma}_{\eta z} \end{bmatrix} .$$

Using Eq. (27), together with Eqs. (12a) and (23), we can derive the element stiffness matrix. This element is free of locking and its stresses calculated from the field-consistent strains are smooth. It is much more difficult to establish field-consistent strains for an 8-node quadrilateral due to incomplete terms to represent strains. Approximate field-consistent approach is a rational means to construct locking-free elements, even though exact field-consistent shear is yet to be established for general quadrilateral.

Problem 20.5. Show that one can express a's of Eq. (20.3:24) in terms of the nodal values of w (Fig. 18.13:3b) as

$$a_{00} = w_9 , \qquad a_{01} = \frac{w_7 - w_5}{2} , \qquad a_{02} = \frac{w_7 - 2w_9 + w_5}{2} ,$$

$$a_{10} = \frac{w_6 - w_8}{2} , \qquad a_{11} = \frac{w_1 - w_2 + w_3 - w_4}{4} ,$$

$$a_{12} = \frac{-w_1 + w_2 + w_3 - w_4}{4} + \frac{w_8 - w_6}{2} ,$$

$$a_{20} = \frac{w_8 - 2w_9 + w_6}{2} , \qquad a_{21} = \frac{-w_1 - w_2 + w_3 + w_4}{4} + \frac{w_5 - w_7}{2} ,$$

$$a_{22} = \frac{w_1 + w_3 + w_2 + w_4}{4} - \frac{w_5 + w_6 + w_7 + w_8}{2} + w_9 .$$

20.4. HYBRID FORMULATIONS FOR PLATES

The hybrid finite element method is an excellent alternative to overcome the compatibility difficulty for thin plate elements. The method is also a good alternative to overcome locking for thick plates. In this section, we shall discuss the hybrid stress method first proposed by Pian (1964) and the more general hybrid formulations for plate analysis.

Assumed Stress Formulation. The functional in Eq. (20.3:1) becomes the hybrid stress model if the transverse shears $\mathbf{Q}$ satisfy *a priori* the moment equilibrium equation (16.15:13). Then Π_{se} and Π_{bfe} becomes

$$(1) \quad \Pi_{se} = \int_{Ae} \left(-\mathbf{Q}^T \boldsymbol{\theta} - \frac{1}{2} \mathbf{Q}^T \mathbf{D}_s^{-1} \mathbf{Q} \right) dx dy + \int_{\partial Ae} w \mathbf{Q}^T \mathbf{n} \, ds,$$

$$(2) \quad \Pi_{bfe} = -\int_{\partial A_{Mne}} \overline{M}_{nn} \theta_n \, ds - \int_{\partial A_{Mse}} \overline{M}_{ns} \theta_s \, ds + \int_{\partial A_{Qe}} \overline{Q} w \, ds,$$

where $\mathbf{n}$ is a unit outward normal to the element boundaries and the tilde over $\mathbf{Q}$ is dropped for simplicity. The hybrid stress functional, like that for the mixed formulation, involves the three fields $\boldsymbol{\theta}$, $\mathbf{Q}$ and w. However, since w is now only needed along the element boundaries, we have more freedom to select the interpolation polynomial for w. Similarly to the mixed formulation, we assume

$$(3) \quad \boldsymbol{\theta} = \mathbf{h}_\theta \mathbf{q}_\theta, \quad \mathbf{Q} = \mathbf{Q}_h + \mathbf{Q}_p = \mathbf{h}_Q \boldsymbol{\beta} + \mathbf{Q}_p, \quad w = \mathbf{h}_w \mathbf{q}_w,$$

where the matrices $\mathbf{h}_\theta$ and $\mathbf{h}_w$ are the same as those in Eq. (20.3:3) except for that $\mathbf{h}_w$ is needed only along ∂A_e, $\mathbf{Q}_h$ and $\mathbf{Q}_p$ are the homogeneous and particular solutions of

$$\frac{\partial Q_x}{\partial x} + \frac{\partial Q_y}{\partial y} = p,$$

and

$$(4) \quad \mathbf{h}_Q = \begin{bmatrix} 1 & 0 & x & y & 0 & \cdots \\ 0 & 1 & -y & 0 & x & \cdots \end{bmatrix}.$$

Within an element, one can always approximate $p(x, y)$ as a polynomial and determine $\mathbf{Q}_p$ accordingly. The expression for $\mathbf{Q}_p$ is not unique, but it generally has no effect on the order of accuracy of the finite element solution. The order of accuracy is determined by the order of complete polynomial for $\mathbf{Q}_h$ or $\mathbf{h}_Q$. A substitution of Eq. (3) into Eq. (20.3:1) with Π_{se} and Π_{bfe} given by Eqs. (1) and (2) gives Π_{Be} in terms of $\mathbf{q}_w$, $\mathbf{q}_\theta$ and $\boldsymbol{\beta}$ (Tong and Rossettos 1977). If $\boldsymbol{\beta}$ is independently assumed for each element, it can be eliminated at the element level based on the stationary condition of Π_{Be}. One can then derive the element matrices in terms of $\mathbf{q}_\theta, \mathbf{q}_w$.

Problem 20.6. Repeat Problem 20.3 by deriving the corresponding hybrid stress elements.

General Hybrid Formulation. Introducing the Lagrangian multipliers $\widehat{w}, \widehat{Q}_n$, one can write Π_{se} of Eq. (20.3:2) in the form (Pian and Tong 1969, 1972, 1980)

$$\Pi_{se} = \int_{\partial Ae} \widehat{Q}_n (\widehat{w} - w) \, ds + \int_{Ae} \left[\mathbf{Q}^T (\mathbf{d}w - \boldsymbol{\theta}) - \frac{1}{2} \mathbf{Q}^T \mathbf{D}_s^{-1} \mathbf{Q} \right] dx dy.$$

The new functional has five dependent fields $\boldsymbol{\theta}$, $\mathbf{Q}$, w, $\widehat{Q}_n$, $\widehat{w}$, of which $\widehat{Q}_n$, $\widehat{w}$ are defined along the element boundaries only. The admissibility requirements are C^0 continuity for $\boldsymbol{\theta}$ and $\widehat{w}$ at inter-element boundaries, smoothness of the field variables within each element and the enforcement of prescribed rigid contraint. The finite element equations can be established by assuming the field variables in terms of interpolated functions with unknown parameters

$$\boldsymbol{\theta} = \mathbf{h}_\theta \mathbf{q}_\theta\,, \quad \mathbf{Q} = \mathbf{h}_Q \boldsymbol{\beta}\,, \quad w = \mathbf{h}_w \mathbf{q}_w\,, \quad \widehat{Q}_n = \mathbf{h}_{\widehat{Q}} \mathbf{q}_{\widehat{Q}}\,, \quad \widehat{w} = \mathbf{h}_{\widehat{w}} \mathbf{q}_{\widehat{w}}\,,$$

where $\mathbf{h}$'s are interpolation matrices and $\mathbf{q}$'s, $\boldsymbol{\beta}$ are unknown parameter column matrices. In the present formulation, there is no restriction on the selection of $\mathbf{h}_Q$ and $\mathbf{h}_w$ while $\mathbf{h}_{\widehat{Q}}$, $\mathbf{h}_{\widehat{w}}$ are assumed along the element boundaries only. The functional is named the *hybrid assumed strain model* (Tang 1983), if $\mathbf{Q}$ is expressed in terms of the transverse shear $\boldsymbol{\gamma}$.

If we set $\bar{\boldsymbol{\gamma}} = \boldsymbol{\gamma} = \mathbf{d}w - \boldsymbol{\theta}$, $\mathbf{Q} = \mathbf{D}_s \boldsymbol{\gamma}$ and $\widehat{Q}_n = \mathbf{n}^T \mathbf{D}_s \boldsymbol{\gamma}$ in Π_{se} above, where $\mathbf{n}$ is a unit outward normal to ∂A_e, the functional becomes

$$(5) \qquad \Pi_{se} = \int_{\partial A_e} \mathbf{n}^T \mathbf{D}_s \boldsymbol{\gamma} (\widehat{w} - w)\, ds + \frac{1}{2} \int_{A_e} \boldsymbol{\gamma}^T \mathbf{D}_s \boldsymbol{\gamma}\, dx dy\,,$$

for the *hybrid assumed displacement formulation* (Tong 1970). The admissibility requirements are C^0 continuity for $\boldsymbol{\theta}$ and $\widehat{w}$ at inter-element boundaries and the enforcement of prescribed rigid contraint(s) for $\boldsymbol{\theta}$ and $\widehat{w}$. There is no restriction on w. The functional has three field variables $\boldsymbol{\theta}$, w and $\widehat{w}$. One can assume

$$\boldsymbol{\theta} = \mathbf{h}_\theta \mathbf{q}_\theta\,, \qquad \widehat{w} = \mathbf{h}_{\widehat{w}} \mathbf{q}_{\widehat{w}}\,, \qquad w = \mathbf{h}_w \mathbf{q}_w$$

to derive the element matrices from Eq. (20.3:1) based on the new Π_{se}. The stability criterion is

$$(6) \qquad\qquad\qquad n_w \geq n_{\widehat{w}}\,,$$

which is similar to Eq. (20.3:9).

Problem 20.7. Repeat Problem 20.3 by deriving the corresponding hybrid displacement elements.

Remarks. One often judges a *thick* plate element based on its performance in *thin* plate applications. The 9-node Lagrangian and the heterosis elements (9-node for $\boldsymbol{\theta}$ and 8-node for w) with selectively reduced integration perform well in examples of clamped and simply supported *square thin plates* with rectangular mesh (Zienkiewics and Taylor 1989, Hughes

1984). The results deteriorate rapidly when the mesh is distorted to include non-rectangular elements. Mixed (field-consistent) and discrete constraint methods are alternatives for eliminating locking for general triangular and quadrilateral elements. The field-consistent and the discretely-constrained elements discussed in the previous section, and the mixed triangular elements developed by Xu (1986), Arnold and Falk (1987), Zienkiewicz and Lefebvre (1988) all perform reasonably well. Locking can be removed using hybrid formulations (Pian 1993). General hybrid approaches have not been fully explored.

The thick plate theory requires three boundary conditions while the thin plate theory needs only two. We may expect the two theories give different deformations for different boundary conditions when the plate is thick (l/t is of order 10, where l is dimension of the plate). Generally the thick plate theory gives considerably larger deflections. For plates with l/t of the order 100 or more, the thick element deflections generally converge to the Kirchhoff plate results for *hard support conditions* (specification of w, M_{nn} and θ_s), but to slightly higher values for *soft support conditions* (prescription of w, M_{nn} and M_{ns}). Babuska and Scapolla (1989) solved a rhombic plate of $l/t = 100$ as a three-dimensional structure with support conditions of the *soft* type and found that the result is very close to the thick plate solution of *soft* support.

20.5. SHELL AS AN ASSEMBLY OF PLATE ELEMENTS

Shell, like plate, is a particular form of three-dimensional solid whose thickness is small as compared with other dimensions. The use of general 3-dimensional finite elements in thin shell applications usually encounter '*locking*' (see Sec. 18.20) unless the element size is of the order of thickness. The Gaussian curvature (i.e., the product of the principal curvature of the midsurface of an unloaded shell) is either zero (as in cylinder) or positive (as in an ellipsoid or spheroid) or negative (as in a hyperboloidal shell). Due to the curvature of the midsurface, the stress resultants can resist pressure normal to the middle surface. In fact, this mechanism is very efficient and is utilized to carry a major part of the external load in engineering applications to shells. In the linearized plate theory, a flat plate supports external loads through bending.

The combination of stretching and bending plus the complexities of the curved geometry make the derivation of two dimensional governing equations for shells difficult and lead to various different shell theories. Modeling the different shell theories results in different shell elements and is a subject of continued research. In this section, we shall consider the approximation of shell as an assembly of flat elements and thus avoid the complication induced by the curvatures of the shell surface. In this model, the flat plate assembly approximates the lateral load carrying mechanism of shells through the interaction of stress resultants between adjacent elements. Through the discrete angle at the junction of neighboring elements, part of the midplane stress resultants of an element is transmitted to its immediate neighbors as transverse shear force to resist the lateral load. As the size of subdivision decreases, the geometry converges. Intuitively, we expect that the flat-element assembly behaves like the shell in the limit.

One advantage of treating shells as an assemblage of flat elements is that *no new elements are needed.* An added advantage is the ease of coupling with edge beams and stiffeners in practical applications. Note that, however, flat elements are not compatible in shell applications if the order of polynomials used for 'in-plane' displacements differs from that for out-plane displacement.

A general shell can always be divided into a triangular mesh. The vertices of a triangle are always co-planar. For meshes involving elements of more than three nodes, the elements may not be flat. Only some shells such as cylinders or spheroids, which can be covered by orthogonal curvilinear coordinates, are representable by flat quadrilaterals. (At the poles of a spheroid, triangles are used. In the finite element method, a triangle can be consiered as a degenerated quadrilateral.) The number of surfaces totally covered by orthogonal curvilinear coordinates is known to be 8. The analysis of such shells (cylinders and spheres) using flat quadrilateral elements has met early success. For more general surfaces, due to the poor bending performance of the early triangular plate elements, shell analysis with triangular elements was not satisfactory. Current improved plate elements can now well represent the behaviors of general shells.

The flat elements in shell applications are subjected to simultaneous in-plane stretching and out-plane bending. Thus, we must include both deformations in the energy expression. Let the local coordinates x and y be in the midplane of the plate element and z be normal to the plane. The functional is the sum of the potential energy of membrane stretching and plate bending, i.e.,

$$(1) \qquad \Pi_e = \Pi_{Be} + \Pi_{Me} \,,$$

where Π_{Be} is the bending potential energy given in Eq. (20.1:10) based on the Kirchhoff theory, or Eq. (20.2:5) for the Reissner–Mindlin theory, and

$$\Pi_{Me} = \Pi_{me} - \Pi_{mfe} \,,$$

where

$$\Pi_{me} = \int_{A_e} \frac{1}{2} t(\mathbf{d}_e \mathbf{u})^T \mathbf{D}_e (\mathbf{d}_e \mathbf{u}) \, dA \,,$$

$$\Pi_{mfe} = \int_{A_e} t\mathbf{u}^T \mathbf{b} \, dA + \int_{\partial A_{\sigma e}} t\mathbf{u}^T \bar{\mathbf{T}} \, ds \,.$$

In the equations above, $\mathbf{u}$ is the in-plane displacement, Π_{Me} is the membrane potential energy of the element, which is the same as Π_e of Eq. (18.14:6) except for the factor t to reflect the thickness of the element, and $\mathbf{D}_e$ is the elastic modulus matrix for plane stress. Note that Π_{me} is the membrane strain energy and Π_{mfe} is the external work associated with membrane action.

Following the earlier derivation, we can write the functional in the matrix form

$$(2) \qquad \Pi_e = \frac{1}{2} \mathbf{q}_M^T \mathbf{k}_M \mathbf{q}_M + \frac{1}{2} \mathbf{q}_B^T \mathbf{k}_B \mathbf{q}_B - \mathbf{q}_M^T \mathbf{f}_M - \mathbf{q}_B^T \mathbf{f}_B \,,$$

where the subscripts M and B denote membrane and bending actions, respectively. The column matrix $\mathbf{q}_M$ is the nodal parameter matrix of the in-plane displacements u and v. The matrix $\mathbf{q}_B$ is the nodal parameter matrix of the out-plane displacement w and its derivatives $w_{,x}$ and $w_{,y}$ for thin plates, or those of w, θ_x and θ_y for thick plates. In subsequent discussion, for expediency, we will also use θ_x and θ_y to denote $w_{,x}$ and $w_{,y}$ for thin plates. The element matrices $\mathbf{k}_M$ and $\mathbf{f}_M$ are $\mathbf{k}$ and $\mathbf{f}$ of Eqs. (18.14:13) and (18.14:14) multiplied by the plate thickness t; while $\mathbf{k}_B$ and $\mathbf{f}_B$ are given in Eqs. (20.1:16) and (20.1:17) for thin plates, and in Eq. (20.3:8) or derived from Eqs. (20.2:6)–(20.2:8) for thick plate. Let us introduce an additional nodal parameter matrix $\mathbf{q}_{z\theta}(= [\theta_{z1} \; \theta_{z2} \ldots]^T)$, which associates with θ_z, the rotation about the z-axis (normal to the midplane of the plate) and define $\mathbf{q}$ as the unknown parameter matrix of the element that

$$(3) \qquad \mathbf{q} = [\mathbf{q}_M^T \quad \mathbf{q}_B^T \quad \mathbf{q}_{z\theta}^T]^T = [\mathbf{q}_M^T \quad \mathbf{q}_w^T \quad \mathbf{q}_\theta^T \quad \mathbf{q}_{z\theta}^T]^T \,.$$

Now each node has six degrees-of-freedom, three translations and three rotations. The rotation about the normal generally does not appear in plate equations but is needed in shell application. We can write Eq. (2) in the standard matrix form

$$(4) \qquad \Pi_e = \frac{1}{2} \mathbf{q}^T \mathbf{k}_{MB} \mathbf{q} - \mathbf{q}^T \mathbf{f}_{MB} \,,$$

where the matrices $\mathbf{k}_{MB}$ and $\mathbf{f}_{MB}$ are defined as

(5)
$$\mathbf{k}_{MB} = \begin{bmatrix} \mathbf{k}_M & \mathbf{0} & \mathbf{0} \\ \mathbf{0} & \mathbf{k}_B & \mathbf{0} \\ \mathbf{0} & \mathbf{0} & \mathbf{0} \end{bmatrix}, \qquad \mathbf{f}_{MB} = \begin{bmatrix} \mathbf{f}_M \\ \mathbf{f}_B \\ \mathbf{0} \end{bmatrix},$$

where the bolded '**0**' denotes a null matrix. Note that *membrane* and *bending actions* within the element are uncoupled and that the rotation θ_z does not contribute to stiffness. This leads to a singular matrix for the system if all elements meeting at a node are coplanar.

Coordinate Transformation. In the previous discussion, x, y are the *in-plane* coordinates and z the *out-plane* coordinate with respect to the midplane of the plate element. All generalized coordinates are therefore also referred to this local coordinate frame. Transformation of the *local generalized coordinates* to a common global coordiante system is necessary before assembling the element. We shall establish the transformation rule below. Let the global coordinates be $\bar{x}, \bar{y}, \bar{z}$ with unit base vectors $\bar{\mathbf{e}}_x, \bar{\mathbf{e}}_y, \bar{\mathbf{e}}_z$, the local cooordinates be x, y, z with unit base vectors $\mathbf{e}_x, \mathbf{e}_y, \mathbf{e}_z$, and the coordinates of node i be specified by the global coordinates $(\bar{x}_i, \bar{y}_i, \bar{z}_i)$. We choose the first node (an arbitrary choice) of the element as the origin of the local coordinates and establish the transformation from the global to the local coordinates in the form (see Sec. 2.8)

$$\begin{bmatrix} x & y & z \end{bmatrix}^T = \mathbf{R}(\bar{\mathbf{x}} - \bar{\mathbf{x}}_1),$$

where

$$\mathbf{R} = \begin{bmatrix} R_{xx} & R_{xy} & R_{xz} \\ R_{yx} & R_{yy} & R_{yz} \\ R_{zx} & R_{zy} & R_{zz} \end{bmatrix}$$

is a rotation matrix with $\mathbf{R}^{-1} = \mathbf{R}^T$. The base vectors of the two coordinate systems are related by

$$\begin{bmatrix} \mathbf{e}_x \\ \mathbf{e}_y \\ \mathbf{e}_z \end{bmatrix} = \mathbf{R} \begin{bmatrix} \bar{\mathbf{e}}_x \\ \bar{\mathbf{e}}_y \\ \bar{\mathbf{e}}_z \end{bmatrix}.$$

Let the x-axis be along the side between node 1 to node 2 (an arbitrary choice also), then

$$\mathbf{e}_x = \bar{x}_{21}\bar{\mathbf{e}}_x + \bar{y}_{21}\bar{\mathbf{e}}_y + \bar{z}_{21}\bar{\mathbf{e}}_z,$$

where

$$\bar{x}_{ji} = (\bar{x}_j - \bar{x}_i)/l_{ji}, \qquad \bar{y}_{ji} = (\bar{y}_j - \bar{y}_i)/l_{ji}, \qquad \bar{z}_{ji} = (\bar{z}_j - \bar{z}_i)/l_{ji},$$

in which $l_{ji} = \sqrt{(\bar{x}_j - \bar{x}_i)^2 + (\bar{y}_j - \bar{y}_i)^2 + (\bar{z}_j - \bar{z}_i)^2}$ is the distance between nodes i and j. Since $\mathbf{e}_x = R_{xx}\mathbf{e}_{\bar{x}} + R_{xy}\mathbf{e}_{\bar{y}} + R_{xz}\mathbf{e}_{\bar{z}}$, one has

$$R_{xx} = \bar{x}_{21}, \qquad R_{xy} = \bar{y}_{21}, \qquad R_{xz} = \bar{z}_{21}.$$

Let $\mathbf{e}$ be a unit vector pointing from node 1 to node 3

$$\mathbf{e} = \bar{x}_{31}\bar{\mathbf{e}}_x + \bar{y}_{31}\bar{\mathbf{e}}_y + \bar{z}_{31}\bar{\mathbf{e}}_z.$$

Assuming nodes 1, 2, 3 are not collinear, we find $\mathbf{e}_z$ from the cross product of $\mathbf{e}_x$ and $\mathbf{e}$,

$$\mathbf{e}_z = \frac{\mathbf{e}_x \times \mathbf{e}}{|\mathbf{e}_x \times \mathbf{e}|} = R_{zx}\bar{\mathbf{e}}_x + R_{zy}\bar{\mathbf{e}}_y + R_{zz}\bar{\mathbf{e}}_z,$$

where

$$R_{zx} = \frac{\bar{y}_{21}\bar{z}_{31} - \bar{y}_{31}\bar{z}_{21}}{A}, \quad R_{zy} = \frac{\bar{z}_{21}\bar{x}_{31} - \bar{z}_{31}\bar{x}_{21}}{A}, \quad R_{zz} = \frac{\bar{x}_{21}\bar{y}_{31} - \bar{x}_{31}\bar{y}_{21}}{A},$$

$$A = \sqrt{(\bar{y}_{21}\bar{z}_{31} - \bar{y}_{31}\bar{z}_{21})^2 + (\bar{z}_{21}\bar{x}_{31} - \bar{z}_{31}\bar{x}_{21})^2 + (\bar{x}_{21}\bar{y}_{31} - \bar{x}_{31}\bar{y}_{21})^2}.$$

Finally, we have

$$\mathbf{e}_y = \mathbf{e}_z \times \mathbf{e}_x = R_{yx}\bar{\mathbf{e}}_x + R_{yy}\bar{\mathbf{e}}_y + R_{yz}\bar{\mathbf{e}}_z,$$

where

$$R_{yx} = R_{zy}R_{xz} - R_{xy}R_{zz}, \qquad R_{yy} = R_{zz}R_{xx} - R_{zx}R_{xz},$$

$$R_{yz} = R_{zx}R_{xy} - R_{xx}R_{zy}.$$

Let $\mathbf{q}_i$ be the generalized coordinate matrix of the ith node defined as

$$\mathbf{q}_i = \begin{bmatrix} u_i & v_i & w_i & \theta_{xi} & \theta_{yi} & \theta_{zi} \end{bmatrix}^T,$$

and $\bar{\mathbf{q}}_i$ the corresponding column matrix referenced to the global coordinate system. The components of the two column matrices are related through rotation:

$$\mathbf{q}_i = \bar{\mathbf{R}}\bar{\mathbf{q}}_i, \qquad \bar{\mathbf{R}} = \begin{bmatrix} \mathbf{R} & \mathbf{0} \\ \mathbf{0} & \mathbf{R}' \end{bmatrix}.$$

where

$$\mathbf{R'} = \begin{bmatrix} R_{yy} & -R_{yx} & -R_{yz} \\ -R_{xy} & R_{xx} & R_{xz} \end{bmatrix}.$$

The matrix $\mathbf{R'}$ is modified from $\mathbf{R}$ to account for mapping two local rotations to three components in the global system and the fact that θ_y, $-\theta_x$ are, respectively, the rotations about the x, y-axes. If $\mathbf{q}$ is arranged in the order that

$$\mathbf{q} = \begin{bmatrix} \mathbf{q}_1 \\ \mathbf{q}_2 \\ \vdots \end{bmatrix} = \begin{bmatrix} \bar{\mathbf{R}} & 0 & \cdots \\ 0 & \bar{\mathbf{R}} & \\ \vdots & & \cdots \end{bmatrix} \begin{bmatrix} \bar{\mathbf{q}}_1 \\ \bar{\mathbf{q}}_2 \\ \vdots \end{bmatrix} = \begin{bmatrix} \bar{\mathbf{R}} & 0 & \cdots \\ 0 & \bar{\mathbf{R}} & \\ \vdots & & \cdots \end{bmatrix} \bar{\mathbf{q}}.$$

When the parameter matrices are expressed in terms of those in the global coordinates, Eq. (4) becomes

(6) $$\Pi_e = \frac{1}{2} \bar{\mathbf{q}}^T \bar{\mathbf{k}}_{MB} \bar{\mathbf{q}} - \bar{\mathbf{q}}^T \bar{\mathbf{f}}_{MB},$$

where

$$\bar{\mathbf{k}}_{MB} = \begin{bmatrix} \bar{\mathbf{R}}^T & 0 & \cdots \\ 0 & \bar{\mathbf{R}}^T & \\ \vdots & & \cdots \end{bmatrix} \mathbf{k}_{MB} \begin{bmatrix} \bar{\mathbf{R}} & 0 & \cdots \\ 0 & \bar{\mathbf{R}} & \\ \vdots & & \cdots \end{bmatrix},$$

$$\bar{\mathbf{f}}_{MB} = \begin{bmatrix} \bar{\mathbf{R}}^T & 0 & \cdots \\ 0 & \bar{\mathbf{R}}^T & \\ \vdots & & \cdots \end{bmatrix} \mathbf{f}_{MB}.$$

The element matrices $\bar{\mathbf{k}}_{MB}$, $\bar{\mathbf{f}}_{MB}$ can be readily assembled. If the global coordinates are not the same for all nodes of the element, a situation occurs often at the boundaries with mixed boundary conditions, one must use the appropriate rotational matrix $\bar{R}$ for the specific nodes involved.

A difficulty arises if all the elements meeting a node are co-planar. This occurs at straight boundaries along the cylindrical axis. Due to zero stiffness for θ_z as shown in Eq. (5), the assembled stiffness matrix is still singular after boundary constraints are imposed. When this occur, the formulation is still valid because the generalized force corresponding to this degree-of-freedom is also zero. However, the singularity of the matrix poses problem in numerical computation. If the global and local coordinate directions are difference, detection of such singularity will be more involved.

Drilling Degree of Freedom. When singularity occurs, one may remove the degrees-of-freedom associated with the normal rotation or add arbitrary stiffness to those degrees-of-freedom. An alternative is to modify the element formulation such that the rotation about a normal to the midplane generates strain whether co-planar or not. Such a rotation is called the *drilling degree-of-freedom*. Abu-Gazaleh (1965) initially proposed the concept. Early results were disappointing (Irons and Ahmad 1980), but it has become a topic of much study (Robinson 1980, Mohr 1982, Bergan and Felippa 1985, Cook 1986, Allman 1988, MacNeal and Harder 1988, Hughes and Brezzi 1989, Iura and Atluri 1992, Chen and Pan 1992, Sze and Ghali 1993). One can use the drilling degree-of-freedom as a tool to improve the performance of low order membrane and solid elements.

For a 4-node quadrilateral, Cook (1986) used drilling degrees-of-freedom θ_z to replace the in-plane displacements normal to the sides of the element at the midside node as follows:

(7)
$$u = \sum_{i=1}^{4} h_i(\xi, \eta) u_i + \sum_{i=1}^{4} h_{i+4}(\xi, \eta)(y_j - y_i)[\theta_{zj} - \theta_{zi}]/8 \,,$$

$$v = \sum_{i=1}^{4} h_i(\xi, \eta) v_i + \sum_{i=1}^{4} h_{i+4}(\xi, \eta)(x_i - x_j)[\theta_{zj} - \theta_{zi}]/8 \,,$$

where h_i are the shape functions for 4-node quadrilateral [Eq. (18.13:2)], h_{i+4} is associated with the midside node of an 8-node quadrilateral between nodes i and j with i, j, k being the cyclic permutations of 1, 2, 3, 4. The shape functions for midside nodes are given in Table 18.3:1. This choice of shape functions for θ_z's is to make the in-plane shear strain constant and consistent with the deformation of in-plane bending for one-dimensional problems. The element now has 3 degrees-of-freedom per node for in-plane displacements (two translations and θ_z) and thus 12 degrees-of-freedom total as opposed to 8 for the regular 4-node element without drilling degrees-of-freedom. In shell applications, three components of rotation at a node are usually needed. Thus the added drilling freedom does not increase the number of degrees-of-freedom per node. While the quadrilateral element with drilling degrees-of-freedom gives good results in plane stress analyses, a corresponding 3-node triangular element performed poorly unless the tangential midside displacements are included. This makes the element equivalent to the 6-node triangle.

Elements with drilling freedoms as defined in Eq. (7) have spurious modes. One is zero translation and constant normal rotation ($u_i = v_i = 0$, $i = 1, 2, 3, 4$, and $\theta_{z1} = \cdots = \theta_{z4}$). This mode is global in a plane. The element also has a second spurious mode $\theta_{z1} = -\theta_{z2} = \theta_{z3} = -\theta_{z4}$). The last mode is non-communicative. There could be even more spurious modes if reduced integration is employed. One way of suppressing the spurious modes is to introduce artificial rotational stiffness by adding

$$(8) \qquad \Pi_d = \frac{1}{2} \int_{A_e} \alpha t G \left[\theta_z - \frac{1}{2} \left(v_{,x} - u_{,y} \right) \right]^2 dA$$

to Eq. (1), where G is the shear modulus and α a small constant. For large value of α, the solution approaches that of an element without drilling freedoms. MacNeal and Harder (1988) gave 10^{-6} as the threshold of α for a spherical shell with the radius to thickness ratio of 250. In practice, one often uses the one-point integration rule to evaluate Π_d. Due to the shell curvature, the drilling degree-of-freedom in a curved shell-element couples the normal rotation to bending strains. The coupling effect on the bending performance requires further studies.

One can use the drilling freedom to improve the in-plane deformation of low order membrane and solid elements in plate and shell applications. One derives the element matrices by including Π_d as defined in Eq. (8) for plane elasticity, or

$$\Pi_d = \frac{G}{2} \int_{V_e} \left\{ \alpha_x \left[\theta_x - \frac{1}{2} \left(w_{,y} - v_{,z} \right) \right]^2 + \alpha_y \left[\theta_y - \frac{1}{2} \left(u_{,z} - w_{,x} \right) \right]^2 \right.$$
$$\left. + \alpha_z \left[\theta_z - \frac{1}{2} (v_{,x} - u_{,y}) \right]^2 \right\} dV$$

for 3-dimensional elasticity, where α_x, α_y, α_z are constants. The optimum values for these constants depend on the aspect ratio of the element. Atluri (1980) introduced variational functionals that include *rotations* as dependent variables. Ibrahimbegovic (1990), Iura and Atluri (1992a), Chen and Pan (1992), and Sze and Ghali (1993) have studies the use of variation functional to formulate elements including drilling freedoms.

Employing drilling freedoms increases the number of degrees-of-freedom per node but improves the accuracy of 3- and 4-node elements. The use of displacements as defined in Eq. (7) (MacNeal 1994) improves the accuracy in strains and provides substantial relief from shear locking, but has little effect on dilatation locking. Pian (1993) and MacNeal (1994) gave comparisons of the performance of various plane elements with and without drilling freedoms.

Four-Node Flat Element Approximation. To approximate a non-coplanar 4-node shell midsurface by a 4-node flat element, one often sets the midplane of the element parallel to the two diagonals of the 4-node midsurface at midway between the diagonals. Thus the nodes of the shell lie on the normal lines through the nodes of the element and are at equal distance to the midplane. Let the coordinates of the nodes of the shell be (x_i, y_i, z_i), then the absolute values of all z_i are equal and the coordinates of the nodes of the element are $(x_i, y_i, 0)$. We assume that node i of the element and node i of the shell move together as a rigid body and the two points have the same rotation parameters. The displacement parameters (u_i', v_i', w_i') of the shell are related to those of the element by

$$\mathbf{u}_i' = \mathbf{u}_i - z_i[\theta_{xi} \ \theta_{yi} \ 0]^T \,, \quad i = 1, \ldots, 4 \,.$$

Defining the nodal parameter matrix as

$$\mathbf{q}_i' = [u_i' \ v_i' \ w_i' \ \theta_{xi} \ \theta_{yi} \ \theta_{zi}]^T \,, \qquad \mathbf{q}_i = [u_i \ v_i \ w_i \ \theta_{xi} \ \theta_{yi} \ \theta_{zi}]^T \,.$$

we have

$$(9) \quad \mathbf{q} = \mathbf{G}\mathbf{q}' \,, \quad \mathbf{q} = \begin{bmatrix} \mathbf{q}_1 \\ \mathbf{q}_2 \\ \vdots \end{bmatrix} \,, \quad \mathbf{q}' = \begin{bmatrix} \mathbf{q}_1' \\ \mathbf{q}_2' \\ \vdots \end{bmatrix} \,, \quad \mathbf{G} = dia[\mathbf{G}_1 \ \mathbf{G}_2 \ \mathbf{G}_3 \ \mathbf{G}_4] \,,$$

where $(G_{14})_i = (G_{25})_i = -z_i$ and all other $(G_{jk})_i = \delta_{jk}$. A substitution of Eq. (9) into Eq. (4) yields

$$\Pi_e = \frac{1}{2} \mathbf{q}'^T \mathbf{k}_{MB}' \mathbf{q}' - \mathbf{q}'^T \mathbf{f}_{MB}' \,,$$

where

$$\mathbf{k}_{MB}' = \mathbf{G}^T \mathbf{k}_{MB} \mathbf{G} \,, \quad \mathbf{f}_{MB}' = \mathbf{G}^T \mathbf{f}_{MB} \,.$$

In this case, we have replace $\mathbf{k}_{MB}$, $\mathbf{f}_{MB}$ with $\mathbf{k}_{MB}'$, $\mathbf{f}_{MB}'$ in Eq. (6). As such a transformation may accidentally excite the bending modes, one may replace the induced moments in $\mathbf{f}_{MB}'$ with vertical couples (the nodal force times z_i at the ith node).

One may improve the flat element approximation by *shallow shell elements* where the element plane is just the projection of the shallow shell. The displacements (u, v, w) and rotations $(\theta_y, -\theta_x, \theta_z)$ are defined as the tangential and normal components of the displacement and rotation of the shell surface. The strains are defined in terms of their derivatives with respect to the coordinates of the projection plane. The in-plane strains involve both the tangential and normal displacements. A formulation of shall shell elements was given by Cowper *et al.* (1970). Shallow shell formulation will introduce membrane locking. Because of the additional complication of shallow shell elements, they are not used often in practice.

Problem 20.8. Use the definition of the nodal parameter matrix $q_i^T = [u_i \ v_i \ w_i \ \theta_{xi} \ \theta_{yi} \ \theta_{zi}]$ for node i, determine the cooresponding $\mathbf{k}$ and $\mathbf{f}$ in terms of components of $\mathbf{k}_b$, $\mathbf{k}_s$, $\mathbf{f}_b$ and $\mathbf{f}_s$ of Sec. 20.2.

Problem 20.9. A common choice of the local coordinates for a flat quadrilateral is to use a bisector of the angles between the element's diagonals as the x- or y-axes. Find the transformation matrix $\mathbf{R}$ (in terms of the nodal coordinates referred to the global system) that transforms the global coordinates to the local ones.

Problem 20.10. Physically, there is a difference between a shell formed by flat surfaces and the approximation of the assembly of flat plate elements (Fig. P20.10). For the shell of flat surfaces, the cross-section is a quadrilateral as shown. The line connecting the top and bottom surfaces along the ζ direction is not generally normal to the midplane. Such differences can be significant for thick shells. Consider the middle element in Fig. P20.10(b): (a) determine the

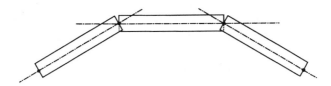

(a) Assembly of flat plate elements

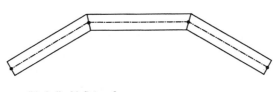

(b) shell with flat surface

P20.10. Difference between shell of flat surfaces and assembly of flat elements.

coordinate transformation from the natural coordinates ξ, ζ to the x, z Cartesian coordinates; (b) find the angle between the ζ- and z-directions; (c) assume that the displacements u, w are bilinear in ξ and ζ, determine the strains e_{xx}, e_{zz}, e_{xz} in terms of ξ, ζ; and (d) find the difference between the strains of the quadrilateral element and those of the flat plate element [Fig. P20.10(a)].

20.6. GENERAL SHELL ELEMENTS

Conceptually, shells can be modeled by curved 3-dimensional elements of reduced dimension in the shell thickness direction. Unfortunately, simple reduction of the thickness dimension leads to overly stiff elements in thin shell applications unless the in-plane dimensions of the element is in the same order as the thickness dimension. To avoid such difficulty, like plates, finite element formulations were introduced to directly model the Kichhoff thin shells and the Reissner-Mindlin moderately thick shells. Reviews of the subject are given by Gilewski and Radwanska (1991) and Ibrahimbegovic (1997). Atluri and Pian (1972) are among the many who developed the early general thin shell elements based on variational approaches. Chang *et al.* (1989) developed a 9-node mixed Lagrange shell element. Ahmad *et al.* (1970) constructed the early curved shell element following the Reissner-Mindlin assumptions for thick plates. The Ahmad element has severe locking problems in thin shell applications.

Finite element method for general thin shells is a difficult subject. Due to bending and stretching coupling, it is harder as compared to plate applications to satisfy the compatibility requirements along the interelement boundaries. The coupling causes further locking, especially for lower order elements. Thus a general shell element is either complicated or simple but only accurate for limited shell configurations.

Direct discretization of 3-dimensional solids in terms of midsurface nodal variables provides great attraction in simplicity. It does not depend on the various forms of shell theories. With attentions to the thickness deformation and transverse shears, one can effectively remove locking in thin shell applications. Elements so derived are called *degenerated solid shell elements*. In subsequent discussion, we shall develop the degeneration process.

Shell Geometry and Coordinates. To account for the double curvatures, plane stress conditions, in-extension of fiber through the shell thickness, we shall use five coordinate systems for different purposes. They are: (1) the local coordinates with Cartesian unit base vectors $\mathbf{e}_x, \mathbf{e}_y, \mathbf{e}_z$ for the geometry description of the element; (2) the natural coordinates with covariant base vectors $\mathbf{g}_\xi, \mathbf{g}_\eta, \mathbf{g}_\zeta$ for strain-displacement relations;

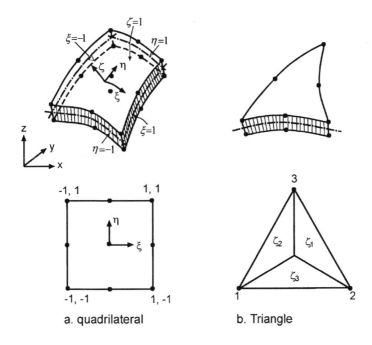

Fig. 20.6:1. Curved shell element.

(3) the fiber coordinates with unit base vectors $\hat{\mathbf{e}}_\xi, \hat{\mathbf{e}}_\eta, \hat{\mathbf{e}}_\varsigma$ for fiber rotations;
(4) the normal coordinates with orthogonal unit base vectors $\bar{\mathbf{e}}_\xi, \bar{\mathbf{e}}_\eta, \bar{\mathbf{e}}_\varsigma$ for defining stress strain relations and enforcing plane stress condition; and
(5) the global coordinates with Cartesian unit base vectors $\tilde{\mathbf{e}}_x, \tilde{\mathbf{e}}_y, \tilde{\mathbf{e}}_z$ for assembling of elements. Using the tensor transformation rules (Chapter 2), one can easily transform quantities from one coordinate system to another. The first and last coordinate systems are generally fixed Cartesian coordinates. Strictly speaking, they can be the same. However, retaining a local system offers some conveniences especially for mixed boundary value problems. The first four systems are local. The second one is usually nonorthogonal for general shells. The third one is always and the fourth is usually orthogonal and curvilinear.

Let x, y, z be the local Cartesian coordinates with unit base vectors $\mathbf{e}_x$, $\mathbf{e}_y$, $\mathbf{e}_z$. The location of a point in the shell (Fig. 20.6:1) is

$$(1) \qquad \mathbf{x} = \mathbf{x}_0(\xi, \eta) + \zeta \mathbf{g}_\zeta(\xi, \eta) = x_0 \mathbf{e}_x + y_0 \mathbf{e}_y + z_0 \mathbf{e}_z + \zeta \mathbf{g}_\zeta(\xi, \eta),$$

where the subscript 0 denotes the value at the reference surface, $\mathbf{x}_0(\xi, \eta)$ is the position vector of the reference surface and $\mathbf{g}_\zeta$ is the covariant base vector in the direction of a fiber through the point at ξ, η on the reference surface. In finite element modeling, the relations between the local Cartesian and the natural coordinates are defined as follows:

$$(2) \qquad \mathbf{x}_0(\xi, \eta) = \sum_i h_i(\xi, \eta)(\mathbf{x}_0)_i \,,$$

$$(3) \qquad \mathbf{g}_\zeta = \sum_i h_i(\xi, \eta)(t\mathbf{s})_i/2 \,,$$

in which h's are the shape functions defined in Chapter 18, i sums over *all the nodes defining the element reference surface*, the subscript i indicates association with node i, e.g., $(\mathbf{x}_0)_i$ is the position vector of node i of the reference surface, and t_i and $\mathbf{s}_i$ are the thickness and the unit vector in the fiber direction at node i defined as

$$(4) \qquad t_i = |\mathbf{x}_i^+ - \mathbf{x}_i^-| \,, \qquad \mathbf{s}_i = \frac{\mathbf{x}_i^+ - \mathbf{x}_i^-}{|\mathbf{x}_i^+ - \mathbf{x}_i^-|} \,,$$

in which $\mathbf{x}_i^+, \mathbf{x}_i^-$ are the position vectors at the top and the bottom surfaces of the shell at node i. The reference surface is usually the neutral surface of the shell. We shall assume the midsurface as the reference surface and leave the derivation for the general case to the readers. Thus the ranges of the natural coordinates ξ, η, ζ are ± 1.

From Eq. (1), we can derive the covariant base vectors [Eq. (2.14:7)] of the natural coordinates of the shell:

$$(5) \qquad \mathbf{g}_\xi = \mathbf{x}_{,\xi} \,, \qquad \mathbf{g}_\eta = \mathbf{x}_{,\eta} \,, \qquad \mathbf{g}_\zeta = \mathbf{x}_{,\zeta} \,,$$

These base vectors are not orthogonal for general shells.

In the *fiber base coordinates*, one of the coordinates is along the fiber direction $\mathbf{g}_\zeta$. The other two are defined to be orthogonal to $\mathbf{g}_\zeta$, otherwise arbitrary. We choose

$$(6) \qquad \widehat{\mathbf{e}}_\zeta = \frac{\mathbf{g}_\zeta}{|\mathbf{g}_\zeta|} \,, \qquad \widehat{\mathbf{e}}_\eta = \frac{\widehat{\mathbf{e}}_\zeta \times \mathbf{g}_\xi}{|\widehat{\mathbf{e}}_\zeta \times \mathbf{g}_\xi|} \,, \qquad \widehat{\mathbf{e}}_\xi = \widehat{\mathbf{e}}_\eta \times \widehat{\mathbf{e}}_\zeta \,,$$

as the unit base vectors where "$\times$" denotes the cross product. Thus, $\widehat{\mathbf{e}}_\zeta = \mathbf{s}_i$ at node i and $\widehat{\mathbf{e}}_\eta$ is tangent to the contravariant base vector $\mathbf{g}^\eta$ of the natural coordinates. If $\mathbf{g}_\xi$, $\mathbf{g}_\eta$ are orthogonal, $\widehat{\mathbf{e}}_\xi = \mathbf{g}_\xi/|\mathbf{g}_\xi|$ and $\widehat{\mathbf{e}}_\eta = \mathbf{g}_\eta/|\mathbf{g}_\eta|$ coincide with $\mathbf{g}_\xi$, $\mathbf{g}_\eta$, respectively. Using $\widehat{\mathbf{e}}_\zeta$ above, one can determine the contravariant base vectors of the natural coordinates (Sec. 2.14):

$$(7) \qquad \mathbf{g}^\xi = \frac{\mathbf{g}_\eta \times \widehat{\mathbf{e}}_\zeta}{|\mathbf{g}_\xi| \, |\mathbf{g}_\eta \times \widehat{\mathbf{e}}_\zeta|}, \qquad \mathbf{g}^\eta = \frac{\widehat{\mathbf{e}}_\zeta \times \mathbf{g}_\xi}{|\mathbf{g}_\eta| \, |\widehat{\mathbf{e}}_\zeta \times \mathbf{g}_\xi|}, \qquad \mathbf{g}^\zeta = \frac{\mathbf{g}_\zeta}{|\mathbf{g}_\zeta|^2} \, .$$

Note that $|\mathbf{g}_\eta \times \widehat{\mathbf{e}}_\zeta|^2 = (\mathbf{g}_\eta \cdot \mathbf{g}_\eta)(\widehat{\mathbf{e}}_\zeta \cdot \widehat{\mathbf{e}}_\zeta) - (\widehat{\mathbf{e}}_\zeta \cdot \mathbf{g}_\eta)^2$.

The *normal base coordinate* system has one axis normal to the surfaces $\zeta = $ constant. The other two base vectors are in the same plane as $\mathbf{g}_\xi$, $\mathbf{g}_\eta$ and are chosen to coincide with the material coordinates of the surfaces $\zeta = $ constant. For isotropic materials, these two base vectors, other than normal to surfaces $\zeta = $ constant, can be arbitrarily chosen. Similar to the fiber base coordinates, we define the unit base vectors as

$$(8) \qquad \bar{\mathbf{e}}_\zeta = \frac{\mathbf{g}_\xi \times \mathbf{g}_\eta}{|\mathbf{g}_\xi \times \mathbf{g}_\eta|}, \qquad \bar{\mathbf{e}}_\eta = \frac{\bar{\mathbf{e}}_\zeta \times \mathbf{g}_\xi}{|\bar{\mathbf{e}}_\zeta \times \mathbf{g}_\xi|}, \qquad \bar{\mathbf{e}}_\xi = \bar{\mathbf{e}}_\eta \times \bar{\mathbf{e}}_\zeta \, .$$

If $\mathbf{g}_\xi$, $\mathbf{g}_\eta$ are orthogonal, then $\bar{\mathbf{e}}_\xi$, $\bar{\mathbf{e}}_\eta$ and $\widehat{\mathbf{e}}_\xi$, $\widehat{\mathbf{e}}_\eta$ coincide.

Covariant Strains in Natural Coordinates. We shall derive the covariant strains in the natural coordinates. Let $\theta_\eta, -\theta_\xi$ represent the rotations of the fiber about the axes along $\widehat{\mathbf{e}}_\xi$, $\widehat{\mathbf{e}}_\eta$, respectively. The displacement vector can be written as

$$(9) \qquad \mathbf{u} = \mathbf{u}_0(\xi, \eta) - t\zeta(\theta_\xi \widehat{\mathbf{e}}_\xi + \theta_\eta \widehat{\mathbf{e}}_\eta)/2 \, ,$$

where $\mathbf{u}_0$ is the displacement of the midsurface. The second term denotes the fiber rotation. Equation (9) allows a fiber to rotate, but not stretch. In finite element modeling, we represent the displacement by

$$(10) \qquad \mathbf{u} = \sum_i h_i(\xi, \eta)(\mathbf{u}_0)_i - \zeta \sum_i h_i(\xi, \eta)[(\theta_\xi \widehat{\mathbf{e}}_\xi + \theta_\eta \widehat{\mathbf{e}}_\eta)t]_i/2 \, ,$$

where the shape functions h's will be different from those in Eq. (2) if the nodes for interpolating displacements differs from those for defining the curved shell surface. The interpolation functions of a 9-node element are given in Table 18.13.1. The first sum in Eq. (10) is the *interpolated displacement* of the midsurface. The second sum is that *induced by fiber rotation*. An alternative is to represent the displacement by

$$(10a) \qquad \mathbf{u} = \sum_i h_i(\xi, \eta)(\mathbf{u}_0)_i - \zeta \sum_i h_i(\xi, \eta)[(\theta_\xi \bar{\mathbf{e}}_\xi + \theta_\eta \bar{\mathbf{e}}_\eta)t]_i/2 \, ,$$

with $\theta_\eta, -\theta_\xi$ denoting the rotations of the normal about the axes along $\bar{\mathbf{e}}_\xi, \bar{\mathbf{e}}_\eta$, respectively. In this case, there is no need for the fiber base coordinates.

From Eq. (10), we can derive the deformation gradient,

$$\frac{\partial \mathbf{u}}{\partial \xi} = \sum_i \frac{\partial h_i}{\partial \xi} \left[\mathbf{u}_0 - \frac{t\zeta}{2} (\theta_\xi \widehat{\mathbf{e}}_\xi + \theta_\eta \widehat{\mathbf{e}}_\eta) \right]_i ,$$

(11)
$$\frac{\partial \mathbf{u}}{\partial \eta} = \sum_i \frac{\partial h_i}{\partial \eta} \left[\mathbf{u}_0 - \frac{t\zeta}{2} (\theta_\xi \widehat{\mathbf{e}}_\xi + \theta_\eta \widehat{\mathbf{e}}_\eta) \right]_i ,$$

$$\frac{\partial \mathbf{u}}{\partial \zeta} = - \sum_i \frac{h_i}{2} [t(\theta_\xi \widehat{\mathbf{e}}_\xi + \theta_\eta \widehat{\mathbf{e}}_\eta)]_i .$$

The covariant in-plane strains are

$$(12) \quad e_{\xi\xi} = \frac{\partial \mathbf{u}}{\partial \xi} \cdot \mathbf{g}_\xi , \quad e_{\eta\eta} = \frac{\partial \mathbf{u}}{\partial \eta} \cdot \mathbf{g}_\eta , \quad e_{\xi\eta} = \frac{1}{2} \left(\frac{\partial \mathbf{u}}{\partial \eta} \cdot \mathbf{g}_\xi + \frac{\partial \mathbf{u}}{\partial \xi} \cdot \mathbf{g}_\eta \right) ,$$

and the covariant transverse shear strains are

$$(13) \quad e_{\xi\zeta} = \frac{1}{2} \left(\frac{\partial \mathbf{u}}{\partial \zeta} \cdot \mathbf{g}_\xi + \frac{\partial \mathbf{u}}{\partial \xi} \cdot \mathbf{g}_\zeta \right) , \quad e_{\eta\zeta} = \frac{1}{2} \left(\frac{\partial \mathbf{u}}{\partial \zeta} \cdot \mathbf{g}_\eta + \frac{\partial \mathbf{u}}{\partial \eta} \cdot \mathbf{g}_\zeta \right) .$$

From Eqs. (10)–(13), one finds the covariant strains in terms of the nodal values of $\mathbf{u}_0$, θ_ξ, θ_η. From Eq. (13), we have

$$e_{\xi\zeta} = \sum_i \frac{h_{i,\xi}}{2} \mathbf{g}_\zeta \cdot (\mathbf{u}_0)_i - \sum_i \frac{h_i}{4} \mathbf{g}_\xi \cdot [(\theta_\xi \widehat{\mathbf{e}}_\xi + \theta_\eta \widehat{\mathbf{e}}_\eta)t]_i$$

$$- \zeta \sum_i \frac{h_{i,\xi}}{4} \mathbf{g}_\zeta \cdot [(\theta_\xi \widehat{\mathbf{e}}_\xi + \theta_\eta \widehat{\mathbf{e}}_\eta)t]_i .$$

The last sum represents the contribution of curvature changes to the transverse shear and is zero for flat elements or only at the nodes in general. A similar expression can be derived using Eq. (10a).

The engineering shear strains can be obtained by multiplying the physical components of the shear strains by 2. Since $\mathbf{g}_\xi$ and $\mathbf{g}_\eta$ are linear in ζ, $e_{\xi\xi}$, $e_{\eta\eta}$, $e_{\xi\eta}$ are quadratic and $e_{\xi\zeta}$, $e_{\eta\zeta}$ are linear in ζ. Note that $e_{\zeta\zeta} = \mathbf{g}_\zeta \cdot \partial \mathbf{u}/\partial \zeta$ is only approximately equal to zero due to the approximation of Eq. (10) or (10a). One often neglects $e_{\zeta\zeta}$ in practical computation. The use of the representation of Eq. (10) or (10a) avoids the explicit calculation of the *Eulerian Christoffel symbols* [Eq. (2.15:4)] in deriving the deformation gradient.

Example-Ring Element. Consider a 2-node ring element of thickness t and radius r_0 over the angle $-\varphi_0 \leq \varphi \leq \varphi_0$. The nodes are at $r = r_0, \varphi_{1,2} =$

$\pm\varphi_0$. The unit base vectors $\hat{\mathbf{e}}_r$, $\hat{\mathbf{e}}_\varphi$ of the fiber base coordinates coincide with those of the cylindrical coordinates at the nodes. The finite element representations of ring and displacements are

$$\mathbf{x} = \mathbf{x}_0 + \zeta t \mathbf{e}_r/2 = (r_0 + \zeta t/2) \sum_{i=1}^{2} (\hat{\mathbf{e}}_r)_i h_i(\xi),$$

$$\mathbf{u} = \sum_{i=1}^{2} h_i(\xi)(\mathbf{u}_0 + \zeta t \theta_\xi \hat{\mathbf{e}}_\varphi/2)_i,$$

where

$$\zeta = 2(r - r_0)/t, \quad \xi = \varphi/\varphi_0, \quad h_1 = (1-\xi)/2, \quad h_2 = (1+\xi)/2.$$

The covariant base vectors and the deformation gradient are

$$\mathbf{g}_\xi = \frac{1}{2}\left(r_0 + \frac{1}{2}\zeta t\right)[(\hat{\mathbf{e}}_r)_2 - (\hat{\mathbf{e}}_r)_1], \quad \mathbf{g}_\zeta = \frac{t}{2}\sum_{i=1}^{2}(\hat{\mathbf{e}}_r)_i h_i(\xi),$$

$$\frac{\partial \mathbf{u}}{\partial \xi} = \frac{(\mathbf{u}_0)_2 - (\mathbf{u}_0)_1}{2} + \frac{(\theta_\xi\hat{\mathbf{e}}_\varphi)_2 - (\theta_\xi\hat{\mathbf{e}}_\varphi)_1}{4}\zeta t, \quad \frac{\partial \mathbf{u}}{\partial \zeta} = \frac{t}{2}\sum_{i=1}^{2}(\theta_\xi\hat{\mathbf{e}}_\varphi)_i h_i,$$

The covariant strains of the natural coordinates are

$$\varepsilon_{\xi\xi} = \frac{1}{4}\left(r_0 + \frac{\zeta t}{2}\right)\left[(\mathbf{u}_0)_2 - (\mathbf{u}_0)_1 + \frac{(\theta_\xi\hat{\mathbf{e}}_\varphi)_2 - (\theta_\xi\hat{\mathbf{e}}_\varphi)_1}{2}\zeta t\right][(\hat{\mathbf{e}}_r)_2 - (\hat{\mathbf{e}}_r)_1],$$

$$\varepsilon_{\xi\zeta} = \frac{t}{8}\Bigg\{\left(r_0 + \frac{\zeta t}{2}\right)[(\hat{\mathbf{e}}_r)_2 - (\hat{\mathbf{e}}_r)_1]\cdot\sum_{i=1}^{2}(\theta_\xi\hat{\mathbf{e}}_\varphi)_i h_i$$

$$+ \left[(\mathbf{u}_0)_2 - (\mathbf{u}_0)_1 + \frac{(\theta_\xi\hat{\mathbf{e}}_\varphi)_2 - (\theta_\xi\hat{\mathbf{e}}_\varphi)_1}{2}\zeta t\right]\cdot\sum_{i=1}^{2}(\hat{\mathbf{e}}_r)_i h_i\Bigg\}.$$

The physical components of the strains [Eq. (4.10:11)] are

$$e_{\xi\xi} = \varepsilon_{\xi\xi}/|\mathbf{g}_\xi|^2, \quad e_{\xi\zeta} = \varepsilon_{\xi\zeta}/(|\mathbf{g}_\xi||\mathbf{g}_\zeta|).$$

Constitutive Equation. A general plane stress constitutive equation can be written with respect to the normal base coordinates in the form

(14) $$\bar{\sigma} = \bar{\mathbf{D}}_{3e}\bar{\varepsilon},$$

where $\bar{\sigma}$, $\bar{\varepsilon}$ are the in-plane stress and strain matrices defined as

(15) $\bar{\sigma} = [\bar{\sigma}_{\xi\xi} \ \bar{\sigma}_{\eta\eta} \ \bar{\sigma}_{\xi\eta} \ \bar{\sigma}_{\xi\zeta} \ \bar{\sigma}_{\eta\zeta}]^T$, $\bar{\varepsilon} = [\bar{\varepsilon}_{\xi\xi} \ \bar{\varepsilon}_{\eta\eta} \ 2\bar{\varepsilon}_{\xi\eta} \ 2\bar{\varepsilon}_{\xi\zeta} \ 2\bar{\varepsilon}_{\eta\zeta}]^T$,

$\bar{\mathbf{D}}_{3e}$ is a 5×5 symmetric elastic modulus matrix for plane stress derived from the 6×6 modulus matrix for 3-dimensional elasticity [Eq. (18.16:4)] by enforcing the zero normal stress condition ($\bar{\sigma}_{\zeta\zeta} = 0$) in the direction normal to the shell surface. We use the subscript 3 to distinct the elastic modulus matrix from the 2-dimensional case. For materials with transverse shear strains $\bar{\varepsilon}_{\xi\zeta}$, $\bar{\varepsilon}_{\eta\zeta}$ decoupled from the rest of the strain components in the constitutive equation, we have

(16)
$$\bar{\mathbf{D}}_{3e} = \begin{bmatrix} \bar{\mathbf{D}}_e & \mathbf{0} \\ \mathbf{0} & \bar{\mathbf{D}}_s \end{bmatrix} ,$$

where $\bar{\mathbf{D}}_e$ is a 3×3 elastic modulus matrix for plane stress [see Eq. (18.14:23) for orthotropic materials], $\bar{\mathbf{D}}_s$ is a 2×2 transverse shear modulus matrix. To attain results consistent with classical bending theory, a shear factor needs to be introduced, which usually amounts to multiplying the transverse shearing moduli by $\kappa = 5/6$. For isotropic materials, $\bar{\mathbf{D}}_s$ is a diagonal matrix given in Eq. (20.2:4).

Strain Transformation. Strain transformation between coordinate systems follows the tensor transformation law (see Sec. 2.5). The physical components of the strain teusor in the normal base coordinates can be expressed as those in the natural coordinates as

(17)
$$\begin{bmatrix} \bar{\varepsilon}_{\xi\xi} & \bar{\varepsilon}_{\xi\eta} & \bar{\varepsilon}_{\xi\zeta} \\ \bar{\varepsilon}_{\xi\eta} & \bar{\varepsilon}_{\eta\eta} & \bar{\varepsilon}_{\eta\zeta} \\ \bar{\varepsilon}_{\xi\zeta} & \bar{\varepsilon}_{\eta\zeta} & \bar{\varepsilon}_{\zeta\zeta} \end{bmatrix} = \bar{\mathbf{R}}^T \begin{bmatrix} \varepsilon_{\xi\xi} & \varepsilon_{\xi\eta} & \varepsilon_{\xi\zeta} \\ \varepsilon_{\xi\eta} & \varepsilon_{\eta\eta} & \varepsilon_{\eta\zeta} \\ \varepsilon_{\xi\zeta} & \varepsilon_{\eta\zeta} & \varepsilon_{\zeta\zeta} \end{bmatrix} \bar{\mathbf{R}} ,$$

(18)
$$\bar{\mathbf{R}} = \begin{bmatrix} \mathbf{g}^\xi \cdot \bar{\mathbf{e}}_\xi & \mathbf{g}^\xi \cdot \bar{\mathbf{e}}_\eta & \mathbf{g}^\xi \cdot \bar{\mathbf{e}}_\zeta \\ \mathbf{g}^\eta \cdot \bar{\mathbf{e}}_\xi & \mathbf{g}^\eta \cdot \bar{\mathbf{e}}_\eta & \mathbf{g}^\eta \cdot \bar{\mathbf{e}}_\zeta \\ \mathbf{g}^\zeta \cdot \bar{\mathbf{e}}_\xi & \mathbf{g}^\zeta \cdot \bar{\mathbf{e}}_\eta & \mathbf{g}^\zeta \cdot \bar{\mathbf{e}}_\zeta \end{bmatrix} ,$$

where ($\cdot$) denotes dot product. From Eqs. (17) and (18) together with Eqs. (12) and (13), we can obtain $\bar{\varepsilon}_{ij}$ in terms of shape functions and nodal values of $\mathbf{u}_0, \theta_\xi, \theta_\eta$. To be consistent with the shell theory, one can neglect $\varepsilon_{\zeta\zeta}$ in Eq. (17). The engineering shears $\bar{\gamma}$'s can be obtained by multiplying the corresponding shear strains $\bar{\varepsilon}$'s by 2. Then we can write the strain-displacement relation in the matrix form

$$(19) \qquad \bar{\varepsilon} = \mathbf{Bq} = \sum_{i=1}^{p} \bar{\mathbf{B}}_i \mathbf{q}_i \,,$$

where $\bar{\mathbf{B}}$ is the strain displacement matrix, p is the number of nodes of the element, the subscript i denotes association with node i and

$$(20) \quad \mathbf{q}_i = [(\mathbf{u}_0)_i^T \quad (\theta_\xi)_i \quad (\theta_\eta)_i]^T \,, \quad \mathbf{q} = [\mathbf{q}_1^T \quad \mathbf{q}_2^T \quad \cdots \quad \mathbf{q}_p^T]^T \,.$$

The calculation of $\bar{\mathbf{B}}$ from Eqs. (1)–(7), (10)–(13), (17) and (18) is tedious but straightforward.

Element Matrices. The shell element shown in Fig. 20.6:1 is a solid whose top and bottom surfaces are curved and whose sides across the thickness are generated by straight lines (fibers). We shall derive the matrices for the shell element from the potential energy given in Eq. (18.16:9), converted to the shell geometry below,

$$(21) \qquad \Pi_e = \int_{A_e} \int_{-1}^{1} \left(\frac{1}{2} \bar{\varepsilon}^T \mathbf{D}_{3e} \bar{\varepsilon} - \mathbf{u}^T \mathbf{b} \right) t J_A d\zeta \, dA$$

$$- \int_{\partial A_{\sigma e}} \int_{-1}^{1} \mathbf{u}^T \bar{\mathbf{T}} t J_s \, d\zeta \, ds + \int_{A_e} p \mathbf{u}_0(\xi, \eta) \cdot \bar{\mathbf{e}}_{\zeta 0} \, dA \,,$$

where zero shear on the top and bottom shell surfaces is assumed and

$$(22) \qquad p = \bar{T}_n^- \, J_A^- - \bar{T}_n^+ \, J_A^+ \cong \bar{T}_n^- - \bar{T}_n^+ \,,$$

in which $\bar{T}_n^\pm$ denote prescribed normal traction on the upper and lower shell surfaces, respectively. The quantities J_A, J_s are, respectively, the *Jacobians* for the volume and the surface integrals normalized by their corresponding values at the midsurface defined as

$$(23) \qquad J_A = \mathbf{g}_\zeta \cdot (\mathbf{g}_\xi \times \mathbf{g}_\eta) / |\mathbf{g}_\zeta \cdot (\mathbf{g}_{\xi 0} \times \mathbf{g}_{\eta 0})| \,,$$

$$(24) \qquad J_s = |\mathbf{g}_\zeta \times \mathbf{g}_\alpha| / |\mathbf{g}_\zeta \times \mathbf{g}_{\alpha 0}| \,,$$

where $\alpha = \xi$ if the boundary is along the curve $\eta = $ constant, and $\alpha = \eta$ if the boundary is along the curve $\xi = $ constant. Both $\mathbf{g}_\xi$ and $\mathbf{g}_\eta$ are functions of ξ, η, ζ and so are J_A and J_s. If the boundary of the shell element is part of cylindrical surface, J_s is 1. In the case of a plate, J_A is also 1. In practice, one may approximate J_A and J_s as unity if the shell is thin.

Using Eqs. (10) or (10a) and (19), from Eq. (21), we can derive the element matrices. The element stiffness matrix is in the form

$$(25) \qquad \Pi_e = \frac{1}{2} \mathbf{q}^T \mathbf{k} \mathbf{q} - \mathbf{q}^T \mathbf{f},$$

where $\mathbf{f}$ involves all integrals linearly in $\mathbf{u}$ and $\mathbf{u}_0$ in Eq. (21) and

$$(26) \qquad \mathbf{k} = \int_{A_2} \int_{-1}^{1} t \mathbf{B}^T \mathbf{D}_{3e} \mathbf{B} J_A \, d\zeta \, dA.$$

Gaussian quadrature is most efficient for integration along the fibers if the shell consists of a single homogeneous layer. Usually a two-point Gauss rule is adequate to sense the bending and membrane actions. For layered laminates of different materials through the thickness, the one or two-point Gauss rule has to be applied to each layer.

Global Coordinates. We must transform $\mathbf{q}$ of the element to the global coordinates before assembling. In Eq. (20), the components of $\mathbf{u}_0$ of node i refers to the local coordinates and θ_ξ, θ_η refer to the fiber base coordinates. Different transformations are needed to transform them to the global system. Let $\tilde{\mathbf{e}}_x, \tilde{\mathbf{e}}_y, \tilde{\mathbf{e}}_z$ be the unit base vectors of the global system and $\tilde{\mathbf{q}}$ the displacement and rotation matrix in global coordinates be defined as

$$(27) \quad \tilde{\mathbf{q}} = [\tilde{\mathbf{q}}_1^T \ \tilde{\mathbf{q}}_2^T \ \cdots \ \tilde{\mathbf{q}}_p^T]^T, \qquad \tilde{\mathbf{q}}_i = [\tilde{\mathbf{u}}_0^T \ \tilde{\boldsymbol{\theta}}^T]_i^T, \qquad \tilde{\boldsymbol{\theta}}_i = [\tilde{\theta}_x \ \tilde{\theta}_y \ \tilde{\theta}_z]_i^T.$$

Note that $\tilde{\mathbf{q}}_i$ is a column matrix of six components and that $\tilde{\theta}_y, -\tilde{\theta}_x$ denote the rotations about the $\tilde{x}, \tilde{y}$-axes. Then

$$(28) \qquad (\mathbf{u}_0)_i = \begin{bmatrix} \mathbf{e}_x \cdot \tilde{\mathbf{e}}_x & \mathbf{e}_x \cdot \tilde{\mathbf{e}}_y & \mathbf{e}_x \cdot \tilde{\mathbf{e}}_z \\ \mathbf{e}_y \cdot \tilde{\mathbf{e}}_x & \mathbf{e}_y \cdot \mathbf{e}_y & \mathbf{e}_y \cdot \mathbf{e}_z \\ \mathbf{e}_z \cdot \tilde{\mathbf{e}}_x & \mathbf{e}_z \cdot \tilde{\mathbf{e}}_y & \mathbf{e}_z \cdot \tilde{\mathbf{e}}_z \end{bmatrix}_i (\tilde{\mathbf{u}}_0)_i,$$

$$(29) \qquad \begin{bmatrix} \theta_\xi \\ \theta_\eta \end{bmatrix}_i = \begin{bmatrix} \hat{\mathbf{e}}_\eta \cdot \tilde{\mathbf{e}}_\eta & -\hat{\mathbf{e}}_\eta \cdot \tilde{\mathbf{e}}_\xi & -\hat{\mathbf{e}}_\eta \cdot \tilde{\mathbf{e}}_\zeta \\ -\hat{\mathbf{e}}_\xi \cdot \tilde{\mathbf{e}}_\eta & \hat{\mathbf{e}}_\xi \cdot \tilde{\mathbf{e}}_\xi & \hat{\mathbf{e}}_\xi \cdot \tilde{\mathbf{e}}_\zeta \end{bmatrix}_i \tilde{\boldsymbol{\theta}}_i.$$

Substituting Eqs. (28) and (29) into Eq. (25), we obtain the element matrices associated with displacements and rotations in global coordinates to be assembled. Note that if we represent the displacement by Eq. (10a), we will have to replace $\hat{\mathbf{e}}_\xi, \hat{\mathbf{e}}_\eta, \hat{\mathbf{e}}_\zeta$ by $\bar{\mathbf{e}}_\xi, \bar{\mathbf{e}}_\eta, \bar{\mathbf{e}}_\zeta$.

Drilling Degree-of-Freedom and Coordinate Transformation. Equation (29) transforms at each node two rotations in local to three rotations in global coordinates. Like plate assembly, the assembled stiffness matrix will be singular, if all elements meeting at a node have the same fiber direction at the common node, which is most likely the case. To avoid this potential singularity, one can modify the element formulation by making the rotations about the nodal fibers contribute to strains using drilling freedoms similar to that of as described in Sec. 20.5. Alternative is to identify a global fiber base coordinates for each node. All rotations at a common node are transformed to the global coordinates before assembling. In this case, there will be only two rotations (or five degrees-of-freedom total) per node. Another alternative is to use the normal base coordinates for rotations as given in Eq. (10a). Due to the approximation of shell geometry by Eq. (1), the normal vectors to different elements at the common node will be usually different. Thus singularity may be avoided.

Remark. A comment on the strain derivation is in order. In literature, the normal and fiber base coordinates are often called *the running local Cartesian coordinates* and the strains referred to these coordinates are simply defined as

$$(30) \qquad \bar{\varepsilon}_{\bar{\xi}\bar{\xi}} = \frac{\partial \bar{u}_{\bar{\xi}}}{\partial \bar{\xi}}, \qquad \bar{\varepsilon}_{\bar{\xi}\bar{\eta}} = \frac{1}{2}\left(\frac{\partial \bar{u}_{\bar{\xi}}}{\partial \bar{\eta}} + \frac{\partial \bar{u}_{\bar{\eta}}}{\partial \bar{\xi}}\right), \ldots, \text{etc}.$$

where $\bar{u}$'s, $\bar{\xi}$'s are the components in the referenced coordinate system. One then expresses $\bar{u}$'s and $\bar{\xi}$'s in terms of their corresponding components in the local fixed Cartesian coordinates based on the coordinate transformation rule and proceeds with the partial differentiation. The transformation does not change the fact that the strains in Eq. (30) are defined based only on partial differentiation rather than on covariant differentiation (Sec. 2.12). Even though the normal and the fiber base coordinates are orthogonal, the *orientation of these running local coordinates changes with location within the element for general shells.* This change contributes to strains in the terms of the *Euclidean Christoffel symbols.* In other words, *strains should be derived from the deformation gradient, $\partial \mathbf{u}/\partial \bar{\xi}, \partial \mathbf{u}/\partial \bar{\eta}$, etc. in order to account for the covariant or contravariant differentiation, rather than just from the partial derivatives of the components of the displacement.*

In the exercises below, we use θ as the circumferential coordinate of the cylindrical coordinates of unit base vectors $\mathbf{e}_r, \mathbf{e}_\theta$, and α for the rotation about the z-axis.

Problem 20.11. Consider a ring element of thickness t and radius r_0 over $-\theta_0 \le \theta \le \theta_0$

$$\mathbf{x} = \mathbf{x}_0 + \zeta t \mathbf{e}_r/2 = r_0(\cos\theta \mathbf{e}_x + \sin\theta \mathbf{e}_y) + \zeta t \mathbf{e}_r/2 = (r_0 + \zeta t/2)\mathbf{e}_r,$$

where

$$\mathbf{e}_r = \cos\theta\mathbf{e}_x + \sin\theta\mathbf{e}_y\,, \qquad \zeta = 2(r - r_0)/t\,, \qquad \xi = \theta/\theta_0\,.$$

The element has three nodes at $\theta_{1,2} = \pm\theta_0, \theta_3 = 0$. Assuming the displacement in the form

$$\mathbf{u} = \sum_{i=1}^{3} h_i(\theta)\left(\mathbf{u}_0 + \frac{t\zeta}{2}\alpha\mathbf{e}_\theta\right)_i\,, \qquad \mathbf{e}_\theta = -\sin\theta\mathbf{e}_x + \cos\theta\mathbf{e}_y\,,$$

$$h_1 = -\frac{\xi(1-\xi)}{2}\,, \qquad h_2\frac{\xi(1+\xi)}{2}\,, \qquad h_3 = 1 - \xi^2\,,$$

show that the covariant base vectors and the metric tensor are

(a) $$\mathbf{g}_\xi = \left(r_0 + \frac{\zeta t}{2}\right)\theta_0\mathbf{e}_\theta\,, \qquad \mathbf{g}_\zeta = \frac{t}{2}\mathbf{e}_r\,,$$

(b) $$\sqrt{g_{\xi\xi}} = \left(r_0 + \frac{\zeta t}{2}\right)\theta_0\,, \qquad \sqrt{g_{\zeta\zeta}} = \frac{t}{2}\,, \qquad g_{\xi\zeta} = 0\,,$$

and that (c) the physical component of the strain in the radial direction is

$$\left(\frac{2}{t}\right)^2 e_{\zeta\zeta} = \left(\frac{2}{t}\right)^2\mathbf{g}_\zeta\cdot\frac{\partial\mathbf{u}}{\partial\zeta} = \frac{(t\alpha)_1}{t}h_1\sin[\theta_0(\xi+1)]$$

$$+ \frac{(t\alpha)_2}{t}h_2\sin[\theta_0(\xi-1)] + \frac{(t\alpha)_3}{t}h_3\sin(\theta_0\xi)\,,$$

which is not zero whereas its representation in the shell theory is zero. This quantity is of the order θ_0^2 for small element (i.e., θ_0 is small). [Hint: Use the following relations:

$$\mathbf{e}_{\theta i}\cdot\mathbf{e}_\theta = \mathbf{e}_{ri}\cdot\mathbf{e}_r = \cos(\theta - \theta_i)\,, \qquad \mathbf{e}_{\theta i}\cdot\mathbf{e}_r = -\mathbf{e}_{ri}\cdot\mathbf{e}_\theta = \sin(\theta - \theta_i)\,,$$

$$\mathbf{u}_{0i} = u_{0i}\mathbf{e}_x + v_{0i}\mathbf{e}_y = u_{r0i}\mathbf{e}_{ri} + u_{\theta0i}\mathbf{e}_{\theta i}\,, \qquad \theta_2 = -\theta_1 = \theta_0, \theta_3 = 0\,.]$$

Problem 20.12. Repeat Problem 20.11 using

$$\mathbf{x} = \sum_{i=1}^{3} h_i(\xi)(\mathbf{x}_0 + \zeta\mathbf{e}_\theta/2)_i\,, \qquad (\mathbf{x}_0)_i = r_0(\mathbf{e}_x\cos\theta_i + \mathbf{e}_y\sin_{\theta_i})\,.$$

Problem 20.13. Let the displacements of the ring discussed above be

$$\mathbf{u} = u_r(\theta)\mathbf{e}_r + u_\theta(\theta)\mathbf{e}_\theta + \alpha(\theta)(r - r_0)\mathbf{e}_\theta\,.$$

(a) Derive the physical strain components $e_{rr}, e_{r\theta}, e_{\theta\theta}$.

(b) Let

$$u_r(\theta)\mathbf{e}_r + u_\theta(\theta)\mathbf{e}_\theta = \sum_{i=1}^{3}(\mathbf{u}_0)h_i(\xi), \qquad \alpha(\theta) = \sum_{i=1}^{3}h_i\alpha_i.$$

Find the differences between strains derived from the displacements above and those from Problem 20.12. Show the difference is of the order of θ_0^2 for small θ_0.

Problem 20.14. Using the tensor transformation rule, determine the components of $\widehat{\mathbf{D}}_e, \widehat{\mathbf{D}}_s$ in the fiber base coordinates in terms of those of $\bar{\mathbf{D}}_e, \bar{\mathbf{D}}_s$ in the normal base coordinates. [Hint: $\hat{\mathbf{e}}^T\widehat{\mathbf{D}}_e\hat{\mathbf{e}} = \bar{\mathbf{e}}^T\bar{\mathbf{D}}_e\bar{\mathbf{e}}$.]

Problem 20.15. Use Eqs. (20.6:17) and (20.6:18) and $\bar{\mathbf{e}}^T\bar{\mathbf{D}}_{3e}\bar{\mathbf{e}} = \mathbf{e}^T\mathbf{D}_{3e}\mathbf{e}$, express the components of $\mathbf{D}_{3e}$ in terms of those of $\bar{\mathbf{D}}_{3e}$.

20.7. LOCKING AND STABLIZATION IN SHELL APPLICATIONS

Doubly curved shell elements can lock in thin shell applications. We illustrate locking potentials in the example as follows. Let us examine the characteristics of $\varepsilon_{\xi\xi}$ and $\varepsilon_{\xi\eta}$ at the midsurface of a 4-node quadrilateral element of constant thickness with $\mathbf{x}_{0i}$ being the position of node i. From Eqs. (20.6:5) and (20.6:11)–(20.6:13), we have

(1) $$\mathbf{g}_\xi = \mathbf{a}_2 + \mathbf{a}_4\eta, \qquad \mathbf{g}_\zeta = \frac{t}{2}\sum_{i=1}^{4}h_i\mathbf{s}_i,$$

(2) $$\frac{\partial\mathbf{u}}{\partial\xi} = \mathbf{b}_2 + \mathbf{b}_4\eta, \qquad \frac{\partial\mathbf{u}}{\partial\zeta} = -\frac{t}{2}\sum_{i=1}^{4}[\theta_\xi\hat{\mathbf{e}}_\xi + \theta_\eta\hat{\mathbf{e}}_\eta]_i h_i,$$

(3a) $$\varepsilon_{\xi\xi} = (\mathbf{a}_2 + \mathbf{a}_4\eta)\cdot(\mathbf{b}_2 + \mathbf{b}_4\eta),$$

(3b) $$\varepsilon_{\xi\zeta} = \frac{t}{4}\left[(\mathbf{b}_2 + \mathbf{b}_4\eta)\cdot\sum_{i=1}^{4}h_i\mathbf{s}_i - (\mathbf{a}_2 + \mathbf{a}_4\eta)\cdot\sum_{i=1}^{4}(\theta_\xi\hat{\mathbf{e}}_\xi + \theta_\eta\hat{\mathbf{e}}_\eta)_i h_i\right],$$

where $\mathbf{s}_i$ is a unit vector in the fiber direction at node i and

(4) $$\mathbf{a}_2 = \frac{-\mathbf{x}_{01} + \mathbf{x}_{02} + \mathbf{x}_{03} - \mathbf{x}_{04}}{4}, \qquad \mathbf{a}_4 = \frac{\mathbf{x}_{01} - \mathbf{x}_{02} + \mathbf{x}_{03} - \mathbf{x}_{04}}{4},$$

(5) $$\mathbf{b}_2 = \frac{-\mathbf{u}_1 + \mathbf{u}_2 + \mathbf{u}_3 - \mathbf{u}_4}{4}, \qquad \mathbf{b}_4 = \frac{\mathbf{u}_1 - \mathbf{u}_2 + \mathbf{u}_3 - \mathbf{u}_4}{4}.$$

In the limit of inextensional bending and zero transverse shear for thin shells, the coefficients of the polynomial for $\varepsilon_{\xi\xi}, \varepsilon_{\xi\zeta}$ vanish. This represents constraints on the parameters appearing in the coefficients. Since

$\varepsilon_{\xi\xi}$ is quadratic in η and independent of ξ and $\varepsilon_{\xi\zeta}$ is quadratic in η but linear in ξ, there are 9 possible constraints (3 from $\varepsilon_{\xi\xi}$ and 6 from $\varepsilon_{\xi\zeta}$) for the parameters. For general shells, it is difficult to identify which of these constraints is field-consistent (Sec. 20.3) and which is not. For flat parallelograms ($\mathbf{a}_4 = 0, \mathbf{s}_1 = \mathbf{s}_2 = \mathbf{s}_3 = \mathbf{s}_4$), one can show that only the transverse shear $\varepsilon_{\xi\zeta}$ has spurious terms which are from the coefficients of ξ and $\xi\eta$ in the last term in Eq. (3b). We define

$$\sum_{i=1}^{4}(\theta_\xi\widehat{\mathbf{e}}_\xi + \theta_\eta\widehat{\mathbf{e}}_\eta)_i h_i = \sum_{i=1}^{4}\mathbf{\Psi}_i h_i = \mathbf{\Theta}_1 + \mathbf{\Theta}_2\xi + \mathbf{\Theta}_3\eta + \mathbf{\Theta}_4\xi\eta\,,$$

where

$$\mathbf{\Psi}_i = (\theta_\xi\widehat{\mathbf{e}}_\xi + \theta_\eta\widehat{\mathbf{e}}_\eta)_i\,,$$

$$\mathbf{\Theta}_1 = \frac{\mathbf{\Psi}_1 + \mathbf{\Psi}_2 + \mathbf{\Psi}_3 + \mathbf{\Psi}_4}{4}\,, \qquad \mathbf{\Theta}_2 = \frac{-\mathbf{\Psi}_1 + \mathbf{\Psi}_2 + \mathbf{\Psi}_3 - \mathbf{\Psi}_4}{4}\,,$$

$$\mathbf{\Theta}_3 = \frac{-\mathbf{\Psi}_1 - \mathbf{\Psi}_2 + \mathbf{\Psi}_3 + \mathbf{\Psi}_4}{4}\,, \qquad \mathbf{\Theta}_4 = \frac{\mathbf{\Psi}_1 - \mathbf{\Psi}_2 + \mathbf{\Psi}_3 - \mathbf{\Psi}_4}{4}\,.$$

The spurious constraints are then,

(6) $$\mathbf{\Theta}_2 \cdot \mathbf{a}_2 \to 0\,, \qquad \mathbf{\Theta}_4 \cdot \mathbf{a}_2 \to 0\,,$$

Similarly we can determine the spurious constraints for $\varepsilon_{\eta\zeta}$. Then the field-consistent transverse shear strains are (Note: $\sum_{i=1}^{4} h_i = \frac{1}{4}$)

(7)
$$\varepsilon_{\xi\zeta} = \frac{t}{4}\left[\frac{\mathbf{b}_2 + \mathbf{b}_4\eta}{4} \cdot \sum_{i=1}^{4}\mathbf{s}_i - \mathbf{a}_2 \cdot (\mathbf{\Theta}_1 + \mathbf{\Theta}_3\eta)\right]\,,$$

$$\varepsilon_{\eta\zeta} = \frac{t}{4}\left[\frac{\mathbf{b}_3 + \mathbf{b}_4\xi}{4} \cdot \sum_{i=1}^{4}\mathbf{s}_i - \mathbf{a}_3 \cdot (\mathbf{\Theta}_1 + \mathbf{\Theta}_2\xi)\right]\,,$$

which are to be used in calculating the shear strain energy to remove locking and in evaluating the transversal shear stresses to avoid linear oscillations within an element. This equivalent to evaluating $\varepsilon_{\xi\zeta}$ at $\xi = 0$ and $\varepsilon_{\eta\zeta}$ at $\eta = 0$.

Literature is full of techniques for locking removal. They include uniformly reduced integration, selectively reduced integration, drilling degrees-of-freedom, mixed formulation etc. No single technique can completely rid of locking for general shell elements. The uniformly one-point integration 4-node and 2×2 reduced integration 9-node quadrilaterals are most frequently used in practice. Reduced integration offers substantial savings in computer time in deriving the element matrices because of reduced integration operations. Further computer savings can be achieved by taking advantages of the symmetric and anti-symmetric properties of $\partial h_i/\partial\xi, \partial h_i/\partial\eta$ at integration points. Choice of an element in applications must balance the computational efficiency against the desired accuracy of the results.

Stabilization. Uniformly reduced integration always and selectively reduced integration usually introduce spurious zero energy modes (kinematic modes). These modes occasionally give spurious solution in static analysis and often give spatially oscillatory solutions in dynamic simulation. The oscillations are called the *hourglass modes*. *Stabilization* schemes have been devised to eliminate the kinematic modes (Hughes and Tezduyar 1981). Belytschko and his associates (Belytschko and Tsay 1983, Belytschko *et al.* 1984, 1989) developed a general stabilization matrix for 4-node quadrilateral plate bending element with one point integration. The stabilization matrix is to be added to the element stiffness matrix. For the nodal parameter matrix

$$(8) \qquad \mathbf{q} = [\mathbf{w}^T \quad \boldsymbol{\theta}_x^T \quad \boldsymbol{\theta}_y^T]^T ,$$

in which

$$(9) \qquad \mathbf{w} = [w_1 \ w_2 \ w_3 \ w_4]^T , \quad \boldsymbol{\theta}_x = [\theta_{x1} \ \cdots]^T , \quad \boldsymbol{\theta}_y = [\theta_{y1} \ \cdots]^T ,$$

with w's (out-plane displacements) and θ's (rotations) being the nodal values, they defined the following column matrices

$$
\begin{aligned}
\mathbf{r} &= [1 \ 1 \ 1 \ 1]^T , \qquad \mathbf{g} = [1 \ -1 \ 1 \ -1]^T , \\
\mathbf{x} &= [x_1 \ x_2 \ x_3 \ x_4]^T , \qquad \mathbf{y} = [y_1 \ y_2 \ y_3 \ y_4]^T \\
\mathbf{b}_1 &= \frac{1}{2} [y_2 - y_4 \quad y_3 - y_1 \quad y_4 - y_2 \quad y_1 - y_3]^T , \\
\mathbf{b}_2 &= -\frac{1}{2} [x_2 - x_4 \quad x_3 - x_1 \quad x_4 - x_2 \quad x_1 - x_3]^T
\end{aligned}
$$

(10)

The rigid body modes are

$$(11) \qquad \mathbf{q} = [\mathbf{r}^T \ \mathbf{0} \ \mathbf{0}]^T , \qquad [\mathbf{x}^T \ \mathbf{r}^T \ \mathbf{0}]^T , \qquad [\mathbf{0} \ \mathbf{y}^T \ \mathbf{r}^T]^T .$$

The kinematic modes are

$$
\begin{aligned}
\mathbf{q} &= [\mathbf{g}^T \ \mathbf{0} \ \mathbf{0}]^T , \qquad [\mathbf{0} \ \mathbf{g}^T \ \mathbf{0}]^T , \\
&\quad [\mathbf{0} \ \mathbf{0} \ \mathbf{g}^T]^T , \qquad [\mathbf{a_1}\mathbf{x}^T + \mathbf{a_2}\mathbf{y}^T \ \mathbf{x}^T \ \mathbf{y}^T]^T ,
\end{aligned}
$$

(12)

where

$$a_1 = \mathbf{r}^T \mathbf{y}/4, \qquad a_2 = \mathbf{r}^T \mathbf{x}/4.$$

The stabilization matrix is in the form

$$(13) \qquad \mathbf{k}_H = \begin{bmatrix} c_1 \boldsymbol{\alpha}\boldsymbol{\alpha}^T & 0 & 0 \\ 0 & c_2 \boldsymbol{\alpha}\boldsymbol{\alpha}^T & 0 \\ 0 & 0 & c_2 \boldsymbol{\alpha}\boldsymbol{\alpha}^T \end{bmatrix},$$

where

$$(14) \qquad \boldsymbol{\alpha} = \mathbf{g} - \frac{1}{\mathbf{x}^T \mathbf{b}_1}[(\mathbf{g}^T\mathbf{x})\mathbf{b}_1 + (\mathbf{g}^T\mathbf{y})\mathbf{b}_2],$$

is chosen to be orthogonal to the rigid body motion. The choice is to assure that the stabilization does not affect the genuine zero strain energy modes. Note that

$$\mathbf{r}^T\mathbf{g} = \mathbf{r}^T\mathbf{b}_1 = \mathbf{r}^T\mathbf{b}_2 = 0, \quad \mathbf{g}^T\mathbf{b}_1 = \mathbf{g}^T\mathbf{b}_2 = \mathbf{y}^T\mathbf{b}_1 = \mathbf{x}^T\mathbf{b}_2 = 0,$$

$$\mathbf{x}^T\mathbf{b}_1 = \mathbf{y}^T\mathbf{b}_2.$$

The constants c's are to be judiciously selected. If they are too large, the element will lock just like the fully integrated element, and whereas too small a value will fail to control oscillations in the solution. Note that $\boldsymbol{\alpha}$ does not control the last kinematic mode of Eq. (12) because $\boldsymbol{\alpha}$ is orthogonal to it. In practice, it does not matter because the last kinematic mode cannot exist in a mesh of two or more elements.

Belytschko and Tsay (1983) proposed to define c's in the form

$$c_1 = c_w \frac{\kappa G t^3}{12(\mathbf{x}^T\mathbf{b}_1)^2}(\mathbf{b}_1^T\mathbf{b}_1 + \mathbf{b}_2^T\mathbf{b}_2),$$

$$c_2 = c_\theta \frac{E t^3}{192(1-\nu^2)\mathbf{x}^T\mathbf{b}_1}(\mathbf{b}_1^T\mathbf{b}_1 + \mathbf{b}_2^T\mathbf{b}_2),$$

where c_w, c_θ are universal constant independent of shape and material properties of the element. The range of c_w is between 0.03 to 0.1. The solution is generally insensitive to c_θ. It can have a value between 0.001 to 10. The first kinematic mode ($\mathbf{w} = \mathbf{g}$), controlled by c_1, is most sensitive.

Remarks. Development of shell elements is conceptually straightforward but practically complicated. The advent of robust simple flat plate elements has made modeling of complex shell structures a routine matter. The trend is toward using shell elements deduced from solid directly. Locking is among the most perplexing problems in applications to general thin shells. It is of a lesser problem for higher order elements, except for the fully integrated 8-node quadrilaterals. General techniques for locking mitigation include reduced integration, mixed and hybrid formulations, addition of drilling freedoms, etc. Reduced integration is very popular because it is simple to implement and offers substantially savings in computer time from reduced integration operations. Locking removal introduces rank deficiency (zero energy mode), which requires stabilization, especially for dynamic simulations. Research is continuing in the area of locking and rank deficiency removal. A simple robust (does not lock, passes the patch test, has no hourglass mode, requires the least number of multiplication, etc.) general shell element is still needed.

In practical applications, the inclination is to use simple elements, especially for dynamic simulations involving large deformation such as metal forming and crashworthiness studies. Choice of element must balance the computational efforts against the desired accuracy of the results. Four-node stabilized quadrilateral elements with one point integration are among the populists.

21

FINITE ELEMENT MODELING OF NONLINEAR ELASTICITY, VISCOELASTICITY, PLASTICITY, VISCOPLASTICITY AND CREEP

Most of the accomplishments of the classical mechanics are based on the principle of superposition, which is rooted in the linearity of the governing equations. Every branch of mechanics is exploring beyond the linear world to examine the effects of nonlinearity. Nonlinear dynamics of the simple Duffing's equation describing the motion of a single particle opened our eyes to the richness of the field. The theory of turbulence in fluid mechanics tells us how deep, controversial, and refractory to research efforts the subject can be. In solid mechanics, the following topics are awaiting for attention: large deformation, nonlinear constitutive equations, nonlinear boundary conditions, mechanical properties of living tissues in biomechanics, new materials of extreme complexity in structural composition, materials in micromachines and nanotechnology, living cells and tissues that cannot be identified as continua in the mathematical sense, micromechanics of heat transfer with large heat flux that occur in computers and electronic communication hardware, and so on. When mechanics has to deal with photonics, DNA, proteins, enzymes, some axioms of the current mechanics have to be changed. Every new change of an axiom leads to a new scientific field. Obviously, nonlinear mechanics is the mechanics of the future. It is obvious also that computational methods have an important role to play.

In Chapter 17, it is proposed that the incremental approach is an efficient way to deal with some nonlinear problems in solid mechanics. In Chapters 18, 19, and 20, the finite element method is singled out for a more detailed discussion in its applications to linear problems. In this final chapter, we discuss the application of finite element method to nonlinear problems in solid mechanics including large deformation, viscoelasticity, plasticity, viscoplasticity, and creep. The purpose is to see some computational features that arise because of nonlinearity.

In incremental approach, small incremental steps are used to linearize the governing equations to be solved by the finite element method. The final solution may involve iterative processes. The computational tools need extreme care when they are used in unknown territories. One of the distinct characteristics of nonlinear problems is that the solution may not be unique. In the nonuniqueness, there lies much of nature's secret. Examples in solid mechanics are the buckling of thin shells, self-equilibrating residual stresses and strains, three-dimensional solutions in bodies with apparently two-dimensional boundary conditions, and many problems in plasticity.

21.1. UPDATED LAGRANGIAN SOLUTION FOR LARGE DEFORMATION

In Chapter 17, it is shown that the updated Lagrangian approach for large deformation problems leads to a set of linear equations for the stress increment rates $\overset{\circ}{\mathbf{T}}, \overset{\circ}{\mathbf{r}}{}^{*}, \overset{\circ}{\mathbf{S}}, \overset{\circ}{\tau}$, the deformation rates $\mathbf{D}, \boldsymbol{\Omega}$ and/or the displacement rate $\mathbf{v}$, depending on the specific formulation. In this section, and this section only, we follow the derivation and notations of Chapter 17, and use boldfaced letters as tensors rather then matrices and follow the tensor operation rules, unless specifically stated otherwise. Note that we use the lower case subscript or superscript n to denote the current configuration. In solving for the rates, the information needed includes the current values of the Cauchy stress $\tau(=\tau_n)$, the increment rate of boundary traction $\overset{\circ}{\mathbf{T}}$ on $S_{\sigma n}$, the increment rate of boundary displacement $\bar{\mathbf{v}}$ on S_{un}, the deformed geometry of the body in D_n, and the constitutive law that relates the desired stress rates to the deformation rates. The constitutive law generally involves the total deformation gradient $\mathbf{F}_0^n$ and the current Cauchy stress τ.

The finite element equations can be based on the integral formulation outlined in Chapter 18. Consider the purely kinematic variational principle of Eq. (17.6:12) in terms of the rate-potential $\overset{\circ}{W}$ defined in Eq. (17.5:25):

$$(1) \qquad \overset{\circ}{\Pi}_3 = \sum_{\text{all elements}} (\overset{\circ}{\Pi}_3)_e \,,$$

where

(2) $\quad (\overset{\circ}{\Pi}_3)_e = \int_{V_{en}} \left[\overset{\circ}{W}(\Delta e) + \frac{1}{2}\boldsymbol{\tau} : (\Delta \mathbf{L} \cdot \Delta \mathbf{L}^T) - \rho_n \Delta \mathbf{b} \cdot \Delta \mathbf{u} \right] dV$

$$- \int_{S_{\sigma n}} \Delta \bar{\mathbf{T}} \cdot \Delta \mathbf{u}\, dA\,,$$

in which

$$\Delta \mathbf{u} = \mathbf{v}\Delta t\,, \qquad \Delta \mathbf{e} = (\boldsymbol{\nabla}_n \Delta \mathbf{u})_s\,,$$

$$\Delta \mathbf{L} = \boldsymbol{\nabla}_n \Delta \mathbf{u}\,, \qquad \Delta \mathbf{b} = \overset{\circ}{\mathbf{b}}\,\Delta t\,, \qquad \Delta \bar{\mathbf{T}} = \overset{\circ}{\bar{\mathbf{T}}}\,\Delta t\,,$$

with $\Delta \mathbf{u}$ being the displacement increment. The prescribed pressure and dead-weight types of boundary traction increment rate $\overset{\circ}{\bar{\mathbf{T}}}$ are given in Eqs. (17.10:2) and (17.10:5), respectively. The prescribed body force increment is given in Eq. (17.10:6). The compatibility requirements are that $\Delta \mathbf{u}$ is continuous and equal to $\Delta \bar{\mathbf{u}}$ on S_{un} as a rigid constraint.

Assuming $\Delta \mathbf{u}$ in the form

$$\Delta \mathbf{u} = \mathbf{h}\Delta \mathbf{q}$$

with $\mathbf{h}$ and $\Delta \mathbf{q}$ being the interpolation matrix and the parameter matrix, respectively, we can derive the element stiffness and force matrices from Eq. (2). The incremental finite element equations are obtained from the stationary condition of $\overset{\circ}{\Pi}_3$ with respect to $\Delta \mathbf{q}$, i.e.,

(3) ▲ $\quad \delta \overset{\circ}{\Pi}_3 = \displaystyle\sum_{\text{all elements}} \left\{ \int_{V_{en}} [\delta \overset{\circ}{W}(\Delta e) + \boldsymbol{\tau} : (\Delta \mathbf{L} \cdot \delta \Delta \mathbf{L}^T) \right.$

$$\left. - \rho_n \Delta \mathbf{b} \cdot \delta \Delta \mathbf{u}]\, dV - \int_{S_{\sigma n}} \Delta \bar{\mathbf{T}} \cdot \delta \Delta \mathbf{u}\, dA \right\} = 0\,,$$

subjected to the rigid constraint $\delta \Delta \mathbf{q} = 0$ over the surface where displacements are prescribed. Oden (1967) gave the early finite element solutions for large deformation of incompressible materials.

From the finite element solution, we can calculate the 2nd Piola–Kichhoff stress increment in the updated Lagrange description [Eq. (17.5:10)]

(4) $$\Delta \mathbf{S} = \frac{\partial \overset{\circ}{W}}{\partial \Delta \mathbf{e}}\,,$$

and the Cauchy stress increment [Eq. (17.9:13)]

(5) ▲ $\quad \Delta \boldsymbol{\tau} = \Delta \mathbf{S} + \Delta \mathbf{L} \cdot \boldsymbol{\tau} + \boldsymbol{\tau} \cdot \Delta \mathbf{L}^T - \text{tr}(\Delta \mathbf{e})\boldsymbol{\tau}\,.$

Note that $\Delta\mathbf{S}$ equals the Truesdell rate of the Cauchy stress. The updated displacements, coordinates and Cauchy stresses, are

(6) $\mathbf{u}_{n+1} = \mathbf{u}_n + \Delta\mathbf{u}, \qquad \mathbf{X}_{n+1} = \mathbf{X}_n + \Delta\mathbf{u} = \mathbf{X}_0 + \mathbf{u}_{n+1},$

$$\boldsymbol{\tau}_{n+1} = \boldsymbol{\tau} + \Delta\boldsymbol{\tau}.$$

Here $\mathbf{u}_n$ is the total displacement vector at the nth state.

One may include $\Delta\boldsymbol{\Omega}(t)$ as an independent field by modifying Eq. (2) as [Eq. (17.6:13)],

(7) $(\overset{\circ}{\Pi}_3)_e = \displaystyle\int_{V_{en}} \left[\overset{\circ}{W}(\Delta e) + \frac{1}{2}\boldsymbol{\tau} : (\Delta\mathbf{L}\cdot\Delta\mathbf{L}^T) + \frac{c}{2}[(\boldsymbol{\nabla}_n\Delta\mathbf{u})_a - \Delta\boldsymbol{\Omega}]^2 \right.$

$$\left. - \rho_n\Delta\mathbf{b}\cdot\Delta\mathbf{u} \right]dV - \int_{S_{\sigma n}} \Delta\bar{\mathbf{T}}\cdot\Delta\mathbf{u}\,dA,$$

where c is a given positive constant of the order of the elastic modulus, and

$$(\boldsymbol{\nabla}_n\Delta\mathbf{u})_a = [\boldsymbol{\nabla}_n\Delta\mathbf{u} - (\boldsymbol{\nabla}_n\Delta\mathbf{u})^T]/2,$$

$$[(\boldsymbol{\nabla}_n\Delta\mathbf{u})_a - \Delta\boldsymbol{\Omega}]^2 = [(\boldsymbol{\nabla}_n\Delta\mathbf{u})_a - \Delta\boldsymbol{\Omega}] : [(\boldsymbol{\nabla}_n\Delta\mathbf{u})_a - \Delta\boldsymbol{\Omega}].$$

Equation (7) is equivalent to the *penalty function approach* in enforcing

(8) $$\Delta\boldsymbol{\Omega} = (\boldsymbol{\nabla}_n\Delta\mathbf{u})_a.$$

If we express the functional in the form

(9) $(\overset{\circ}{\Pi}_3)_e = \displaystyle\int_{V_{en}} \left[\overset{\circ}{P}(\Delta e) + \frac{1}{2}\boldsymbol{\tau} : (\Delta\mathbf{L}\cdot\Delta\mathbf{L}^T - \Delta e\cdot\Delta e) \right.$

$$\left. - \rho_n\Delta\mathbf{b}\cdot\Delta\mathbf{u} \right]dV - \int_{S_{\sigma n}} \Delta\bar{\mathbf{T}}\cdot\Delta\mathbf{u}\,dA,$$

then since [Eq. (17.5:10)]

(10) $$\Delta\sigma_r = \frac{\partial\overset{\circ}{P}}{\partial\Delta e}$$

gives the increment of the corotation stress, which is the Jaumann rate of the Cauchy stress. From Eq. (17.9:12), we have

(11) ▲ $$\Delta\tau = \Delta\sigma_r + \Delta\boldsymbol{\Omega}\cdot\tau - \tau\cdot\Delta\boldsymbol{\Omega}$$

for updating the Cauchy stress.

To improve objectivity in computation, we use the midpoint rule Eqs. (17.9:6) and (17.9:9) that

$$\nabla_{N+1/2} = \varepsilon_I \frac{\partial}{\partial X_I^{N+1/2}},$$

$$\Delta \mathbf{L} = \nabla_{N+1/2} \Delta \mathbf{u},$$

$$\Delta \mathbf{e} = \frac{1}{2} [\nabla_{N+1/2} \Delta \mathbf{u} + (\nabla_{N+1/2} \Delta \mathbf{u})^T],$$

$$\Delta \mathbf{\Omega} = \frac{1}{2} [\nabla_{N+1/2} \Delta \mathbf{u} - (\nabla_{N+1/2} \Delta \mathbf{u})^T]$$

One also updates the total deformation gradient, the related deformation matrices, and the rate-potential using the midpoint rule, e.g., Eq. (17.9:8) for $\mathbf{F}$ and Eq. (17.9:7) or (17.9:11) for $\mathbf{R}$. For the formulation of Eq. (7), we use $\Delta \mathbf{\Omega}$ directly from the finite element solution in Eq. (11).

21.2. INCREMENTAL SOLUTION

Assume that the body force and the prescribed surface traction are in the form[1]

$$\mathbf{b} = p\mathbf{b}_0, \qquad \bar{\mathbf{T}} = p\bar{\mathbf{T}}_0,$$

where p is a loading parameter. Let $\mathbf{q}_n$ be the current solution corresponding to p_n. The assembled and constrained incremental finite element equation derived from Eq. (21.1:3) can be written in the form

(1) ▲ $$\mathbf{K}\Delta \mathbf{q} = \Delta \mathbf{F} = (p_{n+1} - p_n)\mathbf{F}_0 = \Delta p \mathbf{F}_0,$$

where $\mathbf{K}$ and $\Delta \mathbf{F}$, which can be functions of $\mathbf{q}_n$, are the assembled and constrained stiffness matrix and the load increment matrix, respectively. The updated $\mathbf{q}$ is

$$\mathbf{q}_{n+1} = \mathbf{q}_n + \Delta \mathbf{q}.$$

Equation (1) is like a first order ordinary differential equation of p, and can be solved by integration. More accurate $\Delta \mathbf{q}$ (Pian and Tong 1971) can be obtained by the *predictor-corrector integration method* (Ralston, 1965).

If $\mathbf{K}$ becomes singular at any solution stage, say at $p = p_c$, then the numerical integration with respect to p may not be proceeded further. Near

---

[1] A more general form of external load is

$$\mathbf{b} = pc_1 \mathbf{b}_0 + \mathbf{b}', \qquad \bar{\mathbf{T}} = pc_2 \bar{\mathbf{T}}_0 + \bar{\mathbf{T}}',$$

where c_1, c_2 are given constants and $\mathbf{b}'$, $\bar{\mathbf{T}}'$ are given preload. This allows the body force and the surface traction to increase at different ratio.

p_c, a small increment in p can lead to a large increment in $\mathbf{q}$. Thus, near p_c, it is better to carry out the integration with respect to a parameter associated with deformation (Pian and Tong 1971, Zienkiewicz 1971, Bergan 1981). For example, let the new parameter be q_r, a component of $\mathbf{q}$ that changes rapidly with respect to p. Partitioning Eq. (1) in the form

$$
\begin{bmatrix} \mathbf{K}_{11} & K_{21}^T \\ \mathbf{K}_{21} & K_{22} \end{bmatrix} \begin{bmatrix} \Delta \mathbf{q}' \\ \Delta q_r \end{bmatrix} = \Delta p \begin{bmatrix} \mathbf{F}' \\ F_r \end{bmatrix},
$$

we can rewrite Eq. (1) as

$$
(2) \qquad \begin{bmatrix} \mathbf{K}_{11} & -\mathbf{F}' \\ \mathbf{K}_{21} & -F_r \end{bmatrix} \begin{bmatrix} \Delta \mathbf{q}'/\Delta q_r \\ \Delta p/\Delta q_r \end{bmatrix} = \begin{bmatrix} -\mathbf{K}_{21}^T \\ -K_{22} \end{bmatrix},
$$

where $\mathbf{K}_{11}$ is nonsingular and symmetric. Equation (2) can now be integrated with respect to the new parameter q_r (set the sign of Δq_r the same as that of the previous step) to determine $\Delta \mathbf{q}'$ and Δp.

Let $q_r = (q_r)_c$ at $p = p_c$. If $\Delta p/\Delta q_r$ changes sign at $(q_r)_c$, no solution for p exists beyond p_c in the vicinity of $(q_r)_c$ while multiple solutions exist below p_c. Thus, an infinitesimal increment in p beyond p_c can lead to a finite increment in q_r or the collapse of the structure. This is a phenomenon of nonuniqueness in the solution due to nonlinearity. The structure is said to be unstable and is called *buckling* (Fung and Sechler 1960, Timoshenko and Gere 1961). On the other hand, if $\Delta p/\Delta q_r$ is a monotonic function of q_r, the structure is stable and can sustain load beyond p_c. This is called *stable buckling* while q_r changes rapidly in the vicinity of p_c.

The coefficient matrix of $[\Delta \mathbf{q}'/\Delta q_r \ \Delta p/\Delta q_r]^T$ is in general asymmetric. We can, however, solve Eq. (2) as follows to avoid dealing with the asymmetric matrix: let

$$
(3) \qquad \frac{\Delta \mathbf{q}'}{\Delta q_r} = \frac{\Delta \mathbf{q}_1'}{\Delta q_r} + \frac{\Delta p}{\Delta q_r} \frac{\Delta \mathbf{q}_2'}{\Delta q_r},
$$

where

$$
(4) \qquad \mathbf{K}_{11} \frac{\Delta \mathbf{q}_1'}{\Delta q_r} = -\mathbf{K}_{21}^T,
$$

$$
(5) \qquad \mathbf{K}_{11} \frac{\Delta \mathbf{q}_2'}{\Delta q_r} = \mathbf{F}'.
$$

A substitution of Eqs. (3) into Eq. (2) yields

$$
\frac{\Delta p}{\Delta q_r} = \left(K_{22} + \mathbf{K}_{21} \frac{\Delta \mathbf{q}_1'}{\Delta q_r} \right) \Big/ \left(F_r - \mathbf{K}_{21} \frac{\Delta \mathbf{q}_2'}{\Delta q_r} \right).
$$

Thus we can obtain the rate of change with respect to the new parameter q_r without having to deal with asymmetric matrix.

21.3. DYNAMIC SOLUTION

For dynamic simulation, we do not have to solve $\Delta \mathbf{u}$ or $\Delta \mathbf{q}$ of Eq. (21.2:1) if an explicit time integration scheme is used. We can derive the equations of motion using the principle of virtual work discussed in Sec. 18.17, i.e.,

$$(1) \qquad \delta \Pi = \int_{V_{n+1}} [\delta \mathbf{u}^T \ddot{\mathbf{u}} \rho_{n+1} + (\delta \mathbf{e})^T \boldsymbol{\tau}_{n+1} - \rho_{n+1} \delta \mathbf{u}^T \mathbf{b}_{n+1}] \, dV$$

$$- \int_{(\partial V_\sigma)_{n+1}} \delta \mathbf{u}^T \bar{\mathbf{T}}_{n+1} \, dS = 0 \,,$$

in which

$$(2) \qquad \boldsymbol{\tau}_{n+1} = [\tau_{11} \ \tau_{22} \ \cdots \ \tau_{31}]_{n+1}^T \,, \qquad \delta \mathbf{u} = \mathbf{h} \delta \mathbf{q} \,, \qquad \delta \mathbf{e} = \mathbf{B} \delta \mathbf{q} \,,$$

where $\boldsymbol{\tau}_{n+1}$ is the updated Cauchy stress given in Eq. (21.1:6), $\delta \mathbf{e}$ is a strain variation as defined in Eq. (18.16:5), and $\mathbf{h}$ and $\mathbf{B}$ are interpolation matrices relating the parameter $\delta \mathbf{q}$ to the displacement and strain variations as defined in Eq. (18.16:12). The boundary force is calculated from the surface integral in Eq. (1). For pressure surface load

$$(3) \qquad \bar{\mathbf{T}}_{n+1} = -\bar{p}_{n+1} \mathbf{N}_{n+1} \,,$$

and for dead-weight load

$$(4) \qquad \bar{\mathbf{T}}_{n+1} = (\bar{\mathbf{T}}_0)_{n+1} \frac{dS_0}{dS}\bigg|_{n+1} = (\bar{\mathbf{T}}_0)_{n+1} \frac{\sqrt{(\mathbf{N}^T \mathbf{F} \mathbf{F}^T \mathbf{N})_{n+1}}}{J_{n+1}} \,,$$

where $\bar{p}_{n+1}$ is the prescribed pressure referred to the deformed surface in D_{n+1} configuration, and $(\bar{\mathbf{T}}_0)_{n+1}$ is the prescribed traction at the $(n+1)$th stage referred to the underformed surface. The element mass matrix is derived from the first term in the first integral that

$$(5) \qquad \mathbf{m}_{n+1} \ddot{\mathbf{q}} = \left[\int_{(V_e)_{n+1}} \rho_{n+1} \mathbf{h}^T \mathbf{h} \, dV \right] \ddot{\mathbf{q}} \,,$$

which is in the same form as that for small deformation discussed in Sec. 18.17 except that the integration in Eq. (5) is over the deformed element volume. In practice, one often uses the lumped mass matrix by summing all terms of each row to the diagonal.

From Eq. (1), one can calculate the acceleration at stage $n+1$. Using an explicit integration scheme, one determines $\Delta \mathbf{q}$ (i.e., $\Delta \mathbf{u}$) at the $(n+1)$th stage, then the increment $\Delta \boldsymbol{\tau}$ from Eq. (21.1:5) or (21.1:11) and the updated solution from Eq. (21.1:6), and then proceed to the next stage.

21.4. NEWTON-RAPHSON ITERATION METHOD

The Newton–Raphson method (Rafston 1965) is one of the most commonly used methods for solving nonlinear algebraic equations. The method can be used to solve the nonlinear finite element equations of plasticity, viscoplasticity and creep. Almost all solution methods for nonlinear problem involve some sort of iteration. It is no exception for the Newton–Raphson method. The process is illustrated in Fig. 21.4:1. Consider a set of nonlinear algebraic equations in the form

$$(1) \qquad \mathbf{R}(\mathbf{q}) = \boldsymbol{\gamma}(\mathbf{q}) - \mathbf{F} = 0 \,,$$

where $\boldsymbol{\gamma}$ is a nonlinear matrix function of $\mathbf{q}$ and $\mathbf{F}$ is a loading matrix normally independent of $\mathbf{q}$. Assuming that the solution $\mathbf{q}_n$ at $\mathbf{F} = \mathbf{F}_n$ is known, one seeks the solution $\mathbf{q}_{n+1}$ at $\mathbf{F} = \mathbf{F}_{n+1} = \mathbf{F}_n + \Delta \mathbf{F}_n$. Let $\mathbf{q}_{n+1}^i$ be a known approximate solution for $\mathbf{q}_{n+1}$ at the ith iteration, one

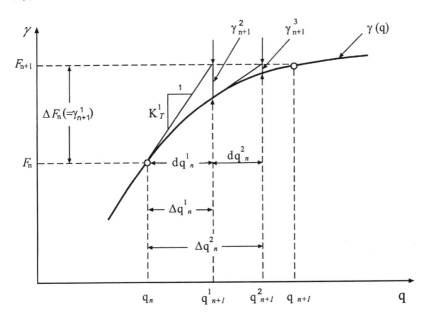

Fig. 21.4:1. Newton-Raphson method.

improves the approximation in the form $\mathbf{q}_{n+1}^{i+1} = \mathbf{q}_{n+1}^{i} + d\mathbf{q}^{i}$. To the first order approximation, Eq. (1) can be written as

(2)
$$\mathbf{K}_T^i d\mathbf{q}^i + \mathbf{R}_n^i = 0 ,$$

where the tangent stiffness matrix $\mathbf{K}_T^i$ and the residual matrix $\mathbf{R}_n^i$ are defined as

$$\mathbf{K}_T^i = \left.\frac{\partial \boldsymbol{\gamma}}{\partial \mathbf{q}}\right|_{\mathbf{q}=\mathbf{q}_{n+1}^i} , \quad \mathbf{R}_n^i = \boldsymbol{\gamma}(\mathbf{q}_{n+1}^i) - \boldsymbol{\gamma}(\mathbf{q}_n) - \Delta\mathbf{F}_n = \boldsymbol{\gamma}(\mathbf{q}_{n+1}^i) - \mathbf{F}_n - \Delta\mathbf{F}_n .$$

One repeats the process until

$$(\mathbf{R}_n^i)^T \mathbf{R}_n^i < e_2 (\Delta\mathbf{F}_n)^T \Delta\mathbf{F}_n , \qquad (i, n \text{ not summed})$$

where e_2 is a chosen small constant to control the accuracy of the solution for $\mathbf{q}_{n+1}$ and $(\mathbf{R}_n^i)^T \mathbf{R}_n^i$ is the *inner product* of the residual and is usually called the *square norm*.

There are variants to the Newton–Raphson method. The most common one is the *modified Newton–Raphson method*, which uses a fixed $\mathbf{K}_T^i$ for several iterations without update. Solving for $d\mathbf{q}^i$ in Eq. (2) requires refactorization of the matrix $\mathbf{K}_T^i$ (see Sec. 18.5), if $\mathbf{K}_T^i$ is updated. Factorization is a time consuming process. Updating $\mathbf{K}_T^i$ less frequently saves computing time. One can even use $\mathbf{K}_T^0$ throughout the entire iteration process of determining $\mathbf{q}_{n+1}$. In this case, $\mathbf{K}_T^0$ is factored only once. Such procedure is likely to converge at a slower rate but the computer time required to reach a converged solution may actually be reduced.

A *linear search procedure* can be used to accelerate the convergence of the Newton–Raphson method. The schematics of the procedure is shown in Fig. 21.4:2. Instead of using $\mathbf{q}_{n+1}^i + d\mathbf{q}^i$, we define

$$\mathbf{q}_{n+1}^{i+1} = \mathbf{q}_{n+1}^i + (1 + \eta) d\mathbf{q}^i$$

as an updated solution. The quantity η is determined by requiring the projection of the residual on the search direction $d\mathbf{q}^i$ to be zero, i.e.,

$$G_i(\eta) = (d\mathbf{q}^i)^T \mathbf{R}[\mathbf{q}_{n+1}^i + (1 + \eta) d\mathbf{q}^i]$$
$$= (d\mathbf{q}^i)^T \{\boldsymbol{\gamma}[\mathbf{q}_{n+1}^i + (1 + \eta) d\mathbf{q}^i] - \mathbf{F}_{n+1}\} = 0 \quad (i \text{ not summed}) .$$

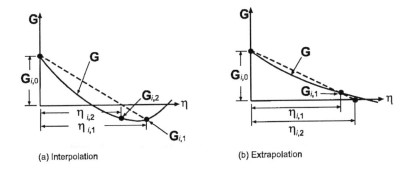

(a) Interpolation   (b) Extrapolation

Fig. 21.4:2. Linear search schematic.

The solution for η can be obtained by iteration with $\eta = 0$ as the initial approximation. The linear search needs additional evaluations of the residual $\mathbf{R}$. One normally undertakes the search only if

$$|G(\eta)| > e_3|(d\mathbf{q}^i)^T[\boldsymbol{\gamma}(\mathbf{q}_{n+1}^i) - \mathbf{F}_{n+1}]|,$$

where e_3 is a number close to 0.5 (Matthies and Strang 1979).

In many problems, for a loading in the form

$$\mathbf{F} = p\mathbf{F}_0, \qquad \Delta\mathbf{F}_n = (p_{n+1} - p_n)\mathbf{F}_0 = \Delta p_n\mathbf{F}_0,$$

no solution exists above a certain maximum value of p. When this occurs, one can recast the formulation to the increment of a single displacement component as shown in Sec. 21.2.

21.5. VISCOELASTICITY

Viscoelasticity is characterized by a constitutive law that the current state of stress depends not only on the current strains, but also on the full history of their development as described in Chapter 15. In the present study, we assume the constitutive equation in the form

$$(1) \qquad \boldsymbol{\sigma} = \mathbf{D}_r(t)\mathbf{e}(0^+) + \int_0^t \mathbf{D}_r(t-\tau)\frac{\partial \mathbf{e}}{\partial \tau}\, d\tau,$$

as given in Eq. (15.1:6), where $\boldsymbol{\sigma}$, $\mathbf{e}$, $\mathbf{D}_r(t)$ are the stress, strain, relaxation matrices, respectively. If the material is inhomogeneous, $\mathbf{D}_r(t)$ will be a function of the spatial coordinates also. Furthermore, in some problems, $\mathbf{D}_r(t)$ may depend on $\boldsymbol{\sigma}$ and/or $\mathbf{e}$. In finite element analysis, we write Eq. (1) in the incremental form

(2) $\sigma_{n+1} - \sigma_n = \dot{\sigma}_n \Delta t = \mathbf{D}_r(0) \Delta \mathbf{e} + [\dot{\mathbf{D}}_r(t_n) \mathbf{e}(0^+) + \mathbf{Q}_n] \Delta t$,

where σ_{n+1}, σ_n denote the stresses at $t = t_{n+1}, t_n$, respectively,

(3) $$\Delta \mathbf{e} = \mathbf{e}_{n+1} - \mathbf{e}_n = \mathbf{B} \Delta \mathbf{q},$$

(4) $$\mathbf{Q}_n = \int_0^{t_n} \dot{\mathbf{D}}_r(t_n - \tau) \frac{\partial \mathbf{e}}{\partial \tau} \, d\tau,$$

in which $\mathbf{B}$ is the strain-displacement matrix relating the generalized coordinates associated with displacements to strains. In Eq. (2), the first term on right most hand side is the elastic solution with $\mathbf{D}_r(0)$ equivalent to the elastic modulus matrix and the second term is the viscoelastic contribution.

The finite element equation of motion can be written in the form [Eq. (18.16:22)]

(5) $$\mathbf{R} = \int_V \mathbf{B}^T \sigma \, dV - \mathbf{F} = 0,$$

where $\mathbf{F}$ is the loading matrix. Substituting Eq. (2) into Eq. (5) and using the strain displacement relation $\Delta \mathbf{e} = \mathbf{B} \Delta \mathbf{q}$, we find

(6) $$\mathbf{K}^0 \Delta \mathbf{q}_n = \mathbf{R}_n,$$

where $\mathbf{K}^0$ is the standard elastic stiffness matrix that

(7) $$\mathbf{K}^0 = \int_V \mathbf{B}^T \mathbf{D}_r(0) \mathbf{B} \, dV$$

and $\mathbf{R}_n$ is the equilibrium residual defined as

(8) $$\mathbf{R}_n = \mathbf{F}_{n+1} - \mathbf{F}_n - \Delta t \int_V \mathbf{B}^T [\dot{\mathbf{D}}_r(t_n) \mathbf{e}(0^+) + \mathbf{Q}_n] \, dV.$$

One can integrate Eq. (6) with initial condition $\mathbf{e}(0^+)$ given. The practical limitation in evaluating Eq. (6) is the requirement of the full history of $\mathbf{e}$ as shown in the integration of $\mathbf{Q}$ in Eq. (4). When t becomes large, this will require extensive computer storage and computation. In practice, often $\mathbf{D}_r$ is an exponentially decaying function of the form

$$\mathbf{D}_r(t) \approx \exp(-t/\tau_0),$$

where τ_0 is a relaxation constant. Equation (4) can be approximated as

(9) $$\mathbf{Q}_n \approx \int_{t_n - a\tau_0}^{t_n} \dot{\mathbf{D}}_r(t_n - \tau) \frac{\partial \mathbf{e}}{\partial \tau} \, d\tau,$$

where a is a constant of order 1. Then it is only necessary to store the recent history of $\mathbf{e}$ from $t_n - a\tau_0$ to t_n.

For linear problems, one can proceed to the next time step after $\Delta\mathbf{q}_n$ is determined from Eq. (6). For nonlinear viscoelastic problems, $\mathbf{B}$ can be a function of $\mathbf{q}$. One may have to use the Newton–Raphson iteration to obtain the updated $\Delta\mathbf{q}_n$.

21.6. PLASTICITY

The linear *incremental plasticity* theory for infinitesimal deformation described in Chapter 6 are summarized below in both matrix and indicial notations:

(1) $\mathbf{e} = \mathbf{e}^e + \mathbf{e}^p$, $e_{ij} = e_{ij}^e + e_{ij}^p$ (elastic and plastic strain-splits),

(2) $\boldsymbol{\tau} = \mathbf{D}_e\mathbf{e}^e = \mathbf{D}_e(\mathbf{e} - \mathbf{e}^p)$,

$\tau_{ij} = D_{ijkl}^e e_{kl}^e = D_{ijkl}^e(e_{kl} - e_{kl}^p)$ (elastic constitutive equation)

(3) $\dot{\mathbf{e}}^p = \dot{\Lambda}\dfrac{\partial h}{\partial \boldsymbol{\tau}} = \dot{\Lambda}h_\tau$, $\dot{e}_{ij}^p = \dot{\Lambda}\dfrac{\partial h}{\partial \tau_{ij}}$ (normality condition flow rule),[2]

(4) $\boldsymbol{\eta} = \boldsymbol{\tau} - \boldsymbol{\alpha}$, $\eta_{ij} = \tau_{ij} - \alpha_{ij}$

(5) $\boldsymbol{\eta}' = \boldsymbol{\tau}' - \boldsymbol{\alpha}'$,

$\eta_{ij}' = \tau_{ij}' - \alpha_{ij}' = \eta_{ij} - \delta_{ij}\eta_{kk}/3$ (deviatoric components)

(6) $\dot{\boldsymbol{\alpha}} = \dot{\Lambda}\mathbf{g}$, $\dot{\alpha}_{ij} = \dot{\Lambda}g_{ij}$ (back stresses)

(7) $\dot{\bar{\varepsilon}}^p = \left(\dfrac{2}{3}\dot{e}_{ij}^p\dot{e}_{ij}^p\right)^{1/2} = \dot{\Lambda}\left(\dfrac{2}{3}\dfrac{\partial h}{\partial \tau_{ij}}\dfrac{\partial h}{\partial \tau_{ij}}\right)^{1/2}$

(equivalent plastic-strain rate),

(8) $h = h(\eta_{ij}, \xi_i)$ (yield potential),

(9) $f(\eta_{ij}, \xi_i) = 0$ (yield surface),

---

[2]In Eq. (21.6:3), h is considered to be a function of 6 stress components, i.e., for $i \neq j$, τ_{ij}, τ_{ji} are the same variable leading to $2e_{ij}^p = \dot{\Lambda}\partial h/\partial \tau_{ij}$ in matrix notation, while h is considered to be a function of 9 stress components, i.e., for $i \neq j$, τ_{ij}, τ_{ji} are the different variables leading to $e_{ij}^p = \dot{\Lambda}\partial h/\partial \tau_{ij}$ in index notation. Note the following relation between tensor and matrix operations: let $(\boldsymbol{\tau})_t$, $(\mathbf{e})_t$ be the stress and strain tensors and $(\boldsymbol{\tau})_m$, $(\mathbf{e})_m$ the stress and strain matrices as defined in Chapter 18, then $(\boldsymbol{\tau})_t : (\mathbf{e})_t = (\boldsymbol{\tau})_m^T(\mathbf{e})_m = (\mathbf{e})_m^T(\boldsymbol{\tau})_m$.

(10) $\dot{\Lambda} = 0$ if $f < 0$ (elastic deformation),

              $\dot{\Lambda} > 0$ if $f = 0$ (plastic deformation),

              $f > 0$ (not allowed),

(11) $$\dot{e}^p_{kk} = \frac{\partial h}{\partial \tau_{kk}} = 0,$$

where σ_{ij}, α_{ij}, e_{ij} are the stress, back stress, and strain tensors, a primed
quantity $(\text{-})'$ denotes the deviatoric component of the corresponding quantity, and ξ_i are internal variables. The *yield potential* h, the *yield surface*
f and the back stress g_{ij} are functions of η_{ij} and *internal variables* ξ_i. The
flow rule is *associated* if $h = f$. The internal variables can be the back
stresses, plastic strains and temperature. Equation (10) written as

$$\dot{\Lambda} \geq 0, \qquad \dot{\Lambda} f = 0,$$

is known as the Kuhn–Tucker form for loading or unloading. In elastic-
plastic analysis, one seeks the solution in terms of displacements and
stresses that satisfy the equations of motion and the boundary conditions
with Eqs. (1)–(11) as constraints.

The plasticity equations are highly nonlinear and can be solved in most
cases only by numerical means. There is a very large literature on the subject (Wilkins 1964, Krieg and Key 1976, Krieg and Krieg 1977, Ortiz and
Popov 1985, Rice and Tracey 1973, Nagtegaal et al. 1974). To construct
the incremental solution, consider the interval (t_n, t_{n+1}), where t is time
for dynamic problems and a loading parameter that characterizes the magnitude of the applied loads or prescribed displacements for static problems.
For simplicity we will just call it time in the subsequent discussion. At t_n
we assume that the total plastic and equivalent plastic strains, the stress
and the back stress, $\mathbf{e}_n$, $\mathbf{e}^p_n$, ε^p_n, τ_n, α_n, are known. We first construct the
updated fields $\mathbf{e}_{n+1}, \mathbf{e}^p_{n+1}, \ldots$, at t_{n+1} in terms of the increment displacement $\Delta \mathbf{u}$ in the manner consistent with the constraints of Eqs. (3)–(10),
and then proceed to the next increment.

The updating process is to determine $\mathbf{e}^p_{n+1}$, ε^p_{n+1} and then derives other
updated fields based on $\mathbf{e}^p_{n+1}$, ε^p_{n+1}. In principle, $\mathbf{e}^p_{n+1}$, ε^p_{n+1} and α_{n+1} can
be determined by integrating the flow and hardening rules over the step
(t_n, t_{n+1}) for a given yield potential h and yield surface f. To illustrate
the update process, we consider an associated von Mises material with
only internal variable ε^p. If the material is isotropic and satisfies a linear

kinematic and nonlinear isotropic hardening law, then

(12) $$\dot{\boldsymbol{\alpha}} = 2c\dot{\Lambda}\boldsymbol{\eta}/3\,,$$

(13) $$h(\boldsymbol{\eta}, \varepsilon^p) = f(\boldsymbol{\eta}, \varepsilon^p) = \frac{1}{2}\boldsymbol{\eta}^T\mathbf{P}\boldsymbol{\eta} - \frac{1}{3}\sigma_Y^2(\varepsilon^p) = 0\,,$$

where σ_Y is the yield stress in unidirectional test, and

(14a) $$\varepsilon^p = \int_0^t \dot{\varepsilon}^p\, dt\,,$$

(14b) $$\mathbf{P} = \frac{1}{3}\begin{bmatrix} 2 & -1 & 0 \\ -1 & 2 & 0 \\ 0 & 0 & 6 \end{bmatrix} \qquad \text{(plane stress)}\,,$$

(14c) $$\mathbf{P} = \begin{bmatrix} \mathbf{P}_1 & \mathbf{0} \\ \mathbf{0} & \mathbf{P}_2 \end{bmatrix}\,, \qquad \mathbf{P}_1 = \frac{1}{3}\begin{bmatrix} 2 & -1 & -1 \\ -1 & 2 & -1 \\ -1 & -1 & 2 \end{bmatrix}\,,$$

$$\mathbf{P}_2 = \begin{bmatrix} 2 & 0 & 0 \\ 0 & 2 & 0 \\ 0 & 0 & 2 \end{bmatrix} \qquad \text{(3-D, plane strain)}\,.$$

Note that $\mathbf{P}$ is the *projection matrix* that makes $\mathbf{P}\boldsymbol{\eta}$ deviatoric.

Following the approach described in Sec. 6.12 and assuming that

$$f(\boldsymbol{\tau}_n + \mathbf{D}_e\Delta\mathbf{e}, \varepsilon_n^p) > 0\,,$$

i.e., yielding will occur, from Eq. (7), we find

(15) $$\dot{\varepsilon}^p = \dot{\Lambda}\sqrt{\frac{2}{3}\boldsymbol{\eta}^T\mathbf{P}\boldsymbol{\eta}} = \frac{2}{3}\dot{\Lambda}\sigma_Y\,.$$

Differentiating Eq. (13) with respect to t gives the consistent equation

(16) $$\boldsymbol{\eta}^T\mathbf{P}\left(\dot{\boldsymbol{\tau}} - \frac{2}{3}c\dot{\Lambda}\boldsymbol{\eta}\right) - \frac{2}{3}\sigma_Y\frac{\partial\sigma_Y}{\partial\varepsilon^p}\dot{\varepsilon}^p = 0\,.$$

Replacing (˙) quantities with $\Delta(\text{-})$ in Eqs. (15) and (16), solving for $\Delta\Lambda$ and substituting it into Eq. (3) and into $\Delta\boldsymbol{\tau} = \mathbf{D}_e(\Delta\mathbf{e} - \Delta\mathbf{e}^p)$, we obtain

(17) $\qquad \mathbf{D}^{ep} = \left\{ \mathbf{I} + \mathbf{D}_e \mathbf{P}\boldsymbol{\eta}(\mathbf{P}\boldsymbol{\eta})^T / \left[\dfrac{4}{9}\sigma_Y^2 \left(c + \dfrac{\partial\sigma_Y}{\partial\varepsilon^P} \right) \right] \right\}^{-1} \mathbf{D}_e \,,$

(18) $\qquad \Delta\boldsymbol{\tau} = \mathbf{D}^{ep}\Delta\mathbf{e} \,,$

(19) $\qquad \Delta\Lambda = \boldsymbol{\eta}^T \mathbf{P}\Delta\boldsymbol{\tau} / \left[\dfrac{4}{9}\sigma_Y^2 \left(c + \dfrac{\partial\sigma_Y}{\partial\varepsilon^P} \right) \right] \,,$

(20) $\qquad \Delta\mathbf{e}^P = \mathbf{P}\boldsymbol{\eta}\Delta\Lambda \,,$

(21) $\qquad \Delta\alpha = \dfrac{2}{3} c\boldsymbol{\eta}\Delta\Lambda \,,$

where $\boldsymbol{\eta} = \boldsymbol{\eta}_n$ and $\sigma_Y = \sigma_Y(\varepsilon_n^P)$. The updated fields are simply

(22) $\qquad \boldsymbol{\tau}_{n+1} = \boldsymbol{\tau}_n + \Delta\boldsymbol{\tau} \,, \qquad \mathbf{e}_{n+1}^p = \mathbf{e}_{n+1}^p + \Delta\mathbf{e}^p \dots .$

This process is called the *tangent modulus approach*.

For ideal plasticity, i.e., $c = \partial\sigma_Y/\partial\varepsilon^P = 0$, we derive $\mathbf{D}^{ep}$ and $\Delta\Lambda$ according to Sec. 6.10 [see Eqs. (6.10:8)–(6.10:13)],

(22a)
$$\mathbf{D}^{ep} = \mathbf{D}_e - \mathbf{D}_e \mathbf{P}\boldsymbol{\tau}\boldsymbol{\tau}^T \mathbf{P}\mathbf{D}_e / (\boldsymbol{\tau}^T \mathbf{P}\mathbf{D}_e \mathbf{P}\boldsymbol{\tau}) \,,$$
$$\Delta\Lambda = \boldsymbol{\tau}^T \mathbf{P}\mathbf{D}_e \Delta\mathbf{e} / (\boldsymbol{\tau}^T \mathbf{P}\mathbf{D}_e \mathbf{P}\boldsymbol{\tau}) \,.$$

The other equations are in the same form as before.

For finite $\Delta\mathbf{u}$, the updated fields satisfy the yield condition with an error of the order $(\Delta\mathbf{u})^2$, i.e., $f(\boldsymbol{\eta}_{n+1}, \varepsilon_{n+1}^p) = O(\Delta\mathbf{u})^2$. We seek improved updated fields in the form (Simo and Taylor 1986):

(23) $\qquad \Delta\mathbf{e} = \mathbf{d}\Delta\mathbf{u} = \mathbf{B}\Delta\mathbf{q} \,,$

(24) $\qquad \Delta\mathbf{e}_{n+a}^p = \Gamma\mathbf{P}\boldsymbol{\eta}_{n+a} \,,$

(25) $\qquad \Delta\alpha_{n+a} = 2c\Gamma\boldsymbol{\eta}_{n+a}/3 \,,$

(26) $\qquad \Delta\varepsilon_{n+a}^p = \sqrt{2/3}\,\Gamma(\boldsymbol{\eta}_{n+a}^T \mathbf{P}\boldsymbol{\eta}_{n+a})^{1/2} \,,$

(27) $\qquad \mathbf{e}_{n+1} = \mathbf{e}_n + \Delta\mathbf{e} = \mathbf{e}_n + \mathbf{B}\Delta\mathbf{q} \,, \qquad \mathbf{e}_{n+a} = \mathbf{e}_n + a\Delta\mathbf{e} \,,$

(28) $\qquad \boldsymbol{\tau}_{n+1} = \boldsymbol{\tau}_n + \mathbf{D}_e(\Delta\mathbf{e} - \Delta\mathbf{e}_{n+a}^p) \,,$

$\qquad\quad\; \boldsymbol{\tau}_{n+a} = \boldsymbol{\tau}_n + a\mathbf{D}_e(\Delta\mathbf{e} - \Delta\mathbf{e}_{n+a}^p) \,,$

(29) $\qquad \alpha_{n+1} = \alpha_n + \Delta\alpha_{n+a} \,, \qquad \alpha_{n+a} = \alpha_n + a\Delta\alpha_{n+a} \,,$

(30) $\mathbf{e}_{n+1}^p = \mathbf{e}_n^p + \Delta\mathbf{e}_{n+a}^p = \mathbf{e}_n^p + \Gamma\mathbf{P}\boldsymbol{\eta}_{n+a}$, $\mathbf{e}_{n+a}^p = \mathbf{e}_n^p + a\Delta\mathbf{e}_{n+a}^p$,

(31) $\varepsilon_{n+a}^p = \varepsilon_n^p + \Delta\varepsilon_{n+a}^p = \varepsilon_n^p + \Gamma(2\boldsymbol{\eta}_{n+a}^T\mathbf{P}\boldsymbol{\eta}_{n+a}/3)^{1/2}$,

$\varepsilon_{n+a}^p = \varepsilon_n^p + a\Delta\varepsilon_{n+a}^p$,

where a is a constant with value commonly $1/2 \le a \le 1$, chosen to improve the accuracy of Eqs. (15)–(17) and Γ is a positive constant to be determined from the yield condition Eq. (13)

(32) $f(\boldsymbol{\eta}_{n+a}, \varepsilon_{n+a}^p) = \dfrac{1}{2}\boldsymbol{\eta}_{n+a}^T\mathbf{P}\boldsymbol{\eta}_{n+a} - \dfrac{1}{3}\sigma_Y^2(\varepsilon_{n+a}^p) = 0$.

In Eq. (32), $\boldsymbol{\eta}_{n+a}$ and ε_{n+a}^p are defined as

(33) $\boldsymbol{\eta}_{n+a} = \boldsymbol{\tau}_n + a\mathbf{D}_e\Delta\mathbf{e} - a\Gamma\mathbf{D}_e\mathbf{P}\boldsymbol{\eta}_{n+a} - \boldsymbol{\alpha}_n - 2ac\Gamma\boldsymbol{\eta}_{n+a}/3$,

(34) $\varepsilon_{n+a}^p = \varepsilon_n^p + a\Gamma\sqrt{2\boldsymbol{\eta}_{n+a}^T\mathbf{P}\boldsymbol{\eta}_{n+a}/3}$.

Solving for $\boldsymbol{\eta}_{n+a}$ from Eq. (33) gives

(35) $\boldsymbol{\eta}_{n+a} = [(1 + 2ac\Gamma/3)\mathbf{I} + a\Gamma\mathbf{D}_e\mathbf{P}]^{-1}(\boldsymbol{\eta}_n + a\mathbf{D}_e\Delta\mathbf{e})$.

A substitution of Eqs. (34) and (35) into Eq. (32) furnishes a nonlinear algebraic equation for Γ to be solved by iteration. One can then update all fields according to Eqs. (27)–(31), which are *second order accurate*, i.e.,

$$f(\boldsymbol{\eta}_{n+1}, \varepsilon_{n+1}^p) = O(\Delta\mathbf{u})^3 .$$

For other yield surfaces, one may not be able to establish a single algebraic equation for Γ from the plastic constrained equations. In this case, one has to solve for Γ, $\boldsymbol{\eta}_{n+a}$, $\mathbf{e}_{n+a}^p$, and ε_{n+a}^p in terms of $\Delta\mathbf{u}$ simultaneously.

For plane strain, $\Delta e_{31} = \Delta e_{32} = \Delta e_{33} = 0$ and $\Delta\mathbf{e}$ is a function of x and y only. For plane stress, $\eta_{31} = \eta_{32} = \eta_{33} = 0$, and $\mathbf{P}$ and $\mathbf{D}_e$ are given in Eqs. (21.6:14b) and (18.14:23). The solution is also a function of x and y only. To assure $(\tau_{33})_{n+1} = 0$, we update e_{33} as

$$(e_{33})_{n+1} = -\frac{\nu}{E}[(\tau_{11})_{n+1} + (\tau_{22})_{n+1}] - [(e_{11}^p)_{n+1} + (e_{22}^p)_{n+1}],$$

for isotropic materials, and according to the appropriate strain-stress relation for general materials.

Isotropic Materials. It can be shown that, for isotropic materials,

(35a) $\mathbf{D}_e\mathbf{P} = \mathbf{P}\mathbf{D}_e$, $\mathbf{D}_e\mathbf{P}\boldsymbol{\Delta} = 2G\mathbf{P}\boldsymbol{\eta}$, $\boldsymbol{\eta}^T\mathbf{P}\mathbf{D}_e\mathbf{P}\boldsymbol{\eta} = 2G\boldsymbol{\eta}^T\mathbf{P}\boldsymbol{\eta}$, if the problem is plane strain or 3 dimensional and,

$$(35b) \qquad \mathbf{D}_e \mathbf{P} \boldsymbol{\eta} = \frac{E}{3(1 - \nu^2)} \begin{bmatrix} (2 - \nu)\eta_{11} - (1 - 2\nu)\eta_{22} \\ (2 - \nu)\eta_{22} - (1 - 2\nu)\eta_{11} \\ 3(1 - \nu)\eta_{12} \end{bmatrix},$$

$$\boldsymbol{\eta}^T \mathbf{P} \mathbf{D}_e \mathbf{P} \boldsymbol{\eta} \neq 2G \boldsymbol{\eta}^T \mathbf{P} \boldsymbol{\eta},$$

if the problem is plane stress. Then,

$$(36) \qquad\qquad\qquad \mathbf{P} = \mathbf{Q} \boldsymbol{\Lambda}_p \mathbf{Q}^T,$$

$$(37) \qquad\qquad\qquad \mathbf{D}_e = \mathbf{Q} \boldsymbol{\Lambda}_d \mathbf{Q}^T,$$

where

$$(38) \qquad\qquad \mathbf{Q} = \frac{1}{\sqrt{2}} \begin{bmatrix} 1 & -1 & 0 \\ 1 & 1 & 0 \\ 0 & 0 & \sqrt{2} \end{bmatrix},$$

$$(39) \qquad\qquad \boldsymbol{\Lambda}_p = \mathrm{dia}[1/3 \ 1 \ 2],$$

$$(40) \qquad\qquad \boldsymbol{\Lambda}_d = \frac{E}{1 - \nu^2} \mathrm{dia} \left[1 + \nu \quad 1 - \nu \quad \frac{1 - \nu}{2} \right],$$

$$(41) \qquad\qquad \boldsymbol{\Lambda}_d \boldsymbol{\Lambda}_p = \frac{E}{1 - \nu^2} \mathrm{dia} \left[\frac{1 + \nu}{3} \quad 1 - \nu \quad 1 - \nu \right],$$

for plane stress, and

$$(42) \qquad \mathbf{Q} = \begin{bmatrix} \mathbf{Q}_1 & \mathbf{0} \\ \mathbf{0} & \mathbf{I} \end{bmatrix}, \qquad \mathbf{Q}_1 = \frac{1}{\sqrt{6}} \begin{bmatrix} \sqrt{3} & -1 & \sqrt{2} \\ -\sqrt{3} & -1 & \sqrt{2} \\ 0 & 2 & \sqrt{2} \end{bmatrix},$$

$$(43) \qquad \boldsymbol{\Lambda}_p = \mathrm{dia}[1 \ 1 \ 0 \ 2 \ 2 \ 2],$$

$$(44) \qquad \boldsymbol{\Lambda}_d = \frac{E}{1 + \nu} \mathrm{dia} \left[1 \ 1 \ \frac{1 + \nu}{1 - 2\nu} \ \frac{1}{2} \ \frac{1}{2} \ \frac{1}{2} \right],$$

$$(45) \qquad \boldsymbol{\Lambda}_d \boldsymbol{\Lambda}_p = \frac{E}{1 + \nu} \mathrm{dia}[1 \ 1 \ 0 \ 1 \ 1 \ 1],$$

for 3-dimensional and plane strain problems, in which dia[...] denotes a diagonal matrix with the diagonal terms given in the bracket and $\mathbf{I}$ is a 3×3 identity matrix. A substitution of Eqs. (36)–(45) into Eq. (35) yields

$$(46) \qquad \boldsymbol{\eta}_{n+a} = \mathbf{Q}[(1 + 2ac\Gamma/3)\mathbf{I} + a\Gamma \boldsymbol{\Lambda}_d \boldsymbol{\Lambda}_p]^{-1} \mathbf{Q}^T (\boldsymbol{\eta}_n + a\mathbf{D}_e \Delta \mathbf{e}),$$

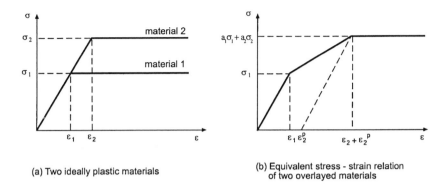

(a) Two ideally plastic materials

(b) Equivalent stress - strain relation
of two overlayed materials

Fig. 21.6:1. Isotropic hardening simulated by two non-hardening materials.

in which $\mathbf{\Lambda}_d\mathbf{\Lambda}_p$ is a diagonal matrix. For pure kinematic hardening materials, i.e., σ_Y is constant, Eq. (32) reduces to a quartic equation in Γ for plane stress. The quartic equation has only one positive root for Γ and can be solved in closed form (Simo and Govindjee 1988).

If the material is also ideal plastic, Eqs. (17) and (19) become

$$(46a) \qquad \mathbf{D}^{ep} = \mathbf{D}_e - \frac{3G}{\sigma_Y^2}\mathbf{P}\tau(\mathbf{P}\tau)^T\,, \qquad \Delta\Lambda = \frac{3}{2\sigma_Y^2}\,\tau^T\mathbf{P}\Delta\mathbf{e}\,,$$

for plane strain and 3 dimensional cases. Other equations can be derived accordingly. For plane stress, $\mathbf{D}_e\mathbf{P}\boldsymbol{\eta}$ in Eq. (22a) is given by Eq. (35b).

Sublayer Hardening Model. An alternative to model isotropic hardening is to use the overlay technique (Pian 1987). In this approach, two or more non-hardening materials are subjected to the same strains. Figure 21.6:1 illustrates a two-nonhardening-material model with the equivalent stress-strain relation in uni-axial tension being

$$\sigma = \begin{cases} E\varepsilon & \text{for} \quad \varepsilon \leq \varepsilon_1 \\ a_2 E(\varepsilon - \varepsilon_1) + \sigma_1 & \text{for} \quad \varepsilon_1 < \varepsilon \leq \varepsilon_2 \\ a_1\sigma_1 + a_2\sigma_2 & \text{for} \quad \varepsilon_2 < \varepsilon\,, \end{cases}$$

in which $a_1, a_2 (= 1 - a_1)$ are the fractions of the two materials, σ_1, σ_2 are their yield stresses and $\varepsilon_1, \varepsilon_2$ are the corresponding strains at the onset of yielding. The yield strain is

$$\varepsilon_1^p = \left.\begin{cases} 0 & \text{for} \quad \varepsilon \leq \varepsilon_1 \\ a_1(\varepsilon - \varepsilon_2) & \text{for} \quad \varepsilon_1 < \varepsilon \end{cases}\right\} \text{ for material 1}\,,$$

$$\varepsilon_2^p = \begin{cases} 0 & \text{for} \quad \varepsilon \leq \varepsilon_2 \\ a_2(\varepsilon - \varepsilon_2) & \text{for} \quad \varepsilon_2 < \varepsilon \end{cases} \quad \text{for material 2}.$$

Dynamic Analysis. In dynamic analysis, $\Delta\mathbf{u} = \mathbf{u}_{n+1} - \mathbf{u}_n$ can be calculated from acceleration in explicit integration. The acceleration at t_{n+1} is derived from the equation of motion based on the principle of virtual work,

(47) ▲ $\quad \delta\Pi = \left[\int_V (\delta\mathbf{u}^T \ddot{\mathbf{u}}\rho + \delta\mathbf{e}^T \boldsymbol{\tau} - \rho\delta\mathbf{u}^T\mathbf{b})dV - \int_{\partial V_\sigma} \delta\mathbf{u}^T\bar{\mathbf{T}}dS \right]_{n+1} = 0$

with $\boldsymbol{\tau}_{n+1}$ from Eq. (28) in terms of $\Delta\mathbf{e}$, $\Delta\mathbf{e}_{n+a}^p$ from Eqs. (23) and (24). From the acceleration, one determines the new $\Delta\mathbf{u}$ and proceeds to the next time increment.

Static Analysis. For static analyses, $\Delta\mathbf{u}$ is not known *a priori* at each increment. It must satisfy the incremental equation of motion, which can be derived from Eq. (47) in the form

(48) $\quad \delta\Delta\Pi = \int_V \left[\delta\mathbf{e}^T \mathbf{D}_e(\Delta\mathbf{e} - \Delta\mathbf{e}_{n+a}^p) - \rho\delta\mathbf{u}^T(\mathbf{b}_{n+1} - \mathbf{b}_n) \right] dV$

$$- \int_{\partial V_\sigma} \delta\mathbf{u}^T(\bar{\mathbf{T}}_{n+1} - \bar{\mathbf{T}}_n) \, dS = 0 \, ,$$

or

(49) $\quad\quad\quad\quad \mathbf{K}\Delta\mathbf{q} = \mathbf{F}_p(\Delta\mathbf{e}_{n+a}^p) + \Delta\mathbf{F} \, ,$

where $\mathbf{F}_p$ is derived from the volume integration of $\delta\mathbf{e}^T\mathbf{D}_e\Delta\mathbf{e}_{n+a}^p$ and $\Delta\mathbf{F}$ from integrals involving the body force increment $(\mathbf{b}_{n+1} - \mathbf{b}_n)$ and the prescribed surface traction increment $(\bar{\mathbf{T}}_{n+1} - \bar{\mathbf{T}}_n)$ and from prescribed boundary displacements by enforcing rigid constraints. Equation (49) differs from the elasticity equation only by the term $\mathbf{F}_p(\Delta\mathbf{e}_{n+a}^p)$, which itself is a function of $\Delta\mathbf{q}$. Recall that $\Delta\mathbf{u} = \mathbf{h}\Delta\mathbf{q}$ and $\Delta\mathbf{e} = \mathbf{B}\Delta\mathbf{q}$.

Equation (49) is to be solved by iterative process together with the plastic constraints of Eqs. (23)–(31), a set of highly nonlinear coupled equations. One approach is to start the iteration by assuming $\Delta\mathbf{u}$ equal to that of the previous step as a predictor, calculating the updated fields according to Eqs. (23)–(31) subjected to the constraint of Eq. (32) for locations where yielding takes place, evaluating $\mathbf{F}_p(\Delta\mathbf{e}_{n+a}^p)$ based on the predicted $\Delta\mathbf{u}$ at the yielded locations, then solving Eq. (49) for the corrected $\Delta\mathbf{u}$. One repeats the process until the estimated error for Eq. (32) is within a specified limit. In this approach, $\mathbf{K}$ is the same as that of the elastic solution and remains constant throughout the entire solution process. All plasticity effects contain in $\mathbf{F}_p(\Delta e_{n+a}^p)$. This iteration process is sometime called the *initial strain approach*, i.e., an initial strain (in terms of $\Delta\mathbf{u}$) is first assumed and corrections are then sought. Various Newton-Raphson schemes have also used to determine $\Delta\mathbf{u}$.

Another approach is to modify Eq. (48) to the form

$$(50) \quad \int_V [\delta e^T \mathbf{D}_e(\Delta e - \Delta e_n^p)] \, dV = \int_v [\delta e^T \mathbf{D}_n^{ep} \Delta e] \, dV$$

$$= \int_V [\delta e \mathbf{D}_e(\Delta e_{n+a}^p - \Delta e_n^p)] \, dV + \int_V \rho \delta \mathbf{u}^T (\mathbf{b}_{n+1} - \mathbf{b}_n) \, dV$$

$$+ \int_{\partial V_\sigma} \delta \mathbf{u}^T (\bar{\mathbf{T}}_{n+1} - \bar{\mathbf{T}}_n) \, dS$$

or

$$(51) \qquad \mathbf{K}_n^{ep} \Delta \mathbf{q} = \mathbf{F}_p'(\Delta e_{n+a}^p - \Delta e_n^p) + \Delta \mathbf{F},$$

where $\mathbf{D}_n^{ep}$ and Δe_n^p are defined in Eqs. (17) and (20) with subscript n denoting the values at t_n, and $\mathbf{F}_p'(\Delta e_{n+a}^p - \Delta e_n^p)$ is derived from the first integral on right most side in Eq. (50). Thus $\mathbf{K}_n^{ep}$ is derived using $\mathbf{D}_n^{ep}$ rather than $\mathbf{D}_e$. For general materials, $\mathbf{D}_n^{ep}$ and e_n^p are derivable from formulae given in Secs. 6.10–6.12. One can use the same iterative method described earlier in this section to establish the solution for $\Delta\mathbf{q}$, i.e., one starts the iteration with assumed $\Delta\mathbf{q}$ to calculate $\mathbf{F}_p'(\Delta e_{n+a}^p - \Delta e_n^p)$ and determine the corrected $\Delta\mathbf{q}$ by solving Eq. (51). The method is called the *modified tangent modulus approach*. The *conventional tangent modulus approach* assumes $\mathbf{F}_p'(\Delta e_{n+a}^p - \Delta e_n^p)$ to be zero. Since $\mathbf{K}_n^{ep}$ depends on the field values, one must refactor $\mathbf{K}_n^{ep}$ in solving for $\Delta\mathbf{q}$, which is a time consuming process. However, the approach often improves the rate of convergence.

One can also update $\mathbf{K}_n^{ep}$ using the current iterated field values during the iteration for a given load increment. This is done in particular if many locations change from loading to unloading or vice versa in the step. Such changes greatly affect the value of $\mathbf{D}^{ep}$ in the evaluation of $\mathbf{K}_n^{ep}$.

Large Deformation. To avoid the confusion between the difference in the definitions of strain matrix and strain tensor, we shall use Cartesian coordinates and the index notation to illustrate the idea. For 3-dimensional problems, a strain matrix is a column matrix with 6 components

[Eq. (18.16:3)],

$$\mathbf{e}^T = [e_{11} \ e_{22} \ e_{33} \ 2e_{12} \ 2e_{23} \ 2e_{31}],$$

whereas a strain tensor has 9 components e_{ij}. If one assumes that the plasticity constraints Eqs. (24)–(26) hold for the incremental plasticity fields, i.e.,

(52)
$$(\Delta e_{ij}^p)_{n+a} = \Gamma \left(\frac{\partial f}{\partial \tau_{ij}} \right)_{n+a} , \qquad (\Delta \alpha_{ij})_{n+a} = \frac{2}{3} c\Gamma \left(\frac{\partial f}{\partial \tau_{ij}} \right)_{n+a} ,$$

$$(\Delta e^p)_{n+a} = \left(\frac{2}{3} \Delta e_{ij}^p \Delta e_{ij}^p \right)^{1/2}_{n+a} ,$$

where $(\Delta e_{ij}^p)_{n+a}$ denotes the value at t_{n+a}, and the strain increments are calculated based on the current configuration,[3] i.e.,

(53) $$\Delta e_{ij} = \Delta D_{ij} = \frac{1}{2} \left[\frac{\partial \Delta u_i}{\partial X_j^n} + \frac{\partial \Delta u_j}{\partial X_i^n} \right] , \qquad \Delta \Omega_{ij} = \frac{1}{2} \left[\frac{\partial \Delta u_i}{\partial X_j^n} - \frac{\partial \Delta u_j}{\partial X_i^n} \right] ,$$

the elastic-plastic analysis carries over for large formation by properly defining the stress measure. For example if the stress increment is based on the Jaumann rate of the Cauchy stress [Eq. (16.9:20) or (17.5:3)], i.e.,

(54)
$$(\Delta \overset{\partial}{\tau}_{ij})_{n+1} = D_{ijkl}^e [\Delta e_{kl} - (\Delta e_{kl}^p)_{n+a}] ,$$

$$(\Delta \overset{\partial}{\tau}_{ij})_{n+a} = a D_{ijkl}^e [\Delta e_{kl} - (\Delta e_{kl}^p)_{n+a}] ,$$

where D_{ijkl}^e is the incremental elastic modulus matrix which can be a function of the deformation gradient and the Cauchy stress at t_n, then the updated Cauchy and back stresses are

(55) $$(\tau_{ij})_{n+1} = (\tau_{ij} + \Delta \Omega_{ik} \tau_{kj} - \tau_{ik} \Delta \Omega_{kj})_n + (\Delta \overset{\partial}{\tau}_{ij})_{n+1} ,$$

(56) $$(\alpha_{ij})_{n+1} = (\alpha_{ij} + \Delta \Omega_{ik} \alpha_{kj} - \alpha_{ik} \Delta \Omega_{kj})_n + (\Delta \alpha_{ij})_{n+1} .$$

If the increment of the 2nd Piola–Kirchhoff stress in the updated Lagrange description is used in the incremental constitutive equation Eq. (54),

---

[3]For dynamic analysis, $\Delta \mathbf{u}$ is calculated from acceleration at each increment. One often defines

$$\Delta e_{ij} = \frac{1}{2} \left[\frac{\partial \Delta u_i}{\partial X_j^{n+1/2}} + \frac{\partial \Delta u_j}{\partial X_i^{n+1/2}} \right] , \qquad \Delta \Omega_{ij} = \frac{1}{2} \left[\frac{\partial \Delta u_i}{\partial X_j^{n+1/2}} - \frac{\partial \Delta u_j}{\partial X_i^{n+1/2}} \right] ,$$

where $X_i^{n+1/2} = X_i^n + \frac{1}{2} \Delta u_i$.

we then calculate the Cauchy and back stress increments according to Eq. (21.1:15).

For finite deformation, solutions obtained by different objective rates can be substantially different. Some of the solutions may be physically unreasonable. For example (Truesdell 1955), when a linear hypoelastic material is subjected to finite simple shear and the Jaumann rate of the Cauchy stress is used in the linear incremental constitutive equation, the shear stress is oscillatory.

For dynamic analysis, the equation of motion has exactly the same for as Eq. (21.3:1), which is reproduced here in index notation

$$(57) \quad \delta\Pi = \left[\int_v (\delta u_i \ddot{u}_i \rho + \delta e_{ij} \tau_{ij} - \rho \delta u_i b_i) \, dV - \int_{\partial V_\sigma} \delta u_i \bar{T}_i \, dS \right]_{n+1} = 0,$$

with the updated Cauchy stress and the Jaumann rate defined in Eqs. (55) and (54). The plasticity effects are contained in the incremental stress-strain relations in terms of $(\Delta e^p_{kl})_{n+a}$ as given in Eq. (54). Thus once $(\Delta \overset{\partial}{\tau}_{ij})_{n+1}$ and $(\Delta \alpha_{ij})_{n+a}$ are determined from the known Δu_i, one uses Eq. (57) to determine the acceleration at t_{n+1} and predict Δu_i for the next increment.

For static analysis, Δu_i is not known *a priori* and must be solved for from the equation of motion together with the plasticity constraints. The equation of motion is the same as Eq. (48) with additional terms to account for large deformation shown below:

$$(58) \quad \delta\Delta\Pi = \int_{V_n} \{\delta e_{ij}[\Delta\Omega_{ik}\tau_{kj} - \tau_{ik}\Delta\Omega_{kj} + (\Delta\overset{\partial}{\tau}_{ij})_{n+1}] - \delta u_i[(\rho b_i)_{n+1}$$

$$- (\rho b_i)_n]\} \, dV - \int_{\partial V_{\sigma n}} \delta u_i [(\bar{T}_i)_{n+1} - (\bar{T}_i)_n] \, dS = 0.$$

It goes without saying that the integrals in Eq. (58) are over the current configuration. The iteration process for small deformation applies to large deformation as well.

The integration can be unstable if the incremental step is too large. Readers are referred to literature for stability criteria (Cormeau[21.3] 1975, Ortiz and Popov[21.3] 1985). A practical criterion for dynamic analysis is to limit the time step below the minimum characteristic dimension of all elements divided by the adiabatic sound speed (Hallquist[21.3] 1998). The characteristic dimension is usually defined as the element volume divided by the largest surface area of the sides of the element.

21.7. VISCOPLASTICITY

For viscoplasticity materials, assuming linear viscosity and small deformation for simplicity, we can write the constitutive relations in the form

(1)
$$\dot{e}^p_{ij} = \frac{H(f)}{\eta} \frac{\partial f(\tau_{ij}, \xi_i)}{\partial \tau_{ij}},$$

(2)
$$\dot{\xi}_i = \frac{H(f)}{\eta} U_i(\tau_{ij}, \xi_i),$$

(3)
$$\dot{\tau}_{ij} = D^e_{ijkl}(\dot{e}_{ij} - \dot{e}^p_{ij}),$$

where η is a known "viscosity" parameter, f is the yield function (only associated material are considered here), U_i is a known function characterizing the internal variable ξ_i, and H is the Heaviside function defined as

(4)
$$H(f) = 0 \quad \text{if} \quad f < 0,$$
$$= f \quad \text{if} \quad f \geq 0.$$

The internal variable can be the equivalent plastic strain and/or the back stress. The finite element equation of motion is derived from

$$\int_V \delta e_{ij} \dot{\tau}_{ij} \, dV - \dot{F}_i = \int_V \delta e_{ij} D^e_{ijkl}(\dot{e}_{kl} - \dot{e}^p_{kl}) \, dV - \dot{F}_i = 0,$$

which leads the finite element equation

(5)
$$\mathbf{K}^0_T \dot{\mathbf{q}} = \mathbf{F}_p(e^p_{ij}) + \dot{\mathbf{F}},$$

where $\mathbf{K}^0_T$ is the stiffness matrix which is the same as elasticity derived from the terms associated with $\dot{e}_{kl}$, $\mathbf{F}_p(\dot{e}^p_{ij})$ is from the terms associated with $\dot{e}^p_{ij}$ which contains all the viscoplasticity effects, and $\dot{\mathbf{F}}$ is the rate of applied load, and is equivalent to $\Delta \mathbf{F}$ of Eq. (21.6:49). Equations (1)–(5) are a set of first order nonlinear differential equations of $\dot{q}_i$, $\dot{e}^p_{kl}$, $\dot{\xi}_i$, which can be integrated with the initial conditions or solved by iteration. Note that, as the stress rate associated with $\dot{e}^p_{kl}$ always points toward the surface of the yield potential $(-D^e_{ijkl}\dot{e}^p_{kl} = -\frac{H(f)}{\eta} D^e_{ijkl} \frac{\partial f}{\partial \tau_{kl}})$ when *yielding occurs*, from Eqs. (1) and (3), the stress state lying outside $(f > 0)$ moves toward the yield surface. The solution approaches the yield surface, i.e., $f(\tau_{ij}, \xi_i) \to 0$ at the steady state.

The main differences between the viscoplasticity and the classical plasticity theories are that the proportional factor $H(f)/\eta$ for the rates of variables in viscoplasticity is specified explicitly in terms of the yield function, and that the time is a physical parameter.

21.8. CREEP

Creep is a time-dependent phenomenon even under a constant load. In infinitesimal deformation, the total strain is split into the instantaneous strain e_{ij}^e and the creep strain e_{ij}^c as

(1) $$e_{ij} = e_{ij}^e + e_{ij}^c .$$

The constitutive law is defined in the form

(2) $$\dot{\tau}_{ij} = D_{ijkl}^e (\dot{e}_{ij} - \dot{e}_{ij}^c)$$

for the instantaneous strain rate and

(3) $$\dot{e}_{ij}^c = \chi_{ij}(\tau_{kl}),$$

for the creep rate, where $\chi_{ij}(\tau_{kl})$ are known functions of the current state of stress. The finite element equation of motion is the same as Eq. (21.7:5) with $\mathbf{F}_p$ derived from terms associated $\dot{e}_{ij}^c$. The creep equations can be solved by iterative methods. (Mendelson et $al.$ 1959, and Greenbaum and Rubinstein 1968).

In this chapter, we have only touched upon briefly the application of finite element method to various nonlinear problems. Readers are referred to literature, which is voluminous, for details. Computational mechanics can provide us tools to explore the enormous richness of nonlinear mechanics: large deformation, composite structures, micro- and nano- technologies, living cells, DNA, proteins, enzymes and beyond.

Problem 21.1. Write the incremental energy function Eq. (17.5:25) in matrix notation

$$\overset{\circ}{W}(\mathbf{D}) = \bar{\mathbf{D}}^T \bar{\mathbf{D}}_e \bar{\mathbf{D}}/2 = \mathbf{D}^T \mathbf{D}_e \mathbf{D}/2,$$

where $\mathbf{D}$ is the symmetric velocity-like gradient referred to the current configuration $\mathbf{x}$ that $\mathbf{D}^T = [D_{11}\ D_{22}\ D_{33}\ 2D_{12}\ 2D_{23}\ 2D_{31}]$, and similarly for $\bar{\mathbf{D}}$ with D's replaced by $\bar{D}'$ s. They are related by the deformation gradient matrix $\mathbf{F}$ that

$$\begin{bmatrix} \bar{D}_{11} & \bar{D}_{12} & \bar{D}_{31} \\ & \bar{D}_{22} & \bar{D}_{23} \\ \text{sym} & & \bar{D}_{33} \end{bmatrix} = \mathbf{F} \begin{bmatrix} D_{11} & D_{12} & D_{31} \\ & D_{22} & D_{23} \\ \text{sym} & & D_{33} \end{bmatrix} \mathbf{F}^T, \qquad \mathbf{F} = \frac{\partial \mathbf{x}}{\partial \mathbf{X}}.$$

The quantities $\bar{\mathbf{D}}_e$, $\mathbf{D}_e$ are incremental elastic modulus matrices. Show that

$$\bar{\mathbf{D}}_e = \frac{4}{J} \left[\frac{\partial W_0}{\partial I_2} (\mathbf{II} - \mathbf{H}) + \frac{\partial W_0}{\partial I_3} \mathbf{A} + \mathbf{G} \right],$$

$$\mathbf{II} - \mathbf{H} = \begin{bmatrix} \mathbf{Q}_1 & \mathbf{0} \\ \mathbf{0} & \mathbf{Q}_2 \end{bmatrix}, \qquad \mathbf{Q}_1 = \begin{bmatrix} 0 & 1 & 1 \\ 1 & 0 & 1 \\ 1 & 1 & 0 \end{bmatrix}, \qquad \mathbf{Q}_2 = -\frac{1}{2} \begin{bmatrix} 1 & 0 & 0 \\ 0 & 1 & 0 \\ 0 & 0 & 1 \end{bmatrix},$$

$$\mathbf{A} = \begin{bmatrix} 0 & C_{33} & C_{22} & 0 & -C_{23} & 0 \\ & 0 & C_{11} & 0 & 0 & -C_{31} \\ & & 0 & -C_{12} & 0 & 0 \\ & & & -C_{33}/2 & C_{31}/2 & C_{23}/2 \\ & \text{sym} & & & -C_{11}/2 & C_{31}/2 \\ & & & & & -C_{22}/2 \end{bmatrix},$$

where $J[= \det(\mathbf{F})]$, C's are the components of the matrix $\mathbf{C}(= \mathbf{F}^T\mathbf{F})$ and, for the energy density function W_0 defined in Eq. (16.11:16a),

$$\mathbf{G} = \frac{\partial^2 W_0}{\partial^2 I_3} I_3 \hat{\mathbf{C}}\hat{\mathbf{C}}^T = b_4 I_3 \hat{\mathbf{C}}\hat{\mathbf{C}}^T , \qquad \hat{\mathbf{C}}^T = [\hat{C}_{11} \ \hat{C}_{22} \ \hat{C}_{33} \ \hat{C}_{12} \ \hat{C}_{23} \ \hat{C}_{31}],$$

where $\hat{C}$'s are the components of the inverse of $\mathbf{C}$. Expressing $\bar{\mathbf{D}}$ in terms of $\mathbf{D}$ in the form $\bar{\mathbf{D}} = \mathbf{R}\mathbf{D}$, one has $\mathbf{D}_e = \mathbf{R}^T\bar{\mathbf{D}}_e\mathbf{R}$. Find $\mathbf{R}$ in terms of the components of $\mathbf{F}$.

BIBLIOGRAPHY

1.1. References for historical remarks.

Todhunter, I. and K. Pearson. A History of the Theory of Elasticity and of the Strength of Materials. Cambridge: University Press, Vol. 1, 1886; Vol. 2, 1893. Reprinted. New York: Dover Publications, 1960. This monumental work demonstrates the amazing vigor with which the theory of elasticity was developed until the time of Lord Kelvin. Originated by the mathematician Todhunter (1820–1884, St. John's College, Cambridge), the book was completed after Todhunter's death by Karl Pearson (1857–1936).

Love, A. E. H. The Mathematical Theory of Elasticity. Cambridge: University Press, 1st ed., 1892, 1893; 4th ed., 1927. Reprinted. New York: Dover Publications, 1944. Note especially the "Historical Introduction," pp. 1–31.

Timoshenko, S. P. History of Strength of Materials. New York: McGraw-Hill, 1953. This book is beautifully written and illustrated.

1.2. Basic references, selected books on solid mechanics.

Atluri, S. N. Structural Integrity and Durability. GA: Technical Science Press, 1997.

Berthelot, J-M. Composite Materials: Mechanical Behavior and Structural Analysis. New York: Springer-Verlag, 1998.

Biezeno, C. B. and R. Grammel. Engineering Dynamics. Vols. 1 and 2. (Translated by M. L. Meyer from the Second German Edition. Glasgow: Blackie, 1955, 1956.

Biot, M. A. Mechanics of Incremental Deformations (Theory of Elasticity and Viscoelasticity of Initially Stressed Solids and Fluids, Including Thermodynamic Foundations and Applications to Finite Strain). New York: Wiley, 1965.

Britvec, S. J. Flexible Space Structures. Stability and Optimization. Switzerland: Birkhauser, 1995.

Carlsson, L. and R. B. Pipes. Experimental Characterization of Advanced Composite Materials, 2nd ed., Lancaster, PA: Technomic. 1997.

Chen, W-F. and A. F. Saleeb, Constitutive Equations for Engineering Materials, Vol. 1: Elasticity and Modeling. New York: Elsevier, 1994.

Chen, W-F., W. O. McCarron and E. Yamaguchi. Constitutive Equations for Engineering Materials, Vol. 2: Plasticity and Modeling. New York: Elsevier, 1994.

Cherepanov, G. P. Methods of Fracture Mechanics: Solid Matter Physics. Solid Mechanics and Its Applications, Vol. 51. Netherland: Kluwer, 1997.

Chou, T. W. Microstructural Design of Fiber Composites. New York: Cambridge University Press, 1992.

Coker, E. G. and L. N. G. Filon. Photoelasticity. Cambridge: University Press, 1931. A classic which presents mathematical solutions as well as experimental results.

Cosserat, E. and F. Cosserat. Theorie des corps deformables. Paris: Hermann, 1909. A remarkable general treatise containing full discussion of curvature and torque stresses.

Davis, R. O. and A. P. S. Selvadurai. Elasticity and Geomechanics. New York: Cambridge UP, 1996.

Dowson, D. History of Tribology. New York: ASME, 1998.

Drumheller, D. S. Introduction to Wave Propagation in Nonlinear Fluids and Solids. New York: Cambridge UP, 1998.

Ellyin, F. Fatigue Damage, Crack Growth and Life Prediction. New York: Chapman and Hall, 1996.

Eringen, A. C. Nonlinear Theory of Continuous Media. New York: McGraw-Hill, 1962.

Felbeck, D. K. and A. G. Atkins. Strength and Fracture of Engineering Solids. Englewood Cliffs, New Jersey: Prentice-Hall, 1996.

Fett, T. and D. Munz. Stress Intensity Factors and Weight Functions. Int. Series Adv. Fracture Comput. Mech. Massachusetts: Billerica, 1997.

Flügge, W. (ed.). Handbook of Engineering Mechanics. New York: McGraw-Hill, 1962.

Föppl, L. Drang und Zwang, Vol. 3. Munich: Leibniz Verlag, 1947.

Fukumoto, Y. (ed). Structural Stability Design: Steel and Composite Structures. New York: Elsevier, 1997.

Goodier, J. N. and N. J. Hoff (eds.). Structural Mechanics. Proc. 1st Symp. Naval Structural Mechanics. New York: Pergamon Press, 1960.

Green, A. E. and W. Zerna. Theoretical Elasticity. London: Oxford University Press, 1968. General tensor notations and convective coordinates are used. The first part concerns finite deformation; the second part is small perturbation about finite deformation; the third part is infinitesimal theory.

Green, A. E. and J. E. Adkins. Large Elastic Deformations and Nonlinear Continuum Mechanics. London: Oxford University Press, 1960. This is an advanced treatise with many explicit solutions.

Herakovich, C. T. Mechanics of Fibrous Composites. New York: Wiley, 1998.

Hyer, M. W. Stress Analysis of Fiber-Reinforced Composite Materials. New York: McGraw-Hill, 1998.

Kalamkarov, A. L. and A. G. Kolpakov. Analysis, Design and Optimization of Composite Structures. United Kingdom: Wiley, 1997.

Magnus, K. and K. Popp. Schwingungen (Vibrations). Stuttgart: BG Teubner, 1997.

McDonald, P. Continuum Mechanics. Boston, Massachusetts: PWS, 1996.

Meirovitch, L. Principles and Techniques of Vibrations. New Jersey: Prentice Hall, 1997.

Meyers, M. A. and K. K. Chawla. Mechanical Behavior of Materials. New Jersey: Prentice Hall, 1999.

Morse, P. M. Vibration and Sound, 2nd ed., New York: McGraw-Hill, 1948.

Muskhelishvili, N. I. Some Basic Problems of the Theory of Elasticity. (Translation of Third Russian Edition by J. R. M. Radok.) Groningen, Netherlands: Noordhoff, 1953.

Nadai, A. Theory of Flow and Fracture of Solids. New York: McGraw-Hill, 1963.

Noll, W. and C. A. Truesdell. The Non-linear Field Theories of Mechanics. Vol. 3, Part 3 of Encyclopedia of Physics, edited by Flügge. Berlin: Springer, 1964. This treatise presents the most advanced picture of continuum mechanics known today.

Papathanasiou, T. D. and D. C. Guell (ed.). Flow-Induced Alignment in Composite Materials. United Kingdom: Wood Head Publishing Ltd., 1997.

Penny, R. K. and D. L. Marriott, Design for Creep. New York: Chapman Hall, 1995.

Prager, W. Introduction to Mechanics of Continua. Boston: Ginn. 1961. An excellent book which treats both solids and fluids.

Prescott, J. Applied Elasticity. New York: Dover Publications, 1946.

Reddy, J. N. Mechanics of Laminated Composite Plates: Theory and Analysis. Boca Raton FL: CRC Press, 1997.

Reissner, E. Selected Works in Applied Mechanics and Mathematics. Sudbury, MA: Jones and Bartlett, 1996.

Rossow, E. C. Analysis and Behavior of Structures. New Jersey: Prentice Hall, 1996.

Sechler, E. E. Elasticity in Engineering. New York: Wiley, 1952. An effective introduction to elasticity with many examples worked out in detail.

Seireg, A. A. and J. Rodriguez. Optimizing the Shape of Mechanical Elements and Structures. New York: Marcel Dekker, 1997.

Sierakowski, R. L. and S. K. Chaturvedi. Dynamic Loading and Characterization of Fiber-Reinforced Composites. New York: Wiley, 1997.

Sokolnikoff, I. S. Mathematical Theory of Elasticity, 2nd ed., New York: McGraw-Hill, 1956. Contains a thorough treatment of the theory of bending and torsion of beams, complex variable technique in two dimensional problems, and variational principles.

Southwell, R. V. Theory of Elasticity. Oxford: The Clarendon Press, 1936, 1941.

Swanson, S. R. Introduction to Design and Analysis with Advanced Composite Materials. New York: Springer-Verlag, 1997.

Timoshenko, S. Vibration Problems in Engineering, 2nd ed., New York: Van Nostrand, 1937.

Timoshenko, S. Strength of Materials, 3rd ed., New York: Van Nostrand, 1955.

Timoshenko, S. and N. Goodier. Theory of Elasticity. New York: McGraw-Hill, 1st ed., 1934; 2nd ed., 1951.

Timoshenko, S. and S. Woinowsky-Krieger. Theory of Plates and Shells. New York: McGraw-Hill, 1st ed., 1940; 2nd ed., 1959.

Timoshenko, S. and J. M. Gere. Theory of Elastic Stability. New York: McGraw-Hill, 1st ed., 1936; 2nd ed., 1961.

Ting, T. C. T. Anisotropic Elasticity: Theory and Applications. New York: Oxford UP, 1996.

Trefftz, E. Mechanik der elastischen Körper. In Handbuch der Physik, H. Geiger and K. Scheel (ed.). Vol. 6. Berlin: Springer, 1928.

Truesdell, C. A. and R. A. Toupin. The Classical Field Theories. Vol. 3, Part 1 of Encyclopedia of Physics, edited by Flügge. Berlin: Springer, 1960. This book explains the development of basic field concepts through exhaustive historic and scientific detail.

Tsai, S. W. Theory of Composite Design. Ohio: Think Composites, 1993.

Voigt, W. Theoretische Studien über die Elasticitätsverhältnisse der Krystalle. Abh. Ges. Wiss. Göttingen, *34* (1887).

Volterra, Vito and Enrico. Sur les distorsions des corps e'lestiques. Paris: Gauthier-Villars, 1960.

von Karman, Theodore and M. A. Biot. Mathematical Methods in Engineering. New York: McGraw-Hill, 1940.

Westergaard, H. M. Theory of Elasticity and Plasticity. Cambridge Massachusetts: Harvard University Press, 1952.

1.3. Biomechanics has a huge literature. Here we list a few for the beginners:

Abé, H., K. Hayashi and M. Sato (eds.). Data Book on Mechanical Properties of Living Cells, Tissues, and Organs. Tokyo: Springer-Verlag, 1996.

Caro, C. G., T. J. Pedley, R. C. Shroter and W. A. Seed. The Mechanics of the Circulation. Oxford University Press, 1978.

Fung, Y. C. Biomechanics. This is a three-volume set published by Springer-Verlag, New York. The volumes are: Biomechanics: Mechanical Properties of Living Tissues. 2nd ed., 1993. Biomechanics: Circulation, 2nd ed., 1997. Biomechanics Motion, Flow, Stress and Growth, 1990.

Mow, Van, C. and W. C. Hayes. Basic Orthopaedic Biomechanics. 2nd ed., Philadelphia: Lippincott–Raven Publishers, 1997.

Sramek, B. Bo, J. Valenta and F. Klimes (eds.). Biomechanics of the Cardiovascular System. Prague, Czechoslovakia Republic: Czech Technical University Press, 1995.

Woo, S. L. Y. and J. A. Buckwalter (eds.). Injury and Repair of the Musculoskeletal Soft Tissues. Park Ridge, IL. American Academy of Orthopedic Surgeons, 1988.

2.1. Tensor analysis has its historical origin in the works on differential geometry by Gauss, Riemann, Beltrami, Christoffel, and others. As a formal system it was first developed and published by Ricci and Levi-Civita.[2.1] A brief historical introduction can be found in Wills.[2.1] An exhaustive presentation is given by Erikson in an appendix to Truesdell and Toupin (Bibliography 1.2).

Brillouin, L. Les Tenseurs en mécanique et en élasticité (Paris, 1938). New York: Dover Publications, 1946.

Jeffreys, H. Cartesian Tensors. Cambridge: University Press, 1931.

Levi-Civita, T. The Absolute Differential Calculus (Translated from Italian by M. Long). London: Blackie, 1927.

Michal, A. D. Matrix and Tensor Calculus, with Applications to Mechanics, Elasticity and Aeronautics. New York: Wiley, 1947.

Ricci, G. and T. Levi-Civita. Méthodes du calcul différentiel absolu et leurs applications. Math. Ann. 54 (1901) 125–201.

Sokolnikoff, I. S. Tensor Analysis, Theory and Applications. New York: Wiley, 1951, 1958.

Synge, J. L. and A. Schild. Tensor Calculus. Toronto: University Press, 1949.

Wills, A. P. Vector Analysis with an Introduction to Tensor Analysis. Englewood Cliffs, New Jersey: Prentice-Hall, Inc., 1938.

3.1. Couple-stress. E. and F. Cosserat (Bibliography 1.2) presents a thorough formulation of the couple-stress theory. Hellinger (Bibliography 10.2) gives the variational principles including couple-stress.

Koiter, W. T. Couple-Stresses in the Theory of Elasticity. Kon. Nederl. Akad. v. Wetenschappen-Amsterdam, Proc. Ser. B 67 (1) (1964).

Kröner, E. On the physical reality of torque stress in continuum mechanics. Int. J. Eng. Sci. 1 (1963) 261–278.

Mindlin, R. D. and H. F. Tiersten. Effects of Couple-Stresses in Linear Elasticity. Arch. Rational Mech. Analysis *11* (1962) 415–448.

Sadowsky, M. A., Y. C. Hsu and M. A. Hussain. Boundary layers in couple-stress elasticity and stiffening of thin layers in shear. Watervliet Arsenal, New York, Report RR-6320, 1963.

4.1. Our presentation of the compatibility conditions follows Sokolnikoff, (Bibliography 1.2), p. 25. The original reference is

Cesaro, E. Sulle formole del Volterra, fondamentali nella teoria delle distorsioni elastiche. Rendiconto dell' Accademia delta Scienze Fisiche e Matematiche (Società Reale di Napoli), 1906, pp. 311–321.

5.1. More rigorous treatment of Gauss' theorem can be found in

Courant, R. Differential and Integral Calculus (Translated by E. J. McShane), New York: Interscience.

Kellogg, O. D. Foundations of Potential Theory. Berlin: Springer, 1929.

6.1. The determination of elastic constants is discussed in the following books.

Born, M. and K. Huang. Dynamical Theory of Crystal Lattices. Oxford: Clarendon Press, 1954.

Duffy, J. and R. D. Mindlin. Stress-Strain Relations of a Granular Medium. J. Appl. Mech. *24* (1957) 585–593,

Huntington, H. B. The Elastic Constants of Crystals. New York: Academic Press, 1958.

Koehler, J. S. Imperfections in Nearly Perfect Crystals. (See p. 197 for influence on elastic constants.) New York: Wiley, 1952.

6.2. Books on mathematical theory of plasticity. See also Bibliography 17.2.

Freudenthal, A. M. and H. Geiringer. The Mathematical Theories of the Inelastic Continuum. Vol. VI of Encyclopedia of Physics. S. Flügge (ed.). Berlin: Springer-Verlag, 1958.

Goodier, J. N. and P. G. Hodge, Jr. Elasticity and Plasticity. New York: Wiley, 1958.

Hill, R. The Mathematical Theory of Plasticity. Oxford: Clarendon Press, 1950.

Hodge, P. G. Plastic Analysis of Structures. New York: McGraw-Hill, 1959.

Khan, A. S. and S. Huang. Continuum Theory of Plasticity. New York: John Wiley and Sons, Inc., 1995.

Lubliner, J. Plasticity Theory. New York: Macmillan Publishing Company, 1990.

Nadai, A. Theory of Flow and Fracture of Solids. 2 volumes, New York: McGraw-Hill, Vol. 1, 1950; Vol. 2, 1963.

Phillips, A. Introduction to Plasticity. New York: Ronald, 1956.

Phillips, A. A review of quasistatic experimental plasticity and viscoplasticity. J. Plasticity. *2* (1986) 315.

Prager, W. and P. G. Hodge, Jr. Theory of Perfectly Plastic Solids. New York: Wiley, 1951.

Prager, W. An Introduction to Plasticity. Reading, Massachusetts: Addison-Wesley, 1959.

Sokolovskij, V. V. Theorie der Plastizität (German Translation of Russian Second Edition). Berlin: Technik Verlag, 1955.

6.3. Plasticity — Constitutive equations. See also Bibliography 17.5.

Boyce, W. E. The bending of a work-hardening circular plate by a uniform transverse load. Quart. Appl. Math. *14* (1957) 277–288.

Bridgman, P. W. The thermodynamics of plastic deformation and generalized entropy. Rev. Mod. Phys. *22* (1950) 56–63.

Budiansky, B. A reassessment of deformation theories of plasticity. J. Appl. Mech. *26* (1959) 259–264.

Crossland, B. The effect of fluid pressure on the shear properties of metals. Proc. Inst. Mech. Eng. *169* (1954) **935**–944.

Drucker, D. C. A more fundamental approach to plastic stress strain relations. Proc. 1st US Nat. Congress Appl. Mech. (Chicago, 1951), 487–491.

Drucker, D. C. Stress-strain relations in the plastic range of metals-experiments and basic concepts. Rheology *1* (1956) 97–119.

Drucker, D. C. A definition of stable inelastic material. J. Appl. Mech. *26* (1959) 101–106.

Hencky, H. Zur Theorie plastischer Deformationen und die hierdurch im Material hervorgerufenen Nach-Spannungen. Z. Ang. Math. Mech. *4* (1924) 323–334.

Hill, R. The mechanics of quasi-static plastic deformation in metals. In Surveys in Mechanics, G. K. Batchelor and R. M. Davies (eds.). Cambridge: University Press, 1956.

Hodge, P. G., Jr. Piecewise linear plasticity. Proc. 9th Int. Congress Appl. Mech. (Brussels, 1956) *8* (1957) 65–72.

Ishlinskii, I. U. General theory of plasticity with linear strain hardening (in Russian). Ukr. Mat. Zh. *6* (1954) 314–324.

Kadashevich, I. and V. V. Novozhilov. The theory of plasticity which takes into account residual microstresses. J. Appl. Math. Mech. (Translation of Prikl. Mat. i. Mekh.) *22* (1959) 104–118.

Koiter, W. T. Stress-strain relations, uniqueness and variational theorems for elastic-plastic materials with a singular yield surface. Quart. Appl. Math. *11* (1953) 350–354.

Lode, W. Versuche über den Einfluss der mittleren Hauptspannung auf das Fliessen der Metalle Eisen, Kupfer, und Nickel. Z. Physik *36* (1926) 913–939.

Naghdi, P. M., F. Essenburg, and W. Koff. An experimental study of initial and subsequent yield surfaces in plasticity. J. Appl. Mech. *25* (1958) 201–209.

Naghdi, P. M. Stress-strain relations in plasticity and thermoplasticity. Article in Lee and Symonds (eds.). Plasticity New York: Pergamon, 1960, pp. 121–169.

Osgood, W. R. Combined-stress tests on 24 ST aluminum alloy tubes. J. Appl. Mech. *14* (1947) 147–153.

Phillips, A. A review of quasistatic experimental plasticity and viscoplasticity. J. Plasticity *2* (1986) 315–328.

Prager, W. The stress-st rain laws of the mathematical theory of plasticity — A survey of recent progress. J. Appl. Mech. *15* (1948) 226–233. Discussions on *16* (1949) 215–218.

Prager, W. The theory of plasticity: A survey of recent achievements (James Clayton Lecture). Proc. Inst. Mech. Eng. *169* (1955) 41–57.

Prager, W. A new method of analyzing stress and strains in work-hardening plastic solids. J. Appl. Mech. *23* (1956) 493–496. Discussions by Budiansky and Hodge *24* (1957) 481–484.

Prager, W. Non-isothermal plastic deformation. Nederl, K. Ak. Wetensch., Proc. *61* (1958) 176–192.

de Saint-Venant, B. Mémoire sur l'établissement des équations différentielles des mouvements intérieurs opérés daus les corps solides ductiles au delà des limites où l'élasticité pourrait les ramener à leur premier état. C. R. Acad. Sci. (Paris), *70* (1870) 473–480.

Shield, R. T. and H. Ziegler. On Prager's hardening rule. Z. Ang. Math. und Phys. (1958) 260–276.

Taylor, G. I. and H. Quinney. The plastic distortion of metals. Phil. Trans. Roy. Soc. (London) Ser. A *230* (1931) 323–262.

Tresca, H. Mémoire sur l'écoulement des corps solides, Mém. prés. par div. Savants *18* (1868) 733–799.

von Mises, R. Mechanik der plastischen Formanderung von Kristallen. Z. Angew. Math. Mech. *8* (1928) 161–185.

von Mises, R. Mechanik der festen Körper im plastisch deformablen Zustand. Göttinger Nachrichten, math.-phys. K1. *1913* (1913) 582–592.

Ziegler, H. A modification of Prager's hardening rule. Quart. Appl. Math. *17* (1959) 55–65.

6.4. Creep. See also Bibliography 14.8.

Andrade, E. N. da C. Creep of metals and recrystallization. Nature *162* (1948) 410.

Andrade, E. N. da C. The flow of metals. J. Iron Steel Inst. *171* (1952) 217–228.

Finnie, I. and W. R. Heller. Creep of Engineering Materials. New York: McGraw-Hill, 1959.

Hoff, N. J. A survey of the theories of creep buckling. Proc. 3rd US Nat. Congress Appl. Mech. p. 29, 1958.

6.5. Plasticity under rapid loading. See also Simmons[1.4].

Bodner, S. R. and P. S. Symonds. Plastic deformations in impact and impulsive loading of beams. In Plasticity. Proc. 2nd Symp. Naval Structural Mechanics. New York: Pergamon, 1960, pp. 488–500.

Fowler, R. G. and J. D. Hood. Very fast dynamical wave phenomenon. Phys. Rev. *128* (3) (November 1962) 991–992.

von Kármán, T. and P. Duwez. The propagation of plastic deformation in solids. J. Appl. Phys. *21* (1950) 987–994.

6.6. Additional plasticity theories (cited in Secs. 6.8–6.13).

Betten, J. Pressure-dependent yield behavior of isotropic and anisotropic materials. In Deformation and Failure of Granular Materials, P. V. Vermeer and H. J. Luger (eds.). Rotterdam: A. R. Balkema, 1982, p. 81.

Betten, J. Application of tensor functions to the formulation of yield criteria for anisotropic materials. Int. J. Plasticity 4 (1988) 29–46.

Chaboche, J. L. Visco-plastic constitutive equations for the description of cyclic and anisotropic behavior of metals. Bull. de l' Acad. Polonaise des Sciences, Ser. Sci. Techn. 25 (1):33, 1977.

Chaboche, J. L. Time independent constitutive theories for cyclic plasticity. Int. J. Plasticity 2 (2) (1986) 149–188.

Dafalias, Y. F. Modeling cyclic plasticity: Simplicity versus sophistication. In Mechanical Engineering Materials, C. S. Desai and R. H. Gallagher (eds.). New York: Wiley, 1984, p. 153.

Dafalias, Y. F. and E. P. Popov. Plastic internal variables formalism of cyclic plasticity. J. Appl. Mech. 43 (1976) 645–651.

Drucker, D. C. and Palgen. On stress-strain relations suitable for cyclic and other loadings. J. Appl. Mech. 48 (1981) 479–491.

Hill, R. A variational principle of maximum plastic work in classical plasticity. Quart. J. Mech. Appl. Math. 1 (1948) 18–28.

Koiter, W. T. General theorems for elastic-plastic solids. Progress in Solid Mechanics, Vol. 1, Sneddon and Hill (eds.). Chapter 4. Amsterdam: North Holland Publishing Co., 1960.

Il'iushin, A. A. On the increment of plastic deformation and the yield surface. PMM 24:663, 1960.

Lee, D. and F. Zavenl, Jr. A generalized strain rate dependent constitutive equation for anisotropic metals. Acta. Metall. 26 (1978) 1771–1780.

Mroz, Z. An attempt to describe the behavior of metals under cyclic loads using a more general work hardening model. Acta. Mech. 7 (1969) 199–212.

Mroz, Z. and N. C. Lind. Simplified theories of cyclic plasticity. Acta. Mech. 22 (1976) 131–152.

Mroz, Z., V. A. Norris and O. C. Zienkiewicz. Application of an anisotropic hardening model in the analyzers of elastoplastic deformation of soils. Geotechnique 29:1, 1979.

Valanis, K. C. A theory of viscoplasticity without a yield surface. Arch. Mech. 23:517, 1971.

Valanis, K. C. Fundamental consequences of a new intrinsic time measure plasticity as a limit of the endochronic theory. Arch. Mech. 32:171, 1980.

Watanabe, O. and S. N. Atluri. Internal time, general internal variable, and multiyield surface theories of plasticity and creep: A unification of concepts. Int. J. Plasticity 2 (1) (1986) 37–57.

6.7. Strain space formulations (cited in Sec. 6.14).

Casey, J. and P. M. Naghdi. On the nonequivalence of the stress space and strain space formulation of plasticity theory. J. Appl. Mech. 50 (1983) 380.

Il'iushin, A. A. On the postulate of plasticity. PMM 25:503, 1961.

Naghdi, P. M. and J. A. Trapp. The significance of formulating plasticity theory with reference to loading surfaces in strain space. Int. J. Eng. Sci. 13:785, 1975.

Naghdi, P. M. and J. A. Trapp. On the nature of normality of plastic strain rate and convexity of yield surfaces in plasticity. J. Appl. Mech. *42* (1975a) 61–66.

Yoder, R. J. and W. D. Iwan. On the formulation of strain-space plasticity with multiple loading surfaces. J. Appl. Mech. *48* (1981) 773–778.

6.8. Large plastic deformation with finite strains.

Atluri, S. N. On the constitutive relations at finite strain: Hypo-elasticity with isotropic or kinematic hardening. Comp. Meth. Appl. Mech. Eng. *43* (1984) 137–171.

Lee, E. H. Elastic-plastic deformation at finite strains. Appl. Mech. *36* (1969) 1–6.

Lee, E. H., R. L. Mallet and T. B. Wertheimer. Stress analysis of anisotropic hardening in finite-deformation plasticity. ASME, J. Appl. Mech. *50* (1983) 554–560.

Naghdi, P. M. and J. A. Trapp. Restrictions on constitutive equations of finitely deformed elastic-plastic materials. Quart. J. Mech. Appl. Math. (1975b) 28:25.

Nemat-Nasser, S. Decomposition of strain measures and their rates in finite deformation elastic-plasticity. Int. J. Solids Structures 15:155, 1979.

6.9. Plastic deformation of crystals.

Hill, R. Continuum micro-mechanics of elastoplastic polycrystals. J. Mech. Phys. Solids *13* (1965) 89–101.

Hill, R. and J. R. Rice. Constitutive analysis of clasto-plastic crystals at arbitrary strain. J. Mech. Phys. Solids *20* (1972) 401–413.

Hill, R., and K. S. Havner. Perspectives in the mechanics of elastoplastic crystals. J. Mech. Phys. Solids *30* (1982) 5–22.

Hutchinson, J. W. Elastic-plastic behavior of polycrystalline metals and composite. Proc. Roy. Soc. (London), A319–347, 1970.

Iwakuma, T. and S. Nemat-Nasser. Finite elastic plastic deformation of polycrystalline metals and composites. Proc. Roy. Soc. (London), Ser. A *394* (1984) 87–119.

Lin, T. H. A physical theory of plasticity and creep. J. Eng. Mater. Tech. 106:290, 1984.

7.1. Linear theory of elasticity. Basic references are listed in Bibliography 1.2. See especially E. Sternberg article in Goodier and Hoff, Structural Mechanics (Bibliography 1.2).

7.2. Torsion and bending. See Basic references, Bibliography 1.2; especially, Biezeno and Grammel, Love, Sechler, Sokolnikoff, Timoshenko, Weber. See Goodier's article in Flügge, Handbook of Engineering Mechanics, Bibliography 1.2. For the important subject of shear center, see the author's book, An Introduction to the Theory of Aeroelasticity (Bibliography 10.5), p. 471–475, where references to work of Goodier, Trefftz, Weinstein, etc. are given. See also the following:

Reissner, E. A note on the shear center problem for shear-deformable plates. Int. J. Solids Structures, *32* (1995) 679–682.

7.3. Stress concentration due to notches or holes. See also Sternberg in Bibliography 8.2.

Koiter, W. T. An infinite row of collinear cracks in an infinite elastic sheet. Ing. Arch. *28* (1959), 168–172.

Ling, C.-B. Collected papers. Institute of Mathematics, Academia Sinica, Taipei, Taiwan, China, 1963. Contains many articles on notches and holes.

Peterson, R. E. Stress Concentration Design Factors. New York: Wiley, 1953.

Williams, M. L. Stress singularities resulting from various boundary conditions in angular corners of plates in extension. J. Appl. Mech. *19* (1952) 526–528.

7.4. Elastic waves. Kolsky[7.4] gives a lucid and attractive introduction. Ewing, Jardetzky and Press[7.4] is comprehensive and contains a large bibliography. Davies,[7.4] and Miklowitz[7.4] also give extensive bibliography. See also Bibliographies 9.2, 11.1 and 11.2.

Brekhovskikh, L. M. Waves in Layered Media. New York: Academic Press, 1960.

Cagniard, L. Reflexion et Refraction des Ondes Seismiques Progressives. Paris: Gauthiers-Villars, 1935.

Davies, R. M. Stress waves in solids, In Surveys in Mechanics. Batchelor and Davies (eds.). Cambridge: University Press, 1956, pp. 64–138.

De Hoop, A. T. Representation Theorems for the Displacement in an Elastic Solid and Their Application to Elastodynamic Diffraction Theory, Thesis, Technische Hogeschool Te Delft 1958.

Ewing, W. M., W. S. Jardetzky and F. Press. Elastic Waves in Layered Media. New York: McGraw-Hill, 1957.

Garvin, W. W. Exact transient solution of the buried line source problem. Proc. Roy. Soc. (London) *234* (1956) 528–541.

Hayes, M. and R. S. Rivlin. Surface waves in deformed elastic materials. Arch. Rational Mech. Analysis *8* (1961) 358–380.

Kolsky, H. Stress Waves in Solids. Oxford: University Press, 1953.

Lamb, H. On the propagation of tremors over the surface of an elastic solid. Phil. Trans. Roy. Soc. (London). Ser. A *203* (1904) 1–42.

Lamb, H. On waves due to a travelling disturbance, with an application to waves in superposed fluids. Phil. Mag. Ser. 6, *13* (1916) 386–399.

Maue, A. W. Die Entspannunaswells bei plotzlichem Einschnitt eines gespannten elastischen Körpers. Z. f. Angew. Math. Mech. *34* (1954) 1–12.

Miklowitz, J. Recent developments in elastic wave propagation. Appl. Mech. Rev. *13* (1960) 865–878.

Pekeris, C. C. The seismic surface pulse. Proc. Nat. Acad. Sci. *41* (1955) 469–480.

Pekeris, C. C. and H. Lifson. Motion of the surface of a uniform elastic half-space produced by a buried pulse. J. Acoustical Soc. Am. *29* (1957) 1233–1238.

8.1. Basic references to Chapters. 8 and 9 are those of Bibliography 1.2. Original references to Lamé, Galerkin, Popkovich, Neuber, Cerrute, Lord Kelvin, and Mindlin can be found in Westergaard[1.2]. References to recent works by Iacovache, Eubanks, Naghdi etc., can be found in Sternberg[8.1] (1960). See also extensive bibliography in Truesdell and Toupin.[1.2]

Almansi, E. Sull'integrazione dell'equazione differenziale, $\Delta^{2n}u = 0$. Annali di matimatica (III) *2*, 1899.

Boussinesq, J. Applications des potentiels à l'étude de l'équilibre et du mouvement des solides elastique. Paris: Gauthier-Villars, 1885.

Biot, M. A. General solutions of the equations of elasticity and consolidation for a porous material. J. Appl. Mech. *23* (1956) 91–96.

Cosserat, E. and F. Cosserat. Sur la théorie de l'elasticité. Annales de la Faculté des Sciences de l'Université de Toulouse *10* (1896).

Dorn, W. S. and A. Schild. A converse to the virtual work theorem for deformable solids. Quart. Appl. Math. *14* (1956) 209–213.

Finzi, B., Integrazione delle equazioni indefinite della meccanica dei sistemi Continui. Rend. Lincei (6) *19* (1934) 578–584, 620–623.

Gurtin, M. E. On Helmholtz's theorem and the completeness of the Papkovich–Neuber stress functions for infinite domains. Arch. Rational Mech. Analysis *9* (1962) 225–233.

Marguerre, K. Ansatze zur Lösung der Grundgleichungen der Elastizitätstheorie, Z. Angew. Math. Mech. *35* (1955) 242–263.

Massonnett, C. General solution of the stress problem of the three-dimensional elasticity (in French). Brussels, Proc. 9th Int. Congress Appl. Mech. *5* (1956) 168–180.

Sternberg, E. and R. A. Eubanks. On stress functions for elastokinetics and the integration of the repeated wave equation. Quart. Appl. Math. *15* (1957) 149–153.

Truesdell, C. Invariant and complete stress functions for general continua. Arch. Rational Mech. Analysis *4* (1959/1960), 1–29.

8.2. Bodies of revolution. See Yu[8.2] and Sterberg's reviews.[8.2]

Mindlin, R. D. Force at a point in the interior of a semi-infinite solid. Physics, *7* (1936) 195–202; Urbana, Ill. Proc. 1st Midwestern Conf. Solid Mech., pp. 56–59, 1953.

Papkovitch, P. F. Solution générale des équations differentielles fondamentales d'élasticité, exprimée par trois fonctions harmoniques. Compt. Rend. Acad. Sci. Paris *195* (1932) 513–515, 754–756.

Sternberg, E. Three-dimensional stress concentrations in the theory of elasticity. Appl. Mech. Rev. *11* (1958) 1.

Yu, Y. Y. Bodies of revolution, Chapter 41 of Handbook of Engineering Mechanics, W. Flügge (ed.). New York: McGraw-Hill, 1962.

9.1. Complex variable technique. See list of Basic References, Bibliography 1.2, especially Muskhelishvili, Green and Zerna, Sokolnikoff, Timoshenko, and Goodier.

Lekhnitsky, S. G. Theory of Elasticity of an Anisotropic Body (in Russian). Moscow, 1950.

Muskhelishvili, N. Investigation of biharmonic boundary value problems and two-dimensional elasticity equations. Math. Ann. *107* (1932) 282–312.

Savin, G. N. Concentration of Stresses around Openings (in Russian). Moscow, 1951.

Shtaerman, I. Y. The Contact Problem of Elasticity Theory (in Russian). Moscow, 1949.

Tong, P. A Hybrid finite element method for damage tolerance analysis. Computers and Structures *19* (1-2) (1984) 263–269.

9.2. Ground shock induced by moving pressure pulse. See also Bibliography 7.4.

Ang, D. D. Transient motion of a line load on the surface of an elastic half-space. Quart. Appl. Math. *18* (1960) 251–256.

Chao, C. C. Dynamical response of an elastic half-space to tangential surface loadings. J. Appl. Mech. *27* (1960) 559–567.

Cole, J. and J. Huth. Stresses produced in a half-plane by moving loads. J. Appl. Mech. *25* (1958) 433–436.

Eringen, A. C. and J. C. Samuels. Impact and moving loads on a slightly curved elastic half-space. J. Appl. Mech. *26* (1959) 491–498.

Morley, L. S. D. Stresses produced in an infinite elastic plate by the application of loads travelling with uniform velocity along the bounding surfaces. British Aeronautical Research Committee R and M 3266, 1962.

Sneddon, I. N. Fourier Transforms. New York: McGraw-Hill, 1951, pp. 445–449.

10.1. There are many books on the calculus of variations.

Bliss, G. A. Calculus of Variations. Mathematical Association of America, 1925. Reprinted, 1944.

Courant, R. and D. Hilbert. Methods of Mathematical Physics, Vol. 1. New York: Interscience Publishers, 1953.

Gelfand, I. M. and S. V. Fomin. Calculus of Variations. Englewood Cliffs, New Jersey: Prentice Hall, Inc., 1963.

Gould, S. H. Variational Methods for Eigenvalue Problems; an Introduction to the Methods of Rayleigh, Ritz, Weinstein, and Aronszajn. Toronto: University Press, 1957.

Lanczos, C. The Variational Principles of Mechanics. Toronto: University Press, 1949.

10.2. Variational principles in elasticity were known to Lagrange, and have been developed steadily. We give some more recent references. In examples of application, Lord Rayleigh's Theory of Sound (Bibliography 1.2) is perhaps the most remarkable contribution. D. Williams' paper[10.2] illustrates the many ways a variational principle may be interpreted and stated. In Chapters 11 and 17 to 21 of the present book, the variational principle is called upon again and again to serve crucial functions.

Gurtin, M. E. A note on the principle of minimum potential energy for linear anisotropic elastic solids. Quart. Appl. Math. *20* (1963) 379–382.

Hellinger, E. Die allgemeinen Ansatze der Mechanik der Kontinua. Articles 30, in F. Klein and C. Müller (eds.). Encylopädie Mathematischen Wissenschaften, mit Einschluss ihrer Anwendungen, Vol. IV/4, Mechanik, 601–694. Teubner: Leipzig, 1914. Hellinger discusses oriented bodies (Cosserat), relativistic thermodynamics, electrodynamics, internal frictions, elastic after-effect, and finite deformation.

Langhaar, H. L. The principles of complementary energy in non-linear elasticity theory. J. Franklin Inst. *256* (1953) 255.

Prager, W. The general variational principle of the theory of structural stability. Quart. Appl. Math. *4* (1947) 378–384.

Reissner, E. On a variational theorem in elasticity. J. Math. Phys. *29* (1950) 90–95.

Reissner, E. On a variational theorem for finite elastic deformations. J. Math. Phys. *32* (1953) 129–135.

Washishu, K. Variational Methods in Elasticity and Plasticity. Oxford: Pergamon Press, 1968.

Williams, D. The relations between the energy theorems applicable in structural theory. Phil. Mag. and J. Sci., 7th Ser. *26* (1938) 617–635.

10.3. Saint-Venant's principle.

Boley, B. A. On a dynamical Saint-Venant principle. ASME Trans. Ser. E, Appl. Mech. pp. 74–78, 1960.

Hoff, N. J. The applicability of Saint-Venant's principle to airplane structures. J. Aero. Sci. *12* (1945) 455–460.

de Saint-Venant, B. Mémoire sur la Torsion des Primes. Mém. des Savants étrangers, Paris, 1855.

Southwell, R. On Castigliano' s theorem of least work and the principle of Saint-Venant. Phil. Mag. Ser. 6, *45* (1923) 193.

Sternberg, E. On Saint-Venant's principle. Quart. Appl. Math. *11* (1954) 393–402.

von Mises, R. On Saint-Venant's principle. Bull. Amer. Math. Soc. *51* (1945) 555.

10.4. Variational principles for plasticity, viscoelasticity, and heat conduction bear close resemblance to those for elasticity. See also Hill (1950) in Bibliography 6.2.

Biot, M. A. Variational and Lagrangian methods in viscoelasticity. Deformation and Flow of Solids, Colloquium Madrid, September 26–30, 1955, Grammel (ed.). Berlin: Springer-Verlag.

Biot, M. A. Variational principles for acoustic-gravity waves. Phys. Fluids *6* (1963) 772–780.

Gurtin, M. E. Variational principles in the linear theory of viscoelasticity. Arch. Rational Mech. Analysis *13* (1963) 179–191.

Herrmann, G. On variational principles in thermoelasticity and heat conduction. Quart. Appl. Math. *21* (1963) 151–155.

Hodge, P. A new interpretation of the plastic minimum principles. Quart. Appl. Math. *19* (1962) 143–144.

Kachanov, L. M. Variational Methods of Solution of Plasticity Problems (Prikl. Mat. i Mekh., May-June 1959, p. 616). PMM — J. Appl. Math. Mech. *3* (1959) 880–883.

Pian, T. H. H. On the variational theorem for creep. J. Aero. Sci. *24* (1957) 846–847.

Sanders, J. L., Jr., H. G. McComb, Jr. and F. R. Schlechte. A variational theorem for creep with applications to plates and columns. NACA Report 1342 (1958).

10.5. Variational principles applied to nonconservative systems are discussed in the following references.

Bolotin, V. V. Nonconservative Problems of the Theory of Elastic Stability. New York: Pergamon Press, 1963.

Fung, Y. C. An Introduction to the Theory of Aeroelasticity. New York: Wiley, 1956; Dover, 1969, 2nd revision, 1993.

10.6. Variational principles used in finite element approach are discussed in the following papers (cited in Sec. 10.11). Details of finite element method are discussed in Chapter 18. See also Bibliography 18.1 to 18.11.

Babuska, I. The finite element method with Lagrange multipliers. Num. Math. *20* (1973) 179–192.

Pian, T. H. H. Derivation of element stiffness matrices by assumed stress distributions. AIAA J. 2(7) (1964) 1333–1336.

Pian, T. H. H. State-of-the-art development of hybrid/mixed finite element methods, Elsevier Science B.V. *21* (1995) 5–20.

Pian, T. H. H. Survey of some recent advances in hybrid/mixed finite element methods. In Modeling and Simulation based engineering, ICES'98, S. N. Atluri and P. E. O'Donoghuie (eds.). pp. 296–301, 1998.

Pian T. H. H. and P. Tong. Basis of finite element methods for solid continua. Int. J. Num. Meth. Eng. *1* (1969) 3–28.

Pian, T. H. H. and P. Tong, Finite element methods in continuum mechanics, Adv. Appl. Mech., C. S. Yih (ed.). *12* (1972) 1–58. San Diego: Academic Press.

Tong, P. and T. H. H. Pian. A variational principle and the convergence of a finite element method based on assumed stress distribution. Int. J. Solids Structures *5* (1969) 463–472.

Tong, P. New Displacement Hybrid Finite Element Models for Solid Continua. Int. J. Num. Meth. Eng. *2* (1970) 73–83.

10.7. Extremum Principles and Limit Analysis (cited in Secs. 10.14–10.15). See also Hill (1950) in Bibliography 6.2.

Drucker, D. C., H. J. Greenberg and W. Prager. The Safety Factor of an Elastic-Plastic Body in Plane Strain. J. of Appl. Mech. 18 (1951) 371–378.

Salencon, J. Applications of the Theory of Plasticity in Soil Mechanics. Wiley, England (1977).

11.1. Waves in beams and plates. See Abramson, Plass, and Ripperger's review.[11.1]

Abramson, H. N., H. J. Plass and E. A. Ripperger. Stress propagation in rods and beams. Adv. Appl. Mech. *5* (1958) 111–194. New York: Academic Press.

Mindlin, R. D. Influence of rotatory inertia and shear on flexural motions of isotropic, elastic plates. J. Appl. Mech. *18* (1951) 31–38.

Robinson, A. Shock transmission in beams. British Aeronautical Research Committee, R and M 2265 (October 1945).

Timoshenko, S. P. On the correction for shear of the differential equation for transverse vibrations of prismatic bars. Phil. Mag. Ser. 6 *41* (1921) 744–746.

Timoshenko, S. P. On the transverse vibrations of bars of uniform cross-section. Phil. Mag. Ser. 6 *43* (1922) 125–131.

11.2. The theory of the Hopkinson's experiments, and the problem of spallation — i.e., a complete or partial separation of a material resulting from tension waves. See also Davies.[7.4]

Butcher, B. M., L. M. Barker, D. E. Munson and C. D. Lundergan. Influence of stress history on time-dependent spall in metals. AIAA J. *2* (1964) 977–990.

Davies, R. M. A critical study of the Hopkinson pressure bar. Phil. Trans. Roy. Soc. (London) *240* (1946–1948) 375–457.

Hopkinson, B. A method of measuring the pressure produced in the detonation of high explosives or by the impact of bullets. Proc. Roy. Soc. (London), Ser. A *213* (1914) 437–456.

Taylor, G. I. The testing of materials at high rates of loading. (James Forrest Lecture.) J. Inst. Civil Eng. *26* (1946) 486–519.

11.3. On generalized coordinates and direct method of solution of variational problems, the following references are recommended.

Collatz, L. The Numerical Treatment of Differential Equations, 3rd ed., Berlin: Springer, 1959.

Duncan, W. J. Galerkin's method in mechanics and differential equations. British Aeronautical Research Committee R and M 1798 (1937).

Kantorovich, L. V. and V. I. Krylov. Approximate Methods of Higher Analysis. (Translation from Russian by C. D. Benster). New York: Interscience Publishers, 1958.

Ritz, W. Über eine neue Methode zur Lösung gewisser Variations probleme der mathematischen Physik. Z. Reine u. Angew. Math. *135* (1909) 1–61.

Temple, G. The accuracy of Rayleigh's method of calculating the natural frequencies of vibrating systems. Proc. Roy. Soc. (London), Ser. A *211* (1952), 204–224.

Trefftz, E. Ein Gegenstück zum Ritzschen Verfahren. Zurich, Proc. 2nd. Int. Congress Appl. Mech., 1926, pp. 131–137.

12.1. Classical thermodynamics, with special reference to the mechanical properties of solids. There is much written on this subject; we list only a few references.

Born, M. Kritische Betrachtungen zur traditionellen Darstellung der Thermodynamik. Physik. Zeitschr. *22* (1921) 218–224, 249–254, 282–386.

Caratheodory, C. Untersuchungen über die Grundlagen der Thermodynamik. Math. Ann. *67* (1909) 355–386.

Gibbs, J. W. On the equilibrium of heterogeneous systems. Trans. Connecticut Acad. *3* (1875) 108–248, (1877) 343–524. The Collected Works of J. Willard Gibbs, Vol. 1, pp. 55–371. New Haven: Yale University Press, 1906. Reprinted 1948.

Lakes, R. Materials with structural hierarchy. Nature *361* (February 1993) 511–515.

Lakes, R. Advances in negative Poisson's ratio materials. Adv. Mater., (1993) 293–295.

Lee, T. and R. S. Lakes. Anisotropic polyurethane foam with Poisson's ratio greater than 1. J. Mater. Sci. *32* (1997) 2397–2401.

Poynting, J. H. and J. J. Thomson. A Text-Book of Physics (Properties of Matter), 4th ed., Chapter 13, Reversible thermal effects accompanying alteration in strains. London: Ch. Griffin, 1907.

Thomson, Sir William (Lord Kelvin). Mathematical and Physical Papers. Cambridge: University Press, Vol. 1, 1882.

Winterbone D. E. Advanced Thermodynamics for Engineers. New York: Wiley, 1997.

13.1. Thermodynamics of irreversible processes.

Biot, M. A. Theory of stress-strain relations in anisotropic viscoelasticity and relaxation phenomena. J. Appl. Phys. *25* (11) (1954) 1385–1391.

Biot, M. A. Variational principles in irreversible thermodynamics with application to viscoelasticity. Phys. Review *97* (6) (1955) 1463–1469.

Biot, M. A. Thermoelasticity and irreversible thermodynamics. J. Appl. Phys. *27* (3) (1956) 240–253.

Biot, M. A. Linear thermodynamics and the mechanics of solids. Proc. 3rd US Nat. Congress Appl. Mech. p. 1–18, 1958.

Biot, M. A. New thermomechanical reciprocity relations with application to thermal stress analysis. J. Aero/Space Sci. *26* (7) (1995) 401–408.

Casimir, H. B. G. On Onsager's principle of microscopic reversibility. Rev. Mod. Phys. *17* (1945) 343–350.

Chandrasekhar, S. Stochastic problems in physics and astronomy. Rev. Mod. Phys. *15* (1943) 1–89, in particular 54–56.

Curie, P. Oeuvres de Pierre Curie. Paris: Gautheir-Villars, 1908.

De Groot, S. R. Thermodynamics of Irreversible Processes. Amsterdam: North-Holland Publishing Co., 1951.

De Groot, S. R. and P. Mazur. Non-Equilibrium Thermodynamics. Amsterdam: North Holland Publishing Co., 1962.

Meixner, J. Die thermodynamische Theorie der Relaxationsercheinumgen und ihr Zusammenhang mit der Nachwirkungstheorie. Kolloid Zeit. *134* (1) (1953) 2–16.

Onsager, L. Reciprocal relations in irreversible processes. I. Phys. Rev. *37* (4) (1931) 405–426. II. Phys. Rev. *38* (12) (1931) 2265–2279.

Onsager, L. and S. Machlup. Fluctuations and irreversible processes. Phys. Rev. *91* (6) (1953) 1505–1512. Part II, 1512–1515.

Prigogine, I. Etude Thermodynamique des Phénomènes Irréversibles. Paris: Dunod., Liège: Ed. Desoer, 1947.

Prigogine, I. Le domaine de validité de la thermodynamique des phenomenes irréversibles. Physica *15* (1949) 272–289 (Netherlands).

Prigogine, I. Introduction to Thermodynamics of Irreversible Processes, 2nd ed., New York: Interscience Publishers, Wiley, 1961.

Prigogine, I. Non-Equilibrium Statistical Mechanics. New York: Interscience Publishers, Wiley, 1962. Extensive use of Feynman diagram.

Voigt, W. Fragen der Krystallphysik. I. Über die rotatorischen Constanten der Wärmeleitung von Apatit und Dolomit. Gött. Nach., 1903, p. 87.

Ziegler, H. Thermodynamik und rheologische Problems. Ing. Arch. *25* (1957) 58.

Ziegler, H. An attempt to generalize Onsager's principle, and its significance for rheological problems. Zeit. f. Ang. Math. u. Phys. *9* (1958) 748–763.

14.1. Thermoelasticity — Basic reference books.

Boley, B. A. and J. H. Weiner. Theory of Thermal Stresses. New York: Wiley, 1960. Dover, New York, 1997.

Carslaw, H. S. and J. C. Jaeger. Conduction of Heat in Solids, 2nd ed., London: Oxford University Press, 1959.

Gatewood, B. E. Thermal Stresses. New York: McGraw-Hill, 1957.

Parkus, H. Instationäre Wärmespannungen. Vienna: Springer, 1959.

Schuh, H. Heat Transfer in Structures. New York: Pergamon Press, 1964.

14.2. Additional references. Goodier's papers[14.2] are recommended for readers who wish to obtain an intuitive physical feeling about thermal stress distributions in common problems.

Goodier, J. N. On the integration of the thermo-elastic equations. Phil. Mag. Ser. 7, *23* (1937) 1017.

Goodier, J. N. Thermal stresses and deformations. J. Appl. Mech. *24* (1957) 467–474.

Hemp, W. S. Fundamental principles and methods of thermo-elasticity. Aircraft Engineering *26* (302) (April 1954) 126–127.

14.3. Aerodynamic heating is a subject of great concern to aerospace engineers. See Hoff (Bibliography 14.6) and the following:

Bisplinghoff, R. L. Some structural and aeroelastic considerations of high-speed flight (19th Wright Brothers Lecture). J. Aero. Sci. *23* (1956) 289–329.

Budiansky, B. and J. Mayers. Influence of aerodynamic heating on the effective torsional stiffness of thin wings. J. Aero. Sci. *23* (1956) 1081–1093.

14.4. Biot's Lagrangian approach to thermomechanical analysis, and other approximate methods of analysis.

Biot, M. A. New methods in heat flow analysis with application to flight structures. J. Aero. Sci. *24* (1957) 857–873.

Biot, M. A. Variational and Lagrangian thermodynamics of thermal convection fundamental shortcomings of the heat-transfer coefficient. J. Aero. Sci. *29* (1) (1962) 105–106.

Biot, M. A. Lagrangian thermodynamics of heat transfer in systems including fluid motion. J. Aero. Sci. *29* (5) (1962) 568–577.

Goodman, T. R. The heating of slabs with arbitrary heat inputs. J. Aero. Sci. *26* (3) (1959) 187–188.

Goodman, T. R. and J. J. Shea. The melting of finite slabs. J. Appl. Mech. *27* (1960) 16–24.

14.5. Physical properties of materials vary with temperature. See various standard handbooks, and the following references.

Barzelay, M. E., K. N. Tong and G. F. Holloway. Effect of pressure on thermal conductance of contact joints. NACA TN 3295 (1955).

Dorn, J. E. (ed.). Mechanical Behavior of Materials at Elevated Temperatures. New York: McGraw-Hill, 1961.

14.6. Thermal stresses at higher temperature and higher stresses usually involve creep problems.

Hoff, N. J. Approximate analysis of structures in the presence of moderately large creep deformations. Quart. Appl. Math. *12* (1954) 49–55.

Hoff, N. J. Buckling and stability. 41st Wright Memoir Lecture, J. Roy. Aero. Soc. *58* (1) (1954) 1–52.

Johnson, A. E. The creep of a nominally isotropic aluminium alloy under combined stress systems at elevated temperature. Metallurgie, *40* (1949) 125–139. Article on Magnesium alloy *42* (1951) 249–262. See also Inst. Mech. Eng. Proc. *164* (4) (1951) 432–447.

Kempner, J. Creep bending and buckling of non-linear viscoelastic columns. NACA TN 3137 (1954).

Mordfin, L. and A. C. Legate. Creep behavior of structural joints of aircraft materials under constant loads and temperatures. NACA TN 3842 (1957).

Odqvist, F. K. G. Influence of Primary Creep on Stresses in Structural Parts. Acta Polytechnica *125* (1953) Mech. Eng. Ser. *2* (9) (also Kungl. Tekniska Hogskolans Handlingar Nr 66) 18 pages.

15.1. Viscoelasticity. Gurtin and Sternberg[15.1] are particularly recommended.

Alfrey, T. Mechanical Behavior of High Polymers. New York: Interscience Publishers, 1948.

Biot, M. A. and H. Odé. On the folding of a viscoelastic medium with adhering layer under compressive initial stress. Quart. Appl. Math. *19* (1962) 351–355.

Bland, D. R. The Theory of Linear Viscoelasticity. New York: Pergamon Press, 1960.

Bodner, S. R. On anomalies in the measurement of the complex modulus. Trans. Soc. Rheology *4* (1960) 141–157.

Coleman, B. D. and W. Noll. On the foundation of linear viscoelasticity. Rev. Mod. Phys. *33* (1961) 239–249.

Drozdov, A. D. Mechanics of Viscoelastic Solids. United Kingdom: Wiley 1998.

Ferry, J. D. Viscoelastic Properties of Polymers. New York: Wiley, 1961.

Gross, B. Mathematical Structures of the Theories of Viscoelasticity. Paris: Hermann, 1953.

Gurtin, M. E. and E. Sternberg. On the linear theory of viscoelasticity. Arch. Rational Mech. Analysis *11* (1962) 291–356.

Naghdi, P. M. and S. A. Murch. On the mechanical behavior of viscoelastic-plastic solids. J. Appl. Mech. *30* (1963) 321–328.

Prager, W. On higher rates of stress and deformation. J. Mech. Phys. Solids, *10* (1962) 133–138.

Schapery, R. A. Approximate methods of transform inversion for viscoelastic stress analysis Proc. 4th US Nat. Congress Appl. Mech. 1962.

15.2. Influence of temperature variations on viscoelasticity.

Koh, S. L. and A. C. Eringen. On the foundations of nonlinear thermo-viscoelasticity. Int. J. Eng. Sci. *1* (1963) 199–229.

Landau, G. H., J. H. Weiner and E. E. Zwicky, Jr. Thermal stress in a viscoelastic-plastic plate with temperature-dependent yield stress. J. Appl. Mech. *27* (1960) 297–302.

Morland, L. W. and E. H. Lee. Stress analysis for linear viscoelastic materials with temperature variation. Trans. Soc. Rheology *4* (1960) 233–263.

Muki, R. and E. Sternberg. On transient thermal stresses in viscoelastic materials with temperature-dependent properties. J. Appl. Mech. *28* (1961) 193–207.

15.3. Reciprocal theorems and their applications.

DiMaggio, F. L. and H. H. Bleich. An application of a dynamic reciprocal theorem. J. Appl. Mech. *26* (1959) 678–679.

Graffi, D. Sul Teoremi di Reciprocità nei Fenomeni Dipendenti dal Temp. Ann. d Mathematica *4* (18) (1939) 173–200.

Greif, R. A. Dynamic Reciprocal Theorem for Thin Shells. J. Appl. Mech. *31* (1964), 724–726.

Knopoff, L. and A. F. Gangi. Seismic Reciprocity. Geophysics *24* (1959) 681–691.

Lamb, H. On reciprocal theorems in dynamics. Proc. London Math. Soc. *19* (1888) 144–151.

Strutt, J. W. (Lord Rayleigh). Some general theorems relating to vibrations. Proc. London Math. Soc. *4* (1873) 357–368.

16. Finite deformations. See Eringen, Green and Zerna, Green and Adkins, Noll and Truesdell, Prager, and Truesdell and Toupin, in the Basic Reference list, Bibliography 1.2. See also Michal (Bibliography 2.1).

Doyle, T. C. and J. L. Erickson. Nonlinear elasticity. Adv. Appl. Mech. *4* (1956) 53–115.

Kappus, R. Zur Elastizitat Theorie endlicher Verschiebungen. Z. f. Ang. Math. u. Mech. *19* (October and December 1939), 271–285, 344–361.

Kirchhoff, G. Über die Gleichungen des Gleichgewichtes eines elastischen Körpers bei nicht unendlich kleinen Verschiebungen seiner Theile. Sitzungsberichte der mathematischnaturwissenschaftlichen Klasse der Akademie der Wissenschaften, Vienna *9* (1852) 762–773.

Noll, W. On the continuity of the solid and fluid states. J. Rational Mech. Analysis *4* (1955) 3–81.

Rivlin, R. S. Mathematics and rheology, the 1958 Bingham medal address. Phys. Today *12* (5) (1959) 32–36.

Truesdell, C. The mechanical foundations of elasticity and fluid dynamics. J. Rational Mech. Analysis *1* (1952) 125–300 and *2* (1953) 593–616.

Truesdell, C. The Principles of Continuum Mechanics. Field Research Laboratory, Socony Mobil Oil Company, Inc. (Dallas, Texas), Colloquium Lectures in Pure and Applied Science, No. 5, 1960. These pungent and informal lecture notes are recommended for beginners and for those who want a taste of the lectures of this gifted speaker.

16.1. The general theory of large deformation and finite strains are discussed in books by Eringen, Green and Zerna, Green and Adkins, Prager and Truesdell and Toupin listed in Bibliography 1.2. Additional basic references (cited in Secs. 16.1–16.10) are the following:

Atluri, S. N. Alternate stress and conjugate strain measures, and mixed variational formulations involving rigid rotations, for computational analyses of finitely deformed solids, with application to plates and shells-1. Theory, Computers and Structures *18* (1) (1984) 93–116.

Atluri, S. N. and A. Cazzani. Rotations in computational solid mechanics. Arch. Comput. Meth. Eng. *2* (1) (1995) 49–138.

Green, A. E. and P. M. Naghdi. A general theory of elastic-plastic continuum. Archives for Rational Mechanics and Analysis, *18* (1965), p. 251.

Lur'e, A. I. Nonlinear Theory of Elasticity. (English translation). New York: North-Holland, 1980.

Rivlin, R. S. and C. Topakoglu. A theorem in the theory of finite elastic deformation. J. Rational Mech. Anal. *3* (1954) 581–589.

Truesdell, C. A first course in rational continuum mechanics. The John Hopkins University, Baltimore, Maryland, 1972.

Truesdell, C. The elements of continuum mechanics. Springer-Verlag, 1985.

Truesdell, C. and W. Noll. The nonlinear field theories of mechanics, Handbuch der physik, Encyclopedia of Physics, Springer-Verlag *3* (3), 1965.

Wang, C. C. and C. Truesdell. Introduction to rational elasticity. Wolters–Noordhoff, Groningen, 1972.

16.2. The constitutive equations of thermoelastic bodies (cited in Sec. 16.11). See also Eringen (1962) in Bibliography 1.2.

Atluri, S. N. On the constitutive relations at finite strain: Hypo-elasticity with isotropic or kinematic hardening. Comp. Meth. Appl. Mech. Eng. *43* (1984) 137–171.

Beatty, M. F. Topics in finite elasticity: Hyperelasticity of rubber, elastomers and biological tissues — With examples. Appl. Mech. Rev. *40* (1987) 1699–1734.

Jones, D. F. and L. R. G. Treloar. The properties of rubber in pure homogeneous strain. J. Phys. *D8* (1975) 1285–1304.

Ling, Y., P. A. Brodsky and W. L. Guo. Finding the constitutive relation for a specific elastomer. J. Electronic Packaging *115* (1993) 329–336.

Ogden, R. W. Recent advances in the phenomenological theory of rubber elasticity. Rubber Chemistry and Technology *59* (1986) 361–383.

Valanis, K. C. and R. F. Landel. The strain-energy function of a hyperelastic material in terms of the extension ratios. J. Appl. Phys. *38* (7) (1967) 2997–3002.

Wang, S. M., T. Y. P. Chang and P. Tong. Nonlinear deformation responses of rubber components by finite element analysis. Theory and Applications. ICES'95, *2* (1995) 3135–3140.

16.3. Variational principles (cited in Sec. 16.13). See also Atluri and Cazzani (1995) in Bibliography 16.1 and Green and Zenna (1968), Green and Adkins (1960) in Bibliography 1.2; Fung (1990, 1993) in Bibliography 1.3.

Koiter, W. T. On the principle of stationary complementary energy in the nonlinear theory of elasticity. SIAM J. Appl. Math. *25* (3) 1973.

Reed, K. W. and S. N. Atluri. On the generalization of certain rate-type constitutive equations for very large strains. Proceeding of the International Conference on Constitutive Laws for Engineering Materials: Theory and Application. University of Arizona. Tucson, 1983, pp. 77–88.

Rivlin, R. S. Large elastic deformations of isotropic materials. VI. Further results in the theory of torsion, shear and flexure. Phil. Trans. Roy. Soc. (London) *242* (1949) 173–195.

Tong, P. An assumed stress hybrid finite element method for an incompressible and near-incompressible material, Int. J. Solids Structures *5* (1969) 455–461.

16.4. The theory of plates and shells (cited in Sec. 16.14). One of the most important problems of plates and shells is stability or buckling. See Timoshenko and Green and Zerna in Bibliography 1.2; also Atluri (1984) in Bibliography 16.2.

Flügge, W. Stresses in Shells. Berlin: Springer, 1960.

Fung, Y. C. and E. E. Sechler. Instability of thin elastic shells. Structural Mechanics, Goodier and Hoff (eds.). New York: Pergamon Press (1960) 115–168.

Koiter, W. T. (ed.). The Theory of Thin Elastic Shells. Symp. IUTAM, Delft, 1959. Amsterdam: North-Holland Publishing Co., 1960.

Mansfield, E. H. The Bending and Stretching of Plates. New York: Pergamon Press, 1964.

von Kármán, T. Festigkeitsprobleme im Maschinenbau. Encyklopädie der mathematischen Wissenschaften, Vol. IV (4) (1910), Chapter 27, p. 349.

Ziegler, H. On the concepts of elastic stability. Adv. Appl. Mech., Vol. IV, 351–403. New York: Academic Press, 1956.

17.1. Incremental approach to solving some nonlinear problems (cited in Secs. 17.1–17.4). See also Biot (1965) in Bibliography 1.2; Pian and Tong (1969) in Bibliography 10.6; Atluri and Cazzani (1995), Truesdell and Noll (1965) in Bibliography 16.1.

Hill, R. Eigen-modal deformations in elastic-plastic continua. J. Mech. Phys. Solids *15* (1967) 371–386.

Iura, M. and S. N. Atluri. Formulation of a membrane finite element with drilling degrees of freedom. Computational Mechanics *9* (1992) 417–428.

Pian, T. H. H. and P. Tong. Variational formulation of finite displacement analysis. High Speed Computing of Elastic Structures. Fraeijs de Veubeke (ed.). University of Liege, Belgium, 1971, pp. 43–63.

17.2. Constitutive laws of incremental approach (cited in Sec. 17.5). See also Truesdel and Noll (1965) in Bibliography 16.1; Atluri (1984) in Bibliography 16.2, Reed and Atluri (1983) in Bibliography 16.3; Hill (1967), Pian and Tong (1971) in Bibliography 17.1.

Atluri, S. N. Rate complementary energy principles, finite strain plasticity problems, and finite elements. Variational Methods in the Mechanics of Solids, S. Nemat-Nasser (ed.). Pergamon Press, 1980, pp. 363–367.

Hill, R. On constitutive inequalities for simple materials — I. J. Mech. Phys. Solids *16* (1968) 229–242.

Rubinstein, R. and S. N. Atluri. Objectivity of incremental constitutive relations over finite time steps in computational finite deformation analyses. Comput. Meth. Appl. Mech. Eng. *36* (1983) 277–290.

17.3. Incremental variational principals (cited in Secs. 17.6–17.7). See also Pian and Tong (1969) in Bibliography 10.6; Atluri and Cazzani (1995) in Bibliography 16.1; Pian and Tong (1971) in Bibliography 17.1.

Atluri, S. N. On rate principles for finite strain analysis of elastic and inelastic nonlinear solids. Recent Research on Mechanical Behavior of Solids (edited by the Committee on Recent Research on Mechanical Behavior of Solids). Tokyo: University of Tokyo Press, 1979, pp. 79–107.

Atluri, S. N. On some new general and complementary energy theorems for the rate problems in finite strain, classical elastoplasticity. J. Structural Mech. *8* (1980a) 61–92.

17.4. Incompressible and merely incompressible materials (cited in Sec. 17.8). See also Tong (1969) in Bibliography 16.3.

Cazzani, A. and S. N. Atluri. Four-node mixed finite elements, using unsymmetric stresses, for linear analysis of membranes. Comput. Mech. *11* (1992) 229–251.

Seki, W. and S. N. Atluri. Analysis of strain localization in strain-softening hyperelastic materials, using assumed stress hybrid elements, Comp. Mech. *14* (6) (1994) 549–585.

17.5. Method of updated solutions (cited in Sec. 17.9). See also Hill (1967), Pian and Tong (1971) in Bibliography 17.1; Rubinstein and Atluri (1983) in 17.2; Atluri (1979, 1980a), Seki and Atluri (1994) in Bibliography 17.3.

Courant, R. and D. Hilbert. Methods of mathematical physics, Vol. I, p. 536, Interscience Publishers, 1953.

Hughes, T. J. R. and J. Winget. Finite rotation effects in numerical integration of rate constitutive equations arising in large-deformation analysis. Int. J. Num. Meth. Eng. 1981.

Key, S. W., C. M. Stone and R. D. Krieg. Dynamic relaxation applied to the quasistatic large deformation, inelastic response of axisymmetric solids. Nonlinear Finite

Element Analysis in Structural Mechanics. W. Wunderlich *et al.* (eds.). Berlin: Springer-Verlag, 1980, pp. 585–621.

Murakawa, H. and S. N. Alturi. Finite elasticity solutions using hybrid finite elements based on a complementary energy principle. ASME J. Appl. Mech. *45* (1978) 539–547.

18.1. Finite element methods. General references. See also Pian (1964, 1995), Pian and Tong (1969), Tong (1970) in Bibliography 10.6; Atluri and Cazzani (1995) in Bibliography 16.1; Atluri (1980a) in Bibliography 17.3.

Argyris, J. H. Energy theorems and structural analysis. Aircraft Eng. *26* (1954); *27* (1955).

Argyris, J. H. Continua and discontinua. Proc. Conf. Matrix Meth. Structural Mech., Air Force Institute of Technology, Wright–Patterson AFB, Ohio, 1966.

Argyris, J. H. and S. Kelsey. Energy theorems and structural analysis. London: Butterworth, 1990.

Besseling, J. F. The complete analogy between the matrix equations and continuous field equations of structural analysis. Colloque International des Techniques de Calcul Analogique et Numerique, Liege, 1963, pp. 223–242.

Bonet, J. and R. D. Wood. Nonlinear Continuum Mechanics for Finite Element Analysis. New York: Cambridge University Press, 1997.

Brebbia, C. A., J. C. F. Telles and L. C. Wrobel. Boundary element techniques. Berlin, Heidelberg, New York, Tokyo: Springer-Verlag, 1984.

Camp, C. V. and G. S. Gipson. Boundary element analysis of non-homogeneous bihamonic phenomena, Springer-Verlag, 1992.

Capon, P. and P. K. Jimack. An adaptive finite element method for the compressible Navier–Stokes equations. Numerical Methods for Fluid Dynamics 5 (1995) 327–334.

Chen, G. and J. Zhou. Boundary element methods. Academic Press, Harcourt Brace Jovanovich Publishers, 1992.

Courant, R. Variational methods for the solution of problems of equilibrium and vibrations. Bull. Am. Math. Soc. *49* (1943) 1–23.

Cruse, T. A. Numerical solutions in three-dimension elastostatics. Int. J. Solids Structures 5 (1969) 1259–1274.

Curse, T. A. and F. J. Rizzo (eds.). Boundary Integral Equation Method: Computational Applications in Applied Mechanics. New York: ASME, 1975.

Fraeij s de Veubeke, B. Upper and lower bounds in matrix structural analysis. Agardograph 72. Oxford: Pergamon, 1964, pp. 165–201.

Gallagher, R. H., J. Padlog and P. P. Bijlaard. Stress analysis of heated complex shapes. ARS J., 1962, pp. 700–707.

Gunzburger, M. D. Finite element methods for viscous incompressible flows. Academic Press, Inc., Harcourt Brace Jovanovich Publishers, 1989.

Hall, W. S. The boundary element method. Kluwer Academic Publishers, 1994.

Hallquist, J. O. LS-TAURUS. An interactive post-processor for the analysis codes LS-NIKE3D, LS-DYNA3D, LS-NIKE3D, and TOPAZ3D. Livermore Software Technology Corporation, Livermore CA, LSTC Report 1001, 1992.

Hallquist, J. O., D. W. Stillman and T. S. Lin. LS-DYNA3D User's manual. Livermore Software Technology Corporation, Livermore CA, LSTC Report 1007, January 1993.

Herrmann, L. R. A bending analysis for plates. Proc. 1st Conf. Matrix Meth. Structural Mech., AFFDL-TR-66-80. Wright–Patterson AFB, 1966, pp. 577–604.

Hughes, T. J. R. The finite element method — Linear static and dynamic finite element analysis. Englewood Cliffs, New Jersey: Prentice-Hall, Inc., 1987.

Hughes, T. and L. Franca. A new finite element formulation for computational fluid dynamics: VII. The Stokes problem with various boundary conditions: Symmetric formulations that converge for all velocity/pressure spaces. Comp. Meth. Appl. Mech. Eng. *65* (1987) 85–96.

Jones, R. E. A generalization of the direct-stiffness method of structural analysis. AIAA J. *2* (5) (1964) 821–826.

Li, L. Y. and P. Bettess. Adaptive finite element methods: A review. Appl. Mech. Rev. *50* (10) (1997) 581–591.

Mackerle, J. A finite element bibliography for biomechanics (1987–1997). Appl. Mech. Rev. *51* (10) (1998) 587–634.

Oden, J. T. A general theory of finite elements. I. Topological considerations. II. Applications. Int. J. Num. Meth. Eng. *1* (1969) 205–221, 247–259.

Prager, W. Variational principles of linear elastostatics for discontinuous displacements, strains and stresses in recent progress in applied mechanics. The Folke–Odquist Volume. B. Broberg, J. Hult and F. Niordson (eds.). Stockholm: Almquist and Wiksell, 1967, pp. 463–474.

Prager, W. Variational principles for elastic plates with relaxed continuity requirements. Int. J. Solids Structures *4* (9) (1968) 837–844.

Shakib, F., T. J. R. Hughes and Z. Johan. A new finite element formulation for computational fluid dynamics: X. The Compressible Euler and Navier–Stokes Equations. Comp. Meth. Appl. Mech. Eng. *89* (1991) 141–219.

Stillman, D. W. and J. O. Hallquist. LS-INGRID: A pre-processor and three-dimensional mesh generator for the programs LS-DYNA3D, LS-NIKE3D, and TOPAZ3D. Livermore Software Technology Corp., Livermore CA, LSTC Report 1019, 1993.

Tezduyar, T., R. Glowinski and J. Liou. Petrov-Galerkin methods on multiply connected domains for the vorticity-stream function formulation of the incompressible Navier–Stokes equations. Int. J. Num. Meth. Fluids *8* (1988) 1269–1290.

Tong, P. Liquid sloshing in an elastic container. AFOSR 66.0943, Ph.D. Thesis, California Institute of Technology, Pasadena, California, 1966.

Tong, P. The finite element method for fluid flow. Matrix Methods of Structural Analysis and Design. Gallagher, Yamada and Oden (eds.). University of Alabama Press, 1970a, pp. 786–808.

Tong, P. and T. H. H. Pian. On the convergence of a finite element method in solving linear elastic problems. Int. J. Solids Structures *3* (1967) 865–879.

Tong, P. and J. N. Rossettos. Finite element method — Basic technique and theory. MIT Press 1977.

Zienkiewicz, O. C. The finite element method in engineering science. New York: McGraw-Hill, 1971.

18.2. Specific references relevant to Secs. 18.1 to 18.4 and 18.6. See also references in Bibliography 18.1

Tong, P. Exact solution of certain problems by the finite element method. AIAA J. *7* (1970b) 178–180.

Turner, M. J., R. W. Clough, H. C. Martin and L. J. Topp. Stiffness and deflection analysis of complex structures. J. Aero. Sci. *23* (9) (1956) 805–823.

18.3. Equation Solving Methods (cited in Sec. 18.5). See also Tong and Rossettos (1977) in Bibliography 18.1.

Belytscho, T., E. J. Plaskacz, J. M. Kennedy and D. M. Greenwell. Finite element analysis on the connection machine. Comp. Meth. Appl. Mech. Eng. *81* (1990) 27–55.

Cuthill, E. and J. McKee. Reducing the bandwidth of sparse symmetric matrices, Proc. 24th Nat, Conf. Assoc. Comp. Mech. ACM Publication, 1969, pp. 157–172.

Chen, P., H. Runesha, D. T. Nguyen, P. Tong and T. Y. P. Chang. Sparse algorithms for indefinite system of linear equations. Comp. Mech. J. *15* (2000) 33–42.

Crane, H. L. Jr., N. E. Gibbs, W. G. Poole, Jr. and P. K. Stockmeyer. Algorithm 508: Matrix bandwidth and profile reduction. ACM Trans. Math. Software, p. 2, 1976.

Duff, I. S. and J. K. Reid. MA47: A Fortran code for direct solution of indefinite sparse symmetric linear systems. RAL Report #95-001, 1995.

George, J. A. and W. H. Liu. Computer solution of large sparse positive definite systems. Englewood Cliffs, New Jersey: Prentice-Hall, 1981.

Gilbert, J. R. and E. Zmijewski. A parallel graph partitioning algorithm for a message passing multiprocessor. Int. J. Parallel Programming *16* (1987) 427–449.

Golub, G. H. and C. F. VanLoan. Matrix computations. Baltimore: The Johns Hopkins University Press, 3rd ed., 1996.

Isaacson, E. and H. B. Keller. Analysis of numerical methods. New York: Wiley, 1966.

John, Z., T. J. R. Hughes, K. K. Mathur and S. L. Johnson. A data parallel finite element method for computational fluid dynamics on the connection machine system. Comp. Meth. Appl. Mech. Eng. *99* (1992) 113–134.

Law, K. H. and D. R. Mackay. A parallel row-oriented sparse solution method for finite element structural analysis. IJNM Eng. *36* (1993) 2895–2919.

Lewis, J. G., B. W. Pfyten and A. Pothen. A fast algorithm for reordering sparse matrices for parallel factorization. SIAM J. Sci. Statist. Comput. *6* (1989) 1146–1173.

Liu, J. W. H. Reordering sparse matrices for parallel elimination. Technical Report 87-01, Computer Science, York University, North York, Ontario, Canada, 1987.

Ng E. and B. W. Peyton. A supermodal Cholesky factorization algorithm for shared-memory multi-processers. SIAM J. Sci. Comput. *14* (4) (1993) 761–769.

Nguyen, D. T., J. Qin, T. Y. P. Chang and P. Tong. Efficient sparse equation solver with unrolling strategies for computational mechanics. CEE Rep#96-001, Old Dominion University, VA, 1996.

Noor, A. K. Parallel processing in finite element structural analysis. Parallel Computations and Their Impact on Mechanics, ASME, 1987, pp. 253–277.

Tong, P. On the numerical problems by the finite element methods. Computer-Aided Engineering, edited by Gladwell, University of Waterloo, Waterloo, Canada, 1970c pp. 539–560.

Zheng, D. and T. Y. P. Chang. Parallel Cholesky method on MIMD with shared memory. Computers and Structures *56* (1) (1985) 25–38.

18.4. General finite element formulation (cited in Sec. 18.7). See also Tong (1984) in Bibliography 9.1.

Tong, P. A hybrid crack element for rectilinear anisotropic material. Int. J. Num. Meth. Eng. *11* (1977) 377–403.

Tong, P., T. H. H. Pian and S. Lasry. A hybrid-element approach to crack problems in plane elasticity. Int. J. Num. Meth. Eng. *7* (1973) 297–308.

18.5. Convergence (cited in Sec. 18.8). See also Tong and Pian (1967), Oden (1969) in Bibliography in Bibliography 18.1

Babuska, I. Error-Bounds for finite element method. Numerische Math. *16* (1971) 322–333.

Oden, J. T. A general theory of finite elements. I. Topological considerations. II. Applications. Int. J. Num. Meth. Eng. *1* (1969) 205–221, 247–259.

Strang, W. G. and G. Fix. An analysis of the finite element method. Englewood Cliffs, New Jersey: Prentice-Hall, 1973.

Szabo, B. A. and G. J. Sahrmann, Hierarchical plate and shell models based on p-extension. Int. J. Num. Meth. Eng. *26* (1988) 1858–1881.

18.6. Two-dimensional shape functions, element matrices, elements of various shapes, plane and axisymmetric problems, rotation, three-dimensional shape functions and the solution of three-dimensional problems (cited in Secs. 18.9–18.16). See also Tong and Rossettos (1977), Zienkiewicz (1971) in Bibliography 18.1.

Coons, S. A. Surfaces for computer aided design of space form. MIT Project MAC, MAC-TR-41, 1967.

Gallagher, R. H. A correlation study of methods of matrix structural analysis. Oxford: Pergamon, 1974.

Gallagher, R. H. Finite element analysis: Fundamentals. Englewood Cliffs, New Jersey: Prentice-Hall, 1975.

Irons, B. M. Numerical integration applied to finite element methods. Conf. Use Digital Comput. Structural Eng., University of Newcastle, 1966.

Irons, B. M. Engineering application of numerical integration in stiffness method. AIAA J. *14* (1966) 2035–2037.

Taig, I. C. Structural analysis by the matrix displacement method. England Electric Aviation Report No. SO17, 1961.

Taylor, R. L., P. J. Beresford and E. L. Wilson. A Noncomforming Element for Solving the Elastic Problem. Int. J. for Num. Meth. in Eng. 10 (1976) 1211–1219.

Zienkiewicz, O. C. and R. L. Taylor. The finite element method. London, New York: McGraw-Hill Book Company, Vol. I, II, 1989.

18.7. Dynamic problems of elastic solids (cited in Sec. 18.17). See also Hughes (1987) in Bibliography 18.1; Tong (1970c) in Bibliography 18.3.

Clough, R. W. and K. J. Bathe. Finite element analysis of dynamic response. Chapter in Advances in Computational Methods in Structural Mechanics and Design. J. T. Oden, R. W. Clough and Y. Yamamoto (eds.). Huntsville: University of Alabama Press, 1972, pp. 153–180.

Dahlquist, G. A special stability problem for linear multi-step methods. BIT *3* (1963) 27–43.

Fried, I. and D. S. Malkus. Finite element mass matrix lumping by numerical integration without convergence rate loss. Int. J. Solids Structures *11* (1976) 461–466.

Gear, C. W. Numerical initial value problems in ordinary differential equations. Englewood Cliffs, New Jersey: Prentice-Hall, 1971.

Geradin, M. A classification and discussion of integration operators for transient structural response. AIAA Paper 74–105, AIAA 12th Aerospace Sciences Meeting, Washington, D. C., 1974.

Goudreau, G. L. Evaluation of discrete methods for the linear dynamic response of elastic and viscoelastic solids. UC SESM Report 69-15, University of California, Berkeley, June 1970.

Goudreau, G. L. and R. L. Taylor. Evaluation of numerical methods in elastodynamics. J. Comput. Meth. Appl. Mech. Eng. *2* (1973) 69–97.

Hilber, H. M., T. J. R. Hughes and R. L. Taylor. Improved numerical dissipation for time integration algorithms in structural dynamics. Earthquake Eng. Structural Dynamics, *5* (1977) 283–292.

Hinton, E., T. Rock and O. C. Zienkiewicz, A note on mass lumping and related processes in the finite element method. Earthquake Eng. Structural Dynamics *4* (1976) 245–249.

Hughes, T. J. R. Analysis of transient algorithms with particular reference to stability behavior. Comp. Meth. Transient Analysis. T. Belytschko and T. J. R. Hughes (eds.). Amsterdam: North-Holland, 1983, pp. 67–155.

Hughes, T. J. R., H. M. Hilber and R. L. Taylor. A reduction scheme for problems of structural dynamics. Int. J. Solids Structures *12* (1976) 749–767.

Krieg, R. D. Unconditional stability in numerical time integration methods. J. Appl. Mech. *40* (1973) 417–421.

Krieg, R. D. and S. W. Key. Transient shell response by numerical time integration. Int. J. Num. Meth. Eng. *7* (1973) 273–286.

Newmark, N. M. A method of computation for structural dynamics. Journal of Engineering Mechanics Division, ASCE, (1959), 67–94.

Park, K. C. Evaluating time integration methods for nonlinear dynamic analysis. Finite Element Analysis of Transient Nonlinear Behavior. T. Belytschko *et al.* (eds.). AMD 14, New York: ASME, 1975, pp. 35–38.

Rossettos, J. N. and H. Weinstock. Finite element three-dimensional dynamic frame model for vehicle crashworthiness prediction. Proc. 3rd Int. Conf. Vehicle System Dynamics. V.P.I., Blacksburg, Virginia, August, 1974.

Tong, P. and J. N. Rossettos. Modular approach to structural simulation for vehicle crashworthiness prediction. Report No., DOT-TSC-NHTSA-74-7, HS-801-475, 1974.

Tong, P., T. H. H. Pian and L. L. Bucciarelli. Mode shapes and frequencies by finite element method using consistent and lumped masses. J. Comp. Structures *1* (1971) 623–638.

Wilson, E. L. A computer program for the dynamic stress analysis of underground structures. SESM Report No. 68-1, Division of Structural Engineering and Structural Mechanics, University of California, Berkeley, 1968.

18.8. Numerical integration methods (cited in Sec. 18.18). See Irons (1966) in Bibliography 18.6.

Cowper, G. R. Gaussian quadrature formulas for triangles. Int. J. Num. Mech. Eng. 7 (1973) 405–408.

Felippa, C. A. Refined finite element analysis of linear and non-linear two-dimensional structures. Structures Materials Research Report No. 66-22, University of California, Berkeley, 1966.

Hammer, P. C., O. P. Marlowe and A. H. Stroud. Numerical integration over simplexes and cones. Math. Tables Aids Comp. 10 (1956) 130–107.

Irons, B. M. Quadrature rules for brick based finite elements. Int. J. Num. Meth. Eng. 3 (1971) 293–294.

18.9. Patch test (cited in Sec. 18.19). See also Hughes (1987) in Bibliography 18.1; Irons (1966) in Bibliography 18.6.

Irons, B. M. Numerical integration applied to finite element methods. Conf. Use Digital Comp. Structural Eng. University of Newcastle, 1966a.

Robinson, J. and S. Blackham. An evaluation of lower order membranes as contained in the MSC/NASTRAN, ASAS, and PAFECFEM Systems. Robinson and Associates, Dorset, England, 1979.

18.10. Locking free elements (cited in Sec. 18.20). See also Hughes (1987); Tong and Rossettos (1977) in Bibliography 18.1; Taylor et at. (1976) in Bibliography 18.6.

Babuska, I. and M. Suri. On locking and robustness in the finite element method. Report BN-1112, Inst. Phys. Sci. Tech. University of Maryland, College Park Campus, 1990.

Barlow, J. More on optimal stress points — Reduced integration, element distortions, and error estimation. Int. J. Num. Meth. Eng. 28 (1989) 1487–1504.

Bazely, G. P., Y. K. Cheung, B. M. Irons and O. C. Zienkiewicz. Triangular elements in plate bending. Conforming and nonconforming solutions. Proc. 1st Conf. Matrix Meth. Structural Mech. pp. 547–576, AFFDLTR-CC-80, Wright–Patterson AFB, OH, 1966.

Belytschko, T., C. S. Tsay and W. K. Liu. A stabilization matrix for the bilinear Mindlin plate element. Comp. Meth. Appl. Meth. Eng. 29 (1981) 313–327.

Belytschko, T. and J. S. J. Ong and W. K. Liu. A consistent control of spurious singer modes in the nine-node Lagrange element for the Laplace and Mindlin plate equations. Comp. Meth. Appl. Mech. Eng. 44 (1984) 269–295.

Doherty, W. P., E. L. Wilson and R. L. Taylor. Stress analysis of axisymmetric solids using higher order quadrilateral finite elements. University of California, Berkeley, Structural Engineering Laboratory Report SESM, 1969.

Hughes, T. J. R. Generalization of selective integration procedures to anisotropic and nonlinear media. Int. J. Num. Meth. Eng. 15 (1980) 1413–1418.

MacNeal, R. H. A simple quadrilateral shell element. Comp. Struct. 8 (1978) 175–183.

MacNeal, R. H. The shape sensitivity of isoparametric elements. Finite Element Methods in the Design Process, Proc. 6th World Congress Finite Element Meth., Banff, Alberta, Canada, 1990.

MacNeal, R. H. Finite elements – Their design and performance. New York, Brazil, Hong Kong: Marcel Dekker Inc., 1994.

Malkus D. S. and T. J. R. Hughes. Mixed finite element methods – Reduced and selective integration techniques: A unification of concepts. Comp. Meth. Appl. Mech. Eng. *15* (1978) 68–81.

Nagtegaal, J. C., D. M. Parks and J. R. Rice. On numerically accurate finite element solutions in the fully plastic range. Comp. Meth. Appl. Mech. Eng. (1974) 153–178.

Prathap, G. The finite element method in structural mechanics, Kluwer Academic Publishers, 1993.

Razzaque, A. Program for triangular bending elements with derivative smoothing. Int. J. Num. Meth. Eng. *6* (1973) 333–343.

Zienkiewicz, O. C., J. Too and R. L. Taylor. Reduced integration technique in general analysis of plates and shells. Int. J. Num. Meth. Eng. *3* (1971) 275–290.

18.11. Spurious modes in reduced integration (cited in Sec. 18.21). See also Zienkiewicz and Taylor (1989) in Bibliography 18.6; MacNeal (1994) in Bibliography 18.10.

Barlow, J. A stiffness matrix for a curved membrane shell, Conf. Recent Adv. Stress Analysis, Royal Aeron. Soc., 1968.

Barlow, J. Optimal stress locations in finite element models, Int. J. Num. Meth. Eng. *10* (1976) 243–245.

Bic'anic', N. and E. Hinton. Spurious modes in two-dimensional isoparametric elements. Int. J. Num. Meth. Eng. *14* (1979) 1545–1547.

Irons, B. M. and T. K. Hellen. On reduced integration in solid isoparametric elements when used in shells with membrane modes. Int. J. Num. Meth. Eng. *10* (1978) 1179–1182.

19.1. Mixed and hybrid formulations. See also Pian (1995, 1998), Tong and Pian (1969) in Bibliography 10.6; Fraeijs de Veubeke (1964) in Bibliography 18.1; Babuska (1971) in Bibliography 18.5.

Brezzi, F. On the existence, uniqueness and approximation of saddle point problems arising from Lagrangian multipliers. RAIRO, 8-R2, 1974, pp. 129–151.

Fraeijs de Veubeke, B. Displacement and equilibrium models in finite element method. Chapter 9 of Stress Analysis. O. C. Zienkiewicz and C. S. Holister (eds.). Wiley, 1965, pp. 145–197.

Fraeijs de Veubeke, B. and O. C. Zienkiewicz. Strain energy bounds in finite element analysis. J. Strain Analysis *2* (1967) 265–271.

Zienkiewicz, O. C., J. P. Vilotte, S. Toyoshima and S. Nakazawa. Iterative method for constrained and mixed approximation. An inexpensive improvement of FEM performance. Comp. Meth. Appl. Mech. Eng. *51* (1985) 3–29.

Zienkiewicz, O. C., K. L. Xi and S. Nakazawa. Iterative solution of mixed problems and stress recovery procedures. Comm. Appl. Num. Meth. Eng. *1* (1985a) 3–9.

19.2. Formulation. See also Tong (1984) in Bibliography 9.1; Pian (1964), Pian and Tong (1969, 1972), Tong (1970), Tong and Pian (1969) in Bibliography 10.6; Atluri and Cazzani (1995) in Bibliography 16.1; Atluri (1980) in Bibliography 17.2; Atluri (1980a) in Bibliography 17.3; Tong and Rossettos (1977) in Bibliography 18.1; Tong *et al.* (1973) in Bibliography 18.4.

Chen, H. S. and C. C. Mei. Oscillations and wave forces in a man-made harbor in the open sea. Proc. 10th Naval Hydrodynamic Symp., Massachusetts: MIT., Cambridge, June 1974.

Pian, T. H. H. and D. P. Chen. Alternative ways for formulation of hybrid elements. Int. J. Num. Meth. Eng. *18* (1982) 1679–1684.

Pian, T. H. H. and K. Sumihara. Rational approach for assumed stress finite elements. Int. J. Num. Meth. Eng. *20* (1984) 1585–1695.

Tong, P. and Y. C. Fung. Slow viscous flow and its application to biomechanics. J. Appl. Mech. *38* (1971) 721–728.

Tong, P. and T. H. H. Pian. On the convergence of the finite element method for problems with singularities. Int. J. Solids Structures *9* (1972) 313–322.

19.3. Hybrid singular elements, displacement element, and alternative formulation. See also Tong (1984) in Bibliography 9.1; Tong and Pian (1972) in Bibliography 10.6; Tong (1977), Tong *et al.* (1973) in Bibliography 18.4.

Barsoum, R. S. On the use of isoparametric finite elements in linear fracture mechanics. Int J. Num. Meth. Eng. *10* (1976) 25–38.

Barsoum, R. S. Triangular quarter point elements as elastic and perfectly elastic crack tip elements. Int. J. Num. Meth. Eng. *11* (1977) 85–98.

Henshell, R. D. and K. G. Shaw. Crack tip elements are unnecessary. Int. J. Num. Meth. Eng. *9* (1975) 495–509.

Jeong, D. Y., J. Zhang, N. Katsube, G. W. Neat, T. H. Flournoy and P. Tong. Corrosion Fatigue Interaction. Proc. 1995 USAF Structural Integrity Program Conf., WL-TR-96-4094, *11* (1996) 1113—1127.

Jirousek, S. J. Hybrid-Treffiz plate bending elements with *p*-method capabilities, Int. J. Num. Meth. Eng. *24* (1987) 1367–1393.

Key, S. W. A convergence investigation of the direct stiffness method. Ph.D. Thesis, University of Washington, Seattle 1966.

Lin, K. Y. and P. Tong. Singular finite elements for the fracture analysis of *v*-notched plate. Int. J. Num. Meth. Eng. *15* (1980) 1343–1354.

Muskhelishvili, N. I. Some basic problems of the theory of elasticity. (Translation by J. R. M. Radok.) Groningen, Netherlands: Noordhoff, 1953.

Orringer, O., K. Y. Lin and P. Tong. *K*-solution with assumed stress hybrid elements. Journal of the Structural Division, ASCE, 103, (1977) 321–324.

Piltner, R. Special finite elements with holes and internal cracks, Int. J. Num. Meth. Eng. *21* (1985) 1471–1485.

Sih, G. C. and H. Liebowitz. Mathematical theories of brittle fracture. Fracture (edited by H. Liebowitz), *11* (1968) 66–188.

19.4. Elements for heterogeneous materials.

Accorsi, M. L. A method for modeling microstructural material discontinuities in a finite element analysis, Int. J. Num. Meth. Eng. *26* (1988) 2187–2197.

Ghosh, S. and S. N. Mukhopadhyay. A two dimensional automatic mesh generator for finite element analysis for random composites. Computers and Structures. *41* (1991) 241–256.

Ghosh, S. and R. L. Mallett. Voronoi cell finite element. Computers and Structures *50* (1994) 33–46.

Ghosh, S. and S. Moorthy. Elastic-plastic analysis of arbitrary heterogeneous materials with the voronoi cell finite element method. Comp. Meth. Appl. Mech. Eng. *121* (1995) 373–409.

Ghosh, S. and Y. Liu. Voronoi cell finite element model based on micropolar theory of thermoelasticity for heterogeneous materials, Int. J. Num. Meth. Eng. *38* (1995) 1361–1398.

Tong, P. and C. C. Mei. Mechanics of Composites of Multiple Scales. Int. J. Comput. Mech. *9* (1992) 195–210.

Zhang, J. and N. Katsube. Problems related to application of eigenstrains in a finite element analysis, Int. J. Num. Meth. Eng. *37* (1994) 3185–3195.

19.5. Element for infinite domain. See also Chen and Mei (1994) in Bibliography 19.2.

Mei, C. C. and H. S. Chen. Hybrid element for water waves. Proc. ASCE Symp. Modeling Techniques. San Francisco, September 1975.

Stoker, J. J. Water waves. New York: Interscience, 1957.

19.6. Incompressible or nearly incompressible materials. See also Tong (1969) in Bibliography 16.3; Hughes (1987) in 18.1; Babuska (1971) in Bibliography 18.5; Malkus and Hughes (1978) in Bibliography 18.10.

Brezzi, F. and J. Pitkaranta. On the stabilization of finite element approximations of the stokes equations. Rept-MAT-A219, Helsinki University of Technology, Institute of Mathematics, Finland, 1984.

Herrmann, L. R. Elasticity equations for nearly incompressible materials by a variational theorem. AIAA J. *3* (1965) 1896–1900.

Oden, J. T. and G. F. Carey. Finite elements — Mathematical aspects, IV. Englewood Cliffs, New Jersey: Prentice-Hall, 1984.

20.1. Bending of thin plates. See also references in Bibliography 16.14; Pian (1964) in Bibliography 10.6; Argyris (1966) in Bibliography 18.1; Tong (1970c) in Bibliography 18.3; Bazeley *et al.* (1966) in Bibliography 18.10.

Bell, K. A Refined Triangular Plate Bending Finite Element. Int. J. Num. Meth. Eng. *1* (1) (1969) 101–122 [also *1* (4) (1969) 395 and *2* (2) (1969) 146–147].

Cheung, Y. K., I. P. King, and O. C. Zienkiewicz. Slab bridges with arbitrary shape and support conditions — A general method of analysis based on finite elements. Proc. Inst. Civl. Eng. *40* (1968) 9–36.

Clough, R. W. and J. L. Tocher. Finite Element Stiffness Matrices for Analysis of Plate Bending. Proc. 1st. Conf. Matrix Meth. Structural Mech., AFFDL-TR-66-80, Wright–Patterson AFB, 1966, pp. 515–546.

Clough, R. W. and C. A. Felippa. A refined quadrilateral element for analysis of plate bending. Proc. 2nd Conf. Matrix Meth. Structural Mech., Air Force Institute of Technology, Wright–Patterson A. F. Base, Ohio, 1968.

Cowper, G. R., E. Kosko, G. M. Lindberg and M. D. Olson. Static and Dynamic Applications of a High-Precision Triangular Plate Element. AIAA J. *7* (10) (1969) 1957–1965.

Dawe, D. J. Parallelogram element in the solution of rhombic cantilever plate problems, J. Strain Analysis *1* (1966) 223–230.

Hinton, E. and H. C. Huang. A family of quadrilateral Mindlin plate elements with substitute shear strain fields. Computers and Structures *23* (1986) 409–431.

Holand, I. and K. Bell (eds.). Finite element methods in stress analysis. Tapir, Technical University of Norway, Trondheim. 1972.

Mau, S. T., T. H. H. Pian and P. Tong. Vibration analysis of laminated plates and shell by a hybrid stress element. AIAA J. *2* (1973) 1450–1452.

Morley, L. S. D. The constant moment plate bending element. J. Strain Analysis *6* (1971) 20–24.

Rossettos, J. N., P. Tong and E. Perl, Finite element analysis of the strength and dynamic behavior of filamentary composite structures. Final Report, NASA grant no. NGR 22-011-073, November, 1972.

Samuelsson, A. The global constant strain condition and the patch test. Energy Methods in Finite Element Analysis. R. Glowinski, E. Y. Rodin and O. C. Zienkiewicz (eds.), Chapter 3, Chichesterm: Wiley, 1979, pp. 49–68.

Specht, B. Modified shape functions for the three node plate bending element passing the patch test. Int. J. Num. Meth. Eng. *26* (1988) 705–715.

Spilker, R. L. and N. I. Munir. The hybrid-stress model for thin plates, Int. J. Num. Meth. Eng. *15* (8) (1980) 1239–1260.

Zienkiewicz, O. C., R. L. Taylor, P. Papadopoulos and E. Onate. Plate bending elements with discrete constraints; new triangular elements. Computers and Structures *35* (1990) 505–522.

20.2. Reissner–Mindlin plates. See also MacNeal (1978), Zienkiewicz *et al.* (1971) in Bibliography 18.10.

Cook, R. D. Concepts and applications of finite element analysis. New York: Wiley, 1974.

Hughes, T. J. R. and M. Cohen, The heterosis finite element for plate bending, Computers and Structures *9* (1978) 445–450.

Hughes, T. J. R., R. L. Taylor and W. Kanoknukulchai, A simple and efficient element for plate bending, Int. J. Num. Meth. Eng. *11* (1977) 1529–1543.

Hughes, T. J. R. and T. E. Tezduyar, Finite elements based upon Mindlin plate theory with particular reference to the four-node bilinear isoparametric element, J. Appl. Mech., 1981, pp. 587–596.

MacNeal, R. H. A simple quadrilateral shell element. Computers and Structures *8* (1978) 175–183.

MacNeal, R. H. Derivation of element stiffness matrices by assumed strain distributions. Nucl. Eng. Design *70* (1982) 3–12.

MacNeal R. H. and R. L. Harder. Eight nodes or nine, Int. J. Num. Meth. Eng. *33* (1992) 1049–1058.

Mindlin, R. D. Influence of rotary inertia and shear in flexural motions of isotropic elastic plates, J. Appl. Mech. *18* (1951) 31–38.

Pawsey, S. F. and R. W. Clough. Improved numerical integration of thick shell finite elements. Int. J. Num. Meth. Eng. *3* (4) (1971) 575–586.

Pian, T. H. H. and P. Tong. Reissner principle in finite element method. Mechanics Today *5* (1980) 377–397.

Pian, T. H. H. and P. Tong. Mixed and hybrid elements. Finite Element Handbook, McGraw Hill, 1987, pp. 2.173–2.203.

Pugh, E. D. L., E. Hinton and O. C. Zienkiewicz. A study of quadrilateral plate bending elements with reduced integration. Int. J. Num. Meth. Eng. *12* (1978) 1059–1079.

Reissner, E. The effect of transverse shear-deformation on the bending of elastic plates, J. Appl. Mech. *12* (A69–A77) 1945; *13* (A252) 1946.

Tong, P. On the numerical problems by the finite element methods. Computer-aided Engineering edited by Gladwall, University of Waterloo, Waterloo, Canada, 1970, pp. 539–560.

Zienkiewicz, O. C. and E. Hinton. Reduced integration, function smoothing and non-conformity in finite element analysis. J. Franklin Inst. *302* (1976) 443–461.

Zienkiewicz, O. C., R. L. Taylor and J. M. Too. Reduced integration technique in general analysis of plates and shells. Int. J. Num. Meth. Eng. *3* (2) (1971) 275–290.

20.3. Mixed functional of Reissner plate theory. See also Pian and Tong (1972), Tong and Pian (1969) in Bibliography 10.6; Herrmann (1966) in Bibliography 18.1; Babuska (1971) in Bibliography 18.5; Brezzi (1974) in Bibliography 19.1; Pian and Tong (1980, 1987) in Bibliography 20.2.

Arnold, D. N. and R. S. Falk. A uniformly accurate finite element method for Mindlin–Reissner plate. IMA Preprint Series No. 307, Institute for Mathematics and its Applications, University of Minnesota, April 1987.

Xu, Z. A simple and efficient triangular finite element for plate bending. Acta Mechanica Sinica, *2* (1986) 185–192.

Zienkiewicz, O. C. and D. Lefebvre. A robust triangular plate bending element of the Reissner–Mindlin type. Int. J. Num. Meth. Eng. *26* (1988) 1169–1184.

20.4. Hybrid formulation for plates. See also Pian (1964), Pian and Tong (1969, 1972), Tong (1970) in Bibliography 10.6; Tong and Rossettos (1977) in Bibliography 18.1; Zienkiewicz and Taylor (1989) in Bibliography 18.6; Pian and Tong (1980) in Bibliography 20.2.

Babuska, I. and T. Scapolla. Benchmark computation and performance evaluation for a rhombic plate bending problem. Int. J. Num. Meth. Eng. *28* (1989) 155–179.

Pian, T. H. H. Some recent advances in assumed stress hybrid finite elements, Symp. Adv. Aerospace Sci. P. Hajela and S. C. Mcintosh Jr. (eds.)., 1993, pp. 397–404.

Tang, L. M., W. J. Chen and Y. X. Liu, String net function approximation and quasi-conforming technique. Hybrid and Mixed Finite Element Methods. S. N. Atluri, R. H. Gallagher and O. C. Zienkiewicz (eds.), New York: Wiley, 1983, pp. 173–188.

20.5. Shells as an assembly of plate elements. See also Iura and Atluri (1992) in Bibliography 17.1; Atluri (1980a) in Bibliography 17.3; MacNeal (1994) in Bibliography 18.10; Pian (1993) in Bibliography 20.4.

Abu-Gazaleh, B. N. Analysis of plate-type prismatic structures. Ph.D. Dissertation, University of California, Berkeley, 1965.

Allman, D. J. A quadrilateral finite element including vertex rotations for plane elasticity analysis, Int. J. Num. Meth. Eng. *26* (1988) 2645–2655.

Atluri, S. N. On some new general and complementary energy principles for the rate problems of finite strain, classical elastoplasticity, J. Structural Mech. *8* (1980) 61–92.

Bergan, P. G. and C. A. Felippa, A triangular membrane element with rotational degrees of freedom, Comp. Meth. Appl. Mech. Eng. *50* (1985) 25–69.

Chen, D. P. and Y. S. Pan, Formulation of hybrid/mixed membrane elements with drilling degrees of freedom, paper presented at ICES'92, Hong Kong, 1992.

Cook, R. D. On the Allman triangle and a related quadrilateral element. Computers and Structures *22* (1986) 1065–1067.

Cowper, G. R., G. M. Lindberg and M. D. Olson, A shallow shell finite element of triangular shape, Int. J. Solids Structures *6* (1970) 1133–1156.

Gilewski, W. and R. Radwanska. A survey of finite element models for the analysis of moderately thick shells, Finite Element Analysis and Design *9* (1991) 1–21.

Hughes, T. J. R. and F. Brezzi, On drilling degrees of freedom, Comp. Meth. Appl. Mech. Eng. *72* (1989) 105–121.

Ibrahimbegovic, A., R. L. Taylor and E. L. Wilson, A robust quadrilateral membrane finite element with drilling degrees of freedom, Int. J. Num. Meth. Eng. *30* (1990) 445–457.

MacNeal, R. H. and R. L. Harder, A refined four-node membrane element with rotational degrees of freedom. Computers and Structures *28* (1988) 75–84.

Mohr, G. A. Finite element formulation by nested interpolations: Application to the drilling freedom problem. Computers and Structures *15* (1982) 185–190.

Robinson, J. Four-node quadrilateral stress membrane element with rotational stiffness, Int. J. Num. Meth. Eng. *16* (1980) 1567–1569.

Sze, K. Y. and A. Ghali, Hybrid plane quadrilateral element with corner rotations, J. Structural Eng. ASCE *119* (1993) 2552–2572.

20.6. General shell elements. See also Gilewski and Radwanska (1991) in Bibliography 20.5.

Ahmad, S., B. M. Irons and O. C. Zienkiewicz. Analysis of thick and thin shell structures by curved finite elements. Int. J. Num. Meth. Eng. *3* (2) (1970) 275–290.

Atluri, S. N. and T. H. H. Pian. Theoretical formulation of finite element methods in linear-elastic analysis of general shells, J. Structural Mech. *1* (1972) 1–41.

Chang, T. Y., A. F. Saleeb and W. Graf. On the mixed formulation of a nine-node Lagrange shell element, Comp. Meth. Appl. Mech. Eng. *73* (1989) 259–281.

Ibrahimbegovic, A. Stress resultant geometrically exact shell theory for finite rotations and its finite element implementation. Appl. Mech. Rev. *50* (4) (1997) 199–226.

Jang, J. and P. M. Pinsky, An assumed covariant strain based nine-node shell element, Int. J. Num. Meth. Eng. *24* (1987) 2389–2411.

Zienkiewicz, O. C., J. Too and R. L. Taylor. Reduced integration technique in general analysis of plates and shells. Int. J. Num. Meth. Eng. *3* (1970) 275–290.

20.7. Locking and Stabilization in shell applications. See also Hughes and Tezduyar (1981) in 20.2.

Belytschko, T. and C. S. Tsay. A stabilization procedure for quadrilateral plate element with one point quadrature. Int. J. Num. Meth. Eng. *19* (1983) 405–419.

Belytschko, T. and J. Lin and C. S. Tsay. Explicit algorithm for nonlinear dynamics of shells, Comp. Meth. Appl. Mech. Eng. *42* (1984) 225–251.

Belytschko, T., B. L. Wong and H. Stolarski. Assumed strain stabilization procedure for the nine-node Lagrange shell element. Int. J. Num. Meth. Eng. *28* (1989) 385–414.

21.1. Updated Lagrangian solution (cited in Secs. 21.1–21.2). See also Bibliography 17.1–17.5; Timoshenko and Gere (1961) in Bibliography 1.2; Fung and Sechler (1960) in Bibliography 16.4; Pian and Tong (1971) in Bibliography 17.1.

Bergan, P. G. Solution by iteration in displacement and load spaces. In Nonlinear Finite Element Analysis in Structural Mechanics. Wunderlich, Stein and Bathe, (eds.).New York: Springer-Verlag (1981) 63–89.

Oden, J. T. Finite plane strain of incompressible elastic solids by the finite element method, Aero. Q. *19* (1967) 254–264.

Zienkiewicz, O. C. Incremental displacement in non-linear analysis. Int. J. Num. Meth. Eng. *3* (1971) 387–392.

21.2. Newton–Raphson iteration. (cited in Sec. 21.4)

Matthies, H. and G. Strang. The solution of nonlinear finite element equations. Int. J. Num. Mech. Eng. *14* (1979) 1613–1626.

Rafston, A. A first course in numerical analysis. New York: McGraw-Hill, 1965.

21.3. Viscoplasticity, plasticity, viscoelasticity, and creep (cited in Secs. 21.5–21.8). See also Bibliographies 6.1–6.9 and 15.1–15.3.

Cormeau, I. Numerical stability in quasi-static elasto/visco-plasticity, Int. J. Num. Meth. Eng. *9* (1975) 109–127.

Greenbaum, G. A. and M. F. Rubinstein. Creep analysis of axisymmetric bodies using finite elements. Nucl. Eng. Design *7* (1968) 379–397.

Hallquist, J. O. LS-DYNA Theoretical Manual, Livermore Software Technology Corporation, Livermore CA, 1998.

Krieg, R. D. and D. B. Krieg. Accuracy of numerical solution methods for the elastic-perfectly plastic model. J. Pressure Vessel Tech., ASME *99* (1977) 510–515.

Krieg, R. D. and S. W. Key. Implementation of a time dependent plasticity theory into structural computer programs. In constitutive equations in visco-plasticity. Computational and Engineering Aspects. Strickline and Saczalski (eds.). ASME, AMD-20 (1976) 125–137.

Mendelson, A. and M. H. Hirschberg and S. S. Manson, A general approach to the practical solution of creep problems. Trans. ASME, J. Basic Eng. *81* (1959) 85–98.

Ortiz, M. and E. P. Popov. Accuracy and stability of integration algorithms for elastic constitutive relations. *21* (1985) 1561–1576.

Pian, T. H. H. Mechanical sublayer model for elastic plastic analyses. Computational Mech. *2* (1987) 26–30.

Rice, J. R. and D. M. Tracey. Computational fracture mechanics. Numerical and Computer Methods in Structural Mechanics. Fenves (ed.). New York: Academic Press, 1973, p. 585.

Simo, J. C. and R. L. Taylor. Algorithm for plane stress elastoplasticity. Int. J. Num. Eng. *22* (1986) 649–670.

Simo, J. C. and S. Govindjee, Exact closed solution of the returning mapping algorithm in plane stress elasto-viscoplasticity, Eng. Computations *5* (1988) 254–258.

Truesdell, C. The Simple Rate Theory of Pure Elasticity. Comm. Pure Appl. Math. 8 (1995) 123.

Wilkins, M. L. Calculation of elastic-plastic flow. Methods of Computational Physics, *3*, Alder, Fernback and Rotenberg (eds.). New York: Academic Press, 1964.

AUTHOR INDEX

SUBJECT INDEX